A TEXTBOOK OF

OPTICS

(For B.Sc. Classes as per UGC Model Syllabus)

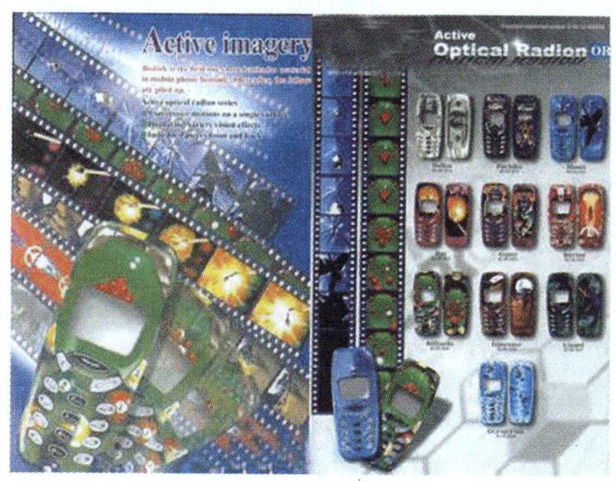

A TEXTBOOK OF
OPTICS

(For B.Sc. Students of Physics)

Dr. N. SUBRAHMANYAM
M.Sc., Ph.D.
Department of Physics
Kirori Mal College, University of Delhi
Delhi

BRIJ LAL
M.Sc.
Reader in Physics
Hindu College, University of Delhi
Delhi

Revised by
Dr. M. N. AVADHANULU
M.Sc.,Ph.D.
Principal
OM College of Engineering, Wardha

S Chand And Company Limited
(ISO 9001 Certified Company)

S Chand And Company Limited

(ISO 9001 Certified Company)

Head Office: D-92, Sector–2, Noida – 201301, U.P. (India), Ph. 91-120-4682700

Registered Office: A-27, 2nd Floor, Mohan Co-operative Industrial Estate, New Delhi – 110 044, Phone: 011-49731800

www.**schandpublishing.com;** e-mail: **info@schandpublishing.com**

Marketing Offices:

Chennai	:	Ph: 23632120; chennai@schandpublishing.com
Guwahati	:	Ph: 2738811, 2735640; guwahati@schandpublishing.com
Hyderabad	:	Ph: 40186018; hyderabad@schandpublishing.com
Jalandhar	:	Ph: 4645630; jalandhar@schandpublishing.com
Kolkata	:	Ph: 23357458, 23353914; kolkata@schandpublishing.com
Lucknow	:	Ph: 4003633; lucknow@schandpublishing.com
Mumbai	:	Ph: 25000297; mumbai@schandpublishing.com
Patna	:	Ph: 4011400; patna@schandpublishing.com

First Edition 1966

Subsequent Editions and Reprints 1994, 95, 96, 97, 98, 99, 2000, 2001, 2002, 2003, 2004 (Twice), 2006, 2007 (Twice), 2008, 2009, 2010 (Twice), 2012

Twenty Fifth Edition 2012; Reprints 2013, 2014, 2015, 2016 (Twice), 2018 (Twice), 2019 (Twice), 2020 (Twice), 2021 (Thrice), 2022 (Four Times), 2024, 2025

Reprint 2026

ISBN: 978-81-219-2611-9 **Product Code:** H6OPT68PHYS10ENAY12O

PRINTED IN INDIA

By Shri Shyamlal Printing Press Private Limited, Plot 20/4, Site-IV, Industrial Area Sahibabad, Ghaziabad – 201 010 and Published by S Chand And Company Limited, A-27, 2nd Floor, Mohan Co-operative Industrial Estate, New Delhi – 110 044.

Books are not paper and words but interaction with thinkers on a one-to-one basis, not of one generation but separated by hundreds and thousands of years.

–Thomas Carlyle

Preface to the Twenty Fifth Revised Edition

The present edition of the book has been thoroughly checked and all the printing mistakes have been removed. The following two chapters, thoroughly revised and rewritten.

1. Polarization (Chapter No. 20)

2. Holography (Chapter No. 23)

The information has been updated in the appendix.

Suggestions for the improvement of the book from the learned faculty members and students shall be highly appreciated and gratefully acknowledged.

Dr. M.N. AVADHANULU
M.Sc., Ph.D.
Reviser

Preface

The past four decades witnessed major inventions in optics which led to a silent revolution in communications and medical fields. It is often said that the near future belongs to photonics, the technology based on the utilization of optical radiation. A good knowledge of optics is essential for following the developments in photonics. This book provides an introduction to optics and is mainly intended for undergraduate students of science and engineering. This book aims to provide the necessary foundation in optics which prepares the student for an intensive study of advanced topics in optics at a later stage. Much of optics requires a good knowledge of mathematics. The traditional approach appears to stress on the derivations, with a lesser attention paid to the physical concepts. As a result, the student feels the subject to be insipid, disconnected and far from inspiring. An attempt is made in this book to balance the requirements of students taking a course in physics and at the same time to make the subject interesting to them. Therefore, the mathematics is kept at the necessary minimum level and concepts are given priority over the derivations. Further, the curriculum for a bachelor's degree in science has been modified by U.G.C. recently which necessitated a radical revision of the contents of the earlier version of this book. Although the book is based on the latest U.G.C. curriculum, additional chapters are included to meet the requirements of other allied courses also. The contents of some of the chapters of this book (Chapters 8,9,10,11,17,18 and 19) are taken as they were in the earlier editions while the rest of the text is reorganized and new chapters are added to suit the revised curriculum.

The material of this book is divided into four major parts, namely Ray Optics, Wave Optics, Quantum Optics and Photonics. The first two chapters of the book serve as a sort of prelude to the book.

Chapter 1 presents the development of various ideas on the nature and behaviour of light in chronological order.

Chapter 2 gives an introduction to the Fermat's principle which is simpler but very effective.

The first part of the book deals with the concepts in Ray Optics which is covered in nine chapters starting from Chapter 3 to Chapter 11.

Chapter 3 reviews the fundamental ideas of reflection and refraction at curved surfaces.

Chapter 4 discusses the thin lenses and the derivation of the lens equation and sets the stage for a better understanding of material in the succeeding three chapters.

Chapter 5 deals with optical systems containing two or more lenses and gives an introduction to the concept of cardinal points.

Chapter 6 applies the concepts developed in the previous chapter to the case of thick lenses.

Chapter 7 introduces to the matrix methods for solving the problems of geometrical optics.

Chapter 8 gives a brief account of dispersion.

Chapter 9 describes the various types of aberrations caused by the lenses and the remedial measures that can be adopted to minimize the aberrations.

Chapter 10 deals with the optical instruments that are commonly encountered.

Chapter 11 describes some of the important experiments conducted to determine the velocity of light and the Doppler effect observed in light.

The next nine chapters, starting from Chapter 12 to chapter 20, are devoted to wave optics, where both scalar wave optics and electromagnetic wave concepts are utilized to discuss the propagation of light and the important phenomena exhibited by light waves.

Chapter 12 deals with the formulation of general wave equation and points out the reason for the discontinuous nature of light waves. The concept of wave trains is introduced and the differences of real light waves from the expected character of harmonic waves are brought out.

Chapter 13 takes into account of the electromagnetic nature of the light waves. The propagation of light and its behaviour at the boundaries of different media are discussed. The electron theory of dispersion is described at this stage.

Chapter 14 introduces the phenomenon of interference and discusses one of the important applications, namely Interferometry.

Chapter 15 gives a detailed account of the interference effects in thin films and the principles of different interferometers are described. The most important application of interference phenomenon in making antireflection coatings, dielectric mirrors and interference filters are discussed.

Chapter 16 explains the concept of coherence and its role in determining the condition of interference.

Chapter 17 deals with the diffraction of Fresnel's class whereas Chapter 18 discusses the Fraunhoffer class of diffraction. The principles and working of different gratings are described.

Chapter 19 discusses the limitation imposed by the diffraction phenomenon and the inability of optical instruments in showing the details in images of objects beyond a certain limit.

Chapter 20 gives a detailed account of the phenomenon of polarization. The different methods of producing and detecting polarized state are discussed. Optical activity and artificial double refraction are discussed.

The third part of the book, starting from Chapter 21 to Chapter 23, deals with quantum optics.

Chapter 21 focuses the failure of classical approach in understanding the interaction of electromagnetic radiation with matter and; step by step describes the evolution of our ideas regarding atomic structure and the mechanism of light emission from atoms. The classical, semi-classical, and wave mechanical models of atom are described. Spontaneous emission is discussed using the Bohr model of atom by incorporating the conclusions drawn from quantum mechanics. Different types of sources of light are discussed.

Chapter 22 looks at the interaction of radiation with matter from quantum mechanical point of view and highlights the role of stimulated emission in producing coherent light. The principles and working of various coherent sources are described.

Chapter 23 provides an introduction to the field of holography.

The fourth part of the book, namely Photonics, contains two chapters, Chapter 24 and Chapter 25.

Chapter 24 deals with light propagation through optical fibres, the light wave-guides. It also describes some of the important applications of optical fibres.

Chapter 25 indicates the methods of generation of coherent light employing the non-linear optical effects.

Chapter 26 is intended to acquaint the reader with the principle of atom laser, the latest invention in the field of matter-wave optics. Atoms are associated with matter waves and lenses, mirrors and beam splitters have all been developed in the past to control the matter waves (atomic beams). Following the development of optical lasers, it is but a logical question to ask whether a source capable of producing an intense, highly directional, and coherent beam of atoms can be fabricated or not. The recent discovery of Bose-Einstein condensation has made the development of atom lasers possible. Physicists are hoping now that the invention of the "atom laser" is going to spark a revolution in the field of atomic optics.

Though this book is intended primarily for undergraduate students, it is hoped to be useful for a wide variety of readers such as those appearing for competitive examinations and as those aiming at obtaining a fair knowledge of the nature and behaviour of light, which gives us the direct intimation of God's subtle splendour.

Dr. M.N. AVADHANULU
M.Sc., Ph.D.
Reviser

Acknowledgements

A large number of works were consulted while preparing the present book. The authors do not claim originality in content or presentation of the material included in this book. The authors record here their indebtedness to the authors of those original works from where much of the information is liberally drawn.

I offer my thanks to Mrs. Nirmala Gupta (CMD), Mr. Navin Joshi, Executive Vice-President and Shri Bhagirath Kaushik, General Manager (S & M), S.Chand & Co. Ltd., New Delhi, for giving the opportunity to revise the book "**Optics**" by Prof. Brij Lal & Prof. Subrahmanyam.

I also thank the editorial staff of S. Chand and Co. Ltd. especially Mr. Shishir Bhatnagar, (Pre-Press Manager) and Mr. Rupesh Kumar Gupta (Subject Editor), editorial contribution wonderful effort in converting the book into **multicolour edition**.

Dr. M.N. AVADHANULU
M.Sc., Ph.D.
Reviser

Contents

Detailed Contents

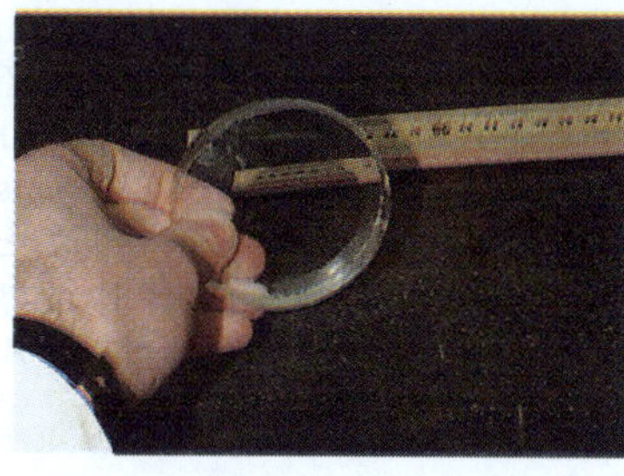

QUANTUM OPTICS

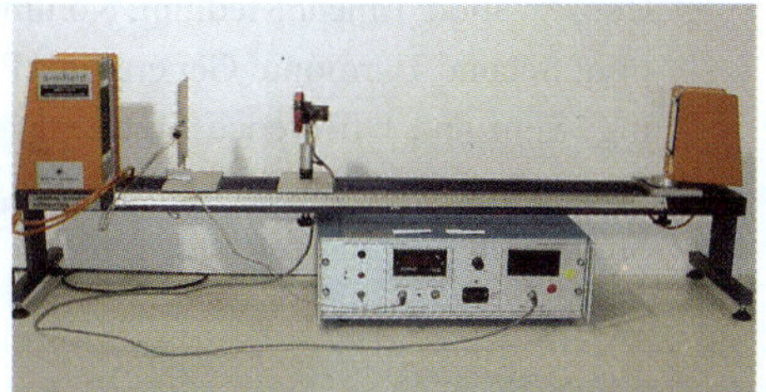

PHOTONICS

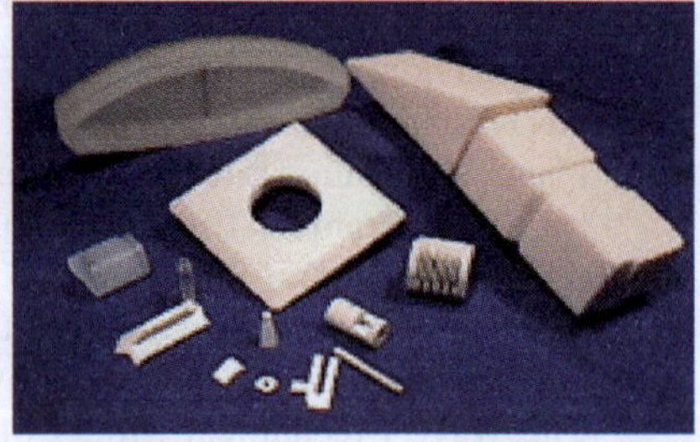

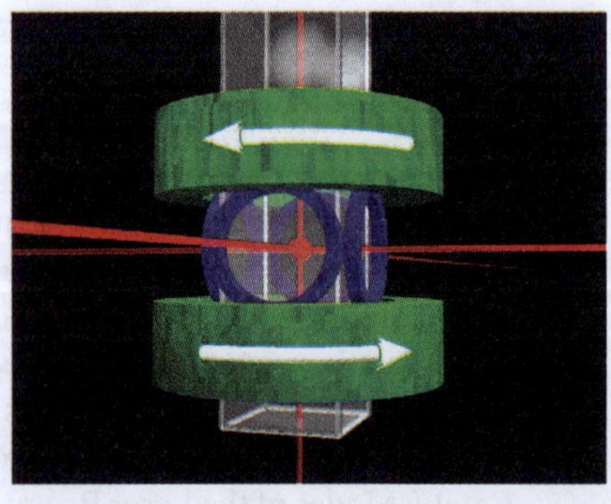

CHAPTER

Light

1.1. INTRODUCTION

We understand the world around us with the help of information reaching our five sense-instruments (organs), namely eyes, ears, nose, tongue and skin. The sense associated with the eyes is known as *vision* (or *sight*). Light is the agent which stimulates our sense of sight.The eyes convert the incoming light into electrical signals and convey them to the brain, which after processing the signals causes images or pictures to be created in our mind. We learn about the properties of atoms and their internal structure through the light emitted by them. We learn about the properties of giant stellar systems by means of the light reaching us from them after traveling for millions of years through the empty space. In fact our world is mostly defined by light. We have always been fascinated by the behaviour of light. Who has not appreciated the charms of rainbow stretching across the vast sky, of the pleasant sight of full moon, of the azure sky with wafting wafers of white clouds, of the glorious saffron coloured rising and setting sun, of the beckoning stars on pitch dark night, of the celestial spectacles of total or partial eclipses of moon and sun? The nature and properties of light have been speculated since ancient times. Nothing has been so elusive or so secretive than light. Man could gain an understanding about the nature of light after many centuries of persistent efforts.

At a Glance

What is light? What is it made up of ? How is it generated? How fast does it travel? How does it propagate across empty space? How does it behave when it comes across an object? How does it interact with matter? These are some of the many questions that arise in our mind. Optics is the branch of physics, which deals with such questions and describes about the phenomena and laws associated with the generation, and propagation of light and its interaction with matter. We briefly answer these questions in this introductory chapter and elaborate them at appropriate places in the book. Let us start with a look at some of the important landmarks in the evolution of our understanding about light.

1.2. BRIEF HISTORY

(A) Development of Geometric Optics:

The Greeks were aware of the rectilinear propagation of light. They knew that when light is reflected from a mirror, the angle of incidence is equal to the angle of reflection. This was stated by Euclid (300 B.C.) in his book Catoptrics. Hero of Alexandria suggested that light traverses the shortest path between two points. They were also aware of refraction of light as it passes from one transparent medium to another. Claudius Ptolemy (130 A.D.) of Alexandria measured the angles of incidence and refraction for several media. Further progress came to a halt with the fall of Roman Empire in 475 A.D. Study of light was again revived in Europe during the thirteenth century. Francis Bacon (1215-1294) suggested the idea of using lenses to improve eyesight. In about 1280, spectacle lenses came into use to correct faulty vision. In 1609 Galileo (1564-1642) devised a practical telescope. Van Leeuwenhoek (1632-1723) developed the first microscope. John Kepler discovered the phenomenon of total internal reflection. In 1621 Willebrod Snell (1591-1626) and independently in 1637 Rene Descartes (1596-1650) discovered the law of refraction. In 1658, Fermat (1601-1655) discovered the principle of least time. According to this principle, light always follows that path which takes it to its destination in the shortest time. He re-derived the law of reflection and refraction applying this principle of least time for the path followed by light. In 1660 the phenomenon of diffraction was noticed by Grimaldi (1618-1663). In 1667 Newton established that white light is composed of seven independent colours. In 1670, Bartholinus (1625-1698) discovered the phenomenon of double refraction.

In 1609, Galileo devised a practical telescope.

In 1675 Isaac Newton (1642-1727) put forward the *corpuscular theory*. According to this theory, a luminous body emits in all directions streams of extremely minute particles, called *corpuscles*. They are supposed to travel through a medium with a tremendous but finite velocity in straight line paths. The particle theory of Newton could explain the straight line propagation of light and that an object casts a sharp shadow; but it failed to explain why the continued loss of particles did not also cause a source of light to lose weight. However, the theory could prove the laws of reflection and refraction of light. Newton predicted that light should travel faster in a denser medium than in a rarer medium. However, the phenomenon of diffraction and Newton's rings could not be explained on the basis of corpuscular theory.

Isaac Newton (1642-1727)

In 1676, Romer (1644-1710) proved that light travels with a finite velocity. Robert Hooke (1635-1703) studied the coloured patterns formed due to thin film interference.

(B) Development of Wave Optics :

In 1678 Huygens (1629-1695), a contemporary of Newton, proposed wave theory of light. According to this theory, light energy is supposed to be transferred from one point to another in the form of waves. Huygens was able to prove the ordinary laws of reflection and refraction. He predicted that light should travel slower in a denser medium than in a rarer medium. He also explained the phenomenon of double refraction by assuming two types of waves. The wave theory was not accepted immediately. The chief reason was that a wave motion needs a medium; but light could travel to us from the sun through the vacuum of space.

In 1803, Thomas Young (1773-1829) demonstrated for the first time the interference of light beams. He also explained Newton's rings and the colours of thin flims on the basis of interference of light waves. Thomas Young provided strong support to the wave theory. In 1808, Malus (1775-1812) discovered the polarization of light. In 1815, Augustin Fresnel (1788-1827) further developed the wave theory and explained the rectilinear propagation of light which has been the chief obstacle in the way of accepting wave theory. He provided a satisfactory explanation of the diffraction phenomenon.

Thomas Young
(1773-1829)

Following Huygens, both Young and Fresnel assumed that light waves are longitudinal. Young and Fresnel conceived of an elastic medium, which was assumed to exist pervading the entire universe, and it was named *luminiferous ether*. The vibrations of the ether propagated as light, just as longitudinal vibrations in air propagate as sound. But the longitudinal wave theory of light could not explain polarization, a property exhibited by transverse waves but not by longitudinal waves. Fresnel and Arago (1786-1853) conducted experiments on superposition of linearly polarized light. Young eventually realized that light is a transverse wave and in 1817 explained the results of Fresnel and Arago's experiments. In 1850, Jean Foucault (1791-1868) established that light travels slower in liquids than in air. This is just opposite to the prediction of Newton's theory.

Finally, the wave model was accepted. The acceptance of the wave theory of light made it obvious that a supporting medium should exist. Subsequently, elastic ether theory was developed during the next ten years. Strange properties were attributed to it. It was assumed to be extremely rigid so that it can support the exceedingly high frequency oscillations of light travelling at a speed of 3×10^8 m/s; yet it does not offer resistance to the motion of celestial bodies through it. Its density was supposed to increase in material substances to account for the lower velocity. In 1823, Fresnel derived expressions for the reflection and transmission coefficients on the basis of ether theory.

(C) Nature of light:

Around 1836, Faraday (1791-1867) showed that a varying magnetic field induces an electromotive force and thus established the intimate connection between electricity and magnetism. Further, Faraday showed that the polarization of light was affected by a strong magnetic field, which was the first hint as to the electromagnetic nature of light. Clerk Maxwell (1831-1879) unified the empirical laws of electricity and magnetism into a coherent theory of electromagnetism. In 1873, Maxwell showed that the speed of electromagnetic waves equals the speed of light. On the strength of this, he made the prediction that light is a high frequency electromagnetic wave. In 1887, Hertz (1857-1894) confirmed Maxwell's theoretical prediction by producing and detecting electromagnetic waves. The electromagnetic waves were initially supposed to be supported by the ether medium. Though electromagnetic theory is capable of explaining the phenomena connected with the propagation of light, it fails to explain the processes of emission and absorption. H.A.Lorentz (1853-1928) assumed that ether is in a state of absolute rest to be the carrier of electromagnetic field.

In 1887, Michelson-Morley performed the famous ether-drift experiment and found that light travels at the same speed irrespective of the position of the earth in its orbit. It led to the conclusion that ether does not exist. Hence, *light is a self-sustaining high frequency electromagnetic wave*. This theory is known as the *Field Theory*.

(D) Development of Quantum Optics:

In 1814 Fraunhofer discovered dark lines in the solar spectrum. In 1861 Bunsen and Kirchhoff attributed them to the absorption of certain wavelengths by the gases in the outer atmosphere of the sun. It was also found that every gaseous chemical element possesses a characteristic line spectrum. The detailed studies of emission and absorption spectra of elements evolved into a separate discipline. In 1900, in order to obtain a correct theoretical expression for the black body radiation, Max Planck (1858-1947) found it necessary to suppose that light is absorbed or emitted in the form of elementary quanta. In 1905, Einstein (1879-1955) made use of the quantum concept to successfully explain the photoelectric emission. According to him, light is a stream of photons. In 1913, applying Planck's quantum hypothesis, Niels Bohr (1885-1962) devised an atomic model for the emission and absorption of light. It successfully explained the simple laws of line spectra of gases. The traditional sources of light produce incoherent light. The first coherent source of light, namely laser was built in 1960. The high power lasers led to a number of nonlinear optical effects such as harmonic generation, frequency mixing etc. Quick developments in holography and fibre optics followed the discovery of lasers.

Max Planck
(1858-1947)

We now visualize a photon as a bundle of electromagnetic radiation that oscillates with a definite frequency and travels through free space with the speed of light. Individual photons carry energy and momentum, so light has particle-like properties. When the number of identical photons is very large, they exhibit the properties of a continuous wave with the same definite frequency and propagation speed as the quantum. The phenomena of interference, diffraction and polarization and propagation of light in space is adequately explained by classical electromagnetic wave theory, whereas the experiments involving interaction of light with matter, such as photoelectric effect are best explained by assuming that light is a particle.

1.3. THE FOUR IMPORTANT THEORIES

Various theories have been put forward about the nature of light. We will make a brief survey of the four important theories which guided the evolution of our understanding of the nature of light. The theories are known as

1. Corpuscular theory
2. Wave theory
3. Electromagnetic theory and
4. Quantum theory.

1.3.1. CORPUSCULAR THEORY

The corpuscular theory was postulated by ancient Greeks and was favoured by Sir Isaac Newton. According to this theory, a luminous body continuously emits tiny, light and elastic particles called *corpuscles* in all directions. These particles or corpuscles are so small that they can readily travel through the interstices of the particles of matter with the velocity of light and they possess the property of reflection from a polished surface or transmission through a transparent medium. When these particles fall on the retina of the eye, they produce the sensation of vision. On the basis of this theory, phenomena like rectilinear propagation, reflection and refraction could be accounted for, satisfactorily. Since the particles are emitted with high speed from a luminous body, they, in the absence of other forces, travel in straight lines according to Newton's second law of motion. This explains rectilinear propagation of light.

1.3.1.1. Reflection of Light on Corpuscular Theory

Let SS' be a reflecting surface and IM the path of a light corpuscle approaching the surface SS'. When the corpuscle comes within a very small distance from the surface (indicated by the dotted line AB) it, according to the theory begins to experience a force of repulsion due to the surface (Fig. 1.1).

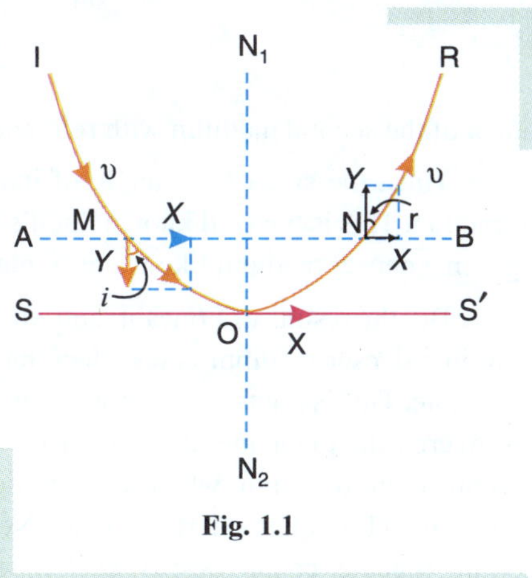

Fig. 1.1

The velocity υ of the corpuscle at M can be resolved into two components x and y parallel and perpendicular to the reflecting surface. The force of repulsion acts perpendicular to the surface SS' and consequently the component y decreases up to O and becomes zero at O the point of incidence on the surface SS'. Beyond O, the perpendicular component of the velocity increases up to N. Its magnitude will be again y at N but in the opposite direction. The parallel component x remains the same throughout. Thus at N, the corpuscle again possesses two components of velocity x and y and the resultant direction of the corpuscle is along NR. The velocity of the corpuscle will be υ. Between the surfaces AB and SS' the path of the corpuscle is convex to the reflecting surface. Beyond the point N, the particle moves unaffected by the presence of the surface SS'.

$$x = \upsilon \sin i = \upsilon \sin r, \qquad \therefore i = r$$

Further, the angles between the incident and the reflected paths of the corpuscles with the normals at M and N are equal. Also, the incident and the reflected path of the corpuscle and the normal lie in the same plane *viz.* the plane of the paper.

1.3.1.2. Refraction of Light on Corpuscular Theory

Newton assumed that when a light corpuscle comes within a very small limiting distance from the refracting surface, it begins to experience a force of attraction towards the surface. Consequently the component of the velocity perpendicular to the surface increases gradually from AB to $A'B'$. SS' is the surface separating the two media (Fig. 1.2) I M is the incident path of the corpuscle travelling in the first medium with a velocity υ and incident at an angle i. AB to $A'B'$ is a narrow region within which the corpuscle experiences a force of attraction. NR is the refracted path of the corpuscle. Let $\upsilon \sin i$ and $\upsilon \cos i$ be the components of the velocity of the corpuscle at M parallel and perpendicular to the surface. The velocity parallel to the surface increases by an amount which is independent of the angle of incidence, but which is

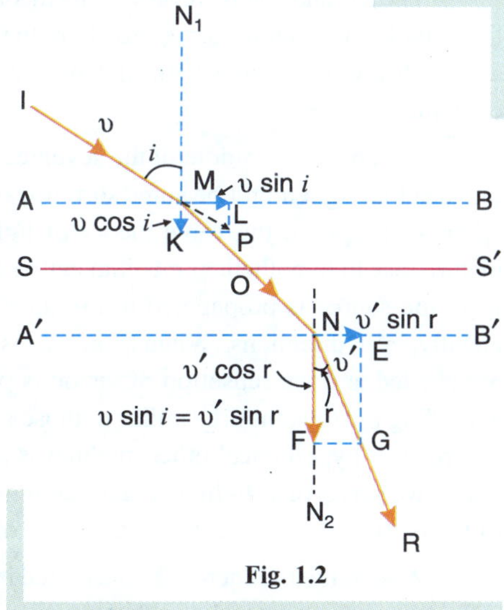

Fig. 1.2

different for different materials. Let υ and υ' be the velocity of the corpuscle in the two media and r the angle of refraction in the second medium.

As the parallel component of the velocity remains the same,

$$\upsilon \sin i = \upsilon' \sin r$$

or

$$\frac{\sin i}{\sin r} = \frac{\upsilon'}{\upsilon}$$

$$= \frac{\text{velocity of light in the second medium}}{\text{velocity of light in the first medium}} = {}_1\mu_2 \text{ (refractive}$$

index of the second medium with reference to the first medium)

Thus, the sine of the angle of incidence bears a constant ratio to the sine of the angle of refraction. This is the well known Snell's law of refraction. If $i > r$, then $\upsilon' > \upsilon$ i.e., the velocity of light in a denser medium like water or glass is greater than that in a rarer medium such as air.

But the results of Foucault and Michelson on the velocity of light showed that the velocity of light in a denser medium is less than that in a rarer medium. Newton's corpuscular theory is thus untenable. This is not the only ground on which Newton's theory is invalid. In the year 1800, Young discovered the phenomenon of interference of light. He experimentally demonstrated that under certain conditions, light when added to light produces darkness. The phenomena belonging to this class cannot be explained, if following Newton, it is supposed that light is material. Two corpuscles coming together cannot destroy each other.

Another case considered by Newton was that of simultaneous reflection and refraction. To explain this he assumed that the particles had fits so that some were in a state favorable to reflection and others were in a condition suitable for transmission. No explanation of interference, diffraction and polarization was attempted because very little was known about these phenomena at the time of Newton. Further, the corpuscular theory has not given any plausible explanation about the origin of the force of repulsion or attraction in a direction normal to the surface.

1.3.2. WAVE THEORY

The test and completeness of any theory consists in its ability to explain the known experimental facts, with a minimum number of hypotheses. From this point of view, the corpuscular theory is above all prejudices and with its help rectilinear propagation, reflection and refraction could be explained.

By about the middle of the seventeenth century, while the corpuscular theory was accepted, the idea that light might be some sort of wave motion had begun to gain ground. In 1679, Christian Huygens proposed the wave theory of light. According to this, a luminous body is a source of disturbance in hypothetical medium called **ether.** This medium pervades all space. The disturbance from the source is propagated in the form of waves through space and the energy is distributed equally, in all directions. When these waves carrying energy are incident on the eye, the optic nerves are excited and the sensation of vision is produced. These vibrations in the hypothetical medium according to Huygens are similar to those produced in solids and liquids. They are of a mechanical nature. The hypothetical ether medium is attributed to the property of transmitting elastic waves, which we perceive as light. Huygens assumed these waves to be longitudinal, in which the vibration of the particles is parallel to the direction of propagation of the wave.

Assuming that energy is transmitted in the form of waves, Huygens could satisfactorily explain reflection, refraction and double refraction noticed in crystals like quartz or calcite. However, the phenomenon of polarization discovered by him could not be explained. It was difficult to conceive unsymmetrical behaviour of longitudinal waves about the axis of propagation. Rectilinear propagation of light also could not be explained on the basis of wave theory. The difficulties mentioned above were overcome, when Fresnel and Young suggested that light wave is transverse and not longitudinal as suggested by Huygens. In a transverse wave, the vibrations of the ether particles take place in a direction perpendicular to the direction of propagation. Fresnel could also explain successfully the rectilinear propagation of light by combining the effect of all the secondary waves starting from the different points of a primary wave front.

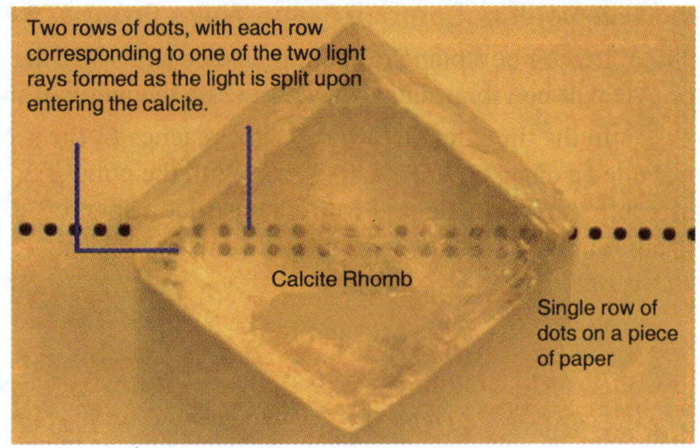

Double refraction.

1.3.2.1. Huygens Principle

Huygens' principle gives a geometrical construction for finding the position of a wave front at a future instant if its position is known at some particular instant. The construction is based on the following two fundamental postulates.

(*i*) Every point on a wave front acts as a 'secondary' source of disturbance. Secondary wavelets spread in all directions from these new sources. The secondary wavelets are spherical and have the same frequency and velocity as the original wave.

(*ii*) The surface, which touches all the wavelets from the secondary sources, gives the new position of the wave front.

We now apply the Huygens' principle to the propagation of spherical wave fronts, of light from a point source, spreading in an isotropic medium.

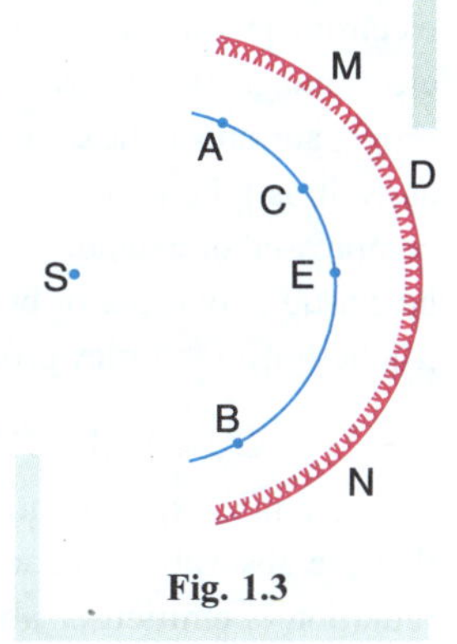

Fig. 1.3

Let S be a point source of light producing spherical wave fronts. Suppose AB is the position of the wave front at some instant, as shown in Fig. 1.3. According to Huygens' principle, every point A,...,C,...., E,....,B on AB is a source of secondary wavelets which advance with the velocity υ. Taking each of these points A,...,C,.... ,E,....,B as a centre, spheres of radii equal to 'υt' are drawn. These small spheres represent the secondary wavelets starting from these points. We now draw a surface MN which touches the small spheres in the forward direction. The tangential surface MN is the new wave front after an interval of t seconds. It is very easy to see that MN is the surface of a sphere having its centre at S.

Huygens' construction is an incomplete concept because it does not explain why there are no backward-going wavelets. In 1882 Kirchhoff eliminated this defect and showed that Huygens' primitive principle was a direct consequence of the differential wave equation. In spite of the defect, Huygens' principle served as very useful guide in explaining the phenomena of interference, diffraction and polarization.

Based on Huygens wave theory and Huygens principle, one can, by constructing the wavefronts, explain satisfactorily reflection and refraction of light. Adopting Fresnel's modification of Huygens principle, rectilinear propagation of light can also be explained.

1.3.2.2. Reflection of a Plane Wave Front at a Plane Surface

Let *XY* be a plane reflecting surface and *AMB* the incident plane wavefront. All the particles on AB will be vibrating in phase. Let *i* be the angle of incidence (Fig. 1.4).

In the time the disturbance at *A* reaches *C*, the secondary waves from the point *B* must have travelled a distance *BD* equal to *AC*. With the point *B* as centre and radius equal to *AC* construct a sphere. From the point *C*, draw tangents *CD* and *CD′*. Then *BD* = *BD′*.

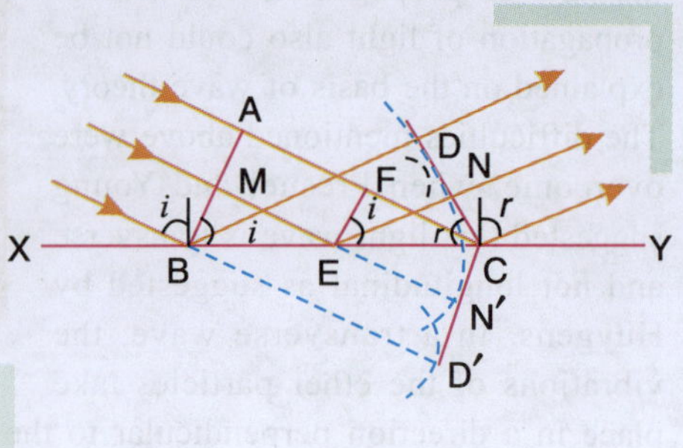

Fig. 1.4

In the Δ^{les} *BAC* and *BDC*

BC is common, $BD = AC$ and

$\angle BAC = \angle BDC = 90°$

∴ The two triangles are congruent.

$\therefore\ \angle ABC = i = \angle BCD = r$

$\therefore\ i = r$

Thus, the angle of incidence is equal to the angle of reflection.

1.3.3. ELECTROMAGNETIC THEORY

In 1862 Maxwell ingeniously synthesized electricity and magnetism and developed equations which succinctly combine the important theories. He showed that electromagnetic waves travel with the speed of light and hence drawn the most important conclusion that light wave itself is an electromagnetic wave. Initially, the existence of ether medium was presumed from propagation of electromagnetic waves in space. However, if light waves which are of very high frequency are to propagate, and at the same time allow a free passage to heavenly bodies, then the ether have to be rigid as well as pliable. It became impossible to visualize the hypothetical solid which could be easily compressed or extended, could permit resistance-free passage of heavenly bodies through it, and yet be elastic to twisting or bending stresses in order to allow propagation of waves. Ultimately, in 1887 Michelson and Morley proved conclusively that there was no ether surrounding the earth or elsewhere.

1.3.4. QUANTUM THEORY

While experimenting on the black body radiation, Max Planck had come to the conclusion that the absorption or radiation of energy is not a continuous process. He postulated that thermal radiation is emitted or absorbed intermittently by indivisible amounts of energy called *quantum*. Each quantum carries an energy *hv* where *h* is a constant now called Planck's constant. Einstein elaborated the quantum concept in an endeavour to account for the phenomenon of photoelectric emission. He postulated that the quanta travel in space as separate entities with the speed 'c'. The quanta are named as photons. The further confirmation for the quantum theory is obtained when Compton effect was discovered in 1923. Compton found that when monochromatic x-rays fell upon matter, the scattered rays contained not only the original x-rays but also x-rays of wavelengths longer than the original. Though the quantum theory explains successfully the interaction of radiation with matter, it cannot account for the phenomenon of polarization, interference and diffraction. The contradictory aspects are reconciled by postulating dual nature of radiation. Accordingly, radiation is viewed as having both the particle as well as wave nature.

1.4. THE SOURCES OF LIGHT

The sun, the stars, lamps give off light. They are called luminous bodies. Other objects moon, mountains, trees etc. are non-luminous. They are visible only when they receive light from some luminous source and they send the light to our eyes.

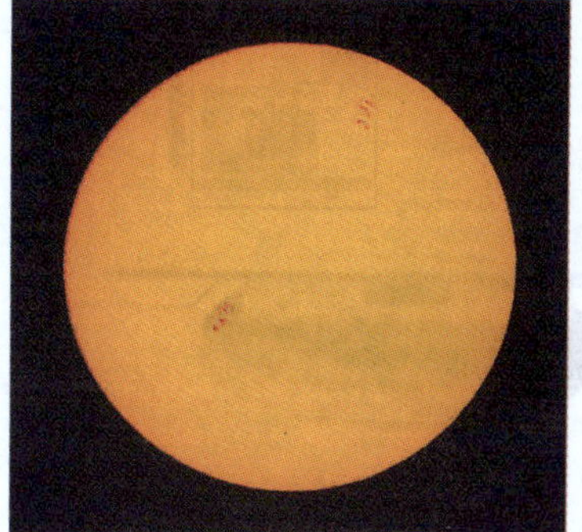

Luminous bodies : Sun, star and lamp.

Non-luminous bodies : Mountain and moon.

Whether a body is luminous or non-luminous depends on the conditions as well as on the material of which it is made. By changing the conditions we can make substances luminous or non-luminous. For e.g. the filament inside the electric bulb is non-luminous unless it is heated by an electric current.

Bodies emit light at the expense of various kinds of energy. The most common is *thermal radiation.* When bodies are heated, to a temperature of 300°C they emit electromagnetic radiation, which lies in infrared region ($\lambda = 5\ \mu m$). They emit light as result of thermal motion of their molecules that is at the expense of their *internal energy*. At a temperature of 800°C, bodies emit visible radiant energy and appear red hot; a larger part of the energy still lies in IR region. At around 3000°C they appear white hot. Such heated materials are known as *incandescent* bodies. Not all sources are incandescent.

Some bodies can emit light, which is not due to transfer of thermal energy into the energy of electromagnetic waves. Emission of light due to supply of energy through processes other than heat is called *luminescence*. There are different kinds of luminescence. Advertisements using the neon and other glow tubes are examples of **electroluminescence**. In this kind of luminescence, charged particles accelerated by an electric field partly transmit their kinetic energy to the atoms of the gas, hen emit light. Many living organisms s fireflies, fish and bacteria emit light due hemical reactions. Such glow is called **emiluminescence**. The cold light emitted by ube lights is a result of **photoluminescence**. The internal surface of the tube light is coated with a phosphor material, which under the action of UV light emits visible light. The TV screens and computer terminal screens glow because of **cathodoluminescence,** which occurs due to bombardment of the screen by high-energy electrons.

Computer screen glows because of cathodoluminescence, which occurs due to bombardment of the screen by high energy electrons.

1.5. PROPERTIES OF LIGHT

Reflection, refraction, dispersion and velocity are the important properties of light. We briefly discuss about them here.

1.5.1. REFLECTION OF LIGHT

When light travelling in a medium encounters a boundary leading to a second

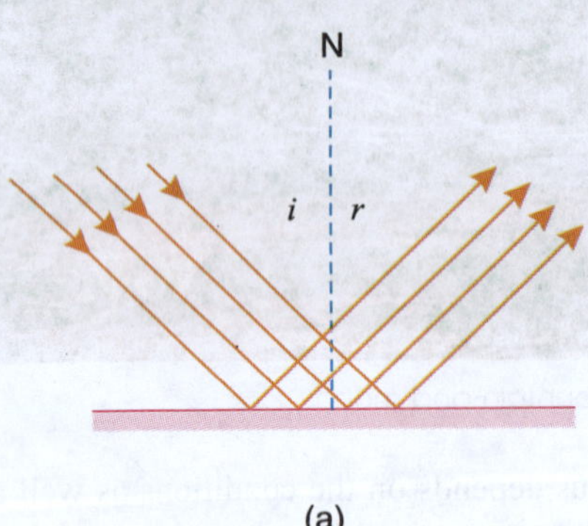

(a)

(b)

Fig. 1.5

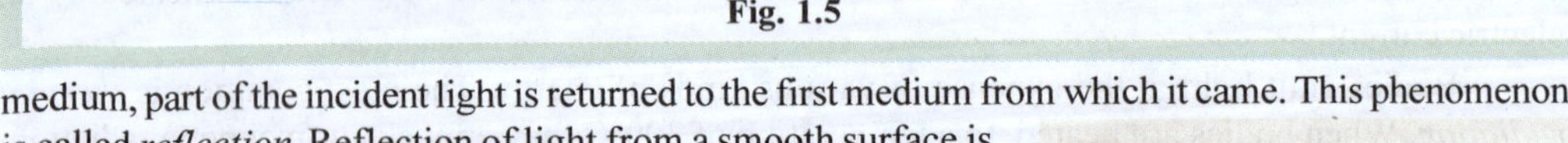

medium, part of the incident light is returned to the first medium from which it came. This phenomenon is called *reflection.* Reflection of light from a smooth surface is called *regular* or *specular reflection.*

Reflection from a rough surface is known as *diffuse reflection.* It is largely by diffuse reflection that we see non-luminous objects around us. The difference between diffuse and specular reflection is a matter of surface roughness. In the study of optics, the term reflection is used to mean specular reflection.

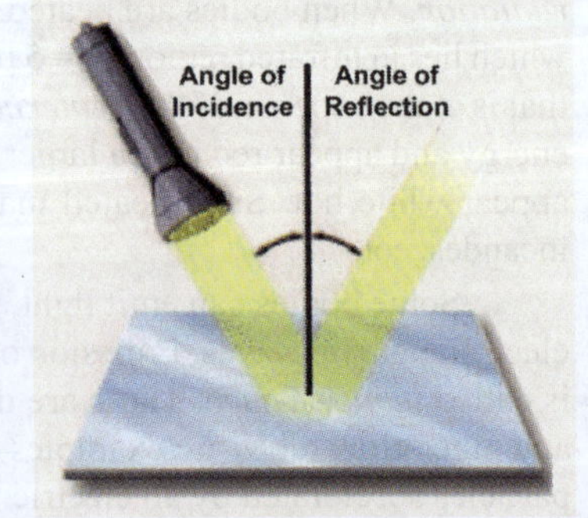

Laws of reflection.

1.5.1.1. Laws of reflection

In Fig. 1.6 the light ray AB, passing through air, is incident on a plane mirror and is reflected via the path BC. The point (B) where the light intersects the surface of the mirror is the *point of incidence*. A line drawn at B, perpendicular to the mirror, is the *surface normal*. The angle subtended by the surface normal and the incident ray is the *angle of incidence, i*. The angle subtended by the surface normal and the reflected ray is the *angle of reflection, r*. For some reason, it is customary to measure the angles from the surface normal toward the ray.

First Law: The incident ray, the reflected ray and the normal at the point of incidence are in the same plane. This plane is called the *plane of incidence*.

Second Law: The angle of reflection is equal to the angle of incidence. Thus, in Fig. 1.6,

$$i = r \qquad (1.1)$$

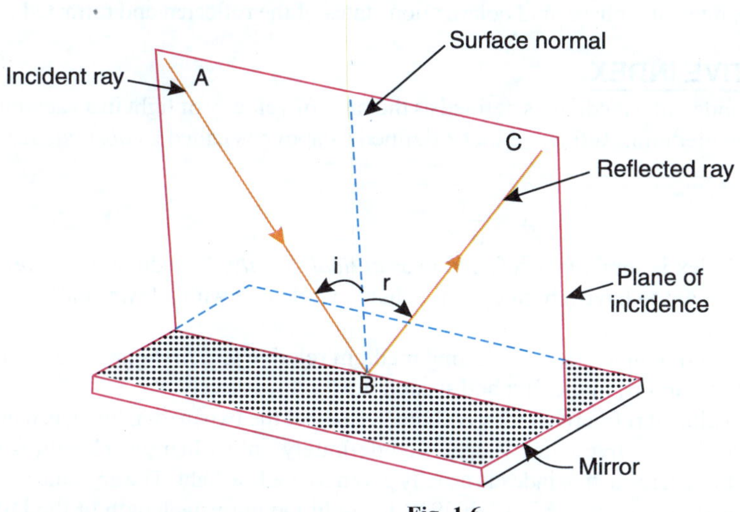

Fig. 1.6

Note — The laws of reflection are obeyed in specular reflection. They do not hold in case of irregular or diffuse reflection.

1.5.2. REFRACTION OF LIGHT

When a ray of light travelling through a transparent medium encounters a boundary leading into another transparent medium, part of the ray is reflected and part of it enters the second medium. The ray that enters the second medium is bent at the boundary and is said to be *refracted*. Thus, refraction means that the light ray follows in the second medium a direction different from its direction in the first medium. The angle '*r*' subtended by the normal and the refracted ray is the *angle of refraction*. This angle is also measured from the surface normal toward the ray.

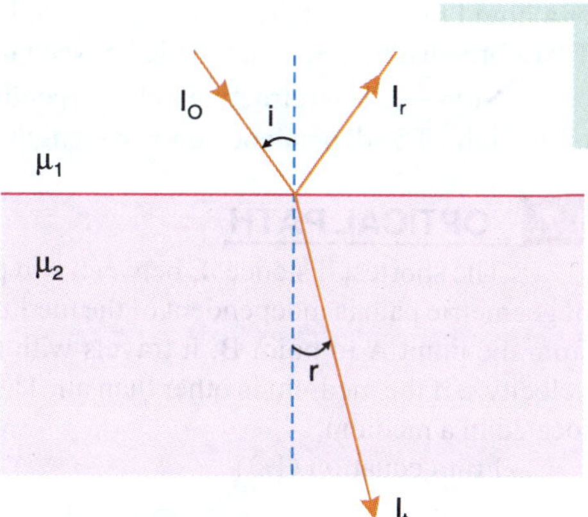

Phenomenon of refraction—A ray obliquely incident on air-glass interface bends toward the normal in glass.

Fig. 1.7

1.5.2.1. Laws of refraction

First Law: The incident ray, the refracted ray and the normal at the point of incidence lie in the same plane.

Second Law: The ratio of the sine of the angle of incidence to the sine of the angle of refraction for any two given media is constant.

$$\frac{\sin i}{\sin r} = \mu \qquad \text{... (1.2)}$$

where μ is called the *refractive index* of the medium.

Note — The laws of reflection and refraction relate only to the directions of the corresponding rays but do not say anything about the intensities of the reflected and refracted rays. These depend on the angle of incidence, the two indexes of refraction, and the polarization of the incident ray. ◆

Note — The laws of reflection and refraction are obtained as experimental results. They can also be derived from Maxwell's equations. The superiority of this treatment is that it enables us to predict the amplitude, intensity, phase, and polarization states of the reflected and refracted rays.

1.6. REFRACTIVE INDEX

The refractive index of a medium is defined as the ratio of velocity of light in a vacuum to the velocity of light in the medium. Refractive index defined as above is called as *absolute refractive index*. Thus,

$$\mu = \frac{c}{\upsilon} \qquad \text{... (1.3)}$$

The refractive index is sometimes referred to as *optical density*. A medium with a relatively high refractive index is said to have a high optical density, while one with a lower index is said to have a low optical density.

The relative refractive index μ_{21} of a second medium relative to a first one is defined as the ratio of speeds of light υ_1 and υ_2 in the first and second media respectively.

The numerical value of refractive index is characteristic of the two media, but it also depends on wavelength. Hence, a refractive index is specified definitely only when wavelength is stated. Unless otherwise mentioned, refractive index is usually given for yellow light. The absolute refractive index for air under standard conditions is 1.0002918 for light having wavelength of the D-line of sodium (5893 Å). Since the refractive index for air is nearly equal to unity, it follows that for a solid or a liquid the absolute refractive index and the refractive index relative to air differ only slightly. Therefore, distinction is not made between them.

Note — The refractive index depends not only on the substance but also on the wavelength of the light. The dependence on wavelength is called **dispersion.** ◆

1.7. OPTICAL PATH

The shortest distance, L between two points A and B is called the *geometric path*. The length of geometric path is independent of the medium that surrounds the path AB. When a light ray travels from the point A to point B, it travels with the velocity 'c' if the medium is air and with a lesser velocity υ if the medium is other than air. Therefore, the light ray takes more time to go from A to B located in a medium.

From equation (1.3)

$$\mu = \frac{c}{\upsilon} = \frac{AB/t}{AB/T} = \frac{T}{t}$$

where t and T are the time taken by the light ray in air and in a medium respectively.

$$\therefore \qquad T = \mu\, t \qquad \text{... (1.4)}$$

The above relation means that a light ray takes μ times more time to cover the distance AB in a medium. To take into account the delay, we use another distance called the *optical length*. If a ray of light travels a distance L in a medium of refractive index μ in a certain interval of time, then it would travel a greater distance Δ in air during the same interval of time. Therefore,

$$\frac{\Delta}{L} = \frac{ct}{\upsilon t} = \mu$$

or $$\Delta = \mu L \quad \text{... (1.5)}$$

i.e., Optical path length = (Refractive index)(Geometric path length)

Thus, ***the optical path length is defined as the product of refractive index and the geometric path length.***

Note that if a ray travels a distance l in a medium of refractive index μ, the optical path length is equal to μl. In a given time light travels the same optical path length in different media. Suppose light travels a distance l_1 in a medium of refractive index μ_1 and a distance l_2 in a medium of refractive index μ_2 in time t. Then

$$\mu_1 l_1 = \mu_2 l_2$$

1.8. DISPERSION

The refractive index is a function of wavelength. Generally, it decreases as the wavelength increases. Consequently, light passing through a material medium will be separated according to wavelength. This is known as chromatic dispersion.

When white light passes through a glass prism, a rainbow of colours emerges.

With the light at normal incidence, blue light would merely lag behind the red light. With the light incident obliquely, the colors become separated in space, both surfaces refracting blue more than red. With a prism the separation is even more pronounced.

Note that dispersion is not proportional to refraction. Refraction and dispersion bear no simple relationship to one another. Some glasses have a high index of refraction and little dispersion; others have just the opposite.

When a ray of white light falls on a glass prism, it splits up into different colours (see Fig.1.8). This display of colours is known as ***spectrum*** of the source of light. This separation of a composite beam into its constituent colours is called *dispersion* and the medium which produces dispersion is known as a *dispersive medium.* When the colours travel through a refracting medium, their velocities are different. Hence the refractive index of the material is different. Out of the seven colours formed by the white light, violet colour suffers greater deviation than red light.

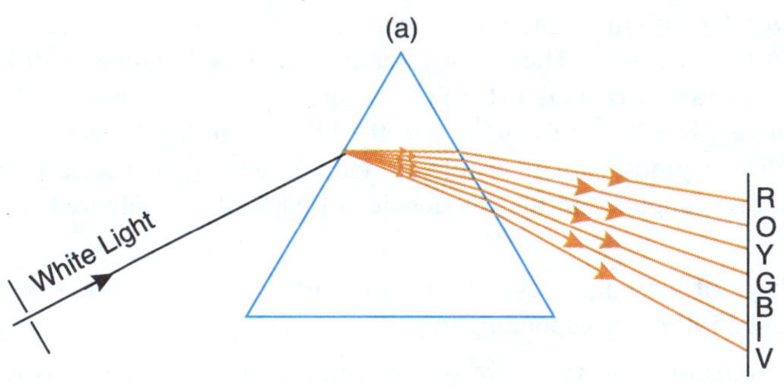

Fig. 1.8

1.9. THE VELOCITY OF LIGHT

The first attempt to measure the velocity of light was made by Galileo in 1667. Two observers were stationed on the tops of two hills about 1.5 km apart. Each observer was provided with one lamp and the experiment was carried out at night. One observer, say A, uncovered his lamp thereby sending a short flash of light to the other observer, say B and noted the instant at which he did so. The other observer B uncovered his lamp as soon as he saw the flash from B's lamp. The interval between these two instants of time noted by A is obviously equal to the time taken by the light to travel the distance AB twice. However, the interval could not be found between the two instants. From this Galileo concluded that velocity of light, if finite, was extremely large.

Eight years later, in 1675, the Danish astronomer, Roemer succeeded in measuring the velocity of light. The satellites of planet Jupiter show eclipses once in every 7 days. He noticed that the time interval between two consecutive eclipses of one of the satellite of Jupiter varied with the relative positions of the earth. This time interval increased as the earth moved away from Jupiter and decreased as it moved toward Jupiter. Roemer correctly attributed this variation to the finite velocity of light and found that light required about 22 minutes to travel a distance equal to the diameter of earth's orbit. The diameter of the earth's orbit was known to be 2.87×10^{11} m.

$$\therefore \qquad c = \frac{2 \times 10^{11}\ \text{m}}{22 \times 60\,\text{s}} = 2.3 \times 10^{8}\ \text{m/s}$$

This value is lower than the actual speed. Regardless of its accuracy it showed that the speed of light was not infinite, as many then believed, but was finite and measurable.

The first laboratory measurement of the speed of light was carried out in 1849 by the French physicist H.L.Fizeau. Later, many experimenters using different methods have measured c. The most precise determination gives a value

$$\therefore \qquad c = 2.997924 \times 10^{8}\ \text{m/s}.$$

which is usually simplified as

$$\mathbf{c = 3 \times 10^{8}\ m/s.} \qquad \text{... (1.7)}$$

1.10. VISIBLE RANGE

The arrangement of the various electromagnetic waves in a continuous sequence of frequencies and wavelengths is called an **electromagnetic spectrum**. The spectrum includes waves covering a broad range of wavelengths (see Fig. 1.9). It is bounded at one end by the gigantic radio waves having wavelengths of a few kilometres and at the other end by γ- rays of tiny wavelengths of the order of 10^{-12}m. **Visible range** is that part of the spectrum constituted by waves, which can be detected by the human eye. It extends from the deepest violet to the deepest red. The limiting range of these waves depends in the individual properties of the eye and varies approximately in the interval λ = 4000Å to λ = 7800Å. The regions flanking the visible range are **infrared (IR)** on the longer wavelength side and **ultraviolet (UV)** on the shorter wavelength side. The IR region lies in the wavelength range 7.8×10^{-7}m to 10^{-3}m and the UV region lies between 4000Å to 10Å. The radiation in these three regions, namely visible, IR and UV, put together is called **optical radiation**. Table - 1 lists the wavelengths and the frequencies of each of the coloured region in the visible region.

The sensitivity of the human eye is a function of wavelength. It has peak **sensitivity** at a wavelength of about 5500 Å corresponding to yellow-green.

The wavelengths of light waves are shorter and therefore smaller units are used to express them. They are usually expressed in angstrom or nanometre or micrometre units.

TABLE-1

Colour	*Vacuum wavelength* (*Å*)	*Frequency* (10^{14}*Hertz*)
Red	7800 - 6200	3.84 - 4.82
Orange	6220 - 5970	4.82 - 5.03
Yellow	5970 - 5770	5.03 - 5.20
Green	5770 - 4920	5.20 - 6.10
Blue	4920 - 4550	6.10 - 6.59
Violet	4550 - 3990	6.59 - 7.69

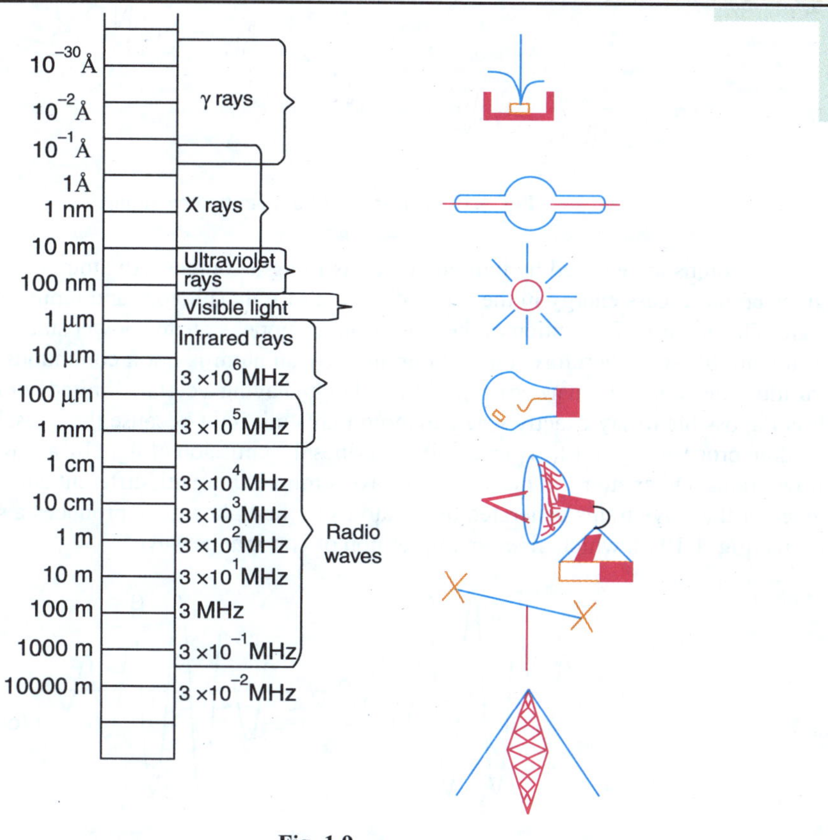

Fig. 1.9

1 angstrom (Å) $= 10^{-10}$ m = 0.1 nm
1 nanometre (nm) $= 10^{-9}$ m = 10 Å
1 micrometre (μm) $= 10^{-6}$ m = 10,000 Å = 1000 nm

1.11. PHOTONS

According to quantum theory, light is considered as a stream of special particles, namely **photons**. Photons do not have rest mass and travel with a velocity equal to the speed of light in a vacuum. The fundamental characteristics of a photon are its energy E and momentum p.

$$E = h\nu \qquad \text{... (1.8)}$$

and
$$p = \frac{h\nu}{c} = \hbar k \qquad \text{... (1.9)}$$

The momentum vector p of a photon has a direction coinciding with that of the wave vector *k*. *k* has a magnitude 2 π/λ and a direction coinciding with that of wave velocity. Photon mass $m = \frac{h\nu}{c^2}$ is the mass of the electromagnetic field and is not associated with a rest mass because photons at rest do not exist.

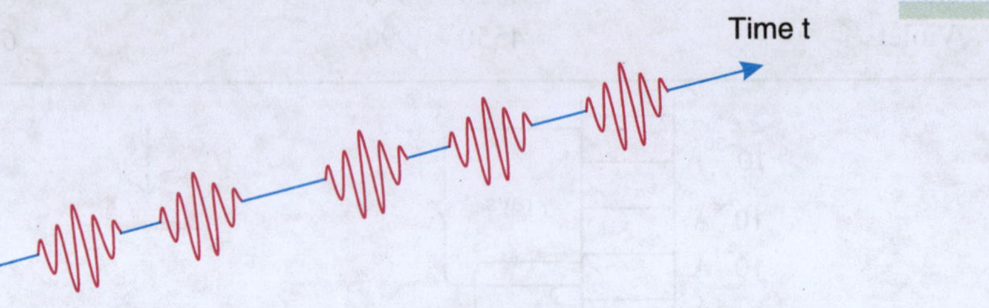

Fig. 1.10 (a) Bursts of light emitted by atoms.

Photons are emitted by individual atoms in light sources. An atom leaving an excited state gives up the excess energy in the form of a burst of light (photon) and jumps to the lower normal state. The process of transition of the atom from an upper state to a lower state lasts for a brief time of about 10^{-8}sec. Therefore, the light emitted by an atom is not a continuous harmonic wave of infinite extension but is a *wave train* of finite length having a certain limited number of oscillations. It is impossible to say exactly when an atom may emit light because the emission is completely a random process. Fig. 1.10(a) pictorially describes the emission of light by a single atom in terms of wave trains. Other atoms in the source behave similarly but with different emission times. Adding together the wave trains generated by all atoms in the light source produces a succession of wave trains (Fig. 1.10 b), which gives an impression of continuous wave.

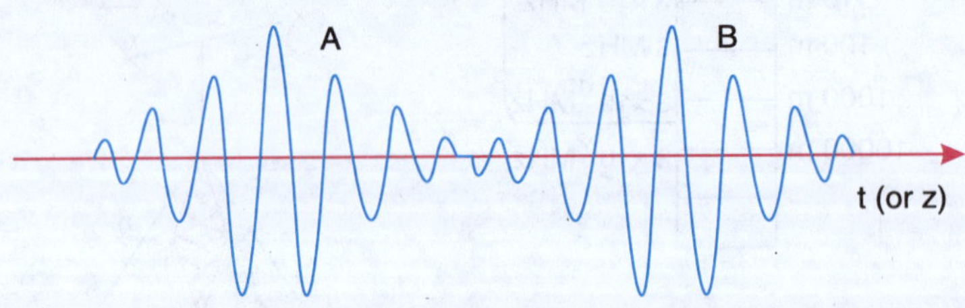

Fig. 1.10 (b) A succession of wave trains gives an impression of continuous wave.

1.12. THE DUAL NATURE

On one hand light appears to be a continuous electromagnetic wave of frequency ν and on the other hand it appears to be a collection of photons having energy E and momentum p. It has been found that neither of the models can separately explain *all* the experimental facts. A particle is precisely localized in space whereas a continuous wave cannot be attributed to a particular location in space. Thus, the corpuscular nature and wave nature appear to be mutually exclusive. However, the experimental evidence points out that light behaves both as a continuous wave and a particle. Therefore, we say that light exhibits **dual nature**. The reason for the wave-particle dualism may be understood as follows. At the lower frequency end of the electromagnetic spectrum are radio waves whose wavelengths are so large that they spread over a very large volume of space (see Fig. 1.11). Therefore, the energy available at any point is insignificantly small and their particle nature cannot be observed. On the higher frequency side of the spectrum, we have x-rays and γ-rays.Their wavelengths are so short that the wave energy is concentrated in a point of very small dimension and the particle nature is readily experienced whereas the wave nature is less noticeable. The visible region represents the transition region in which both the aspects of light can be observed.

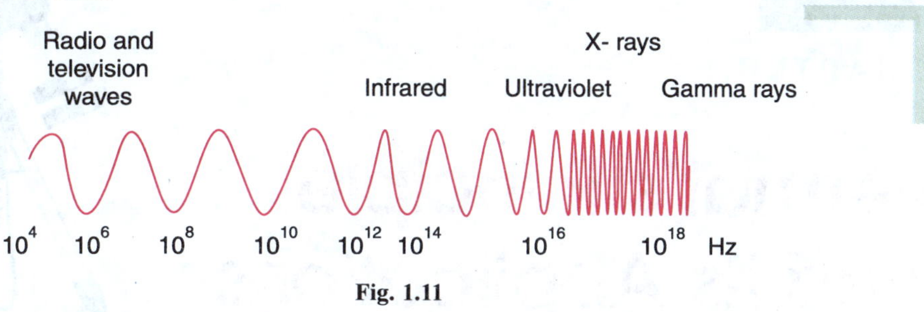

Fig. 1.11

In photon description the light intensity is given by $I = N\,h\nu$ and in the wave description it is given by $I = |E|^2$. It implies that the square of the amplitude of a light wave at a point in space is proportional to the number of photons arriving at that point. In other words the amplitude of a light wave determines the probability that a photon can be found at a particular point in space. Probability to observe photons is thus proportional to $|E|^2$. Thus, the ultimate connection between the wave behaviour and the particle behaviour are interpreted.

To sum up, light *rays* consist of streams of *photons*, which are basically *electromagnetic wave trains*. We use all the three descriptions namely rays, waves and photons appropriately to describe the behaviour of light. When light is refracted or reflected by optical elements such as lenses, we mostly use the ray description; when light propagates through space or any medium we use the electromagnetic wave description and whenever light interacts with matter we make use of the photon description.

QUESTIONS

1. What is meant by reflection?
2. State and explain the law of reflection.
3. What do you mean by refraction of light?
4. What is absolute refractive index of a medium?
5. What is Snell's law?
6. What is meant by optical path? How is it different from geometrical path length?
7. What are the four important theories of light?
8. Explain briefly the corpuscular theory and its limitations?
9. Describe the Huygens' wave theory?
10. Why light is classified in the category of electromagnetic waves?
11. Explain what the visible region means?
12. What is photon theory of light?
13. Explain the dual nature of light?

CHAPTER

Fermat's Principle and its Applications

2.1. INTRODUCTION

Fermat's principle is a remarkable and important principle in optics, which explains why light propagates along straight lines and the laws of reflection and refraction. It forms the basis of geometrical optics. It is not so much a computational device as it is a concise way of thinking about the propagation of light. It is a statement about the grand scheme of things without any concern for the contributing mechanisms.

Refraction at the water surface gives the broken pencil effect.

At a Glance

- Introduction
- Fermat's Principle of Least Time
- Rectilinear Propagation of Light
- Reversibility of Light Rays
- Laws of Reflection
- Laws of Refraction
- Parabolic Mirror
- Elliptical Mirror - Optical Path Stationary
- Law of Refraction at a Spherical Refracting Surface
- The Thin Lens Formula

2.2. FERMAT'S PRINCIPLE OF LEAST TIME

Hero of Alexandria postulated that reflected light travels from one point to another along the shortest path. It is clear from Fig. 2.1 that ABC, which complies with the laws of reflection, is shorter than any other imaginable path from the point A to C, for example the path ADC. The length of ABC equals the length of the line AC_1, whereas the length of ADC actually equals the length of the broken line ADC_1 (C_1 is the mirror image of the point C).

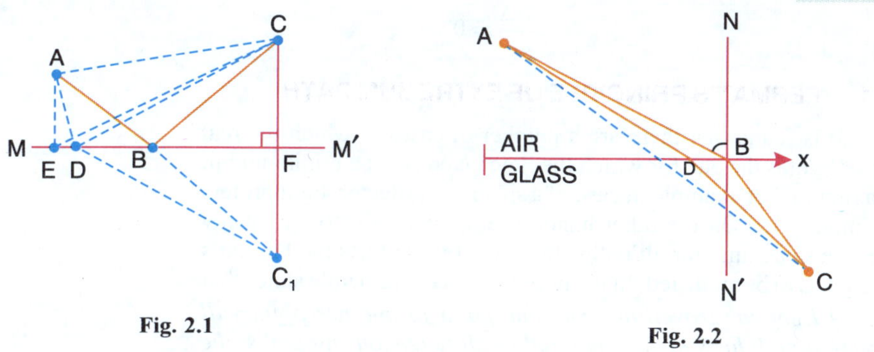

Fig. 2.1

Fig. 2.2

It is obvious from Fig. 2.2 that the refraction of light does not obey the principle of the shortest path. The paths ABC, ADC etc. are longer than the path AC. Taking this fact into consideration, Fermat suggested in 1650 that the *principle of shortest path* be replaced with the *principle of least time*. The principle of least time states that

When a light ray travels between two points P and Q, it follows, out of all possible paths from P to Q, a path, which requires the least time.

Suppose P and Q are two points (Fig. 2.3) in the same medium. Light needs the time dt to travel the distance dL in the medium.

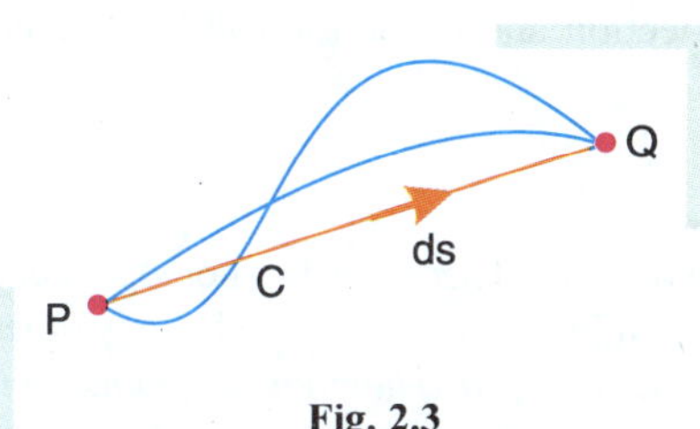

Fig. 2.3

$$dt = \frac{dL}{\upsilon} \tag{2.1}$$

The time required to travel the total distance PQ is

$$t = \int_P^Q \frac{dL}{\upsilon} \tag{2.2}$$

As $\upsilon = \frac{c}{\mu}$, we can rewrite equ.(2.2) as

$$t = \frac{1}{c}\int_P^Q \mu \, dL \tag{2.3}$$

The quantity $\int_P^Q \mu \, dL$ is the ***optical path length***, Δ.

If we consider different paths from P to Q, then Fermat's principle states that

$$\frac{dt}{ds} = 0 \tag{2.4}$$

Pierre de Fermat
(1601 – 1665)

where ds is a parameter that expresses the difference between any two given paths under comparison.

Equ. (2.3) may be rewritten as

$$t = \frac{\Delta}{c} \tag{2.5}$$

According to equ. (2.5), the time of traverse is proportional to the optical path length, Δ. We therefore restate the Fermat's principle as follows:

Light travels along a path having the minimum optical path length.

Thus, the condition may now be expressed as

$$\frac{d\Delta}{ds} = 0 \tag{2.6}$$

2.2.1. FERMAT'S PRINCIPLE OF EXTREMUM PATH

It is found that there are a number of cases in which the real path of light is the one for which the time taken is not a minimum but a maximum. For example in case of a spherical reflector, light prefers maximum time. On the other hand, in elliptical reflectors light ray takes the same time for all paths. In view of these facts the Fermat's principle is to be modified. In its modern form the principle states that

A light ray travelling from one point to another point will traverse a path for which, compared to all neighbouring paths, the time required is a minimum or a maximum or stationary.

Fig. 2.4

This is known as *"Fermat's principle of extremum path"* or *"Fermat's principle of stationary time"*.

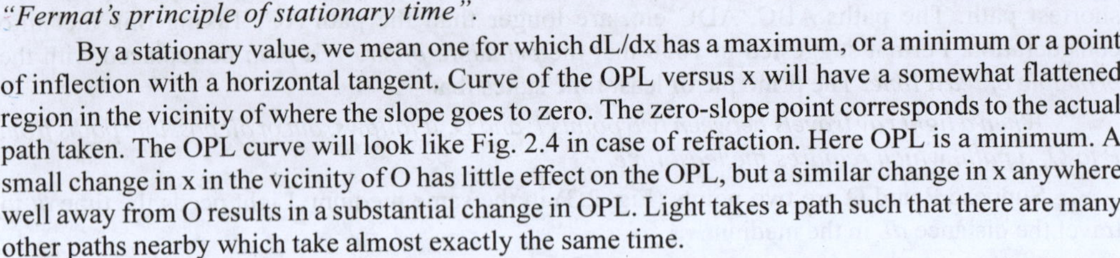

By a stationary value, we mean one for which dL/dx has a maximum, or a minimum or a point of inflection with a horizontal tangent. Curve of the OPL versus x will have a somewhat flattened region in the vicinity of where the slope goes to zero. The zero-slope point corresponds to the actual path taken. The OPL curve will look like Fig. 2.4 in case of refraction. Here OPL is a minimum. A small change in x in the vicinity of O has little effect on the OPL, but a similar change in x anywhere well away from O results in a substantial change in OPL. Light takes a path such that there are many other paths nearby which take almost exactly the same time.

The fundamental laws of rectilinear propagation, reflection and refraction can be derived from Fermat's principle.

2.3. RECTILINEAR PROPAGATION OF LIGHT

A straight line is the path of shortest distance between two points. Therefore, in a homogeneous medium the time taken by a ray of light along a straight line is a minimum as compared to any other path. It is obvious that *the paths are straight lines, when light rays travel in a homogeneous medium.*

2.4. REVERSIBILITY OF LIGHT RAYS

The optical path that is a minimum when light ray travels from point P to point Q is also a minimum when the ray travels in the opposite direction. Therefore, a ray that has traveled from P to Q will cover the same path in the opposite direction from Q to P. Thus, the reversibility of light rays follows from Fermat's principle.

2.5. LAWS OF REFLECTION

We wish to find the path that a light ray takes in going from A to C after first striking the mirror MN such that the time taken is a minimum.

One way would be to go to the mirror quickly and then to go to C along the path ADC. Though the path AD is short, the path DC is longer. In order to find the path, for which the time would be shortest, a geometrical trick is used. Point I is located behind the mirror, such that CF = IF. In the Δ^{le} CDI, $\angle CFD = 90^{o}$, CF = FI and DC = DI. Therefore, the sum of the distances AD + DC,

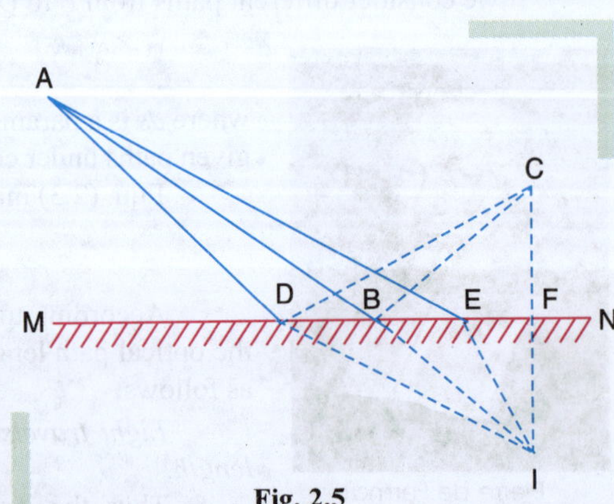

Fig. 2.5

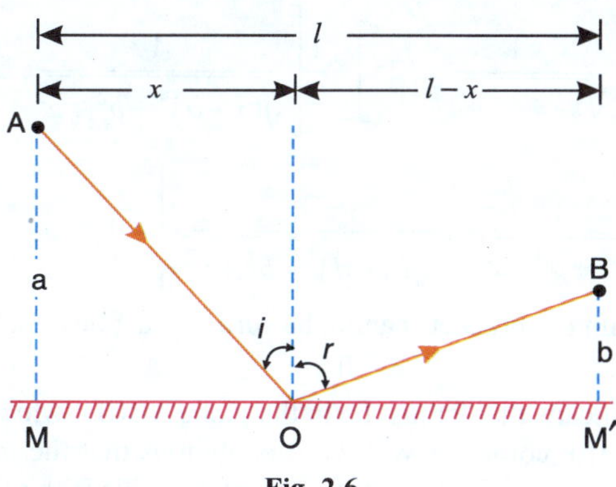

Fig. 2.6

which is proportional to the time taken for light, is also the sum of the two lengths AD + DI. Then the problem consists in finding the path that a ray from A takes in going to I in the shortest time.

Is it path ADI? As AD and ID are two sides of the triangle AID, their sum is greater than the length of the third side AI. So the answer is no. Perhaps it is the path AEI? The answer is again no for exactly the same reason: AE + EI > AI. Proceeding in this way we arrive at the only possible answer- the ray must travel along path ABI because this is the shortest route. This means that in going from A to C, the ray is reflected at B.

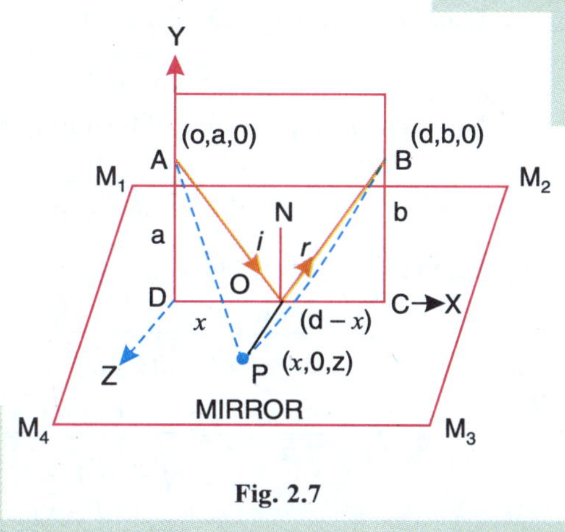

Fig. 2.7

Let us suppose that a ray from A strike the mirror at O before being reflected to B, as in Fig. 2.6. The basic method is, first, to obtain an expression for the path length L traveled by the ray from A to B, in terms of *x, l, a and b*. Secondly, minimize L by altering the position of point O. Thirdly, substitute the angles *i* and *r* into the resulting expression. The minimum path length may now be found by differentiating L with respect to x and equating the resulting expression to zero.

First law of reflection:

Consider a plane mirror. A and B are two points above the mirror plane and are located in a plane ABCD normal to the mirror plane. Light coming from point A is reflected toward B (Fig. 2.7). Suppose the light ray passes through a point P. It means that the light is incident along AP and is reflected along PB. Thus, APB is a most general conceivable path from A to B.

Draw a plane ABCD normal to the mirror plane $M_1M_2M_3M_4$. Take the origin of coordinates at D. Let DC and DA be the x and y axes. Let DA = a, CB = b and DC = d. The point P has general coordinates (x,o,z). If now AP + PB = L, we get

$$L = \left[(x-0)^2 + (0-a)^2 + (z-0)^2\right]^{1/2} + [(x-d)^2 + (0-b)^2 + (z-0)^2]^{1/2}$$

or $$L = \sqrt{x^2 + a^2 + z^2} + \sqrt{(x-d)^2 + b^2 + z^2} \tag{2.7}$$

We now apply Fermat's principle to get the actual path. The path APB can be varied by varying x and z. We obtain the minimum value of L, that is the shortest optical path by taking the derivative of L with respect to *z* and setting the derivative equal to zero. Thus,

$$\left(\frac{\partial L}{\partial z}\right)_x = \frac{1}{2}\left[\frac{1}{\sqrt{x^2+a^2+z^2}}\cdot 2z\right] + \frac{1}{2}\left[\frac{1}{\sqrt{(x-d)^2+b^2+z^2}}\cdot 2z\right] = 0$$

$$\therefore \qquad z\left[\frac{1}{\sqrt{x^2+a^2+z^2}} + \frac{1}{\sqrt{(x-d)^2+b^2+z^2}}\right] = 0$$

As the factor within the brackets cannot be zero, z must be equal to zero.

$$\therefore \qquad z = 0$$

The above result means that P must lie in the plane ABCD, which is normal to the mirror. O is such a position for P. If P coincides with O, it is obvious that the incident ray AO, the surface normal ON and the reflected ray OB lie in the same plane. This is the first law of reflection.

Second law of reflection :

Using z = 0 into equ. (2.7), we get

$$L = \sqrt{x^2+a^2} + \sqrt{(x-d)^2+b^2} \qquad (2.8)$$

Now taking the derivative of L with respect to x, we get

$$\frac{dL}{dx} = \frac{1}{2}\frac{1}{\sqrt{x^2+a^2}}\cdot 2x + \frac{1}{2}\frac{1}{\sqrt{(x-d)^2+b^2}}\cdot 2(x-d)$$

Setting $\frac{dL}{dx} = 0$, we obtain

$$\frac{x}{\sqrt{x^2+a^2}} + \frac{(x-d)}{\sqrt{(x-d)^2+b^2}} = 0$$

$$\therefore \qquad \frac{x}{\sqrt{x^2+a^2}} = \frac{-(x-d)}{\sqrt{(x-d)^2+b^2}}$$

or

$$\frac{x}{\sqrt{x^2+a^2}} = \frac{(d-x)}{\sqrt{(x-d)^2+b^2}}$$

Looking at triangles AOD and BOC

$$\sin i = \frac{x}{\sqrt{x^2+a^2}} \text{ and } \sin r = \frac{(d-x)}{\sqrt{(x-d)^2+b^2}}$$

$$\therefore \qquad \sin i = \sin r$$

$$\therefore \qquad i = r \qquad (2.9)$$

This is the second law of reflection.

2.6. LAWS OF REFRACTION

We now apply Fermat's principle for refraction phenomenon. We have a boundary PQ between two media. Light goes from A to C. If the speed of light on both sides of the boundary were the same, the path from A to C would be a straight line. The speed above PQ and that below PQ are different and we assume that the medium above PQ is air and below PQ is water.

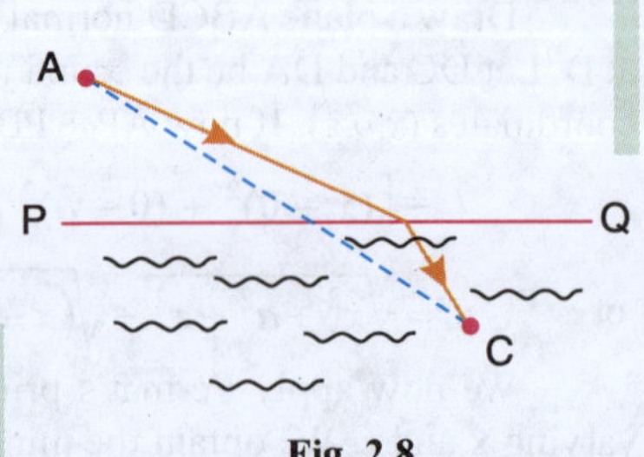

Fig. 2.8

In his book, Feynman has given an apt analogy which

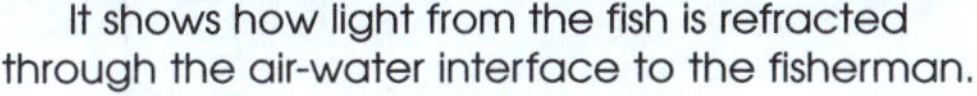
It shows how light from the fish is refracted through the air-water interface to the fisherman.

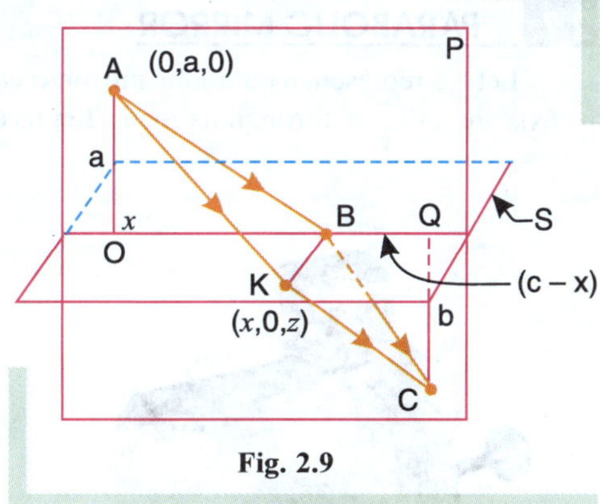

Fig. 2.9

we cite here. "To illustrate that the best thing to do is not just to go in a straight line, let us imagine that a beautiful girl has fallen out of a boat, and she is screaming for help in the water at point C. We are at point A on land and we see the accident, and we can run and can also swim. But we can run faster than we can swim. What do we do? Do we go in a straight line? By using a little intelligence we would realize that it would be advantageous to travel a greater distance on land in order to decrease the distance in the water, because we go so much slower in the water."

Consider a plane surface S separating two media. Let A and C be two points lying in the two different media. We must find the path from A to C, which can be covered by light faster than it can cover any other hypothetical path. Clearly, the path must consist of two straight lines namely AB in medium 1 and BC in medium 2; the point B in the plane S has to be found.

Let K be any general point on the surface S. It has the coordinates (x,0,z). The optical path through the point K (x,0,z) is given by

$$\Delta = (AK)\,\mu_1 + (KC)\mu_2$$

$$= \mu_1\left(x^2 + a^2 + z^2\right)^{1/2} + \mu_2\left[(x-c)^2 + b^2 + z^2\right]^{1/2} \tag{2.10}$$

Now apply Fermat's condition to get the actual path.

$$\left(\frac{\partial \Delta}{\partial z}\right)_x = 0$$

$$\therefore \qquad z = 0$$

It means that the incident ray, the refracted ray and the normal to the plane PQRS lie in the same plane. This is the first law of refraction.

Applying the second condition that $\left(\frac{\partial \Delta}{\partial z}\right)_x = 0$, we get

$$\left(\frac{\partial \Delta}{\partial z}\right)_x = \frac{\mu_1}{2}\left[\frac{2x}{\sqrt{x^2 + a^2 + z^2}}\right] + \frac{\mu_2}{2}\left[\frac{2(x-c)}{\sqrt{(x-c)^2 + b^2 + z^2}}\right] = 0$$

$$\frac{\mu_1 x}{\sqrt{x^2 + a^2 + z^2}} = \frac{\mu_2\,(c-x)}{\sqrt{(x-c)^2 + b^2 + z^2}}$$

$$\mu_1 \sin i = \mu_2 \sin r$$

$$\therefore \qquad \frac{\sin i}{\sin r} = \frac{\mu_2}{\mu_1} \tag{2.11}$$

This is *Snell's law*.

2.7. PARABOLIC MIRROR

Let RS represent a parabolic mirror. It can be shown that rays incident on the mirror parallel to its axis are reflected through its focus. Let us consider the ray path be PQ′F, which is not parallel to

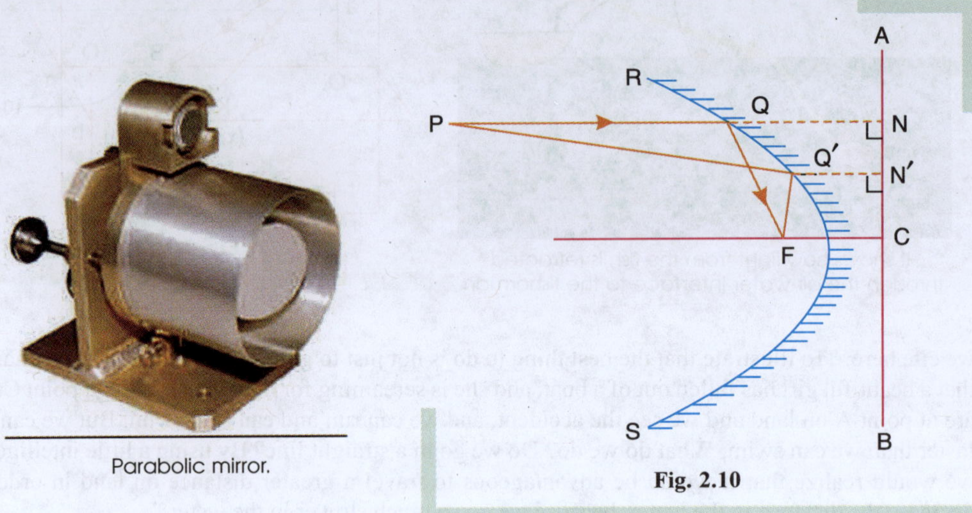

Fig. 2.10

the axis of the mirror. According to Fermat's principle the ray path will correspond to a minimum. From the point Q′ we drop a perpendicular Q′N′ on the directrix AB. From the definition of parabola it follows that Q′N′ = Q′F.

Thus, $PQ' + Q'F = PQ' + Q'N'$

Let N be the foot of the perpendicular drawn from point P on AB. Then for $PQ' + Q'N'$ to be a minimum, the point Q′ should lie on the straight line PN. The actual ray will be PQ + QF where PQ is parallel to the axis. Therefore, all rays parallel to the axis will pass through F and conversely all rays emanating from the point F will become parallel to the axis after reflection.

2.8. ELLIPTICAL MIRROR - OPTICAL PATH STATIONARY

Let us consider reflection from the inner surface of an ellipsoid of revolution. F_1 and F_2 are the foci of the ellipsoid. According to the definition of an ellipse, the paths F_1OF_2, $F_1O'F_2$, $F_1O''F_2$

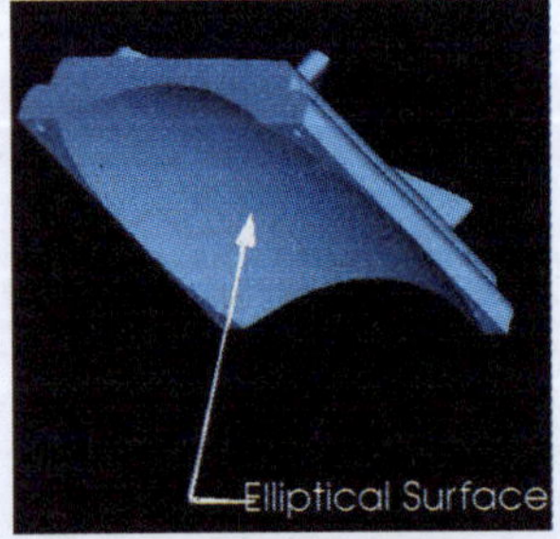

Elliptical mirror.

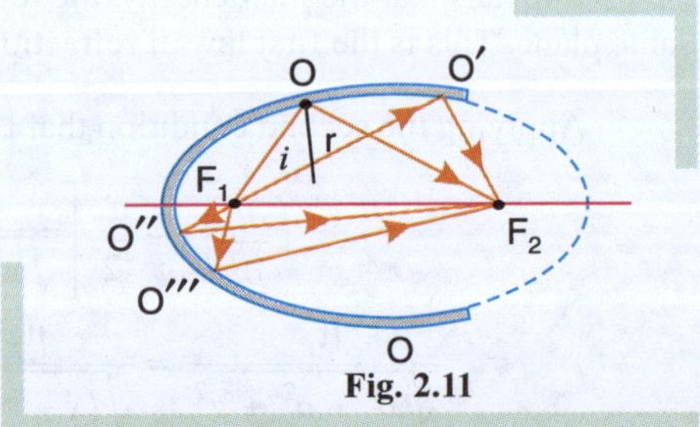

Fig. 2.11

etc. are identical in length. Hence all the rays leaving focus F_1 and arriving after reflection at focus F_2 take same time. In this case the optical path is **stationary**. It is also a geometrical property of the ellipse that $i = r$ for any location of O. All optical paths from F_1 to F_2 via a reflection are therefore precisely equal. None is a maximum. The optical path length is clearly stationary with respect to variations.

2.8.1. OPTICAL PATH MINIMUM

If we replace the surface of the ellipsoid with surface MN having a smaller curvature and oriented so that a ray leaving point F_1 arrives at point F_2 after being reflected from MN, then the path F_1OF_2 will be a minimum.

Let MN be a curved mirror. Let F_1OF_2 be the actual path of a ray from F_1 reaching F_2 after reflection at this mirror. Let M′ be a point on the mirror near O, so that $F_1M'F_2$ be a neighbouring path. Let F_1M' intercept the ellipse at M″.

$$F_1O + OF_2 = F_1M'' + M''F_2 = \text{constant}$$

(Property of the ellipse)

Fig. 2.12

In the neighbouring path $F_1M'F_2$ is given by

$$F_1M' = F_1M'' + M''M'$$

The difference between the actual path F_1OF_2 and the neighbouring path $F_1M'F_2$ is Δ.

$$\begin{aligned}\Delta &= F_1OF_2 - F_1M'F_2\\ &= (F_1O + OF_2) - (F_1M' + M'F_2)\\ &= F_1M'' + M''F_2 - F_1M'' - M''M' - M'F_2\\ &= M''F_2 - (M''M' + M'F_2)\end{aligned} \qquad (2.12a)$$

This quantity is always negative. In triangle $M'F_2M''$

$$M''M' + M'F_2 > M''F_2$$

∴ The actual path is the **minimum** (shortest) among the neighbouring paths.

2.8.2. OPTICAL PATH MAXIMUM

In case of the surface NN′ having a curvature greater than that of ellipsoid, path F_1OF_2 will be a maximum. Now, we have to extend F_1M' to meet the ellipse at M″. Now $F_1O + OF_2 = F_1M'' + M''F_2$

The difference between the actual path F_1OF_2 and the neighbouring path $F_1M'F_2$ is Δ.

$$\begin{aligned}\Delta &= F_1OF_2 - F_1M'F_2\\ &= (F_1O + OF_2) - (F_1M' + M'F_2)\\ &= F_1M'' + M''F_2 - (F_1M'' - M''M') + M'F_2\\ &= M''F_2 - (M'F_2 - M'M'')\end{aligned} \qquad (2.12b)$$

which is always positive. The actual path is the **maximum** (longest) among the neighbouring paths.

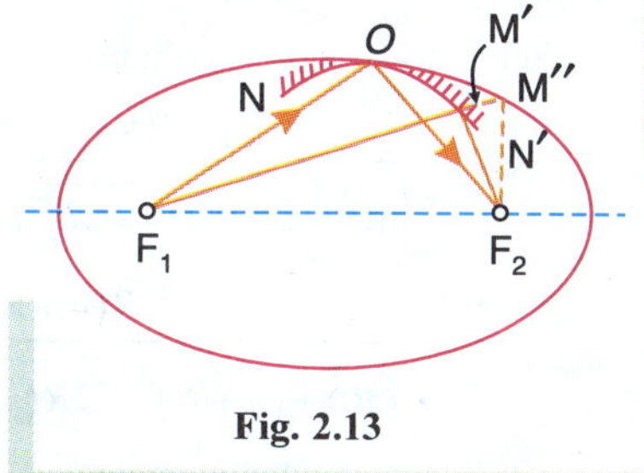

Fig. 2.13

Obviously, the optical path is stationary among the neighbouring paths, if the reflecting surface coincides with the locus surface $F_1O + OF_2 = \text{constant}$.

2.9. LAW OF REFRACTION AT A SPHERICAL REFRACTING SURFACE

Fermat's principle can be applied to a spherical refracting surface and the formula can be derived without resorting to the law of refraction.

Fig. 2.14 shows a spherical refracting surface of radius of curvature R. Let a ray of light OP be incident on the spherical surface. The ray refracts at the surface and reaches the point I. Let OM

= u and MI = υ and MC = PC = R. According to Fermat's principle, the optical path length connecting points O and I must be a minimum in comparison with all neighbouring paths of the same general character. If all rays from O are to reach I, it follows that the optical path length must be the same for each ray.

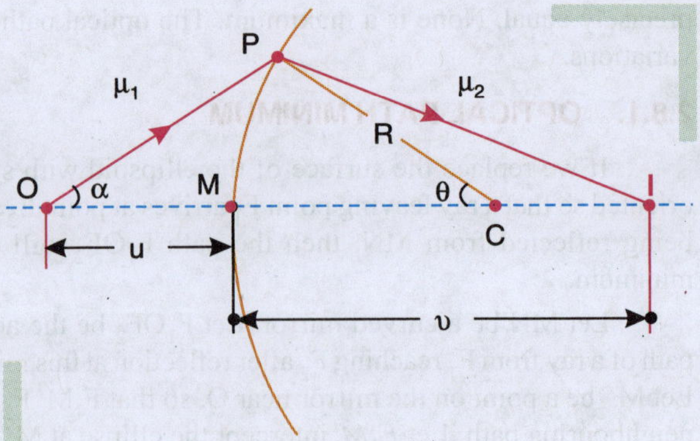

The axial ray OPI that connects the object point O with the image point I has the optical length

$$L = \mu_1 OP + \mu_2 PI \qquad (2.13)$$

where μ_1 is the refractive index of the medium 1 and μ_2 is the refractive of index of the medium 2.

Fig. 2.14

From ΔOPC, it is seen that

$$OP^2 = OC^2 + CP^2 - 2\, OC{\cdot}CP \cos\theta$$

$$\therefore \quad OP = [OC^2 + CP^2 - 2\, OC{\cdot}CP \cos\theta]^{1/2}$$

But $\quad OC = (OM+MC)$

$$\therefore \quad OP = \left[(u+R)^2 + R^2 - 2(u+R)R\cos\theta\right]^{1/2} \qquad (2.14)$$

From ΔPIC, we have

$$PI = \left[IC^2 + CP^2 + 2CI \cdot PC \cos\theta\right]^{1/2}$$

But IC = $(\upsilon - R)$

$$\therefore \quad PI = \left[(\upsilon\text{-R})^2 + R^2 + 2(\upsilon - R)R\cos\theta\right]^{1/2} \qquad (2.15)$$

The total optical path

$$L = \mu_1\left[(u+R)^2 + R^2 - 2(u+R)R\cos\theta\right]^{1/2} + \mu_2\left[(\upsilon\text{-R})^2 + R^2 + 2(\upsilon - R)R\cos\theta\right]^{1/2} \qquad (2.16)$$

Setting $\frac{dL}{d\theta} = 0$, we obtain

$$\frac{\mu_1 R(u+R)\sin\theta}{\left[R^2 + (u+R)^2 - 2R(u+R)\cos\theta\right]^{1/2}} = \frac{\mu_2 R(\upsilon - R)\sin\theta}{\left[R^2 + (\upsilon - R)^2 + 2R(\upsilon - R)\cos\theta\right]^{1/2}}$$

$$\therefore \quad \frac{\mu_1 R(u+R)}{\left[R^2 + (u+R)^2 - 2R(u+R)\cos\theta\right]^{1/2}} = \frac{\mu_2 R(\upsilon - R)}{\left[R^2 + (\upsilon - R)^2 + 2R(\upsilon - R)\cos\theta\right]^{1/2}}$$

When θ is small, cos θ ≅ 1.

$$\frac{\mu_1 R(u+R)}{\left[R^2 + (u+R)^2 - 2R(u+R)\right]^{1/2}} = \frac{\mu_2 R(\upsilon - R)}{\left[R^2 + (\upsilon - R)^2 + 2R(\upsilon - R)\right]^{1/2}}$$

$$\therefore \quad \frac{\mu_1 R(u+R)}{u} = \frac{\mu_2 R(\upsilon - R)}{\upsilon} \qquad (2.17)$$

By dividing the above equation with R^2, we get

$$\mu_1 \cdot \frac{u+R}{uR} = \mu_2 \cdot \frac{\upsilon - R}{\upsilon R}$$

or
$$\frac{\mu_1}{R} + \frac{\mu_1}{u} = \frac{\mu_2}{R} - \frac{\mu_2}{\upsilon}$$

$$\frac{\mu_1}{u} + \frac{\mu_2}{\upsilon} = \frac{\mu_2 - \mu_1}{R} \tag{2.18}$$

Using sign convention that υ and R are positive and u is negative and taking the medium to the left of the refracting surface as air (μ_1 =1 and μ_2= μ), the above equation reduces to

$$\frac{\mu}{\upsilon} - \frac{1}{u} = \frac{\mu - 1}{R} \tag{2.19}$$

2.10. THE THIN LENS FORMULA

Fermat's principle can be applied to the case of a thin lens also and the thin lens formula can be derived without resorting to the law of refraction.

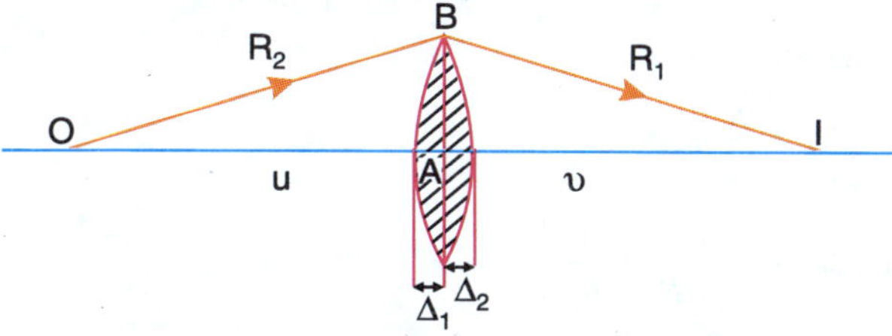

Fig. 2.15

We consider two paths through the lens - one a straight-line path OAI connecting O and I and the other is the one touching the edge B, that is the ray OBI.

The time required to cover OAI is

$$T_1 = \left[u + \mu(\Delta_1 + \Delta_2) + \upsilon\right]/c \tag{2.20}$$

The time required to cover OBI is

$$T_2 = \left[\sqrt{(u+\Delta_1)^2 + h^2} + \sqrt{(\upsilon+\Delta_2)^2 + h^2}\right]/c \tag{2.21}$$

Equating T_1 and T_2, we get

$$u + \mu(\Delta_1 + \Delta_2) + \upsilon = \left[\sqrt{(u+\Delta_1)^2 + h^2} + \sqrt{(\upsilon+\Delta_2)^2 + h^2}\right] \tag{2.22}$$

We now use the paraxial approximation.

$$h << (u+\Delta_1) \quad \text{and} \quad h << (\upsilon + \Delta_2)$$

$$\sqrt{(u+\Delta_1)^2 + h^2} = (u+\Delta_1)\sqrt{1 + \left(\frac{h}{u+\Delta_1}\right)^2}$$

$$= (u+\Delta_1)\left[1 + \tfrac{1}{2}\left(\frac{h}{u+\Delta_1}\right)^2\right] = u + \Delta_1 + \frac{h^2}{2(u+\Delta_1)} \tag{2.23}$$

Similarly, $$\sqrt{(\upsilon+\Delta_2)^2+h^2}=\upsilon+\Delta_2+\frac{h^2}{2(\upsilon+\Delta_2)} \quad (2.24)$$

Equation (2.22) becomes

$$u+\mu(\Delta_1+\Delta_2)+\upsilon=u+\Delta_1+\frac{h^2}{2(u+\Delta_1)}+\upsilon+\Delta_2+\frac{h^2}{2(\upsilon+\Delta_2)}$$

$$(\mu-1)(\Delta_1+\Delta_2)=\frac{h^2}{2}\left[\frac{1}{u+\Delta_1}+\frac{1}{\upsilon+\Delta_2}\right] \quad (2.25)$$

But $\Delta_1 << u$ and $\Delta_2 << \upsilon$

$$\therefore \quad (\mu-1)(\Delta_1+\Delta_2)=\frac{h^2}{2}\left[\frac{1}{u}+\frac{1}{\upsilon}\right] \quad (2.26)$$

We can write $$\Delta_1=R_1-\sqrt{R_1^2-h^2}=R_1-R_1\sqrt{1-\left(\frac{h}{R_1}\right)^2}$$

$$=R_1-R_1\left[1-\frac{1}{2}\left(\frac{h}{R_1}\right)^2\right]=\frac{h^2}{2R_1}$$

Similarly, it can be shown that

$$\Delta_2=\frac{h^2}{2R_2}$$

$$\Delta_1+\Delta_2=\frac{h^2}{2}\left[\frac{1}{R_1}+\frac{1}{R_2}\right] \quad (2.27)$$

$$\therefore \quad \frac{1}{u}+\frac{1}{\upsilon}=(\mu-1)\left[\frac{1}{R_1}+\frac{1}{R_2}\right] \quad (2.28)$$

Using sign convention that u and R_2 are negative and; υ and R_1 are positive, the above equation reduces to

$$\frac{1}{\upsilon}-\frac{1}{u}=(\mu-1)\left[\frac{1}{R_1}-\frac{1}{R_2}\right] \quad (2.29)$$

Equ. (2.29) is the thin lens formula.

QUESTIONS

1. State and explain Fermat's principle of extremum path and use it to deduce the laws of reflection and refraction of light. **(Nagpur, 2004; Awadh, 2001; Garhwal, 2002)**
2. State and explain Fermat's principle of stationary time. Derive the laws of refraction using this principle. Give an example where the path of light is a relative maximum rather than a minimum.
3. State and explain Fermat's principle of extremum path and analyze a case where the actual path of light may be a maximum. Use Fermat's principle to deduce Snell's law of refraction. **(Kovempu, 2005)**
4. "Fermat's law is a law of extremum path, not only minimum path". Analyze a case where the actual path of light may be a minimum.

RAY OPTICS

Ray optics is the simplest theory of light. Here light is described in terms of rays, and we ignore the wave and photon character of the light. We also do not bother about the nature of light, but are only interested to understand how it behaves on a large scale. The basic concept here is that a light ray travels along a well-defined path. The optics of light rays involves only geometric considerations and the laws are formulated using the language of geometry. Therefore, ray optics is also called **geometrical optics**. The ray description is adequate when the wavelengths involved are very small compared with the dimensions of the objects. Ray optics is just an approximation, but it is of great importance technically. It is concerned with the location and direction of light rays. Hence it is useful in studying image formation by mirrors, lenses and the working of many optical instruments and devices. However, many aspects of the behaviour of light cannot be explained on the basis of ray theory; for example we cannot explain the phenomenon of interference, diffraction, and polarization using ray concept. G.Kirchhoff, A.Sommerfeld and others showed that the ray model of propagation light could be derived from the basic electromagnetic theory.

Ray optics is based mainly on the following three simple laws:

THE THREE LAWS

1. Law of rectilinear propagation of light:

It is a common sight that an object kept in the path of light coming from a point source produces a sharp shadow. Similarly, on misty mornings sunlight passing through the dense foliage of trees appears as light shafts. There are many such instances where light appears to exhibit *rectilinear propagation*. The law of rectilinear propagation states that *in an optically homogeneous medium light propagates in a straight line*. A medium is said to be *optically homogeneous* if its refractive index is everywhere the same. This law is approximate and when light passes through very small openings, deviations from a straight line are observed.

2. Law of independence of light rays:

Light rays do not disturb one another when they intersect. If several rays are passing through a medium simultaneously in different directions, then the path of any ray is the same as it would be if all others were absent. The intersection of rays does not hinder the rays from propagating independently of each other. Rays of light always preserve their individuality. The rays pass through each other simultaneously without mutual interaction. For example when we view an object, light rays passing in other directions does not obstruct the light that comes to us from the object.

3. Law of reversibility of path:

If the path of a light ray is reversed, it will exactly retrace its path, irrespective of the number of reflections and refractions.

CHAPTER

Reflection and Refraction

3.1. INTRODUCTION

Optics is regarded by many as a branch of physics, which has out-lived its purpose. This is not true. Since the invention of camera, and subsequently movies and TV, optics has been playing a key role in the entertainment industry. The transmission and storage of images is central to many fields, such as remote sensing, and space exploration. Optics has regained its status in the eyes of public with the advent of optical fibres and their application in communications.

Reflection, refraction, and dispersion are fundamental processes of optics. These can be understood using either ray concept or wave concept. In this and next few chapters we apply the ray concept to explain the principles and working of optical systems and instruments.

3.2. LIGHT RAYS

The conception of energy travelling out from the source along rays can be used to explain the rectilinear propagation of light. We may define a *light ray* as a narrow stream of radiation that travels in a line, never diverging or converging. A light ray is represented by a line drawn in the direction in which the light energy is travelling. We can think of a broad beam of light as a bundle of parallel rays. Ray concept was used to describe light long before its wave nature was firmly established. The ray model is highly useful in studying reflection, refraction and formation of images by mirrors and lenses.

At a Glance

Note — Light rays as such *do not exist*, nor can they be isolated experimentally; they exist only in theory. If we attempt to isolate a ray by means of a series of small holes, all that we can isolate are pencils of light. If the holes are made very small (of a few wavelengths size), light ceases to act as a ray and behaves more like a wave. ◆

3.3. REFLECTION AT PLANE SURFACES (MIRRORS)

Plane mirror.

Any smooth surface acts as a mirror. A mirror may be *plane* or *curved*. Mirrors were usually made in the past, by coating glass with silver. Nowadays, they are made by depositing in vacuum a thin film of aluminum on a polished surface. The reflecting film is protected by deposition of a thin layer of silicon monoxide or magnesium fluoride over it. The plane mirrors used as looking glass are coated at the back so that the reflecting layer is completely protected. The mirrors used for technological purposes are coated on the front surface so that losses in energy due to transmission through the substrate material are reduced.

Mirrors redirect light rays, thereby forming an *image* of some *object*.

3.3.1. OBJECT AND IMAGE

An *object* is anything from which light rays radiate. This light could be emitted by the object itself or reflected from it. We refer normally to two kinds of objects- a *point object*, which has no physical extent, and an *extended or distributed object*, which has a length, width and height.

When light rays proceeding from an object are reflected at a mirror, they *converge to* or *appear to diverge from* a position *different* from that of the original object and give the impression of object being there. The *apparent object* as seen by the observer is called the *image* of the object.

The image formed by an optical component may be real or virtual. The eye sees a virtual image as well as it sees a real image. A *real image* contains light energy; the image can be received and seen on a screen. And also, the real image can be photographed by simply placing a photographic film (or plate) at the position of the image. Real images are *essential* for photography.

On the other hand, a *virtual image* cannot be received on a screen. The light rays never actually pass through the virtual image; hence, the image cannot be received on a screen. For the same reason, it cannot be recorded on a photographic plate placed at its position. Virtual image can be photographed only with the help of a converging optical system, which uses the virtual image as a *virtual object* and forms its real image.

3.3.2. IMAGE OF A POINT OBJECT

Image formation by mirrors involves only the law of reflection.

Fig. 3.1 shows a point object O located at a distance u in front (to the left) of a plane mirror. u is called the *object distance*. The ray OA is incident normally on the mirror and is reflected along its original path. The ray OB makes an angle i and is reflected at an equal angle with the normal. When we extend the two reflected rays backward, they intersect at point I, at a distance υ behind the mirror. To the observer the reflected rays appear to come from I lying behind the mirror. Thus, I is the image of the object point O. The image is *virtual* because no rays actually come from it. We call υ the *image distance*. Let us draw the line OI perpendicular to the mirror. The two triangles OAB and IAB are congruent, so O and I are at equal distances from

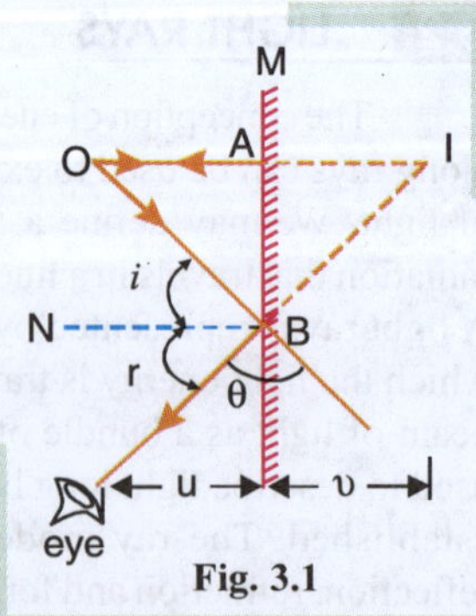

Fig. 3.1

the mirror, and u and v have equal magnitudes. It means that the image point I is located exactly opposite the object point O and is as far behind the mirror as the object point is from the front of the mirror. The object distance u is positive because the object point is on the incoming side of the reflecting surface. The image distance v is negative because the image point I is not on the left side of the mirror. The object and the image distances are related by

$$u = -v \tag{3.1}$$

From Fig. 3.1 it is easy to see that the incident ray is deviated through an angle $\theta = 180^\circ - 2\,i$.

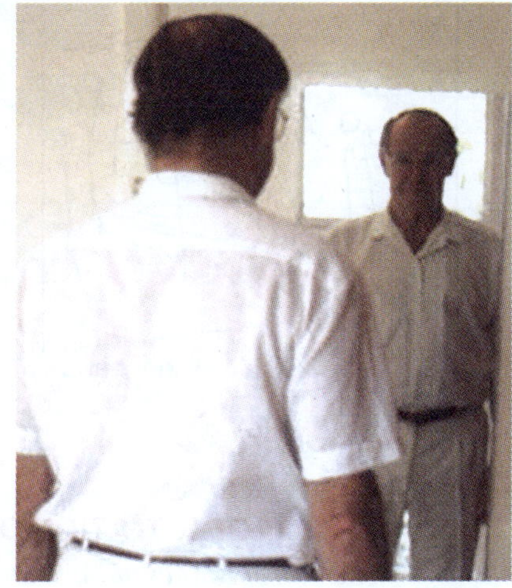

Image formation by mirror.

3.3.3. IMAGE OF AN EXTENDED OBJECT

We now consider an *extended object* with finite size. For sake of simplicity, we consider a linear object such as an arrow with its height oriented parallel to the mirror.

Fig. 3.2 shows an arrow AB of height h. The image formed by such an extended object is an extended image; to each point on the object, there corresponds a point on the image. Two of the rays from B are shown. The ray BD is incident normally on the mirror and is reflected along the path DB. The ray BC, incident on the mirror at an angle i, is reflected at an equal angle along CQ. These two (in fact *all*) rays from B appear to diverge from the image point B′ after reflection. The image of the arrow is the line A′B′, with height h'. Other points of the object AB have image points between A′ and B′. The triangles ABC and A′B′C are congruent, so the object AB and image A′B′ have the same size and orientation, and $h = h'$. Thus, the image is *erect*.

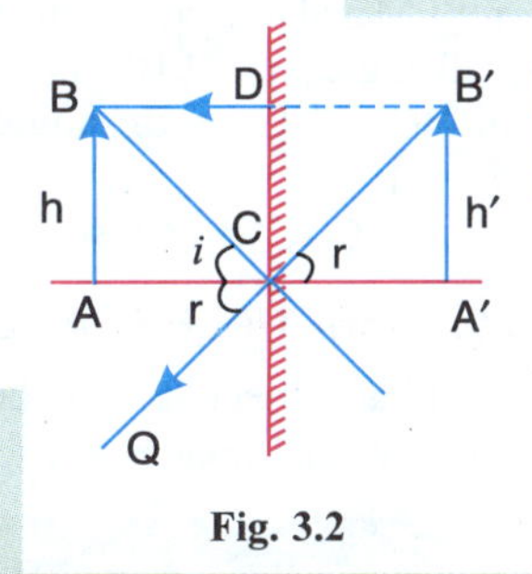

Fig. 3.2

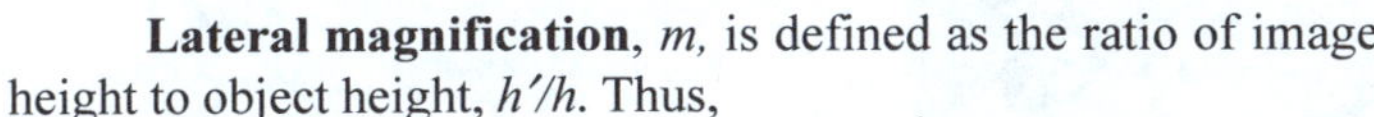

Lateral magnification, m, is defined as the ratio of image height to object height, h'/h. Thus,

$$m = \frac{h'}{h} \tag{3.2}$$

In case of a plane mirror, $m = +1$.

Note — The lateral magnification for a plane mirror is always +1.

3.3.4. IMAGE REVERSAL

When we consider a three-dimensional object, a three-dimensional image is formed by the mirror (See Fig. 3.3). Three fingers of a hand form the object (Fig. 3.3b). The images of two fingers that lie in a plane parallel to the mirror are just like their objects and are not reversed at all. The image of the finger that points toward the mirror is reversed front to back (Observe the arrows associated with the fingers). Looking at Fig. 3.3(a) it is also observed that a mirror reverses right and left. Then, obviously, the question arises in our mind – why does the mirror reverses right and left; and not up and down. Following the paths of rays, it is more accurate to say that a mirror reverses front to back than that it reverses right to left.

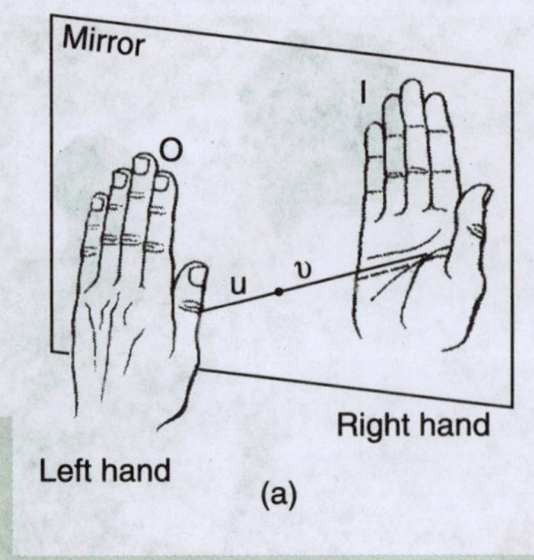

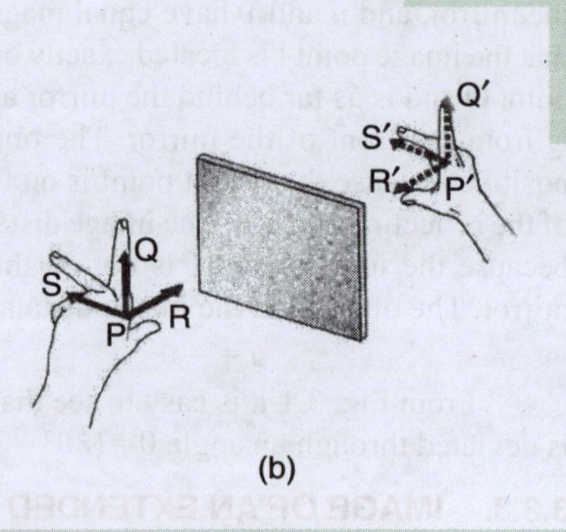

Fig. 3.3

3.3.5. EFFECT OF ROTATION OF MIRROR ON THE REFLECTED RAY

When *a mirror rotates through an angle α, the reflected beam will move through double the angle, that is 2α*.

Fig. 3.4 shows a ray AB being incident at an angle i on a mirror LM and reflected along BC. Now, the mirror is rotated through a small angle α (=∠LBL′) because of which the ray AB is reflected along BC′. ∠CBC′ is the angle through which the reflected ray gets displaced.

$$\angle CBC' = \angle ABC' - \angle ABC = 2(i + \alpha) - 2\,i = 2\alpha.$$

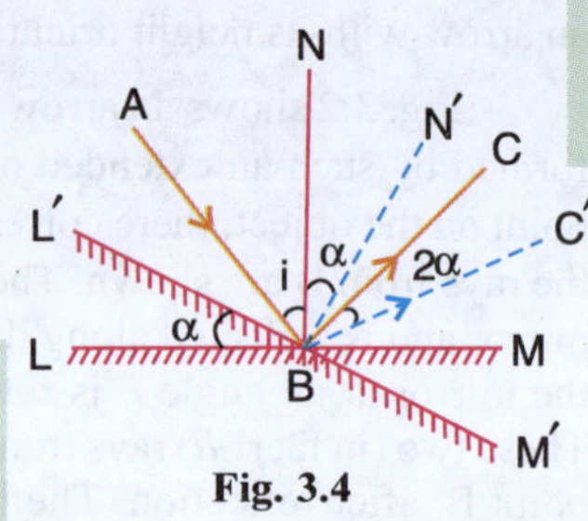

Fig. 3.4

3.3.6. MULTIPLE PLANE MIRRORS

If an object is held between two or more plane mirrors, multiple virtual images are formed because of multiple reflections of light from the mirrors. The number of images formed depends upon the angle between the mirrors. Let the angle between the mirrors be α, then the number of images, N is given by

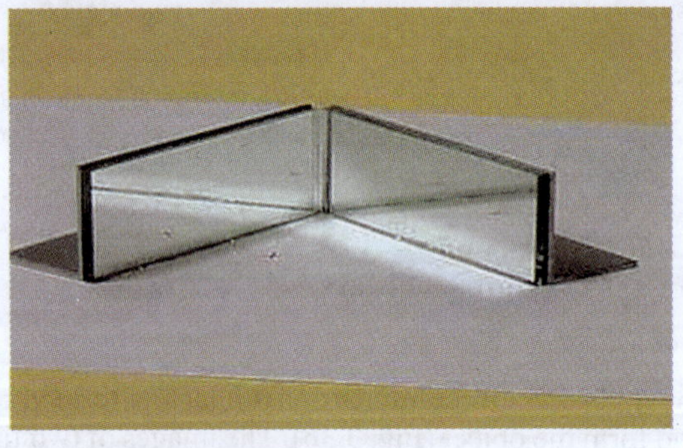

Two plane mirrors touching at one edge.

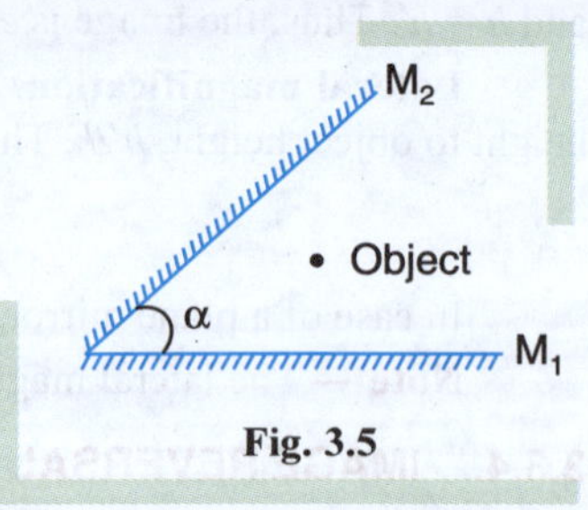

Fig. 3.5

$$N = \frac{360^o}{\alpha} - 1 \qquad (3.3)$$

For a pair of parallel mirrors α =0°, N is infinity and therefore, infinite number of images will be formed.

3.3.7. PROPERTIES OF IMAGES IN PLANE MIRRORS

We now summarize the properties of images in plane mirrors.

(*a*) Image formed by a plane mirror is virtual and erect.
(*b*) The image is as far behind as the object is in front of it.
(*c*) The right side of the object is transformed into left side of the image.
(*d*) Magnification produced by a plane mirror is unity. Hence, the image is as large as the object.
(*e*) When a plane mirror is rotated through a certain angle, the reflected ray turns through twice the angle.
(*f*) If an object is held between two or more plane mirrors, multiple virtual images are formed.

The number of images is given by

$$N = \frac{360^o}{\alpha} - 1$$

where α is the angle between the mirrors.

3.3.8. APPLICATIONS

Plane mirrors are used as choppers, beam deflectors, image rotators and scanners. They are also used to amplify and measure the slight rotations of laboratory apparatus such as galvanometers, torsion pendulums etc.

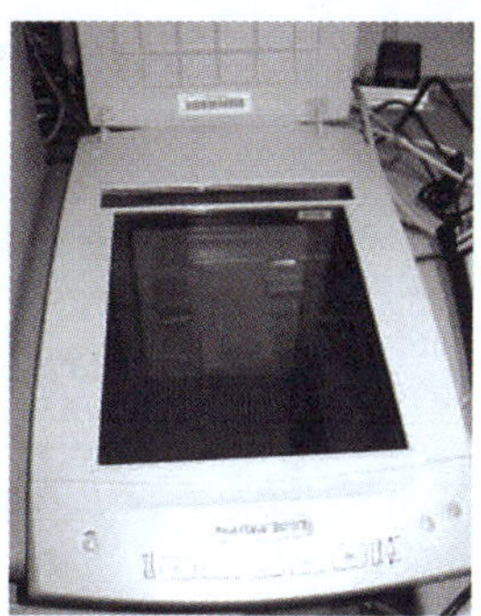

Plane mirror is used in scanner.

Caution—
1. *Light will be reflected from a mirror only if the average depth of the surface irregularities of the reflector is much less than the wavelength of incident light.*
2. *The size of the reflector must be much larger than the wavelength of the incident light; otherwise the light will be scattered in all directions.*

3.4. REFLECTION AT SPHERICAL MIRRORS

Mirrors need not have to be always flat; they may be curved. Spherical mirrors are of the simple type among the curved mirrors. A **spherical mirror** is a segment of a spherical surface and usually has circular edges. There are two types of spherical mirrors- *concave* and *convex* mirrors. When the reflection takes place from the inner surface of the spherical segment, then the mirror is called a **concave** mirror. If light is reflected from the outer bulging surface of the spherical segment, then the mirror is called a **convex** mirror.

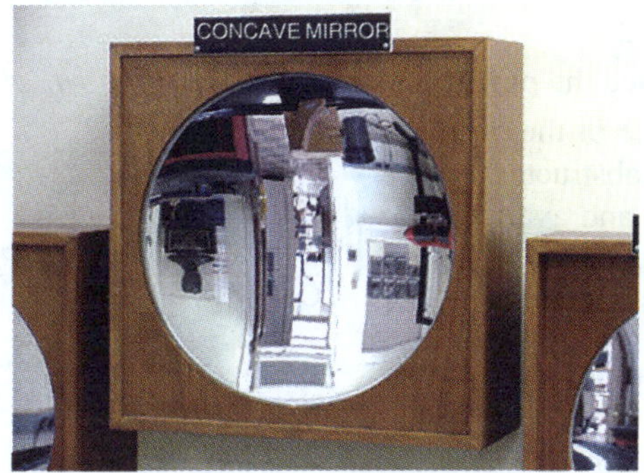

Cancave mirror

Convex mirror

The laws of reflection are equally applicable to the curved mirrors. Using these laws, we can find out the nature and position of the image when the object is situated at different distances from the mirror; we can also determine the relation between the distance of the object and image and the focal length of the mirror. In order to obtain the above information, we have to first acquaint with the basic terms and definitions used in connection with the curved mirrors.

3.4.1. BASIC TERMS

The basic terms associated with spherical mirrors are as follows:

(*i*) The **centre of curvature,** C is the centre of the sphere of which the mirror is a small segment;

(*ii*) The **vertex (or pole),** P is the midpoint of the mirror;

(*iii*) The **radius of curvature,** R is the radius of the sphere of which the mirror is a small section;

(*iv*) The **principal axis** is the line passing through the centre of curvature C and the vertex P. It extends indefinitely in both directions (See Fig. 3.6).

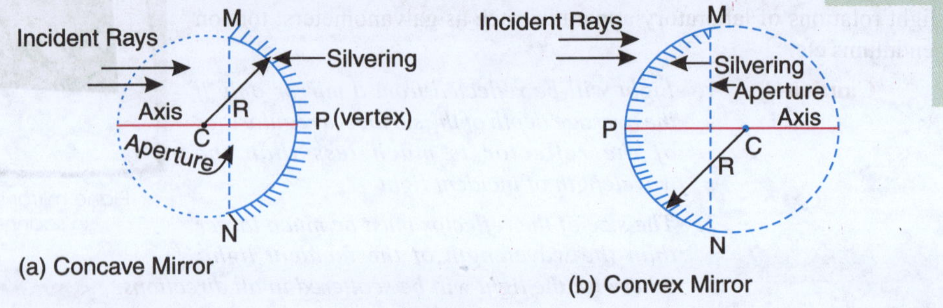

Fig. 3.6

(*v*) With a concave mirror, all rays parallel to the principal axis pass through a single point F after reflection. F is called the **principal focus.** The parallel rays converge to F after reflection (See Fig. 3.9). However, in case of a convex mirror, the reflected rays *appear* to emerge from F behind the mirror. Thus, the reflected rays *appear to diverge* from the principal focus.

(*vi*) **Focal plane** is a plane passing through the point F and perpendicular to the principal axis.

(*vii*) The **focal length,** f of the mirror is the distance, PF, between the vertex P and the principal focus F. It will be shown later that $f = R/2$.

(*viii*) The reciprocal of the focal length is called the **power** of the mirror. Thus, $P = 1/f$.

(*ix*) The diameter MN of the circular outline of the mirror is called the **aperture** of the mirror. The unobstructed surface area of the mirror, having circular rim and available for reflection, represents the aperture (See Fig. 3.6). It is measured in terms of the angle θ, subtended by the extremities of the mirror at the centre of curvature, as shown in Fig. 3.6(c). The aperture determines the amount of light energy that is received by the mirror. In other words, it determines the light gathering capability of the mirror.

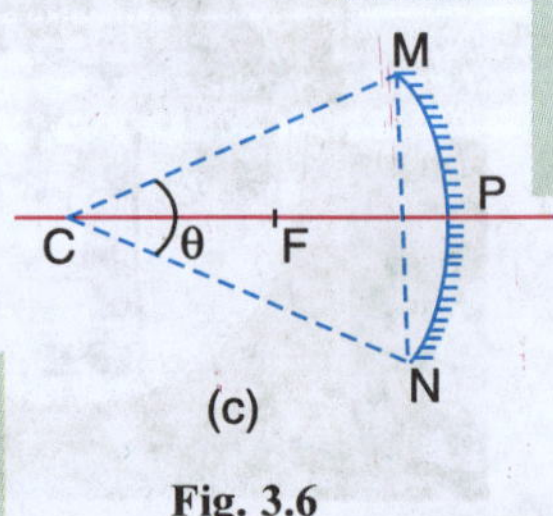

Fig. 3.6

3.4.2. PARAXIAL RAYS AND PARAXIAL APPROXIMATION

A ray is said to be *paraxial* if the angle θ between the ray and the principal axis (or symmetry axis) is small. Only rays inclined at less than about 10° to the principal axis are considered as paraxial rays and they lie close to the axis throughout the distance from object to image. As the angles are small, and when expressed in radians, we can set the cosines equal to unity and the sines and tangents equal to the angles. This is known as the *small angle approximation.* Accordingly, we can set

$\cos\theta \cong 1$ for $\cos\theta$, $\sin\theta \cong \theta$ for $\sin\theta$ and $\tan\theta \cong \theta$ for $\tan\theta$.

This is also known as **paraxial approximation**. As we are approximating sin θ to θ, paraxial approximation is also called as the **first-order theory**. A small area in the immediate surrounding of the optical axis is known as the **paraxial region**. Gauss, in 1841, was the first to give a systematic analysis of the formation of images under paraxial approximation. The results are known as *first-order*, paraxial or **Gaussian optics**. If we restrict the rays to the paraxial rays, a good image is formed with monochromatic light by any optical system. Therefore, we restrict ourselves to paraxial rays in our study of mirrors and lenses.

3.4.3. SIGN CONVENTION

In order to specify the position of the object and the image, we need a reference point and sign convention. There are many sign conventions in use in different books, which cause a lot of confusion in the minds of the students. We adopt the following sign convention known as *Cartesian sign convention* in our study. This is a relatively simpler convention, widely used and easy to remember. We use this convention uniformly in the present and next chapters.

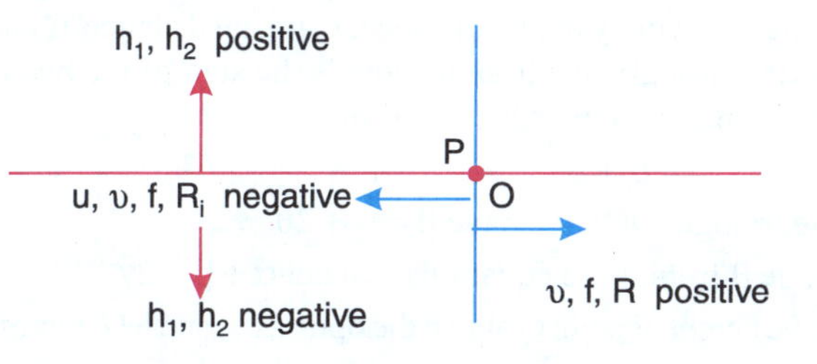

Fig. 3.7

(*a*) The diagrams are drawn with the incident light travelling from left to right.

(*b*) The distances are measured by taking the vertex as the origin.

(*c*) The distances measured in the direction of the incident light are considered positive while those measured in the direction opposite to the incident light are taken as negative. (All quantities measured to the right of P are positive and all those to its left are negative.)

(*d*) Heights measured upward and perpendicular to the principal axis are taken as positive while those measured downward are considered negative.

(*e*) Angles measured clockwise, beginning at the optic axis, are negative. Angles measured counterclockwise are positive.

In order to derive the formula for spherical mirrors (and lenses), we first draw a diagram and treat all the distances and angles as positive, and from the geometry of the diagram deduce an equation relating to the quantities. When the final equation is obtained, we apply the above sign convention and change the signs in front of quantities that are negative in the figure. This procedure gives us the *algebraic equation*, which can be applied to any possible case.

3.4.4. SPHERICAL MIRROR EQUATION

We now consider the image formation by a spherical mirror. Fig. 3.8 shows a concave mirror with radius of curvature R facing the incident light. The *centre of curvature* of the surface is at C and the *vertex* of the mirror is at P. The line CP is the *optic axis*. Point O is an object point that lies on the optic axis and we assume that the distance from O to P is greater than R.

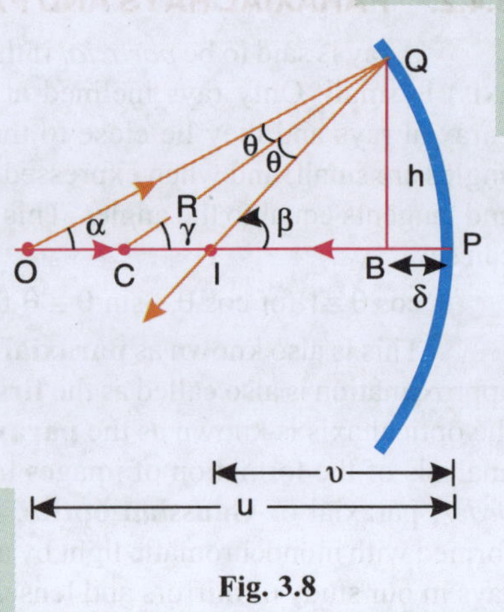

Fig. 3.8

A ray from O, at an angle α with the axis, strikes the mirror at Q and is reflected, the incident and reflected rays making equal angles θ with the normal CQ. A second ray from O, passing through C, strikes the mirror normally and is reflected back on itself. The two reflected rays intersect at I, forming a real image of O. The image is *real* since the light energy passes through the image position.

Spherical mirror.

Let us now determine the location of the image point I. The basic equation governing spherical mirrors relates three lengths, namely object distance, *u*, image distance, υ, and the radius of curvature of the mirror, R. An exterior angle of a triangle equals the sum of the two opposite interior angles. Applying this theorem to the triangle OQC, we have

$$\gamma = \alpha + \theta \tag{3.4}$$

Applying to the triangle OQI, we have $\beta = \alpha + 2\theta$ (3.5)

Eliminating angle θ from these equations, we get $\alpha + \beta = 2\gamma$ (3.6)

Let h represent the height of point Q above the optic axis, and let δ represent the short distance BP.

From $\Delta^{le}OQB$, $\tan\alpha = \dfrac{h}{u - \delta}$

From $\Delta^{le}IQB$, $\tan\beta = \dfrac{h}{\upsilon - \delta}$ and

From $\Delta^{le}CQB$, $\tan\gamma = \dfrac{h}{R - \delta}$

In the paraxial ray approximation, the angles α, β and γ are very small and we can replace *tan α* by *α* and so on. When α is small we can also neglect the distance *x* in comparison to *u*, υ and R.

$$\tan\alpha \cong \alpha = \frac{h}{u} \quad ; \quad \tan\beta \cong \beta = \frac{h}{\upsilon} \quad \text{and} \quad \tan\gamma \cong \gamma = \frac{h}{R}$$

Using the values of α, β and γ into the equation (3.6) and on simplification, we get the *spherical mirror equation*

$$\frac{1}{u} + \frac{1}{\upsilon} = \frac{2}{R} \tag{3.7}$$

The above equation represents the *numerical relationship* between *u*, υ and R. We now assign the signs to these quantities as per the convention mentioned earlier. The pole P is taken as the origin and the line CP as x-axis. The rays are incident from left to right. We take the direction from left to right as the positive x-direction. The points I,C, and O are situated to the left of the origin P in the figure. Therefore, the quantities u, υ and R are negative. Putting the signs in Eq. (3.7), we get

$$\frac{1}{-u}+\frac{1}{-\upsilon}=\frac{2}{-R}$$

or
$$\frac{1}{u}+\frac{1}{\upsilon}=\frac{2}{R} \qquad (3.8)$$

Eq. (3.8) is known as the *spherical mirror equation*. It is the algebraic relationship which incorporated the signs. Though it is derived for a special situation shown in Fig.3.8, it is valid in all other situations with a spherical mirror.

Equation (3.8) is an approximation valid for paraxial rays only. Since the angles α, β, and γ do not appear in this equation, we conclude that, all paraxial rays passing through object point O will go through the same image point I after they are reflected at the surface. Since all such reflected light rays converge on the image point, a concave mirror is also called a **converging mirror**.

Concave mirror is also called a converging mirror.

In an actual case the rays can be made more and more paraxial by putting a circular diaphragm in front of the mirror, centered about the vertex P. As a result, the image will become sharper. But, as the diaphragm *decreases the aperture* of the mirror, the image will become fainter because less total light energy will be reflected from the mirror. A compromise must often be made between image quality and image brightness.

Note— ***(i)*** The initial equation (3.7) is a *numerical relationship* between the various distances for a particular type of spherical mirror. We then adopted a sign convention for the distances, which enabled us derive an *algebraic relationship*. Such an algebraic relation (3.8) obtained with the sign convention adopted by us, will serve for all types of mirrors used in all possible ways. Any sign convention that is different than that adopted by us is likely to give a different algebraic relationship.

(ii) We have to bear in mind that u,υ, and R are algebraic quantities and have a sign, numerical value and a unit. We have to assign the proper sign and value for these parameters while working out the numerical problems.

(iii) For larger values of α, the reflected rays pass through different points of the axis. Eq. (3.7) is to be corrected and a first correction to it gives

$$\frac{1}{u}+\frac{1}{\upsilon}=\frac{2}{R}+\frac{h^2}{R}\left(\frac{1}{R}-\frac{1}{u}\right)^2$$

which shows that υ depends on the distance h and therefore there is no unique point image. ◆

3.4.4.1. Focal Point and Focal Length

If the object O in Fig. 3.8 is taken at a very large distance from the mirror ($u = \infty$), the incoming rays become almost parallel. Then, the image distance υ is given by

$$\frac{1}{\infty}+\frac{1}{\upsilon}=\frac{2}{R}. \text{ Therefore, } \upsilon=\frac{R}{2}.$$

The incident parallel rays converge, after reflection from the mirror, to a common point F at a distance R/2 from the vertex (see Fig. 3.9a). The point F at which the incident parallel rays converge is called **focal point**. The distance from the vertex to the focal point is called **focal length** and is denoted by f. *Focal length* is defined as the image distance when the object is at infinity. Thus, $\upsilon = f$ and

$$f=\frac{R}{2} \tag{3.9}$$

Thus, the focal length of a concave spherical mirror is half the radius of curvature.

Now let us look at the opposite situation. If the object is placed at the focal point F, the object distance u = R/2. The image distance is given by

$$\frac{2}{R}+\frac{1}{\upsilon}=\frac{2}{R} \quad \text{or} \quad \frac{1}{\upsilon}=0, \quad \text{therefore } \upsilon=\infty.$$

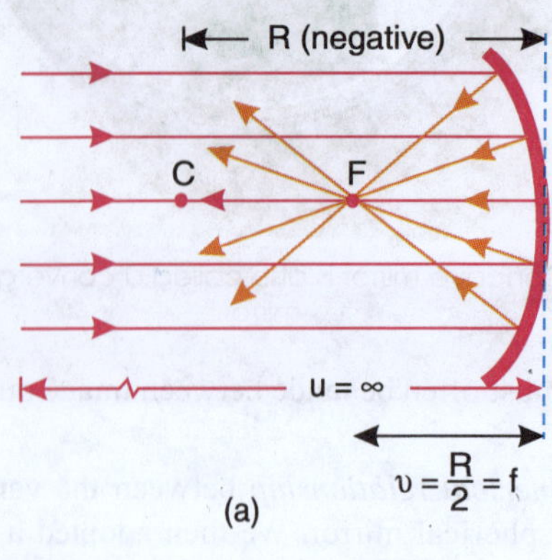

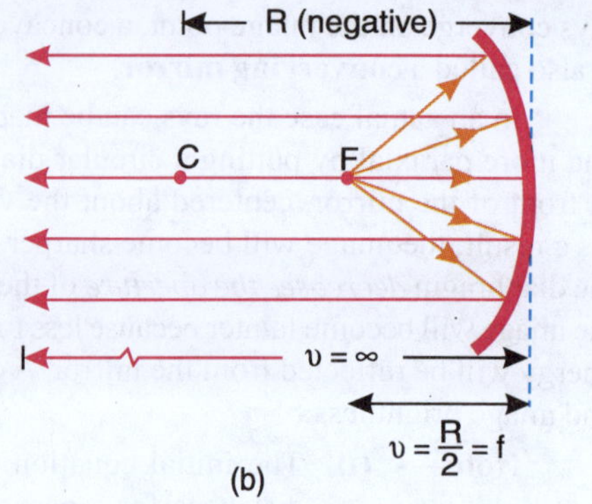

Fig. 3.9

With the object at the focal point, the reflected rays are parallel to the optic axis, so the image is at infinity (see Fig. 3.9b).

We usually express the relationship between object and image distances in terms of the focal length f:

$$\frac{1}{u}+\frac{1}{\upsilon}=\frac{1}{f} \tag{3.10}$$

This is known as the **Gauss formula** for a spherical mirror. Since f can be determined experimentally by observing the point of convergence of rays that are parallel to the principal axis, it is not necessary to know the radius of R in order to apply Eq. (3.10).

Face warped by the spherical surface.

Mathematically we can interchange u and υ and still satisfy the mirror equation. It illustrates the principle of *optical reversibility*, which states that if any ray is reversed, it will retrace its path through an optical system.

3.4.5. CONJUGATE POINTS

If a point object placed at a distance u from the mirror gives an image at a distance υ from the mirror, then an object at a distance υ from the mirror will give an image at a distance u from the mirror. Thus, the position of the object and the image are mutually interchangeable. Two points, which can interchange their positions, are known as **conjugate points.** Therefore, the two points at distances u and υ are conjugate points. The centre of curvature is a self-conjugate point. Planes on which the conjugate points lie are **conjugate planes** and the distances from the vertex to these planes are **conjugate distances**.

3.4.6. EXTENDED OBJECT

Now suppose we have an object with finite size, represented by the arrow AB in Fig. 3.10, perpendicular to the optic axis. A real and inverted image A'B' is formed between C and F.

Δ^{les} ABC and A′B′C are similar and hence

$$\frac{AB}{A'B'} = \frac{CB}{CB'} \tag{3.11}$$

Also, Δ^{les} ABP and A′B′P are similar.

Hence $$\frac{AB}{A'B'} = \frac{PB}{PB'}. \tag{3.12}$$

From equations (3.11) and (3.12), we get

$$\frac{CB}{CB'} = \frac{PB}{PB'}$$

or $$\frac{PB - PC}{PC - PB'} = \frac{PB}{PB'}$$

∴ $$\frac{u - R}{R - \upsilon} = \frac{u}{\upsilon}$$

∴ $$uR + \upsilon R = 2u\upsilon$$

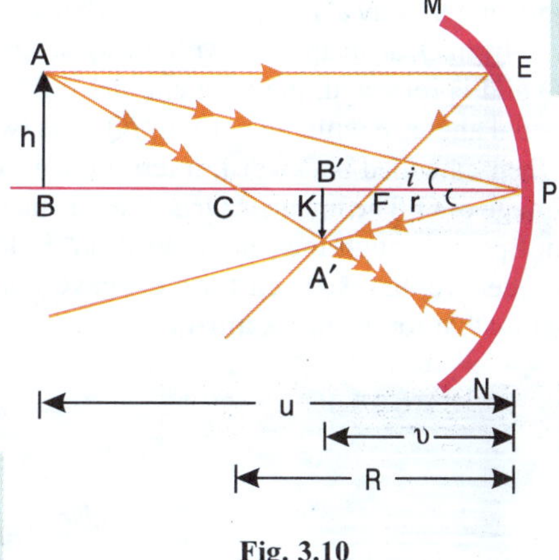

Fig. 3.10

Dividing throughout by $u\upsilon R$, we get

$$\frac{1}{u} + \frac{1}{\upsilon} = \frac{2}{R} \tag{3.13}$$

Using the sign convention, u = – ve, R = – ve and υ = –ve into the above equation, we get

$$\frac{1}{-u} + \frac{1}{-\upsilon} = \frac{2}{-R}$$

∴ $$\frac{1}{u} + \frac{1}{\upsilon} = \frac{2}{R} \tag{3.14}$$

Using the relation (3.9) into (3.14), we obtain

$$\frac{1}{u} + \frac{1}{\upsilon} = \frac{1}{f} \tag{3.15}$$

This relation is the *Gauss' formula.*

3.4.7. LATERAL MAGNIFICATION

The ratio of the transverse dimensions of the final image formed by an optical system to the corresponding dimension of the object is defined as the *transverse* or *lateral magnification.* It is noted that the object and image have different sizes and that they have opposite orientations. Because triangles ABP and A′ B′ P are similar, we have

$$\frac{h}{u} = -\frac{h'}{\upsilon}$$

$$\therefore \quad m = \frac{h'}{h} = -\frac{\upsilon}{u} \tag{3.16}$$

Note— The term magnification is generally used in a wider sense. Magnification does not mean that the image is enlarged. The image formed by an optical system (mirror, lens etc.) may be larger than, smaller than, or of the same size as the object. ◀

Convex mirror

3.4.8. CONVEX MIRROR

Fig. 3.11(a) shows the formation of image of a point object by a convex mirror.

A ray from O, incident along the axis, strikes the mirror normally at P and is reflected back on itself. A second ray from O, at an angle α with the axis, strikes the mirror at Q and is reflected, the incident and reflected rays making equal angles θ with the normal CQ. The two reflected rays, when extended backward, intersect at I, forming a virtual image of O. The image is *virtual* since the light energy does not pass through the image position. Following the same procedure that we used for a concave mirror, it is easy to show that for a convex mirror.

$$\frac{1}{u} + \frac{1}{\upsilon} = \frac{2}{R}.$$

Fig. 3.11 (a)

Fig. 3.11 (b) shows the formation image with an extended object. The object AB is in front of the mirror and a virtual and erect image A′ B′ is formed behind the mirror. The triangles ABC and A′B′C are similar. Hence we have

$$\frac{AB}{A'B'} = \frac{CB}{CB'} \tag{3.17}$$

Also, the triangles ABP and A′ B′ P are similar. Hence we have

$$\frac{AB}{A'B'} = \frac{PB}{PB'} \qquad (3.18)$$

From (3.17) and (3.18), we get

$$\frac{CB}{CB'} = \frac{PB}{PB'} \quad \text{or} \quad \frac{PB+PC}{PC-PB'} = \frac{PB}{PB'}$$

$$\therefore \qquad \frac{u+R}{R-\upsilon} = \frac{u}{\upsilon}$$

or $$\upsilon R - uR = -2u\upsilon$$

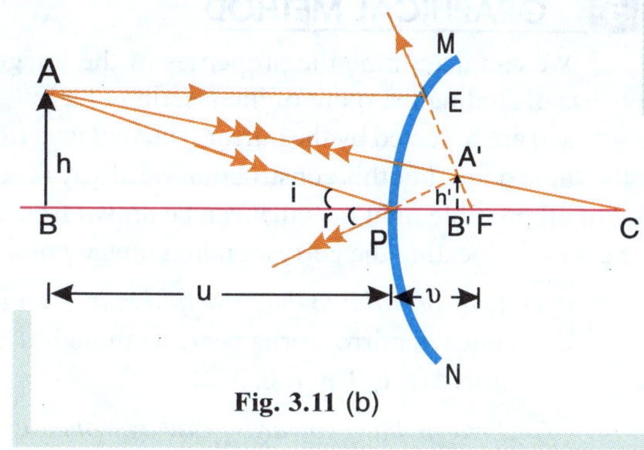

Fig. 3.11 (b)

Dividing the above equation throughout by $u\upsilon R$, we obtain

$$\frac{1}{u} - \frac{1}{\upsilon} = -\frac{2}{R} \qquad (3.19)$$

Using the sign convention, $u = -ve$, υ = +ve and R = +ve, we get

$$\frac{1}{-u} - \frac{1}{\upsilon} = -\frac{2}{R}$$

or $$\frac{1}{u} + \frac{1}{\upsilon} = \frac{2}{R} \qquad (3.20)$$

$$\frac{1}{u} + \frac{1}{\upsilon} = \frac{1}{f} \qquad (3.21)$$

Note— Numerical relations (3.13) and (3.19) for concave and convex mirrors are different but the algebraic relationships (3.14) and (3.20) have the same form.

3.4.9. SPHERICAL MIRROR EQUATION APPLIED TO A PLANE MIRROR

In case of a plane mirror, we may treat that R tends towards infinity. Using this value of R into the mirror equation (3.20), we get

$$\frac{1}{u} + \frac{1}{\upsilon} = \frac{2}{\infty} = 0$$

$$\therefore \qquad u = -\upsilon$$

This result confirms our conclusion (eq.3.1) about the object-image relationship in case of a plane mirror.

3.4.10. LATERAL MAGNIFICATION

It is seen from Fig. 3.11(b) that the object and image have different sizes but both are erect. Because triangles ABP and A′ B′ P are similar, we have

$$\frac{h}{u} = \frac{h'}{\upsilon}$$

$$\therefore \qquad m = \frac{h'}{h} = \frac{\upsilon}{u} \qquad (3.22)$$

We, thus, find that the basic relationships for image formation by a spherical mirror are valid for both concave and convex mirrors, provided that we use the sign rules consistently.

3.5. GRAPHICAL METHOD

We can determine the properties of the image by a simple *graphical* method. This method consists of finding the point of intersection of a few particular rays that diverge from a point of the object and are reflected by the mirror. Then *all* rays from this point that strike the mirror will intersect at the same point. For this construction we always choose an object point that is *not* on the optic axis. Out of all possible light rays that can be drawn from a point on an object, there are four rays, which are useful in locating the corresponding image point. These are called **principal rays** or **easy rays**.

(*a*) *A ray parallel to the principal axis*, after reflection, passes through the focus F in case of concave mirror; or appears as though it came from the focal point F in case of convex mirror (Ray 1 in Fig. 3.12).

(*b*) *A ray heading towards or away from the centre of curvature* C is reflected back along its original path (Ray 2 in Fig. 3.12). Because the radius and tangent at the point of incidence are at 90^0 to one another, any ray incident normally on a mirror is reflected along the same path.)

(*c*) *A ray incident at vertex* P at an angle *i* is reflected at an angle *i* ((Ray 3 in Fig.3.12).

(*d*) *A ray through the focal point* F is reflected parallel to the axis (Ray 1 in Fig.3.12).

Normally, any of the two rays mentioned above are sufficient to locate the image of a linear object. However, a third ray is often used as a safeguard against errors.

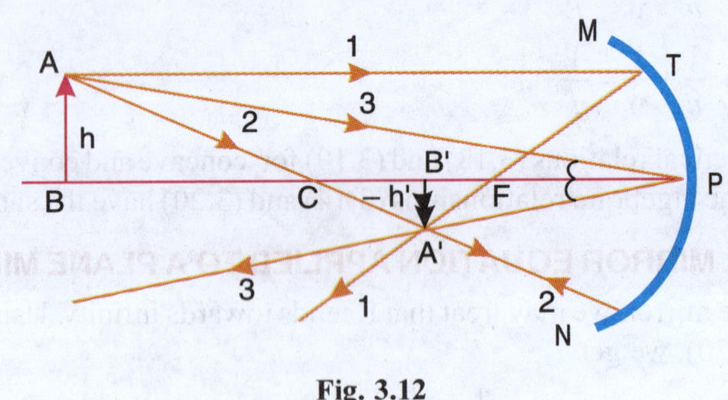

Fig. 3.12

Fig. 3.12 demonstrates how these rays are used to locate the image formed by a concave mirror.

3.5.1. EFFECT OF ALTERING THE OBJECT DISTANCE

We will now consider the location of an image graphically when the object lies at various distances from the mirror. In each case we shall find out the position, nature and relative size of the image.

(*i*) Object at infinity:

The graphic construction is shown in Fig. 3.13 (a). Since the object is at infinity, the incident rays become parallel. Two such rays are shown in the figure, one along the principal axis and the other parallel to it. The ray parallel to optic axis is incident on the mirror at N and the reflected ray passes through F. The ray along the principal axis is reflected at P and retraces

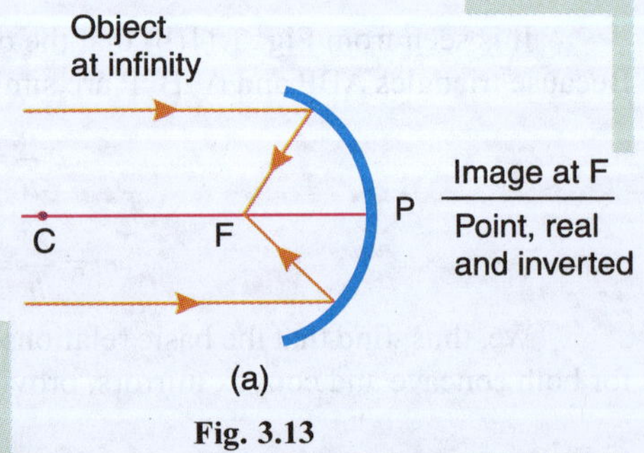

Fig. 3.13

its path. The two reflected rays intersect at F and the image is formed at point F.

Here, the object distance $u = \infty$ and image distance $\upsilon = f$.

Magnification $m = \frac{\upsilon}{u} = \frac{f}{\infty} = 0$ (i.e. the image is very small in size).

***(ii)* Object between infinity and point C:**

It is seen from the Fig. 3.13 (b) that the image has finite dimensions. It is real, inverted and smaller in size than the object. It lies between C and F.

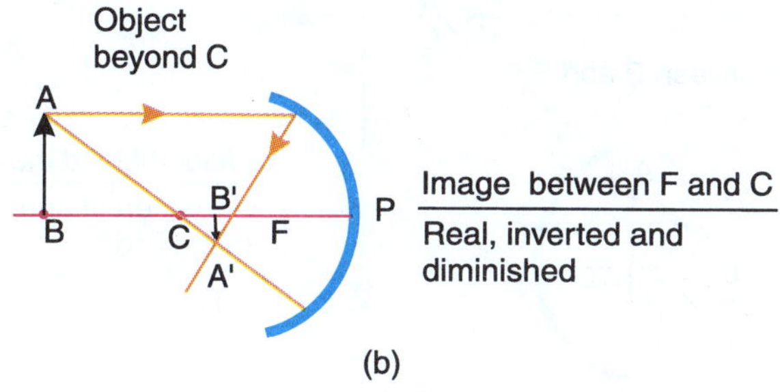

(b)

Fig. 3.13

Magnification $m = \frac{\upsilon}{u} < 1$

***(iii)* Object at C:**

In this case, the image lies at the centre of curvature. It is real and inverted and of the same size as the object.

Magnification $m = \frac{\upsilon}{u} = 1$

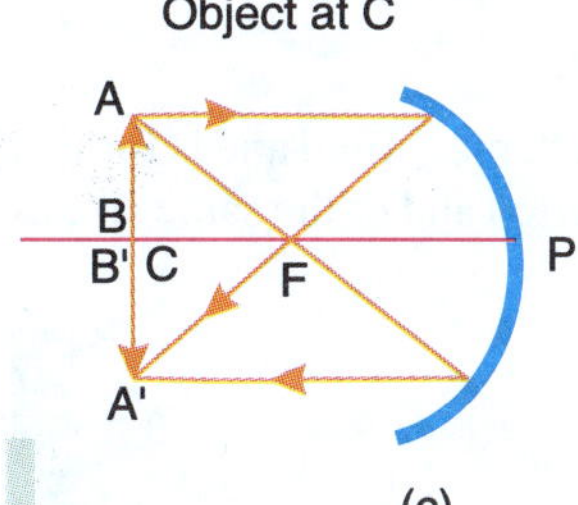

(c)

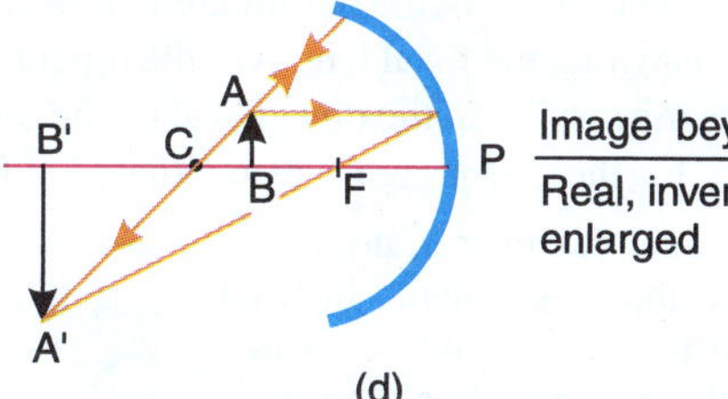

(d)

Fig. 3.13

***(iv)* Object between C and F:**

The image in this case lies between C and infinity. It is inverted, and real, as shown in Fig.3.13(d).

Magnification $m = \frac{\upsilon}{u} > 1$

Therefore, the image is larger in size than the object.

***(v)* Object at Focus F:**

The image forms at infinity, as shown in Fig. 3.13 (e).

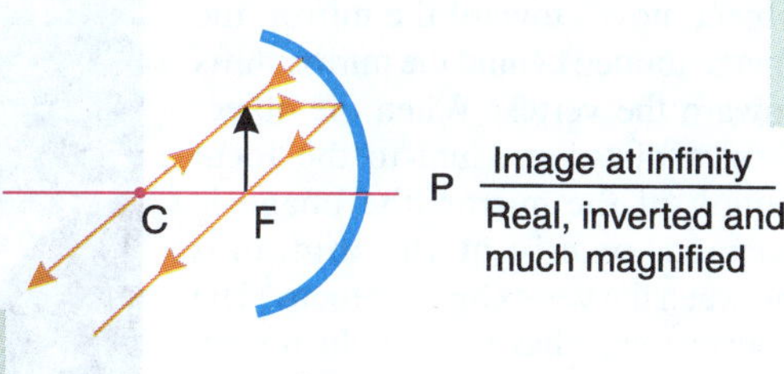

(e)

Fig. 3.13

Magnification $m = \frac{v}{u} = \frac{\infty}{f} = \infty$

(*vi*) Object between F and P:

In this case, the image is virtual and erect. It is magnified and lies on the other side of the mirror, as shown in Fig. 3.13 (f).

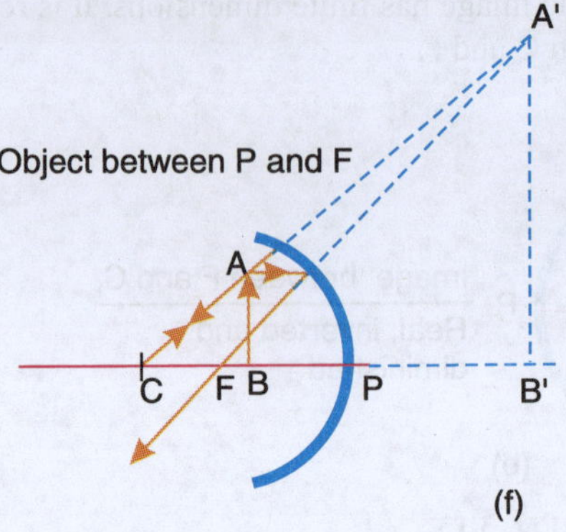

(f)

Fig. 3.13

A concave mirror may form a real or virtual image of an object depending upon the distance of the object from the mirror. The reflected rays actually pass through the image points in case of a *real image* and the image can be projected onto a screen. Note that as the object moves toward the mirror, the image recedes away from the mirror and finally appears behind the mirror. As the object approaches the focus of the mirror, the image recedes further from the mirror and when the object is at focus, the image is at infinity. When the object is within the focal length of the mirror, the image is virtual, erect, magnified and is formed behind the mirror. When the object is at C, the image formed is real, inverted and of the same size as the object. Further the image is also situated at C.

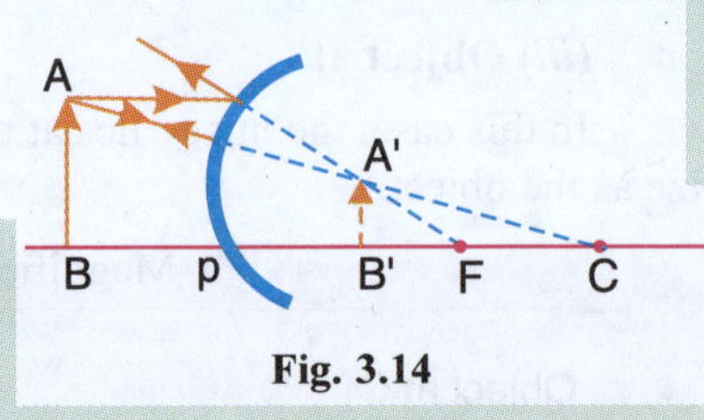

Fig. 3.14

In a convex mirror, no matter where the object is situated in front of the mirror, the image is formed behind the mirror. The image is virtual, erect, diminished and is always formed between the vertex P and the focus F (see Fig.3.14). As the object moves toward the mirror, the image formed behind the mirror shifts toward the vertex. When the object is at a distance equal to the focal length of the mirror, the image is formed exactly at the midpoint between the vertex and the focus. The convex mirror has a wide field of view and hence is used as a driving mirror.

Convex mirror is used as a driving mirror.

Caution— In actual practice the paraxial condition is difficult to ensure. The presence of large angle rays, falling on a spherical mirror of wide aperture, will not converge to or diverge from a single point. Hence the image of an extended object appears blurred. The lack of image sharpness caused by the spherical mirror surface is called the **spherical aberration**. The rays reflected by the mirror intersect along a conical surface and produce a bright curve, called the **reflection caustic curve** (see Fig.3.15). Such caustic curves are seen under appropriate illumination condition on the surface of milk or tea in a cup. ◆

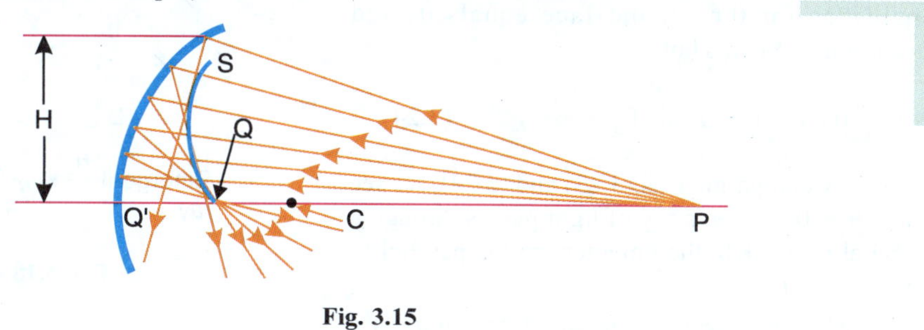

Fig. 3.15

3.6. ASPHERIC MIRRORS

Spherical aberration cannot be eliminated completely, but by proper design of the surface it can be suppressed for certain positions, called **anastigmatic**. Anastigmatic positions can be modified by changing the shape of the surface. An **elliptical mirror** is anastigmatic for an object placed at one focus of the ellipse whose image falls exactly at the other focus (See Fig. 3.16). The two foci thus are conjugate. Such mirrors are used for optical pumping of ruby laser.

Similarly, a **parabolic mirror** produces no aberration for rays that are parallel to the principal axis (See Fig. 3.17). All these rays will pass through the focus of the parabolic mirror. Such mirrors are used in astronomical telescopes, search lights, and microwave antennas.

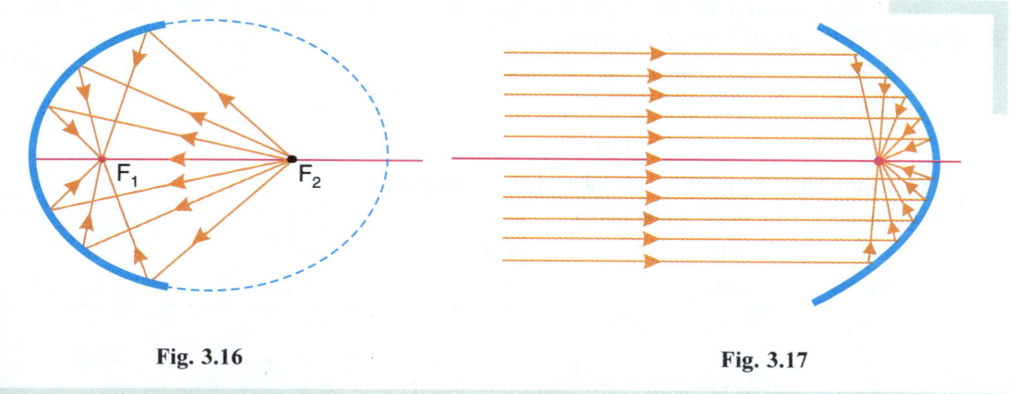

Fig. 3.16 Fig. 3.17

3.7. REFRACTION OF LIGHT

3.7.1. REFRACTION THROUGH A GLASS SLAB

A light ray that travels through a parallel-sided transparent slab (e.g. plate glass window) is refracted at both faces. The two refractions at the parallel surfaces result in a 'sideways' displacement, but do not change the direction of the ray (Fig. 3.18). To show that there is no change in direction, we can apply Snell's law at both faces.

At the first face, the Snell's law takes the form

$$\mu_1 \sin \theta_1 = \mu_2 \sin \theta_2 \qquad (3.23)$$

Because the faces are parallel, the angle of incidence for the second face equals θ_2 and according to Snell's law

$$\mu_2 \sin \theta_2 = \mu_1 \sin \theta_3 \qquad (3.24)$$

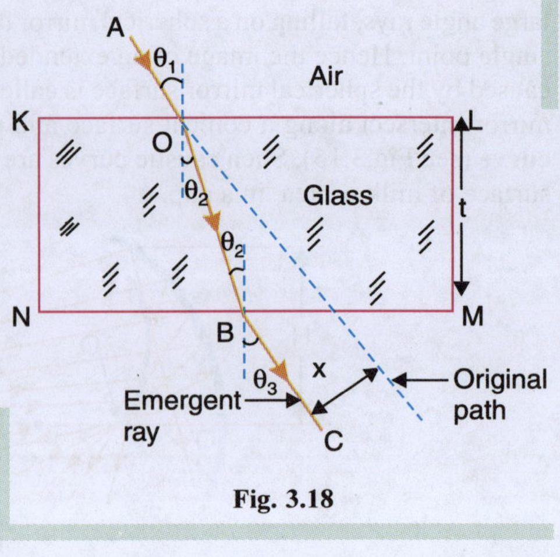

Fig. 3.18

By comparing Eqs.(3.23) and (3.24), we see that $\theta_1 = \theta_3$. When a ray of light passes through a parallel-sided slab, the emergent ray is parallel to the incident ray.

The net effect of the two refractions is a parallel displacement of the ray. The distance x between the incident and emergent ray is given by

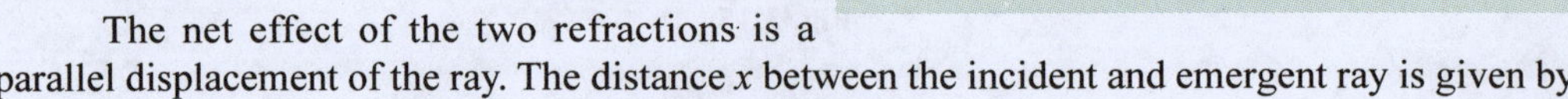

$$x = \frac{t \cdot \sin(\theta_1 - \theta_2)}{\cos \theta_2} \qquad (3.25a)$$

where t is the thickness of the glass slab.

It can also be shown that $x = t\theta\left(\frac{\mu - 1}{\mu}\right)$ (3.25b)

for small angles θ and θ measured in radians.

For an ordinary windowpane the displacement is generally not more than a few millimeters.

3.7.2. REFRACTION THROUGH A COMPOUND SLAB

For a single parallel-sided slab, the emergent beam will be parallel to the incident beam. If r is the angle of refraction, the Snell's gives

$$\frac{\sin i}{\sin r} = {}_1\mu_2 \text{ and } \frac{\sin r}{\sin i} = {}_2\mu_1 \qquad (3.26)$$

$$ {}_1\mu_2 \cdot {}_2\mu_1 = 1 \qquad (3.27)$$

For a number of parallel-sided slabs, placed one upon the other (Fig.3.19), the Snell's law gives

$$\frac{\sin i}{\sin r_1} = {}_1\mu_2, \frac{\sin r_1}{\sin r_2} = {}_2\mu_3, \text{ and } \frac{\sin r_2}{\sin i} = {}_3\mu_1$$

$$\therefore \quad {}_1\mu_2 \cdot {}_2\mu_3 \cdot {}_3\mu_1 = 1$$

$$\therefore \quad {}_2\mu_3 = \frac{1}{{}_1\mu_2 \cdot {}_3\mu_1} = \frac{{}_1\mu_3}{{}_1\mu_2} \qquad (3.28)$$

Note— The visibility of a transparent medium is due to the difference in its refractive index from that of the surrounding medium. A piece of transparent glass placed in water is visible because of the difference between their refractive indexes.

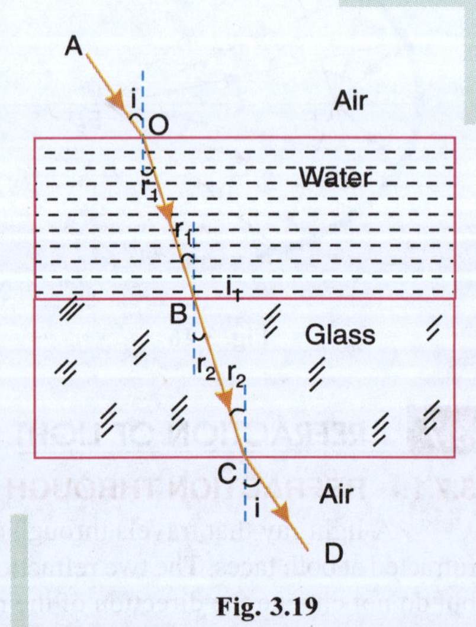

Fig. 3.19

3.7.3. APPARENT DEPTH

Let an object lie below a refracting medium of refractive index μ_2 (Fig. 3.20) and is seen almost normally in a medium of refractive index $\mu_{1.}$

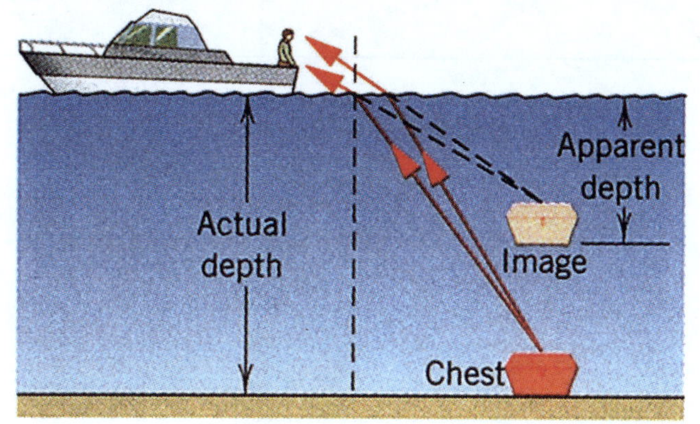

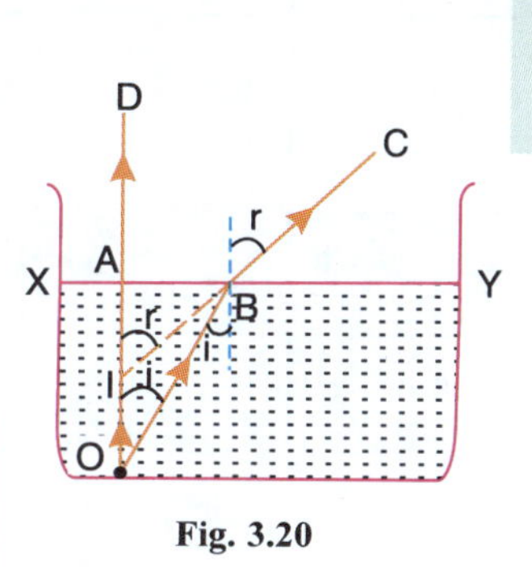

Fig. 3.20

According to Snell's law, we have

$$\mu_2 \sin i = \mu_1 \sin r$$

or
$$\mu_2 \frac{AB}{OB} = \mu_1 \frac{AB}{IB}$$

Let $OB = u$ and $IB = \upsilon$. Then $\mu_2 \upsilon = \mu_1 u$

or
$$\upsilon = \frac{\mu_1}{\mu_2} \mu$$

If $\mu_2 > \mu_{1,}$ $\upsilon < u$, and the image I of an object O appears to be raised through a distance OI. This distance is given by

$$OI = t\left(1 - \frac{\mu_1}{\mu_2}\right) \qquad (3.29)$$

where t is the thickness of the medium.

In case of viewing other than vertically, the position of the image changes with the angle, which the line of vision makes with the surface of separation of the two media. As the angle increases, the apparent depth decreases. Also, the object appears curved when viewed obliquely through the refracting medium.

3.8. TOTAL INTERNAL REFLECTION

Light is partially reflected and partially transmitted at an interface between two materials with different indexes of refraction. Under certain circumstances, however, all of the light can be reflected back from the interface, with none of it being transmitted, even though the second material is transparent. Fig.3.21 shows several rays radiating from a point source P in material having a refractive index μ_2. The rays strike the surface of a second material having refractive index μ_1 ($\mu_2 > \mu_1$). From Snell's law of refraction

$$\sin \theta_2 = \frac{\mu_2}{\mu_1} \sin \theta_1 \qquad (3.30)$$

Because μ_2 / μ_1 is greater than unity, $\sin \theta_2 > \sin \theta_1$. Therefore, the ray is bent away from normal. Thus there must be some value of θ_1 less than 90° for which Snell's law gives $\sin \theta_2 = 1$ and

$\theta_2 = 90^o$. Ray 3 in the diagram emerges just grazing the surface at an angle of refraction of 90°.

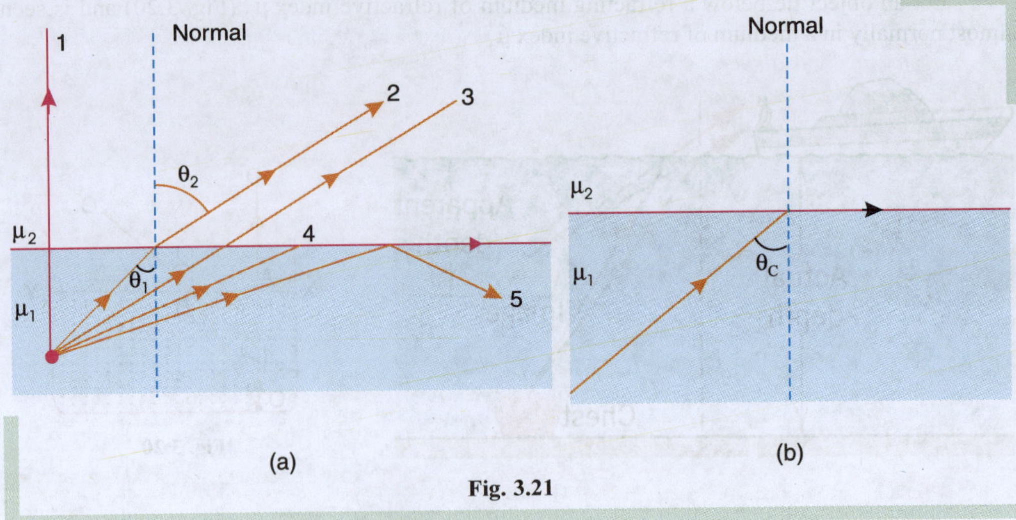

Fig. 3.21

The angle of incidence, for which the refracted ray emerges tangent to the surface, is called the **critical angle**, denoted by θ_{crit}. If the angle of incidence is *greater than* the critical angle, the sine of the angle of refraction, as computed by Snell's law, would have to be greater than unity, which is impossible. Beyond the critical angle, the ray *cannot pass* into the upper material; it is trapped in the lower material and is completely reflected at the boundary surface. This is called **total internal reflection**.

This is an excellent experiment to show total internal reflection.

We can find the critical angle for two given materials by setting $\theta_2 = 90^o$ in Snell's law. We then have

$$\sin \theta_{crit} = \frac{\mu_1}{\mu_2} \qquad (3.31)$$

Light propagating within the glass ($\mu \cong 1.52$) will be totally reflected if it strikes the glass-air surface at an angle of about 41° or more.

When a beam of light enters at one end of a transparent rod, the light can be totally reflected internally, if the index of refraction of the rod is greater than that of the surrounding material. The light is trapped with in the rod even if the rod is curved. Such a rod is called a *light pipe*.

3.9. REFLECTING PRISMS

Prisms perform different roles in optics. We broadly classify them into two major groups as *reflecting prisms* and *dispersing prisms.* We first study the reflecting prisms. Reflecting prisms make use of *total internal reflection* and are employed for the purpose of changing either the direction of light or the orientation of the image or both. Because the critical angle for glass is slightly less than 45°, it is possible to use a prism with angles of 45°-45°-90° as a totally reflecting surface. As reflectors, totally reflecting prisms have some advantages over metallic surfaces such as coated glass mirrors. While no metallic surface reflects 100% of the light incident on it, light can be totally reflected by a prism. The reflecting properties of a prism have the additional advantage of being permanent and unaffected by tarnishing.

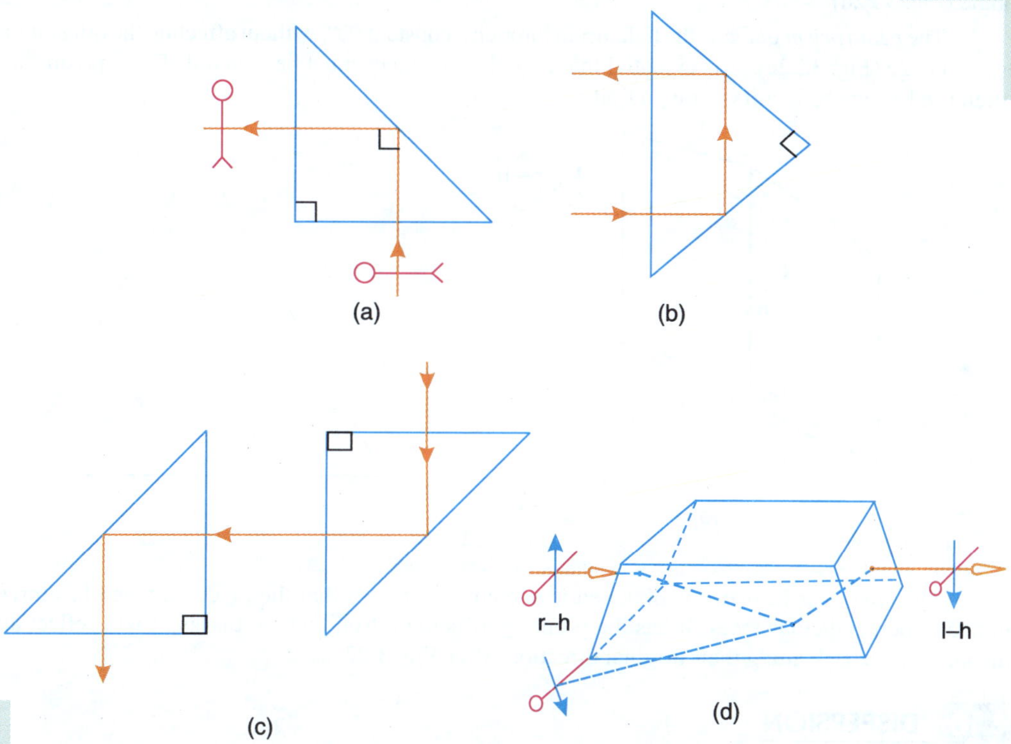

(a) (b)

(c) r–h l–h (d)

Fig. 3.22

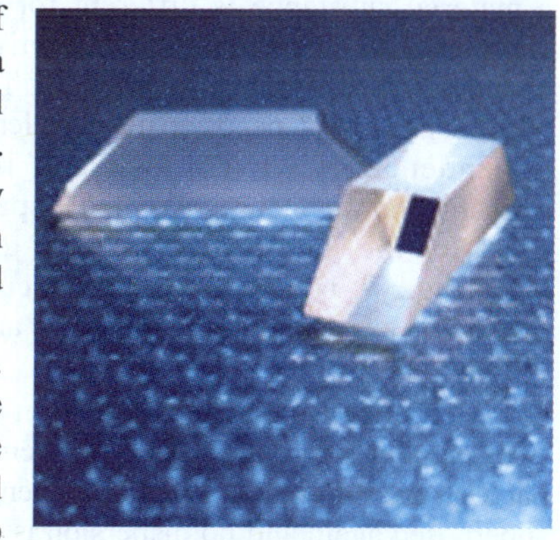

Dove Prism

A light beam is incident on the prism in such a way that at least one internal reflection takes place within the prism and emerges without dispersion. Some of the commonly used reflecting prisms are discussed here.

The *right-angled prism* is a 45°-45°-90° prism (Fig. 3.22 a). The light is reflected at the hypotenuse and rays are deviated normal to the incident face by 90°. It is easy to see that the top and bottom of the image have been interchanged but the right and left sides have not. It means that the top surface acts like a plane mirror.

The *Porro prism* is physically the same as the right angle prism but is used in a different orientation. Light enters and leaves at right angles to the hypotenuse and is totally reflected at each of the shorter faces (Fig. 3.22 b). The total change of direction of the rays is 180^0. Thus, if enters right-handed, it leaves right-handed. Binoculars use combinations of two Porro prisms (Fig. 3.22c).

The *Dove prism* is a truncated version of the right-angle prism, used almost exclusively in collimated light. As the prism is rotated about the line of sight, the image rotates through twice the angle (Fig. 3.22d).

The *penta prism* deflects the light beam through a constant 90°, without affecting the orientation of the image (Fig. 3.22e). It is essential that two of its surfaces must be silvered. These prisms are often used as end reflectors in range finders.

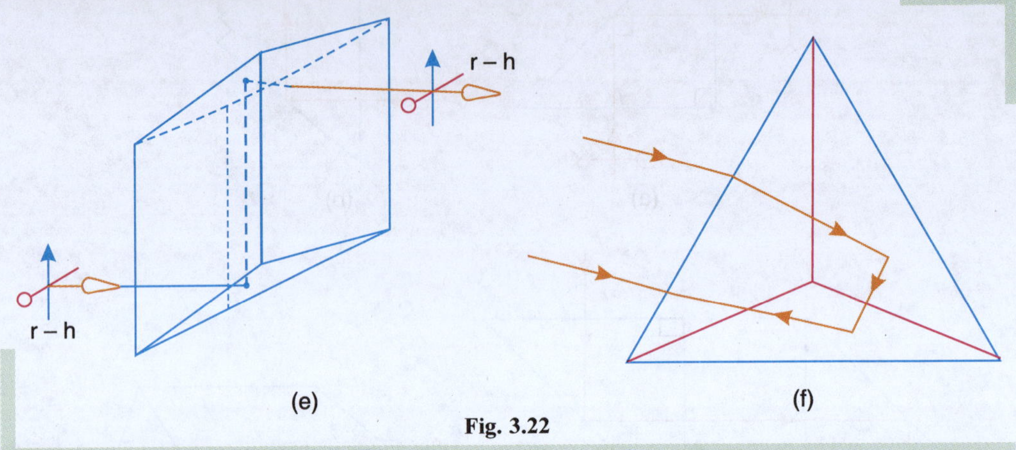

Fig. 3.22

The *corner-cube prism* is obtained if one cuts a cube so that the piece removed has three mutually perpendicular faces. It has the property of being retrodirective; that is, it will reflect all incoming rays back along their original directions (See Fig. 3.22f).

3.10. DISPERSION

Ordinary white light is a mixture of waves with wavelengths extending throughout the visible spectrum. The speed of light in vacuum is the same for all wavelengths, but the speed in a material substance is different for different wavelengths. Therefore, the refractive index changes as a function of wavelength. The angle of refraction also depends on the wavelength. The dependence of wave speed and index of refraction on wavelength is called **dispersion**. A graph of the index of refraction, μ, versus wavelength, λ, is called a **dispersion curve**. The dispersion curves for some materials are shown in Fig. 3.23.

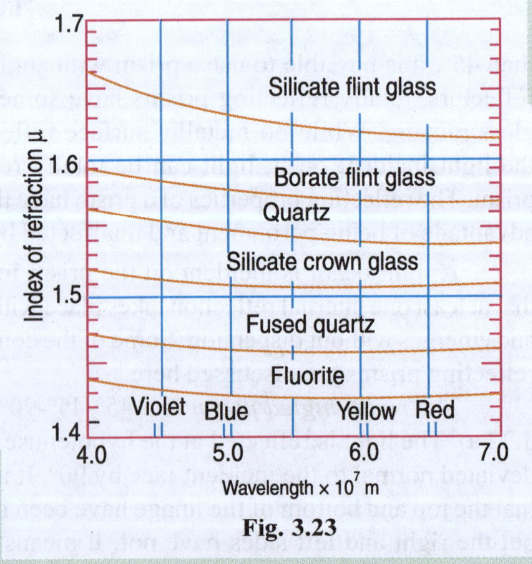

Fig. 3.23

If a piece of glass is in the form of a plate with parallel sides, the rays which emerge are parallel, the different colours (wavelengths) are superposed again and no dispersion is observed except at the very edges of the image (Fig. 3.24a). The effect is not normally noticeable. But if the light passes through a prism, the different wavelengths in white light are refracted through different angles; and the emerging rays are not parallel for the different colours (Fig. 3.24b) and the dispersion is clearly noticeable. The light spreads out into a fan-shaped beam, showing a band of colours. The deviation (change of direction) produced by the prism increases with increasing frequency or decreasing

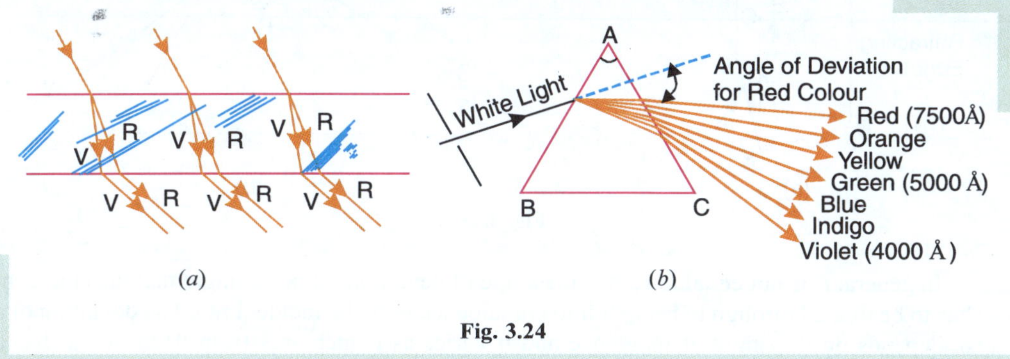

Fig. 3.24

wavelength. Violet light is deviated most, and red is deviated least; other colours are in intermediate positions. The light is said to be *dispersed* into a **spectrum**. The effect is known as **chromatic dispersion**. A spectacular example of the dispersion of light is the rainbow.

The amount of dispersion depends on the difference between the refractive indexes for violet and for red light. The angle subtended by the directions of the two colours is called the **angular dispersion**.

3.11. DISPERSIVE PRISMS

If the light contains more than one wavelength, a refracting prism will disperse the light. Prisms that separate light according to wavelengths are known as **dispersive prisms**. The majority

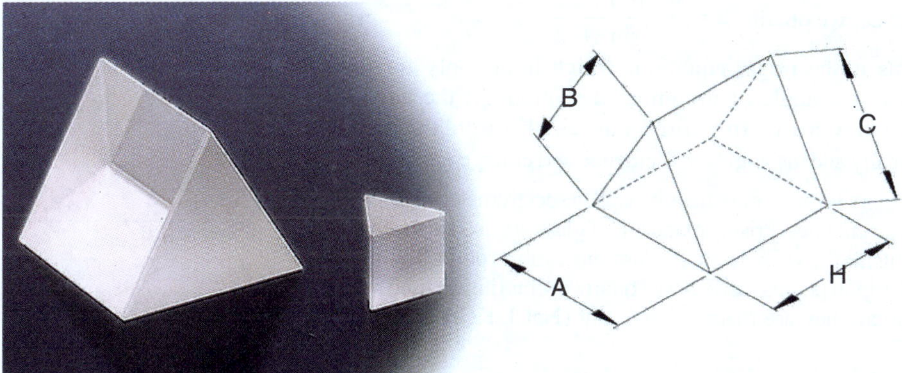

Dispersive prisms are mainly used in spectrometers to separate closely adjacent lines.

of dispersive prisms are right prisms on triangular bases. Such a prism consists of two refracting surfaces inclined at some angle to one another (see Fig. 3.25). In Fig. 3.25 (a) the two refracting faces are labeled ABED and ACFD. The angle between them is called the *refracting angle* A, or simply the *angle of the prism* and AD is called the refracting edge. The optics of the prism can be understood by considering a triangular section cut parallel to ABC and DEF (see Fig. 3.25 b).

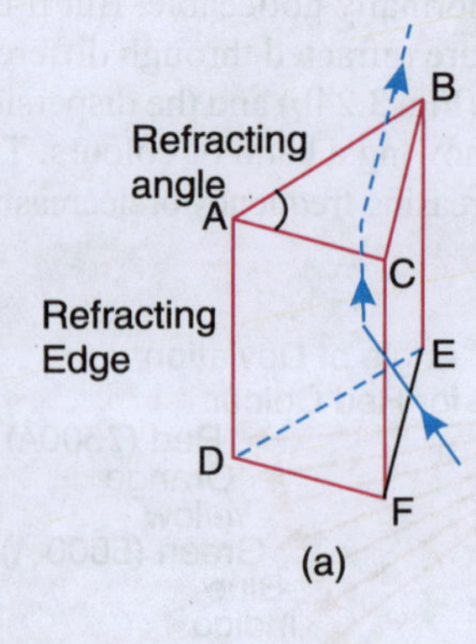

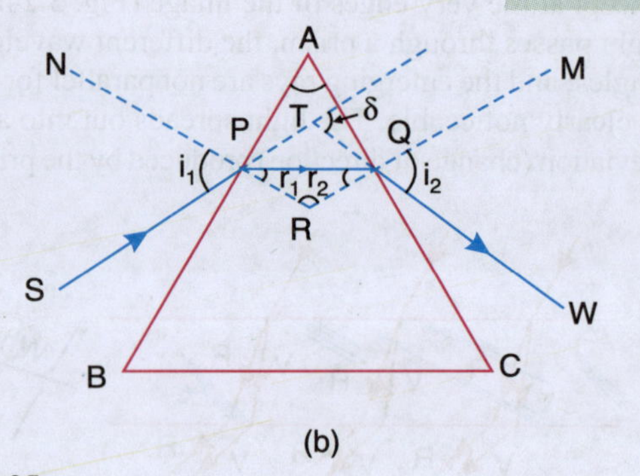

Fig. 3.25

In general i_1 is not equal to i_2. δ is the angle of deviation : it is the angle that the emergent ray has to be rotated through to bring it into coincidence with the incident ray. The deviation of a ray,δ, depends on the angle of incidence and it varies as i_1 increases from 0° to 90°. It has a minimum, δ_m, at one value of i_1. When this occurs the ray travels through the prism symmetrically, that is $i_1 = i_2 = i$. Fig. 3.25 (b) is drawn when minimum deviation occurs. NR and MR are normals at the point of incidence P and the point of emergence Q respectively. The internal angles of quadrilateral APRQ add to 360°.

$$\angle A + \angle PRQ + 180^\circ = 360^\circ$$

$\therefore$ $\angle PRQ = 180^\circ - \angle A$

From the triangle PRQ, $\angle PRQ = 180^\circ - 2r \qquad (r_1 = r_2 = r, \text{ say})$

$\therefore$ $A = 2r$

From the triangle PTQ $\delta_m = 2(i - r) = 2i - A$

$\therefore$ $i = (A + \delta_m)/2$

Now Snell's law gives $\sin i = \mu \sin r$

Hence, we obtain $$\mu = \frac{\sin\,[(A + \delta_m)/2]}{\sin A/2} \qquad (3.32)$$

This is the prism equation, which holds only for light deflected at the angle of minimum deviation. At the angle of minimum deviation the light passes through the prism symmetrically and the angles of incidence and emergence are equal.

Dispersive prisms are mainly used in spectrometers to separate closely adjacent lines. Prisms made out of glass are used in the visible region. Outside the visible region, glass shows absorption. Therefore, in the UV region prisms made out of quartz or fluorite are used and in the IR region, they are made of rock salt (NaCl, KCl) or sapphire (Al_2O_3).

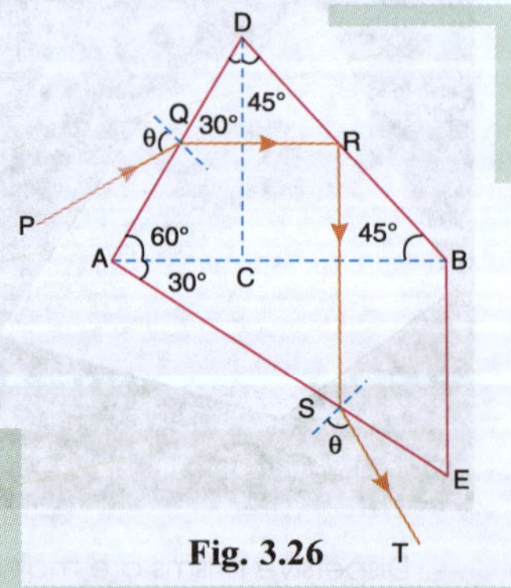

Fig. 3.26

3.11.1. CONSTANT DEVIATION DISPERSING PRISM

Constant deviation dispersing prisms are widely used in spectroscopy. It may be regarded as made up of two 30°-60°-30° prisms and one 45°-45°-90° prism (see Fig. 3.26). Suppose that in the position shown a monochromatic ray of wavelength λ traverses the component prism DAE symmetrically, thereafter to be reflected at 45° from the face AB. The ray will then traverse prism CDB symmetrically having experienced a total deviation of 90°. The ray can be thought of as having passed through an ordinary 60° prism at minimum deviation. All other wavelengths present in the beam will emerge at other angles. If the prism is rotated slightly about an axis normal to the paper, the incoming beam will have a new incident angle. Now, a different wavelength component undergoes a minimum deviation, which is again 90°- hence the name constant deviation. With a prism of this sort, one can conveniently set up the light source and viewing system at a fixed angle, i.e. 90° and then simply rotate the prism to look at a particular wavelength.

3.12. REFRACTION AT SPHERICAL SURFACES

Images can be formed by reflection as well as by refraction. We study here the refraction at a spherical surface, that is, at a spherical interface between two transparent materials having different indices of refraction. Fig. 3.27 shows what happens when a real object is brought successively closer to a convex refracting surface.

When the object distance is large, the rays passing through the refracting surface come together and form a real image. If the object is moved closer to the surface to the position F_1 the emergent light is parallel and the image forms at infinity. If the object is brought even closer, the rays emergent on the right hand side of the surface do not come together. Instead, the light appears to come from a point I, lying to the left side of the surface and the image is a virtual image. Now we derive the relationships between the various distances in two different situations.

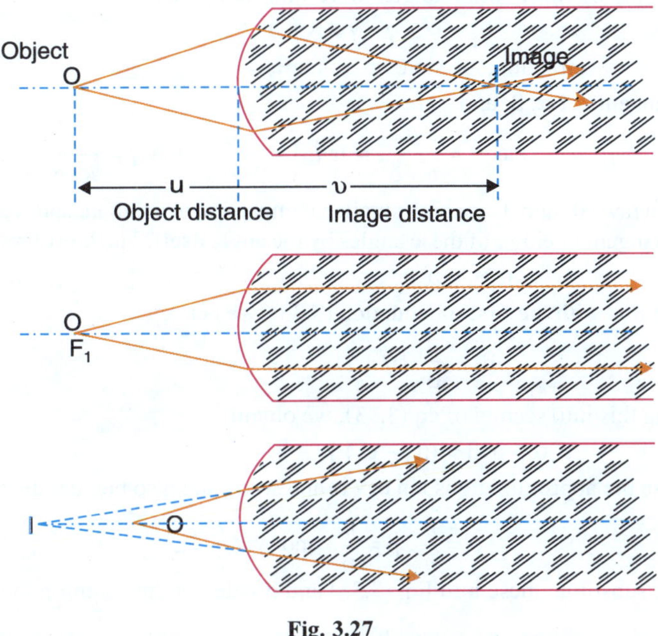

Fig. 3.27

1. (a) Refraction at a convex surface forms a real image:

In Fig.3.28, a convex surface XY with radius R forms an interface between two materials with different indexes of refraction μ_1 and μ_2. Let C be the centre of curvature and P the pole of the spherical surface. Consider a point object O lying on the principal axis in the rarer medium of refractive index μ_1 at a distance u from the vertex in front of the convex surface; the distance u being greater than the radius of curvature R of the surface. Ray OP travelling along the principal axis strikes the vertex P and passes into the second medium without deviation. Ray OA, making an angle α with the axis, is incident on the surface at an angle θ_1 with the normal and is refracted along AI at an angle θ_2. These rays intersect at I at a distance υ to the right of the vertex. The figure is drawn for the case $\mu_1 < \mu_2$.

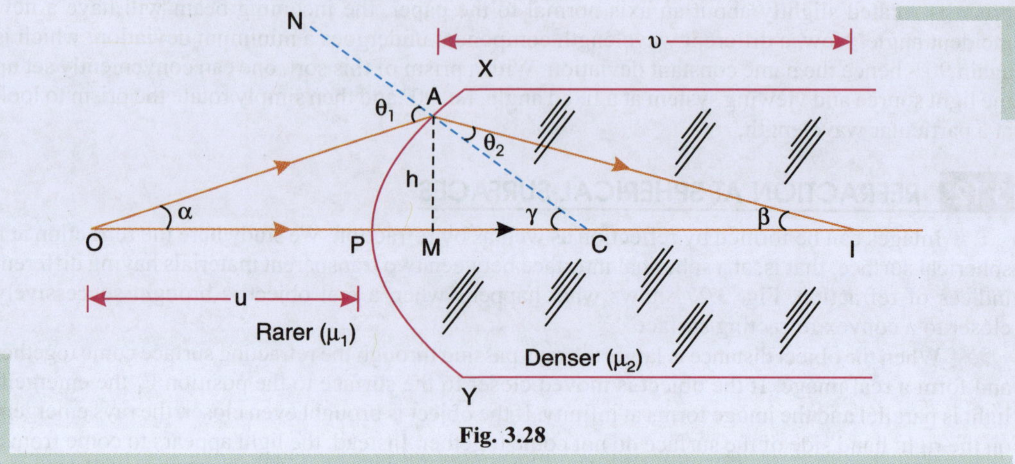

Fig. 3.28

We use the theorem that an exterior angle of a triangle equals the sum of the two opposite interior angles. Applying this to the triangles OAC and ACI gives

$$\theta_1 = \alpha + \gamma, \quad \gamma = \beta + \theta_2 \tag{3.33}$$

From the law of refraction, $\mu_1 \sin \theta_1 = \mu_2 \sin \theta_2$

Also the tangents of α, β, and γ are

$$\tan \alpha = \frac{h}{u+\delta}, = \tan \beta = \frac{h}{\upsilon - \delta}, \tan \gamma = \frac{h}{R-\delta} \tag{3.34}$$

For paraxial rays, θ_1 and θ_2 are both small in comparison to a radian, and we may approximate both the sine and tangent of either of these angles by the angle itself. The law of refraction then gives

$$\mu_1\theta_1 = \mu_2\theta_2 \tag{3.35}$$

Combining this with the first of equation (3.33), we get

$$\theta_2 = \frac{\mu_1}{\mu_2}(\alpha + \gamma)$$

Substituting this into second of eq.(3.33), we obtain

$$\mu_1\alpha + \mu_2\beta = (\mu_2 - \mu_1)\,\gamma \tag{3.36}$$

Now we use the approximations $\tan \alpha = \alpha$ and so on and also neglect the small distance δ.

Thus, $$\alpha = \frac{h}{u}, \ \beta = \frac{h}{\upsilon}, \text{ and } \gamma = \frac{h}{R}. \tag{3.37}$$

Finally, we substitute these into Eq. (3.35) and divide out the common factor h. We obtain

$$\frac{\mu_1}{u} + \frac{\mu_2}{\upsilon} = \frac{\mu_2 - \mu_1}{R} \tag{3.38}$$

We now apply the sign convention: u is -ve, υ is +ve and R is +ve. It gives us

$$\frac{\mu_2}{\upsilon} - \frac{\mu_1}{u} = \frac{\mu_2 - \mu_1}{R} \tag{3.39}$$

This is the **Gaussian relation** for a single spherical surface. Though this relation is derived for a convex surface and for a real object and real image, it is equally valid for all other conditions, as shown in the following sections.

If the first medium is air and the second medium is of refractive index μ, then $\mu_1 = 1$ and $\mu_2 = \mu$.

$$\frac{\mu}{\upsilon} - \frac{1}{u} = \frac{\mu - 1}{R} \tag{3.40}$$

1. (b). Refraction at a convex surface forms a virtual image:

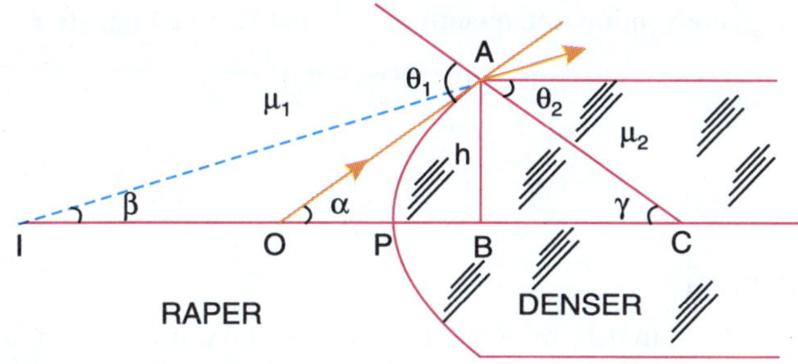

Fig. 3.29

When the point object O lies on the principal axis close to the vertex of the convex refracting surface, it forms a virtual image at I (See Fig. 3.29).

In Δ^{le} OAC $\quad \theta_1 = \alpha + \gamma \quad$ and

In Δ^{le} IAC $\quad \theta_2 = \beta + \gamma$

Using these in the Snell's equation (3.35), we get

$$\mu_1(\alpha+\gamma) = \mu_2(\beta+\gamma)$$
$$\mu_1\alpha - \mu_2\beta = (\mu_2 - \mu_1)\gamma$$

Substituting the values of α, β, and γ, and dividing throughout by h, we get

$$\frac{-\mu_2}{\upsilon} + \frac{\mu_1}{u} = \frac{(\mu_2 - \mu_1)}{R} \tag{3.41}$$

According to sign convention both u and υ are negative and R is positive.

$$\therefore \quad \frac{-\mu_2}{-\upsilon} + \frac{\mu_1}{-u} = \frac{(\mu_2 - \mu_1)}{R}$$

or

$$\frac{\mu_2}{\upsilon} - \frac{\mu_1}{u} = \frac{\mu_2 - \mu_1}{R} \tag{3.42}$$

2. Refraction at a concave surface:

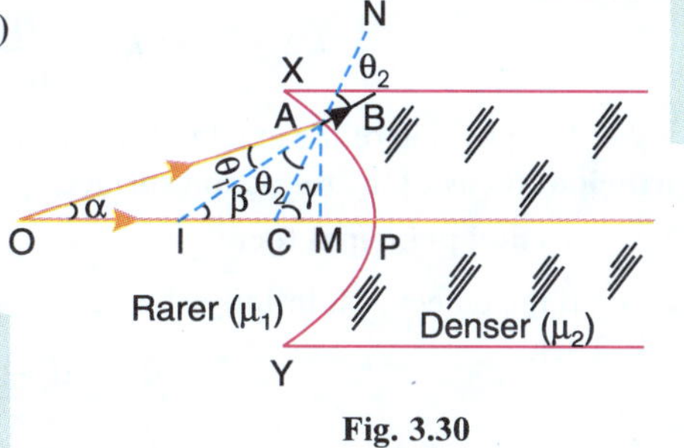

Fig. 3.30

Let O be the point object placed on the principal axis. The ray OA incident on the surface at angle θ_1 is refracted along AB at an angle θ_2 (See Fig.3.30). The ray of light starting from O and travelling along the principal axis passes undeviated through vertex P. The two refracted rays do not meet at any point but appear to come from point I on the principal axis. Therefore, I is the virtual image of the object O.

Referring to the Fig. 3.30, in Δ^{le} OAC $\quad \theta_1 = \gamma - \alpha$ and

In Δ^{le} IAC $\quad \theta_2 = \gamma - \beta$

Using these in the Snell's equation (3.35), we get

$$\mu_1(\gamma - \alpha) = \mu_2(\gamma - \beta)$$
$$-\mu_1\alpha + \mu_2\beta = (\mu_2 - \mu_1)\gamma$$

Substituting the values of α,β, and γ, and dividing throughout by *h*, we get

$$\frac{\mu_2}{\upsilon} - \frac{\mu_1}{u} = \frac{(\mu_2 - \mu_1)}{R} \qquad (3.43)$$

According to sign convention, the quantities *u*, υ and R are all negative.

$$\therefore \qquad \frac{\mu_2}{-\upsilon} - \frac{\mu_1}{-u} = \frac{(\mu_2 - \mu_1)}{-R}$$

or

$$\frac{\mu_2}{\upsilon} - \frac{\mu_1}{u} = \frac{\mu_2 - \mu_1}{R} \qquad (3.44)$$

3.12.1. PRINCIPAL FOCI

We can use equ. (3.44) to determine the first and second principal foci. F_1 is the position of the object for which the image lies at infinity, while F_2 is the position of the image for which the object lies at infinity (See Fig. 3.31).

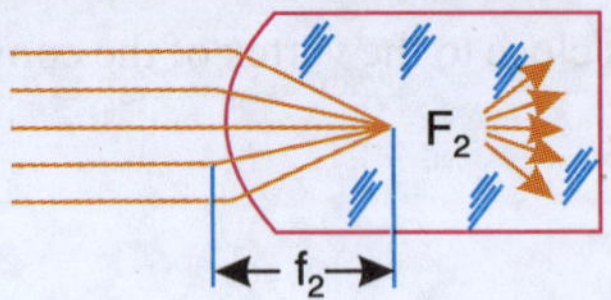

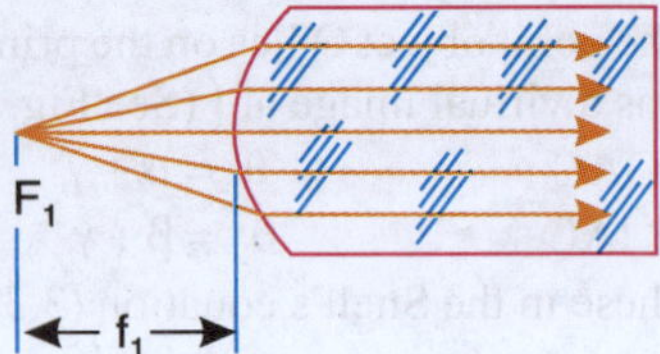

Fig. 3.31

First principal focus:

If the object is placed at F_1 such that the refracted rays become parallel to the principal axis, $\upsilon = \infty$. The object distance *u* is denoted by f_1 and is called the **first principal focal length** of the refracting surface.

$$\frac{\mu_2}{\infty} - \frac{\mu_1}{f_1} = \frac{\mu_2 - \mu_1}{R}$$

$$\therefore \qquad f_1 = -\frac{\mu_1 R}{\mu_2 - \mu_1} \qquad (3.45)$$

The position of the axial point object whose image is formed at infinity is known as the **first principal focus** of the refracting surface.

Second principal focus:

If the object is at infinity, then $u = \infty$ and $1/u = 0$. Let f_2 be the value of υ corresponding to u = ∞. Then

$$\frac{\mu_2}{f_2} = \frac{\mu_2 - \mu_1}{R}$$

or

$$f_2 = \frac{\mu_2 R}{\mu_2 - \mu_1} \qquad (3.46)$$

This value of $\upsilon = f_2$ is termed as the **second principal focal length** and the position of the image corresponding to the object at infinity is called the **second principal focus** of the refracting surface.

It is easy to see that **the first focal length f_1 for the spherical refracting surface is not equal to the second focal length f_2.**

In case of a convex surface R is positive and f_1 is negative and f_2 is positive. On the other hand, for a concave surface R is negative and hence f_1 is positive and f_2 is negative. When f_1 is positive the system is called **convergent**, and when f_1 is negative it is called **divergent**.

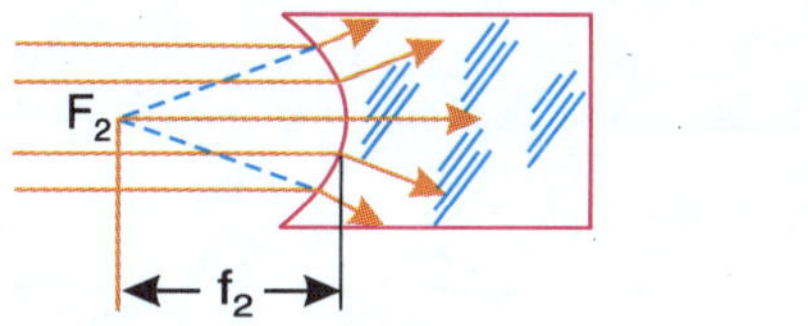

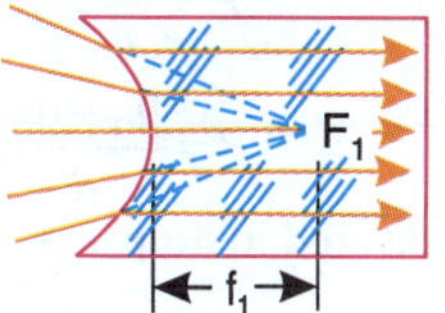

Fig. 3.32

The following points are interesting and are to be remembered.

If *parallel light* is incident on a *convex surface*, the light *converges* at the *second focal point*, F_2 (See Fig. 3.31). If the *convex surface renders* the incident *light into* a *parallel beam*, the light comes from the *first focal point*, F_1. The same applies to a concave surface also. In case of a convex surface, the second focal point F_2 is in the medium at right side while the first focal point F_1 is in air at left side to the surface (See Fig. 3.32). The reverse is true in case of a concave surface.

3.12.2. VARIOUS RELATIONS INVOLVING f_1 AND f_2

(*i*) The sum of the two focal lengths equals the radius of curvature.

$$f_1 + f_2 = \left[\frac{-\mu_1}{\mu_2 - \mu_1}\right]R + \left[\frac{\mu_2}{\mu_2 - \mu_1}\right]R = R \tag{3.47}$$

(*ii*) We see that using relation (3.45) into spherical mirror equation (3.39) or (3.44) we get

$$\frac{\mu_1}{u} - \frac{\mu_2}{\upsilon} = \frac{\mu_1}{f_1} \tag{3.48}$$

This is known as **Gauss' formula for a spherical surface.**

(*iii*) We may write

$$\frac{f_1}{\mu_1} = \frac{-R}{\mu_2 - \mu_1}, \quad \frac{f_2}{\mu_2} = \frac{R}{\mu_2 - \mu_1}$$

Adding,

$$\frac{f_1}{\mu_1} + \frac{f_2}{\mu_2} = 0 \quad \text{or} \quad \frac{f_1}{f_2} = -\frac{\mu_1}{\mu_2} \tag{3.49}$$

As μ_1 and μ_2 are always positive and unequal, f_1 and f_2 are always of opposite sign and are unequal. The ratio of focal lengths f_1/f_2 is equal to the ratio μ_1/μ_2 of the corresponding refractive indices.

(*iv*) We may write $\frac{\mu_2}{\upsilon} - \frac{\mu_1}{u} = \frac{\mu_2 - \mu_1}{R}$

as

$$\frac{\mu_2 R}{\upsilon(\mu_2 - \mu_1)} - \frac{\mu_1 R}{u(\mu_2 - \mu_1)} = 1$$

∴

$$\frac{f_2}{\upsilon} + \frac{f_1}{u} = 1 \tag{3.50}$$

3.12.3. POWER

The **refracting power** of a spherical surface is defined as the inverse of the *reduced* focal length. Any distance divided by the refractive index of the space in which it is measured is called a

reduced distance. Therefore, the ratios f_1/μ_1 and f_2/μ_2 are the reduced focal lengths. Since $\frac{f_1}{\mu_1} = -\frac{f_2}{\mu_2}$, the refracting power is given by

$$P = \frac{\mu_1}{f_1} = -\frac{\mu_2}{f_2} \tag{3.51}$$

$$\therefore \quad P = \frac{\mu_2 - \mu_1}{R} \tag{3.52}$$

Power is measured in **diopters**.

3.13. LATERAL MAGNIFICATION

AB is a small linear object situated on the axis far away from P in the medium of refractive index μ_1 (See Fig. 3.33). The image of the point A is formed at a point A′ on the axis. To find the image of the point B, take a ray BQ parallel to the principal axis, which after refraction passes through the second focus F_2. Another ray BC passes through the centre of curvature without deviation. The two rays intersect at the point B′, which is the image of B. Therefore, A′B′ is the linear image of the object AB.

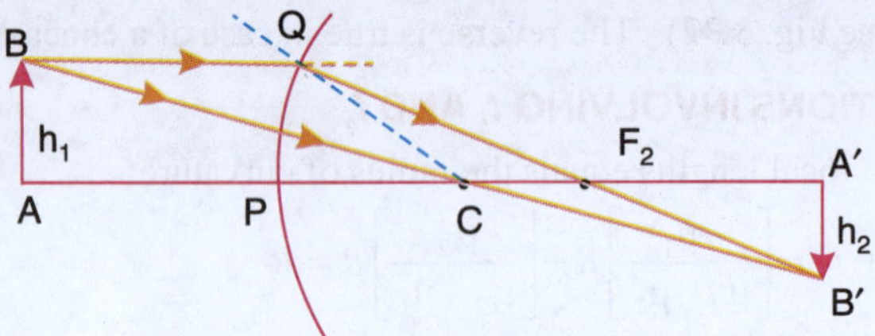

Fig. 3.33

$$\text{Linear Magnification} = m = \frac{A'B'}{AB} = \frac{h_2}{h_1}$$

Δ^{les} ABC and A′B′C are similar. $\therefore \frac{A'B'}{AB} = \frac{CA'}{CA}$

$$\frac{-h_2}{h_1} = \frac{PA' - PC}{PA + PC} = \frac{+\upsilon - (+R)}{-u + (+R)} = \frac{\upsilon - R}{-u + R}$$

or

$$\frac{h_2}{h_1} = \frac{\upsilon - R}{u - R} \tag{3.53}$$

We have

$$\frac{\mu_2}{\upsilon} - \frac{\mu_1}{u} = \frac{\mu_2 - \mu_1}{R} = \frac{\mu_2}{R} - \frac{\mu_1}{R}$$

$$\frac{\mu_2}{\upsilon} - \frac{\mu_2}{R} = \frac{\mu_1}{u} - \frac{\mu_1}{R}$$

$$\frac{\mu_2[R - \upsilon]}{\upsilon R} = \frac{\mu_1[R - u]}{uR}$$

$$\frac{\mu_2[\upsilon - R]}{\upsilon} = \frac{\mu_1[u - R]}{u}$$

$$\frac{[\upsilon - R]}{[u - R]} = \frac{\mu_1}{\mu_2} \cdot \frac{\upsilon}{u} \tag{3.54}$$

But $$m = \frac{h_2}{h_1} = \frac{\upsilon - R}{u - R}$$

$\therefore$ $$m = \frac{\mu_1}{\mu_2} \cdot \frac{\upsilon}{u} \qquad (3.55)$$

From the above equation it follows that when υ is positive, the image is real and inverted; and when υ is negative, the image is virtual and erect.

3.14. LONGITUDINAL MAGNIFICATION

In general objects are not linear but are distributed or extended. Therefore, in addition to lateral magnification, longitudinal magnification also occurs. The *ratio of a short length of the image dυ, measured along the axis, to the conjugate length du of the object*, is known as **longitudinal or axial magnification**. Thus,

Longitudinal magnification $$L = \frac{d\upsilon}{du} \qquad (3.56)$$

But $$\frac{\mu_2}{\upsilon} - \frac{\mu_1}{u} = \frac{\mu_2 - \mu_1}{R}$$

Differentiating the above expression, we get

$$-\frac{\mu_2}{\upsilon^2} \cdot \frac{d\upsilon}{du} + \frac{\mu_1}{u^2} = 0$$

or $$\frac{d\upsilon}{du} = \frac{\mu_1}{u^2} \cdot \frac{\upsilon^2}{\mu_2} = \frac{\mu_1}{\mu_2} \cdot \frac{\upsilon^2}{u^2}$$

$\therefore$ $$L = \frac{\mu_1}{\mu_2} \cdot \frac{\upsilon^2}{u^2} \qquad (3.57)$$

We can rewrite the above expression (3.57) as

$$L = \left[\frac{\mu_1 \upsilon}{\mu_1 \upsilon}\right]^2 \cdot \frac{\mu_2}{\mu_1}$$

$\therefore$ $$L = [m]^2 \cdot \frac{\mu_2}{\mu_1} \qquad (3.58)$$

Thus, *the longitudinal magnification is the square of lateral magnification.* The longitudinal magnification is always positive. If the object extends from left to right, the image likewise extends from left to right.

3.15. SMITH-HELMHOLTZ EQUATION AND LAGRANGE LAW

Referring to Fig. 3.34, A′B′ is the real inverted image of the object AB. The incident ray AQ after refraction passes through A′. Let the refracted ray QA′ be inclined at an angle θ_2 to the axis in the paraxial region.

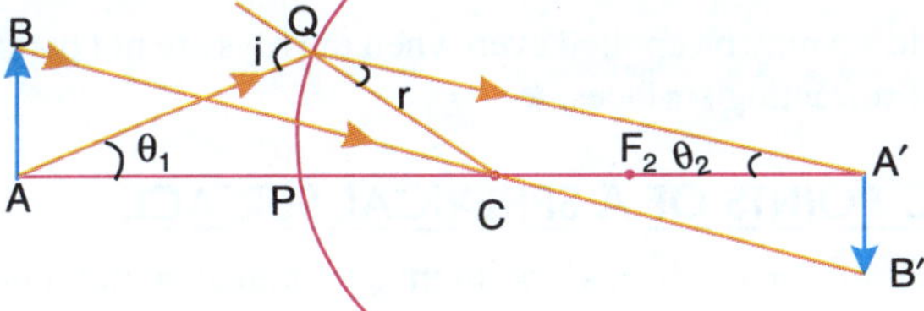

Fig. 3.34

Angular magnification $\alpha = \dfrac{\tan\theta_2}{\tan\theta_1}$

or $$\alpha = \frac{QP/\upsilon}{AP/u} = \frac{u}{\upsilon} \tag{3.59}$$

Linear magnification $m = \dfrac{h_2}{h_1} = \dfrac{\mu_1}{\mu_2}\cdot\dfrac{\upsilon}{u}$

$$\therefore \quad \frac{h_2}{h_1}\cdot\frac{u}{\upsilon} = \frac{\mu_1}{\mu_2}$$

or $$\frac{h_2}{h_1}\cdot\frac{\tan\theta_2}{\tan\theta_1} = \frac{\mu_1}{\mu_2}$$

Thus, linear magnification × angular magnification = μ_1 / μ_2. (3.60)

$$\therefore \quad \mu_1 h_1 \tan\theta_1 = \mu_2 h_2 \tan\theta_2 \tag{3.61}$$

This is known as *Smith-Helmholtz equation.*

Further, for paraxial rays, $\tan\theta_1 \cong \theta_1$ and $\tan\theta_2 \cong \theta_2$. Therefore, the above relation may be expressed as

$$\mu_1 h_1 \theta_1 = \mu_2 h_2 \theta_2 \tag{3.62}$$

This is known as **Langrange's law.**

3.16. ABBE'S SINE CONDITION

Referring to Fig. 3.34, let i and r be the angles of incidence and refraction for the rays AQ and QA′.

Now $$\frac{\sin i}{\sin\theta_1} = \frac{AC}{QC} \text{ and } \frac{\sin r}{\sin\theta_2} = \frac{CA'}{QC}$$

$$\frac{\sin i}{\sin r} = \frac{AC}{CA'}\cdot\frac{\sin\theta_1}{\sin\theta_2} \tag{3.63}$$

But $$\mu_1 \sin i = \mu_2 \sin r$$

$$\therefore \quad \mu_1 \cdot AC \sin\theta_1 = \mu_2 \cdot CA' \sin\theta_2 \tag{3.64}$$

Δ^{les}ABC and A′B′C are similar.

$$\frac{CA'}{AC} = \frac{A'B'}{AB} = \frac{h_2}{h_1}$$

$$\mu_1 h_1 \sin\theta_1 = \mu_2 h_2 \sin\theta_2 \tag{3.65}$$

The above condition is known as *Abbe's sine condition*. The linear magnification will be constant if the ratio $\sin\theta_1 / \sin\theta_2$ is constant. We shall use this condition for calculating the resolving power of a microscope.

Note— The sine condition may be applied even when the rays are not paraxial and is applicable to any system of refracting surfaces. ♦

3.17. APLANATIC POINTS OF A SPHERICAL SURFACE

A spherical refracting surface forms a point image of an axial point object only in the case of paraxial rays. When this condition is not satisfied, the refracted rays will not have a common point of intersection. The image of the axial point object will not be a point image. This departure from the ideal point-image is termed as *spherical aberration*. However, this defect may be overcome by the use of an aplanatic surface. An image-forming surface, which is free from spherical aberration and coma, is known as an *aplanatic surface*. It is a spherical surface and is, with respect to two conjugate points, such that all the rays originating from one of the two points are reflected or refracted through the other point. Therefore, the surface forms a geometrical point image of a point object on its axis, however big the aperture of the surface may be. These conjugate points are known as *aplanatic foci* or *aplanatic points*.

Fig. 3.35 shows a spherical surface APB, having C as the centre of its curvature and P as its pole. Let the refractive index of the medium of the sphere be μ_1 and that of the medium outside the sphere be μ_2 where $\mu_1 > \mu_2$. Consider any axial wide-angled ray OA from an axial point object O striking the surface at point A. The refracted ray bends away from the normal and the virtual image is formed at point I. It will now be shown that, provided object O is so placed that OC = (μ_2 / μ_1) R, then all rays originating from O and striking the surface APB will seem to come from I. That is, I is a perfect point image of the point object O and free from spherical aberration.

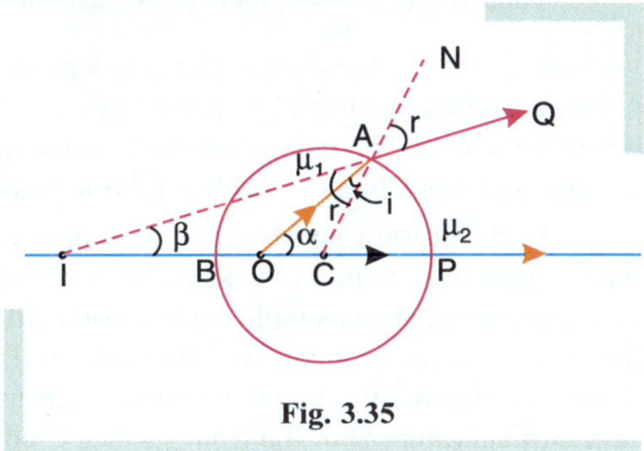

Fig. 3.35

Referring to the Fig. 3.35, the ray OA is incident on the sphere at A and is bent away from the normal and is refracted along AQ. Another ray OC, incident normally at P, passes straight without any deviation. The two rays AQ and OCP appear to diverge from I, which is the virtual image of O.

Now $\angle OAC = i$, $\angle NAQ = \angle IAC = r$, $\angle AOC = \alpha$, and $\angle AIC = \beta$

In Δ^{le} AOC,

$$\frac{\sin i}{\sin \alpha} = \frac{CO}{CA} = \frac{CO}{R}$$

The object is so placed that $CO = \frac{\mu_2}{\mu_1} R$.

$$\frac{\sin i}{\sin \alpha} = \frac{\mu_2 R}{\mu_1 R} = \frac{\mu_2}{\mu_1} \quad (3.66)$$

From Snell's law
$$\frac{\sin i}{\sin r} = \frac{\mu_2}{\mu_1} \quad (3.67)$$

$$\frac{\sin i}{\sin \alpha} = \frac{\sin i}{\sin r} \quad (3.68)$$

$$\sin \alpha = \sin r$$

$\therefore$
$$i = r$$

In the Δ^{le} IAO, exterior $\angle AOC = \beta + \angle IAO$

$\therefore$
$$\alpha = \beta + r - i. \quad \text{But } \alpha = r$$

$\therefore$
$$\beta = i$$

From Δ^{le} IAC,
$$\frac{\sin \beta}{\sin r} = \frac{CA}{CI}$$

Since $\beta = i$, and $CA = R$,
$$\frac{\sin i}{\sin r} = \frac{R}{CI} \quad (3.69)$$

Using equ.(3.67) into the above relation (3.69), we get

$$\frac{\mu_2}{\mu_1} = \frac{R}{CI}$$

$$\therefore \qquad CI = \frac{\mu_1 R}{\mu_2} \tag{3.70}$$

Thus, if $CO = \frac{\mu_2}{\mu_1} R$, then the corresponding image lies at a distance CI such that $CI = \frac{\mu_1 R}{\mu_2}$ and vice versa. Hence, the distance CI is independent of the position of A. In other words, all rays from point object O, however large an angle they make with the axis, appear to diverge from point I after emerging from the boundary between the two media. Therefore, there is no spherical aberration for this particular position of the object. Point O and I are the aplanatic points of the refracting surface.

Such aplanatic surfaces are made use of in microscope objectives where it is essential that as wide a pencil of light as possible from each point of the extremely small object should enter the objective; otherwise the greatly magnified image would be too faintly illuminated to be seen distinctly. The objective is so designed that the centre of curvature O of the front surface C coincides with the aplanatic point of the outer surface D (See Fig. 3.36). Hence, if a point object is placed at O, then rays from it enter the lens normally and suffer no refraction or deviation at the first surface. Therefore, the refraction at the outer surface D is without any spherical aberration.

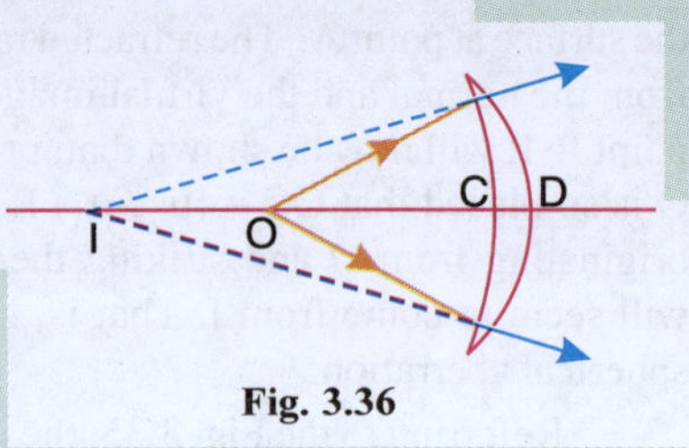

Fig. 3.36

WORKED OUT EXAMPLES

Example 3.1: A boy 1.5 m tall stands in front of a vertical plane mirror. How tall must be a vertical mirror and how high must its lower edge be above the floor if he is to be able to see his entire length? Assume his eyes are 10 cm below the top of his head.

Solution: Fig. 3.37 shows the boy JK standing in front of the mirror AB. It also shows the paths of rays that leave his head and enter his eyes after reflection from the mirror at point A; and the rays that leave his feet and enter his eyes reflected at the point B. The mirror need to be of length L = AB. From the Fig. 3.37, it is easy to see that

$$CG = \tfrac{1}{2} JG$$

$$\therefore \qquad L = CD = CG + GD$$

$$\therefore \qquad = \tfrac{1}{2} JG + \tfrac{1}{2} GK$$

$$= \tfrac{1}{2}(JG + GK) = \tfrac{1}{2} JK.$$

$$L = 1.5m/2 = 0.75m$$

Fig. 3.37

The distance of the bottom edge of the mirror above floor

= BE = GK/2 = (JK – JG)/2 = (1.5 – 0.1)/2 = **0.7 m.**

Note that *the horizontal distance of the boy from the mirror makes no difference.*

Example 3.2: A point light source lies on the principal axis of a concave spherical mirror with radius of curvature 160 cm. Its vertical image appears to be at a distance of 70 cm from it. Determine the location of the light source.

Solution: $\frac{1}{u}+\frac{1}{\upsilon}=\frac{2}{R}$ Here υ = 70 cm, R = –160 cm

$$\frac{1}{u}=\frac{2}{R}-\frac{1}{\upsilon}$$

$$\therefore \quad \frac{1}{u}=\frac{2}{-160\,\text{cm}}-\frac{1}{70\,\text{cm}}=-\frac{15}{560\,\text{cm}}$$

$$\therefore \quad u=-\frac{560}{15}\,\text{cm}=-37\text{ cm}$$

The light is at a distance of 37 cm in front (to the left of the vertex) of the mirror.

Example 3.3: A point source of light is located 20 cm in front of a convex mirror with f = 15 cm. Determine the position and character of the image point.

Solution: $\frac{1}{u}+\frac{1}{\upsilon}=\frac{1}{f}$ Here u = – 20 cm, f = 15 cm.

$$\frac{1}{\upsilon}=\frac{1}{f}-\frac{1}{u}=\frac{1}{15\,\text{cm}}-\frac{1}{-20\,\text{cm}}=\frac{35}{300\,\text{cm}}=\frac{7}{60\,\text{cm}}$$

$$\upsilon=8.6\text{ cm}$$

As υ is positive, the image is located behind (to the right side of the vertex of) the mirror. Hence, the image is *virtual*.

Example 3.4: A concave spherical surface of radius of curvature 100 cm separates two media of refractive indices 1.50 and 4/3. An object is kept in the first medium at a distance of 30 cm from the surface. Calculate the position of the image.

Solution: μ_1= 1.50, μ_2 = 4/3, u = – 30 cm, R = – 100 cm

$$\frac{\mu_2}{\upsilon}-\frac{\mu_1}{u}=\frac{\mu_2-\mu_1}{R}$$

$$\frac{4/3}{\upsilon}-\frac{1.50}{-30\,\text{cm}}=\frac{4/3-1.50}{-100\,\text{cm}}$$

$$\boldsymbol{\upsilon=-27.58\text{ cm}}$$

Example 3.5: Show that for refraction at a concave spherical surface (separating glass-air medium), the distance of the object should be greater than three times the radius of curvature of the refracting surface for the image to be real.

Solution: μ_1= 1.50, μ_2 = 1, u and R are negative.

$$\frac{\mu_2}{\upsilon}-\frac{\mu_1}{u}=\frac{\mu_2-\mu_1}{R}$$

$$\frac{1}{\upsilon}-\frac{1.50}{-u}=\frac{1-1.50}{-R},\quad \frac{1}{\upsilon}+\frac{3}{2u}=\frac{1}{2R}$$

$$\therefore \quad \frac{1}{\upsilon}=\frac{1}{2R}-\frac{3}{2u}$$

For υ to be positive $\frac{1}{2R}>\frac{3}{2u}$ or $u>3R$

Example 3.6: A convex surface of radius of curvature 40 cm separates two media of refractive indices 4/3 and 1.50. An object is kept in the first medium at a distance of 20 cm from the surface. Calculate the position of the image.

Solution: $\mu_1 = 4/3,\ \mu_2 = 1.50,\ u = -20$ cm, R = 40 cm

$$\frac{\mu_2}{\upsilon} - \frac{\mu_1}{u} = \frac{\mu_2 - \mu_1}{R}$$

$$\frac{1.50}{\upsilon} - \frac{4/3}{-20\text{ cm}} = \frac{1.50 - 4/3}{40\text{ cm}}$$

$$\boldsymbol{\upsilon = -24\text{ cm}}$$

Example 3.7: A convex refracting surface of radius of curvature 15 cm separates two media of refractive indices 4/3 and 1.50. An object is kept in the first medium at a distance of 240 cm from the surface. Calculate the position of the image.

Solution: $\mu_1 = 4/3,\ \mu_2 = 1.50,\ u = -240$ cm, R = 15 cm

$$\frac{\mu_2}{\upsilon} - \frac{\mu_1}{u} = \frac{\mu_2 - \mu_1}{R}$$

$$\frac{1.50}{\upsilon} - \frac{4/3}{-240\text{ cm}} = \frac{1.50 - 4/3}{15\text{ cm}}$$

$$\boldsymbol{\upsilon = 270\text{ cm}}$$

A real image forms in the second medium at a distance of 270 cm from the refracting surface.

Example 3.8: The eye can be regarded as a single spherical refracting surface of radius of curvature of cornea 7.8 mm, separating two media of refractive indices 1.00 and 1.34. Calculate the distance from the refracting surface at which a parallel beam of light will come to focus.

Solution: $\mu_1 = 1.00,\ \mu_2 = 1.34,\ u = -\infty$, R = 0.78 cm

$$\frac{\mu_2}{\upsilon} - \frac{\mu_1}{u} = \frac{\mu_2 - \mu_1}{R}$$

$$\frac{1.34}{\upsilon} - \frac{1.00}{-\infty} = \frac{1.34 - 1.00}{0.78\text{ cm}}$$

$$\boldsymbol{\upsilon = 3.075\text{ cm}}$$

A real image forms in the second medium at a distance of 3.075 cm from the refracting surface.

Example 3.9: A glass dumbbell of length 50 cm and refractive index 1.50 has ends of 5 cm radius of curvature. Find the position of the image formed due to refraction at one end only, when a point object is situated in air at a distance of 20 cm from the end of the dumbbell along the axis.

Solution:

$$\frac{\mu_2}{\upsilon} - \frac{\mu_1}{u} = \frac{\mu_2 - \mu_1}{R}$$

Here $\mu_1 = 1.50,\ \mu_2 = 1,\ u = -20$ cm, R = 5 cm

$$\frac{1.5}{\upsilon} - \frac{1.00}{-20\text{ cm}} = \frac{1.50 - 1.00}{5\text{ cm}}$$

$$\boldsymbol{\upsilon = 30\text{ cm}}$$

Since υ is positive, the image is formed 30 cm to the right of the vertex as shown in Fig. 3.38.

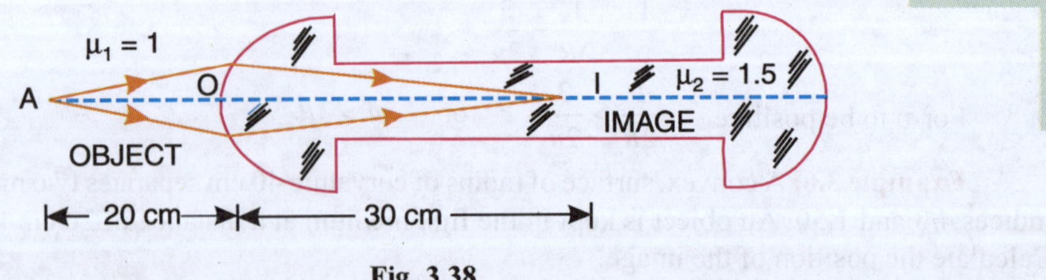

Fig. 3.38

Example 3.10: In the above example, if the object is 5 cm from the dumbbell, what is the position of the image?

Solution:
$$\frac{1.5}{\upsilon} - \frac{1.00}{-5\text{ cm}} = \frac{1.50-1.00}{5\text{ cm}}$$
$$\boldsymbol{\upsilon = -15\text{ cm}}$$

Therefore, the image is formed to the left of the vertex, as shown in Fig. 3.39.

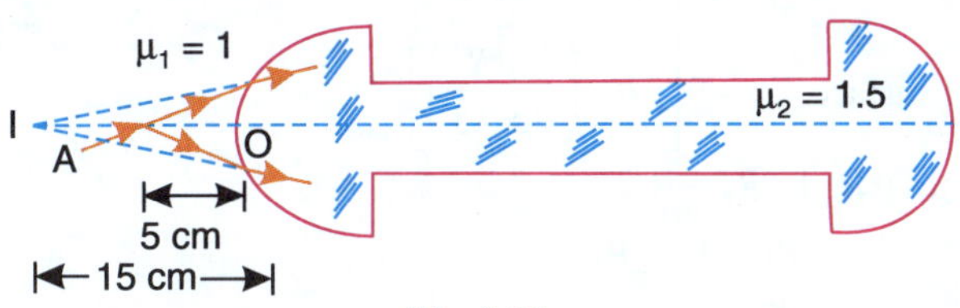

Fig. 3.39

Example 3.11: A small filament is at the centre of a hollow glass sphere of inner and outer radii 8 cm and 9 cm respectively. The refractive index of glass is 1.50. Calculate the position of the image of the filament when viewed from outside the sphere.

Solution: For refraction at the first surface, $\mu_1 = 1$, $\mu_2 = 1.50$, $u = -8$ cm, $R = -8$ cm

$$\frac{\mu_2}{\upsilon} - \frac{\mu_1}{u} = \frac{\mu_2 - \mu_1}{R}$$
$$\frac{1.5}{\upsilon'} - \frac{1.00}{-8\text{ cm}} = \frac{1.50-1.00}{-8\text{ cm}}$$
$$\boldsymbol{\upsilon' = -8\text{ cm}}$$

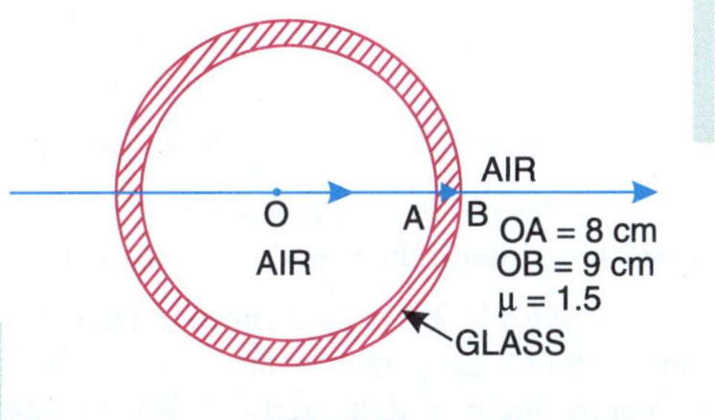

Fig. 3.40

It means that due to the first surface the image is formed at the centre of the sphere.

For the second surface, $\mu_1 = 1.50$, $\mu_2 = 1$, $u = -9$ cm, $R = -9$ cm

$$\frac{\mu_2}{\upsilon} - \frac{\mu_1}{u} = \frac{\mu_2 - \mu_1}{R}$$
$$\frac{1}{\upsilon} - \frac{1.50}{-9\text{ cm}} = \frac{1-1.50}{-8\text{ cm}}$$
$$\boldsymbol{\upsilon = -9\text{ cm}}$$

Hence, the final image is formed at the centre of the sphere.

Example 3.12: A glass rod has ends as shown in Fig. 3.41. The refractive index of glass is μ. The object point O is at a distance 2R from the surface of larger radius of curvature. The distance between the apexes of the ends is 3R. Show that the image point I is formed at a distance of $\frac{(9-4\mu)R}{(10\mu-9)(\mu-2)}$ from the right hand vertex.

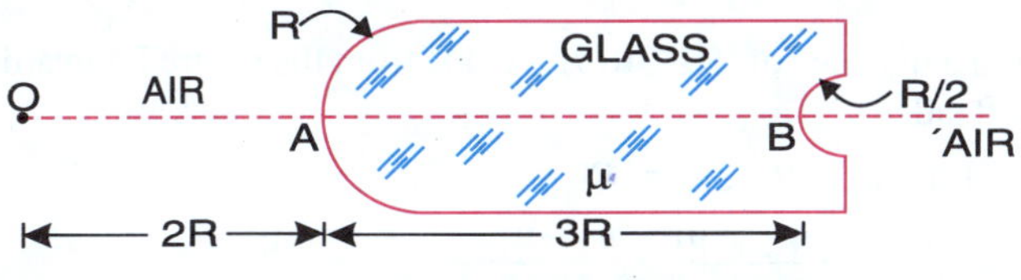

Fig. 3.41

Solution: For the first surface $\frac{\mu_2}{\upsilon'} - \frac{\mu_1}{u} = \frac{\mu_2 - \mu_1}{R}$

$\mu_1 = 1,\ \mu_2 = \mu,\ u = -2R,\ R_1 = R$

$$\frac{\mu}{\upsilon'} - \frac{1}{-2R} = \frac{\mu - 1}{R} \quad \text{or} \quad \frac{\mu}{\upsilon'} = \frac{2\mu - 3}{2R}$$

$$\therefore \quad \upsilon' = \frac{2\mu R}{2\mu - 3}$$

For the second surface, $u = -\left[3R - \left(\frac{2\mu R}{2\mu - 3}\right)\right]$

Here $\mu_1 = \mu,\ \mu_2 = 1,\ R_2 = R/2$

$$\frac{\mu_2}{\upsilon} - \frac{\mu_1}{u} = \frac{\mu_2 - \mu_1}{R}$$

$$\frac{1}{\upsilon} + \frac{\mu}{\left[3R - \left(\frac{2\mu R}{2\mu - 3}\right)\right]} = \frac{1 - \mu}{R/2}$$

$$\therefore \quad \frac{1}{\upsilon} = \frac{2 - 2\mu}{R} - \frac{\mu(2\mu - 3)}{(6\mu R - 9R - 2\mu R)}$$

$$\therefore \quad \frac{1}{\upsilon} = \frac{-10\mu^2 + 29\mu - 18}{R(4\mu - 9)}$$

$$\therefore \quad \upsilon = \frac{(9 - 4\mu)R}{(10\mu - 9)(\mu - 2)}$$

In this problem, the image will be real if the refractive index of glass is between 2 and 2.25. It will be virtual if the refractive index is less than 2 or is more than 2.25.

Example 3.13: Two concave refracting surfaces of equal radii of curvature and refractive index 1.5 face each other in air as shown in Fig. 3.42. A point object is placed midway in between the centre and one of the vertices. What is the separation between the images of O formed by each refracting surface?

Solution: We have for the image formed by the surface on the right side,

$\mu_1 = 1,\ \mu_2 = 1.5,\ u = -R/2,\ R_1 = -R$

$$\frac{\mu_2}{\upsilon} - \frac{\mu_1}{u} = \frac{\mu_2 - \mu_1}{R_1}$$

$$\frac{1.5}{\upsilon} - \frac{2}{-R} = \frac{1.50 - 1.00}{-R}$$

$$\frac{1.5}{\upsilon} = \frac{1}{-2R} - \frac{2}{R} = \frac{-5}{2R}$$

$$\therefore \quad \upsilon = -\frac{3R}{5}$$

The image is at a distance of (R – 3R/5) = 0.4 R from the centre P towards the right side. For the surface on the left side,

$\mu_1 = 1,\ \mu_2 = 1.5,\ u = 3R/2,\ R_2 = R$

$$\frac{\mu_2}{\upsilon} - \frac{\mu_1}{u} = \frac{\mu_2 - \mu_1}{R_2}$$

$$\frac{1.5}{\upsilon} - \frac{2}{-3R} = \frac{1.50 - 1.00}{R}$$

$$\frac{1.5}{\upsilon} = \frac{1}{2R} + \frac{2}{3R} = \frac{7}{6R}$$

$$\therefore \quad \upsilon = \frac{9R}{7}$$

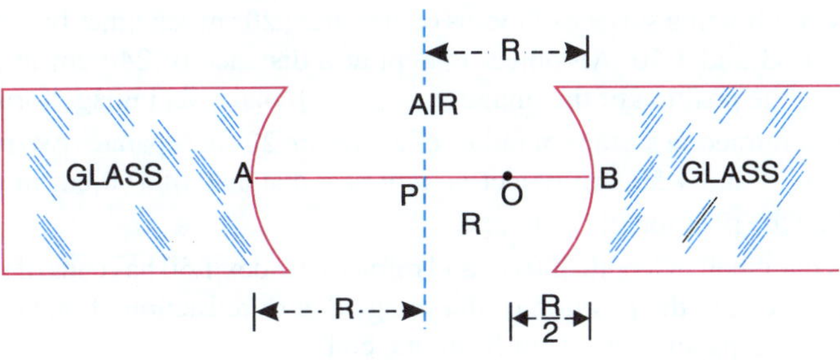

Fig. 3.42

Hence, the image is at a distance of (9R / 7–R) = 2R / 7 from the centre P towards the right hand side. The distance between the two images $= 0.4R - 2R/7 = 0.114R$

QUESTIONS

1. Discuss refraction at a convex surface when the image is virtual. Show that for air medium
2. Prove that in the case of refraction at a concave surface
$$\frac{\mu}{\upsilon} - \frac{1}{u} = \frac{u-1}{R}$$
3. Discuss refraction at a convex surface when the image formed is real. Derive the necessary formula.
$$\frac{\mu_2}{\upsilon} - \frac{\mu_1}{u} = \frac{u_2 - u_1}{R}$$
4. Explain the terms first principal focus and second principal focus and show that
$$f_1\upsilon + f_2u = u\upsilon$$
5. Calculate the transverse and the longitudinal magnifications for refraction at a spherical surface.
6. Derive Helmholtz's relation as used for a number of coaxial refracting surfaces.
7. What do you mean by aplanatic points and aplanatic surface? What are the advantages of aplanatic surfaces? **(Nagpur, 2004, 2005; Madhurai Kamaraj, 2003)**
8. Obtain an expression for refraction at a single spherical surface. Hence derive the lens-maker's formula.
9. Derive an expression connecting the object and image distances, when refraction of light takes place at a spherical surface separating two media.
10. State and explain Lagrange's equation for magnification. **(Nagpur, 2005)**

PROBLEMS FOR PRACTICE

1. A light spot is 150 cm away from a convex spherical mirror with radius of curvature 72 cm. Compute the image distance. [**Ans:** – 29 cm]

2. An object is placed before a concave mirror normal to its principal axis so that its magnification is 1.2. After the object has been moved further from the mirror by 25 cm, the magnification becomes 0.4. Calculate the focal length and radius of curvature of the mirror. **[Ans:** f = 15 cm; R = 30 cm]

3. A convex refracting surface of radius of curvature 50 cm separates two media of refractive indices 1.33 and 1.50. An object is placed at a distance of 100 cm in the first medium. Calculate the position of the image. **[Ans:** A virtual image forms at $\upsilon = -150$ cm.]

4. A convex refracting surface of radius of curvature 20 cm separates two media of refractive indices 1.33 and 1.50. An object is kept at a distance of 240 cm in the first medium. Calculate the position of the image. **[Ans:** a real image forms at $\upsilon = 540$ cm]

5. A convex refracting surface of radius of curvature 20 cm separates two media of refractive indices 1.33 and 1.50. An object is kept at a distance of 160 cm in the first medium. Calculate the position of the image. **[Ans:** $\upsilon = \infty$]

6. A glass dumbbell of length 200 cm and refractive index 1.50 has ends of radius of curvature 10 cm. Calculate the position of the image due to refraction at one end only, when the object is at a distance of 40 cm from one end. **[Ans:** 60 cm]

7. A glass dumbbell of length one metre and refractive index 1.5 has ends of radius of curvature 5 cm. Calculate the position of the image due to refraction at one end only, when the object is at a distance of 50 cm from one end. **[Ans:** 18.75 cm]

8. Obtain the condition for the formation of a real image when paraxial rays of light are refracted at a concave spherical water-air interface. Radius of curvature of the surface is R. **[Ans:** $u > 4R$]

9. A small filament is at the centre of a hollow glass sphere of inner and outer radii 4 cm and 4.5 cm respectively. The refractive index of glass is 1.50. Calculate the position of the image of the filament when viewed from outside the sphere. **[Ans:** At the centre]

CHAPTER

Lenses

4.1. INTRODUCTION

A lens is an image-forming device. It forms an image by refraction of light at its two bounding surfaces. *In general, a lens is made of glass and is bounded by two regular curved surfaces; or by one spherical surface and a plane surface.* Spherical surfaces are easy to make. Therefore, most lenses are made of spherical surfaces and have a wide range of curvatures. Other transparent materials such as quartz, fused silica and plastics are also used in making lenses. A single lens with two refracting surfaces is a **simple lens**. We study here the behaviour of a simple lens, with a view to gain familiarity with lens systems. We use the ray concept to understand the behaviour of light passing through lenses and derive the relationship between focal length of the lens, object distance and image distance.

4.2. LENSES

Lenses are mainly of two types- *convex lens* and *concave* lens. A **convex lens** is thicker at the center than at

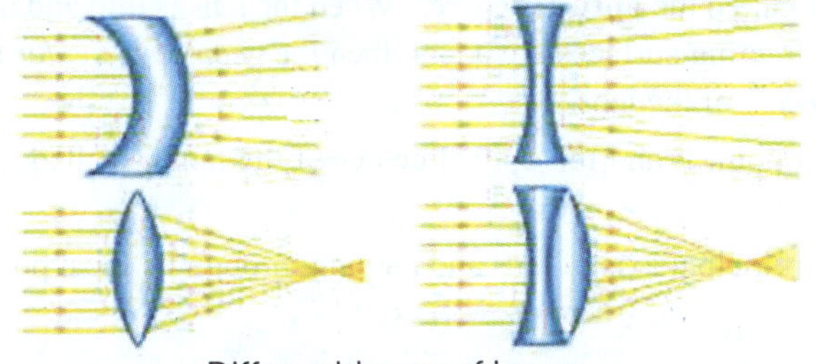

Different types of lenses.

At a Glance

- Introduction
- Lenses
- Terminology
- Conjugate Points, Planes and Distances
- Image Tracing
- Location of the Image
- Sign Convention
- Thin Lens
- Lens Equation
- Lens Maker's Equation
- Newton's Lens Equation
- Magnification
- Smallest Separation of Object and Real Image
- Displacement of Lens when Object and Screen are Fixed
- Deviation by a Thin Lens
- Power
- Equivalent Focal Length of Two Thin Lenses

the edges while a **concave lens** is thinner at the center than at the edges. A convex lens is a *converging* lens since a parallel beam of light, after refraction, converges to a point, F. A concave lens is called a *diverging lens* since rays coming parallel to the principal axis, after refraction, diverge out and seem to come from a point, F. Within these two categories there is a variety of simple lenses; some of the standard forms are shown in Fig. 4.1.

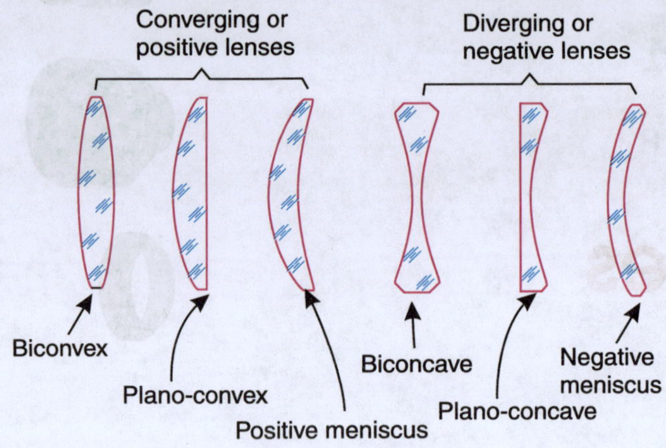

Fig. 4.1

4.3. TERMINOLOGY

We first acquaint with the terminology and the sign convention associated with lenses.

- A lens has two curved surfaces, each surface having a **curvature**.
- The length of the radius of curvature of surface is called the **radius of curvature**, R.
- The reciprocal of the length of the radius of curvature is known as the **curvature** C ($C = 1/R$). A lens has two centers of curvature and two radii of curvature, one for each refracting surface.
- The line joining the centers of curvature of the two curved surfaces is called the **principal axis** or simply **axis** of the lens.
- The points where the principal axis intersects the two refracting surfaces are called the **front vertex** and the **back vertex**.
- The point F to which a set of rays parallel to the principal axis is caused to converge (in case of convex lens) or appear to diverge (in case of concave lens) is the **principal focus**.
- For every lens, there is a point on the principal axis for which the rays passing through it are not deviated by the lens. Any ray passing through it emerges in a direction parallel to the incident ray. Such a point is called the **optical centre.** When the lens is thin and the radii of curvature of the two refracting surfaces are equal, then the *geometric center* of the lens becomes the *optical center* of the lens.
- The distance between the focal point F and the optical center of the lens is called the **focal length** of the lens.
- The plane perpendicular to the principal axis of lens and passing through its focal point is known as the **focal plane**.

- When a point object or a linear object is placed on one side of a convex lens beyond the focal plane, an image is formed on the opposite side, as shown in Fig.4.3. The distance from the front vertex to the object is called the **object distance u** and the distance from the back vertex to the image is the **image distance, υ.**
- The **power** of a lens is the reciprocal of its focal length.

4.4. CONJUGATE POINTS, PLANES AND DISTANCES

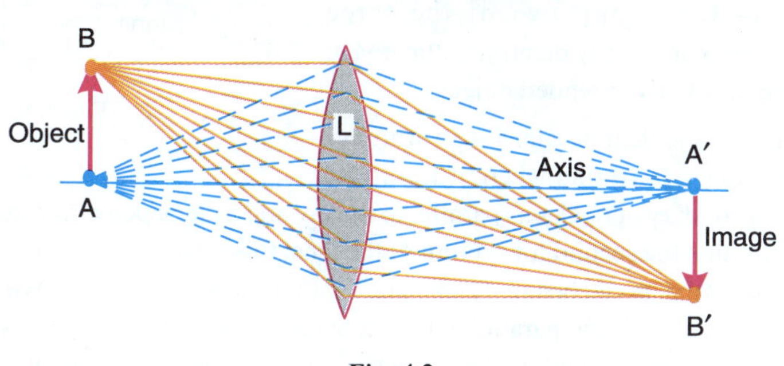

Fig. 4.2

In Fig.4.2 all the rays coming from an *object point* B are brought to a focus at B′. Similarly, the rays from point A are brought to a focus at A′. A′ B′ is the *image* of the linear object AB. It follows from the principle of reversibility of light rays that the object and image positions are interchangeable. Thus, when the image A′ B′ becomes the object and AB becomes its image. The *object and image are therefore conjugate.* Any pair of object and image points A and A′, as in Fig.4.2, are **conjugate points**. Planes on which the conjugate points lie are **conjugate planes** and the distances from the vertex to these planes are **conjugate distances.**

4.5. IMAGE TRACING

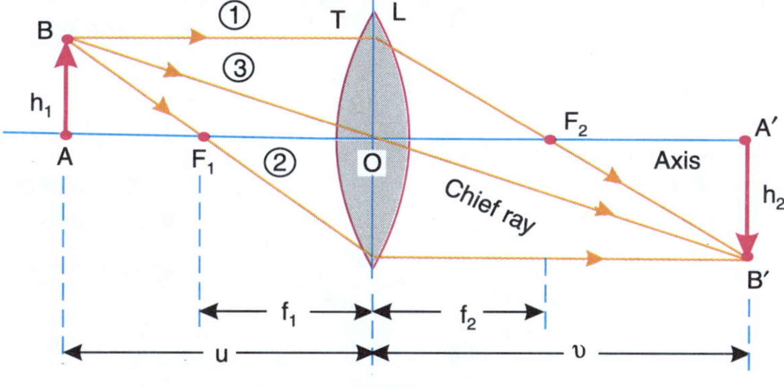

Fig. 4.3

We may use graphical ray tracing to determine the position of the image formed by a lens. To find the image, we take the help of characteristic rays shown in Fig. 4.3.

1. One is the ray parallel to the principal axis, which after refraction, passes through focal point F_2.

2. Second ray is the ray that passes through the first focal point F_1 of the lens; after refraction, it travels parallel to the principal axis.
3. The third ray, usually called *chief ray* goes through the optical centre of the lens and emerges without deviation.

Thus, we have three characteristic rays whose paths are known. Using any two of the three characteristic rays, we can readily determine the image of any object-point or of any extended object.

Fig. 4.4

Suppose a ray is incident on a lens at an arbitrary angle, as shown in Fig.4.4. To find where it forms the image, we draw a secondary optical axis parallel to the ray and a plane perpendicular to the principal axis and passing through the second focal point of the lens; this plane is the second focal plane. It is readily seen that just as a paraxial beam of light parallel to the principal optical axis converges at the focal point F_2, so the beam of light parallel to the secondary optical axis converges at a point on the focal plane. Thus, light rays incident on the surface of a convex lens at a particular angle to its *axis*, are redirected towards a particular position in the *focal plane* of the lens. We can therefore regard a convex lens as an *angle-to-position converter* for light.

4.6. LOCATION OF THE IMAGE

A convex lens produces a real or virtual image depending on the location of the object. A concave lens *always* produces virtual images of real objects. Fig. 4.5 illustrates the behaviour of a convex lens pictorially.

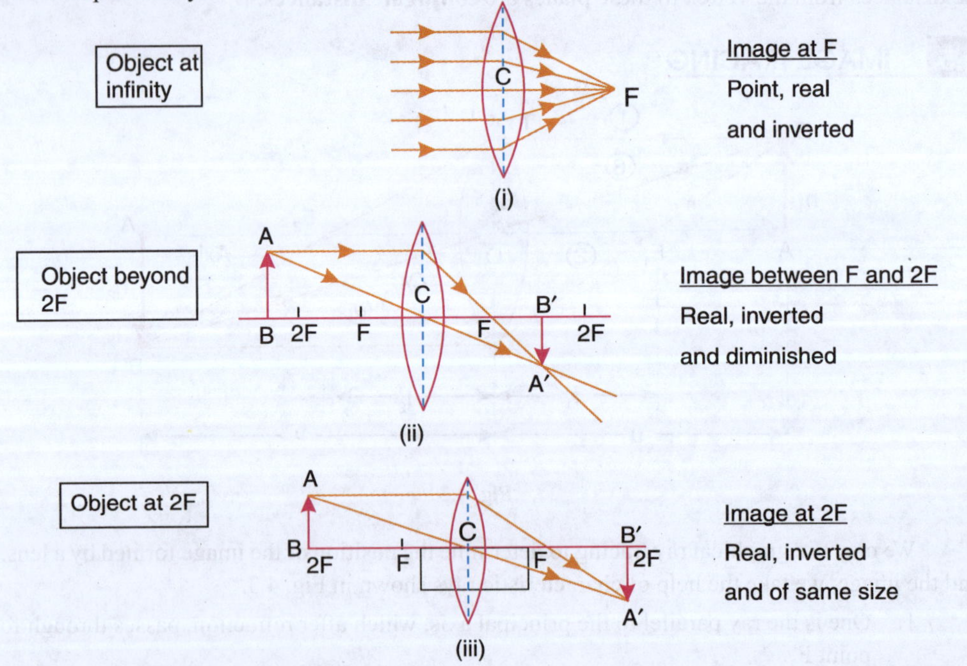

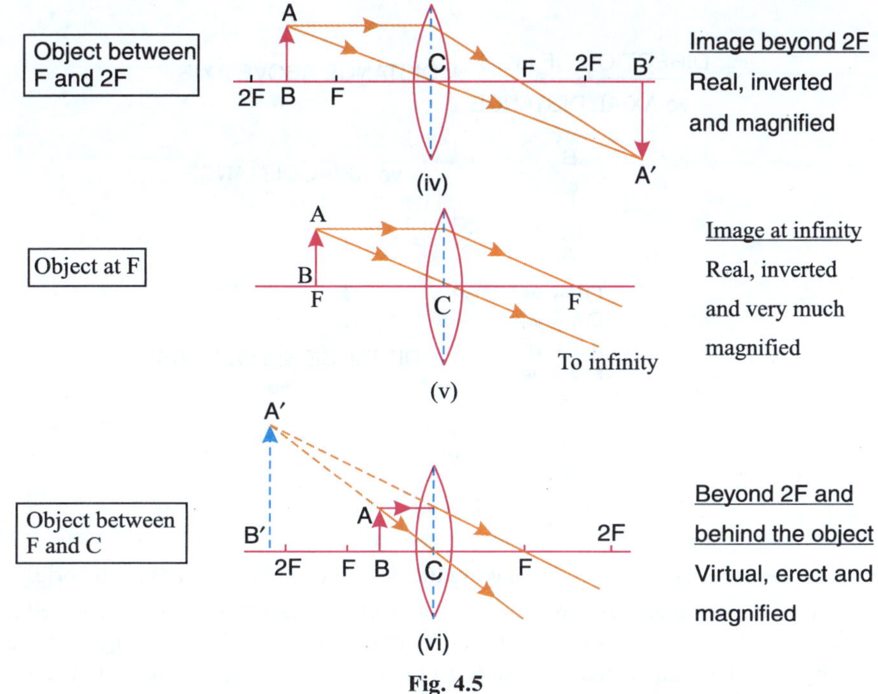

Fig. 4.5

As the object approaches the lens, it is seen that the real image moves away from it.

(*i*) When the object is very far away, the image is just to the right of the focal plane. The image is real, inverted, and smaller in size than the object (m < 1).

(*ii*) As the object approaches the lens, the image moves away to the right of the focal plane getting larger and larger. The image is real and inverted. This is the configuration for cameras and eyeballs.

(*iii*) When the object is at 2F, the image is real, inverted and of the same size as the object (m = 1). This is the configuration of a photocopier.

(*iv*) When the object is in between 2F and F, the image is enlarged (m > 1), real and inverted. This configuration corresponds to the film projector.

(*v*) When the object is precisely at F, there is no image as the emerging rays are parallel and in effect the image is at infinity.

(*vi*) With the object closer in than one F, the image reappears. It is virtual, erect and enlarged (m > 1). This is the configuration of the magnifying glass.

It may be noted that as an image-forming device, a convex lens is similar to a convex mirror. On the other hand, a concave lens resembles a convex mirror.

4.7. SIGN CONVENTION

The following convention of the signs is adopted for obtaining the relation between these quantities.

(*a*) The diagrams are drawn showing the incident light travelling from left to right.

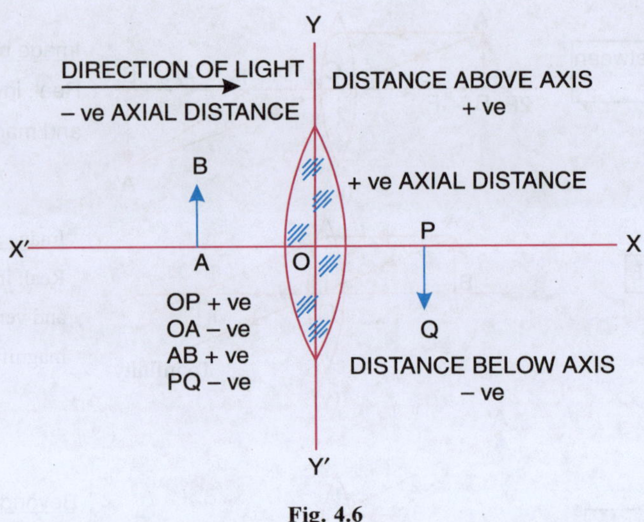

Fig. 4.6

(*b*) The distances are measured by taking the optical centre of the lens as the origin.

(*c*) The distances measured in the direction of the incident light are considered positive while those measured in the direction opposite to the incident light are taken as negative. (All quantities measured to the right of P are positive and all those to its left are negative.)

(*d*) Heights measured upward and perpendicular to the principal axis are taken as positive while those measured downward are considered negative.

(*e*) The angle made by a ray with the principal axis is taken to be positive if the ray has to be rotated anti-clockwise to become coincident with the axis, otherwise it is negative.

4.8. THIN LENS

Lenses are broadly classified into *thin* and *thick* lenses. *A lens is said to be* **thin** *if the thickness of the lens can be neglected when compared to the lengths of the radii of curvature of its two refracting surfaces, and to the distances of the objects and images from it.* No lens is actually a thin lens. Yet many simple lenses commonly used can be treated as equivalent to a thin lens.

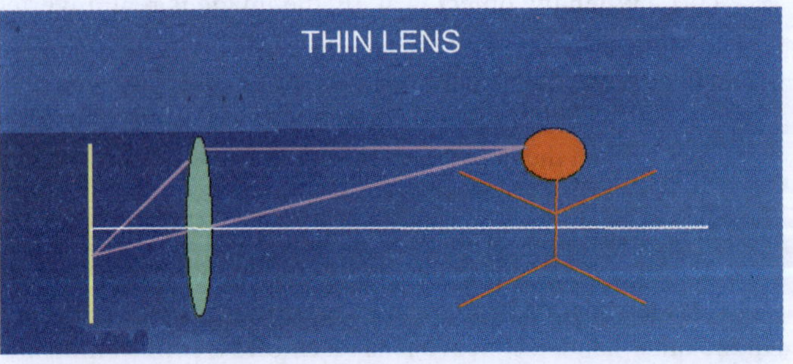

4.9. LENS EQUATION

A lens forms an image by refraction of light at its two bounding surfaces. Each surface acts as an image-forming component, and contributes to the final image formed by the lens. If we know the focal length of a lens and the position of an object, the position of the image can be determined either by using graphical construction (ray diagram) or using mathematical relation. It is not always convenient to draw a ray diagram. We study here how the position of the image is determined using mathematical equation.

Now let us consider a *thin* convex lens XY with optical center at C, as shown in Fig. 4.7. Let the absolute refractive index of the lens materials be μ_2; and the lens be surrounded on all sides by air or any other rarer medium of refractive index μ_1. The centers of curvatures of the two refracting surfaces of the lens are C_1 and C_2. Let the radius of the curvature of the surfaces be R_1 and R_2 respectively. Consider a point object 'O' situated on the principal axis of the lens.

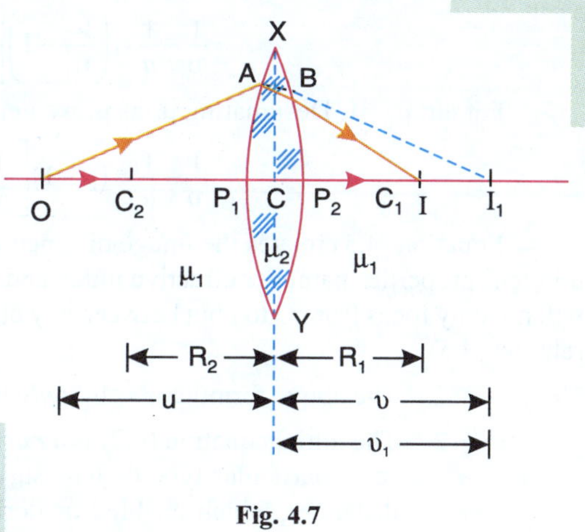

Fig. 4.7

A ray of light leaving the axial point-object O and travelling along the principal axis passes without deviation. Another ray OA strikes the first surface at A and is refracted in a direction BI_1. The ray is further refracted by the second surface in a direction BI and meets the ray along the principal axis at I. Therefore, I is the final point-image of the object O formed after refraction by the two surfaces of the lens.

By considering refraction at spherical surfaces we can derive an equation that describes image formation by a lens. The basic idea is that the image formed by the first refracting surface acts as a *virtual object* for the second refracting surface. We can apply **Gauss formula**

$$\frac{\mu_1}{u}+\frac{\mu_2}{\upsilon}=\frac{\mu_2-\mu_1}{R} \tag{4.1}$$

for refraction at the first surface.

The first surface, XAP_1Y, forms a *real* point-image, I_1. We replace υ by υ_1 and R by R_1 in equ.(4.1) and apply the sign convention. We see that u is – ve. Therefore, the equation (4.1) is written for the first surface as

$$\frac{\mu_2}{\upsilon_1}-\frac{\mu_1}{u}=\frac{\mu_2-\mu_1}{R_1} \tag{4.2}$$

In the Fig. 4.7, the light leaving the first surface would reach I_1 if the second surface XBP_2Y did not intervene. When the second surface is present, the light does not reach I_1. I_1 becomes the virtual object for the second surface and the second surface forms the image at I, which is the final image. The position of the final image is again found using Gauss formula using $u = \upsilon$ and $R = R_2$. Thus, we get

$$\frac{\mu_2}{\upsilon_1}+\frac{\mu_1}{\upsilon}=\frac{\mu_2-\mu_1}{R_2}$$

Applying the sign convention, we get

$$-\frac{\mu_2}{\upsilon_1}+\frac{\mu_1}{\upsilon}=-\frac{\mu_2-\mu_1}{R_2} \tag{4.3}$$

Adding equations (4.2) and (4.3) , we obtain

$$\mu_1\left(\frac{1}{\upsilon}-\frac{1}{u}\right)=(\mu_2-\mu_1)\left[\frac{1}{R_1}-\frac{1}{R_2}\right]$$

$$\left(\frac{1}{\upsilon}-\frac{1}{u}\right)=\left(\frac{\mu_2-\mu_1}{\mu_1}\right)\left[\frac{1}{R_1}-\frac{1}{R_2}\right]$$

$$\frac{1}{\upsilon}-\frac{1}{u}=\left(\frac{\mu_2}{\mu_1}-1\right)\left[\frac{1}{R_1}-\frac{1}{R_2}\right] \tag{4.4}$$

For air, $\mu_1 = 1$. Designating μ_2 as μ, we get

$$\frac{1}{\upsilon}-\frac{1}{u}=(\mu-1)\left[\frac{1}{R_1}-\frac{1}{R_2}\right] \tag{4.5}$$

Equation (4.5) relates the image distance υ of a thin lens to the object distance u and to the thin lens properties namely, refractive index and the radii of curvature. It is to be noted that a lens will not only focus from O to I but between any other pairs of points, as long as the points satisfy the relation (4.5).

Note.—
1. The above formula is valid *only* for nearly paraxial rays and strictly a *thin* lens.
2. The initial equation (4.2) is a *numerical relationship* between the various distances for a particular type of lens surface. We then adopted a sign convention for the distances, which enabled us derive an *algebraic relationship*. Such an algebraic relation (4.5) obtained with the sign convention will serve for all types of lenses used in all possible ways. Any sign convention that is different than that adopted by us is likely to give a different algebraic relationship.
3. u, υ, f and R are algebraic quantities each having a sign, numerical value and a unit. While solving the numerical problems, we have to assign the proper signs and values for these three parameters. ♦

4.10. LENS MAKER'S EQUATION

If the object is at infinity, the image is formed at the *principal focus* of the lens. When $u = \infty, \frac{1}{u} = 0$ and $\upsilon = f$. Equ.(4.5) becomes

$$\frac{1}{f}-\frac{1}{\infty}=(\mu-1)\left[\frac{1}{R_1}-\frac{1}{R_2}\right]$$

$$\therefore \qquad \frac{1}{f}=(\mu-1)\left[\frac{1}{R_1}-\frac{1}{R_2}\right] \tag{4.6}$$

Equn (4.6) is known as the *lens makers' formula*, since it enables one to calculate f from the known properties of the lens. It can also be used to determine the values of R_1 and R_2 needed for a desired focal length of a lens of a given index of refraction.

If the lens is turned around, R_1 and R_2 are interchanged and the sign of each is reversed. Consequently, there is no change in f. Therefore, *for a thin lens, the focal length is independent of the order of the surfaces.*

It is clear from the lens makers' formula that to get a short focal length, the lens has to have a surface of smaller radius of curvature R and be made of a material with a high refractive index.

Comparing equations (4.5) and (4.6), we see that

$$\frac{1}{\upsilon}-\frac{1}{u}=\frac{1}{f} \tag{4.7}$$

The above equation is known as the **Gauss' formula for a lens**. This is a more useful formula than equ.(4.5), because if we know the focal length of the lens, we can relate the position of the object and its image without necessarily knowing the index of refraction or the radii of the lens.

Note—
1. The μ in equ.(4.6) stands for the refractive index of the lens material with respect to the surrounding medium. If the lens is placed in a medium such as water, then the value of μ should be the refractive index of glass with respect to water. That is,

$$\mu = \mu_{wg} \quad \text{or} \quad \mu = \mu_g / \mu_w$$

2. Another sign convention is there in which the distances of the ***real*** objects or images from the lens would be considered as *positive* and those of ***virtual*** images would be regarded as *negative*. Though that sign convention is simple, it is found inadequate in many respects. When that sign convention is used, we get the following formula for the focal length of a lens. $\frac{1}{u}+\frac{1}{\upsilon}=\frac{1}{f}$

4.10.1. POSITIONS OF THE PRINCIPAL FOCI

Each individual surface of the lens has its own focal points and planes and the lens as a whole has its own *pair* of focal points and focal planes. The focal points and focal planes of the lens are known as principal focal points and principal focal planes. We are interested in knowing the locations of these principal focal points and principal focal planes. They are obtained as follows.

(*i*) If a point object is placed on the principal axis such that the rays refracted by the lens are parallel to the axis, then the position of the point object is called the **first principal focus** F_1 of the lens(see Fig.4.8a). The distance at the first principal focus from the optical center C of the lens is called the **first principal focal length,** f_1. We can find f_1 as follows.

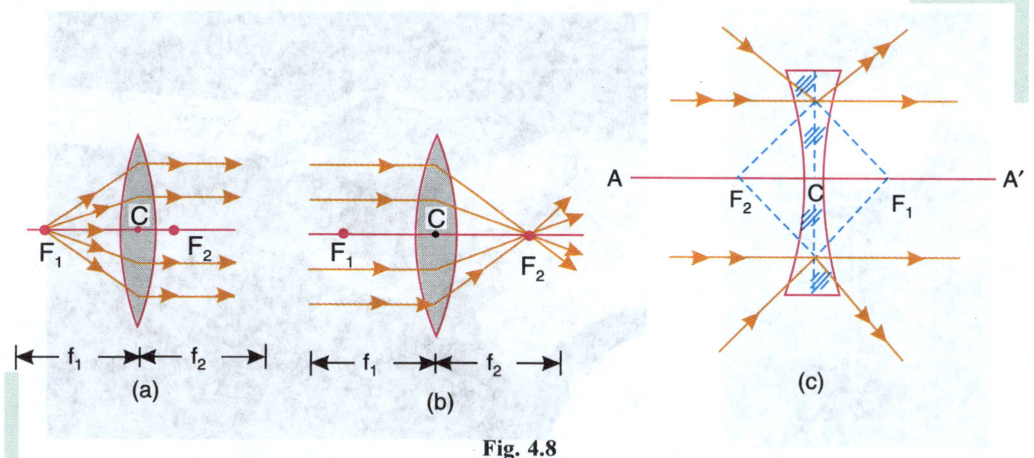

Fig. 4.8

Using $u = f_1$, and $\upsilon = \infty$ into equ.(4.6), we get

$$\frac{1}{\infty}-\left(-\frac{1}{f_1}\right)=(\mu-1)\left[\frac{1}{R_1}-\frac{1}{R_2}\right]$$

or
$$\frac{1}{f_1}=(\mu-1)\left[\frac{1}{R_1}-\frac{1}{R_2}\right] \tag{4.8}$$

The plane perpendicular to the axis and passing through the first focal point is known as the **first principal focal plane**.

(*ii*) If the object is situated at infinity, the position of the image on the axis is known as the **second principal focus** F_2 (see Fig.4.8b). The distance of the second principal focus from the optical center C is called the **second principal focal length,** f_2.

Using $u = \infty$ and $\upsilon = f_2$ into equ. (4.5), we get

$$\frac{1}{f_2}-\frac{1}{\infty}=(\mu-1)\left[\frac{1}{R_1}-\frac{1}{R_2}\right]$$

$$\frac{1}{f_2} = (\mu - 1)\left[\frac{1}{R_1} - \frac{1}{R_2}\right] \tag{4.9}$$

The plane perpendicular to the axis and passing through the second focal point is known as the **second principal focal plane**.

It follows from equ.(4.8) and (4.9) that

$$f_1 = f_2 \tag{4.10}$$

Thus, *every thin lens in air has two focal points (F_1 and F_2), one on each side of the lens and equidistant from the centre.*

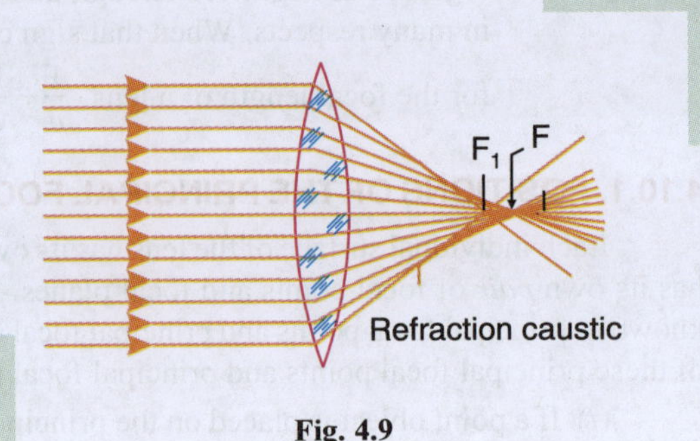

Fig. 4.9

It will be seen that the second focal length (f_2) of a converging lens is positive and the first (f_1) negative, while for a diverging lens the reverse is true (see Fig.4.8c). *The two focal lengths of thin lens in air are numerically equal.* Frequently, one refers simply to "focal length" of a lens; it will be assumed that this always

Converging lens

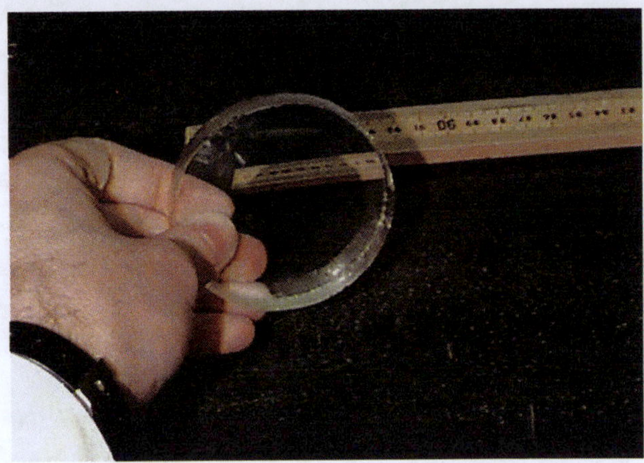

Diverging lens

refers to the *second focal length* so that the focal length of a converging lens is positive, while for a diverging lens it is negative.

Note — For lenses having a large diameter or aperture, the image of a point is not a point. Incident rays that are parallel to the principal axis intersect at different points, depending on their distance from that axis. The refracted rays intersect over a conical surface called the **refraction caustic.**

4.11. NEWTON'S LENS EQUATION

The lens-maker's formula (4.6) tells the lens-maker what is needed to know to grind a lens of a given power: the refractive index of the glass and the two radii of curvature. Now we obtain a lens equation for the practical applications. It relates the focal length to the object and image distances. This is also known as *lens users' formula* or *Newton's lens equation.*

Let h_1 be the height of the object and h_2 be the image height. From the similar triangles to the left of the lens we find that

$$\frac{h_1}{x_1} = \frac{h_2}{f_1}$$

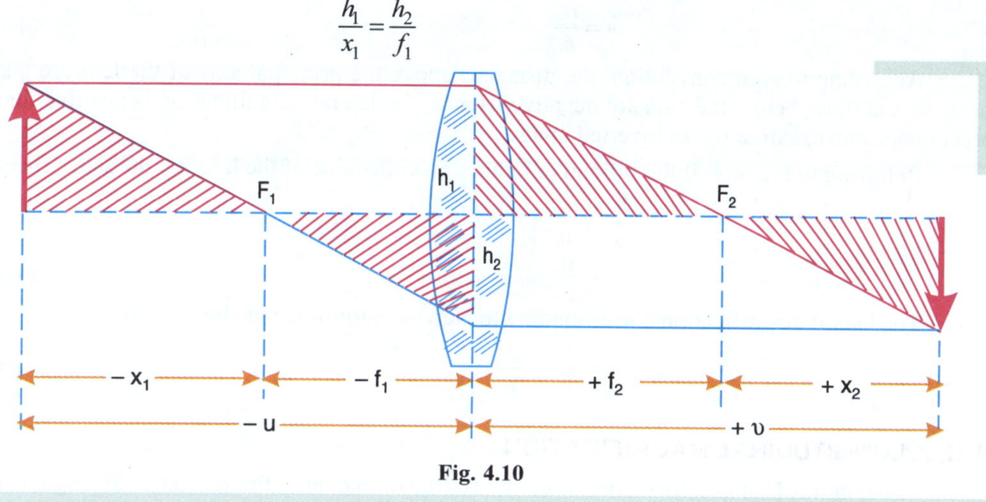

Fig. 4.10

To the right of the lens we have

$$\frac{h_1}{f_2} = \frac{h_2}{x_2}$$

Rearranging both equations and combining them by eliminating h_1/h_2 gives

$$\frac{x_1}{f_1} = \frac{f_2}{x_2}$$

or

$$x_1 x_2 = f_1 f_2 \qquad (4.11)$$

This is known as **Newton's lens equation**. The terms x_1 and x_2 are *extra focal* object and image distances respectively. They are *measured from the foci rather than from the lens*. Consequently, Newton's equation can be used with both "thin" and "thick" lenses.

When the medium is the same on both sides of the lens, the equ.(4.11) reduces to

$$x_1 x_2 = f^2 \qquad (4.12)$$

Note— In the Gaussian formula the object and image distances are measured from the lens centre, while in the Newton's formula they are measured from focal points.

4.12. MAGNIFICATION

Lenses are used to form a magnified image of an object. Magnification is defined as

$$\text{Magnification, } m = \frac{\text{size of the image}}{\text{size of the object}}$$

The image of a three-dimensional object will occupy a three-dimensional region of space. The lens can affect the transverse, longitudinal and angular dimensions of the image. Since the image and object relationship is expressible in terms of axial, lateral and angular distances, we distinguish three types of magnification, namely *lateral magnification, longitudinal magnification and angular magnification.*

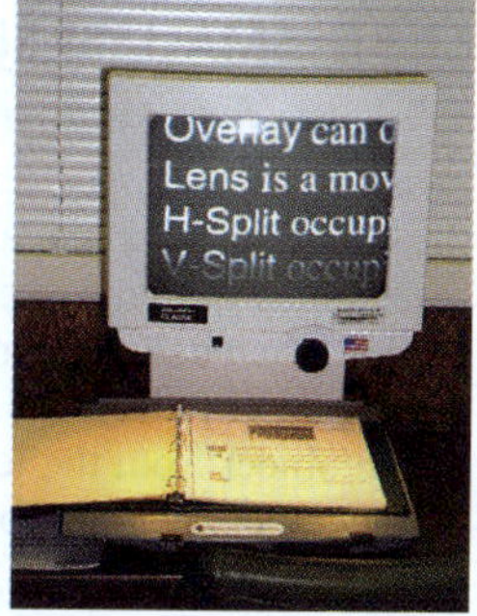

Classic video magnifier.

4.12.1. LATERAL MAGNIFICATION

Lateral or **transverse magnification** of a lens is defined as the ratio of the length of the image to the length of the object size, both the lengths being measured perpendicular to the principal axis. Thus,

$$m = \frac{h_2}{h_1} \tag{4.13}$$

According to sign convention, the distances above the principal axis of the lens are taken positive and those below the axis are negative. Hence, the lateral magnification is positive for an erect image and negative for an inverted image.

Referring to Fig. 4.4, if the refractive indices on either side of the lens are the same, then

$$\frac{h_1}{u} = \frac{h_2}{\upsilon}$$

$$\therefore \quad m = \frac{\upsilon}{u} \tag{4.14}$$

The lateral magnification corresponding to Newton's formula may be written as

$$m = \frac{h_2}{h_1} = \frac{f_1}{x_1} = \frac{x_2}{f_2} \tag{4.15}$$

4.12.2. LONGITUDINAL MAGNIFICATION

The **longitudinal magnification** is defined as the ratio of an infinitesimal axial length in the region of the image to the corresponding length in the region of the object. Thus, if dx_i and dx_o are the extensions of the object and the image along the principal axis respectively, then the longitudinal magnification of the lens is given by

$$m_L = \frac{dx_i}{dx_o} \tag{4.16}$$

Differentiating equ.(4.12), we get

$$m_L = -\frac{f^2}{x_o^2} = -m^2 \tag{4.17}$$

4.12.3. ANGULAR MAGNIFICATION

Angular magnification is defined as the ratio of slopes of emergent ray and conjugate incident ray with the principal axis. If the incident ray and the conjugate emergent ray make angles θ_1 and θ_2 with the principal axis respectively, then the angular magnification of the lens is given by

$$\gamma = \frac{\tan\ \theta_2}{\tan\ \theta_1} \tag{4.18}$$

4.13. SMALLEST SEPARATION OF OBJECT AND REAL IMAGE

From Fig. 4.7 we find that it is not always possible for a convex lens to form a real image of the object on a screen, although the object and the screen may both may be at a greater distance from the lens than its focal length. A convex lens forms a real or a virtual image depending on the distance of the object from the lens. We now show that the distance between an object and a screen must be equal to or greater than '4*f*' if a real image is to be formed with a convex lens.

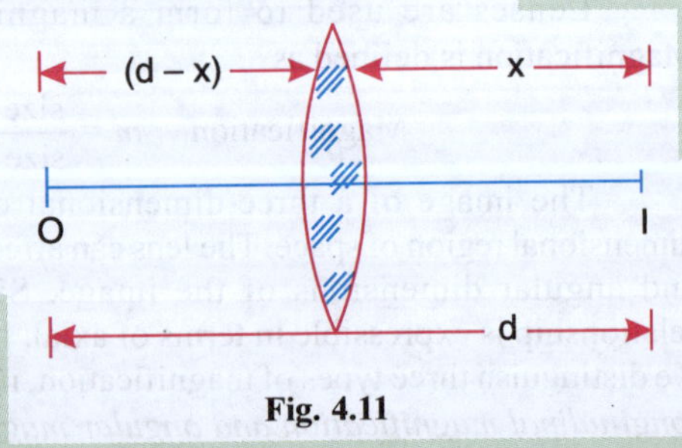

Fig. 4.11

Suppose I is the real image of a point object O in a converging lens. If the image distance is *x*, and the distance OI = *d*, the object distance = [*d* – *x*] (See Fig. 4.11). Thus, $\upsilon = x$, and $u = (d - x)$. Substituting these in the lens equation, we get

$$\frac{1}{x} + \frac{1}{d-x} = \frac{1}{f}$$

$$\therefore \qquad \frac{d}{x(d-x)} = \frac{1}{f}$$

$$\therefore \qquad x^2 - xd + fd = 0$$

If the image is to be real, the roots of this quadratic equation for x must be real.

$$\therefore \qquad d^2 - 4fd > 0$$

or $$d > 4f. \tag{4.19}$$

Hence, the *minimum* distance between an object and its real image for a convex lens must be $4f$ where f is the focal length of the lens.

4.14. DISPLACEMENT OF LENS WHEN OBJECT AND SCREEN ARE FIXED

Let us consider a linear object, O, placed in front of a convex lens, A, which gives rise to an image on the screen at I (See Fig. 4.12). Since the image distance AI (= υ) is greater than the object distance AO (= u), the image is larger than the object. If the object and the screen are kept fixed at O and I respectively, and the lens is moved from A to position B, then another clear image is obtained. This time the image is smaller than the object.

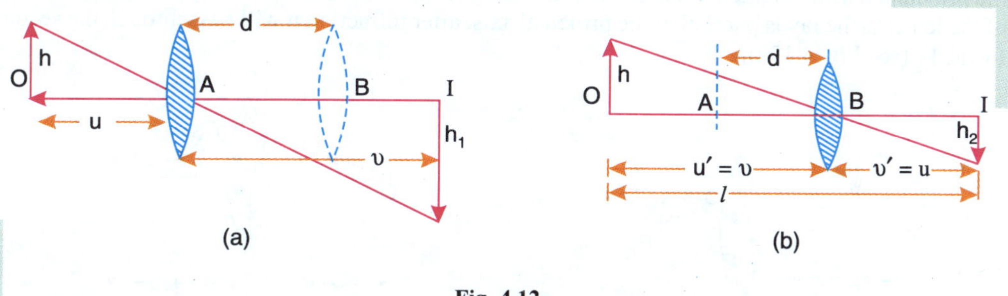

Fig. 4.12

Since O and I are conjugate, OB = IA and IB = OA. If we designate the displacement AB of the lens by d and the constant distance OI by l, then OA+BI = $l - d$.

$\therefore$ OA = $(l - d)/2$.

And AI = AB + BI = OA + AB = $(l-d)/2 + d = (l+d)/2$.

But OA = u and AI = υ for the lens position A. Substituting for OA and AI in the lens formula, we get

$$\frac{2}{l+d} - \frac{2}{l-d} = \frac{1}{f}$$

$$\therefore \qquad \frac{4l}{l^2 - d^2} = \frac{1}{f}$$

$$\therefore \qquad f = \frac{l^2 - d^2}{4l} \tag{4.20}$$

Thus, if the displacement d of the lens and the distance l between the object and the screen are measured, the focal length f of the lens can be found from the above equation (4.20). This gives us a very useful method of measuring the focal length of a lens.

(*ii*) **Magnification.** When the lens in position A, the lateral magnification, m_1 of the object is

$$m_1 = \frac{\upsilon}{u} = \frac{AI}{OA}$$

As AI > OA, the image, h_1, is bigger than the object, h.

$$\frac{h_1}{h} = \frac{AI}{OA} \tag{4.21}$$

When the lens is in position B, the image is smaller than the object. The lateral magnification is given by

$$m_2 = \frac{BI}{OB} \quad \therefore \quad \frac{h_2}{h} = \frac{BI}{OB} \tag{4.22}$$

As AI = OB and OA = BI, we get from equn. (4.21) and (4.22)

$$\frac{h}{h_1} = \frac{h_2}{h}$$

$$\therefore \quad h^2 = h_1 h_2$$

$$\text{or} \quad h = \sqrt{h_1 h_2} \tag{4.23}$$

It means that the size of the object is equal to the geometric mean of the two images. This method of measuring the size of the object is useful when the object is inaccessible.

4.15. DEVIATION BY A THIN LENS

A lens may be considered to be made up of a large number of prisms placed one above the other. As the action of the lens is to *deviate* the incident rays of light, it is necessary to find the deviation produced by a particular section of the lens. Let a ray of monochromatic light parallel to the principal axis be incident on a thin lens at a height h above the axis and let f be the focal length of the lens. As the ray is parallel to the principal axis, after refraction it will pass through the second focus, F_2 (see Fig. 4.13a).

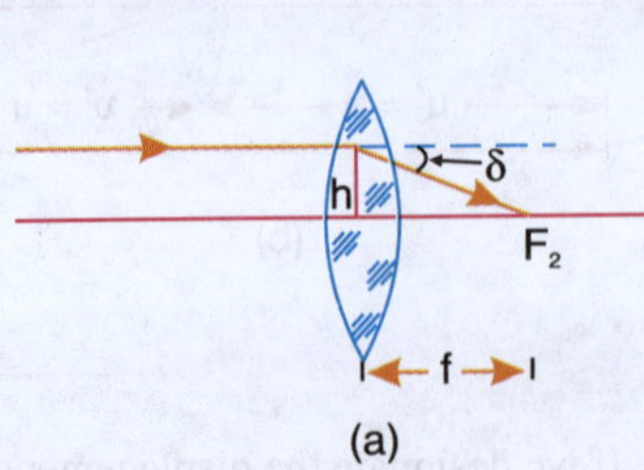

(b)

Fig. 4.13

The deviation suffered by the ray is given by

$$\tan \delta = \frac{h}{f}$$

In the paraxial region δ being small, $\tan \delta \approx \delta$.

$$\therefore \quad \delta = \frac{h}{f}$$

Next, consider a luminous point object O and its corresponding image I (see Fig. 4.13b). The deviation suffered by the ray OA incident at A is given by

$$\delta = \angle AOL + \angle AIL$$

$$\delta = \frac{h}{-u} + \frac{h}{+\upsilon} = h\left[\frac{1}{\upsilon} - \frac{1}{u}\right] = +h\left[\frac{1}{f}\right]$$

$$\therefore \quad \delta = \frac{h}{f} \tag{4.24}$$

This shows that the deviation produced by a lens is *independent* of the position of the object.

Equ. (4.24) is valid for all rays incident at the same point A of a lens, whatever may be the direction of incidence. All the rays suffer the same deviation in refraction through the lens.

4.16. POWER

The **power** of a lens is the measure of its ability to produce convergence of a parallel beam of light. A convex lens of large focal length produces a small converging effect and a convex lens of small focal length produces a large converging effect. Due to this reason, the power of a *convex lens* is taken as *positive* and a convex lens of small focal length has high power. On the other hand, a *concave lens* produces divergence. Therefore, its power is taken as *negative*.

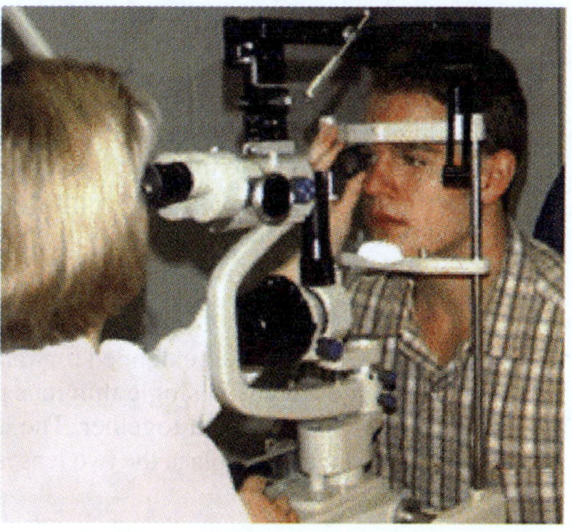

Power of a lens.

The unit in which the power of a lens is measured is called a **diopter** (D). A convex lens of focal length one metre has a power = +1 diopter and a convex lens of focal length 2 m has a power = + 0.5 diopter.

Mathematically, $$\text{Power} = \frac{1}{\text{Focal length in metres}} \quad (4.25)$$

The concept power is useful because it allows us to work out the effective focal length of a combination of lenses very easily. The power of a pair of lenses, of focal lengths f_1 and f_2, placed in contact is simply the sum of their individual powers.

$$\frac{1}{f} = \frac{1}{f_1} + \frac{1}{f_2}$$

$$\therefore \quad P = P_1 + P_2 \quad (4.26)$$

where P_1 and P_2 are the powers of the two lenses and P is the equivalent power.

4.17. EQUIVALENT FOCAL LENGTH OF TWO THIN LENSES

When two thin lenses are arranged coaxially, the image formed by the first lens system becomes the object for a second lens system and the two systems act as a single optical system forming the final image from the original object.

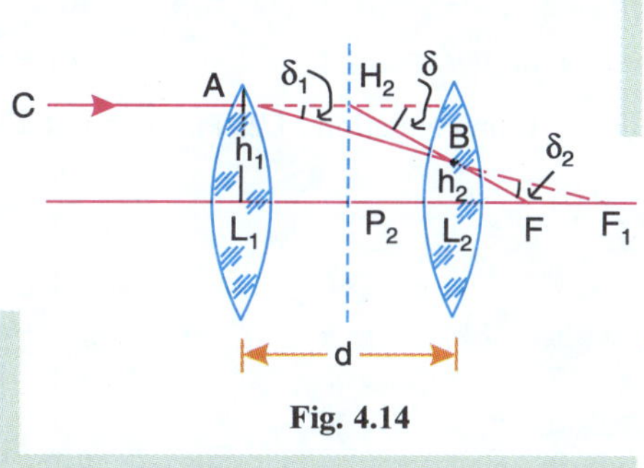

Fig. 4.14

Let us consider a simple optical system that consists of two thin lenses L_1 and L_2, placed on a common axis and separated by a distance *d*, as shown in Fig. 4.14. The lenses are separated by a distance and have focal lengths f_1 and f_2. We are interested to know how the combination works. We find that two lenses, separated by a finite distance, can be replaced by a single thin lens called an **equivalent lens**. The equivalent lens, when placed at a suitable fixed point, will produce an image of the same size as that produced by the combination of the two lenses. The focal length of equivalent lens is called **equivalent focal length**. We now derive an expression for the equivalent focal length, *f*, of the combination of two lenses.

Let a ray CA of monochromatic light parallel to the principal axis be incident on the first lens L_1 at a height h_1 above the axis. The ray CA is deviated through an angle δ_1 by the lens L_1. The incident ray CA, after refraction, is directed toward F_1, which is the *second principal focus of lens* L_1. Then the deviation produced by the first lens is given by

$$\delta_1 = \frac{h_1}{f_1}$$

The emergent ray, AB, from the first lens is incident on the lens L_2 at a height h_2. On refraction, the ray is deviated through an angle δ_2 by the lens L_2 and meets the principal axis at F. Since the incident ray CA is parallel to the principal axis and after refraction through the optical system meets the axis at F. F must be the *second principal focus of the combined lens system*. The deviation produced by the second lens is given by

$$\delta_2 = \frac{h_2}{f_2}$$

If the incident ray CA is extended forward and the final emergent ray BF backward, they meet at a point H_2. It is clear that a single thin lens placed at P_2 will produce the same deviation as that produced by the two lenses put together. The lens of focal length P_2F placed at P_2 is termed as the *equivalent lens*, which can replace the two lenses L_1 and L_2. The deviation produced by the equivalent lens is

$$\delta = \frac{h_1}{f}$$

4.17.1. FOCAL LENGTH OF THE EQUIVALENT LENS

Deviation produced by the first lens L_1 is $\delta_1 = h_1 / f_1$

Deviation produced by the second lens L_2 is $\delta_2 = h_2 / f_2$

But $$\delta = \delta_1 + \delta_2$$

$$\therefore \quad \frac{h_1}{f} = \frac{h_1}{f_1} + \frac{h_2}{f_2} \tag{4.27}$$

The Δ^{les} AL_1F_1 and BL_2F_1 are similar.

$$\therefore \quad \frac{AL_1}{L_1F_1} = \frac{BL_2}{L_2F_1}$$

or $$\frac{h_1}{f_1} = \frac{h_2}{f_1 - d} \tag{4.28}$$

or $$h_2 = \frac{h_1(f_1 - d)}{f_1} \tag{4.29}$$

Using equation (4.29) into equ.(4.27), we get

$$\frac{h_1}{f} = \frac{h_1}{f_1} + \frac{h_1(f_1 - d)}{f_1 f_2}$$

$$\frac{1}{f} = \frac{1}{f_1} + \frac{f_1 - d}{f_1 f_2}$$

$$\frac{1}{f} = \frac{1}{f_1} + \frac{1}{f_2} - \frac{d}{f_1 f_2} \tag{4.30}$$

Therefore, the equivalent focal length is given by

$$f = \frac{f_1 f_2}{f_1 + f_2 - d} \tag{4.31}$$

or $$f = \frac{-f_1 f_2}{\Delta} \tag{4.32}$$

where $\Delta = d - (f_1 + f_2)$ and is known as the **optical interval** between the two lenses. It is numerically equal to the distance between the second principal focus of the first lens and the first principal focus of the second lens.

From Fig. 4.14, it may be seen that the rays converge at F and appear to have come from a plane EP_2. Plane EP_2 is therefore, the position of a single lens, the effect of which is the same that of the two lenses, L_1 and L_2. The distance L_2F is the *equivalent focal length.* The equivalent focal length is independent of the direction from which the light enters the system. The behaviour of the system cannot be reproduced in all respects by the equivalent thin lens placed in a fixed position. It will have to be placed in the position H_2P_2 when light enters from left and in position H_1P_1 when light enters from the right.

Note— 1. Equ. (4.31) shows that if the distance d between the two convex lenses exceeds the sum of their focal lengths $(f_1 + f_2)$, the system becomes divergent, because of negative focal length.

2. If the medium between the two convex lenses is other than air, then the equ.(4.31) for equivalent focal length would become

$$f = \frac{f_1\, f_2}{f_1 + f_2 - d/\mu}$$

where μ is the refractive index of the medium.

4.17.2. DISTANCE OF EQUIVALENT LENS FROM L_2

Let us say the plane EP_2 is located at a distance of L_2P_2 from the second lens L_2. Now consider the similar Δ^{les} EP_2F and BL_2F.

$$\frac{P_2F}{L_2F} = \frac{EP_2}{BL_2}$$

From the Fig. 4.15 it is seen that $P_2F = f$, $L_2F = f - L_2P_2$ and $EP_2 = AL_1 = h_1$

$$\frac{h_1}{h_2} = \frac{f}{f - L_2P_2} \qquad (4.33)$$

Comparing equations (4.33) and (4.28), we obtain

$$\frac{f}{f - L_2P_2} = \frac{f_1}{f_1 - d}$$

$$ff_1 - fd = ff_1 + (L_2P_2)f_1$$

or

$$L_2P_2 = \frac{fd}{f_1} \qquad (4.34)$$

4.17.3. DISTANCE OF EQUIVALENT LENS FROM L_1

The distance of equivalent lens from L_1 is given by

$$L_1P_2 = (d - L_2P_2) = d - \frac{fd}{f_1} = d\left(1 - \frac{f}{f_1}\right) \qquad (4.35)$$

If we consider a parallel beam of light to be incident from right and falls first on lens L_2, the equivalent lens moves to the left, as shown in Fig. 4.15.

The equivalent lens is now at a distance of

$$L_1P_1 = \frac{fd}{f_2} \qquad (4.36)$$

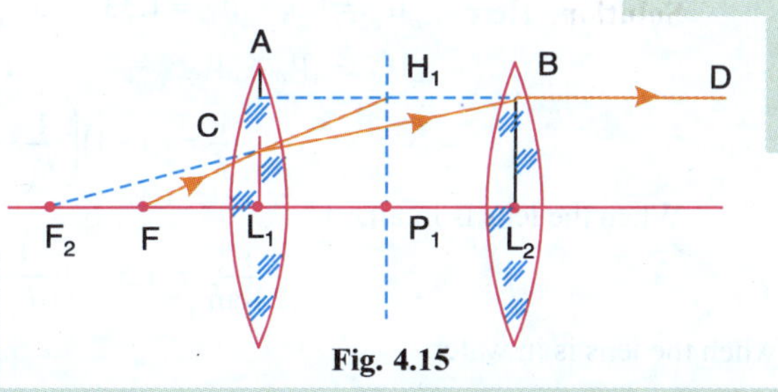

Fig. 4.15

from the lens L_1. These two positions of the equivalent lens are called the **principal planes**.

4.17.4. POWER

When two thin lenses of focal lengths f_1 and f_2 are placed coaxial and separated by a distance d, the *equivalent focal length* is given by

$$\frac{1}{f} = \frac{1}{f_1} + \frac{1}{f_2} - \frac{d}{f_1 f_2}$$

$$\therefore \quad P = P_1 + P_2 - d \cdot P_1 P_2 \qquad (4.37)$$

WORKED OUT PROBLEMS

Example 4.1: A sunshine record globe of 30 cm diameter is made of glass (μ = 1.5). A ray enters the globe parallel to the axis. Find the position where the ray crosses the axis.

Solution. A sunshine recorder globe is a sphere of glass (Fig. 4.16).

The radius of curvature of the first surface is +15 cm and u = ∞.

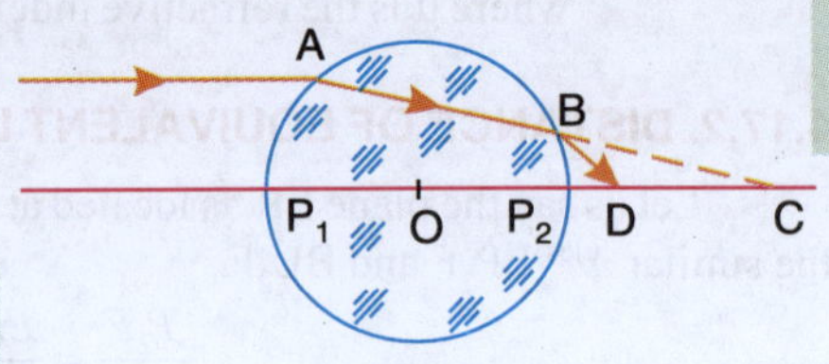

Fig. 4.16

$$\frac{\mu}{\upsilon_1} - \frac{1}{u} = \frac{(\mu - 1)}{R}$$

$$\frac{1.5}{\upsilon_1} - \frac{1}{\infty} = \frac{(1.5 - 1)}{15 \text{ cm}}$$

or $$\upsilon_1 = 45 \text{ cm}$$

$$\therefore \quad P_1C = 45 \text{ cm.}$$

For the second surface, point C is the virtual object.

$$\therefore \quad CP_2 = 45 \text{ cm} - 30 \text{ cm} = 15 \text{ cm.}$$

At the second surface, the ray passes from a denser to a rarer medium and $R = -15$ cm.

$$\frac{1/\mu}{\upsilon} - \frac{1}{u} = \frac{(1/\mu - 1)}{R}$$

$$\frac{1/1.5}{\upsilon} - \frac{1}{15cm} = \frac{(1/1.5 - 1)}{-15cm}$$

$$\therefore \quad \upsilon = 7.5 \text{ cm}$$

Therefore, the ray crosses the axis at a point D which is 5 cm from the point P_2 of the globe.

Example 4.2: A convex lens of focal length 24 cm (μ = 1.5) is totally immersed in water (μ = 1.33). Find its focal length in water.

Solution. Here $_a\mu_g = 1.5, \; _a\mu_w = 1.33 \quad \therefore \; _w\mu_a = 1/1.33$

$$_w\mu_g = \; _w\mu_a \times \; _a\mu_g = 1.5/1.33 = 1.125$$

$$\frac{1}{f} = (\mu - 1)\left(\frac{1}{R_1} - \frac{1}{R_2}\right)$$

When the lens is in air,

$$\frac{1}{24\,\text{cm}} = (1.5 - 1)\left(\frac{1}{R_1} - \frac{1}{R_2}\right) \qquad (i)$$

when the lens is in water,

$$\frac{1}{f} = (1.125-1)\left(\frac{1}{R_1} - \frac{1}{R_2}\right) \quad (ii)$$

Dividing the expression (*i*) by (*ii*), we get

$$\frac{f}{24\,\text{cm}} = \frac{0.5}{0.125}$$

$$f = 96 \text{ cm.}$$

QUESTIONS

1. Prove that in the case of a thin convex lens
$$\frac{1}{f} = \frac{1}{\upsilon} - \frac{1}{u} = (\mu-1)\left(\frac{1}{R_1} - \frac{1}{R_2}\right)$$
2. Show that the least possible distance between an object and its real image in a convex lens is four times the focal length of the lens.
3. Deduce for a thin lens an equation connecting the focal length, the radii of curvature of the surfaces and the refractive indices of the material of the lens and surrounding the medium.
4. Show that the deviation produced by a thin lens is independent of the position of the object.
5. Two thin convex lenses of focal length f_1 cm and f_2 cm are coaxial and separated by d. show that the equivalent focal length f of the combination is given by the relation
$$f = \frac{f_1 f_2}{f_1 + f_2 - d}$$
6. Derive an expression for the equivalent focal length of a system of two thin lenses separated by a finite distance when the space between them is filled with a medium of refractive index μ ($\mu > 1$).
7. Calculate the equivalent focal length of two thin co-axial lenses separated by a finite distance.
8. Prove that for a combination of two thin lenses of focal lengths f_1 and f_2 separated by a distance d, the focal length of the combination is given by
$$\frac{1}{f} = \frac{1}{f_1} + \frac{1}{f_2} - \frac{d}{f_1 f_2}$$
9. What is an equivalent lens? In what respect it is called an equivalent lens?

   Derive an expression for the power of an equivalent lens corresponding to two thin lenses of known power kept coaxially in air separated by a certain distance. Also find an expression for its position from any of the two lenses.
10. What do you understand by the power of a lens? Calculate the power of two thin lenses of focal length f_1 and f_2 separated by a distance d.

PROBLEMS FOR PRACTICE

Plano-convex lens.

1. Calculate the focal length of a plano-convex lens for which the radius of the curved surface is 40 cm. ($\mu = 1.5$). **[Ans:** $f = 40$ cm]
2. Find the focal length of a plano-convex lens, the radius of the curved surface being 10 cm ($\mu = 1.5$). **[Ans:** $f = 20$ cm]
3. A sunshine recorder globe of 10 cm diameter is made of glass of refractive index 1.5. A ray of light enters it parallel to the axis. Find the position where the ray meets the axis. **[Ans:** 2.5 cm from the second surface]
4. A convex lens of focal length 24 cm ($\mu = 1.5$) is totally immersed in water ($\mu = 1.33$). Find the focal length of the lens in water. **[Ans:** $f = 96$ cm]
5. The two surfaces of a double concave lens are of radii of curvature 10 and 30 cm. Find its focal length in water. ($\mu_{water} = 1.33$, $\mu_{glass} = 1.5$). **[Ans:** $f = 60$ cm]
6. A concavo-convex lens has a refractive index of 1.5 and the radii of curvature of its surfaces are 15 cm and 30 cm. The concave surface is upwards and it is filled with a liquid of refractive index 1.6. Calculate the focal length of the liquid-glass combination. **[Ans:** $f = 27.27$cm]
7. Two convex lenses of focal length 10 cm and 20 cm are placed at 5 cm apart in air, find the focal length of equivalent lens. **(Nagpur, 2005)**
8. Calculate the Focal length of a double convex lens for which the radius of curvature of each surface is 30 cm and Refractive Index of glass is 1.5. **(Nagpur, 2004)**

CHAPTER

Optical System and Cardinal Points

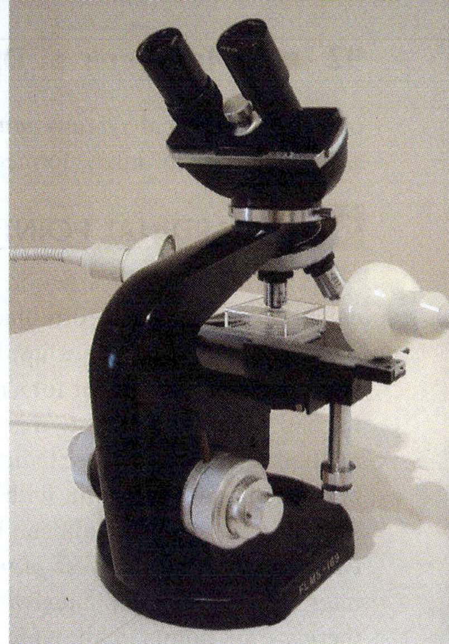

5.1. INTRODUCTION

Single lenses are rarely used for image formation, as they suffer from various defects. In optical instruments such as cameras, microscopes, telescopes etc., a collection of lenses are employed for forming images of

In telescope, a group of lenses are employed for forming images of objects.

At a Glance

- Introduction
- Cardinal Points
- Construction of the Image Using Cardinal Points
- Newton's Formula
- Relationship Between f_1 and f_2
- Relationship Between f_1 and f_2 and μ_1 and μ_2
- Gaussian Formula
- The Three Magnifications and Their Inter-Relationships
- Nodal Slide
- Cardinal Points of a Coaxial System of Two Thin Lenses

objects. **An optical system consists of a number of lenses placed apart, and having a common principal axis.** The image formed by such a coaxial optical system is good and almost free of aberrations.

5.2. CARDINAL POINTS

In the case of refraction through a thin lens, the thickness of the lens has been neglected in calculating the various formulae. It is then immaterial from which point of the lens the distances are measured. But we cannot apply the above approximation for an optical system consisting of a combination of lenses (or for a *thick* lens). One way of calculating the position and size of the image formed by an optical system is to consider refraction at each surface of a lens successively, but the method proves to be more tedious. In 1841, Gauss showed that any number of coaxial lenses could be treated as a single unit, without the necessity of treating the single surfaces of lenses separately. The lens makers' formula can be applied to the system provided the distances are measured from two *hypothetical parallel planes*, fixed with reference to the refracting system. The points of intersection of these planes with the axis are called the **principal points** or **Gauss points**. In fact there are six points in all, which characterize an optical system. They are

(*i*) two focal points,

(*ii*) two principal points and

(*iii*) two nodal points.

These six points are known as **cardinal points** of an optical system. The planes passing through these points and which are perpendicular to the principal axis are known as **cardinal planes**. The cardinal points and cardinal planes are intrinsic properties of a particular optical system and determine the image forming properties of the system. If these are known, one can find the image of any object without making a detailed study of the passage of the rays through the system. It is not necessary to consider the refraction of the rays at the various surfaces.

We describe here how to locate the cardinal points and planes for a coaxial optical system.

5.2.1. PRINCIPAL POINTS AND PRINCIPAL PLANES

Consider an optical system having its principal focal points F_1 and F_2. A ray OA travelling parallel to the principal axis and incident at A is brought to focus at F_2 in the **image space** of the optical system as shown in Fig. 5.1(a). The actual ray is refracted at each surface of the optical system and follows the path $OABF_2$. If we extend the incident ray OA forward and the emergent ray BF_2 backward, they meet each other within the optical system at H_2. Now, we can describe the refraction of the incident ray OA in terms of a *single refraction* at a plane passing through H_2. A plane drawn through the point H_2 and perpendicular to the axis may be regarded as the surface at which refraction takes place. This plane is called the *principal plane* of the optical system. Thus, the four consecutive deviations of the light rays caused by the four surfaces of the optical system are equivalent to a single refraction at H_2, taking place at the *principal plane*. We now define the *principal plane* of an optical system as *the loci where we assume refraction to occur without reference as to where the refraction actually*

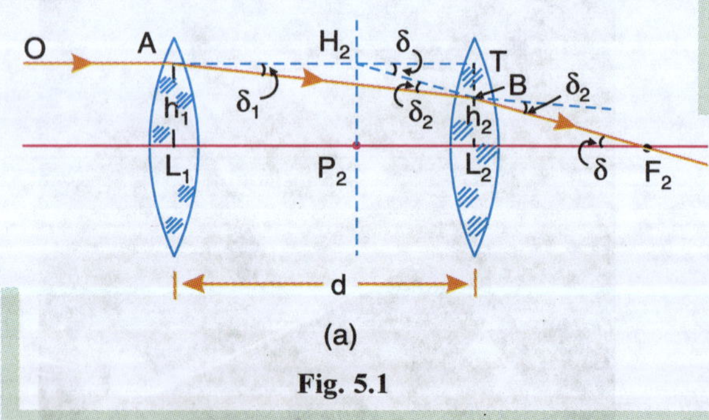

Fig. 5.1

occurs. H_2P_2 is the principal plane in the image space and is called the **second principal plane**. The point P_2, at which the second principal plane intersects the axis, is called the **second principal point**.

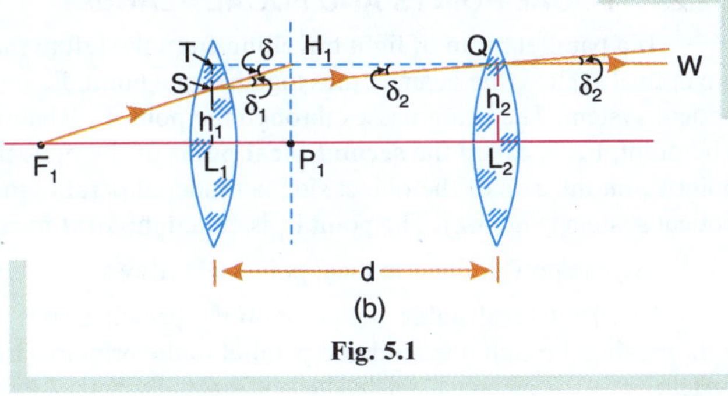

(b)

Fig. 5.1

By adopting similar procedure, as shown in Fig. 5.1(b), we can locate the principal plane H_1P_1 and principal point P_1 in the **object space**. Consider the ray F_1S passing through the first principal focus F_1 such that after refraction it emerges along QW parallel to the axis at the same height as that of the ray OA (see Fig. 5.1a). The rays F_1S and QW when produced intersect at H_1. A plane perpendicular to the axis and passing through H_1 is called the **first principal plane**. The point of intersection, P_1, of the first principal plane with the axis is called the **first principal point**.

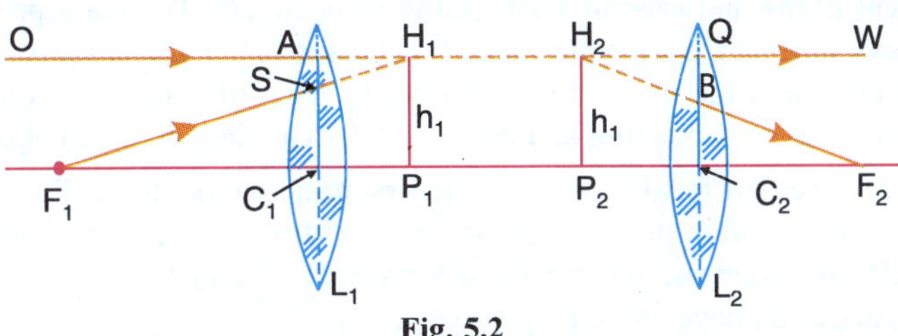

Fig. 5.2

It is seen from Fig. 5.2 that two incident rays are directed towards H_1 and after refraction seem to come from H_2. Therefore, H_2 is the image of H_1. Thus, H_1 and H_2 are the *conjugate points* and the planes H_1P_1 and H_2P_2 are a pair of *conjugate planes*. It is also seen that

$$H_2P_2 = H_1P_1.$$

Hence, the lateral magnification of the planes is +1.Thus, the first and second principal planes are planes of unit magnification and are therefore called *unit planes* and the points P_1 and P_2 are called *unit points*.

Note— The principal planes are *conceptual* planes and do not have physical existence within the optical system. ◆

5.2.1.1. Some Remarkable Features of Principal Planes

1. Even a complex optical system has *only two principal planes*.
2. Between H_1 and H_2 all rays are *parallel* to the principal axis.
3. The location of the principal planes is characteristic of a given optical system. Their positions do not change with the object and image distance used.
4. The principal planes are conjugate to each other. An object in the first principal plane is imaged in the second principal plane with *unit magnification.* Any ray directed towards a point on the first principal plane emerge from the lens as if it originated at the corresponding point (at the same distance above or below the axis) on the second principal plane. Hence, the name *unit planes*.
5. The principal points H_1 and H_2 provide a set of references from which several system parameters are measured.

5.2.2. FOCAL POINTS AND FOCAL PLANES

If a parallel beam of light travelling from the left to the right (in object space) is incident on the optical system, the beam comes together at a point, F_2 on the other side (in image space) of the optical system. The beam passes through the point F_2 whatever may be its path inside the system. The point, F_2, is called the **second focal point** of the optical system. A beam of light passing the point F_1 on the axis of the object side is rendered parallel to the axis after emergence through the optical system (Fig. 5.2). The point F_1 is called the **first focal point**.

We can now define the focal points as follows:

The **first focal point** is a point on the principal axis of optical system such that a beam of light passing through it is rendered parallel to the principal axis after refraction through the optical system.

The **second focal point** is a point on the principal axis of the optical system such that a beam of light travelling parallel to the principal axis of the optical system, after refraction through the system, passes through it.

The planes passing through the principal focal points F_1 and F_2 and perpendicular to the axis are called **first focal plane** and **second focal plane** respectively. The main property of the focal planes is that the rays starting from a point in the focal plane in the object space correspond to a set of conjugate parallel rays in the image space. Similarly, a set of parallel rays in the object space corresponds to a set of rays intersecting at a point in the focal plane in the image space.

The distance of the first focal point from the first principal point, i.e.,$F_1 P_1$, is called the **first focal length,** f_1, of the optical system and the distance of the second focal point from the second principal point, F_2P_2, is called the **second focal length,** f_2 . f_1 and f_2 are also known as the focal lengths in object space and image space respectively.

When the medium is same on the two sides of the optical system $f_1 = f_2$ (numerically).

5.2.3. NODAL POINTS AND NODAL PLANES

Nodal points are points on the principal axis of the optical system where light rays, ***without refraction,*** intersect the optic axis. In a *thin lens* the nodal point is the centre of the lens. Light passing through the centre of a thin lens does not deviate. In an optical system the centre separates into two nodal points. The planes passing through the nodal points and perpendicular to the principal axis are called the **nodal planes**. Whereas the principal planes are planes where all refraction are assumed to occur, the nodal planes are planes where refraction does *not* take place. Fig. 5.3 represents an optical system with the help of its cardinal planes. It is seen from the Fig.5.3 that a ray of light, AN_1, directed towards one of the nodal points, N_1, after refraction through the optical system, along N_1N_2, emerges out from the second nodal point, N_2, in a direction, N_2R, parallel to the incident ray. The distances of the nodal points are measured from the focal points.

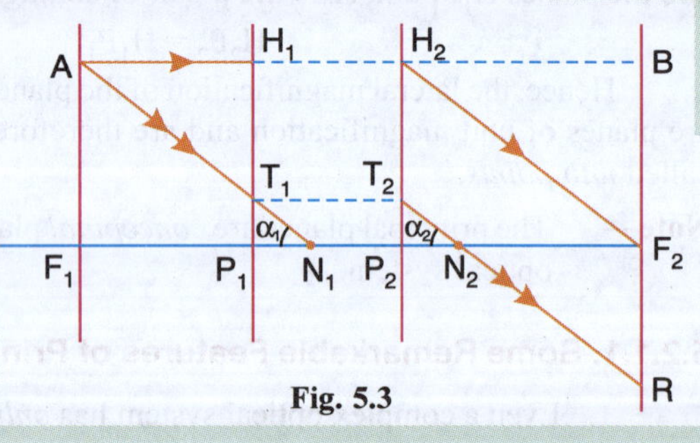

Fig. 5.3

5.2.3.1. The nodal points are a pair of conjugate points on the axis having unit positive angular magnification

Let H_1P_1 and H_2P_2 be the first and second principal planes of an optical system. Let AF_1 and BF_2 be its first and second focal planes respectively (Refer to Fig.5.3). Consider a point A situated on the first focal plane. From A draw a ray AH_1 parallel to the axis. The conjugate ray will proceed from H_2, a point in the second principal plane such that $H_2P_2 = H_1P_1$ and will pass through the second focus.

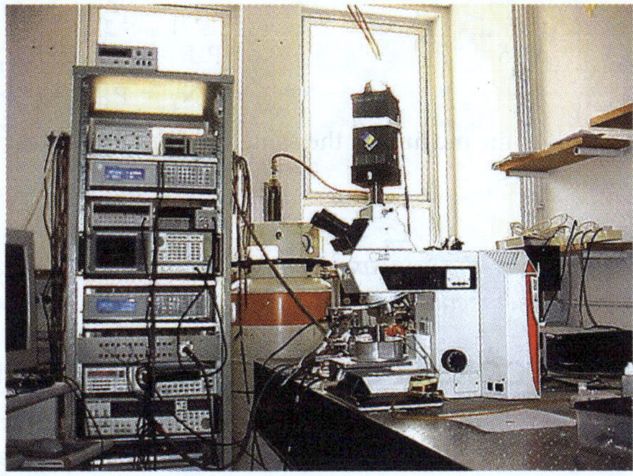
Magneto-optical system.

Take another ray AT_1 parallel to the emergent ray H_2F_2 and incident on the first principal plane at T_1. It will emerge out from T_2, a point on the second principal plane such that $T_2P_2 = T_1P_1$, and will proceed parallel to the ray H_2F_2. The points of intersection of the incident ray AT_1 and the conjugate emergent ray T_2R with the axis give the positions of the nodal points. It is clear that the two points N_1 and N_2 are a pair of conjugate points and the incident ray AN_1 is parallel to the conjugate emergent ray T_2R.

Further $$\tan \alpha_1 = \tan \alpha_2$$

The ratio $\dfrac{\tan \alpha_2}{\tan \alpha_1} = \gamma$ represents the angular magnification.

$$\therefore \qquad \frac{\tan \alpha_2}{\tan \alpha_1} = 1 \qquad (5.1)$$

Therefore, *the points N_1 and N_2 are a pair of conjugate points on the axis having unit positive angular magnification.*

5.2.3.2. The distance between two nodal points is always equal to the distance between two principal points.

Referring to Fig.5.3, we see that in the right angled Δ^{les} $T_1P_1N_1$ and $T_2P_2N_2$

$$T_1P_1 = T_2P_2$$

$$\angle T_1 N_1P_1 = \angle T_2 N_2P_2 = \alpha$$

Therefore, the two Δ^{les} are congruent.

$$\therefore \qquad P_1N_1 = P_2N_2$$

Adding N_1P_2 to both the sides, we get

$$\therefore \qquad P_1N_1 + N_1P_2 = P_2N_2 + N_1P_2$$

$$\therefore \qquad P_1P_2 = N_1N_2 \qquad (5.2)$$

Thus, *the distance between the principal points N_1 and N_2 is equal to the distance between the principal points P_1 and P_2.*

5.2.3.3. The nodal points N_1 and N_2 coincide with the principal points P_1 and P_2 respectively whenever the refractive indices on either side of the lens are the same.

Now consider the two right angled Δ^{les} AF_1N_1 and $H_2P_2F_2$ in Fig.5.3.

$$AF_1 = H_2P_2$$

$$\angle A N_1F_1 = \angle H_2F_2 P_2$$

$\therefore$ The two Δ^{les} are congruent.

$$F_1N_1 = P_2F_2$$

But $$F_1N_1 = F_1P_1 + P_1 N_1$$

$$\therefore \qquad F_1P_1 + P_1 N_1 = P_2F_2$$

$$\therefore \qquad P_1 N_1 = P_2F_2 - F_1P_1$$

Also $\qquad P_2F_2 = +f_2$ and $P_1F_1 = -f_1$

$$\therefore \qquad P_1N_1 = P_2N_2 = (f_1 + f_2)$$

As the medium is the same, say air, on both the sides of the system

$$f_2 = -f_1$$

$$\therefore \qquad P_1N_1 = P_2N_2 = 0 \qquad (5.3)$$

Thus, the principal points coincide with the nodal points when the optical system is situated in the same medium.

5.3. CONSTRUCTION OF THE IMAGE USING CARDINAL POINTS

From the knowledge of the cardinal points of an optical system, the image corresponding to any object placed on the principal axis of the system can be constructed. It is not necessary to know the position and curvatures of the refracting surfaces or the nature of the intermediate media. Only knowledge of cardinal points and cardinal planes is sufficient.

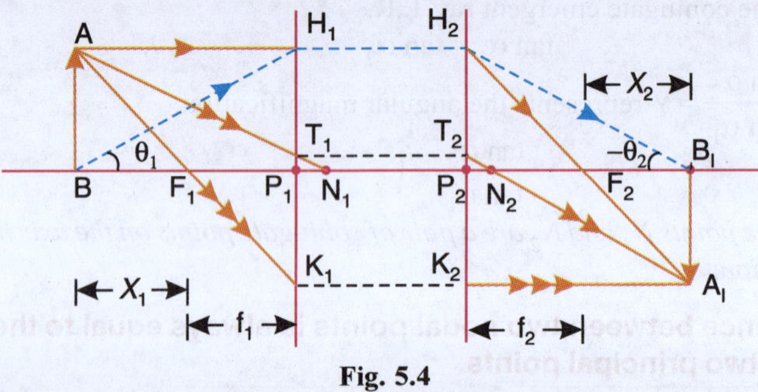

Fig. 5.4

Let F_1, F_2 be the principal foci, P_1, P_2 the principal points and N_1, N_2 the nodal points of the optical system, shown in Fig. 5.4. AB is a linear object on the axis. In order to find the image of the point A we make the following construction.

(1) A ray AH_1 is drawn parallel to the axis touching the first principal plane at H_1. The conjugate ray will proceed from H_2 a point on the second principal plane such that $H_2P_2 = H_1P_1$ and will pass through the second principal focus F_2.

(2) A second ray AF_1K_1 is drawn passing through the first principal focus F_1 and touching the first principal plane at K_1. Its conjugate ray will proceed from K_2 such that $K_2P_2 = K_1P_1$ and it will be parallel to the axis.

(3) A third ray $A_1T_1N_1$ is drawn which is directed toward the first nodal point N_1. This ray passes after refraction through N_2 in a direction parallel to AN_1.

The point of intersection of any of the above two refracted rays will give the image of A. Let it be A_1. From A_1, if a perpendicular is drawn on to the axis, it gives the image A_1B_1 of the object AB.

5.4. NEWTON'S FORMULA

Referring to the Fig. 5.4, it is seen that Δ^{les} ABF_1 and $F_1K_1P_1$ are similar.

$$\therefore \qquad \frac{K_1P_1}{AB} = \frac{P_1F_1}{BF_1} \qquad \text{But, } K_1P_1 = A_1B_1$$

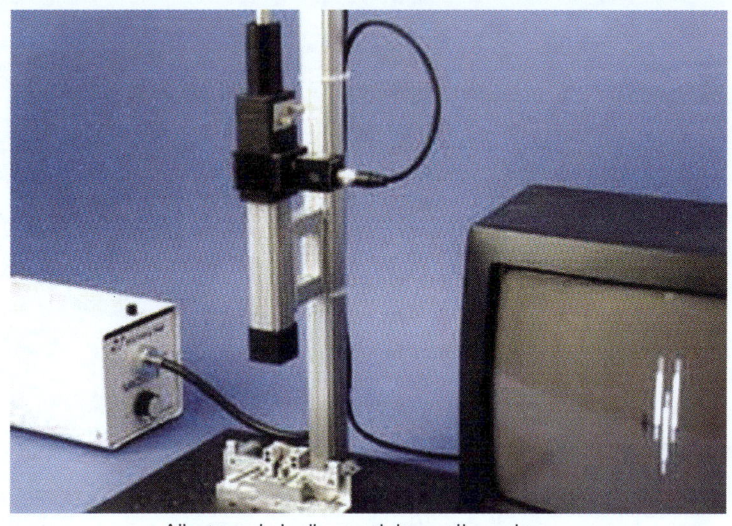

Alignment studio modular optic system.

$$\therefore \qquad \frac{A_1B_1}{AB} = \frac{f_1}{x_1} \tag{5.4}$$

Further, $\Delta^{les} A_1B_1F_2$ and $H_2P_2F_2$ are similar.

$$\frac{A_1B_1}{H_2P_2} = \frac{B_1F_2}{P_2F_2} \qquad \text{But, } H_2P_2 = AB$$

$$\therefore \qquad \frac{A_1B_1}{AB} = \frac{x_2}{f_2} \tag{5.5}$$

From equations (5.4) and (5.5), we get

$$\frac{h_2}{h_1} = \frac{f_1}{x_1} = \frac{x_2}{f_2} \tag{5.6}$$

or

$$x_1x_2 = f_1f_2 \tag{5.7}$$

This is the **Newton's formula**. In the foregoing discussion, the distances of the image and the object have been measured from their respective foci. But it is sometimes convenient to measure the conjugate distances from the principal points.

5.5. RELATIONSHIP BETWEEN f_1 AND f_2

Referring to Fig. 5.5 and using the sign convention,

$$P_1B = H_1A = -u, \; P_2B_1 = K_2A_1 = +v$$

Also, $AB = P_1H_1 = P_2H_2 = +h_1, \quad A_1B_1 = K_1P_1 = K_2P_2 = -h_2$

$$K_1H_1 = K_1P_1 + P_1H_1 = -h_2 + h_1 \quad \text{and} \quad K_2H_2 = K_2P_2 + P_2H_2 = -h_2 + h_1$$

Δ^{les} $K_1F_1P_1$ and K_1AH_1 are similar.

$$\therefore \qquad \frac{P_1F_1}{H_1A} = \frac{K_1P_1}{K_1H_1}$$

$$\frac{-f_1}{-u} = \frac{-h_2}{-h_2 + h_1}$$

or
$$\frac{f_1}{u} = \frac{-h_2}{-h_2 + h_1} \tag{5.8}$$

Δ^{les} $H_2P_2F_2$ and $H_2K_2A_1$ are similar.

$$\therefore \quad \frac{P_2F_2}{K_2A_1} = \frac{P_2H_2}{K_2H_2}$$

$$\frac{f_2}{\upsilon} = \frac{h_2}{-h_2 + h_1} \tag{5.9}$$

Adding the equations (5.8) and (5.9), we get

$$\frac{f_1}{u} + \frac{f_2}{\upsilon} = \frac{-h_2 + h_1}{-h_2 + h_1}$$

$$\therefore \quad \frac{f_1}{u} + \frac{f_2}{\upsilon} = 1 \tag{5.10}$$

Equ. (5.10) can be rewritten as

$$\frac{\upsilon}{u} = \frac{(\upsilon - f_2)}{f_1} \tag{5.11}$$

When the system is situated in air $f_2 = -f_1 = f$.

$$\therefore \quad f_1 = -f \text{ and } f_2 = f \tag{5.12}$$

5.6. RELATIONSHIP BETWEEN f_1 AND f_2 AND μ_1 AND μ_2

In Fig. 5.4, the conjugate rays BH_1 and H_2B_1 make angles θ_1 and θ_2 respectively with the principal axis. We have

$$\frac{\tan \theta_1}{\tan (-\theta_2)} = \frac{h_1/(-u)}{h_1/\upsilon} = -\frac{\upsilon}{u} \tag{5.13}$$

From the equ.(5.5), we have

$$\frac{h_2}{h_1} = \frac{x_2}{f_2} = \frac{\upsilon - f_2}{f_2} \quad \because \quad x_2 = \upsilon - f_2 \tag{5.14}$$

Using equ.(5.11) into equ.(5.14), we get

$$\frac{\tan \theta_1}{\tan \theta_2} = -\frac{h_2}{h_1} \cdot \frac{f_2}{f_1}$$

In the paraxial approximation, the above expression reduces to

$$\therefore \quad \frac{\theta_1}{\theta_2} = -\frac{h_2}{h_1} \cdot \frac{f_2}{f_1} \tag{5.15}$$

According to the Lagrange law, we have $\mu_1 h_1 \theta_1 = \mu_2 h_2 \theta_2$. [Refer to equ.(3.62)]

$$\frac{\theta_1}{\theta_2} = \frac{\mu_2 h_2}{\mu_1 h_1} \tag{5.16}$$

Comparing equs.(5.15) and (5.16), we obtain

$$\frac{f_1}{f_2} = -\frac{\mu_1}{\mu_2} \tag{5.17}$$

When the system is situated in air, $f_2 = -f_1 = f$

$$\therefore \quad f_1 = -f \text{ and } f_2 = f \tag{5.18}$$

5.7. GAUSSIAN FORMULA

Using the result (5.18) into (5.10), we get

$$\frac{-f}{u}+\frac{f}{\upsilon}=1$$

$$\therefore \quad \frac{1}{\upsilon}-\frac{1}{u}=\frac{1}{f} \tag{5.19}$$

This is the **Gauss' formula.**

Thus, the simple formula for a thin lens can also be used for a thick lens or for a compound lens system, provided the distances are measured from the corresponding points. A similar equation can be obtained for a divergent system.

5.8. THE THREE MAGNIFICATIONS AND THEIR INTER-RELATIONSHIPS

(i) Lateral magnification is given by $m=\frac{h_2}{h_1}$ (5.20)

(ii) Axial magnification is given by $m_L=\frac{d\upsilon}{du}=\frac{\Delta x_2}{\Delta x_1}$ (5.21)

Differentiating Newton's formula (5.7), we get

$$x_1\Delta x_2+\Delta x_2 x_1=0$$

$$m_L=\frac{\Delta x_2}{\Delta x_1}=-\frac{x_2}{x_1}$$

Using equ.(5.6) and (5.20) into the above expression, we get

$$m_L=-\frac{mf_2}{f_1/m}=-m^2\cdot\frac{f_2}{f_1} \tag{5.22}$$

Using equ.(5.17), equ.(5.22) may also be written as

$$m_L=m^2\cdot\frac{\mu_2}{\mu_1} \tag{5.23}$$

(iii) Angular magnification $\alpha=\frac{\theta_2}{\theta_1}$

In case of paraxial approximation, the Lagrange law gives

$$\frac{\theta_2}{\theta_1}=\frac{\mu_1 h_1}{\mu_2 h_2}=\frac{\mu_1}{\mu_2}\cdot\frac{1}{m}$$

$$\alpha\cdot m=\frac{\mu_1}{\mu_2} \tag{5.24}$$

Combining equations (5.23) and (5.24), we obtain the relationship between the linear, axial and angular magnifications.

$$\alpha\cdot m_L=m \tag{5.25}$$

When $\mu_1=\mu_2$, equ.(5.23) gives that $m_L=m^2$. Therefore, $\alpha\cdot m=1$ (5.26)

Equ. (5.26) shows that the angular magnification and lateral magnification are inversely proportional to each other.

5.9. NODAL SLIDE

A **nodal slide** is a particular type of horizontal metal support for a lens system, which is capable of rotation about a vertical axis (Fig. 5.5).

2nd generation nodal slide.

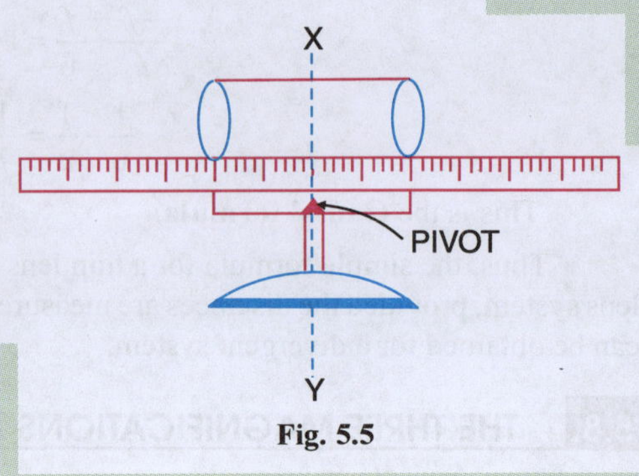

Fig. 5.5

The nodal slide provides a convenient method for locating the focal and nodal points and for determining the focal length of a lens system.

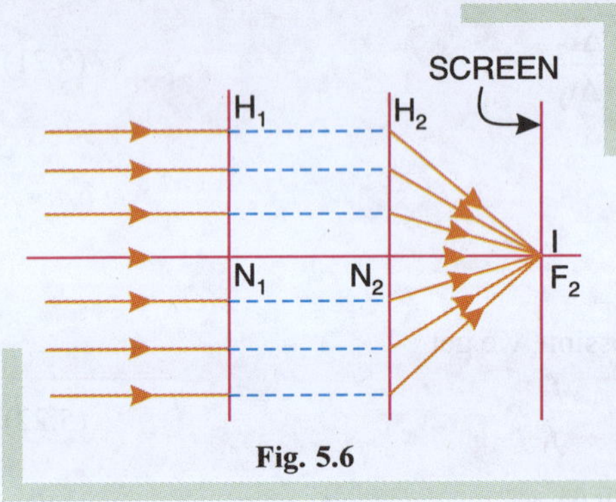

Fig. 5.6

Principle: Any optical system has two nodal points N_1 and N_2. An incident light ray directed towards N_1, after refraction through the system, proceeds from N_2 in a direction parallel to the incident ray. The method of locating the nodal points with the help of the nodal slide involves the following principle.

" If a parallel beam of light is incident on a convergent lens system, it forms an image on a screen held at its second focal plane. When the lens system is rotated through a small angle about a vertical axis through its second nodal point, the image does not shift laterally and remains stationary."

Let us suppose a beam of parallel rays is incident on a coaxial lens system. The beam passes through the system and converges to the second focus F_2 and a real image is formed on the screen.

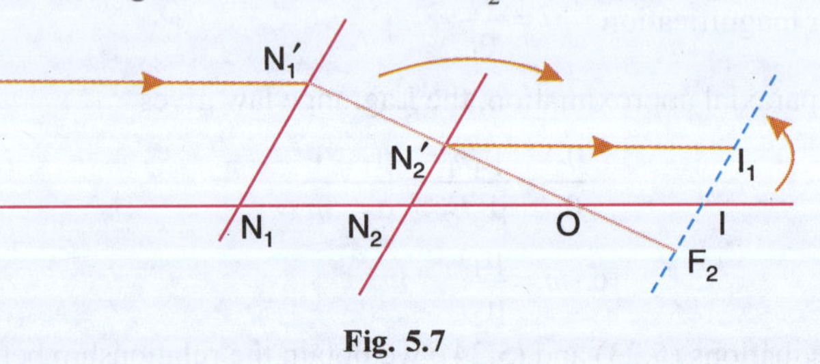

Fig. 5.7

Now let the system be rotated about a perpendicular axis through O, which lies between N_2 and F_2. Due to this rotation, the nodal points N_1 and N_2 shift to the positions N_1' and N_2' respectively. A ray incident at N_1' travels along $N_2'I_1$ parallel to the incident ray (see Fig.5.7). Since the incident beam is parallel, the image lies on the second focal plane. The point of intersection of the ray $N_2'I_1$ with the focal plane gives the new position of the image. Thus, when the axis of rotation lies between N_2 and F_2, a slight rotation of the system changes the position of the image.

Next let us say the system is rotated through a small angle about an axis passing through N_2. Then N_1 shifts to N_1' while N_2 remains fixed. A parallel ray incident at N_1' on passing through N_2 follows the path N_2I. Therefore, the image remains stationary, as seen from Fig. 5.8.

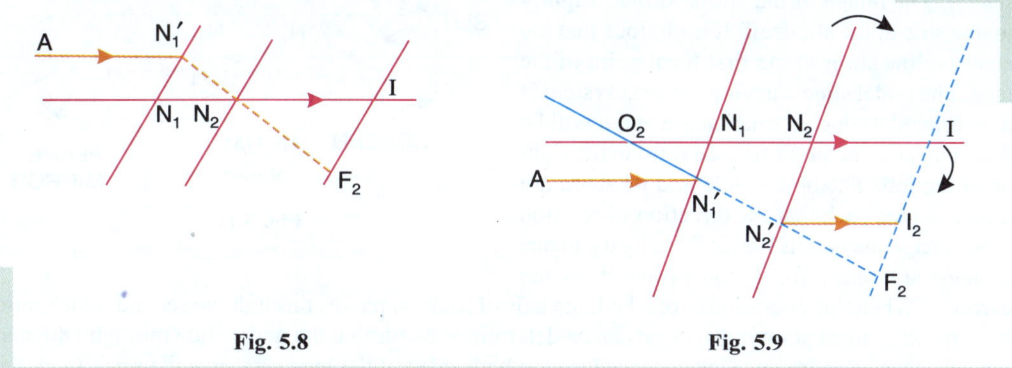

Fig. 5.8

Fig. 5.9

If the axis of rotation lies before N_1, as in Fig. 5.9, any small amount of rotation displaces N_1 and N_2 and consequently the image position changes.

Thus, the position of the axis of rotation for which there is no displacement of the image can be found. It gives the second nodal point. As the media on both sides of the system being the same, the second nodal point coincides with second principal point.

5.9.1. DETERMINATION OF NODAL POINTS

The experimental arrangement consists of an optical bench on which four uprights are kept. They carry a plane mirror, the nodal slide, and a screen provided with a slit fitted with cross wires and lamp housing (see Fig. 5.10). Light from the lamp passes through the slit and is incident on the lens system. It is rendered parallel and on passing through the lens system it is reflected back by the vertical plane mirror. The reflected light once again passes through the lens system and is brought to a focus in the plane of the stilt, as shown in Fig. 5.11.

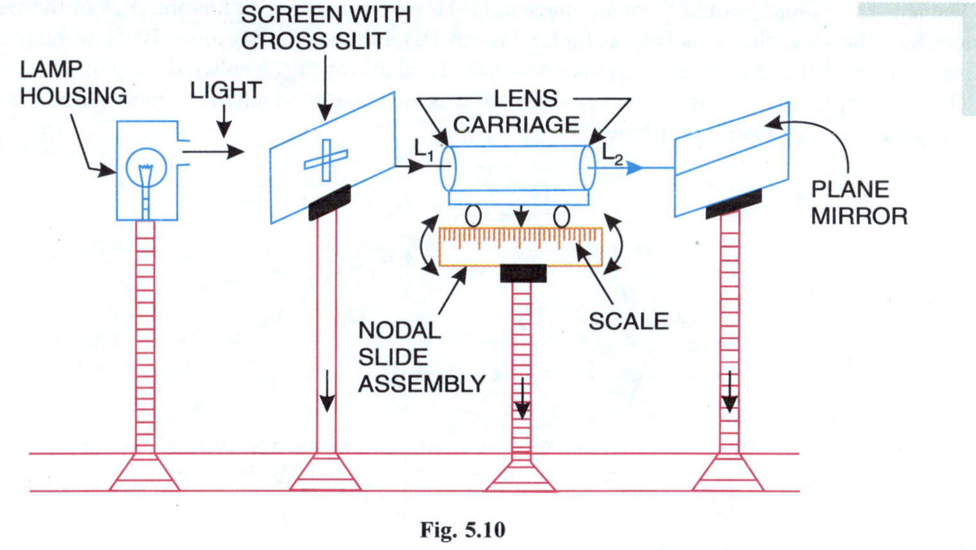

Fig. 5.10

The distance between the optical system and the screen is adjusted in such a way that a well-defined image of the slit is obtained on the screen. The image of the slit is formed slightly to one side of the slit itself. It is obvious that the centre of the slit is at the first focal point of the lens. The nodal slide carrying the lens system is now rotated through a small angle and it will be found that the image shifts side ward to the right or to the left. The nodal slide and its stand are then adjusted such that the direction of rotation of the image changes its sign and finally the image remains stationary for a slight rotation of the carriage. When this condition is reached, the axis of rotation passes through the second nodal point N_2. The other focal point nodal point can be determined by turning the nodal slide through 180° and repeating the experiment. Since the medium on both sides of the lens system is the same (air), the nodal points are also the principal points. The distance between the screen and the axis of rotation for the stationary image is an accurate measure of the first focal length of the lens system.

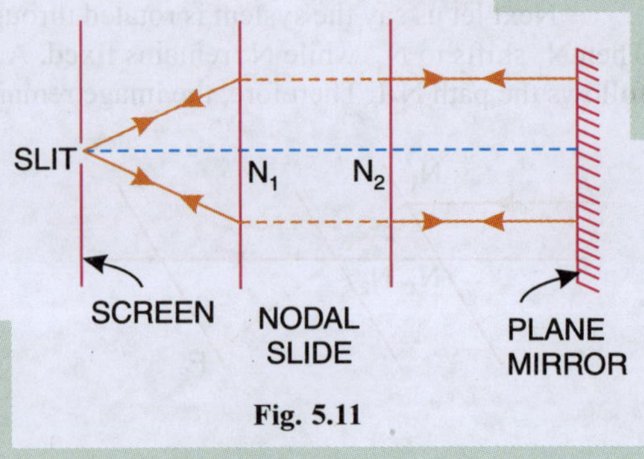

Fig. 5.11

5.10. CARDINAL POINTS OF A COAXIAL SYSTEM OF TWO THIN LENSES

We determine the cardinal points of a coaxial optical system by assuming first that the object at infinity and then the case of object located on the principal axis at a certain distance from the system. We shall find that the computations would yield identical results in both the cases. We shall also observe that the results are identical with those obtained in the deviation method used in Chapter- 4.

5.10.1. OBJECT AT INFINITY

We now consider an object located at infinity, as shown in Fig. 5.12.

AB is a ray of light coming from an object situated at a very large distance, such that $u_1 = \infty$. The lens L_1, if alone, would form an image at G. However, because of the presence of the second lens L_2, G becomes the virtual object for L_2. The ray BD, instead of going along BDG, refracts along the path DF_2. When the ray AB is produced forward and the ray DF_2 backward, they intersect at H_1. The plane H_1P_1 normal to the axis may considered as the plane at which the refraction occurred and this plane is called **principal plane.**

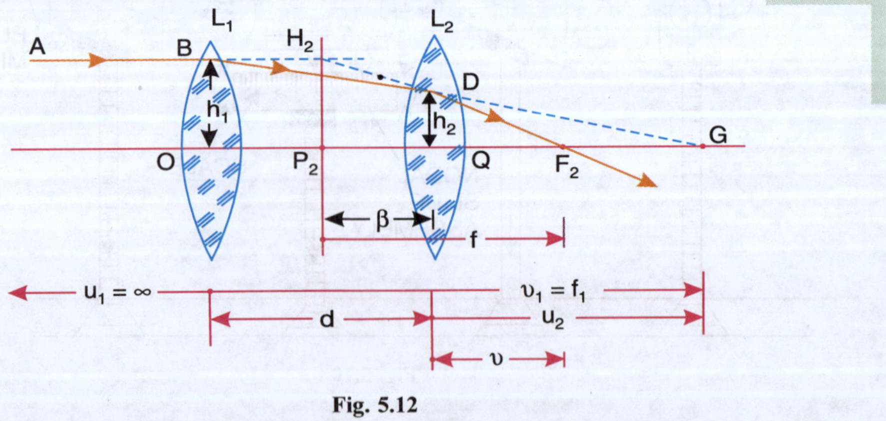

Fig. 5.12

5.10.1.1. Focal Length of the System

Now, we can write the expression for the refraction taking place at the surface of first lens as follows.

$$\frac{1}{\upsilon_1}-\frac{1}{u_1}=\frac{1}{f_1} \quad \therefore \quad \frac{1}{OG}-\frac{1}{u_1}=\frac{1}{f_1}$$

As $u_1=\infty$, we obtain $OG=f_1$

The equation for the refraction at the second lens may be written as

$$\frac{1}{\upsilon}-\frac{1}{u_2}=\frac{1}{f_2} \quad \therefore \quad \frac{1}{QF_2}-\frac{1}{QG}=\frac{1}{f_2}$$

or $$\frac{1}{QF_2}=\frac{1}{f_2}+\frac{1}{f_1-d} \quad \therefore \quad \frac{1}{QF_2}=\frac{f_1+f_2-d}{f_2(f_1-d)} \tag{5.27}$$

The Δ^{les} BOG and DQG are similar and also the Δ^{les} CP_1F_2 and DQF_2 are similar.

$$\frac{BO}{OG}=\frac{DQ}{-QG} \quad \therefore \quad \frac{h_1}{f_1}=\frac{h_2}{-(f_1-d)} \tag{5.28}$$

$$\frac{CP_1}{P_1F_2}=\frac{DQ}{-QF_2} \quad \therefore \quad \frac{h_1}{f}=\frac{h_2}{QF_2} \tag{5.29}$$

From equs.(5.28), (5.29) and (5.27), we get

$$\frac{h_1}{h_2}=\frac{f_1}{-(f_1-d)}=\frac{f(f_1+f_2-d)}{f_2(f_1-d)}$$

or $$\frac{1}{f}=\frac{f_1+f_2-d}{f_1f_2} \tag{5.30}$$

$\therefore$ $$\frac{1}{f}=\frac{1}{f_1}+\frac{1}{f_2}-\frac{d}{f_1f_2} \tag{5.31}$$

which is the same as equ.(4.30). Because the location of the focal point F_2 is determined by QF_2, which is known from the equation, the position of the principal plane P_1 is specified by the value of f calculated from the equ.(5.31).

5.10.1.2. Cardinal Points

(i) Second Principal point : Let us say the second principal plane H_2P_2 is located at a distance of $L_2P_2=\beta$ from the second lens L_2. According to sign convention β would be *negative* as it is measured toward the left of the lens.

$\therefore$ $$QF_2=f-(-\beta)=f+\beta$$

We can determine β using the equation for f into the above relation. Thus,

$$f+\beta=\frac{f_2(f_1-d)}{f_1+f_2-d}$$

$$\beta=-f+\frac{f_2(f_1-d)}{f_1+f_2-d}=\frac{-f_1f_2}{\Delta}+\frac{f_2f_1-f_2d}{\Delta}$$

where $\Delta=f_1+f_2-d$.

$\therefore$ $$\beta=-f_2\frac{d}{\Delta} \tag{5.32}$$

or $$\beta=-\frac{f_2d}{f_1+f_2-d} \tag{5.33}$$

But from equ.(5.30), we have $f_1 + f_2 - d = \dfrac{f_1 f_2}{f}$.

$$\therefore \qquad \beta = -\frac{f\,d}{f_1} \tag{5.34}$$

This is identical to (4.34).

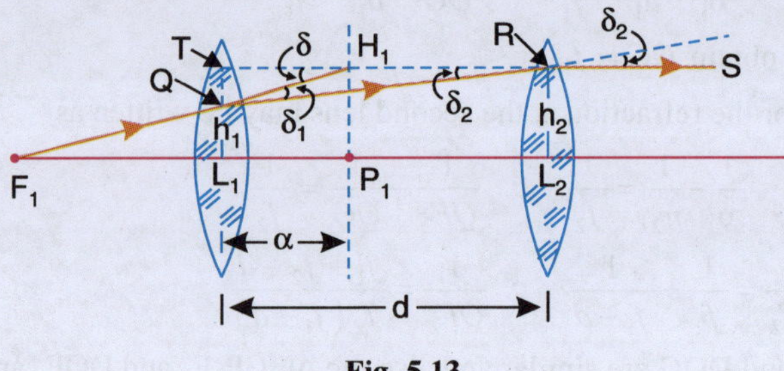

Fig. 5.13

(ii) First Principal Point: By considering a ray of light parallel to the axis and incident on the second lens L_2 from the right side (See Fig. 5.13), we can show that the distance of first principal plane, $L_1P_1 = \alpha$ from the first lens L_1 is given by

$$\alpha = f_1 \frac{d}{\Delta} \tag{5.35}$$

Also,

$$\alpha = +\frac{f\,d}{f_2} \tag{5.36}$$

This is the same as the result (4.35).

Note— In a combination of two lenses, the sequence of the principal planes is in the reverse order- H_2P_2 is to the left of the centre and H_1P_1 is to the right. Compare this result with that obtained for a thick lens in § 6.2.4 and § 6.3.1.2. ◆

(iii) Second Focal Point: Referring to Fig. 5.12, the distance of the second focal point F_2 from the second lens L_2 is given by

$$L_2F_2 = P_2F_2 - P_2L_2$$

$$= f - (-L_2P_2) = f + \beta$$

$$= f + \left(-\frac{f\,d}{f_1}\right)$$

$$\therefore \qquad L_2F_2 = f\left(1 - \frac{d}{f_1}\right) \tag{5.37}$$

(iv) First Focal Point: The distance of the first focal point F_1 from the first lens L_1 is given by

$$L_1F_1 = P_1F_1 - P_1L_1$$

$$= -f - (-L_1P_1) = -f + \alpha$$

$$= -f + \left(\frac{f\,d}{f_2}\right)$$

$$\therefore \qquad L_1F_1 = -f\left(1 - \frac{d}{f_2}\right) \tag{5.38}$$

(v) and (vi) Nodal Points: As the optical system is considered to be located in air, P_1 and P_2 are also the positions of nodal points N_1 and N_2 respectively.

5.10.2. AN AXIAL POINT-OBJECT

Let us consider a point-object O is placed at a distance u from the first lens and the final image is formed at I. The first image due to the first lens is formed at I′. Application of Gauss formula gives

$$\frac{1}{\upsilon'} - \frac{1}{u} = \frac{1}{f_1} \quad \text{or} \quad \frac{1}{\upsilon'} = \frac{1}{f_1} + \frac{1}{u} = \frac{u + f_1}{f_1 u}$$

$$\therefore \qquad \upsilon' = \frac{f_1 u}{u + f_1} \qquad (5.39)$$

The image I′ acts as a *virtual object* for the second lens and the final image is formed at I. The object distance for the second lens is $\upsilon' - d$.

$$\therefore \qquad \frac{1}{\upsilon} - \frac{1}{\upsilon' - d} = \frac{1}{f_2} \quad \text{or} \quad \frac{1}{\upsilon' - d} = \frac{1}{\upsilon} - \frac{1}{f_2} = \frac{f_2 - \upsilon}{\upsilon f_2}$$

$$\therefore \qquad \upsilon' - d = \frac{\upsilon f_2}{f_2 - \upsilon} \qquad (5.40)$$

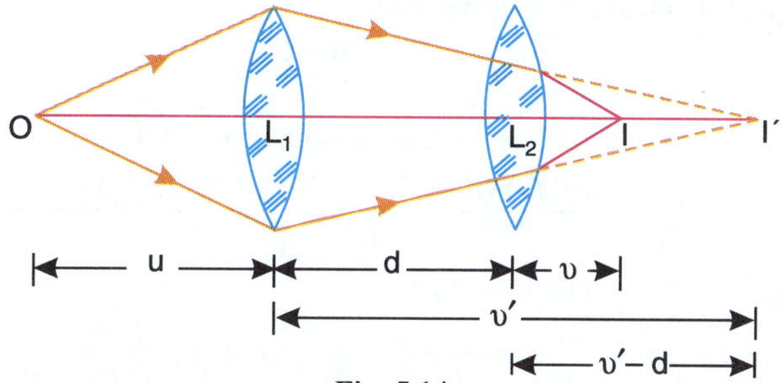

Fig. 5.14

Using the equ.(5.39) into equ.(5.40), we get

$$\frac{f_1 u}{u + f_1} - d = \frac{\upsilon f_2}{f_2 - \upsilon}$$

Multiplying the above equation by $(u + f_1)(f_2 - \upsilon)$,

$$f_1 u (f_2 - \upsilon) - d(\upsilon + f_1)(f_2 - \upsilon) = \upsilon f_2 (u + f_1)$$

Rearranging the terms, we get

$$u\upsilon (d - f_1 - f_2) + u(f_1 f_2 - d f_2) + \upsilon (d f_1 - f_1 f_2) - d f_1 f_2 = 0 \qquad (5.41)$$

This equation can be written in the form

$$u\upsilon A + uB + \upsilon C + D = 0$$

where A,B, C and D are coefficients.

$$u\upsilon + u\frac{B}{A} + \upsilon\frac{C}{A} + \frac{D}{A} = 0 \qquad (5.42)$$

Suppose the focal length of the equivalent lens is f and the reduced object-distance $U = u - \alpha$ and the reduced image-distance $V = u - \beta$. Here α represents the distance of the first lens surface from the first principal plane and β represents the distance of second surface of the second lens from the second principal plane. Then,

$$\frac{1}{V} - \frac{1}{U} = \frac{1}{f} \qquad (5.43)$$

$$\frac{1}{\upsilon - \beta} - \frac{1}{u - \alpha} = \frac{1}{f} \qquad (5.44)$$

Multiplying the above expression by $(u - \alpha)(\upsilon - \beta) f$, we obtain

$$(u - \alpha) f - (\upsilon - \beta) f = (u - \alpha)(\upsilon - \beta)$$

Simplifying the above expression and rearranging the terms, we get

$$u\upsilon + u(-\beta - f) + \upsilon(-\alpha + f) + (\alpha\beta - \beta f + \alpha f) = 0 \qquad (5.45)$$

Comparing equs. (5.42) and (5.45), we have

$$-\beta - f = \frac{B}{A} \qquad (5.46)$$

$$-\alpha + f = \frac{C}{A} \qquad (5.47)$$

$$\alpha\beta - \beta f + \alpha f = \frac{D}{A} \qquad (5.48)$$

Multiplying (5.46) and (5.47), we get

$$\alpha\beta - \beta f + \alpha f - f^2 = \frac{B \times C}{A^2} \qquad (5.49)$$

Subtracting equ.(5.47) from (5.49), we obtain

$$f^2 = \frac{D}{A} - \frac{B \times C}{A^2} = \frac{DA - BC}{A^2} \qquad (5.50)$$

Substituting the values of coefficients A,B,C and D from (5.41), we obtain

$$f^2 = \frac{(-df_1 f_2)(d - f_1 - f_2) - (f_1 f_2 - df_2)(df_1 - f_1 f_2)}{(d - f_1 - f_2)^2}$$

On simplifying the above equation we get

$$f^2 = \frac{f_1^2 f_2^2}{(f_1 + f_2 - d)^2} \qquad \therefore \qquad f = \frac{f_1 f_2}{(f_1 + f_2 - d)} \qquad (5.51)$$

or

$$\frac{1}{f} = \frac{1}{f_1} + \frac{1}{f_2} - \frac{d}{f_1 f_2} \qquad (5.52)$$

From equ.(5.47), $\alpha = -\frac{C}{A} + f$

$$\therefore \qquad \alpha = -\frac{df_1 - f_1 f_2}{d - f_1 - f_2} + \frac{f_1 f_2}{f_1 + f_2 - d}$$

$$= \frac{df_1}{f_1 + f_2 - d} \qquad (5.53a)$$

But from equ. (5.51), we find that $f_1 + f_2 - d = \frac{f_1 f_2}{f}$

$$\therefore \qquad \alpha = \frac{+df}{f_2} \qquad (5.53)$$

This result is identical to (5.36).

A lens combination is characterized by two principal planes, namely first and second principal planes. α gives the position of the first principal plane; as it is a positive quantity, the plane is to the right of the first lens.

From equ. (5.46), we see that

$$\beta = -\left(\frac{B}{A} + f\right) \quad \text{or} \quad \beta = -\frac{(f_1 f_2 - df_2)}{d - f_1 - f_2} - \frac{f_1 f_2}{f_1 + f_2 - d}$$

$$\therefore \qquad \beta = \frac{-df_2}{f_1 + f_2 - d}$$

or $$\beta = \frac{-d f}{f_1} \tag{5.54}$$

This result is identical to (5.34).

β gives the position of the second principal plane; as it is a negative quantity, the plane is to the left of the first lens.

The positions of focal points are given by expressions (5.37) and (5.38).

WORKED OUT EXAMPLES

Example 5.1: A coaxial lens system placed in air has two lenses of focal lengths 3F and F separated by a distance 2F. Find the positions of the cardinal points.

Solution: $f_1 = 3F,\ f_2 = F,\quad d = 2F$

$$f = \frac{f_1 f_2}{f_1 + f_2 - d}$$

$$f = \frac{3F \times F}{3F + F - 2F} = \frac{3}{2}F$$

$$\alpha = +\frac{df}{f_2} = \frac{2F \cdot (3F/2)}{F} = +3F$$

Therefore, the first principal point P_1 is at a distance 3F to the right of the first lens, as shown in Fig. 5.15.

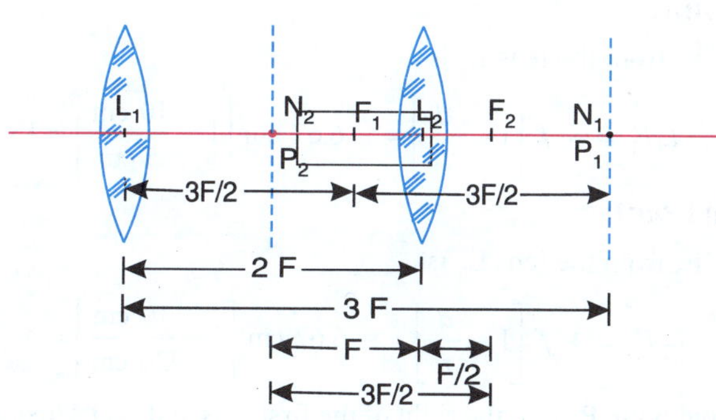

Fig. 5.15

$$\beta = -\frac{df}{f_1} = \frac{-2F \cdot 3F/2}{3F} = -F$$

The second principal point P_2 is at a distance F to the left of the second lens.

The first focal point F_1 is at a distance 3F – 3F/2 = 3F/2 from the first lens and F_2 is at a distance 3F/2 – F = F/2 from the second lens. As the medium on the two sides of the lens system is the same, the nodal points N_1 and N_2 coincide with P_1 and P_2.

Example 5.2: Two thin convex lenses of focal lengths 20 cm and 5 cm are kept coaxially separated by a distance of 10 cm. Plot the positions of the cardinal points for the combination.

Solution: Given that $f_1 = 20$ cm, $f_2 = 5$ cm and $d = 10$ cm

$$f = \frac{f_1 f_2}{f_1 + f_2 - d} = \frac{20\text{ cm} \times 5\text{ cm}}{20\text{ cm} + 5\text{ cm} - 10\text{ cm}} = 6.67\text{ cm}$$

First Principal Point:

$$\alpha = \frac{fd}{f_2} = \frac{10 \text{ cm} \times 6.67 \text{ cm}}{5 \text{ cm}} = 13.33 \text{ cm}$$

Second Principal Point:

$$\beta = \frac{fd}{f_1} = -\frac{10 \text{ cm} \times 6.67 \text{ cm}}{20 \text{ cm}} = -3.33 \text{ cm}$$

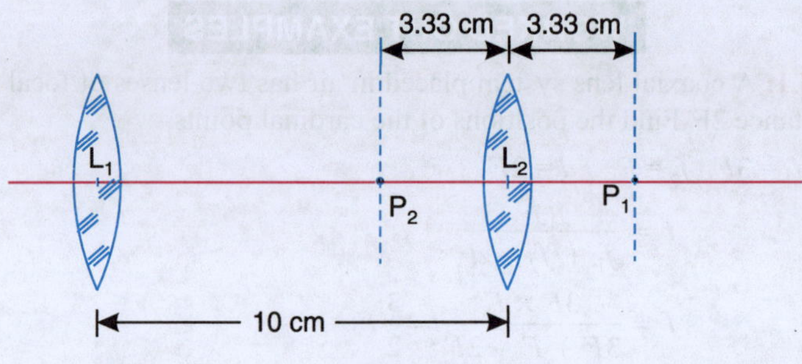

Fig. 5.16

Nodal Points:

As the system is situated in air, the nodal points are same as the principal points.

First Focal Point:

The distance of F_1 from the lens L_1 is

$$L_1F_1 = -f\left[1 - \frac{d}{f_2}\right] = -6.67 \text{ cm}\left[1 - \frac{10 \text{ cm}}{5 \text{ cm}}\right] = 13.33 \text{ cm}$$

Second Focal Point:

The distance of F_2 from the lens L_2 is

$$L_2F_2 = +f\left[1 - \frac{d}{f_1}\right] = -6.67 \text{ cm}\left[1 - \frac{10 \text{ cm}}{20 \text{ cm}}\right] = +3.33 \text{ cm}$$

The first principal point P_1 is to the right of the first lens and is at a distance of 13.33 cm from it. The second principal point P_2 is to the left of the second lens and is at a distance of 3.33 cm from it. The cardinal points are plotted in Fig. 5.16.

Example 5.3: Two identical thin convex lenses of focal lengths 8 cm each are coaxial and 4 cm apart. Find the equivalent focal length and the positions of the principal points. Also, find the position of the object for which the image is formed at infinity.

Solution: Here, $f_1 = f_2 = 8$ cm, $d = 4$ cm

$$f = \frac{f_1f_2}{f_1 + f_2 - d} = \frac{8 \text{ cm} \times 8 \text{ cm}}{8 \text{ cm} + 8 \text{ cm} - 4 \text{ cm}} = \frac{64}{12} \text{ cm} = 5.33 \text{ cm}$$

$$\alpha = +\frac{d\ f}{f_2} = \frac{4 \text{ cm} \times 5.33 \text{ cm}}{8 \text{ cm}} = 2.66 \text{ cm}$$

$$\beta = -\frac{d\ f}{f_1} = -\frac{4 \text{ cm} \times 5.33 \text{ cm}}{8 \text{ cm}} = -2.66 \text{ cm}$$

The points are plotted in Fig. 5.17. For the final image to be formed at infinity

$$\frac{1}{V} - \frac{1}{U} = \frac{1}{f}$$

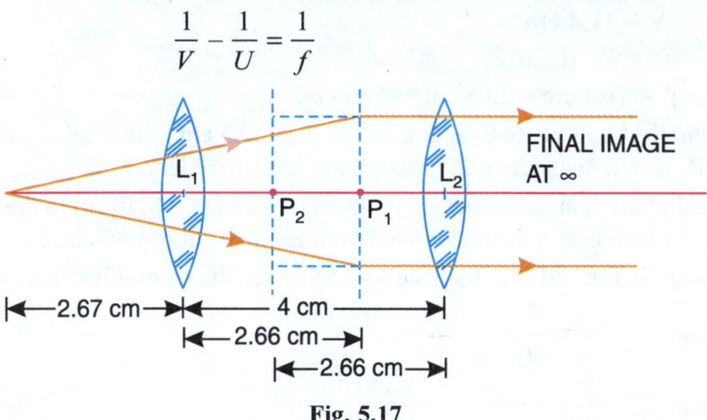

Fig. 5.17

As V = ∞ (given), f = 5.33 cm, we obtain U = – f = – 5.33 cm. But U = $u - \alpha$.

∴ $u = U + \alpha$ = – 5.33 cm + (+ 2.66 cm) = **– 2.67 cm.**

Therefore, the object is at a distance of 2.67 cm to the left of the first lens.

Example 5.4: Two thin convergent lenses each of 20 cm focal length are set coaxially 5 cm apart. An image of upright pole 200 m distant and 10 m high is formed by the combination. Find the position of the unit and focal planes and the image. Also find the size of the image.

Fig. 5.18

Solution: Here, $f_1 = f_2 = 20$ cm, $d = 5$ cm

$$f = \frac{f_1 f_2}{f_1 + f_2 - d} = \frac{20 \text{ cm} \times 20 \text{ cm}}{20 \text{ cm} + 20 \text{ cm} - 5 \text{ cm}} = \frac{80}{7} \text{ cm} = 11.43 \text{ cm}$$

$$\alpha = + \frac{d\,f}{f_2} = \frac{5 \text{ cm} \times 11.43 \text{ cm}}{20 \text{ cm}} = 2.85 \text{ cm}$$

$$\beta = - \frac{d\,f}{f_1} = \frac{5 \text{ cm} \times 11.43 \text{ cm}}{20 \text{ cm}} = - 2.85 \text{ cm}$$

We have u = – 20000 cm, α = 2.85 cm.

$U = u - \alpha$ = – 20000 cm – 2.85 cm = – 20002.85 cm

$$\frac{1}{V} - \frac{1}{U} = \frac{1}{f}$$

$$\therefore \quad \frac{1}{V} = \frac{1}{f} + \frac{1}{U} = \frac{1}{11.43 \text{ cm}} + \frac{1}{-20002.85 \text{ cm}}$$

or $\quad$ **V = 11.4 cm**

But $\quad V = \upsilon - \beta$

$\therefore \quad \upsilon = V + \beta = 11.4 \text{ cm} - 2.85 \text{ cm} = \mathbf{8.55 \text{ cm.}}$

Therefore, the first unit plane P_1 is at a distance of 2.85 cm to the right of the lens L_1 and the second unit plane P_2 is at a distance of 2.85 cm to the left of the lens L_2.

The first focal plane is at a distance of 11.43 cm – 2.85 cm = 8.58 cm to the left of the lens L_1 and the second focal plane is at a distance of 8.58 cm to the right of the lens L_2.

The final image is formed at a distance of 8.55 cm to the right of the lens L_2.

Magnification, $\quad m = \frac{I}{O} = \frac{V}{U}$

As O = 1000 cm, $\quad \frac{I}{1000 \text{ cm}} = -\frac{11.4 \text{ cm}}{20002.85 \text{ cm}}$

$\therefore \quad$ **I = – 0.57 cm.**

Example 5.5: A thin convex and a thin concave lens each of 50 cm focal length are coaxially situated and separated by 10 cm. Find the position and nature of the final image formed of an object placed 20 cm from the convex lens (i.e. 30 cm from the concave lens).

Solution: Here, $f_1 = +50$ cm, $f_2 = -50$ cm, $d = 10$ cm

$$f = \frac{f_1 f_2}{f_1 + f_2 - d} = \frac{-50 \text{ cm} \times 50 \text{ cm}}{50 \text{ cm} - 50 \text{ cm} - 10 \text{ cm}} = +250 \text{ cm}$$

$$\alpha = +\frac{d\,f}{f_2} = \frac{10 \text{ cm} \times 250 \text{ cm}}{-50 \text{ cm}} = -50 \text{ cm}$$

$$\beta = +\frac{d\,f}{f_1} = -\frac{10 \text{ cm} \times 250 \text{ cm}}{50 \text{ cm}} = -50 \text{ cm}$$

Fig. 5.19

U = 30 cm, f = + 250 cm.

$$\frac{1}{V} - \frac{1}{U} = \frac{1}{f}$$

$$\therefore \quad \frac{1}{V} = \frac{1}{f} + \frac{1}{U} = \frac{1}{30 \text{ cm}} + \frac{1}{250 \text{ cm}} = 26.8 \text{ cm.}$$

The image is formed at a distance 26.8 cm from P_2. Therefore, the image is at a distance 13.21 cm to the left of the convex lens. The image is virtual and erect.

Example 5.6: An optical system consists of two thin lenses, a convex lens of focal length 20 cm and a concave lens of focal length 10 cm and separated by a distance of 8 cm. An object 1 cm in height is placed at a distance 50 cm from the convex lens. Find the position and size of the image.

Solution: Here, $f_1 = +20$ cm, $f_2 = -10$ cm, $d = 8$ cm

$$f = \frac{f_1 f_2}{f_1 + f_2 - d} = \frac{-20 \text{ cm} \times 10 \text{ cm}}{20 \text{ cm} - 10 \text{ cm} - 8 \text{ cm}} = -100 \text{ cm}$$

$$\alpha = +\frac{d\,f}{f_2} = \frac{8 \text{ cm} \times (-100 \text{ cm})}{-10 \text{ cm}} = +80 \text{ cm}$$

The first principal point P_1 is at a distance 80 cm to the right of the convex lens.

$$\beta = -\frac{d\,f}{f_1} = -\frac{-8 \text{ cm} \times (-100 \text{ cm})}{20 \text{ cm}} = +40 \text{cm}$$

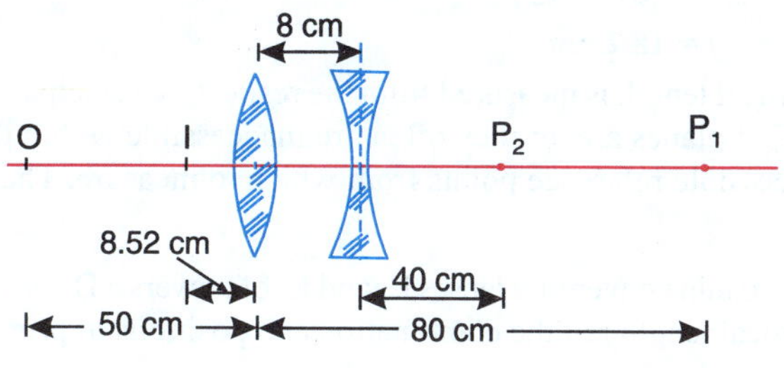

Fig. 5.20

The second principal point P_2 is at a distance of 40 cm to the right of the concave lens.

For the combined system, distance of the object from P_1

$$U = -(50 + 80) \text{ cm} = -130 \text{ cm.}$$

$$\frac{1}{V} - \frac{1}{U} = \frac{1}{f}$$

$$\therefore \quad \frac{1}{V} = \frac{1}{f} + \frac{1}{U} = -\frac{1}{100 \text{ cm}} - \frac{1}{130 \text{ cm}} = -56.5 \text{ cm}$$

The image is at a distance of 56.5 cm to the left of P_2, i.e. 8.52 cm to the left of the convex lens.

$$\frac{I}{O} = \frac{V}{U}$$

$$\therefore \quad \frac{I}{1 \text{ cm}} = \frac{56.5 \text{ cm}}{130 \text{ cm}} \quad \text{or} \quad \mathbf{I = 0.435 \text{ cm}}$$

The image is virtual, erect and diminished.

Example 5.7: A concavo-convex lens has an index of refraction 1.5 and the radii of curvature of its surface are 10 cm and 20 cm. The concave surface is upwards and it is filled with an oil of refractive index 1.6. Calculate the focal length of the oil-glass combination.

Solution: For the glass lens

$$\frac{1}{f_1} = (\mu - 1)\left[\frac{1}{R_1} + \frac{1}{R_2}\right]$$

Here $R_1 = 10$ cm, $R_2 = 20$ cm and $\mu = 1.5$

$$\frac{1}{f_1} = (1.5 - 1)\left[\frac{1}{10 \text{ cm}} + \frac{1}{20 \text{ cm}}\right] = \frac{1}{40 \text{ cm}}$$

For the oil lens

$$\frac{1}{f_2} = (\mu - 1)\left[\frac{1}{R_1} + \frac{1}{R_2}\right]$$

Here $R_1 = 20$ cm, $R_2 = \infty$ and $\mu = 1.6$

$$\frac{1}{f_2} = (1.6 - 1)\left[\frac{1}{20 \text{ cm}} - \frac{1}{\infty}\right] = \frac{3}{100 \text{ cm}}$$

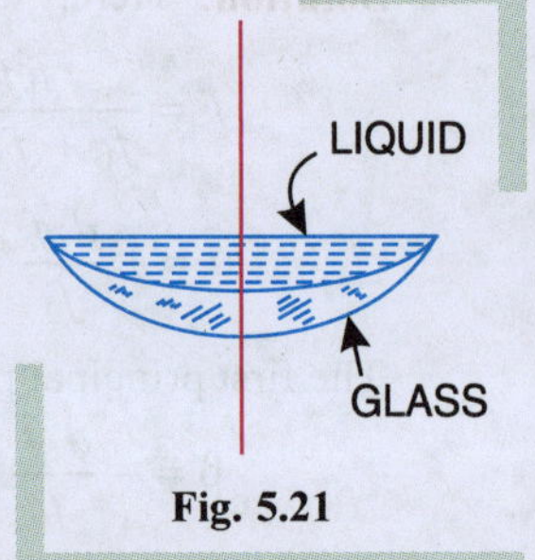

Fig. 5.21

For the oil-glass combination

$$\frac{1}{f} = \frac{1}{f_1} + \frac{1}{f_2} = \frac{1}{40 \text{ cm}} + \frac{3}{100 \text{ cm}} = \frac{11}{200 \text{ cm}}$$

$\therefore$ $\quad$ $\mathbf{f = 18.2}$ **cm.**

The equivalent focal length is measured from the respective principal plane. Often, we do not know where the principal planes are, or they often are inaccessible within the lens or lens system. But we need easily accessible reference points from which to measure. These reference points are the vertices of the lens.

Example 5.8: Two thin converging lenses L_1 and L_2 of powers 5 D and 4 D are placed coaxially 10 cm apart. Find the focal length f of the combination and positions of principal points P_1 and P_2.

(Purvanchal 2000)

Solution: The focal length of the first lens, $f_1 = \frac{1}{P} = \frac{1}{5D} = \frac{1 \text{ m}}{5} = 20$ cm

The focal length of the second lens, $\quad f_2 = \frac{1}{P} = \frac{1}{4D} = \frac{1 \text{ m}}{4} = 25$ cm

The equivalent focal length of the coaxial system is given by

$$f = \frac{f_1 f_2}{f_1 + f_2 - d}$$

$$\therefore \quad f = \frac{20 \text{ cm} \times 25 \text{ cm}}{20 \text{ cm} + 25 \text{ cm} - 10 \text{ cm}} = 14.29 \text{ cm}$$

$$\alpha = +\frac{d\ f}{f_1} = -\frac{10 \text{ cm} \times (14.29 \text{ cm})}{25 \text{ cm}} = 5.7 \text{ cm}$$

The first principal point P_1 is at a distance 5.7 cm to the right of the first lens.

$$\beta = -\frac{d\ f}{f_1} = -\frac{-10 \text{ cm} \times 14.29 \text{ cm}}{20 \text{ cm}} = -7.14 \text{ cm}$$

The second principal point P_2 is at a distance of 40 cm to the left of the second lens.

Example 5.9: Two thin convex lenses having focal lengths 20 cm and 5.0 cm are coaxial and separated by a distance of 10 cm. Find the equivalent focal length and power of the combination.

(Purvanchal 2001)

Solution: The equivalent focal length of the coaxial system is given by

$$f = \frac{f_1 f_2}{f_1 + f_2 - d}$$

$$f = \frac{20 \text{ cm} \times 5 \text{ cm}}{20 \text{ cm} + 5 \text{ cm} - 10 \text{ cm}} = 6.67 \text{ cm}$$

The power of the combination is given by

$$P = \frac{1}{f(m)} = \frac{1}{0.0667\,\text{m}} = 15\,D$$

Example 5.10: Two thin convex lenses of focal lengths 30 cm and 10 cm are separated by a distance of 25 cm in air. Calculate the positions of the cardinal points. **(Agra 2001)**

Solution: Given that $f_1 = +\,30$ cm, $f_2 = +\,10$ cm and $d = 25$ cm

$$f = \frac{f_1 f_2}{f_1 + f_2 - d} = \frac{30\text{ cm} \times 10\text{ cm}}{30\text{ cm} + 10\text{ cm} - 25\text{ cm}} = +\,20\text{ cm}$$

First Principal Point:

$$\alpha = \frac{fd}{f_2} = \frac{20\text{ cm} \times 25\text{ cm}}{10\text{ cm}} = +\,50\text{ cm}$$

Second Principal Point:

$$\beta = -\frac{fd}{f_1} = -\frac{20\text{ cm} \times 25\text{ cm}}{30\text{ cm}} = -\,16.7\text{ cm}$$

First Focal Point:

The distance of F_1 from the lens L_1 is

$$L_1F_1 = -f\left[1 - \frac{d}{f_2}\right] = -\,20\text{ cm}\left[1 - \frac{25\text{ cm}}{10\text{ cm}}\right] = +\,30\text{ cm}$$

Second Focal Point:

The distance of F_2 from the lens L_2 is

$$L_2F_2 = f\left[1 - \frac{d}{f_1}\right] = +\,20\text{ cm}\left[1 - \frac{25\text{ cm}}{30\text{ cm}}\right] = +\,3.3\text{ cm}$$

The first principal point P_1 is to the right of the first lens and is at a distance of 50 cm from it. The second principal point P_2 is to the left of the second lens and is at a distance of – 16.7 cm from it. The first focal point F_1 is to the right of the first lens and is at a distance of 30 cm from it. The second focal point F_2 is to the right of the second lens and is at a distance of + 3.3 cm from it.

Nodal Points:

As the system is situated in air, the nodal points are same as the principal points. The cardinal points are plotted in Fig. 5.22.

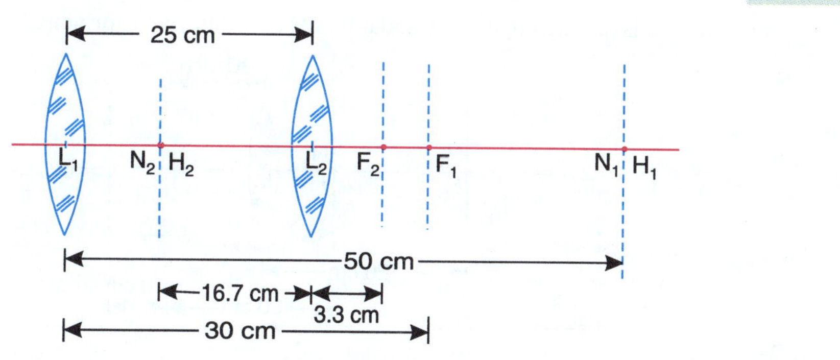

Fig. 5.22

Example 5.11: Two lenses of powers + 2.5 D and – 5 D are kept coaxially separated by 30 cm. Find the positions of cardinal points. **(Allahabad 2002)**

Solution: The focal length of the first lens, $f_1 = \frac{1}{P} = \frac{1}{2.5\ D} = \frac{1\ \text{m}}{2.5} = 40\ \text{cm}$

The focal length of the second lens, $f_2 = \frac{1}{P} = -\frac{1}{5\ D} = -\frac{1\ \text{m}}{5} = -20\ \text{cm}$

The equivalent focal length of the coaxial system is given by

$$f = \frac{f_1 f_2}{f_1 + f_2 - d}$$

$$\therefore \quad f = \frac{40\ \text{cm} \times (-20\ \text{cm})}{40\ \text{cm} - 20\ \text{cm} - 30\ \text{cm}} = +80\ \text{cm}$$

$$\alpha = +\frac{d\,f}{f_2} = -\frac{(30\ \text{cm}) \times (80\ \text{cm})}{-20\ \text{cm}} = -120\ \text{cm}$$

The first principal point P_1 is at a distance 120 cm to the left of the first lens.

$$\beta = -\frac{d\,f}{f_1} = -\frac{-30\ \text{cm} \times 80\ \text{cm}}{40\ \text{cm}} = -60\ \text{cm}$$

The second principal point P_2 is at a distance of 60 cm to the left of the second lens.

First Focal Point:

The distance of F_1 from the lens L_1 is

$$L_1F_1 = -f\left[1 - \frac{d}{f_2}\right] = -80\ \text{cm}\left[1 - \frac{30\ \text{cm}}{(-20\ \text{cm})}\right] = -200\ \text{cm}$$

Second Focal Point:

The distance of F_2 from the lens L_2 is

$$L_2F_2 = +f\left[1 - \frac{d}{f_1}\right] = 80\ \text{cm}\left[1 - \frac{30\ \text{cm}}{40\ \text{cm}}\right] = +20\ \text{cm}$$

The first focal point F_1 is to the left of the first lens and is at a distance of 200 cm from it. The second focal point F_2 is to the right of the second lens and is at a distance of 20 cm from it. The cardinal points are plotted in Fig. 5.23.

Nodal Points:

As the system is situated in air, the nodal points are same as the principal points.

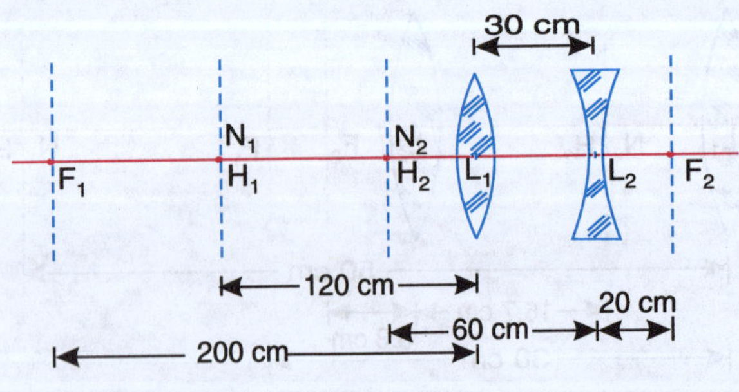

Fig. 5.23

Example 5.12: A convex lens of 10 cm focal length is placed in air from a concave lens of focal length 20 cm, at a distance of 5 cm. Find the distance between the two principal points of the combination. **(Garhwal 2002)**

Solution: Given that $f_1 = +10$ cm, $f_2 = -20$ cm and $d = 5$ cm

The equivalent focal length of the coaxial system is given by

$$f = \frac{f_1 f_2}{f_1 + f_2 - d}$$

$$\therefore \qquad f = \frac{10 \text{ cm} \times (-20 \text{ cm})}{10 \text{ cm} - 20 \text{ cm} - 5 \text{ cm}} = 13.3 \text{ cm}$$

$$\alpha = +\frac{d\,f}{f_2} = \frac{(5 \text{ cm}) \times (13.3 \text{ cm})}{-20 \text{ cm}} = -3.32 \text{ cm}$$

The first principal point P_1 is at a distance 3.32 cm to the left of the first lens.

$$\beta = -\frac{d\,f}{f_1} = \frac{(5 \text{ cm}) \times (13.3 \text{ cm})}{10 \text{ cm}} = -6.65 \text{ cm}$$

The second principal point P_2 is at a distance of 6.65 cm to the left of the second lens.

The distance between the two principal points of the combination

$$= \beta - \alpha - d = [-6.65 - (-3.32) - (-5)] \text{ cm} = 1.67 \text{ cm}$$

QUESTIONS

1. Explain the term cardinal points with reference to a coaxial system. **(Nagpur, 2005)**
2. What are the principal points and principal planes? Show that the principal planes are the planes of unit linear magnification. **(Nagpur, 2004)**
3. What are nodal points and nodal planes? Give their properties. Show that the nodal planes are planes of unit angular magnification.
4. Derive Newton's formula for a convergent system of lenses forming a real image.
5. Describe the construction and working of a nodal slide and show how the nodal points of a system can be located with its help.
6. Show that for a co-axial lens system, $x x' = f f'$ where x and x' are the respective distances of the object and the image from the first and the second focal points and f and f' are the two focal lengths. What form does the expression take when the media on the two sides of the system are the same?
7. Define cardinal points of a system of co-axial lenses. Describe how you would determine experimentally the principal planes of a combination of two thin lenses separated by a distance. **(Nagpur, 2004)**
8. What are the properties of cardinal points of a co-axial lens system? Plot the cardinal points of a Huygens eyepiece. How can they be determined experimentally?
9. (a) Explain the term cardinal points with reference to a co-axial lens system.

   (b) What do you understand by the term power of a lens?
10. Two thin convex lenses of focal length f_1 cm and f_2 cm are coaxial and separated by d. Show that the equivalent focal length f of the combination is given by the relation

$$f = \frac{f_1 f_2}{f_1 + f_2 - d}$$

11. Define cardinal points of a coaxial lens system. Show how the principal planes of such a system can be located using the theory of the method of deviation or otherwise. State the sign convention you use.
12. Derive an expression for the equivalent focal length of a system of two thin lenses separated by a finite distance when the space between them is filled with a medium of refractive index μ ($\mu > 1$).
13. Derive expressions for the equivalent focal length and the positions of principal points and focal points of a coaxial system of two thin lenses separated by a finite distance. **(Meerut, 2005; Nagpur, 2004; Kurukshetra, 2002)**
14. Derive Lagrange's Equation for Magnification. **(Nagpur, 2004)**
15. Obtain condition of achromatism for a coaxial system of two thin lenses separated by a distance *d*. **(Meerut, 2005)**
16. Explain the meaning of the term Achromatism? Find the condition of achromatism of two lenses in contact. **(Garhwal, 2005)**
17. Derive the condition for achromatism for two thin convex lenses in contact. **(Gulbarga, 2005)**
18. Obtain the condition of achromatism of two coaxial thin lenses in contact. **(RTMNU, 2010)**

PROBLEMS FOR PRACTICE

1. Two thin convex lenses of focal length 12 cm and 6 cm are placed coaxially 8 cm apart. Determine: (*a*) Focal length of equivalent thin lens.
   (*b*) Position of principal points. **(Nagpur, 2004)**
2. Two convex lenses of focal length 10 cm and 20 cm are placed 5 cm apart in air. Find equivalent focal length of the combination. **(Nagpur, 2004)**
3. The focal length of an achromatic doublet is 0.9 m. and the dispersive powers of the two lenses are 0.024 and 0.036. Calculate their focal lengths. **(Bangalore, 2005)**
4. Calculate the focal length of a convex lens of crown glass (dispersive power 0.012) and a concave lens of flint glass (dispersive power 0.020) so that when placed in contact they form an achromatic converging combination of focal length 30 cm. **(Kovempu, 2005)**
5. A convergent doublet of s separated lenses, corrected for spherical aberration, has an equivalent focal length of 10 cm. The lenses of the doublet are separated by 2 cm. What are the focal lengths of its component lenses? **(Garhwal, 2005)**
6. Two thin convex lenses of focal length 50 cm and 20 cm are placed coaxially in air and are separated by a distance of 30 cm. Determine the position of the cardinal points. Find the position of the image and its magnification of an object placed 25 cm in front of the first lens. **(Garhwal, 2005)**
7. Two thin lenses of focal lengths f_1 and f_2 separated by a distance '*d*' have an equivalent focal length 50 cm. The combination satisfied the conditions for no chromatic aberration and minimum spherical aberration. Find the values of f_1, f_2 and *d*. Assume that both the lenses are of the same material. **(Madurai Kamaraj, 2003)**
8. A thin convex and concave lenses of focal lengths 50cm each are co-axially separated by a distance of 10 cm. Find the position of principal points. **(RTMNU, 2010)**

CHAPTER

Thick Lenses

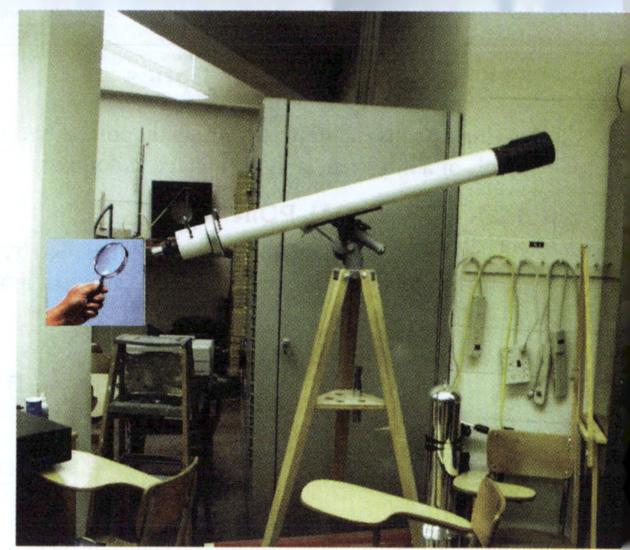

6.1. THICK LENS

A **thick lens** is a physically large lens having two spherical surfaces separated by a distance, which is not negligible in comparison to the radii of curvature of the spherical surfaces. In other words, a thick lens is a lens whose thickness cannot be treated as small in comparison to its focal length. We can determine the position and size of the image of an object by considering refraction at each surface successively, but the procedure is comparatively tedious. The position and size of the image can be determined *directly* if we know the cardinal points of the thick lens.

At a Glance

- Thick Lens
- Cardinal Points of a Thick Lens
- Thick Lens Equation
- Behaviour of Lens as Thickness Increases
- Glass Sphere as a Lens
- Combination of Two Thick Lenses
- Principal Planes in a Two-Lens System Move Out When the Lenses are Separated
- Applications of Lens Combinations

6.2. CARDINAL POINTS OF A THICK LENS

As in the case of an optical system consisting of two or more lenses, a thick lens is also characterized by six

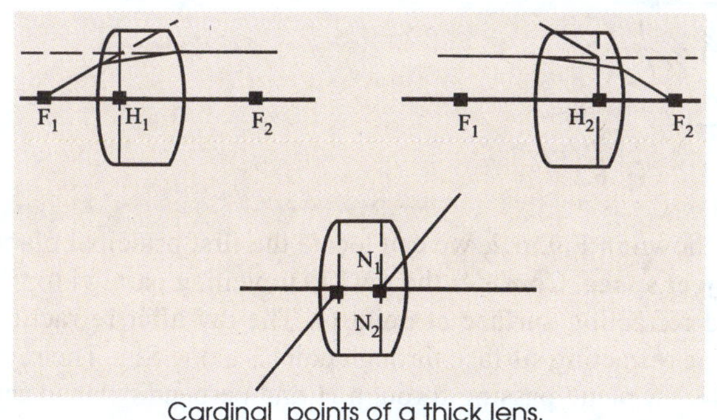

Cardinal points of a thick lens.

cardinal points, namely two principal points, two focal points and two nodal points. The cardinal points of a thick lens may be located as follows:

6.2.1. PRINCIPAL POINTS

Fig. 6.1 shows a thick lens, AB, bounded by two spherical surfaces separated by a distance. Let a ray OA travelling parallel to the principal axis be incident on it. After refraction through the

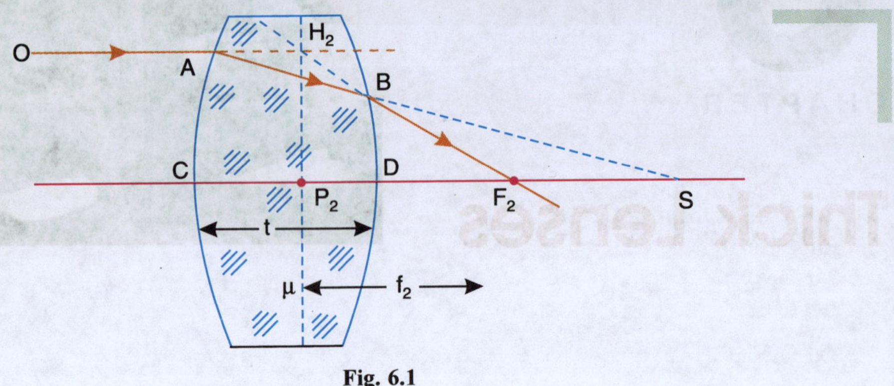

Fig. 6.1

lens, the ray OA passes through the focal point F_2 in the **image space** of the lens. The actual ray is refracted at each surface of the lens and follows the path $OABF_2$. If we extend the ray OA forward and the ray BF_2 backward, they meet each other within the lens at H_2. Now, we can describe the refraction of the incident ray OA in terms of a *single refraction* at a plane passing through H_2. A plane drawn through the point H_2 and perpendicular to the axis may be regarded as the surface at which refraction takes place. Such a plane H_2P_2 is called a **principal plane.** The point P_2 is called the **second principal point**. Thus, the two consecutive deviations of the light rays caused by the two surfaces at A and B of the thick lens are equivalent to a single refraction at H_2, taking place at the *principal plane*. H_2P_2 is the principal plane in the image space and is called the **second principal plane**.

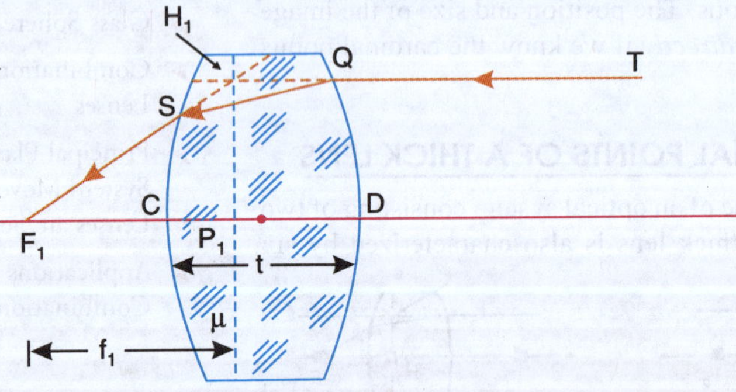

Fig. 6.2

By adopting similar procedure, as shown in Fig. 6.2, we can locate the first principal plane H_1P_1 and first principal point P_1 in the **object space**. Consider the ray TQ travelling parallel to the principal axis and incident on the second refracting surface at point Q. The ray after refraction through the lens along TS, emerges from the refracting surface through point S along SF_1. The rays F_1S and TQ when produced intersect at H_1. A plane passing through H_1 and perpendicular to the principal axis is called the **first principal plane**. The point of intersection, P_1, of the first principal plane, H_1P_1, with the principal axis is called the **first principal point**.

It is seen from Fig. 6.5 that the two incident rays AB and F_1S are directed towards H_1 and after refraction seem to come from H_2. Therefore, H_2 is the image of H_1. Thus, H_1 and H_2 are the *conjugate points* and the planes H_1P_1 and H_2P_2 are a pair of *conjugate planes*. It is also seen that

$$H_2P_2 = H_1P_1.$$

Hence, the lateral magnification of the planes is +1. Thus, the first and second principal planes are planes of unit magnification and are therefore called *unit planes* and the points P_1 and P_2 are called *unit points*.

Note— The principal planes are *conceptual* planes and do not have physical existence within the lens. ◆

6.2.2. FOCAL POINTS

The **first focal point** is a point on the principal axis of the lens such that a beam of light passing through it is rendered parallel to the principal axis after refraction through the lens. The **second focal point** is a point on the principal axis of the lens such that a beam of light travelling parallel to the principal axis of the lens, after refraction through the lens, passes through it.

The planes passing through the principal focal points F_1 and F_2 and perpendicular to the axis are called **first focal plane** and **second focal plane** respectively. The main property of the focal planes is that the rays starting from a point in the focal plane in the object space correspond to a set of conjugate parallel rays in the image space. Similarly, a set of parallel rays in the object space corresponds to a set of rays intersecting at a point in the focal plane in the image space.

The distance of the first focal point from the first principal point, i.e., F_1P_1, is called the **first focal length,** f_1, of the lens and the distance of the second focal point from the second principal point, F_2P_2, is called the **second focal length,** f_2. f_1 and f_2 are also known as the focal lengths in object space and image space respectively.

When the medium is same on the two sides of the lens $f_1 = f_2$ (numerically).

6.2.3. NODAL POINTS

Nodal points are points on the principal axis of the lens where light rays, ***without refraction,*** intersect the optic axis. The planes passing through the nodal points and perpendicular to the principal

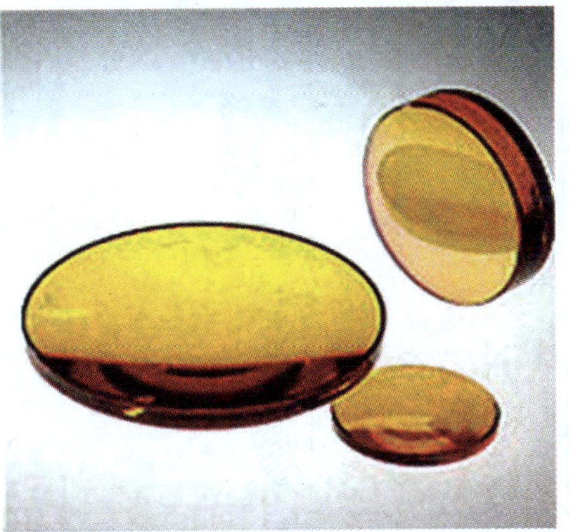

Meniscus lenses, plano-convex lenses.

axis are called the **nodal planes**. Whereas the principal planes are planes where all refraction are assumed to occur, the nodal planes are planes where refraction does *not* take place. Fig. 6.3 represents a thick lens with the help of its cardinal planes. It is seen from the Fig. 6.3 that a ray of light directed towards one of the nodal points, N_1, after refraction through the lens emerges out from the second nodal point, N_2, in a direction parallel to the incident ray. The distances of the nodal points are measured from the focal points. The cardinal points and cardinal planes of a thick lens are shown in Fig. 6.3.

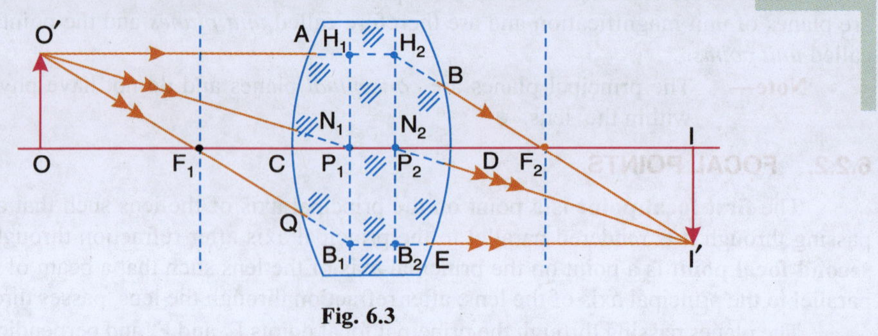

Fig. 6.3

The formation of the image of an object OO′ is also illustrated in Fig. 6.3. An object OO′ is placed on the axis of a thick lens. The paraxial incident ray O′A travels parallel to the axis and is incident on the lens at A. The ray after refraction through the lens passes through the second focal point F_2. Another paraxial ray O′F_1Q starting from O′ passes through the first focal point F_1 and after refraction through the lens emerges at E parallel to the principal axis. Another paraxial ray O′N_1 starting from O′ passes through the first nodal point N_1 and emerges at along N_2D after refraction through the lens. The three emergent rays intersect at I′ which is the image of O′. The line I I′ drawn perpendicular on the axis represents the image of the object OO′.

Note— The position of principal planes depends on the shape of the lens. They are shown in Fig. 6.4.

In *symmetrical lenses*, the principal planes are *symmetrically* located and lie *within* the lens. In a plano-convex and even more so in a meniscus, the two principal planes move to one side and may even move outside of the lens, towards the side of *steeper* curvature.

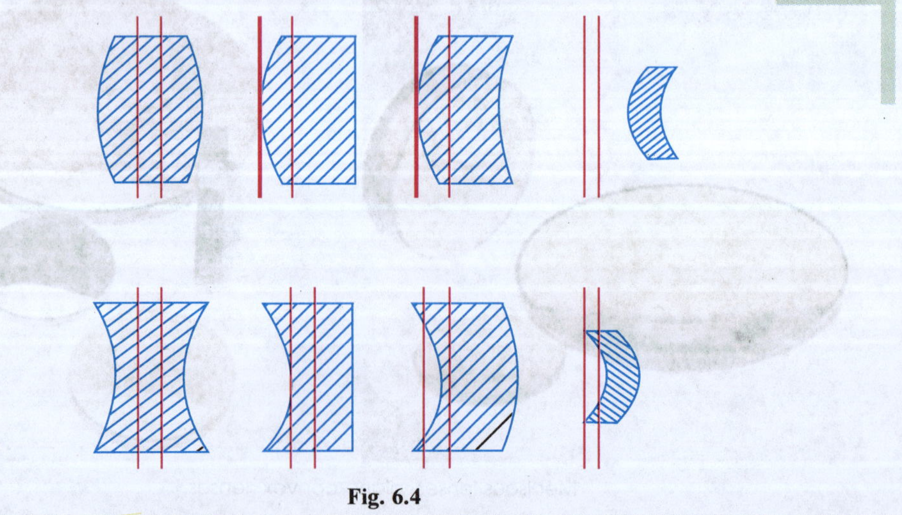
Fig. 6.4

6.3. THICK LENS EQUATION

We determine the cardinal points of a thick lens by assuming first that the object at infinity and then for the case of object located on the principal axis at a certain distance from the system. We shall find that the computations would yield identical results in both the cases.

6.3.1. EXTENDED OBJECT AT INFINITY

Let AB be a ray parallel to the axis and coming from infinity. After refracting along BG, it emerges at G at the second surface and goes along GF_2. Let H_1P_1 and H_2P_2 be the first and the second principal planes. Let R_1 be the radius of curvature of the first surface and R_2 be the radius of curvature of the second surface. Let μ be the refractive index of the lens material and let t be the thickness of the lens. Let f_1 and f_2 be the first and second focal lengths of the lens respectively and let the medium on either side of the lens be air.

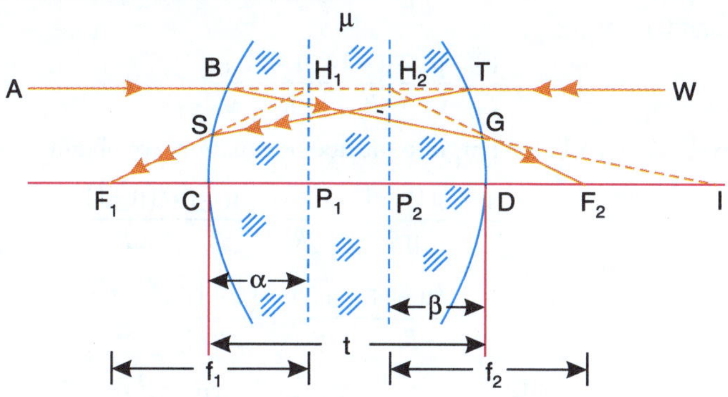

Fig. 6.5

6.3.1.1. Focal length of a thick lens

Let $P_1F_1 = -f_1$ and $P_2F_2 = f_2$. But $f_2 = -f_1$. We designate the numerical value by f.

Let I be the position of the image formed by refraction at the first surface BC when the object is lying at infinity. Thus,

$$CI = \upsilon_1$$

Since $u = \infty$, we can write for refraction at the first surface

$$\frac{\mu}{\upsilon_1} - \frac{1}{\infty} = \frac{\mu - 1}{R_1} \quad \text{or} \quad \frac{1}{\upsilon_1} = \frac{\mu - 1}{\mu R_1} \tag{6.1}$$

As the second surface refracted the ray BG along F_2 and formed the final image at F_2, we can write for refraction at the second surface

$$\frac{1}{DF_2} - \frac{\mu}{DI} = \frac{\mu - 1}{-R_2} \tag{6.2}$$

The pairs of Δ^{les} $H_2P_2F_2$, GDF_2 and BCI, GDI are similar. Therefore,

$$\frac{P_2F_2}{DF_2} = \frac{H_2P_2}{GD} = \frac{BC}{GD} = \frac{CI}{DI}$$

$$\therefore \quad \frac{DF_2}{P_2F_2} = \frac{DI}{CI} \tag{6.3}$$

or

$$\frac{1}{f} = \frac{1}{DF_2} \cdot \frac{DI}{\upsilon_1} \tag{6.4}$$

Multiplying equ.(6.2) by DI, we get

$$\frac{DI}{DF_2}-\mu=\frac{\mu-1}{-R_2}\cdot DI$$

$$\therefore \qquad \frac{DI}{DF_2}=\mu+\frac{\mu-1}{-R_2}\cdot DI$$

Substituting the above expression in (6.4), we obtain

$$\frac{1}{f}=\frac{1}{\upsilon_1}\left[\mu-\frac{\mu-1}{R_2}\cdot DI\right]$$

$$=\frac{1}{\upsilon_1}\left[\mu-\frac{\mu-1}{R_2}\cdot(CI-CD)\right]$$

As CD = t and CI = υ_1,

$$\frac{1}{f}=\frac{\mu}{\upsilon_1}-\frac{\mu-1}{R_2}+\frac{\mu-1}{R_2}\cdot\frac{t}{\upsilon_1} \qquad (6.5)$$

Using the expression (6.1) for $1/\upsilon_1$ into the above equation, we obtain

$$\frac{1}{f}=\frac{\mu(\mu-1)}{\mu R_1}-\frac{\mu-1}{R_2}+\frac{\mu-1}{R_2}\cdot\frac{t(\mu-1)}{\mu R_1}$$

$$=\frac{(\mu-1)}{R_1}-\frac{\mu-1}{R_2}+\frac{(\mu-1)^2 t}{\mu R_1 R_2}$$

$$\frac{1}{f}=(\mu-1)\left[\frac{1}{R_1}-\frac{1}{R_2}\right]+\frac{(\mu-1)^2 t}{\mu R_1 R_2}$$

or

$$\frac{1}{f}=(\mu-1)\left[\left(\frac{1}{R_1}-\frac{1}{R_2}\right)+\frac{(\mu-1)\,t}{\mu R_1 R_2}\right] \qquad (6.6)$$

On putting $t = 0$, equ.(6.6) reduces to equation for a thin lens

$$\frac{1}{f_t}=(\mu-1)\left[\frac{1}{R_1}-\frac{1}{R_2}\right]$$

Therefore, we can write equ.(6.6) as also

$$\frac{1}{f}=\frac{1}{f_t}+\frac{(\mu-1)^2 t}{\mu R_1 R_2} \qquad (6.7)$$

6.3.1.2. Cardinal points of a thick lens

When a beam of light passes through a thick lens or a combination of lenses, we encounter problems in measuring the object distance, the image distance and focal length. These distances *cannot* be measured from the centre of the lens system. They are therefore measured from the *principal planes*.

(i) Second Focal Point: The distance of the second focal point F_2 from the second surface of the lens is DF_2. From the relation (6.3), we have

$$DF_2=P_2F_2\,\frac{DI}{CI}=f\cdot\frac{DI}{CI}$$

$$=f\cdot\frac{CI-CD}{CI}=f\cdot\frac{CI-t}{CI}$$

$$\therefore \qquad DF_2 = f\left(1-\frac{t}{CI}\right) = f\left(1-\frac{t}{\upsilon_1}\right)$$

Using the value of υ_1 from (6.1) into the above equation, we get

$$DF_2 = f\left[1-\frac{(\mu-1)\,t}{\mu R_1}\right] \qquad (6.8)$$

(ii) Second Principal Plane: The second principal plane is located at a distance $P_2D = \beta$ from the second refracting surface. From the Fig. 6.5, it is seen that the pairs of Δ^{les} $H_2P_2F_2$, GDF_2 and BCI, GDI are similar.

$$\therefore \qquad \frac{P_2D}{P_2F_2} = \frac{H_2G}{H_2F_2} = \frac{BG}{GI} = \frac{CD}{CI} = \frac{t}{\upsilon_1} \qquad (6.9)$$

$$\therefore \qquad P_2D = P_2F_2 \cdot \frac{t}{\upsilon_1}$$

$$\therefore \qquad -\beta = f_2 \cdot t \cdot \frac{(\mu-1)}{\mu R_1} \qquad \text{But } f_2 = f.$$

$$\therefore \qquad \beta = -f \cdot \frac{(\mu-1)\,t}{\mu R_1} \qquad (6.10)$$

(iii) First Focal Point: If we consider the incident ray WT coming from the right, then the first focal point F_1 from the first surface C is CF_1. Obviously, the positions of R_1 and R_2 interchange and all the signs become opposite. The distance of the first focal point is then given by

$$CF_1 = -f\left[1+\frac{(\mu-1)\,t}{\mu R_2}\right] \qquad (6.11)$$

(iv) First Principal Plane: The first principal plane is located at a distance $CP_1 = \alpha$ from the first refracting surface. Consider the ray WT coming from the right side parallel to the axis and incident on the second surface. The surfaces TGD and BSC successively refract it. The final ray emerges at the first surface and goes through the first focal point F_1. Following arguments similar to those used for deriving an expression for β, we obtain for the distance of the first principal plane

$$\alpha = -f \cdot t \cdot \frac{(\mu-1)}{\mu\, R_2} \qquad (6.12)$$

(v) and (vi) Nodal Points: As the medium on either side of the lens is air, the nodal points N_1 and N_2 will coincide with the principal points H_1 and H_2 respectively.

6.3.1.3. Power

The power of a thick lens is given by

$$P = \frac{1}{f} = \frac{(\mu-1)}{R_1} - \frac{(\mu-1)}{R_2} + \frac{(\mu-1)^2}{R_1R_2} \cdot \frac{t}{\mu}$$

The power of the first refracting surface is

$$P_1 = \frac{(\mu-1)}{R_1}$$

and the power of the second refracting surface is

$$P_2 = -\frac{(\mu-1)}{R_2}$$

$$\therefore \qquad P = P_1 + P_2 - P_1P_2 \cdot \frac{t}{\mu} \qquad (6.13)$$

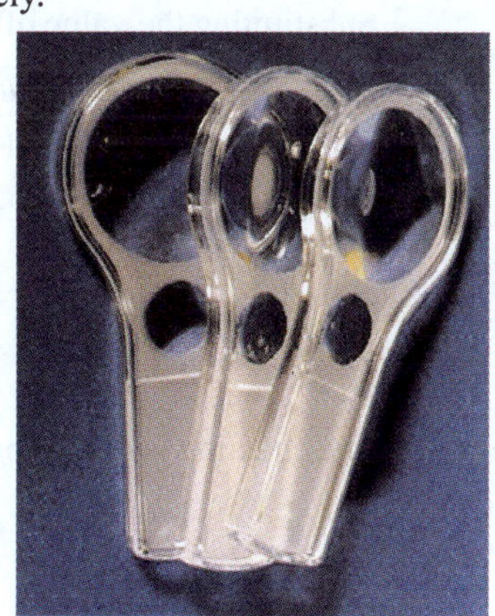
Power of a thick lens.

6.3.2. AN AXIAL POINT-OBJECT

Now let us consider the case of a point object located on the principal axis of the thick lens.

6.3.2.1. Focal length

Consider a lens of thickness t and of refractive index μ, placed in air. The radii of curvature are R_1 and R_2. A point object O is situated on the axis at a distance u from the first refracting surface and forms an image I′ at a distance υ_1 from P.

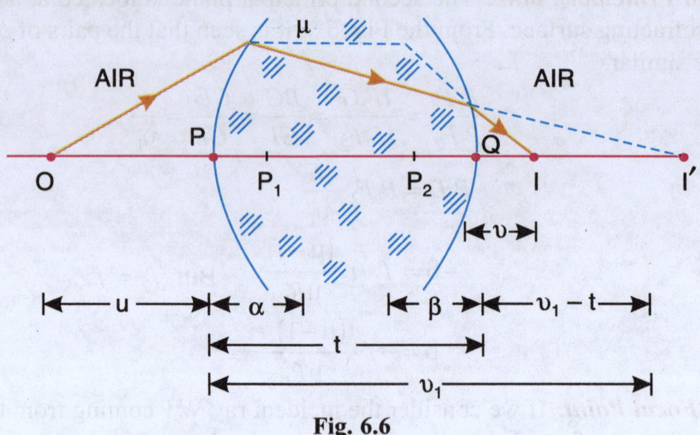

Fig. 6.6

$$\frac{\mu}{\upsilon_1} - \frac{1}{u} = \frac{\mu - 1}{R_1}$$

$$\therefore \qquad \upsilon_1 = \frac{R_1 u \mu}{R_1 + u(\mu - 1)} \qquad (6.14)$$

The image formed by the first surface acts as the object for the second surface and the final image is formed at I.

$$\frac{1/\mu}{\upsilon} - \frac{1}{(\upsilon_1 - t)} = \frac{1/\mu - 1}{R_2}$$

$$(\upsilon_1 - t) = \frac{\mu \upsilon R_2}{R_2 + \upsilon(\mu - 1)} \qquad (6.15)$$

Substituting the value of υ_1 from equ. (6.14) in equ.(6.15), we get

$$\frac{R_1 u \mu}{R_1 + u(\mu - 1)} - t = \frac{\mu \upsilon R_2}{R_2 + \upsilon(\mu - 1)}$$

On simplification of the above expression, we obtain

$$u\upsilon\left[\mu(\mu-1)(R_1 - R_2) - (\mu-1)^2 t\right] + u\left[\mu R_1 R_2 - tR_2(\mu - 1)\right]$$

$$+ \upsilon\left[-\mu R_1 R_2 - tR_1(\mu - 1)\right] - tR_1 R_2 = 0 \qquad (6.16)$$

The equation is of the form

$$u\,\upsilon\,A + u\,B + \upsilon\,C + D = 0$$

or

$$u\,\upsilon + u\frac{B}{A} + \upsilon\frac{C}{A} + \frac{D}{A} = 0 \qquad (6.17)$$

where A, B, C, and D are coefficients.

Let us take V = υ – β and U = u – α and the focal length = f

$$\frac{1}{\upsilon-\beta}-\frac{1}{u-\alpha}=\frac{1}{f} \tag{6.18}$$

Simplifying and rearranging the terms, we obtain

$$u\,\upsilon+(-\beta-f)\;u+(-\alpha+f)\;\upsilon+(-\beta f+\alpha f+\alpha\,\beta)=0 \tag{6.19}$$

Comparing equ. (6.17) and (6.19), we have

$$-\beta-f=\frac{B}{A} \tag{6.20}$$

$$-\alpha+f=\frac{C}{A} \tag{6.21}$$

and
$$-\beta f+\alpha f+\alpha\,\beta=\frac{D}{A} \tag{6.22}$$

From equ.(6.20) to (6.22), we get

$$f^2=\frac{D}{A}-\frac{BC}{A^2}=\frac{AD-BC}{A^2}$$

Substituting the values of A, B, C and D in the above expression and after simplification, we get

$$f=\frac{\mu^2R_1^2R_2^2}{\left[\mu(\mu-1)(R_1-R_2)-(\mu-1)^2\,t\right]^2}$$

$$f=\pm\frac{\mu R_1R_2}{\left[\mu(\mu-1)(R_1-R_2)-(\mu-1)^2\,t\right]}$$

$$\therefore\qquad \frac{1}{f}=(\mu-1)\left[\frac{1}{R_1}-\frac{1}{R_2}+\frac{(\mu-1)\;t}{\mu R_1R_2}\right] \tag{6.23}$$

For a thin lens, t = 0 and from equ.(6.23), we see that

$$\therefore\qquad \frac{1}{f_t}=(\mu-1)\left[\frac{1}{R_1}-\frac{1}{R_2}\right] \tag{6.24}$$

which is the same as equ. (4.8).

6.3.2.2. Position of the Principal Points

The first principal point P_1 is at a distance α from P. From equ.(6.21),

$$-\alpha+f=\frac{C}{A}$$

$$\alpha=f-\frac{C}{A}$$

$$\alpha=\frac{+t\;R_1}{\mu(R_1-R_2)-t\;(\mu-1)} \tag{6.25}$$

The second principal point P_2 at a distance β from Q. We have

$$-\beta - f = \frac{B}{A}$$

$$\therefore \qquad \beta = -\left(\frac{B}{A} + f\right)$$

Substituting the values of B, A, and f and after simplification we get

$$\beta = \frac{+t R_2}{\mu(R_1 - R_2) - t(\mu - 1)} \qquad (6.26)$$

It is easy to see that

$$\frac{\alpha}{\beta} = \frac{R_1}{R_2}$$

Equ.(6.18) may now be written as

$$\frac{1}{V} - \frac{1}{U} = \frac{1}{f} \qquad (6.27)$$

This is a relation similar to that of a thin lens; with the difference that the object distance is measured from the first principal plane and the image distance is measured from the second principal plane.

6.4. BEHAVIOUR OF LENS AS THICKNESS INCREASES

In case of a double convex lens, R_1 is positive and R_2 is negative. When we take these signs into consideration, the thick lens equation may be rewritten as

$$\frac{1}{f} = (\mu - 1)\left[\frac{1}{R'} + \frac{1}{R''} - \frac{(\mu - 1)\, t}{\mu R' R''}\right] \qquad (6.28)$$

Equ.(6.28) indicates that the value of $1/f$ decreases with the increasing thickness. As t increases, a stage will be reached when $1/f$ approaches zero and beyond that it becomes negative. It then acts as a divergent lens. The **critical thickness, t_c,** beyond which a thick convergent lens changes into a divergent lens is given by

$$\left[\frac{1}{R'} + \frac{1}{R''} - \frac{(\mu - 1)\, t_c}{\mu R' R''}\right] = 0$$

$$\therefore \qquad t_c = \frac{\mu\,(R' + R'')}{\mu - 1} \qquad (6.29)$$

Divergent image of a person standing in front of an aluminized Fresnel lens.

For a lens for which R′ = R″ (= R), the above equation reduces to

$$t_c = \frac{2\mu R}{\mu - 1} \qquad (6.30)$$

6.5. GLASS SPHERE AS A LENS

Let us consider a glass sphere of radius R and refractive index, μ placed in air (See Fig. 6.7). It may be treated as a *thick lens* with O_1 and O_2 as poles of two refracting surfaces. The focal length of a thick lens in air is given by

$$\frac{1}{f}=(\mu-1)\left[\frac{1}{R_1}-\frac{1}{R_2}+\frac{(\mu-1)\,t}{\mu R_1 R_2}\right]$$

In the case of a sphere-lens, we have $R_1 = +R$, $R_2 = -R$ and $t = 2R$.

$$\therefore \quad \frac{1}{f}=(\mu-1)\left[\frac{1}{R}+\frac{1}{R}-\frac{(\mu-1)\,(2R)}{\mu R^2}\right]$$

$$=(\mu-1)\frac{2}{R}\left[1-\frac{\mu-1}{\mu}\right]$$

$$=\frac{2(\mu-1)}{\mu R}$$

$$\therefore \quad f=\frac{\mu R}{2(\mu-1)} \qquad (6.31)$$

The relation (6.31) gives the focal length of a sphere-lens.

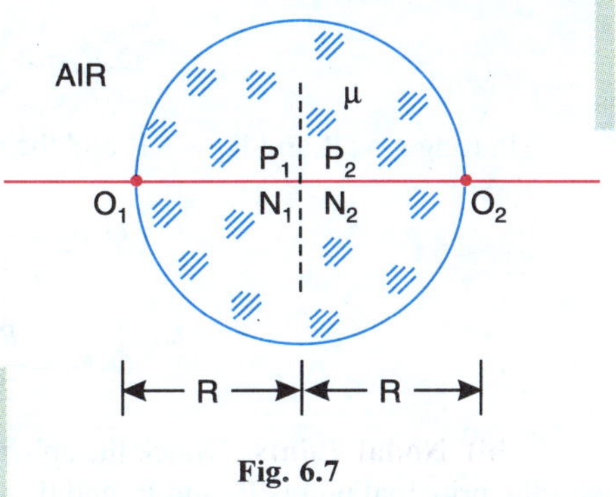

Fig. 6.7

6.5.1. CARDINAL POINTS

(i) Principal Points: The distance of the first principal point P_1 from the first surface of the lens is given by

$$O_1P_1=-f\,\frac{(\mu-1)t}{\mu R_2}$$

Using $t = 2R$ and $R_2 = -R$ and the relation (6.31) for f into the above equation, we get

$$O_1P_1=-\frac{\mu R}{2(\mu-1)}\cdot\frac{(\mu-1)(2R)}{\mu(-R)}=+R \qquad (6.32)$$

It means that the first principal point P_1 is at a distance on the right of O_1, that is, at the centre of the sphere (See Fig. 6.7).

The distance of the second principal point P_2 from the rear surface is given by

$$O_2P_2=-f\,\frac{(\mu-1)t}{\mu R_2}$$

Again putting $t = 2R$ and $R_1 = +R$ and the relation (6.29) for f into the above equation, we get

$$O_2P_2=-\frac{\mu R}{2(\mu-1)}\cdot\frac{(\mu-1)(2R)}{\mu(R)}=-R \qquad (6.33)$$

It means that the second principal point P_2 is at a distance on the left of O_2, that is, at the centre of the sphere (See Fig. 6.7).

(ii) Focal Points: The distance of the *first focal point* F_1 from the first surface is given by

$$O_1F_1=-f\left[1+\frac{(\mu-1)t}{\mu R_2}\right]$$

Putting $t = 2R$ and $R_2 = -R$ and the relation (6.29) for f into the above equation, we get

$$O_1F_1=-\frac{\mu R}{2(\mu-1)}\left[1+\frac{(\mu-1)2R}{\mu(-R)}\right]=-\frac{\mu R}{2(\mu-1)}+R$$

$$\therefore \qquad O_1F_1 = -\frac{R(2-\mu)}{2(\mu-1)} \qquad (6.34)$$

The distance of the *second focal point* F_2 from the second surface is given by

$$O_2F_2 = +f\left[1-\frac{(\mu-1)t}{\mu R_1}\right]$$

Putting $t = 2R$ and $R_1 = +R$ and the relation (6.24) for f into the above equation, we get

$$O_2F_2 = \frac{\mu R}{2(\mu-1)}\left[1-\frac{(\mu-1)2R}{\mu(R)}\right] = \frac{\mu R}{2(\mu-1)} - R$$

$$\therefore \qquad O_2F_2 = \frac{R(2-\mu)}{2(\mu-1)} \qquad (6.35)$$

(iii) Nodal Points : Since the sphere is located in air, the nodal points N_1 and N_2 coincide with the principal points P_1 and P_2 and thus, lie at the centre of the sphere, as shown in Fig. 6.7. It may be noted that the centre of the sphere is also the optical centre of the sphere-lens because the nodal points are coinciding with the centre.

6.6. COMBINATION OF TWO THICK LENSES

6.6.1. EXTENDED OBJECT AT INFINITY

As already mentioned in § 4.17, the image formed by the first optical system becomes the object for a second system and the two systems act as one forming the final image from the original object. We have to find the cardinal points of the combined system.

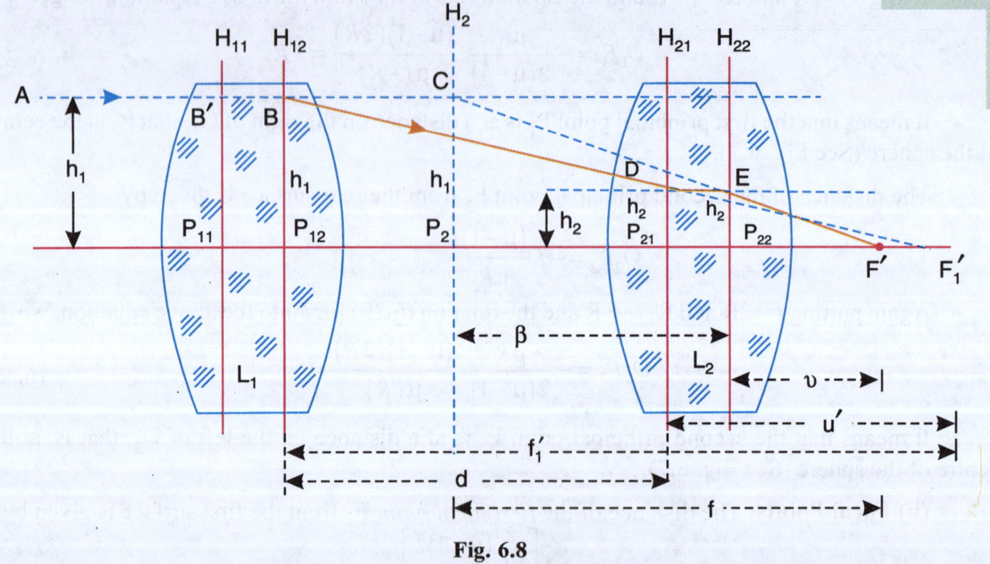

Fig. 6.8

Principal points P_{11} and P_{12} and the focal points F_1 and F_1' represent the first system. Principal points P_{21} and P_{22} and the focal points F_2 and F_2' represent the second system. The separation of the two systems is specified by the separation of their adjacent principal planes P_{12} and P_{21}, which is say d. d is positive when measured to the right from P_{12} to P_{21}. Principal points P_1 (not shown in diagram) and P_2 and the focal points F (again, not shown in Fig. 6.8) and F′ represent the combined system.

Consider a ray AB′ parallel to the axis and at height h_1. It meets the P_{11} plane at B′. It emerges from the first lens at B at the same height h_1 and would have followed the path BF_1'. Thus, BF_1' is the emergent ray from the first system. Ray BF_1' meets the plane P_{21} at D, which has height h_2. F_1' forms the object point for the second system. The second lens deviates the ray BF_1' along EF′ which is the final emergent ray. Then F′ will be the *second focal point* of the combined system.

If the ray AB is produced forward and the ray EF′ backward, they intersect at C. The plane CP_2 *is the principal plane of the equivalent lens.* $P_2F' = f$ is the *focal length* of *the equivalent lens.* Δ^{les} BF_1' P_{12} and EF_1' P_{22} are similar. We have

$$\frac{BP_{12}}{EP_{22}} = \frac{P_{12}F_1'}{P_{22}F_1'} \quad \therefore \quad \frac{h_1}{h_2} = \frac{f_1'}{f_1' - d} \tag{6.36}$$

Also $\Delta^{les}CP_2F'$ and $E\,P_{22}F'$ are similar.

$$\frac{CP_2}{EP_{22}} = \frac{P_2F'}{P_{22}F'} \quad \therefore \quad \frac{h_1}{h_2} = \frac{f}{f - \beta} \tag{6.37}$$

From the equ. (6.36) and (6.37), we obtain

$$\frac{f}{f-\beta} = \frac{f_1'}{f_1' - d} \quad \text{or} \quad \frac{1}{f-\beta} = \frac{f_1'}{f(f_1' - d)} \tag{6.38a}$$

Further, $$\frac{1}{\upsilon} - \frac{1}{u'} = \frac{1}{P_{22}F'} - \frac{1}{P_{21}F_1'} = \frac{1}{f_2} \quad \text{or} \quad \frac{1}{f-\beta} - \frac{1}{f_1' - d} = \frac{1}{f_2'}$$

$$\frac{1}{f-\beta} = \frac{1}{f_1' - d} + \frac{1}{f_2'} \tag{6.38b}$$

Equating (6.38a) and (6.38b), we get

$$\frac{f_1'}{f(f_1' - d)} = \frac{1}{f_1' - d} + \frac{1}{f_2'}$$

$$\therefore \quad f_1' = f\left[\frac{f_1' - d}{f_2'} + 1\right]$$

or $$f = \frac{f_1' f_2'}{f_1' + f_2' - d} \tag{6.39a}$$

$$\therefore \quad \frac{1}{f} = \frac{1}{f_1'} + \frac{1}{f_2'} - \frac{d}{f_1' f_2'} \tag{6.39b}$$

6.6.2. METHOD OF CONSTANT DEVIATION

We shall now apply the method of deviation to determine the equivalent focal length of the combination of two coaxial thick lenses. For the sake of simplicity, we represent the lens system by its two principal planes.

Referring to the Fig. 6.9, we see that there is a deviation of path of the incident ray OA due to the presence of optical system and the ray subsequently followed the path B A′ C. The angular deviation is δ, which is given by

$$\delta = \angle BAA' + \angle BA'A = \angle BOO' + \angle BO'O$$

$$\delta = \frac{h}{-u} + \frac{h}{\upsilon} = h\left(\frac{1}{\upsilon} - \frac{1}{u}\right) = \frac{1}{f_2}$$

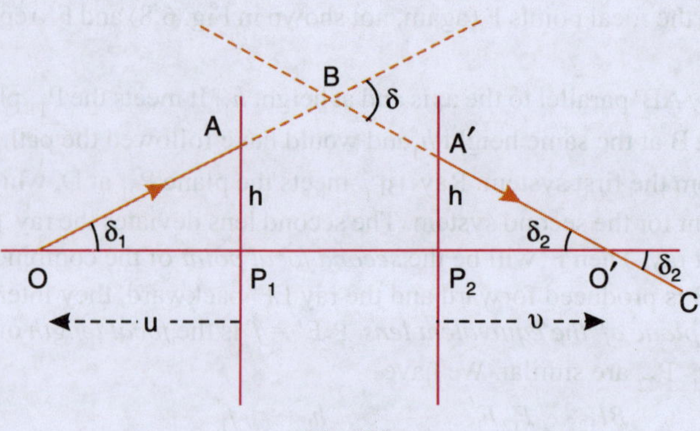

Fig. 6.9

It is easy to see that the deviation is independent of u; it is proportional to the height where the ray meets the principal plane and inversely proportional to the focal length f_2.

We can now apply the above result to the combination of lenses. Referring to the Fig. 6.10, the ray AB is parallel to the principal axis, and the first lens causes a deviation of $\delta_1 = \dfrac{h_1}{f_1'}$ and the second lens causes a deviation of $\delta_2 = \dfrac{h_2}{f_2'}$.

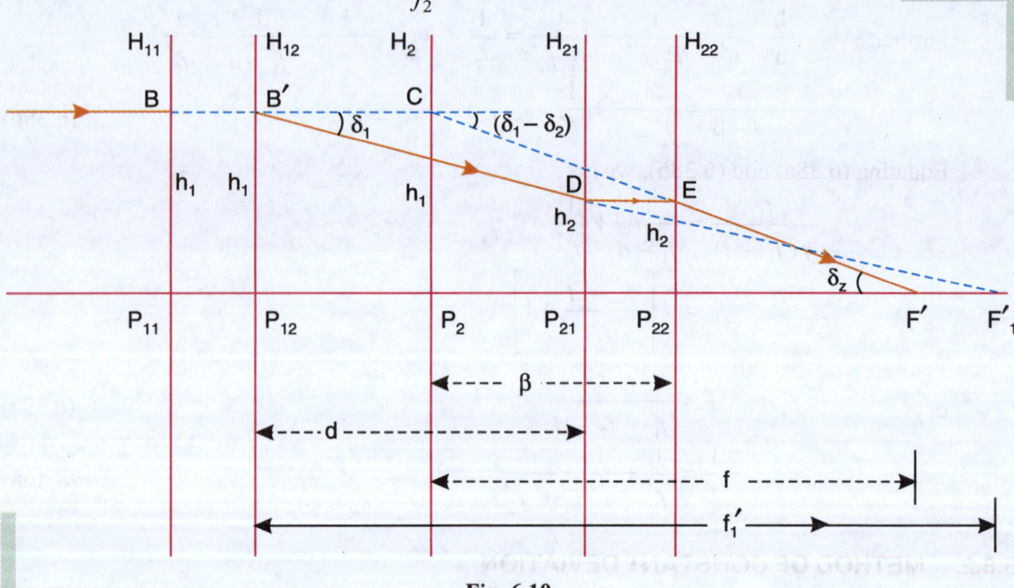

Fig. 6.10

The deviation due to the total system is $\delta = \delta_1 + \delta_2 = \dfrac{h_1}{f}$

$$\therefore \qquad \frac{h_1}{f} = \frac{h_1}{f_1'} + \frac{h_2}{f_2'} \quad \text{or} \quad h_1\left(\frac{1}{f} - \frac{1}{f_1'}\right) = \frac{h_2}{f_2'}$$

$$\therefore \qquad \frac{h_1}{h_2} = \frac{1}{f_2'\left(\dfrac{1}{f} - \dfrac{1}{f_1'}\right)} \qquad (6.40)$$

The Δ^{les} $B'P_{12}F_1'$ and $DP_{21}F_1'$ are similar. Therefore,

$$\frac{B'P_{12}}{DP_{21}} = \frac{P_{12}F_1'}{P_{21}F_1'} \quad \text{or} \quad \frac{h_1}{h_2} = \frac{f_1'}{\left(f_1' - d\right)} \tag{6.41}$$

Equating (6.40) and (6.41), we find that

$$\frac{f_1'}{\left(f_1' - d\right)} = \frac{1}{f_2'\left(\frac{1}{f} - \frac{1}{f_1'}\right)}$$

On simplifying the above expression, we get

$$\frac{1}{f} = \frac{f_1' + f_2' - d}{f_1' f_2'} \tag{6.42}$$

Eq. (6.42) is same as eq. (6.39b).

Considering the similar $\Delta^{les}CP_2F'$ and $EP_{22}F'$ we get

$$\frac{h_1}{h_2} = \frac{f}{f - \beta} \tag{6.43}$$

Equating (6.41) and (6.43), we find that

$$\frac{f}{f - \beta} = \frac{f_1'}{f_1' - d}$$

We get the following result after simplification of the above equation.

$$\beta = +\frac{f\,d}{f_1'} \tag{6.44}$$

6.7. PRINCIPAL PLANES IN A TWO-LENS SYSTEM MOVE OUT WHEN THE LENSES ARE SEPARATED

It is interesting to note how the positions of principal planes change when two positive lenses of equal power are first in contact and then are gradually separated.

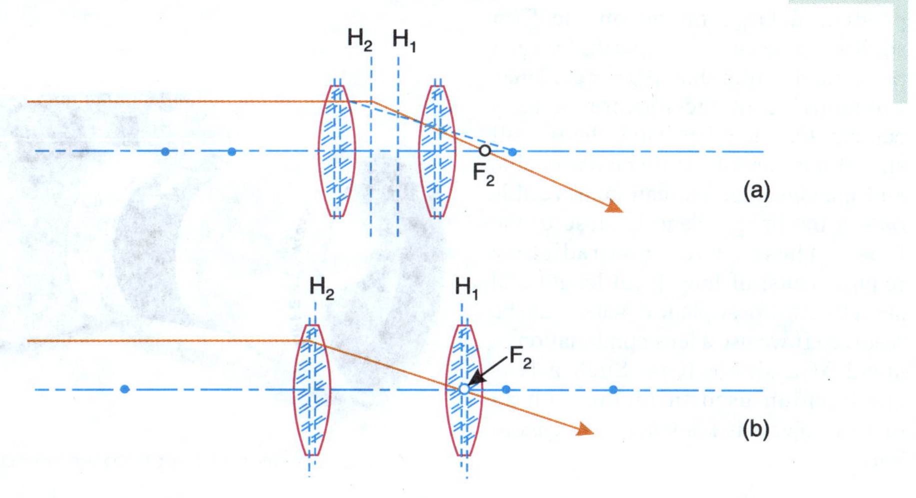

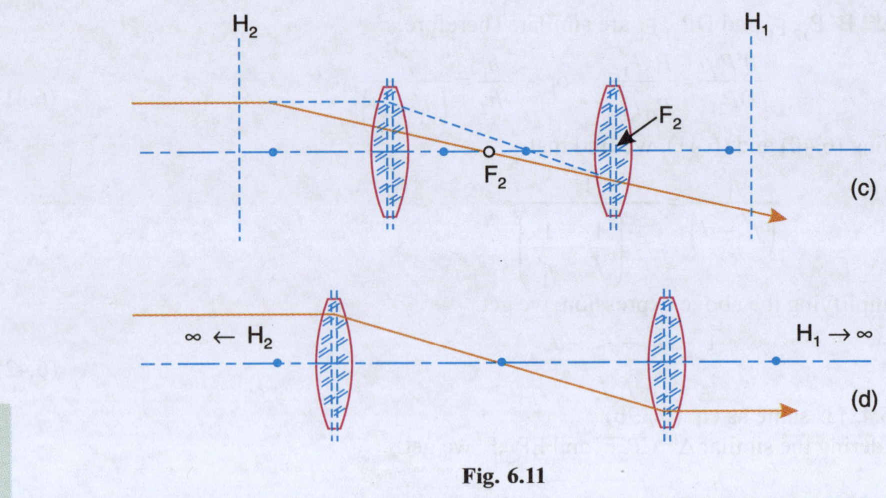

Fig. 6.11

When the lenses are in contact the two principal planes are close together within the system. As the lenses are separated, the principal planes also separate. When the distance between the lenses is equal to their focal length, the principal planes coincide with the lenses. When the distance is larger than the focal length, the planes have moved outside the lenses. And when the distance is equal to 2f the planes have moved out to infinity.

6.8. APPLICATIONS OF LENS COMBINATIONS

Optical instruments employ a variety of lens combinations to achieve the desired functions. We cannot study all the different combinations here. We study to two typical applications, telephoto lens and telescopic combinations and take up some other combinations, eyepieces, at a later stage.

6.8.1. TELEPHOTO LENS

To photograph a distant object, a camera should have a convex lens of long focal length. The image size for a distant object is directly proportional to the focal length of the lens. Therefore, to produce a large image on the film surface, the distance between the lens and photographic film should be large. Thus, a magnified image requires a long camera. On the other hand, the overall size of the camera should be reasonable and manageable. We can achieve this *only if* the image plane is close to the lens. These two contradictory requirements, of long focal length and short lens-to-focal-plane distance, can be resolved if we use a lens combination in stead of a single lens. Such a lens combination used in photographing distant objects is known as a *telephoto lens*.

One application of telephoto lens in camera.

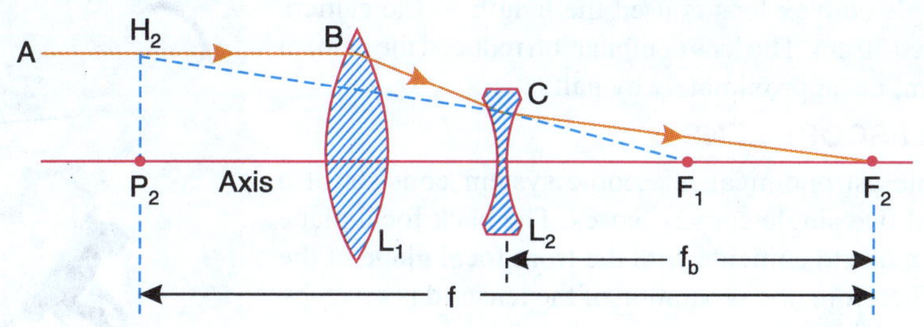

Fig. 6.12

A **telephoto lens** consists of a convex lens of focal length f_1 and a concave lens of focal length f_2 separated by a distance slightly greater than $f_1 + f_2$. The concave lens L_2 is placed in the original position of the lens in the camera whereas the convex lens L_1 is in front of L_2, as shown in Fig. 6.12. The combination of these two lenses acts as single convex lens of large focal length. The equivalent focal length of the combination is given by

$$\frac{1}{f} = \frac{1}{f_1} + \frac{1}{f_2} - \frac{d}{f_1 f_2}$$

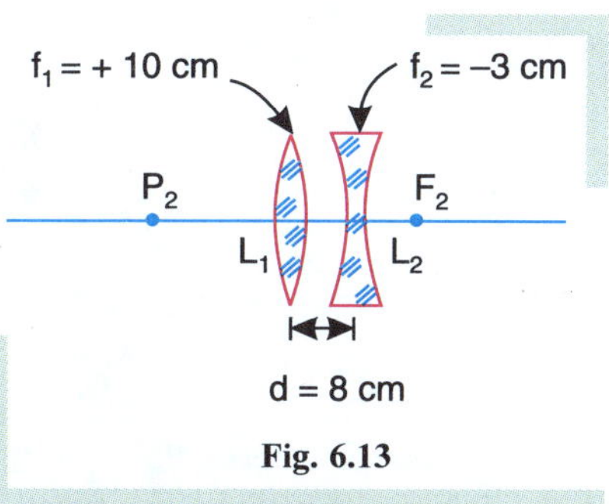

Fig. 6.13

The focal length f of the system is measured from the second principal plane H_2P_2 and the lens to focal plane distance, L_2F_2 is measured from L_2 to the photographic film. The distance from the first lens to the photographic film, L_1F_2, determines the overall length of the camera. What we require is shorter L_1F_2 and longer P_2F_2. The position of the second principal plane is determined by the intersection of a ray, AB, parallel to the axis in the object space with its conjugate ray, CF_2, in the final image space. It is seen that the second principal plane H_2P_2 lies well in front of the first lens L_2. Therefore, the equivalent focal length of the system, $f = P_2F_2$, is quite large. The lens-to-focal-plane distance is reduced because the back focal length, $f_b = L_2F_2$, is shorter. Thus, the overall length of the photo lens plus camera is made shorter.

To fully appreciate the function of the lens combination in the telephoto lens, let us consider the following example. Fig. 6.13 shows a telephoto lens consisting of a convex lens of focal length 10 cm and a concave lens of focal length (–3 cm), separated by a distance 8 cm.

The focal length of the combination, $f = \frac{f_1 f_2}{f_1 + f_2 - d} = \frac{10\,\text{cm} \times (-3\,\text{cm})}{10\,\text{cm} - 3\,\text{cm} - 8\,\text{cm}} = +30\,\text{cm}$

$\therefore$ $f = P_2F_2 =$ **30 cm.**

$$\beta = -\frac{fd}{f_1} = -\frac{30\,\text{cm} \times 8\,\text{cm}}{10\,\text{cm}} = -24\,\text{cm}$$

$\therefore$ $L_2P_2 = -\beta = 24$ cm

$\therefore$ Lens-to-focal-plane distance $L_2F_2 = P_2F_2 - P_2L_2 = (30 - 24)$ cm = **6 cm.**

$$\alpha = \frac{fd}{f_2} = -\frac{30\,\text{cm} \times 8\,\text{cm}}{-3\,\text{cm}} = -80\,\text{cm}$$

$\therefore$ $L_1P_1 =$ **–80 cm.**

Overall length of the camera $L_1F_2 = d + L_2F_2 = (8 + 6)$ cm = **14 cm.**

If a single convex lens is used the length of the camera would have been 30 cm. The lens combination reduced the camera length to 14 cm, i.e. approximately by half.

6.8.2. TELESCOPE LENS

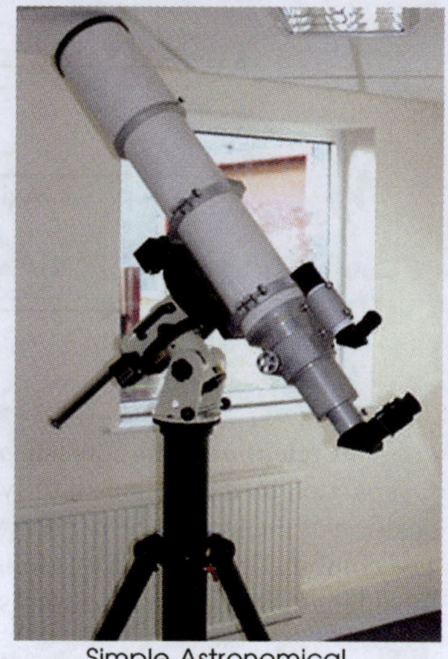
Simple Astronomical Telescope

A simple astronomical telescopic system consists of a combination of two single convex lenses. The back focal plane of first lens is made to coincide with the front focal plane of the second lens. Therefore, the separation of the lenses d is given by $f_1 + f_2$. Consequently,

$$\Delta = f_1 + f_2 - d = 0$$

and the equivalent focal length of the combination, $f = \dfrac{f_1 f_2}{\Delta} = \infty$. Further, $\alpha = \dfrac{fd}{f_2} = \infty$ and $\beta = -fd/f_1 = \infty$. Thus, all the cardinal planes (focal and principal planes) are at infinity (see Fig. 6.11d). It means that a telescope gives images at infinity for distant objects; this is the condition appropriate for viewing by a relaxed eye.

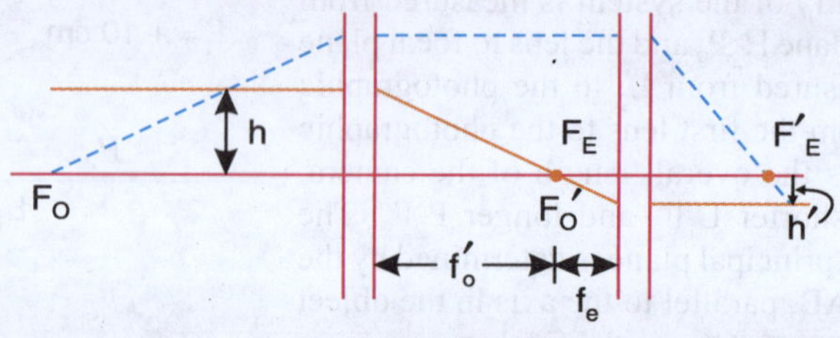

Fig. 6.14

Telescopes perform two functions. One is to produce an enlarged image of a distant object, and the other is to gather more light from the object than is possible with the unaided eye.

From the Fig. 6.14 it is apparent that the linear magnification h′/ h is a constant for all object and image positions in a telescopic system.

WORKED OUT PROBLEMS

Example 6.1: A convex lens of thickness 4 cm has radii of curvature 6 cm and 10 cm. Find the focal length and the positions of the focal points and the principal points. The refractive index of lens material is $\mu = 1.5$.

Solution: Here $R_1 = 6$ cm, $R_2 = -10$ cm and $t = 4$ cm.

$$f = \frac{-\mu R_1 R_2}{\mu(\mu-1)(R_1 - R_2) - (\mu-1)^2 t}$$

$$= \frac{-1.5 \times 6 \text{ cm} \times (-10 \text{ cm})}{1.5(1.5-1)(6 \text{ cm} + 10 \text{ cm}) - (1.5-1)^2 \times 4 \text{ cm}}$$

$f = 8.18$ cm.

$$\alpha = \frac{R_1 t}{\mu(R_1 - R_2) - t(\mu - 1)}$$

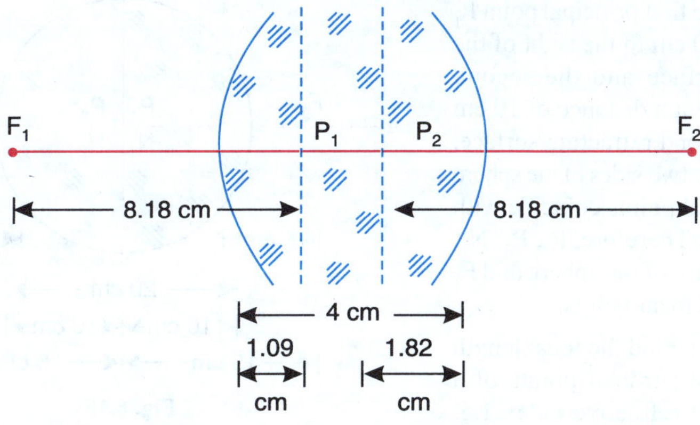

Fig. 6.15

$$= \frac{6\,\text{cm} \times 4\,\text{cm}}{1.5\left[6\,\text{cm} - (-10\,\text{cm})\right] - 4\,\text{cm}\,(1.5-1)}$$

$= 1.09$ cm.

$$\beta = \frac{R_2 t}{\mu(R_1 - R_2) - t(\mu - 1)}$$

$$= \frac{4\,\text{cm}(-10\,\text{cm})}{1.5\left[6\,\text{cm} - (-10\,\text{cm})\right] - 4\,\text{cm}\,(1.5-1)}$$

$= -1.82$ cm.

In Fig. 6.15, P_1 and P_2 show the positions of the principal points and F_1 and F_2 show the positions of the focal points. As the medium on the two sides of the lens is the same, the nodal points coincide with the principal points.

Example 6.2: Determine the position of the focal points, principal points and nodal points in the case of a sphere of radius 10 cm and $\mu = 1.5$. Indicate their positions in a diagram.

Solution: Here $R_1 = 10$ cm, $R_2 = -10$ cm and $t = 20$ cm.

$$f = \frac{-\mu R_1 R_2}{\mu(\mu-1)(R_1 - R_2) - (\mu-1)^2 t}$$

$$= \frac{-1.5 \times 10\text{ cm} \times (-10\text{ cm})}{1.5(1.5\text{-}1)(10\text{ cm} + 10\text{ cm}) - (1.5-1)^2 \times 20\text{ cm}}$$

$$\boldsymbol{f = 15\text{ cm}}$$

$$\alpha = \frac{R_1 t}{\mu(R_1 - R_2) - t(\mu - 1)}$$

$$= \frac{10\,\text{cm} \times 20\,\text{cm}}{1.5[10\,\text{cm} - (-10\,\text{cm})] - 20\,\text{cm}\,(1.5 - 1)}$$

$= +\,10$ cm

$$\beta = \frac{R_2 t}{\mu(R_1 - R_2) - t(\mu - 1)}$$

$$= \frac{20\,\text{cm}(-10\,\text{cm})}{1.5\left[10\,\text{cm} - (-10\,\text{cm})\right] - 20\,\text{cm}\,(1.5-1)} = -10\text{ cm}.$$

Therefore, the first principal point P_1 is at a distance of 10 cm to the right of the first refracting surface and the second principal point P_2 is at a distance of 10 cm to the left of the second refracting surface. As the medium on the two sides of the sphere is the same, the nodal points coincide with the principal points. Therefore, P_1, P_2, N_1, and N_2 lie at the centre of the sphere and F_1 and F_2 represent the focal points.

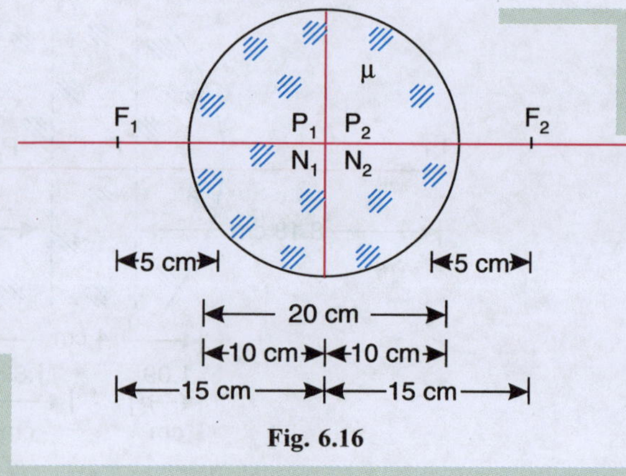

Fig. 6.16

Example 6.3: Find the focal length and the positions of cardinal points of a pano-convex lens of refractive index 1.5, the radius of curvature of the curved surface being 20 cm and thickness 1.0 cm. **(Kurukshetra, 2001)**

Solution: Given $R_1 = \infty$, $R_2 = -20$ cm, $t = 1$ cm and $\mu = 1.5$.

***(i)* Focal Length:** The focal length of a lens of thickness t is given by

$$\frac{1}{f} = (\mu - 1)\left[\frac{1}{R_1} - \frac{1}{R_2} + \frac{(\mu-1)}{\mu}\cdot\frac{t}{R_1 R_2}\right]$$

$$\frac{1}{f} = (1.5-1)\left[\frac{1}{\infty} - \frac{1}{-20\,\text{cm}} + \frac{(1.5-1)}{1.5}\cdot\frac{1\,\text{cm}}{\infty(-20\,\text{cm})}\right]$$

$$\frac{1}{f} = (1.5-1)\left[\frac{1}{20\,\text{cm}}\right] = \frac{0.5}{20\,\text{cm}}$$

$$f = +40\,\text{cm}$$

Plano-convex lens.

***(ii)* Second Focal Point:**

$$DF_2 = f\left[1 - \frac{(\mu-1)\,t}{\mu R_1}\right] = 40\,\text{cm}\left[1 - \frac{(1.5-1)\,1\,\text{cm}}{1.5(\infty)}\right] = \mathbf{+\,40\ cm}$$

***(iii)* Second Principal Plane:**

$$\beta = -f\cdot\frac{(\mu-1)\,t}{\mu R_1} = -40\ \text{cm}\ \frac{1.5-1}{1.5(\infty)} = \mathbf{0}$$

***(iv)* First Focal Point:**

$$CF_1 = -f\left[1 + \frac{(\mu-1)\,t}{\mu R_2}\right] = -40\,\text{cm}\left[1 + \frac{(1.5-1)\,1\,\text{cm}}{1.5(-20\,\text{cm})}\right] = \mathbf{-\,39.3\ cm}$$

***(v)* First Principal Plane:**

$$\alpha = -f\cdot t\cdot\frac{(\mu-1)}{\mu R_2} = -40\,\text{cm}.1\,\text{cm}.\frac{(1.5-1)}{1.5(-20\,\text{cm})} = \mathbf{0.66\ cm}$$

***(vi)* and *(vii)* Nodal Points:** The nodal points, N_1 and N_2 coincide with the principal points, P_1 and P_2 respectively.

Example 6.4: The radii of curvature of a convex meniscus lens are 15 cm and 10 cm and its thickness is 2 cm. Calculate its focal length and the positions of cardinal points. If an object is placed at a distance of 100 cm from this lens, calculate the position and magnification of the image.

Solution: A meniscus lens has both the surfaces curved in the same direction. In a convex meniscus lens $R_1 > R_2$. Given $R_1 = 15$, $R_2 = 10$ cm, $t = 2$ cm and $\mu = 1.5$ (assumed).

***(i)* Focal Length:** The focal length of a lens of thickness t is given by

$$\frac{1}{f} = (\mu - 1)\left[\frac{1}{R_1} - \frac{1}{R_2} + \frac{(\mu-1)}{\mu}\cdot\frac{t}{R_1R_2}\right]$$

$$\frac{1}{f} = (1.5-1)\left[\frac{1}{-15\,\text{cm}} - \frac{1}{-10\,\text{cm}} + \frac{(1.5-1)}{1.5}\cdot\frac{2\,\text{cm}}{(-15\,\text{cm})(-10\,\text{cm})}\right]$$

$$\frac{1}{f} = (1.5-1)\left[-\frac{1}{15\,\text{cm}} + \frac{1}{10\,\text{cm}} + \frac{1}{225\,\text{cm}}\right] = \frac{17}{900\,\text{cm}}$$

$$f = \frac{900}{17}\,\text{cm} = 52.9\,\text{cm}$$

***(ii)* Second Focal Point:**

$$DF_2 = f\left[1 - \frac{(\mu-1)\,t}{\mu R_1}\right] = 52.9\,\text{cm}\left[1 - \frac{(1.5-1)\,2\,\text{cm}}{1.5(-15\,\text{cm})}\right] = \mathbf{+\,55.3\ cm}$$

***(iii)* Second Principal Plane:**

$$\beta = -f\cdot\frac{(\mu-1)\,t}{\mu R_1} = -52.9\,\text{cm}\;\frac{(1.5-1)2\,\text{cm}}{1.5(-15\,\text{cm})} = \mathbf{2.35cm}$$

***(iv)* First Focal Point:**

$$CF_1 = -f\left[1 + \frac{(\mu-1)\,t}{\mu R_2}\right] = -52.9\,\text{cm}\left[1 + \frac{(1.5-1)\,2\,\text{cm}}{1.5(-10\,\text{cm})}\right] = \mathbf{-\,49.4\ cm}$$

***(v)* First Principal Plane:**

$$\alpha = -f\cdot t\cdot\frac{(\mu-1)}{\mu\,R_2} = -52.9\,\text{cm}\,(2\,\text{cm})\frac{(1.5-1)}{1.5(-10\,\text{cm})} = \mathbf{-\,3.53\ cm}$$

***(vi)* and *(vii)* Nodal Points:** The nodal points, N_1 and N_2 coincide with the principal points, P_1 and P_2 respectively.

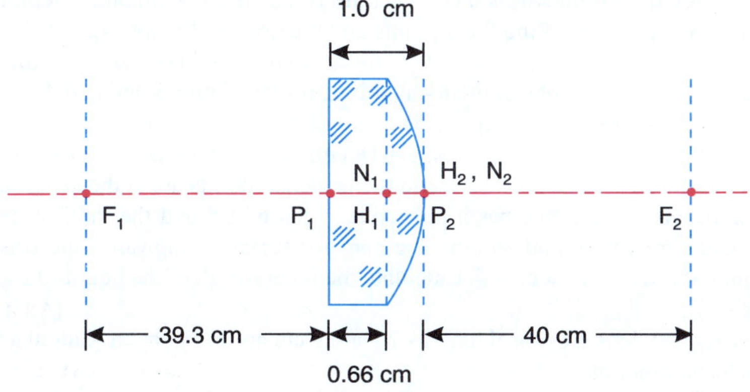

Fig. 6.17

From Fig. 6.17, we find that $-U = OP_1 = OC + CP_1 = 100\text{ cm} + 3.53\text{ cm} = 103.53\text{ cm}$

Or $U = -\,103.53$ cm

We have $$\frac{1}{f} = \frac{1}{V} - \frac{1}{U}$$

$$\therefore \quad \frac{1}{V}=\frac{1}{f}+\frac{1}{U}=\frac{1}{52.9\,\text{cm}}-\frac{1}{103.53\,\text{cm}}=\mathbf{109.3\ cm.}$$

The image lies at a distance 109.3 cm to the right of the second principal point P_2.

The distance of the image from the second surface is

$$DI = DP_2+P_2I = 2.35\text{cm} + 109.3\text{cm} = \mathbf{111.65\ cm}$$

Magnification, $m=\dfrac{V}{U}=\dfrac{111.65\,\text{cm}}{-103.53\,\text{cm}}=-1.08$

QUESTIONS

1. What are cardinal points of a thick lens or a coaxial combination of lenses? Define them, state their characteristics and show them diagrammatically for a lens system. Explain image formation with the help of these points in the case of a coaxial lens system. **(Delhi, 2000)**
2. Calculate the focal length of a thick convex lens of radii of curvature R_1 and R_2 and thickness t. Calculate the position of the principal points.
3. Define cardinal points of a system of coaxial lenses. Show that the principal planes are conjugate planes of +1 magnification. **(Udaipur, 2000)**
4. Show that the focal length of a thick lens is given by

$$\frac{1}{f}=(\mu-1)\left[\frac{1}{R_1}-\frac{1}{R_2}+\frac{(\mu-1)}{\mu}\cdot\frac{d}{R_1R_2}\right]$$

Here μ is the refractive index of the material of the lens, R1 and R2 are the radii of curvatures of the lens surfaces and d is the lens thickness along the axis. Discuss how the converging and diverging properties of such lenses vary with thickness. **(Bombay, 2000)**

5. Describe nodal assembly method to locate cardinal points of a lens system experimentally. Explain its working with particular reference to the characteristics of nodal points. **(Mysore, 2001)**

PROBLEMS FOR PRACTICE

1. A convex lens of thickness 6 cm has radii of curvature 8 cm and 10 cm. Find the focal length and position of the focal points and the principal points. (μ = 1.5). [**Ans:** f = 10 cm; α = + 2cm; β = – 2.5 cm]
2. Determine the positions of the focal points, principal points and nodal points in the case of sphere of radius 12 cm. (μ = 1.5). [**Ans:** f = 18 cm; α = + 12cm; β = –12 cm.The nodal points and the principal points are at the centre of the sphere.]
3. A concavo-convex lens has a refractive index of 1.5 and the radii of curvature of its surfaces are 15 cm and 30 cm. The concave surface is upwards and it is filled with a liquid of refractive index 1.6. Calculate the focal length of the liquid-glass combination. [**Ans:** f = 27.27cm]
4. Two convex lenses of focal lengths 20 and 5 cm are 10 cm apart. Calculate the power of such combination. [**Ans:** P = 15 diopters]
5. Two thin convex lenses having focal lengths 6 cm and 2 cm are co-axial and separated by a distance of 4 cm. Calculate the combined focal length and the positions of the principal planes. [**Ans:** f = 3 cm; α = + 6cm; β = – 2 cm]
6. Find the focal length of a plano convex lens of radius of curved surface being 30 cm and refractive index of glass μ = 1.50. **(Nagpur, 2004)**

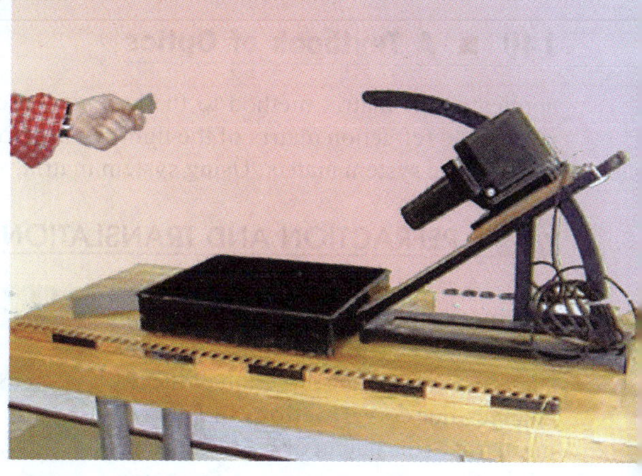

7 CHAPTER

Matrix Methods

7.1. INTRODUCTION

Optical systems are, in general, made up of a large number of refracting surfaces. In order to obtain the position of the final image due to an optical system, one has to calculate step by step the position of the image due to each surface and consider this image as an object for the next surface. Such a step by step analysis becomes lengthy and tedious. In order to solve such problems easily and more efficiently K.Hallback has introduced in 1964 matrix methods in the study of geometrical optics. The matrix method is less cumbersome and above that, it is more amenable to computer use. We study in this chapter the

At a Glance

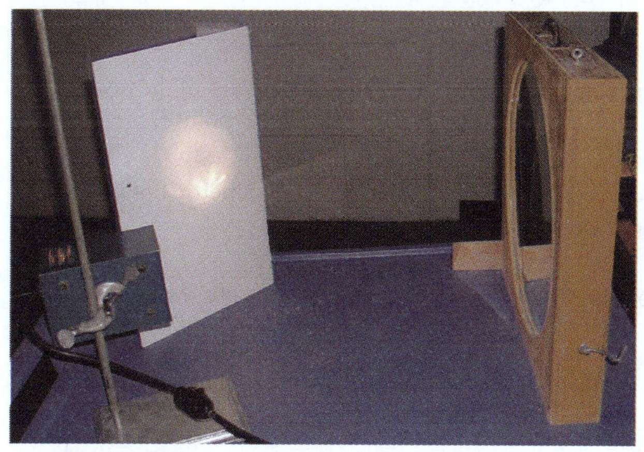

Matrix method in the study of geometrical Optics.

application of matrix method to the case of lenses. In this method we determine the translation matrix and refraction matrix of the light ray in an optical system. By multiplying these matrices we arrive at the system matrix. Using system matrix, we can find out the properties of the system.

7.2. REFRACTION AND TRANSLATION

A ray of light propagating through a cylindrically symmetric optical system undergoes two operations. At each surface boundary the direction of the ray changes due to **refraction**. In between the surfaces the height of the ray changes while the direction remains the same. This is known as **translation**. Therefore, to fully describe the ray propagation through an optical system, we make use of two operators, one for the refraction process and the other for the translation process. These two operators are known as **refraction matrix** and the **translation matrix** respectively.

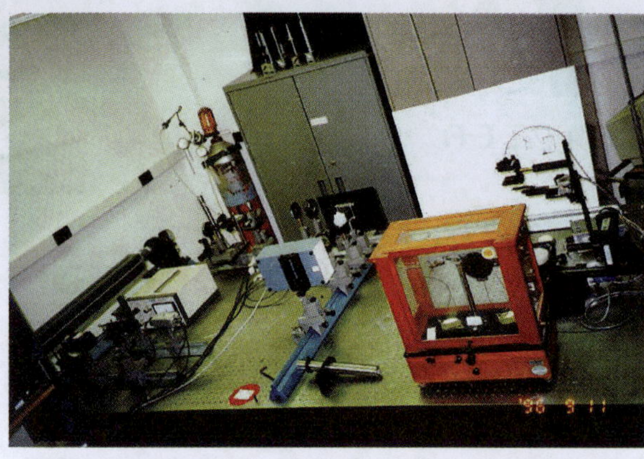

A complete bistatic and fully polarimetric optical system.

There is a point to point correspondence between object space and the image space. Every point (x, y, z) in object space gets transformed into the point (x', y', z') in the image space. The transformation is linear in case of paraxial ray approximation and hence matrix operation can be used for the purpose.

Fig. 7.1

A ray is specified by its distance from the axis of the optical system and the angle that it makes with the axis. Let us consider a ray inclined to the axis and passing through the points A and B (see Fig. 7.1). The point A is at a distance x_1 from the axis. Also the ray is inclined at an angle α_1 at A with the z-axis, which is chosen as the axis of the optical system. If the ray makes an angle θ with the x-axis, then

$$\lambda_1 = \mu \cos \theta \qquad (7.1)$$

where λ_1 is known as the optical direction cosine of the ray at A. The coordinates of the ray at A are specified by either (x_1, α_1) or (λ_1, x_1). Similarly, the coordinates of the ray at point B are given by either (x_2, α_2) or (λ_2, x_2).

Now let us consider a refracting curved surface of a cylindrically symmetric optical system (Fig.7.2), the axis of symmetry being the z-axis.

The incident ray is specified by the column matrix $\begin{bmatrix} \lambda_1 \\ x_1 \end{bmatrix}$ where λ_1 and x_1 represent the coordinates of A. The corresponding image ray is specified by $\begin{bmatrix} \lambda_2 \\ x_2 \end{bmatrix}$. It means that there is a matrix, which operates on the column vector $\begin{bmatrix} \lambda_1 \\ x_1 \end{bmatrix}$ specifying the object to produce the column vector $\begin{bmatrix} \lambda_2 \\ x_2 \end{bmatrix}$ representing the image. We may now write that

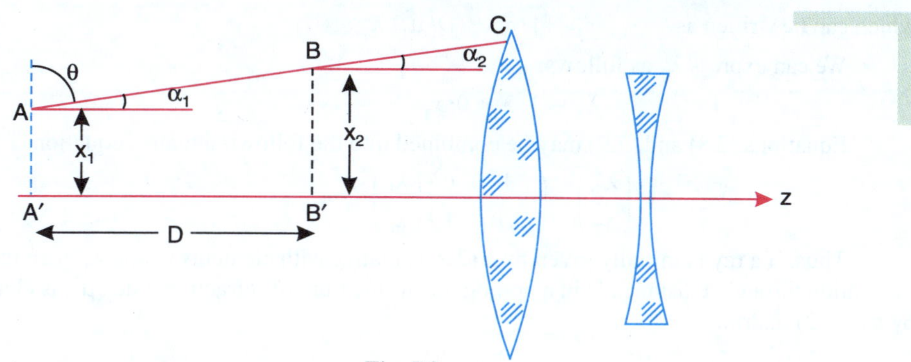

Fig. 7.2

$$\begin{bmatrix} \lambda_2 \\ x_2 \end{bmatrix} = [R] \begin{bmatrix} \lambda_1 \\ x_1 \end{bmatrix}$$

or
$$[R] \begin{bmatrix} \lambda_1 \\ x_1 \end{bmatrix} = \begin{bmatrix} \lambda_2 \\ x_2 \end{bmatrix} \tag{7.2}$$

The matrix $[R]$ is called the *refraction matrix*. Note that the refraction matrix operating on the initial *coordinates gives the coordinates* of the refracted ray.

In the same way we can represent the effect of translation as

$$\begin{bmatrix} \lambda_2 \\ x_2 \end{bmatrix} = [T] \begin{bmatrix} \lambda_1 \\ x_1 \end{bmatrix}$$

or
$$[T] \begin{bmatrix} \lambda_1 \\ x_1 \end{bmatrix} = \begin{bmatrix} \lambda_2 \\ x_2 \end{bmatrix} \tag{7.3}$$

The matrix $[T]$ is called the *translation matrix*.

7.3. TRANSLATION MATRIX

Let us consider a paraxial ray traveling in a homogeneous medium of refractive index μ_1 which is at a distance of x_1 from the z-axis (axis of symmetry) as shown in Fig.7.1. Let A be a point on the ray. Let it be at a distance x_1 from the z-axis and be inclined at an angle α_1 with the z-axis. Thus, (x_1, α_1) are the coordinates of the ray at point A. Let (x_2, α_2) be the coordinates of the ray at B through which the ray passes at a later instant. In traveling from A to B the ray undergoes *translation*. Since the medium is homogeneous, the ray travels in a straight line. Therefore, as is seen from Fig. 7.1,

$$\alpha_1 = \alpha_2 \tag{7.4}$$

and
$$x_2 = x_1 + D \tan \alpha_1 \approx x_1 + D\alpha_1 \tag{7.5}$$

Since the medium in which the light travels is the same, $\mu_1 = \mu_2$. It follows that

$$\mu_1\alpha_1 = \mu_2\alpha_2 \tag{7.6}$$

or
$$\lambda_1 = \lambda_2$$

where
$$\lambda_1 = \mu_1\alpha_1 \text{ and } \lambda_2 = \mu_2\alpha_2. \tag{7.7}$$

Substituting the value of α_1 from equation (7.7) into equation (7.5), we get

$$x_2 = x_1 + \mathrm{D}\,(\lambda_1/\,\mu_1)$$

which can be written as $$x_2 = 1 \cdot x_1 + (D/\mu_1)\,\lambda_1 \tag{7.8}$$

We can express λ_2 as follows:

$$\lambda_2 = 1 \cdot \lambda_1 + 0 \cdot x_1 \tag{7.9}$$

Equations (7.8) and (7.9) may be combined into the following matrix equation.

$$\begin{pmatrix} \lambda_2 \\ x_2 \end{pmatrix} = \begin{pmatrix} 1 & 0 \\ D/\mu_1 & 1 \end{pmatrix} \begin{pmatrix} \lambda_1 \\ x_1 \end{pmatrix} \tag{7.10}$$

Thus, if a ray is initially given by a (2 × 1) matrix with elements x_1 and λ_1, then the effect of translation through a distance D in a homogeneous medium of refractive index μ_1, is characterized by a (2 × 2) matrix.

$$T = \begin{pmatrix} 1 & 0 \\ D/\mu_1 & 1 \end{pmatrix} \tag{7.11}$$

The matrix is called *translation matrix*. It may be noted that

$$\det T = \begin{pmatrix} 1 & 0 \\ D/\mu_1 & 1 \end{pmatrix} = 1. \tag{7.12}$$

7.4. REFRACTION MATRIX

Let us consider a convex spherical surface of radius of curvature R separating two media of refracting index μ_1 and μ_2. Let a ray AB be incident on the surface SS_1 at a point B and be refracted

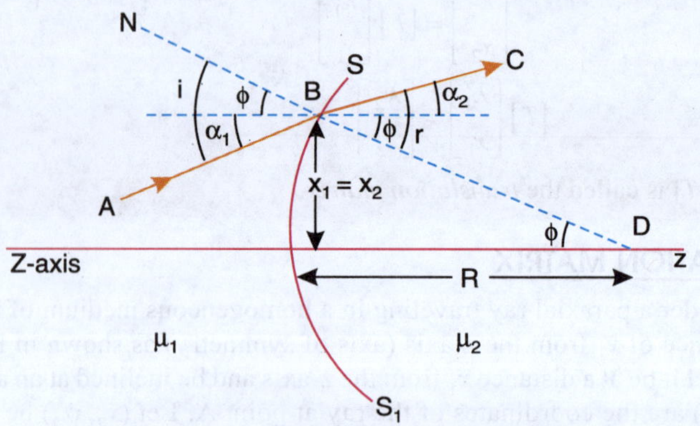

The refraction of a ray at a spherical surface.

Fig. 7.3

along BC as shown in Fig.7.3. If i and r be the angles of incidence and refraction with the normal to the surface, then according to the Snell's law,

$$\mu_1 \sin i = \mu_2 \sin r \tag{7.13}$$

In case of paraxial rays, we approximate $\sin i \approx i$ and $\sin r \approx r$

$$\therefore \qquad \mu_1 i = \mu_2 r \tag{7.14}$$

From Fig. 7.3, we find that

$$i = \phi + \alpha_1 \quad \text{and } r = \phi + \alpha_2 \tag{7.15}$$

Substituting the above values in equ.(7.14), we obtain

$$\mu_1(\phi+\alpha_1)=\mu_2(\phi+\alpha_2)$$

$$\mu_1\alpha_2=\mu_1\alpha_1-(\mu_2-\mu_1)\,\phi \tag{7.16}$$

If the point is at a distance x_1 from the axis of symmetry (z-axis), then

$$\tan\phi=\frac{x_1}{R}$$

$$\phi\approx\frac{x_1}{R}\qquad\text{(as }\phi\text{ is small)} \tag{7.17}$$

Substituting the above value in equ.(7.16), we get

$$\mu_2\alpha_2=\mu_1\alpha_1-(\mu_2-\mu_1)\,\frac{x_1}{R}$$

$$\lambda_2=\lambda_1-\left(\frac{\mu_2-\mu_1}{R}\right)x_1\ (\text{as } \lambda_2=\mu_2\,\alpha_2 \text{ and } \lambda_1=\mu_1\,\alpha_1)$$

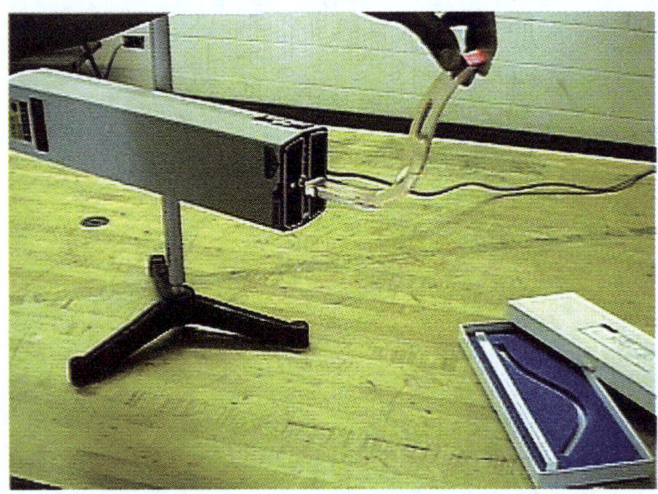

Computer optics Inc. custom designed lens.

$$\lambda_2=\lambda_1-Px_1 \tag{7.18}$$

where $P=\left(\frac{\mu_2-\mu_1}{R}\right)$ is known as *power* of the refracting surface.

Since the distance of the ray at P before and after refraction is the same,

so $$x_2=x_1$$

$$x_2=0\cdot\lambda_1+x_1 \tag{7.19}$$

From equ. (7.18) and (7.19)

$$\begin{pmatrix}\lambda_2\\ x_2\end{pmatrix}=\begin{pmatrix}1 & -P\\ 0 & 1\end{pmatrix}\begin{pmatrix}\lambda_1\\ x_1\end{pmatrix} \tag{7.20}$$

Therefore, the refraction through a spherical surface can be characterized by 2×2 matrix

$$R=\begin{pmatrix}1 & -P\\ 0 & 1\end{pmatrix} \tag{7.21}$$

which is known as the *refraction matrix*.

Again it may be noted that

$$\det R = \begin{vmatrix} 1 & -P \\ 0 & 1 \end{vmatrix} = 1. \tag{7.22}$$

7.5. SYSTEM MATRIX

Let us consider the example of a double convex lens (Fig.7.4). When a ray passes through a lens, refraction occurs twice, at each curved surface, and translation occurs once, between the two curved surfaces. Hence, we require two refraction matrices $[R_1]$ and $[R_2]$ and one translation matrix, $[T_{21}]$. Multiplication of these matrices, written from right to left, leads to the **system matrix**, [S]. The system matrix indicates how a ray of light is affected in passing through the optical system. Thus

$$[R_2]\,[T_{21}]\,[R_1] = [S] \tag{7.23}$$

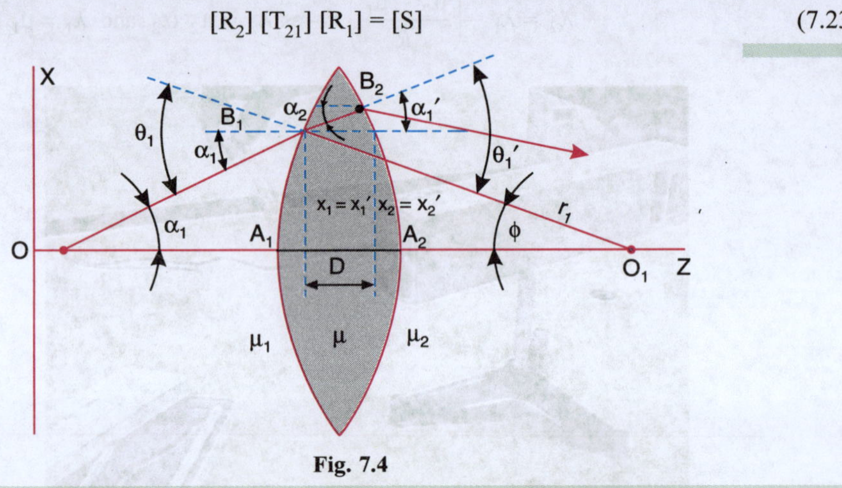

Fig. 7.4

The passage of the ray through the first refracting surface is described by

$$\begin{pmatrix} \lambda_1' \\ x_1' \end{pmatrix} = \begin{pmatrix} 1 & -P_1 \\ 0 & 1 \end{pmatrix} \begin{pmatrix} \lambda_1 \\ x_1 \end{pmatrix} \tag{7.24}$$

where P_1 is the refracting power of the first surface. Similarly, the propagation of the ray from the first to the second refracting surface inside the lens is described by

$$\begin{pmatrix} \lambda_2 \\ x_2 \end{pmatrix} = \begin{pmatrix} 1 & 0 \\ D/\mu_1 & 1 \end{pmatrix} \begin{pmatrix} \lambda_1' \\ x_1' \end{pmatrix} \tag{7.25}$$

The refraction at the second surface may be written as

$$\begin{pmatrix} \lambda_2' \\ x_2' \end{pmatrix} = \begin{pmatrix} 1 & -P_2 \\ 0 & 1 \end{pmatrix} \begin{pmatrix} \lambda_2 \\ x_2 \end{pmatrix} \tag{7.26}$$

where P_2 is the power of the second refracting surface. Combining the above equations (7.24),(7.25) and (7.26) we get the relation between the characteristics of the ray incident on the lens and the ray refracted by the lens. Thus, we get

$$\begin{pmatrix} \lambda_2' \\ x_2' \end{pmatrix} = [R_2]\,[T_{21}]\,[R_1] \begin{pmatrix} \lambda_1 \\ x_1 \end{pmatrix}$$

or

$$\begin{pmatrix} \lambda_2' \\ x_2' \end{pmatrix} = \begin{pmatrix} 1 & -P_2 \\ 0 & 1 \end{pmatrix} \begin{pmatrix} 1 & 0 \\ D/\mu_1 & 1 \end{pmatrix} \begin{pmatrix} 1 & -P_1 \\ 0 & 1 \end{pmatrix} \begin{pmatrix} \lambda_1 \\ x_1 \end{pmatrix} \tag{7.27}$$

or $$\begin{pmatrix} \lambda_2' \\ x_2' \end{pmatrix} = [S] \begin{pmatrix} \lambda_1 \\ x_1 \end{pmatrix}$$

$$\therefore \quad [S] = [R_2]\,[T_{21}]\,[R_1] = \begin{pmatrix} 1 & -P_2 \\ 0 & 1 \end{pmatrix} \begin{pmatrix} 1 & 0 \\ D/\mu_1 & 1 \end{pmatrix} \begin{pmatrix} 1 & -P_1 \\ 0 & 1 \end{pmatrix} \tag{7.28}$$

$$= \begin{pmatrix} b & -a \\ -d & c \end{pmatrix}$$

where a, b, c, and d are known as **Gauss constants**. Three of the four constants are independent and usually a, b, and c are chosen as the independent constants.

The matrix $$[S] = \begin{pmatrix} b & -a \\ -d & c \end{pmatrix} \tag{7.29}$$

is called **system matrix.**

It follows from equ.(7.28) that

$$a = P_1 + P_2 - P_1 P_2 D/\mu_1 \tag{7.30a}$$

$$b = 1 - P_2 D/\mu_1 \tag{7.30b}$$

$$c = 1 - P_1 D/\mu_1 \tag{7.30c}$$

$$d = -D/\mu_1 \tag{7.30d}$$

Generally, an optical system consists of a number of refracting surfaces. So we get translation and refraction matrices for every surface. By multiplying all the individual matrices in proper order we get the final system matrix.

Since the system matrix is the product of translation and refraction matrices and as the determinants of these matrices are unity, the determinant of the system matrix is also unity.

Thus $$\det S = \begin{vmatrix} b & -a \\ -d & c \end{vmatrix} = bc - ad = 1 \tag{7.31}$$

7.6. POSITION OF THE IMAGE PLANE

Let us consider an object plane 00' at a distance $-D_1$ from the refracting surface of an optical system as shown in Fig. 7.5. According to sign convention the distance towards the left of the refracting system are taken as negative while on the right side they arc taken as positive. The image lane I I′ is at a distance D_2.

Let a ray O′ A start from the point O′ of the object plane and strikes the first refracting surface at point A. It emerges from the last refracting surface along BI′ from B.

Let (λ_1, x_1), (λ_1', x_1'), (λ_2', x_2'), (λ_2, x_2) be the co-ordinates of point O′, A, B and I′ respectively.

Between O′ and A, there is only translation

Hence

$$\begin{pmatrix} \lambda_1' \\ x_1' \end{pmatrix} = \begin{pmatrix} 1 & 0 \\ -D_1 & 1 \end{pmatrix} \begin{pmatrix} \lambda_1 \\ x_1 \end{pmatrix} \tag{7.32}$$

Between A and B there will be refraction followed by translation and the refraction at B.

Hence it can be represented by system matrix.

$$\begin{pmatrix} \lambda_2' \\ x_2' \end{pmatrix} = \begin{pmatrix} b & -a \\ -d & c \end{pmatrix} \begin{pmatrix} \lambda_1' \\ x_1' \end{pmatrix} \tag{7.33}$$

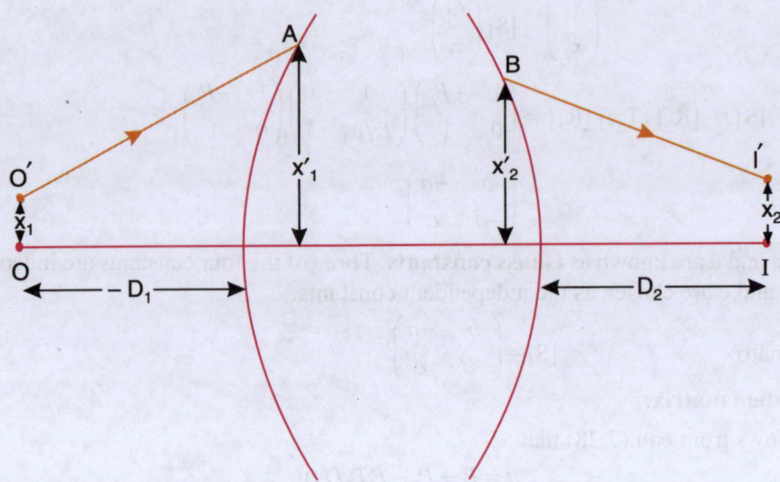

The object point O is at a distance $(-D_1)$ from the first refracting surface. The paraxial image is assumed to be formed at a distance D_2 from the last refracting surface.

Fig. 7.5

between B and I′, there will be translation

$$\begin{pmatrix}\lambda_2\\ x_2\end{pmatrix}=\begin{pmatrix}1 & 0\\ -D_2 & 1\end{pmatrix}\begin{pmatrix}\lambda_2'\\ x_2'\end{pmatrix} \tag{7.34}$$

The values of the matrix $\begin{pmatrix}\lambda_2\\ x_2\end{pmatrix}$ in terms of matrix $\begin{pmatrix}\lambda_1\\ x_1\end{pmatrix}$ can be obtained by multiplying equation (7.32), (7.33) and (7.34).

i.e.,

$$\begin{pmatrix}\lambda_2\\ x_2\end{pmatrix}=\begin{pmatrix}1 & 0\\ -D_2 & 1\end{pmatrix}\begin{pmatrix}b & -a\\ -d & c\end{pmatrix}\begin{pmatrix}1 & 0\\ -D_1 & 1\end{pmatrix}\begin{pmatrix}\lambda_1\\ x_1\end{pmatrix} \tag{7.35}$$

Multiplying the right hand side in proper order we get

$$\begin{pmatrix}\lambda_2\\ x_2\end{pmatrix}=\begin{pmatrix}b+aD_1 & -a\\ bD_2+aD_1D_2-cD_1-d & c-aD_2\end{pmatrix}\begin{pmatrix}\lambda_1\\ x_1\end{pmatrix} \tag{7.36}$$

$$\begin{pmatrix}\lambda_2\\ x_2\end{pmatrix}=\begin{pmatrix}(b+aD_1)\lambda_1-ax_1\\ (bD_2+aD_1D_2-cD_1-d)\lambda_1+(c-aD_2)x_1\end{pmatrix} \tag{7.37}$$

Equating the corresponding elements we have

$$x_2=(bD_2+aD_1D_2-cD_1-d)\lambda_1+(c-aD_2)x_1 \tag{7.38}$$

When the object is point size and lying on the axis, the image is also a point lying on the axis. So $x_1 = 0$ and $x_2 = 0$. Therefore, equ.(7.38) reduces to

$$0=(bD_2+aD_1D_2-cD_1-d)\lambda_1$$

But $\lambda_1 \neq 0$

$\therefore$

$$bD_2 + aD_1D_2 - D_1 - d = 0 \tag{7.39}$$

This is the condition for image plane. Now the system matrix for image plane can be written as

$$\begin{pmatrix} \lambda_2 \\ x_2 \end{pmatrix} = \begin{pmatrix} b + aD_1 & -a \\ 0 & c - aD_2 \end{pmatrix} \begin{pmatrix} \lambda_1 \\ x_1 \end{pmatrix} \tag{7.40}$$

If $x_1 \neq 0$ and $x_2 \neq 0$, then using (7.39) in equ.(7.38), we get

$$x_2 = (c - aD_2)\, x_1 \tag{7.41}$$

7.7. MAGNIFICATION

The magnification of a system is defined as $m = \frac{x_2}{x_1}$. From equ. (7.41), we obtain that

$$\frac{x_2}{x_1} = m = (c - aD_2)$$

Mapping with procedural and texture magnification.

From equation (7.40), we get
$$\begin{vmatrix} b + aD_1 & -a \\ 0 & c - aD_2 \end{vmatrix} = 1 \tag{7.42}$$

$$(b + aD_1)(c - aD_2) = 1$$

$$(b + aD_1) = \frac{1}{c - aD_2} = \frac{1}{m} \tag{7.43}$$

In terms of magnification of the system, the system matrix for image plane is given by

$$\begin{pmatrix} \lambda_2 \\ x_2 \end{pmatrix} = \begin{pmatrix} \frac{1}{m} & -a \\ 0 & m \end{pmatrix} \begin{pmatrix} \lambda_1 \\ x_1 \end{pmatrix} \tag{7.44}$$

7.8. SYSTEM MATRIX FOR THICK LENS

Now let us consider a thick lens of thickness 't' bounded by two curved surfaces and refractive index μ with respect to air. Let R_1, R_2 be the radii of curvature of two refracting surfaces. Suppose a ray OA strikes the first surface at point A as shown in Fig. 7.6 and emerges at B. (λ_1, x_1), (λ_2, x_2) are the coordinates at A and B. When a ray is propagating through the thick lens first it undergoes refraction at A and translation between A and B and finally refraction at B. The entire process of image formation is represented by a series of matrix operations as follows:

$$\begin{pmatrix} \lambda_2 \\ x_2 \end{pmatrix} = \begin{pmatrix} 1 & -P_2 \\ 0 & 1 \end{pmatrix} \begin{pmatrix} 1 & 0 \\ \frac{t}{\mu} & 1 \end{pmatrix} \begin{pmatrix} 1 & -P_1 \\ 0 & 1 \end{pmatrix} \begin{pmatrix} \lambda_1 \\ x_1 \end{pmatrix} \tag{7.45}$$

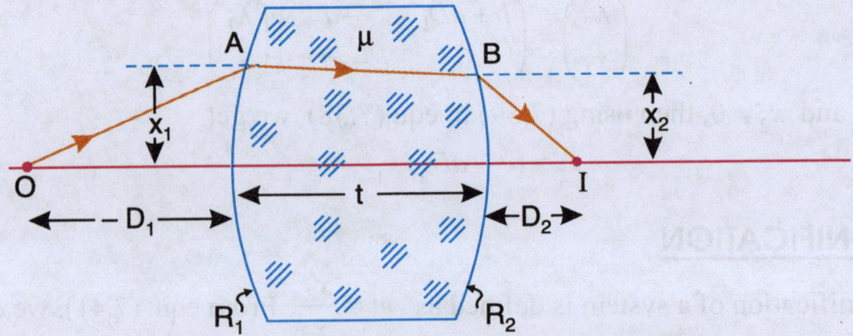

A paraxial ray passing through a thick lens of thickness '*t*'

Fig. 7.6

where $$P_1 = \frac{(\mu - 1)}{R_1}$$

and $$P_2 = \left(\frac{1-\mu}{R_2}\right) = -\frac{(\mu - 1)}{R_2}$$

Now the system matrix is given by

$$S = \begin{pmatrix} 1 & -P_2 \\ 0 & 1 \end{pmatrix} \begin{pmatrix} 1 & 0 \\ \dfrac{t}{\mu} & 1 \end{pmatrix} \begin{pmatrix} 1 & -P_1 \\ 0 & 1 \end{pmatrix}$$

$$\text{S} = \begin{pmatrix} 1 - P_2 t/\mu & -P_1 - P_2(1 - P_1 t/\mu) \\ t/\mu & 1 - P_1 t/\mu \end{pmatrix} \tag{7.46}$$

7.8.1. FOCAL LENGTH OF THICK LENS

Comparing the system matrix for thick lens with the general matrix system matrix $\begin{pmatrix} b & -a \\ -d & c \end{pmatrix}$,

we obtain the Gaussian constants as follows.

$$a = P_1 + P_2(1 - P_1 t/\mu) = P_1 + P_2 - P_1 P_2 t/\mu$$

$$b = 1 - P_2 t/\mu$$

$$c = 1 - P_1 t/\mu$$

$$d = -t/\mu$$

For the image plane, $$bD_2 + aD_1 D_2 - cD_1 - d = 0$$

Substituting the values of a, b,c, and d into the above equation, we obtain

$$(1 - P_2 t/\mu)\, D_2 + P_1 + P_2 - P_1 P_2 t/\mu\, (D_1\ D_2 - (1 - P_1 t/\mu)\, D_1 - (-\,t/\mu) = 0$$

or $$P_1 + P_2 - \frac{P_1 P_2 t}{\mu} = \frac{1}{D_1} + \frac{1}{D_2} + \frac{P_2 t}{\mu D_1} - \frac{P_1 t}{\mu D_2} - \frac{t}{\mu D_1}$$

But D_1 or $D_2 >> t$.

$$\therefore \quad P_1 + P_2 - \frac{P_1 P_2 t}{\mu} = \frac{1}{D_1} - \frac{1}{D_2} \tag{7.47}$$

From the Fig.7.6, it is readily seen that $D_1 = -\,u$ and $D_2 = \upsilon$.

$$\therefore \qquad \frac{1}{D_1} - \frac{1}{D_2} = \frac{1}{\upsilon} + \frac{1}{u}$$

But
$$\frac{1}{\upsilon} + \frac{1}{u} = \frac{1}{f}$$

$$\therefore \qquad P_1 + P_2 - \frac{P_1 P_2 t}{\mu} = \frac{1}{f}$$

Using the expressions for P_1 and P_2 into the above equation, we obtain

$$\frac{(\mu-1)}{R_1} - \frac{(\mu-1)}{R_2} - \frac{(\mu-1)}{R_1}\frac{[-(\mu-1)]}{R_2}\frac{t}{\mu} = \frac{1}{f}$$

or
$$\frac{1}{f} = (\mu-1)\left[\frac{1}{R_1} - \frac{1}{R_2} + \frac{(\mu-1)}{R_1 R_2}\cdot\frac{t}{\mu}\right] \qquad (7.48)$$

7.9. SYSTEM MATRIX FOR THIN LENS

The system matrix for thick lens is given by equ.(7.46).

$$S = \begin{pmatrix} 1 - P_2 t/\mu & -P_1 - P_2(1 - P_1 t/\mu) \\ t/\mu & 1 - P_1 t/\mu \end{pmatrix}$$

In the case of thin lens, the thickness may be assumed to be negligibly small ($t \to 0$)and hence the system matrix takes the form

$$S = \begin{pmatrix} 1 & -P_1 - P_2 \\ 0 & 1 \end{pmatrix} \qquad (7.49)$$

7.9.1. LENS MAKER'S FORMULA OR FOCAL LENGTH OF THIN LENS

System matrix for thin lens is given by

$$S = \begin{pmatrix} b & -a \\ -d & c \end{pmatrix} = \begin{pmatrix} 1 & -P_1 - P_2 \\ 0 & 1 \end{pmatrix}$$

From equ.(7.47), we find the Gaussian constants as

$$a = P_1 + P_2, \quad b = 1, \quad c = 1, \text{ and } d = 0$$

For the image plane, $bD_2 + aD_1D_2 - cD_1 - d = 0$

Substituting the values of a, b,c, and d in the above equation, we obtain

$$D_2 + (P_1 + P_2)\, D_1 D_2 - D_1 = 0$$

or
$$(P_1 + P_2) D_1 D_2 = D_1 - D_2$$

Dividing the above equation with D_1D_2, we get

$$(P_1 + P_2) = \frac{1}{D_2} - \frac{1}{D_1}$$

Using the expressions for P_1 and P_2 into the above equation, we obtain

$$\left(\frac{\mu-1}{R_1}\right) - \left(\frac{\mu-1}{R_2}\right) = \frac{1}{D_2} - \frac{1}{D_1} \qquad (7.50)$$

From the Fig. 7.6, it is readily seen that $D_1 = -u$ and $D_2 = \upsilon$.

$$\therefore \qquad (\mu-1)\left[\frac{1}{R_1} - \frac{1}{R_2}\right] = \frac{1}{\upsilon} + \frac{1}{u}$$

But $$\frac{1}{\upsilon}+\frac{1}{u}=\frac{1}{f}$$

where f is the focal length of the lens.

$$\therefore \qquad (\mu-1)\left[\frac{1}{R_1}-\frac{1}{R_2}\right]=\frac{1}{f} \qquad (7.51)$$

Equation (7.5) gives the focal length or ***Lens Maker's Formula*** for thin lens.

7.10. CARDINAL POINTS OF AN OPTICAL SYSTEM

The cardinal points of an optical system are,

(*i*) two principal or unit points,

(*ii*) two focal points, and

(*iii*) two nodal points.

The position and size of the image of an object placed in front of a coaxial system of lenses (number of lenses having a common principal axis) can be determined with the help of the *cardinal points*. The cardinal points and planes of an optical system are shown in Fig. 7.7.

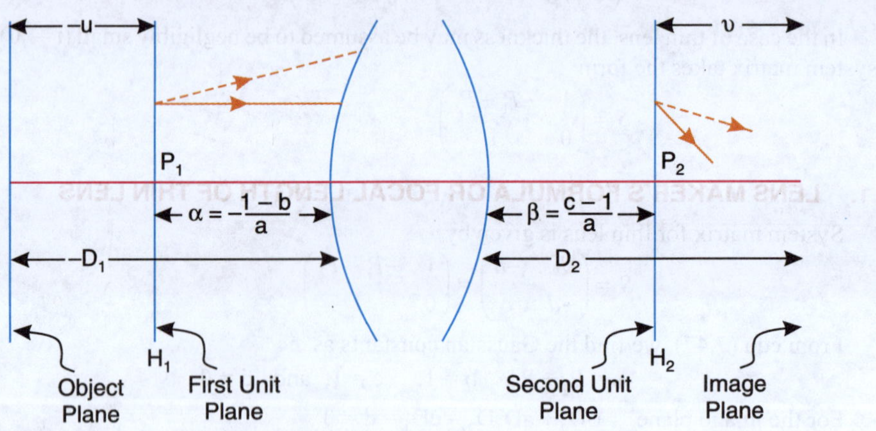

P_1H_1 and P_2H_2 are the two unit planes. A ray emanating at any height from the first unit plane will cross the second unit plane at the same height.

Fig. 7.7

7.10.1. PRINCIPAL OR UNIT POINTS

The principal or unit points (P_1 and P_2 in **Fig.7.7**) are a pair of conjugate points on the principal axis for which linear transverse magnification is unity and positive. The planes passing through these points and perpendicular to the principal axis are called principal planes or unit planes.

Location of Unit Planes:

The unit planes are two conjugate planes, one of which lies in object plane and the other in image plane. If a ray strikes the first unit plane (in object side) at a certain height, it emerges out from the second unit plane (on the image side) at the same height.

Let α and β be the respective distances of the first and second unit planes from the refracting surfaces of a lens system as shown in Fig.7.7. For these planes magnification is unity, i.e. m = 1. From equ.(7.43), we have

$$(b+aD_1) = \frac{1}{c-aD_2} = \frac{1}{m}$$

Using $D_1 = \alpha$ and $D_2 = \beta$ in the above equation, we get

$$(b+a\alpha) = 1 \qquad \text{or} \qquad \alpha = \frac{1-b}{a} \tag{7.52}$$

and

$$c - a\beta = 1 \qquad \text{or} \qquad \beta = \frac{c-1}{a} \tag{7.53}$$

Thus, the unit planes are completely determined by the elements of the system matrix S.

It is usually convenient to measure distances from the unit planes. Let u be the distance of the object plane from the first unit plane and υ be the distance of image plane from the second unit plane. Then

$$D_1 = u + \alpha \ = u + \frac{1-b}{a} \tag{7.54}$$

$$D_2 = U + \beta = \upsilon + \frac{c-1}{a} \tag{7.55}$$

Now the image plane condition is that

$$bD_2 + aD_1D_2 - cD_1 - d = 0$$

$$D_2\,(aD_1 + b) = d + cD_1$$

or

$$D_2 = \frac{d + cD_1}{(aD_1 + b)} \tag{7.56}$$

Substituting the values of D_2 and D_1 in the above equation, we get

$$\upsilon + \frac{c-1}{a} = \frac{d + c(u + \frac{1-b}{a})}{b + a(u + \frac{1-b}{a})}$$

Simplifying the above relation we get

$$\upsilon = \frac{ad - bc + c(au+1)}{a(1+au)} - \frac{c-1}{a} \tag{7.57}$$

We know that $S = \begin{pmatrix} b & -a \\ -d & c \end{pmatrix}$ and det S =1

Therefore, bc – ad =1

or ad – bc = –1

Further simplification of equ.(7.57) using the above result yields

$$U = \frac{au}{a(1+au)} = \frac{u}{1+au} \tag{7.58}$$

$$\frac{1}{\upsilon} = \frac{1+au}{u} = \frac{1}{u} + a$$

or

$$\frac{1}{\upsilon} - \frac{1}{u} = a \tag{7.59}$$

Comparing above equation with the relation $\frac{1}{\upsilon}-\frac{1}{u}=\frac{1}{f}$ we see that $\frac{1}{a}$ represents the focal length of the system, provided the distances are measured from the two unit planes.

7.10.2. UNIT PLANES FOR THICK LENS

The *system matrix* for thick lens is given by

$$S=\begin{pmatrix} b & -a \\ -d & c \end{pmatrix}=\begin{pmatrix} 1-P_2t/\mu & -P_1-P_2(1-P_1t/\mu) \\ \frac{t}{\mu} & 1-P_1t/\mu \end{pmatrix}$$

Comparing with the system matrix

$$a = P_1+P_2(1-P_1t/\mu)$$
$$b = 1-P_2t/\mu$$
$$c = 1-P_1t/\mu$$
$$d = -\frac{t}{\mu}$$

The location of the unit planes are expressed as

$$\alpha=\frac{1-b}{a}=\frac{1-(1-P_2\frac{t}{\mu})}{P_1+P_2(1-P_1\frac{t}{\mu})}=\frac{P_2\frac{t}{\mu}}{P_1+P_2(1-P_1\frac{t}{\mu})} \quad (7.60)$$

For a bi-convex lens $|R_1|=|R_2|=R$ (say) and $P_1=P_2=\frac{\mu-1}{R}$

$$\therefore \quad \alpha=\frac{t}{\mu\left\{1+\left(1-\frac{t}{\mu}\frac{\mu-1}{R}\right)\right\}}=\frac{t}{2\mu-\frac{t(\mu-1)}{R}} \quad (7.61)$$

When $t \ll R$

$$\alpha=\frac{t}{2\mu} \quad (7.62)$$

further

$$\beta=\frac{c-1}{a}=\frac{1-P_1\frac{t}{\mu}-1}{P_1+P_2(1-P_1\frac{t}{\mu})} \quad (7.63)$$

Substituting the values of P_1 and P_2 for a biconvex lens in equ.(7.63) and simplifying the expression, we get

$$\beta=\frac{-t}{2\mu-t\frac{(\mu-1)}{R}}\approx\frac{-t}{2\mu} \quad (7.64)$$

Thus, the first unit plane lies at the right at a distance t /2μ from first refracting surface and second unit plane to the left of second refracting surface at the same distance of t /2μ as shown in Fig.7.7.

60 mm bi-convex lens.

7.10.3. NODAL PLANES

Nodal points are a pair of conjugate points on the axis, which have a relative angular magnification of unity. According to the property of nodal points, a ray striking the first nodal point at an angle α emerges from the second nodal point at the same angle. The planes passing through these points and normal to the axis are known as nodal planes.

To determine the position of the nodal points, we consider two axial points N_1 and N_2 at a distance γ and δ from two refracting surfaces of thick lens as shown in Fig. 7.8.

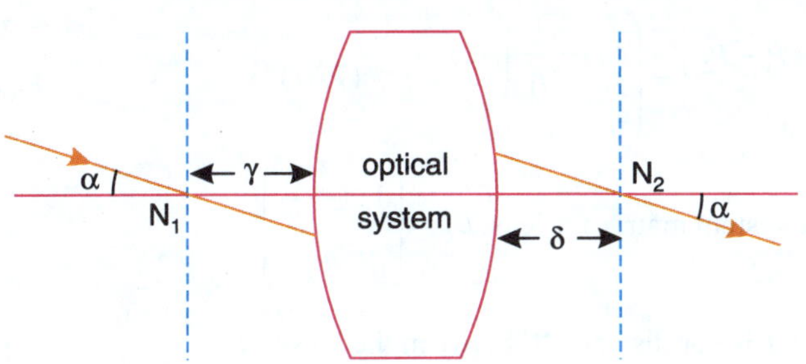

Fig. 7.8

Since the medium on either side of the system is the same, $\lambda_1 = \lambda_2$

Further we are considering the object point on the axis of the system. Therefore,

$$x_1 = 0 \text{ and } x_2 = 0$$

From equation (7.40) we have

$$\begin{pmatrix} \lambda_2 \\ x_2 \end{pmatrix} = \begin{pmatrix} b + aD_1 & -a \\ 0 & c - aD_2 \end{pmatrix} \begin{pmatrix} \lambda_1 \\ x_1 \end{pmatrix}$$

Taking $x_1 = x_2 = 0$, $D_1 = \gamma$ and $D_2 = \delta$, the above matrix becomes

$$\begin{pmatrix} \lambda_2 \\ 0 \end{pmatrix} = \begin{pmatrix} b + a\gamma & -a \\ 0 & c - a\delta \end{pmatrix} \begin{pmatrix} \lambda_1 \\ 0 \end{pmatrix}$$

$$\therefore \quad \lambda_2 = (b + a\gamma)\, \lambda_1 = \lambda_1.$$

$$\therefore \quad (b + a\gamma) = 1$$

or

$$\gamma = \frac{1 - b}{a} \tag{7.65}$$

Further, we have

$$(b + aD_1) = \frac{1}{c - aD_2} = \frac{1}{m}$$

As m = 1,

$$c - a\delta = 1$$

$$\delta = \frac{c - 1}{a} \tag{7.66}$$

Comparing equations (7.65) and (7.66) with equ.(7.52) and (7.53), we find that

$$\gamma = \alpha$$

and

$$\delta = \beta$$

Thus, when the medium on either side of the optical system is the same, the nodal points coincide with the principal planes.

7.11. SYSTEM MATRIX FOR TWO THIN LENSES

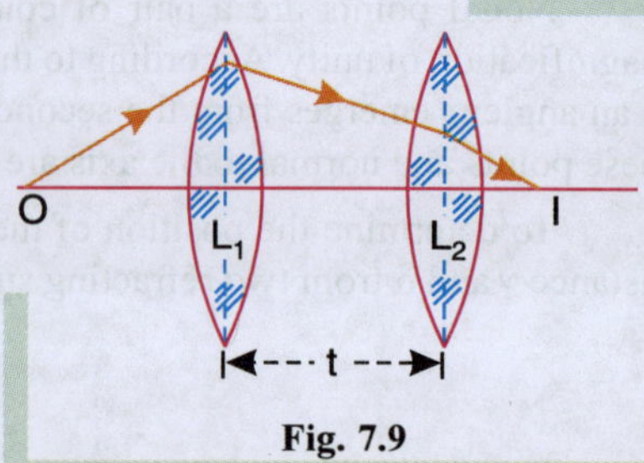

Fig. 7.9

Let us now consider two thin lenses of focal lengths f_1 and f_2 placed coaxially and separated by a distance 't'. The system matrix can be written by combining the system matrices of individual lenses and the translation matrix for the distance of separation between them.

System matrix for lens

$$L_1 = \begin{pmatrix} 1 & -P_1 - P_2 \\ 0 & 1 \end{pmatrix} = \begin{pmatrix} 1 & -\frac{1}{f_1} \\ 0 & 1 \end{pmatrix} \quad (7.67)$$

Similarly, the system matrix for lens $L_2 = \begin{pmatrix} 1 & -\frac{1}{f_2} \\ 0 & 1 \end{pmatrix}$ (7.68)

Translation matrix for distance 't' between the lenses $T = \begin{pmatrix} 1 & 0 \\ t & 1 \end{pmatrix}$ (as $\mu = 1$) (7.69)

Now the total system matrix is given by

$$S = \begin{pmatrix} 1 & -\frac{1}{f_2} \\ 0 & 1 \end{pmatrix} \begin{pmatrix} 1 & 0 \\ t & 1 \end{pmatrix} \begin{pmatrix} 1 & -\frac{1}{f_1} \\ 0 & 1 \end{pmatrix}$$

$$S = \begin{pmatrix} 1 - \frac{t}{f_2} & \frac{-1}{f_1} - \frac{1}{f_2} + \frac{t}{f_1 f_2} \\ t & 1 - \frac{t}{f_1} \end{pmatrix} \quad (7.70)$$

7.11.1. EFFECTIVE FOCAL LENGTH

Comparing the above matrix in equ.(7.70) with the general system matrix $\begin{pmatrix} b & -a \\ -d & c \end{pmatrix}$, we have

$$-a = \frac{-1}{f_1} - \frac{1}{f_2} + \frac{t}{f_1 f_2}$$

or

$$a = \frac{1}{f_1} + \frac{1}{f_2} - \frac{t}{f_1 f_2}$$

We know that the element 'a' in the system matrix represents the inverse of focal length F of the system. Therefore,

$$\frac{1}{F} = \frac{1}{f_1} + \frac{1}{f_2} - \frac{t}{f_1 f_2} \quad (7.71)$$

F represents the effective focal length of two thin lenses separated by a distance t.

7.11.2. POSITIONS OF THE PRINCIPAL PLANES

The comparison of the system matrix (7.70) with the general system matrix gives that

$$b = 1 - \frac{t}{f_2}$$

$$c = 1 - \frac{t}{f_1}$$

$$d = -t.$$

The positions of the principal planes are given by

$$\alpha = \frac{1-b}{a} = \frac{1-b}{F} = \frac{1-[1-t/f_2]}{F} = \frac{tF}{f_2} \quad (7.72)$$

$$\beta = \frac{c-1}{a} = \frac{[1-t/f_1]-1}{F} = -\frac{tF}{f_1} \quad (7.73)$$

WORKED OUT EXAMPLES

Example 7.1: Obtain the system matrix for a thin lens placed in air and made of refractive index 1.5 and radii of curvature 50 cm each.

Solution: System matrix for a thick lens is given by

$$S = \begin{pmatrix} 1 & -P_1 - P_2 \\ 0 & 1 \end{pmatrix}$$

The powers of two surfaces are equal. Hence

$$P_1 = P_2 = \left(\frac{\mu-1}{R_1}\right) = \left(\frac{\mu-1}{|R_2|}\right) = \frac{1.5-1}{50} = 0.01$$

$$\therefore \qquad S = \begin{pmatrix} 1 & -0.01-0.01 \\ 0 & 1 \end{pmatrix} = \begin{pmatrix} 1 & -0.02 \\ 0 & 1 \end{pmatrix}$$

Further $\qquad f = \dfrac{1}{0.02} = 50 \text{ cm.}$

Example 7.2: The focal lengths of two thin convex lenses are 20 cm and 10 cm. They are at a distance of 10 cm on the axis of symmetry. If a point object of length 1 cm is placed at a distance of 20 cm from the first lens, find the position of the image and magnification.

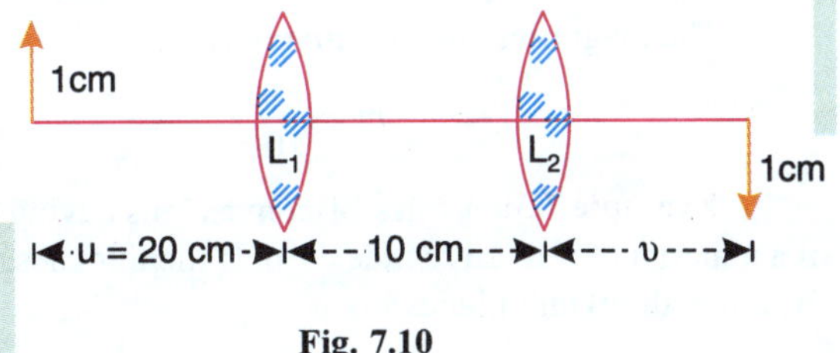

Fig. 7.10

Solution: The situation is as shown in Fig. 7.10.

Let the image be formed at a distance of υ cm from the second lens. The system matrix can be obtained in the following way.

The translation of a matrix between object and lens L_1 is given by

$$\begin{pmatrix} 1 & 0 \\ d & 1 \end{pmatrix} = \begin{pmatrix} 1 & 0 \\ 20 & 1 \end{pmatrix}$$

The refraction matrix for lens L_1 is given by

$$\begin{pmatrix} 1 & \frac{-1}{f_1} \\ 0 & 1 \end{pmatrix} = \begin{pmatrix} 1 & -\frac{1}{20} \\ 0 & 1 \end{pmatrix}$$

The translation matrix between the lens L_1 and L_2 is given by

$$\begin{pmatrix} 1 & 0 \\ d & 1 \end{pmatrix} = \begin{pmatrix} 1 & 0 \\ 10 & 1 \end{pmatrix}$$

The refraction matrix lens L_2 is given by

$$\begin{pmatrix} 1 & \dfrac{-1}{f_2} \\ 0 & 1 \end{pmatrix} = \begin{pmatrix} 1 & -\dfrac{1}{10} \\ 0 & 1 \end{pmatrix}$$

The translation matrix between L_2 and image is given by

$$\begin{pmatrix} 1 & 0 \\ d & 1 \end{pmatrix} = \begin{pmatrix} 1 & 0 \\ \upsilon & 1 \end{pmatrix}$$

Now the system matrix is given by

$$S = \begin{pmatrix} 1 & 0 \\ \upsilon & 1 \end{pmatrix} \begin{pmatrix} 1 & -\dfrac{1}{10} \\ 0 & 1 \end{pmatrix} \begin{pmatrix} 1 & 0 \\ 10 & 1 \end{pmatrix} \begin{pmatrix} 1 & -\dfrac{1}{20} \\ 0 & 1 \end{pmatrix} \begin{pmatrix} 1 & 0 \\ 20 & 1 \end{pmatrix}$$

$$= \begin{pmatrix} -2 & \dfrac{-1}{10} \\ -2\upsilon + 20 & \dfrac{-\upsilon}{10} + \dfrac{1}{2} \end{pmatrix}$$

For image plane we must have

$$-2\upsilon + 20 = 0$$

or $\quad U = 10$ cm.

Hence the image will be formed at a distance 10 cm from the second lens.

The magnification m is given by

$$m = \frac{-\upsilon}{10} + \frac{1}{2} \approx \frac{-1}{2}.$$

Example 7.3: A thick bi-convex lens having radii of curvature + 6 cm and −6 cm is made up of a material of refractive index 1.6. If the thickness of the lens is 2 cm. Find (a) System matrix (b) focal length (c) unit planes.

Solution: Radii of curvature of lens $|R_1| = |R_2| = 6$ cm

Refracting powers of the surfaces

$$P_1 = P_2 = \left(\frac{\mu - 1}{R}\right) = \frac{1.6 - 1}{6} = 0.1 \text{ cm}^{-1}$$

System matrix of the lens

$$S = \begin{pmatrix} 1 - P_2 t/\mu & -P_1 - P_2(1 - P_1 t/\mu) \\ \dfrac{t}{\mu} & 1 - P_1 t/\mu \end{pmatrix}$$

$$= \begin{pmatrix} 1 - \dfrac{0.1 \times 2}{1.6} & -0.1 - 0.1(1 - \dfrac{2}{1.6} \times 0.1) \\ \dfrac{2}{1.6} & 1 - \dfrac{0.1 \times 2}{1.6} \end{pmatrix}$$

$$= \begin{pmatrix} 0.875 & -0.1875 \\ 1.25 & 0.875 \end{pmatrix}$$

Comparing with the general system matrix,

$$a = 0.1875;\ b = 0.875;\ c = 0.875;\ d = -1.25$$

Focal length of the lens $f = \frac{1}{a} = \frac{1}{0.1875} = 53$ cm

Distances of unit planes

$$\alpha = \frac{1-b}{a} = \frac{1-0.875}{0.1875} = 0.67 \text{ cm}$$

$$\beta = \frac{c-1}{a} = \frac{0.875-1}{0.1875} = -0.67 \text{ cm}$$

Thus, first unit plane is at a distance of 0.67 cm to the right of the first refracting surface, where as the second unit plane is at a distance of 0.67 cm to the left of the second refracting surface.

Example 7.4: Two thin converging lenses of powers 5 and 4 diopters are placed co-axially 10 cm apart. Find the focal length of the combination.

Solution: Power of a lens $P = \frac{1}{f}$

$$f_1 = \frac{1}{P_1} = \frac{1}{4} = 0.25 \text{ m} = 25 \text{ cm}$$

$$f_2 = \frac{1}{P_2} = \frac{1}{5} = 0.20 \text{ m} = 20 \text{ cm}$$

$$\frac{1}{F} = \frac{1}{f_1} + \frac{1}{f_2} - \frac{t}{f_1 f_2}$$

$$F = \frac{f_1 f_2}{f_1 + f_2 - t} = \frac{25 \times 20}{25 + 20 - 10} = 14.3 \text{ cm}$$

Example 7.5: Find the position of the paraxial point of a lens in the form of a sphere of glass of radius 5 cm and refractive index 1.5 for the Fig. 7.11 is shown below.

Solution:

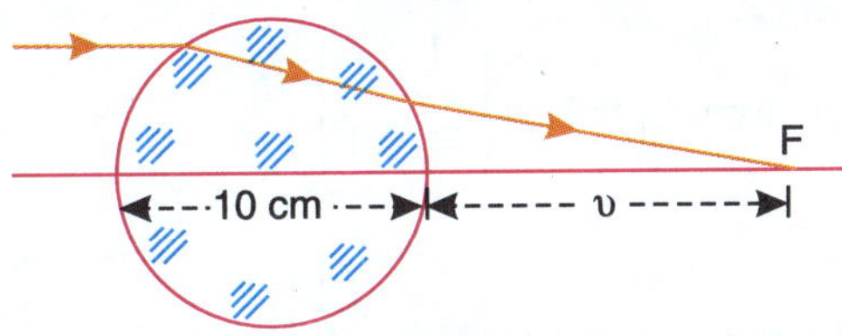

Fig. 7.11

The system matrix can be obtained by multiplying the translation and refraction matrices in proper order.

$$S = \begin{pmatrix} 1 & 0 \\ \upsilon & 1 \end{pmatrix} \begin{pmatrix} 1 & -\frac{0.5}{5} \\ 0 & 1 \end{pmatrix} \begin{pmatrix} 1 & 0 \\ \frac{10}{1.5} & 0 \end{pmatrix} \begin{pmatrix} 1 & -\frac{0.5}{5} \\ 0 & 1 \end{pmatrix} = \begin{pmatrix} \frac{1}{3} & \frac{-2}{15} \\ \frac{\upsilon}{3} + \frac{2\upsilon}{3} & \frac{-2\upsilon}{15} + \frac{1}{3} \end{pmatrix}$$

$$\begin{pmatrix} \lambda_2 \\ x_2 \end{pmatrix} = \begin{pmatrix} \frac{1}{3} & \frac{-2}{15} \\ \frac{\upsilon}{3} + \frac{2\upsilon}{3} & \frac{-2\upsilon}{15} + \frac{1}{3} \end{pmatrix} \begin{pmatrix} \lambda_1 \\ x_1 \end{pmatrix}$$

or
$$x_2 = \left(\frac{\upsilon}{3} + \frac{2\upsilon}{3}\right)\lambda_1 + \left(\frac{-2\upsilon}{15} + \frac{1}{3}\right)x_1$$

Here $x_2 = 0$ and $\lambda_2 = 0$

$$\left(\frac{-2\upsilon}{15} + \frac{1}{3}\right)x_1 = 0$$

$$U = \mathbf{2.5\ cm.}$$

Example 7.6: Find the focal length, and unit planes of Ramsden's eye piece.

Solution: Ramsden's eye piece consists of two plano convex lenses each of focal length f separated by a distance of $\frac{2f}{3}$. The Ramsden's eye piece is as shown in Fig. 7.12.

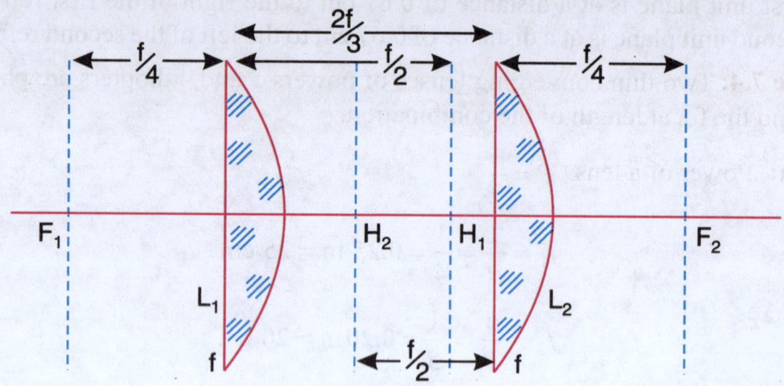

Fig. 7.12

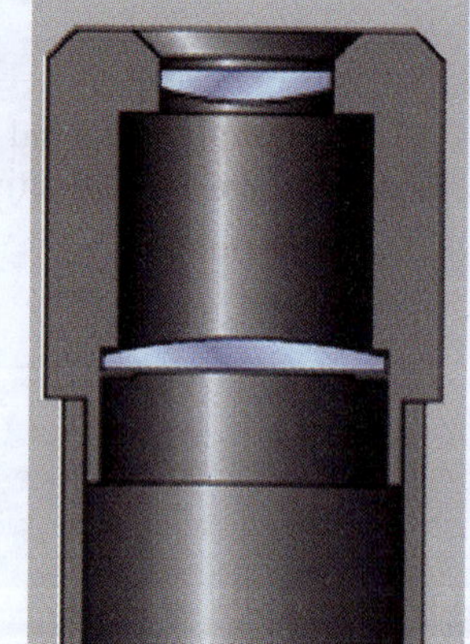
Ramsden's eye piece

The system matrix for the lens $L_1 = \begin{pmatrix} 1 & \frac{-1}{f} \\ 0 & 1 \end{pmatrix}$

The translation matrix from L_1 to $L_2 = \begin{pmatrix} 1 & 0 \\ \frac{2f}{3} & 1 \end{pmatrix}$

The system matrix for the second lens $= \begin{pmatrix} 1 & \frac{-1}{f} \\ 0 & 1 \end{pmatrix}$

The system matrix of Ramsden's eye piece

$$S = \begin{pmatrix} 1 & \frac{-1}{f} \\ 0 & 1 \end{pmatrix}\begin{pmatrix} 1 & 0 \\ \frac{2f}{3} & 1 \end{pmatrix}\begin{pmatrix} 1 & \frac{-1}{f} \\ 0 & 1 \end{pmatrix} = \begin{pmatrix} \frac{1}{3} & \frac{-4}{3f} \\ \frac{2f}{3} & \frac{1}{3} \end{pmatrix}$$

Comparing this with the general system matrix, Gaussian constants are

$$a = \frac{4}{3f};\ b = \frac{1}{3};\ c = \frac{1}{3};\ d = \frac{-2f}{3}$$

The combined focal length of the system F is given by

$$F = \frac{1}{a} = \frac{3f}{4}$$

The distances of the unit planes

$$\alpha = \frac{1-b}{a} = \frac{1-(1/3)}{(4/3f)} = \frac{2/3}{4/3} xf = \frac{1}{2} f$$

$$\beta = \frac{c-1}{a} = \frac{(1/3)-1}{4/3} = \frac{-1}{2} f$$

Nodal planes coincide with unit planes.

Example 7.7: Find the focal length, unit planes of Huygens's eye piece.

Solution: Huygens's eye piece consists of two Plano convex lenses each of focal lengths 3f and f separated by a distance of 2f. The Huygens's eye piece is as shown in Fig. 7.13.

The system matrix for the lens $L_1 = \begin{pmatrix} 1 & \frac{-1}{3f} \\ 0 & 1 \end{pmatrix}$

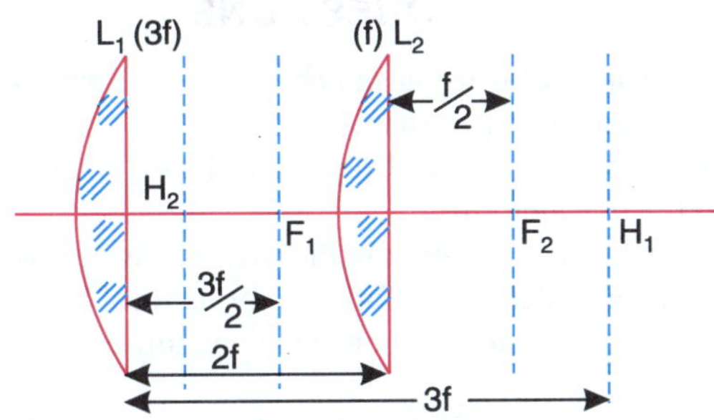

Fig. 7.13

The translation matrix from L_1 to $L_2 = \begin{pmatrix} 1 & 0 \\ 2f & 1 \end{pmatrix}$

The system matrix for the second lens $= \begin{pmatrix} 1 & \frac{-1}{f} \\ 0 & 1 \end{pmatrix}$

The system matrix of Ramsden's eye piece

$$S = \begin{pmatrix} 1 & \frac{-1}{f} \\ 0 & 1 \end{pmatrix} \begin{pmatrix} 1 & 0 \\ 2f & 1 \end{pmatrix} \begin{pmatrix} 1 & \frac{-1}{3f} \\ 0 & 1 \end{pmatrix} = \begin{pmatrix} 1 & \frac{-2}{3f} \\ 2f & \frac{1}{3} \end{pmatrix}$$

Huygens's eye piece.

Comparing this with the general system matrix, Gaussian constants are

$$a = \frac{2}{3f}\ ; b = -1\ ; c = \frac{1}{3}\ ; d = -2f$$

The combined focal length of the system F is given by

$$F = \frac{1}{a} = \frac{3f}{2}$$

The distances of the unit planes

$$\alpha = \frac{1-b}{a} = \frac{1+1}{2} x3f = 3f$$

$$\beta = \frac{c-1}{a} = \frac{(1/3)-1}{2} x3f = f$$

Nodal planes coincide with unit planes.

QUESTIONS

1. What is the advantage of using matrix method in paraxial optics? Obtain the *system matrix* for a general optical system.
2. Determine the matrices that will represent the effect of translation and refraction in a medium of refractive index μ.
3. What is System matrix? Obtain it in the case of a system of two thin lenses separated by a distance and hence derive the formula for its focal length.
4. Formulate the equations for the image and magnification of an optical system using the matrix method?
5. What are unit planes and nodal planes of a kens system? Find the positions of the unit planes in a thick lens and hence deduce the expression for its focal length.
6. Obtain the system matrix for the thick lens and drive the lens formula for thin lens.
7. Show that if the distances are measured from the unit planes, the focal length of the system is given by the reciprocal of the element of the system matrix.
8. Using matrix method obtain the expression for equivalent focal length and position of cardinal points of Ramsden's eye piece?
9. Using matrix method obtain the expression for equivalent focal length and position of cardinal points of Huygens's eye piece?
10. Explain how translation and refraction matrices are formed and hence build up the system matrix for a thick lens.
11. Define Translation matrix and Refraction Matrix of an optical system.

(A.P.University, 2010)

PROBLEMS FOR PRACTICE

1. The radius of curvature of the surfaces of a double convex lens are $R_1 = R_2 = 50$ cm. The refractive index of the material of the lens is 1.5. Find the optical power of the lens. **[Ans: 2 diopters]**
2. Consider a system of two thin lenses of focal lengths 10 & 30 cm separated by a distance of 20 cm in air. (a) Determine the system matrix elements (b) the positions of the unit planes.
3. Consider a thick lens of the form shown in Fig. 7.14. The radii of curvature of the first and second surfaces are

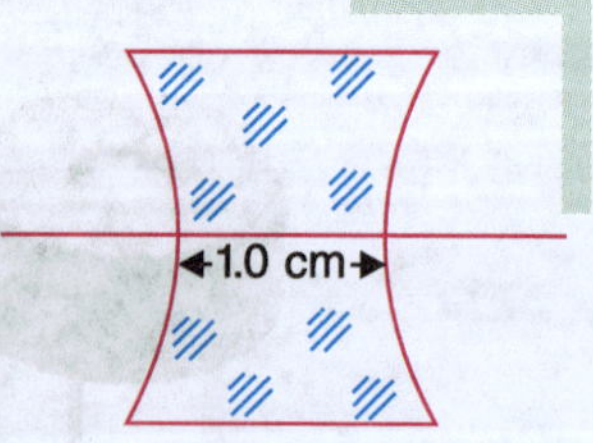

Fig. 7.14

−10 cm and 20 cm. respectively and the thickness of the lens is 1.0 cm. The refractive index of the material of the lens is 1.5. Determine the positions of the principal planes.

[Ans: $d_1 = 20/91$ cm, $d_2 = 40/91$ cm]

4. Consider a thick equi-convex lens (made of a material of refractive index 1.5) of the type shown in Fig. 7.15 the magnitude of the radii of curvature of the two surfaces is 4 cm, the thickness of the lens is 1 cm and the lens is placed in air. Obtain the system matrix and determine the focal length and the positions of unit planes.

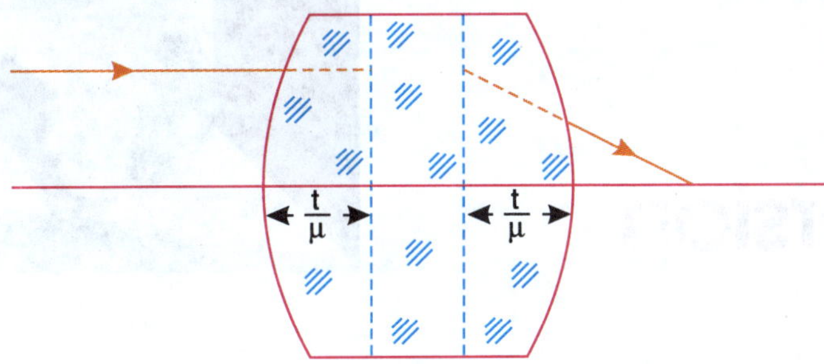

Fig. 7.15

[Ans: $\begin{pmatrix} 0.9167 & -0.240 \\ 0.6667 & 0.9167 \end{pmatrix}$, f = 4.2 cm, α = 0.35 cm, β = – 0.35 cm]

5. The focal length of each lens of Ramsden's eye piece is 4 cm by matrix formulation. Find the focal length and determine the positions of the cardinal points.

[Ans: 2 cm, –2cm, –1cm, 1cm, & F = 3cm]

6. A lens of thickness 2cm and refractive index 1.5 has radii of curvature +8cm and –8cm. Find the system matrix and focal length of thick lens. **(A.P.University, 2010)**

CHAPTER

Dispersion

8.1. DISPERSION BY A PRISM

A beam of white light, when it passes through a prism is split up into its constituent colours. Different colours suffer different amounts of deviation. The spread of colours is called **dispersion of light**. The image thus formed on a screen is called a **spectrum**. The medium, which produces dispersion, is called a **dispersive medium**.

At a Glance

- Dispersion by a Prism
- Refraction Through a Prism
- Angular Dispersion
- Dispersive Power
- Angular and Chromatic Dispersions
- Achromatic Combination of Prisms – Deviation without Dispersion
- Dispersion Without Deviation
- Direct Vision Spectroscope

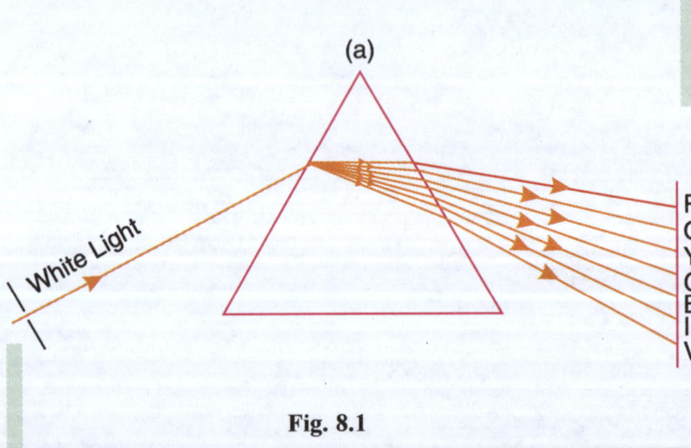

Fig. 8.1

The spectrum consists of visible and invisible regions. In the visible region the order of the colours is from violet to red. The principal colours are given by the word **VIBGYOR** (Violet, Indigo, Blue, Green, Yellow,

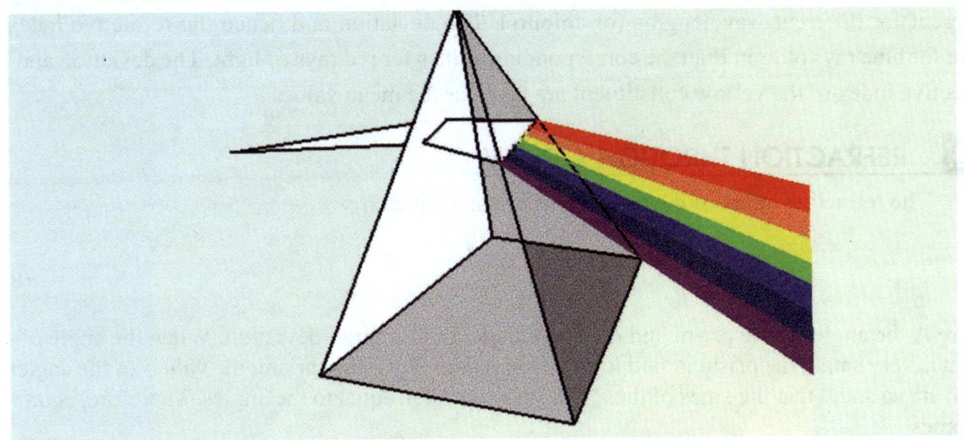

Light splitting up into its constituent colours.

Orange and Red). The deviation produced for the violet rays of light is maximum and for red rays of light it is minimum. Fig. 8.1 represents the dispersion of a white ray of light by a prism in the visible region. The region of the spectrum of wavelengths shorter than violet is called **ultra-violet** and the region of wavelengths longer than red is called **infrared**. In the present chapter, the discussion relates only to the visible region of the spectrum.

It is evident from Fig. 8.2 that the refractive index for the material of a prism (or a lens) is

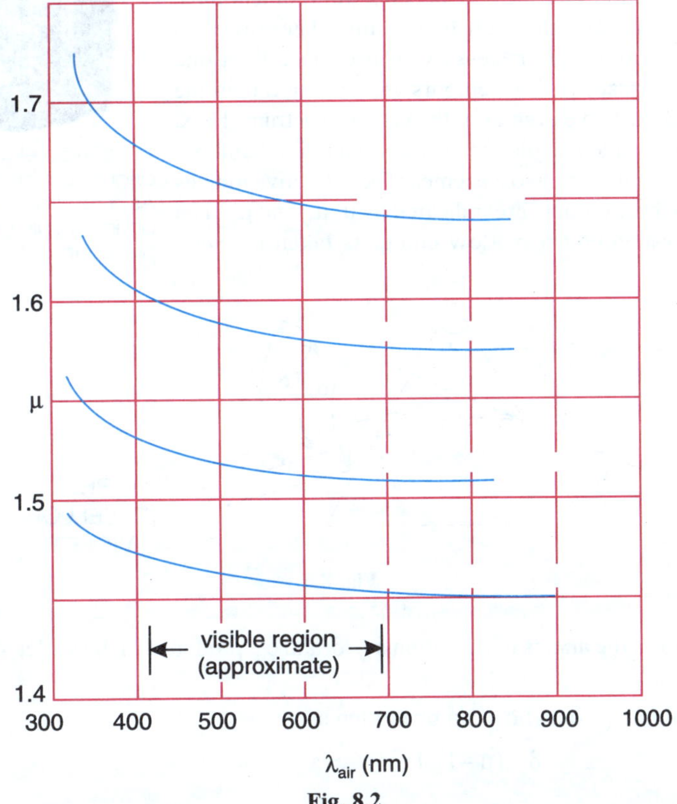

Fig. 8.2

different for different wavelengths (or colours). The deviation and hence the refractive index is more for blue rays of light than the corresponding values for red rays of light. The deviation and the refractive index of the yellow constituent are taken as the mean values.

8.2. REFRACTION THROUGH A PRISM

The refractive index of the material of prism is given by

$$\mu = \frac{\sin \frac{A+\delta_m}{2}}{\sin A/2} \tag{8.1}$$

where A the angle of the prism and δ_m is the angle of minimum deviation. When the angle of the prism is very small, the prism is said to be a **thin prism**. For a thin prism, the values of the angles A and δ are so small that the sines of the angles may be taken equal to the angles. Therefore, equ.(8.1) becomes

$$\mu = \frac{A+\delta_m}{A}$$

or $$\delta_m = (\mu - 1)A \tag{8.2}$$

8.3. ANGULAR DISPERSION

The refraction and dispersion does not bear a simple relationship to one another. It is known that some glasses have a high index of refraction and little dispersion while others have low index of refraction and high dispersion. Therefore, to characterize a material, it is necessary to use more than one refractive index. In practice three colours are chosen; one in the middle and two at the two extremes of the visible spectrum. Thus, the yellow colour is taken as the mean colour and violet and red colours are considered at the two extremes. The refractive indices corresponding to these colours are indicated by μ, μ_V and μ_R. The deviation δ corresponding to yellow colour is taken as *mean deviation*.

The curved electrostat has good angular dispersion characteristics. As a result the speaker provides an incredibly clear, natural and pleasant sound image.

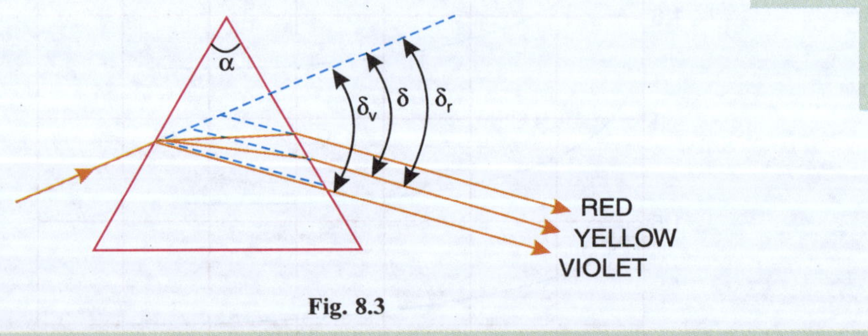

Fig. 8.3

Fig. 8.3 shows the angles of deviation δ_V, δ and δ_R produced in the violet, mean yellow and red rays of light.

The deviation δ_V, δ and δ_R can be written as

$$\delta = (\mu - 1)A \tag{8.3a}$$

$$\delta_V = (\mu_V - 1)A \tag{8.3b}$$

$$\delta_R = (\mu_R - 1)A \tag{8.3c}$$

The total angle through which the spectrum is spread is called as the **angular dispersion**. It is measured by the difference in deviation between the two extreme colours of the spectrum. Therefore, the angular dispersion is given by

$$\theta = \delta_V - \delta_R = (\mu_V - \mu_R)A \tag{8.4}$$

Thus, the angular dispersion depends on the nature of the material of the prism and upon the angle of the prism.

8.4. DISPERSIVE POWER

Dispersive power indicates the ability of the material of the prism to disperse the light rays. It is defined as the ratio of the angular dispersion to the deviation of the mean ray.

$$\text{Dispersive power, } \omega = \frac{\delta_V - \delta_R}{\delta} \tag{8.5}$$

Using the set of equations (8.3a,8.3b and 8.3c), we can rewrite equ.(8.5) as

$$\omega = \frac{\mu_V - \mu_R}{\mu - 1} \tag{8.6}$$

This is the expression for *dispersive power* used mainly in ***geometric optics*** and stated in terms of mean μ and refractive indices for violet and red light. The difference $(\mu_V - \mu_R)$ is known as the *mean dispersion*. It is seen that the dispersive power is independent of the angle of the prism and the angle of incidence.

Note— There is no universal choice about colours chosen for defining the dispersive power. In case of viewing with the aid of eyes, the range violet to red is more useful, whereas in photography the range violet to green is most useful.

Flint glass bellflower.

Crown glass.

It is also customary to express dispersive power more precisely with reference to C, F and D Fraunhofer lines (dark lines) in the solar spectrum. The F, D and C lines lie in the blue, yellow and red regions of the spectrum and their wavelengths are 4861Å and 5893Å and 6563Å respectively. (1Å = 1 Angstrom unit = 10^{-10}m). Thus, the dispersive power may be expressed as

$$\omega = \frac{\mu_F - \mu_C}{\mu_D - 1}$$

The inverse of dispersive power is called as **Abbe's number** or **V-number** and is denoted by V.

$$V = \frac{\mu_D - 1}{\mu_F - \mu_C} \tag{8.7}$$

A small Abbe's number means high dispersion. Glasses of low dispersion, having V-number above 55, are called **crowns;** and glasses of high dispersion with V-number below 55 are called **flints**.

8.5. ANGULAR AND CHROMATIC DISPERSIONS

Now we note here the definition of dispersive power as used in ***wave theory*** of light.

The variation of refractive index μ of a given medium with wavelength is called dispersion. If μ varies rapidly with wavelength, the medium is said to be highly dispersive.

The refractive index μ of the material of a prism is given by

$$\mu = \frac{\sin \frac{A+\theta}{2}}{\sin A / 2} \tag{8.8}$$

where A is the angle of the prism and θ is the angle of minimum deviation.

$$\sin \frac{A+\theta}{2} = \mu \sin \frac{A}{2}$$

$$\therefore \qquad \cos \frac{A+\theta}{2} = \sqrt{1-\mu^2 \sin^2 \frac{A}{2}}$$

If $d\theta$ is the difference in the angle of deviation between two spectral lines of wavelengths λ and $\lambda + d\lambda$, then $\frac{d\theta}{d\lambda}$ is called the *angular dispersion* between the wavelengths λ and $\lambda + d\lambda$.

Differentiating equation (8.8) with respect to μ on the left-hand side and θ on the right hand side, we obtain

$$d\mu = \frac{\frac{1}{2} \cos \frac{A+\theta}{2}}{\sin A / 2} . d\theta$$

Dividing by $d\lambda$

$$\frac{d\mu}{d\lambda} = \frac{\cos \frac{A+\theta}{2}}{2 \sin A / 2} . \frac{d\theta}{d\lambda}$$

$$\therefore \qquad \frac{d\theta}{d\lambda} = \frac{2 \sin \frac{A}{2}}{\cos \frac{A+\theta}{2}} . \frac{d\mu}{d\lambda}$$

In chromatic dispersion, source light is broken up into its component parts.

or
$$\frac{d\theta}{d\lambda} = \frac{2\sin\frac{A}{2}}{\sqrt{1-\mu^2\sin^2\frac{A}{2}}}\cdot\frac{d\mu}{d\lambda} \tag{8.9}$$

$\frac{d\mu}{d\lambda}$ is called **chromatic dispersion** of the material of the prism. It depends on the angle of the prism and refractive index of the material of the prism. In general $\frac{d\mu}{d\lambda}$ is not constant and decreases as λ increases. Using a spectrometer and the given prism, a graph is drawn between μ and λ (μ along the y-axis and λ along the x-axis). The tangent to the curve at any point measures the chromatic dispersion $d\mu/d\lambda$ for that particular wavelength. Substituting this value of $d\mu/d\lambda$ in equation (8.9), $d\theta/d\lambda$ can be calculated.

8.6. ACHROMATIC COMBINATION OF PRISMS — DEVIATION WITHOUT DISPERSION

An achromatic prism is a combination of two appropriate prisms so constructed that it shows no colours. It is noted that the dispersive powers of different materials are different. Flint glasses have higher dispersive power than crown glasses. Hence, it is possible to combine two prisms of *different* materials and specified angles such that a ray of white light may pass through the combination *without dispersion*, though it may suffer deviation. Such a combination is called an *achromatic combination*.

Fig. 8.4 shows an arrangement of two prisms, which does not cause dispersion though it may produce deviation. Let the first prism be made of crown and the second of flint. Let δ_V and δ_R be the deviations and μ_V and μ_R the refractive indices for the violet and red rays in case of crown glass prism. Let the angle of the prism be A. The angular dispersion produced by the crown glass prism is

Fig. 8.4

$$\delta_V - \delta_R = (\mu_V - \mu_R)A \tag{8.10}$$

Let the refracting angle of the second prism of flint glass be A′. It is arranged in such a way that it produces a deviation in a direction opposite to that of the first prism. Let δ_V' and δ_R' be the deviations and μ_V' and μ_R' the refractive indices for the violet and red rays. The angular dispersion produced by the flint glass prism is

$$\delta_V' - \delta_R' = \left(\mu_V' - \mu_R'\right)A' \tag{8.11}$$

For achromatism the net angular dispersion produced by the two prisms together must be zero.

$$\therefore \quad (\mu_V - \mu_R)A + \left(\mu_V' - \mu_R'\right)A' = 0 \tag{8.12}$$

This is the condition for achromatism.

$$\therefore \quad A = -\left(\frac{\mu_V' - \mu_R'}{\mu_V - \mu_R}\right)A' \tag{8.13}$$

The above expression gives the relation between the angles of the two prisms in order that the combination may function as an achromatic combination. The negative sign shows that the two prisms should be placed with their refracting angles in opposite directions. Equ.(8.12) may be written as

$$\frac{(\mu_V - \mu_R)}{\mu - 1} \cdot (\mu - 1) A + \frac{(\mu_V' - \mu_R')}{\mu - 1} \cdot (\mu' - 1) A' = 0$$

Using equ. (8.6) we can write the above equation as

$$\omega\delta = -\omega'\delta' \tag{8.14}$$

where ω and ω′ are the dispersive powers of the two prisms and δ and δ′ their mean deviations. Equ.(8.14) indicates that the dispersive powers of the prism should be in the inverse ratio of their mean deviation. It is found that when a crown glass prism of refracting angle 60° is combined with a flint glass prism of an angle of 29°17′, the combination becomes achromatic. Though the combination is achromatic, yet it produces deviation.

8.7. DISPERSION WITHOUT DEVIATION

Achromatic prisms are not of much practical use. A more useful combination of prisms is the one that produces *dispersion but no deviation*. It is called a *direct vision prism*.

It consists of two prisms of different dispersive powers and different angles and combined in such a way that the combination does not produce deviation of the mean rays. If the angles of the crown and flint glass prisms are so adjusted that the deviation produced for the mean rays by the first prism is equal and opposite to that produced by the second prism, then the final beam will be parallel to the incident beam. However, such a combination of two prisms produces dispersion of the incident beam without deviation.

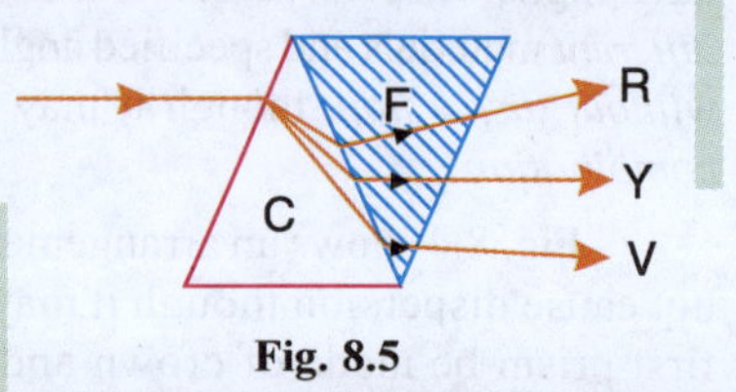

Fig. 8.5

Let us consider two thin prisms, one made of crown glass and the other of flint glass. Let A and A′ be the refracting angles of the prisms and μ and μ′ the refractive indices. Let δ and δ′ be the deviations produced for the mean ray by the two prisms respectively. The deviations are given by

$$\delta = (\mu - 1) A$$

$$\delta' = (\mu' - 1) A'$$

The net deviation produced by the prism combination is to be zero.

That is, $\delta + \delta' = 0$

Or $(\mu - 1) A + (\mu' - 1) A' = 0$

$$\therefore \quad A = -\left(\frac{\mu' - 1}{\mu - 1}\right) A' \tag{8.15}$$

The negative sign shows that the refracting angles of the two prisms are in opposite direction.

We calculate the net angular dispersion produced by the combination of prisms as follows:

The total deviation for the violet rays is given by

$$\delta_V = (\mu_V - 1) A + (\mu_V' - 1) A'$$

and the total deviation for the red rays is given by

$$\delta_R = (\mu_R - 1) A + (\mu_R' - 1) A'$$

Hence the total dispersion is equal to

$$\delta_V - \delta_R = (\mu_V - \mu_R)A + (\mu_V' - \mu_R')A'$$

Using equ.(8.15) into the above equation, we obtain

$$\delta_V - \delta_R = (\mu_V - \mu_R)A + (\mu_V' - \mu_R')\left(-\frac{\mu - 1}{\mu' - 1}\right)A$$

or $$\delta_V - \delta_R = \left[\frac{(\mu_V - \mu_R)}{(\mu - 1)} - \frac{(\mu_V' - \mu_R')}{\mu' - 1}\right](\mu - 1)A$$

It follows from equ.(8.6) that

$$\delta_V - \delta_R = (\omega - \omega')(\mu - 1)A \tag{8.16}$$

Thus the resultant dispersion is equal to the difference between the two dispersions.

8.8. DIRECT VISION SPECTROSCOPE

A direct vision spectroscope (Fig. 8.6) consists of three crown and two flint glass prisms of suitable refracting angles. The prisms are fixed in a metal tube. A collimating lens is fixed at one end of the tube and a telescope at the other. The angles of the prisms are such that the total deviation produced for the mean rays is zero. The refracting angles of crown and flint glass prisms are in opposite directions.

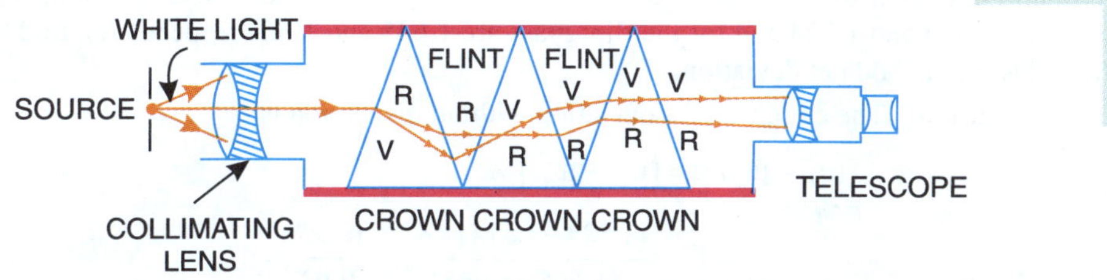

Fig. 8.6

Thus, the deviation produced by the crown glass prisms in one direction is equal and opposite to that produced by the flint glass prisms in the opposite direction. There will be resultant dispersion and dispersed beam is almost parallel to the incident beam. The prisms are cemented together with canada balsam (μ =1.54) to minimize the reflection losses at the interfaces. A spectroscope of this type is very handy and is used to study qualitatively the spectra of different sources of light. Use of more than two prisms increases the resolving power of the instrument, i.e., the spectral lines will appear well separated from one another.

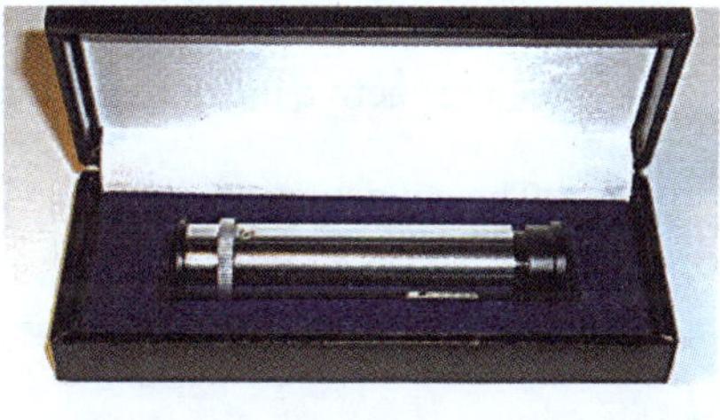

Direct vision spectroscope is used as a small instrument for visually observing the spectra of relatively bright light sources.

WORKED OUT EXAMPLES

Example 8.1: Caculate the dispersive power for crown and flint glass from the following data:

	C	D	F
Crown	1.5145	1.5170	1.5230
Flint	1.6444	1.6520	1.6637

Solution:

$$\omega_1 = \frac{\mu_F - \mu_C}{\mu_D - 1} = \frac{1.5230 - 1.5145}{1.5170 - 1} = 0.01644$$

$$\omega_2 = \frac{\mu_F - \mu_C}{\mu_D - 1} = \frac{1.6637 - 1.6444}{1.6520 - 1} = 0.02961$$

Example 8.2: For a crown and flint glass for C and F lines $\mu_C = 1.515$, and $\mu_F = 1.523$ and $\mu_C = 1.644$, $\mu_F = 1.664$ respectively. Calculate the angle of flint glass prism which may be combined with crown glass prism having refracting angle 20° so that the combination is achromatic for C and F rays.

Solution: The condition for the combination to be achromatic is that

$$(\mu_F - \mu_C)\, A = \left(\mu_F' - \mu_C'\right) A'$$

$$(1.523 - 1.515) \times 20^\circ = (1.664 - 1.644)\, \alpha'$$

$$\therefore \qquad \alpha' = \frac{0.008}{0.02} \times 20^\circ = \mathbf{8^\circ}.$$

Example 8.3: A crown glass prism of refracting angle 8° is combined with a flint glass prism to obtain deviation without dispersion. If the refractive indices for red and violet rays for crown glass are 1.514 and 1.524 and for the flint glass are 1.645 and 1.665 respectively, find the angle of flint glass prism and net deviation.

Solution: The condition for deviation without dispersion is

$$(\mu_V - \mu_R)\, A = \left(\mu_V' - \mu_R'\right) A'$$

$$\therefore \qquad A' = \frac{(1.524 - 1.514) \times 8^\circ}{(1.665 - 1.645)} = \frac{0.08^\circ}{0.02} = 4^\circ$$

Mean refractive index for crown glass $\mu = \frac{1.514 + 1.524}{2} = 1.519$

Mean refractive index for flint glass $\mu = \frac{1.645 + 1.665}{2} = 1.655$

$$\therefore \text{ The net deviation } (\delta - \delta') = (\mu - 1)\, A - (\mu' - 1)\, A'$$

$$= 0.519 \times 8^\circ - 0.655 \times 4^\circ$$

$$= \mathbf{1.53^\circ}.$$

QUESTIONS

1. Define angular dispersion and dispersive power. Derive the condition to produce dispersion without deviation in a combination of prisms.
2. Can we have dispersion without deviation with a pair of prisms of the same material but of different angles?
3. What do you mean by the terms angular dispersion, angular deviation, and dispersive power of the material? How does dispersive power depend on the refractive index?
4. Derive the condition for the combination of two thin prisms to produce mean deviation without net dispersion. Derive an expression for the net mean deviation.

(Madhurai Kamaraj, 2003)

5. Explain the principle, construction and working of direct vision spectroscope and derive an expression for the net dispersion produced by it.
6. How can one achieve deviation without dispersion using a pair of prisms? Why we cannot have deviation by combination of two prisms made of the same material but of different angles?
7. What is meant by dispersion? **(Madhurai Kamaraj, 2003)**

PROBLEMS FOR PRACTICE

1. Find the angle of a prism of dispersive power 0.021 and refractive index 1.62 to form an achromatic combination with a prism of angle 4.2° and dispersive power 0.045 having refractive index 1.65. Find the resultant deviation. [**Ans: A =11.25°; δ = 3.1°**]
2. If white light is incident at angle of 30° on a slab of glass that for blue light has a refractive index of 1.7 and for red light of 1.6, what is the angular dispersion between blue and red inside the glass? [**Ans: 1.1°**]
3. The dispersive powers of crown and flint glasses are 0.03 and 0.05 respectively. If the difference in the refractive indices of blue and red colours is 0.015 for crown glass and 0.022 for flint glass, calculate the angles of the two prisms for a deviation of 2° without dispersion. [**Ans:** A = 10°; A′ = 6.8°]

CHAPTER

Lens Aberrations

9.1. INTRODUCTION

One of the basic problems of lenses is the imperfect quality of the images. The simple equations derived and discussed in the earlier chapters, connecting object and image distances, focal length etc are based on the assumption that the angles made by the rays of light with the axis are small and *paraxial approximation* may be made. In actual practice the objects are bigger and a lens is required to produce a bright and magnified image. We are therefore required to take into consideration the wide-angle rays from the centre of the object and also the upper and lower parts of the object and falling near the top and bottom of the lens. These rays are known as *peripheral or marginal rays*. In general, peripheral rays of light do not meet at a single point after refraction through the lens. Secondly, the refractive index and hence the focal length of a lens is different for different wavelengths of light. For a given lens, the refractive index for violet light is more than that for red light. Thus, if the light coming from an object point is not monochromatic, the lens forms a number of coloured images. These images, even though formed by paraxial rays, are at different positions and are of different sizes.

9.2. ABERRATIONS

The departures of real images from the ideal images, in respect of the actual size, shape, and position, are called **aberrations**. In other words, an aberration is any failure of

At a Glance

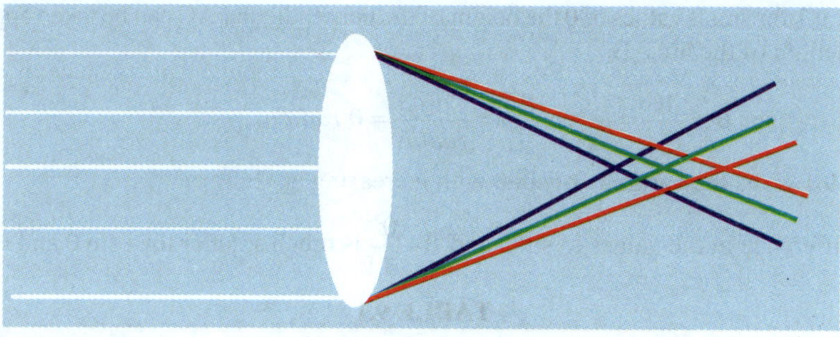

Chromatic aberration

a mirror or a lens to behave precisely according to the simple *formulae* we have derived. Aberrations are only due to inherent shortcomings of a lens and not caused by faulty construction of the lens, such as irregularities in its surfaces. *They are inevitable consequences of the laws of refraction at spherical surfaces*.

Aberrations are divided broadly into two categories- *monochromatic aberrations* and *chromatic aberrations*. The defects due to wide-angle incidence and peripheral incidence, which occur even with monochromatic light, are called **monochromatic aberrations**. Aberrations that occur due to dispersion of light are called **chromatic aberrations**. Chromatic aberration occurs with light that contains atleast two wavelengths. Monochromatic aberrations are again divided into five types:

1. Spherical aberration
2. Coma
3. Astigmatism
4. Curvature of field
5. Distortion.

The deviations from the actual size, shape and position of an image as calculated from the earlier simple equations are called the aberrations produced by a lens. The aberrations produced by the variation of refractive index with wavelength of light are called chromatic aberrations. The other aberrations are caused even if monochromatic light is used and they are called monochromatic aberrations. Lens aberrations are just the consequence of the refraction laws at the spherical surfaces and not due to defective construction of a lens such as the surfaces and not due to defective construction of a lens such as the surfaces being not spherical etc.

9.3. FIRST ORDER THEORY

To understand satisfactorily the theory of lens aberrations, it is necessary to start with the expansion of the sines of angles into a power series. According to Maclaurin's theorem the expansion of sin θ is given by

$$\sin\theta = \theta - \frac{\theta^3}{3!} + \frac{\theta^5}{5!} - \frac{\theta^7}{7!} + ---- \qquad (9.1)$$

When the value of θ is small, the series is a rapidly converging one i.e., the value of any term is smaller than the preceding one. In case the slope angle is small, sin θ = θ approximately. The equations developed on the basis that the sines of the angles are equal to the angles form the basis of the **first order theory**.

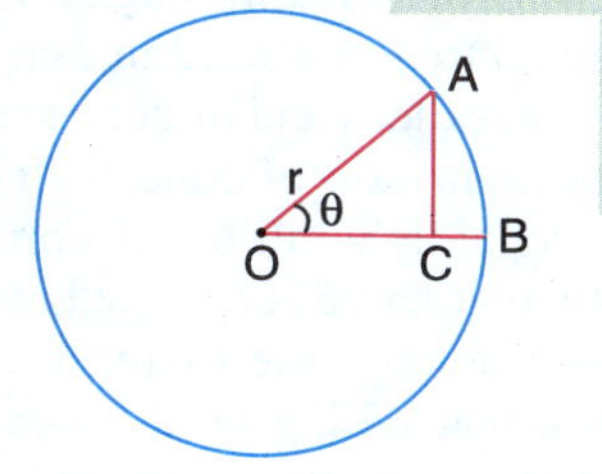

Fig. 9.1

In Fig. 9.1 for small values of θ the height of the perpendicular AC can be taken approximately equal to the length of the arc AB.

$$\sin\ \theta = \frac{AC}{OA} = \frac{AC}{r} = \frac{AB}{r} = \frac{arc}{radius} = \theta\ radians \qquad (9.2)$$

Table 9.1 gives the variation of sin θ with increasing angle.

The difference in the values of sin θ and $\theta - \frac{\theta^3}{3\,!}$ is much smaller than sin θ and θ.

TABLE 9.1

Angle (degrees)	sin θ	θ (radians)	$\theta - \frac{\theta^3}{3\,!}$
10	0.1736481	0.1745329	0.1736458
20	0.3420201	0.3490658	0.3419770
30	0.50000000	0.5235988	0.4996742
40	0.6427876	0.6981316	0.6414228

9.4. THIRD ORDER THEORY

If, in the formulae for reflection and refraction at spherical surfaces the first two terms of the series are replaced for values of sines of angles, the results obtained represent the **third order theory**. The formulae thus obtained give a fairly accurate account of the principal aberrations. In the third order theory, the aberration of a ray of light, viz., its deviations from the path obtained from Gauss formulae, is denoted by five sums called the **Seidel sums**. A lens will be free from all the aberrations, if all the five sums are equal to zero. But in practice, no optical system can be made to satisfy all the conditions at the same time. Let S_1, S_2 etc denote the five seidel sums. Then spherical aberration is absent if $S_1 = 0$; coma is absent if $S_1 = 0$ and $S_2 = 0$; astigmatism and curvature of the field are absent if $S_1 = 0$, $S_2 = 0$, $S_3 = 0$ and $S_4 = 0$. Finally if S_5 is also equal to zero the image of an axial object will be free from distortion as well. These five defects of an image are called the **monochromatic aberrations**.

9.5. SPHERICAL ABERRATION

A lens may be regarded as made up of a large number of prisms, of increasing angles from the centre to outward in case of a convex lens and of decreasing angles in case of concave lens. A ray of light falling on a prism of larger angle is deviated more towards the base of the prism than that falling on a prism of small angle. Therefore, peripheral light rays passing through a lens farther away from the axis are refracted more and come to focus closer to the

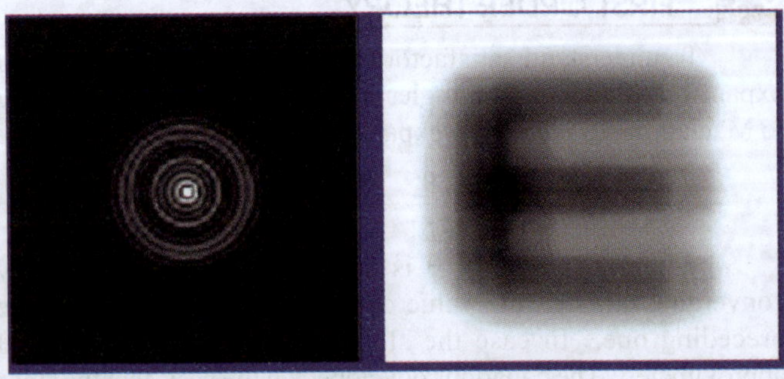

Spherical Aberration : Causes halos around points of light.

lens. Paraxial rays passing through the lens close to the axis are refracted less and come to focus farther from the lens. Therefore, rays passing through different zones of a lens surface come to different foci. An image formed by paraxial rays will be surrounded by a diffuse halo formed by peripheral rays and consequently the image is blurred. This phenomenon is known as ***spherical aberration.***

The presence of spherical aberration in the image formed by a single lens is illustrated in Fig. 9.2.O is a point object on the axis of the lens and I_p and I_m are the images formed by the paraxial

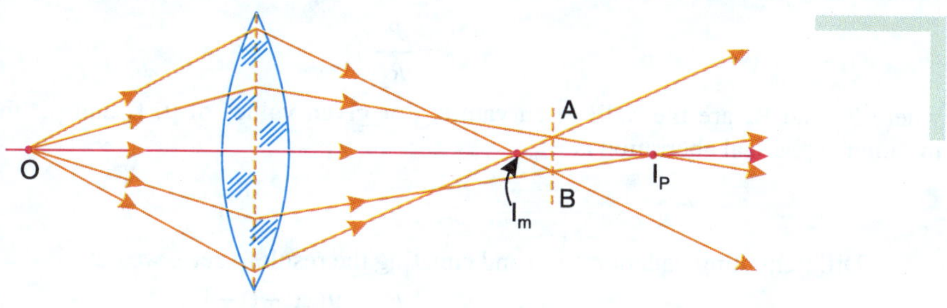

Fig. 9.2

and marginal rays respectively. It is clear from the figure that the paraxial rays of light form the image at a longer distance from the lens than the marginal rays. The image is not sharp at any point on the axis. However, if the screen is placed perpendicular to the axis at AB, the image appears to be a circular patch of diameter AB. At positions on the two sides of AB, the image patch has a larger diameter. This patch of diameter AB is called the **circle of least confusion**, which corresponds, to the position of the best image. The distance I_m I_p measures the **longitudinal spherical aberration**. The radius of the circle of least confusion measures the **lateral spherical aberration**. When the aperture of the lens is relatively large compared to the focal length of the lens, the cones of the rays of light refracted through the different zones of the lens surface are not brought to focus at the same point I_m and the axial rays come to focus at a farther point I_p. Thus, for an object point O on the axis, the image extends over the length I_m I_p. This effect is called *spherical aberration* and arises due to the fact that different annular zones have different focal lengths. The spherical aberration produced by a concave lens is illustrated in Fig. 9.3.

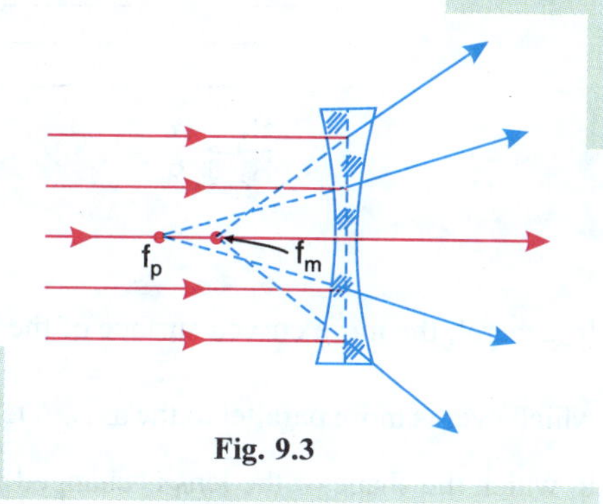

Fig. 9.3

The spherical aberration produced by a lens depends on the distance of the object point and varies approximately as the square of the distance of the object ray above the axis of the lens. The spherical aberration produced by a convex lens is positive and that produced by a concave lens is negative.

9.5.1. REDUCING SPHERICAL ABERRATION

Spherical aberration produced by lenses is minimized or eliminated by the following methods.

1. Spherical aberration can be minimized by using **stops**, which reduce the effective lens aperture. The stop used can be such as to permit either the axial rays of light or the marginal rays of light. However, as the amount of light passing through the lens is reduced, correspondingly the image appears less bright.

2. The longitudinal spherical aberration produced by a thin lens for a parallel incident beam is given by

$$x = \frac{\rho_2}{f_2}\left[\frac{k^2\mu^2 + k\left(\mu + 2\mu^2 + 2\mu^3\right) + \mu^3 - 2\mu^2 - 2}{2\mu(\mu-1)^2(1-k)^2}\right] \quad (9.3)$$

where x is the longitudinal spherical aberration, ρ is the radius of the lens aperture and f_2 is the second principal focal length.

$$k = \frac{R_1}{R_2}$$

where R_1 and R_2 are the radii of curvature. For given values of μ, f_2 and ρ, the condition for minimum spherical aberration is

$$\frac{dx}{dk} = 0.$$

Differentiating equation (9.3) and equating the result to zero, we get

$$k = \frac{R_1}{R_2} = \frac{\mu(2\mu-1)-4}{\mu(2\mu+1)} \quad (9.4)$$

From equation (9.4), for a lens whose material has a refractive index $\mu = 1.5$, $k = -\frac{1}{6}$. Thus, the lens, which produces minimum spherical aberration, is biconcave and the radius of curvature of the surface facing the incident light is one-sixth the radius of curvature of the other face.

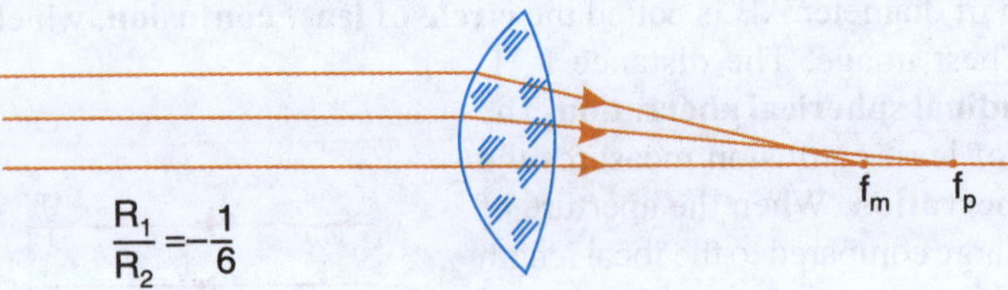

Fig. 9.4

In general, the more curved surface of the lens should face the incident or emergent beam of light which ever is more parallel to the axis. A lens whose $\left(\frac{R_1}{R_2} = -\frac{1}{6}\right)$ is called a **cross lens**. The process in which the shape of the lens is changed without changing the focal length of the lens is called bending of the lens for minimum spherical aberration. A crossed lens is shown in Fig. 9.4.

It is clear from the figure that the deviation produced by the two surfaces is the same and the axial and marginal rays of light come to focus with minimum of spherical aberration. However, it should be noted that the spherical aberration cannot be completely eliminated in a lens with spherical surfaces. For a lens of refractive index 1.5, focal length 100 cm and radius of the lens aperture 10 cm, the longitudinal spherical aberration is1.07 cm for $k = -1/6$. For the same values of focal length and radius of the lens aperture, if the values of μ and k are 2 and +1/5, the longitudinal spherical aberration reduces to 0.44cm.

Plano-convex lens

3. Plano-convex lenses are used in optical instruments so as to reduce the spherical aberration. When the curved surface of the lens faces the incident or emergent light whichever is more parallel to the axis, the spherical aberration is minimum. The spherical aberration in a crossed lens $\frac{R_1}{R_2} = -\frac{1}{6}$ is only 8% less than that of a plano-convex lens having the same focal length and radius of the lens aperture. This is the reason why plano-convex lenses are generally used in place of crossed lenses without increasing the spherical aberration appreciably. Fig. 9.5 represents the variation of longitudinal spherical aberration with the radius of the lens aperture for lenses of the same focal length and refractive index.

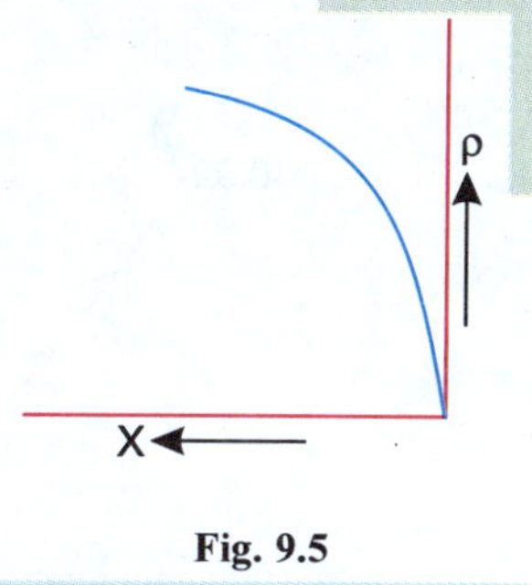

Fig. 9.5

The spherical aberration will, however, be very large if the plane surface faces the incident light. The spherical aberration is a result of larger deviation of the marginal rays than the paraxial rays. If the deviation of the marginal rays of light is made minimum, the focus f_m for a parallel incident beam will shift towards f_p, the focus for the paraxial rays of light and the spherical aberration will be minimum.

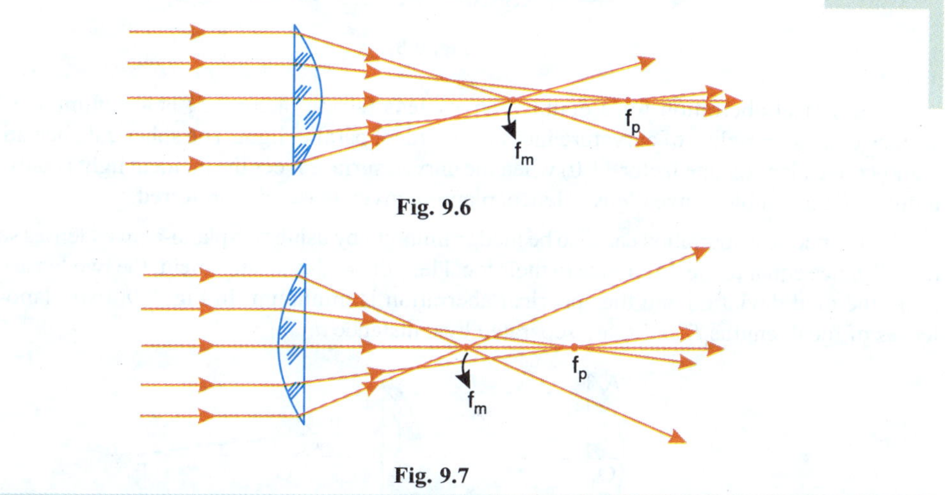

Fig. 9.6

Fig. 9.7

As the deviation is a minimum in a prism, when the angles of incidence and emergence are equal, similarly in a lens also, spherical aberration can be minimized if the total deviation produced by a lens is equally shared by the two surfaces. In a plano-convex lens, when the plane surface faces the parallel beam of light, the deviation is produced only at the curved surface and hence the longitudinal spherical aberration (Fig. 9.6) is more than when the curved surface faces the incident light (Fig. 9.7). In the latter case the spherical aberration is less than the former, because the total deviation in the second case is divided between the two surfaces. Thus, the spherical aberration produced by a single lens can be minimized by choosing proper radii of curvature. The shape factor q of lens is given by

$$q = \frac{R_1 + R_2}{R_2 - R_1} \tag{9.5}$$

Fig. 9.8 represents lenses of the same values of f, μ and ρ but of different shape factors calculated according to equation (9.5). The longitudinal spherical aberration is represented along the y-axis and the shape factor q along the x-axis.

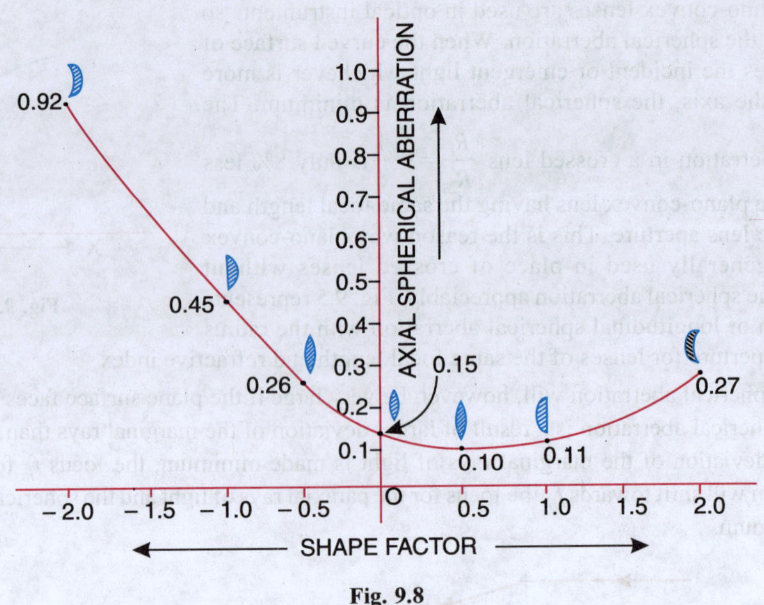

Fig. 9.8

Spherical aberration for a double convex lens (shape factor 0.5) is a minimum when the surface of smaller radius of curvature faces the incident parallel light. The spherical aberration for a plano-convex lens (shape factor +1.0) when the curved surface faces the incident light is only slightly more than the double convex lens. Hence, plano- convex lenses are preferred.

4. Spherical aberration can also be made minimum by using two plano-convex lenses separated by a distance equal to the difference in their focal length. In this arrangement, the two lenses equally share the total deviation and the spherical aberration is minimum. In Fig. 9.9, two plano-convex lenses of focal lengths f_1 and f_2 are separated by a distance d.

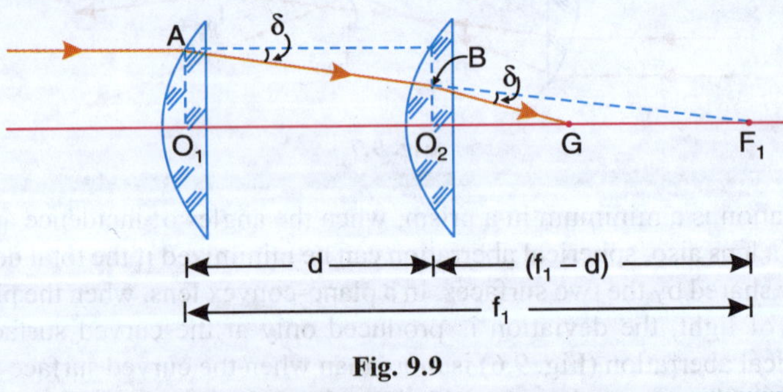

Fig. 9.9

Let δ be the angle of deviation produced by each lens (see Fig. 9.9).

$$\angle BF_1G = \delta,\ \angle F_1BG = \delta$$

and from $\Delta^{le}BGF_1$, $BG = GF_1$ or $O_2G = GF_1$ (approximately)

$$\therefore \qquad O_2\ G = \tfrac{1}{2}(O_2F_1) = \tfrac{1}{2}(f_1 - d)$$

For the second lens, F_1 is the virtual object and G is the real image. Substituting these values of object and image distances in the formula

$$\frac{1}{\upsilon}-\frac{1}{u}=\frac{1}{f_2} \qquad (u, \upsilon, f_2 \text{ are all } +ve)$$

$$\frac{1}{O_2G}-\frac{1}{O_2F_1}=\frac{1}{f_2}$$

$$\frac{2}{(f_1-d)}-\frac{1}{(f_1-d)}=\frac{1}{f_2}$$

$$\frac{1}{(f_1-d)}=\frac{1}{f_2}$$

$$f_2 = f_1 - d \quad \text{or} \quad d = f_1 - f_2 \tag{9.6}$$

Thus, **the condition for minimum spherical aberration is that the distance between the two lenses is equal to the difference in their focal lengths.**

5. Spherical aberration for a convex lens is +ve and that for a concave lens is –ve.By a suitable combination of convex and concave lenses spherical aberration can be made minimum.

6. Spherical aberration may be minimized by using axial-GRIN lenses (See § 9.32).

9.6. COMA

The effect of rays from an object point not situated on the axis of the lens results in an aberration called **coma**. Comatic aberration is similar to spherical aberration in that both are due to the failure of the lens to bring all rays from a point object to focus at the same point. Spherical aberration refers to object points situated on the axis whereas comatic aberration refers to object points situated off the axis. In the case of spherical aberration, the image is a circle of varying diameter along the axis and in the case of comatic aberration the image is comet-shaped and hence the name coma.

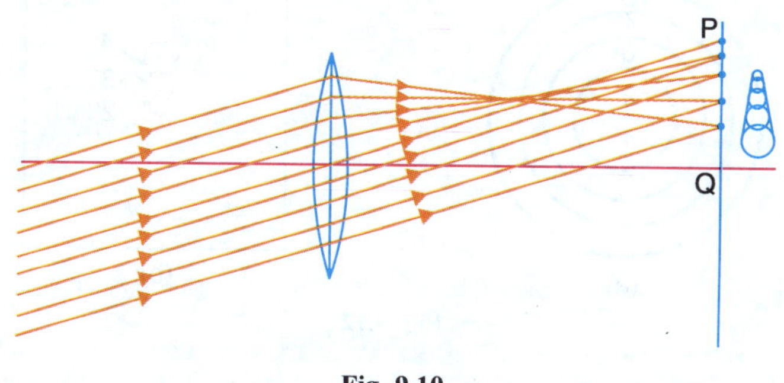

Fig. 9.10

Fig. 9.10 illustrates the effect of coma. The resultant image of a distant point off the axis is shown on the right side of the figure. The rays of light in the tangential plane are represented in the figure.

Fig. 9.11 illustrates the presence of coma in the image due to a point object, O, situated of the axis of the lens. Rays of light getting refracted through the centre of the lens (ray1) meet the screen XY at the point P. Rays 2,2; 3,3 etc getting refracted through the outer zones of the lens come to focus at points Q,R,S, etc, nearer the lens and overlapping circular patches of gradually increasing diameter are formed on the screen. The resultant image of the point is comet-shaped as indicated on the right side of the figure.

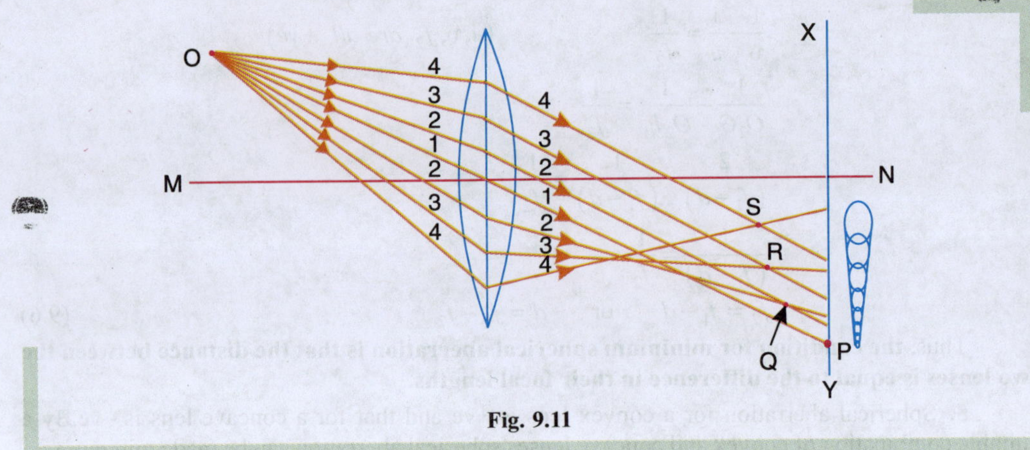

Fig. 9.11

Let 1, 2, 3 etc be the various zones of the lens (Fig. 9.12a). Rays of light getting refracted through these different zones give rise to circular patches of light 1′, 2′, 3′, etc. The screen is placed perpendicular to the axis of the lens and at the position where the central rays come to focus [Fig. 9.12 (b)]. Like spherical aberration, comatic aberration produced by a single lens can also be corrected by properly choosing the radii of curvature of the lens surface. Coma can be altogether eliminated for a given pair of object and image points whereas spherical aberration cannot be completely corrected. Further, a lens corrected for coma will not be free from spherical aberration and the one corrected for spherical aberration will not be free from coma. Use of a stop or a diaphragm at the proper position eliminates coma.

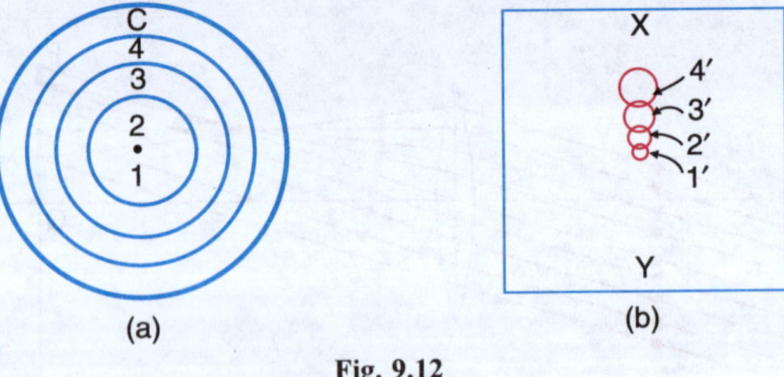

Fig. 9.12

Coma is the result of varying magnification for rays refracted through different zones of the lens. For example, in Fig. 9.11, rays of light getting refracted through the outer zones come to focus at points nearer the lens. Hence the magnification of the image due to the outer zones is larger than the inner zones and in this case coma is said to be positive. On the other hand if the magnification produced in an image due to the outer zones is smaller, coma is said to be negative.

According to Abbe, a German optician, coma can be eliminated if a lens satisfies the Abbe's sine condition viz.

$$\mu_1 y_1 \sin\theta_1 = \mu_2 y_2 \sin\theta_2 \qquad (9.7)$$

where μ_1, y_1 and θ_1 refer to the refractive index, height of the object above the axis and the slope angle of the incident ray of light respectively. Similarly μ_2, y_2 and θ_2 refer to the corresponding quantities in the image space. The magnification of the image is given by $\frac{y_2}{y_1}$.

$$\frac{y_2}{y_1} = \frac{\mu_1 \sin \theta_1}{\mu_2 \sin \theta_2} \tag{9.8}$$

Elimination of coma is possible if the lateral magnification y_2 / y_1 is the same for all rays of light, irrespective of the slope angles θ_1 and θ_2. Thus, coma can be eliminated if, $\sin \theta_1 / \sin \theta_2$ is a constant because μ_1/μ_2 is constant. A lens that satisfies the above condition is called an *aplanatic lens*.

9.6.1. APLANATIC POINTS

A spherical lens, which is free from the defects of spherical aberration and coma is called an **aplanatic lens**. A pair of conjugate points free from spherical aberration and coma is called aplanatic points. Fig. 9.13 illustrates the property of an aplantic lens. Let O be the center of curvature of the lens of refractive index μ and radius of curvature R. P is a point on the axis of the lens such that PO = R/μ. It can be shown that all rays passing through the point P appear to diverge through the point Q irrespective of the slope angle made by the incident rays. PA is the incident ray and AC is the refracted ray. The ray AC appears to diverge from the point Q which is the image of P. Let *i* and *r* be the angles of incidence and refraction and α and β the slope angles made by the incident and refracted rays.

Aplanatic lens.

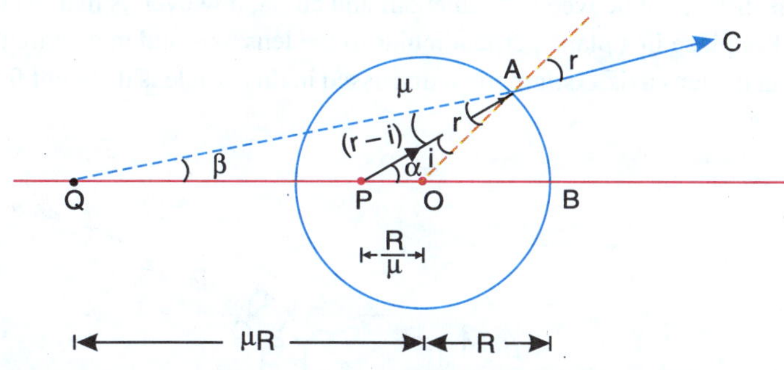

Fig. 9.13

Then,
$$\frac{\sin i}{\sin r} = \frac{1}{\mu} \tag{9.9}$$

In the Δ *APO*,
$$\frac{\sin i}{\sin \alpha} = \frac{PO}{R} = \frac{R}{\mu R} = \frac{1}{\mu} \tag{9.10}$$

From (9.9) and (9.10)
$$\frac{\sin i}{\sin r} = \frac{\sin i}{\sin \alpha}$$

∴
$$\angle \alpha = \angle r \tag{9.11}$$

In the Δ *APQ*,
$$\angle \alpha = \beta + (r - i) \tag{9.12}$$

Substituting $\angle \alpha = \angle r$ in equation (9.12), we get
$$r = \beta + r - i \text{ or } i = \beta$$

In the Δ *AOQ*

$$\frac{\sin r}{\sin \beta} = \frac{\sin r}{\sin i} = \mu = \frac{OQ}{OA} = \frac{OQ}{R}$$

$$\therefore \qquad OQ = \mu.R$$

Thus, if the distance of the object point P is R /μ from the centre of curvature, then distance of the image point Q is μR irrespective of the slope angles α and β. The object and image distances of the conjugate points that satisfy the Abbe condition (9.8) can be seen from the Fig. 9.5 as

$$BP = R + \frac{R}{\mu} \quad \text{and} \quad BQ = R + \mu R. \qquad (9.13)$$

The object and image points which are at distances given by equ. (9.13) from the pole of the spherical surface are free from spherical aberration as well as coma. Such points are known as *aplanatic points.*

An aplanatic lens is mostly used as the front lens of a high power microscope objective called the oil immersion objective(§ 9.14). As it is not possible to place an object inside a solid spherical lens, the lens is ground a little and the object to be examined is embedded in between a drop of oil and the lens surface. The oil chosen is such that it has the same refractive index as that of the lens.

9.7. ASTIGMATISM

Astigmatism, similar to coma, is the aberration in the image formed by a lens, of object points off the axis. The difference between astigmatism and coma, however, is that in coma the spreading of the image takes place in a plane perpendicular to the lens axis and in astigmatism the spreading takes place along the lens axis. Astigmatism discussed in this article is different from the one treated in defective vision.

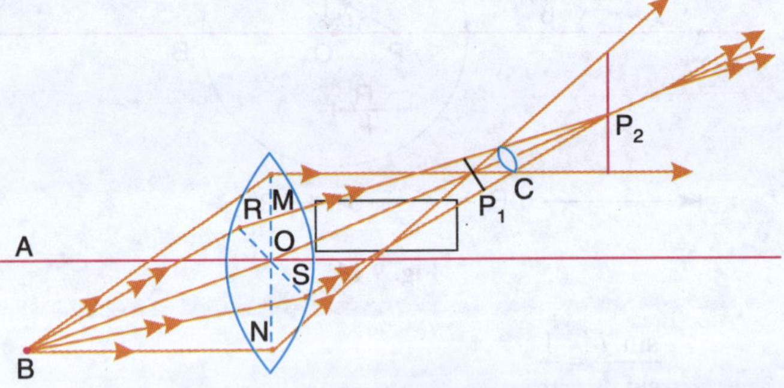

Fig. 9.14

Fig. 9.14 illustrates the defect of astigmatism in the image of a point B situated off the axis. Two portions of the cone of rays of light diverging from the point B are taken. The cone of the rays of light refracted through the tangential (vertical) plane BMN comes to focus at point P_1 nearer the lens and the cone of rays refracted through the sagittal (horizontal) plane BRS comes to focus at the point P_2 away from the lens. All rays pass through a horizontal line passing through P_1 called the primary image and also through a vertical line passing through P_2

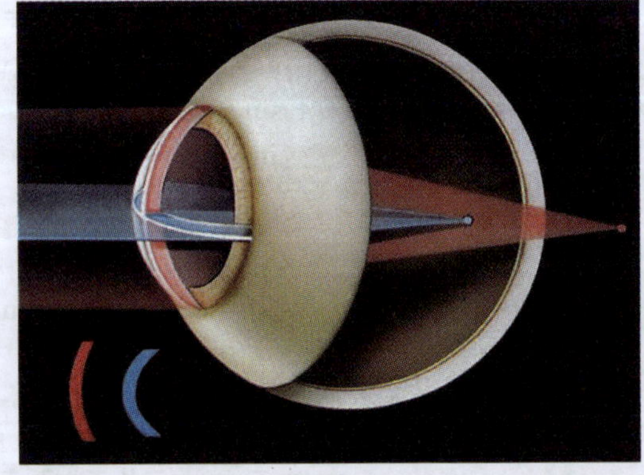

Astigmatism.

called the secondary image. The refracted beam has an elliptical cross-section, which ends to a horizontal line at P_1 and a vertical line at P_2. The cross section of the refracted beam is circular at some point between the primary and the secondary images and this is called the *circle of least confusion*. If a screen is held perpendicular to the refracted beam between the points P_1 and P_2, the shape of the image at different positions is as shown in Fig. 9.15.

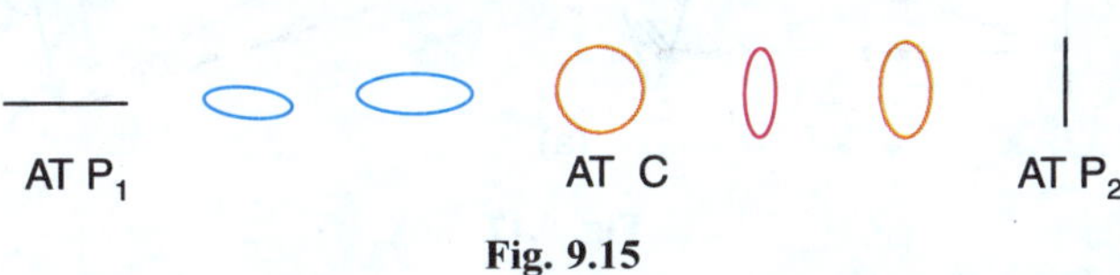

Fig. 9.15

The locus of the primary images of all points in the object plane gives the surface of revolution about the lens axis and is called the *primary image surface*. The locus of the secondary images gives the *secondary image surface*. The surface of best focus is given by the locus of the circles or least confusion. The primary and the secondary image surfaces and the surface of best focus are illustrated in Fig.9.16. P_1 and P_2 are the images of the object point B. TPN and SPR are the first and the second image surfaces and KPL is the surface of best focus. The three surfaces touch at the point P on the axis. Generally, the surface of best focus is not plane but curved as shown. This defect is called the *curvature of the field*. The shape of the image surfaces depends on the shape of the lens and the position of the stops. If the primary image surface is to the left of the secondary image surface, astigmatism is said to be positive, otherwise negative. By using a convex and a concave lens of suitable focal lengths and separated by a distance, it is possible to minimize the astigmatic difference and such a lens combination is called an **anastigmat**.

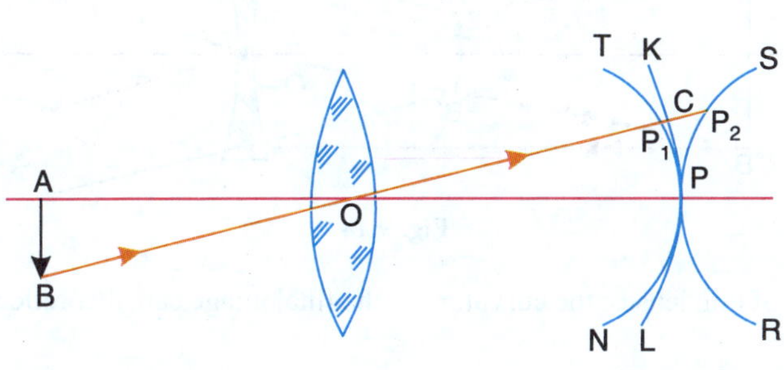

Fig. 9.16

9.8. CURVATURE OF THE FIELD

The image of an extended object due to a single lens is not a flat one but it will be a curved surface. The central portion of the image nearer the axis is in focus but the outer regions of the image away from the axis are blurred. This defect is called the **curvature of the field**.

This defect is due to the fact that the paraxial focal length is greater than the marginal focal length. This aberration is present even if the aperture of the lens is reduced by a suitable stop, usually employed to reduce spherical aberration, coma and astigmatism. Fig. 9.17 illustrates the presence of curvature of the field in the image formed by a convex lens. A real image formed by a convex lens curves towards the lens [Fig. 9.17(a)] and a virtual image curves away from the lens [Fig. 9.17(b)]. Fig. 9.18 represents the curvature of the field present in the image formed by a curvature of the field present in the image formed by a concave lens.

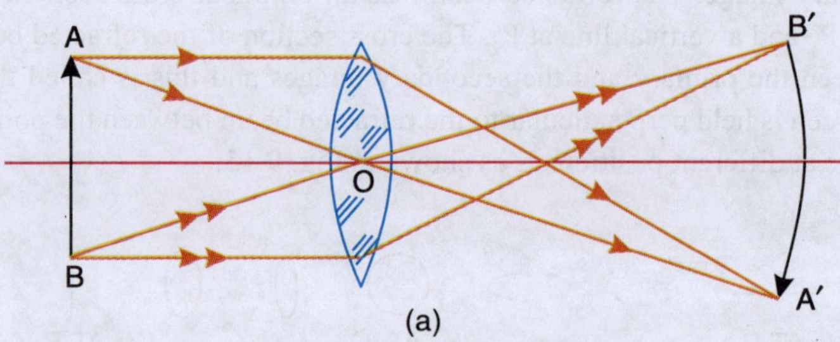

(a)

Fig. 9.17

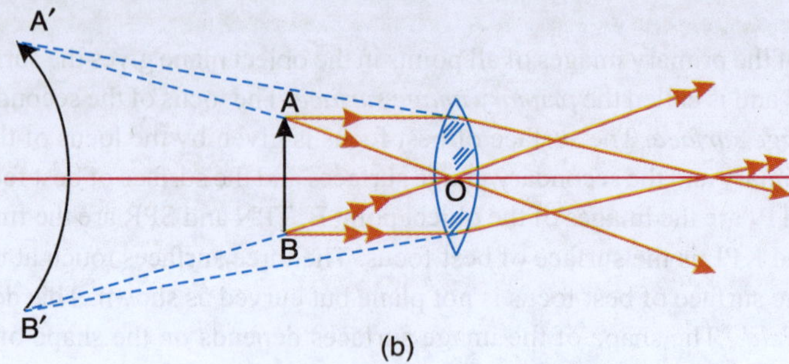

(b)

Fig. 9.17

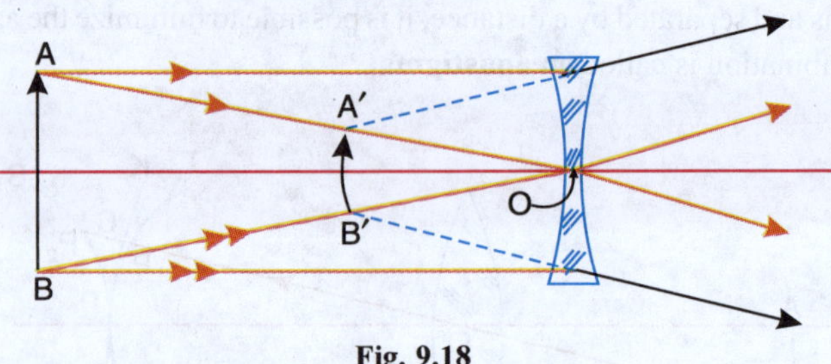

Fig. 9.18

For a system of thin lenses, the curvature of the final image can, theoretically, be given by the expression

$$\frac{1}{R} = \sum \frac{1}{\mu_n f_n} \tag{9.14}$$

where R is the radius of curvature of the final image, μ_n and f_n are the refractive index and focal length of the nth lens. For the image to be flat, R must be infinity.

$$\therefore \quad \frac{1}{R} = \sum \frac{1}{\mu_n f_n} = \frac{1}{\infty} = 0$$

Correspondingly, the condition for two lenses placed in air, reduces to

$$\frac{1}{\mu_1 f_1} + \frac{1}{\mu_2 f_2} = 0 \tag{9.15}$$

This is known as Petzwal's condition for no curvature. This condition holds good whether the lenses are seperated by a distance or placed in contact. As the refractive indices μ_1 and μ_2 are positive, the above condition will be satisfied if the lenses are of opposite sign. If one of the lenses is convex, the other must be concave.

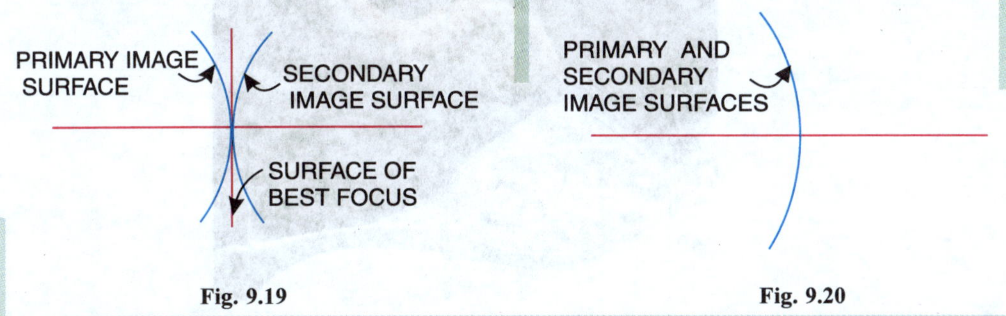

Fig. 9.19

Fig. 9.20

Astigmatism and coma are completely eliminated if the primary and secondary image surfaces are coincident and plane. In this case, the surface of best focus will also be a plane one. But this cannot be achieved with a single lens. Astigmatism or curvature of the field can be minimized by introducing suitable stops on the lens axis. If the primary and the secondary image surfaces are made to have equal and opposite curvatures (Fig. 9.19), the surface of best focus will be plane and midway between them. Astigmatism will, however, be present. Astigmatism can be eliminated by having the same curvature for the primary and secondary image surfaces. In this case curvature of the field will be present (Fig. 9.20). Correction for coma is more important than astigmatism for object points having comparatively small angular distances from the axis. Hence, telescope objectives, whose field of view is small, are corrected for coma rather than for astigmatism. On the other hand a camera lens of wide field has to be necessarily corrected for astigmatism.

9.9. DISTORTION

The failure of a lens to form a point image due to a point object is due to the presence of spherical aberration, coma and astigmatism. The variation in the magnification produced by a lens

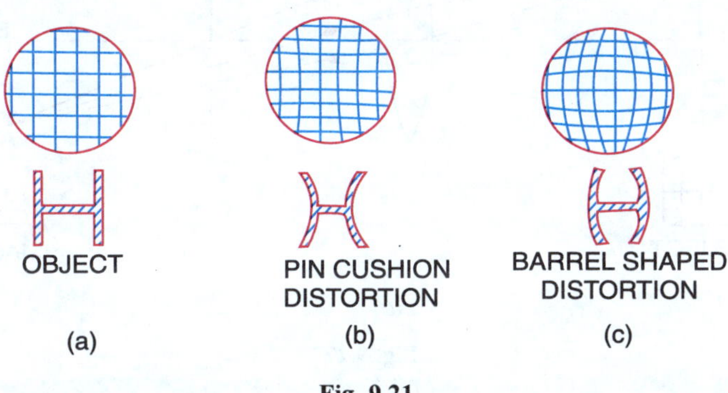

Fig. 9.21

for different axial distances results in an aberration called **distortion**. This aberration is not due to the lack of sharpness in the '**image**'. Distortion is of two types viz., (a) pin–cushion distortion and (b) barrel-shaped distortion. In pin-cushion distortion, the magnification increases with increasing axial distance and the image of an object [Fig. 9.21(a)] appears as shown in Fig.9.21(b). On the other hand, if the magnification decreases with increasing axial distance, it results in barrel-shaped distortion and the image appears as shown in Fig. 9.21(c).

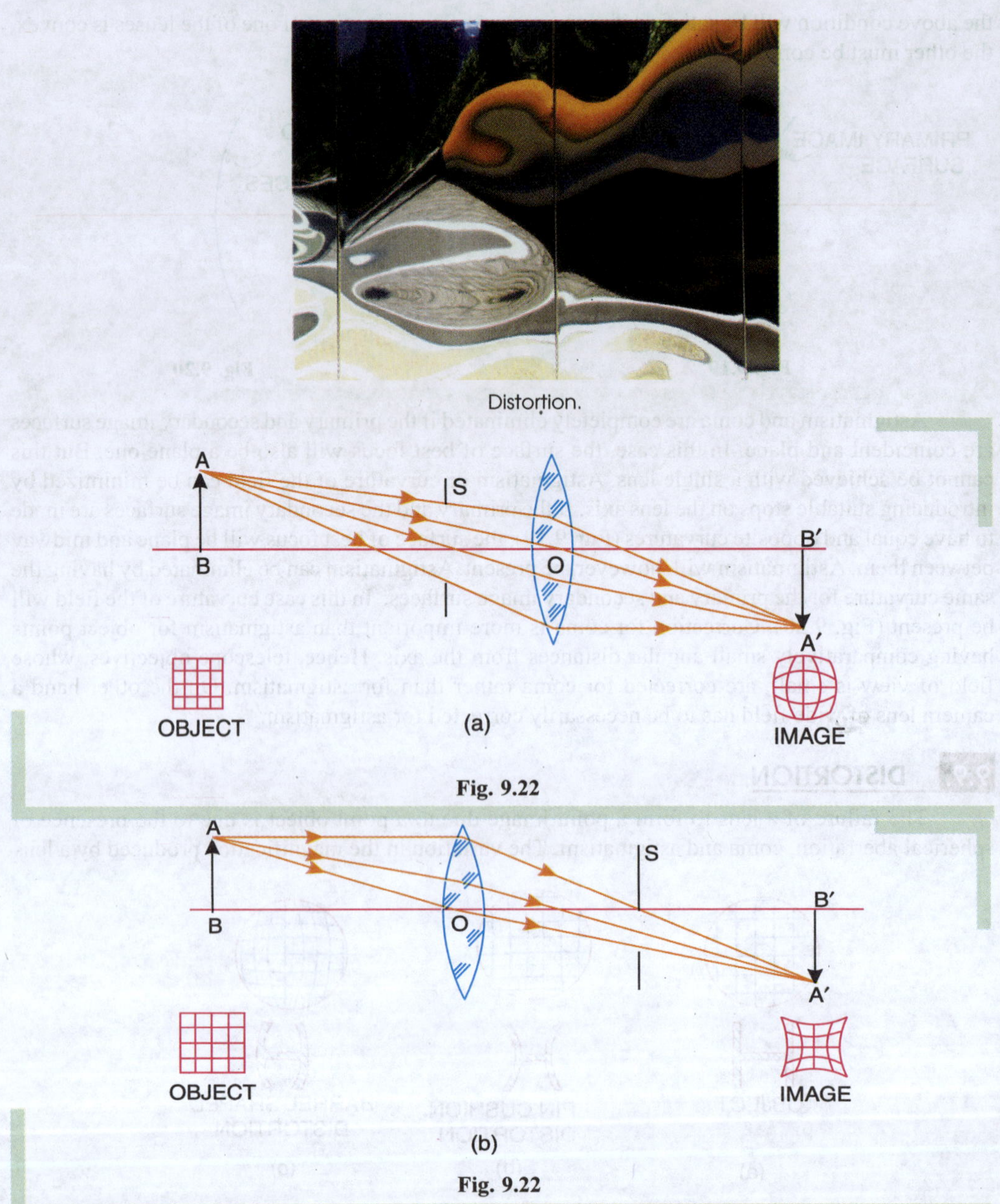

Distortion.

(a)

Fig. 9.22

(b)

Fig. 9.22

In the case of optical instruments intended mainly for visual observation, a little amount of distortion may be present but it must be completely eliminated in a photographic camera lens, where the magnification of the various regions of the object must be the same. In the absence of stops, which limit the cone of rays or light striking the lens, a single lens is free from distortion. But, if stops are used, the resulting image is distorted. If a stop is placed before the lens the distortion is barrel-shaped [Fig. 9.22 (a)] and if a stop is placed after the lens, the distortion is pin-cushion type [Fig. 9.22 (b)]. To eliminate distortion a stop is placed in between two symmetrical lenses, so that the pin-cushion distortion produced by the first lens is compensated by the barrel-shaped distortion produced by the second lens [Fig. 9.22 (c)]. Projection and camera-lenses are constructed in this way.

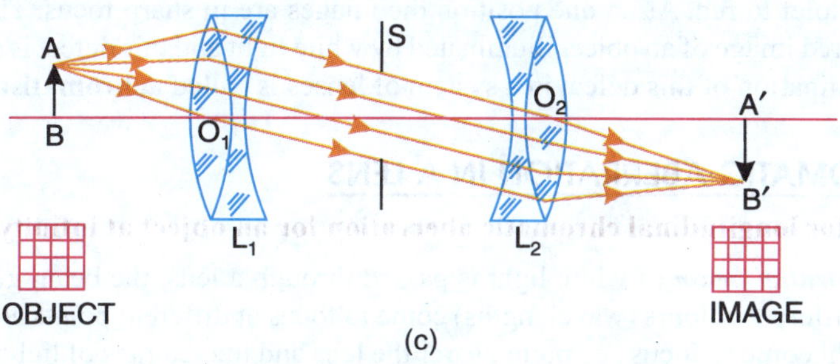

Fig. 9.22

9.10. CHROMATIC ABERRATION

The refractive index of the material of a lens is different for different wavelengths of light. Hence the focal length of a lens is different for different wavelengths. Further, as the magnification of the image is dependent on the focal length of a lens, the size of the image is different for different wavelengths (colours). The variation of the image distance from the lens with refractive index measures **axial or longitudinal chromatic aberration** and the variation in the size of the image measures **lateral chromatic aberration**. Fig. 9.23 illustrates chromatic aberration present in an image formed by a single lens L. AB is an object placed in front of the lens. A′B′ and A″B″ are the violet and the red images. The violet image is formed nearer the lens than the red image. The monochromatic aberrations are assumed to be absent in this case. The distance x measures the axial or longitudinal

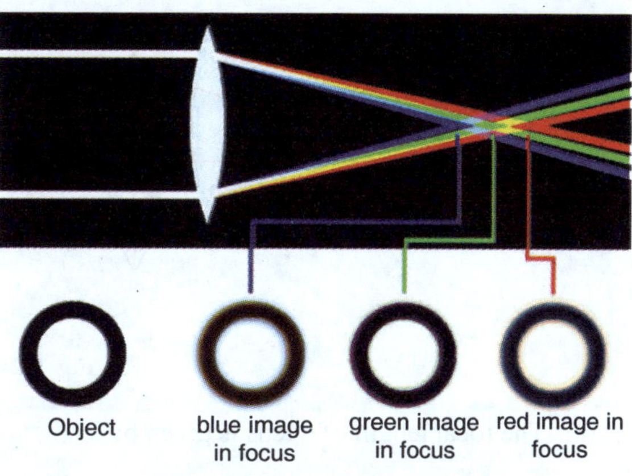

Axial chromatic aberration.

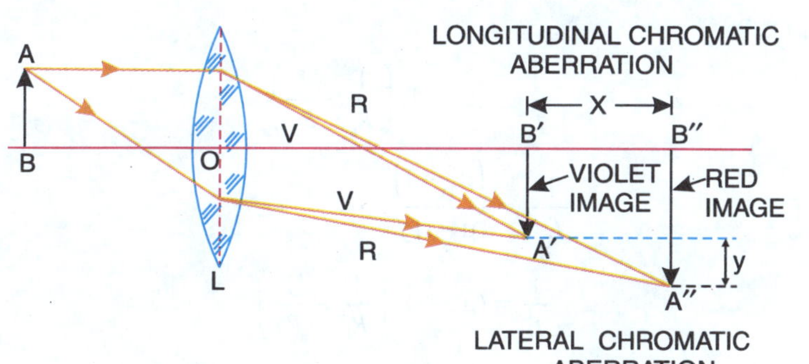

Fig. 9.23

chromatic aberration and the distance y measures the lateral chromatic aberration. The images of intermediate colours between violet and red lie in between the images A′B′ and A″B″ and their size increases from violet to red. At no one position the images are in sharp focus. Thus, a single lens produces a coloured image of an object illuminated by white light and this defect is called **chromatic aberration.** Elimination of this defect in a system of lenses is called **achromatism.**

9.11. CHROMATIC ABERRATION IN A LENS

A. Expression for longitudinal chromatic aberration for an object at infinity:

When a ***parallel beam*** of white light is passed through a lens, the beam gets dispersed and rays of light of different colours (wavelengths) come to focus at different points along the axis. The violet rays of light come to focus at a point nearer the lens and the red rays of light at a farther point (Fig. 9.24). f_υ is the focus for the violet rays and f_r is the focus for the red rays. The colours in between violet and red come to focus between f_υ and f_r. The distance $(f_r - f_\upsilon) = x$ is called the **longitudinal** or **axial chromatic aberration**.

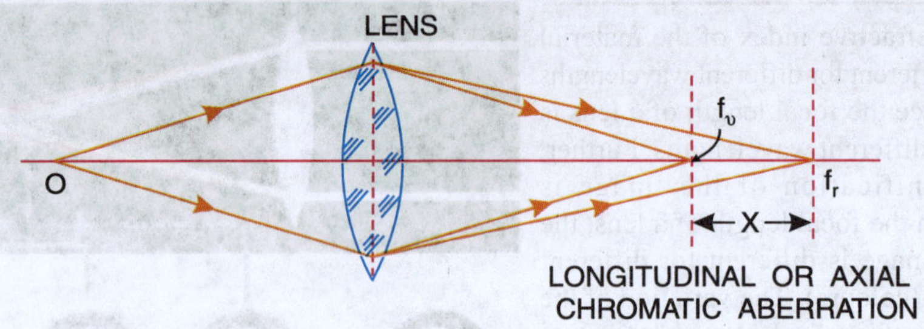

Fig. 9.24

The focal length of a lens is given by

$$\frac{1}{f} = (\mu - 1)\left(\frac{1}{R_1} - \frac{1}{R_2}\right) \qquad (9.16)$$

or

$$\left(\frac{1}{R_1} - \frac{1}{R_2}\right) = \frac{1}{f(\mu - 1)} \qquad (9.17)$$

Similarly

$$\frac{1}{f_\upsilon} = (\mu_\upsilon - 1)\left(\frac{1}{R_1} - \frac{1}{R_2}\right) \qquad (9.18)$$

$$\frac{1}{f_\upsilon} = \frac{(\mu_\upsilon - 1)}{(\mu - 1)f} \qquad (9.19)$$

and

$$\frac{1}{f_r} = (\mu_r - 1)\left(\frac{1}{R_1} - \frac{1}{R_2}\right) \qquad (9.20)$$

$$\frac{1}{f_r} = \frac{(\mu_r - 1)}{(\mu - 1)f} \qquad (9.21)$$

From equations (9.18) and (9.20), we get

$$\frac{1}{f_\upsilon}-\frac{1}{f_r}=\frac{1}{(\mu-1)f}(\mu_\upsilon-1-\mu_r+1)$$

$$\frac{f_r-f_\upsilon}{f_\upsilon f_r}=\frac{(\mu_\upsilon-\mu_r)}{(\mu-1)f}$$

Taking $f_\upsilon f_r=f^2$ (where f is the mean focal length), we can write

$$\frac{f_r-f_\upsilon}{f^2}=\frac{\mu_\upsilon-\mu_r}{(\mu-1)f}$$

or
$$f_r-f_\upsilon=\frac{(\mu_\upsilon-\mu_r)f}{(\mu-1)}$$

$\therefore$
$$f_r-f_\upsilon=\omega.f \tag{9.22}$$

where $\omega=\left[\frac{\mu_\upsilon-\mu_r}{\mu-1}\right]$ is known as the dispersive power of the material.

Thus, **the axial chromatic aberration for a thin lens for an object at infinity is equal to the product of the dispersive power of the material of the lens and f is the mean focal length.** Further, it is clear, from the above, that *a single lens cannot form an image free from chromatic aberration.*

B. Expression for longitudinal chromatic aberration for an object at finite distance:

Let us consider a point object illuminated by white light and situated on the axis of the lens. Coloured images are formed along the axis. The violet image is nearest the lens and the red image is the farthest. In between these two images (Fig. 9.25), if a screen is placed at the position XY, the image of least chromatic aberration is formed.

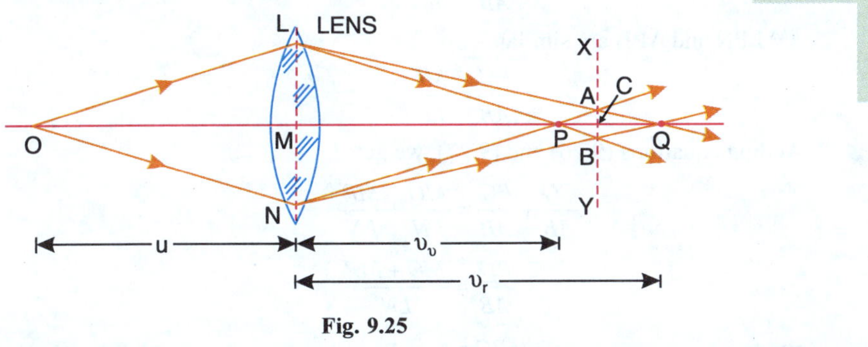

Fig. 9.25

Let u be the distance of the object point and υ_υ and υ_r the distances of the violet and red images on the axis of the lens. If f_υ and f_r are the focal lengths for the violet and red rays of light, then,

$$\frac{1}{\upsilon_\upsilon}-\frac{1}{u}=\frac{1}{f_\upsilon} \tag{9.23}$$

and
$$\frac{1}{\upsilon_r}-\frac{1}{u}=\frac{1}{f_r} \tag{9.24}$$

Subtracting (9.24) from (9.23), we get

$$\frac{1}{\upsilon_\upsilon}-\frac{1}{\upsilon_r}=\frac{1}{f_\upsilon}-\frac{1}{f_r}$$

$$\frac{\upsilon_r-\upsilon_\upsilon}{\upsilon_\upsilon\ \upsilon_r}=\frac{f_r-f_\upsilon}{f_\upsilon\ .f_r}$$

Taking $\upsilon_\upsilon \upsilon_r = \upsilon^2$ and $f_\upsilon . f_r = f^2$

$$\frac{\upsilon_r - \upsilon_\upsilon}{\upsilon^2} = \frac{f_r - f_\upsilon}{f^2}.$$

But $f_r - f_\upsilon = \omega . f$ (see equ. 9.22)

$$\frac{\upsilon_r - \upsilon_\upsilon}{\upsilon^2} = \frac{\omega . f}{f^2} = \frac{\omega}{f}$$

or
$$\upsilon_r - \upsilon_\upsilon = \frac{\omega . \upsilon^2}{f} \tag{9.25}$$

It is clearly seen in this case that *the longitudinal chromatic aberration depends on the distance of the image and hence on the distance of the object from the lens, in addition to its dependence on the dispersive power and focal length of the lens.*

9.12. CIRCLE OF LEAST CHROMATIC ABERRATION

Sometimes it becomes necessary to know the diameter of the circle of least chromatic aberration. It can be calculated as follows. Let O be the source of light on the axis (Fig.9.25). P is the image for violet and Q that for red rays. AB is the diameter of the circle of least chromatic aberration,d. Let LN be the aperture of the lens, D and υ be the mean distance of the image.

Δ^{les} LQN and AQB are similar.

$$\frac{CQ}{AB} = \frac{MQ}{LN} \tag{9.26}$$

Δ^{les} LPN and APB are similar.

$$\frac{PC}{AB} = \frac{MP}{LN} \tag{9.27}$$

Adding equations (9.26) and (9.27), we get

$$\frac{CQ}{AB} + \frac{PC}{AB} = \frac{MQ}{LN} + \frac{MP}{LN}$$

∴
$$\frac{PQ}{AB} = \frac{MQ + MP}{LN} \tag{9.28}$$

But
$$PQ = \upsilon_r - \upsilon_\upsilon$$

And
$$MQ + MP = \upsilon_r + \upsilon_\upsilon = 2\upsilon \text{ approximately}$$

And substituting these values in equation (9.28), we get

$$\frac{\upsilon_r - \upsilon_\upsilon}{d} = \frac{2\upsilon}{D} \quad (\because AB = d \text{ and } LN = D)$$

∴
$$d = D \cdot \frac{\upsilon_r - \upsilon_\upsilon}{2\upsilon}$$

But, from equation (9.25), we have $\upsilon_r - \upsilon_\upsilon = \dfrac{\omega . \upsilon^2}{f}$

∴
$$d = D . \frac{\omega . \upsilon^2}{f . 2\upsilon} = \frac{1}{2} D\omega \cdot \frac{\upsilon}{f} \tag{9.29}$$

If incident light is a parallel beam, then

$$\upsilon = f \text{ and } d = \frac{1}{2}D\omega.\frac{f}{f} = \frac{1}{2}D\omega \tag{9.30}$$

In this case, *diameter of the circle of least chromatic aberration depends on the diameter of the lens aperture and the dispersive power of the material, but is independent of the focal length of the lens.*

9.13. ACHROMATIC LENSES

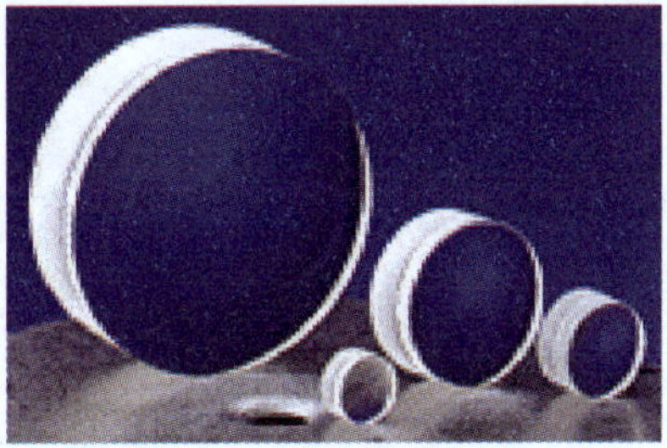
Achromatic lenses.

Using a suitable combination of two or more lenses of different materials, the chromatic aberration can be minimized. A combination of two lenses is usually called an "achromatic combination, an achromat or an achromatic doublet". In practice, it is possible to correct the chromatic aberration for only two colours. The achromatism would be ideal if the images of different colours are formed in the same position and are of the same size, thus the longitudinal and lateral chromatic aberrations being totally eliminated. With a convex lens, the violet rays of light come to focus at a point nearer the lens and with a concave lens and a converging incident beam of light the red rays of light meet the axis at a point nearer the lens. Hence, to have an achromatic combination of two lenses, one of the lenses should be convex (crown) and the other concave (flint). Therefore, an achromatic doublet is made using two thin lenses, a convex lens of crown glass of smaller focal length and a concave lens of flint glass of larger focal length.

1. Condition for achromatism of two lenses placed in contact:

An achromat is made by placing in contact, lenses of different materials and of suitable focal lengths, such that the focal length of the combination is the same for both the extreme colours. It means that the focal length of the achromat is independent of the refractive index.

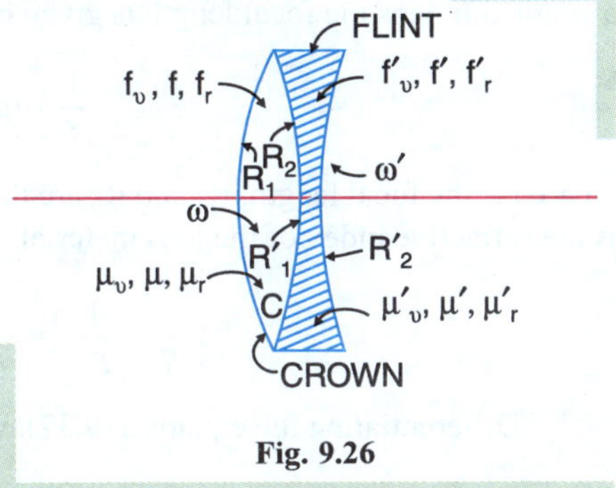

Fig. 9.26

Let the lenses of the doublet have mean focal lengths f and f' and mean refractive indices of their materials μ and μ' respectively. For the first lens for violet rays, we have

$$\frac{1}{f_\upsilon} = (\mu_\upsilon - 1)\left(\frac{1}{R_1} - \frac{1}{R_2}\right) = \left(\frac{\mu_\upsilon - 1}{\mu - 1}\right)\cdot\frac{1}{f} \tag{9.31}$$

Similarly for red rays, we write

$$\frac{1}{f_r} = \left(\frac{\mu_r - 1}{\mu - 1}\right)\cdot\frac{1}{f} \tag{9.32}$$

For the second lens of mean focal length f' and refractive index μ', we can write

$$\frac{1}{f'_\upsilon} = \left(\frac{\mu'_\upsilon - 1}{\mu' - 1}\right)\cdot\frac{1}{f'} \tag{9.33}$$

and

$$\frac{1}{f'_r} = \left(\frac{\mu'_r - 1}{\mu' - 1}\right)\cdot\frac{1}{f'} \tag{9.34}$$

Let F_υ and F_r be the focal lengths of the combination of the doublet for violet and red rays of light. Then

$$\frac{1}{F_\upsilon}=\frac{1}{f_\upsilon}+\frac{1}{f'_\upsilon}=\frac{(\mu_\upsilon-1)}{(\mu-1)f}+\frac{(\mu'_\upsilon-1)}{(\mu'-1)f'}$$

$$\frac{1}{F_r}=\frac{1}{f_r}+\frac{1}{f'_r}=\frac{(\mu_r-1)}{(\mu-1)f}+\frac{(\mu'_r-1)}{(\mu'-1)f'}$$

And for the combination to be achromatic, the focal lengths F_υ and F_r must be equal.

$$F_\upsilon=F_r \quad \text{or} \quad \frac{1}{F_\upsilon}=\frac{1}{F_r}$$

$$\therefore \quad \frac{(\mu_\upsilon-1)}{(\mu-1)f}+\frac{(\mu'_\upsilon-1)}{(\mu'-1)f'}=\frac{(\mu_r-1)}{(\mu-1)f}+\frac{(\mu'_r-1)}{(\mu'-1)f'}$$

$$\text{or} \quad \frac{(\mu_\upsilon-\mu_r)}{(\mu-1)f}+\frac{\mu'_\upsilon-\mu'_r}{(\mu'-1)f'}=0$$

$$\frac{\omega}{f}+\frac{\omega'}{f'}=0 \tag{9.35}$$

$$\frac{\omega}{f}=-\frac{\omega'}{f'}$$

$$\therefore \quad f'=-f.\frac{\omega'}{\omega} \tag{9.36}$$

Since, ω and ω' are positive quantities, f' is negative if f is positive i.e, if the crown glass lens is convex, then the flint glass lens is concave. The ratio of the dispersive powers of the materials of the lenses is equal to the ratio of the focal length of the two lenses.

Alternative Method:

The above condition $\frac{\omega}{f}+\frac{\omega'}{f'}=0$ can be obtained more easily by the method of calculus. For a single thin lens the focal length is given by

$$\frac{1}{f}=(\mu-1)\left(\frac{1}{R_1}-\frac{1}{R_2}\right) \tag{9.37}$$

where f is the focal length, R_1 and R_2 are the radii of curvature of the two surfaces of the lens and μ is the refractive index of the lens material. We may write equ.(9.37) as

$$\left(\frac{1}{R_1}-\frac{1}{R_2}\right)=\frac{1}{f(\mu-1)} \tag{9.38}$$

Differentiating the equation (9.37), we get

$$-\frac{df}{f^2}=d\mu\left(\frac{1}{R_1}-\frac{1}{R_2}\right)$$

Using equ.(9.38) into the above equation, we get

$$-\frac{df}{f^2}=\frac{d\mu}{(\mu-1)f}$$

We designate the small change df in f by δf corresponding to a change $d\mu$ in refractive index for two colours by $\delta\mu$.

But $\frac{\delta\mu}{(\mu-1)} = \omega$ where ω is the dispersive power of the lens between the two colours.

$$\therefore \qquad -\frac{\delta f}{f^2} = \frac{\omega}{f} \tag{9.39}$$

The –ve sign in (9.39) indicates that the focal length decreases with increase in refractive index. If f and f' are the focal lengths of the two lenses of different materials (crown and flint) placed in contact, then

$$\frac{1}{F} = \frac{1}{f} + \frac{1}{f'} \tag{9.40}$$

where F is the combined focal length for the mean rays. Differentiating equation (9.40), and using equ.(9.39), we obtain

$$-\frac{\delta F}{F^2} = -\frac{\delta f}{f^2} - \frac{\delta f'}{f'^2} = \frac{\omega}{f} + \frac{\omega'}{f'}$$

If the combination is to be achromatic then $\frac{1}{F}$ should be the same for the violet and red or $\delta\left(\frac{1}{F}\right)$ must be equal to zero.

$$-\frac{\delta F}{F^2} = 0$$

$$\therefore \qquad \frac{\omega}{f} + \frac{\omega'}{f'} = 0$$

$$\therefore \qquad \frac{f}{f'} = -\frac{\omega}{\omega'} \tag{9.41}$$

We draw the following conclusions from the above equation (9.41).

1. ω and ω′ are positive quantities; therefore f and f' must have opposite signs. It means that one lens must be convex and the other concave.
2. If the two lenses are made of the same material, ω = ω′ and it follows from equ.(9.40) that

   $$\frac{1}{F} = \frac{1}{f} + \frac{1}{f'} = 0 \text{ or } F = \infty.$$

   It means that the combination acts not as a lens but as a plane glass sheet. Therefore the two lenses should be of different materials.
3. The convex lens should be made of a material of smaller dispersive power if the lens combination is to act as a converging system. The dispersive power of crown glass is smaller than that of flint glass. Therefore, a convex lens of crown glass in contact with a concave lens of flint glass is used to make an achromatic doublet.

Achromatic doublet is used in telescope.

Achromats are used in telescopes and cameras. The achromatic doublets used in telescopes are achromatized for blue and red colours while those used in cameras are achromatized for violet and green colours.

2. Condition for achromatism of two thin lenses separated by a finite distance:

If the two lenses are of the same material, their achromatic combination is possible only if they are separated by a finite distance.

Let f_1 and f_2 be the focal lengths of two lenses placed coaxial and separated by a distance (Fig. 9.27). The mean focal length F of the combination is given by

$$\frac{1}{F}=\frac{1}{f_1}+\frac{1}{f_2}-\frac{x}{f_1 f_2} \qquad (9.42)$$

Fig. 9.27

Differentiating the above equation, we obtain

$$-\frac{\delta F}{F^2}=-\frac{\delta f_1}{f_1^2}-\frac{\delta f_2}{f_2^2}+\frac{x}{f_2}\left(-\frac{\delta f_1}{f_1^2}\right)-\frac{x}{f_1}\left(-\frac{\delta f_2}{f_2^2}\right) \qquad (9.43)$$

It follows from equ. (9.39) that

$$\left(-\frac{\delta f_1}{f_1^2}=\frac{\omega_1}{f_1}\right) \text{ and } \left(-\frac{\delta f_2}{f_2^2}=\frac{\omega_2}{f_2}\right)$$

where ω_1 and ω_2 are the dispersive powers of the first and second lens respectively. Using these into (9.43), we get

$$-\frac{\delta F}{F^2}=\frac{\omega_1}{f_1}+\frac{\omega_2}{f_2}-\frac{x}{f_2}\left(\frac{\omega_1}{f_1}\right)-\frac{x}{f_1}\left(\frac{\omega_2}{f_2}\right)$$

or

$$-\frac{\delta F}{F^2}=\frac{\omega_1 f_2+\omega_2 f_1}{f_1 f_2}-\frac{x(\omega_1+\omega_2)}{f_1 f_2} \qquad (9.44)$$

For an achromatic combination, F should be the same for violet and red colours. That is change in F must be zero. Thus,

$$\delta F = 0$$

$$\therefore \quad \frac{\omega_1 f_2+\omega_2 f_1}{f_1 f_2}-\frac{x(\omega_1+\omega_2)}{f_1 f_2}=0$$

$$(\omega_1 f_2+\omega_2 f_1)-x(\omega_1+\omega_2)=0$$

or

$$x=\frac{\omega_1 f_2+\omega_2 f_1}{\omega_1+\omega_2} \qquad (9.45)$$

When the two lenses are of the same material, $\omega_1 = \omega_2$. Then equ. (9.45) becomes

$$x=\frac{f_1+f_2}{2} \qquad (9.46)$$

Thus, the condition for achromatism of two thin co-axial lenses made of same material and separated by a distance is that the distance between the two lenses must be equal to the mean focal length of the two lenses. This principle is used in Huygens eyepiece in which the focal lengths of the components are $3f$ and f and they are separated by a distance

$$x = \frac{3f + f}{2} = 2f$$

Equ. (9.46) does not involve dispersive power and hence the combination has the same focal length for all colours. Therefore, when a ray of white light parallel to the axis passes through the lens system, all the colours suffer the same amount of deviation. Hence the images of all colours appear to be of the same size. However, the positions of different colours are different.

9.14. OIL-IMMERSION OBJECTIVE OF HIGH POWER MICROSCOPE

The high-power microscopes have an overall magnification of about 500. The objective in these microscopes is required to produce a magnification of the order of 50. Hence the objective must satisfy the following conditions.

(*i*) The image produced by the objective should be free from chromatic aberration, spherical aberration and coma. The rest of the aberrations are unimportant because the field of view of a microscope is very small.

Analysing concrete with a high power microscope.

(*ii*) Sufficient light should enter through the lens as the object to be viewed is very small and is placed very close to the lens.

Oil immersion lenses make use of aplanatic surfaces for obtaining a very wide-angled pencil of light from each point of the object. If an object is placed at one aplanatic point of a surface, an image free from spherical aberration and coma is formed at the second aplanatic point (refer to § 9.6.1). In case of a single spherical surface, these points are at distances R/μ and μR from the centre of curvature.

An object cannot be placed inside a glass lens. However, we can create an equivalent situation. The object is immersed in a film of oil, having the same refractive index as the material of the lens

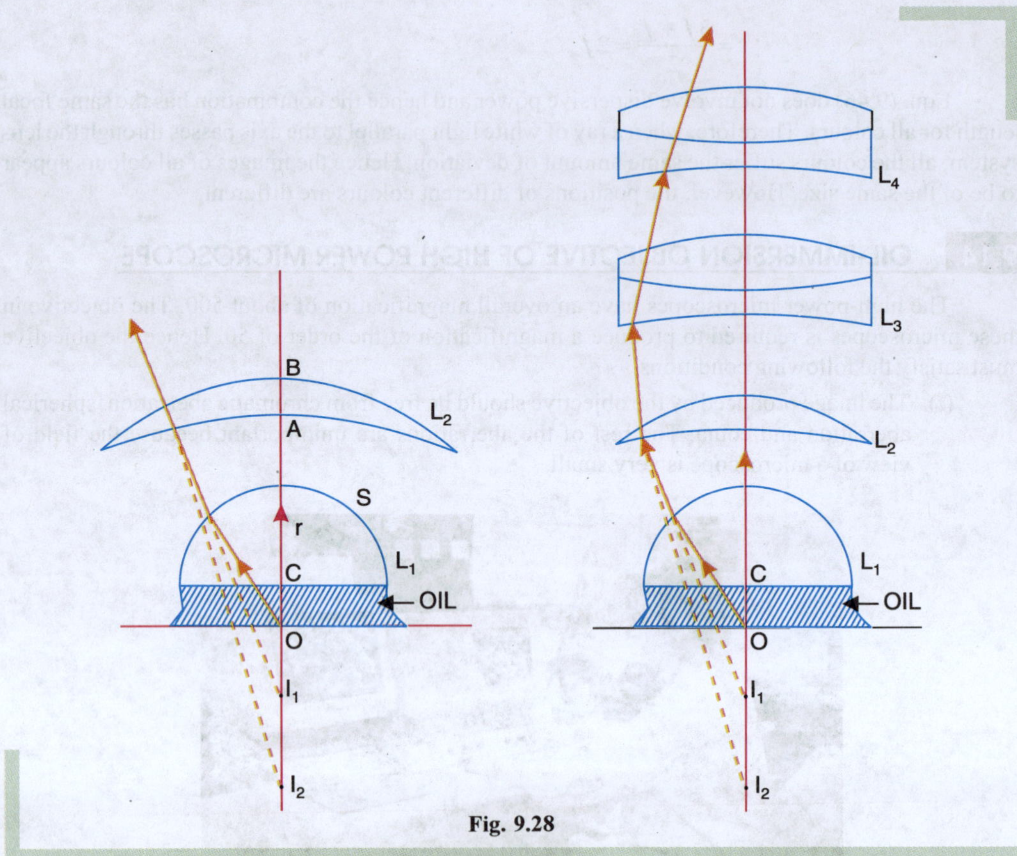

Fig. 9.28

and the front surface of the lens is brought in contact with oil as shown in Fig. 9.28. This forms the oil-immersion objective lens of a microscope. The oil-immersion objective is in practice made of a combination of lenses as shown in Fig. 9.28. The lower-most lens L_1 that is close to the object O is a plano-hemispherical convex lens. The flat surface of the lens is kept in contact with cedar-wood oil whose refractive index is the same as that of L_1. The object O is immersed in the oil and for all practical purposes may be treated to be within the lens. The rays of light starting from O do not suffer refraction till they reach the spherical surface of the lens L_1. The depth of the oil is so adjusted that the distance of O is R/μ from the centre of curvature C of the lens L_1. Thus, O is an aplanatic point for the lens. Hence all rays starting from it after refraction through the lens appear to come from the other aplanatic point I_1, where $CI_1 = \mu R$. The image I_1 is free from spherical aberration.

The position of the meniscus lens L_2 is so adjusted that the image I_1 lies at the centre of curvature of surface A of L_2. Therefore, the rays enter L_2 without deviation. I_1 is also the aplanatic point for the second surface B of L_2, which consequently forms a virtual image of I_1 at I_2. To sum up, the points O and I_1 are aplanatic points for the spherical surface S of the lens L_1. The points I_1 and I_2 are the aplanatic points for the second surface B of lens L_2.

Thus, the original wide-angled pencil of light starting from object O is changed into a narrow pencil diverging from point image I_2. The lenses L_3 and L_4 form the final image almost free from spherical aberration, chromatic aberration and coma.

9.15. ACHROMATISM OF TELESCOPE OBJECTIVE

A telescope is used for visual observation of distant objects. The important visible portion of the spectrum lies between red and blue colours. Therefore, the telescope objective must be achromatized between these colours, both for longitudinal and lateral aberrations.

Two lenses of different materials form an achromatic combination for two colours when they satisfy the following condition.

$$\frac{\omega_1}{f_1}+\frac{\omega_2}{f_2}=0 \quad \text{or} \quad \frac{\omega_1}{\omega_2}=-\frac{f_1}{f_2}$$

where f_1 and f_2 are the focal lengths of the lenses, and ω_1 and ω_2 are the dispersive powers of their materials. The focal length F of the combination is given by

$$\frac{1}{F}=\frac{1}{f_1}+\frac{1}{f_2}$$

An achromatic telescope objective of a given focal length can be designed with the help of the above two equations.

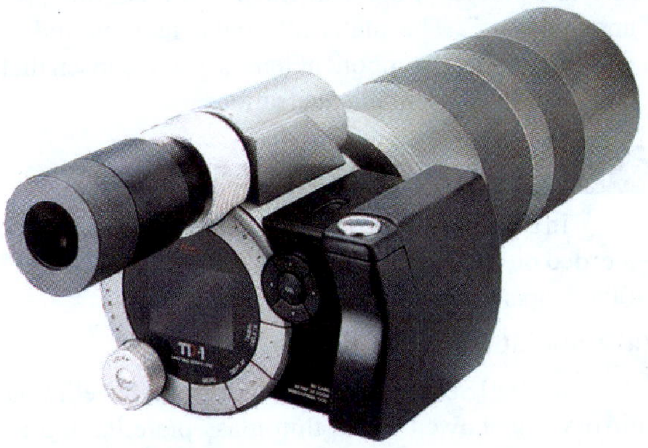

Telescope objective must be achromatized between red and blue colours for both longitudinal and lateral aberrations.

(1) As ω_1 and ω_2 are positive quantities, f_1 and f_2 should have opposite signs. Therefore, one lens must be convex and the other concave.

(2) As the telescope objective is a converging lens system, the focal length f_1 of the convex lens should be smaller than the focal length f_2 of the concave lens. If f_1 is to be smaller than f_2, then ω_1 should be less than ω_2. Thus, the convex lens should be made of glass of smaller dispersive power. The dispersive power of crown glass is smaller than that of flint glass. Therefore, the achromatic combination must be made of a convex lens of crown glass and a concave lens of flint glass. The radii of curvatures R_1 and R_2 of the convex lens and R_1' and R_2' of the concave lenses can be calculated using the following equations.

$$\frac{1}{f_1}=(\mu_1-1)\left(\frac{1}{R_1}-\frac{1}{R_2}\right)$$

and

$$\frac{1}{f_2}=(\mu_2-1)\left(\frac{1}{R_1'}-\frac{1}{R_2'}\right)$$

The above relations give different sets of values of R_1 and R_2 (or of R_1' and R_2') for the same values of μ_1 (or of μ_2). In practice, the objective is made plano-convex lens because it has minimum spherical aberration when convex side faces the incident light. Hence we make one face of the concave lens plane (i.e. $R_2'=\infty$) and then calculate the radius of curvature R_1' of the other face. The surfaces of the lenses in contact are made of the same radius of curvature ($R_2=R_1'$). The lenses are cemented together by canada balsam to avoid loss of light by reflection at the interface. Using the value of R_2, the radius of curvature R_1 of the first face of the convex lens can then be determined.

9.16. ACHROMATISM OF A CAMERA LENS

The lenses used in cameras are achromatic doublets which consist of two lenses, one convex and the other concave, of different materials placed in contact so as to satisfy the following condition for two specified colours;

$$\frac{\omega_1}{f_1} + \frac{\omega_2}{f_2} = 0$$

Usually photographic plates are most sensitive between violet and green colours. Therefore, the camera lens-doublet is achromatized for these two colours. In the achromatic doublet the equality of focal length is achieved for only two colours. In high quality cameras it is required to achieve equality of focal lengths for three or more colours. Then we use three or more component lenses of different materials placed in contact. Such combinations are called "apochromats". The materials, focal lengths and radii of curvature of these component lenses are so chosen that apochromat is free from spherical aberration, coma etc monochromatic aberrations.

Apochromat is a combination of three or more component lenses of different materials.

9.17. CORRECTOR PLATES

In case of large telescopes, large spherical mirrors are used to form images and the images are recorded on a curved film. The systems are free from astigmatism and coma etc. However, because of large aperture spherical aberration occurs.

(a) Schmidt corrector plate:

In 1929 Schmidt designed a system that eliminates spherical aberration in such large spherical mirrors. He showed that a thin glass plate having a shallow toroidal surface and held before the spherical mirror could help minimizing the spherical aberration in a mirror. Such plates are known as **corrector plates.** Light rays passing through the outer regions of the corrector plate would be deviated by just the amount needed to be sharply focussed on the curved film.

18" Schmidt Corrector Plate

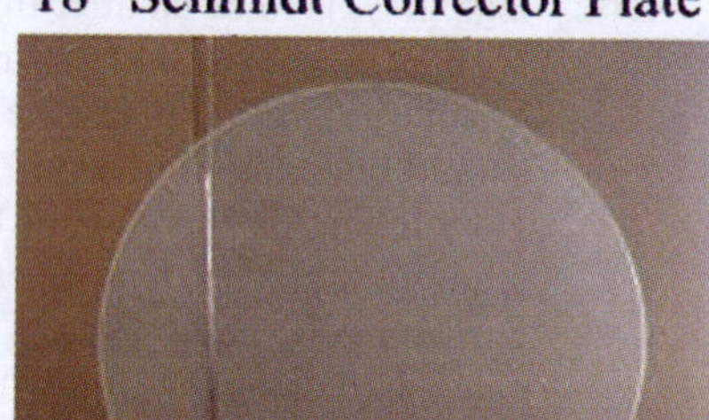

Fig. 9.29 shows the **Schmidt corrector plate** P used with large aperture reflector telescopes. One of the surfaces of the corrector plate is plane and the second surface is convex towards the axis and concave at the ends. The axial rays are refracted towards the axis and the marginal rays are refracted outwards. All the refracted rays come to focus at F.

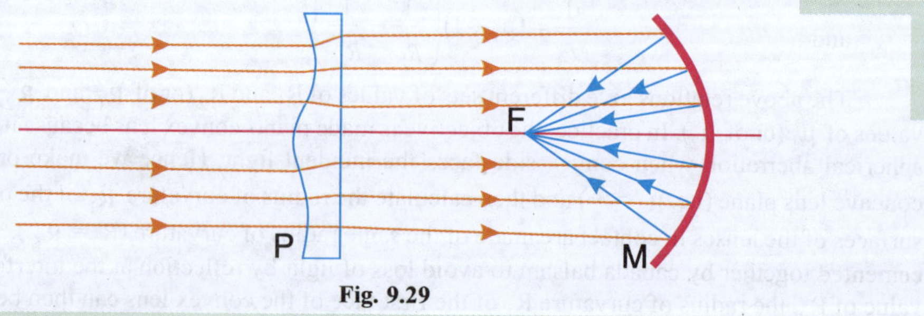

Fig. 9.29

(b) Maksutov corrector:

The grinding of one of the surfaces of a Schmidt corrector is complicated and difficult. The difficulty is avoided in the Maksutov corrector. The Maksutov corrector consists of a thick meniscus lens as shown in Fig. 9.30. In this corrector both the surfaces are spherical but of unequal curvature. The radius of curvature of the surface facing the incident parallel beam is less than the second surface. The course of a parallel beam of light due to refraction through the lens and reflection from, the mirror is indicated in the figure (Fig. 9.30). The parallel beam comes to focus at F. A lens of the type of Maksutov corrector is simple and cheap to manufacture.

Maksutov corrector.

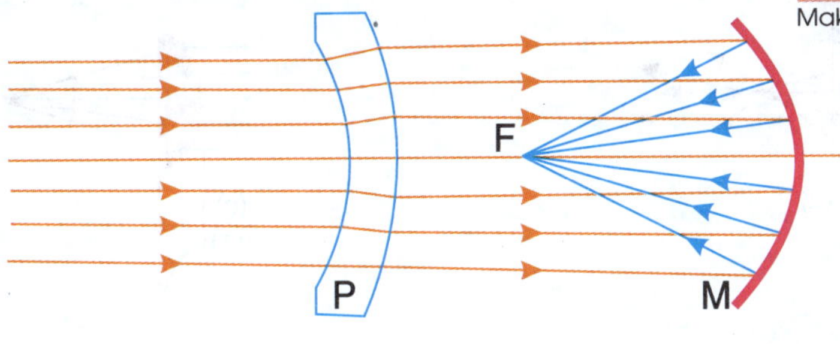

Fig. 9.30

9.18. CONCLUSION

It is clear from the previous articles that no single lens is free from all the aberrations. It is also not possible to minimize all the aberrations simultaneously in a lens system. A lens system needed for a precision work must consist of a number of lens components so that any aberration produced by some of them is cancelled by the rest. A high degree of correction for the aberrations can be achieved by increasing the number of lens components. These individual lenses chosen must be of required radii of curvature, focal lengths and dispersive powers. However, it is not always necessary to eliminate all the aberrations simultaneously in all the optical instruments. Different instruments need relative correction for different aberrations. Correction for spherical aberration, coma and axial chromatic aberration is essential in the case of telescope objectives, which cover a small angular field. On the other hand, a photographic objective, which covers a wide field of view, has to be corrected for astigmatism, distortion and curvature of the field. In general, images of high quality are produced only by lenses corrected for small angular field and small aperture.

9.19. GRADIENT-INDEX LENSES

In an ordinary homogeneous lens, the refractive index is the same throughout. Refraction takes place *only* at the surfaces of the lens. When light propagates through an inhomogeneous medium, it slows down in optically dense regions and speeds up in less dense regions. Lenses can be made from optically inhomogeneous materials, which are characterized by a gradient in the index of refraction. Such lenses are known as **gradient-index** or **GRIN** lenses. In GRIN lenses refraction of light takes place

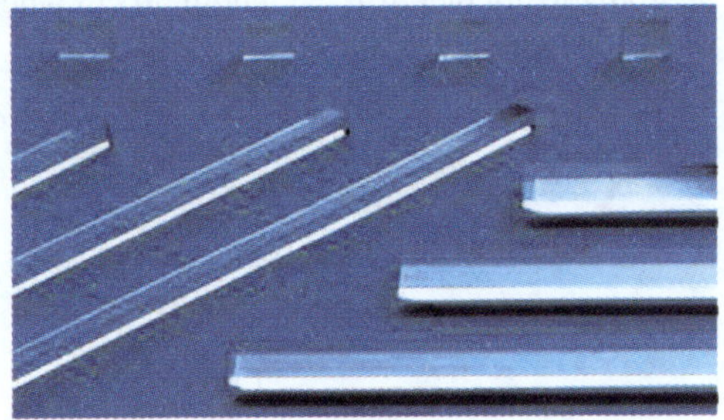

Gradient index lenses are produced by silver-ion or lithium-ion exchange in a special glass

within the lens also. They give us many advantages. The GRIN lenses can replace aspherical lenses, which are hard to prepare.

GRIN lenses can be made in three forms. (i) They can be made with cylindrical symmetry where the index of refraction varies as a function of distance from the optic axis.

(ii) They can be made to have an axial gradient. Here the refractive index varies as a function of distance along the axis. The surfaces of constant index are plane and normal to the axis. Axial refractive gradients are particularly useful for the correction of spherical aberration.

(iii) They can be made to have spherical symmetry, where the refractive index varies symmetrically about a point. The surfaces of constant index are spheres. An example of such a lens is the crystalline lens of the human eye.

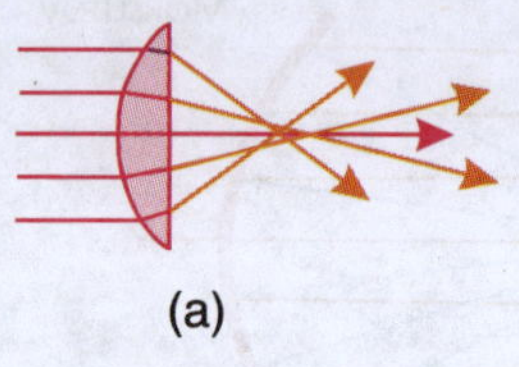

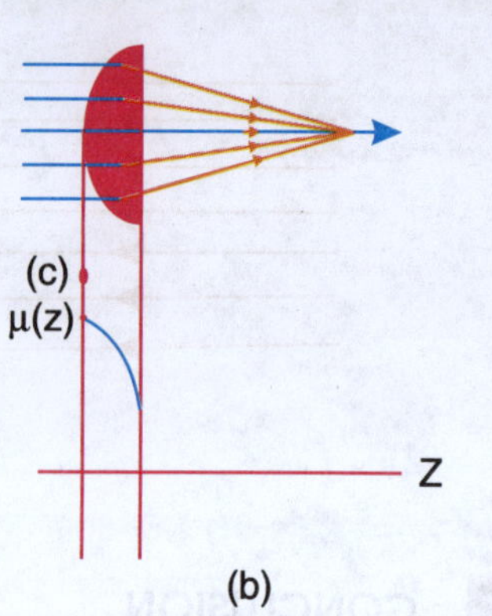

Fig. 9.31

Spherical aberration occurs in an ordinary convex lens because the edges of the lens refract the marginal rays very strongly. If we now consider a grin-lens containing an axial gradient and the refractive index is highest near the front surface of the lens, then the marginal rays are bent less. With a properly chosen gradient profile, all rays passing through the lens come together at the same point. Thus, spherical aberration is eliminated when GRIN lenses are used, as shown in Fig. 9.31.

Further, introduction of a GRIN element into a compound lens simplifies the system by greatly reducing the number of elements but without sacrificing performance.

WORKED OUT PROBLEMS

Examples 9.1. Calculate the focal lengths of the crown and flint glass lenses that form an achromatic doublet of focal length 50 cm when placed in contact. The refractive indices for the C and F lines are given below:

Crown	$\mu_C = 1.515$	$\mu_F = 1.523$
Flint	$\mu_C = 1.644$	$\mu_F = 1.664$

Solution: $$\omega_1 = \frac{1.523 - 1.515}{\frac{1.523 + 1.515}{2} - 1} = \frac{0.008}{0.519} = \frac{8}{519}$$

$$\omega_2 = \frac{1.664 - 1.644}{\frac{1.664 + 1.644}{2} - 1} = \frac{0.020}{0.654} = \frac{10}{327}$$

For achromatism

$$\frac{\omega_1}{f_1}+\frac{\omega_2}{f_2}=0 \quad \text{or} \quad \frac{\omega_1}{f_1}=\frac{-\omega_2}{f_2}$$

$$\therefore \quad \frac{1}{f_2}=\frac{\omega_1}{\omega_2 f_1}=\frac{8\times327}{519\times10 f_1} \quad (1)$$

$$\frac{1}{F}=\frac{1}{f_1}+\frac{1}{f_2}=\frac{1}{50}$$

$$\therefore \quad \frac{1}{f_1}-\frac{8\times327}{512\times10f}=\frac{1}{50} \quad \text{or} \quad \frac{1}{f_1}\left[1-\frac{8\times327}{519\times10}\right]=\frac{1}{50}$$

f_1 **= 24.79cm (crown glass, convex lens)**

From equation (1) $f_2=-\dfrac{519\times10\times24.79}{8\times327}$

= – 49.19 cm **(Flint glass, concave lens)**

Example 9.2: Two glasses have dispersive powers in the ratio 2:3. These glasses are to be used in the manufacture of an achromatic objective of focal length 20 cm. What are the focal lengths of the lenses?

Solution: For an achromatic objective,

$$\frac{\omega_1}{f_1}+\frac{\omega_2}{f_2}=0 \quad \therefore \quad \frac{1}{f_2}=-\frac{\omega_1}{\omega_2 f_1}$$

Also $\dfrac{1}{F}=\dfrac{1}{f_1}+\dfrac{1}{f_2}=\dfrac{1}{50}$

Here $\dfrac{\omega_1}{\omega_2}=\dfrac{2}{3}$ and F = 20 cm

$$\therefore \quad \frac{1}{f_2}=-\frac{2}{3f_1}$$

$$\frac{1}{20}=\frac{1}{f_1}+\frac{1}{f_2}=\frac{1}{f_1}-\frac{2}{3f_1}=\frac{1}{3f_1}$$

f_1 = 20/3 = **6.67 cm**

$$\frac{1}{f_2}=-\frac{2\times3}{3\times20}=-\frac{1}{10} \quad \text{or} \quad f_2=-10 \text{ cm}$$

Example 9.3: Calculate the focal lengths of a convex lens of crown glass (dispersive power 0.012) and a concave lens of flint glass (dispersive power 0.020) so that when placed in contact they form an achromatic converging combination of focal length 30 cm.

Solution:

For achromatism,

$$\frac{\omega_1}{f_1}+\frac{\omega_2}{f_2}=0$$

Here, ω_1= 0.012 and ω_2= 0.020

$$\frac{0.012}{f_1}+\frac{0.020}{f_2}=0 \quad \therefore \frac{1}{f_2}=-\frac{0.6}{f_1}$$

Also

$$\frac{1}{F} = \frac{1}{f_1} + \frac{1}{f_2} \qquad \text{But} \qquad F = +30 \text{ cm}$$

$$\frac{1}{30 \text{ cm}} = \frac{1}{f_1} - \frac{0.6}{f_1} \qquad f_1 = 30 \text{ cm} \times 0.4 = 12 \text{ cm.}$$

$$\frac{1}{f_2} = -\frac{0.6}{12\text{cm}} \qquad f_2 = -20 \text{ cm}$$

The focal length of the convex lens is 12 cm and the focal length of the concave lens is 20 cm.

Example 9.4: It is desired to make a converging achromatic lens of mean focal length 30cm by using two lenses of materials A and B. If the dispersive powers of A and B are in the ratio 1:2, find the focal length of each lens.

Solution:

Given that $\frac{\omega_1}{\omega_2} = \frac{1}{2}$

We know that $\frac{\omega_1}{f_1} + \frac{\omega_2}{f_2} = 0 \quad \therefore \quad \frac{1}{f_2} = -\frac{\omega_1}{\omega_2 f_1} \quad \therefore \quad \frac{1}{f_2} = -\frac{1}{2f_1}$ (1)

$$\frac{1}{F} = \frac{1}{f_1} + \frac{1}{f_2} = \frac{1}{30} \text{ or } \frac{1}{f_1} - \frac{1}{2f_1} = \frac{1}{30} \quad \therefore f_1 = 15 \text{ cm}$$

From the equation (1) $f_2 = -2f_1 = -2 \times 15 \text{ cm} = -30 \text{ cm.}$

Example 9.5: The dispersive powers of crown and flint glass are 0.015 and 0.030 respectively. The refractive indices for the mean rays are 1.52 and 1.65. If one of the surfaces of the flint glass lens is plane, calculate the radii of curvature of the other surfaces of the two lenses which form an achromatic doublet of focal length 80 cm.

Solution:

Given that $\omega_1 = 0.015, \omega_2 = 0.030$, and $\mu_1 = 1.52, \mu_2 = 1.65$

$$\frac{\omega_1}{f_1} + \frac{\omega_2}{f_2} = 0 \quad \therefore \frac{1}{f_2} = -\frac{\omega_1}{\omega_2 f_1} \quad \therefore \frac{1}{f_2} = -\frac{0.015}{0.030 f_1} \quad \therefore \frac{1}{f_2} = -\frac{1}{2f_1}$$

But $\frac{1}{F} = \frac{1}{f_1} + \frac{1}{f_2} = \frac{1}{80}$ or $\frac{1}{f_1} - \frac{1}{2f_1} = \frac{1}{80}$ or $f_1 = 40 \text{ cm}$

$\therefore \quad f_2 = -2f_1 = -80 \text{ cm.}$

For the flint glass lens,

$$\frac{1}{f_2} = (\mu_2 - 1)\left(\frac{1}{R_1'} - \frac{1}{R_2'}\right)$$

Here, $R_2' = \infty \quad \therefore \; -\frac{1}{80 \text{ cm}} = (1.65 - 1)\left(\frac{1}{R_1'}\right)$ or $R_1' = -52 \text{ cm.}$

For the crown glass lens, $R_2 = R_1' = -52$ cm and

$$\frac{1}{f_1} = (\mu - 1)\left(\frac{1}{R_1} - \frac{1}{R_2}\right) \qquad \therefore \frac{1}{40 \text{cm}} = (1.52 - 1)\left(\frac{1}{R_1} - \frac{1}{-52 \text{ cm}}\right)$$

$\therefore \quad \mathbf{R_1 = 34.67 \text{ cm}}$

For the crown glass lens, $R_1 = 34.67$cm, $R_2 = -52$cm

For the flint glass lens, $R_1' = -52$cm, $R_2' = \infty$.

Example 9.6: Two lenses of focal lengths 8 cm and 4 cm are placed at a certain distance apart. Calculate the distance between the lenses if they form an achromatic combination.

Solution:

Here given that $f_1 = 8\text{cm}$, $f_2 = 4\text{cm}$, $x = ?$

We know that $x = \dfrac{f_1 + f_2}{2}$ $\therefore$ $x = \dfrac{8\text{cm} + 4\text{cm}}{2} = 6\,\text{cm}$.

Example 9.7: Two thin lenses of focal lengths f_1 and f_2 separated by a distance have an equivalent focal length 50cm. The combination satisfies the conditions for no chromatic aberration and minimum spherical aberration. Find the values of f_1, f_2 and d. Assume that both the lenses are of the same material.

Solution:

We know that $$\frac{1}{F} = \frac{1}{f_1} + \frac{1}{f_2} - \frac{x}{f_1 f_2} = \frac{f_1 + f_2 - x}{f_1 f_2} \quad \ldots(1)$$

Here F = 50cm and for no chromatic aberration

$$x = \frac{f_1 + f_2}{2} \quad \ldots.(2)$$

For minimum spherical aberration, $x = f_1 - f_2$ ...(3)

From (2) and (3), we get $f_1 = \dfrac{3x}{2}$ and $f_2 = \dfrac{x}{2}$

Substituting these values of f_1 and f_2 in equation (1), we obtain $\dfrac{1}{50\,\text{cm}} = \dfrac{\dfrac{3x}{2} + \dfrac{x}{2} - x}{\dfrac{3x}{2}\left(\dfrac{x}{2}\right)}$

$$\therefore \quad x = \frac{200\,\text{cm}}{3} = 66.67\,\text{cm}$$

$$\therefore \quad f_1 = \frac{3x}{2} = 100\,\text{cm}$$

and $$f_2 = \frac{x}{2} = 33.33\ \text{cm}$$

Example 9.8: A converging achromat of 40cm focal length is to be constructed out of a thin crown glass and a thin flint glass lens, the surfaces in contact having a common radius of cruvature of 25cm. Calculate the radius of curvature of the second surface of each lens, given that the values of the dispersive powers and refractive indices are 0.017 and 1.5 for crown glass, and 0.034 and 1.7 for flint glass.

Solution:

Given that $\omega_1 = 0.017$, $\omega_2 = 0.034$ and $\mu = 1.5$, $\mu_2 = 1.7$

We have $\dfrac{\omega_1}{f_1} + \dfrac{\omega_2}{f_2} = 0$ $\quad\therefore\quad \dfrac{1}{f_2} = -\dfrac{\omega_1}{\omega_2 f_1}$ $\quad\therefore\quad \dfrac{1}{f_2} = -\dfrac{0.017}{0.034 f_1} = -\dfrac{1}{2 f_1}$

But $\dfrac{1}{F} = \dfrac{1}{f_1} + \dfrac{1}{f_2} = \dfrac{1}{40}$ or $\dfrac{1}{f_1} - \dfrac{1}{2f_1} = \dfrac{1}{40}$ $\quad\therefore\quad f_1 = 20$ cm.

$$\therefore \quad f_2 = -2f_1 = -40\ \text{cm}$$

For the crown lens

$$\frac{1}{f_1}=(\mu_1-1)\left(\frac{1}{R_1}-\frac{1}{R_2}\right) \quad \text{or} \quad \frac{1}{20\,\text{cm}}=(1.5-1)\left(\frac{1}{R_1}-\frac{1}{-25\,\text{cm}}\right)$$

$$R_1=\frac{50}{3}\,\text{cm}=16.67\,\text{cm}$$

For the flint glass lens

$$\frac{1}{f_2}=(\mu_2-1)\left(\frac{1}{R_1'}-\frac{1}{R_2'}\right) \quad \text{or} \quad -\frac{1}{40\,\text{cm}}=(1.7-1)\left(-\frac{1}{25\,\text{cm}}-\frac{1}{R_2'}\right)$$

$$R_2' = -233.3\,\text{cm}$$

For crown glass lens, $R_1 = 16.67\text{cm}$, $R_2 = -25$ cm.

For flint glass lens, $R_1 = 25\text{cm}$, $R_2 = -233.33\text{cm}$.

Example 9.9: An achromatic doublet of focal length 20 cm is to be formed out of a combination of crown and flint glasses. The radius of curvature of the faces in contact is 15 cm. Calculate the radius of curvature of the other faces, given the dispersive powers of the crown and flint glasses as 0.02 and 0.04 and their respective refractive indices are 1.52 and 1.65.

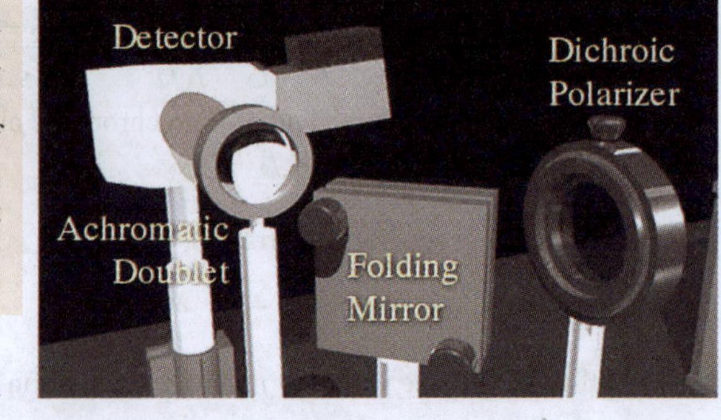

Solution:

Given that $\omega_1 = 0.02, \omega_2 = 0.04$ and $\mu_1 = 1.52, \mu_2 = 1.65$

Also $F = 20$cm.

We have $\dfrac{\omega_1}{f_1}+\dfrac{\omega_2}{f_2}=0 \quad \therefore \dfrac{1}{f_2}=-\dfrac{\omega_1}{\omega_2 f_1} \quad \therefore \dfrac{1}{f_2}=-\dfrac{0.02}{0.04 f_1}=-\dfrac{1}{2f_1}$

But $\dfrac{1}{F}=\dfrac{1}{f_1}+\dfrac{1}{f_2}=\dfrac{1}{20\,\text{cm}}$ or $\dfrac{1}{f_1}-\dfrac{1}{2f_1}=\dfrac{1}{20\,\text{cm}} \quad \therefore f_1=10$ cm.

$\therefore \quad f_2=-2f_1=-20\,\text{cm}$

For the crown lens

$$\frac{1}{f_1}=(\mu_1-1)\left(\frac{1}{R_1}-\frac{1}{R_2}\right)$$

Here $R_2 = -15$ cm, $\quad \therefore \dfrac{1}{10\,\text{cm}}=(1.52-1)\left(\dfrac{1}{R_1}-\dfrac{1}{-15\,\text{cm}}\right) \quad \therefore R_1 = 7.96$ cm

In the second lens

$$\frac{1}{f_2}=(\mu_2-1)\left(\frac{1}{R_1'}-\frac{1}{R_2'}\right) \quad \therefore -\frac{1}{20\,\text{cm}}=(1.65-1)\left(-\frac{1}{15\,\text{cm}}-\frac{1}{R_2'}\right)$$

$\therefore \quad R_2' = 97.5$ cm

Example 9.10: An object of a telescope is an achromat of focal length 90cm. The magnitudes of the dispersive powers of the two lenses are 0.024 and 0.036. Calculate their focal lengths.

Solution:

Here given that $\omega_1 = 0.024$, $\omega_2 = 0.036$ and F = 90cm.

$$\frac{\omega_1}{f_1} + \frac{\omega_2}{f_2} = 0 \qquad \therefore \quad \frac{1}{f_2} = -\frac{\omega_1}{\omega_2 f_1} = \frac{1}{f_2} = -\frac{0.024}{0.036 f_1} = -\frac{2}{3 f_1}$$

Also $$\frac{1}{F} = \frac{1}{f_1} + \frac{1}{f_2} = \frac{1}{90\,\text{cm}} \text{ or } \frac{1}{f_1} - \frac{2}{3f_1} = \frac{1}{90\,\text{cm}} \text{ or } f_1 = 30 \text{ cm}$$

$\therefore$ $$f_2 = -\frac{3}{2} f_1 = -45 \text{ cm}$$

The object glass is made of a combination of convex lens of focal length 30 cm and a concave lens of focal length 45 cm.

Example 9.11: The focal length of an achromatic combination of two lenses in contact is 150cm. If the dispersive powers of the materials of the two lenses are 0.018 and 0.027, calculate the focal lengths of the two lenses.

Solution:

Here given that $\omega_1 = 0.018$, $\omega_2 = 0.027$ and F = 150cm.

We have that

$$\frac{\omega_1}{f_1} + \frac{\omega_2}{f_2} = 0 \qquad \frac{0.018}{f_1} + \frac{0.027}{f_2} = 0 \qquad \frac{1}{f_2} = -\frac{2}{3f_1}$$

Also $$\frac{1}{F} = \frac{1}{f_1} + \frac{1}{f_2} = \frac{1}{150\,\text{cm}} \text{ or } \frac{1}{f_1} - \frac{2}{3f_1} = \frac{1}{150\,\text{cm}} \qquad \therefore f_1 = 50\,\text{cm}$$

$\therefore$ $$f_2 = -\frac{3}{2} f_1 = -\frac{3 \times 50 \text{ cm}}{2} = -75 \text{ cm}$$

QUESTIONS

1. What are the common defects in the images produced by a single lens? How can these defects be removed?
2. What to you mean by spherical and chromatic aberration of a lens? Explain how they are caused. How would you correct for chromatic aberration in the case of a lens system in contact? **(Nagpur, 2005)**
3. Enumerate the various defects of image formation by an optical system. Explain the methods of removing them.
4. Obtain an expression for the dispersive power of a lens. Find the condition for achromatism of a combination of two thin coaxial lenses (i) when in contact and (ii) when separated by a distance.
5. Explain the defects astigmatism, coma and curvature. Explain how they may be minimized.
6. Discuss the monochromatic aberrations in the image formed by a single lens with respect to a certain position of an object.
7. What is the difference between pincushion distortion and barrel-shaped distortion? How can these defects be removed?

8. Show that the diameter of the circle of least confusion depends on the diameter of the lens aperture and the dispersive power of the material. Also show that it is independent of the focal length of the lens.
9. Discuss the common defects to which a lens system is liable. Describe how these defects have been removed in an eyepiece.
10. What are aplanatic surfaces? Show that a spherical refracting surface is aplanatic with respect to a certain position of an object.
11. Explain what is meant by chromatic aberration in lenses. Derive the condition for the elimination of this defect in telescope objectives.
12. Enumerate the chief defects of spherical lenses. Find the condition for achromatism of two thin lenses when they are in contact.
13. Explain the meaning of chromatic aberration in lenses. Derive the condition for the achromatism of two thin lenses separated by a distance. Give the name and construction of the eyepiece, which satisfies the condition.
14. Why is spherical aberration large in the case of a microscope objective even though its aperture is small?
15. Briefly explain the principal defects present in the images formed by spherical lenses. How can these defects be corrected?
16. What is chromatic aberration of a lens? Explain clearly how it is removed in a photographic objective.
17. What is spherical aberration and how does it effect the definition of the image formed by a lens? How can it be minimized in the case of common eyepieces?
18. Explain fully how the errors due to spherical aberration have been overcome in the construction of objective glasses of high power microscope.
19. What are spherical and chromatic aberrations for a thin lens? Describe without mathematical details, how these defects can be eliminated?
20. Discuss the various defects from which images formed by a single lens suffer. Indicate how these defects may be minimized?
21. Describe with suitable diagrams, the lateral and longitudinal chromatic aberrations. How will you achromatize a system of two thin lenses (a) of different materials and (b) of the same material. **(Nagpur, 2004)**
22. Explain how chromatic aberration may be removed in the case of a combination of two lenses of the same material? Is such a combination of lenses truly achromatic?
23. Obtain an expression for the longitudinal chromatic aberration of a lens. A system of lenses consists of two thin convex lenses separated by a finite distance on a common axis. Derive the conditions under which the system is free from chromatic and spherical aberrations.

24. Briefly explain spherical and chromatic aberrations in optical images? What is an achromatic lens system and how is it obtained in practice? Obtain conditions for achromatism with two thin lenses (a) when they are in contact and (b) when they are separated. **(Garhwal, 2005)**

25. Two thin lenses of the same materiel and mounted coaxial are to have minimum spherical and chromatic aberration. Find the ratio of their focal lengths and the separation between them in terms of their focal lengths.

26. What is chromatic aberration? Derive the condition for achromatism in an achromatic doublet. Discuss the validity of the condition for the choice of the lenses.

27. Distinguish between lateral and longitudinal spherical aberration. Mention two methods for minimizing spherical aberration. **(Nagpur, 2004)**

28. (*i*) Define aberration in lenses.
(*ii*) Explain longitudinal and lateral spherical aberration in lens with neat diagram.
(*iii*) How spherical aberration can be removed by using a planoconvex lens?
(*iv*) Distinguish between chromatic and monochromatic aberrations. **(Nagpur, 2004)**

29. (*i*) What are the types and subtypes of aberrations?
(ii) With neat ray diagram explain spherical aberration in lens. **(Nagpur, 2004)**

30. What is meniscus lens? **(Nagpur, 2004)**

31. What is spherical aberration? How it can be eliminated? Obtain the condition for minimizing spherical aberration in case of two coaxial lenses separated by a distance. **(A.P.University, 2010)**

32. What is Coma? How is it eliminated? **(A.P.University, 2010)**

33. Give the methods to reduce spherical aberration. **(RTMNU, 2010)**

PROBLEMS FOR PRACTICE

1. The dispersive powers for crown and flint glass are 0.016 and 0.032 respectively. Calculate the focal lengths of the lens made of crown and flint glass, which form an achromatic doublet of focal length 20cm when placed in contact.

[**Ans:** $f_1 = +10$ cm, $f_2 = -20$ cm]

2. Two lenses of focal lengths 4.5cm and 1.5cm are placed at a certain distance apart. Calculate the distance between the lenses if they form an achromatic combination.

[**Ans:** $d = 3$cm.]

3. A converging lens of focal length 35.0 and achromatic for the c and f lines is to be made with two lenses. One of hard crown and the other of dense flint glass placed in contact. Find the focal lengths of these lenses. The following data for the glasses are given

Hard crown ; $\mu_D = 1.5175$; $\mu_F - \mu_C = 0.00856$
Dense flint ; $\mu_D = 1.6264$; $\mu_F - \mu_C = 0.01722$

(μ_D is the refractive index for the mean ray.)

[**Ans:** Crown glass convex lens, f = + 13.94 cm; Flint glass concave lens, $f = -24.67$ cm.]

4. The dispersive powers for crown and flint glass are 0.015 and 0.030 respectively. Calculate the focal lengths of the lenses (made of crown and flint glass) which form an achromatic doublet of focal length 60 cm when placed contact. [**Ans:** 30 cm and – 60 cm.]

5. The object glass of a telescope is an achromat of focal length 90 cm. If the magnitude of the dispersive powers of the two lenses are 0.024 and 0.036, calculate their focal lengths. [**Ans:** 30 cm and – 45 cm]

6. A double convex lens has radii of curvature of 40cm and 10cm. Find the longitudinal chromatic aberration for the object at infinite. Given $\mu_v = 1.5230$ and $\mu_R = 1.5145$.

(A.P.University, 2010)

10 CHAPTER

Optical Instruments

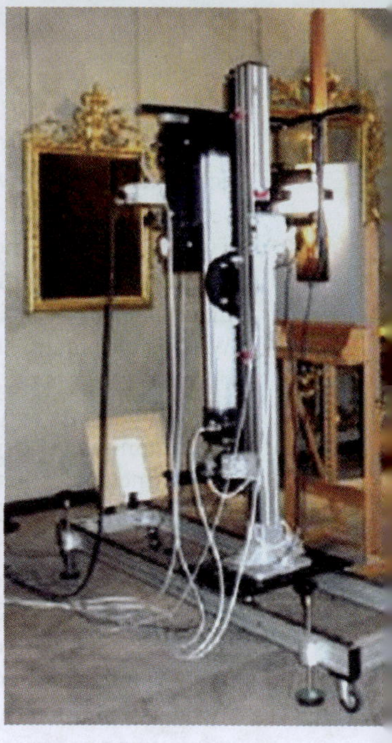

At a Glance

10.1. INTRODUCTION

The optical instruments can be broadly divided into two groups. One group forms *real images* of an object, which is projected onto a screen or photographic plate. The image can be viewed simultaneously by many observers. Projectors belong to this category. The other group forms *virtual images* of an object and only one observer can see the image. The virtual image formed by the instrument is transformed by the eye into a real image. Therefore, the optical instruments forming a virtual image are often called *optical aids*. The spectacles, microscopes and telescopes belong to this group. We study about some of the important optical instruments in this chapter.

10.2. THE EYE

The observer's eye is an essential part of all optical instruments. A sectional view of the human eye is shown in Fig. 10.1. It is nearly spherical in shape and about 2.5 cm in diameter. It has a tough outer skin called *sclera,* which protects the eye and gives the necessary stiffness. The sclera is lined inside with vascular tissue, which consists of blood vessels feeding the eye. At the front of the eye, the sclera extends into a thin convex transparent membrane called the *cornea*, which acts as an entrance lens for the eye. Back of the cornea is the *iris,* which is a pigmented muscular ring. The iris can be of different colours, which explains why people have blue, green, brown or black eyes. The iris contains an aperture with variable diameter called the *pupil*, which opens and closes to adapt to changing light intensity. Behind the iris is a *crystalline lens*, which is a biconvex lens made by Nature herself. The lens contains a fibrous jelly, hard at the centre and progressively softer at the outer portions. The crystalline lens is held in place by ligaments that attach it to the ciliary muscle, which encircles it.

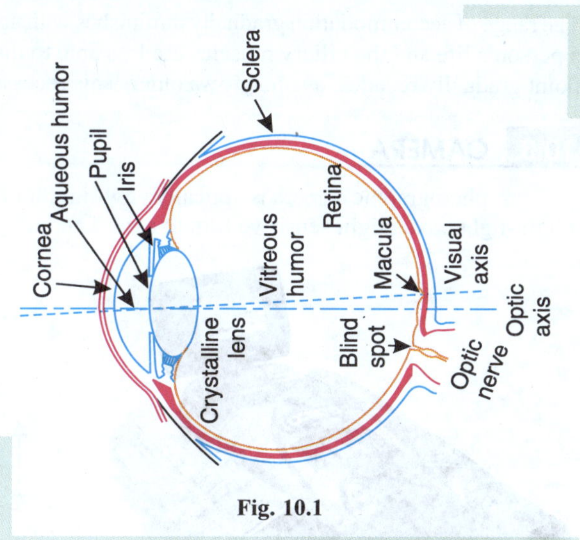

Fig. 10.1

The chamber in front of the lens and behind the cornea contains a liquid called *aqueous humor*. Behind the lens, the eye is filled with a thin water jelly called *vitreous humor*. The indices of refraction of both the aqueous humor and vitreous humor are about 1.336. The crystalline lens, while not homogeneous, has an average index of 1.437. Most of the refraction of light entering the eye occurs at the outer surface of the cornea. The chamber holding the vitreous humor is lined with a sensory layer called the *retina.* The retina has the shape of a hemisphere and contains light receptors called *rods* and *cones*. The human eye has a total of 125 million rods and 6.5 million cones. They sense the image and transmit it via the *optic nerve* to the brain. At the point where the optic nerve enters the eye, there are no rods or cones and this portion of the retina is called the *blind spot*. In contrast, vision is most acute in a small central region at the axis of the eyeball called the *fovea centralis*. Here, cones responsible for colour vision are concentrated. The remaining areas of the retina are occupied mainly by rods.

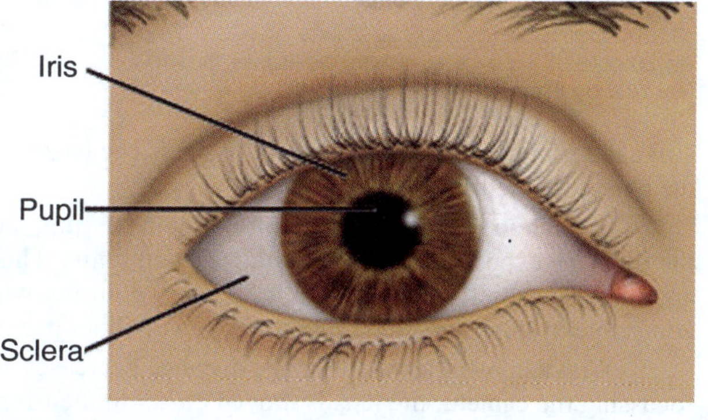

Refraction at the cornea and the surfaces of the lens produces a real and inverted image of the object on the retina. The optic nerve sends a signal to the brain, which makes the corrections necessary for us to see objects in their natural positions.

For an object to be seen sharply, the image must be formed exactly at the location of the retina. The eye adjusts to different object distances by changing the focal length f of its lens; the lens-to-retina distance does not change. The focal length of the eye is adjusted by varying the radii of curvature of the crystalline lens, with the help of ciliary muscle. This process is called *accommodation*.

The eye accommodates involuntarily, but its ability to accommodate is not unlimited. We describe the limited ability of the eye in terms of the *far point* and the *near point* of the eye. The far point of a normal eye is infinity. The ciliary muscle is fully relaxed in this position. The nearest point on which the eye is focussed with the ciliary muscle fully contracted is the near point of the eye. The minimum distance at which an eye can see objects distinctly and without getting tired is called the *normal viewing distance or normal distance of distinct vision* (NDDV); it is 25 cm for a normal eye.

The range of accommodation gradually diminishes with age because the crystalline lens grows through a person's life and the ciliary muscles are less able to distort a larger lens. For this reason, the near point gradually recedes, as one grows older. This recession of the near point is called *presbyopia.*

10.3. CAMERA

A photographic camera is optically very similar to the human eye. It consists of a convex lens, a light-tight box, a light sensitive film to record an image and a shutter to let the light from the lens

A photographic camera consists of a convex lens, a light-tight box, a light sensitive film to record an image.

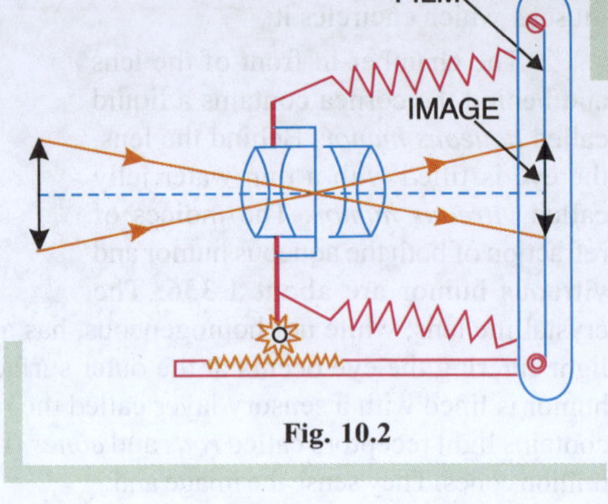

Fig. 10.2

strike the film for a prescribed length of time. The lens forms a reduced real image of the object on the photographic film.

In order to record the image properly on the film, the total light energy per unit area reaching the film (the exposure) must fall within certain limits. This is controlled by the shutter and the lens aperture. The shutter controls the time interval during which light enters the lens. This is usually adjustable in steps corresponding to factors of about two, often from 1s to1/1000s.

With a convex lens, the image distance decreases as the object distance increases. Hence in focussing the camera, the lens is moved closer to the film for a distant object and farther from the film for a nearby object. This is achieved by turning the lens in a threaded mount.

The intensity of light reaching the film is proportional to the area viewed by the camera lens and to the effective area of the lens. It is calculated that the intensity of light reaching the film with a particular lens is proportional to D^2/f^2.

The light gathering capability of a lens is commonly expressed in terms of the *f-number* of the lens. It is given by

$$f - \text{number} = \frac{\text{focal length}}{\text{diameter of the stop}} = \frac{f}{D}$$

Therefore, the light intensity reaching the film is inversely proportional to the square of the *f*-number.

Suppose the focal length of the lens is 11 cm and the diameter of the stop is 1 cm, then the *f*-number is 11 and $D = f/11$. The diameters of the stop of a camera are usually marked so that the exposure time is doubled between each position such as

$f/2, f/2.8, f/4, f/5.6, f/8, f/11, f/16$ and so on.

The larger numbers represent smaller apertures and exposures, and each step corresponds to a factor of two in intensity. The actual exposure is proportional to both to the aperture area and the time of exposure. Thus, $f/4$ and 1/500 s, $f/5.6$ and 1/250 s and f/8 and 1/125 s all correspond to the same exposure.

10.3.1. CAMERA LENSES

The simplest type of a camera lens is the *biconvex* lens. It suffers from every type of aberration. It was used in the earliest stages. With the advancement in technology, lens combinations are used. They are classified as normal, wide-angle, telephoto, zoom etc.

Telephoto lens.

(*i*) The *normal lens* has a large aperture to reduce exposure time, an angular field of view of about 50° and produces an undistorted image.

(*ii*) A *telephoto lens* has a long focal length, gives a small angle of view and a large image of distant object.

(*iii*) A *wide-angle lens* is a lens of short focal length and gives a small image and a wide angle of view.

(*iv*) A *zoom lens* consists of a large number of individual lenses to decrease aberrations. The focal length of an ordinary camera may be increased up to six times without moving the image out of focus.

10.4. SIZE OF AN OBJECT

We often equate the *size of the object* to the linear dimension of the object, that is the height of the object. In practice, our feeling of the size of an object is actually derived from the size of its image cast on the retina. The size of the retinal image depends upon the **visual angle**, which is the angle subtended by the object at the eye. Therefore, it is known as the **angular size**.

Consider the objects A_1 and A_2 placed in front of the eye (Fig. 10.3). The objects are of different sizes, but they subtend equal angles at the eye and their images formed on the retina are of the same size. If k is the distance between the retina and the lens,

$$\frac{h_1}{k} = \theta_1 \quad \text{or} \quad h_1 = k\theta_1$$

Thus $$h_1 \propto \theta_1 \tag{10.1}$$

Fig. 10.3

Thus, the size of the image on the retina is proportional to the angle θ.

Consequently, we bring objects close to our eyes to see them in more detail. This action makes the subtended angle and the retinal image as large as possible. As the object is brought closer to the eye (A_2 taken to the position A_3), the image on the retina becomes larger and larger on account of increase in the visual angle. However, the object cannot be brought nearer to the eye beyond the least distance of distinct vision. The distance of distinct vision for a normal eye, NDDV, is 25 cm. If the object is brought nearer than this distance the eye cannot accommodate and in the process will have to strain itself.

An optical instrument forms the virtual image of the object, which acts as an object for the eye and subtends a much greater visual angle. A large image of a small object is thus formed on the retina. Thus we see small objects which cannot be seen by the naked eye. The instruments used for increasing visual angle are called **magnifying instruments**.

10.5. THE SIMPLE MAGNIFIER

A converging lens can be used to form a virtual image that is larger and farther from the eye than the object itself, as shown in Fig.10.4. Then the object may be moved closer to the eye and the *angular size* of the image may be substantially larger than the angular size of the object held at 25 cm without the lens. A lens used in this way is called a **magnifier** or a **simple microscope.** The virtual image is most comfortable to view when it is placed at infinity, as the muscle of the eye is relaxed.

Simple magnifier.

In Fig. 10.4 (a) is shown an object held at the near point (NDDV) where it subtends an angle θ at the eye. In Fig. 10.4 (b), a magnifier is held in front of the eye forming an image at infinity and the angle θ′ subtended at the magnifier.

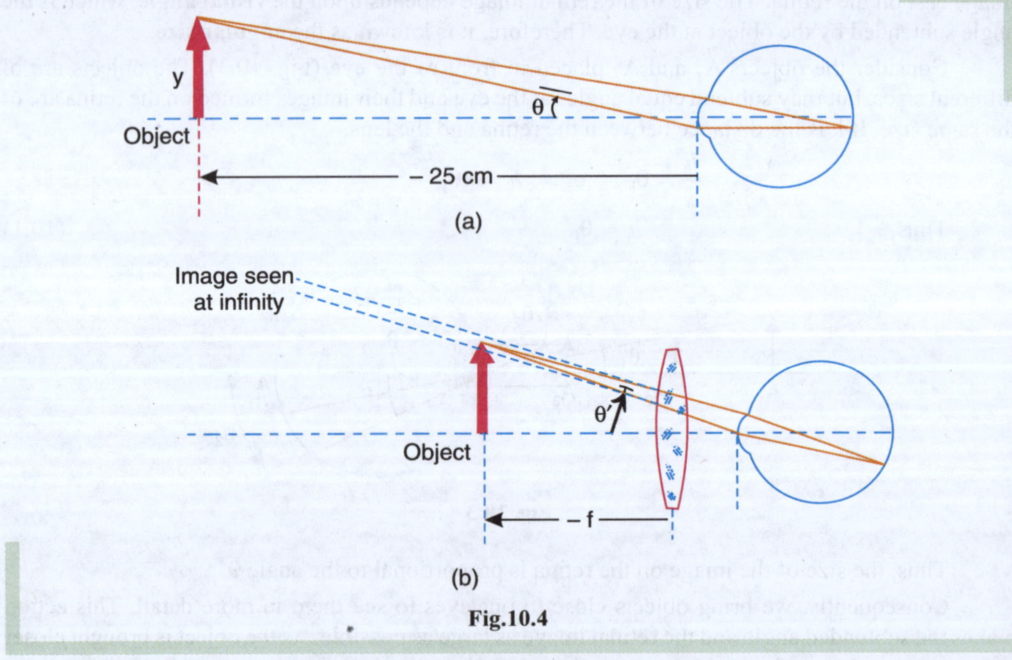

Fig.10.4

$$\text{Angular magnification } M = \frac{\theta}{\theta'}$$

Note— **Angular** magnification is not the same as lateral magnification. Angular magnification is the ratio of the angular size of an image to the angular size of the corresponding object. On the other hand, lateral magnification is the ratio of the height of an image to the height of the corresponding object. The difference can be understood this way. In Fig. 10.3 the angular magnification is of about 2 times, since the object subtends an angle about twice larger than that without the lens. Hence the object looks about two times larger to the eye. However the lateral magnification is infinite, because the virtual image is at infinity. It does not mean that the object is infinitely large. When dealing with a magnifier M is useful but m is not.

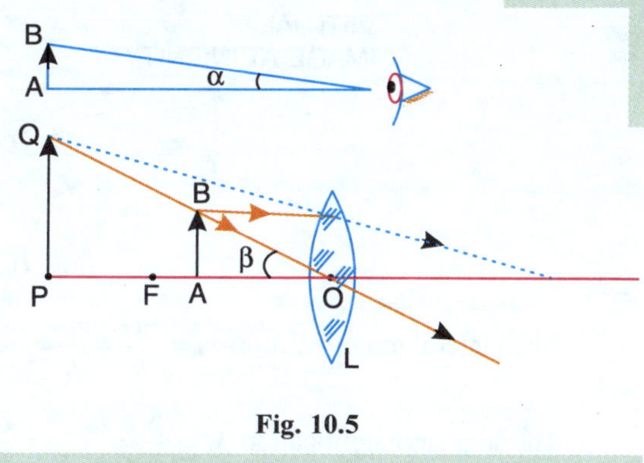

Fig. 10.5

Suppose an object of length AB = h is placed at the near point A and viewed by the eye (Fig. 10.5). Then the visual angle is α.

$$\tan \alpha = \frac{h}{D} \tag{10.1}$$

We can calculate how much the lens magnifies. Suppose that a convex lens L of focal length f is used as a magnifying glass and the object is placed between O and F, such that the magnified erect image PQ is formed. If the observer's eye is close to the lens, then the distance OP is equal to the least distance of distinct vision. Here, the visual angle is given by

$$\tan \beta = \frac{PQ}{D} \tag{10.2}$$

Then $$\frac{\tan \beta}{\tan \alpha} \cong \frac{\beta}{\alpha} = \frac{PQ}{h}$$

But $$\frac{PQ}{h} = m = \frac{\upsilon}{u} = \frac{D}{u}$$

Hence, angular magnification $$M = \frac{\beta}{\alpha} = \frac{D}{u} \tag{10.3}$$

We know that

$$\frac{1}{\upsilon} - \frac{1}{u} = \frac{1}{-D} - \frac{1}{-u} = \frac{1}{f}$$ (as per sign convention both D and u are negative)

$$\therefore \quad -1 + \frac{D}{u} = \frac{D}{f} \quad \text{or} \quad \frac{D}{u} = 1 + \frac{D}{f} \tag{10.4}$$

Using the result (10.4) into equ. (10.3), we obtain

$$M = 1 + \frac{D}{f} \tag{10.5}$$

To see the image in the above case, the observer has to *accommodate* his eye. **Accommodation** means that the image is seen at the distance of most distinct vision, i.e. 25 cm.

When the object is placed at the focus of the lens, the image is formed at infinity, as shown in Fig. 10.6. In this case, the image is seen *without accommodation.*

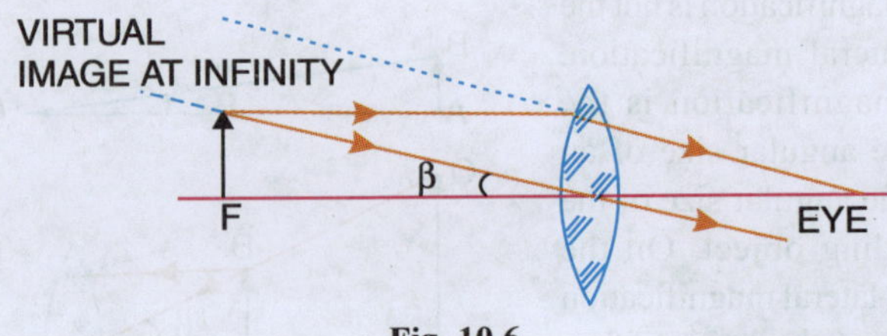

Fig. 10.6

Then lateral magnification $m = \frac{PQ}{h} = \frac{\infty}{h} = \infty.$

The angular magnification $M = \frac{\beta}{\alpha} = \frac{h/f}{h/D} = \frac{D}{f}$ (10.6)

From equ. (10.6), it may seem that we can make the angular magnification as large as we like by decreasing the focal length. In fact, the aberrations of a simple lens set a limit to M of about 3× to 4×. Angular magnification may be made as great as 20× by correcting for aberrations. When greater magnification is required we use a compound microscope.

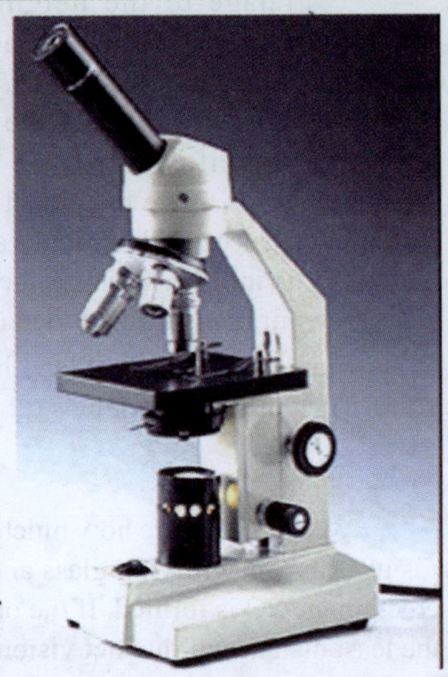

Compound microscope is used when greater magnification is required.

10.6. FIELD OF VIEW

Let us consider the situation in Fig. 10.7 (*a*). Light is passing through the aperture AB and reaching the eye E. The eye can view objects in the angle range PEQ. This is called the **field of view.** If a concave lens is added in plane AB, light from a wider field P′Q′ can reach the eye. We say *the field of view is enlarged.* On the other hand, if a convex lens is placed in plane AB would narrow the field of view. In compound systems such as telescopes and microscopes, the situation is slightly different and we discuss about it at the appropriate places.

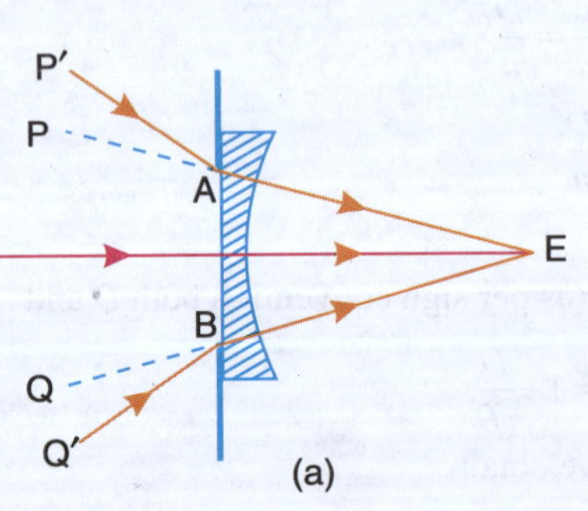

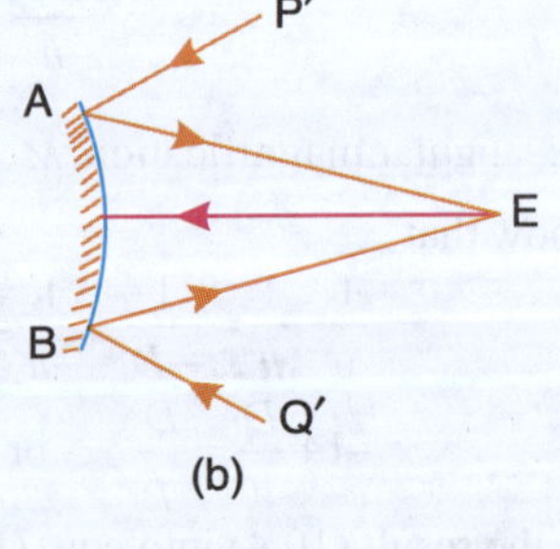

Fig. 10.7

10.7. STOPS AND PUPILS

When an optical instrument is used to study an object, light reflected from the object enters the instrument through its lenses. The light passing through the instrument is limited in cross-section either by the finite diameter of the lenses or diaphragms called *stops*. The amount of light that reaches the image plane of the instrument depends on the brightness of the object and also on the dimensions of the stops located in the path of light. The pupil of the observer's eye should be placed at the exit pupil of a system in order to have a comfortable viewing of the image. Therefore, in an optical system stops and pupils are as much important as lenses or mirrors.

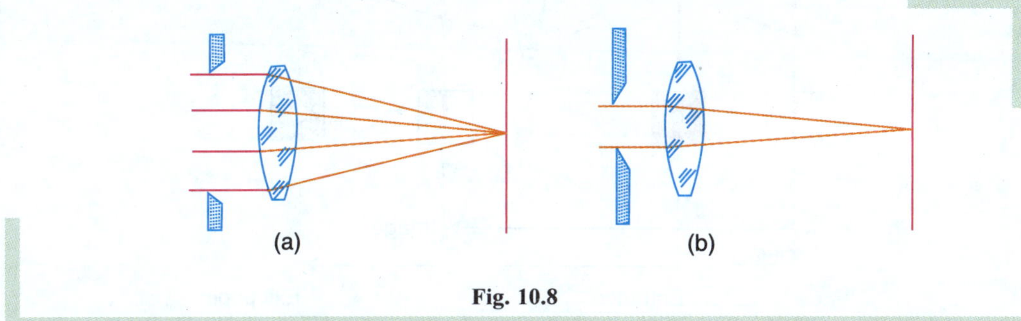

Fig. 10.8

Aperture stop: An **aperture stop** is a circular opening in an opaque screen. It limits the cross-section of the light that forms the image. Stops are used to restrict the light entering the instrument (Fig.10.8a) in order to obtain a good image. A smaller aperture makes the final image less bright than before; but the size of the image remains the same. The *clear* diameter of the lens admitting the light into the instrument acts as an *aperture* while the peripheral rim acts as a *stop.*

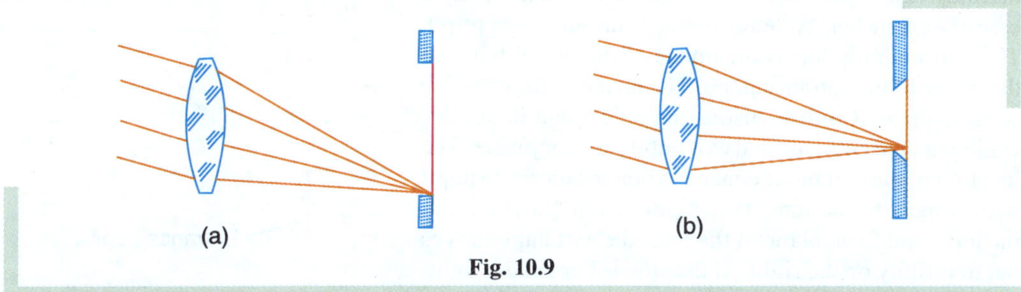

Fig. 10.9

Field stop: It is also necessary to limit the size of the image if good images are required. For this purpose, we use a diaphragm known as the **field stop**. The field stop limits the field of view of the instrument. As an example, the film or plate holder acts as the field stop in a camera (Fig. 10.9). As the field stop is made smaller, the field of view becomes narrower but the amount of light admitted to the image space remains constant. A field stop is often placed in the image plane.

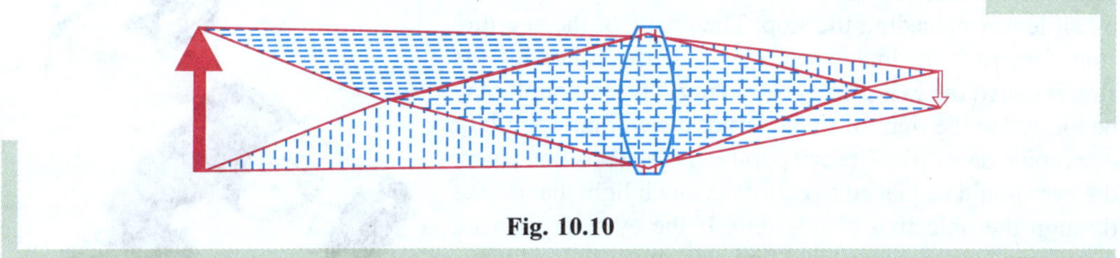

Fig. 10.10

Pupils: While a stop is a real object, a pupil is more conceptual. A pupil may be a real physical stop; or it may be an image of a stop. A **pupil** is defined as the cross-section of a bundle where light from all parts of the object passes through, completely mixed and without preferential spatial separation, as shown in Fig. 10.10.

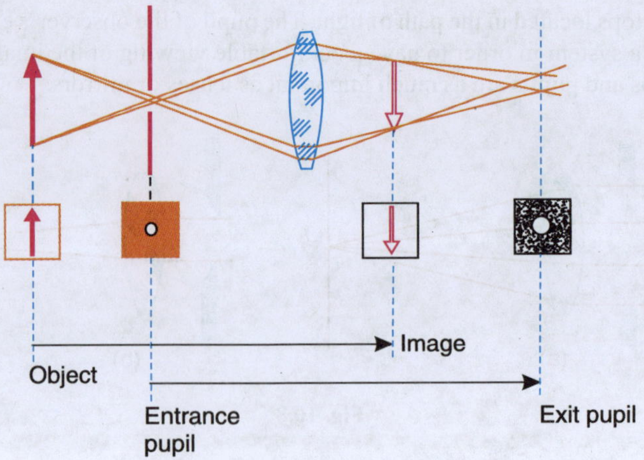

Fig. 10.11

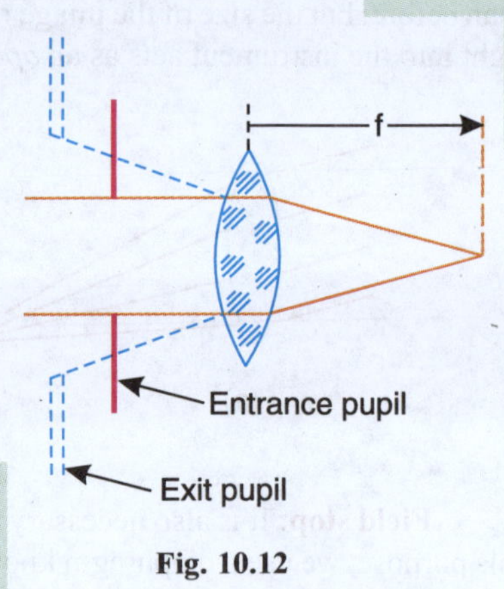

Fig. 10.12

If we place a stop between the object and a lens, that stop will become the pupil (Fig. 10.11). In this simple case of a single lens system, the stop is the **entrance pupil.** The lens not only forms the image of the object but also the image of the entrance pupil. The image of the entrance pupil is the **exit pupil**. Just as the object and image are conjugate, the entrance and exit pupils are conjugate. The location of the exit pupil depends upon the entrance pupil with respect to the lens. If the entrance pupil is kept at the left-hand focal plane of the lens, the exit pupil moves out to infinity on the right. If the stop is kept at the right-hand focal plane, then the exit pupil moves to infinity on the left. If the stop is kept within the focal length of the lens, the entrance pupil and and exit pupil lie on the same side of the lens, as shown in Fig. 10.12.

Combination of lenses: The *entrance pupil* of a combination of lenses is the image of the aperture stop formed by all lenses preceding the stop. The image of the aperture stop, formed in the image space by all lenses following the stop is called the **exit pupil** or **eye ring**. The exit pupil may be located to the right of to the left of the entrance pupil, or even coincide with it. The exit pupil is the best position where the eye should be placed to collect as much light that passes through the objective as possible. If the eye *were* placed elsewhere the observer would find a narrower field of view. A complex optical system consists of a combination of lenses and stops. There may be several diaphragms in addition to the aperture and field stops. The image of the field stop, formed by the lens preceding it, is called the ***entrance window***. The image of the entrance window, as formed by the complete system, is called the ***exit window***.

Astronomical telescope is mainly used in night.

In visual instruments, the observer's eye is positioned at the centre of the exit pupil. The pupil of the eye varies from 2mm to 8 mm depending on the level of the illumination. An astronomical telescope mainly used at nights might have an exit pupil of atleast 8 mm.

10.8. OBJECTIVE AND EYEPIECE

The magnifying power of simple microscope can be increased by decreasing the focal length of the lens. However, the focal length of a lens cannot be decreased beyond a certain limit. Moreover, the lens of small focal length has a smaller diameter because the curvature of the surface is large and the field of view is small.

Therefore, to increase the magnifying power, two separate lenses are used. The lens near the object is called the **objective**, which forms a real image of an object under examination (Fig. 10.13). The lens used to enlarge this image further to form a final image and which is then viewed by the eye is called an **eyepiece** or **ocular**. The objective of an ordinary microscope is usually an achromat with a magnification of about ×5 and an eyepiece usually consists of multiple lenses.

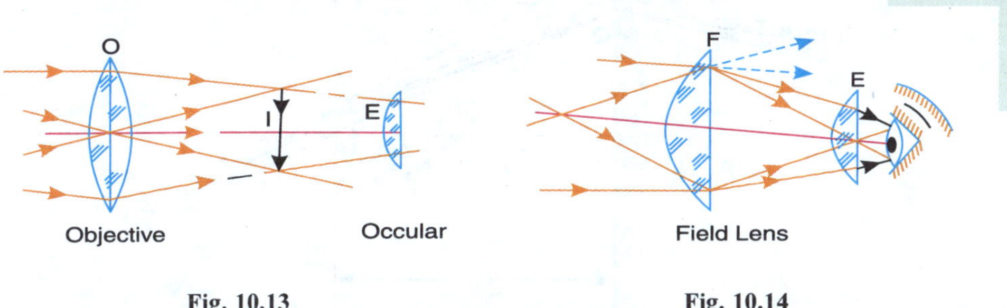

Fig. 10.13　　Fig. 10.14

An optical instrument is required to produce a magnified image free from aberrations and a bright image covering a wide field of view. If a single lens is used as an eyepiece, the final image will suffer from spherical and chromatic aberrations. Another drawback is the small field of view, which becomes lesser and lesser as the magnification of the instrument is increased. The rays passing through the outer portions of the image are refracted through the peripheral portions of the eye lens and they cannot simultaneously enter the small aperture of the pupil of the eye placed close to the eye lens (Fig. 10.13). Hence, only that part of the image, which is nearer to the axis, will be seen. Therefore, the final image will cover a small field of view. The field of view will progressively decrease as the distance between the objective and ocular is increased. The distance is varied in order to increase the magnification. In other words, the greater the magnifying power of the instrument, the smaller the field of view. It becomes therefore expedient to place an additional lens in the eyepiece to cause all the rays from the image to enter the eye lens. This extra lens is called a **field lens**. The function of the field lens is to gather in more of the rays from the objective toward the axis of the eyepiece (See Fig. 10.14). The field lens and the eye lens together constitute

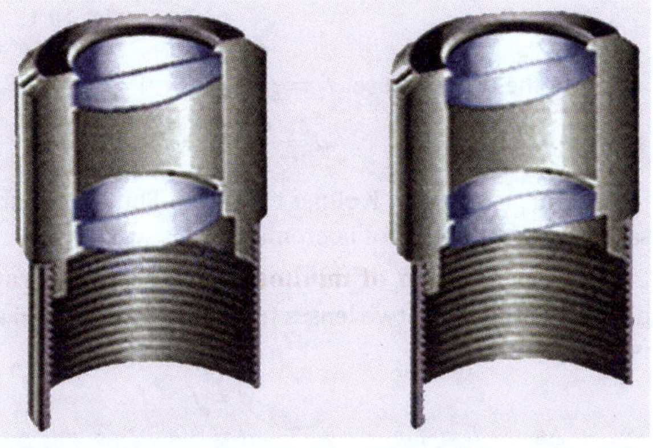

Huygen's and Ramsden's eyepieces.

an ocular or eyepiece. The two lenses are made and kept in such a way that their combination is achromatic and free from spherical aberrations. Two common eyepieces are the Huygens and the Ramsden types.

10.9. KELLNER'S EYEPIECE

It consists of two plano-convex lenses of equal focal lengths. The distance between the two lenses is equal to the focal length of either lens. The convex surfaces of both the lenses face the incident light.

***(i)* Condition for achromatism:** The distance between the two lenses for achromatism is

$$d = \frac{f_1 + f_2}{2} \tag{10.7}$$

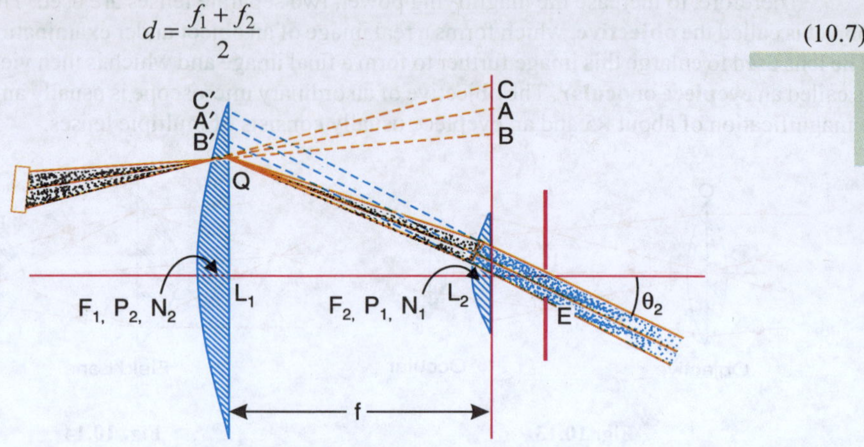

Kellner's eyepiece and its cardinal points.

Fig. 10.15

In the present case $f_1 = f_2 = f\,(say)$

$$\therefore \quad d = \frac{f+f}{2} = f \tag{10.8}$$

This is true for Kellner eyepiece. Thus, Kellner eyepiece satisfies the condition of achromatism.

***(ii)* Condition of minimum spherical aberration:** The distance between the two lenses for minimum spherical aberration is

$$d = f_1 - f_2 \tag{10.9}$$

As $\quad f_1 = f_2 \qquad d = 0.$

But the distance between the two lenses in Kellner eyepiece is f. Thus, the eyepiece does not satisfy the condition of minimum spherical aberration.

***(iii)* Equivalent focal length:** The equivalent focal length of the eyepiece is given by

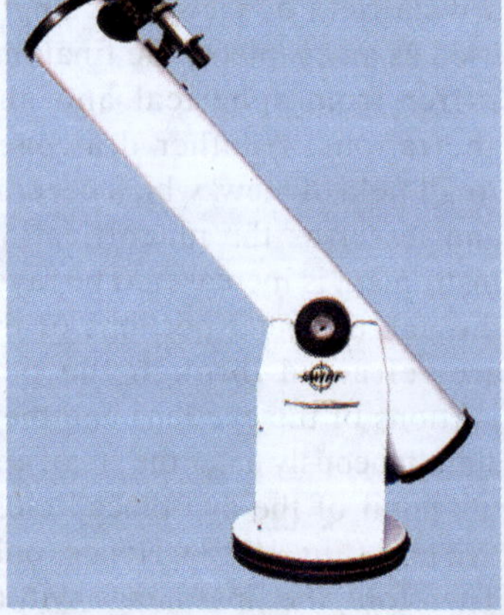

Kellner's eyepiece.

$$F = \frac{f_1 f_2}{f_1 + f_2 - d} = \frac{f\,f}{f + f - f} = f \tag{10.10}$$

The equivalent focal length of the eyepiece is thus equal to the focal length of either lens.

***(iv)* Principal points:** The distance of the first principal point P_1 from the lens L_1 is given by

$$\alpha_1 = \frac{d\,F}{f_2} = \frac{f\,f}{f} = f \tag{10.11}$$

That is, the first principal point lies at a distance f to the right of the field lens. The distance of the second principal point, P_2 from the eye lens L_2 is given by

$$\alpha_2 = -\frac{dF}{f_1} = -\frac{f\,f}{f} = -f \tag{10.12}$$

That is, the second principal point lies at a distance of f to the left of the eye lens.

(v) Focal points: The distance of the first focal point F_1 from the field lens L_1 is

$$\beta_1 = -F\left(1-\frac{d}{f_2}\right) = -f\left(1-\frac{f}{f}\right) = 0 \tag{10.13}$$

That is, the first focal point coincides with the pole of the field lens.

The distance of the second focal point F_2 from the eye-lens L_2 is

$$\beta_2 = F\left(1-\frac{d}{f_1}\right) = f\left(1-\frac{f}{f}\right) = 0 \tag{10.14}$$

That is, the second focal point coincides with the pole of the eye lens.

(vi) Nodal points: As the lenses are in the air medium, the nodal points N_1 and N_2 coincide with the principal points P_1 and P_2 respectively.

Position of cross-wires:

An eyepiece when adjusted for normal vision forms the final image at infinity. For this the objective should form the image at first focal plane. Hence the cross-wire should be placed on the plane surface of the field lens, so that they may be in focus with the final image.

Merits and demerits:

1. The field of view of the eyepiece is very large. Therefore, it may be used in the microscope and high power telescopes.
2. As the eyepiece does not satisfy the condition of minimum spherical aberration, the defect is present.
3. The cross-wires are permanently marked on the field lens. Any scratch or dust particle present on the field lens also come into sharp focus and as a result the final image is spoiled.
4. The defect of distortion is produced.

10.10. HUYGENS EYEPIECE

In the Huygens eyepiece a converging beam enters the field lens and forms a virtual image before the eye lens. The need for a converging beam implies that this eyepiece does not act like a simple magnifier.

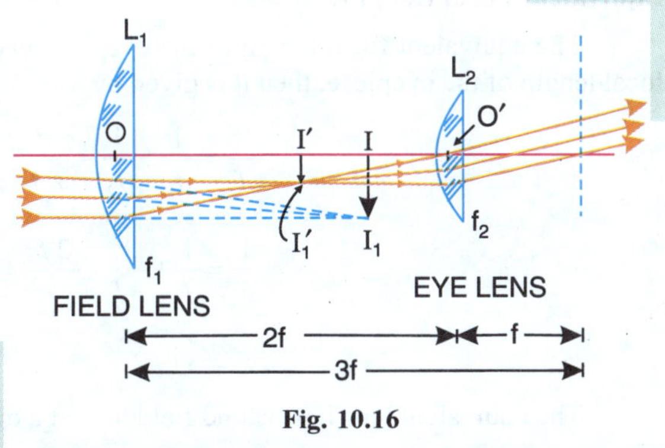

Fig. 10.16

Construction:

Huygens' eyepiece consists of two lenses having focal lengths in the ratio 3:1 and the distance between them is equal to the difference in their focal lengths. The focal lengths and the positions of the two lenses are such that each lens produces an equal deviation of the ray and the system is achromatic. This eyepiece is free from chromatic as well as spherical aberrations as it satisfies the two conditions simultaneously.

1. The lens combination acts as an achromatic system if $D = \frac{f_1 + f_2}{2}$ where D is the distance between the lenses and f_1 and f_2 are the focal lengths of the two lenses.
2. The lenses produce equal deviation of the incident ray when the distance between the two lenses is equal to $(f_1 - f_2)$.

The field and the eye lenses used are plano-convex and are placed with their convex surfaces towards the incident ray. In this way, the total deviation due to the combination is divided into four parts which makes the combination to have minimum spherical aberration.

If spherical and chromatic aberrations are to be minimised simultaneously, the following condition is to be satisfied.

$$D = \frac{f_1 + f_2}{2} \qquad \text{Chromatic aberration}$$

$$D = f_1 - f_2 \qquad \text{Spherical aberration}$$

Combining the two conditions, we obtain

$$\frac{f_1 + f_2}{2} = f_2 - f_1$$

or

$$f_2 = 3f_1$$

$$\therefore \quad D = 3f_1 - f_1 = 2f_1 \tag{10.15}$$

To satisfy the conditions for minimum chromatic and spherical aberrations, the focal length of the field lens should be three times the focal length of the eye lens and the distance between them should be equal to twice the focal length of the eye lens. Huygens' eyepiece is constructed on this principle.

Theory:

The objective forms an image, which serves as a virtual object for the field lens. The field lens forms real inverte image $I'I'_1$. If this image is situated at the principal focus of the eye lens, then the final image is at infinity.

Equivalent Focal Length:

The equivalent focal length of the eyepiece can be found as follows. If F is the equivalent focal length of the eyepiece, then it is given by

$$\frac{1}{F} = \frac{1}{f_1} + \frac{1}{f_2} - \frac{d}{f_1 f_2}$$

$$\frac{1}{F} = \frac{1}{f} + \frac{1}{3f} - \frac{2f}{3f^2} = \frac{2}{3f}$$

$$\therefore \quad F = 3f/2 \tag{10.16}$$

The equivalent lens lies behind field lens at a distance of

$$x = \frac{d \times F}{f} = \frac{2f \times 3f/2}{f} = 3f \qquad (10.17)$$

In other words, the equivalent lens is at a distance of $3f - 2f = f$ behind the eye lens.

Position of Cross-wires:

It is noted above that the principal focus of the equivalent lens lies $f/2$ ahead of the eye lens. It is here that the image due to objective must be formed in order that the final image is at infinity. The rays coming from the objective are, however, intercepted by the field lens, thus displacing the image to position I_1. It is here that the cross-wires should be placed.

We further noted above that the image I I_1 is at a distance of $f/2$ from the eye lens or $3f/2$ from the field lens. Hence, for field lens $u = -3f/2$. Then

$$\frac{1}{u} + \frac{1}{\upsilon} = \frac{1}{f}$$

$$\therefore \qquad -\frac{2}{3f} + \frac{1}{\upsilon} = \frac{1}{3f}$$

$$\text{or} \qquad v = f \qquad (10.18)$$

In other words, I I_1 lies midway between the two lenses and so fixes the position of the cross-wires or scale, if used. Since the image formed by the objective lies behind the field lens (instead of in the front) this eyepiece is sometimes referred to as a *negative eyepiece*. It may be noted that the image of the scale is formed by the eye lens whereas the final image is produced by both the lenses. Hence the image and the scale would not be magnified equally and so the measurements will not be reliable. The image of the cross-wire or scale would have all the defects of an image formed by a single lens. Hence in instruments using Huygens' eyepieces, scale is not used except when magnification is low.

Merits and Demerits:

(*i*) The Huygens' eyepiece is fully free from chromatic aberration because the distance between the lenses is equal to half the sum of their focal lengths.

(*ii*) Spherical aberration is also minimum because the distance between the two lenses is equal to the difference of their focal lengths.

(*iii*) The field of view of this eyepiece is smaller than that of Ramsden's eyepiece.

10.10.1. CARDINAL POINTS OF HUYGENS EYEPIECE

Position of Principal Points:

The equivalent lens should be placed at a distance α from the field lens, which is given by

$$\alpha = \frac{d \times F}{f} = \frac{2f \times 3f/2}{f} = 3f \qquad (10.19)$$

Since the distance d between the field lens L_1 and the eye lens L_2 is $2f$, the position of equivalent lens is $3f - 2f = f$, i.e.,it should be placed away from the eye lens, as shown by dotted line at P_1 in Fig. 10.17.

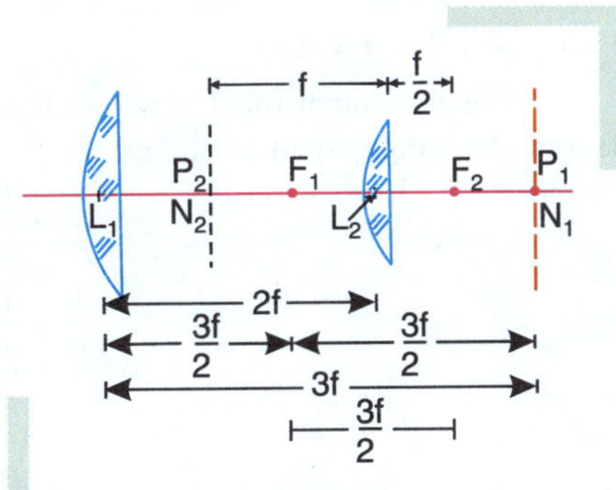

Fig. 10.17

The first principal point P_1 lies at a distance $\alpha = 3f$ from the field lens. The second principal point P_2 lies at a distance β from the eye lens towards the field lens and is given by

$$\beta = -\frac{d \times F}{f} = \frac{-2f \times \frac{3}{2}f}{3f} = -f \tag{10.20}$$

Position of the Focal Points:

The first focal point F_1 lies at a distance $3f/2$ from the first principal point P_1, i.e., at a distance $\left\{\frac{3f}{2} - f\right\} = \frac{f}{2}$ from the eye lens on the side of the field lens.

The second focal point lies at a distance $3f/2$ from the second principal point P_2, i.e., at a distance $\left\{\frac{3f}{2} - f\right\} = \frac{f}{2}$ from the eye lens away from the field lens.

10.11. RAMSDEN EYEPIECE

Ramsden's eyepiece consists of two plano-convex lenses each of focal length f separated by a distance equal to $(2/3)f$. The lenses are kept with their curved surfaces facing each other, as shown in Fig.10.18, thereby reducing spherical aberration. The field lens is a little larger than the intermediate image and is placed close to this image to allow as much light as possible to pass through it. The eye lens has a smaller diameter but carries out the actual magnification.

Theory:

The objective forms the real inverted image I of a distant object. This serves as an object for the field lens, which gives rise to a virtual image I_1. I_1 in turn serves as an object for the eye lens, which gives the final image at infinity, because I_1 is made to lie at its principal focus.

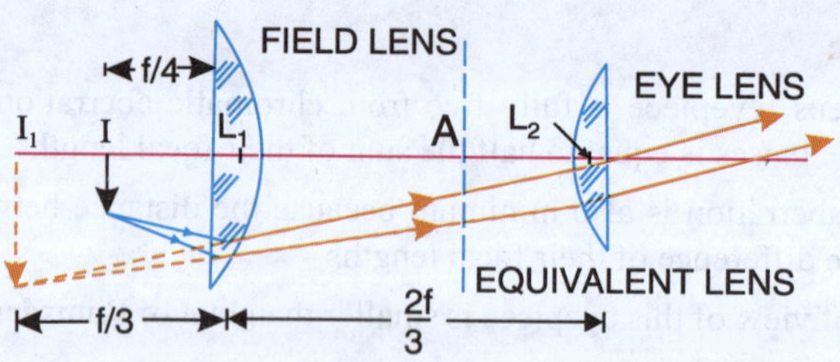

Fig. 10.18

Equivalent focal length :

The equivalent focal length of the eyepiece can be found as follows. If F denotes the focal length of the equivalent lens, then

$$\frac{1}{F} = \frac{1}{f_1} + \frac{1}{f_2} - \frac{d}{f_1 f_2}$$

$$\therefore \quad \frac{1}{F} = \frac{1}{f} + \frac{1}{f} - \frac{(2/3)f}{f^2}$$

$$= \frac{2}{f} - \frac{2}{3f}$$

$$= \frac{4}{3f}$$

$$\therefore \qquad F = 3f/4 \qquad (10.21)$$

The equivalent lens of focal length $3f/4$ must be placed behind the field lens at a distance

$$\alpha = \frac{F \times d}{f} = \frac{(3/4)f \times (2/3)f}{f} = \frac{f}{2} \qquad (10.22)$$

Thus the equivalent lens lies between the field lens and the eye lens.

Position of the cross-wires:

The cross-wires should be placed at the position of I. Now, the position of I relative to field lens can be found as follows. If the final image is to be formed at infinity, the image I should lie in the focal plane of the equivalent lens. In other words, the distance AI should be equal to $F = 3f/4$. Since $AL_1 = f/2$, $I_4 = f/4$. Therefore, the objective should produce the image at a distance of $f/4$ in front of the field lens. A fine scale may be placed here if it is desired to measure the size of the image. Since the scale and image would be magnified equally, the measurement would be trustworthy.

Merits and Demerits:

(*i*) The field of view of this eyepiece is fairly wide.

(*ii*) It is not entirely free from chromatic aberration since the distance between the two lenses is not equal to half the sum of their focal lengths. However, chromatic aberration is minimised by using an achromatic combination both for the field lens and the eye lens.

(*iii*) Spherical aberration is minimised by using two plano-convex lenses thereby spreading deviation over four surfaces.

Ramsden's eyepiece is used practically in all instruments where measurements of the size of the final image are to be made. This eyepiece is sometimes referred to as *positive eyepiece* because a real image is formed by the objective in front of the field lens, which further acts as a real object for the eye lens.

10.11.1. CARDINAL POINTS OF RAMSDEN EYEPIECE

Position of the principal planes:

The first principal plane of the equivalent lens lies at P_1 behind the field lens at a distance α given by the relation

$$\alpha = \frac{F \times d}{f} = \frac{(3/4)f \times (2/3)f}{f} = \frac{f}{2} \qquad (10.23)$$

Thus, the equivalent lens is placed at P_1 in between the field lens F and the eye lens E at a distance $f/2$ behind the field lens, as shown in Fig. 10.19. The point P_1 where the principal plane cuts the axis is the first principal point.

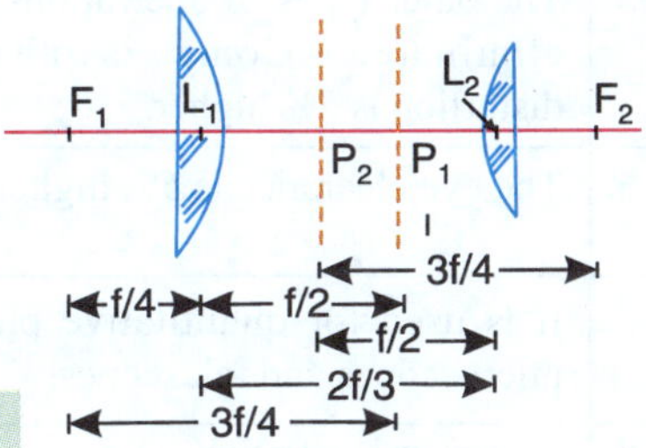

Fig. 10.19

The second principal plane is at a distance β from the eye lens L_2.

$$\beta = -\frac{d \times F}{f_1} = \frac{-\frac{2}{3}f \times \frac{3}{4}f}{f} = -\frac{f}{2} \qquad (10.24)$$

The second principal plane lies behind the eye lens L_1 at a distance $f/2$. The point P_2 where the principal plane cuts the axis is the second principal point.

Nodal Points:

As the lens system is situated in air, the nodal points coincide with the principal points. Thus, P_1 and P_2 are the nodal points, N_1 and N_2 respectively.

Position of Focal Points :

The equivalent focal length of the eyepiece is $3f/4$. Hence the first focal point F_1 lies at a distance $3f/4$ from the first principal point P_1. That is at a distance of $\left\{\frac{3}{4}f - \frac{1}{2}f\right\} = \frac{1}{4}f$ to the left of the field lens L_1.

The second focal point F_2 lies at a distance $3f/4$ from the second principal point P_2 i.e., at a distance $f/4$ behind the eye lens L_2 away from the field lens.

10.12. COMPARISON OF RAMSDEN EYEPIECE WITH HUYGENS EYEPIECE

	Ramsden Eyepiece	*Huygens Eyepiece*
1.	Ramsden's eyepiece is a positive eyepiece. The image formed by the objective lies in front of the field lens. Therefore, cross-wires can be used.	Huygens' eyepiece is a negative eyepiece. The image formed by the objective lies in between the two lenses. Therefore, cross-wires cannot be used.
2.	The condition for minimum spherical aberration is not satisfied. But by spreading the deviations over four surfaces, spherical aberration is minimized.	The condition for minimum spherical aberration is satisfied.
3.	It does not satisfy the condition for achromatism but can be made achromatic by using an achromatic doublet as the eye lens.	It satisfies the condition for achromatism.
4.	It is achromatic for only two chosen colours.	It is achromatic for all colours.
5.	The other types of aberrations are better eliminated. Coma is absent and distortion is 5% higher.	Other aberrations like pincushion distortion are not eliminated.
6.	The eye clearance is 5% higher.	The eye clearance is too small and less comfortable.
7.	It is used for quantitative purposes in microscopes and telescopes.	It is used for qualitative purposes in microscopes and telescopes.
8.	Its power is positive.	Its power is positive.
9.	The two principal planes are crossed.	The two principal planes are crossed.
10.	It can be used as a simple microscope because the first principal plane lies to the left of the field lens and the focal plane is real.	It cannot be used as simple microscope because the first focal plane lies to the right of the field lens and the focal plane is virtual.
11.	The nodal points coincide with the principal points.	The nodal points coincide with the principal points.

10.13. GAUSS EYEPIECE

Gauss eyepiece is a modification of Ramsden eyepiece. The field lens and the eye lens are two plano-convex lenses of equal focal length and separated by a distance equal to two third of the focal length of either. To illuminate the field of view a glass plate G is held at an angle of 45° to the axis of the lens system (Fig.10. 20). S is a source of light. Light reflected from G illuminates the field of view. The cross-wire C is kept at a distance $\frac{1}{4}f$ in front of the field lens. This eyepiece is used in the telescope of a spectrometer.

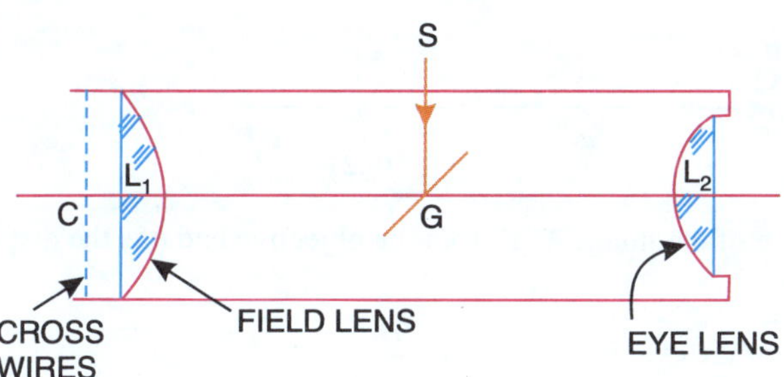

Fig. 10.20

10.14. COMPOUND MICROSCOPE

A simple magnifying lens is not useful where large magnification is required. A highly magnified image must be produced in two stages. A compound microscope is used for the purpose. Around 1590, Janssen, a Dutch spectacle maker, invented the compound microscope. Galileo announced his invention of a compound microscope in 1610.

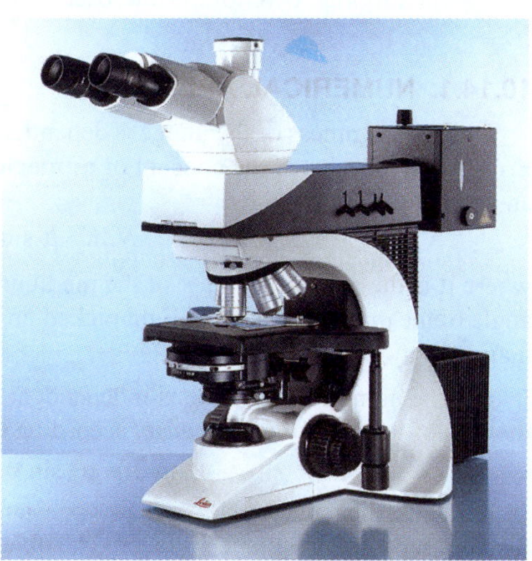

Compound microscope

A compound microscope has two lenses, an objective and an eyepiece. The object AB is placed at a distance slightly greater than the focal length of the objective. The objective forms a real, inverted, magnified image A′B′ at A′. The eyepiece is adjusted so that the distance of A′ from it is less than its focal length. The eyepiece further magnifies the image A′B′ and forms the final image PQ that is magnified virtual. Both the objective and the eyepiece of an actual microscope are highly corrected compound lenses with several optical elements, but for simplicity we show them here as simple thin lenses.

Magnifying power: The overall angular magnification of the compound microscope is the product of two factors. The first factor is the lateral magnification m of the objective, which determines the linear size of the real image A′B′ and the second factor is the angular magnification M_e of the eyepiece. When the microscope is in normal use, the final image PQ is formed at the near point, i.e. at D = 25cm from the eye.

Magnification of the objective $m = \frac{A'B'}{AB} = -\frac{\upsilon}{u}$

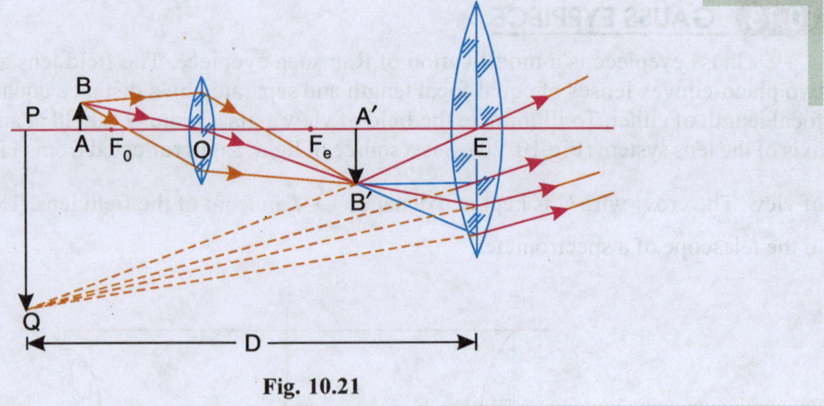

Fig. 10.21

where υ is the distance of the image A′B′ from the objective and u is the distance of the object AB from the objective.

Magnification of the eyepiece $M_e = \left(1 + \frac{D}{f_e}\right)$ (10.25)

where f_e is the focal length of the eyepiece.

Magnifying Power of the microscope $M = mM_e = -\frac{v}{u}\left(1 + \frac{D}{f_e}\right) = -\frac{v}{u}\left(1 + \frac{25cm}{f_e}\right)$ (10.26)

10.14.1. NUMERICAL APERTURE

The brightness of the image is dependent on the amount of light gathered in by the objective. Ernst Abbe introduced the concept of **numerical aperture** to specify the light-gathering ability of an objective. It is given by

$$N.A. = \mu \sin \theta_{max} \qquad (10.27)$$

where μ is the refractive index of the medium adjacent to the objective lens, and θ_{max} is the semi-angle of the maximum cone of light picked up by that lens. In other words, it is the angle made by a marginal ray with the axis.

The numerical aperture is independent of any refraction at plane surfaces at right angles to the axis of the instrument because, according to the law of refraction at plane surfaces

$$\mu \sin \theta = \mu_1 \sin \theta_1 = \text{constant} \qquad (10.28)$$

Thus, the value of N.A remains constant as the beam travels from one medium to another if they are separated by plane surfaces. The object is generally placed on a glass slide embedded in a material such as canada balsam and protected by a thin cover glass. If the medium between the cover glass and the front lens of the objective is air, which is, of lower refractive index than the cover glass.

Then, $\quad$ N.A. $= \mu \sin \theta = \mu_1 \sin \theta_1$

For air $\mu_1 = 1$, $\theta_1 = 90°$ and μ for glass = 1.5

∴ $\quad$ N.A. $= 1.5 \sin \theta = 1 \times \sin 90° = 1$

$\theta = 42°$ $\quad$ i.e. most of the rays are lost by total internal reflection.

If we allow for some working distance between the cover glass and the objective, the maximum N.A is not reached and the maximum possible value will be 0.95.

If the space between the cover glass and the objective is filled with oil of the same refractive index as that of cover glass, the aperture is increased.

$$\text{N.A.} = \mu \sin 90^\circ = 1.5 \times 1 = 1.5$$

The angle of 90° is not reached due to the thickness of cover glass and the working distance. It has been found that the practical value for $\theta = 67.5^\circ$

$$\text{N.A.} = \mu \sin 67.5^\circ = 1.5 \times \sin 67.5^\circ = 1.40$$

Hence for a dry objective, $(\text{N.A})_{max} = 0.95$ and for an oil immersion objective, $(\text{N.A})_{max} = 1.40$.

The magnification and the numerical aperture are engraved on the barrel of the objective.

10.15. TELESCOPES

Jan Lippershey invented the first telescope in 1608. In the year 1609 Galileo designed a telescope with which he made astounding astronomical observations. Newton designed a reflecting telescope because of the difficulty in manufacturing lenses free from chromatic aberrations. An astronomical telescope is used to view heavenly bodies. The telescope that uses a lens as an objective is called a **refracting telescope**. However, many telescopes use a curved mirror as an objective. Such telescopes are known as **reflecting telescopes**.

105 mm refracting astronomical telescope.

10.15.1. REFRACTING ASTRONOMICAL TELESCOPE

The simple astronomical telescope consists of two lenses (Fig. 10.22). The objective is a convex lens of long focal length and forms a real image of the distant object. The eyepiece is a convex lens of small focal length. It may be noted that the optical system of a telescope is similar to that of a microscope. The objective lens forms a real and reduced image AB of the object. The eyepiece enlarges this image and forms a magnified virtual image at infinity. As the object viewed by the telescope lies far away, the first image is formed nearly at the second focal point of the objective lens. The telescope is in *normal adjustment* when the image is formed at infinity. If the final image formed by the eyepiece lies at infinity, the first image must also be at the first focal point of the eyepiece. The distance between the objective and eyepiece is therefore the sum of their focal lengths, $F + f$.

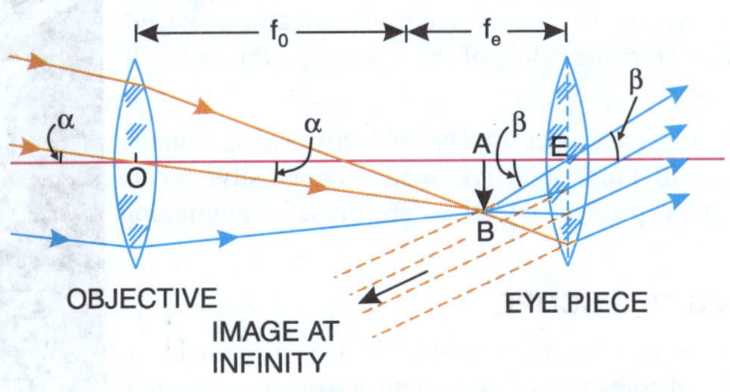

Fig. 10.22

To calculate the angular magnification M of the telescope, it is assumed that the eye is very close to the eyepiece. Moreover, the distance between the objective and the eyepiece is very small compared to the distance of the object from the objective or eyepiece. The angle α subtended at the unaided eye by the object can be taken as the angle subtended by the object at the objective.

$$M = \frac{\beta}{\alpha}$$

In the ΔABE, $\tan\beta = \frac{AB}{AE}$ or $\beta = \frac{AB}{AE}$

In the ΔOAB $\tan\alpha = \frac{AB}{OA}$ or $\alpha = \frac{AB}{OA}$

$$M = \frac{\beta}{\alpha} = -\frac{AB/AE}{AB/OA} = -\frac{OA}{AE} = -\frac{F}{f} \quad (10.29)$$

Thus, the magnification is equal to the ratio of the focal length of the objective and the eyepiece. The negative sign shows that the final image is inverted. To increase the magnification of the telescope, the objective should be of large focal length and the eyepiece should be of short focal length. By contrast, we may note that a microscope should have an objective of short focal length. An inverted image is not of any particular disadvantage for astronomical observations.

10.16. REFLECTING TELESCOPE

In the reflecting telescope the objective lens is replaced by a concave mirror. In large telescopes this has many advantages. The objective of the telescope should have a large aperture if fine details and a bright image are required. In the case of a refracting telescope the objective is a combination of lenses. But objective of very large aperture cannot be manufactured. Therefore, in very large telescopes the objective is a paraboloidal mirror instead of a converging lens. Mirrors are inherently free from chromatic and spherical aberrations. Spherical aberration in the case of a mirror can be avoided by the use of a paraboloidal mirror. The material of the mirror need not be transparent. Moreover, the mounting of a mirror is easier than that of a lens because the lens can be supported only at the rim. Because the image is formed in a region traversed by incoming rays, this image can be observed directly with an eyepiece by blocking part of the incoming beam. When a telescope is used for photography the eyepiece is removed and photographic film or an electronic detector is placed at the position of the real image. Most of the telescopes for research do not use an eyepiece.

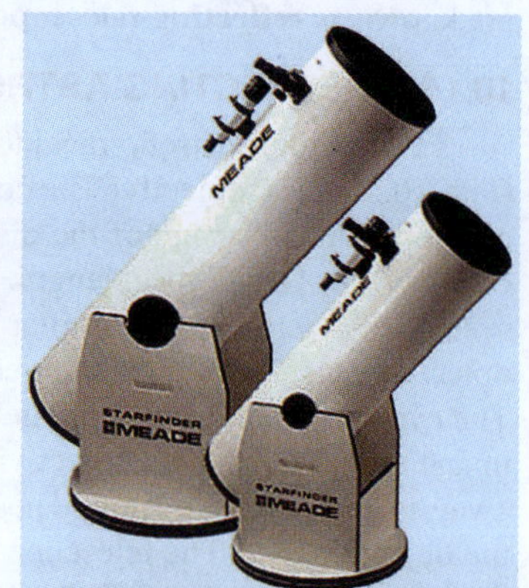

Reflecting telescope.

The early reflecting telescopes used had objectives of speculum, an alloy of copper and tin. Later objectives made of glass silvered on its surface were used. At present objectives are made of aluminum coated Pyrex.

10.16.1. NEWTON'S TELESCOPE

The objective is a large concave spherical mirror made of speculum metal (alloy of copper and tin). A Ramsden eyepiece is used

Newton's telescope.

with this telescope. The rays from the distant star are reflected from the concave spherical mirror A of large aperture and long focal length. Then they are reflected to one side at right angles to the axis, by means of a small plane mirror B. The final image is seen through the eyepiece E (Fig. 10.23).

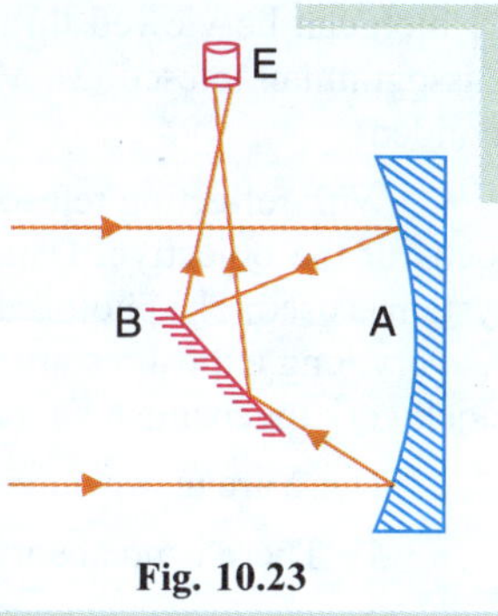

Fig. 10.23

10.16.2. OTHER REFLECTING TELESCOPES

Herschel's telescope is similar to the Newton's telescope but the axis of the mirror A is inclined to the incident rays from the distant star. This avoids the use of the mirror. The final image is formed on one side and is seen through the eyepiece E (Fig. 10.24).

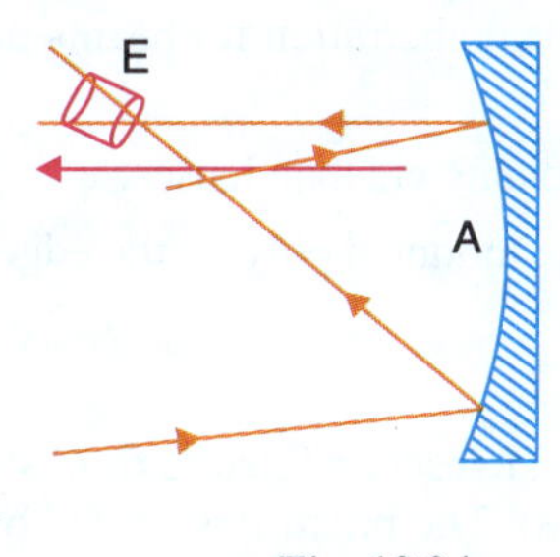

Fig. 10.24

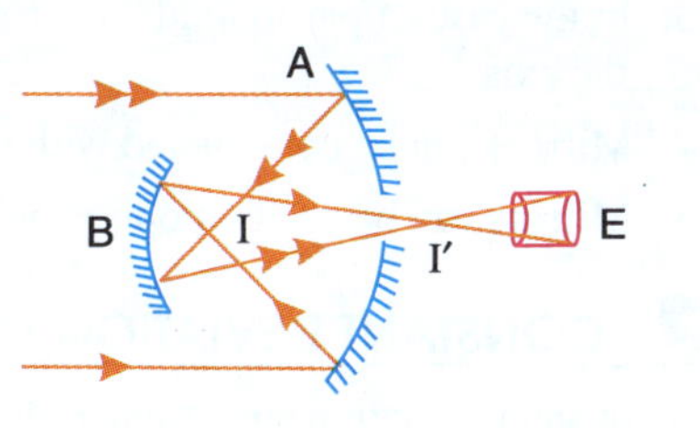

Fig. 10.25

In **Gregory's telescope** the objective A has small central aperture (Fig. 10.25). The rays from the distant star are reflected by the objective A. These rays form an image I at the point which lies between the focus and the centre of curvature of the concave mirror A. These rays after reflection from the mirror B form the final image I′ which is viewed by the eyepiece E. I′ is the secondary image of I.

Cassegrain's telescope has a large spherical mirror A, having an aperture in the centre just similar to Gregory's telescope. The rays from the distant star after reflection from the mirror A fall on the convex mirror B and are allowed to converge at I. The final image is formed at I′, which can be viewed through the eyepiece. The 500 cm Mount Polomar telescope is Cassegrain type (Fig. 10.26).

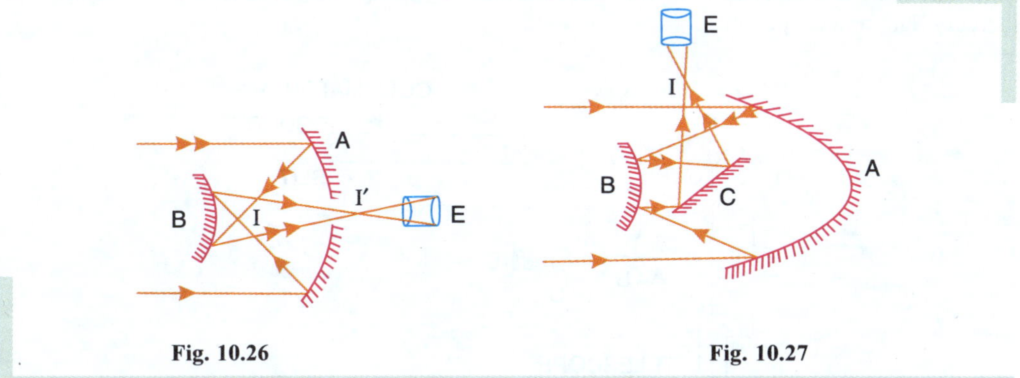

Fig. 10.26

Fig. 10.27

Coude's telescope has a large paraboloidal mirror A. B is a convex mirror and C is a plane mirror (Fig. 10.27). The rays from the distant star after reflection from the mirror A fall on the convex mirror B and after falling on the mirror C converge to a point I. The final image is formed at

I which can be viewed through the eyepiece. The telescope is a combination of Newtonian and Cassegrainian telescopes. Moreover, the necessity of the aperture in the centre of the mirror A is avoided.

With reflecting telescopes, visual observation may be made in the Newtonian manner at the focus of the objective. Usually to direct photography for spectroscopic observations Cassegrain system is used. The photographs are taken by placing the plate at the principal focus of the objective. As very long exposures are required the telescope is mounted so as to turn about the polar axis by clockwork mechanism so that the star observed is always in the centre of the field.

There are many points in favour of reflecting telescopes:

1. There is no absorption of light as in thick lenses.
2. The mirrors are free from chromatic aberration.
3. In the case of paroboloidal mirror, there is no spherical aberration for beams parallel to the axis.
4. Mirrors can be constructed with considerably large diameters than lenses.
5. Mirrors can be easily mounted whereas lenses can be mounted only on the edges or rim.

10.17. CONSTANT DEVIATION SPECTROMETER

A constant deviation spectrometer has a collimator and a telescope fixed to rigid stand such that they are mutually perpendicular to one another (Fig.10.28). The prism P is placed on a table which can be rotate about a vertical axis with the help of a drum D. A is a spring which allows the lever B to move when the drum D is rotated. The drum D has a graduated scale, which is calibrated and the index C shows the wavelengths of the different lines directly.

To start with, the slit is illuminated with sodium light. The drum D is rotated such that the index C is on a mark showing the reading 5890Å. The prism P is placed on the table and rotated slowly with hand such that the yellow line D_1 of sodium light is seen in the field of view of the telescope and the line coincides with the centre of the cross wires. The prism is clamped in this position and sodium light is removed. The source whose wavelengths for different lines are to be determined, is placed before the slit. The drum D is rotated to bring the lines at the centre of the cross wires in turn and the corresponding readings as shown by the index C are read. Therefore, with the help of this spectrometer, wavelengths of spectral lines can be determined directly and in a short time. The scale is usually calibrated and standardized by using lines of known wavelengths from mercury light or a copper arc.

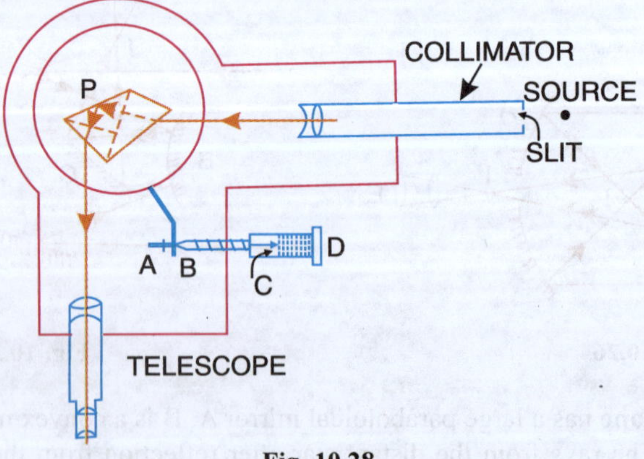

Fig. 10.28

The constant deviation spectrometer is helpful in finding any impurity in different salts and also for radical analysis. The substance is allowed to emit radiation in a carbon arc and the wavelengths of different spectral lines are found. From the standard charts of wavelengths of the spectral lines of different substances, the nature of the radicals present in a salt can be found. If by any chance, the prism is disturbed, it can be reset by using a spectral line of known wavelength.

10.18. PULFRICH REFRACTOMETER

It is used to measure the refractive indices of solids and liquids. It consists of a right angled prism A having its two faces perfectly plane. One of the faces is horizontal and the other is vertical

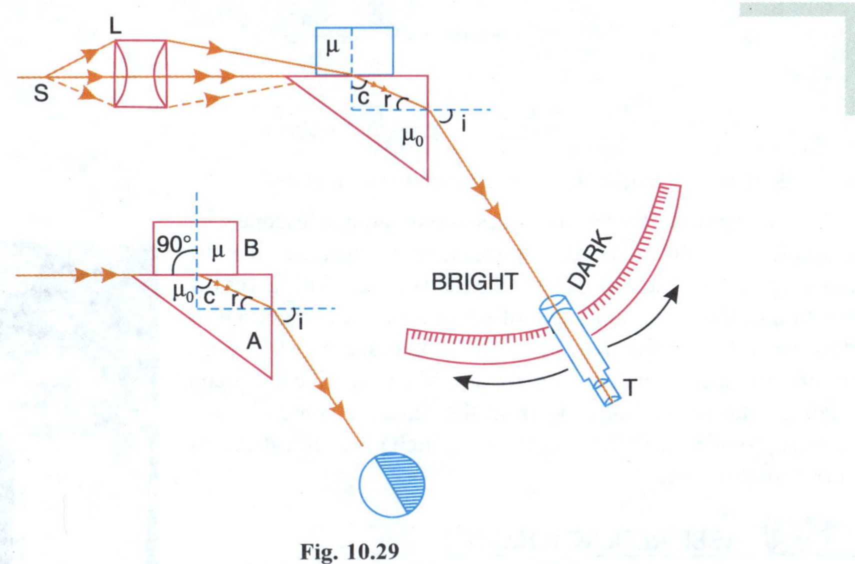

Fig. 10.29

(Fig. 10.29). The solid B whose refractive index is to be determined is taken having two faces cut perpendicular to one another. The faces are made perfectly plane.

The solid B is placed on the prism A. To bring perfect contact between the horizontal faces of A and B, a few drops of a liquid whose refractive index is greater than the solid is used. In the case of a liquid whose refractive index is to be determined, a glass cell is used.

Light is incident in a direction parallel to the horizontal surface so that light entering the prism A is incident at the critical angle c with the normal. Finally it emerges from the prism at an angle i.

Let the refractive index of the solid be μ_0.

Here $\mu_0. > \mu$ and c is the critical angle

$$\sin c = \frac{\mu}{\mu_0} \qquad (10.30)$$

also

$$\frac{\sin i}{\sin r} = \mu_0$$

Pulfrich refractometer.

$$r + c = \frac{\pi}{2} \qquad r = \frac{\pi}{2} - c$$

$$\frac{\sin i}{\sin(\frac{\pi}{2} - c)} = \frac{\sin i}{\cos c} = \mu_0$$

$$\cos c = \frac{\sin i}{\mu_0} \tag{10.31}$$

Squaring and adding equations (10.30) and (10.31), we get

$$\sin^2 c + \cos^2 c = \frac{\mu^2}{\mu_0^{\ 2}} + \frac{\sin^2 i}{\mu_0^{\ 2}}$$

$$\mu = \sqrt{\mu_0^{\ 2} - \sin^2 i} \tag{10.32}$$

Knowing μ_0 and i, the value of μ can be calculated.

To measure i, the apparatus has a collimating telescope with a circular scale attached to it. The telescope is adjusted so that the cross wires lie on the dark edge of the field of view. This gives the position of the minimum angle of emergence i. the, material of the prism A has a refractive index of 1.74. In modern instruments, the circular scale is calibrated in terms of the refractive index and readings can be read directly from the scale. In some cases, a table is provided with the instruments which give the value of μ corresponding to i.

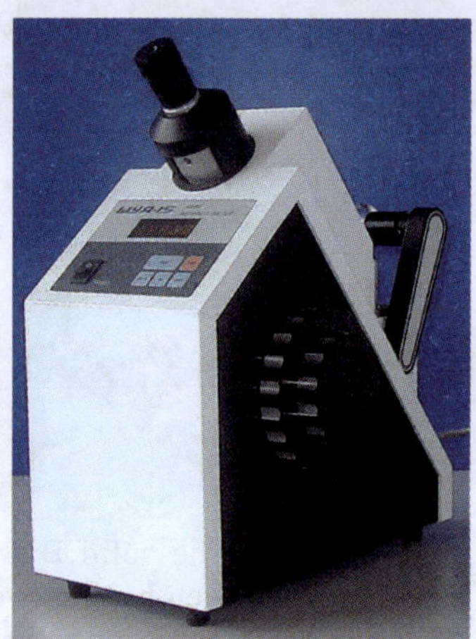

Abbe refractometer.

10.19. ABBE REFRACTOMETER

Abbe refractometer is used to measure the refractive index of liquids. It consists of two prisms A and B. The prisms have angles 30^0, 60^0 and 90^0. The liquid whose refractive index is to be determined is placed between the two prisms (Fig. 10.30).

Light is reflected from the glass plate or a ground glass surface P into the prism A. the emergent light is received by the telescope. The rays of light are incident on the hypotenuse face of the prism at various angles if incidence. At a particular angle of incidence, the ray passes along the plane of the liquid and just passes into the second prism. These are the rays, which have an angle of incidence equal to the critical angle. The theory is just similar to that of Pulfrich refractometer. The telescope is set to a position where the cross wires lie on the dark edge of the field of view. The reading is noted. The scale attached to the telescope is calibrated to give the refractive index of the liquid directly.

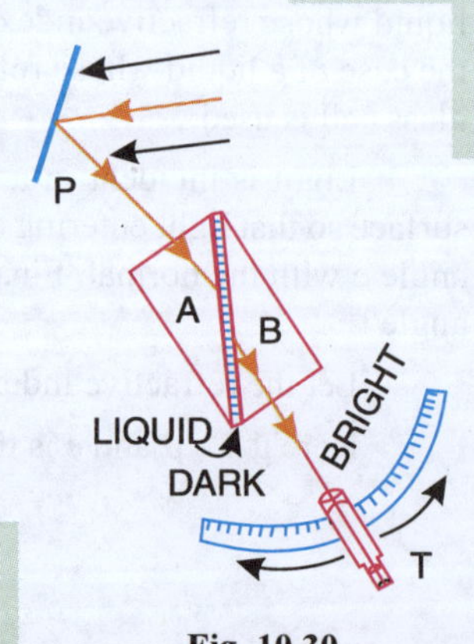

Fig. 10.30

10.20. PRISM BINOCULARS

Prism binoculars consists of two astronomical telescopes each containing two totally reflecting right angled prisms. On this way, the distance between the objectives of the telescope is made larger than the distance between the two eyes, thus increasing the field of view and the stereoscopic effect pertaining to distance. Moreover the effective length is made 1/3 of the length of the telescope to be used for the same magnification and the final image is also erect.

Prism binoculars.

The prism A is placed with its refractive edge horizontal while prism B is placed with its refracting edge vertical. In an astronomical telescope the final formed is inverted.

The objective and the eyepiece are convex lenses. The prism A turns the image in the vertical direction upside down (Fig. 10.31). The prism B turns the image in the horizontal direction. The image produced after reflecting is erect and magnified. The optical path is about three times the length of the binoculars. Therefore, it works as an astronomical telescope equal to three times its length.

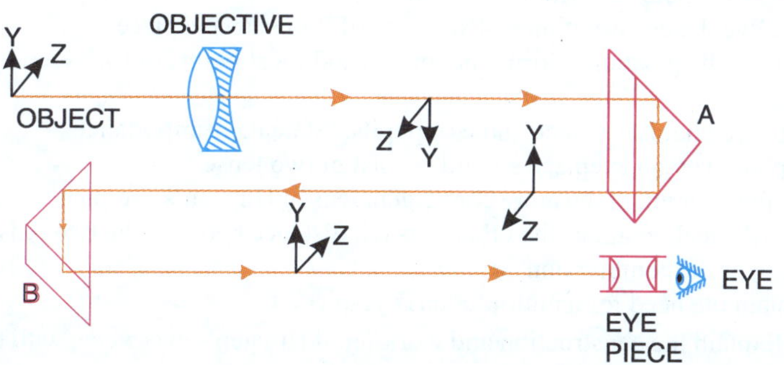

Fig. 10.31

QUESTIONS

1. What do you understand by visual angle and angular magnification? Give the construction and working of a simple microscope. Calculate its magnifying power.
2. Describe a simple photographic camera. What do you understand by *f*-number? Give the construction of a telephoto lens and explain its working.
3. What are the advantages of a compound microscope over a simple microscope? Calculate its magnifying power.
4. Explain the terms: entrance pupil and exit pupil. **(Nagpur, 2004)**
5. What are the functions of a field lens used in an eyepiece? Give the construction of a Huygens eyepiece and calculate the positions of the cardinal points.
6. What is an eyepiece and what is its advantage over a single lens?
7. Give the construction and working of a Ramsden eyepiece. How are chromatic and spherical aberrations minimized in this eyepiece?

8. Explain the construction of a Huygens eyepiece. Why cannot a cross-wire be used with it?
9. Give the name and construction of the eyepiece, which satisfies the condition for achromatism.
10. Describe and point out the respective merits of Ramsden and Huygens eyepieces.
11. Explain the principle of an oil-immersion objective.
12. Explain what do you understand by spherical and chromatic aberrations. Describe how these are minimized in a Huygens eyepiece.
13. Give the construction and working of a Ramsden eyepiece. How are chromatic and spherical aberrations minimized in this eyepiece? Indicate in a diagram the position of the cardinal points. Make the necessary calculations. **(Nagpur, 2004)**
14. Explain with the help of a neat diagram the construction and working of a Huygens eyepiece and clearly indicate the positions of its cardinal points. Why is it referred to as a theoretically perfect but a negative eyepiece?
15. What is an eyepiece and what is its advantage over a single lens? Give with a neat sketch the theory of Ramsden eyepiece. What types of eyepieces would you recommend for
    (*i*) low power microscope
    (*ii*) spectrometer telescope ?
16. (*i*) Give the construction and working of Ramsden eyepiece
    (*ii*) Find the position of principal points and focal points of Ramsden eyepiece. **(Nagpur, 2005)**
17. Explain what is entrance and exit pupils. State their importance. **(Nagpur, 2005)**
18. Explain why an eyepiece should consist of two lenses? **(Nagpur, 2005)**
19. The focal length of the more convergent lens of Huygen's eye-piece is 0.5 cm. Calculate the (*i*) focal length of the other lens (*ii*) distance between the two lenses and (*iii*) focal length of the combination **(Nagpur, 2004)**
20. Explain the need for a multiple lens eye-piece. **(Nagpur, 2004)**
21. (*i*) Explain the construction and working of Huygen's eye-piece, with the help of a neat diagram.
    (*ii*) Find the positions of principal points and focal points.
    (*iii*) Why is it called a negative eye-piece? **(RTMNU, 2010)**
22. The focal length of the more convergent lens of a Huygen's eye-piece is 0.5cm. Calculate the focal length of the eye-piece. **(RTMNU, 2010)**
23. Give the construction and working of Telephoto lens. **(RTMNU, 2010)**

11
CHAPTER

Velocity of Light

11.1. INTRODUCTION

Determination of velocity of light is of great importance. Einstein has shown that the energy released by the nucleus of an atom or otherwise is given by $E = mc^2$; where m is the decrease in mass and c is the velocity of light.

But, before the 17th century, it was thought that velocity of light is infinite. The fact that the flash of lightening is seen instantaneously and the sound is heard after some time shows that velocity of light is greater than the velocity of sound.

The first attempt to find the velocity of light was made by Galileo in 1600.

11.2. GALILEO'S EXPERIMENT

Two observers were stationed at a distance of a few kilometers. One observer uncovered his lamp and the second observer uncovered his lamp after seeing light from the lamp of the first observer. The first observer tried to measure the time interval between the uncovering of his lamp and the light seen from the lamp of the second observer. If the distance between the two observers = x, then $c = \frac{2x}{t}$. But Galileo failed to find the velocity of light as the time interval t was very small and could not be measured accurately.

At a Glance

The first successful attempt was made by Romer, a Danish astronomer in 1676.

11.3. ROMER'S ASTRONOMICAL METHOD

Romer observed the eclipses of the Jupiter's satellites at times when earth was at different positions with respect to Jupiter. He found that while the earth, in its orbital motion round the sun, receding from Jupiter, the mean period between two successive eclipses of a particular satellite is longer than that when the earth is moving nearer the Jupiter. This anomaly formed the basis for the calculation of velocity of light. He explained when the earth is receding from Jupiter; light has to travel a greater distance at each successive disappearance of satellite whereas when the earth is approaching the Jupiter, light has to travel a shorter distance at each successive disappearance of the satellite.

Jupiter has a number of satellites or moons revolving round it. Jupiter makes a complete revolution around the sun in 11.86 years whereas the earth completes one revolution in one year. It is assumed for the sake of simplicity that the orbit of the earth and Jupiter are circular.

The satellites which revolve round the Jupiter have their periods lying between 11 hours 58 minutes for the satellite nearest the planet and 16 days, 16 hours, 32 minutes and 11 seconds for the most remote satellite.

As the satellites revolve in orbits nearly parallel to the plane of the Jupiter's orbit, each satellite, once in every revolution, enters the shadow of the Jupiter and so becomes eclipsed (Fig 11.1). Romer studied the eclipse of the innermost satellite of Jupiter. At some time, Jupiter J_1 and the earth E_1 are on the same side of the sun and are in conjunction. If light were transmitted instantaneously, the actual time of eclipse and its observation on the earth should be the same. If light has a velocity c, then light from the satellite at the time of eclipse has to travel a distance J_1E_1 before reaching the earth. Thus the eclipse will be observed at a time J_1E_1/c later on the earth. If the actual time of the eclipse is T_0 and T_ϕ is the time when it is observed on the earth then

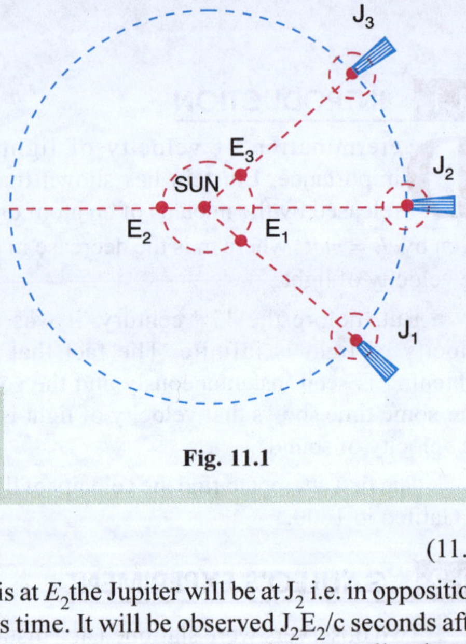

J3
E3
J2
SUN
E2
E1
J1

Fig. 11.1

$$T_1 = T_0 + J_1E_1/c \tag{11.1}$$

After a lapse of 0.545 of a year, when the earth is at E_2 the Jupiter will be at J_2 i.e. in opposition. Let the eclipse of the same satellite be observed at this time. It will be observed J_2E_2/c seconds after its actual occurrence. If t is the time for one revolution of the satellite the time actually elapsed between the first and the nth eclipse will be $= (n-1)\,t$. If T_2 is the time of observation of the eclipse on the earth, then

$$T_2 = T_0 + (n-1)\,t + \frac{J_2E_2}{c} \tag{11.2}$$

Subtracting (11.1) from (11.2),we get

$$T_2 - T_1 = (n-1)\,t + \frac{J_2E_2 - J_1E_1}{c} \tag{11.3}$$

Let $$T = T_2 - T_1 = (n-1)\,t + \frac{J_2E_2 - J_1E_1}{c}$$

$$= (n-1)\,t + \frac{d}{c}$$

where d is the diameter of the earth's orbit round the sun

$$T = (n-1)t + d/c \qquad (11.4)$$

Similarly, it can be proved that the time interval between the first and the nth eclipse observed on the earth when the earth moves from E_2 to E_3 will be

$$T' = (n-1)t - d/c \qquad (11.5)$$

i.e., when the earth moves from opposition to conjunction.

Subtracting (11.5) from (11.4),we obtain

$$T - T' = 2d/c$$

Romer observed that the time interval T was 32 minutes 52 seconds more than the time interval T′. Taking the diameter of the earth's orbit as 185.6×10^6 miles and the time interval T – T′ =1972 seconds, Romer calculated the value of c as

$$c = \frac{2 \times 185.6 \times 10^6}{1972} \text{ miles/second}$$

$$c = 186{,}000 \text{ miles/second (approx.)}$$

This method is not very accurate due to the following reasons: (i) orbits are not circular but they are elliptical (ii) correct value of the diameter of the earth is not known (iii) it may not be exact time when the eclipse occurs.

11.4. BRADLEY'S ABERRATION METHOD

This method is based on the principle of relative velocity. When rain is falling vertically downward, a person standing on the road holds his umbrella vertically, but when he moves on the road, he holds his umbrella in a direction making some angle with the vertical.

This fact has been used in the measurement of the velocity of light by Bradley.

In 1729, Bradley observed that the angular elevation of a 'fixed' star varied slightly according to the position of the earth in its orbit around the sun. One day he was being rowed across a stream in a boat and noticed that the boat drifted slightly downstream. He at once thought that the difference between the actual and the observed angular elevation of the star was due to a combination of the velocity of the earth in its orbit and that of the velocity of light.

James Bradley (1693-1762)

If the earth were stationary, the telescope would have to point in the direction AS analogous to umbrella when the man is stationary to observe the star [Fig. 11.2 (*i*)]. But since the earth is moving with a velocity of 18.5 miles/second around the sun, the telescope would have to be directed along AS′ to observe the star (analogous to umbrella when the man is moving in rain) as shown in Fig. 11.2 (*ii*).

The direction of AS′ is that of the relative velocity of light with respect to the earth. The angle between the true and the apparent direction of the star is α.

Here $\qquad \tan\alpha = \upsilon/c$

Since the angle is small, $\qquad \alpha = \upsilon/c$

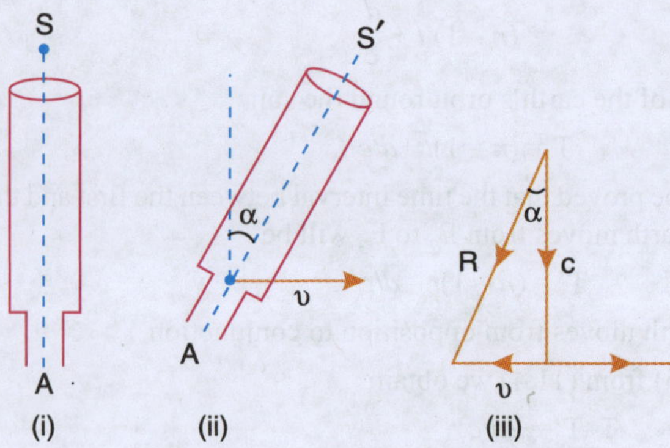

(*i*) When the earth is stationary; (*ii*) When the earth is moving in its orbit;
(*iii*) Relative velocity of light with respect to the velocity of the earth.

Fig. 11.2

It will be clear from the Fig. 11.3, that when the earth is at E_1 it is moving towards the star, and at E_3 it is moving away from the star. In these cases for the purpose of receiving light from the star, it is stationary and there is no aberration. The aberration is maximum at E_2 or E_4 because the earth will be moving normal to true direction of the star (analogous to the man moving normal to the direction of the rain).

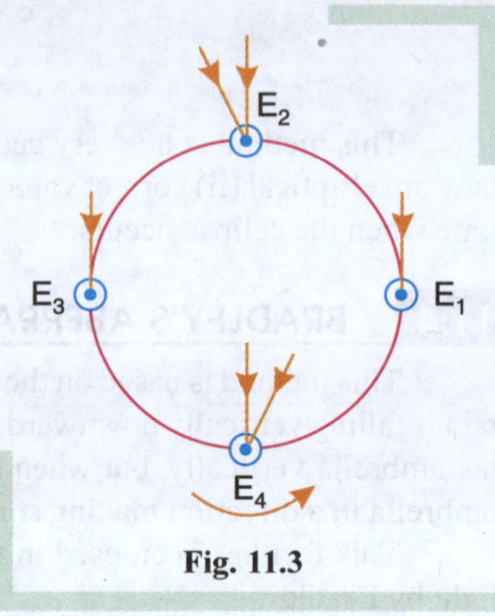

Fig. 11.3

The direction of the motion of the earth is continually changing and the earth is moving in diametrically opposite directions at the spring and autumn equinoxes.Therfore, on observing the position of the star near the zenith at these times, through a telescope it will be seen to occupy different positions with respect to the horizon. At the spring equinox, $\alpha = \upsilon/c$ and at the autumn equinox, $\alpha = \upsilon/c$.

From the two positions of the telescope, the aberration can be calculated and it is equal to half the angle between the two positions of the telescope.

The angle α was found to be 20.24″ of an arc and since υ = 28.5 miles /second the value of the velocity of light c was calculated as 186, 230 miles/second. However, aberration is independent of the distance of the star from the earth and should not be confused with parallax.

11.5. FIZEAU'S METHOD

The first terrestrial method for determining the velocity of light was performed by Fizeau in 1849.The experimental arrangement is shown in Fig. 11.4.

Armand Fizeau (1819-1896)

A bright source S emits light which after passing through a lens L and after reflection from the plane P is converged to a point H. The point H lies at the focus of the lens L_1 and in a space between the two teeth of the wheel W. Therefore, the light after passing through L_1 is rendered parallel and after traveling a distance of few miles is allowed to fall on the lens L_2. The light after refraction through the lens L_2 is brought to focus at A which is also the pole of the concave mirror M.The radius of curvature of the mirror M is equal to twice the focal length of the lens L_2. Due to this ,the rays are reflected along their original path. An image of S can thus be observed by the eye placed behind the eyepiece E (Fig. 11.4).

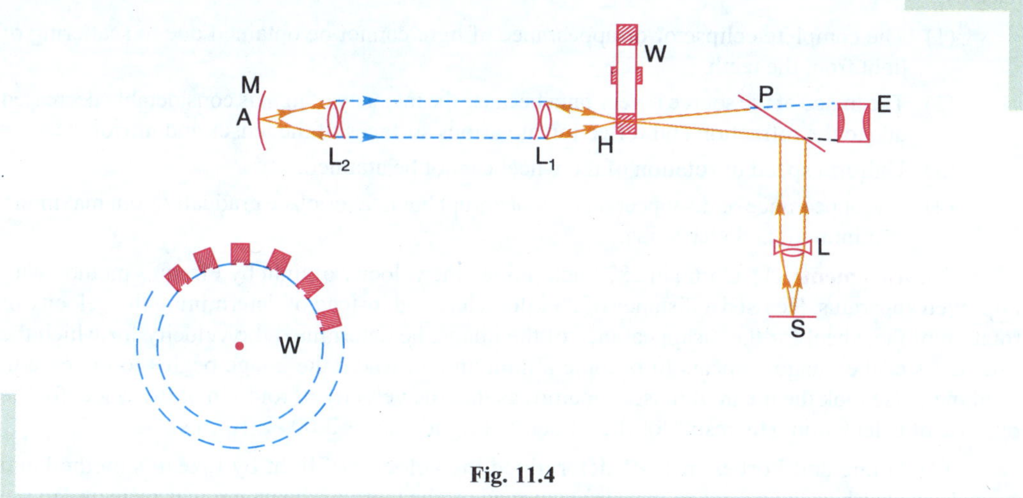

Fig. 11.4

The rim of the toothed wheel W is at H and the wheel is rotated about a horizontal axis. This is the important path of Fizeau's experiment. The teeth and spaces of the wheel are of equal width.Fizeau used a wheel with 720 teeth.

Working: When the wheel W is rotated, an image of the source is observed through E as long as light passes through the space of the wheel towards E.The wheel is rotated at such a speed that the rays passing the space fall on the tooth on their return. When this is adjusted, then no image of the source S is seen by the eye. If the speed of the rotation of the wheel is doubled, again the image of the source S is seen by the eye because the rays on their return journey again pass through the next space. When the speed of the wheel is three times the first, no image of the source is seen by the eye because the rays on the return fall on the next tooth. In this way, if the speed of the wheel is continuously increased, the image of the source S will be alternately blocked and seen.

If the distance between H and A is equal to d and the wheel makes n rotations per second when the first eclipse is observed, then the time taken by light to travel from H to A and back is

$$t = 2d / c$$

If the wheel has 'm' teeth, m spaces and in the time t, the wheel moves from the centre of the space to the centre of the tooth, then

$$t = 1/2\, m\, n$$

$$2d / c = 1/2\, m\, n$$

and

$$c = 4\, m\, n\, d \qquad (11.6)$$

If n_2 is the number of rotations per second of the wheel when the second eclipse is observed then $c = 4\, m\, nd/3$, because $n_2 = 3\, n$. In Fizeau's actual experiment

$$n = 12.6 \text{ rotations/second}$$

$$m = 720$$

$$d = 8633 \text{ meters}$$

$$c = 4\, m\, n\, d$$

$$c = 3.13 \times 10^8 \text{ m/s}$$

The main advantage of this method is that the principle involved is simple and Fizeau actually took the idea from the experiment attempted by Galileo (covering and uncovering the lamp).

It should be remembered that Fizeau's experiment is not free from criticism due to the following reasons:

(1) The complete eclipse or disappearance of light cannot be obtained due to scattering of light from the teeth.

(2) The image of the source is very faint because the intensity of light is considerably decreased due to the refraction and reflection at various surfaces of the lenses and mirror.

(3) Uniform speed of rotation of the wheel cannot be attained.

(4) The appearance or disappearance is not abrupt but it takes place gradually from maximum to minimum and vice versa.

Improvement: (1) Cornu in 1874 determined the velocity of light by Fizeau's method with improved apparatus. He used a distance of 23 kilometers and instead of determining the velocity of rotation of the wheel for the disappearance of the image, he determined the velocity for which the brightness of the image, appears to become minimum and where the image begins to increase in brightness. He took the mean of these velocities as the true velocity of rotation of the wheel for the purpose of calculation. His result for the velocity of light in air is 3.004×10^8m/s.

(2) Young and Forbes in 1881 determined the velocity of light by Fizeau's method and beveled the teeth of the wheel so that light stopped by the teeth of the wheel was reflected to the sides. Morever, they used the silvered plates having a small aperture for viewing the image instead of the glass plate P. They calculated the value of c as 3.013×10^8 m/s.

(3) Perrotin used Fizeau's method and kept the distance equal to 40 kilometers and calculated the value of the velocity of light in air as 2.999×10^8m/s.

11.6. FOUCAULT'S ROTATING MIRROR METHOD

In 1862, Foucault designed an apparatus for measurement of velocity of light. It requires much shorter distance than Fizeau's method and it is a modification of Fizeau's method.

Foucault used a rotating mirror instead of a toothed wheel.

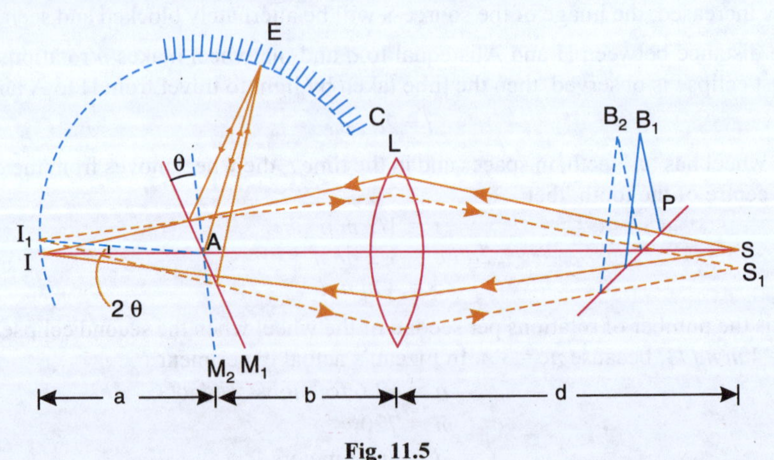

Fig. 11.5

Light from a strong source S falls on an achromatic lens L, after passing through the glass plate P (Fig. 11.5). The light after passing through the lens L will converge at the point I.

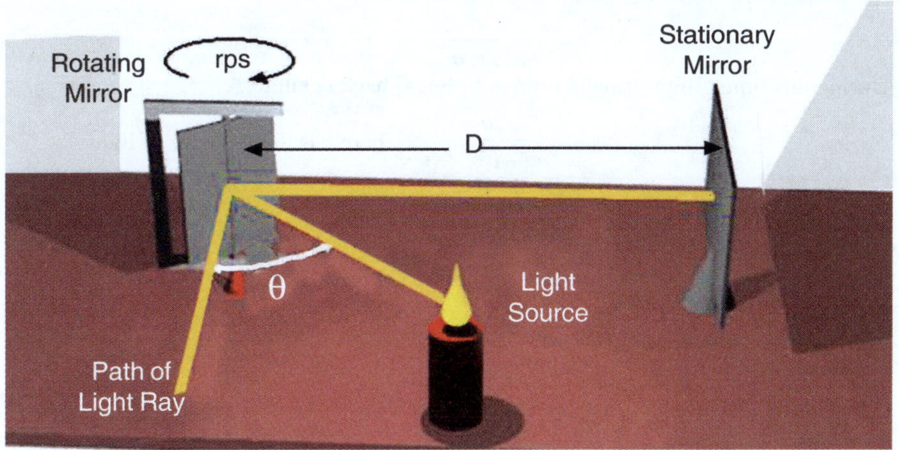

Velocity of light demonstration by the foucault rotating mirror method. This is a simplified version of setup.

If a mirror M_1 is placed at A, light after reflection from the surface of the mirror M_1, converges at the pole of concave mirror C, whose distance from A is adjusted equal to its radius of curvature. Light is reflected back from C along its original path and finally the image is formed on the source S. As there is a half silvered glass plate P inclined at an angle of 45° to the axis of the lens, the image is formed at B_1, which can be viewed with the help of a micrometer eyepiece.

Suppose, the plane mirror M_1, is rotated rapidly at a uniform speed about an axis passing through A, the rays after reflecting from the concave mirror C, find the plane mirror displaced by an angle θ to a new position M_2. The image is now observed at B_2. The displacement B_1B_2 of the image is measured and the velocity of light is calculated.

Theory: Consider a point E on the concave mirror from which light is reflected back to the rotating mirror. When the plane mirror is at the position M_1, the image of S is at I such that AE = AI = *a*, where *a* is the radius of curvature of the concave mirror. When the plane mirror is rotating with a uniform speed and is at a position M_2, the image of S will be formed at I and the light appears to diverge from I_1 instead of I.

If the mirror turned through an angle θ, the reflected ray is turned through an angle 2 θ.

$$I I_1 = a \times 2\,\theta$$

As S and S_1 are conjugate points with respect to I and I_1, for the lens L

$$\frac{SS_1}{d} = \frac{II_1}{(a+b)}$$

$$S_1S = \frac{a \times 2\theta . d}{(a+b)}$$

Suppose $S_1S = B_1B_2 = x$

Then

$$x = \frac{2a\theta\, d}{(a+b)}$$

and $$\theta = \frac{x(a+b)}{2ad} \qquad (11.7)$$

If the plane mirror makes *n* rotations per second, then the time taken by the plane mirror to rotate through an angle θ is

$$t = \frac{\theta}{2\pi\, n}$$

During this time t, light travels from A to E and back again to A.

$$\therefore \qquad t = \frac{2a}{c} \qquad \text{where } c \text{ is the velocity of light.}$$

$$\frac{2a}{c} = \frac{\theta}{2\pi\, n}$$

$$\theta = \frac{4\pi\, n\, a}{c} \qquad (11.8)$$

From equ. (11.7) and (11.8), we obtain

$$\frac{x(a+b)}{2ad} = \frac{4\pi\, n\, a}{c}$$

$$\therefore \qquad c = \frac{8\pi\, n\, a^2 d}{x(a+b)} \qquad (11.9)$$

As *n, a, d, x* and b are measurable quantities, c can be calculated.

The displacement of the image in the Foucault's experiment was only 0.70 mm and could be measured with the help of micrometer eyepiece. The value of c found by Foucault was 2.96×10^8 m/s.

The main disadvantage of the Foucault's method is that the image obtained is not very bright due to reflection and refraction of light at various surfaces.

Foucault observed the velocity of light in water by placing a long cylindrical pipe of water between the concave mirror and the plane mirror. He found that the displacement *x* of the image when water was used, was greater for the same speed of rotation of the mirror than when air was used.

Therefore, according to the relation $c = \frac{8\pi\, n\, a^2 d}{x(a+b)}$, the value of c in water should be less than the value of c in air.

Newton's corpuscular theory of light predicted that the velocity of light in water should be more than that in air. The result of Foucault gave a death blow to the corpuscular theory of light. The wave theory of light predicted that the velocity of light in water should be less than in air. Thus, Foucault's experiment justified the validity of the wave theory of light.

11.7. MICHELSON'S METHOD (ROTATING MIRROR NULL METHOD)

A. A. Michelson, an American Physicist, spent many years of his life in measuring the velocity of light. The method devised by him in the year 1926 at Mount Wilson observatory is considered accurate.

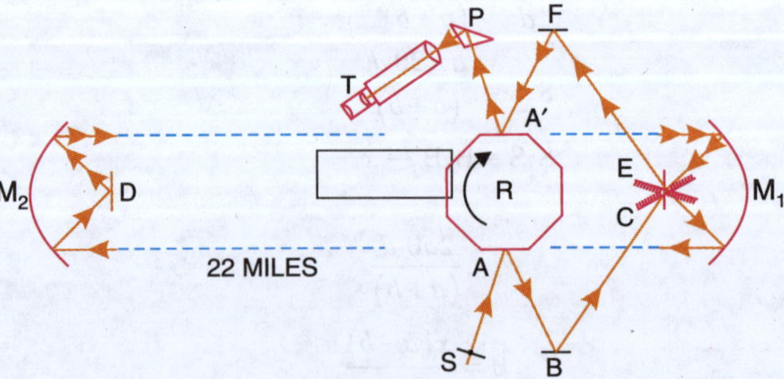

Fig. 11.6

A.A. Michelson (America's first Nobel Prize winner in Physics)

The form of apparatus designed by him is shown in Fig 11.6. Light from an arc after passing through a narrow slit S is reflected from one face of the octagonal rotating mirror R. Then it is reflected from the small fixed mirrors B and C to a large concave mirror M_1.The point on the mirror C from where light is reflected is the focus of the concave mirror M_1.The light then travels as a parallel beam to another concave mirror M_2, and it is reflected to a plane mirror D at the focus of the concave mirror M_2.The light is then reflected back to the concave mirror M_2, travels back to M_1 and is therefore reflected to a plane mirror F and is then incident on the face A′ of the octagonal mirror opposite to A. The final image is viewed through a micrometer eyepiece T with the help of a total reflecting prism P. When the rotating octagonal mirror is stationary, the image is seen by the light reflected from the top surface A′. When it is rotated the image disappears. The speed of rotation of R is suitably adjusted so that the image reappears and is seen in the same position as when R is stationary. The light reflected from A now arrives at A′ in the time taken by R to rotate through 45° or 1/8 of a revolution so that the next face is present at A. For this critical speed, the beam is reflected from the next face exactly along the same path as when the mirror is stationary.

If the distance traveled by light in its journey from A to A′ is equal to d, then

$$c = d/t$$

If R makes one revolution per second, then

$$t = \frac{1}{8n}$$

$$c = \frac{d}{1/8n} = 8nd \qquad (11.10)$$

Thus c can be calculated.

Michelson set up M_1 at an observing station on Mt.Wilson and M_2 at Mt.St.Antonio in California, at a distance of 22 miles. The velocity of revolving mirror R was measured by stroboscopic comparison with an electrically maintained tuning fork. He obtained the value of c = 2.99797×10^8m /s.

11.8. KERR CELL METHOD (A LABORATORY METHOD FOR FINDING THE VELOCITY OF LIGHT)

Kerr cell method is an improvement over the Fizeau's toothed wheel apparatus. It was first devised by Gaviola in 1925. In place of toothed wheel, electro optic shutter is used which is capable of chopping a beam of light several hundred times more rapidly than can be done by the toothed wheel.

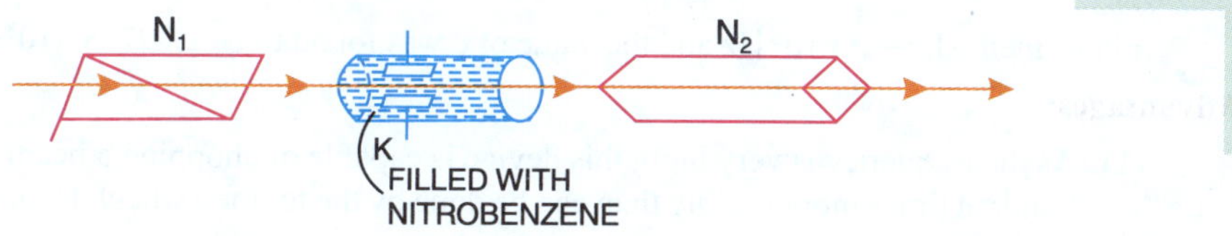

Fig. 11.7

Electro optic Shutter

It consists of a Kerr cell K placed between two crossed Nicol prisms N_1 and N_2. Kerr cell

is a small glass container having two electrodes filled with nitrobenzene (Fig. 11.7). *When a high voltage is applied to the electrodes of K, the light is transmitted through the system. On the other hand, when the field is switched off, light is stopped and not transmitted through the system.*

Thus, by using an electrical oscillator which supplies high frequency voltage, the beam of light can be interrupted at the rate of many millions of times in one second.

Karolus and Mittestadt at Leipzig in 1926 designated an arrangement similar to Fizeau as shown in Fig. 11.8.

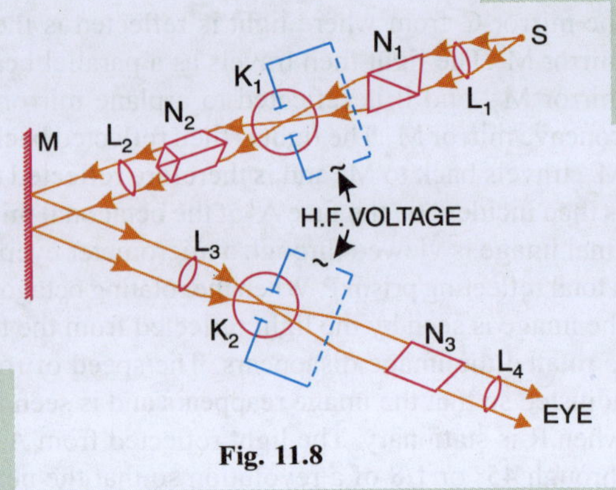

Fig. 11.8

Light from a source S after passing through the lens L_1 is made to pass through the nicol prism N_1 and it becomes plane polarized. This beam of light is focused at the centre of the Kerr cell K_1 and falls on the nicol prism N_2. This beam of light is rendered parallel by the lens L_2 and after reflection from the plane mirror M is allowed to fall on the Lens L_3 which concentrates the beam in the middle of the Kerr cell K_2. Finally the beam passes through the nicol prism N_3 and lens L_4.

As the two nicol prisms N_1 and N_2 are crossed, no light reaches the mirror M and the eye. When a high frequency voltage is applied to the Kerr cell K_1 the beam reaches the mirror M and is reflected. As this reflected beam reaches N_2 after passing through K_2, it is again obstructed by N_3 and no light is observed by the eye. It should be remembered that N_2 and N_3 are crossed.

Suppose a high frequency oscillatory voltage is applied to the Kerr cells, K_1 and K_2 simultaneously such that there is no phase difference in their voltages. This means that the two cells will act as two shutters and are allowed to open and close simultaneously. Further, consider the light passing through the cell K_1 , when the voltage is maximum and reaching the cell K_2, after some time when the voltage across the cell K_2 is minimum .Then no light reaches the eye.

Thus we find similar to Fizeau's method, the arrangement N_1, K_1 and N_2 acts as a space for the passage of light while the arrangement of K_2 and N_3 acts as a toothed wheel for the obstruction of the light.

If the distance between the two cells is d and f is the frequency of the H.F. voltage, then time taken by light to travel from K_1 to K_2 is

$$t = \frac{1}{4f} \qquad \text{But, } c = \frac{d}{t}$$

$$\therefore \quad c = \frac{d}{1/4f} = 4fd \tag{11.11}$$

In this method, $f = 3 \times 10^6$ Hz and the value of c was found to be $2.99778 \times 10^8 \pm 20 \times 10^3$ m/s.

Advantages:

(1) As the frequency is very high, this device is capable of chopping a beam of light several hundred times more rapidly than can be done by the toothed wheel. Hence a shorter base line can be used.

(2) The apparatus can be set up in a laboratory.

(3) The accurate frequency of the high frequency oscillator is known.

11.9. ANDERSON'S METHOD

In 1941, W.C.Anderson made use of one Kerr cell and divided the transmitted light pulses in to two beams by means of a half silvered glass plate. His arrangement is shown in Fig. 11.9. Light from a source S is allowed to pass through a modulated Kerr cell and is split up into two beams by the half silvered glass plate M_1. One beam of the light goes towards M_2 and is reflected back, and the other beam goes along a longer path towards M_3, M_4, M_5 and M_6 and after reflection from M_6 retraces its path and finally the two beams reach the photoelectric cell P.

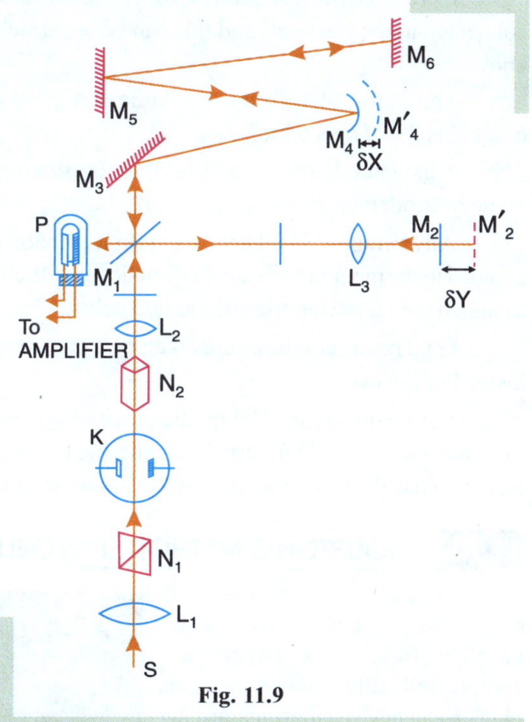

Fig. 11.9

If the longer path exceeds the shorter path by $\frac{(2n+1)\lambda}{2}$, the resultant of the two modulated waves reaching the photo cell will give a constant intensity. The amplifier connected to the photocell will give zero response in this case.

After this, light after reflection from M_3 is allowed to fall normally on M'_4 and it retraces its own path towards M_1. This is done for cutting the light to go to M_5 and M_6. The position of mirror M_2 is adjusted so that again there is zero response in the amplifier connected to the photocell. If the extra path (M_4 to M_6 and back) is a whole number of λ, no change in the response of the photo cell is observed.

In the actual experiment the extra path was found to be equal to 11λ.

Here
x = distance from M_1 to M_4 and back
y = distance from M_1 to M_2 and back
S = distance from M_4 to M_6 and back
δx = the distance between M_4 and M'_4
δy = the distance between M_2 and M'_4

In the first case, the path difference is

$$(x + S) - y = (2n + 1)\frac{\lambda}{2} \quad (11.12)$$

and in the second case the path difference is

$$(x + 2\,\delta x) - (y + 2\,\delta y) = \frac{\lambda}{2} \quad (11.13)$$

Subtracting equ. (11.13) from (11.12),we get

$$S + 2\,\delta y - 2\,\delta x = n\,\lambda \quad (11.14)$$

It was found that $n = 11$

∴ λ can be known from equation (11.14).

$$c = \frac{\lambda}{T} \quad (11.15)$$

where $$T = \frac{1}{f}.$$

Here f is the frequency of H.F.oscillator connected to the Kerr cell and this can be accurately known.

The value of c found by Anderson by this method is 299,776 ± 6 km/s.

This value is the mean of 2895 observations made by Anderson.

Photo cell

Advantages: (1) The use of two Kerr cells is avoided in the method because it is practically difficult to match the characteristics of the two cells.

(2) The observations are taken by a photo cell instead of the eye.

Source of error: The main source of error in this method is the difficulty in arranging for both beams to use the same position of the photoelectric cell.

11.10. HOUSTON'S METHOD (PIEZOELECTRIC GRATING METHOD)

When a quartz crystal is subjected to a high frequency electric field it is alternately compressed and extended and thus high frequency oscillations are set up. This effect is known as *piezo electric effect*. When these high frequency oscillations are set up in the form of pulses in a quartz crystal, it will be crossed by several parallel nodal planes. In such a case the intensity and refractive index at nodal planes is different from those at other planes and thus the crystal will act as a diffraction grating.

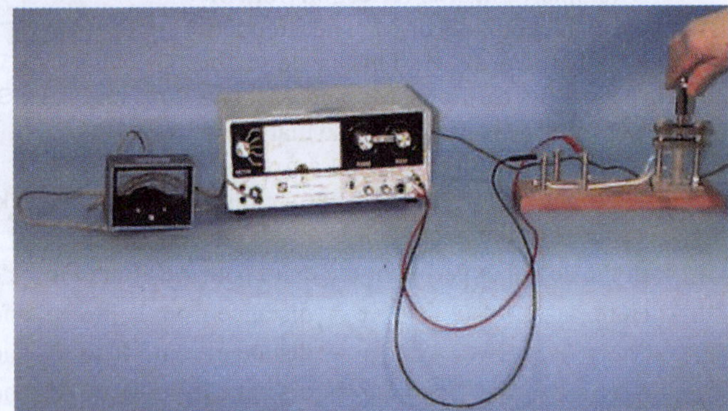

When a pressure is applied to a piezo-electric crystal in a sudden pulse, a current pulse is given off.

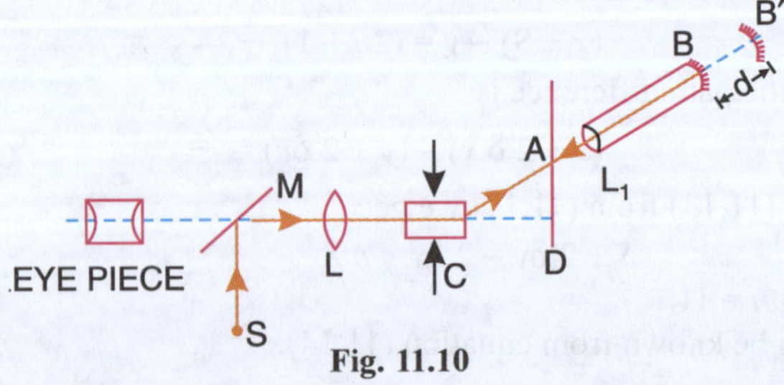

Fig. 11.10

Monochromatic light from a source S is allowed to fall on the quartz crystal C (whose faces are cut parallel) after reflection from the glass plate M. The lens L is used to make the beam parallel

(Fig 11.10). As mentioned, the quartz crystal will act as a grating. Therefore, light after passing through the crystal passes through the slit A of the screen. Finally it is brought to the focus by the lens L_1. Convex lens L_1 and concave mirror B are fixed at the two ends of a tube such that the length of the tube is equal to the focal length of the lens L_1 and also equal to the radius of curvature of the concave mirror B.

The light retraces its own path and on arriving at the quartz crystal C. If it finds the crystal exactly in the same condition as on its forward journey, then the image of S is seen by the eye. If on the other hand the light on arriving at the crystal C finds that the crystal is not exactly in the same condition as on its forward journey, no image of S is seen by the eye.

Therefore, the quartz crystal C acts similar to a toothed wheel in Fizeau's method or Kerr cell in Anderson's method.

Let the frequency of oscillations be *f*. Then the rate of formation of the grating is $2f$ per second. Therefore, when the image is seen by the eye, the time taken by light to go from C to B and back is a simple multiple of $\frac{1}{2f}$.

The tube containing L_1 and B is moved away from the screen along CB. At first it is observed that the image disappears. It again reappears when the distance BB′ and back (i.e., 2d) is travelled by light in the time $\frac{1}{2f}$ second. Several such points can be located when the tube L_1B is moved away from the screen which is *d* cm apart from each other. For accuracy the mean of all these distances is taken and let it be equal to *d*.

$$c = \frac{2d}{t}$$

But
$$t = \frac{1}{2f}.$$

Therefore
$$c = \frac{2d}{1} \times \frac{2f}{1} = 4df$$

In the actual experiment performed by Houston, the value of *f* was equal to 115×10^6 Hz. The value of c calculated by Houston in vacuum was 299,782 ± 9 km/s, which agrees with the results obtained by other methods.

11.11. THE DOPPLER EFFECT

Christian Johann Doppler stated in 1842 that a change of frequency of the wavemotion should be observed when a source of sound or light was moving. ***The apparent change in the observed frequency of a wave due to the relative motion between the source of waves and the observer is known as the Doppler effect.*** The Doppler effect occurs for all types of waves whenever there is relative motion between the source of waves and the observer.

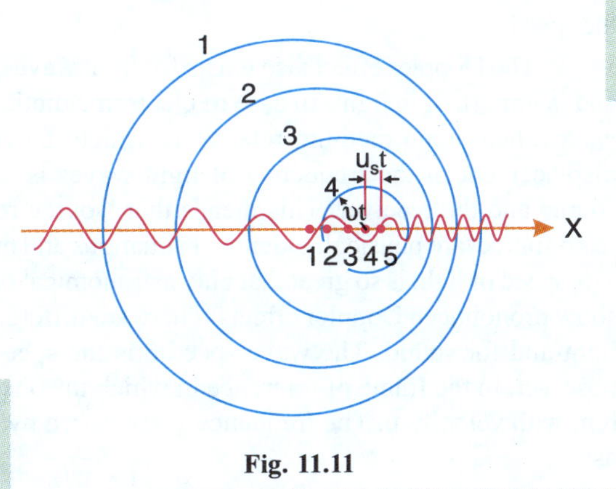

Fig. 11.11

As a train approaches and passes by, there is a distinct change in the sound pitch or frequency of its whistle. The sound gets shriller (higher frequency) as the source approaches a stationary observer and becomes lower in pitch (lower frequency) as it recedes. The reason is simple. The source (here the train) radiates spherical waves; but since the source is moving, it emits each successive wave at a different location (Fig. 11.11). As the train approaches the stationary observer, the waves moving in the same direction as the source are *bunched up* in front of the source and towards the observer causing a decrease in wavelength and increase in frequency of the sound heard. As the train moves away from the observer, the sound waves are *spread out* behind the source causing an increase in wavelength and a decrease in frequency of the sound heard. Doppler predicted that the colour of a luminous body must change in the same way by relative motion of the body and the observer.

Christian Johann Doppler (1803-1853)

In the general case where both the source and observer are moving, we obtain the following expression for the *apparent frequency* of the sound heard by the observer.

$$\nu' = \frac{\upsilon + u_o}{\upsilon - u_s}\,\nu \tag{11.16}$$

where υ is the velocity of sound in air,

u_o is the velocity of the observer,

u_s is the velocity of the source of sound and

ν is the actual frequency of the source.

Equ. (11.16) is the classical expression for the Doppler shift and includes all possibilities for collinear motion of source and observer relative to the medium. It is seen that the change in frequency due to the Doppler effect is determined by the velocities of the source and the receiver relative to the medium. It is not just the relative motion of source and observer that determines the frequency change. In fact, even when the relative motion is the same, we obtain different quantitative results, depending on whether the source or the observer is moving. This difference occurs because u_o and u_s are measured relative to the medium in which the sound wave is propagated and because the medium determines the speed.

Doppler effect

The Doppler effect also exists for light waves and is important in light. In case of electromagnetic waves there is no medium relative to which the velocity can be defined and hence the Doppler displacement of the frequency of light waves is determined only by the relative velocity of the source and the receiver. This means that 'source receding from observer' and 'observer receding from source' are physically identical situations and must exhibit exactly the same Doppler frequency. The speed of light is so great that only astronomical or atomic sources which have very high velocities show pronounced Doppler effects. There are differences, however, in the Doppler effect formula for light and for sound. The wave speed c is the speed of light and is the same for both source and observer. In the frame of reference in which the observer is at rest, the source is moving away from him with velocity υ. The frequency ν' measured by the observer is given by the theory of relativity as

$$\nu' = \nu\sqrt{\frac{1 - \upsilon/c}{1 + \upsilon/c}} \tag{11.17}$$

where υ is the relative velocity of source and observer. When υ is positive, the source moves away from the observer and $v' < v$; when υ is negative, the source moves towards the observer and $v' > v$. This is known as the **longitudinal Doppler effect.** It may be noted that the qualitative effect is the same as for sound although the quantitative relationship is different. If a source moves towards the observer the light appears more violet, and if it moves away it appears more red.

There is a further consequence of relativity on the Doppler effect. Whereas in the classical case there is no frequency shift if the relative velocity of source and receiver is perpendicular to the line joining them (for example, when the source travels along a circle at whose centre is located the receiver), that is no longer true in the relativistic case. In addition to the longitudinal effect, a **transverse Doppler effect** exists for light waves. The transverse Doppler effect is a strictly relativistic phenomenon and is a consequence of time dilation. It consists in a reduction in the frequency picked up by the receiver observed when the vector of the relative velocity is directed at right angles to the straight line passing through the receiver and the source. In this case, the frequency v in the frame of the source is associated with the frequency v' in the frame of the receiver by the relation

$$v' = v\sqrt{1 - \upsilon^2 / c^2} \approx v\left[1 - \frac{1}{2}.\frac{\upsilon^2}{c^2}\right] \tag{11.18}$$

The relative change in frequency in the transverse Doppler effect, given by

$$\frac{\Delta v}{v} = -\frac{1}{2}\frac{\upsilon^2}{c^2} \tag{11.19}$$

is proportional to the square of the ratio υ/c and is consequently considerably smaller than in the longitudinal effect for which the relative change in the frequency is proportional to the first power of υ/c. The existence of the transverse Doppler effect was proved experimentally by the American physicists H.E. Ives and G.R. Stilwell in 1938.

The longitudinal Doppler effect is used to determine the speeds at which luminous heavenly bodies are moving towards us or receding from us. Analysis of the spectra of light from distant stars shows shifts in wavelength compared to spectra of the same elements from the stationary light sources here on the earth. These can be interpreted as Doppler shifts due to motion of the stars. The shift is nearly always toward the longer wavelength or red end of the spectrum and is therefore called the **red shift**. Such Doppler red shift shows that the galaxies are moving away from us, the recession velocity being greater for the more distant galaxies. These observations are the basis of the concept of "expanding universe". According to "Big Bang" theory, all the matter in the Universe was in one big mass and a big bang occurred in the remote past. Since then the bodies have been moving away. Since we can calculate the expansion rate of the galaxies, it is estimated that the expansion of the Universe began about 15 billion years ago.

The thermal motion of the molecules of a luminous gas, owing to the Doppler effect, leads to **broadening of the spectral lines**. As a result of the chaotic nature of the thermal motion, all the directions of the molecular velocities relative to a spectrograph are equally probable. Therefore, the radiation registered by the instrument contains all the frequencies in the interval from $v_o(1 - \upsilon/c)$ to $v_o(1 + \upsilon/c)$, where v_o is the frequency emitted by the molecules, and υ is the velocity of thermal motion. The width of a recorded spectral line is thus $2v_o\upsilon / c$. The magnitude of the Doppler broadening of spectral lines makes it possible to assess the velocity of thermal motion of the molecules and consequently, the temperature of a luminous gas.

The Doppler effect provides a convenient means of tracking a satellite, of measuring the speed of an aeroplane, or the speed of an automobile. The satellite emits a radio signal of constant frequency v. The frequency of the signal received on earth decreases as the satellite is passing over. If the received signal is combined with a constant signal generated in the receiver, it produces beats. The beat frequency produces an audible note whose pitch decreases as the satellite passes overhead. To measure the speed of speeding cars, a generator of electromagnetic waves is located usually in a patrol car stationed at the side of the road. The wave is reflected from a moving car, which thus acts as a moving source. The reflected wave is Doppler-shifted in frequency. Measurement of the frequency shift using beats permits the measurement of the speed of the car.

QUESTIONS

1. Give a brief account of the methods for finding the velocity of light and give the details of the method which you consider most accurate.
2. Describe in detail Fizeau's method for finding the velocity of light. What are the chief difficulties met with, in carrying out this experiment ?
3. Describe Foucault's method for finding the velocity of light. How does this method justify the correctness of the wave theory of light ?
4. Describe the Kerr cell method finding the velocity of light in the laboratory. What are the advantages of this method over other methods ?
5. Describe and explain Michelson's rotating mirror method for finding the velocity of light and compare its merits and demerits with those other methods.
6. Light travels to a target and back in 0.5 s, in carbon disulphide. Calculate the distance of the target if the refractive index of the carbon disulphide is 1.46.
7. Describe a modern method for measuring the velocity of light.
8. Describe Anderson's method for determining the velocity of light. What are the merits and demerits of the method?
9. Calculate the time taken by a beam of light to travel through a glass pane of thickness 1 mm. Refractive index of glass is 1.50 and velocity of light through air is 3×10^8 m/s.
10. A certain monochromatic radiation has a wavelength of 5000 Å in water. What is the wavelength in (i) vacuum and (ii) in carbon disulphide? μ for water is 1.333 and μ for carbon disulphide = 1.628.

WAVE OPTICS

Huygens (1629-1695) proposed the wave theory of light in 1678. According to this theory, light energy is supposed to be transferred from one point to another in the form of waves. He had suggested a simple method to explain the propagation of light waves from one point to another point in a medium. The method is now known as **Huygens' principle**. His contemporaries raised certain objections regarding the hypothesis of wave nature of light. One of the objections was that if light were a wave motion, one should see light around corners just as we hear sound round the corners, though we are not in line with the source of sound. Second objection was that any wave motion requires a supporting medium for propagation; but light is known to travel through a vacuum. Thomas Young explained the colours exhibited by thin films like soap bubbles basing on the wave theory, where he applied the principle of superposition of waves. For the first time he measured the wavelength of light waves using the double slit experiment. Fresnel developed a mathematical theory, which not only removed the defects of the simple Huygens' principle but also explained the diffraction of light as well as rectilinear propagation of light. Huygens, Young and Fresnel assumed that light waves are longitudinal. Young and Fresnel conceived of an all pervading elastic medium, which enables light propagation to occur. It was assumed to exist in the entire universe, and it was named ***luminiferous ether***. The vibrations of the ether propagated as light, just as longitudinal vibrations in air propagate as sound. But the longitudinal wave theory of light could not explain polarization of light. Young eventually realized that light is a transverse wave. Subsequently, elastic ether theory was developed during the next ten years. Strange properties were attributed to it. It was assumed to be extremely rigid so that it can support the exceedingly high frequency oscillations of light travelling at a speed of 3×10^8 m/s; yet it does not offer resistance to the motion of celestial bodies through it. In spite of the success of the wave theory to explain many optical phenomena, the basic question as to what light is still remained unanswered. This simple wave theory in which light is described by a single scalar function is known as **wave optics** or more precisely **scalar wave optics.** This theory is sufficient to explain reflection, transmission, interference, diffraction, Fourier optics, and Holography etc phenomena.

Maxwell in 1873 made a brilliant guess that light is an **electromagnetic wave** of high frequency. It is described by the same theoretical principles that govern all forms of electromagnetic radiation. Light propagates in the form of two mutually coupled **vector waves**, an electric field wave and a magnetic field wave. The electromagnetic theory of light encompasses wave optics, which in turn encompasses geometrical optics. Electromagnetic theory was highly successful in explaining the propagation of light waves and related phenomena.

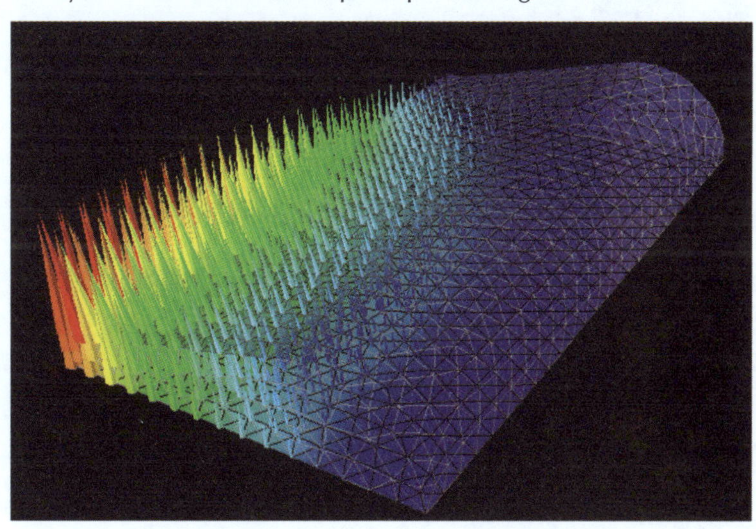

High frequency electromagnetic waves.

Since the electromagnetic theory takes vector nature into account, it easily explains the polarization of light. The polarization of light plays an important role in the interaction of light with matter.

1. The amount of light reflected at the boundary between two materials depends on polarization state of incident wave.
2. The amount of light absorbed by certain materials is polarization dependent.
3. Light scattering from matter is in general polarization sensitive.
4. The refractive index of anisotropic materials depends on the polarization.
5. Optically active materials have the natural ability to rotate the polarization state of the light.

The electromagnetic theory is highly useful in the study of **guided wave optics** and **integrated optics.**

However, the wave theory could not explain the absorption and the emission process of light, and certain aspects of interaction of light with matter.

12

CHAPTER

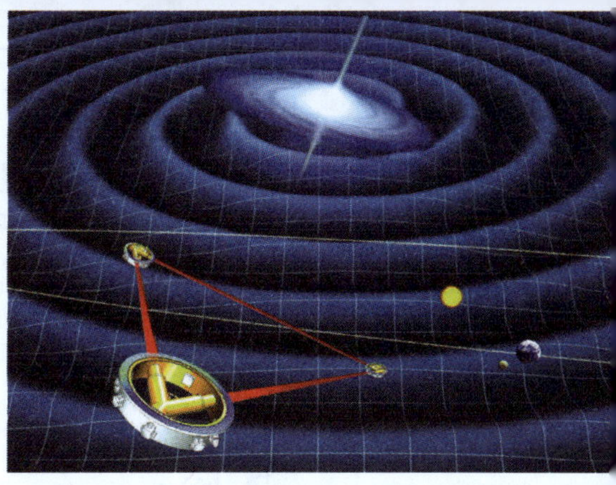

Waves and Wave Packets

12.1. OSCILLATIONS

All motions occurring in nature can be broadly categorized into three types, namely translational, rotational and oscillatory motions. A body undergoes oscillatory motion if the force acting on it is not constant but varies during the motion. In the oscillatory motion the body is disturbed from its equilibrium position and is subjected to a restoring force. The oscillatory motion is periodic and repeats itself over and over in equal intervals of time. The oscillatory motion in which the force is directly proportional to the displacement of the body is called **simple harmonic vibration.** In practice, the amplitude of vibration in simple harmonic motion does not remain constant but becomes progressively smaller. Such a motion is said to be **damped**; the motion dies out gradually transferring mechanical energy into thermal energy by the action of the frictional forces. Oscillations are sustained if the body is disturbed repeatedly. In order to keep the body in continuous oscillatory motion, some external periodic force must be used. The frequency of this force is called the *forcing frequency* and the oscillations are said to be **forced oscillations**. When the forcing frequency is equal to the natural frequency, **resonance** is said to occur. In the state of resonance, energy is transferred to the body from the source of the external periodic force.

At a Glance

- Oscillations
- Waves
- Travelling Waves
- Examples of Waves
- Characteristics of a Wave
- Mathematical Representation of Travelling Waves
- General Wave Equation
- Phase Velocity
- Complex Representation of a Plane Wave
- Light Sources Emit Wave Packets
- Wave Packet and Bandwidth
- Fourier Series and Transforms
- Wave Packet and bandwidth Theorem
- Group Velocity
- Real Light Waves

12.2. WAVES

When a disturbance passes through a medium, a series of points are affected. A local displacement from equilibrium caused in one part of the medium is transmitted successively to the next by interaction among the particles, and such displacements together make up a *wave*. Simple harmonic vibration of particles in the medium generates a **simple harmonic wave**. A **wave** is any disturbance, which travels through the medium due to the repeated periodic motion of the particles (of the medium) about their mean position.

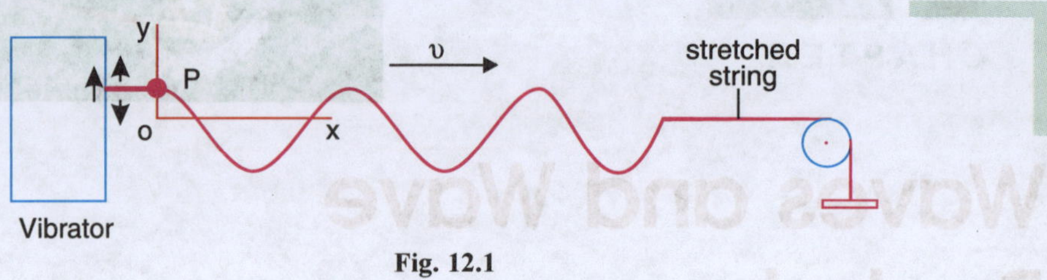

Fig. 12.1

A simple method of generating a wave is shown in Fig.12.1. A string is connected to a blade. When the blade is set into vibration, the blade oscillates vertically with simple harmonic motion. As a result each particle of the string such as P, oscillates vertically in the *y* direction with simple harmonic motion. Each particle vibrates with a frequency equal to the frequency of vibration of the blade. Consequently, a wave travelling to the right is set up on the string. Though the particle oscillates in the *y* direction, the wave travels in the *x* direction with a speed υ. It may be noted that the wave is a **one-dimensional wave**.

Light waves are three dimensional waves.

We are familiar with waves on water surface. When a pebble is thrown into sill water of a pond, ripples are produced, which spread out slowly outwards in ever-widening circles. Water particles do not travel along with the wave but vibrate only up and down about their equilibrium position. We identify the wave motion with the help of crests and troughs travelling away from the centre of disturbance. These waves are **two-dimensional waves**.

A point source of light emits light waves in all directions, which spread out uniformly in the form of ever-increasing concentric spheres with a velocity of 3×10^8m/s (see Fig. 12.2). The light waves are **three-dimensional waves**.

Travelling wave.

12.3. TRAVELLING WAVES

Waves such as those we see on the surface of water, which move away from the centre of disturbance are called **travelling waves** or **progressive waves**. A progressive wave consists of a sequence of waveforms. They transfer energy outwards from the source. The amplitude A of a travelling wave produced by a point source gradually decreases in moving away from the source, as follows.

$$A \propto \frac{1}{r}$$

where r is the distance of the point of observation from the position of the point source. Hence the intensity of the *spherical waves* falls off as $1/r^2$.

12.3.1. WAVE FRONT AND THE RAY

Waves start from a source and spread out into new and new regions of space. Ripples on a water surface start from the point of disturbance and expand in the form of circles. The circles are in fact the crests of the waves. All the particles located at the crest will be in the same state of oscillation and hence in the same phase. The continuous locus of all particles, which are in the same phase, is called a **wave front** (Fig. 12.2). A surface, which passes through these points and completely surrounds the source, is called the **wave surface**.

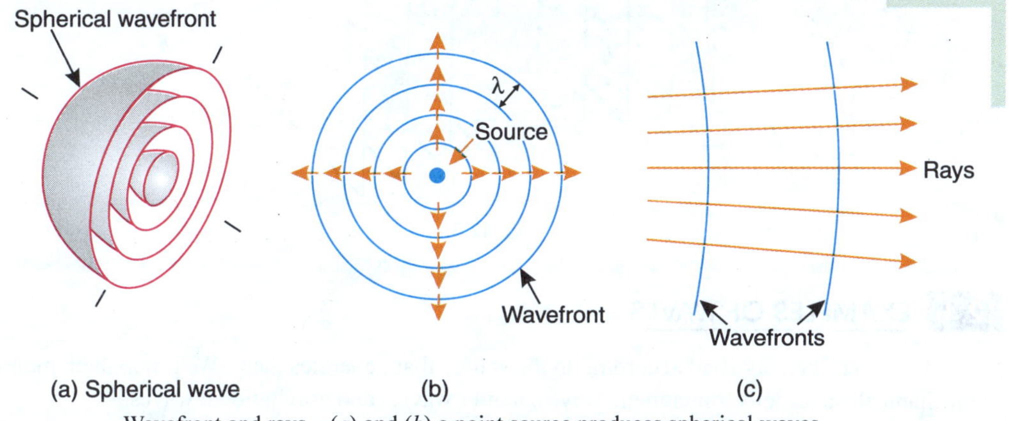

Wavefront and rays—(*a*) and (*b*) a point source produces spherical waves.
(*c*) at large distance from the source, the wavefronts tend to be parallel planes.

Fig. 12.2

In case of water ripples the wave fronts are *circles* whereas in the case of a point light source they are *spheres*. In diagrams, wave fronts are often shown connecting crests or troughs, but in fact any surface connecting points of equal phase is a wave front. The propagation of the wave is visualized by the advancing wave front and stationary wave surfaces behind it. The wave surfaces are separated by one wavelength, as shown in Fig. 12.2 (b).

At considerable distances from the source, a *spherical wave front* becomes very large and a small portion of it may be considered to be nearly planar (see Fig.12.2c). In that case, it is called a *plane wave front.* If a point source of light is placed at the primary focus of a convex lens, the spherical wave front produced by the light source is transformed into a plane wave front. In case of *plane waves*, the amplitude and hence the *intensity remains constant over a long distance* (see Fig. 12.3).

It is sometimes more convenient to describe the wave propagation in terms of rays instead of wave fronts. A line drawn perpendicular to a wave front is called a **ray**. It shows the direction along which a wave front moves (see Fig.12.2c).

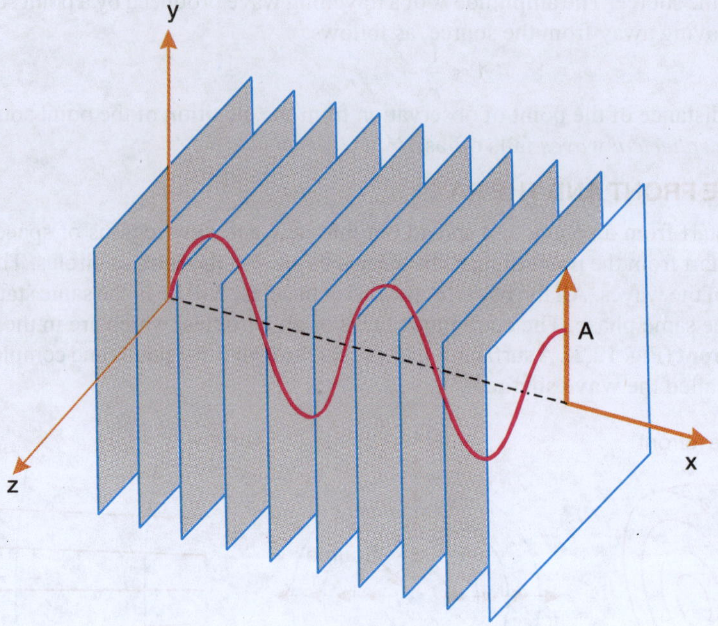

Fig. 12.3

12.4. EXAMPLES OF WAVES

Waves can be classified according to the source that generates them. We group them mainly as mechanical waves, electromagnetic waves, matter waves, and gravitational waves.

Mechanical waves: Mechanical waves or elastic waves are governed by Newton's laws and require a material medium for their propagation. Sound waves, seismic waves, water waves in bodies of water such as ocean, river, and ponds are examples of mechanical waves.

Electromagnetic waves: Visible light, radio waves, microwaves, x-rays and γ-rays belong to this category. Electromagnetic waves consist of oscillating electric and magnetic fields and do not require material medium for their propagation. They all travel in free space with the same speed 'c'.

Matter waves: Atomic particles exhibit wave properties under certain conditions. The laws of quantum mechanics govern such *matter waves*.

Gravitational waves: It is suggested that the cosmic bodies such as galaxies, stars produce gravitational waves and interact with each other through these waves. The gravitational waves are believed to propagate with the velocity of light.

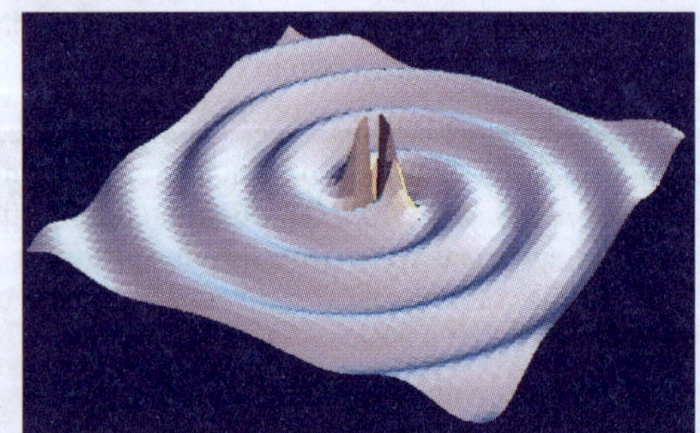

Representation of gravitational wave caused by two massive objects that are orbiting each other and are obout to collide.

12.5. CHARACTERISTICS OF A WAVE

If a *snapshot* of a progressive wave is taken at any instant, we observe a wave profile as in Fig.12.4. It consists of a sequence of waveforms.

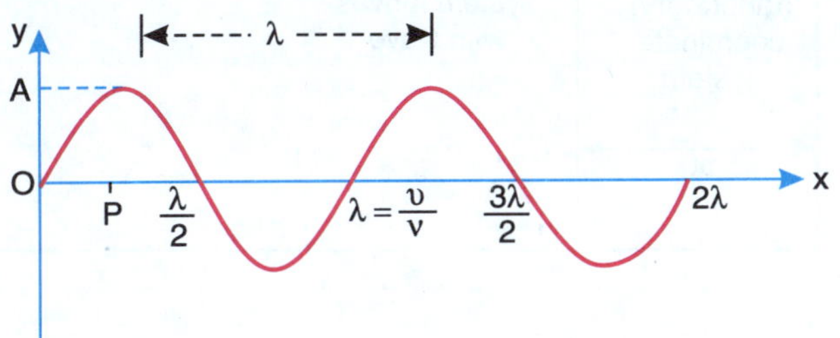

Fig. 12.4

Any wave is characterized by the following parameters:

(*a*) Time Period, T: If a point is chosen and the wave profile is observed as it passes this point, then the profile is seen to repeat at equal intervals of time. This repeat time is known as the *time period* of the wave.

(*b*) Wavelength, λ: The distance between the corresponding points, such as two successive crests, in successive waveforms is called the *wavelength.* It represents the spatial period of the wave.

(*c*) Amplitude, A: The maximum displacement in a waveform is known as the *amplitude.*

(*d*) Velocity, υ: Each time the source (of disturbance) vibrates once, the wave moves forward at a distance λ. If there are ν vibrations in one second, the wave moves forward at a distance of 'νλ'. The distance that the wave moves in one second is the *velocity* of the wave, υ. Thus,

$$\upsilon = \nu\lambda \tag{12.1}$$

where

$$\nu = \frac{1}{T} \tag{12.2}$$

(*e*) Phase angle, ϕ: The extent of displacement of particles in the medium and the direction of their displacement change from point to point along the wave. The quantity, which represents the displacement, is called the *phase of the vibration*, ϕ. The phase may be expressed in terms of degrees or radians; or as the ratio of time t to the time period T; or as the ratio of the distance x to the wavelength, λ. The ratios t/T and x/λ are fractional numbers and have a maximum value of 1. When expressed in terms of radians (or degrees), the maximum value that the phase can take is 2π radians (or 360°).

(*f*) Intensity, I: The energy transferred on an average by a wave in unit time, through a unit area perpendicular to its propagation direction, is known as the *intensity* of the wave. It is established that the intensity of a wave is directly proportional to the square of the amplitude of the wave. Thus,

$$I \propto |A|^2 \tag{12.3}$$

12.6. MATHEMATICAL REPRESENTATION OF TRAVELLING WAVES

We obtain the entire picture of wave motion only when we consider the harmonic motion of a series of points in the medium. As the oscillations are communicated from point to point, the points in the medium will be in different states of oscillation at different times. The displacement of a particle in the medium is, therefore, a function of *space coordinates* as well as a function of *time*. We denote the displacement by y. Thus,

$$y = f(x,t)$$

The displacement y is called the **wave function.**

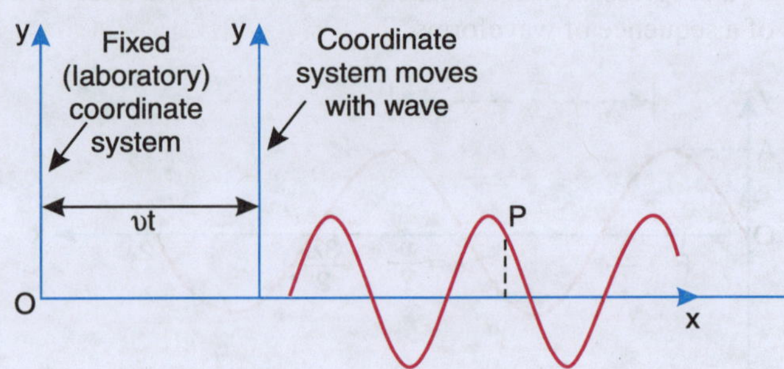

Fig. 12.5

Let us consider the case of a one-dimensional wave moving along +x-axis, as in Fig. 12.5.

We first consider the displacement as a *function of time*, at the position $x = 0$. Then,

$$y = f(t)$$

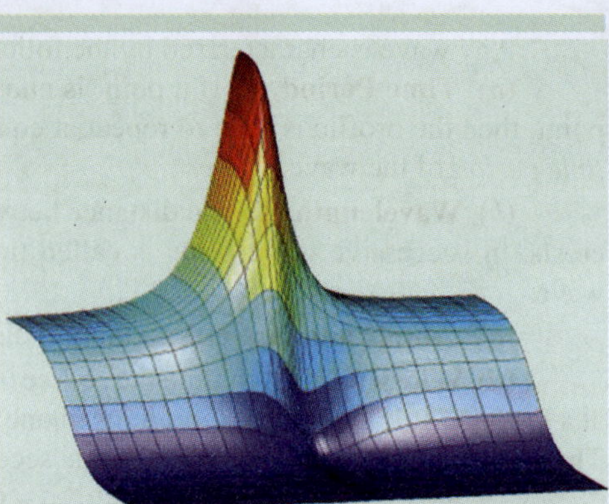

Travelling wave solution.

1. Since the oscillations are sinusoidal, we can describe the displacement y in terms of *time* as

$$y = A \sin \omega t$$

or $$y = A \sin 2\,\pi \nu t \quad (12.4)$$

The wave is travelling forward to the right with a velocity, say υ. Then after time t, the wave has moved through the distance $x = \upsilon t$. The displacement at x can be represented by

$$y = f(x - \upsilon\, t) \quad (12.5)$$

$$\upsilon = \nu\lambda$$

Also $$\upsilon = \frac{x}{t}. \text{ Therefore, } \nu\lambda = x/t$$

or $$\nu = \frac{x}{\lambda t}. \quad (12.6)$$

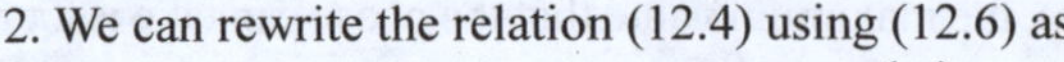

2. We can rewrite the relation (12.4) using (12.6) as

$$y = A \sin 2\pi\left(\frac{x}{\lambda}\right) \quad (12.7)$$

This describes the displacement in terms of *space*.

Using the equations (12.5) and (12.7), we can describe the displacement of any point on a harmonic wave in terms of both space and time as

$$y = A \sin\left[\frac{2\pi}{\lambda}(x - \upsilon\, t)\right] \quad (12.8)$$

This equation gives the relationship between the space and time dependence of disturbances in a medium. It is seen from the above that the wave is periodic in both space and time.

The equation (12.8) may be rewritten as

$$y = A \sin k\,(x - \upsilon\, t) \quad (12.9)$$

where $$k = 2\pi/\lambda.$$

k is known as **propagation constant** or **propagation number**.

The equation (12.9) may further be rewritten as

$$y = A \sin(kx - \omega t) \tag{12.10}$$

where $\omega = 2\pi v$.

Equ.(12.10) represents a **progressive** or **travelling wave**.

1. The wave is said to be **monochromatic** because it has a single frequency, ν.
2. It is an undamped wave since its **amplitude A** is constant along the direction of propagation. It is a **plane wave,** since the amplitude is constant everywhere.
3. It represents a continuous train of waves stretching from $x = -\infty$ to $x = +\infty$. The disturbance is sinusoidal and continues forever.
4. It is a mathematically **idealized wave**. Such ideal waves **do not occur** in nature.

For many purposes the light disturbance at any point can be represented by the single scalar quantity "y". It is assumed that the variations of y are propagated in the form of a wave motion, and equ.(12.10) represents the light wave.

12.7. GENERAL WAVE EQUATION

To know how the displacement y varies as a function of space, x and time, t we have to do partial differentiation of y with respect to x and t.

$$\frac{\partial y}{\partial x} = \frac{2\pi}{\lambda} A \cos\left[\frac{2\pi}{\lambda}(x - \upsilon t)\right] \tag{12.11}$$

$$\frac{\partial y}{\partial t} = -\frac{2\pi\upsilon}{\lambda} A \cos\left[\frac{2\pi}{\lambda}(x - \upsilon t)\right] \tag{12.12}$$

Combining both these equations and eliminating equal factors, we get

$$\frac{\partial y}{\partial x} = -\frac{1}{\upsilon}\frac{\partial y}{\partial t} \tag{12.13}$$

If we take the second derivatives, it will hold for any sinusoidal wave, independent of the direction of travel, either $-x$ or $+x$.

$$\frac{\partial^2 y}{\partial x^2} = \frac{1}{\upsilon^2}\frac{\partial^2 y}{\partial t^2} \quad \text{or} \quad \frac{\partial^2 y}{\partial t^2} = \upsilon^2 \frac{\partial^2 y}{\partial x^2} \tag{12.14}$$

We replace y by the more general term ξ, which stands for any disturbance.

$$\frac{\partial^2 \xi}{\partial t^2} = \upsilon^2 \frac{\partial^2 \xi}{\partial x^2} \tag{12.15}$$

This is the *one-dimensional wave equation.* It connects the variations of ξ in space and time to the velocity of propagation of the wave.

If we are to include waves propagating in any direction, we need to extend the right hand term to the y and z axes, and replace it by

$$\frac{\partial^2 \xi}{\partial x^2} + \frac{\partial^2 \xi}{\partial y^2} + \frac{\partial^2 \xi}{\partial z^2}$$

Using the Laplacian operator $\nabla^2 = \frac{\partial^2}{\partial x^2} + \frac{\partial^2}{\partial y^2} + \frac{\partial^2}{\partial z^2}$, we can write the equation as

$$\frac{\partial^2 \xi}{\partial t^2} = \upsilon^2 \nabla^2 \xi \tag{12.16}$$

This is the *general three-dimensional wave equation.*

12.8. PHASE VELOCITY

A monochromatic wave train is an *infinite sequence* of waves in time and space of crests and troughs. Following equation (12.10), the equation of a harmonic wave propagating along the x-axis has the following form

$$y = A \sin\left[(kx - \omega t) + \phi\right] \tag{12.17}$$

where ϕ is the initial phase of the wave which is determined by our choice of the beginning of counting x and t. Let us fix a value of the phase by assuming that

$$\left[(kx - \omega t) + \phi\right] = \text{constant} \tag{12.18}$$

This expression determines the relation between the time t and the place where the phase has a fixed value. The value of $\frac{dx}{dt}$ calculated from (12.18) gives the velocity with which the given value of the phase propagates.

$$k\frac{dx}{dt} - \omega = 0$$

$$\frac{dx}{dt} = \frac{\omega}{k} = \frac{2\pi\nu}{2\pi/\lambda} = \nu\lambda = \upsilon \tag{12.19}$$

Thus, the velocity of wave propagation 'υ' is the velocity of phase propagation and it is therefore called the **phase velocity**. The *phase velocity* υ of a wave is the velocity with which the wave front moves forward. It is the same as the velocity of propagation of the wave. When the waves are travelling through a non-dispersing medium, the common velocity of the waves is the phase velocity.

12.9. COMPLEX REPRESENTATION OF A PLANE WAVE

An expression similar to equ.(12.10) can be written in terms of cosine as

$$y = A \cos(kx - \omega t) \tag{12.20}$$

We can express equ.(12.10) and (12.20) in a single equation, using Euler's formula.

$$e^{i\theta} = \cos\theta + i\sin\theta$$

$$\therefore \quad y = Ae^{i(kx - \omega t)} \tag{12.21a}$$

or

$$y = Ae^{-i(\omega t - kx)} \tag{12.21b}$$

The advantage of the above complex representation is as follows. The complex quantity used to represent the wave may be split into its space and time parts to give

$$y = Ae^{-i\omega t}e^{ikx}$$

It is seen to consist of a complex amplitude $\psi = Ae^{ikx}$ and a harmonic factor $e^{-i\omega t}$.

The modulus and argument of the complex amplitude give the amplitude and phase of the disturbance. When adding a number of waves of the same frequency, $e^{-i\omega t}$ is a common factor and the sum of the complex amplitudes gives the complex amplitude of the resultant. Frequently, we will be interested only in the resultant intensity and it is not necessary to know both the modulus and the argument. The resultant intensity A^2 is obviously given by the product of ψ and its complex conjugate ψ^* and this is the usual method of finding the intensity when an expression for ψ has been obtained.

12.10. LIGHT SOURCES EMIT WAVE PACKETS

A harmonic wave is of *infinite extension* and is strictly *monochromatic* (see Fig.12.6). In a harmonic wave, there is a definite relationship between the phase of the wave at a given time and at a certain time after; and also the phase at a given point and another point at a certain distance away. However, a source of light does not emit a continuous train of waves but can be said to emit a succession of wave trains of limited length and there will be no fixed phase relationship between successive wave trains.

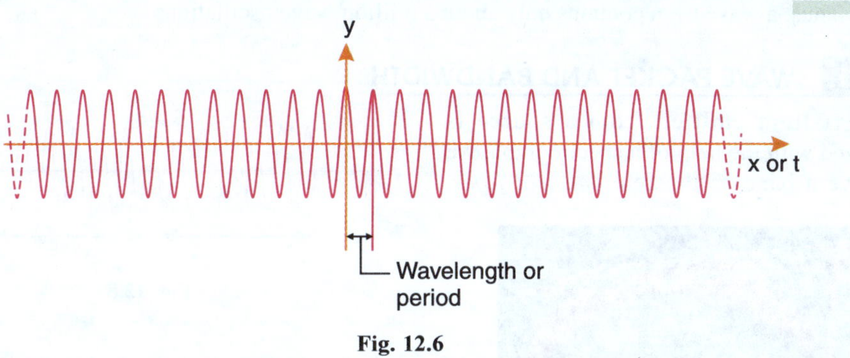

Fig. 12.6

Bohr explained for the first time how light is emitted from a light source. According to his theory, atoms are the emitters of light in any light source and light is emitted when excited atoms pass from an excited state to a lower energy state. An atom starts emitting a light wave as it leaves the excited state and ceases emission as soon as it reaches the lower energy state. Thus, an emission event produces a *light burst* (or a photon). Each light burst occurs over a period of about 10^{-8}s only and each is a **wave train** containing only a certain limited number of wave oscillations in it. The same atom again receives energy after some time and jumps into excited state and subsequently emits another burst of light. This is true for all atoms in the light source. The emission events occur quite randomly. Each atom in the source acts independently and different atoms emit wave trains at different instants and their combination in millions and millions constitutes the light radiation from the light source. In order to appreciate some of the peculiarities of natural light, the following fact is to be well understood. ***The light emitted by an ordinary light source is not an infinitely long, simple harmonic wave but it is a jumble of finite wave trains*** (see Fig.12.7).

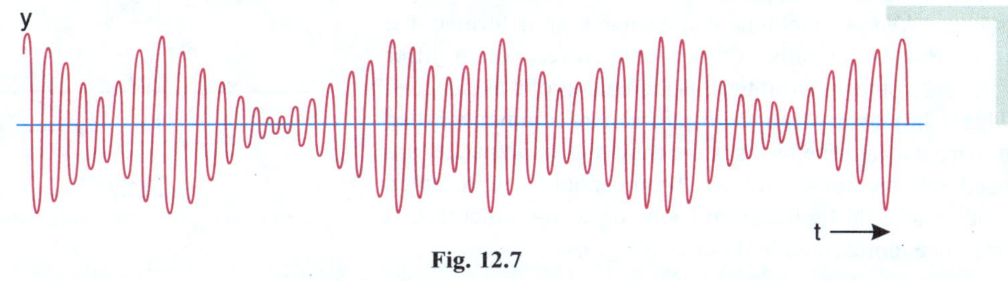

Fig. 12.7

If a wave train (see Fig. 12.8) lasts for a time interval Δt, then the length of the wave train in a vacuum is

$$l = (x_2 - x_1) = c\Delta t \qquad (12.22)$$

where c is the velocity of light in a vacuum. If we take $\Delta t = 10^{-8}$ s,

$$l = (3 \times 10^8 \text{m/s})(10^{-8}\text{s}) = 3 \text{ m}.$$

The number of wave oscillations present in a wave train is

$$N = \frac{l}{\lambda} \qquad (12.23)$$

where λ is the wavelength. Assuming that $\lambda = 5000 \text{ Å} = 5 \times 10^{-7}$m,

$$N = \frac{3\text{m}}{5\times10^{-7}\text{m}} = 6\times10^6$$

Thus, a wave train contains only about a million wave oscillations.

12.11. WAVE PACKET AND BANDWIDTH

If light emitted from a source is analyzed with the help of a spectrograph, we observe a band of discrete *spectral lines*.

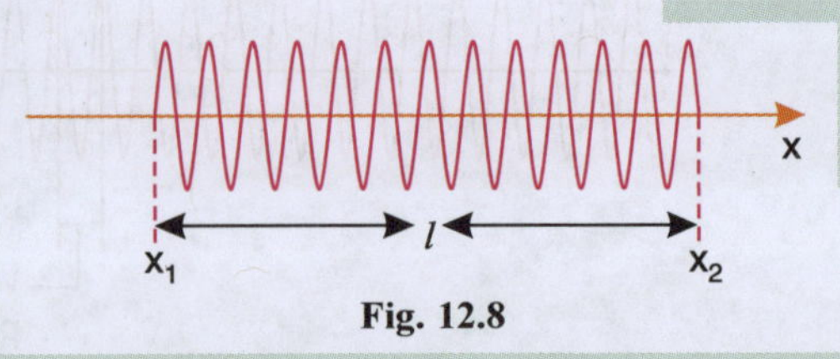

Fig. 12.8

Wave packet.

Obviously the wave packets, emitted by the atoms in the light source, form the spectral lines. These spectral lines are produced by light of specific wavelengths. A wave train (Fig. 12.8) is not a harmonic wave. Therefore, the wavelength of a wave train is *not* sharply defined, but consists of a continuous spread of wavelengths over a range, $\Delta\lambda$, centered on a wavelength λ_o (see Fig. 12.11 b). The spread of wavelengths, $\Delta\lambda$, is called the **band width.** The *bandwidth* is the wavelength interval, which contains the major portion of the energy of the wave packet. We may conclude that a wave train is formed due to addition of a number of harmonic waves, which differ from each other by infinitesimal increments of wavelengths. This is the reason why a wave train is more commonly called a *wave packet.* The terms wave train and wave packet are used synonymously. A wave train cannot be represented mathematically by the simple sine or cosine functions. It has to be represented by Fourier integrals.

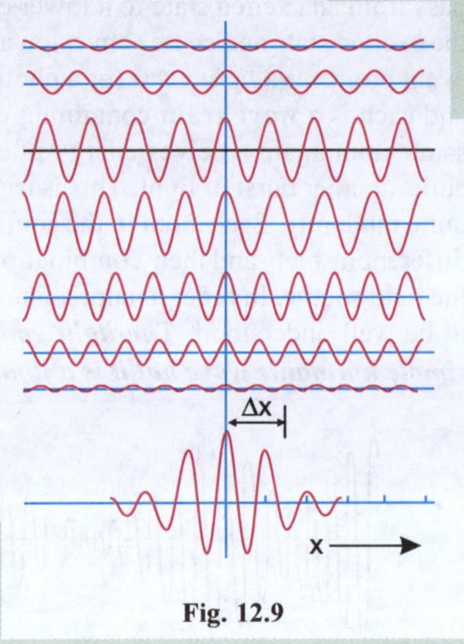

Fig. 12.9

12.12. FOURIER SERIES AND TRANSFORMS

Any periodic function may be represented as a sum of sine and cosine functions. A non-periodic function can also be represented in a similar way, provided

(*i*) it is piecewise continuous and

(*ii*) the integral $\int_{-\infty}^{\infty} f(t)dt$ does exist and has a finite value.

A non-periodic function may be expanded as

$$f(t) = \int_{-\infty}^{\infty} \left[\frac{1}{2\pi} \int_{-\infty}^{\infty} f(t) e^{-i\omega t} dt \right] e^{i\omega t} d\omega \tag{12.25}$$

The function inside the square bracket can be represented as

$$F(\omega)=\frac{1}{2\pi}\int_{-\infty}^{\infty} f(t)e^{-i\omega t}\,dt \tag{12.26}$$

Thus, we can write

$$f(t)=\int_{-\infty}^{\infty} F(\omega)e^{i\omega t}\,d\omega \tag{12.27}$$

The two functions $f(t)$ and $F(\omega)$ are related to each other by equations (12.26) and (12.27) are called a pair of Fourier transforms. Function $F(\omega)$ is the **Fourier transform** of $f(t)$ and function $f(t)$ is the **inverse Fourier transform** of function $F(\omega)$.

Joseph Fourier (1768-1830)

Equation (12.26) and (12.27) can be written as

$$f(t)=\frac{1}{\sqrt{2\pi}}\int_{-\infty}^{\infty} F(\omega)e^{i\omega t}\,d\omega \quad \text{and} \quad F(\omega)=\frac{1}{\sqrt{2\pi}}\int_{-\infty}^{\infty} f(t)e^{-i\omega t}\,dt$$

The factor $\frac{1}{\sqrt{2\pi}}$ is introduced for sake of symmetry and is arbitrary. We can omit this factor and write

$$f(t)=\int_{-\infty}^{\infty} F(\omega)e^{i\omega t}\,d\omega \tag{12.28}$$

and

$$F(\omega)=\int_{-\infty}^{\infty} f(t)e^{-i\omega t}\,dt \tag{12.29}$$

12.13. WAVE PACKET AND BANDWIDTH THEOREM

Let $F(t)$ be the light disturbance at a point at time t, due to a wave train. We assume F to be zero for $|t| \geq t_0$, and express it as a Fourier integral

$$F(t)=\int_{-\infty}^{\infty} f(\nu)e^{-2\pi i\nu t}\,d\nu \tag{12.30}$$

where by Fourier transform theorem

$$f(\nu)=\int_{-\infty}^{\infty} F(t)e^{2\pi i\nu t}\,dt \tag{12.31}$$

Equ.(12.31) represents a harmonic wave of infinite extension in time. Now suppose the wave trains are all of duration Δt, during which $F(t)$ is simply periodic with frequency ν_o. That is

$$\left.\begin{aligned} F(t) &= f_0 e^{-2\pi i\nu_0 t} && \text{when } |t| \leq \frac{\Delta t}{2} \\ &= 0 && \text{when } |t| > \frac{\Delta t}{2} \end{aligned}\right\} \tag{12.32}$$

where f_0 is constant. Equ.(12.32) represents a wave packet. Then from (12.31) and (12.32), we get

$$f(\nu)=f_0\int_{-\Delta t/2}^{\Delta t/2} e^{2\pi i(\nu-\nu_0)t}\,dt$$

$$= f_0\left|\frac{e^{2\pi i(\nu-\nu_0)t}}{2\pi i(\nu-\nu_0)}\right|_{\frac{-\Delta t}{2}}^{\frac{\Delta t}{2}}$$

Since $e^{ix} - e^{-ix} = 2\,i \sin x$, we write the above equation as

$$f(\nu) = f_o \frac{\left[e^{2\pi\, i\, (\nu-\nu_o)\Delta t/2} - e^{-\,2\pi\, i(\nu-\nu_o)\Delta t/2}\right]}{2\pi\, i(\nu-\nu_o)}$$

$$= f_0 \left|\frac{2i \sin\ [\pi(\nu-\nu_0)\Delta t]}{2\pi\ i(\nu-\nu_0)\Delta t}\right| \Delta t$$

$$= f_0\,\Delta t\left[\frac{\sin\{\pi(\nu-\nu_0)\Delta t\}}{\pi(\nu-\nu_0)\Delta t}\right] \qquad (12.33)$$

The relative intensities of the Fourier components of the wave train is given by the square of the function $\left[\frac{\sin\{\pi(\nu-\nu_0)\Delta t}{\pi(\nu-\nu_0)\Delta t}\right]$. A plot of $\left[\frac{\sin\{\pi(\nu-\nu_0)\,\Delta t}{\pi(\nu-\nu_0)\,\Delta t}\right]^2$ is shown in Fig. 12.10. It may be seen from the plot that most of the power is in the central maximum and the intensity falls to zero at frequency values $\nu_0 - \frac{\Delta\nu}{2}$ and $\nu_0 + \frac{\Delta\nu}{2}$. It is clear that the bandwidth of the wave packet is $\Delta\nu$ and is given by

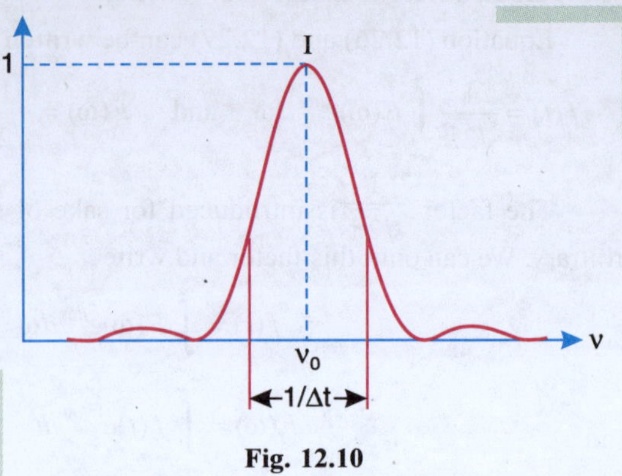

Fig. 12.10

$$\Delta\nu \sim \frac{1}{\Delta t} \qquad (12.34)$$

Thus, *the effective frequency range of the Fourier spectrum is of the order of the reciprocal of the duration of a single wave train.*

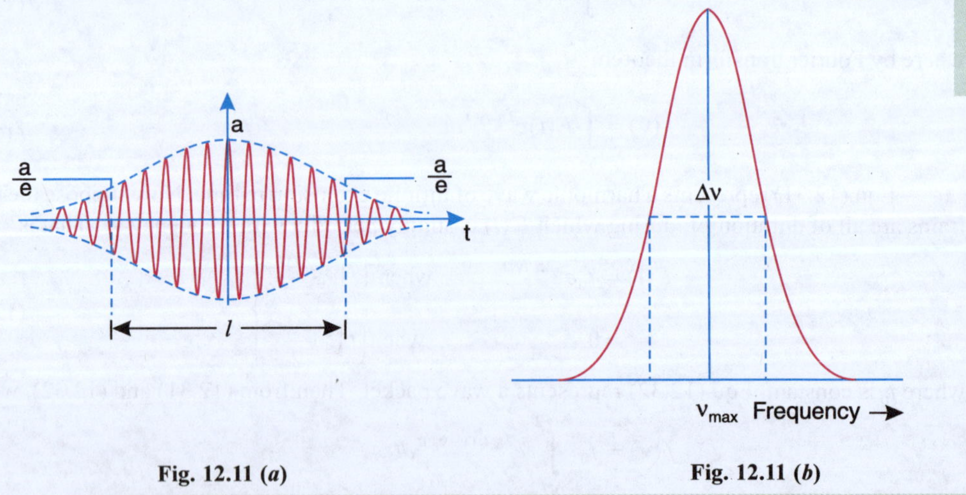

Fig. 12.11 (*a*)

Fig. 12.11 (*b*)

A source of light is unlikely to emit a train of waves that is sharply limited, as shown in Fig.12.8. It is more likely to emit a train of waves whose amplitude increases gradually to a maximum and then decreases gradually, as illustrated in Fig. 12.11(*a*), which will have a spectral line profile of the form shown in Fig.12.11 (b). As t or $\nu \to \infty$, both the envelope of the wave train and the intensity distribution curve tend to zero. Such wave trains are more generally called **wave packets**. We define the width of the wave packet by taking the value of t when the curve falls to 1/e of its maximum. Thus, the width of the wave packet shown in Fig. 12.11 (a) is taken to be equal to l. Similarly, we define the **spectral line width** by taking the value of t when the curve (see Fig.12.11 b) falls to 1/e of its maximum and is denoted by $\Delta\nu$.

12.14. GROUP VELOCITY

A light beam thus consists of a stream of wave packets and the wave packets are formed as a result of superposition of waves of a number of frequencies. When a wave packet propagates through a medium, it cannot preserve its shape over a reasonable propagation distance. The component waves get dispersed, as each component wave has its own propagation velocity in the medium. Such a medium in which the wave packet loses its initial shape is called a **dispersive** medium. Most media in nature are dispersive. If a wave packet travels through a medium without changing its shape over a long distance, then the medium is said to be a **non-dispersive** medium. The wave packet generally has the maximum amplitude at a particular value of x and the velocity of this maximum amplitude point is called the **group velocity** (see Fig. 12.12). Thus, the velocity at which a wave packet (or a pulse) travels is the group velocity of the wave packet. This velocity also represents the velocity with which energy of the wave packet is transmitted.

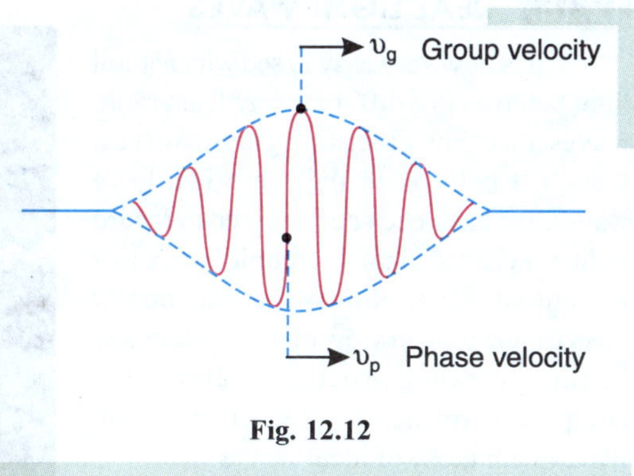

Fig. 12.12

Let each component wave in the wave packet has its own phase velocity, $\upsilon = \nu\lambda$. The wave packet has amplitude that is large in a small region and very small outside it. The amplitude of the wave packet varies with x and t. Such variation of amplitude is called the *modulation* of the wave. The velocity of propagation of the modulation is known as the *group velocity*, υ_g. It is given by

$$\upsilon_g = \frac{d\omega}{dk} \qquad (12.35)$$

$$\upsilon_g = \frac{d(\upsilon k)}{dk} = \upsilon + k\frac{d\upsilon}{dk}$$

We further write

$$\frac{d\upsilon}{dk} = \frac{d\upsilon}{d\lambda}\frac{d\lambda}{dk}$$

But

$$\lambda = \frac{2\pi}{k}$$

Differentiating the above expression, we get

$$\frac{d\lambda}{dk} = -\frac{2\pi}{k^2} = -\frac{\lambda}{k}$$

$\therefore$

$$k\frac{d\upsilon}{dk} = -\lambda\frac{d\upsilon}{d\lambda}$$

$$\upsilon_g = \upsilon - \lambda\frac{d\upsilon}{d\lambda} \qquad (12.36)$$

This is the relation that connects phase velocity and group velocity.

The velocity with which the wave packet advances in the medium is the *group velocity* υ_g. Phase velocity is a characteristic of an individual wave whereas group velocity characterizes a group of waves. Group velocity will be the same as phase velocity if all the constituent waves travel with the same velocity. It means in a non-dispersive medium,

$$\upsilon_g = \upsilon. \tag{12.37}$$

However, the waves of different wavelengths travel in a medium with different velocities. Therefore, the group velocity is in general less than the phase velocity.

12.15. REAL LIGHT WAVES

It is now very easy to see why natural light behaves in a different way from radio waves or other electromagnetic waves, though it belongs to the same family of waves. We have been accustomed to regard light waves as ideal harmonic waves of infinite extension. Now we have to modify this visualization in view of the discreteness in the emission process of light. We compare here the features of real light waves with those of ideal waves.

Real light waves.

(*a*) Real light waves are of limited extension:

Ideal waves are of infinite extension in both space and time and are of constant amplitude.

Light emitted from common sources is in the form of wave trains (or wave packets). The amplitude varies from one end of the wave packet to the other end. A jumble of such wave packets constitutes the real light wave.

(*b*) Real light waves are not monochromatic:

Ideal waves are harmonic and possess a single frequency. Hence, they are strictly monochromatic.

In contrast, the wave trains emitted by a light source are not harmonic but are pulses of short duration. Such non-harmonic waves may be regarded as arising due to the superposition of a series of harmonic waves having a range of frequencies $\Delta\nu$ centred about a central frequency ν_o. The degree of monochromaticity of a source is given by

$$\xi = \frac{\Delta\nu}{\nu_o} \tag{12.38}$$

where $\Delta\nu$ is the band width. When $\Delta\nu/\nu_o = 0$, the radiation is ideally monochromatic. If $\Delta\nu/\nu_o \ll 1$, the radiation is **quasi-monochromatic**.

(*c*) Real light waves are non-directional :

In case of real light waves, there is no definite direction of propagation as light is emitted randomly and in all possible directions. Therefore the light is divergent and its intensity diminishes at large distance from the source.

(*d*) Real light waves are incoherent :

Coherence means the coordinated motion of several waves. When two or more waves are coherent, they will maintain a fixed and predictable phase relationship with each other. Monochromatic plane waves are ideally coherent.

As the emission acts occur without any coordination in the source, the resulting wave trains will not have any correlation in their phase values. The phases of wave trains vary at random from one wave train to another wave train and fluctuate irregularly at a rate of about 10^8 times per second. Consequently, the real light waves are incoherent.

(*e*) Real light waves are unpolarized:

Light waves belong to the category of transverse waves. In ideal transverse waves, the vibrations are perpendicular to the direction of propagation and are confined to a plane perpendicular to it. Therefore, the waves are polarized.

In case of real light waves, each wave train taken alone is polarized. However, owing to the haphazardness in the acts of emission of wave trains by atoms, the different wave trains possess different orientations of planes of polarization. The radiation consists of wave trains with planes of vibration distributed in all possible directions about the direction of propagation. Therefore, the real light is highly disordered and unpolarized.

WORKED OUT EXAMPLES

Example 12.1: A particle performs harmonic oscillations with a period of 2.0 sec. The amplitude of the oscillations is 10 cm. Find the displacement, velocity and acceleration of the particle in 0.2 sec after it passes through its equilibrium position, if the beginning of the oscillations coincides with the equilibrium position.

Given: Time period T = 2.0 sec, Amplitude A=10 cm = 0.10 m, instant of time t = 0.2 sec

The initial phase angle $\alpha = 0$

Solution

The displacement, $y = A \sin(\omega t + \alpha) = A \sin \omega t$ (as $\alpha = 0$) $= \text{A}\sin \dfrac{2\pi}{T} t$

$$\therefore \quad y = (0.10\text{m}) \sin \frac{2\times 3.14}{2}\times 0.2 = 0.10 \sin 36° = 0.059 \text{ m}$$

The velocity of the particle is

$$\upsilon = \frac{dy}{dx} = \omega A \cos \omega t = \frac{2\pi}{T} A \cos \frac{2\pi}{T} t = \frac{2\times 3.142}{2s} \times 0.10 \text{ m} \times \cos 36° = 0.25 \text{ m/sec}.$$

The acceleration of the particle is a

$$a = \frac{d^2 y}{dy^2} = -\omega^2 A \sin \omega t = -\omega^2 y$$

$$\therefore \quad |a| = \frac{4\pi^2}{T^2} y = \frac{4(3.142)^2}{(2)^2 s^2} \times 0.059 \text{ m} = 0.57 \text{ m/s}^2$$

Example 12.2: The equation of a wave traveling on a string is given by $y = 2 \sin \pi (0.5x - 200t)$, where x and y are in cm and t in seconds. Find the amplitude, wavelength, frequency, period and velocity of propagation.

Solution:

The equation of harmonic motion is $y = A \sin \dfrac{2\pi}{\lambda}(\upsilon t - x)$

or

$$y = A \sin 2\pi \left(\nu t - \frac{x}{\lambda}\right) \tag{1}$$

The given equation is $y = 2 \sin \pi (0.5x - 200t)$

It may be written as

$$y = 2 \sin 2\pi \left(100t - \frac{x}{4}\right) \qquad (2)$$

Comparing equation (2) with equation (1), we get

Amplitude A = **2 cm,** Wavelength λ = **4 cm,** Frequency = **100 Hz,**

Time period $T = \frac{1}{100 \text{ Hz}}$ and velocity υ = (100Hz) (4cm) (10^{-2}m/cm) = **4 m/s**

QUESTIONS

1. What is an oscillatory motion?
2. What are the conditions for appearance of oscillations?
3. What are the parameters that characterize an oscillatory motion?
4. Explain the terms: displacement and phase.
5. Discribe the oscillatory motion mathematically?
6. What is meant by simple harmonic motion (SHM)?
7. Explain the characteristics of SHM?
8. What are meant by free oscillations? What is meant by natural frequency?
9. What is meant by wave motion?
10. What is propagated in wave motion?
11. Summarize the characteristics of wave motion.
12. What is the difference between oscillatory motion and wave motion?

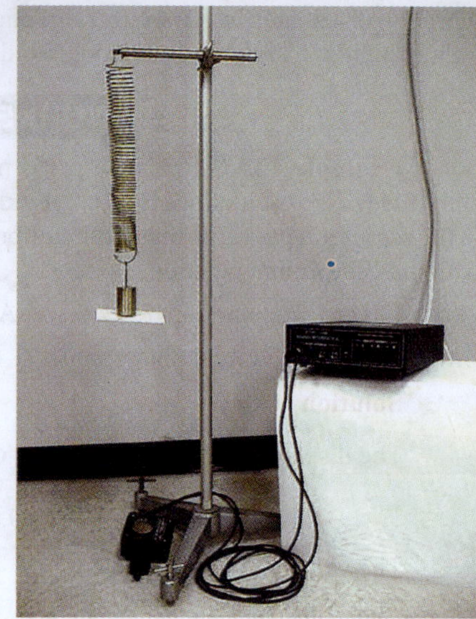

Simple harmonic motion

13. What are the different types of wave motion?
14. What are the parameters that characterize a wave?
15. Describe mathematically a simple harmonic wave traveling in the positive X-direction.
16. Obtain the relation between wave velocity and the particle velocity.
17. Write down the differential equation of simple harmonic wave.
18. What do you understand by the term "phase velocity"?
19. Explain the following: (*a*) wave surface (*b*) wave front and (*c*) ray.
20. How are the following produced: (*a*) spherical waves (*b*) plane waves?
21. What is a wave front? Mention different types. **(Kovempu, 2005)**

13 CHAPTER

Propagation of Light Waves

13.1. INTRODUCTION

Propagation of light is an electromagnetic phenomenon. Light waves are known to behave like any other electromagnetic wave. The physical laws describing propagation, reflection, refraction, attenuation of electromagnetic waves accurately describe the behaviour of light waves. In the electromagnetic theory of light, the mechanical displacement of the medium is replaced by a variation of the electric field at the corresponding point. The light wave consists of varying electric and magnetic fields and can be described by two vectors, the amplitude of electric field strength, **E**, and the amplitude of the magnetic field strength, **H**. These vectors oscillate at right angles to each other and to the direction of propagation. They cannot be separated. Thus, a beam of light is a traveling configuration of electric and magnetic fields. The electromagnetic theory was developed by James Clerk Maxwell in 1873. The first and foremost outcome of Maxwell's equations was the prediction of existence of electromagnetic waves and the brilliant prediction that light waves are electromagnetic waves. Starting from the Maxwell's equations, the propagation of light waves in space and materials can be easily explained and all the fundamental laws of optics such as law of reflection, Snell's law of refraction etc can be derived. Expressions to calculate the reflectivity, transmissivity and absorptivity of material media can be obtained from these equations. Further, the electromagnetic

At a Glance

theory explains the origin of refractive index and the dispersion properties of optical media. In this chapter we briefly acquaint ourselves with the propagation of light waves and their interaction with material media.

13.2. MAXWELL'S EQUATIONS

James Clerk Maxwell
(1831–1909)

Maxwell unified the theories of electricity and magnetism by way of deducing four very important equations which combine the experimental observations reported by Gauss, Ampere, and Faraday with his concept of displacement current. The equations encapsulate the connection between the electric field and electric charge and between the magnetic field and electric current. The Maxwell's equations also define the bilateral coupling between the electric and magnetic field quantities. They along with some auxiliary equations form the fundamental tenets of electromagnetic theory. When the charge and current sources vary with time, the electric and magnetic fields become interconnected and the coupling between them produces electromagnetic waves capable of traveling through free space and in material media. In all, there are four Maxwell's equations. These equations cannot be derived since they are the fundamental axioms or postulates of electrodynamics, obtained with the help of generalization of experimental results.

The Maxwell's equations are expressed in differential form and integral form in the following way:

Law	Differential Form	Integral Form	
Gauss's law	$\nabla \cdot \mathbf{D} = \rho$	$\oint \mathbf{D} \cdot d\mathbf{S} = \int \rho \, dV$	(13.1)
Faraday's law	$\nabla \times \mathbf{E} = -\dfrac{\partial \mathbf{B}}{\partial t}$	$\oint \mathbf{E} \cdot d\mathbf{l} = -\int_S \dfrac{\partial \mathbf{B}}{\partial t} \cdot d\mathbf{S}$	(13.2)
Gauss's law for magnetism	$\nabla \cdot \mathbf{B} = 0$	$\oint \mathbf{B} \cdot d\mathbf{S} = 0$	(13.3)
Ampere's law	$\nabla \times \mathbf{H} = \mathbf{J} + \dfrac{\partial \mathbf{D}}{\partial t}$	$\oint \mathbf{H} \cdot d\mathbf{l} = \int \left[\dfrac{\partial \mathbf{D}}{\partial t} + \mathbf{J} \right] \cdot d\mathbf{S}$	(13.4)

Maxwell's equations contain only the first derivatives of fields **E** and **B** with respect to time and space coordinates and the first powers of densities ρ and J of electric charges and currents. Therefore, these equations are linear and the fields obey superposition principle.

13.2.1. PHYSICAL SIGNIFICANCE OF MAXWELL'S EQUATIONS

The physical significance of Maxwell's equations can be readily interpreted from their mathematical statement in the integral form.

(1) Maxwell's first equation (13.1) shows that the total electric flux density **D** through the surface enclosing a volume is equal to the charge density ρ within the volume. It means that a charge distribution generates a steady electric field.

(2) The second equation shows that the e.m.f. around a closed path is equal to the time derivative of the magnetic flux density through the surface bounded by the path. It means that an electric field can also be generated by a time-varying magnetic field.

(3) Maxwell's third equation tells us that the net magnetic flux through a closed surface is zero. It implies that magnetic poles do not exist separately in the way as electric charges do. Thus, in other words, magnetic monopoles do not exist.

(4) Maxwell's fourth equation shows that the magneto motive force around a closed path is equal to the conduction current plus the time derivative of the electric flux density D through any surface bounded by the path. The time derivative of the electric flux density $\partial D / \partial t$ is called displacement current. Thus this equation means that a magnetic field is generated by a time-varying electric field.

13.2.2. ELECTROMAGNETIC WAVES

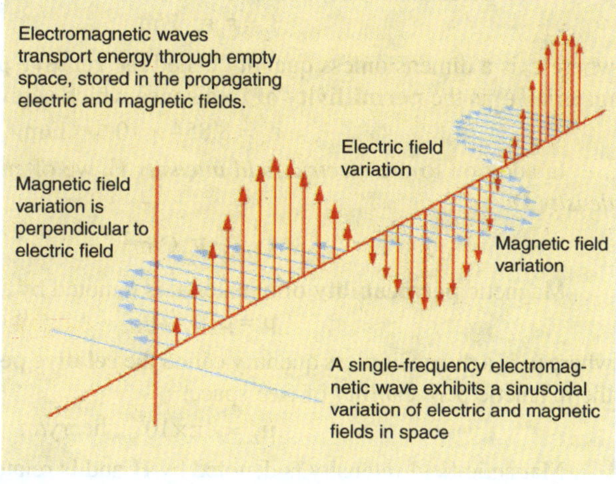

Maxwell showed that by combining the four equations a *wave equation* was obtained which described the propagation of waves. The time variation of a magnetic field induces an electric field, while a variation of an electric field, in its turn, induces a magnetic field and electromagnetic fields can exist independently, without electric charges and currents. The continuous inter-conversion of the fields preserves them and an electromagnetic perturbation propagates in space. Such fields are called **electromagnetic waves**. The generation of electromagnetic wave does not require any medium. **The electromagnetic waves propagate through space entirely on their own.** Maxwell's theory placed no restriction on possible wavelengths for the electromagnetic radiation. The vector cross products in Maxwell's second and fourth equations imply that the two fields, **E** and **H** are normal to each other and also normal to the direction of propagation. In a vacuum, these waves always propagate at a velocity equal to the velocity of light c.

The vectors **E** and **B** in an electromagnetic wave always oscillate in phase (see Fig. 13.1). The instantaneous values of **E** and **B** at any point are connected through the relation

$$\sqrt{\varepsilon}\, E = \sqrt{\mu}\, H$$

This means that **E** and **H** (or **B**) simultaneously attain their maximum values, vanish etc.

The directions of E and H in a uniform plane wave.

Fig. 13.1

Note that electromagnetic waves are, in general, neither longitudinal nor transverse. In some types of electromagnetic waves, the electric vector **E** is at right angles to the ray while the magnetic vector **H** is not. Waves of this type are called **transverse electric waves** or TE waves. Since **H** vector is not normal to the ray, it must have a component along the ray direction. In another type of waves the magnetic vector is at right angles to the ray while the electric vector **E** is not. Waves of this type are called **transverse magnetic waves** or TM waves. If both the vectors **E** and **H** are at right angles to the ray, the waves are called **transverse electromagnetic waves** or TEM waves.

13.3. CONSTITUTIVE RELATIONS

The electric and magnetic properties of a medium are described by three quantities: relative permittivity,ε_r, the relative permeability,μ_r, and the conductivity,σ. If Maxwell's equations are to be extended to material media, they should be supplemented with relations which would contain these quantities characterizing individual properties of the medium. Such relations are called **constitutive relations** and also known as *material equations.*

The **permittivity** of a dielectric material is denoted by

$$\varepsilon = \varepsilon_o \varepsilon_r \ \text{F/m} \tag{13.5}$$

where ε_r is a dimensionless quantity called the **relative permittivity** or **dielectric constant** of the material. ε_o is the **permittivity of free space** which is given by

$$\varepsilon_o = 8.854 \times 10^{-12} \ \text{F/m}$$

In addition to the *electric field intensity* **E**, we often use the related quantity, the *electric flux density* **D**.

$$\mathbf{D} = \varepsilon \mathbf{E} \ \text{C/m}^2 \tag{13.6}$$

Magnetic **permeability** of a material is denoted by

$$\mu = \mu_o \mu_r \ \text{H/m} \tag{13.7}$$

where μ_r is a dimensionless quantity called the **relative permeability** of the material and μ_o is called the magnetic permeability of free space.

$$\mu_o = 4\pi \times 10^{-7} \ \text{henry/m}$$

Magnetic field intensity is denoted by **H** and is related to magnetic induction **B** through

$$\mathbf{B} = \mu \ \mathbf{H} \tag{13.8}$$

The point form of Ohm's law describes the flow of current in a material and it is given by

$$\mathbf{J} = \sigma \ \mathbf{E} \tag{13.9}$$

where J is the current density and σ the electrical conductivity of the medium.

We usually assume ε_r,μ_r, and σ are not functions of time and that the medium is linear, homogeneous, isotropic; ε, μ, and σ are therefore constant and uniform throughout the medium. A medium is a **homogeneous medium** when the quantities ε, μ, and σ are constant throughout the medium. The medium is **isotropic** if ε is a scalar constant, so that **D** and **E** have everywhere the same direction. Material equations have the simplest form for sufficiently weak electric fields. Equations (13.6), (13.8), and (13.9) hold good for the case of isotropic non-ferroelectric and non-ferromagnetic media.

When the relations (13.6), (13.8), and (13.9) are inserted in the Maxwell's equations (13.1) to (13.4) we get the following differential equations relating the electric and magnetic field strengths **E** and **H**. If they are then solved as simultaneous equations, they will determine the laws which both **E** and **H** must obey.

$$\nabla \cdot \mathbf{E} = \frac{\rho}{\varepsilon} \tag{13.10a}$$

$$\nabla \times \mathbf{E} = -\mu \frac{\partial \mathbf{H}}{\partial t} \tag{13.10b}$$

$$\nabla . \mathbf{H} = 0 \tag{13.10c}$$

$$\nabla \times \mathbf{H} = J + \varepsilon \frac{\partial \mathbf{E}}{\partial t} \tag{13.10d}$$

13.4. WAVE EQUATION FOR FREE-SPACE

Maxwell's equations are coupled partial differential equations in **E** and **H** which cannot be solved in general. In order to simplify the equations, we should uncouple the set and obtain differential equations in **E** or **B** alone. In order to understand the nature of waves, we consider *free space*, which is a large empty volume of space. Free space is a perfect dielectric and does not contain charges ($\rho = 0$) and there are no conduction currents ($J = 0$) flowing in it. Maxwell's equations for free space (or a dielectric medium) become

$$\nabla \cdot \mathbf{E} = 0 \qquad (13.11a) \qquad\qquad \nabla \times \mathbf{E} = -\frac{\partial \mathbf{B}}{\partial t} \qquad (13.11b)$$

$$\nabla \cdot \mathbf{B} = 0 \qquad (13.11c) \qquad\qquad \nabla \times \mathbf{H} = \frac{\partial \mathbf{D}}{\partial t} \qquad (13.11d)$$

In order to describe the propagation of an electromagnetic wave in free space, we need to derive wave equation for **E** and **B** and then solve them to obtain explicit expressions for **E** and **B** as functions of (x,y,z). We start by taking curl of both sides of equ.(13.11b) and obtain

$$\nabla \times (\nabla \times \mathbf{E}) = -\nabla\left(\frac{\partial \mathbf{B}}{\partial t}\right) = -\frac{\partial}{\partial t}(\nabla \times \mathbf{B}) = -\mu_0 \frac{\partial}{\partial t}(\nabla \times \mathbf{H})$$

Using equ.(13.11d) into the above equation, we get

$$\nabla \times (\nabla \times \mathbf{E}) = -\mu_o \varepsilon_o \frac{\partial^2 \mathbf{E}}{\partial t^2}$$

But $\qquad \nabla \times (\nabla \times \mathbf{E}) = \nabla(\nabla \cdot \mathbf{E}) - \nabla^2 \mathbf{E}$

But $\qquad \nabla \cdot \mathbf{E} = 0$

∴ $\qquad \nabla \times (\nabla \times \mathbf{E}) = -\nabla^2 \mathbf{E}$

∴
$$\nabla^2 \mathbf{E} = \mu_o \varepsilon_o \frac{\partial^2 \mathbf{E}}{\partial t^2} \qquad (13.12a)$$

This is the law that **E** must obey.

A similar procedure for **B** gives us

∴
$$\nabla^2 \mathbf{B} = \mu_o \varepsilon_o \frac{\partial^2 \mathbf{B}}{\partial t^2} \qquad (13.12b)$$

This is the law that **B** must obey.

Equations (13.12a) and (13.12b) are very much similar to the general wave equation (12.16), and constitute **wave equations**. They are *three-dimensional vector wave equations* and describe the propagation of an electromagnetic wave through a uniform medium. Since the wave equations for **E** and **B** are of the same form, their solutions will also have the same form. The solution to (13.12) leads to waves that can exist in free space. Even though the electric and magnetic fields of the waves start out on charges and currents, they detach themselves from them and move through free space as independent entities.

13.4.1. VELOCITY OF THE ELECTROMAGNETIC WAVE

The propagation characteristics of the electromagnetic wave are contained in the solution of equ.(13.12). To bring out the characteristics, we compare it with the general wave equation (12.16).

$$\nabla^2 \xi = \frac{1}{\upsilon^2} \frac{\partial^2 \xi}{\partial t^2}$$

The comparison gives

$$\upsilon^2 = \frac{1}{\mu_o \ \varepsilon_o}$$

or
$$\upsilon = \frac{1}{\sqrt{\mu_o \ \varepsilon_o}} \qquad (13.13)$$

Substituting the values of μ_o and ε_o, we find that

$$\frac{1}{\sqrt{\mu_o\varepsilon_o}} = \frac{1}{\sqrt{(4\pi\times10^{-7}\ wb.A.m^2)\left(8.9\times10^{-12}\ C^2/N.m^2\right)}} = 3.0\times10^8 \text{ m/s.}$$

$$\therefore \qquad \frac{1}{\sqrt{\mu_o\varepsilon_o}} = c \text{, the velocity of light} \qquad (13.14)$$

Obviously, electromagnetic waves travel with the velocity of light in free space.

13.4.2. RELATION BETWEEN THE REFRACTIVE INDEX AND RELATIVE PERMITTIVITY OF A MEDIUM

In case of a medium other than vacuum, we have to use $\varepsilon_o\varepsilon_r$ (= ε) and $\mu_o\ \mu_r$ (= μ), instead of ε_o and μ_o in equ.(13.12). We then get

$$\upsilon = \frac{1}{\sqrt{\mu_o\mu_r\varepsilon_o\varepsilon_r}} = \frac{1}{\sqrt{\mu_o\varepsilon_o}}\cdot\frac{1}{\sqrt{\mu_r\varepsilon_r}} = \frac{c}{\sqrt{\mu_r\varepsilon_r}}$$

or $$\frac{c}{\upsilon} = \sqrt{\mu_r\varepsilon_r}$$

But $$\frac{c}{\upsilon} = n \qquad \text{(refractive index of the medium)}$$

$$\therefore \qquad n = \sqrt{\mu_r\varepsilon_r}$$

For a non-magnetic medium μ_r = 1. Therefore,

$$n = \sqrt{\varepsilon_r} \qquad (13.15)$$

or $$n^2 = \varepsilon_r$$

13.5. UNIFORM PLANE WAVES

When energy is emitted by a source, it expands outwardly from the source in the form of spherical waves. The spherical wave travels at the same speed in all directions and therefore expands at the same rate. To an observer very far away from the source, the wave front of the spherical wave appears approximately planar. A plane wave is the simplest example of wave motion. In a plane wave, the electric and magnetic intensities are of constant value over any plane perpendicular to the direction of propagation. Such a plane is a surface of equal phase. A **plane wave** is thus a wave for which the phase has the same value at all points on an infinite plane. As the amplitude is also constant over the plane surface, it is called a **uniform plane wave**. Plane waves vary only in the direction of propagation and are uniform in planes normal to the direction of propagation. In this chapter we confine ourselves to the plane wave propagation in unbounded media. Plane wave propagation can be described by Cartesian coordinates, which are easier to work with mathematically than the spherical coordinates needed for describing propagation of a spherical wave.

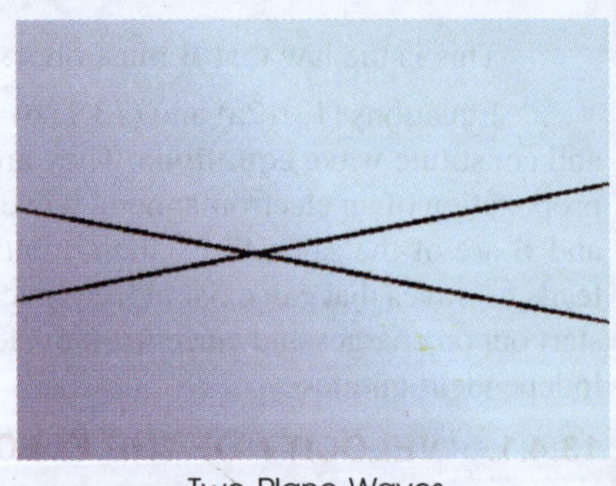

Two Plane Waves

13.5.1. THE TRANSVERSE NATURE OF PLANE WAVES

Let us assume that the plane waves are traveling along the z-direction and hence **E** is constant over any given plane parallel to the xy-plane. Similarly **H** is constant over the xy-plane. We then have

$$\frac{\partial \mathbf{E}}{\partial x} = \frac{\partial \mathbf{E}}{\partial y} = 0 \text{ and } \frac{\partial \mathbf{H}}{\partial x} = \frac{\partial \mathbf{H}}{\partial y} = 0 \tag{13.16}$$

These relations imply that

$$\frac{\partial E_x}{\partial x} = 0, \frac{\partial E_y}{\partial x} = 0, \frac{\partial E_z}{\partial x} = 0 \text{ and } \frac{\partial E_x}{\partial y} = 0, \frac{\partial E_y}{\partial y} = 0, \frac{\partial E_z}{\partial y} = 0 \tag{13.17a}$$

$$\frac{\partial H_x}{\partial x} = 0, \frac{\partial H_y}{\partial x} = 0, \frac{\partial H_z}{\partial x} = 0 \text{ and } \frac{\partial H_x}{\partial y} = 0, \frac{\partial H_y}{\partial y} = 0, \frac{\partial H_z}{\partial y} = 0 \tag{13.17b}$$

According to Maxwell' equation (13.11a), we have

$$\nabla . \mathbf{E} = \mathbf{0}$$

which means that

$$\frac{\partial E_x}{\partial x} + \frac{\partial E_y}{\partial y} + \frac{\partial E_z}{\partial z} = 0$$

Since $\frac{\partial E_x}{\partial x} = 0$ and $\frac{\partial E_y}{\partial y} = 0$, then $\frac{\partial E_z}{\partial z} = 0.$ (13.18a)

It means that E_z is independent of z and has the same value all along the z-axis. Wave motion consists of changing values of **E** along the direction of propagation. As E_z is constant along z-direction, it does not contribute to the wave motion and therefore E_z must be zero. $E_z = 0$ implies that **E** lies in a plane perpendicular to the direction of propagation of the wave. Hence, the electric wave is a *transverse* wave.

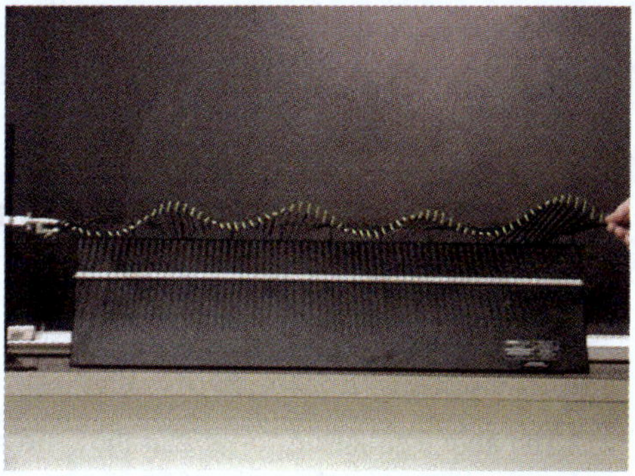

Transverse wave.

Again, according to Maxwell' equation (13.11c), we have

$$\nabla \cdot \mathbf{B} = \mathbf{0} \text{ and hence } \nabla \cdot \mathbf{H} = \mathbf{0}$$

which means that

$$\frac{\partial H_x}{\partial x} + \frac{\partial H_y}{\partial y} + \frac{\partial H_z}{\partial z} = 0$$

Since $\frac{\partial H_x}{\partial x} = 0$ and $\frac{\partial H_y}{\partial y} = 0$, then $\frac{\partial H_z}{\partial z} = 0.$ (13.18b)

Thus, H_z is independent of z and has the same value all along the z-axis. As H_z is constant along z-direction, it does not contribute to the wave motion and therefore H_z must be zero. $H_z = 0$ implies that **H** lies in a plane perpendicular to the direction of propagation of the wave. Hence, the magnetic wave also is a *transverse* wave.

13.5.2. RELATION BETWEEN E AND H IN A UNIFORM PLANE WAVE

Now we assume that the variations of **E** are simple harmonic and that **E** is parallel to the y-axis. Then $E_z = 0$ and from equ.(13.12a), we may write the solution for a wave traveling in the positive z-direction,

$$E_y = E_1 \, e^{-i(\omega t - k z)} \tag{13.19a}$$

or in a simpler form as $E_y = E_1 \cos\left[\omega(t - z/c)\right]$ (13.19b)

Maxwell's equation (13.11b) can be written into the following three scalar equations.

$$\frac{\partial E_z}{\partial y} - \frac{\partial E_y}{\partial z} = -\frac{\partial B_x}{\partial t} \qquad \frac{\partial E_x}{\partial z} - \frac{\partial E_z}{\partial x} = -\frac{\partial B_y}{\partial t} \qquad \frac{\partial E_y}{\partial x} - \frac{\partial E_x}{\partial y} = -\frac{\partial B_z}{\partial t} \tag{13.20}$$

Using the results of equ.(13.17a) into the first relation of (13.20), we get

$$\frac{\partial E_y}{\partial z} = \frac{\partial B_x}{\partial t} \tag{13.21}$$

The magnetic flux density B_x associated with the electric field E_y can be found by integrating equ.(13.21). Thus,

$$B_x = \int \frac{\partial E_y}{\partial z}\, dt = -\frac{E_1 \omega}{c} \int \sin\left[\omega(t - z/c)\right] dt$$

$$= \frac{1}{c} E_1 \cos\left[\omega(t - z/c)\right]$$

The constant of integration is omitted since we are not interested in a constant component of the field.

$$\therefore \quad B_x = \frac{1}{c} E_y \qquad (13.22)$$

or

$$H_x = \frac{1}{\mu_o c} E_y = \frac{\sqrt{\mu_o \varepsilon_o}}{\mu_o} E_y$$

$$\therefore \quad H_x = \sqrt{\frac{\varepsilon_o}{\mu_o}}\, E_y \qquad (13.23)$$

Since E_y and H_x differ only by a scalar and have the same time dependence, **E** and **H** are in phase at all points in space and are mutually perpendicular. Their cross product **E × H** points in the direction of propagation denoted by the vector **k**. The three vectors are oriented as in a left-handed co-ordinate system, as depicted in Fig. 13.1, where *k* points along *z*-direction.

13.5.3. CHARACTERISTIC IMPEDANCE

Equ.(13.23) may be rewritten as

$$\frac{E_y}{H_x} = \sqrt{\frac{\mu_o}{\varepsilon_o}}$$

The above equation states that the ratio of the amplitudes of the vectors **E** and **H** always equal to the square root of the ratio of μ_o and ε_o. The ratio has the dimensions of impedance and hence it is called the **characteristic impedance** or **intrinsic impedance** of free space. It is denoted by η_o. We generalize equ.(13.23) by writing it as

$$\frac{\mathbf{E}}{\mathbf{H}} = \eta_o = \sqrt{\frac{\mu_0}{\varepsilon_0}} \qquad (13.24)$$

Using the values of μ_o and ε_o into the above equation, we get

$$\eta_o = \sqrt{\frac{4\pi \quad \times \quad 10^{-7}\ H/m}{10^{-19}\ /36\pi\ F/m}} = 120\,\pi \ = 377\ \text{ohms} \qquad (13.25)$$

In a dielectric material, $\eta = 377/\varepsilon_r^{1/2} = 377/n$, where *n* is the index of refraction of the material.

13.6. WAVE POLARIZATION

The polarization of an electromagnetic wave is taken as the direction of the **E** vector. It describes *the shape and locus of the tip of the* **E** *vector* (in the plane orthogonal to the direction of propagation) *at a given point in space as a function of time*. The uniform plane wave described by equ.(13.19) has only y-component of electric field. The direction of the field vector at some point in space and time lies along a line in a plane perpendicular to the direction of propagation.

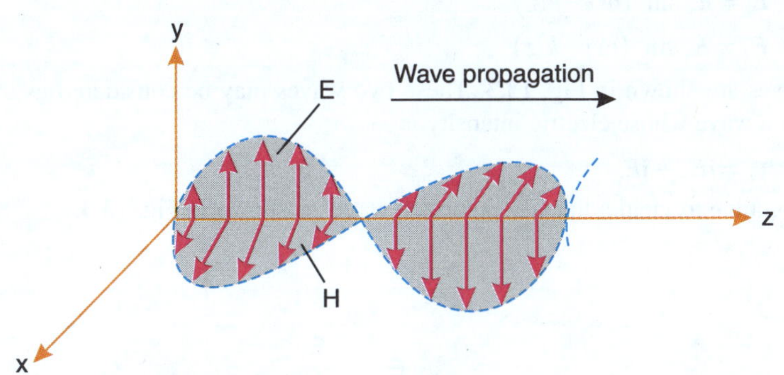

(*a*) The linearly polarized uniform plane wave (*a*) as a function of distance at $t = 0$

Fig. 13.2

If we take a cross-sectional view of the wave in the xy-plane at z = 0, we observe the magnitude of the field vectors varying sinusoidally, as shown in Fig. 13.2 (*b*). Note that the field vectors reverse direction with each half-cycle but that this variation is always along a line (the ± y direction for **E** and the ± x direction for **H**). Since the direction of the field vectors at some point in space and every point in time lies along a line in a plane perpendicular to the direction of propagation, the wave is said to be linearly polarized. Thus, the uniform plane waves are **linearly polarized**.

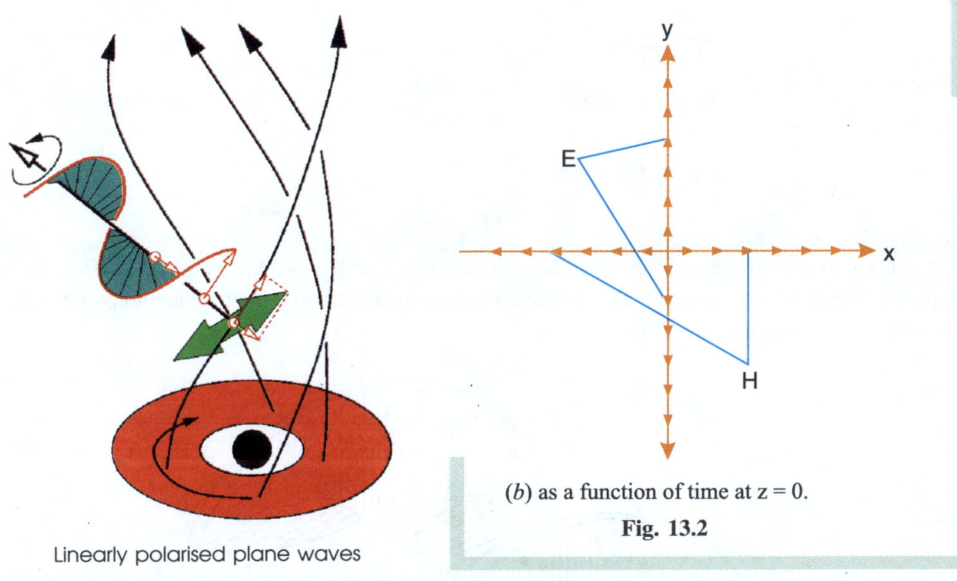

Linearly polarised plane waves

(*b*) as a function of time at z = 0.

Fig. 13.2

In general, waves can have either E_y or E_x components when travelling in z-direction. If a wave has E_y-component, then the wave is polarized along y-direction and if the wave has E_x-component, then it is polarized along x-direction. If two coherent and linearly polarized waves polarized in mutually perpendicular directions propagate in the same direction, they combine to give a resultant wave whose state of polarization depends on their phase difference.

Case 1: If the waves are of equal or of unequal intensity ($E_y = E_x$ or $E_y \neq E_x$) and are in-phase, they may be represented by

$$E_1 = E_y \sin\ (\omega t - k z)$$

and $$E_2 = E_x \sin\ (\omega t - k z)$$

The waves are shown in Fig. 13.3. These two waves may be considered as the rectangular components of a wave whose electric intensity is

$$\mathbf{E} = \mathbf{i}E_x + \mathbf{j}E_y \tag{13.26}$$

They combine to yield a linearly polarized wave, as shown in Fig. 13.3.

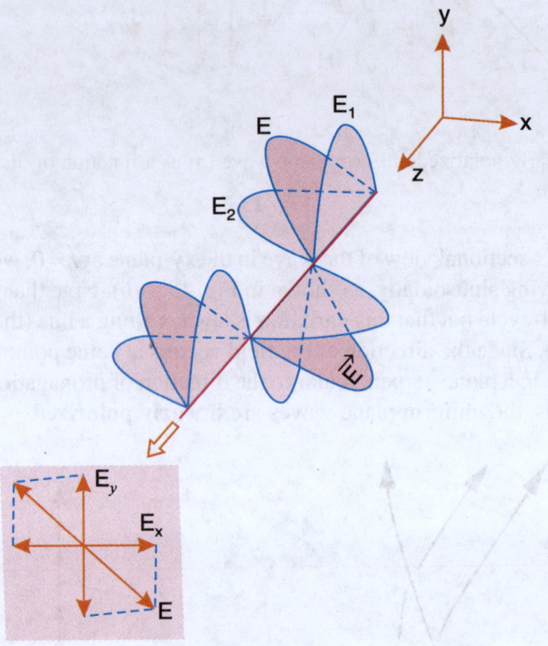

Fig. 13.3

Case 2: The waves may be of unequal intensity ($E_y \neq E_x$) and have a phase difference of 90°, as shown in Fig.13.4. It is seen that one wave is equal to zero at any given time and place, where the other wave is a maximum.

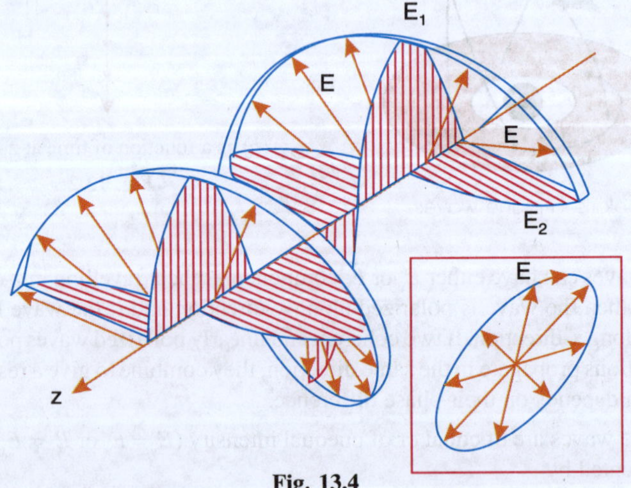

Fig. 13.4

At any given place, as time is varied, the tip of the **E** vector traces an ellipse (see inset in Fig.13.4) during the course of one cycle and the resultant wave is **elliptical polarized.** At any given instant of time the **E** vector describes an elliptical helix in space (see Fig.13.4). If **E** vector rotates in clockwise direction it is referred as right hand elliptically polarized wave since if one places the fingers of the right hand in the direction of the rotation, the thumb will point in the direction of propagation. If **E** rotates counter clockwise it is referred as left hand elliptical polarization.

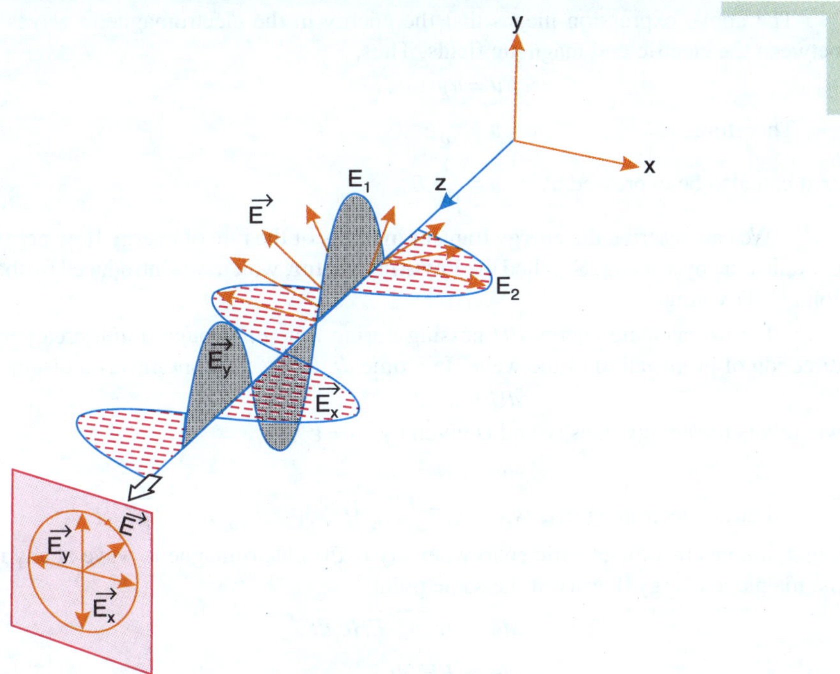

Two perpendicular waves with a 90° phase difference produce circularly polarized electromagnetic wave.

Fig. 13.5

Case 3: In the particular case where the waves have equal intensity and maintain a phase difference of 90°, the tip of the **E** vector of the resultant wave traces out a circle and the wave is referred to as **circularly polarized** wave. At a given instant **E** and **H** describe the form of regular helixes, the two being everywhere 90° apart. If **E** vector rotates in clockwise direction it is referred as *right hand circularly polarized* wave and if it rotates counter clockwise it is referred as *left hand circular polarization.*

13.7. ENERGY DENSITY, THE POYNTING VECTOR AND INTENSITY

One of the important characteristics of electromagnetic waves is that it transports energy from one region to another region. The *energy density*, *u*, of a wave is the radiant energy per unit volume. The electromagnetic wave consists of electric field and magnetic field which independently can store energy. The energy density stored in the electric field **E,** say existing between the plates of a charged capacitor, is given by

$$u_E = \frac{\varepsilon_o}{2} E^2 \tag{13.27a}$$

Similarly the energy density stored in a magnetic field, say produced by a current carrying loop, is given by

$$u_B = \frac{1}{2\mu_o} B^2 \tag{13.27b}$$

According to (13.22), the vectors E and B are related through

$$\boldsymbol{E} = c\,\boldsymbol{B}$$

$$\therefore \quad u_E = \frac{\varepsilon_o}{2} E^2 = \frac{\varepsilon_o}{2} c^2 B^2 = \frac{1}{2\mu_o} B^2 = u_B$$

The above expression means that the energy in the electromagnetic waves is stored equally between the electric and magnetic fields. Thus,

$$u = u_E + u_B$$

Therefore, $$u = \varepsilon_o E^2 \tag{13.28a}$$

or it can also be expressed as $$u = \frac{1}{\mu_o} B^2. \tag{13.28b}$$

We can describe the energy transfer in terms of the rate of energy flow per unit area or power per unit area, by a vector, **S**, called the **Poynting vector**, which was introduced by the British physicist John H. Poynting.

Let us calculate energy dU passing during time dt through a unit area perpendicular to the direction of propagation of the wave. In a time dt, the wave front moves a distance $dz = c.dt$.

$$dU = uc\,dt$$

where u is the energy density and is given by $u = \varepsilon_o E^2$.

$$\therefore \quad du = \varepsilon_o E^2 c\,dt$$

For an electromagnetic wave $\varepsilon_o E^2 = \mu_o H^2$ or $\sqrt{\varepsilon_o}\,E = \sqrt{\mu_o}\,H$

which implies that the electric energy density in the electromagnetic wave at any instant is equal to the magnetic energy density at the same point.

$$du = \sqrt{\varepsilon_o \mu_o}\,EHc\,dt$$

or $$du = EH\,dt$$

The term EH represents the magnitude of energy flux density vector. The corresponding vector **S** is then

$$\boldsymbol{S} = \boldsymbol{E} \times \boldsymbol{H} \tag{13.29}$$

$$\boldsymbol{S} = \frac{1}{\mu_o} \boldsymbol{E} \times \boldsymbol{B} \tag{13.30a}$$

or $$\boldsymbol{S} = c^2 \varepsilon_o\,(\mathbf{E} \times \mathbf{B}) \tag{13.30b}$$

S is known as the *Poynting vector* has dimensions of energy per unit time per unit area.

A plane electromagnetic wave traveling in the positive z-direction is described by the expression (13.19b).

$$\therefore \quad S = \frac{1}{\mu_o} E_o B_o \cos^2(\omega t - kz)$$

The electromagnetic power P that the plane wave delivers to a receiver oriented perpendicular to z-axis is given by

$$P = SA = \frac{1}{\mu_o} E_o B_o A \cos^2(\omega t - kz)$$

$$P = \frac{1}{\mu_o c} E_o^2 A \cos^2(\omega t - kz) \tag{13.31}$$

The **intensity** of the electromagnetic wave is defined as the *average power per unit area.* If T is the observation time, then the average power is

$$P_{ave} = \frac{1}{T}\int_o^T P\,dt = \frac{1}{\mu_o c} E_o^2 A \cdot \frac{1}{T}\int_o^T \left[\cos^2(\omega t - kz)\right] dt = \frac{1}{2\mu_o c} E_o^2 A$$

$$\therefore \qquad I = \frac{P_{ave}}{A} = \frac{1}{2\mu_o c} E_o^2$$

or
$$I = \tfrac{1}{2} c\varepsilon_o E_o^2 \qquad \text{W/m}^2 \tag{13.32a}$$

Or equivalently,
$$I = \frac{c}{2\mu_o} B_o^2 \tag{13.32b}$$

From equ.(13.32a) we find that

$$I \propto |E_o|^2 \tag{13.33}$$

It means that the *intensity of the light wave is proportional to the square of the amplitude of the electric vector.*

13.8. RADIATION PRESSURE AND MOMENTUM

Electromagnetic waves transport energy. Therefore, they carry momentum and exhibit a force in the direction of propagation. The momentum is a property of the field alone and is not associated with any moving mass. The momentum density, that is momentum *dp* per volume dV is given by

$$\frac{dp}{dV} = \frac{EB}{\mu_o c^2} = \frac{S}{c^2} \tag{13.34}$$

The volume dV occupied by an electromagnetic wave that passes through an area A in time *dt* is

$$dV = Acdt$$

Therefore, the momentum flow rate per unit area is given by

$$\frac{1}{A}\frac{dp}{dt} = \frac{S}{c}$$

This is the momentum transferred per unit surface area per unit time. The average rate of momentum transfer per unit area is given by

$$\frac{1}{A}\frac{dp}{dt} = \frac{S_{ave}}{c} = \frac{I}{c}$$

This momentum is responsible for the **radiation pressure**.

When the surface, on which light is incident, is a **perfect absorber**, the momentum of the wave is transferred to the absorbing surface. Assuming normal incidence of the wave on the surface, the radiation pressure is the average value of dp/dt divided by the absorbing area. Therefore,

$$\wp = \frac{S_{ave}}{c} = \frac{I}{c} \tag{13.35a}$$

If the surface is a **perfect reflector**, the light incident normally on it with a velocity + c is reflected with a velocity – c. As a result, the change in momentum will be equal to 2*p*.

$$\wp = \frac{2S_{ave}}{c} = \frac{2I}{c} \tag{13.35b}$$

13.9. LIGHT WAVES AT BOUNDARIES

We have studied so far the propagation of a light wave in free space. When light proceeds from one medium into another (e.g., from air into a solid substance), several things happen. Some of the light radiation may be transmitted through the medium, some is absorbed, and some is reflected at the interface between the two media. Obviously, the intensity I_0 of the beam incident on the surface of the medium must equal the sum of the intensities of the transmitted, absorbed and reflected beams. Thus,

$$I_o = I_T + I_A + I_R \tag{13.36}$$

where I_T, I_A and I_R are the intensities of the transmitted, absorbed and reflected beams respectively. Equ.(13.36) may be rewritten as

$$1 = \frac{I_T}{I_o} + \frac{I_A}{I_o} + \frac{I_R}{I_o}$$

or

$$T + A + R = 1 \tag{13.37a}$$

where $T = I_T / I_o$ is the fraction of the light transmitted by the material and is called the **transmittivity** or **transmittance**, $A = I_A / I_o$ is the fraction of light absorbed by the material and is called **absorptivity**, and $R = I_R / I_o$ is the fraction of light reflected by the material and is called the **reflectivity** or **reflectance** of the material. Glass and other transparent materials reflect only a small portion of the incident radiation and transmit the remainder. Absorption of light in the medium will be negligible. Therefore, we consider here only reflection and transmission of light. In such cases equ.(13.37a) reduces to

$$T + R = 1 \tag{13.37b}$$

We now develop the familiar laws of optics by means of Maxwell's equations.

13.9.1. BOUNDARY CONDITIONS

Maxwell's equations in the differential form presume that all quantities vary continuously in space and time. When the waves appear at the interface between two different materials, Maxwell's equations require that certain relationships must exist between the fields on the two sides of the interface. These are known as *boundary conditions*. They are obtained as follows.

At the boundary between two different media, we resolve the total electric field **E** into a component tangent to the boundary surface, E_t and a component normal to the boundary surface, E_n, as shown in Fig. 13.6 (a).

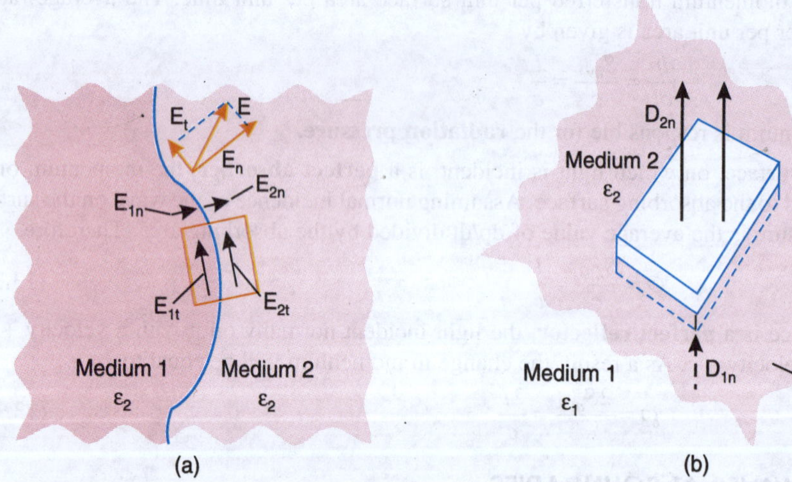

Electric field components at a boundary.

Fig. 13.6

1. The components of the electric field vector **E** that are tangent to the surface of a boundary must be continuous across the boundary. It means that the tangential components of **E** are equal on the two sides of the interface.

$$E_{2t} = E_{1t} \tag{13.38a}$$

2. The difference between the normal components of **D** is equal to the free surface charge density at the boundary.

$$D_{2n} - D_{1n} = \rho_s \tag{13.38b}$$

Again at the boundary between two different media, we resolve the total magnetic flux density **B** into a component tangential to the boundary surface, B_t and a component normal to the boundary surface, B_n, as shown in Fig. 13.7.

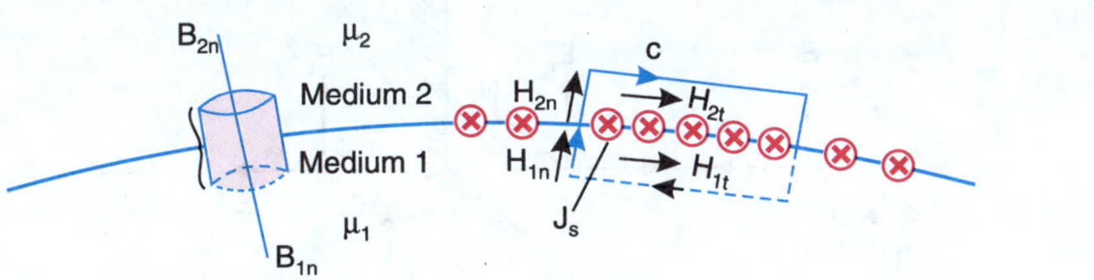

Magnetic field components at a boundary.

Fig. 13.7

3. At the interface between two different materials, the normal component of **B** is continuous across the interface. It means that the normal components of **B** are equal on the two sides of the interface.

$$B_{2n} = B_{1n} \tag{13.38c}$$

4. The difference between the tangential components of **H** is equal to the current density.

$$H_{2t} - H_{1t} = J_s \tag{13.38d}$$

The boundary conditions at the interface between two dielectric materials, where there is no free charge and no current, are;

1. Tangential components of **E** and **H** are continuous across the interface. That is

$$E_{2t} = E_{1t} \tag{13.39a}$$

$$H_{2t} = H_{1t} \tag{13.39b}$$

2. Normal components of **D** and **B** are continuous across the interface.

$$D_{2n} = D_{1n} \tag{13.39c}$$

$$B_{2n} = B_{1n} \tag{13.39d}$$

These boundary conditions are used to write equations relating the fields on each side of the interface.

13.10. WAVE INCIDENT NORMALLY ON BOUNDARY

Let us consider a plane surface separating two unlike media, for instance air and glass. Since the dielectric constant of one material is different from that of the other, the same equations cannot describe the wave on the two sides of the boundary. Accordingly, an additional expression representing a reflected wave must be added. Thus, three distinct wave systems exist at every boundary of media. These three are known as the **incident**, the **reflected** and the **refracted waves**. It is our problem to find relations between the amplitude and phase of these three waves which satisfy the general boundary conditions.

Fig. 13.8 shows a plane wave traveling in the z-direction and incident **normally** on a boundary parallel to the plane z = 0 which is the plane separating the two semi-infinite perfect dielectrics 1 and 2. At the plane z = 0, a part of the energy is reflected back in medium 1 along the normal, giving a reflected wave with P_r directed opposite to P_i, and the remaining energy is transmitted as a wave

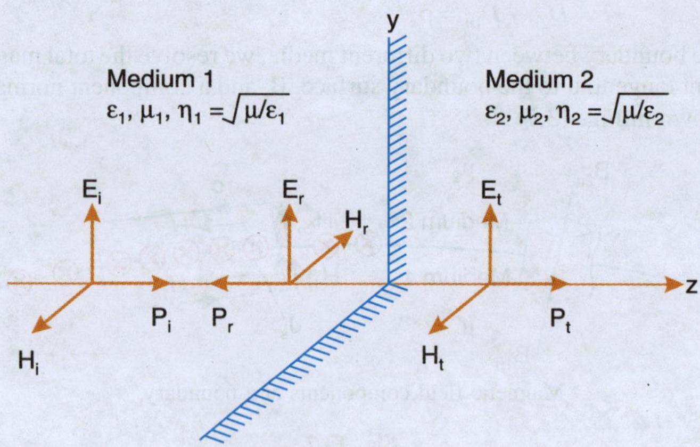

The boundary between two dielectric media is the *xy* plane. Medium 1 contains the incident and reflected waves, medium 2 the transmitted wave.

Fig. 13.8

along the normal in medium 2. Thus, P_i is in the same direction as E_i. E_i is the electric field strength of the incident wave, E_r the electric field strength of the reflected wave and E_t the electric field strength of the transmitted wave propagated in the second medium. Similarly, H_i, H_r and H_t represent the strengths of incident, reflected and transmitted magnetic fields. These three waves have to satisfy the conditions of continuity for the tangential components of E and H at the interface. Accordingly, the sum of the tangential components of E_i and E_r just to the left of the interface must be equal to the tangential component of E_t just to the right of the interface; a similar situation holds for H. Thus,

$$\left.\begin{array}{l}\text{Tangential component of electric field}\\ \text{in medium 1}\end{array}\right\} = \left\{\begin{array}{l}\text{Tangential component of electric field}\\ \text{in medium 2}\end{array}\right.$$

That is,

$$E_i + E_r = E_t \tag{13.40}$$

Similarly, applying the boundary condition for magnetic fields and noting that $\sigma = 0$ for perfect dielectrics, we get

$$H_i + H_r = H_t \tag{13.41}$$

Also from equation (13.24)

$$E_i = \eta_1 H_i \tag{13.42a}$$

$$E_r = -\eta_1 H_r \tag{13.42b}$$

and

$$E_t = \eta_2 H_t \tag{13.42c}$$

Substituting these values in equation (13.4) we get

$$\therefore \quad \frac{1}{\eta_1}(E_i - E_r) = \frac{1}{\eta_2}(E_i + E_r)$$

or

$$\eta_2 (E_i - E_r) = \eta_1 (E_i + E_r)$$

or

$$E_i (\eta_2 - \eta_1) = E_r (\eta_2 + \eta_1)$$

Therefore, the **reflection coefficient** is given by

$$\Gamma = \frac{E_r}{E_i} = \frac{\eta_2 - \eta_1}{\eta_2 + \eta_1} \tag{13.43}$$

The **transmission coefficient** is given by

$$\tau = \frac{E_t}{E_i} = \frac{E_i + E_r}{E_i} = 1 + \frac{E_r}{E_i}$$

$$= 1 + \frac{\eta_2 - \eta_1}{\eta_2 + \eta_1}$$

or
$$\tau = \frac{2\eta_2}{\eta_2 + \eta_1} \tag{13.44}$$

If $\Gamma = 0$, it follows from equ.(13.43) that there will be no reflection of the wave. This implies that $\eta_2 = \eta_1$ or there is no change of medium. From equ.(13.43) and (13.44), it can be seen that Γ and τ are interrelated by the formula

$$\tau = 1 + \Gamma \tag{13.45}$$

For nonmagnetic media $\eta_1 = \sqrt{\frac{\mu_1}{\varepsilon_1}}$ and $\eta_2 = \sqrt{\frac{\mu_2}{\varepsilon_2}}$ and using these values into (13.43) and (13.44), we respectively get

$$\Gamma = \frac{\sqrt{\varepsilon_1} - \sqrt{\varepsilon_2}}{\sqrt{\varepsilon_1} + \sqrt{\varepsilon_2}} = \frac{n_1 - n_2}{n_1 + n_2} \tag{13.46a}$$

and
$$\tau = \frac{2\sqrt{\varepsilon_1}}{\sqrt{\varepsilon_1} + \sqrt{\varepsilon_2}} = \frac{2n_1}{n_1 + n_2} \tag{13.46b}$$

The reflection and transmission coefficients are useful in finding the fields. Knowing the magnitude and direction of the **E** vector and intrinsic impedance of the material, the magnitude and direction of the **H** vector can be found.

13.11. WAVE INCIDENT OBLIQUELY ON BOUNDARY

A light wave incident on the boundary separating two different optical media is known to obey the law of reflection, Snell's law of refraction, total internal reflection and Brewster's law etc. These laws can be obtained without reference to the polarization of the wave. We first obtain the first two laws before the polarization of the waves is taken into consideration.

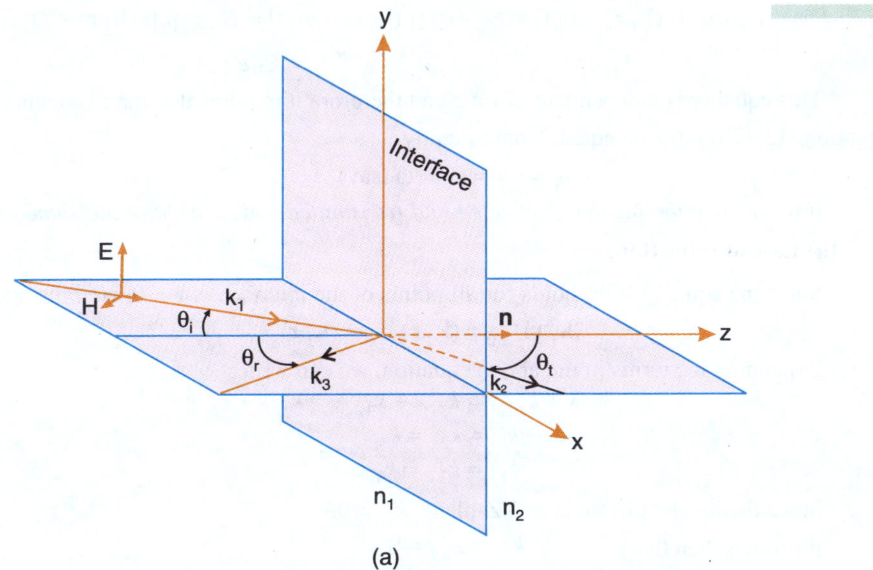

Reflection and refraction of a plane wave at a plane dielectric interface

Fig. 13.9

Fig. 13.9 shows a plane polarized electromagnetic wave incident on an interface of two lossless dielectric media. The interface is represented by z = 0. Let (ε_1, μ_1) and (ε_2, μ_2) represent the permittivity and permeability of the media. Let a plane wave with wave vector $\mathbf{k_1}$, and frequency ω_1 be incident from medium 1 at point O on the interface. The wave is partly reflected and partly refracted (transmitted). Let E_i, E_t, and E_r denote the electric fields associated with the incident, transmitted and reflected waves respectively. Let the reflected and transmitted waves have wave vectors $\mathbf{k_3}$ and $\mathbf{k_2}$ and frequencies ω_3 and ω_2 respectively. Let **n** be the unit vector normal to the interface and directed from medium 1 into medium 2. The electric field vectors may be represented by the following equations.

(b)

Fig. 13.9

$$\left.\begin{aligned} &\text{Incident wave:} && E_i = E_1 \exp\left[i(\mathbf{k_1}\cdot r) - \omega_1 t\right] \quad (a) \\ &\text{Transmitted wave:} && E_t = E_2 \exp\left[i(\mathbf{k_2}\cdot r) - \omega_2 t\right] \quad (b) \\ &\text{Reflected wave:} && E_r = E_3 \exp\left[i(\mathbf{k_3}\cdot r) - \omega_3 t\right] \quad (c) \end{aligned}\right\} \quad (13.47)$$

Since the fields must satisfy Maxwell's equations, we get

$$k_1^2 = \omega_1^2\varepsilon_1\mu_1,\ k_2^2 = \omega_2^2\varepsilon_2\mu_2 \text{ and } k_3^2 = \omega_3^2\varepsilon_1\mu_1 \quad (13.47a)$$

Further, the fields have to satisfy the boundary conditions at the interface. These boundary conditions are to be satisfied at all points on the interface and at all times. We apply the boundary condition (13.40) which states that the tangential component of the electric field must be continuous across the interface between the media. Hence

$$(E_i)_{\text{tangential}} + (E_r)_{\text{tangential}} = (E_t)_{\text{tangential}}$$

$$E_{1t} \exp\left[i(\mathbf{k_1}.\mathbf{r}) - \omega_1 t\right] + E_{3t} \exp\left[i(\mathbf{k_3}.\mathbf{r}) - \omega_3 t\right] = E_{2t} \exp\left[i(\mathbf{k_2}.\mathbf{r}) - \omega_2 t\right]$$

$$E_{1t}\, e^{i(k_1.r)} e^{-i\omega_1 t} + E_{3t} e^{i(k_3.r)} e^{-i\omega_3 t} = E_{2t} e^{i(k_2.r)} e^{-i\omega_2 t} \quad (13.47b)$$

This equality is independent of time and therefore it requires that the exponents in the above equation (13.47b) must be equal. Consequently,

$$\omega_1 = \omega_2 = \omega_3 = \omega \text{ (say)} \quad (13.48)$$

It means that *the incident, reflected and transmitted waves all have the same frequency.*

(a) Law of reflection:

Since the equ. (13.47b) holds for all points of the interface at z = 0, we must have

$$(\mathbf{k_1}.\mathbf{r})_{z=0} = (\mathbf{k_3}.\mathbf{r})_{z=0} = (\mathbf{k_2}.\mathbf{r})_{z=0} \quad (13.48a)$$

Expanding the terms in the above equation, we can write

$$k_{1x}x + k_{1y}\ y = k_{3x}x + k_{3y}\ \ y = k_{2x}x + k_{2y}y$$

$$\therefore \quad k_{1z} = k_{2z} = k_{3z}$$

and $$k_{1y} = k_{2y} = k_{3y}$$

Since the incident beam is in XZ-plane, $k_{1y} = 0$.

It follows then that $$k_{2y} = k_{3y} = 0 \quad (13.49)$$

It means that the propagation vectors $\mathbf{k_3}$ and $\mathbf{k_2}$ also lie in XZ-plane. As normal **n** is along z-axis, we conclude that all the three propagation vectors and normal to the interface lie in the same plane. Thus, the incident, transmitted and reflected waves all lie in the same plane.

We have from the Fig. 13.9 that

$$(\boldsymbol{k_1}.\mathbf{r}) = k_1\,(x \sin\theta_i + z\cos\theta_i) \quad (a) \tag{13.50}$$

$$(\boldsymbol{k_2}.\mathbf{r}) = k_2\,(x \sin\theta_t + z\cos\theta_t) \quad (b)$$

$$(\boldsymbol{k_3}.\mathbf{r}) = k_3\,(x \sin\theta_r + z\cos\theta_r) \quad (c)$$

From equn. (13.48a) $(\boldsymbol{k_1}.\mathbf{r})_{z=0} = (\boldsymbol{k_3}.\mathbf{r})_{z=0}$

and using (13.50), we get

$$k_1 x \sin\theta_i = k_3 x \sin\,\theta_r$$

or

$$k_1 \sin\,\theta_i = k_3 \sin\,\theta_r \tag{13.51}$$

This is known as **phase matching condition**.

It is seen from equ. (13.47a) that the magnitudes of the propagation vectors are given by

$$k_1 = k_3 = \omega\sqrt{(\mu_o\ \varepsilon_1)} \quad \text{and} \quad k_2 = \omega\sqrt{(\mu_o\ \varepsilon_2)}$$

Since $k_1 = k_3$, we deduce from equ.(13.51) that

$$k_1 \sin\theta_i = k_1 \sin\theta_r$$

$$\therefore \quad \theta_i = \theta_r \tag{13.52}$$

This is the **law of reflection**, which states that the angle of reflection is equal to the angle of incidence.

(b) Snell's law:

Further from equn. (13.48a) $(\boldsymbol{k_1}.r)_{z=0} = (\boldsymbol{k_2}.r)_{z=0}$ and equn. (13.50), we get

$$k_1 x \sin\theta_i = k_2 x \sin\theta_t$$

$$\therefore \quad k_1 \sin\,\theta_i = k_2 \sin\,\theta_t$$

or

$$\frac{\sin\theta_i}{\sin\theta_t} = \frac{k_2}{k_1} = \frac{\sqrt{\varepsilon_2}}{\sqrt{\varepsilon_1}}$$

$$\frac{\sin\theta_i}{\sin\theta_t} = \frac{n_2}{n_1} \tag{13.53}$$

where n_1 and n_2 represent the refractive indices of media 1 and 2 respectively.

Equ.(13.53) is the **Snell's law of refraction**.

(c) Reflection coefficients:

For normal incidence, the reflection coefficient Γ and transmission coefficient τ of a boundary between two different media are independent of the polarization of the incident wave, because the electric and magnetic fields of a normally incident plane wave are both always tangential to the boundary regardless of the wave polarization. This is not the case for oblique incidence. In general, the light waves coming from common sources are unpolarized and can be considered as the superposition of two orthogonally polarized waves, one with its electric field parallel to the plane of incidence – and it is called **parallel polarization**- and another with its electric field perpendicular to the plane of incidence- and it is called **perpendicular polarization**. These two polarization configurations are shown in Fig. 13.10.

We now consider a plane polarized electromagnetic wave incident on an interface of two dielectric materials and derive expressions for the reflection and transmission coefficients in both cases of polarization.

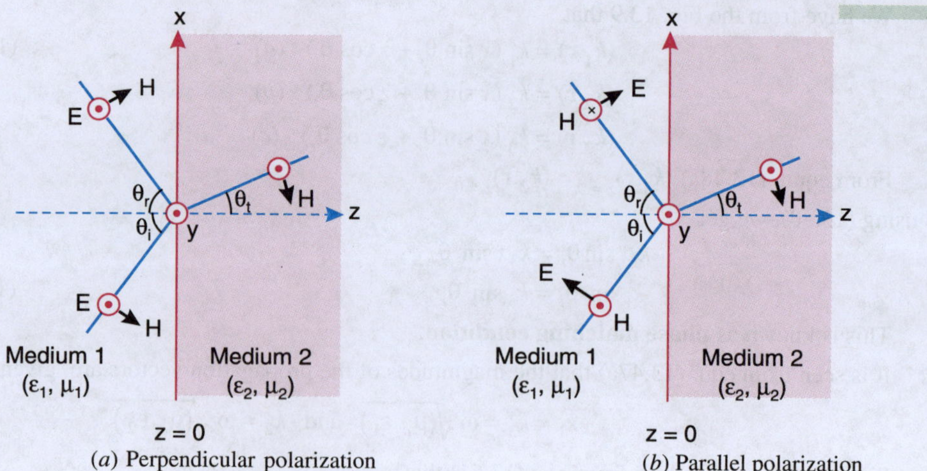

(*a*) Perpendicular polarization (*b*) Parallel polarization

The plane of incidence is the plane containing the direction of wave travel, $\hat{\mathbf{k}}$, and the surface normal to the boundary, which in the present case is the plane of the paper. A wave is (*a*) perpendicularly polarized when its **E** is perpendicular to the plane of incidence and (*b*) parallel polarized when its **E** lies in the plane of incidence.

Fig. 13.10

13.11.1. PERPENDICULAR POLARIZATION (E–NORMAL TO THE PLANE OF INCIDENCE)

Let us first consider the case of perpendicular polarization where the electric vector is perpendicular to the plane of incidence, as shown in Fig. 13.11.

Continuity of the tangential components of H at the boundary is expressed by

$$H_\perp^i \cos\theta_i - H_\perp^r \cos\theta_i = H_\perp^t \cos\theta_t \tag{13.53}$$

Since in the electromagnetic wave **E** and **H** are at right angles and $E = \eta\, H$, the above equation (13.53) may be written as

$$\frac{E_\perp^i}{\eta_1}\cos\theta_i - \frac{E_\perp^r}{\eta_1}\cos\theta_i = \frac{E_\perp^t}{\eta_2}\cos\theta_t \qquad \text{at } z = 0$$

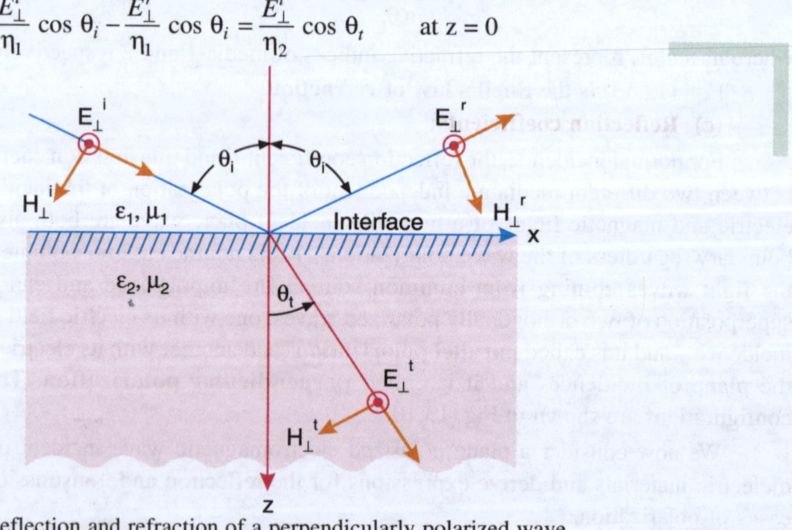

Reflection and refraction of a perpendicularly polarized wave.

Fig. 13.11

$$\cos\theta_i\,(E_\perp^i - E_\perp^r) = \frac{\eta_1}{\eta_2} E_\perp^t \cos\theta_t \tag{13.54}$$

Similarly, continuity of tangential components of E gives

$$E_\perp^i + E_\perp^r = E_\perp^t \quad \text{at } z = 0 \tag{13.55}$$

$$\frac{\eta_2\cos\theta_i}{\eta_1\cos\theta_t}(E_\perp^i - E_\perp^r) = E_\perp^t$$

$$\therefore \quad \frac{\eta_2\cos\theta_i}{\eta_1\cos\theta_t}(E_\perp^i - E_\perp^r) = E_\perp^i + E_\perp^r$$

or

$$-E_\perp^r\left[1+\frac{\eta_2\cos\theta_i}{\eta_1\cos\theta_t}\right] = E_\perp^i\left[1-\frac{\eta_2\cos\theta_i}{\eta_1\cos\theta_t}\right] \tag{13.56}$$

Therefore, the *reflection coefficient* is

$$\Gamma_\perp = \frac{E_\perp^r}{E_\perp^i} = -\frac{1-\dfrac{\eta_2\cos\theta_i}{\eta_1\cos\theta_t}}{1+\dfrac{\eta_2\cos\theta_i}{\eta_1\cos\theta_t}}$$

$$\Gamma_\perp = \frac{E_\perp^r}{E_\perp^i} = \frac{\eta_2\cos\theta_i - \eta_1\cos\theta_t}{\eta_2\cos\theta_i + \eta_1\cos\theta_t} \tag{13.57a}$$

The *transmission coefficient* is

$$\tau_\perp = \frac{E_\perp^t}{E_\perp^i} = \frac{E_\perp^i + E_\perp^r}{E_\perp^i} = 1 + \frac{E_\perp^r}{E_\perp^i}$$

$$= 1 + \frac{\eta_2\cos\theta_i - \eta_1\cos\theta_t}{\eta_2\cos\theta_i + \eta_1\cos\theta_t}$$

or

$$\tau_\perp = \frac{2\eta_2\cos\theta_i}{\eta_2\cos\theta_i + \eta_1\cos\theta_t} \tag{13.57b}$$

Equ.(13.57a) for reflection coefficient can be expressed in terms of permittivity and refractive index as follows.

$$\Gamma_\perp = \frac{E_\perp^r}{E_\perp^i} = \frac{\sqrt{\varepsilon_1}\cos\theta_i - \sqrt{\varepsilon_2}\cos\theta_t}{\sqrt{\varepsilon_1}\cos\theta_i + \sqrt{\varepsilon_2}\cos\theta_t} = \frac{n_1\cos\theta_i - n_2\cos\theta_t}{n_1\cos\theta_i + n_2\cos\theta_t} \tag{13.58a}$$

Similarly, equ.(13.57b) for transmission coefficient can be expressed in terms of permittivity and refractive index as follows.

$$\tau_\perp = \frac{E_\perp^t}{E_\perp^i} = \frac{2\sqrt{\varepsilon_1}\cos\theta_i}{\sqrt{\varepsilon_1}\cos\theta_i + \sqrt{\varepsilon_2}\cos\theta_t} = \frac{2n_1\cos\theta_i}{n_1\cos\theta_i + n_2\cos\theta_t} \tag{13.58b}$$

These two coefficients $\Gamma_\perp$ and $\tau_\perp$ are known as the **Fresnel reflection and transmission coefficients for perpendicular polarization.** They are related by

$$\tau_\perp = 1 + \Gamma_\perp \tag{13.59}$$

Using Snell's law into (13.58a) and (13.58b), we can show that

$$\Gamma_\perp \approx \frac{\sin\theta_t\cos\theta_i - \sin\theta_i\cos\theta_t}{\sin\theta_t\cos\theta_i + \sin\theta_i\cos\theta_t} = -\frac{\sin(\theta_i - \theta_t)}{\sin(\theta_i + \theta_t)} \tag{13.60a}$$

and

$$\tau_\perp \approx \frac{2\sin\theta_t\cos\theta_i}{\sin(\theta_i + \theta_t)} \tag{13.60b}$$

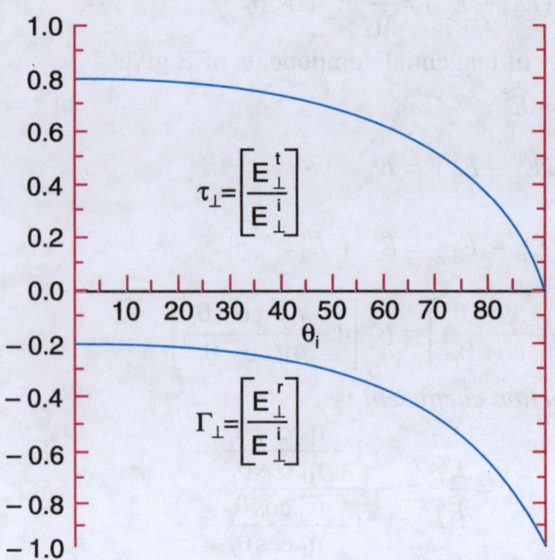

The ratios $\Gamma_\perp$ and $\tau_\perp$ as functions of the angle of incidence θ_i. This corresponds to light incident in air on a glass with $n = 1.5$. The wave is polarized with its E vector normal to the plane of incidence.

Fig. 13.12

Fig. 13.12 shows the angular variations of the magnitudes of $\Gamma_\perp$ and $\tau_\perp$.

13.11.2. PARALLEL POLARIZATION (E-PARALLEL TO THE PLANE OF INCIDENCE)

For this polarization the total **H** field is parallel to the boundary surface. The tangential component of **H** is continuous across the boundary, because surface currents cannot flow in a dielectric.

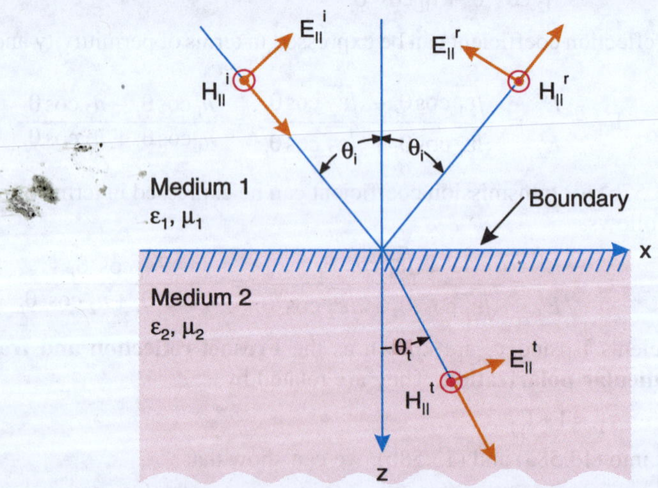

Reflection and refraction of a parallel-polarized wave. Medium 1 and 2 are lossless, nonferromagnetic dielectrics.

Fig. 13.13

Therefore,

$$H_{11}^i + H_{11}^r = H_{11}^t \qquad \text{at } z = 0$$

Since **E** and **H** are related by $\eta = \sqrt{\mu/\varepsilon}$, the above equation can be written as

$$\sqrt{\frac{\varepsilon_1}{\mu_1}} E_{11}^i + \sqrt{\frac{\varepsilon_1}{\mu_1}} E_{11}^r = \sqrt{\frac{\varepsilon_2}{\mu_2}} E_{11}^t \tag{13.61}$$

Since $\mu_2 = \mu_1 = \mu_o$, equ.(13.61) becomes

$$E_{11}^i + E_{11}^r = \sqrt{\frac{\varepsilon_2}{\varepsilon_1}} E_{11}^t \tag{13.62}$$

The tangential components of **E** must also be continuous across the boundary. We get

$$E_{11}^i \cos\theta_i - E_{11}^r \cos\theta_i = E_{11}^t \cos\theta_t \qquad \text{at } z = 0 \tag{13.63}$$

By using equ.(13.62) into (13.63) , we get

$$E_{11}^i \cos\theta_i - E_{11}^r \cos\theta_i = \sqrt{\frac{\varepsilon_1}{\varepsilon_2}} (E_{11}^i + E_{11}^r) \cos\theta_t$$

$$E_{11}^i \left(\sqrt{\frac{\varepsilon_1}{\varepsilon_2}} \cos\theta_t - \cos\theta_i \right) = -E_{11}^r \left(\cos\theta_i + \sqrt{\frac{\varepsilon_1}{\varepsilon_2}} \cos\theta_t \right)$$

$$E_{11}^r = E_{11}^i \frac{\left(\cos\theta_i - \sqrt{\frac{\varepsilon_1}{\varepsilon_2}} \cos\theta_t \right)}{\left(\cos\theta_i + \sqrt{\frac{\varepsilon_1}{\varepsilon_2}} \cos\theta_t \right)} = E_{11}^i \frac{\left(\sqrt{\frac{\varepsilon_2}{\varepsilon_1}} \cos\theta_i - \cos\theta_t \right)}{\left(\sqrt{\frac{\varepsilon_2}{\varepsilon_1}} \cos\theta_i + \cos\theta_t \right)}$$

We therefore get for the reflected field

$$E_{11}^r = E_{11}^i \left[\frac{\sqrt{\varepsilon_2/\varepsilon_1}\ \cos\theta_i - \cos\theta_t}{\sqrt{\varepsilon_2/\varepsilon_1}\ \cos\theta_i + \cos\theta_t} \right] \tag{13.64}$$

Therefore the reflection coefficient is given by

$$\Gamma_{11} = -\frac{E_{11}^r}{E_{11}^i} = -\frac{\sqrt{\varepsilon_2/\varepsilon_1} \cos\theta_i - \cos\theta_t}{\sqrt{\varepsilon_2/\varepsilon_1} \cos\theta_i + \cos\theta_t} = \frac{n_1 \cos\theta_t - n_2 \cos\theta_i}{n_1 \cos\theta_t + n_2 \cos\theta_i} \tag{13.65}$$

By using equ.(13.61) into (13.63), we get

$$E_{11}^i \cos\theta_i - \left(\sqrt{\varepsilon_2/\varepsilon_1} E_{11}^t - E_{11}^i \right) \cos\theta_i = E_{11}^t \cos\theta_t$$

$$2E_{11}^i \cos\theta_i = E_{11}^t \left(\sqrt{\varepsilon_2/\varepsilon_1}\ \cos\theta_i + \cos\theta_t \right)$$

We thus get for the transmitted field

$$E_{11}^t = E_{11}^i \left[\frac{2\cos\theta_i}{\sqrt{\varepsilon_2/\varepsilon_1}\ \cos\theta_i + \cos\theta_t} \right] \tag{13.66}$$

Therefore, the transmission coefficient is given by

$$\tau_{11} = \frac{E_{11}^t}{E_{11}^i} = \frac{2\cos\theta_i}{\sqrt{\varepsilon_2/\varepsilon_1} \cos\theta_i + \cos\theta_t} = \frac{2n_1 \cos\theta_i}{n_1 \cos\theta_t + n_2 \cos\theta_i} \tag{13.67}$$

Γ_{11} and τ_{11} are known as the **Fresnel's reflection and transmission coefficients for parallel polarization.** The relationship between reflection and transmission coefficients can be obtained from (13.65) and (13.67). Thus,

$$\tau_{11} = (1+\Gamma_{11})\frac{\cos\theta_i}{\cos\theta_t} \quad (13.68a)$$

and
$$1-\Gamma_{11} = \frac{\eta_1}{\eta_2}\tau_{11} \quad (13.68b)$$

Fig. 13.14 shows the angular variations of the magnitudes of $\Gamma_{||}$ and $\tau_{||}$.

Using Snell's law into (13.65) and (13.67), we get

$$\Gamma_{11} = \frac{\tan\ (\theta_i - \theta_t)}{\tan\ (\theta_i + \theta_t)} \quad (13.69a)$$

$$\tau_{11} = \frac{2\sin\theta_t\cos\theta_i}{\sin(\theta_i+\theta_t)\cos(\theta_i-\theta_t)} \quad (13.69b)$$

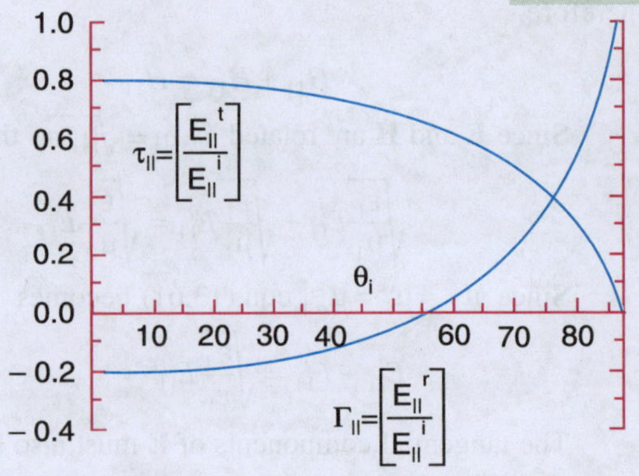

The ratios $\Gamma_{||}$ and τ_{11} as functions of the angle of incidence θ_i corresponding to light incident in air on a glass with $n = 1.5$. The wave is polarized with its E vector *parallel* to the plane of incidence.

Fig. 13.14

1. No reflection:

It can be readily seen from equ,(13.65) and (13.67) that

$$\Gamma_{11} = 0 \text{ and } \tau_{11} = 1$$

when $n_2 = n_1$ and $\theta_1 = \theta_2$. It means that there is *no reflection* when the second medium has the same refractive index as that of the first medium, and all the light is transmitted.

13.12. REFLECTANCE AND TRANSMITTANCE

Let us consider a beam of light incident at an angle on a surface, as shown in Fig.13.15. It illuminates an area A of the surface. The cross-sectional areas of the incident, reflected and transmitted beams are respectively A cos θ_i, A cos θ_r and A cos θ_t. Then the incident power is I_i A cos θ_i, the reflected power is I_rA cos θ_r and the transmitted power is I_t cos θ_t. The **reflectance** of the surface is defined as *the ratio of the reflected power to the incident power.* Thus,

$$R = \frac{I_r A\ \cos\ \theta_r}{I_i A\ \cos\ \theta_i} = \frac{I_r}{I_i}$$

where $I_r = \frac{1}{2}\upsilon_i\varepsilon_i E_r^2$ and $I_i = \frac{1}{2}\upsilon_r\varepsilon_r E_i^2$. Note that the incident and reflected waves are in the same medium and therefore υ and ε are equal for these two waves.

$$\therefore \quad R = \frac{E_r^2}{E_i^2} = \Gamma^2 \quad (13.70a)$$

The **transmittance** is defined as *the ratio of the transmitted power to the incident power.*

$$T = \frac{I_t A\ \cos\ \theta_t}{I_i A\ \cos\ \theta_i} = \frac{n_t \cos\ \theta_t}{n_i \cos\ \theta_i}\left(\frac{E_t^2}{E_i^2}\right)$$

$$\therefore \quad T = \frac{n_t \cos\ \theta_t}{n_i \cos\ \theta_i}\tau^2 \quad (13.70b)$$

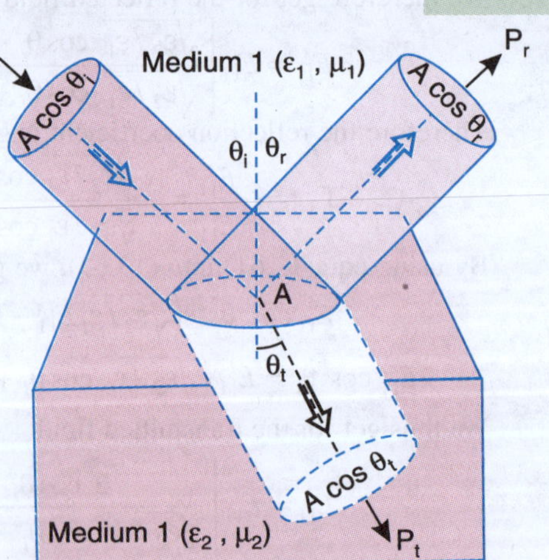

Reflection and transmission of an incident circular beam illuminating a spot of size *A* on the interface.

Fig. 13.15

We may write the equ.(13.69) in the component forms as follows.

$$R_{11} = \Gamma_{11}^2 \quad (13.71a)$$

and
$$R_{\perp} = \Gamma_{\perp}^2 \quad (13.71b)$$

$$T_{11} = \frac{n_t \cos\theta_t}{n_i \cos\theta_i} \tau_{11}^{\ 2} \tag{13.72a}$$

and
$$T_{\perp} = \frac{n_t \cos\theta_t}{n_i \cos\theta_i} \tau_{\perp}^{\ 2} \tag{13.72b}$$

The variations of these components R_{11} and T_{11} and $R_{\perp}$ and $T_{\perp}$ with incident angle are shown in Fig. 13.16 (*a*) and (*b*) respectively.

For the case of normal incidence of the wave on the interface, the above equations reduce to

$$R = R_{11} = R_{\perp} = \left[\frac{n_t - n_i}{n_t + n_i}\right]^2 \tag{13.73}$$

and
$$T = T_{11} = T_{\perp} = \frac{4 n_t n_i}{(n_t + n_i)^2} \tag{13.74}$$

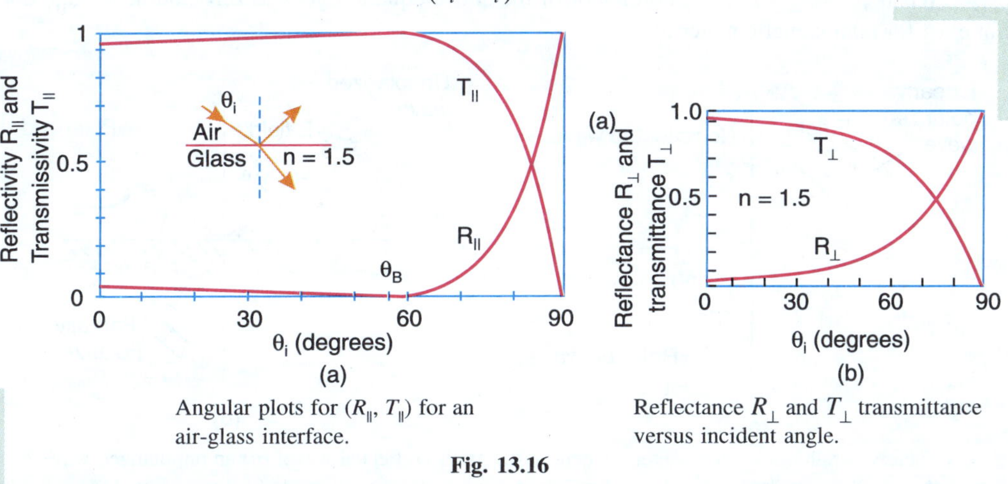

Angular plots for ($R_{\|}$, $T_{\|}$) for an air-glass interface.

Reflectance $R_{\perp}$ and $T_{\perp}$ transmittance versus incident angle.

Fig. 13.16

We summarize in Table-1 the amplitude reflection and transmission coefficients for the two cases of polarization.

TABLE -1

Property	Normal Incidence $\theta_1 = \theta_2 = 0$	Perpendicular Polarization	Parallel Polarization
Reflection coefficient	$\Gamma = \frac{\eta_2 - \eta_1}{\eta_2 + \eta_1}$	$\Gamma_{\perp} = \frac{\eta_2 \cos\theta_i - \eta_1 \cos\theta_t}{\eta_2 \cos\theta_i + \eta_1 \cos\theta_t}$	$\Gamma_{\|} = \frac{\eta_2 \cos\theta_t - \eta_1 \cos\theta_i}{\eta_2 \cos\theta_t + \eta_1 \cos\theta_i}$
Transmission coefficient	$\tau = \frac{2\eta_2}{\eta_2 + \eta_1}$	$\tau_{\perp} = \frac{2\eta_2 \cos\theta_i}{\eta_2 \cos\theta_i + \eta_1 \cos\theta_t}$	$\tau_{\|} = \frac{2\eta_2 \cos\theta_i}{\eta_2 \cos\theta_t + \eta_1 \cos\theta_i}$
Relation of Γ to τ	$\tau = 1 + \Gamma$	$\tau_{\perp} = 1 + \Gamma_{\perp}$	$\tau_{\|} = (1 + \Gamma_{\|}) \frac{\cos\theta_i}{\cos\theta_t}$
Reflectivity	$R = \lvert\Gamma\rvert^2$	$R_{\perp} = \lvert\Gamma_{\perp}\rvert^2$	$R_{\|} = \lvert\Gamma_{\|}\rvert^2$
Transmissivity	$T = \lvert\tau\rvert^2 \left(\frac{\eta_1}{\eta_2}\right)$	$T_{\perp} = \lvert\tau_{\perp}\rvert^2 \frac{\eta_1 \cos\theta_t}{\eta_2 \cos\theta_i}$	$T_{\|} = \lvert\tau_{\|}\rvert^2 \frac{\eta_1 \cos\theta_t}{\eta_2 \cos\theta_i}$
Relation of *R* to *T*	$T = 1 - R$	$T_{\perp} = 1 - R_{\perp}$	$T_{\|} = 1 - R_{\|}$

Note: (1) $\sin\theta_t = \sqrt{\mu_1 \varepsilon_1 / \mu_2 \varepsilon_2} \sin\theta_i$; (2) $\eta_1 = \sqrt{\mu_1/\varepsilon_1}$; (3) $\eta_2 = \sqrt{\mu_2/\varepsilon_2}$; for nonmagnetic media, $\eta_2/\eta_1 = \eta_1/\eta_2$.

13.13. BREWSTER'S LAW

The Brewster angle θ_B may be defined as the incidence angle θ_i at which the Fresnel reflection coefficient $\Gamma = 0$.

Perpendicular polarization:

For perpendicular polarization, the Brewster angle can be obtained by setting the numerator of the expression (13.57a) equal to zero. It gives

$$\eta_2 \cos\theta_i = \eta_1 \cos\theta_t$$

$$\sin\theta_{B\perp} = \sqrt{\frac{1-(\mu_1\varepsilon_2/\mu_2\varepsilon_1)}{1-(\mu_1/\mu_2)^2}} \quad (13.75)$$

When $\mu_1 = \mu_2 = 1$ the denominator of the above equation goes to zero and hence $\theta_{B\perp}$ does not exist for nonmagnetic materials.

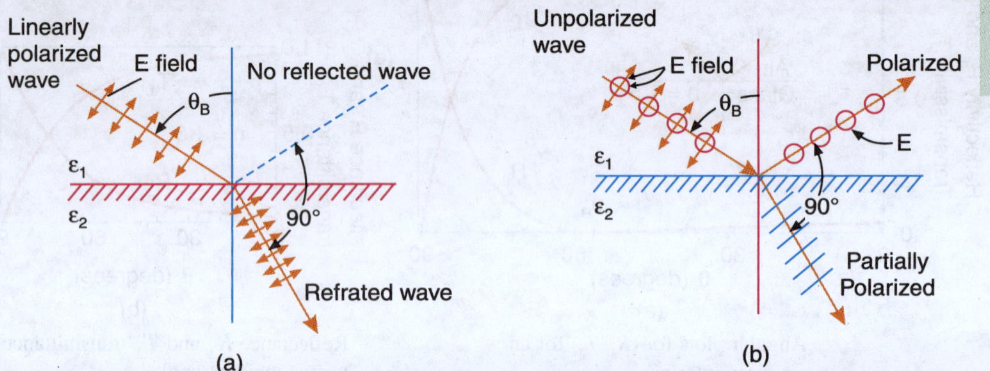

A linearly parallel-polarized wave, incident at θ_B, has no reflected wave; (*b*) an unpolarized wave, such as ordinary light, has a polarized reflected wave and a partially polarized transmitted wave.

Fig. 13.17

Parallel polarization:

For parallel polarization, the Brewster angle can be obtained by setting the numerator of the expression (13.65) equal to zero. It gives

$$\sqrt{\frac{\varepsilon_2}{\varepsilon_1}}\cos\theta_i = \cos\theta_t \quad (13.76)$$

$$\theta_B = \sin^{-1}\left(\sqrt{\frac{\varepsilon_2}{\varepsilon_1+\varepsilon_2}}\right) \quad (13.77)$$

or

$$\theta_B = \tan^{-1}\left(\sqrt{\frac{\varepsilon_2}{\varepsilon_1}}\right) \quad (13.78)$$

This expression is called **Brewster's law.**

If a plane wave of mixed polarization or an unpolarized wave is incident on a plane interface at the Brewster's angle, then the only portion of wave with electric vector perpendicular to the plane of the incidence will be reflected. Thus, the reflected radiation is completely linearly polarized. Even an unpolarized wave is reflected at angles other than the Brewster's angle, the reflected wave is predominantly polarized perpendicular to the plane of incidence.

13.14. TOTAL REFLECTION

When an electromagnetic wave passes from a denser medium to the rarer medium, we have

$$\frac{\sin\,\theta_1}{\sin\,\theta_2}=\frac{\upsilon_1}{\upsilon_2}=n<1.\text{ It means }\sin\,\theta_2>\sin\,\theta_1 \tag{13.79}$$

From equ.(13.60a) we have

$$E_\perp^r=\frac{\sin\,\theta_t\cos\,\theta_i-\sin\,\theta_i\cos\,\theta_t}{\sin\,\theta_t\cos\,\theta_i+\sin\,\theta_i\cos\,\theta_t}E_\perp^i \tag{13.80}$$

From the condition shown above, if $\upsilon_2 > \upsilon_1$, the law of refraction leads to a value of $\sin\theta_t > 1$. Consequently, $\cos\theta_t$ becomes a complex quantity with a phase shift in the incident wave and there is also attenuation. If we eliminate θ_t and remember that $n < 1$, then

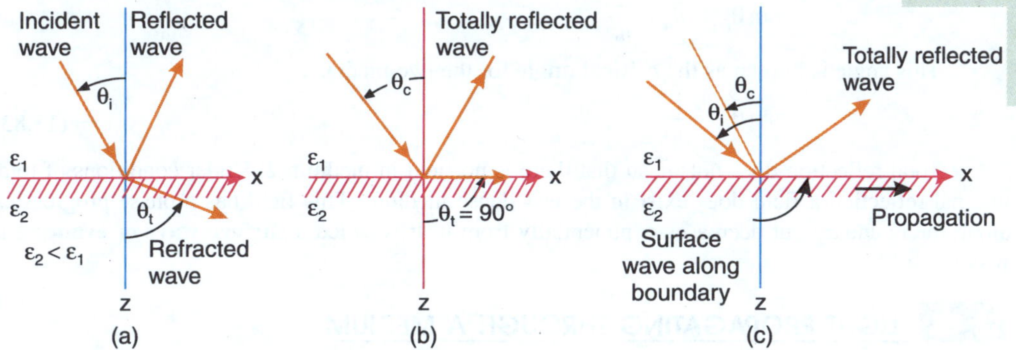

(*a*) If medium 1 has a permittivity larger than medium 2 ($\varepsilon_1 > \varepsilon_2$), a wave incident from medium 1 will refract into medium 2 at a larger angle; (*b*) because θ_2 cannot exceed 90°, the incident wave is completely reflected back into medium 1 when $\theta_i \geq \theta_c$; (*c*) surface wave along boundary, when $\theta_i > \theta_c$.

Fig. 13.18

$$E_\perp^r=\frac{\cos\,\theta_i-i\sqrt{\sin^2\theta_i-n^2}}{\cos\,\theta_i+i\sqrt{\sin^2\theta_i-n^2}}E_\perp^i=\frac{\sqrt{a^2+b^2}e^{-i\phi}}{\sqrt{a^2+b^2}e^{i\phi}}E_\perp^i=E_\perp^i e^{-2i\phi} \tag{13.81}$$

where $a=\cos\,\theta_i$, $b=\sqrt{\sin^2\theta_i-n^2}$ and $\tan\phi=b/a$.

Equ.(13.81) shows that the amplitude of the reflected wave is equal to that of the incident wave but there is a phase change. It means that the incident wave is totally internally reflected back into the denser medium.

Equ.(13.64) gives the electric component in the plane of incidence.

$$E_{11}^r=E_{11}^i\left[\frac{\sqrt{\varepsilon_2/\varepsilon_1}\,\cos\,\theta_i-\cos\,\theta_t}{\sqrt{\varepsilon_2/\varepsilon_1}\,\cos\,\theta_i+\cos\,\theta_t}\right] \tag{13.82}$$

$$E_{11}^r=E_{11}^i\frac{c-i\,d}{c+i\,d}=E_{11}^i\frac{\sqrt{c^2+d^2}e^{-i\psi}}{\sqrt{c^2+d^2}e^{i\psi}}=E_{11}^i e^{-2i\psi}$$

Thus, the amplitude of the reflected wave is equal to that of the incident wave but there is a phase change. This phenomenon is known as *total internal reflection*. Because of phase change the reflected wave will be elliptically polarized wave.

The difference of phase between the two components is given by

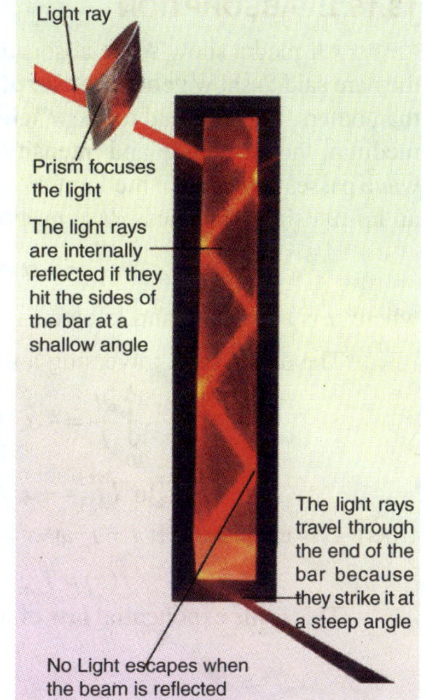

Total Internal Reflection.

$$\tan(\phi-\psi)=\frac{\tan\phi-\tan\psi}{1+\tan\phi\ \ \tan\psi}=-\frac{\cos\theta_i\left(\sin^2\theta_i-n^2\right)^{1/2}}{\sin^2\theta_i}$$

If the phase difference is zero, that is $(\phi-\psi)=0$, then for $\theta_i\neq\pi/2$, we get

$$\sin^2\theta_i=n^2 \text{ or } \sin\theta_i=n$$

where n is the refractive index of the second medium with respect to the first medium. If the second medium be air, then n represents the refractive index of air with respect to the first medium.

$$\therefore \qquad \sin\theta_i={}^1n_a=\frac{1}{n_1}$$

This angle is known as the **critical angle** for the two media.

$$\therefore \qquad \sin\theta_c=\frac{1}{n_1} \tag{13.83}$$

Total reflection does not mean that there is no field in medium 2. Under conditions of total internal reflection a field does exist in the less dense medium. This field has a phase progressing along the boundary but decreases exponentially from it. It is called a **surface wave** or **evanescent wave.**

13.15. LIGHT PROPAGATING THROUGH A MEDIUM

When a beam of light is propagated in a material medium its velocity is less than the velocity in a vacuum and its intensity gradually decreases as it progresses through the medium. The velocity of light in a material medium varies with the wavelength, and this variation is referred to as **dispersion**. As the beam traverses the medium, its intensity decreases gradually. The reduction in the intensity of a beam traversing a medium is partly due to **scattering** and partly due to **absorption**. In the true absorption the light disappears, the energy being converted into heat. The phenomenon of scattering, absorption and dispersion are inseparably connected.

13.15.1. ABSORPTION

All media show some absorption. Some media absorb all wavelengths more or less equally; they are said to show **general absorption**. Others absorb some wavelengths very much more strongly than others; they are said to show **selective absorption**. If we assume plane waes propagating in the medium, the amplitude and intensity of plane waves will not be constant but will decrease as the wave passes through the medium. For plane waves, the fraction dI/I of the intensity lost in traversing an infinitesimal thickness dx is proportional to dx, so that

$$\frac{dI}{I}=-\alpha\,dx \tag{13.84}$$

where α is a constant and known as the *absorption coefficient*.

The decrease in traversing a finite thickness x, is

$$\int_0^x\frac{dI}{I}=-\alpha\int_0^x dx$$

$$[\ln I]_0^x=-\alpha\,x+C$$

where C is a constant. If $I=I_o$ at $x=0$, $C=\ln I_o$ and one has

$$I(x)=I_oe^{-\alpha x} \tag{13.85}$$

This is the exponential **law of absorption** discovered by Bouguer and Lambert independently.

13.15.2. SCATTERING

Tyndall observed that when a beam of light passes through a transparent substance, solid, liquid or gaseous, a certain portion of light is scattered in all directions. This is called the **Tyndall effect**. Tyndall thought that the faint blue light emerging sideways out of the beam was due to scattering of light by small particles present in the medium. Rayleigh recognized that the scattering was due to individual molecules rather than due to other extraneous particles and such scattering accounts for the blue of the sky. The molecular scattering is hence called the **Rayleigh scattering**. For Rayleigh scattering to take place the molecules or group of molecules should be much smaller in linear dimension than the wavelength of light.

Tyndal effect is seen here using a laser pointer.

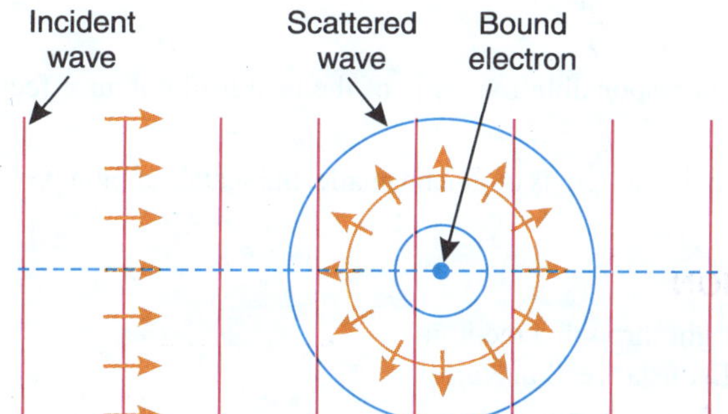

Scattering of electromagnetic radiation by a bound electron.

Fig. 13.19

13.15.2.1. Rayleigh Scattering

A molecule contains an equal number of positive and negative charges. When a molecule is put into a static electric field it suffers some distortion, the positively charged nuclei being attracted towards the negative pole of the field, the electrons to the positive pole. The separation of charge centers causes an *induced electric dipole moment* to be set up in the molecule and the molecule is said to be **polarized**. The size of the induced dipole, μ, depends both on the magnitude of the applied field, E, and on the ease with which the molecule can be distorted. We may write

$$\mu = \alpha E \tag{13.86}$$

where α is the polarizability of the molecule. In case of majority of the molecules, the polarizability is anisotropic. That is the electrons forming the bond are more easily displaced by an electric field applied along the bond axis than one across this direction.

When a sample of such molecules is subjected to a beam of light of frequency ν the electric field experienced by each molecule varies according to the equation

$$E = E_o \sin 2\pi \nu t$$

and thus the induced dipole also undergoes oscillations of frequency ν;

$$\mu = \alpha E = \alpha E_o \sin 2\pi \nu t \tag{13.87}$$

Thus, the light filed will make the induced dipole oscillate. Such an oscillating dipole emits light of its own oscillation frequency. In other words, light does not simply pass through transparent matter; instead it forms dipoles, these dipoles radiate and this re-radiation appears as light. The light is seen in any direction perpendicular to the axis of the dipole. This is the same as with a dipole antenna, which radiates in all directions except in the direction of its own length. This is the explanation of **Rayleigh scattering**. The amplitude of the scattered light is inversely proportional to the square of wavelength and since $I \propto E^2$, the intensity is inversely proportional to the fourth power of the wavelength. Thus,

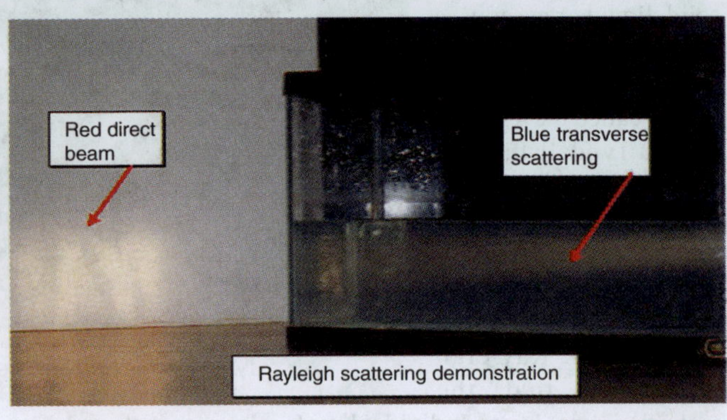

$$I \propto \frac{1}{\lambda^4} \tag{13.88}$$

Light scattering is responsible for some of the beautiful colour effects such as blue sky and red sunset.

However, if the incident light is monochromatic, the scattered light is observed to be unchanged in frequency.

13.15.3. DISPERSION

The passage of light through a medium causes an electrical disturbance that varies sinusoidally with time. As a consequence, any charged particle in the medium is subjected to a fluctuating electrical field and is set into forced vibrations. Since atoms consist of positive nuclei surrounded by negatively charge clouds of electrons, these differently charged parts of atom will tend to be displaced slightly as a light wave passes. The electrons cannot leave the atom because of the intense local electric field, but they get slightly displaced to one side in a direction opposite to that of the field, as shown in Fig.13.20a. The nucleus being relatively heavy and thus having a larger inertia will not be moved as much. The electrons being lighter will be forced back and forth in response to the vibrating field. The result is a rapidly oscillating atomic dipole, as shown in Fig. 13.20b.

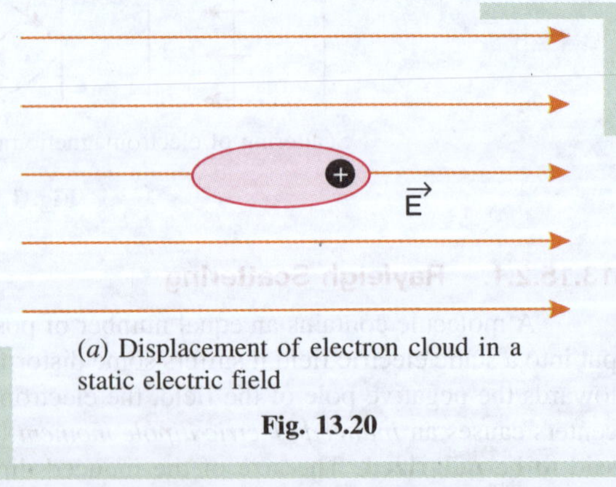

(*a*) Displacement of electron cloud in a static electric field

Fig. 13.20

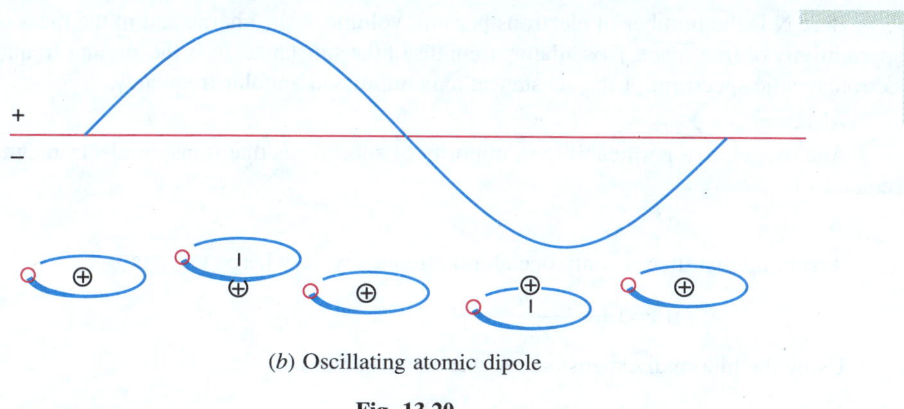

(*b*) Oscillating atomic dipole

Fig. 13.20

Thus, when an electromagnetic wave is incident on an atom or a molecule, the periodic electric force of the wave sets the bound charges into vibratory motion. The frequency with which these charges are forced to vibrate is equal to the frequency of the wave. The phase of this motion as compared to the impressed electric force will depend on the impressed frequency. It will vary with the difference between the impressed frequency and the natural frequency of the charges.

Crystal dispersion.

Dispersion can be explained with the concept of secondary waves that are produced by the induced oscillations of the bound charges. When a beam of light propagates through a transparent medium (solid or liquid), the amount of lateral scattering is very small. The scattered waves traveling in a lateral direction produce destructive interference.

However, the secondary waves traveling in the same direction as the incident beam superimpose on one another. The resultant vibration will depend on the phase difference between the primary and the secondary waves. This superimposition changes the phase of the primary waves and this is equivalent to a change in the wave velocity. Wave velocity is defined as the speed at which a condition of equal phases is propagated. Hence, the variation in phase due to interference, changes the velocity of the wave through the medium. The phase of the oscillations and hence that of the secondary waves depends upon the impressed frequency. It is clear, therefore that the velocity of light in the medium varies with the frequency of light.

Also, refractive index depends upon the velocity of light in the medium. Therefore, refractive index of the medium varies with frequency (wavelength) of light.

13.16. CAUCHY'S DISPERSION FORMULA

The relative permittivity of the medium in the case of dynamic polarizability is given by,

$$\varepsilon_r = 1 + \chi_e = 1 + \frac{Ne^2}{\varepsilon_o m} \sum_i \frac{f_i}{\omega_i^2 - \omega^2} \tag{13.89}$$

Here N is the number of electrons per unit volume, e the charge and m the mass of electron, ε_o permittivity of free space, f_i oscillator strengths of the substance, ω_i is the angular frequency of the electromagnetic spectrum of the substance, ω is impressed angular frequency.

Also, $\sum f_i = 1$

And the relative permeability in minority of substances that transmit electromagnetic waves is equal to 1.

$$\therefore \quad \mu^2 = \varepsilon_r$$

Assuming that there is only one atomic frequency ω_o where $\omega \ll \omega_o$

$$\therefore \quad \mu^2 = 1 + \frac{Ne^2}{\varepsilon_0 m(\omega_o^2 - \omega^2)} \tag{13.90}$$

Using the binomial expansion, we write equ(13.90) as

$$\mu = \left[1 + \frac{Ne^2}{\varepsilon_0 m(\omega_0^2 - \omega^2)}\right]^{\frac{1}{2}}$$

or

$$\mu = 1 + \frac{Ne^2}{2\varepsilon_0 m(\omega_o^2 - \omega^2)} = 1 + \left(\frac{Ne^2}{2\varepsilon_0 m\omega_o^2}\right)\left(1 - \frac{\omega^2}{\omega_o^2}\right)^{-1}$$

$$= 1 + \left(\frac{Ne^2}{2\varepsilon_0 m\omega_o^2}\right)\left(1 + \frac{\omega^2}{\omega_o^2}\right)$$

As $\omega = \frac{2\pi c}{\lambda}$ and $\omega_0 = \frac{2\pi c}{\lambda_o}$,

$$\mu = 1 + \frac{Ne^2}{2\,\varepsilon_0\, m\left(\frac{2\pi c}{\lambda_o}\right)^2}\left(1 + \frac{\lambda_o^2}{\lambda^2}\right)$$

$$\mu = 1 + \frac{Ne^2\lambda_o^2}{8\pi^2\varepsilon_0 mc^2} + \frac{Ne^2\lambda_o^4}{8\pi^2\varepsilon_0 mc^2\lambda^2} \tag{13.91}$$

Taking
$$1 + \frac{Ne^2\lambda_o^4}{8\pi^2\varepsilon_o mc^2} = B \tag{13.92}$$

and
$$\frac{Ne^2\lambda_o^2}{8\pi^2\varepsilon_o mc^2} = A \tag{13.93}$$

$$\mu = A + \frac{B}{\lambda^2} \tag{13.94}$$

Equation (13.94) represents *Cauchy's dispersion formula*. A and B are called *Cauchy's constants*. The values of A and B depend on the medium. From eq. (13.94) it is evident that the refractive index of the medium decreases with increase in wavelength of light.

If a graph is plotted between μ and $\frac{1}{\lambda^2}$ it will be a straight line. The intercept OP on the Y-axis gives the value of A. The slope of the line PC gives the value of B.

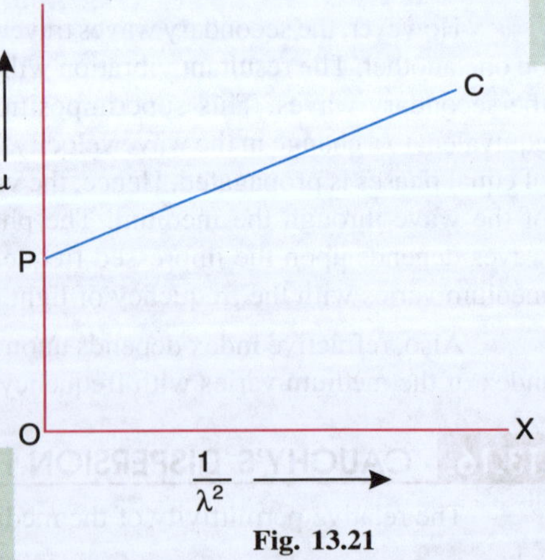

Fig. 13.21

13.17. DISPERSIVE POWER

According to Cauchy's formula $$\mu = A + \frac{B}{\lambda^2} \tag{13.95}$$

Differentiating equation (13.95), we get

$$\frac{d\mu}{d\lambda} = -\frac{2B}{\lambda^3} \tag{13.96}$$

$\frac{d\mu}{d\lambda}$ is the *dispersive power* of the medium. Therefore, dispersive power is inversely proportional to cube of the wavelength of light. At $\lambda = 4000$Å (violet) the dispersive power is about eight times the dispersive power at $\lambda = 8000$Å (red). It means that the spectral lines are more dispersed near the violet end of the spectrum than that at the red end.

13.18. ANOMALOUS DISPERSION

Cauchy's dispersion formula is applicable to the normal dispersion of many transparent media in the visible region of the spectrum. In general, the refractive index decreases with increase in wavelength. But, over small wavelength regions, the refractive index increases with increase in wavelength accompanied by increase absorption of the radiation.

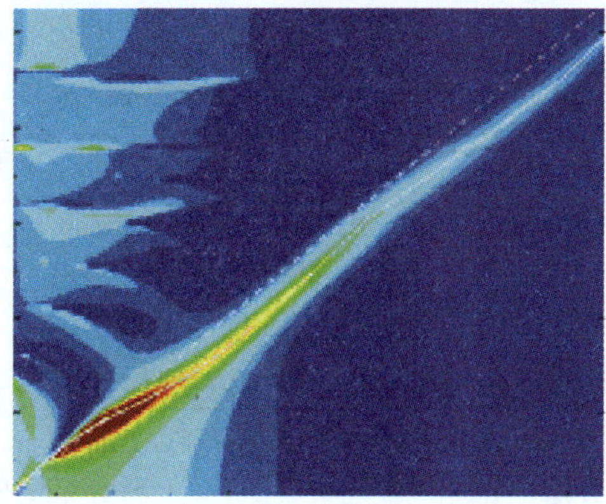

Anomalous dispersion in strongly scattering media.

Increase in refractive index with increase in wavelength is termed as **anomalous dispersion.** It is noticed that anomalous dispersion is present at those wavelengths corresponding to the absorption bands of the medium. For example, in the case of sodium vapour in the visible region, anomalous dispersion is noticed at wavelengths 5890 Å and 5896 Å.

Anomalous dispersion is shown in Fig. 13.22.

The curves, AB, CD, and EF show normal dispersion. However, at the absorption wavelengths λ_1 and λ_2, the refractive index increases with wavelength. BC and DE represent anomalous dispersion. Iodine vapour has an absorption band in the visible region. When a hollow prism containing iodine vapour is used, red rays are deviated more than violet rays.

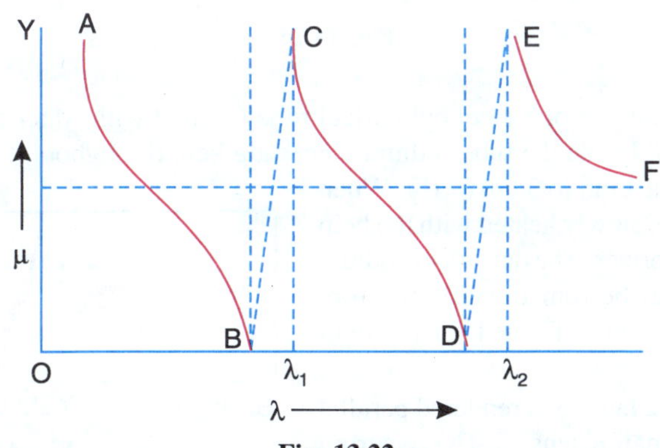

Fig. 13.22

Sellmeier's formula for the refractive index of a medium is given by,

$$\mu^2 = 1 + \sum_p \frac{A_p \lambda^2}{\lambda^2 - \lambda_p^2} \tag{13.97}$$

Here A_p is proportional to the number of particles per unit volume vibrating with a natural frequency corresponding to the wavelength λ_p and λ is the wavelength of incident radiation. If the medium has two absorption wavelengths corresponding to λ_o and λ_1, the Sellmeier's relation will be,

$$\mu^2 = 1 + \frac{A_o \lambda^2}{\lambda^2 - \lambda_o^2} + \frac{A_1 \lambda^2}{\lambda^2 - \lambda_o^2} \tag{13.98}$$

From this relation it is clear that as μ approaches λ_o or λ_1 from the shorter wavelength side, μ tends to $-\infty$ at resonance. Similarly from the longer wavelength side μ tends to $+\infty$ at resonance. The curves obtained between μ and λ are shown in Fig. 13.23.

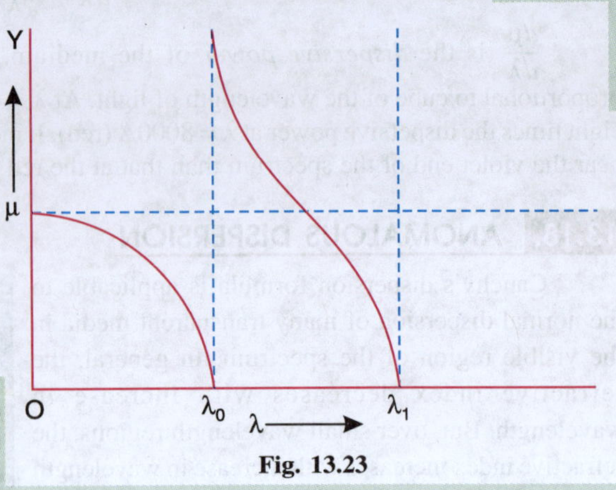

Fig. 13.23

Cauchy's dispersion formula is applicable for normal dispersion where as Sellmeier's formula explains both normal and anomalous dispersion.

13.19. WOODS'S EXPERIMENT

Anomalous dispersion of sodium vapour in the neighborhood of D_1 and D_2 lines can be demonstrated with the help of Wood's experiment.

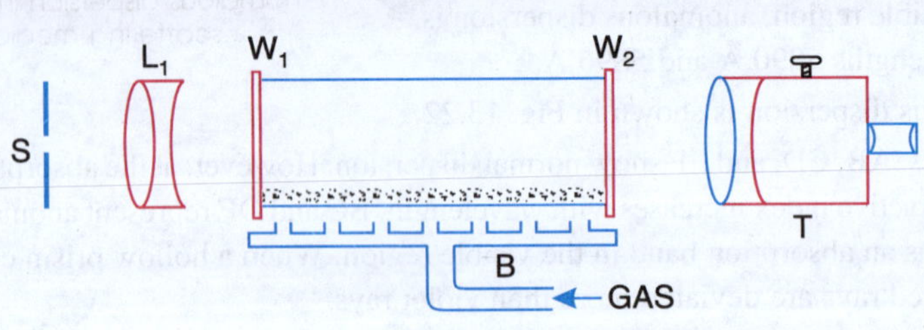

Fig. 13.24

The apparatus consists of a steel cylindrical tube C fitted with water cooled glass windows W_1 and W_2 (Fig.13.24). Inside the tube sodium pieces are kept throughout its length. The pressure inside the tube is kept at 1 cm to 3 cm of Hg. When the sodium in the tube is slowly heated with the help of burners, sodium vaporizes. The density of sodium vapour decreases from bottom towards the top. Practically a sodium prism is formed for the light passing through the tube.

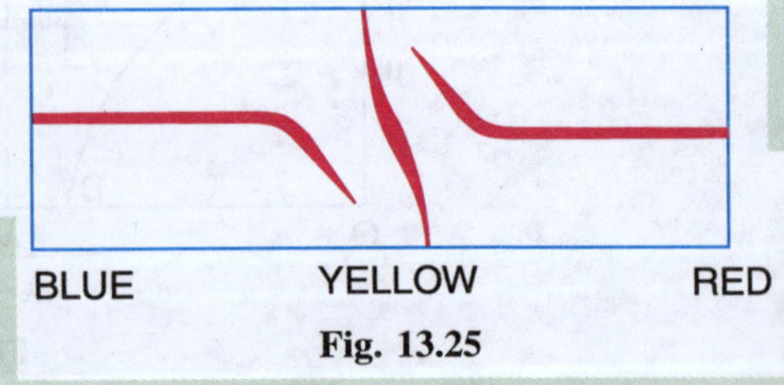

Fig. 13.25

Light from an arc lamp S is rendered parallel with the help of a collimating lens L_1. The spectrum is obtained on a photographic plate. It is noticed that the spectrum is continuous except at the positions corresponding to D_1 and D_2 lines (Fig. 13.25).

13.20. ELECTRON THEORY OF DISPERSION

A complete theory of dispersion based on electromagnetic theory was propounded by Lorentz. He assumed that the atomic dipoles in a medium execute forced oscillations under the influence of electric field due to the incident electromagnetic radiation.

If E is the electric field at any instant, the force on the electron is = eE. In addition, the electron also experiences;

(a) an elastic force proportional to the displacement $= -\alpha\xi$.

(b) a frictional damping force proportional to the velocity $= -\beta\frac{dx}{dt}$

If the mass of the electron is m, the total force

$$F = eE - \alpha x - \beta\frac{dx}{dt} \qquad (13.99)$$

But $$F = m\frac{d^2x}{dt^2} \quad \therefore \quad m\frac{d^2x}{dt^2} = eE - \alpha x - \beta\frac{dx}{dt}$$

or $$\frac{d^2x}{dt^2} + \left(\frac{\beta}{m}\right)\frac{dx}{dt} + \left(\frac{\alpha}{m}\right)x = \left(\frac{eE}{m}\right) \qquad (13.100)$$

In the absence of the external electric field, E = 0.

$$\frac{d^2x}{dt^2} + \left(\frac{\beta}{m}\right)\frac{dx}{dt} + \left(\frac{\alpha}{m}\right)x = 0 \qquad (13.101)$$

The equation is similar to the equation, $\frac{d^2x}{dt^2} + 2b\frac{dx}{dt} + k^2x = 0$

The general solution of the equation is $x = e^{-bt}[Ae^{mt} + Be^{-mt}]$

Where $m = \sqrt{b^2 - k^2}$, $b = \frac{\beta}{2m}$ and $k^2 = \frac{\alpha}{m}$, $\omega_0 = \sqrt{\frac{\alpha}{m} - \frac{\beta^2}{4m^2}}$

The natural frequency of the system is $\nu_0 = \frac{1}{2\pi}\sqrt{\frac{\alpha}{m} - \frac{\beta^2}{4m^2}}$

If damping is low, $\beta = 0$.

$$\therefore \quad \nu_0 = \frac{1}{2\pi}\sqrt{\frac{\alpha}{m}} \qquad (13.102)$$

$$\omega_0^2 = \frac{\alpha}{m} \qquad (13.103)$$

Further, taking $\left(\frac{\beta}{m}\right) = g_0$ and substituting these values in eq (13.100), we get,

$$\frac{d^2x}{dt^2} + g_0\frac{dx}{dt} + \omega_0^2 x = \left(\frac{eE}{m}\right) \qquad (13.104)$$

The polarization P of the medium is equal to the dipole moment per unit volume. If the number of dispersion electrons per unit volume is N, then, P = Nex

$$\therefore \quad \frac{dP}{dt} = (Ne)\frac{dx}{dt} \text{ and } \frac{d^2P}{dt^2} = (Ne)\frac{d^2x}{dt^2}$$

∴ Multiplying eq. (13.104) by Ne and substituting these values we get,

$$\frac{d^2P}{dt^2} + g_0\frac{dP}{dt} + \omega_0^2 P = Ne\left(\frac{eE}{m}\right)$$

or
$$\frac{d^2P}{dt^2} + g_0 \frac{dP}{dt} + \omega_0^2 P = \left(\frac{Ne^2}{m}\right) E \tag{13.105}$$

The incident radiation corresponds to the electric filed is, $E = E_0 e^{j\omega t}$

The polarization at any instant is given by $P = P_0 e^{j\omega t}$

$$\therefore \quad \frac{dP}{dt} = i\omega P_0 \, e^{j\omega t} = i\omega P \text{ and } \frac{d^2P}{dt^2} = -\omega^2 P$$

Substituting these values in equation (13.105), we get,

$$-\omega^2 P + g_o i\omega P + \omega_o^2 P = \left(\frac{Ne^2}{m}\right) E$$

$$P = \left(\frac{Ne^2}{m}\right)\left[\frac{E}{\omega_0^2 - \omega^2 + ig_0\omega}\right] \tag{13.106}$$

But dielectric constant $K = 1 + 4\pi\left(\frac{P}{E}\right)$ or $\left(\frac{4\pi P}{E}\right) = K - 1$

The expression $\left(\frac{4\pi P}{E}\right)$ is a complex number. Therefore, the dielectric constant K and refractive index μ are complex numbers. Using μ^* as complex refractive index we have,

$$K - 1 = \mu^{*2} - 1$$

$$\therefore \quad \left(\frac{4\pi P}{E}\right) = \mu^{*2} - 1 \tag{13.107}$$

From equ.(13.106), we write

$$4\pi\left(\frac{P}{E}\right) = 4\pi\left(\frac{Ne^2}{m}\right)\left[\frac{1}{\omega_0^2 - \omega^2 + ig_o\omega}\right]$$

$$\mu^{*2} - 1 = 4\pi\left(\frac{Ne^2}{m}\right)\left[\frac{1}{\omega_0^2 - \omega^2 + ig_o\omega}\right] \tag{13.108}$$

Equation (13.108) is obtained on the assumption that all the particles have the same natural frequency ω_o i.e., the medium has only one absorption line. In practice, however, there may be large number of absorption wavelengths for a given medium. Equation (13.108) has to be modified for a general case.

$$\mu^{*2} - 1 = 4\pi\left(\frac{Ne^2}{m}\right)\sum_p\left[\frac{f_p}{\omega_0^2 - \omega^2 + ig_p\omega}\right] \tag{13.109}$$

Here μ^{*2} is complex.

This is the general dispersion formula. Here f represents the number of oscillators in each molecule. It may be noted that in equation (13.109) the refractive index is dependent on frequency i.e., wavelength.

Special Case. In the case of transparent substance, absorption is negligible i.e., $g_p\omega$ is negligibly small as compared to $\omega_p^2 - \omega^2$. Therefore from equation (13.109),

$$\mu^2 - 1 = 4\pi\left(\frac{Ne^2}{m}\right)\sum_p\left[\frac{f_p}{\omega_0^2 - \omega^2}\right] \tag{13.110}$$

Here μ is real.

In vacuum, $\lambda = \dfrac{2\pi c}{\omega}$ and $\lambda_p = \dfrac{2\pi c}{\omega_p}$

$$\therefore \quad \omega_p^2 - \omega^2 = 4\pi^2 c^2 \left[\frac{\lambda^2 - \lambda_p^2}{\lambda^2 \lambda_p^2}\right] \qquad (13.111)$$

Substituting the value in equation (13.110), we obtain

$$\mu^2 - 1 = 4\pi\left(\frac{Ne^2}{m}\right)\sum_p \left[\frac{f_p \lambda_p^2 \lambda^2}{(\lambda^2 - \lambda_p^2)\, 4\pi^2 c^2}\right]$$

$$\mu^2 - 1 = \sum_p \left(\frac{Ne^2 f_p \lambda_p^2}{m\pi c^2}\right)\left[\frac{\lambda^2}{(\lambda^2 - \lambda_p^2)}\right]$$

$$\left(\frac{Ne^2 f_p \lambda_p^2}{m\pi c^2}\right) = A_p \qquad (13.112)$$

$$\mu^2 - 1 = \sum_p \frac{A_p \lambda^2}{\lambda^2 - \lambda_p^2}$$

$$\therefore \quad \mu^2 = 1 + \sum_p \frac{A_p \lambda^2}{\lambda^2 - \lambda_p^2} \qquad (13.113)$$

Equation (13.113) represents **Sellmeier's dispersion formula**.

WORKED OUT EXAMPLES

Example 13.1: The wave function of a light wave is $E(z,t) = 10^3 \sin \pi (3\times10^6 x - 9\times10^{14} t)$

(*a*) determine the speed, wavelength, frequency and period of the wave,

(*b*) determine the magnetic field associated with the wave.

Solution: (*i*) $E(z,t) = 10^3 \sin \pi (3\times10^6 x - 9\times10^{14} t)$

$$= 10^3 \sin 3\times10^6 \pi \, (x - 3\times10^8 t) \qquad (1)$$

The equation is similar to the general wave equation,

$$E(z,t) = E_o \sin k\,(x - vt) \qquad (2)$$

Comparing (1) with (2), we get,

$$v = 3\times10^8 \text{ m/s}, \text{ and } k = 3\times10^6 \pi \text{ /m}$$

$$\therefore \quad \lambda = \frac{2\pi}{k} = \frac{2\pi}{3\times10^6 \pi} = 6.666\times10^{-7} \text{ m} = 6666 \text{ Å}$$

$$\nu = \frac{\upsilon}{\lambda} = \frac{3\times10^8 \text{ m/s}}{6.666\times10^{-7} \text{ m}} = 4.5\times10^{14} \text{ Hz}$$

$$T = \frac{1}{\nu} = \frac{1}{4.5\times10^{14} \text{ Hz}} = 2.2\times10^{-15} \text{ s}$$

(*ii*) The light wave propagates in the z- direction while the E- vector oscillates along x- direction in xz- plane. Since in a EM wave, the magnetic field **B** is normal to both **E** and wave propagation directions, it should be in the yz- plane.

Thus, $B_x = 0 \,;\, B_z = 0$

$$B = B_y\,(z,t) = B(z,t)$$

As $E = cB,\ B = \dfrac{E}{c}$

$$\therefore\quad B(z,t) = \frac{10^3 \sin \pi \left(3\times 10^6 y - 9\times 10^{14} t\right)}{c} = \frac{10^3 \sin \pi \left(3\times 10^6 y - 9\times 10^{14} t\right)}{3\times 10^8}$$

$$= 3.33\times 10^{-6} \sin \pi \left(3\times 10^6\, y - 9\times 10^{14}\, t\right)$$

Example 13.2: A electromagnetic wave moving through free space has an electric field given by, $E = 100 \sin \left[8\pi\times 10^{14}\left(t - \dfrac{z}{3\times 10^8}\right)\right]$. Calculate the corresponding intensity.

Solution: $E = 100 \sin \left[8\pi\times 10^{14}\left(t - \dfrac{z}{3\times 10^8}\right)\right]$; $\therefore\ E_o = 100$ V/m

$$\text{Intensity,}\quad I = \left(\frac{c\varepsilon_o}{2}\right)E_o^2 = \frac{\left(3\times 10^8\right)\left(8.85\times 10^{-12}\ \dfrac{\text{C}^2}{\text{Nm}^2}\right)(100\ \text{V/m})^2}{2} = \mathbf{13.3\ W/m^2}$$

Example 13.3: Determine the reflection coefficient and the transmission coefficient of an electric field wave traveling in air and incident normally on a boundary between air and dielectric having permeability = μ_o and permittivity, $\varepsilon_r = 4$.

Solution: We know that, $\eta_1 = \sqrt{\dfrac{\mu_o}{\varepsilon_o}}$ and $\eta_2 = \sqrt{\dfrac{\mu_o}{\varepsilon_o\varepsilon_r}}$

But it is given that, $\varepsilon_r = 4$. $\quad\therefore\ \eta_2 = \dfrac{1}{2}\sqrt{\dfrac{\mu_o}{\varepsilon_o}} = \dfrac{1}{2}\eta_1$

Reflection coefficient is given by, $\Gamma = \dfrac{\eta_2 - \eta_1}{\eta_2 + \eta_1}$

Putting the values of η_1 and η_2, we have, $\Gamma = \dfrac{\dfrac{\eta_1}{2} - \eta_1}{\dfrac{\eta_1}{2} + \eta_1} = -\dfrac{1}{3} = -0.33$

Transmission coefficient is given by, $\tau = \dfrac{2\eta_2}{\eta_2 + \eta_1} = \dfrac{2\times\dfrac{\eta_1}{2}}{\dfrac{\eta_1}{2} + \eta_1} = \dfrac{2}{3} = 0.66$

Example 13.4: Given two dielectric media, medium 1 is free space and medium 2 has $\varepsilon_2 = 4\varepsilon_o$ and $\mu = \mu_o$. Determine reflection coefficient for oblique incidence $\theta_1 = 30^o$, for

(*a*) Perpendicular polarization,

(*b*) Parallel polarization

Solution: (*a*) *Perpendicular polarization*

Reflection coefficient is given by, $\Gamma = \dfrac{\sqrt{\varepsilon_o}\cos\theta_1 - 2\sqrt{\varepsilon_o}\cos\theta_2}{\sqrt{\varepsilon_o}\cos\theta_1 + 2\sqrt{\varepsilon_o}\cos\theta_2}$

$$= \frac{\cos\theta_1 - 2\cos\theta_2}{\cos\theta_1 + 2\cos\theta_2}$$

We also know that, $\frac{\sin\theta_1}{\sin\theta_2} = \sqrt{\frac{\varepsilon_2}{\varepsilon_1}} = \sqrt{\frac{4\varepsilon_o}{\varepsilon_o}} = 2$

Hence, $\sin\theta_2 = \frac{1}{2}\sin\theta_1 = \frac{1}{2}\sin 30^o = \frac{1}{4}$

Therefore, $\Gamma = \frac{\cos 30° - 2\times 0.968}{\cos 30° + 2\times 0.968} = \frac{0.866 - 1.936}{0.866 + 1.936}$

$= -0.381$

(*b*) *Parallel polarization*

$$\Gamma = \frac{2\cos\theta_1 - \cos\theta_2}{2\cos\theta_1 + \cos\theta_2}$$

$$= \frac{2\times 0.866 - 0.968}{2\times 0.866 + 0.968}$$

$= 0.283$

Example 13.5: A uniform plane electromagnetic wave is incident at angle θ_1 at the surface of discontinuity between two homogeneous isotropic dielectrics with permittivity ε_1 and ε_2. ε_2 being the permittivity of the dielectric into which the wave gets refracted at an angle θ_2. If E_i, E_r and E_t are the electric intensities respectively of the incident, reflected and transmitted waves, show that the reflection coefficient for parallel polarization is given by,

$$\frac{E_r}{E_i} = \frac{\tan(\theta_1 - \theta_2)}{\tan(\theta_1 + \theta_2)}$$

Solution: We know that, $\frac{E_r}{E_i} = \frac{\sqrt{\varepsilon_2}\cos\theta_1 - \sqrt{\varepsilon_1}\cos\theta_2}{\sqrt{\varepsilon_2}\cos\theta_1 + \sqrt{\varepsilon_1}\cos\theta_2}$

Dividing numerator and denominator by $\sqrt{\varepsilon_2}$, we get,

$$\frac{E_r}{E_i} = \frac{\cos\theta_1 - \sqrt{\frac{\varepsilon_1}{\varepsilon_2}}\cos\theta_2}{\cos\theta_1 + \sqrt{\frac{\varepsilon_1}{\varepsilon_2}}\cos\theta_2}$$

$$= \frac{\cos\theta_1 - \frac{\sin\theta_1}{\sin\theta_2}\cos\theta_2}{\cos\theta_1 + \frac{\sin\theta_1}{\sin\theta_2}\cos\theta_2}$$

$$= \frac{\sin\theta_2\cos\theta_1 - \sin\theta_1\cos\theta_2}{\cos\theta_1\sin\theta_2 + \sin\theta_1\cos\theta_2}$$

Dividing again numerator and denominator by $\cos\theta_1$, and $\cos\theta_2$, we get,

$$\frac{E_r}{E_i} = \frac{\tan\theta_2 - \tan\theta_1}{\tan\theta_2 + \tan\theta_1}$$

Dividing again numerator and denominator by $1 + (\tan\theta_1 \tan\theta_2)$ we get,

$$\frac{E_r}{E_i} = \frac{\tan(\theta_2 - \theta_1)}{\tan(\theta_2 + \theta_1)}\text{; Hence proved.}$$

QUESTIONS

1. State and explain Maxwell's equations for electromagnetic field. Starting from Maxwell's equations, deduce the wave equation for a plane wave in free space.
2. Show that the ratio of the electric and magnetic fields of a uniform plane wave is constant depending upon the medium.
3. What is meant by the Poynting vector? Derive Poynting vector from Maxwell's equations and explain its physical significance.
4. Solve the Maxwell's equations for perfect dielectric containing no charge and no conduction current.
5. Prove Snell's law of reflection and refraction taking the oblique incidence of electromagnetic wave on an interface between two dielectric media.
6. What is Brewster's angle? In which case does it have a value? Find its value in terms of the parameters of the media.
7. Discuss briefly the salient features of total internal reflection. Does the critical angle θ_c depend on the polarization of the **E**-vector?
8. Define Brewster's angle. Does the angle depend on the nature of polarization of **E**-vector?
9. Obtain an expression for the reflection coefficient for perpendicularly polarized wave when it is incident at an angle θ_i on the surface.
10. Give an account of electromagnetic theory of dispersion.
11. What is normal and anomalous dispersion? Derive Sellmeier's equation for the refractive index of a dielectric medium.
12. Solve the wave equations to show that the phase velocity of plane monochromatic waves in a non conducting media is given by $1/\sqrt{\mu\varepsilon}$ and hence, find an expression for index of refraction.
13. State Maxwell's equations in their general /integral form. Derive their form for harmonically varying fields.
14. show that the speed of propagation of electromagnetic wave in free space is given by $c = 1/\sqrt{\mu_0 \varepsilon_0}$
15. Establish Maxwell's equations for electromagnetic fields and obtain an expression for pointing vector. **(Meerut, 2000)**
16. Write down Maxwell's field equations in differential and explain their physical meaning. Prove pointing theorem relating, the flow of energy at a point in space in an electromagnetic field. **(Gulbarga, 2005)**
17. Obtain Maxwells equations and deduce an expression for the velocity of propagation of plane electromagnetic wave in a medium of dielectric constant ε and permeability μ. **(Patna, 2000)**
18. Show that the following electromagnetic fields satisfy Maxwell's equations: $E_x = E_y = 0$ and $E_z = \cos(y - ct)$, $B_x = \cos(y - ct)$, $B_y = B_z = 0$.

19. Obtain an expression for: (*i*) the electric vector, (*ii*) the pointing vector, (*iii*) the field energy density and (*iv*) the momentum density in the field for a plane wave given by $\mathbf{B} = B_0 \cos \frac{2\pi}{\lambda}(z-ct)\,y$ moving along the z-direction in free space. **[Andhra 2001]**

PROBLEMS FOR PRACTICE

1. A plane electromagnetic wave in free space has an average pointing vector of 1 watt/m^2. Find the average energy density. [**Ans**: 3.33×10^{-7} J/m^3]

2. Find the velocity of a plane wave in a lossless medium having a relative permittivity of 5 and relative permeability of 1. [**Ans**: 1.34×10^8 m/sec]

3. A plane sinusoidal linearly polarized electromagnetic wave of wavelength $\lambda = 5 \times 10^{-7}$m travels in vacuum in the direction of the X-axis. The average intensity of the wave per unit area is 0.1 W/m^2 and plane of the vibration of the electric field is parallel to the Y-axis. Write the equations describing the electric and magnetic fields of the wave.

[**Ans:** $E_y = \sqrt{24\ \ \pi} \cos\left[4\pi \times 10^6 (x-ct)\right] N/\text{coulomb},\ B_z = \left(E_y/c\right) Tesla$]

4. What must be the strength of uniform electric field if it is to have the same energy density as that possessed by a 400 gauss magnetic field [1 W/m^2 =10^{4}gausses, ε_0= 8.85 × 10^{12} F /m and $\mu_0 = 4\pi \times 10^{-7}$ H/m].

5. Find the Brewster angle for a parallel-polarized wave traveling from air into glass for which ε_r =5.0 **(Ans: 65.91°)**

6. In free space, E(z,t) =1.0 sin ($\omega t - \beta z)a_x$ V/m. Show that the average power crossing a circular disk of radius 15.5 m in a z = const plane is 1 W.

7. In free space, $H(z,t) = 1.33 \times 10^{-1} \cos(4 \times 10^7 t - \beta z)\, a_x$ A/m. Obtain an expression for $E(z,t)$. Find β and λ. **(E_0 = 50 V/m, 4/3 rad/m, 15 πm.)**

8. A plane electromagnetic wave is traveling in an unbounded lossless dielectric medium having $\mu_r = 1$ and $\varepsilon_r = 3$ and has peak electric field intensity of 6 V/m. Find

(*a*) velocity of the wave

(*b*) the intrinsic impedance of the medium

(*c*) the peak value of the magnetic intensity

(Ans: 1.73 × 10^8 m/s; 217.66 ohm; 2.76 × 10^{-2} A/m)

14
CHAPTER

Interference

14.1. INTRODUCTION

In 1680 Christian Huygens suggested that light propagates in the form of waves. Huygens did not know anything about the nature of the light wave; whether it is a transverse wave or a longitudinal wave; he had no knowledge about the speed of light or its wavelength. The wave theory of light received the first experimental evidence in 1801 from the interference experiments conducted by Thomas Young. Using the principle of superposition and Huygens' wave concept, Young explained the interference effects observed in various instances such as double slit experiment, the striking colours observed on oil slicks exposed to sunlight and Newton's rings.

14.2. LIGHT WAVES

A light wave is a harmonic electromagnetic wave consisting of periodically varying electric and magnetic fields oscillating at right angles to each other and also to the direction of propagation of the wave (see Fig.14.1a). The

Light wave.

At a Glance

electric field in the wave is defined by the electric field strength vector **E** and the magnetic field by the vector of magnetic induction **B**. Vectors **E** and **B** are of equal importance to the wave. However, a light wave is often represented by the **E** wave (see Fig.14.1b), since many of the effects of light such as photoelectric effect, photochemical and physiological actions are found to be mostly due to the electric vector **E**. The vector **E** is often referred to as the **light vector** or **optical vector.** The electric field is known as **optical field**, radiation field, wave field or light field. The magnetic field is implied to be oscillating in a plane normal to the plane of the electric field oscillations and is not shown specifically in the diagrams.

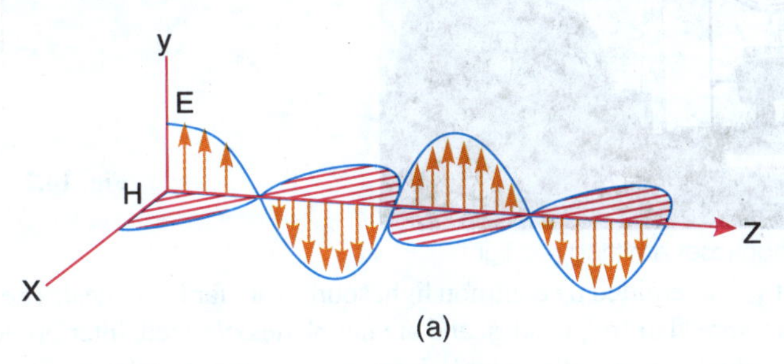

(a)

Fig. 14.1

The light wave depicted in Fig. 14.1(b) is mathematically represented by the expression

$$E = E_0 \sin\left(k z - \omega t\right) \tag{14.1}$$

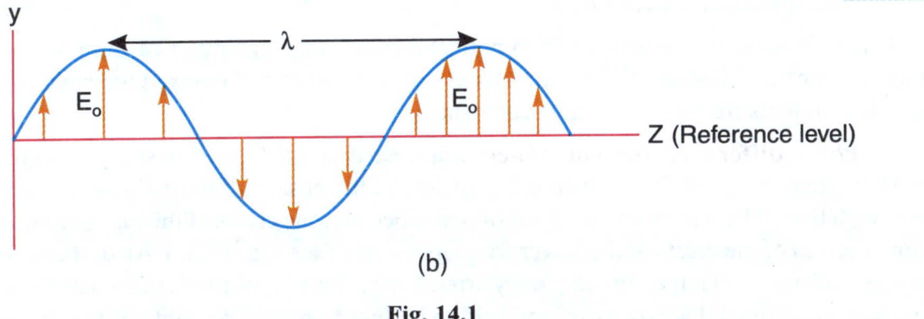

(b)

Fig. 14.1

In fact, the wave shown in Fig. 14.1 (b) and represented by equ. (14.1) is an **ideal** electromagnetic wave. **No real light source emits such perfect waves.**

1. We note the following features about such a mathematical wave.

(*i*) The wave has a *single* definite frequency ν (= ω / 2π). In optics, waves having a single frequency and wavelength are called **monochromatic** waves.

(*ii*) It is a *harmonic* wave and is of **infinite extension** consisting of a continuous **train of waves.** At any instant the wave extends from $z = -\infty$ to $z = +\infty$, and at any point whose position corresponds to a particular value of z the wave continues from $t = -\infty$ to $t = +\infty$.

(*iii*) The amplitude of the wave E_o stays constant as the wave propagates through air. Hence it is a plane wave and its wavefront is normal to the z-axis.

(*iv*) The electric vector **E** of the wave oscillates always parallel to a fixed direction in space, i.e. y-direction. In other words, the E-vibrations are confined to yz – plane. Therefore the wave is **plane polarized** (or **linearly polarized**).

Helium neon laser emitting red light.

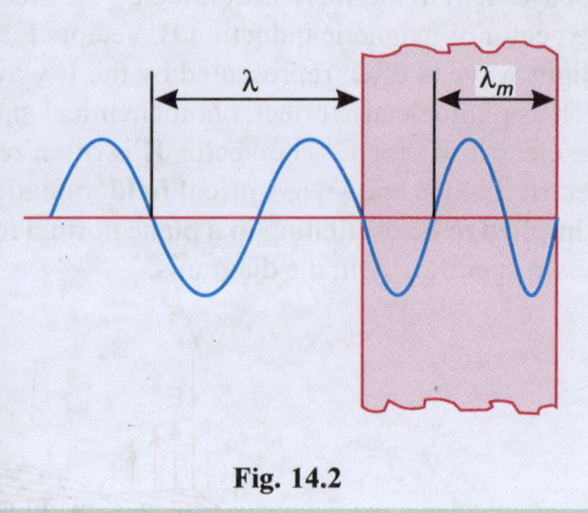

Fig. 14.2

The real light waves emitted by common light sources are far from ideal. They are wave trains of limited length, have a spread of frequencies, and are not plane polarized. Interference effects are easily observed when sinusoidal waves with a single frequency, ν, and wavelength, λ, overlap. While it is fairly easy to produce sound waves and r.f. waves of single frequency, common sources of light *do not emit* monochromatic light. The common light sources emit a continuous distribution of wavelengths. The sodium lamp used in optics laboratories is a fairly monochromatic source of light, which emits light in the yellow region at the wavelength 5893 Å. The most nearly monochromatic source is the helium-neon laser that emits red light at 6328 Å.

2. A light wave travels *slower* in an optical medium than in air or a vacuum. It travels with a velocity υ, which is less than 'c'. The wavelength of light wave *decreases in the medium*, as shown in Fig. 14.2 while its frequency remains constant.

3. Phase difference and coherence: Suppose two waves are passing through a point in space. If the frequencies of the two waves are different, the phase difference between the vibrations changes with time. The waves will drift out of phase because the crests of the higher frequency wave will arrive ahead of the crests of the lower frequency wave (see Fig. 14.3c). Also, if one (or both) of the waves undergoes changes in frequency irregularly, the phase difference changes irregularly. Under these conditions the two waves are said to be **incoherent**. The light emitted by most of the light sources is incoherent as the frequency of light changes abruptly and irregularly, though we can

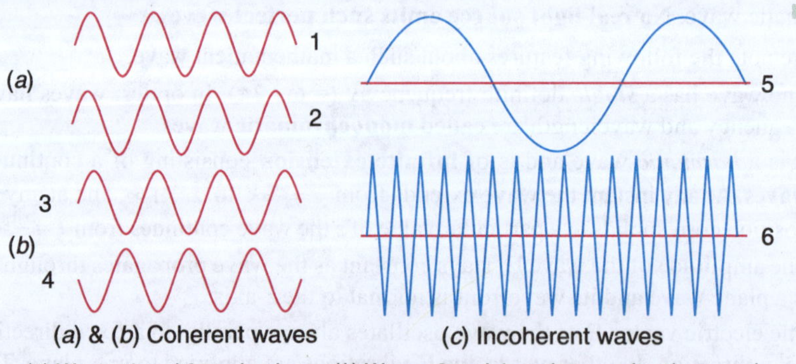

(*a*) Waves 1 and 2 are in-phase, (*b*) Waves 3 and 4 are in opposite phase and (*c*) Waves 5 and 6 are incoherent.

Fig. 14.3

think of an average frequency associated with the light wave. On the other hand, if we consider two waves of same frequency, they may differ in the amplitudes but they maintain a predictable phase relationship (see Fig. 14.3a). The difference in their phases may have any value from zero radians to a maximum of 2π radians; but the phase difference remains constant. Thus, two or more waves of the *same* frequency can maintain the same phase or constant phase difference over a distance and time. Such waves are said to be **coherent waves.**

When coherent waves rise or fall together, reaching the crest (or trough) at the same time, they are said to be **in phase,** as shown in Fig. 14.3 (a). The phase difference will be 0 or 2π radians, which remains constant as the waves propagate in space. The path difference between the waves will then be zero or an integral multiple of a wavelength, λ. Such waves move through space with a *crest-to-crest correspondence.* When one wave reaches its crest while the other falls to its trough, then the phase difference between the waves is π and the waves are said to be in **opposite phase** and the phase difference remains 180° (see Fig. 14.3b). They are inverted with respect to each other everywhere. Then the path difference between the waves will be $\lambda / 2$ or an odd integral multiple of $\lambda / 2$.

The waves may have also any *constant phase difference* other than zero or π radians.

If two (or more) waves maintain a constant phase difference over a long distance and time, then they are said to be coherent.

Two waves of different frequency (see Fig. 14.3c) can never maintain a constant phase difference, because their phase difference goes on fluctuating and changes arbitrarily. They are said to be **incoherent**, as the phase difference between the waves fluctuates with time.

Thus, **coherence** means the coordinated motion of several waves in a medium maintaining a fixed and predictable phase relationship over a length of time.

The waves shown in Fig. 14.3 (a) and (b) are coherent waves. Sources, which produce coherent waves, are called **coherent sources**.

In the study of interference, we focus our attention mainly on the two specific types of disposition of the waves, namely in phase and opposite phase dispositions, at the point of observation.

4. Optical path and Phase change: Optical path length indicates the number of light waves that fit into that path. Thus, $\Delta = N\lambda$, where Δ is the optical path length and N is an integer or a mixed fraction. Optical path is related to the geometric path and the relation may be found as follows. The distance traversed by light in a medium of refractive index μ in time t is given by

$$L = \upsilon\, t \tag{14.2}$$

where υ is the velocity of light in the medium. The distance travelled by light in a vacuum in the same time t, is $\Delta = c\, t = c \cdot \dfrac{L}{\upsilon} = \mu\, L$

The distance L is called the *geometric path length* (GPL). Δ is the equivalent distance in a vacuum and is called *optical path length* (OPL). Thus,

$$O.P.L. = \mu \times G.P.L.$$

or

$$\Delta = \mu\, L \tag{14.3}$$

The above result means that *more number of waveforms is accommodated along the optical path than in the corresponding geometrical path.* Therefore, ***in the study of interference we always must calculate the optical paths travelled by light rays.***

(a) Effect of Optical path: Optical path determines the **phase** of a light wave arriving at a point. We know that if a wave covers in air a distance of one wavelength, i.e. 1λ, its phase changes by 2π radians. Therefore, we compute that if a wave travels a distance L in air, its phase change is given by

$$\delta = \frac{2\pi L}{\lambda} \tag{14.4}$$

When the wave travels the distance L in a medium, then

$$\delta = \frac{2\pi\Delta}{\lambda} = \frac{2\pi\mu L}{\lambda} \tag{14.5}$$

where Δ is the optical path or optical path difference.

Comparing equns. (14.4) and (14.5), we find that a light path of geometric length L in a medium of refractive index μ produces the same phase change as a light path of length μL in a vacuum.

(b) Effect of reflection: The process of reflection also affects the phase of a light wave. When light is incident on a surface, part of the light gets reflected while a major portion may be transmitted or absorbed. The quantity characterizing the reflectivity of a surface is called the **reflection coefficient,** ρ. ρ depends on the nature of surface, the angle of incidence of light and many other factors. Augustin Jean Fresnel, about 150 years ago, derived a set of expressions which allow us to calculate the amount of light reflected and transmitted at an interface. For normal incidence it was shown that

$$\rho = \frac{\mu_1 - \mu_2}{\mu_1 + \mu_2} \tag{14.6}$$

It may be seen that ρ is positive when $\mu_2 < \mu_1$. It implies that the oscillations in the incident and reflected waves occur in the same phase. On the other hand, when $\mu_2 > \mu_1$, ρ is *negative* signifying that the *oscillations in the incident and reflected waves are in opposite phase*. Hence, we draw the following conclusions:

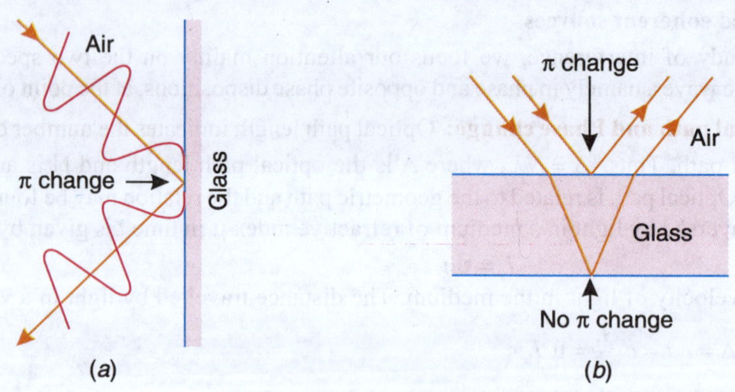

Phase change due to reflection.

Fig. 14.4

(*i*) A light wave travelling from a rarer medium (μ_1) to a denser medium (μ_2) undergoes a phase change of π radians when it gets reflected at the boundary of denser medium, as shown in Fig. 14.4 (a). The wave loses a half-wave on reflection at the boundary of rarer-to-denser medium.

(*ii*) A light wave travelling from a denser medium (μ_2) to a rarer medium (μ_1) does not undergo a change in phase on reflection at the boundary of denser-to-rarer medium (Fig. 14.4b). Therefore, the change in path is zero.

14.3. SUPERPOSITION OF WAVES

Frequently it is necessary to find the resultant disturbance at a point when a number of disturbances arrive simultaneously. According to the **principle of superposition—**

when two or more waves overlap, the resultant displacement at any point and at any instant may be found by adding the instantaneous displacements that would be produced at the point by the individual waves if each were present alone.

It means that the resultant is simply the sum of the disturbances. The principle of superposition applies to electromagnetic waves also and is the most important principle in wave optics. In case of electromagnetic waves, the term *displacement* refers to the amplitude of the electric field vector.

Interference is an important consequence of superposition of *coherent* waves.

14.4. INTERFERENCE

If two or more light waves of the same frequency overlap at a point, the resultant effect depends on the *phases* of the waves as well as their *amplitudes*. The resultant wave at *any point* at any instant of time is governed by the **principle of superposition.** The combined effect at each point of the region of superposition is obtained by adding algebraically the amplitudes of the individual waves. Let us assume here that the component waves are of the *same amplitude*.

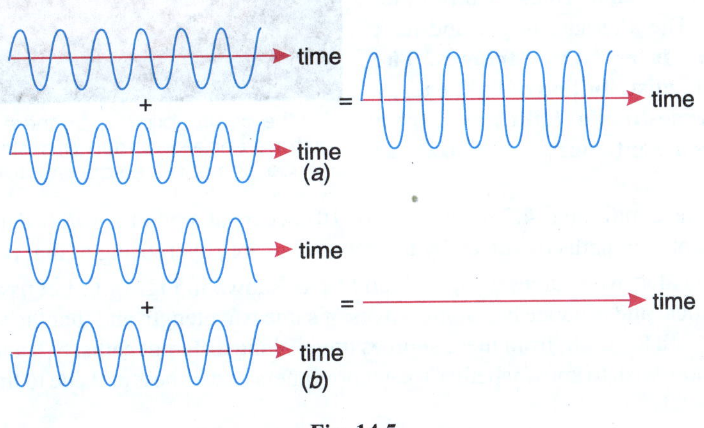

Fig. 14.5

At certain points, the two waves may be *in phase*. The amplitude of the resultant wave will then be equal to the sum of the amplitudes of the two waves, as shown in Fig.14.5(a). Thus, the amplitude of the resultant wave

$$A_R = A + A = 2A. \tag{14.7}$$

Hence, the intensity of the resultant wave is

$$I_R \propto A_R^2 = 2^2 A^2 = 2^2 I. \tag{14.8}$$

It is obvious that the resultant intensity is greater than the sum of the intensities due to individual waves.

$$I_R > I + I = 2I \tag{14.9}$$

Therefore, the interference produced at these points is known as **constructive interference**. A **stationary bright band** of light is observed at points of constructive interference.

At certain other points, the two waves may be *in opposite phase*. The amplitude of the resultant wave will then be equal to the sum of the amplitudes of the two waves, as shown in Fig.14.4 (b). Thus, the amplitude of the resultant wave

$$A_R = A - A = 0. \tag{14.10}$$

Hence, the intensity of the resultant wave is

$$I_R \propto 0^2 = 0. \tag{14.11}$$

It is obvious that the resultant intensity is less than the sum of the intensities due to individual waves.

$$I_R < 2I \tag{14.12}$$

Therefore, the interference produced at these points is known as **destructive interference.** A **stationary dark band** of light is observed at points of destructive interference. Thus, we see that a redistribution of energy took place in the region.

Thus, when two or more coherent waves of light are superposed, the resultant effect is **brightness** in certain regions and **darkness** at other regions. The regions of brightness and darkness alternate and may take the form of straight bands, or circular rings or any other complex shape. The alternate bright and dark bands are called **interference fringes. The phenomenon of redistribution of light energy due to the superposition of light waves from two or more coherent sources is known as interference.**

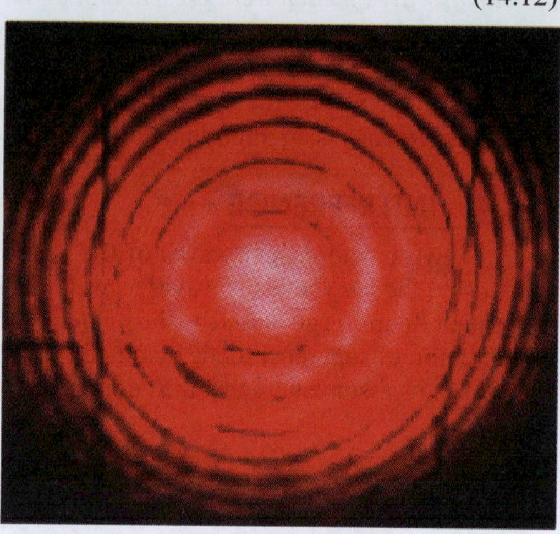

In the central bright spot, there is constructive interference and then a destructive interference ring and then constructive, and so on.

Whether the condition (14.7) occurs or (14.10) occurs at a point is *solely* determined by the difference in the optical paths traversed by the waves that are superposing at that point.

Let us consider two sources of light S_1 and S_2, as shown in Fig. 14.6. Let us assume that the sources are identical and produce harmonic waves of same wavelength and that the waves are in the same phase at S_1 and S_2. Light from these sources travel along different paths, S_1P and S_2P, and meet at a point P. We now wish to know whether we get brightness or darkness at P due to the superposition of waves.

Referring to Fig. 14.6, we find that the waves move along the geometric paths $S_1P = r_1$ and $S_2P = r_2$, which are different in length. Also, the media through which the two waves travelled, may be different. As a result, the optical path lengths are different. If μ_1 is the refractive index of the medium in which the ray S_1P travelled, the corresponding optical path length is $\mu_1 r_1$. Similarly, if μ_2 is the refractive index of the medium in which the ray S_2P travelled, the corresponding optical path length is $\mu_2 r_2$. These optical paths accommodate different number of waveforms along their lengths. The optical path difference between the waves at the point P is $(\mu_2 r_2 - \mu_1 r_1)$. It may come to a few full waves or a mixed fraction of waves. It means that though the waves started with the same phase, they may arrive at P with different phases because they travelled along different optical path lengths.

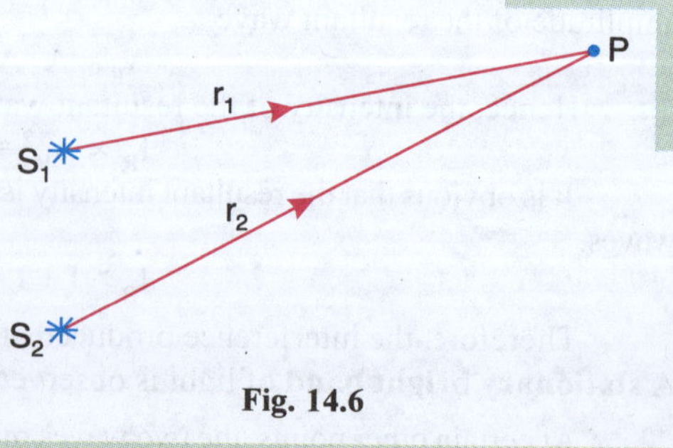

Fig. 14.6

If the optical path difference $\Delta = (\mu_2 r_2 - \mu_1 r_1)$ is equal to zero or an integral multiple of wavelength λ, then the waves arrive in phase at P and superpose with crest-to-crest correspondence. That is, if

$$\Delta = m\lambda \tag{14.13}$$

where m is an integer and takes values, m = 0,1,2,3,4,5,......,then the waves are in phase (see Fig.14.5a) and their overlapping at P produces constructive interference or brightness.

On the other hand, if the optical path difference $\Delta = (\mu_2 r_2 - \mu_1 r_1)$ is equal to an odd integral multiple of half-wavelength, $\lambda/2$, then the waves arrive out of phase at P and superpose with crest-to-trough correspondence. That is, if

$$\Delta = (2m+1)\frac{\lambda}{2} \tag{14.14}$$

where m is an integer and takes values, m = 0,1,2,3,4,5,.......,then the waves are inverted with respect to each other (see Fig. 14.5b) and their overlapping at P produces destructive interference or darkness.

The regions of brightness and darkness are also known as regions of **maxima** and **minima**.

14.4.1. THEORY OF INTERFERENCE

(a) Analytical Method: Let us assume that the electric field components of the two waves arriving at point P vary with time as

$$E_A = E_1 \sin \omega t \tag{14.15}$$

and

$$E_B = E_2 \sin (\omega t + \delta) \tag{14.16}$$

where δ is the phase difference between them. According to Young's principle of superposition, the resultant electric field at the point P due to the simultaneous action of the two waves is given by

$$E_R = E_A + E_B \tag{14.17}$$

$$= E_1 \sin \omega t + E_2 \sin (\omega t + \delta)$$

$$= E_1 \sin \omega t + E_2 (\sin \omega t \cos \delta + \cos \omega t \sin \delta)$$

$$= (E_1 + E_2 \cos \delta) \sin \omega t + E_2 \sin \delta \cos \omega t \tag{14.18}$$

Equ. (14.18) shows that ***the superposition of two sinusoidal waves having the same frequency but with a phase difference produces a sinusoidal wave with the same frequency but with a different amplitude E.***

Let

$$E_1 + E_2 \cos \delta = E \cos \phi \tag{14.19}$$

and

$$E_2 \sin \delta = E \sin \phi \tag{14.20}$$

where E is the amplitude of the resultant wave and ϕ is the new initial phase angle. In order to solve for E and ϕ, we square the equ.(14.19) and (14.20) and add them.

$$(E_1 + E_2 \cos \delta)^2 + E_2^2 \sin^2 \delta = E^2 \left(\cos^2 \phi + \sin^2 \phi\right)$$

or

$$E^2 = E_1^2 + E_2^2 \cos^2 \delta + 2E_1E_2 \cos \delta + E_2^2 \sin^2 \delta$$

or

$$E^2 = E_1^2 + E_2^2 + 2E_1E_2 \cos \delta \tag{14.21}$$

Thus, it is seen that the square of the amplitude of the resultant wave is not a simple sum of the squares of the amplitudes of the superposing waves, there is an additional term which is known as the *interference term*.

(b) Phasor diagram *and* phasor addition: A wave may be viewed either sideways or end-on. In sideways view, as the wave travels through a distance λ, the phase angle changes from 0 to 2π radians. In the end-on view we find a point on the profile of the wave oscillating linearly. The two perspectives may be combined as follows. A circle having a radius OA equal to the amplitude of the wave motions is drawn (see Fig. 14.7). Now consider a point P on the circumference of the circle; Q is the projection of P on the vertical axis. As P moves around the circumference with constant angular velocity ω, Q oscillates vertically. This is the end-on view. The Fig. 14.7 is called the **phasor diagram**. OP is called a **rotating vector** or a **phasor**. It means that the *length* of a phasor is proportional to the *amplitude* of the sinusoidal wave and the *projection* of a phasor on the vertical axis is proportional to the *instantaneous value* of the alternating quantity.

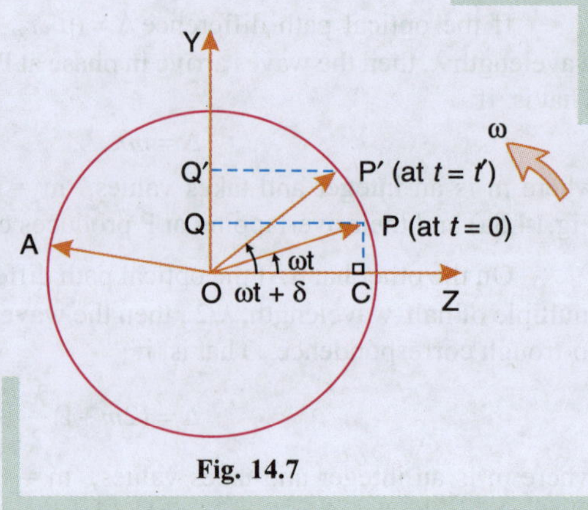

Fig. 14.7

Phasor representation may be used to *add sinusoidal functions* with a phase difference. The wave E_A is represented with a vector of amplitude E_1 rotating about the origin in a counter clockwise direction with an angular frequency ω (See Fig. 14.8 a). As the phasor E_1 rotates, the projection E_A oscillates along the vertical axis. The second wave, E_B, has amplitude E_2 and angular frequency ω but its phase is δ with respect to wave E_A. It is also shown in Fig. 14.8 (a). The resultant E_R is the sum of E_A and E_B obtained by drawing the phasors end to end, by placing the foot of one arrow at the head of the other (as in Fig. 14.8b), maintaining the proper phase difference. The whole assembly rotates counterclockwise about the origin. The sum of the projections on the vertical axis at any time gives the instantaneous value of the total field at a point. The amplitude E_R of the resultant sinusoidal wave at P is the *vector sum* of the other two phasors, as shown in Fig. 14.8 (b).

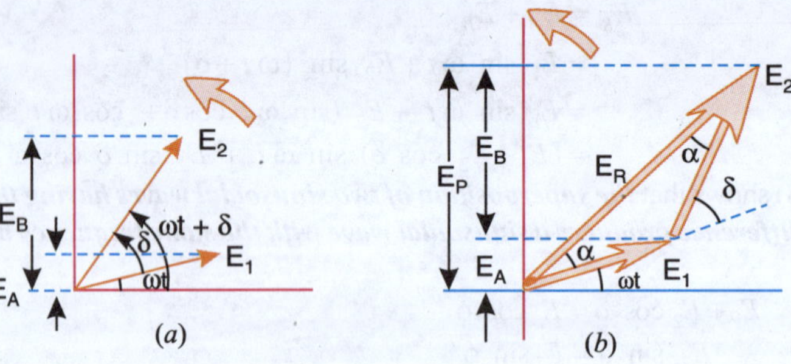

Fig. 14.8

To find E_R, we use the law of cosines.

$$E_R^2 = E_1^2 + E_2^2 - 2E_1E_2 cos(\pi\text{-}\delta)$$

or

$$E_R^2 = E_1^2 + E_2^2 + 2E_1E_2 \cos \delta$$

This is the same as the equ.(14.21). This method is particularly convenient when several wave amplitudes have to be added.

14.4.2. INTENSITY DISTRIBUTION

The intensity of a light wave is given by the square of its amplitude.

$$I = \tfrac{1}{2}\varepsilon_o cE^2 \propto E^2$$

Using this relation into (14.21), we get

$$I = I_1 + I_2 + 2\sqrt{I_1 I_2}\cos\delta \tag{14.22}$$

We see that the resultant intensity at P on the screen is not just the sum of the intensities due to the separate waves. The term $2\sqrt{I_1 I_2}\cos\delta$ is known as the **interference term**. Whenever the phase difference between the waves is zero, i.e. $\delta = 0$, we have maximum amount of light. Thus,

$$I_{max} = I_1 + I_2 + 2\sqrt{I_1 I_2} \tag{14.23}$$

When $I_1 = I_2 = I_o$

$$I_{max} = 4I_o \tag{14.23a}$$

It means that the resultant intensity I will be *more than the sum* of the intensities due to the two sources.

When the phase difference is $\delta = 180°$, $\cos 180° = -1$ and we have a minimum amount of light.

$$I_{min} = I_1 + I_2 - 2\sqrt{I_1 I_2} \tag{14.24}$$

which, when $I_1 = I_2$, becomes

$$I_{min} = 0 \tag{14.24a}$$

It means that the resultant intensity I will be *less than the sum* of the intensities due to the two sources.

At points that lie between the maxima and minima, when $I_1 = I_2 = I_0$, we get

$$I = I_o + I_o + 2I_o\cos\delta$$

$$= 2I_o(1 + \cos\delta)$$

Then using the identity $1 + \cos\delta = 2\cos^2\left(\frac{1}{2}\delta\right)$, we get

$$I = 4I_o\cos^2\left(\tfrac{1}{2}\delta\right) \tag{14.25}$$

Equ. (14.25) shows that the intensity varies along the screen in accordance with the ***law of cosine square.*** Fig. 14.9 shows the variation of intensity as a function of phase angle δ.

It is seen from the plot that the intensity varies from zero at the fringe minima to $4I_o$ at the fringe maxima.

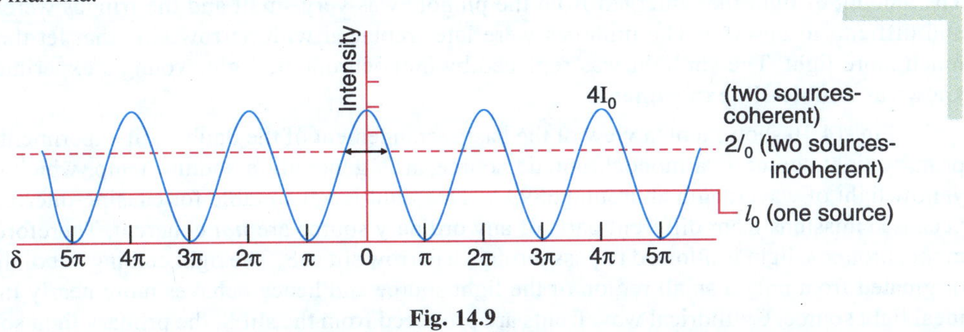

Fig. 14.9

14.4.3. SUPERPOSITION OF INCOHERENT WAVES

Incoherent waves are the waves that do not maintain a constant phase difference. Then the phase of the waves fluctuate irregularly with time and independently of each other. In case of light waves the phase fluctuates randomly at a rate of about 10^8 per second. Light detectors such as human eye, photographic film etc cannot respond to such rapid changes. The detected intensity is always the average intensity, averaged over a time interval which is very much larger than the time of fluctuation. Thus,

$$I_{ave} = I_1 + I_2 + 2\sqrt{I_1 I_2}\ \langle\cos\delta\rangle$$

The average value of the cosine over a large time interval will be zero and hence the interference term becomes zero. Therefore, the average intensity of the resultant wave is

$$I_{ave} = I_1 + I_2$$

If $I_1 = I_2$, then $$I_{ave} = 2I \tag{14.26}$$

It implies that the superposition of incoherent waves does not produce interference but gives a uniform illumination. The average intensity at any point is simply equal to the sum of the intensities of the component waves.

14.4.4. SUPERPOSITION OF MANY COHERENT WAVES

The result (14.23) may be written as

$$I_{max} = 2^2 I_o$$

which gives the resultant intensity when two coherent waves superpose. The resultant maximum intensity due to N coherent waves will be therefore

$$I_{max} = N^2 I_o \tag{14.27a}$$

and the minimum intensity $$I_{min} = 0 \tag{14.27b}$$

where N represents the number of coherent waves superposing at a point.

14.5. YOUNG'S DOUBLE SLIT EXPERIMENT – WAVEFRONT DIVISION

As early as in 1665 Grimaldi attempted to produce interference between two beams of light. He directed sunlight into a dark room through two pinholes in a screen, with an expectation that bright and dark bands would be observed in the area where the beams overlap on each other. He observed uniform illumination instead. In 1801, about one hundred thirty six years later, Thomas Young gave the first demonstration of the interference of light waves. Young admitted the sunlight through a single pinhole and then directed the emerging light onto two pinholes. Finally the light was received on a screen. The spherical waves emerging from the pinholes interfered with each other and a few coloured fringes were observed on the screen. The amount of light that emerged from the pinhole was very small and the fringes were faint and difficult to observe. The pinholes were later replaced with narrow slits that let through much more light. The sunlight was replaced by monochromatic light. Young's experiment is known as **double-slit experiment**.

Fig. 14.10 shows a plan view of the basic arrangement of the double slit experiment. The primary light source is a monochromatic source; it is generally a sodium lamp, which emits yellow light of wavelength at around 5893Å. This light is not suitable for causing interference because emissions from different parts of any ordinary source are ***not*** coherent. Therefore, the monochromatic light is allowed to pass through a narrow slit at S. The light coming out of the slit originated from only a small region of the light source and hence behaves more nearly like an ideal light source. Cylindrical wavefronts are produced from the slit S, the primary light source, which fall on the two narrow closely spaced slits, S_1 and S_2 as shown in Fig. 14.10. The slits at S_1 and S_2 are very narrow. The cylindrical waves emerging from the slits overlap. If the slits are *equidistant* from S, the *phase* of the wave at S_1 will be the same as the *phase* of the wave at S_2. Further, waves leaving S_1 and S_2 are therefore always *in phase*. Hence, sources S_1 and S_2 act as *secondary coherent sources*. The waves leaving from S_1 and S_2 interfere and produce alternate bright and dark bands on the screen at T.

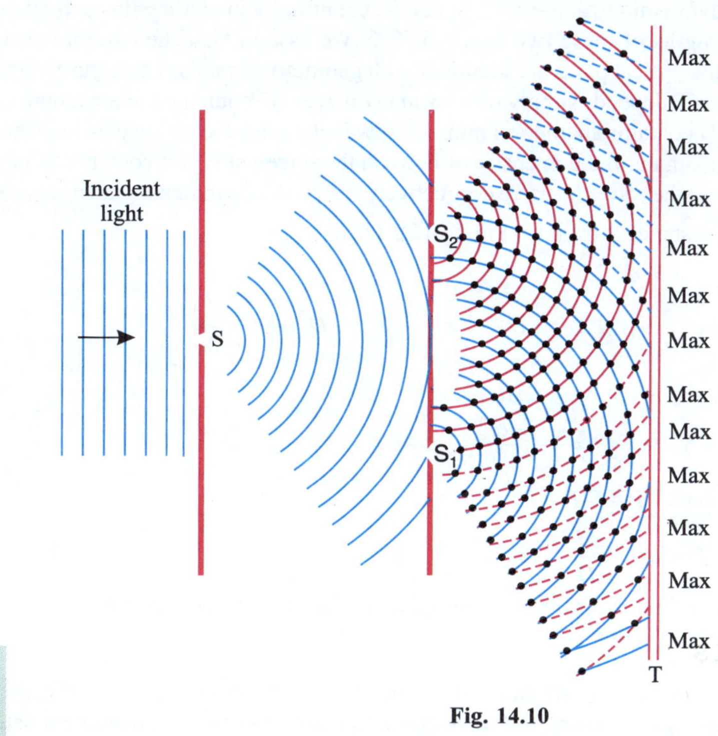

Fig. 14.10

14.5.1. OPTICAL PATH DIFFERENCE BETWEEN THE WAVES AT P:

Let P be an arbitrary point on screen T, which is at a distance D from the double slits. Let θ be the angle between MP and the horizontal line MO. Let S_1N be a normal on to the line S_2P. The distances

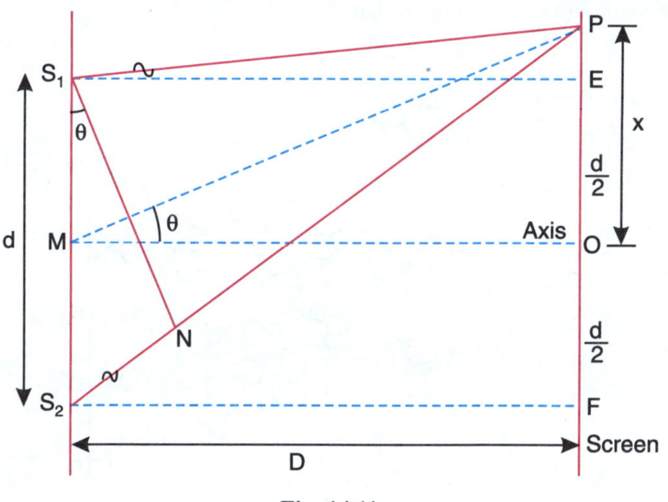

Fig. 14.11

PS_1 and PN are equal. The waves emitted at the slits, S_1 and S_2 are initially in phase with each other. The difference in the path lengths of these two waves is S_2N. We assume that the experiment is carried out in air. Therefore, the optical paths are identical with geometrical paths. The nature of the interference of the two waves at P depends simply on how many waves are contained in the length of the path difference S_2N. If S_2N contains an integral number of wavelengths, the two waves interfere constructively, producing a maximum in the intensity of light on the screen at P. If it contains an odd number of half-wavelengths, then the waves interfere destructively and produce a minimum intensity at P.

Let the point P be at a distance x from O (Fig.14.11). Then

PE $= x - d/2$ and PF $= x + d/2$.

$$(S_2P)^2 - (S_1P)^2 = \left[D^2 + \left(x + \frac{d}{2}\right)^2\right] - \left[D^2 + \left(x - \frac{d}{2}\right)^2\right]$$

$$(S_2P)^2 - (S_1P)^2 = 2xd$$

$$S_2P - S_1P = \frac{2xd}{S_2P + S_1P}$$

We can approximate that $S_2P \cong S_1P \cong D$.

$$\therefore \text{Path difference} = S_2P - S_1P = \frac{xd}{D} \qquad (14.28)$$

We now find out the conditions for observing bright and dark fringes on the screen.

14.5.2. BRIGHT FRINGES

Bright fringes occur wherever the waves from S_1 and S_2 interfere constructively. The first time this occurs is at O, the axial point. There, the waves from S_1 and S_2 travel the same optical path length to O and arrive in phase. The next bright fringe occurs when the wave from S_2 travels one complete wavelength further the wave from S_1. In general constructive interference occurs if S_1P and S_2P differ by a whole number of wavelengths.

The condition for finding a bright fringe at P is that

$$S_2P - S_1P = m\lambda$$

Using the equation (14.28), it means that

$$\frac{xd}{D} = m\lambda \qquad (14.29)$$

where m is called the **order of the fringe**.

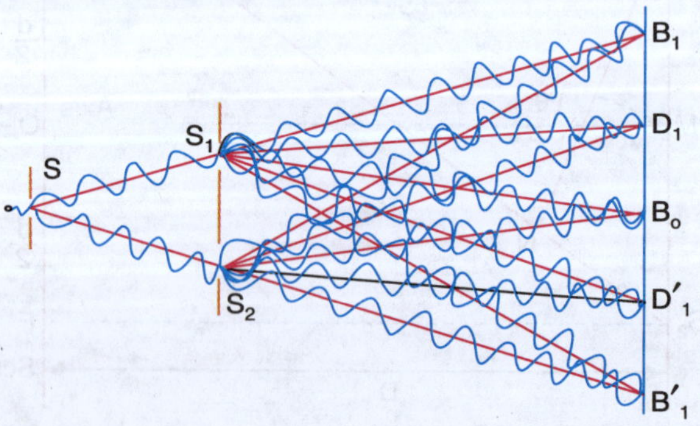

Fig. 14.12

The bright fringe B_0 (at O), corresponding to m = 0, is called the *zero-order* fringe. It means the path difference between the two waves reaching at O is zero. Fringe at B_1 is the *first-order bright fringe* from the axis corresponding to m = 1; the path difference between the two waves reaching at B_1 is one λ. The *second order bright fringe* (m = 2) will be located where the path difference is 2λ and so on.

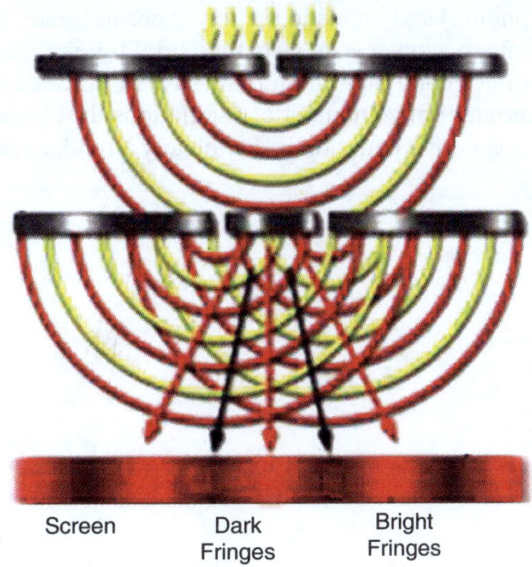

14.5.3. DARK FRINGES

The first dark fringe occurs when $(S_2P - S_1P)$ is equal to λ / 2. The waves are now in opposite phase at P. The second dark fringe occurs when $(S_2P - S_1P)$ equals 3λ / 2. The mth dark fringe occurs when

$$(S_2P - S_1P) = (2m + 1)\,\lambda / 2$$

The condition for finding a dark fringe is $\dfrac{xd}{D} = (2m+1)\dfrac{\lambda}{2}$ (14.30)

The *first-order dark fringe* D_1 (Fig. 14.12) from the axis corresponds to m = 0, where the path difference between the two waves is λ/2. The second order dark fringe (m = 1) will be produced where the path difference is 3λ/2 and so on.

14.5.4. SEPARATION BETWEEN NEIGHBOURING BRIGHT FRINGES

The mth order fringe occurs when $x_m = \dfrac{m\lambda\, D}{d}$

and the (m+1)[th] order fringe occurs when $x_{m+1} = \dfrac{(m+1)\lambda\, D}{d}$

The fringe separation, β is given by $\beta = x_{m+1} - x_m = \dfrac{\lambda D}{d}$ (14.31)

The same result will be obtained for dark fringes. Thus, the distance between any two consecutive bright or dark fringes is known as the ***fringe width*** and is the *same* everywhere on the screen. Further, the width of the bright fringe is equal to the width of the dark fringe. Therefore, the alternate bright and dark fringes are *parallel*. From the equ.(14.31), we find the following:

(*i*) The fringe width β is independent of the *order* of the fringe.It is directly proportional to the wavelength of light, i.e.$\beta \propto \lambda$. The fringes produced by red light are less closer compared to those produced by blue light.

(*ii*) The width of the fringe is *directly proportional* to the distance of the screen from the two slits, $\beta \propto D$. The farther the screen, the wider is the fringe separation.

(*iii*) The width of the fringe is *inversely proportional* to the distance between the two slits. The closer are the slits, the wider will be the fringes.

14.6. COHERENCE

Interference fringes did not appear on the screen in the experiment of Grimaldi as he did not keep the slit S before the double slit arrangement. He obtained only a uniform illumination. It was so because the beams arriving at the screen were not coherent and the phase difference between them varied with time in a haphazard way. The reason for the lack of coherence lies in the very process of light emission. In ordinary sources of visible light, individual atoms are responsible for the emission of light. An atom leaving an excited state gives up the excess energy in the form of a burst of light (photon) and jumps to the lower normal state. The process of transition of the atom from an upper state to a lower state lasts for a brief time of about 10^{-8}sec. Therefore, the light emitted by an atom is not a continuous harmonic wave of infinite extension but is a *wave train* of finite length having a certain limited number of oscillations. It is impossible to say exactly when an atom may emit light because the emission is completely a random process.

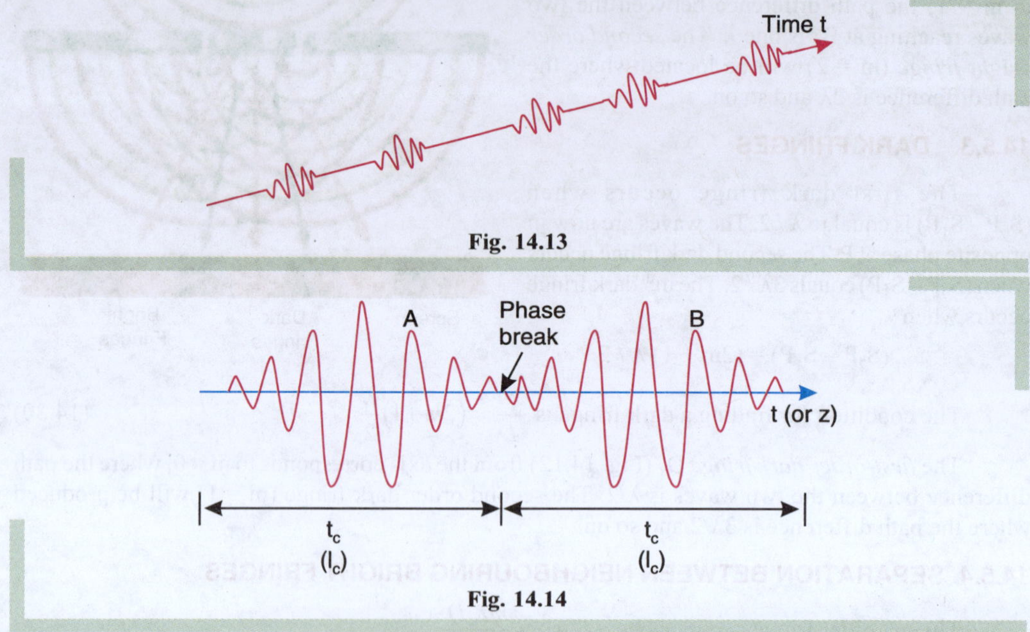

Fig. 14.13

Fig. 14.14

Fig. 14.13 pictorially describes the emission of light by a single atom in terms of wave trains. Other atoms in the source behave similarly but with different emission times. Adding together the wave trains generated by all atoms in the light source produces a succession of wave trains which have their phases distributed randomly. In passing from one wave train to the next there is an abrupt change in phase. Therefore, it is not possible to relate the phase at a point in wave train B to a point in wave train A. The phase of the wave train from an atom would remain constant with respect to the phase of the wave train from another atom only for about 10^{-8} sec. It implies that the two wave trains can be coherent for a maximum time of about 10^{-8} sec. Therefore, light from conventional sources is characterized by two important parameters, namely coherence time and coherence length.

Coherence time: It is *the average time during which the wave remains sinusoidal and phase of the wave packet can be predicted reliably.*

Coherence length: It is *the length of the wave packet over which it may be assumed to be sinusoidal and has predictable phase.*

Light from a sodium discharge lamp has a coherence length of about 2 to 3 cm, while the coherence length of white light is a fraction of a cm. In the double slit experiment, the presence of slit S ensures that the same group of wave trains are incident on slits S_1 and S_2. When the phase of the wave changes at S this change is communicated simultaneously to S_1 and S_2. Therefore, the waves emerging from S_1 and S_2 will be coherent with respect to each other and a *stationary* interference pattern is produced on the screen.

14.7. CONDITIONS FOR INTERFERENCE

We may now summarize the conditions that are to be fulfilled in order to observe a distinct well-defined interference pattern.

(A) Conditions for sustained interference:

(*i*) **The waves from the two sources must be of the same frequency.**

If the light waves differ in frequency, the phase difference fluctuates irregularly with time. Consequently, the intensity at any point fluctuates with time and we will not observe steady interference.

(*ii*) **The two light waves must be coherent.**

If the light waves are coherent, then they maintain a fixed phase difference over a time and space. Hence, a stationary interference pattern will be observed.

(*iii*) **The path difference between the overlapping waves must be less than the coherence length of the waves.**

We have already learnt that light is emitted in the form of wave trains and a finite coherence length characterizes them. If we consider two interfering wave trains, having constant phase difference, as in Fig. 14.15, the interference effects occur due to parts QR of wave 1 and ST of wave 2.

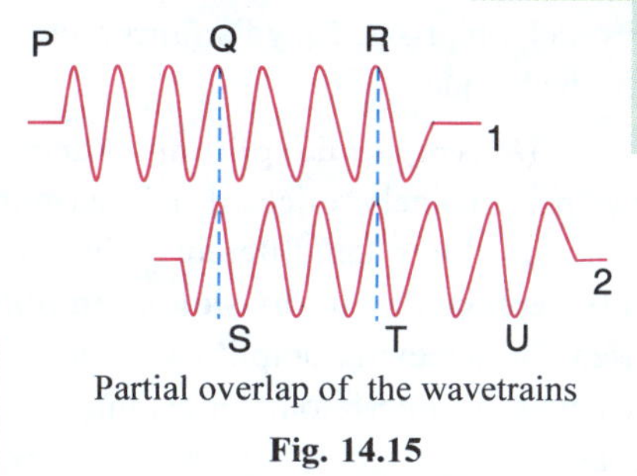

Partial overlap of the wavetrains

Fig. 14.15

For the parts PQ and TU interference will not occur. Therefore, the interference pattern does not appear distinctly. When the entire wave train PR overlaps on the wave train SU, interference pattern will be distinct. On the other hand, when the path difference between the waves 1 and 2 becomes very large, the wave trains arrive at different times and do not overlap on each other. Therefore, in such cases interference does not take place. The interference pattern completely vanishes if the path difference is equal to the coherence length. It is hence required that

$$\Delta < l_{coh} \tag{14.32}$$

(*iv*) **If the two sets of waves are plane polarized, their planes of polarization must be the same.** Waves polarized in perpendicular planes cannot produce interference effects.

(B) Condition for formation of distinct fringe pattern:

(v) **The two coherent sources must lie close to each other in order to discern the fringe pattern.** If the sources are far apart, the fringe width will be very small and fringes are not seen separately.

(*vi*) **The distance of the screen from the two sources must be large.**

(*vii*) **The vector sum of the overlapping electric field vectors should be zero in the dark regions** for obtaining distinct bright and dark fringes. The sum will be zero only if the vectors are anti-parallel and have the same magnitude.

14.8. TECHNIQUES OF OBTAINING INTERFERENCE

The phase relation between the waves emitted by two independent light sources rapidly changes with time and therefore they can *never* be coherent, though the sources are identical in all respects. However, if two sources are derived from a single source by some device, then any phase change occurring in one source is simultaneously accompanied by the same phase change in the other source. Therefore, the phase difference between the waves emerging from the two sources remains constant and the sources are *coherent*. The techniques used for creating coherent sources of light can be divided into the following two broad classes.

(*a*) **Wavefront splitting:** One of the methods consists in dividing a light wavefront, emerging from a narrow slit, by passing it through two slits closely spaced side by side. The two parts of the same wavefront travel through different paths and reunite on a screen to produce fringe pattern. This is known as **interference due to division of wavefront.** This method is useful only with *narrow* sources. Young's double slit, Fresnel's double mirror, Fresnel's biprism, Lloyd's mirror, etc employ this technique.

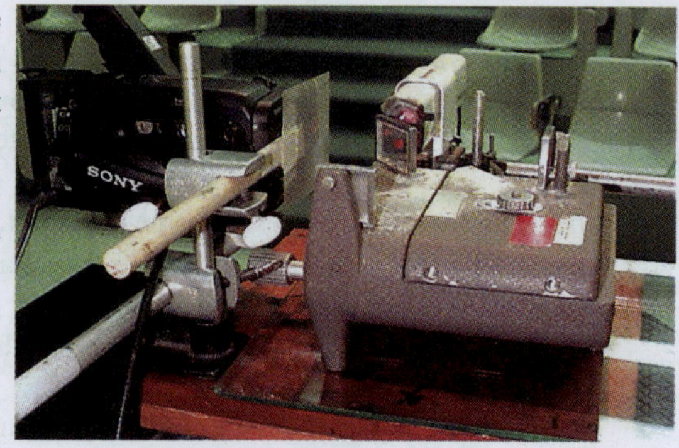

Michelson's Interferometer.

(*b*) **Amplitude splitting:** Alternately, the amplitude (intensity) of a light wave is divided into two parts, namely reflected and transmitted components, by partial reflection at a surface. The two parts travel through different paths and reunite to produce interference fringes. This is known as **interference due to division of amplitude.** Optical elements such as beam splitters, mirrors are used for achieving amplitude division. Interference in thin films (wedge, Newton's rings etc), Michelson's interferometer etc interferometers utilize this method. This method requires *extended* source.

14.9. FRESNEL BIPRISM

Fresnel used a biprism to show interference phenomenon. The biprism consists of two prisms of very small refracting angles joined base to base. In practice, a thin glass plate is taken and one of its faces is ground and polished till a prism (Fig. 14.16 a) is formed with an obtuse angle of about 179° and two side angles of the order of 30'.

When a light ray is incident on an ordinary prism, the ray is bent through an angle called the *angle of deviation*. As a result, the ray emerging out of the prism appears to have emanated from a source S′ located at a small distance above the real source, as shown in Fig.14.16(b). We say that the prism produced a *virtual image* of the source. A biprism, in the same way, creates two virtual sources S_1 and S_2, as seen in Fig.14.16(c). These two virtual sources are images of the same source S produced by refraction and are hence *coherent*.

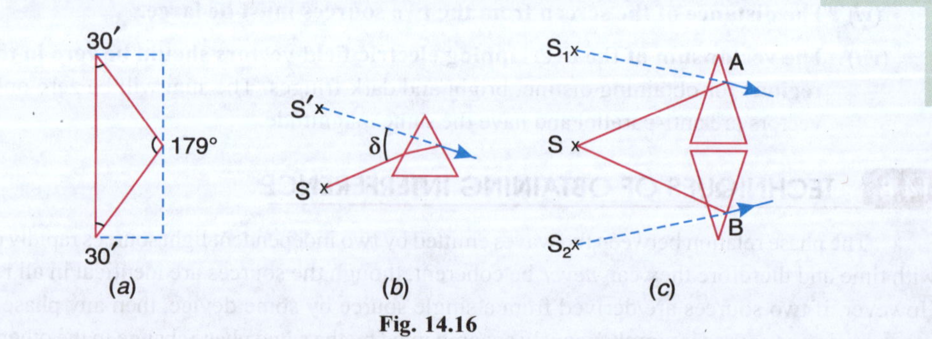

Fig. 14.16

14.9.1. EXPERIMENTAL ARRANGEMENT

The biprism is mounted suitably on an optical bench. An optical bench consists of two horizontal long rods, which are kept strictly parallel to each other and at the same level. The rods carry uprights on which the optical components are positioned. A monochromatic light source such as sodium vapour lamp illuminates a vertical slit S. Therefore, the slit S acts as a narrow linear monochromatic light source. The biprism is placed in such a way that its refracting edge is parallel to the length of the slit S. A single cylindrical wavefront impinges on both prisms. The top portion of wavefront is refracted downward and appears to have emanated from the virtual image S_1. The lower segment, falling on the lower part of the biprism, is refracted upward and appears to have emanated from the virtual source S_2. The virtual sources S_1 and S_2 are coherent (see Fig.14.17), and hence the light waves are in a position to interfere in the region beyond the biprism. If a screen is held there, interference fringes are seen. In order to observe fringes, a micrometer eyepiece is used.

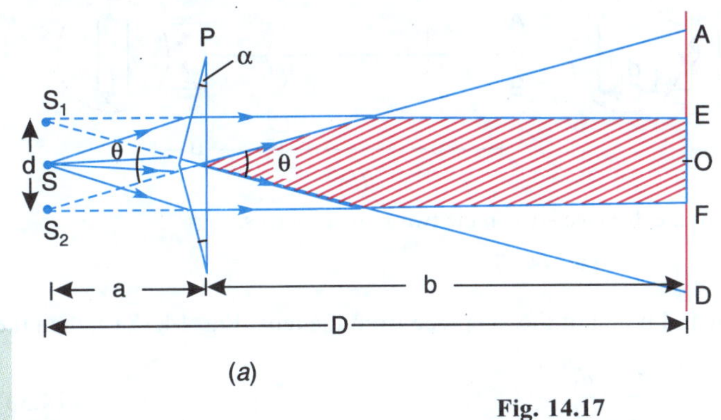

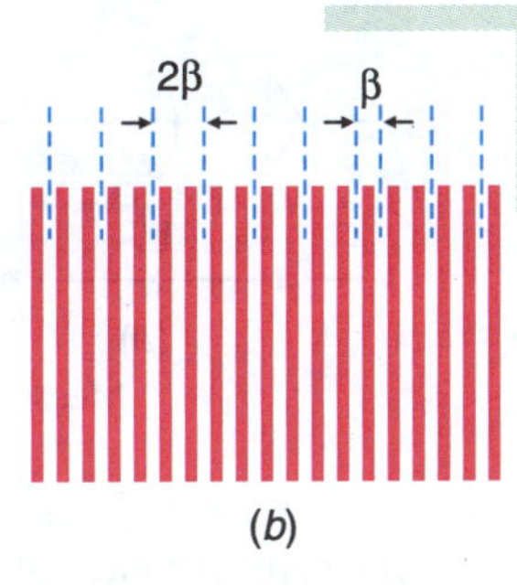

Fig. 14.17

Theory:

The theory of the interference and fringe formation in case of Fresnel biprism is the same as described in § 14.6 for the double-slit. As the point O is equidistant from S_1 and S_2, the central bright fringe of maximum intensity occurs there. On both sides of O, alternate bright and dark fringes, as shown in Fig.14.17(b), are produced. The width of the dark or bright fringe is given by equ.(14.31).

$$\beta = \frac{\lambda D}{d}$$

where D(= a + b) is the distance of the sources from the eye-piece.

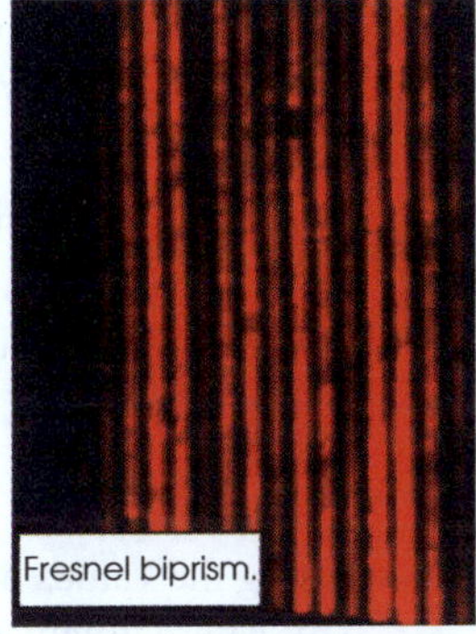

Dark or bright fringes.

14.9.2. DETERMINATION OF WAVELENGTH OF LIGHT

The wavelength of the light can be determined using the equ.(14.31). For using the relation, the values of β, D and d are to be measured. These measurements are done as follows.

Adjustments:

A narrow adjustable slit S, the biprism, and a micrometer eyepiece are mounted on the uprights and are adjusted to be at the same height and in a straight line. The slit is made vertical and parallel to the refracting edge of the biprism by rotating it in its own plane. It is illuminated with the light from the monochromatic source. The biprism is moved along the optical bench till, on looking through it along the axis of the optical bench, two equally bright vertical slit images are seen. Then the eyepiece is moved till the fringes appear in the focal plane of the eyepiece.

(*i*) Determination of fringe width β: When the fringes are observed in the field view of the eyepiece, the vertical cross-wire is made to coincide with the centre of one of the bright fringes. The position of the eyepiece is read on the scale, say x_o. The micrometer screw of the eyepiece is moved slowly and the number of the bright fringes N that pass across the cross-wire is counted. The position of the cross-wire is again read, say x_N. The fringe width is then given by $\beta = \frac{x_N - x_o}{N}$

(*ii*) Determination of 'd': (*a*) A convex lens of short focal length is placed between the slit and the eyepiece without disturbing their positions. The lens is moved back and forth near the biprism till a sharp pair of images of the slit is obtained in the field view of the eyepiece. The distance between the images is measured. Let it be denoted by d_1.

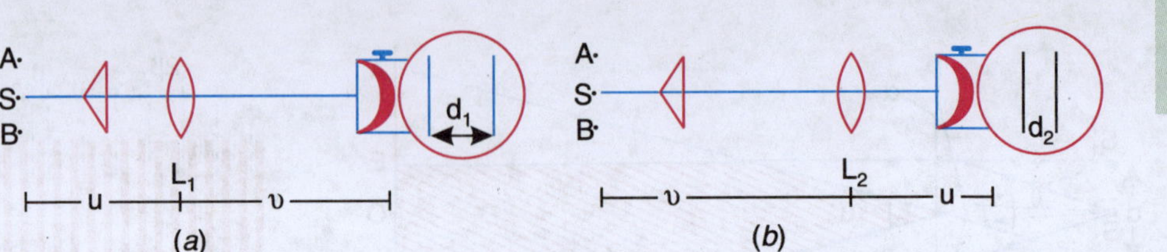

Measurement of the distance etween the two virtual sources.

Fig. 14.18

If u is the distance of the slit and υ that of the eyepiece from the lens (Fig. 14.18a.), then the magnification is

$$\frac{\upsilon}{u} = \frac{d_1}{d} \tag{14.33}$$

The lens is then moved to a position nearer to the eyepiece, where again a pair of images of the slit is seen. The distance between the two sharp images is again measured. Let it be d_2. Again magnification is given by

$$\frac{u}{\upsilon} = \frac{d_2}{d} \tag{14.34}$$

Note that the magnification in one position is the reciprocal of the magnification in the other position.

Multiplying the equations (14.33) and (14.34), we obtain

$$\frac{d_1 d_2}{d^2} = 1$$

$$d = \sqrt{d_1 d_2} \tag{14.35}$$

Using the values of β, d and D in the equation (14.31), the wavelength λ can be computed.

(b) Alternatively, the value of d can be determined as follows. The deviation δ produced in the path of a ray by a thin prism is given by

$$\delta = (\mu - 1)\alpha$$

where α is the refracting angle of the prism. From the Fig. 14.17, it is seen that $\delta = \theta/2$. Since d is very small, we can also write $d = a\theta$.

$$\therefore \quad \frac{\theta}{2} = \frac{d}{2\,a} = (\mu - 1)\alpha$$

$$\therefore \quad d = 2a(\mu - 1)\alpha \tag{14.36}$$

14.9.3. INTERFERENCE FRINGES WITH WHITE LIGHT

In the biprism experiment if the slit is illuminated by white light, the interference pattern consists of a central **white fringe** flanked on its both sides by a few coloured fringes and general illumination beyond the fringes. The central white fringe is the *zero-order* fringe.

With monochromatic light all the bright fringes are of the same colour and it is not possible to locate the zero-order fringe. Therefore, in order to locate the zero order fringe the biprism is to be illuminated by white light.

14.9.4. LATERAL DISPLACEMENT OF FRINGES

The biprism experiment can be used to determine the thickness of a given thin sheet of transparent material such as glass or mica. If a thin transparent sheet is introduced in the path of one of the two interfering beams, the fringe system gets displaced towards the beam in whose path the sheet is introduced. By measuring the amount of displacement, the thickness of the sheet can be determined.

Suppose S_1 and S_2 are the virtual coherent monochromatic sources. The point O is equidistant from S_1 and S_2, where we obtain the *central bright fringe*. Therefore, the optical path $S_1O = S_2O$. Let a transparent plate G of thickness t and refractive index μ be introduced in the path of one of the beams (see Fig. 14.19). The optical path lengths S_1O and S_2O are now not equal and the *central bright fringe* shifts to P from O. The light waves from S_1 to P travel partly in air and partly in the sheet G; the distance travelled in air is $(S_1P - t)$ and that in the sheet is t.

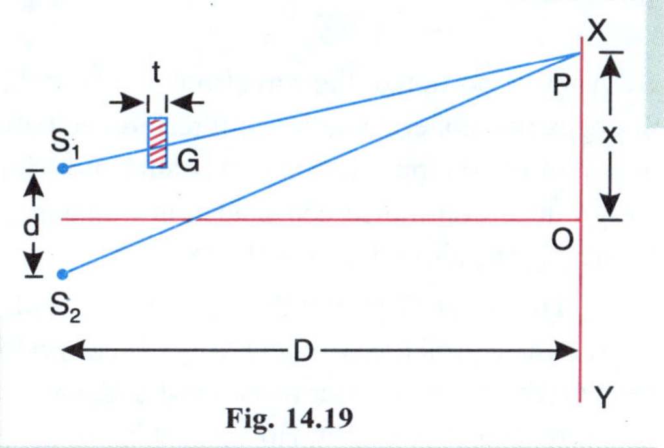

Fig. 14.19

The optical path $\Delta_{S_1P} = (S_1P - t) + \mu\, t = S_1P + (\mu - 1)\, t$

The optical path $\Delta_{S_2P} = S_2P$

The optical path difference at P is $\Delta_{S_1P} - \Delta_{S_2P} = 0$, since in the presence of the thin sheet, the optical path lengths S_1P and S_2P are *equal* and central zero fringe is obtained at P.

$$\therefore \quad \Delta_{S_1P} = \Delta_{S_2P}$$

$$[S_1P + (\mu - 1)\, t] = S_2P$$

$$\therefore \quad S_2P - S_1P = (\mu - 1)\, t$$

But according to the relation (14.28), $S_2P - S_1P = \dfrac{x\,d}{D}$

where x is the **lateral shift** of the central fringe due to the introduction of the thin sheet.

$$\therefore \quad (\mu - 1)\, t = \frac{xd}{D}$$

Hence, the thickness of the sheet is $t = \dfrac{xd}{D(\mu - 1)}$ (14.37)

14.10. LLOYD'S SINGLE MIRROR

In 1834, Lloyd devised an interesting method of producing interference, using a single mirror and using almost grazing incidence. The Lloyd's mirror consists of a plane mirror about 30 cm in length and 6 to 8 cm in breadth (see Fig. 14.20). It is polished on the front surface and blackened at the back to avoid multiple reflections. A cylindrical wavefront coming from a narrow slit S_1 falls on the mirror which reflects a portion of the incident wavefront, giving rise to a virtual image of the slit

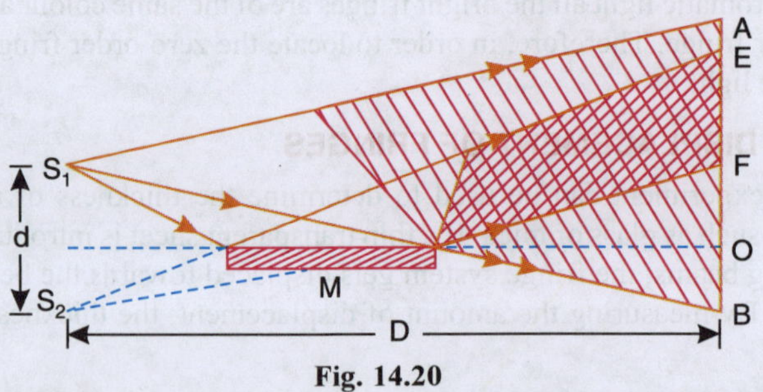

Fig. 14.20

S_2. Another portion of the wavefront proceeds directly from the slit S_1 to the screen. The slits S_1 and S_2 act as two coherent sources. Interference between direct and reflected waves occurs within the region of overlapping of the two beams and fringes are produced on the screen placed at a distance D from S_1 in the shaded portion EF.

Lloyd's mirror - white light.

The point O is equidistant from S_1 and S_2. Therefore, central (zero-order) fringe is expected to lie at O (the perpendicular bisector of S_1S_2) and it is also expected to be bright. However it is not usually seen since the point O lies outside the region of interference (only the direct light and not the reflected light reaches O).

By moving the screen nearer to the mirror such that it comes into contact with the mirror, the point O can be just brought into the region of interference. With white light the central fringe at O is expected to be white but in practice it is *dark*. The occurrence of dark fringe can be understood taking into the consideration of the phase change of π that light suffers when reflected from the mirror. The phase change leads to a path difference of $\lambda/2$ and hence destructive interference occurs there.

14.10.1. DETERMINATION OF WAVELENGTH

The fringe width is given by equ. (14.31). Thus,

$$\beta = \frac{\lambda D}{d}$$

Measuring β,D and d, the wavelength λ can be determined.

Comparison between the fringes produced by biprism and Lloyd's mirror:

1. In biprism the complete set of fringes is obtained. In Lloyd's mirror a few fringes on one side of the central fringe are observed, the central fringe being itself invisible.
2. In biprism the central fringe is bright whereas in case of Lloyd's mirror, it is dark.
3. The central fringe is less sharp in biprism than that in Lloyd's mirror.

14.11. FRESNEL'S DOUBLE MIRROR

Fresnel's double mirror is an arrangement for obtaining two coherent sources by using the phenomenon of reflection. It consists of two plane mirrors inclined to each other at a very small angle, as shown in Fig. 14.21.The mirrors are silvered on their front surfaces and are arranged at nearly 180° such that their surfaces are nearly coplanar.

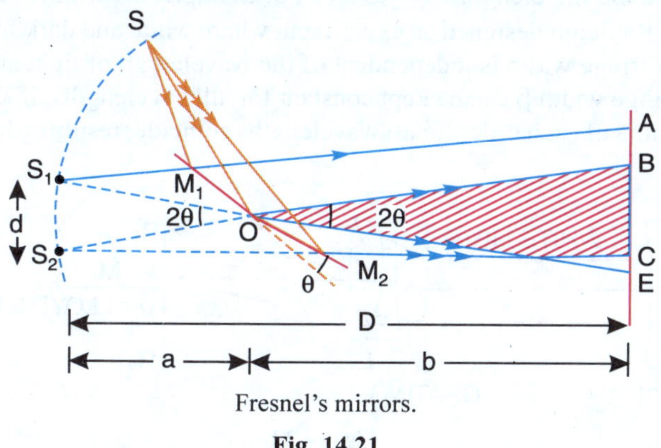

Fresnel's mirrors.

Fig. 14.21

A narrow slit S is placed parallel to the line of intersection of the mirror surfaces and is illuminated with monochromatic light. One portion of the cylindrical wavefront coming from slit S is reflected from the first mirror and another portion of the wavefront is reflected from the second mirror. After reflection, the light appears to diverge from S_1 and S_2, which are the virtual images of S. As the images S_1 and S_2 of the slit are derived from the same source S, they behave as two coherent sources, placed at a distance d apart. The waves diverging from S_1 and S_2 overlap and interference fringes are produced in the overlapping region EF on the screen. The fringes are of equal width.

Fringe width:

It is seen from the geometry of the figure (Fig. 14.21) that $OS_1 = OS_2 = OS$. That is, S_1, S_2 and S lie on a circle with O as a centre. Let 'a' be the distance of the sources and 'b' be the distance of the screen from O. Then the fringe width is given by

$$\beta = \frac{\lambda D}{d} = \frac{(a+b)\lambda}{d} \tag{14.38}$$

OE and OB are the reflected rays from OM_1 and OM_2 respectively, corresponding to the incident ray SO. Therefore, the angle between OE and OB is twice the angle between the mirrors. Hence, $\angle S_1OS_2 = \angle BOE = 2\theta$. Now,

$$\text{Arc } S_1S_2 = a \times 2\theta$$

$$\therefore \quad d = a \times 2\theta$$

Using the above result into equ.(14.38), we get

$$\beta = \frac{(a+b)}{2a\theta}\lambda \tag{14.39}$$

Comparison between the fringes produced by biprism and double mirror :

The fringes in both cases are similar in appearance. However, the double mirror fringes are narrower than the biprism fringes.

14.12. ACHROMATIC FRINGES

A system of white and dark fringes, without any colours, obtained by white light is called *achromatic fringes*.

When the slit is illuminated by white light in any interference experiment, we obtain a central white fringe flanked by a few coloured fringes. Coloured fringes are obtained because the fringe width is dependent on the wavelength $(\beta = \lambda D / d)$. For example, the width of the red fringe is more than the blue fringe. Rayleigh designed an experiment where white and dark fringes were obtained. It can be done if the fringe width is independent of the wavelength of light and is the same for all wavelengths. The fringe width β can be kept constant for all wavelengths, if λ / d is the same in all cases. Then the maxima of each order for all wavelengths coincide, resulting in achromatic fringes.

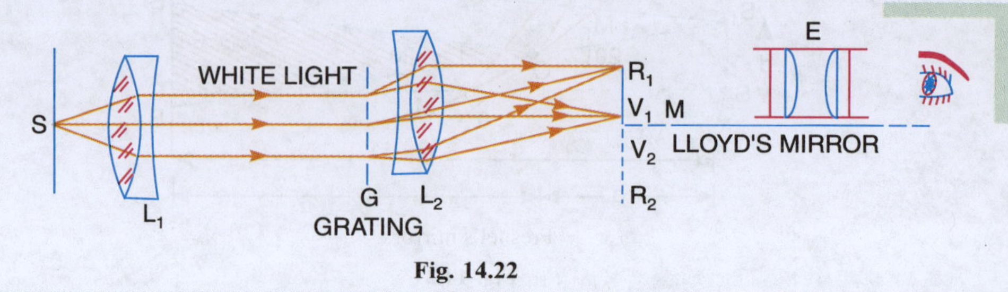

Fig. 14.22

In practice, achromatic fringes may be obtained as follows. S is a narrow source of white light at the focal plane of the converging lens L_1. A grating G having 800 to 1200 lines per cm is placed normal to light emerging from L_1. Another achromatic lens L_2 is used to form the second order spectrum on an opaque screen with a narrow opening in it. The narrow opening is adjusted so that only the first order spectrum is allowed to pass through it. The violet end is nearer to the highly polished Lloyd's mirror M than the red end. The position of M is adjusted such that V_2 and R_2 are the images of V_1 and R_1. Interference occurs between the beams from V_1R_1 and those from V_2R_2. The violet fringes are produced by V_1 and V_2 while red fringes are produced by R_1 and R_2.

Suppose $V_1V_2 = d_1$ and $R_1R_2 = d_2$.

If $\frac{\lambda_V}{d_1} = \frac{\lambda_R}{d_2}$, the fringe width β will be the same and interference fringes due to different colours will overlap and white achromatic fringes are produced in the field of view. The white and dark fringes are seen through the eyepiece or can be projected on a screen.

14.13. NON-LOCALIZED FRINGES

Point sources produce fringes, which can be seen at different distances from the source. As the screen is moved far from the source, the fringe spacing increases and conversely, when the screen is moved nearer to the source, the fringes come closer. Therefore, we say the fringes are **non-localized**. Narrow sources produce non-localized fringes.

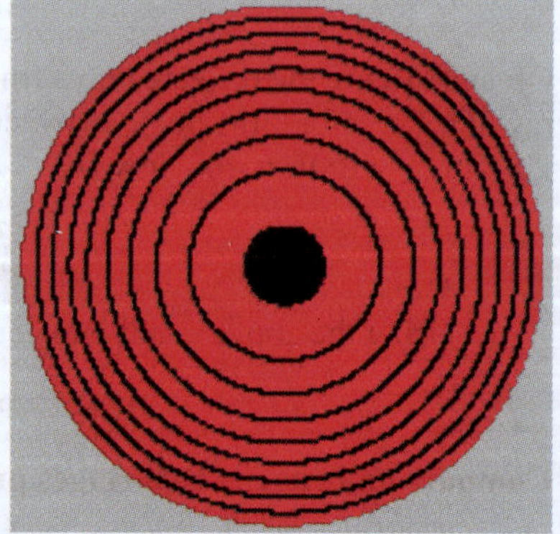

Non-localized fringes.

14.14. VISIBILITY OF FRINGES

The contrast of the interference fringes can be quantitatively described by the parameter called *visibility*. Visibility is defined as

$$V = \frac{I_{max} - I_{min}}{I_{max} + I_{min}} \tag{14.40}$$

the value of visibility varies between 0 and 1. When the fringes are of maximum intensity in the bright areas and totally dark in the dark areas, the visibility is equal to 1. As the phase difference increases, the coherence between the light waves decreases and the visibility is reduced. Finally, when the coherence between the two light waves disappears, I_{max} and I_{min} become equal and the visibility goes to zero. Then fringes are not observed and instead we observe uniform illumination.

From equ.(14.23) and (14.24), the visibility of the fringes produced by two beams can be expressed as

$$V = \frac{(E_1 + E_2)^2 - (E_1 - E_2)^2}{(E_1 + E_2)^2 + (E_1 - E_2)^2} = \frac{2E_1E_2}{E_1^2 + E_2^2} = \frac{2\sqrt{I_1 I_2}}{I_1 + I_2} = \frac{2\sqrt{I_1 / I_2}}{1 + I_1 / I_2} \tag{14.41}$$

It is seen from the above equation that the closer the intensities of the two waves, the higher is visibility of the fringes. When $I_1 = I_2$, V = 1. It may be noted that V is always equal to 1 when monochromatic light is used.

14.15. FRINGE PATTERN WITH WHITE LIGHT

When the slit is illuminated with white light in any interference experiment, we obtain a central white fringe flanked by a few coloured fringes. At the centre of the screen O, there is zero path difference and the bright fringes produced by all the colours there, add to each other. As a result a white fringe is produced at O. At other places away from O, the bright fringes in each order separate, because the fringe width is dependent on the wavelength ($\beta = \lambda D / d$). For example, the width of the red fringe is more than the blue fringe. Hence coloured fringes are produced on either side of the central white fringe.

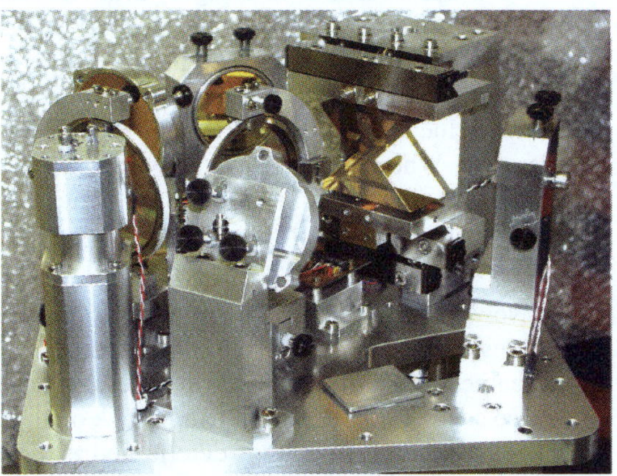

Interferometer is based on the principle of interference of light

14.16. INTERFEROMETRY

Instruments based on the principle of interference of light are known as **interferometers**. The instruments designed by Jamin and Rayleigh are used to determine the refractive index of gases and are known as *refractometers*.

14.16.1. JAMIN'S REFRACTOMETER

Jamin's refractometer consists of two exactly identical and optically plane glass blocks M_1 and M_2 (cut from the same block) and silvered on their back surfaces and arranged with their faces parallel to each other, as shown in Fig. 14.23. Light from an extended monochromatic source S incident on M_1 at an angle 45° is broken into two parallel rays (1) and (2) by reflection at the upper and lower surfaces of M_1. These two rays combine after suffering reflections at the two surfaces of M_2 as shown and form interference fringes as observed in a telescope. If the plates M_1 and M_2 are parallel, the light paths will be identical.

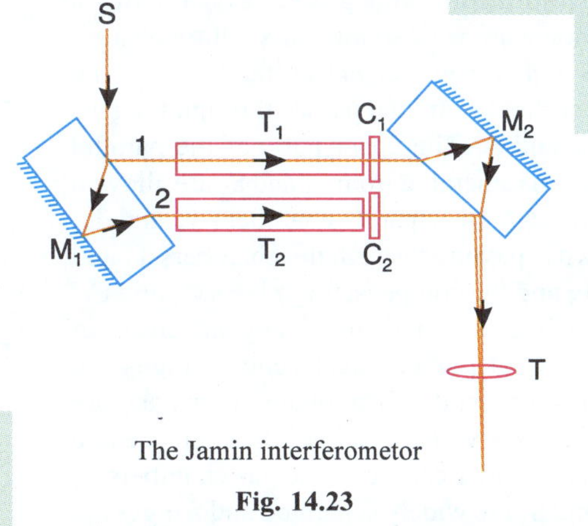

The Jamin interferometer

Fig. 14.23

Measurement of refractive index :

To measure the refractive index of a gas at a given temperature and pressure, two similar evacuated tubes T_1 and T_2 are placed in the paths of the two parallel beams. The gas whose refractive index is to be measured is then slowly allowed to enter into one of the tubes. As the gas enters, the optical path of the beam passing through the tube increases. The fringes therefore move past the cross-wire of the telescope. The fringes are counted until the gas entering the tube attains the given pressure and temperature.

If l is the length of the tube and μ the refractive index of the gas, then the change in the optical path difference due to the presence of the gas is $(\mu - 1)\, l$. If N be the number of fringes passed across the field of view, then

$$(\mu - 1)\, l = N\lambda \tag{14.42}$$

The refractive index, μ can be calculated using the above expression.

Compensator:

To avoid counting of fringes, a device called the "Jamin's compensator" is used. The compensator consists of two equally thick glass plates C_1 and C_2 cut from the same piece of glass and inclined at a small angle. The plates can be rotated together about a horizontal axis and the rotation is read on a divided circle D. One plate of the compensator is placed in the path of each beam. When the plates are equally inclined to the incident beam, the optical paths through the plates are the same. When the plates are rotated, the angles of incidence of the two beams change and a relative phase difference is introduced which varies as the compensator is rotated. By using monochromatic light and observing the passage of fringes across the field of view as the compensator is rotated, the scale can be calibrated to read the optical path difference in terms of wavelength.

Now, in the actual experiment, white light is used and the central achromatic fringe is adjusted on the cross wire by adjusting the compensator. The gas is then introduced in one of the tubes at the given temperature and pressure which results in shifting of the fringes. The compensator is now rotated so as to bring the central fringe back on the cross wire again. The change in the optical path difference due to this rotation is determined as the compensator is already calibrated. This must be equal to $(\mu - 1)\, l$, from which μ can be calculated.

14.16.2. RAYLEIGH REFRACTOMETER

The Rayleigh refractometer is mainly used to determine the refractive indices of inert gases and slight variation in the refractive index of solutions and gases. Light from a monochromatic source passes through a slit S and is then incident on the lens L_1. The parallel beam then passes through the slits S_1 and S_2. The upper parts of the parallel beams emerging from S_1 and S_2 are allowed through the separate chambers T_1 and T_2. After passing through the chambers T_1 and T_2 and the compensating plates C_1 and C_2, the beams are recombined by the lens L_2 in its focal plane. Interference fringes are obtained in the focal plane of lens and are observed with the help of the eyepiece or a telescope. Because of the gas chambers S_1 and S_2 are widely separated and fringes are closely spaced. The lens L_2 forms a *second*

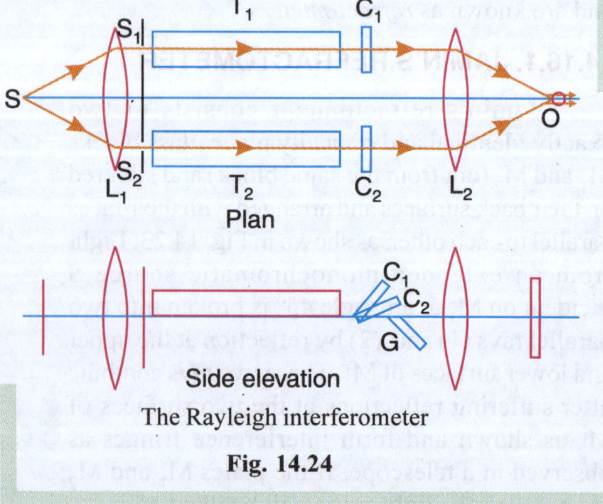

The Rayleigh interferometer

Fig. 14.24

system of interference pattern, which is due to the superposition of the lower parts of the two beams emerging from S_1 and S_2 and passing beneath the chambers T_1 and T_2. The upper edge of the lower fringe system is made to coincide with the lower edge of the upper system with the help of an inclined thick plate P (not shown in the figure) held across the lower parts of the slits. The centres of the two sets of interference patterns coincide and the fringe spacing in the two sets is the same. The first fringe system shifts with the change of gas pressure etc whereas the second fringe system remains *stationary*.

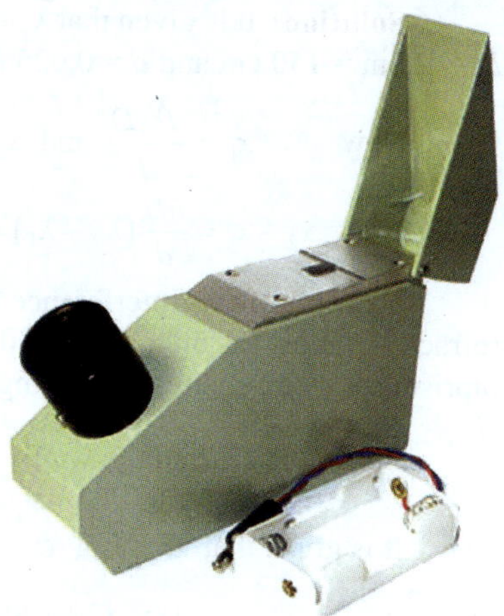

It is used to determine the refractive index of inert gases.

The circular disc D attached to the compensating plates C_1 and C_2 is previously calibrated in terms of wavelength and refractive index. Initially, both the tubes T_1 and T_2 are evacuated. Using white light, the central white fringe is observed in the field of view of the eyepiece. The gas at a known pressure and temperature is introduced into the tube T_1. The central white fringe shifts from the field of view.

By rotating the circular disc D, which in turn displaces the compensating plates C_1 and C_2, the central white fringe is brought back to the centre of the field of view. The number of wavelengths graduated on the circular disc is noted. The change in the path difference is

$$(\mu - 1)\, l = N\lambda$$

where N is the number of fringes that crossed the field of view. Knowing l, N and λ, the refractive index μ can be calculated.

WORKED OUT PROBLEMS

Example 14.1: Green light of wavelength 5100 Å from a narrow slit is incident on a double slit. If the overall separation of 10 fringes on a screen 200 cm away is 2 cm, find the slit separation.

Solution: The fringe width $\beta = \dfrac{\lambda D}{d}$

It is given that $D = 200$ cm, $\lambda = 5100 \times 10^{-8}$cm and $10\,\beta = 2$ cm.

$\therefore \qquad \beta = 0.2$ cm.

The slit separation $d = \dfrac{\lambda D}{\beta} = \dfrac{5100 \times 10^{-8}\, cm \times 200 cm}{0.2 cm} =$ **0.05 cm.**

Example 14.2: Two coherent sources are 0.18 mm apart and the fringes are observed on a screen 80 cm away. It is found that with a certain monochromatic source of light, the fourth bright fringe is situated at a distance of 10.8 mm from the central fringe. Calculate the wavelength of light.

Solution: The distance of the nth fringe from the central fringe is $x = \dfrac{n\lambda D}{d}$.

It is given that D = 80 cm, d = 0.18 mm = 0.018 cm, x = 10.8 mm = 1.08 cm and n = 4.

$\therefore \qquad \lambda = \dfrac{xd}{n\,D} = \dfrac{1.08\,\text{cm} \times 0.018\,\text{cm}}{4 \times 80\,\text{cm}} = 6075 \times 10^{-8}\text{cm} =$ **6075 Å.**

Example 14.3: A light source emits light of two wavelengths 4300Å and 5100Å. The source is used in a double slit experiment. The distance between the sources and the screen is 1.5 m and the distance between the slits is 0.025mm. Calculate the separation between the third order bright fringes due to these two wavelengths.

Solution: It is given that $\lambda_1 = 4300\text{Å} = 4300 \times 10^{-8}\text{cm}$, $\lambda_2 = 5100\text{Å} = 5100 \times 10^{-8}\text{cm}$, $n = 3$, $D = 1.5\text{ m} = 150\text{ cm}$ and $d = 0.025\text{mm} = 0.0025\text{cm}$.

Now $x_1 = \dfrac{n\lambda_1 D}{d}$ and $x_2 = \dfrac{n\lambda_2 D}{d}$

$$\therefore \quad x_2 - x_1 = \frac{nD}{d}(\lambda_2 - \lambda_1) = \frac{3 \times 150\text{ cm}}{0.0025\text{ cm}}(5100 - 4300) \times 10^{-8}\text{ cm} = \mathbf{1.44\ cm}$$

Example 14.4: Interference fringes are observed with a biprism of refracting angle 1° and refractive index 1.5 on a screen 80 cm away from it. If the distance between the source and the biprism is 20 cm, calculate the fringe width when the wavelength of light used is 6900Å.

Solution: The fringe width $\beta = \dfrac{\lambda D}{d}$ and $d = 2(\mu - 1)\alpha a$

It is given that $\mu = 1.5$, $\alpha = 1° = \dfrac{\pi}{180°}$, $a = 20$ cm and $b = 80$ cm, $\lambda = 6900 \times 10^{-8}$cm

$\therefore \quad D = (20 + 80)\text{ cm} = 100\text{ cm}.$

$$\beta = \frac{\lambda D}{2(\mu - 1)\alpha a} = \frac{(6900 \times 10^{-8}\text{ cm}) \times 100\text{ cm}}{2(1.5 - 1) \times (\pi/180) \times 20\text{ cm}} = \mathbf{0.02\ cm}$$

Example 14.5: In a biprism experiment the eyepiece is placed at a distance of 1.2 m from the source. The distance between the virtual sources was found to be 7.5×10^{-4}m. Find the wavelength of light, if the eyepiece is to be moved transversely through a distance of 1.888cm for 20 fringes.

Solution: The fringe width $\beta = \dfrac{\lambda D}{d}$; But $\beta = \dfrac{l}{n}$ $\therefore\ \lambda = \dfrac{ld}{nD}$

It is given that $l = 1.888$cm $= 0.01888$m, $d = 7.5 \times 10^{-4}$m, $n = 20$ and D = 1.2m.

$$\therefore \quad \lambda = \frac{0.01888\text{m} \times 7.5 \times 10^{-4}\text{m}}{20 \times 1.2\text{m}} = 5900 \times 10^{-10}\text{ m} = \mathbf{5900Å}$$

Example 14.6: A thin sheet of a transparent material ($\mu = 1.60$) is placed in the path of one of the interfering beams in a biprism experiment using sodium light, $\lambda = 5890$ Å. The central fringe shifts to a position originally occupied by the 12th bright fringe. Calculate the thickness of the sheet.

Solution: The thickness of the sheet, $t = \dfrac{n\lambda}{\mu - 1}$

It is given that $\mu = 1.60$, $n = 12$, $\lambda = 5890$ Å.

$$\therefore \quad t = \frac{12 \times 5890 \times 10^{-10}\text{ m}}{1.60 - 1} = 1.18 \times 10^{-5}\text{ m} = \mathbf{0.12\ \mu m}$$

Example 14.7: When a thin sheet of transparent material of thickness 6.3×10^{-4}cm is introduced in the path of one of the interfering beams, the central fringe shifts to a position occupied by the sixth fringe. If $\lambda = 5460$ Å, find the refractive index of the sheet.

Solution: $(\mu - 1)\,t = n\lambda$ $\quad\therefore$ The refractive index of the sheet $\mu = \dfrac{n\lambda}{t} + 1$.

It is given that $t = 6.3 \times 10^{-4}$cm, $n = 6$ and $\lambda = 5460$ Å $= 5460 \times 10^{-8}$cm.

$$\therefore \quad \mu = \frac{6 \times 5460 \times 10^{-8}\text{ cm}}{6.3 \times 10^{-4}\text{ cm}} + 1 = 1.52$$

Example 14.8: In Lloyd's single mirror interference experiment, the slit source is at a distance of 2 mm from the plane of the mirror. The screen is kept at a distance of 1.5 m from the source. Calculate the fringe width. Wavelength of light is 5890 Å.

Solution: The fringe width $\beta = \frac{\lambda D}{d}$

It is given that λ = 5890 Å = 5890×10^{-10}m, D = 1.5 m and $d/2$ = 2mm. $\therefore d = 4 \times 10^{-3}$m.

$$\beta = \frac{5890 \times 10^{-10}\,\text{m} \times 1.5\text{m}}{4 \times 10^{-3}\,\text{m}} = 22\text{mm}$$

QUESTIONS

1. What are the conditions necessary for observing interference fringes?
2. Why is the condition of coherence necessary to observe interference fringes?
3. Is it possible to observe interference fringes with light emanating from two independent sources? If not, why?
4. How can coherent sources be obtained in practice?
5. Is it necessary that the interfering waves should have the same frequency? If so, why?
6. Is it necessary that the interfering waves should have equal amplitudes? Explain.
7. What are coherent sources? How are they realized in practice? Describe a method for determining the refractive index of a gas using the interference phenomenon. **(Madhurai Kamaraj, 2003)**
8. How would you determine the wavelength of light with the Lloyd's mirror experiment? In what respect do the fringes in this case differ from those obtained with Fresnel's biprism? How would you obtain achromatic fringes with this arrangement?
9. Describe Fresnel's biprism. Explain how the wavelength of light can be determined with its help.
10. What are coherent sources? Explain the importance of such sources in interference phenomenon. Two coherent sources form interference fringes. Obtain an expression for the distance between two consecutive bright fringes. **(Nagpur, 2004)**
11. What are coherent sources? Discuss why two independent sources of light of the same wavelength cannot produce interference fringes? Give a diagram showing clearly how coherent sources are produced in a biprism. Derive the formula for the fringe width in the biprism experiment.
12. What is meant by interference of light? State the fundamental conditions for the production of interference fringes.
13. State the basic conditions for the phenomenon of interference of light. Briefly discuss the effect of introducing a thin plate in the path of one of the interference beams in a biprism experiment. Deduce an expression for the displacement of the fringes. Show how this method is used for finding the thickness of a mica sheet.
14. Discuss the conditions for interference. Describe Young's experiment and derive an expression for (i) intensity at a point on the screen and (ii) fringe width. **(Punjab, 2005; Nagpur 2004)**
15. Derive an expression for the resultant intensity when two coherent beams of light are superposed.
    What is the visibility of fringes:-
    (*a*) for two slits of equal intensities?
    (*b*) if intensity of one slit is 4 times the other?
    What will be the intensity when the two sources are incoherent?
16. Show that the distance between the two virtual coherent sources in Fresnel's biprism arrangement is $2d(n-1)\theta$ where d is the distance between the source and the biprism, θ is the angle of the biprism and n is the refractive index of the material of the prism.

17. State the principle of superposition of light waves.
18. Why two independent sources can not produce observable interference pattern?
19. How are the fringes of equal inclination obtained? **(Nagpur, 2005)**
20. Derive the equation for optical path difference between two reflected rays from a thin film. **(Nagpur, 2005)**
21. Explain optical path of light in denser medium. **(Nagpur, 2005)**
22. Draw a well-labelled diagram of Rayleigh interferometer. **(Nagpur, 2004)**
23. What are the conditions for sustained interference pattern? **(Kovempu, 2005)**
24. How are coherent sources formed in the Biprism? On what factors does the separation between coherent sources depend?
25. State and explain conditions for the interference of light. **(Garhwal, 2005)**
26. Calculate the displacement of fringes when a thin transparent film is introduced in the path of one of the interfering beams in biprism.
27. State the principle of superposition of light waves.
28. Explain the phase change when light is reflected from a denser surface. **(RTMNU, 2010)**
29. Explain the conditions for interference of light. **(A.P.University, 2010)**
30. (*i*) Explain lateral displacement of fringes in double-slit experiment.
    (*ii*) How is it useful to determine the thickness of a sample? **(RTMNU, 2010)**

PROBLEMS FOR PRACTICE

1. Two coherent sources, whose intensity ratio is 9:4, produce interference fringes. Deduce the ratio of maximum to minimum intensity of the fringe system. **(Purvanchal, 2002)** [**Ans:** 25:1]
2. Interference fringes are formed by a biprism whose acute angle is 20' and refractive index is 1.5. The slit is at 10 cm from the biprism and is illuminated by light of wavelength 6000 Å. Find the fringe width on a screen placed at a distance of one metre from the biprism. **(Lucknow, 2000)** [**Ans:** 1.1 mm]
3. A biprism is placed 5 cm from a slit illuminated by sodium light ($\lambda = 5890$ Å). The width of the fringes obtained on a screen 75 cm from the biprism is 9.424×10^{-2} cm. What is the distance between the two coherent sources? **(Kanpur, 2002)** [**Ans:** 0.5 mm]
4. A biprism forms interference fringes with monochromatic light of wavelength 5450 Å. On introducing a thin glass plate ($\mu = 1.5$) in the path of one of the interfering beams, the central fringe shifts to the position previously occupied by the third bright fringe. Does fringe width change? Find the thickness of the plate. **(Awadh, 2001)** [**Ans:** 0.03 μm]
5. A parallel beam of light of wavelength 5890 Å is incident on a thin glass plate of refraction in glass is 60º, calculate the smallest thickness of the plate, which will appear dark by reflected light. **(Nagpur, 2005)**
6. A Biprism of obtuse angle 176º is made of glass of Refractive index 1.5. A slit illuminated with monochromatic light is placed 20 cm. behind and the width of interference fringes formed on a screen 80 cm in front of Biprism is found to be 9.25×10^{-3} cm. Calculate the wavelength of light. **(Lucknow, 2004)**
7. A thin mica sheet ($\mu = 1.6$) of 7×10^{-4} cm thickness introduced in the path of one of the interfering beams in a biprism arrangement shifts the central fringe to a position occupied by the 7th bright fringe from the center. Find the wavelength of the light used. **(Garhwal, 2005)**
8. In a double slit experiment, when a thin plate of transparent material is introduced in the path of one of the interfering beams, the central fringe is displaced by 3.6 fringes. Calculate the thickness of the plate. Given $\mu = 1.4$, $\lambda = 5500$Å. **(RTMNU, 2010)**

15 CHAPTER

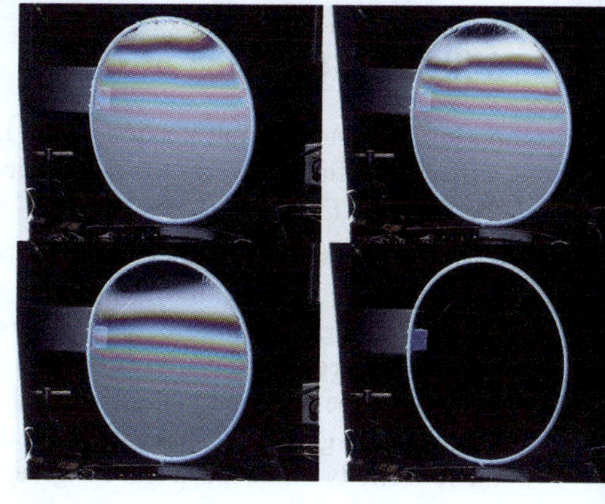

Interference In Thin Films

15.1. THIN FILM

An optical medium is called a **thin film** when its thickness is about the order of 1 wavelength of light in visible region. Thus, a film of thickness in the range 0.5 μm to 10 μm may be considered as a thin film. A thin film may be a thin sheet of transparent material such as glass, mica, an air film enclosed between two transparent plates or a soap bubble. When light is incident on such a

Soap bubble.

film, a small part of it gets reflected from the top surface and a major part is transmitted into the film. Again, a small part of the transmitted component is reflected back into the film by the bottom surface and the rest of it emerges

At a Glance

- Thin Film
- Plane Parallel Film
- Interference Due to Transmitted Light
- Haidinger Fringes
- Variable Thickness (Wedge-Shaped) Film
- Newton's Rings
- Michelson's Interferometer
- Applications of Michelson Interferometer
- Twyman and Green Interferometer
- Mach-Zehnder Interferometer
- Multiple Beam Interference
- Fabry-Perot Interferometer and Etalon
- Lummer and Gehrcke Plate
- Applications of Thin Film Interference
- Antireflection Coatings
- Dielectric Mirrors
- Interference Filters

out of the film. A small portion of the light thus gets reflected partially several times in succession within the film (see Fig. 15.1).

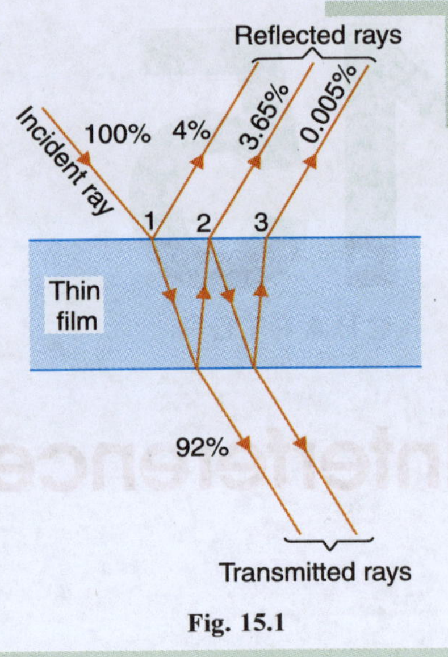

Fig. 15.1

In transparent thin films, the two bounding surfaces strongly transmit light and only weakly reflect the incident light. Therefore, only the first reflection at the top surface and the first reflection at the bottom surface will be of appreciable strength. For example, if we consider a glass plate, having a refractive index 1.52, the reflectivity of the top surface is given by

$$r = \left[\frac{1.52-1}{1.52+1}\right]^2 = 0.042$$

It means that about 4% of the incident light is reflected by the top surface of the glass plate, while 96% of it is transmitted into the plate. Out of the light reaching the bottom surface, again 3.8% is reflected and 92% is transmitted out of the plate. Then, again out of the 3.8% of the light 0.15% is reflected at the inner boundary of the top surface and about 3.65% is transmitted out into the air. After two reflections, the intensity will become insignificantly small. At each reflection, the intensity and hence the *amplitude of light wave is divided* into a reflected component and a refracted component. The reflected and refracted components travel along different paths and subsequently overlap to produce interference. Therefore, the interference in thin films is called interference by division of amplitude. Newton and Robert Hooke first observed the thin film interference. However, Thomas Young gave the correct explanation of the phenomena. A thin film may be uniform or non-uniform in its structure. However, as long as its thickness lies within the specified limits, interference of light occurs.

15.2. PLANE PARALLEL FILM

A transparent thin film of uniform thickness bounded by two parallel surfaces is known as a *plane parallel thin film.*

When light is incident on a parallel thin film, a small portion of it gets reflected from the top surface and a major portion is transmitted into the film. Again, a small part of the transmitted component is reflected back into the film by the bottom surface and the rest of it is transmitted from the lower surface of the film. Thin films transmit incident light strongly and reflect only weakly. After two reflections, the intensities of reflected rays drop to a negligible strength. Therefore, we consider the first two reflected rays only (see Fig. 15.2). These two rays are derived from the same incident ray but appear to come from two sources located below the film. The sources are virtual coherent sources. The reflected waves 1 and 2 travel along parallel paths and interfere at infinity. This is a case of *two-beam* interference.

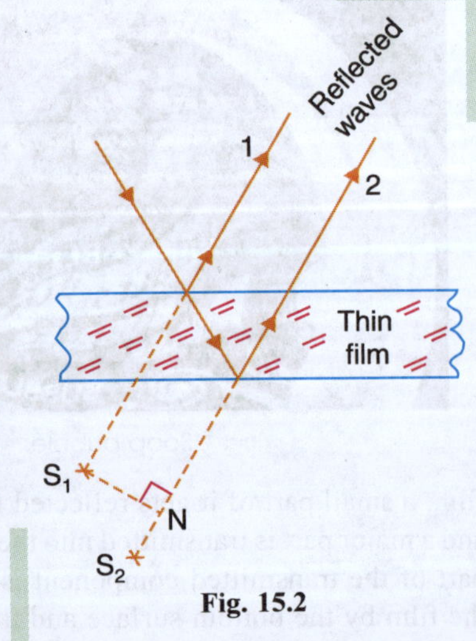

Fig. 15.2

The condition for maxima and minima can be deduced once we have calculated the optical path difference between the two rays at the point of their meeting.

15.2.1. INTERFERENCE DUE TO REFLECTED LIGHT

Let us consider a transparent film of uniform thickness 't' bounded by two parallel surfaces as shown in Fig. 15.3. Let the refractive index of the material be μ. The film is surrounded by air on both the sides. Let us consider plane waves from a monochromatic source falling on the thin film at an angle of incidence *'i'*. Part of a ray such as AB is reflected along BC, and part of it is transmitted

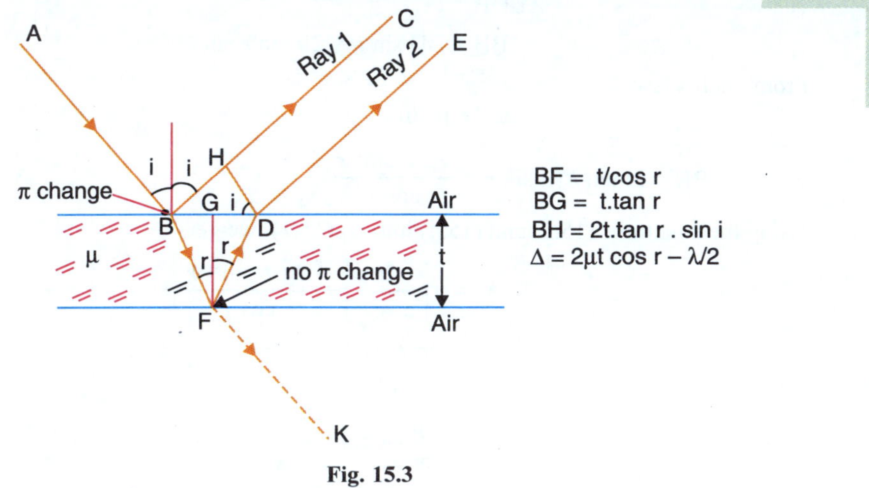

Fig. 15.3

into the film along BF. The transmitted ray BF makes an angle 'r' with the normal to the surface at the point G. The ray BF is in turn partly reflected back into the film along FD while a major part refracts into the surrounding medium along FK. Part of the reflected ray FD is transmitted at the upper surface and travels along DE. Since the film boundaries are parallel, the reflected rays BC and DE will be parallel to each other. The waves travelling along the paths BC and BFDE are derived from a single incident wave AB. Therefore they are coherent and can produce interference if they are made to overlap by a condensing lens or the eye.

(*i*) Geometrical Path Difference:

Let DH be normal to BC. From points H and D onwards, the rays HC and DE travel equal path. The ray BH travels in air while the ray BD travels in the film of refractive index μ along the path BF and FD. The geometric path difference between the two rays is

$$BF + FD - BH.$$

(*ii*) Optical Path Difference:

Optical path difference $\Delta_a = \mu L$

$$\therefore \quad \Delta_a = \mu\,(BF+FD) - 1(BH) \tag{15.1}$$

∴ In the ΔBFD, ∠BFG = ∠GFD = ∠r

$$BF = FD$$

$$BF = \frac{FG}{\cos r} = \frac{t}{\cos r}$$

$$\therefore \quad BF + FD = \frac{2t}{\cos r} \tag{15.2}$$

Also, $BG = GD$

$$\therefore \quad BD = 2BG$$

$$BG = FG \tan r = t \tan r$$

$\therefore$ $$BD = 2\,t\,\tan r$$

In the Δ^{le} BHD $$\angle HBD = (90 - i)$$

$$\angle BHD = 90^0$$

$\therefore$ $$\angle BDH = i$$

$\therefore$ $$BH = BD \sin i = 2\,t \tan r \sin i \qquad (15.3)$$

From Snell's law,

$$\sin i = \mu \sin r$$

$\therefore$ $$BH = 2t \tan r\,(\mu \sin r) = \frac{2\mu\, t\, \sin^2 r}{\cos r} \qquad (15.4)$$

Using the equations (15.2) and (15.4) into eq.(15.1), we get

$$\Delta_a = \mu\left[\frac{2t}{\cos r}\right] - \left[\frac{2\mu\, t\, \sin^2 r}{\cos r}\right]$$

$$= \frac{2\mu\, t}{\cos r}\left[1 - \sin^2 r\right]$$

$$= \frac{2\mu\, t}{\cos r}\cos^2 r$$

$\therefore$ $$\Delta_a = 2\mu\, t\, \cos r \qquad (15.5)$$

(*iii*) Correction on account of phase change at reflection:

When a ray is reflected at the boundary of a rarer to denser medium, a path change of λ/2 occurs for the ray BC (see Fig.15.3). There is no path difference due to transmission at D. Including the change in path difference due to reflection, the true path difference

$$\Delta_t = 2\,\mu\, t \cos r - \frac{\lambda}{2} \qquad (15.6)$$

15.2.2. CONDITIONS FOR MAXIMA (BRIGHTNESS) AND MINIMA (DARKNESS)

Maxima occur when the optical path difference Δ = m λ. If the difference in the optical path between the two rays is equal to an *integral number of full waves*, then the rays meet each other in phase. The crests of one wave falls on the crests of the others and the waves *interfere constructively*. Thus, when

$$2\mu\, t\, \cos r - \frac{\lambda}{2} = m\lambda \qquad (15.7)$$

the reflected rays undergo constructive interference to produce brightness or maxima at the point of their meeting.

$$2\mu\, t\, \cos r = m\lambda + \lambda/2$$

or $$2\mu\, t\, \cos r = (2m+1)\lambda\,/2 \qquad \textbf{Condition for Brightness} \qquad (15.8)$$

Minima occur when the optical path difference is $\Delta = (2m+1)\,\lambda\,/\,2$. If the difference in the optical path between the two rays is equal to an *odd integral number of half-waves*, then the rays meet each other in opposite phase. The crests of one wave falls on the troughs of the others and *the waves interfere destructively*. Thus, when

$$2\mu\, t\, \cos r - \lambda\,/2 = (2m+1)\,\lambda\,/2 \qquad (15.9)$$

the reflected rays undergo destructive interference to produce darkness. Equ.(15.9) may be rewritten as

$$2\mu\, t \cos r = (m+1)\,\lambda$$

The phase relationship of the interfering waves does not change if one full wave is added to or subtracted from any of the interfering waves. Therefore (m + 1)λ can as well be replaced by mλ for simplicity in expression. Thus,

$$2\mu\, t \cos r = m\lambda \qquad \textbf{Condition for Darkness} \qquad (15.10)$$

15.2.3. SOME IMPORTANT POINTS

(a) It is seen that the conditions of interference depend on four parameters, namely μ, t, λ and r. In the case of constant thickness (parallel) film, (μt) is constant. When a parallel beam of light is incident on such a film, r also remains constant. Then the interference conditions solely depend on the wavelength λ.

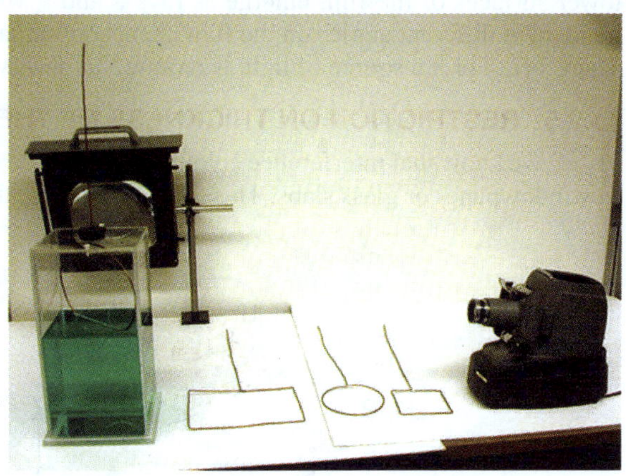

Thin film interference -soap films.

(b) When monochromatic light falls on a parallel beam, the whole film will appear *uniformly* dark or bright. If the condition of constructive interference is satisfied, the film will show intense colour corresponding to the incident light.

(c) If a parallel beam of white light falls on a parallel film, those wavelengths for which the path difference is mλ, will be absent from the reflected light. The other colours will be reflected. Therefore, the film will appear uniformly coloured with one colour being absent.

15.2.4. NARROW LIGHT SOURCE VERSUS EXTENDED LIGHT SOURCE

In case of Fresnel's biprism and Lloyd's mirror, interference fringes were produced by two coherent sources. The initial source is narrow. The fringes obtained on a screen are viewed with an eyepiece. In case of a thin film, a narrow source limits the area of the film that can be viewed. Consider a thin film illuminated by a narrow source of light S (Fig. 15.4). The ray 1 produces

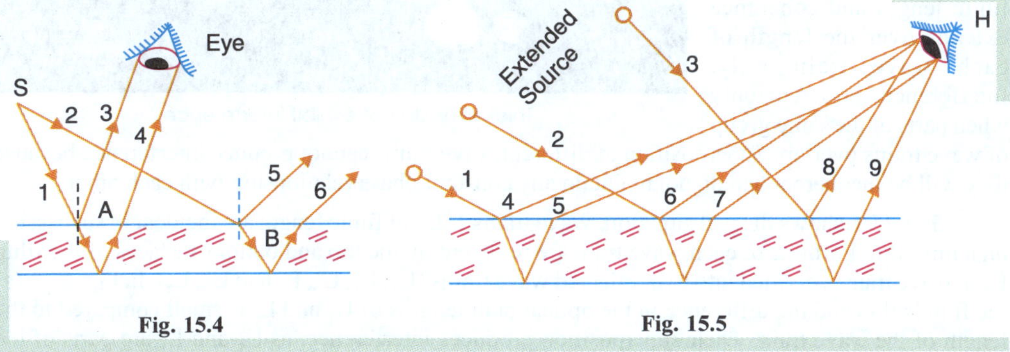

Fig. 15.4

Fig. 15.5

interference fringes because rays 3 and 4 reach the eye. The ray 2 is incident on the surface of the film at a different angle and is reflected along 5 and 6. The rays 5 and 6 do not reach the eye. Similar is the case for other rays incident at different angles on the film surface. The reflected rays do not reach the eye. Thus, *only* the portion A of the film is visible to the eye.

If an extended (or broad) light source is used to illuminate the film, as in Fig.15.5, a larger area of the film surface is observed. The ray 1 after reflection from the upper and the lower surface of the film emerges as rays 4 and 5, which reach the eye. Ray 2 from some other point of the source after reflection from the upper and lower surfaces of the film emerge as rays 6 and 7 which also reach the eye. Also, ray 3 from some other point of the source after reflection from the upper and lower surfaces of the film emerge as rays 8 and 9 which also reach the eye. Therefore, the rays incident at different angles on the film are accommodated by the eye and the field of view is large. Therefore, a broad source of light is required to observe interference in thin films.

15.2.5. RESTRICTION ON THICKNESS OF THE FILM

We know that interference colours are observed only in thin films but not in thick plates such as windowpanes or glass slabs. This is due to the fact that light waves can interfere only when both

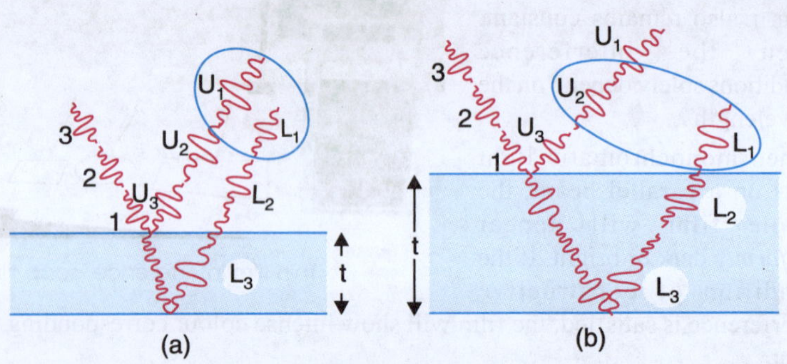

Fig. 15.6

the conditions of temporal and spatial coherence are satisfied. In Fig. 15.3 we have assumed that a monochromatic wave of infinite length is incident on the film. In reality, the incident light consists of wave trains of finite length and coherence extends over the length of each wave train only. Interference can occur only when parts of the same group of wave trains overlap. Superposition of different wave trains cannot produce interference because they will be incoherent and do not maintain any constant phase relationship with each other.

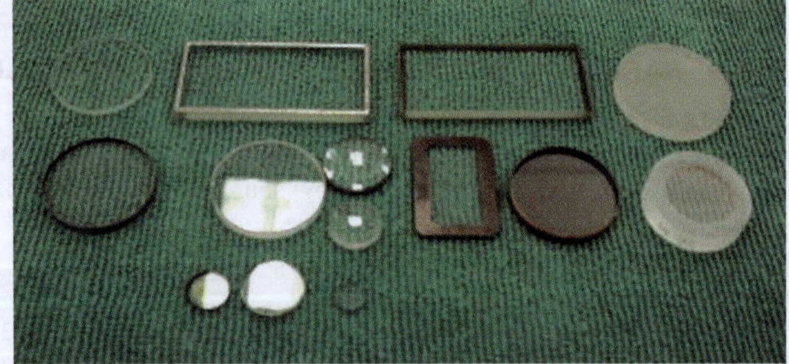

Thick films do not exhibit interference.

Fig. 15.6 shows the real situation. Wave trains 1,2,3 of finite length are incident in succession on a thin film. Portions of each wave train are reflected by the top and bottom surfaces of the film. Each wave train is divided into two reflected wave trains (U_1, L_1, U_2, L_2 and U_3, L_3). In Fig. 15.6 (a) the film is thin and the difference in the optical path lengths of U_1 and L_1 is small compared to the length of the wave train. Their superposition produces interference, as U_1 and L_1 are parts of the same wave train 1 and hence are coherent. In Fig.15.6 (b) the film is thicker and the optical path difference between U_1 and L_1 is large than the coherence length. Consequently, superposition takes place between parts of different wave trains, U_2 and L_1 and U_3 and L_2. Therefore interference does not take place.

It implies that interference occurs *only when* the optical path difference, Δ, between the superposing waves is less than the coherence length (see § 16.3).

i.e., $\Delta << l_{coh}$ (15.11)

∴ $(2\mu t \cos r - \lambda/2) << l_{coh}$ (15.12)

But $$l_{coh} = \frac{\lambda^2}{\Delta\lambda}$$ (Refer to equ. 16.11)

∴ $(2\mu t \cos r - \lambda/2) < \lambda^2/\Delta\lambda$

Rearranging the terms, we obtain

$$t < \frac{\lambda\left[\frac{\lambda}{\Delta\lambda} + \frac{1}{2}\right]}{2\mu \cos r} \tag{15.13}$$

$\lambda/\Delta\lambda >> 1/2$ and for normal incidence $\cos r = 1$.

∴ $$t < \frac{\lambda^2}{2\mu\Delta\lambda} \tag{15.14}$$

The above equation indicates that *interference in thin film will be observed if the thickness of the film is less than the coherence length of the incident light waves.* Normally, the coherence length of the light from ordinary sources is of the order of a fraction of a millimeter. Therefore, interference is seen with the films of thickness of the order of a few hundred microns only. It is because of this reason that thick films do not exhibit interference.

15.3. INTERFERENCE DUE TO TRANSMITTED LIGHT

Consider a thin transparent film of thickness t and refractive index μ. A ray SA after refraction goes along AB. At B it is partly reflected along BC and partly refracted along BR. The ray BC, after reflection at C, finally emerges along DQ. Here at B and C reflection takes place at the rarer medium. Therefore, no phase change occurs. Draw BM normal to CD and DN normal to BR. The optical path difference between DQ and BR is given by

$$\Delta = \mu(BC + CD) - BN$$

Also, $\mu = \frac{\sin i}{\sin r} = \frac{BN}{MD}$ or $BN = \mu.MD$

In Fig.15.7, ∠BPC = r and CP = BC = CD

∴ BC + CD = PD

∴ Δ = μ(PD) – μ (MD) = μ(PD – MD) = μ.PM

In ΔBPM, $\cos r = \frac{PM}{BP}$ or PM = BP. $\cos r$

But BP = 2 t

∴ PM = 2 $t \cos r$

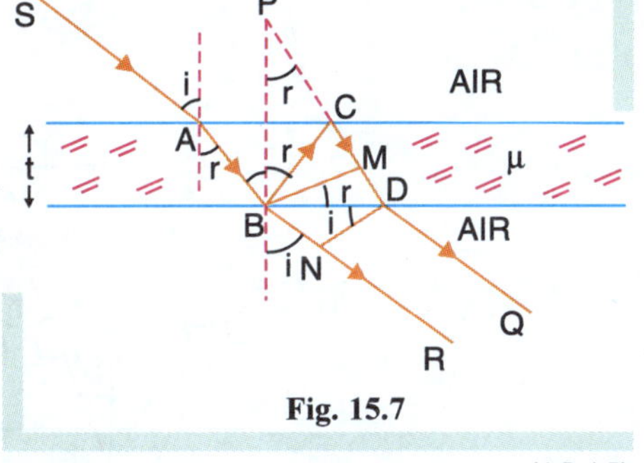

Fig. 15.7

∴ $$\Delta = \mu.PM = 2\mu t \cos r \tag{15.15}$$

Bright Fringes:

When the optical path difference $\Delta = m\lambda$, bright fringe occurs.

∴ $$2\mu t \cos r = m\lambda \tag{15.16}$$

where m = 0,1,2,3,....

Dark Fringes:

When the optical path difference $\Delta = (2m + 1)\,\lambda/2$, dark fringe occurs.

$$\therefore \qquad 2\mu\, t\, \cos r = \frac{(2m+1)}{2}\lambda \qquad (15.17)$$

where m = 0,1,2,3,….

In case of transmitted light, the fringes are less distinct because the difference in amplitudes of BR and DQ is very large. However, when the angle of incidence is nearly 45° the fringes are more distinct.

15.4. HAIDINGER FRINGES

In thin films interference fringes are produced due to the path difference $2\mu\, t \cos r$ between the overlapping rays. For a given film the path difference may arise due to (*i*) the angle of refraction *r* inside the film or (*ii*) the change in thickness. We can express the change in path difference by differentiating the expression $2\,\mu\, t \cos r$.

$$\text{Change in path difference, } \delta(\Delta) = 2\mu\, t \cdot \delta(\cos\, r) + 2\mu\, \cos\, r\, (\delta\, t) \qquad (15.18)$$

When the film is of **uniform** (constant) thickness, the change in path difference is only due to the change in *r*. If the thickness of the film is large, the path difference will change appreciably

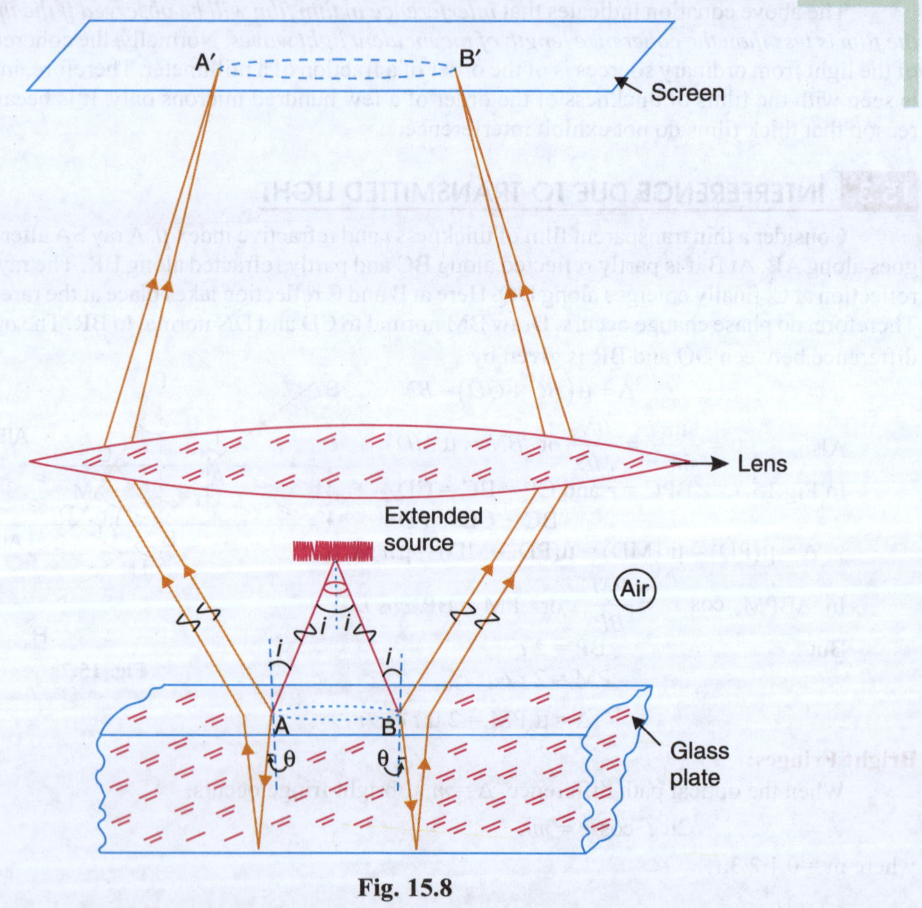

Fig. 15.8

even when r changes in a small way. Fringes are produced in this case due to the superposition of rays, which are equally inclined to the normal. These fringes are called **fringes of equal inclination**. The fringes of equal inclination are known as **Haidinger fringes.** In this case all the pairs of interfering rays of equal inclination pass through the plate as a parallel beam and hence meet at infinity. The other pairs of differrent inclination meet at different points at infinity. Therefore, they can be located with a telescope focussed to infinity. The fringes are therefore said to be *localized at infinity*.

To produce Haidinger fringes, the source must be an extended source, the film thickness must be appreciably large and the observing instrument is to be focussed for parallel rays.

Fig. 15.8 shows the formation of Haidinger fringes. Let us consider that a thin plate is illuminated by an extended monochromatic light source. A lens is arranged parallel to the plate and a screen is kept in the focal plane of the lens. Light from the extended source is incident on the plate in diverse directions. The waves propagating parallel to the plane of the page and falling on the plate at an angle i at points A and B get reflected from the top and bottom surfaces of the plate. The reflected pairs of waves will meet at points A′ and B′ respectively on the screen due to the focussing action of the lens. Depending on their path difference, the reflected waves produce either brightness or darkness on the screen. In fact the waves incident at the top surface of the plate at an angle i travel along the generators of a cone as shown in Fig. 15.8. Each pair of parallel reflected waves interfere at diametrically opposite points. Thus, a circular fringe is produced. Similarly, the waves incident at a different angle will produce a collection of identical points arranged along a circle of another radius. As a result, a system of alternating bright and dark circular fringes with a common centre will be observed on the screen.

Each fringe is characterized by a particular value of m. Bright fringes are produced when the condition $2\,\mu\,t\cos r = m\lambda$ is satisfied; and dark fringes are produced where the condition

$2\,\mu\,t\cos r = (2m+1)\,\lambda/2$ is satisfied. The parallel pairs of reflected rays meet only at infinity; therefore a lens is used to focus them. Accordingly, these fringes of equal inclination are said to be localized at infinity.

15.5. VARIABLE THICKNESS (WEDGE-SHAPED) FILM

Let us now study the interference of light in a film of varying thickness. *A thin film having zero thickness at one end and progressively increasing to a particular thickness at the other end is called a wedge.* A thin wedge of air film can be formed by two glass slides resting on each other at one edge and separated by a thin spacer at the opposite edge.

The arrangement for observing the interference pattern in a wedge shaped film is shown in Fig. 15.9(a). The wedge angle is usually very small and of the order of a fraction of a degree. When a parallel beam of *monochromatic* light illuminates the wedge from above, the rays reflected from its two bounding surfaces will not be parallel. They appear to diverge from a point near the film. The path difference between the rays reflected from the upper and lower surfaces of the air film varies along its length due to variation in film thickness. Therefore, alternate bright and dark fringes are observed on its top surface (see Fig. 15.9b). The fringes are localized at the top surface of the film.

When the light is incident on the wedge from above, it gets partly reflected from the glass-to-air boundary at the top of the air film. Part of the light is transmitted through the air film and gets reflected partly at the air-to-glass boundary, as shown in Fig. 15.10. The two rays BC and DE, thus reflected from the top and bottom of the air film, are coherent as they are derived from the same ray AB through *division of amplitude*. The rays are close enough if the thickness of the film is of the order of a wavelength of light. For small film thickness the rays interfere producing darkness or brightness depending on the phase difference. The thickness of the glass plates is large compared

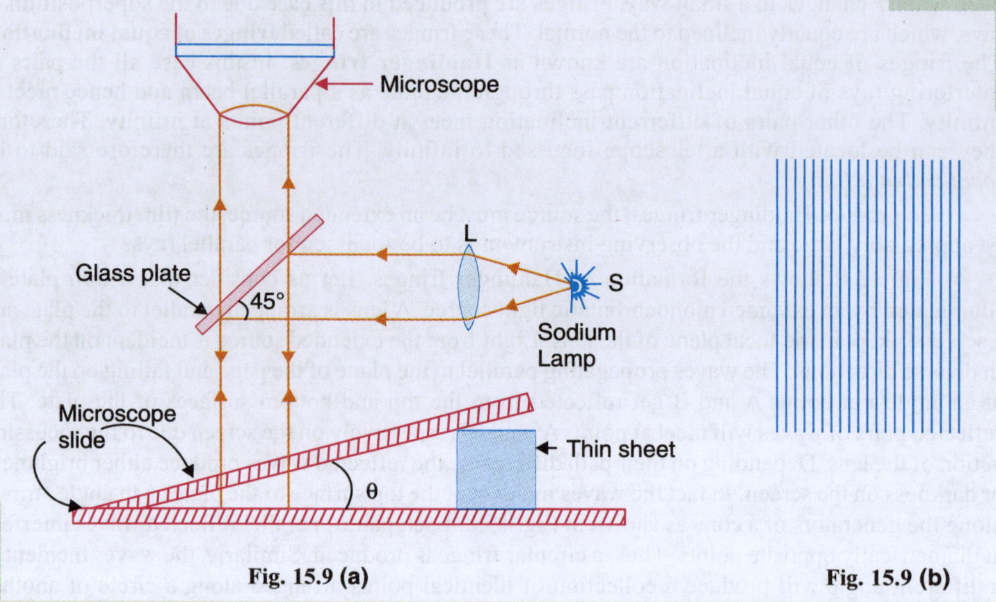

Fig. 15.9 (a) Fig. 15.9 (b)

with the wavelength of the incident light. Hence, the observed interference effects are entirely due to the wedge-shaped air film.

The optical difference between the two rays BC and DE is given by

$$\Delta = 2\mu\, t \cos r - \lambda/2$$

where λ/2 takes account the gain of half–wave due to the abrupt jump of π radians in the phase of the wave reflected from the bottom boundary of air – to – glass.

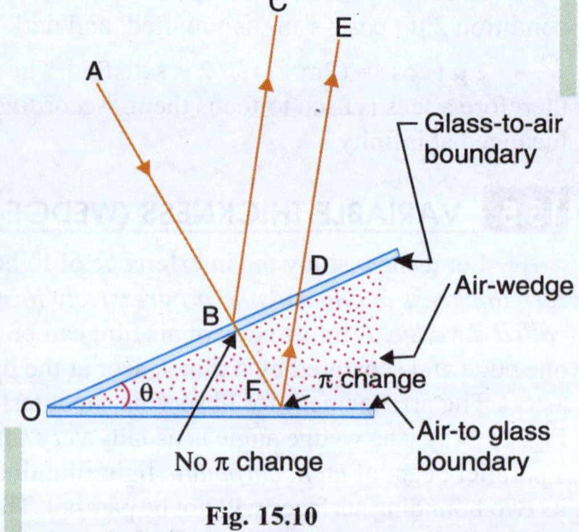

Fig. 15.10

Maxima occur when the optical path difference Δ = m λ. If the difference in the optical path between the two rays is equal to an *integral number of full waves*, then the rays meet each other in phase. The crests of one wave falls on the crests of the others and the waves *interfere constructively*. This needs that

$$\Delta = 2\mu\, t \cos r - \lambda/2$$

Minima occur when the optical path difference is Δ = (2 m + 1) λ/2. If the difference in the optical path between the two rays is equal to an *odd integral number of half-waves*, then the rays meet each other in opposite phase. The crests of one wave falls on the troughs of the others and *the waves interfere destructively*. It needs that

$$2\mu\, t \cos r = m\lambda.$$

Referring to Fig.15.11, let us say a dark fringe occurs at A where the relation

$$2\mu\, t \cos r = m\lambda$$

is satisfied. If normal incidence is assumed, *cos r = 1* and if the thickness of air film at A is denoted by t_1, then at A

$$2\mu\, t_1 = m\lambda \qquad (15.19)$$

The next dark fringe will occur, say, at C where the thickness CL = t_2. Then at C

$$2\mu\, t_2 = (m+1)\,\lambda \qquad (15.20)$$

Subtracting eq. (15.19) from eq. (15.20), we get

$$2\mu\,(t_2 - t_1) = \lambda \qquad (15.21)$$

Fig. 15.11

But $(t_2 - t_1) = BC$

$\therefore \quad 2\mu(BC) = \lambda$

or $$BC = \frac{\lambda}{2\mu} \qquad (15.22)$$

From the Δ^{le}ABC, $\angle CAB = \theta$ and BC = AB tan θ

$$\therefore \quad (AB)\tan\theta = \frac{\lambda}{2\mu} \qquad (15.23)$$

AB is the distance between successive dark fringes and it also equals the separation of the successive bright fringes. It is, therefore, called the **fringe width,β**. That is AB = β. We may write eq. (15.23) as

$$\beta = \frac{\lambda}{2\mu\tan\theta} \qquad (15.24)$$

For small values of θ, tan θ ≈ θ.

$$\therefore \quad \beta = \frac{\lambda}{2\mu\theta} \qquad (15.25)$$

As the quantities on the right side of the above equation are all constant, β is constant for a given wedge angle. According to equ.(15.25), an increase in the angle θ makes the fringes move closer. At an angle θ ≈ 1°, the interference pattern vanishes. On the other hand, if θ is gradually decreased, the fringe separation increases, and ultimately the fringes disappear as the faces of the film become parallel.

The interference pattern has the following salient features.

(*i*) Fringe at the apex is dark.

(*ii*) Fringes are straight and parallel.

(*iii*) Fringes are equidistant.

(*iv*) Fringes are localized.

(*v*) Fringes are of equal thickness.

***(i)* Fringe at the apex is dark:** At the apex, the two glass slides are in contact with each other. Therefore, the thickness of the air film at the contact edge is negligible (t ≅ 0). The optical path difference there becomes

$$\Delta = 2\mu t - \lambda/2 = 0 - \lambda/2 = -\lambda/2 \qquad (15.26)$$

It implies that a path difference of λ/2 or a phase difference of π occurs between the reflected waves at the edge. The two waves interference destructively. Therefore, the fringe at the apex is always dark (See Fig. 15.12).

***(ii)* Straight and parallel fringes:** Each fringe in the pattern is produced by the interference of rays reflected from sections of the wedge having the same thickness. The locus of points having the same thickness lie along lines parallel to the contact edge. Therefore, the fringes are straight. Since the fringes are equidistant [see (iii)], they will be parallel (See Fig. 15.12).

***(iii)* Equidistant fringes:** The fringe width β is given by

$$\beta \approx \lambda/2\theta \qquad (15.27)$$

where λ is the wavelength of the incident light and θ is the angle of the wedge. As the quantities λ and θ are constants, β is constant for a given wedge angle. Therefore, the fringes are equidistant (see Fig.15.12).

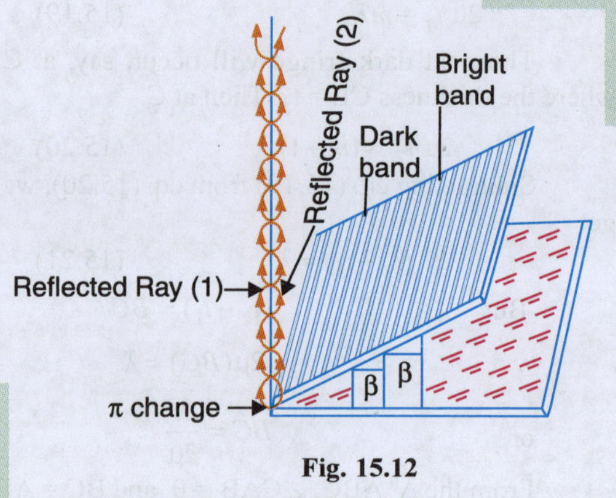

Fig. 15.12

***(iv)* Localized fringes:** The fringes form very close to the top surface of the wedge and can be seen with a microscope.

***(v)* Fringes of equal thickness:** In thin films of thickness of the order of a few λ, the rays from various parts of the film have almost the same inclination and hence the path difference between the overlapping waves changes mainly due to change of thickness. The fringes produced in such cases are mainly due to the variation in thickness of the film. Each fringe will be the locus of points of the same thickness. Such fringes are called **fringes of equal thickness.**

15.5.1. DETERMINATION OF THE WEDGE ANGLE

The wedge angle θ can be experimentally determined with the help of a travelling microscope. Using the microscope the positions of dark fringes at two distant points Q and R are noted (Fig. 15.13). Let the distance OQ be x_1 and OR be x_2. Let the thickness of the wedge be t_1 at Q and t_2 at R.

Fig. 15.13

The dark fringe at Q is given by

$$2\mu t_1 = m\lambda \qquad (15.28)$$

But as θ is very small, we can write

$$t_1 = x_1 \tan\theta \cong x_1\theta$$

$$\therefore \quad 2\mu\, x_1\, \theta = m\lambda \qquad (15.29)$$

We can write similarly for the dark fringe at R as

$$2\mu\, x_2\, \theta = (m+N)\lambda \qquad (15.30)$$

where N is the number of dark fringes lying between the positions Q and R. Subtracting equ.(15.29) from equ.(15.30), we get

$$2\mu(x_2 - x_1)\theta = N\lambda$$

$$\therefore \quad \theta = \frac{N\lambda}{2\mu(x_2 - x_1)} \qquad (15.31)$$

In case of air μ = 1 and the above relation reduces to

$$\theta = \frac{N\lambda}{2(x_2 - x_1)} \qquad (15.32)$$

15.5.2. DETERMINATION OF THE THICKNESS OF THE SPACER

The thickness of the spacer used to form the wedge shaped air film between the glass slides can be determined from the above measurements. If 't' is the thickness of the spacer (foil or wire) used, we can write from Fig.15.13 that

$$t = l \tan \theta \cong l\,\theta \tag{15.33}$$

where l is the length of the air wedge. Using the equ.(15.32) into equ.(15.33), we obtain

$$\therefore \qquad t = \frac{lN\lambda}{2(x_2 - x_1)} \tag{15.34}$$

15.5.3. FIZEAU FRINGES

If a parallel beam of light is incident perpendicularly or nearly perpendicular on a variable thickness film, then dark and bright fringes are seen in reflected light. These fringes are fringes of equal thickness, because each fringe corresponds to lines of equal optical thickness. These fringes or localized fringes and are observed at the top of the film. These localized fringes of equal thickness are known as *Fizeau fringes*. Contours following lines of equal optical thickness are seen if the area is large. The fringes may be obtained in case of thick films also if the source is small.

15.5.4. COLOURS IN THIN FILMS

The colours exhibited in reflection by thin films of oil, mica, soap bubbles and coatings of oxides on heated metals etc are due to interference of light from an extended source such as sky. Thomas Young explained the origin of colours in thin films. It may be understood as follows. The films are usually observed by reflected light. The eye looking at the thin film receives light waves reflected from the top and bottom surfaces of the film. The reflected rays are very close to each other and are in a position to interfere. The optical path difference between the interfering rays is $\Delta = 2\mu t \cos r - \lambda/2$. It is seen that the path difference depends upon the thickness t of the film, the wavelength λ and the angle r, which is related to the angle of incidence of light on the film. White light consists of a range of wavelengths and for specific values of t and r, waves of only certain wavelengths (colours) constructively interfere. Therefore, only those colours are present in the reflected light. The other wavelengths interfere destructively and hence are absent from the reflected light. Hence, the film at a particular point appears coloured. As the thickness and the angle of incidence vary from point to point, different colours are intensified at different places. The colours seen are not isolated colours, as at each place there is a mixture of colours. The composition of colours is different at different places and contours of impressive hues are observed over the entire surface of the film.

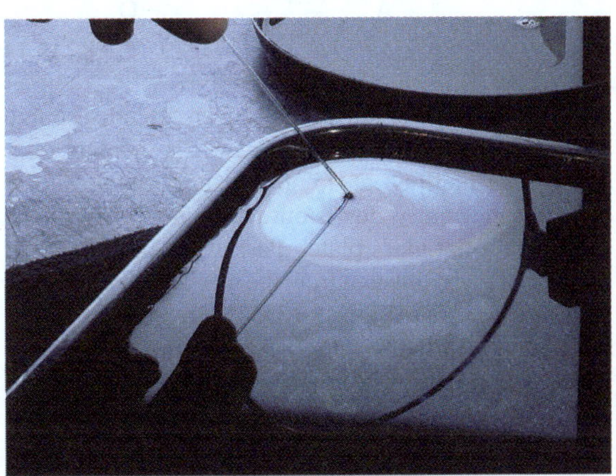

Colours in thin films of oil.

15.6. NEWTON'S RINGS

Newton's rings are an example of fringes of equal thickness. Newton's rings are formed when a plano-convex lens P of a large radius of curvature placed on a sheet of plane glass AB is illuminated from the top with monochromatic light (Fig. 15.14). The combination forms a thin circular air film of variable thickness in all directions around the point of contact of the lens and the glass plate. The locus of all points corresponding to specific thickness of air film falls on a circle whose centre is at O. Consequently, interference fringes are observed in the form of a series of concentric rings with their centre at O. Newton originally observed these concentric circular fringes and hence they are called **Newton's rings**.

The experimental arrangement for observing Newton's rings is shown in Fig. 15.14.

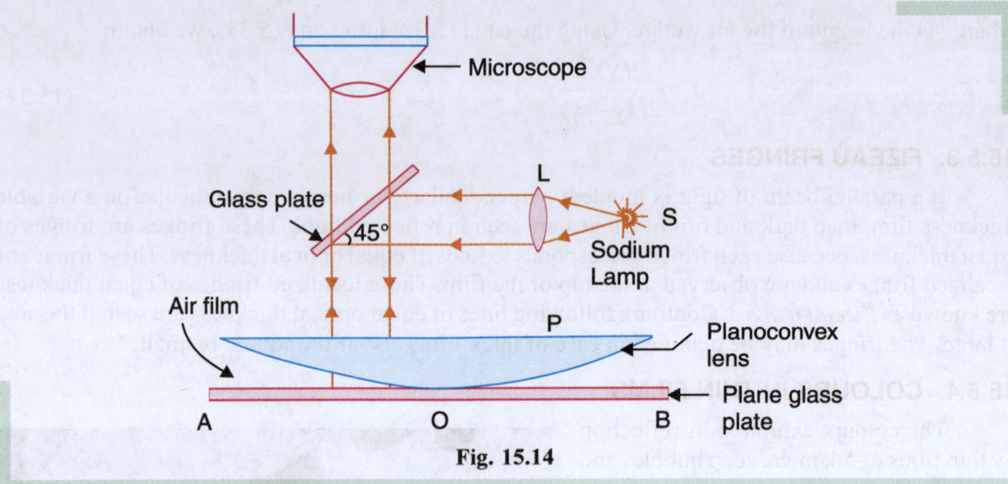

Fig. 15.14

Monochromatic light from an extended source S is rendered parallel by a lens L. It is incident on a glass plate inclined at 45° to the horizontal, and is reflected normally down onto a plano-convex lens placed on a flat glass plate. Part of the light incident on the system is reflected from the glass-to-air boundary, say from point D (Fig. 15.15). The remainder of the light is transmitted through the air film. It is again reflected from the air-to-glass boundary, say from point J. The two rays reflected from the top and bottom of the air film are derived through division of amplitude from the same incident ray CD and are therefore coherent. The rays 1 and 2 are close to each other and interfere to produce darkness or brightness. The condition of brightness or darkness depends on the path difference between the two reflected light rays, which in turn depends on the thickness of the air film at the point of incidence.

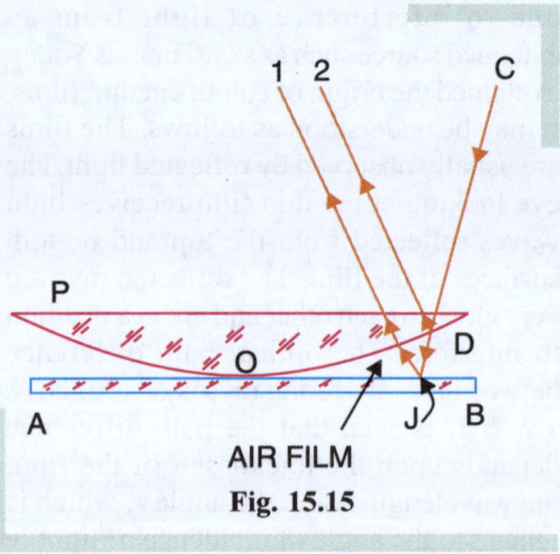

Fig. 15.15

15.6.1. CONDITION FOR BRIGHT AND DARK RINGS

The optical path difference between the rays is given by $\Delta = 2\mu t \cos r - \lambda/2$. Since $\mu = 1$ for air and $\cos r = 1$ for normal incidence of light,

$$\Delta = 2t - \lambda/2 \tag{15.35}$$

Intensity maxima occur when the optical path difference $\Delta = m\lambda$. If the difference in the optical path between the two rays is equal to an *integral number of full waves*, then the rays meet each other in phase. The crests of one wave falls on the crests of the other and the waves *interfere constructively*. Thus, if $2t - \lambda/2 = m\lambda$

$$\boxed{2t = (2m+1)\lambda/2} \qquad (15.36)$$

bright fringe is obtained.

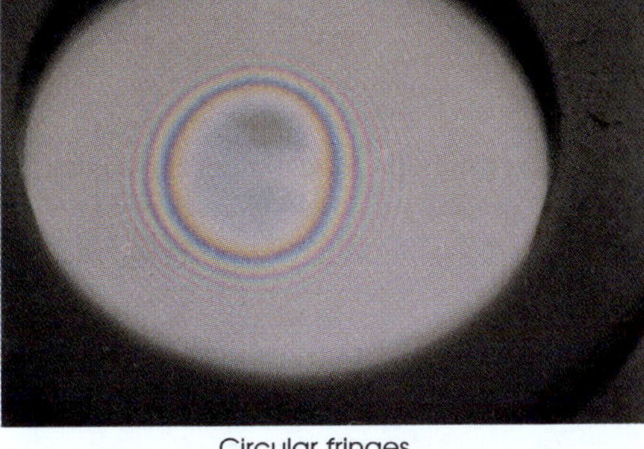

Circular fringes.

Intensity minima occur when the optical path difference is $\Delta = (2\ m + 1)\ \lambda/2$. If the difference in the optical path between the two rays is equal to an *odd integral number of half-waves*, then the rays meet each other in opposite phase. The crests of one wave fall on the troughs of the other and *the waves interfere destructively*.

Hence, if $\quad 2t - \lambda/2 = (2m + 1)\ \lambda/2$

$$\boxed{2t = m\lambda} \qquad (15.37)$$

dark fringe is produced.

15.6.2. CIRCULAR FRINGES

In Newton's ring arrangement, a thin air film is enclosed between a plano-convex lens and a glass plate. The thickness of the air film at the point of contact is zero and gradually increases as we move outward. The locus of points where the air film has the same thickness then fall on a circle whose centre is the point of contact. Thus, the thickness of air film is constant at points on any circle having the point of lens–glass plate contact as the centre. The fringes are therefore circular.

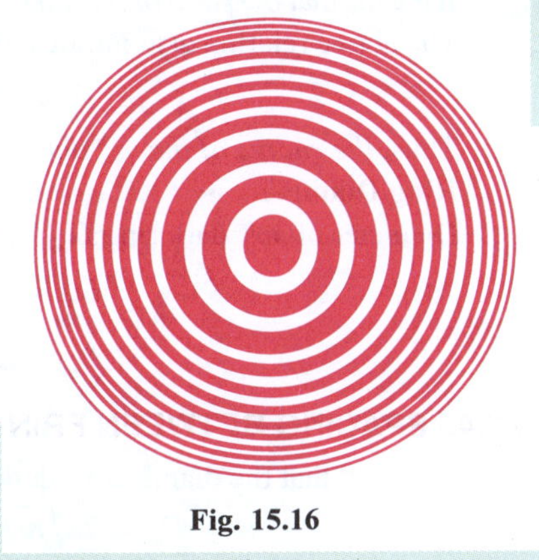

Fig. 15.16

15.6.3. RADII OF DARK FRINGES

Let R be the radius of curvature of the lens (Fig.15.17). Let a dark fringe be located at Q. Let the thickness of the air film at Q be PQ = t. Let the radius of the circular fringe at Q be OQ= r_m. By the Pythagorus theorem,

Fig. 15.17

$$PM^2 = PN^2 + MN^2$$

$$\therefore \quad R^2 = r_m^2 + (R-t)^2$$

or $$r_m^2 = 2Rt - t^2 \qquad (15.38)$$

As R >> t, 2Rt >> t^2.

$$\therefore \quad r_m^2 \cong 2Rt \qquad (15.39)$$

The condition for darkness at Q is that

$$2t = m\lambda$$

$$\therefore \quad r_m^2 \cong m\lambda R$$

$$r_m = \sqrt{m\lambda R} \qquad (15.40)$$

The radii of dark fringes can be found by inserting values 1,2,3, ...for m. Thus,

$$r_1 = \sqrt{1\lambda R} \quad \text{or} \quad r_1 \propto \sqrt{1}$$

$$r_2 = \sqrt{2\lambda R} \quad \text{or} \quad r_2 \propto \sqrt{2}$$

$$r_3 = \sqrt{3\lambda R} \quad \text{or} \quad r_3 \propto \sqrt{3} \quad \text{and so on}$$

It means that *the radii of the dark rings are proportional to under root of the natural numbers.* The above relation also implies that

$$r_m \propto \sqrt{\lambda}$$

Thus, *the radius of the mth dark ring is proportional to under root of wavelength.*

Ring Diameter:

Diameter of mth dark ring $D_m = 2r_m$

$$D_m = 2\sqrt{2Rt}$$

$$D_m = 2\sqrt{m\lambda R} \qquad (15.41)$$

15.6.4. SPACING BETWEEN FRINGES

It is seen that the diameter of dark rings is given by

$$D_m = 2\sqrt{m\lambda R}$$

where m = 1,2,3, ...

The diameters of dark rings are proportional to the square root of the natural numbers. Therefore, the diameter of the ring does not increase in the same proportion as the order of the ring, for example, if m increases as 1,2,3,4, ...the diameters are

$$D_1 = 2\sqrt{\lambda R}$$

$$D_2 = 2\,(1.4)\sqrt{\lambda R}$$

$$D_3 = 2\,(1.7)\sqrt{\lambda R}$$

$$D_4 = 2\,(2)\sqrt{\lambda R} \quad \text{and so on.}$$

Therefore, the rings get closer and closer, as m increases. This is why the rings are not evenly spaced.

15.6.5. FRINGES OF EQUAL THICKNESS

Newton's rings are formed as result of interference between light waves reflected from the top and bottom surfaces of a thin air film enclosed between a plano-convex lens and a plane glass plate.

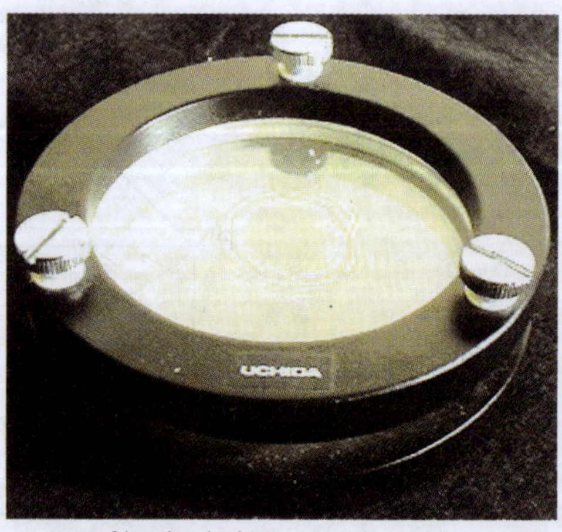

Newton's rings arrangement.

The occurrence of alternate bright and dark rings depends on the optical path difference arising between the reflected rays. If the light falls normally on the air film the optical path difference between the waves reflected from the two surfaces of the film is

$$\Delta = 2t - \lambda/2$$

It is seen that the path difference between the reflected rays arises due to the variation in the thickness 't' of the air film. Reflected light will be of minimum intensity for those thickness for which the path difference is m λ and maximum intensity for those thickness for which the path difference is $(2m+1)\lambda/2$. Thus, each maxima and minima is a locus of constant film thickness. Therefore, the fringes are known as fringes of equal thickness.

15.6.6. DARK CENTRAL SPOT

The central spot is dark as seen by reflection. Newton's rings are produced due to superposition of light rays reflected from the top and bottom surfaces of a thin air film enclosed between a plano-convex lens and a plane glass plate. The occurrence of brightness or darkness depends on the optical path difference arising between the reflected rays. The optical path difference is given by $\Delta = 2t - \lambda/2$.

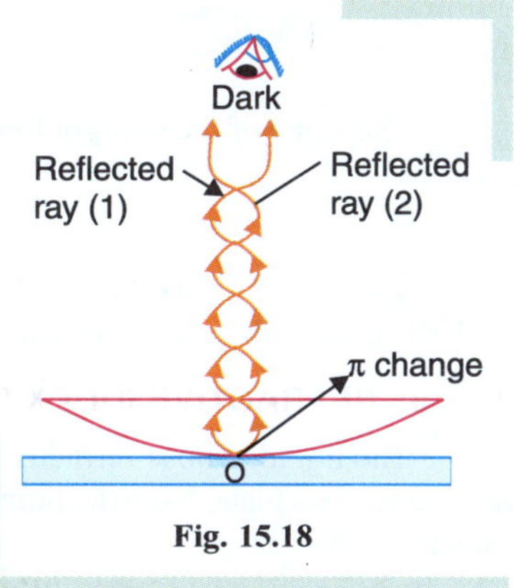

Fig. 15.18

At the point of contact 'O' of the lens and glass plate (Fig.15.18), the thickness of air film is negligibly small compared to a wavelength of light.

$$\therefore \quad t \cong 0$$

$$\therefore \quad \Delta \cong \lambda / 2$$

The wave reflected from the lower surface of the air film suffers a phase change of π while the wave reflected from the upper surface of the film does not suffer such change.

Thus, the superposing waves are out of step by $\lambda/2$ which is equivalent to a phase difference of 180° (or π rad). Thus the two interfering waves at the centre are opposite in phase and produce a dark spot.

15.6.7. DETERMINATION OF WAVELENGTH OF LIGHT

A plano-convex lens of large radius of curvature (about 100 cm) and a flat glass plate are cleaned. The lens is kept with its convex face on the glass plate and they are held in position with the help of a metal ring arrangement. The system is held under a low power travelling microscope kept before a sodium vapour lamp. It is arranged that the yellow light coming from the sodium lamp falls on a glass plate held at 45° light beam. The light is turned through 90° and is incident normally on the lens-plate system. The microscope is adjusted till the circular rings came into focus. The centre of the cross-wire is made to come into focus on the centre of the dark spot, which is at the centre of the circular ring system. Now, turning the screw the microscope is moved on the carriage slowly towards one side, say right side. As the cross-wires move in the field of view, dark rings are counted. The movement is stopped when the 22nd dark ring is reached. Then the microscope is moved in the opposite direction and stopped at the 20th or 19th dark ring. The vertical cross-wire is made tangential to

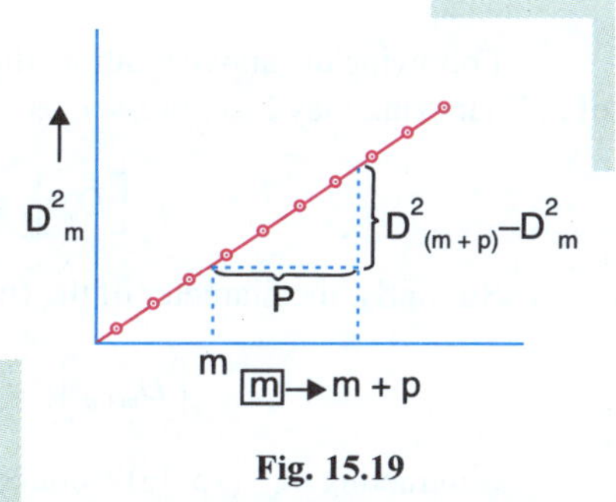

Fig. 15.19

the 19th ring and the reading is noted with the help of the scale graduated on the carriage. Thus, starting from the 19th ring, the tangential positions of the 18th, 17th, 16th,.....,5th dark rings are noted down. Now, the microscope is moved quickly to the left side of the ring system and it is stopped at the 5th dark ring. The cross-wire is again made tangential to the 5th dark ring and its position is noted. The difference between the readings on right and left sides of the 5th dark ring gives its diameter value. The procedure is repeated till 19th ring is reached and its reading is noted. From the value of the diameters the squares of the diameters are calculated. A graph is plotted between D_m^2 and the ring number '*m*'. A straight line would be obtained, as shown in Fig. 15.19.

We have

$$D_m^2 = 4m\lambda R \tag{15.42}$$

For the (m+p)th ring,

$$D_{m+p}^2 = 4(m+p)\lambda R \tag{15.43}$$

$$\therefore \quad D_{m+p}^2 - D_m^2 = 4p\lambda R$$

$$\lambda = \frac{D_{m+p}^2 - D_m^2}{4pR} \tag{15.44}$$

The slope of the straight line (Fig.15.18) gives the value of 4λR. Thus,

$$\lambda = \frac{Slope}{4R} \tag{15.45}$$

The radius of curvature R of the lens may be determined using a spherometer and λ is computed with the help of the above equation.

15.6.8. REFRACTIVE INDEX OF A LIQUID

The liquid, whose refractive index is to be determined, is filled in the gap between the lens and plane glass plate. Now the liquid film substitutes the air film. The condition for interference may then be written as

$$2\mu t \cos r = m\lambda \qquad \text{Darkness}$$

where μ is the refractive index of the liquid. For normal incidence the equation becomes

$$2\mu t = m\lambda$$

As $t = \frac{r^2}{2R}$,

$$\frac{2\mu\, r^2}{2R} = m\lambda$$

Or

$$r^2 = \frac{m\lambda R}{\mu}$$

$$\therefore \quad D^2 = \frac{4m\lambda R}{\mu}$$

Refractive index detector.

Following the above relation, the diameter of mth dark ring may be expressed as

$$\left[D_m^2\right]_L = \frac{4m\lambda R}{\mu} \tag{15.46}$$

Similarly, the diameter of the (m+p)th ring is given by

$$\left[D_{m+p}^2\right]_L = \frac{4(m+p)\lambda R}{\mu} \tag{15.47}$$

Subtracting eq. (15.46) from eq. (15.47), we get

$$\left[D_{m+p}^2\right]_L - \left[D_m^2\right]_L = \frac{4p\lambda R}{\mu} \tag{15.48}$$

But we know that

$$\left(D_{m+p}^2\right)_{air} - \left(D_m^2\right)_{air} = 4p\lambda R \tag{15.49}$$

$$\therefore \qquad \mu = \frac{\left(D_{m+p}^2\right)_{air} - \left(D_m^2\right)_{air}}{\left(D_{m+p}^2\right)_{liq} - \left(D_m^2\right)_{liq.}} \tag{15.50}$$

15.6.9. NEWTON'S RINGS IN TRANSMITTED LIGHT

Newton's rings in transmitted light may be observed with the arrangement made as in Fig. 15.20. The condition for maxima or bright rings is

$$2\mu\, t \cos r = m\lambda$$

and for dark rings $\quad 2\mu\, t \cos r = (2m+1)\lambda/2$

As μ = 1 for air and r = 0 for normal observation, the above expressions may be simplified to

For bright fringes $\quad 2t = \mu\lambda$

and for dark rings $\quad 2t = (2m+1)\lambda/2$

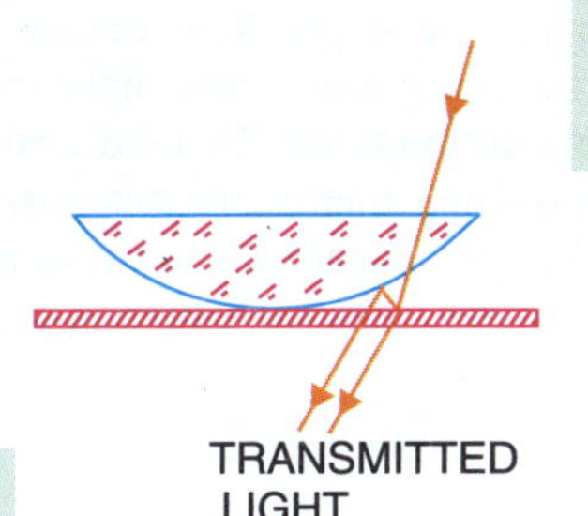

Fig. 15.20

As $t = \dfrac{r^2}{2R}$, the radius for the bright ring is given by $r_m^2 = m\lambda R$ (15.51)

and the radius for dark rings is given by $r_m^2 = (2m+1)\lambda R/2$ (15.52)

15.6.10. NEWTON'S RINGS FORMED BY TWO CURVED SURFACES

Case 1: Lower surface concave:

Let us consider two curved surfaces of radii of curvature R_1 and R_2 in contact at the point O. A thin air film is enclosed between the two surfaces. The dark and bright rings are formed and can be viewed with a travelling microscope. Suppose the radius of the mth dark ring is r. The thickness of the air film at P is

$$PQ = PT - QT$$

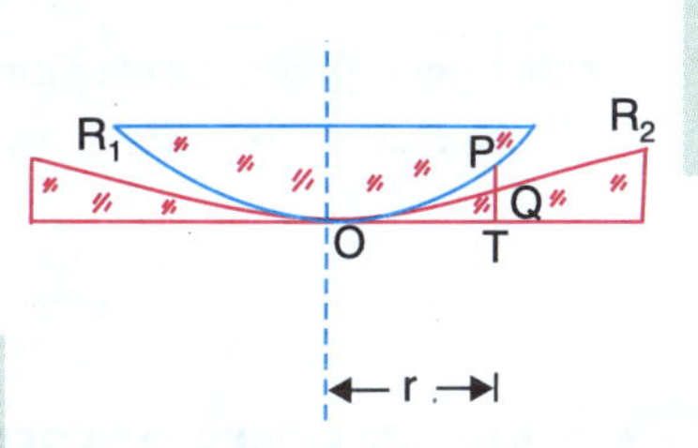

Fig. 15.21

From geometry, $PT = \dfrac{r^2}{2R_1}$ and $QT = \dfrac{r^2}{2R_2}$

$$\therefore \qquad PQ = \frac{r^2}{2R_1} - \frac{r^2}{2R_2} \tag{15.53}$$

But PQ = t. The condition for dark rings in reflected light is given by $2\mu\, t \cos r = m\lambda$. As μ = 1 and cos r = 1 for normal incidence, the above condition reduces to $2t = m\lambda$.

$$\therefore \qquad 2\left(\frac{r^2}{2R_1} - \frac{r^2}{2R_2}\right) = m\lambda$$

$$r^2\left(\frac{1}{R_1}-\frac{1}{R_2}\right)=m\lambda \quad \text{where } m = 0,1,2,3,\ldots.. \tag{15.54}$$

For bright fringes the condition is $2\mu\, t\, \cos r = (2m+1)\lambda/2$

which reduces to $2t = (2m+1)\lambda/2$

or
$$r^2\left(\frac{1}{R_1}-\frac{1}{R_2}\right)=\frac{(2m+1)\lambda}{2} \quad \text{where } m = 0,1,2,3,\ldots.. \tag{15.55}$$

Case 2: Lower surface convex:

Let us consider two curved surfaces of radii of curvature R_1 and R_2 in contact at the point O. A thin air film is enclosed between the two surfaces. The dark and bright rings are formed and can be viewed with a travelling microscope. Suppose the radius of the mth dark ring is r. The thickness of the air film at P is

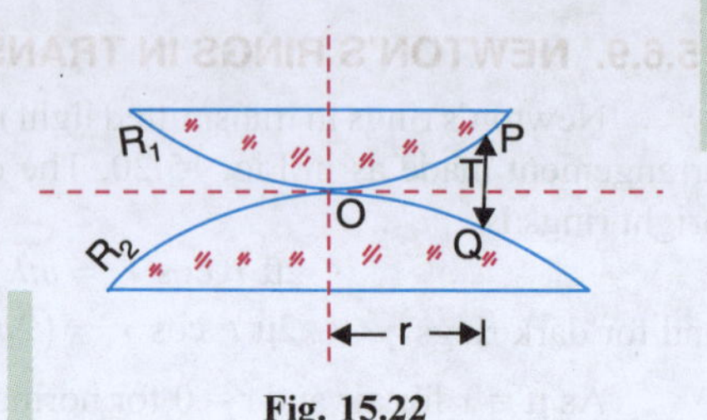

Fig. 15.22

$$PQ = PT + QT$$

From geometry $PT=\dfrac{r^2}{2R_1}$ and $QT=\dfrac{r^2}{2R_2}$

$$\therefore \quad PQ=\frac{r^2}{2R_1}+\frac{r^2}{2R_2}$$

But PQ = t. The condition for dark rings in reflected light is given by $2\mu\, t \cos r = m\lambda$. As $\mu = 1$ and $\cos r = 1$ for normal incidence, the above condition reduces to $2t = m\lambda$

$$\therefore \quad 2\left(\frac{r^2}{2R_1}+\frac{r^2}{2R_2}\right)=m\lambda$$

$$r^2\left(\frac{1}{R_1}+\frac{1}{R_2}\right)=m\lambda \quad \text{where } m = 0,1,2,3,\ldots \tag{15.56}$$

For bright fringes the condition is $2\mu\, t\, \cos r = (2m+1)\lambda/2$

which reduces to $2t = (2m+1)\lambda/2$

or
$$r^2\left(\frac{1}{R_1}+\frac{1}{R_2}\right)=\frac{(2m+1)\lambda}{2} \quad \text{where } m = 0,1,2,3,\ldots \tag{15.57}$$

15.7. MICHELSON'S INTERFEROMETER

An interferometer is an instrument in which the phenomenon of interference is used to make precise measurements of wavelengths or distances.

15.7.1. PRINCIPLE

In Michelson interferometer, a beam of light from an extended source is divided into two parts of equal intensities by partial reflection and refraction. These beams travel in two mutually perpendicular directions and come together after reflection from plane mirrors. The beams overlap on each other and produce interference fringes.

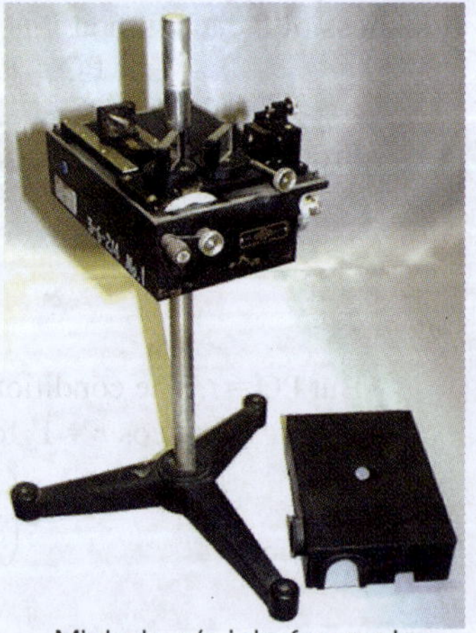

Michelson's interferometer.

15.7.2. CONSTRUCTION

The schematic of a simple Michelson interferometer is shown in Fig.15.23. It consists of a beam splitter G_1, a compensating plate G_2, and two plane mirrors M_1 and M_2. The beam splitter G_1 is a partially silvered plane parallel glass plate. The compensating plate G_2 is a simple plane parallel glass plate having the same thickness as G_1. The two plates G_1 and G_2 are held parallel to each other and are inclined at an angle of 45° with respect to the mirror M_2. The mirror M_1 is mounted on a carriage and can be moved exactly parallel to itself with the help of a micrometer screw. The distance through which the mirror M_1 is moved can be read with the help of a graduated drum attached to the screw. Displacements of the order of 0.1 μm (1000 Å) can be easily read. The plane mirrors M_1 and M_2 can be made perfectly perpendicular with the help of the fine screws attached to them. The interference bands are observed in the field of view of the telescope T.

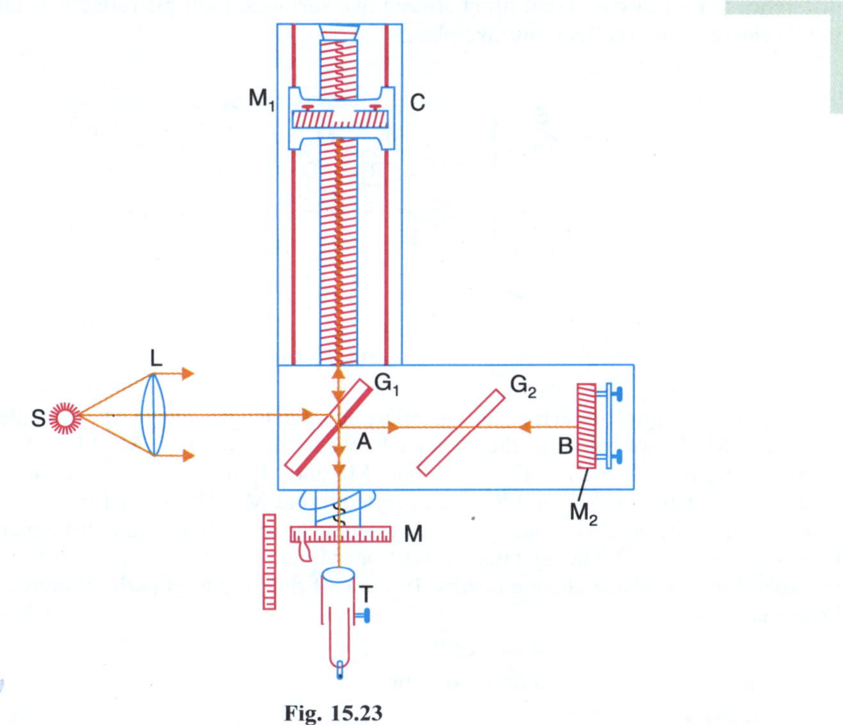

Fig. 15.23

15.7.3. WORKING

Monochromatic light from an extended source S is rendered parallel by means of a collimating lens L and is made incident on the beam splitter G_1. It is partly reflected at the back surface of G_1 along AC and partly transmitted along AB. The beam AC travels normally towards the plane mirror M_1 and is reflected back along the same path and comes out along AT. The transmitted beam travels toward the mirror M_2 and is reflected along the same path. It is reflected at the back surface of G_1 and proceeds along AT. The two beams received along AT are produced from a single source through division of amplitude and are hence *coherent*. The superposition of these beams leads to interference and produces interference fringes.

From the Fig. 15.23 it is clearly seen that a light ray starting from the source S and undergoing reflection at the mirror M_1 passes through the glass plate G_1 three times. On the other hand, in the absence of plate G_2, the ray reflected at M_2 travels through the glass plate G_1 only once. For compensating this path difference, a compensating plate G_2 of the same thickness is inserted into the path AB and is held exactly parallel to G_1.

If we look into the instrument from T, we see mirror M_1 and in addition we see a virtual image, M'_2, of mirror M_2. Depending on the positions of the mirrors, image M'_2 may be in front of, or behind, or exactly coincident with mirror M_1.

15.7.4. CIRCULAR FRINGES

Circular fringes are produced with monochromatic light when the mirrors M_1 and M_2 are exactly perpendicular to each other. The origin of the circular fringes can be understood as follows.

If we look into the instrument from T, we see mirror M_1 directly, and in addition we will see the virtual image M'_2 of mirror M_2 formed by reflection in the glass plate G_1 (Fig 15.24). It means that one of the interfering beams come from M_1 and the other beam appears to come from the virtual image M'_2. The situation is similar to an air film enclosed between mirrors M_1 and M'_2 with the difference that in case of a real film between two surfaces, multiple reflections take place, whereas in this case only two reflections take place.

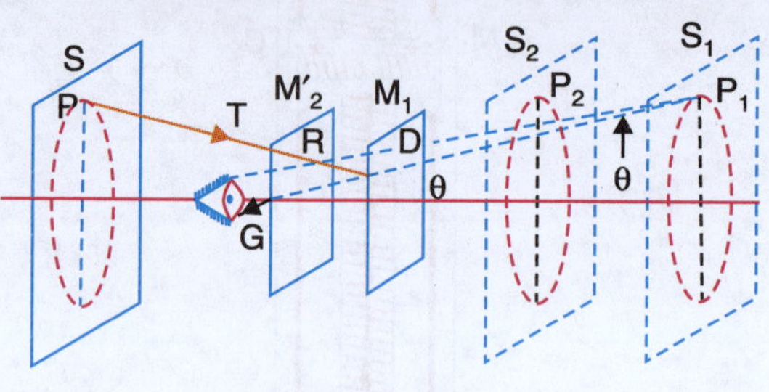

Fig. 15.24

If the two arms of the interferometer are equal in length, image M'_2 coincides with mirror M_1. If M'_2 and M_1 do not coincide, the distance between them is finite, $M'_2\ M_1 = d$. Now if a light ray comes from a point S and is reflected by both M'_2 and M_1, the observer will see two virtual images-S_1 due to reflection at M'_2 and S_2 due to reflection at M_1. The virtual images are separated by a distance 2d. If the observer looks into the system at an angle θ, the path difference between the two beams will be 2d cos θ. The light that comes from M_2 and goes to T undergoes rare-to-dense reflection and therefore a π-phase change occurs. In view of this, the total path difference between the two beams is given by

$$\Delta = 2d \cos \theta + \lambda / 2.$$

The condition for obtaining brightness

$$2d \cos \theta + \lambda / 2 = m\lambda$$

where m = 0,1,2, ...

For a given mirror separation d, a given wavelength λ and order m, angle θ is constant. This means that the fringes are of circular shape. They are called fringes of equal inclination.

In case the mirror M_1 coincides with the virtual image M'_2, d = 0. The path difference between the interfering beams will be λ / 2. Consequently, we obtain a minimum at the coincidence position and the centre of the field will be dark, as shown in Fig. 15.25 (a).

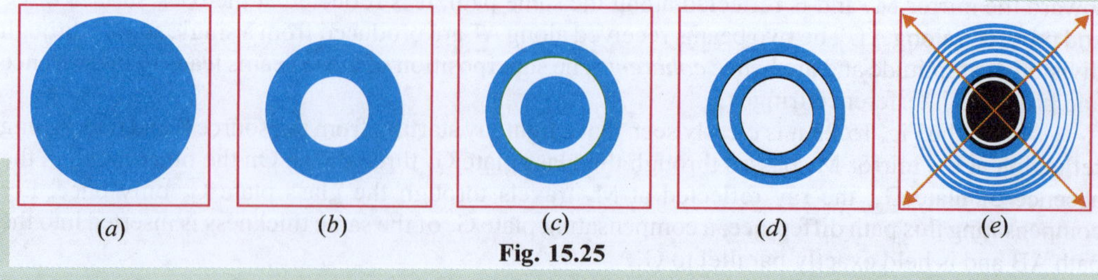

Fig. 15.25

If one of the mirrors is now moved through a distance $\lambda/4$, the path difference changes by $\lambda/2$ and therefore a maximum is obtained. By moving the mirror through another $\lambda/4$, a minimum is obtained; moving it by another $\lambda/4$ again a maximum is obtained and so on. Therefore, a new ring appears in the centre of the field each time the mirror is moved through $\lambda/2$. As d increases new rings appear in the centre faster than rings already present disappear in the periphery; and the field becomes more crowded with thinner rings (Fig. 15.25e). Conversely, as d is made smaller, the rings contract and disappear in the centre.

15.7.5. LOCALIZED FRINGES

When the two mirrors are tilted, they are not exactly perpendicular to each other and therefore the mirror M_1 and the virtual image M'_2 are not parallel. In this case the air path between them is wedge-shaped and the fringes appear to be straight. If one of the mirrors is moved, the fringes move across the field. The position of any particular bright fringe is taken up by the one next to it. The fringes can be counted as they pass a reference mark. If m fringes move across the field of view when M_1 moves through a distance d, then

$$d = m\,\lambda/2$$

or

$$\lambda = \frac{2d}{m} \qquad (15.58)$$

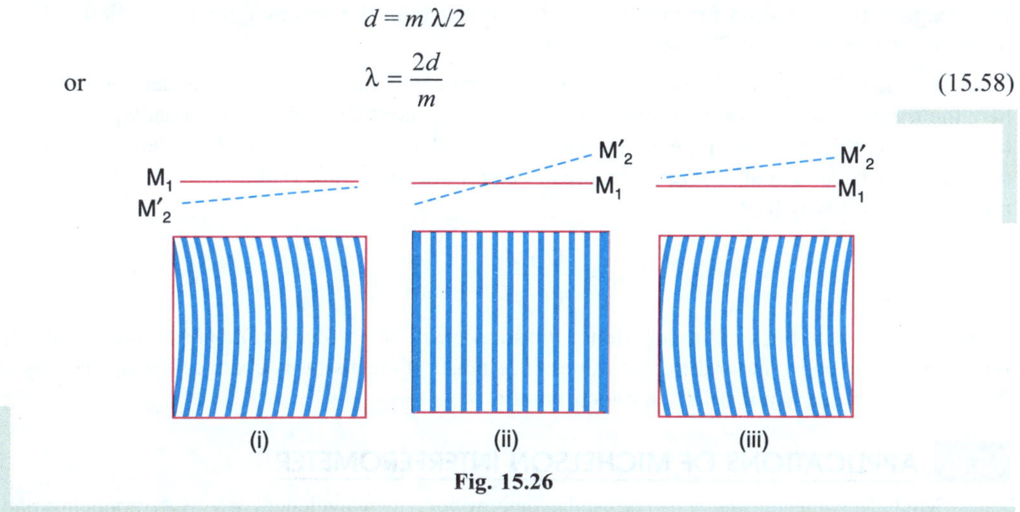

Fig. 15.26

15.7.6. WHITE LIGHT FRINGES

Instead of a monochromatic source, if a white light source is used, a few coloured fringes with a central dark fringe can be observed. In observing these fringes, the mirrors are slightly tilted as for localised fringes and position of M_1 is found where it intersects M'_2. This position is often difficult to find with white light. The position can best be located with monochromatic light when the fringes become straight. Then a very slow motion of M_1 in this region using white light will bring these fringes into view, when a central dark fringe is surrounded by 8 to 10 coloured fringes on either side are observed. These fringes are useful for the determination of zero path difference.

White light fringes.

15.7.7. VISIBILITY OF FRINGES

In case of Michelson interferometer, the intensity is given by

$$I = 4A^2 \cos^2 \frac{\delta}{2}$$

Here $$\delta = \frac{2\pi}{\lambda}(2d \cos \theta)$$

where d is the distance between M_1 and M_2'. The intensity is maximum when δ is an integral multiple of 2π. The intensity is zero when δ is an odd multiple of π. When a monochromatic source of light is used, the minimum intensity of the fringes is zero. The visibility of fringes in the case of a Michelson interferometer is

$$V = \frac{I_{max} - I_{min}}{I_{max} + I_{min}}$$

for monochromatic light, $I_{min} = 0$ and therefore, $V = 1$

However, if the source of light is not strictly monochromatic, but contains two nearby wavelengths, the condition for maximum intensity for both the wavelengths is satisfied only for particular values of path difference (2d cos θ).

As the value of d is altered, the two wavelengths do coincide over a considerable range and here the fringe visibility is a maximum. For values of d other than maximum intensity positions for both wavelengths, the two fringe patterns will be complementary, provided the intensities for both the wavelengths are equal. If intensities are not equal, the minimum visibility will not be zero. The minimum visibility will be

$$V_{min} = \frac{A_1^2 - A_2^2}{A_1^2 + A_2^2}$$

where A_1 and A_2 are the amplitudes. Hence the source will be perfectly monochromatic if visibility is maximum and constant for different values of 2d cos θ. If the visibility changes with the change of 2d cos θ, the source is not strictly monochromatic.

15.8. APPLICATIONS OF MICHELSON INTERFEROMETER

Michelson interferometer can be used to determine (i) the wavelength of a given monochromatic source of light (ii) the difference between the two neighbouring wavelengths or resolution of the spectral lines, (iii) refractive index and thickness of various thin transparent materials and (iv) for measurement of the standard metre in terms of the wavelength of light.

15.8.1. MEASUREMENT OF WAVELENGTH

Michelson interferometer is used to determine the wavelength of light from a monochromatic source. The monochromatic source is kept at S. If the mirrors M_1 and M_2 are exactly perpendicular, circular fringes are obtained. If the mirror M_1 is moved forward or backward, the circular fringes appear or disappear at the centre. Now, as the mirror is moved through a known distance d and the number of fringes disappearing at the centre is counted. Suppose d_1 is the initial thickness of the air film between the mirror M_1 and the image of M_2 corresponding to the bright fringe of order m_1 and d_2 is the final thickness of the air film corresponding to a bright fringe of order m_n in the same position. Then,

$$2d_1 = m_1\lambda$$

and $$2d_2 = m_n\lambda$$

By subtraction, we get $2(d_2 - d_1) = (m_n - m_1)\lambda$

$\therefore$ $2d = N\lambda$ where $(d_2 - d_1) = d$ and $(m_n - m_1) = N$

$$\therefore \quad \lambda = \frac{2d}{N} \tag{15.59}$$

15.8.2. DETERMINATION OF THE DIFFERENCE IN THE WAVELENGTH OF TWO WAVES

If a source of light consists of two wavelengths λ_1 and λ_2, which differ slightly, then two sets of fringes corresponding to the two wavelengths are produced in a Michelson interferometer. By adjusting the position of the mirror M_1 of the interferometer, the position is found when the fringes are very bright. In this position, the bright fringe due to λ_1 coincides with the bright fringes due to λ_2. When the mirror M_1 is moved, the two sets of fringes get out of step because their wavelengths are different. When the mirror M_1 has been moved through a certain distance, the bright fringe due to one set will coincide with the dark fringe due to the other set and no fringes will be seen in this case. Again by moving the mirror M_1, a position is reached when a bright fringe of one set falls on the bright fringe of the other and the fringes are again distinct. This is possible when the mth order of the longer wavelength coincides with the (m + 1)th order of the shorter wavelength.

Let m_1 and m_2 be the changes in the order at the centre of the field when the mirror M_1 is displaced through a distance d between two consecutive positions of maximum distinctness of the fringes.

$$\therefore \quad 2d = m_1 \lambda_1 = m_2 \lambda_2$$

If λ_1 is greater than λ_2,

$$m_2 = m_1 + 1$$

$$\therefore \quad 2d = m_1 \lambda_1 = (m_1 + 1)\lambda_2$$

$$\therefore \quad m_1 = \frac{\lambda_2}{\lambda_1 - \lambda_2}$$

$$\therefore \quad 2d = \frac{\lambda_1 \lambda_2}{\lambda_1 - \lambda_2}$$

or

$$\lambda_1 - \lambda_2 = \frac{\lambda_1 \lambda_2}{2d}$$

Taking λ as the mean wavelength of the two wavelengths λ_1 and λ_2, the small difference $\Delta\lambda$ is given by

$$\Delta\lambda = \lambda_1 - \lambda_2 = \frac{\lambda^2}{2d} \tag{15.60}$$

15.8.3. THICKNESS OF A THIN TRANSPARENT SHEET

Let a transparent sheet of thickness t and refractive index μ be inserted in the path of one of the interfering beams of Michelson interferometer. The optical path of that beam increases because of the sheet. It becomes μt instead of t. The increase in the optical path is $(\mu t - t)$ or $(\mu - 1)t$. Since the beam traverses the medium twice, the extra path difference between the two interfering beams is $2(\mu - 1)t$. If m is the number of fringes by which the fringe system is displaced, then

$$2(\mu - 1)t = m\lambda$$

When monochromatic light is used, it is difficult to distinguish the sudden shift of fringes when the thin sheet is inserted. It is also not possible to count the number of fringes shifted. The difficulty is overcome by using white light first to locate the central dark fringe and it is made to coincide with the cross-wire of the telescope. The thin sheet is then introduced into the path of the beam. Position of mirror M_1 is adjusted till again a dark fringe of zero path difference coincides with the cross-wire of the telescope. The distance *d* through which the mirror is moved is noted. The white light is now replaced with the monochromatic light and the mirror M_1 is moved back slowly and the number of fringes contained in *d* is found. The thickness t is obtained from the relation

$$t = \frac{m\lambda}{2(\mu - 1)} \qquad (15.61)$$

15.8.4. DETERMINATION OF THE REFRACTIVE INDEX OF GASES

When a tube containing a gas is introduced in the path of the beam going towards M_1, a path difference equal to $2(\mu - 1)\,l$ is introduced between the two interfering beams. Here, μ is the refractive index of the gas and *l* is the length of the tube. If *m* fringes cross the centre of the field of view, then $2(\mu - 1)\,l = m\lambda$. Knowing *l*, *m*, and λ, μ an be calculated.

In the path of the rays going towards M_1, a tube containing air at atmospheric pressure is introduced and the fringes are obtained in the centre of the field of view. In that case, refractive index of the air at various pressures can be determined. Let the length of the tube be *l* and let it contain air at atmospheric pressure. The tube is completely evacuated and *m* fringes cross the centre of the field of view. The path difference introduced between the two interfering beams is $2(\mu - 1)\,l$.

$$\therefore \quad 2(\mu - 1)\,l = m\lambda$$

$$\therefore \quad \mu = \frac{m\lambda}{2l} + 1$$

15.8.5. STANDARDISATION OF THE METRE

The experiment to measure the standard metre in terms of the wavelength of the cadmium red line was first performed by Michelson and Benoit in 1895. It is not possible to count the number of fringes which cross the field of view when one of the mirrors of the Michelson interferometer is moved through whole length of one metre. Moreover, for a path difference of more than 20 cm, it is not possible to obtain the fringes. Therefore, the mirror must not be moved through a distance of more than 10cm. In practice nine **etalons** were used, each being twice the length of the preceding etalon. The length of the shortest etalon used is 0.390625 mm and of the longest was 10 cm. The experiment is divided into two main parts.

(*i*) The number of wavelengths of the monochromatic cadmium light is counted for the shortest etalon.

(*ii*) The length of the second etalon is compared with the shorter etalon and the process is repeated until the number of wavelengths for a length of 10 cm-etalon is known. From this 10 cm-etalon, the number of wavelengths for a length of one metre in terms of the wavelength of cadmium red line is known. This acts as a standard metre because, even if the original standard metre is destroyed, the standard metre can be formed again from the knowledge of the number of wavelengths. The standard metre is represented in terms of the wavelengths of red, green and blue lines of cadmium.

Etalon

An etalon is a substandard for length. It consists of two mirrors, which are plane-parallel and silvered on their front faces. The distance between their surfaces is *l* (Fig. 15.27). The mirrors can be made perfectly parallel by means of screws attached to them.

Experiment:

(*i*) The Michelson interferometer is used as shown in Fig. 15.28. Light from the source S

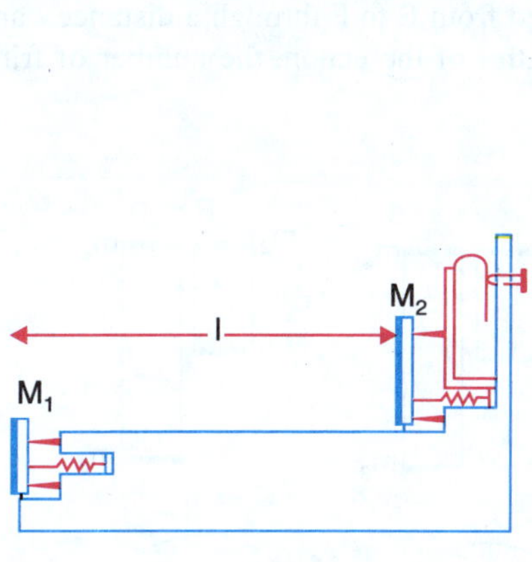

Fig. 15.27

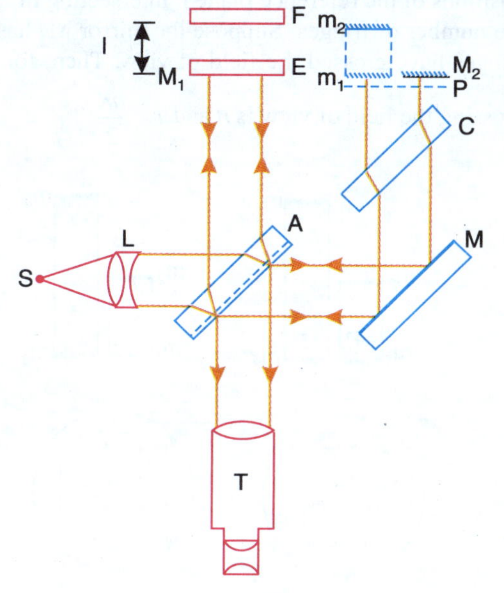

Fig. 15.28

after passing through the lens L is incident on the glass plate A. One portion of the beam is reflected towards M_1 while the other part after reflection from M falls on the mirrors m_1 and m_2 of the shortest etalon and the fixed mirror M_2. The centre of m_2 lies in a horizontal plane parallel and above the plane containing the centres of m_1 and M_2. P is the reference plane and is the image of M_1. The mirrors m_1 and m_2 are adjusted such that their planes are parallel to the reference plane P. Circular fringes are visible in the field of view of both the mirrors m_1 and m_2 when seen through the telescope T having a small aperture. Then the mirror M_1 is adjusted such that the reference plane P makes a small angle with m_1 and m_2. When the reference plane P intersects the plane of m_1 just in the middle, straight-line fringes are obtained with white light, as shown in Fig.15.29 (*a*). The fringes are due to the wedge-shaped film between P and m_1.

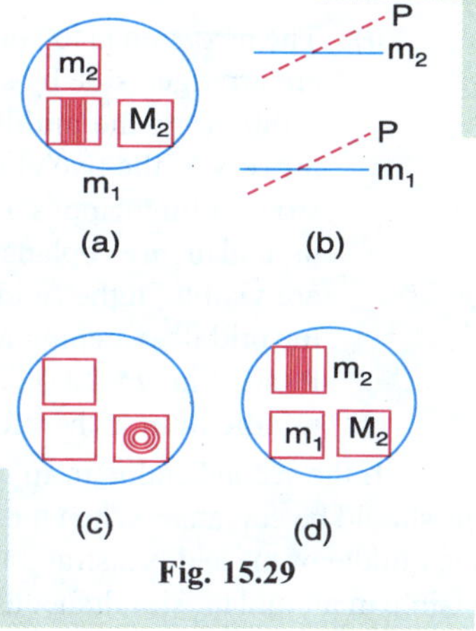

Fig. 15.29

White light is replaced by monochromatic cadmium light and the fixed mirror M_2 is adjusted to be perfectly parallel to M_1 so that circular fringes are visible in the field of view. It should be remembered that with white light, when the mirror M_1 is at E, straight-line fringes are visible in m_1 and with monochromatic cadmium red light circular fringes are visible in M_2 in the field of view of the telescope. The circular fringes are formed at infinity.

The mirror M_1 is moved and the number of circular fringes that cross the field of view are counted. When the plane of reference P intersects the middle of m_2, straight-line fringes will be seen in m_1 with white light (See Fig. 15.29d). White light is used only to note the initial and the final positions of the reference plane P intersecting m_1 and m_2, whereas cadmium light is used to count the number of fringes. Suppose the mirror M_1 has moved from E to F through a distance l and n fringes have crossed the field of view. Then, for a length l of the etalon, the number of fringes crossing the field of view is n and $l = \frac{n\lambda}{2}$.

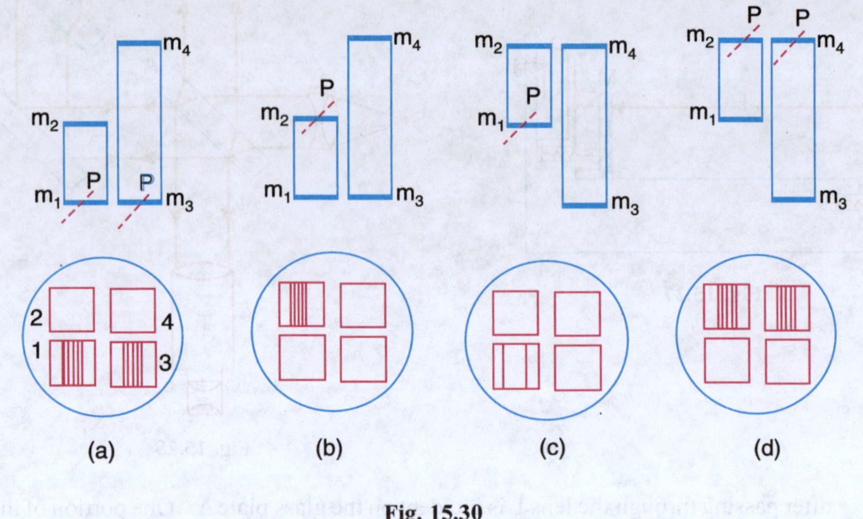

Fig. 15.30

(ii) The next step is to compare the shortest etalon with the next etalon. These two etalons are arranged side by side. First the mirror M_1 is adjusted such that the reference plane P intersects the middle of the mirrors m_1 and m_3. Here m_1 and m_3 are the two lower mirrors of the two etalons whereas m_2 and m_4 are the upper mirrors. With white light straight line fringes are visible in m_1 and m_3, as shown in Fig. 15.30 (a). It shows that m_1 and m_3 are coplanar. After this, the mirror M_1 is moved such that straight-line fringes are visible in the field of view of the upper mirror m_2 and the central dark fringe is in the middle. Keeping M_1 fixed, the etalon m_1m_2 is moved backwards until m_1 intersects P (see Fig. 15.30c). The etalon is adjusted such that the central dark fringe is in the middle of m_1. Therefore, the etalon has moved through a distance equal to its length.

If the second etalon m_3m_4 is exactly twice the length of the etalon m_1m_2, the mirrors m_2 and m_4 should be coplanar. When the mirror M_1 is moved such that the reference plane P just intersects the middle of m_2 and m_4, straight line fringes are produced (see Fig. 15.30d). If the fringes are not visible in m_2 and m_4 simultaneously, the mirrors m_2 and m_4 are not coplanar and the etalon m_3 m_4 is not exactly twice the length of the etalon m_1m_2. The etalon m_3m_4 is compared with etalon m_1m_2. Similarly, the etalon m_3m_4 is compared with the next etalon and the process is repeated. The 10 cm etalon is taken as a substandard and is compared directly with a prototype standard metre.

The final results with cadmium red, green and blue lines are:

a. Cadmium red line, $\lambda_R = 6438.4722$ Å and 1 metre = 1,553,163.5 λ_R.
b. Cadmium green line, $\lambda_G = 5085.8240$ Å and 1 metre = 1,966,249.7 λ_G.
c. Cadmium blue line, $\lambda_B = 4799.9107$ Å and 1 metre = 2,083,372.1 λ_B.

The green spectral line of mercury as emitted by the single isotope Hg (198) is considerably sharper than the cadmium red line. Its wavelength is 5460.7532 Å. It is also used to calibrate the standard metre.

15.9. TWYMAN AND GREEN INTERFEROMETER

Twyman and Green interferometer has been designed for testing the optical homogeneity of prisms, lenses and glass plates. It is also used for testing the rulings of gratings for the absence of ghosts.

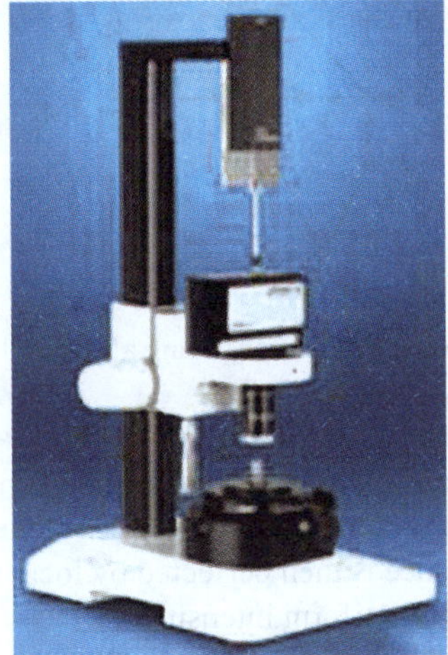

Twyman and Green Interferometer

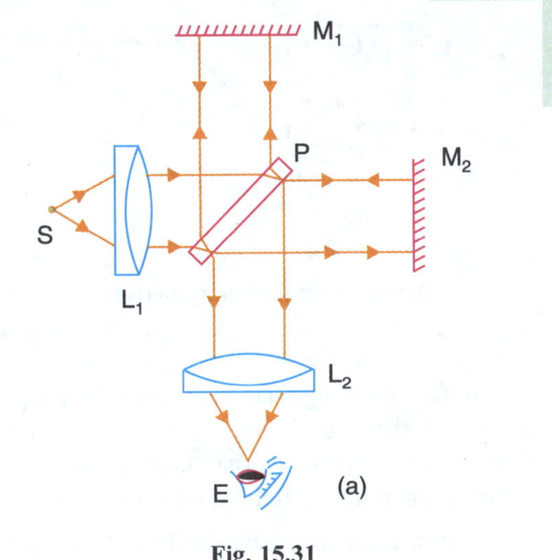

Fig. 15.31

This interferometer resembles the Michelson interferometer in which the extended source is replaced by a monochromatic point source S at the focus of a well-corrected lens L_1. Single plane wavefront emerging from L_1, after partial reflection at the half-silvered plate P set at 45°, gives rise to two plane wavefronts which fall normally on plane mirrors M_1 and M_2 set perpendicular to each other. The plane wavefronts reflected at M_1 and M_2 are superposed at P and focussed by a second well-corrected lens L_2 at E where the observer's eye is situated. When M_1 and M_2 are exactly perpendicular, the superposed plane wavefronts are exactly parallel. Therefore, the phase difference between the superposed disturbances is the same at every point in the field of view. The field is therefore, of uniform intensity depending on the difference between the paths PM_1P and PM_2P. The intensity is a maximum when this path difference is $n\lambda$, where $n = 0,1,2,...$

In order to adjust the instrument, an illuminated pinhole in a screen is placed at S in the focal plane of L_1, and also a screen is placed at E in the focal plane of L_2. The distance PM_1 is made equal to PM_2. Now, two images of the pinhole due to reflections at M_1 and M_2 are received on the screen alongside the pinhole itself. Similarly two images are seen on the screen at E. The mirrors M_1 and M_2 adjusted by tilting screws until the two images at S coincide with pinhole and at the same time the two images at E coincide. On removing the screen and using a monochromatic point source at S, no fringes should be seen. If otherwise, the mirrors are further adjusted until all types of fringes disappear.

Now, suppose after adjusting the interferometer the plane mirror M_2 is replaced by an optically perfect prism plus a plane mirror adjusted as in Fig. 15.31(b) or a perfect lens plus a spherical mirror adjusted as in Fig. 15.31(c). The field of view will remain of uniform intensity. If however the prism or the lens is not optically perfect, the wavefront returning to P will no longer be plane and the phase difference between the superposed disturbances will vary across the field of view and so fringes will be seen. Since fringes are foci of equal phase difference, the observed fringes are contours of the deformed wavefront. Thus, the imperfections of the prism or the lens are found in terms of the wavefront deformation.

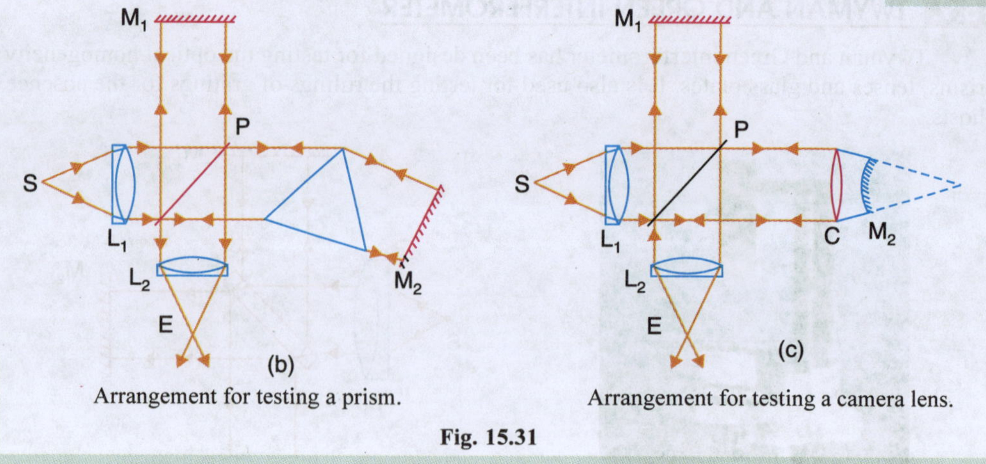

Arrangement for testing a prism.

Arrangement for testing a camera lens.

Fig. 15.31

In fact, the fringes are not in focus in any particular plane but if the eye is focussed on a prism face as seen through the lens L_2, the fringes are observed coincident with the prism face. These fringes are marked on the prism face by a pencil. The prism face is then perfected by local figuring and polishing, until the field of view in the interferometer is of uniform intensity.

In case of a lens, the lens under test is mounted on a nodal slide, which automatically maintains the centre of curvature of M_2 at the focus of the lens. If the lens is not perfect, the fringes will be seen which can be marked and corrected as in case of prism.

15.10. MACH-ZEHNDER INTERFEROMETER

This interferometer is used to study slight changes in refractive index of various gases over a considerable region. Its principle is similar to that of Jamin's interferometer. It consists of two beam splitters B_1 and B_2 and two totally reflecting mirrors M_1 and M_2 (Fig. 15.32). Two similar tubes T_1 and T_2 are placed in the two arms of interferometer and are evacuated. While T_1 remains evacuated, gas is admitted slowly into T_2. The number of fringes, that cross through the cetre of the field of view of the telescope, is counted. If the length of the tube is l and N fringes cross the field of view when the refractive index changes from μ_1 to μ_2, then

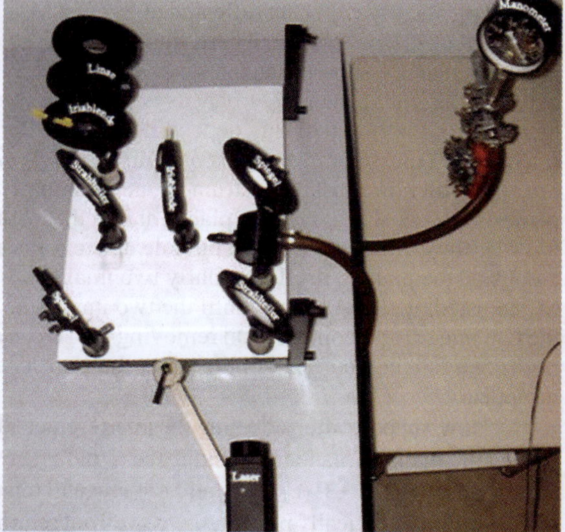

Mach-Zehnder Interferometer.

$$(\mu_2 l - \mu_1 l) = N\lambda$$

$$(\mu_2 - \mu_1)\ l = N\lambda$$

or $$\Delta\mu = \frac{N\lambda}{l} \tag{15.62}$$

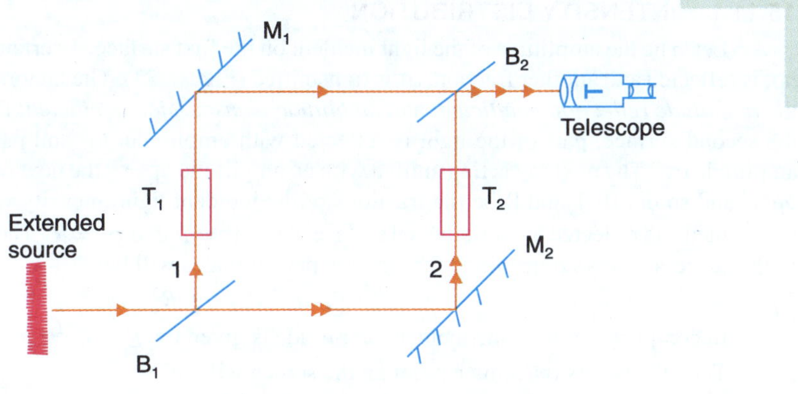

Fig. 15.32

Using the above expression the change in refractive index can be determined. Since the tube can be very long, a small change in refractive index can be measured accurately.

This interferometer is particularly useful in studying the flow pattern in wind tunnels. In this application only one tube is used, which is the test chamber, such as a wind tunnel or shock tube. Since some wind tunnels are several meters in diameter, the interferometer is usually very large. It has been found that the value of (μ – 1) is directly proportional to the air pressure P at a given temperature.

$$(\mu-1)=\frac{2}{3}P \tag{15.63}$$

The variation in air density in the tunnel (test chamber) therefore causes nonuniform distribution of refractive index. The direct beam and the beam passing through test chamber on combining produce fringe pattern.

15.11. MULTIPLE BEAM INTERFERENCE

We assumed in § 15.1 that the high order reflections occurring at interfaces of thin film are negligible. However, if for any reason the reflectance of the interfaces is not negligible, then the higher order reflections are to be taken into account. When the reflected or transmitted beams meet, multiple beam interference takes place. We are specifically interested in the fringes associated with an air space between two reflecting surfaces. Usually, these surfaces consist of metal films deposited on glass plates.

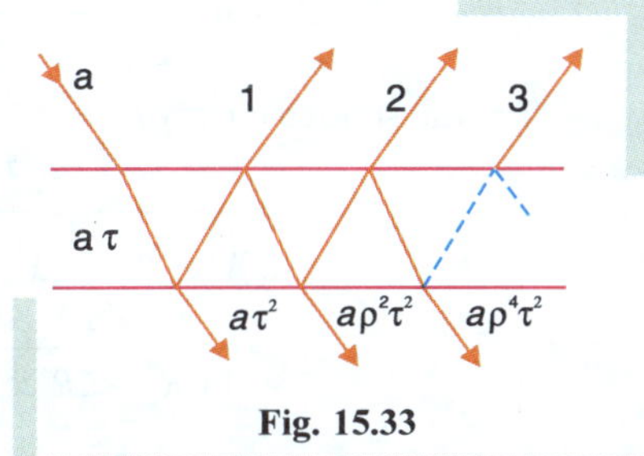

Fig. 15.33

Let us consider the reflected rays 1,2,3, etc as shown in Fig. 15.33. The amplitude of the incident ray is a. Let ρ be the reflection coefficient, τ the transmission coefficient.

The amplitude coefficient of reflection is

$$\rho=\frac{\text{amplitude of the reflected wave}}{\text{amplitude of the incident wave}} \tag{15.64}$$

If the film does not absorb light, the amplitudes of the reflected and transmitted waves are $a\rho$ and $a(1-\rho)$ respectively.

15.11.1. INTENSITY DISTRIBUTION

Let a be the amplitude of the light incident on the first surface. A certain fraction of this light, $a\rho$, is reflected and another fraction, $a\tau$ is transmitted (Fig. 15 33). The factors ρ and τ are known as the *amplitude reflection coefficient* and *amplitude transmission coefficient* respectively. Again, at the second surface, part of the light is reflected with amplitude $a\rho^2$ and part is transmitted with amplitude $a\tau^2$. The next ray is transmitted with an amplitude $a\rho^2\tau^2$, the next one with after that with $a\rho^4\tau^2$ and so on. If T and R be the fractions of the incident light intensity which are respectively transmitted and reflected at each silvered surface, then, $\tau^2 = T$ and $\rho^2 = R$. Therefore, the amplitudes of the successive rays transmitted through the pair of plates will be

$$a\,T,\; a\,T\,R,\; a\,T\,R^2, \ldots\ldots$$

In complex notation, the incident amplitude is given by $E = ae^{i\omega t}$.

Then the waves reaching a point on the screen will be

$$E_1 = a\,T\,e^{i\omega t}$$

$$E_2 = a\,T\,R\,e^{i(\omega t - \delta)}$$

$$E_3 = a\,T\,R^2\,e^{i(\omega t - 2\delta)}, \text{ and so on.}$$

$$\therefore \quad E_N = aR^{(N-1)}\,T\,e^{i\,[\omega t - (N-1)\delta]}$$

By the principle of superposition, the resultant amplitude is given by

$$A = a\,T + a\,T\,R\,e^{-i\delta} + a\,T\,R^2 e^{-2i\delta} + a\,T\,R^3 e^{-3i\delta} + \ldots.$$

$$= a\,T\left[1 + Re^{-i\delta} + R^2 e^{-2i\delta} + R^3 e^{-3i\delta} + \ldots..\right]$$

Using the expression for sum of the terms of a geometrical progression, we get

$$A = a\frac{1 - R^N e^{-iN\delta}}{1 - \text{Re}^{-i\delta}}$$

When the number of terms in the above expression approaches infinity, the term $R^N e^{-iN\delta}$ tends to zero, and the transmitted amplitude reduces to

$$A = a\,T\left[\frac{1}{1 - Re^{-i\delta}}\right] \tag{15.65}$$

The complex conjugate of A is given by

$$A^* = a\,T\left[\frac{1}{1 - Re^{+i\delta}}\right] \tag{15.66}$$

The transmitted energy $I_T = AA^*$

$$= \frac{a^2T^2}{\left(1 - Re^{-i\delta}\right)\left(1 - Re^{+i\delta}\right)} = \frac{a^2T^2}{1 + R^2 - R\left(e^{i\delta} + e^{-i\delta}\right)}$$

$$= \frac{a^2T^2}{1 + R^2 - 2R\cos\delta} = \frac{a^2T^2}{(1-R)^2 + 2R(1 - \cos\delta)}$$

$$= \frac{a^2T^2}{(1-R)^2 + 4R\sin^2\frac{\delta}{2}} = \frac{a^2T^2}{(1-R)^2}\left[\frac{1}{1 + \frac{4R}{(1-R)^2}\sin^2\frac{\delta}{2}}\right] \tag{15.67}$$

The intensity will be maximum when $\sin^2\frac{\delta}{2} = 0$, *i.e.*, $\delta = 2m\pi$, where m = 0,1,2,3,—Thus,

$$I_{max} = \frac{a^2T^2}{(1-R)^2} \tag{15.68}$$

The intensity will be a minimum, when $\sin^2\frac{\delta}{2} = 1$ .i.e., $\delta = (2m+1)\pi$ where m = 0,1,2,3... Thus,

$$I_{min} = \frac{a^2T^2}{(1-R)^2} \cdot \frac{1}{1+\frac{4R}{(1-R)^2}} = \frac{a^2T^2}{(1+R)^2} \tag{15.69}$$

We can now rewrite the equ. (15.67) as

$$I_T = \frac{I_{max}}{1+\frac{4R}{(1-R)^2}\sin^2\frac{\delta}{2}} \tag{15.70}$$

Similarly, the interference intensity from the reflected light beams can be shown to be

$$I_R = \frac{4R\sin^2\left(\frac{\delta}{2}\right)I_{max}}{\left(1-R^2\right)+4R\sin^2\left(\frac{\delta}{2}\right)} \tag{15.71}$$

15.11.2. COEFFICIENT OF FINESSE

We now introduce a quantity F, which is called the *coefficient of finesse*. It is defined as

$$F = \frac{4R}{(1-R)^2} \tag{15.72}$$

Then the relative inteference intensity distribution can be expressed as

$$\frac{I_T}{I_{max}} = \frac{1}{1+F\sin^2\left(\frac{\delta}{2}\right)} \tag{15.73}$$

and

$$\frac{I_R}{I_{max}} = \frac{F\sin^2\left(\frac{\delta}{2}\right)}{1+F\sin^2\left(\frac{\delta}{2}\right)} \tag{15.74}$$

15.11.3. VISIBILITY OF FRINGES

The visibility of fringes is given by $V = \frac{I_{max}-I_{min}}{I_{max}+I_{min}}$

Substituting the values of I_{max} and I_{min}, we get $V = \frac{2R}{1+R^2}$ (15.75)

Equ. (15.75) shows that the visibility of fringes is a function of reflectivity only. The visibility of fringes increases with increase in the value of R. V reaches the value 0.8 when R = 0.5 and approaches unity as R approaches 1. Thus the higher the reflectivity, the greater is the contrast of the fringes.

15.11.4. SHARPNESS OF THE FRINGES

If a plot is drawn for I against δ at different values of R, we obtain a set of curves, as shown in Fig. 15.34.

It is noted from the graphs that the intensity falls off on both sides of the maximum at higher values of R. The sharpness of a fringe is measured by the half-width of the curve. The **half-width** is the width of the I-δ curve at the position where $I = \frac{1}{2} I_{max}$. We have that

$$I = \frac{I_{max}}{1 + \frac{4R}{(1-R)^2}\sin^2\frac{\delta}{2}}$$

$$\therefore \quad \frac{1}{2} = \frac{1}{1 + \frac{4R}{(1-R)^2}\sin^2\frac{\delta}{2}}$$

$$\text{or} \quad 1 + \frac{4R}{(1-R)^2}\sin^2\frac{\delta}{2} = 2$$

$$\text{or} \quad \sin^2\frac{\delta}{2} = \frac{(1-R)^2}{4R}$$

$$\therefore \quad \delta = 2\sin^{-1}\left(\frac{1-R}{2\sqrt{R}}\right) \qquad (15.76)$$

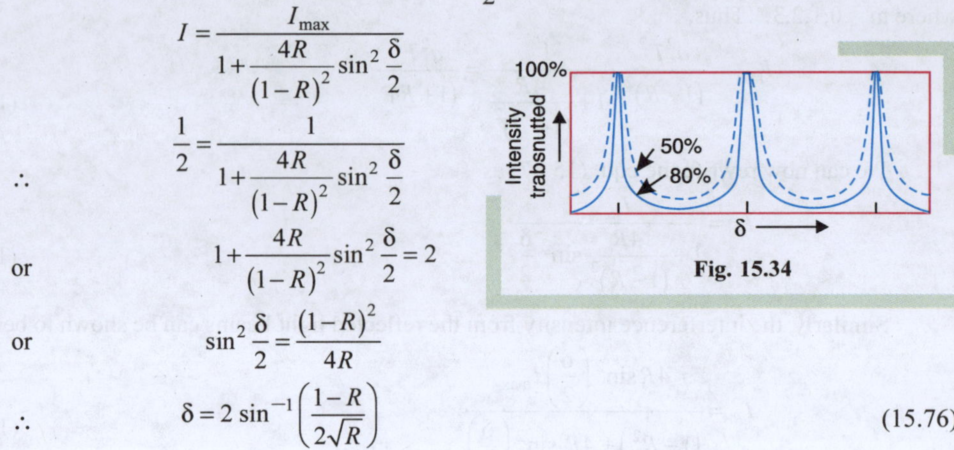

Fig. 15.34

15.12. FABRY-PEROT INTERFEROMETER AND ETALON

The Fabry-Perot interferometer is a high resolving power instrument, which makes use of the 'fringes of equal inclination', produced by the transmitted light after multiple reflections in an air film between two parallel highly reflecting glass plates.

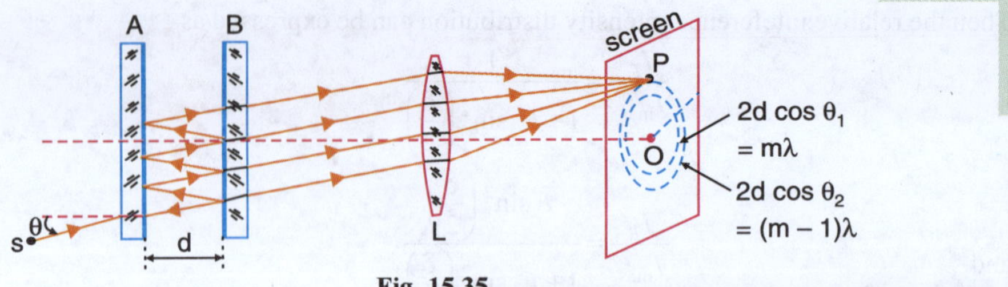

Fig. 15.35

The interferometer consists of two optically plane glass plates A and B with their inner surfaces silvered, and placed accurately parallel to each other. Screws are provided to secure parallelism if disturbed. This system is difficult to manufacture and is no more in use. Instead an etalon which is much more easily manufactured is used. The etalon consists of two semi-silvered plates rigidly held parallel at a fixed distance apart. The reflectance of the two surfaces can be as high as 90 to 99.9%. Although both reflected and transmitted beams interfere

Fabry-Perot Interferometer.

with each other, the Fabry-Perot interferometer is usually used in the transmissive mode.

S is a broad source of monochromatic light and L_1 a convex lens (not shown in Fig. 15.35) which makes the rays parallel. An incident ray suffers a large number of internal reflections successively at the two silvered surfaces, as shown in Fig.15.35. At each reflection a small fraction of the light is transmitted also. Thus, each incident ray produces a group of coherent and parallel, transmitted rays with a constant path difference between any two successive rays. A second convex lens L brings these rays together to a point in its focal plane where they interfere. Hence the rays from all points of the source produce an interference pattern on a screen placed in the focal plane of L.

15.12.1. FORMATION OF FRINGES

Let d be the separation between the two silvered surfaces and θ the inclination of a particular ray with the normal to the plates. The path difference between any two successive transmitted rays corresponding to the incident ray is $2d \cos \theta$. The condition for these rays to produce maximum intensity is given by

$$2d \cos \theta = m \lambda$$

where m is an integer. The locus of points in the source, which give rays of constant inclination, θ is a circle. Hence, with an extended source, the interference pattern consists of a system of bright concentric rings on a dark background, each ring corresponding to a particular value of θ.

15.12.2. DETERMINATION OF WAVELENGTH

When the reflecting surfaces A and B of the interferometer are adjusted exactly parallel, circular fringes are obtained. Let m be the order of the bright fringe at the centre of the fringe system. As at the centre $\theta = 0$, we have

$$2t = m\lambda$$

If the movable plate is moved a distance $\lambda/2$, 2t changes by λ and hence a bright fringe of the next order appears at the centre. If the movable plate is moved from the position x_1 to x_2 and the number of fringes appearing at the centre during this movement is N, then

$$N \cdot \frac{\lambda}{2} = x_2 - x_1$$

or
$$\lambda = \frac{2(x_2 - x_1)}{N} \tag{15.77}$$

Measuring x_2, x_1 and N, we can determine the value of λ.

15.12.3. MEASUREMENT OF DIFFERENCE IN WAVELENGTH

The light emitted by a source may consist of two or more wavelengths, as D_1 and D_2 lines in case of sodium. Separate fringe patterns corresponding to the two wavelengths are not produced in Michelson interferometer. Hence, Michelson interferometer is not suitable to study the *fine structure* of spectral lines. On the other hand, in Fabry-Perot interferometer, each wavelength produces its own ring pattern and the patterns are separated from each other. Therefore, Fabry-Perot interferometer is suitable to study the fine structure of spectral lines.

Difference in wavelengths can be found using coincidence method. Let λ_1 and λ_2 be two very close wavelengths in the incident light. Let us assume that $\lambda_1 > \lambda_2$. Initially, the two plates of the interferometer are brought into contact. Then the rings due to λ_1 and λ_2 coincide partially. Then the movable plate is slowly moved away such that the ring systems separate and maximum discordance occurs. Then the rings due to λ_2 are half way between those due to λ_1. Let t_1 be the separation between the plates when maximum discordance occurs. At the centre

$$2t_1 = m_1\lambda_1 = \left(m_1 + \tfrac{1}{2}\lambda_2\right) \tag{15.78}$$

or $$m_1(\lambda_1 - \lambda_2) = \lambda_2 / 2$$

$$\therefore \quad m_1 = \frac{\lambda_2}{2(\lambda_1 - \lambda_2)} \tag{15.79}$$

Using the value of m_1 in equ.(15.78), we get $2t_1 = \frac{\lambda_2}{2(\lambda_1 - \lambda_2)}\lambda_1$

$$\therefore \quad \lambda_1 - \lambda_2 = \frac{\lambda_1 \lambda_2}{4t_1} \cong \frac{\lambda_{mean}^2}{4t_1} \tag{15.80}$$

(since $\lambda_1\lambda_2 = \lambda_{mean}{}^2$ as $\lambda_1 - \lambda_2$ is very small).

When the separation between the plates is further increased, the ring systems coincide again and then separate out and maximum discordance occurs once again. If t_2 is the thickness now,

$$2t_2 = m_2\lambda_1 = \left(m_2 + \tfrac{3}{2}\lambda_2\right) \tag{15.81}$$

From equ.(15.81) and (15.78), we get

$$2(t_2 - t_1) = (m_2 - m_1)\lambda_1 = (m_2 - m_1)\lambda_2 + \lambda_2 \tag{15.82}$$

or $$(m_2 - m_1) = \frac{\lambda_2}{\lambda_1 - \lambda_2}$$

Using the above expression into equ.(15.82), we obtain

$$2(t_2 - t_1) = \frac{\lambda_1\lambda_2}{(\lambda_1 - \lambda_2)}$$

or $$\lambda_1 - \lambda_2 = \frac{\lambda_1\lambda_2}{2(t_2 - t_1)} \cong \frac{\lambda_{mean}^2}{2(t_2 - t_1)} \tag{15.83}$$

15.13. LUMMER AND GEHRCKE PLATE

It consists of a plane parallel glass plate of about 10 cm long and a few mm thick. A prism C is cemented at one end. A beam of parallel light enters the prism C and after reflection falls on the plate (Fig. 15.36). The angle of incidence at which the ray strikes the face A of the plate is slightly less than the critical angle for the material of the plate. The beam is reflected up and down between the faces A and B of the plate. At each reflection, a beam leaves the plate. The path difference is the same between the beams. They are received by the telescope and interference fringes are viewed in the field of view. It is used for observing the fine structure of spectral lines and Zeeman effect.

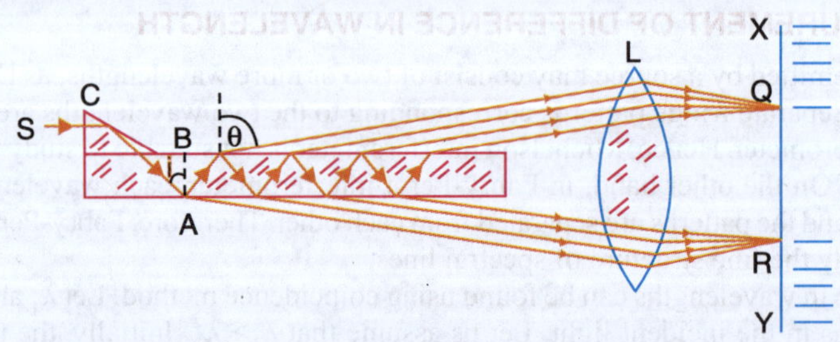

Fig. 15.36

15.14. APPLICATIONS OF THIN FILM INTERFERENCE

The application of interference phenomenon is wide and varied. Interference is used for making precision measurements. For example, the wavelength of light can be measured with accuracy up to eight significant digits. Therefore, interferometers are used to determine and redefine the length of a standard metre. Standard metre was defined formerly as the distance between two marks on a platinum-iridium bar. According to the modern definition the standard metre is a length which contains exactly 1,650,763.73 wavelengths of orange-red light emitted by krypton-86. Another interesting application is in astronomy where double slit interference is used to determine the angular separation of double stars and the diameter of fixed stars. We discuss here a few selected applications.

15.14.1. MEASUREMENT OF SMALL DISPLACEMENTS

The interference phenomenon is used to determine small displacements such as those produced by compression or elongation of a metal rod, and thermal expansion of crystals etc. The order of crystal expansion is quite small and it can be conveniently measured by the interference methods. We study here how Fizeau adopted Newton's rings to study the thermal expansion of a crystal.

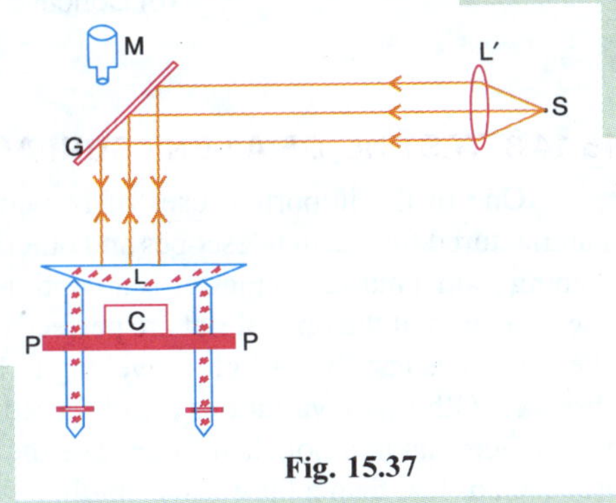

Fig. 15.37

The crystal under test is placed on a metal disc supported by three screws, and a lens is kept on the ends of the screws, as shown in Fig. 15.37. A thin film of air is left between the crystal and the lens. The position of the lens can be adjusted with the help of the screws. When monochromatic light is made incident on the lens crystal combination, Newton's rings are produced which can be examined by means of a travelling microscope. When the crystal is heated, the thickness of the air film between the lens decreases due to thermal expansion of the crystal. As a result, Newton's rings undergo expansion. The expanding rings are counted with the help of the microscope. The distance x between the lens and the plate at any instant is given by

$$2\mu\left(x+\frac{r_m^2}{2R}\right)=m\lambda \tag{15.84}$$

where r_m is the radius of the mth dark ring seen by reflected light, R the radius of curvature of the lower surface of the lens and μ is the refractive index of air between the lens and the plate.

15.14.2. TESTING OF SURFACE FINISH

In modern technology interference is widely used for estimating the quality of a surface finish. Machine components retain surface irregularities left after machining. The extent of suitability of the component for a particular application depends on the irregularities which act as sources of stress leading to fatigue cracks. The surfaces of components, which are going to be subjected to high stresses and load reversals, are therefore required to have a high surface finish. The smoothness of the surface may be quickly inspected visually by keeping an optical flat on the component at an angle and illuminating it with monochromatic light. The air wedge formed therein produces straight and equidistant bands if the surface of the component is smooth. If the bands are curved towards the contact edge the surface is concave and if the fringes curve away, it is convex, as shown in Fig.15.38.

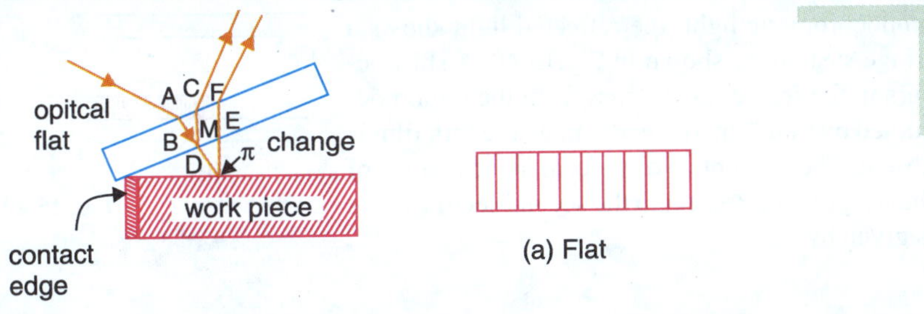

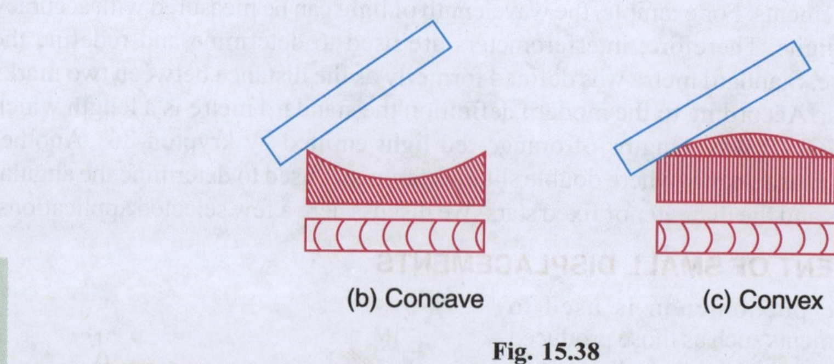
(b) Concave (c) Convex

Fig. 15.38

15.14.3. TESTING OF A LENS SURFACE

One of the important uses of Newton's rings is in the testing of the optical components manufactured for use in telescopes and other instruments. The grinding of a lens surface is tested by keeping it on a master. A master is an optical flat which is a cylindrical disc made of fused quartz. The two faces of the optical flat are perfectly parallel to each other. The departure from the flatness of each face is less than a light wavelength. If a lens is ground perfectly, a circular fringe pattern is observed. Otherwise variations are observed (Fig. 15.39*b*)which give an indication of how the lens must be ground and polished to remove the imperfections. High quality lenses are ground with a precision of less than a light wavelength.

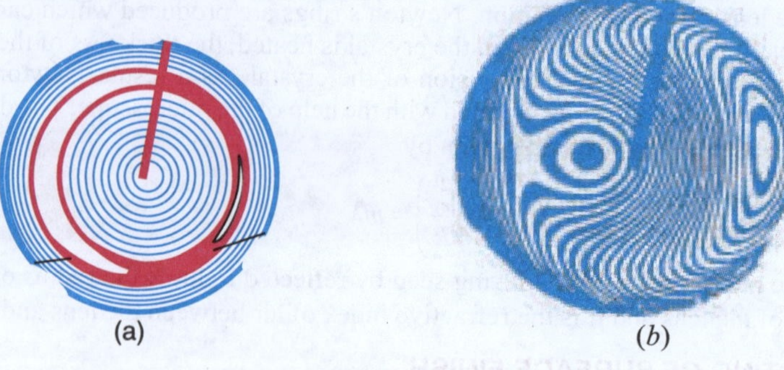
(a) (*b*)

Fig. 15.39

15.14.4. THICKNESS OF A THIN FILM COATING

Dielectric and metallic thin films are often coated on optical components, solar cells etc. One of the methods of determination of thickness of such thin films is based on multiple beam interference. A partially coated substrate is used for the determination. The surfaces of the substrate and the thin film on it are coated with a transparent metallic film of uniform thickness. A glass plate is also coated on one of its surfaces with the transparent metallic film. When the substrate and the glass plate are placed in contact and examined under monochromatic light, the reflected light shows a fringe system, as shown in Fig.15.40. A shift occurs in the fringes as we pass from the region occupied by thin film to the region where thin film is absent. The amount of displacement of one set of the fringes with respect to the second set of fringes is given by

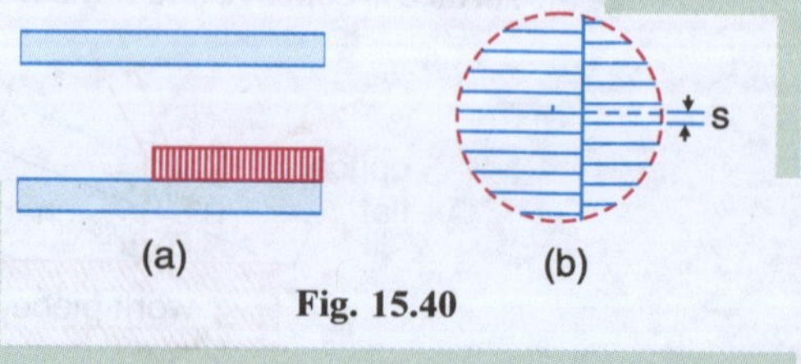

(a) (b)

Fig. 15.40

$$s = 2t \quad \text{or} \quad t = s/2$$

where t is the thickness of the thin film. By measuring 's', t can be calculated.

15.15. ANTIREFLECTION COATINGS

One of the most important applications of thin film interference is in producing antireflection coatings. Optical instruments such as cameras and telescopes use multi-component glass lenses. It is noted that part of the light incident on a glass surface is reflected and that much amount is subtracted from the transmitted light. When the number of reflections are large, the quality of the image produced by the device will be poor. Alexander Smakula discovered in 1935 that the reflections from a surface can be reduced by coating the surface with a thin transparent dielectric film.

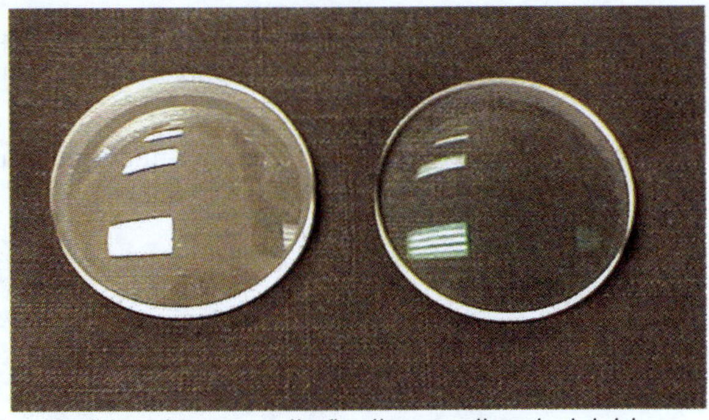

Left lens has no antireflection coatings but right lens has antireflection coatings.

A transparent thin film coated on a surface with a view to suppress the surface reflections is called an antireflection (AR) coating or a nonreflecting film.

A thin film can act as an AR coating if it meets the following two conditions:

(*i*) **Phase condition:** The waves reflected from the top and bottom surfaces of the thin film are in *opposite phase* such that their overlapping leads to destructive interference, and

(*ii*) **Amplitude condition:** The waves have *equal* amplitudes.

The above conditions enable us determine respectively (a) the required thickness of the film and (b) the refractive index of the material to be used for forming the film.

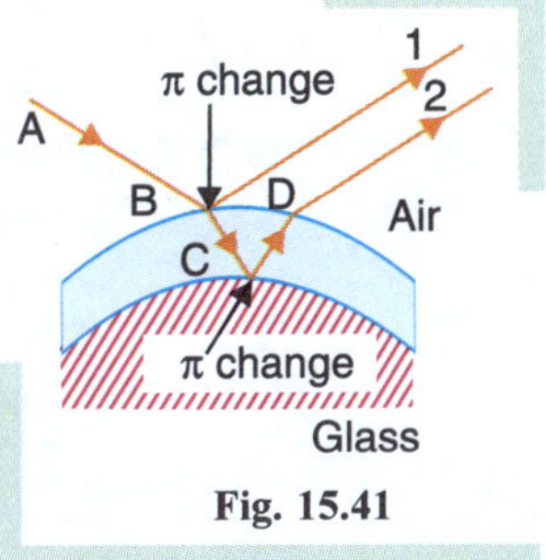

Fig. 15.41

(*i*) **Phase condition and minimum thickness of the film:** Let the thickness of the film be t and the refractive index of the film-material be μ_f .The phase condition requires that the waves (ray 1 and ray 2) reflected from the top and bottom surfaces of thin film be 180° out of phase. It requires that the optical path difference between the two rays must equal one half-wave or an odd number of half-waves. Referring to Fig.15.41, the optical path difference between ray 1 and ray 2 is

$$\Delta = 2\mu_f t \cos r - \lambda/2 - \lambda/2$$

The first $\lambda/2$ corresponds to the π change at the top surface of the film (air-to-film boundary) and the second $\lambda/2$ to the π change that occurs at the film-to glass boundary. If we assume normal incidence of light, $\cos r = 1$ and the above equation reduces to

$$\Delta = 2\mu_f t - \lambda = 2\mu_f t$$

We wrote the above equality remembering that an *addition of a full wave or subtraction of a full wave from a train of waves does not affect the original phase relation.* The ray 1and ray 2 interfere destructively if the optical path difference satisfies the condition that

$$\Delta = (2m+1)\lambda/2$$

Thus, it requires that $2\mu_f t = (2m+1)\lambda/2$

For the film to be transparent, its thickness should be a minimum, which happens when m = 0.

$$2\mu_f t_{min} = \lambda/2$$

$$t_{min} = \frac{\lambda}{4\mu_f} \qquad (15.85)$$

It means that the optical thickness of the AR coating should be of one-quarter wavelength. Such quarter-wavelength coatings suppress the reflections and cause the light to pass into the transmitted component.

(ii) **Amplitude condition :** The amplitude condition requires that the amplitudes of reflected rays, ray 1 and ray 2 are equal. That is,

$$E_1 = E_2 \qquad (15.86)$$

It requires that

$$\left[\frac{\mu_f - \mu_a}{\mu_f + \mu_a}\right]^2 = \left[\frac{\mu_g - \mu_f}{\mu_g + \mu_f}\right]^2 \qquad (15.87)$$

where μ_a, μ_f, and μ_g are the refractive indices of air, thin film and glass substrate respectively. As $\mu_a = 1$, the above expression may be rewritten as

$$\left[\frac{\mu_f - 1}{\mu_f + 1}\right]^2 = \left[\frac{\mu_g - \mu_f}{\mu_g + \mu_f}\right]^2$$

Expanding the above equation, we get

$$\frac{\mu_f^2 - 2\mu_f + 1}{\mu_f^2 + 2\mu_f + 1} = \frac{\mu_g^2 - 2\mu_g\mu_f + \mu_f^2}{\mu_g^2 + 2\mu_g\mu_f + \mu_f^2}$$

$$4\mu_f^3\mu_g + 4\mu_f\mu_g = 4\mu_f^3 + 4\mu_f\mu_g^2$$

Dividing by 4 μ_f and rearranging the terms

$$\mu_f^2 - \mu_g\mu_f^2 + \mu_g^2 - \mu_g = 0$$

$$\mu_f^2 = \mu_g\left(1 + \mu_f^2 - \mu_g\right)$$

$$\therefore \qquad \mu_f^2 \cong \mu_g$$

$$\therefore \qquad \mu_f = \sqrt{\mu_g} \qquad (15.88)$$

It implies that the refractive index of thin film should be less than that of the substrate and possibly nearer to its square root.

In case of glass, if we take $\mu_g = 1.5$, $\mu_f = \sqrt{\mu_g} = 1.22$.

The materials which have refractive index nearer to this value are magnesium fluoride, MgF_2 (μ = 1.38) and cryolite, $3NaF.AlF_3$ (μ =1.36). Apart from the refractive index, the material should possess some more additional properties. The film should adhere well, should be durable, scratch proof and insoluble in ordinary solvents. MgF_2 and cryolite satisfy these requirements. However, among the two, magnesium fluoride is cheaper and is hence widely used as AR coating.

It may be noted that the condition (15.85) is satisfied only at one particular wavelength. The wavelength normally chosen is 5500 Å for which the eye is most sensitive. This wavelength is located in the yellow-green portion of the spectrum. Consequently, the reflection of red and violet light will be larger when white light is incident on the component. Hence, the component shows *purple hue* in reflected light.

15.15.1. MULTILAYER AR COATINGS

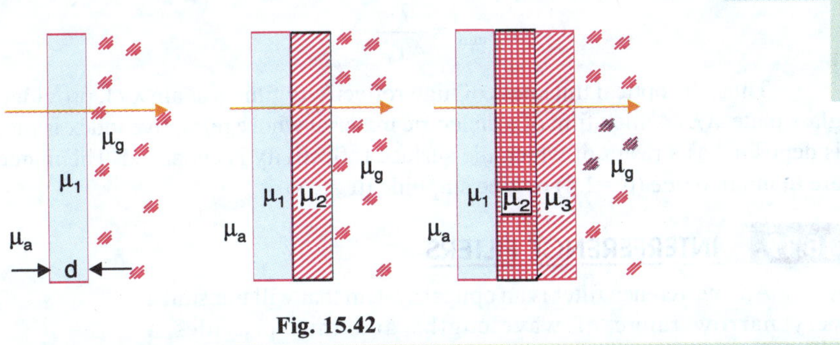

Fig. 15.42

A single layer AR coating is effective only at one particular wavelength. A much wider coverage across the spectrum is possible with multiple coatings, called *multilayers*. In practice three layer coatings are widely used and are highly effective over most of the visible spectrum. The central layer is half-wave(λ / 2) thick and is of high refractive index materials such as zirconium dioxide (ZrO_2, μ =2.1). The outside layer is of magnesium fluoride having λ / 4 thickness and the layer adjacent to the substrate is again a λ /4 thick coating of cesium fluoride (CeF_3, μ =1.63) or aluminium oxide (Al_2O_3, μ=1.76). Some of the antireflection coatings use up to 100 layers of alternating high and low refractive index materials.

15.16. DIELECTRIC MIRRORS

Another important application of the thin film interference phenomenon is in *increasing* the reflectivity of a substrate. If the refractive index of the thin film coated on glass is higher than that of

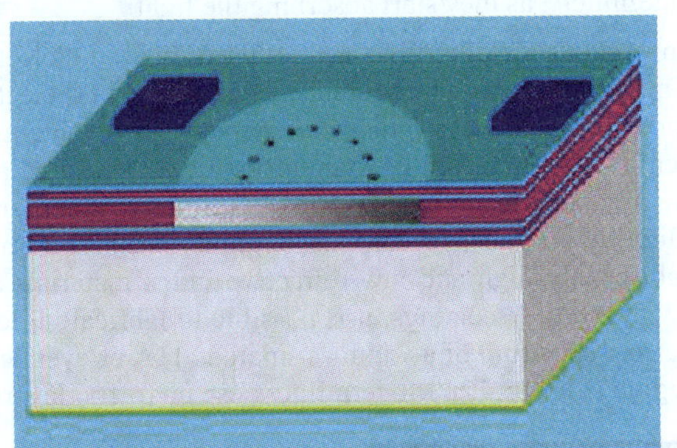

Two dielectric mirrors separated by an air gap.

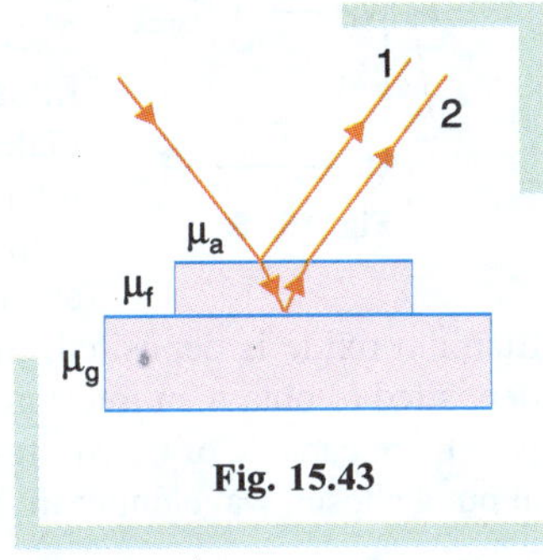

Fig. 15.43

glass ($\mu_f > \mu_g$), then the reflectivity of glass surface increases. Referring to Fig. 15.43, the optical path between rays 1 and 2 is given by

$$\Delta = 2\mu_f t - \lambda / 2$$

where normal incidence of light is assumed. Ray 1 and 2 should constructively interfere if the reflection is to be more from the surface. Therefore, the condition that $\Delta = m\lambda$ has to be satisfied. Thus, the condition for more reflection is

$$2\mu_f t - \lambda/2 = m\lambda$$

or
$$2\mu_f t = (2m+1)\lambda/2$$

$m = 0$ gives the minimum thickness of the coating.

$$\therefore \quad t_{min} = \frac{\lambda}{4\mu_f} \tag{15.89}$$

Thus, the optical thickness of high reflectivity film is again $\lambda/4$, provided $\mu_f > \mu_g$. Thus, on a glass plate, a $\lambda/4$ thick film of a dielectric material whose refractive index is more than that of glass is deposited as a result of which the surface reflectivity is enhanced. The materials generally used are titanium oxide (μ =2.8) or zinc sulphide (μ =2.3).

15.17. INTERFERENCE FILTERS

An interference filter is an optical system that will transmit a very narrow range of wavelengths and thus provides a monochromatic beam of light.

Interference filters are fabricated earlier as follows. A thin metallic film, usually of aluminium or silver, is deposited on a glass substrate by vacuum deposition technique. Then a thin layer of cryolite is deposited over this. The structure is again covered by another metallic film. Another plate is placed over it to protect the thin film structure. The filter is shown in Fig. 15.44. By varying the thickness of the dielectric film, any particular wavelength can be filtered out. However, the filtered light will have a narrow spectrum centered on the chosen wavelength. By increasing the reflectivity of the surfaces, the transmitted spectrum can be made narrower. But it is not possible to increase the thickness of metallic films indefinitely, as they start absorbing the light.

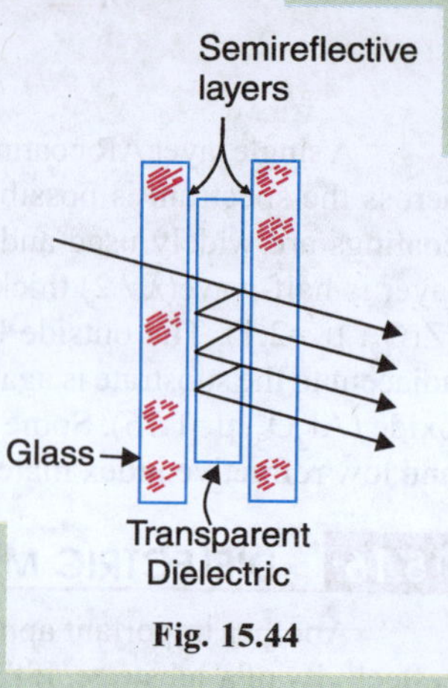

Fig. 15.44

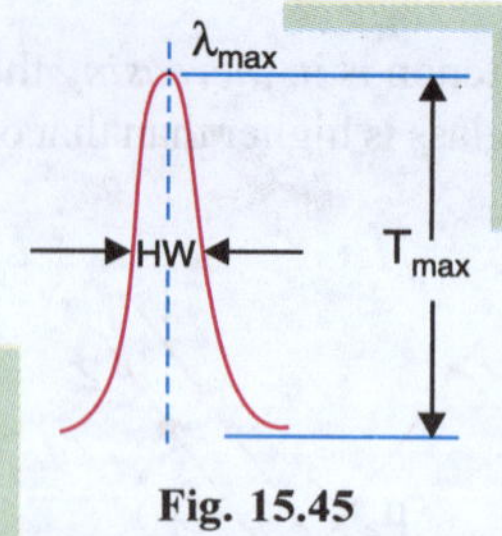

Fig. 15.45

In modern versions metallic films are not used; instead dielectric films are used. In an all dielectric interference filter, layers of dielectric materials of appropriate refractive indices are deposited. To obtain an interference filter, a $\lambda/4$ thick film of titanium oxide is deposited and then over it a film of dielectric material with lower refractive index, such as magnesium fluoride is deposited. On this, again a $\lambda/4$ thick film of titanium oxide is deposited. In this way alternately high and low refractive index materials are deposited to obtain an interference filter. With multiple coatings, it is possible to fabricate filters, which are capable of transmitting a very narrow spectrum of a width as small as 11Å or even less, about a chosen wavelength in the visible region (Fig. 15.45). Modern filters use up to 100 layers.

WORKED OUT PROBLEMS

Example 15.1: A soap film 5×10^{-5} cm thick is viewed at an angle of 35° to the normal. Find the wavelengths of light in the visible spectrum which will be absent from the reflected light ($\mu = 1.33$).

Solution: Let i be the angle of incidence and r be the angle of refraction. Then

$$\mu = \frac{\sin i}{\sin r} \quad \therefore \quad 1.33 = \frac{\sin 35°}{\sin r} \quad \therefore r = 25.5° \text{ and } \cos r = 0.90$$

The condition for destructive interference is $2\mu\, t \cos r = m\lambda$.

Using different values for m in the above relation, we get following values for wavelengths.

When m = 1, $\lambda_1 = 2 \times 1.33 \times 5 \times 10^{-5}$ cm $\times 0.90 = 12.0 \times 10^{-5}$ cm =**120μm.**

When m = 2, $\lambda_2 = (2 \times 1.33 \times 5 \times 10^{-5}$ cm $\times 0.90) \div 2 = 6.0 \times 10^{-5}$ cm = **6000 Å.**

When m = 3, $\lambda_3 = (2 \times 1.33 \times 5 \times 10^{-5}$ cm $\times 0.90) \div 3 = 4.0 \times 10^{-5}$ cm = **4000 Å.**

When m = 4, $\lambda_4 = (2 \times 1.33 \times 5 \times 10^{-5}$ cm $\times 0.90) \div 4 = 3.0 \times 10^{-5}$ cm = **3000 Å.**

Out of the above wavelengths, λ_2 = 6000 Å, and λ_3 = 4000 Å lie in the visible region. Therefore, these two wavelengths are absent in the reflected light.

Example 15.2: A glass wedge of angle 0.01 radian is illuminated by monochromatic light of wavelength 6000 Å falling normally on it. At what distance from the edge of the wedge will the 10th fringe be observed by reflected light?

Solution: Given that θ = 0.01 rad, m = 10, $\lambda = 6000 \times 10^{-8}$ cm.

The condition for dark fringe is $2\,t = m\,\lambda$

The angle of the wedge $\theta = \frac{t}{x}$ or $t = \theta x$

$$\therefore \quad 2\,\theta\, x = m\,\lambda$$

or
$$x = \frac{m\lambda}{2\theta} = \frac{10 \times 6000 \times 10^{-8}\ \text{cm}}{2 \times 0.01} = \mathbf{3\ mm}$$

Example 15.3: A beam of monochromatic light of wavelength 5.82×10^{-7} m falls normally on a glass wedge with the wedge angle of 20 seconds of an arc. If the refractive index of glass is 1.5, find the number of dark fringes per cm of the wedge length.

Solution: Given wedge angle $\theta = 20'' = \frac{20 \times \pi}{60 \times 60 \times 180}$ radians, $\lambda = 5.82 \times 10^{-7}$ m, μ = 1.5.

$$\text{Fringe width } \beta = \frac{\lambda}{2\mu\theta} = \frac{5.82 \times 10^{-7}\ \text{m} \times 60 \times 60 \times 180}{2 \times 1.5 \times 20 \times \pi} = 2\ \text{mm}.$$

$$\text{Number of fringes per cm} = \frac{1}{0.2\ \text{cm}} = \mathbf{5\ per\ cm}.$$

Example 15.4: A thin equiconvex lens of focal length 4 m and refractive index 1.50 rests on and in contact with an optical flat, and using light of wavelength 5460Å, Newton's rings are viewed normally by reflection. What is the diameter of the 5th bright ring?

Solution: Given that m =5, λ = 5460Å = 5460×10^{-10} m, f = 4m, μ =1.5

We know that $\frac{1}{f} = (\mu - 1)\left[\frac{1}{R_1} - \frac{1}{R_2}\right]$. Here R_1 = R and R_2 = – R

$$\therefore \quad \frac{1}{f} = (\mu - 1)\left[\frac{2}{R}\right]$$

$$\therefore \quad \frac{1}{4\text{m}} = 0.5\left[\frac{2}{R}\right] \text{ or } R = 4\ \text{m}.$$

The diameter of the mth bright ring is given by

$$D_m = \sqrt{2(2m-1)\lambda R} = \sqrt{2(2 \times 5 - 1) \times 5460 \times 10^{-10}\ \text{m} \times 4\text{m}} = \mathbf{6.2\ mm}$$

Example 15.5: Newton's rings are observed in reflected light of λ = 5.9 × 10 $^{-5}$ cm. The diameter of the 10th dark ring is 0.5 cm. Find the radius of curvature of the lens and the thickness of the air film.

Solution: Given that λ = 5.9 × 10 $^{-5}$ cm, m = 10.

The radius of mth dark ring is given by

$$\therefore \quad R = \frac{r^2}{m\lambda} = \frac{(0.5\,\text{cm})^2}{10 \times 5.9 \times 10^{-5}\,\text{cm}} = 106\text{ cm} = \mathbf{1.06m}$$

The thickness of air film is given by

$$\therefore \quad t = \frac{m\lambda}{2} = \frac{10 \times 5.9 \times 10^{-5}\,\text{cm}}{2} = \mathbf{2.95\mu m.}$$

Example 15.6: In a Newton's rings experiment, the diameter of 10th dark ring due to wavelength 6000 Å in air is 0.5 cm. Find the radius of curvature of the lens.

Solution: Radius of curvature, $R = \frac{(D/2)^2}{m\lambda} = \frac{(0.5 \times 10^{-2}/2)^2\,\text{m}^2}{10 \times 6000 \times 10^{-10}\,\text{m}} = \mathbf{1.04m}$

Example 15.7: In a Newton's rings experiment the diameter of the 15th ring was found to be 0.59 cm and that of the 5th ring was 0.336 cm. If the radius of the plano-convex lens is 100 cm, calculate the wavelength of light used.

Solution: $\lambda = \frac{D_{m+p}^2 - D_m^2}{4pR} = \frac{D_{15}^2 - D_5^2}{4 \times 10 \times R} = \frac{(5.9 - 3.36)^2 \times 10^{-6}\,\text{m}^2}{4 \times 10 \times 1\text{m}} = \mathbf{5880\ Å.}$

Example 15.8: In a Newton's rings experiment the diameter of 10th ring changes from 1.40 to 1.27 cm when a drop of liquid is introduced between the lens and the glass plate. Calculate the refractive index of the liquid.

Solution: $\mu = \frac{\left(D_m^2\right)_{air}}{\left(D_m^2\right)_{liq.}} = \frac{(1.40\,\text{cm})^2}{(1.27\,\text{cm})^2} = \mathbf{1.215}$

Example 15.9: In a Michelson interferometer 200 fringes cross the field of view when the movable mirror is moved through 0.0589 mm. Calculate the wavelength of light used.

Solution: $\lambda = \frac{2d}{m} = \frac{2 \times 0.0589 \times 10^{-3}\,\text{m}}{200} = \mathbf{5890\ Å}$

Example 15.10: In an experiment for determining the refractive index of gas using Michelson interferometer a shift of 140 fringes is observed, when all the gas is removed from the tube. If the wavelength of light used is 5460 Å and the length of the tube is 20 cm, calculate the refractive index of the gas.

Solution: Given that n = 140, λ = 5460 Å = 5460 × 10 $^{-10}$m, l = 20 cm = 0.2 m.

$$\mu = 1 + \left(\frac{n\lambda}{2l}\right) = 1 + \left[\frac{140 \times 5460 \times 10^{-10}\,\text{m}}{2 \times 0.2\,\text{m}}\right] = \mathbf{1.00019.}$$

Example 15.11: A glass microscope lens (μ = 1.50) is coated with magnesium fluoride (μ = 1.38) film to increase the transmission of normally incident yellow light (λ = 5800 Å). What minimum film thickness should be deposited on the lens?

Solution: Given that μ_g = 1.50, μ_f = 1.38, λ = 5800 Å.

Minimum film thickness $t_{min} = \frac{\lambda}{4\mu_f} = \frac{5800}{4 \times 1.38}\text{ Å} = \mathbf{1050\ Å}$

QUESTIONS

1. How will you find the wavelength of monochromatic light by using Michelson's Interferometer ? **(Nagpur, 2005)**
2. Draw a well labeled diagram of Michelson Interferometer. **(Nagpur, 2005)**
3. (*i*) Describe Michelson Interferometer with a neat diagram.
   (*ii*) How it is used to find difference between two close wavelengths?
4. (*i*) Explain the formation of circular fringes in Michelson Interferometer.
   (*ii*) Describe how you will use Michelson Interferometer to determine the difference between two wavelengths very close to each other.
5. With the help of a neat diagram, explain the construction of Fabry-Perot Interferometer.
6. Explain the construction of Michelson's interferometer. How it is used to determine the difference in wave lengths between two closely spaced spectral lines. **(Bangalore, 2005)**
7. Give the theory of Newton's rings. **(Kovempu, 2005)**
8. Describe the working of a Michelson interferometer. State the condition for obtaining white light fringes. Show with necessary theory how this interferometer can be used to measure wavelength of light. **(Agra, 2005)**
9. How is the wavelength of sodium light determined by Newton's rings method? Derive the formula used. Why are the rings circular? **(Meerut, 2005)**
10. Describe the construction and working of Michelson's interferometer. How will you use it to determine the difference between wavelength of two D-lines of sodium? **(Meerut, 2005)**
11. What will happen if wedge shaped film is placed in white light? **(Meerut, 2005)**
12. Soap bubble or a thin film of oil spread over the surface of water appears coloured in sunlight. Why? **(Lucknow, 2004)**
13. How would you obtain Newton's rings with bright center? **(Lucknow, 2004)**
14. Explain the construction and working of a Fabry-Perot interferometer. **(Lucknow 2004)**
15. Light containing two wavelengths λ_1 and λ_e falls normally on a plano convex lens of radius of curvature R, resting on a plane glass plate. If the nth dark ring due to λ_1 coincides with the ($_n$+1) th dark ring due to λ_e, show that the radius of the $_n$th dark ring of λ_1 is :
$$\sqrt{\frac{\lambda_1\lambda_e R}{\lambda_1-\lambda_e}}$$
(Lucknow, 2004)
16. Describe a Michelson interferometer. How can it be used, for measuring the wavelength of monochromatic light? **(Garhwal, 2005)**
17. What are Newton's rings? Describe an experiment to determine the radius of curvature of a plae convex lens. **(Gulbarga, 2005)**
18. Describe Michelson interferometer and explain the formation of fringes in it. **(Punjab, 2005)**
19. What will happen if a transparent thin sheet is introduced in the path of one of the interfering beams? **(Punjab, 2005)**
20. What is the effect of phase change at each reflection in a Fabry – Perot interferometer? **(Punjab, 2005)**
21. Describe with theory, the Newton's rings experiment to determine the wavelength of monochromatic source of radiation. **(A.P.University, 2010)**
22. What happens to the rings pattern when a liquid is introduced between the plano-convex lens and plate in Newton's rings experiment? **(RTMNU, 2010)**
23. Explain the complementary nature of fringes due to reflected and transmitted light in thin films. **(RTMNU, 2010)**
24. Describe the principle and working of Fabry-Parot interferometer. **(GNDU, Amritsar, 2010)**

25. Describe the construction and working Fabry – Perot interferometer. How is it used to determine wavelength of light? How is it superior to Michelson's interferometer? **(RTMNU, Nagpur, 2010)**
26. Obtain the conditions for the observing the interference fringes. **(GNDU, Amritsar, 2010)**
27. Discuss applications of Michelson's interferometer. **(GNDU, Amritsar, 2010)**
28. Show that the separation between two successive bright Newton's rings is given by:

$$S = \frac{\sqrt{2\lambda R}}{\sqrt{2n+1}+\sqrt{2n-1}}$$ **(RTMNU, 2010)**

PROBLEMS FOR PRACTICE

1. In Michelson Interferometer 100 fringes cross the field of view when movable mirror is displaced through 2.948×10^{-3} cm. Calculate the wave length of monochromatic light used. **(Nagpur, 2005)**
2. When the movable mirror of Michelson's interferometer is shifted by 0.0589 mm, a shift of 200 fringes is obtained. Find the wavelength of light. **(Nagpur, 2005)**
3. By how much distance the movable mirror or a Michelson Interferometer be moved to obtain consecutive positions of maximum distinctness for sodium D-lines λ_1 = 5890 Å and λ_2 = 5896 Å. **(Nagpur, 2004)**
4. Calculate the distance between two successive position of a movable mirror of a Michelson Interferometer giving maximum visibility fringes for sodium D-lines of wavelength 5890 Å and 5896 Å.
5. Calculate the minimum thickness of a half wave plate of calcite for wavelength 5890 Å. Given : For Calcite $\mu_o = 1.658$, $\mu_o = 1.486$. **(Nagpur, 2004)**
6. Newton's rings are formed in a reflected light of wave length 590 nm. The diameter of the 10th dark ring is 0.5×10^{-2} m. Find the radius of curvature of the lens. **(Bangalore, 2005)**
7. A shift of 100 circular fringes is observed when the movable mirror of the Michelson interferometer is shifted by 0.0295 mm. Calculate the wavelength of light. **(Kovempu, 2005)**
8. A soap film of RI 1.33 and thickness 1.5×10^{-4} cm is illuminated by white incident at an angle of 60°. The light reflected by it is examined by a spectroscope in which is found a dark band corresponding to a wavelength of 5×10^{-5} cm. Calculate the order of interference of the dark band. **(Gulbarga, 2005)**
9. A shift of 200 fringes is observed when movable mirror of F-P interferometer is shifted by 0.0295 mm. Calculate the wavelength used. **(Punjab, 2005)**
10. In Michelson interferometer when the mirror M_1 is moved maximum visibility is observed for the position 0.03 cm. Calculate the difference between the wavelengths, if mean wavelength is 5893Å.
11. In a Fabry-Perot interferometer, the separation between the plates is 4×10^{-4} cm. Light of wavelength 5000Å falls normally on the plates. Find the order of the plates. Find the order of the maximum at the centre. **(RTMNU, 2010)**
12. The Michelson's interferometer experiment is performed with a source which consists of two wavelengths 4882Å. Through what distance does the mirror have to be moved between two positions of the disappearance of the fringes? **(GNDU, Amritsar, 2010)**
13. In a Newton's rings experiment, the diameter of 3rd and 23rd dark tings are 0.2cm and 0.6cm respectively. If the radius of curvature of plane convex lens is 92 cm, find the wavelength of light. **(A.P.University, 2010)**

16

CHAPTER

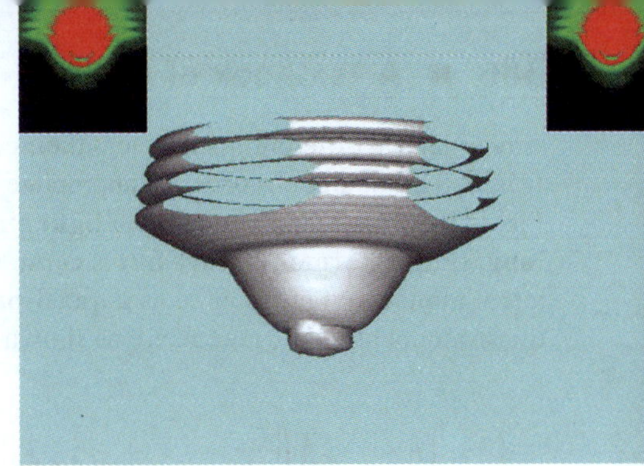

Coherence

16.1. INTRODUCTION

We have so far assumed naively that light sources emit *perfect harmonic waves.* In an ideal harmonic wave there exists a definite relationship between the phase of the wave at a given time and at a certain time later; and also at a given point and at a certain distance away. In reality, light sources do not emit perfectly harmonic waves. Even a very best practical *monochromatic* source emits a finite range of wavelengths and the light waves are quasi-monochromatic. If ***it were not so***, the light waves would have been ideally coherent and interference would be observed at all times.

In practice, light is emitted from a light source when excited atoms pass from the upper excited state to a lower energy state. The atom gives up the excess energy in the form of a photon. The process of transition from upper state to a lower state lasts for a brief time of about 10^{-8}s. It means that an atom starts emitting a light wave as it leaves the excited state and ceases emission as soon as it reaches the lower energy state. Therefore, an emission event produces a light burst. Each light burst occurs over a period of 10^{-8}s only, during which period a train of finite length having a certain limited number of wave oscillations is generated. Such a light burst is known as a **wave train** or a **wave packet.** After some time the atom again receives energy and jumps into excited state and subsequently emits another burst of light. These emission events occur quite randomly. Each atom in the source acts independently and different atoms emit

At a Glance

wave trains at different instants and their combination in millions and millions constitutes the light from a light source. In order to appreciate some of the peculiarities of natural light, the following fact is to be well understood. **The light emitted by an ordinary light source is not an infinitely long, simple harmonic wave but is composed of a jumble of finite wave trains**. We therefore call a real monochromatic source as a **quasi-monochromatic** source. The wave trains issuing out of a quasi-monochromatic source are as shown in Fig. 16.1.

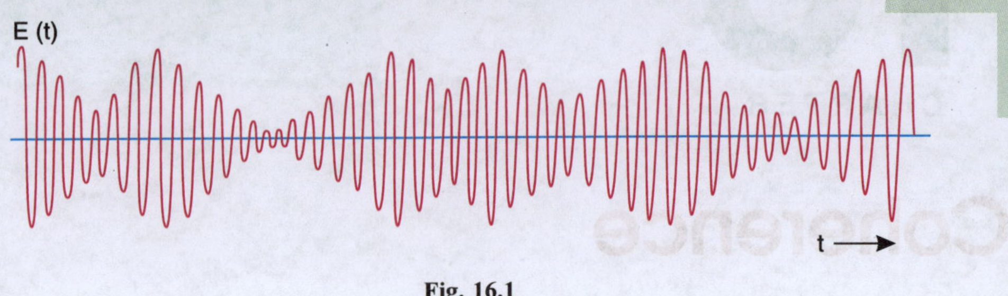

Fig. 16.1

16.2. WAVE TRAIN

Fig. 16.2 shows a wave train generated by an atom.

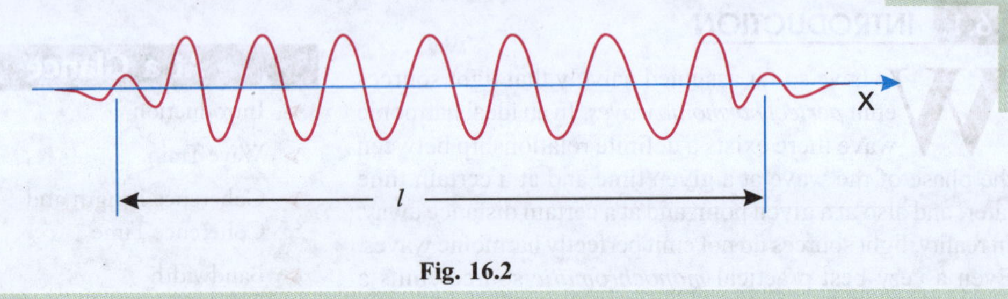

Fig. 16.2

If such a wave train lasts for a time interval Δt, then the length of the wave train in a vacuum is

$$l = c\Delta t \tag{16.1}$$

where c is the velocity of light in a vacuum.

For example, if $\Delta t = 10^{-8}$s, and $c = 3\times10^8$m/s, then $l = (3\times10^8\text{m/s})(10^{-8}\text{s}) = 3$ m.

The number of oscillations present in the wave train is

$$N = \frac{l}{\lambda} \tag{16.2}$$

where λ is the wavelength. If we assume $\lambda = 5000\text{Å} = 5\times10^{-7}$m, then

$$N = \frac{3\text{ m}}{5\times10^{-7}\text{ m}} = \mathbf{6\times10^6}$$

Thus, a wave train contains about a million wave oscillations in it.

Adding together the wave packets generated by all atoms in the light source, one finds a succession of wave trains, as shown in Fig. 16.3. In passing from one wave train to the next, there is an abrupt change in phase and also in plane of polarization. It is not possible to relate the phase at a point in wave train Q to a point in wave train P.

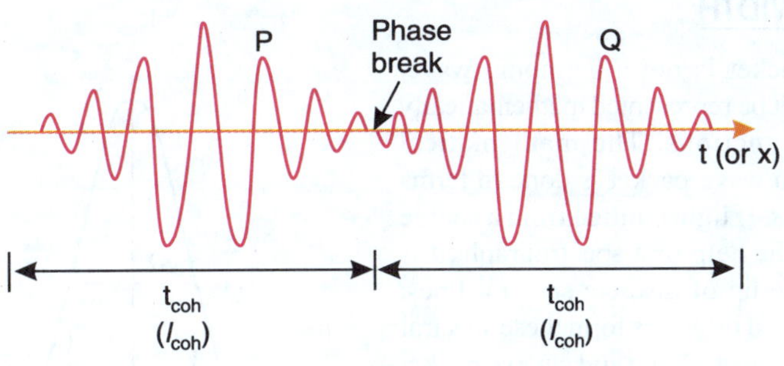

Fig. 16.3

Consequently there is no correlation between the phase of different wave trains. Each wave train has a sustained phase for only about 10^{-8} s, after which a new wave train is emitted with a totally random phase which also lasts only for about 10^{-8} s. The phase of the wave train from one atom will remain constant with respect to the phase of the wave train from another atom for utmost 10^{-8} s. It means that the wave trains can be coherent for a maximum of 10^{-8} s only. If two light waves overlap, sustained interference is not observed since the phase relationship between the waves changes rapidly, nearly at the rate of 10^8 times per second.

16.3. COHERENCE LENGTH AND COHERENCE TIME

The wave train, shown in Fig. 16.2, appears fairly sinusoidal for some number of oscillations between abrupt changes of frequency and phase. The length of the wave train over which it may be assumed to have a fairly sinusoidal character and predictable phase is known as **coherence length**. We denote it by $l_{coh.}$ We may consider coherence length as approximately equal to the length of the wave train, $c\Delta t$, over which its phase is predictable. The time interval during which the phase of the wave train can be predicted reliably is called **coherence time**. It is the time, Δt, during which the phase of the wave train does not become randomized but undergoes change in a regular systematic way. Coherence time is denoted by t_{coh}. We can therefore write

$$l_{coh} = c\,\Delta\,t \tag{16.3}$$

and

$$t_{coh} = \Delta\,t \tag{16.4}$$

$\therefore$

$$l_{coh} = c\,t_{coh} \tag{16.5}$$

A wave train consists of a group of waves, which have a continuous spread of wavelengths over a finite range $\Delta\lambda_o$ centered on a wavelength λ_o. According to Fourier analysis the frequency bandwidth $\Delta\nu$ is given by

$$\Delta\nu = \frac{1}{\Delta t}$$

where Δt is the average lifetime of the excited state of the atom. However, Δt is the time during which a wave train is radiated by the atom and corresponds to the coherence time, t_{coh}, of the wave train.

$\therefore$

$$\Delta\nu = \frac{1}{\Delta t} = \frac{1}{t_{coh}} \tag{16.6}$$

Using the relation (16.5) into equ. (16.6), we get

$$\Delta\nu = \frac{c}{l_{coh}} \tag{16.7}$$

16.4. BANDWIDTH

A wave packet is not a harmonic wave. Therefore, it cannot be represented mathematically by simple sine functions. The mathematical representation of a wave packet is done in terms of Fourier integrals. If light emitted from a source is analyzed with the help of a spectrograph, it is known to be made up of discrete spectral lines. Wave packets emitted by atoms form these spectral lines. Therefore, a spectral line and a wave packet are equivalent descriptions. The wavelength of a wave packet or a spectral line is not precisely defined. There is a continuous spread of wavelengths over a finite range, $\Delta\lambda$, centered on a wavelength λ_o. The maximum intensity of the wave packet occurs at λ_o and the intensity falls off rapidly on either side of λ_o, as shown in Fig. 16.4. The spread of wavelengths is called the *bandwidth*.

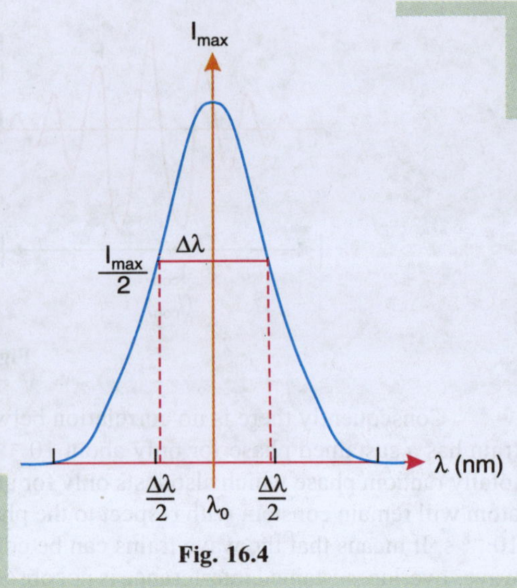

Fig. 16.4

The bandwidth is the wavelength interval from $\lambda_o - \Delta\lambda/2$ to $\lambda_o + \Delta\lambda/2$ which contains the major portion of the energy of the wave packet. In practice a source, which is said to produce line spectrum, produces a number of sharp wavelength distributions.

16.5. RELATION BETWEEN COHERENCE LENGTH AND BANDWIDTH

The frequency and wavelength of a light wave are related through the equation

$$\nu = \frac{c}{\lambda} \tag{16.8}$$

where λ_o is the vacuum wavelength.

Differentiating equ.(16.8) on both sides, we get

$$\Delta\nu = -\frac{c}{\lambda^2}\Delta\lambda \tag{16.9}$$

Using the relation (16.7) into equ.(16.9), we obtain

$$\therefore \quad \frac{c}{l_{coh}} = -\frac{c}{\lambda^2}\Delta\lambda \tag{16.10}$$

Rearranging the terms, we get

$$l_{coh} = \frac{\lambda^2}{\Delta\lambda} \tag{16.11}$$

The minus sign has no significance and hence is ignored. Equ.(16.11) means that the coherence length (the length of the wave packet) and the bandwidth of the wave packet are related to each other. The longer the wave packet, the narrower will be the bandwidth (see Fig. 16.5). In the limiting case, when the wave is infinitely long, we obtain monochromatic radiation of frequency ν_o (wavelength λ_o).

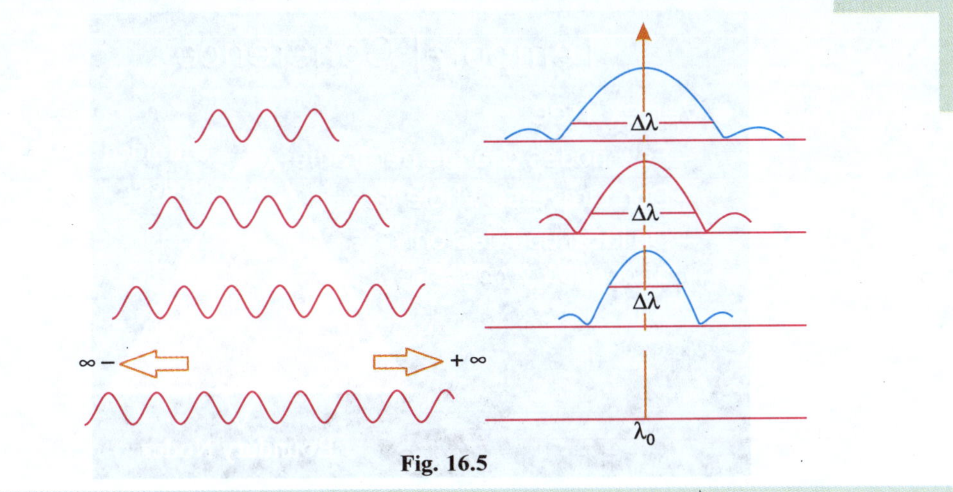

Fig. 16.5

Form the equ.(16.2), the coherence length may be defined as the product of the number of wave oscillations N contained in the wave train and of the wavelength, λ. Thus,

$$l_{coh} = N\lambda \qquad (16.12)$$

Equating (16.11) and (16.12), we get

$$N = \frac{\lambda}{\Delta\lambda}$$

$$\therefore \qquad \frac{\Delta\lambda}{\lambda} = \frac{1}{N} \qquad (16.13)$$

Equ.(16.13) shows that the larger the number of wave oscillations in a wave packet, the smaller is the bandwidth. In the limiting case, when N is infinitely large, that is when the wave packet is infinitely long, the wave will be monochromatic having a precisely defined wavelength. The dependence of bandwidth on the length of the wave packet is schematically shown in Fig. 16.5.

16.6. COHERENCE

Coherence is an important property of light. It refers to the connection between the phase of light waves at one point and time, and the phase of the light waves at another point and time. Coherence effects are mainly divided into two categories: *temporal and spatial*. The temporal coherence is related directly to the finite bandwidth of the source, whereas the spatial coherence is related to the finite size of the source.

16.6.1. TEMPORAL COHERENCE

Temporal coherence is also known as longitudinal coherence. Let a point source of quasi-monochromatic light S (Fig.16.6) emit light in all directions. Let us consider light travelling along the line SP_1P_2. The phase relationship between the points P_1 and P_2 depends on the distance P_1P_2 and the coherence length of the light beam. The electric fields at P_1 and P_2 will be correlated in phase when a single wave train extends over greater length than the distance P_1P_2; that is if the distance P_1P_2 is less than the coherence length l_{coh}. Then, the waves are correlated in their rising and falling and they will preserve a constant phase difference.

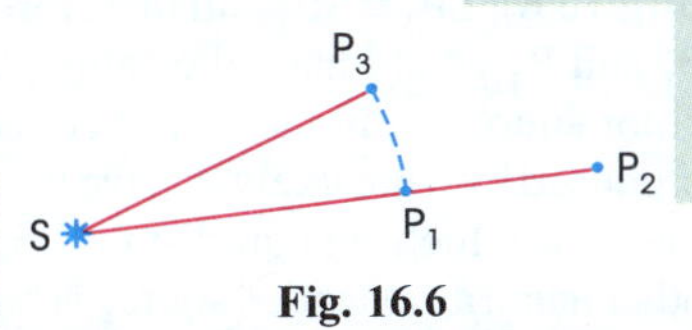

Fig. 16.6

The points P_1 and P_2 would not have any phase relationship if the longitudinal distance P_1P_2 is greater than

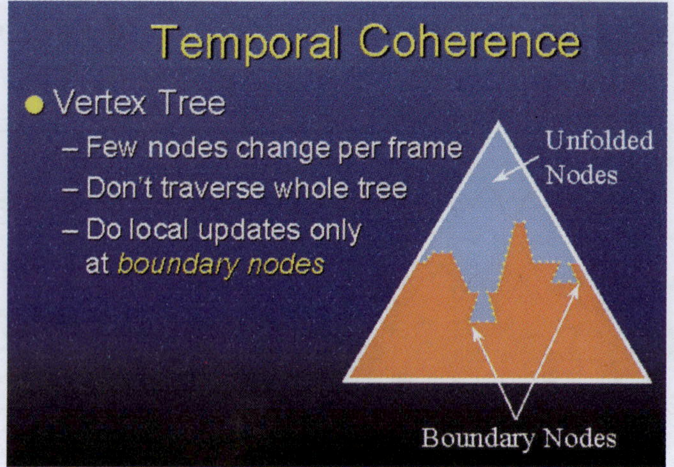

l_{coh}, since in such a case many wave trains would span the distance. It means different independent wave trains would be at P_1 and P_2 at any instant and therefore the phase at the two points would be independent of each other. The degree to which a correlation exists is known as the amount of *longitudinal coherence.*

16.6.1.1. Monochromaticity

Form equ.(16.11) and Fig.16.5 we conclude that temporal coherence is indicative of **monochromaticity** of the source. An ideally monochromatic source is an absolutely coherent source. The degree of monochromaticity of a source is given by

$$\xi = \frac{\Delta\nu}{\nu_o} \tag{16.14}$$

When the ratio $\Delta\nu / \nu_o = 0$, the light wave is ideally monochromatic.

16.6.1.2. Purity of spectral line

The width of a spectral line is given by $\Delta\lambda$ (see Fig. 16.4). It is seen from equ.(16.11) that it is related to the temporal coherence. Thus,

$$\Delta\lambda = \frac{\lambda^2}{l_{coh}} \tag{16.15}$$

16.6.2. SPATIAL COHERENCE

Spatial coherence refers to the continuity and uniformity of a wave in a direction perpendicular to the direction of propagation. If the phase difference for any two fixed points in a plane normal to the wave propagation does not vary with time, then the wave is said to exhibit *spatial coherence*. It is also known as *lateral coherence*. Again looking at the point source S (Fig.16.6), $SP_1 = SP_3$ and therefore, the fields at points P_1 and P_3 would have the same phase. Thus, an ideal point source exhibits spatial coherence, as the waves produced by it are likely to have the same phase at points in space, which are equidistant from the source. On the other hand, an extended source is bound to exhibit lesser

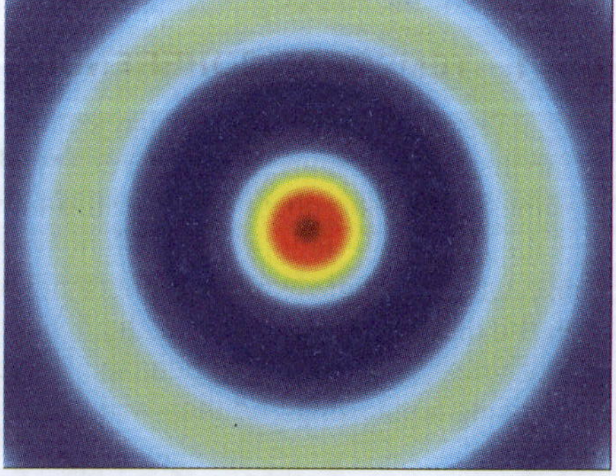
Spatial Coherence.

lateral spatial coherence. Two points on the source separated by a lateral distance greater than one wavelength will behave quite independently. Therefore, correlation is absent between the phases of the waves emitted by them. The degree of contrast of the interference fringes produced by a source is a measure of the degree of the spatial coherence of its waves. The higher the contrast, the better is the spatial coherence.

16.7. DETERMINATION OF COHERENCE LENGTH

The coherence length can be measured by means of Michelson interferometer. In a Michelson interferometer, a light beam from the source S is incident on a semi-silvered glass plate G (see Fig.16.7) and gets divided into two components; one component is reflected, 1,and the other,2, is transmitted. These two beams, 1 and 2, are reflected back at mirrors M_1 and M_2 respectively and are received by the telescope where interference fringes are produced. It is obvious that the beams produce stationary interference only if they are coherent.

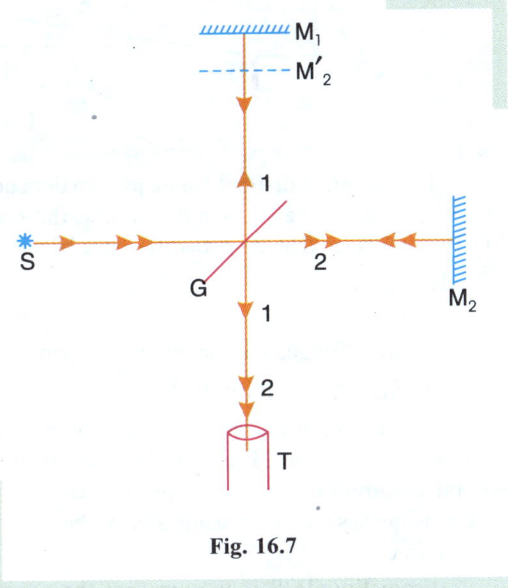

Fig. 16.7

Let M_2' be the image of M_2 formed by G. If the reflecting surfaces M_1 and M_2' (the image of M_2) are separated by a distance d, then $2d$ will be the path difference between the interfering waves. The condition of fixed phase relationship between the two waves, 1 and 2, will be satisfied if

$$2d << l_{coh}$$

In such a case distinct interference fringes will be seen. If, however,

$$2d >> l_{coh}$$

then the phases of the two waves are not correlated and interference fringes will not be seen. To determine the coherence length of waves emitted by a light source, the distance d between the mirrors M_1 and M_2'(the image of M_2) is varied by moving one of the mirrors. As the distance varies, the contrast of the fringes decreases and ultimately they disappear. The path difference $2d$ at the particular stage where the fringes disappear gives us the coherence length.

The light from a sodium lamp has coherence length of the order of 1 mm, that of green mercury line is about 1 cm, neon red line 3 cm, red cadmium line 30 cm, orange krypton line 80 cm and that of a commercial He-Ne laser is about 15m. The coherence length of light from some of the lasers goes up to a few km.

16.8. CONDITION FOR SPATIAL COHERENCE

The degree of spatial coherence of a beam of light can be deduced from the contrast of the fringes produced by it. The broader the source of the light, the lesser is the degree of coherence. In Young's double slit experiment, if the slits S_1 and S_2 are directly illuminated by a source, interference fringes are not observed. Instead the screen is uniformly illuminated. The absence of fringes indicates that the light issuing from the slits do not possess spatial coherence. If a narrow slit is introduced before the double slit, light passing through the narrow slit S illuminates S_1 and S_2. The waves emerging from them, having been derived through wave front division, are coherent and stationary interference pattern will be observed on the screen. If the width of the slit S is gradually increased the contrast of fringe pattern decreases and fringes disappear. When the slit S is wider, S_1 and S_2 receive waves from different parts of S which do not maintain coherence. When S is narrow, it ensures that the wave trains incident on slits S_1 and S_2 originate from a small region of the source and hence they have spatial coherence.

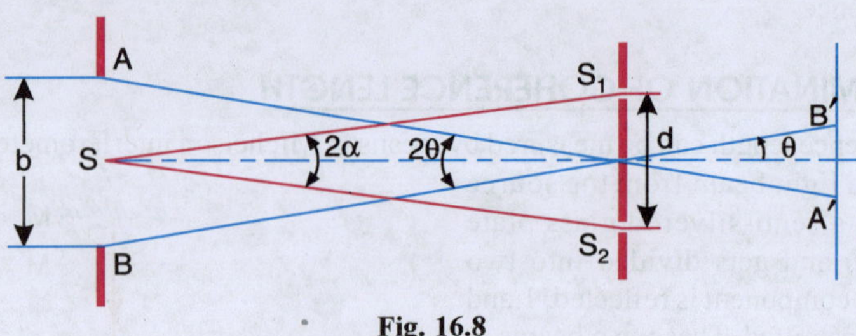

Fig. 16.8

Formation of distinct fringe pattern depends on two parameters in the double slit experiment. One is the size of the slit S and second is the separation, d, between the two slits S_1 and S_2. From Fig.16.9, it is seen that the path difference between the two waves passing at the edges A and B of the slit S is

$$\Delta = b \sin \alpha$$

Distinct fringes will be obtained when $\Delta << \frac{1}{2}\lambda$

Therefore, it requires that $$b \sin \alpha << \frac{1}{2}\lambda \quad (16.16)$$

This equation determines the size of the source, which is spatially coherent to produce fringes of satisfactory contrast. Equ.(16.16) shows that the smaller is the size of the source, the better is the spatial coherence. Spatial coherence of waves decreases with increasing size of the light source.

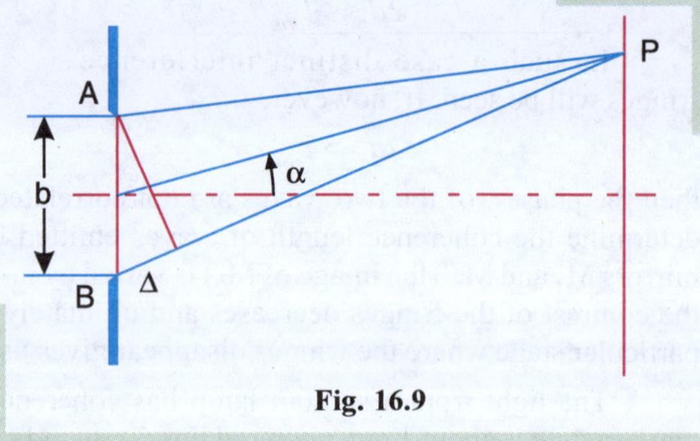

Fig. 16.9

From Fig. 16.8 it is seen that the edges A and B of the slit S subtend an angle 2θ at the centre of the double slit and the double slit subtends an angle 2α at the centre of S. Waves from A produce interference pattern which has its centre at A′. Similarly, waves from B produce fringe pattern with its centre at B′. The patterns cancel each other if the maximum of one falls on the minimum of the other. The condition for no fringes is that

$$2\theta = \frac{\lambda}{2d} \quad (16.17)$$

When the slit S is made wider, the edges A and B move away from each other. As a result the patterns at A′ and B′ also move away. When the maxima of one pattern fall on the maxima of other, the fringes appear. Therefore, the condition for fringes is that

$$2\theta = \frac{\lambda}{d}$$

or $$d = \frac{\lambda}{2\theta} \quad (16.18)$$

In other words, there is restriction imposed on the separation of the slits S_1 and S_2. d must be very small compared to $\lambda / 2\theta$ in order to maintain spatial coherence and obtain distinct fringe pattern on the screen.

Light produced by lasers is highly coherent. Therefore, when a laser is used as the source, interference fringes can be obtained without the aid of slit S.

WORKED OUT PROBLEMS

Example 16.1: A sodium atom radiates for 4×10^{-12} s. What is the coherence length of light from a sodium lamp?

Solution: It is given that coherence time $t_{coh} = 4\times 10^{-12}$ s.

$$\text{Coherence length } l_{coh} = c\, t_{coh} = (3\times 10^{8} \text{ m/s})(4\times 10^{-12} \text{ s})$$

$$= 12\times 10^{-4} \text{ m} = \mathbf{1.2\ mm.}$$

Example 16.2: Calculate the coherence length for CO_2 laser whose line width is 1×10^{-5} nm at IR emission wavelength of 10.6 μm.

Solution: Coherence length, $l_{coh} = \dfrac{\lambda^2}{\Delta\lambda} = \dfrac{(10.6\times 10^{-6})^2 \text{ m}^2}{1\times 10^{-5}\times 10^{-9} \text{ m}} = \mathbf{11.2\ km}.$

Example 16.3: Compute the coherence length of yellow light with 5893Å in 10^{-12} second pulse duration. Find also the bandwidth.

Solution: Coherence length is given by $l_{coh} = ct_{coh}$.

∴ $$l_{coh} = (3\times 10^{8} \text{ m/s})(10^{-12}\text{s}) = \mathbf{0.3\ mm}$$

$$\text{Bandwidth is given by } \Delta\lambda = \frac{\lambda^2}{l_{coh}} = \frac{(5893\times 10^{-10} \text{ m})^2}{3\times 10^{-4} \text{ m}} = \mathbf{11.6Å}$$

QUESTIONS

1. Explain the terms coherence length and coherence time for a light wave. Derive an expression for the coherence length of a wave train that has a frequency bandwidth $\Delta\nu$.
2. Explain in brief:
   (*i*) Coherence length (*ii*) spatial coherence (*iii*) temporal coherence.
   Obtain an expression for coherence length.
3. Explain in brief the terms: (*i*) Temporal Coherence (*ii*) Spatial Coherence. **(Punjab 2005)**
4. Distinguish between spatial and temporal coherence. What are coherence length and coherence time? Why is it impossible to observe interference between light waves emitted by independent sources?
5. The light of wavelength 6000Å has wave trains 13.2×10^{-6} m. Calculate the coherent time. **(GNDU, Amritsar, 2010)**
6. Determine what emission frequency width would be required to have a temporal coherence length of 10m at a source wavelength of 488 nm? **(GNDU, Amritsar, 2010)**

PROBLEMS FOR PRACTICE

1. Calculate the frequency bandwidth for white light (frequency range 4×10^{14}Hz to 7.5×10^{14}Hz). Also find (i) coherence time and (ii) coherence length of white light.
   [**Ans:** $\Delta\nu = 3.5\times 10^{14}$Hz; $t_{coh} = 2.8\times 10^{-15}$s; $l_{coh} = 8.6\times 10^{-7}$m]
2. A quasi-monochromatic source emits radiations of mean wavelength $\lambda = 5461$Å and has a bandwidth $\Delta\nu = 10^9$Hz. Calculate coherence time, coherence length and frequency stability. [**Ans:** $t_{coh} = 10^{-9}$s; $l_{coh} = 0.3$ m; $\Delta\nu/\nu = 1.8\times 10^{-6}$]
3. An optical filter has a line width of 1.5 nm and mean wavelength 550 nm. With white light incident on the filter, calculate (i) coherence length and (ii) the number of wavelengths in the wave train. [**Ans:** $l_{coh} = 2\times 10^{-4}$m; N = 366]

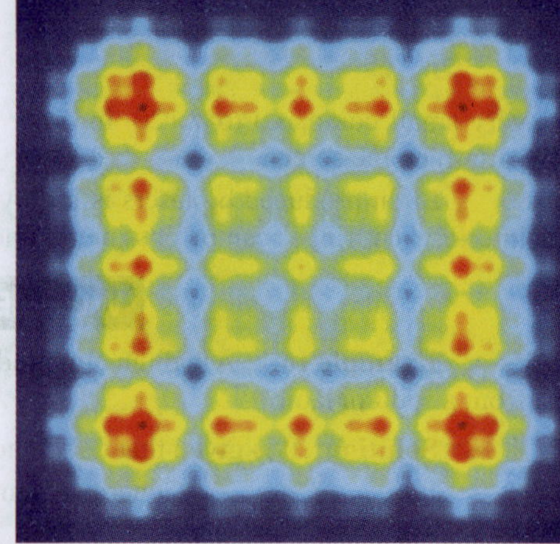

17 CHAPTER

Fresnel Diffraction

17.1. INTRODUCTION

When waves encounter obstacles (or openings), they bend round the edges of the obstacles, if the dimensions of the obstacles are comparable to the wavelength of the waves. The bending of waves around the edges of an obstacle is called *diffraction*.

Fig. 17.1 illustrates the passage of waves through an opening. When the opening is large compared to the wavelength, the waves do not bend round the edges. When the opening is small, the bending round the edges is noticeable. When the opening is very small, the waves spread over the

Light diffraction by a razor blade.

At a Glance

entire surface behind the opening. The opening acts an independent source of waves, which propagate in all directions. The diffraction effect is observable quite close to the opening when the size of the opening is very small. When the opening is large, diffraction effect is observed at greater distances from the opening. In general diffraction of waves becomes noticeable only when the size of the obstacle is comparable to a wavelength.

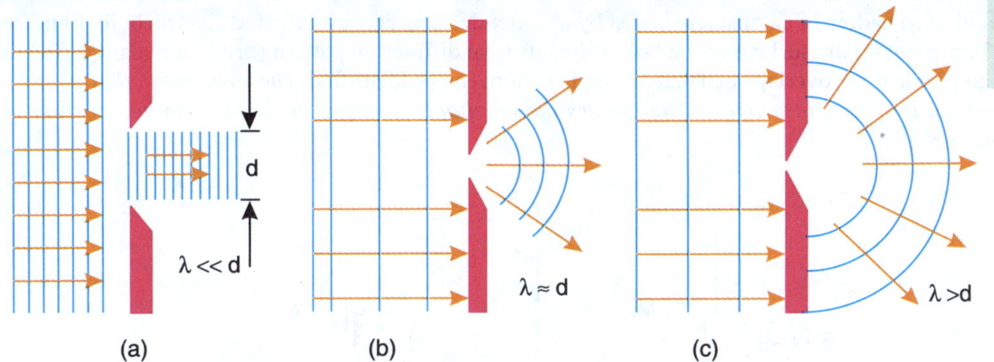

Diffraction–(*a*) A plane wave does not bend at the slit if the opening $d \gg \lambda$. (*b*) Bending is perceptible when $\lambda \approx d$ (*c*). When $\lambda > d$, the bending takes place to such an extent that light can be perceived in a direction normal to the ray propagation suggesting that the opening acts as a point source.

Fig. 17.1

It is a matter of common experience that the path of light entering a dark room through a hole in the window illuminated by sunlight is straight. Similarly, if an opaque obstacle is placed in the path of light, a sharp shadow is cast on the screen, indicating thereby that light travels in straight lines. Rectilinear propagation of light can be easily explained on the basis of Newton's corpuscular theory. But it has been observed that when a beam of light passes through a small opening (a small circular hole or a narrow slit) it spreads to some extent into the region of the geometrical shadow also. If light energy is propagated in the form of waves, then similar to sound waves, one would expect bending of a beam of light round the edges of an opaque obstacle or illumination of the geometrical shadow.

However, diffraction phenomenon is not readily apparent in case of light waves. It becomes significant when the aperture size is of the order of one wavelength wide. Diffraction and interference are basically equivalent.

17.2. HUYGENS-FRESNEL THEORY

According to Huygen's wave theory of light, each progressive wave produces secondary waves, the envelope of which forms the secondary wave front. In Fig. 17.2 (a), S is a source of monochromatic light and MN is a small aperture. XY is the screen placed in the path of light. AB is the illuminated portion of the screen and above A and below B is the region of the geometrical shadow. Considering MN as the primary wavefront, according to Huygen's construction, if secondary wave fronts are drawn, one would expect encroachment of light

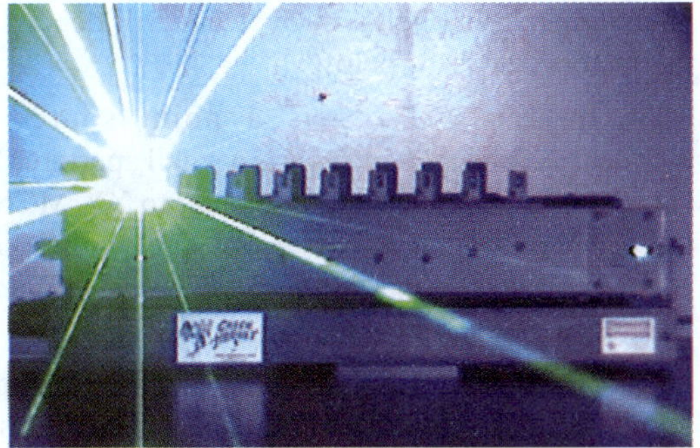

Diffraction effect.

in the geometrical shadow. Thus, the shadows formed by small obstacles are not sharp. This bending of light round the edges of an obstacle or the encroachment of light within the geometrical shadow is known as *diffraction*. Similarly, if an opaque obstacle MN is placed in the path of light [Fig.17.2 (b)], there should be illumination in the geometrical shadow region AB also. But the illumination in the geometrical shadow of an obstacle is not commonly observed because the light sources are not point sources and secondly the obstacles used are of very large size compared to the wavelength of light. If a shadow of an obstacle is cast by an extended source, say a frosted electric bulb, light from every point on the surface of the bulb forms its own diffraction pattern (bright and dark diffraction bands) and these overlap such that no single pattern can be identified. *The term diffraction is referred to such problems in which one considers the resultant effect produced by a limited portion of a wavefront.*

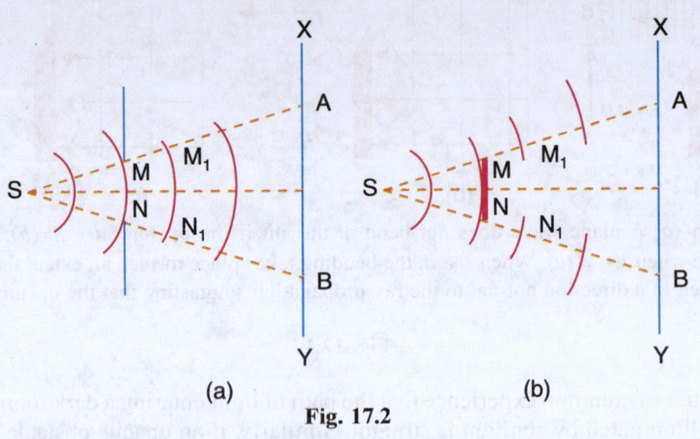

Fig. 17.2

Diffraction phenomena are part of our common experience. The luminous border that surrounds the profile of a mountain just before the sun rises behind it, the light streaks that one sees while looking at a strong source of light with half shut eyes and the colored spectra (arranged in the form of a cross) that one sees while viewing a distant source of light through a fine piece of cloth are all examples of diffraction effects.

Augustine Jean Fresnel in 1815, combined in a striking manner Huygens' wavelets with the principle of interference and could satisfactorily explain the bending of light round obstacles and also the rectilinear propagation of light.

17.3. FRESNEL'S ASSUMPTIONS

According to Fresnel, the resultant effect at an external point due to a wavefront will depend on the factors discussed below:

In Fig. 17.3, S is a point source of monochromatic light and MN is a small aperture. XY is the screen and SO is perpendicular to XY. MCN is the incident spherical wavefront due to the point source S. To obtain the resultant effect at a point P on the screen, Fresnel assumed the following:

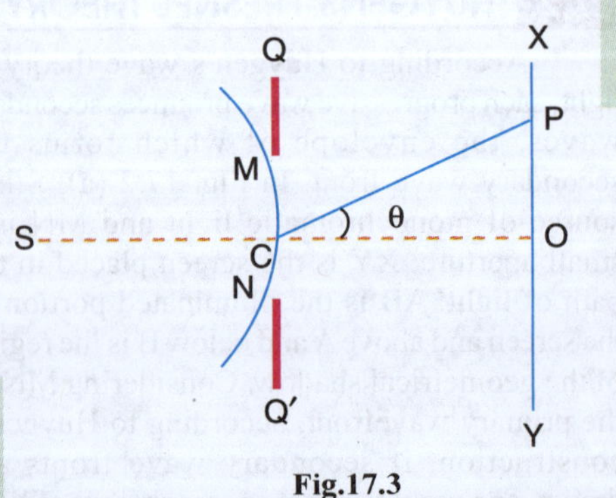

Fig.17.3

(1) A wave front can be divided into a large number of strips or **zones** called Fresnel's zones of

small area and the resultant effect at any point will depend on the combined effect of all the secondary waves emanating from the various zones;

(2) The effect at a point due to any particular zone will depend on the distance of the point from the zone;

Augustin Jean Fresnel (1788-1827)

(3) The effect at P will also depend on the obliquity of the point with reference to the zone under consideration, e.g. due to the part of the wavefront at C, the effect will be a maximum at O and decreases with increasing obliquity. It is a maximum in a direction radially outwards from C and it decreases in the opposite direction. The effect at a point due to the obliquity factor is proportional to $(1+ \cos\theta)$ where $\angle PCO = \theta$. Considering an elementary wavefront at C, the effect is maximum at O because $\theta = 0$ and $\cos\theta = 1$. Similarly, in a direction tangential to the primary wavefront at C (along CQ) the resultant effect is one half of that along CO because $\theta = 90^\circ$ and $\cos 90^\circ = 0$. In the direction CS, the resultant effect is zero since $\theta = 180^\circ$ and $\cos 180^\circ = -1$ and $1 + \cos 180^\circ = 1-1 = 0$. This property of the secondary waves eliminates one of the difficulties experienced with the simpler form of Huygens principle viz., that if the secondary waves spread out in all directions from each point on the primary wavefront, they should give a wave traveling forward as well as backward. Now, as the amplitude at the rear of the wave is zero there will evidently be no back wave.

17.4. RECTILINEAR PROPAGATION OF LIGHT

ABCD is a plane wavefront perpendicular to the plane of the paper [Fig. 17.4 (a)] and P is an external point at a distance b perpendicular to ABCD. To find the resultant intensity at P due to the wavefront ABCD, Fresnel's method consists in dividing the wavefront into a number of *half period elements* or *zones* called Fresnel's zones and to find the effect of all the zones at the point P.

If spheres are constructed with P as centre and radii equal to $b+\lambda/2$, $b+2\lambda/2$, $b+3\lambda/2$ etc., they will cut out circular areas of radii OM_1, OM_2, OM_3, etc., on the wave front. These circular zones are called **half period zones** or **half period elements**. Each zone differs from its neighbour by a phase difference of π or path difference of $\lambda/2$. Thus the secondary waves starting from the point O and M_1 and reaching P will have a phase difference of π or a path difference $\lambda/2$. A Fresnel half period zone with respect to an actual point P is a thin annular zone of the primary wavefront in which the secondary waves from any two corresponding points of neighboring zones differ in path by $\lambda/2$.

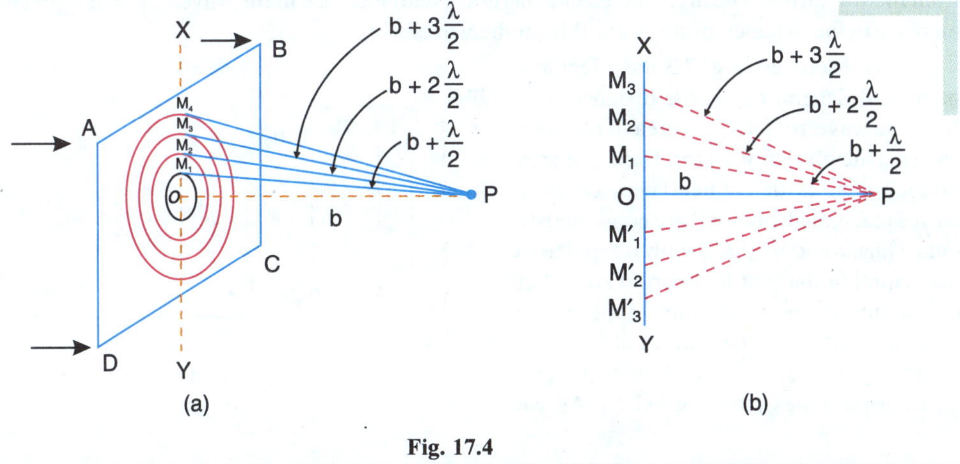

Fig. 17.4

In Fig. 17.4 (b) O is the pole of the wavefront XY with reference to the external point P. OP is perpendicular to XY. In Fig.17.4 (c) 1, 2, 3 etc are the half period zones constructed on the primary wavefront XY. OM_1 is the radius of the first zone. OM_2 is the radius of the second zone and so on. P is the point at which the resultant intensity has to be calculated.

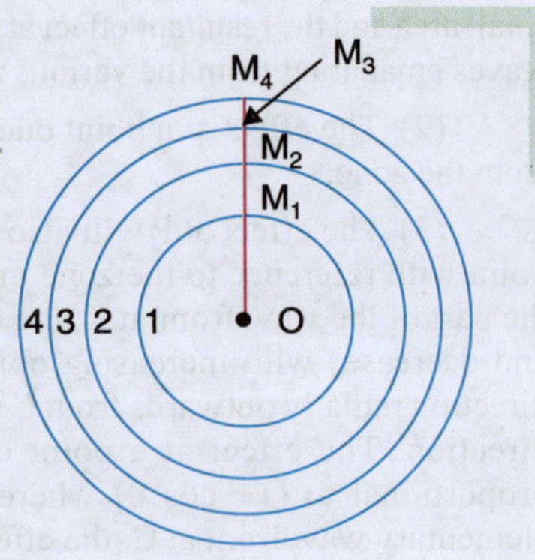

Fig. 17.4 (c)

$$OP = b,\ OM_1 = r_1,\ OM_2 = r_2,\ OM_3 = r_3, \text{ etc.}$$

And

$$M_1P = b+\frac{\lambda}{2}, \quad M_2P = b+\frac{2\lambda}{2}, \quad M_3P = b+\frac{3\lambda}{2} \text{ etc.}$$

The area of the first half period zone is

$$\pi\, OM_1^2 = \pi\,[M_1P^2 - OP^2] = \pi\left[\left(b+\frac{\lambda}{2}\right)^2 - b^2\right]$$

$$= \pi\left[b\lambda + \frac{\lambda^2}{4}\right] = \pi\, b\lambda \qquad (17.1)$$

As λ is small, λ^2 term is neglected.

The radius of the first half period zone is, $r_1 = OM_1 = \sqrt{b\lambda}$

The radius of the second half period zone is, $OM_2 = [M_2P^2 - OP^2]^{1/2}$

$$= \left[(b+\lambda)^2 - b^2\right]^{1/2}$$

$$= \sqrt{2b\lambda}$$

The area of the second half period zone, $= \pi\left[OM_2^2 - OM_1^2\right]$

$$= \pi\,[2b\lambda - b\lambda]$$

$$= \pi\, b\lambda \qquad (17.2)$$

Thus, the area of each half period zone is equal to $\pi\, b\, \lambda$. Also the radii of the 1st, 2nd, 3rd, etc half period zones are $\sqrt{1b\lambda}, \sqrt{2b\lambda}, \sqrt{3b\lambda}$ etc. Therefore, the radii are proportional to the square roots of the natural numbers. However, it should be remembered that the areas of the zones are not constant but are dependent on - (i) λ , the wavelength of light and (ii) b, the distance of the point from the wavefront. The area of the zone increases with increase in the wavelength of light and with increase in the distance of the point P from the wavefront.

As discussed in §17.3 the effect at a point will depend on (i) the distance of P from the wavefront, (ii) the area of the zone and (iii) the obliquity factor. Here, the area of each zone is the same. The secondary waves reaching the point P are continuously out of phase and in phase with reference to the central or the first half period zone. Let m_1, m_2, m_3, etc represent the amplitudes of vibration of the ether particles at P due to secondary waves from the 1st , 2nd, 3rd, etc. half period zones (see Fig. 17.5). As we

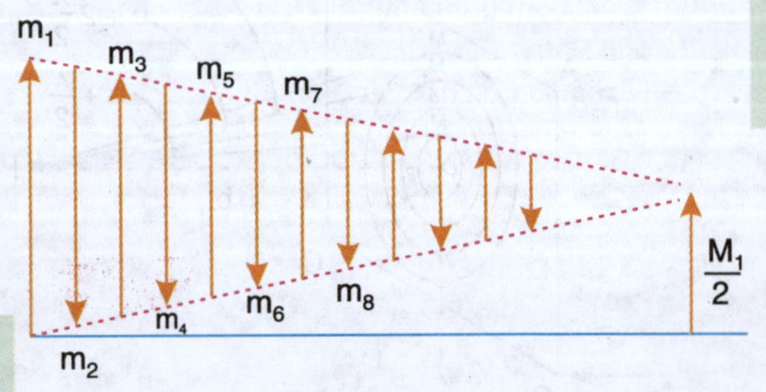

Fig. 17.5

consider the zones outwards from O, the obliquity increases and hence the quantities m_1, m_2, m_3 etc are of continuously decreasing order. Thus, m_1 is slightly greater than m_2; m_2 is slightly greater than m_3 and so on. Due to the phase difference of π between any two consecutive zones, if the displacements of the ether particles due to odd numbered zones is in the positive direction, then due to the even numbered zones the displacement will be in the negative direction at the same instant. As the amplitudes are of gradually decreasing magnitude, the amplitude of vibration at P due to any zone can be approximately taken as the mean of the amplitudes due to the zones preceding and succeeding it.

e.g. $$m_2 = \frac{m_1 + m_3}{2}$$

The resultant amplitude at P at any instant is given by,

$$A = m_1 - m_2 + m_3 - m_4 ...... + m_n \text{ if } n \text{ is odd.}$$

(If n is even, the last quantity is $- m_n$).

∴ $$A = \frac{m_1}{2} + \left[\frac{m_1}{2} - m_2 + \frac{m_3}{2}\right] + \left[\frac{m_3}{2} - m_4 + \frac{m_5}{2}\right] + ...$$

But $$m_2 = \frac{m_2}{2} + \frac{m_3}{2} \quad and \quad m_4 = \frac{m_3}{2} + \frac{m_5}{2}$$

$$A = \frac{m_1}{2} + \frac{m_n}{2} ....if \ n \text{ is odd.}$$

$$A = \frac{m_1}{2} + \frac{m_{n-1}}{2} - m_n ....if \ n \text{ is even.}$$

If the whole wave front ABCD is unobstructed, the number of half period zones that can be constructed with reference to the point P is infinite i.e. $n \to \infty$. As the amplitudes are of gradually diminishing order, m_n and m_{n-1} tend to be zero.

Therefore, the resultant amplitude at P due to the whole wavefront $= A = \frac{m_1}{2}$. (17.3)

The intensity at a point is proportional to the square of the amplitude.

∴ $$I \propto \frac{m_1^2}{4} \quad (17.4)$$

Thus, the intensity at P is only one–fourth of that due to the first half period zone alone. Here, only half the area of the first half period zone is effective in producing the illumination at the point P. A small obstacle of the size of half the area of the first half period zone placed at O will screen the effect of the whole wavefront and the intensity at P due to the rest of the wavefront will be zero. While considering the rectilinear propagation of light, the size of the obstacle used is far greater than the area of the first half period zone and hence the bending effect of light round corners (diffraction effects) cannot be noticed. In the case of sound waves, the wavelengths are far greater than the wavelength of light, and hence the area of the first half period zone for a plane wavefront of sound is very large. If the effect of sound at a point beyond an obstacle is to be shadowed, an obstacle of very large size has to be used to get no sound effect. If the size of the obstacles placed in the path of light is comparable to the wavelength of light, then it is possible to observe illumination in the region of the geometrical shadow also. Thus, rectilinear propagation of light is only approximately true.

17.5. ZONE PLATE

A zone plate is a specially constructed screen such that light is obstructed from every alternate zone. It can be designed so as to cut off light due to the even numbered zones or that due to the odd numbered zones. The correctness of Fresnel's method in dividing a wavefront into half period zones can be verified with its help.

To construct a zone plate, concentric circles are drawn on white paper such that the radii are proportional to the square roots of the natural numbers (as shown in §17.4, the radii are proportional to the square roots of the natural numbers). The odd numbered zones (i.e. 1st, 3rd, 5th, etc) are covered with black ink and a reduced photograph is taken. The drawing appears as shown in Fig. 17.6 (b). The negative of the photograph will be as shown in Fig. 17.6 (a). In the developed negative, the odd zones are transparent to incident light and the even zones will cut off light.

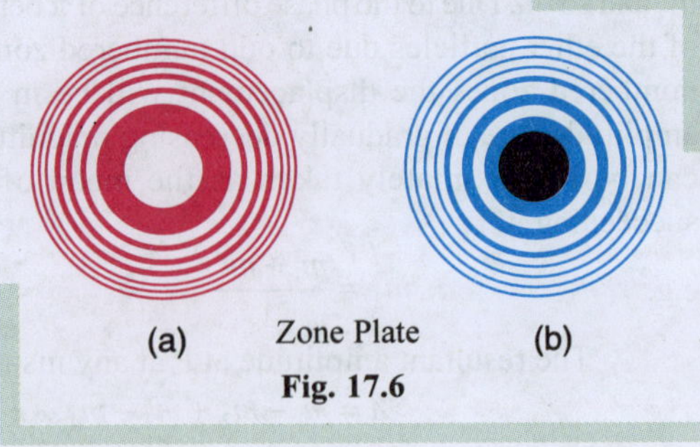

Fig. 17.6

If such a plate is held perpendicular to an incident beam of light and a screen is moved on the other side to get the image, it will be observed that maximum brightness is possible at some position of the screen say b cm from the zone plate (Fig. 17.7) XO is the upper half of the incident plane wavefront. P is the point at which the light intensity is to be considered. The distance of the point P from the wavefront is b. $OM_1(=r_1), OM_2(=r_2)$ etc are the radii of the zones.

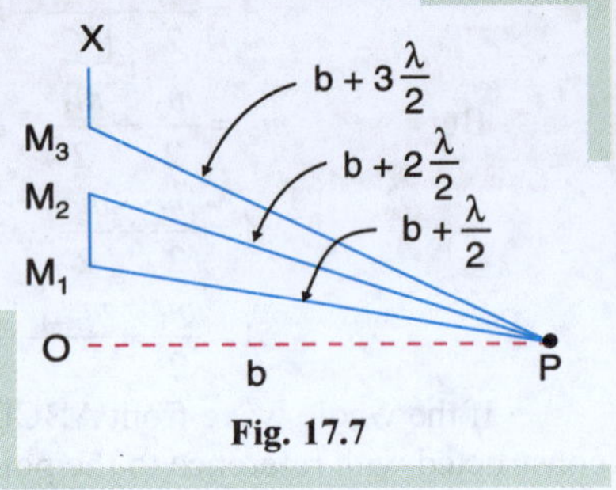

Fig. 17.7

$r_1 = \sqrt{b\lambda}$ and $r_2 = \sqrt{2b\lambda}$ where λ is the wavelength of light.

$$r_n = \sqrt{nb\lambda} \quad \text{or} \quad b = \frac{r_n^2}{n\lambda} \tag{17.5}$$

If the source is at a large distance from the zone plate, a bright spot will be obtained at P. As the distance of the source is large, the incident wavefront can be taken as a plane one with respect to the small area of the zone plate. The even numbered zones cut off the light and hence resultant amplitude at $P = A = m_1 + m_2 + m_3 + ...$ etc. In this case the focal length of the zone plate f_n is given by

$$f_n = b = \frac{r_n^2}{n\lambda} \tag{17.6}$$

Thus, a zone plate has different foci for different wavelengths. The radius of the nth zone increases with increasing value of λ. *It is very interesting to note that as the even numbered zones are opaque, the intensity at P is much greater than that when the whole wavefront is exposed to the point P.*

In the first case the resultant amplitude is given by

$$A = m_1 + m_3 + m_5 + ...m_n .... \quad (n \text{ is odd})$$

When the whole wavefront is unobstructed, the amplitude is given by

$$A = m_1 - m_2 + m_3 - m_4 ...... + m_n$$

$$= \frac{m_1}{2} \quad (\text{if } n \text{ is very large and } n \text{ is odd}).$$

If a parallel beam of white light is incident on the zone plate, different colours come to focus at different points along the line OP. Thus, the function of a *zone plate is similar to that of a convex (converging) lens* and a formula connecting the distance of the object and image points can be obtained for a zone plate also.

17.5.1. ACTION OF A ZONE PLATE FOR AN INCIDENT SPHERICAL WAVE FRONT

Let XY represent the section of the zone plate perpendicular to the plane of the paper. S is a point source of light, P is the position of the screen for a bright image, 'a' is the distance of the source from the zone plate and b is the distance of the screen from the plate. OM_1, OM_2, OM_3, (r_1, r_2, r_3) etc are the radii of the 1st, 2nd, 3rd etc. half period zones. The position of the screen is such that from one zone to the next there is an increasing path difference of $\frac{\lambda}{2}$.

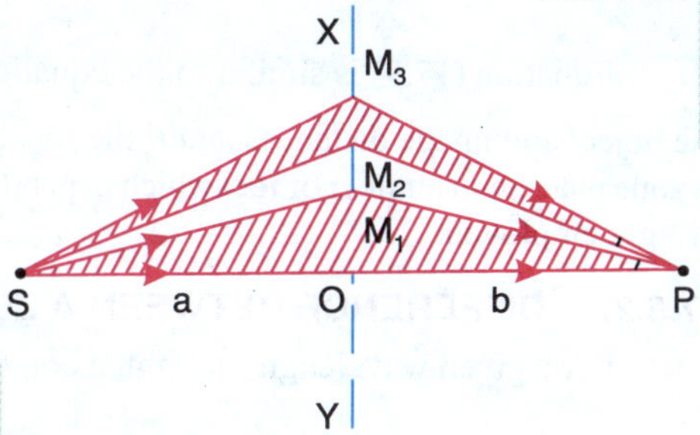

Fig. 17.8

Thus,

$$SO + OP = a + b$$

$$SM_1 + M_1P = a + b + \frac{\lambda}{2} \tag{17.7}$$

$$SM_2 + M_2P = a + b + \frac{2\lambda}{2} \quad \text{and so on.}$$

From the $\Delta\, SM_1O$

$$SM_1 = \left(SO^2 + OM_1^2\right)^{1/2}$$

$$= \left(a^2 + r_1^2\right)^{1/2}$$

Similarly from the ΔOM_1P

$$M_1P = \left(OP^2 + OM_1^2\right)^{1/2}$$

$$= \left(b^2 + r_1^2\right)^{1/2}$$

Substituting the values of SM_1 and M_1P in equation (17.7), we get

$$(a^2 + r_1^2)^{1/2} + \left(b^2 + r_1^2\right)^{1/2} = a + b + \frac{\lambda}{2}$$

$$a\left(1 + \frac{r_1^2}{a^2}\right)^{1/2} + b\left(1 + \frac{r_1^2}{b^2}\right)^{1/2} = a + b + \frac{\lambda}{2}$$

$$a + \frac{r_1^2}{2a} + b + \frac{r_1^2}{2b} = a + b + \frac{\lambda}{2}$$

$$\frac{r_1^2}{2}\left[\frac{1}{a} + \frac{1}{b}\right] = \frac{\lambda}{2}$$

$$r_1^2\left(\frac{1}{a} + \frac{1}{b}\right) = \lambda$$

Similarly for r_n i.e. the radius of the nth zone, the relation can be written as

$$r_n^2\left[\frac{1}{a} + \frac{1}{b}\right] = n\lambda \tag{17.8}$$

Applying the sign convention,

$$\frac{1}{b}-\frac{1}{a}=\frac{n\lambda}{r^2}=\frac{1}{f_n} \quad (17.9)$$

or
$$f_n=\frac{r_n^2}{n\lambda}$$

Equation (17.9) is similar to the equation $\left(\frac{1}{\upsilon}-\frac{1}{u}=\frac{1}{f}\right)$ in the case of lenses with a and b as the object and image distances and f_n the focal length. Thus, a zone plate acts as a converging lens. A zone plate has a number of foci which depend on the number of zones used as well as the wavelength of light employed.

17.5.2. DIFFERENCE BETWEEN A ZONE PLATE AND A CONVEX LENS

For a given wavelength of light, a convex lens has only one focal length given by

$$\frac{1}{f}=(\mu-1)\left(\frac{1}{R_1}-\frac{1}{R_2}\right)$$

where f is the focal length of the lens, μ is the refractive index of the material of the lens and R_1 and R_2 are the radii of curvature. In a convex lens, the violet rays of light come to focus nearer the lens than the red rays of light because for a given material the refractive index for violet rays of light is more than for red rays of light.

In the case of a zone plate, there are a number of foci between the point O and P (Fig. 17.8). Each focus corresponds to the position where, with reference to P an odd number of half period elements can be constructed on each zone. As the screen is moved nearer the zone plate, the area of the half period elements decreases and more half period elements can be present on each zone. If P_m is the position on the image when $(2_m - 1)$ half period elements can be present on each zone, f_m the focal length of the zone plate is given by

$$f_m=\frac{r_n^2}{(2m-1)n\lambda} \quad (17.10)$$

Putting m = 1, 2, 3.., etc., the different positions of the screen for a bright image can be obtained. In equation (17.10), r_n is the radius of the nth zone of the wavefront, λ is the wavelength of light and (2m –1) is the number of odd half period elements present on each zone. For example, if the position of the screen is such that with reference to the point P, three half period elements can be constructed on each zone, then the focal length of the zone plate f_3 is given by

$$f_3=\frac{r_n^2}{5n\lambda}$$

With the decrease in the focal length of the zone plate, the brightness of the image decreases. Let the first zone contain only one period element. Then, the amplitude at P due to this zone is m_1. If the first zone contains three half period elements for a particular position of the screen, then the amplitude at P due to the first zone

$$=m_1-m_2+m_3=\frac{m_1}{2}+\left[\frac{m_1}{2}-m_2+\frac{m_3}{2}\right]+\frac{m_3}{2}=\frac{m_1}{2}+\frac{m_2}{2}$$

But, $\frac{m_1+m_3}{2}$ is less than m_1, because $m_1 > m_3$. Further, in a zone plate (for the same number of odd half period elements contained in each zone) the focal length for violet light is more than for red light, which is reverse in the case of a convex lens.

Comparison between a zone plate and a convex lens:

1. Both the zone plate and convex lens form a real image of the object and the equations connecting the conjugate distances are similar.

2. The focal lengths of both depend on the wavelength, λ and hence suffer from chromatic aberration. The chromatic aberration in a zone plate is much more severe than in a convex lens.

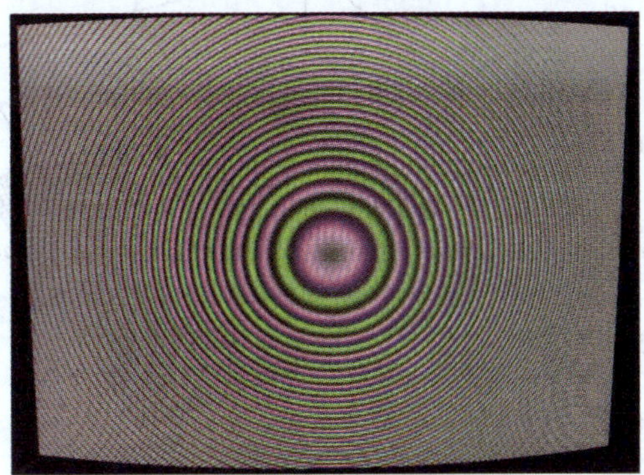

Comparison between a zone plate and a convex lens.

3. A zone plate acts simultaneously as a convex lens and as a concave lens. In addition to a real image, a virtual image is also formed simultaneously.
   A convex lens forms only a real image.
4. In case of zone plate the image is formed by the diffraction phenomenon.
   In case of a convex lens the image is formed due to refraction of light.
5. The zone plate has got multiple foci on either side of the plate. Hence, the intensity of the image formed will be much less.
   Convex lens has only one focus. As all the light is focused at one point, the intensity of the image will be more.
6. In a zone plate, waves reaching the image point through any two alternate zones differ in path by λ and in phase by 2π.
   In case of a convex lens all the rays reaching the image point have zero path or phase difference.
7. A zone plate can be used over a wide range of wavelengths from microwaves to x-rays. Glass lens cannot be used beyond the visible region.

17.6. DISTINCTION BETWEEN INTERFERENCE AND DIFFRACTION

The main differences between interference and diffraction are as follows:

INTERFERENCE	DIFFRACTION
1. Interference is the result of interaction of light coming from different wave fronts originating from the source.	1. Diffraction is the result of interaction of light coming from different parts of the same wavefront.
2. Interference fringes may or may not be of the same width.	2. Diffraction fringes are not of the same width.
3. Regions of minimum intensity are perfectly dark.	3. Regions of minimum intensity are not perfectly dark.
4. All bright bands are of same intensity.	4. The different maxima are of varying intensities with maximum intensity for central maximum.

17.7. FRESNEL AND FRAUNHOFFER TYPES OF DIFFRACTION

The diffraction phenomena are broadly classified into two types: Fresnel diffraction and Fraunhoffer diffraction.

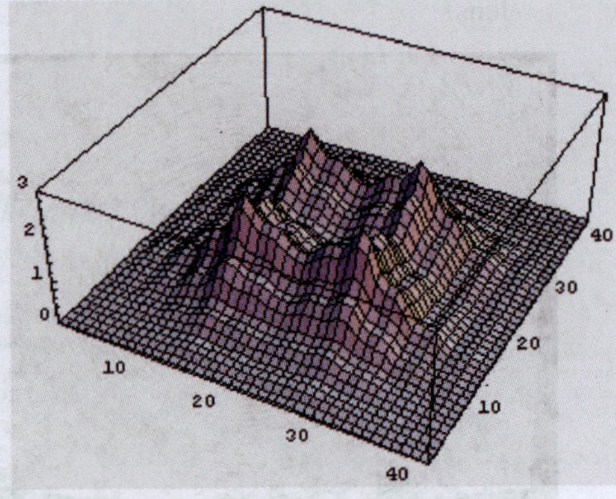

Surface of Fresnel diffraction.

1. Fresnel diffraction: In this type of diffraction, the source of light and the screen are effectively at finite distances from the obstacle (Fig. 17.9a). Observation of Fresnel diffraction phenomenon does not require any lenses. The incident wave front is not planar. As a result, the phase of secondary wavelets is not the same at all points in the plane of the obstacle. The resultant amplitude at any point of the screen is obtained by the mutual interference of secondary wavelets from different elements of unblocked portions of wave front. It is experimentally simple but the analysis proves to be very complex.

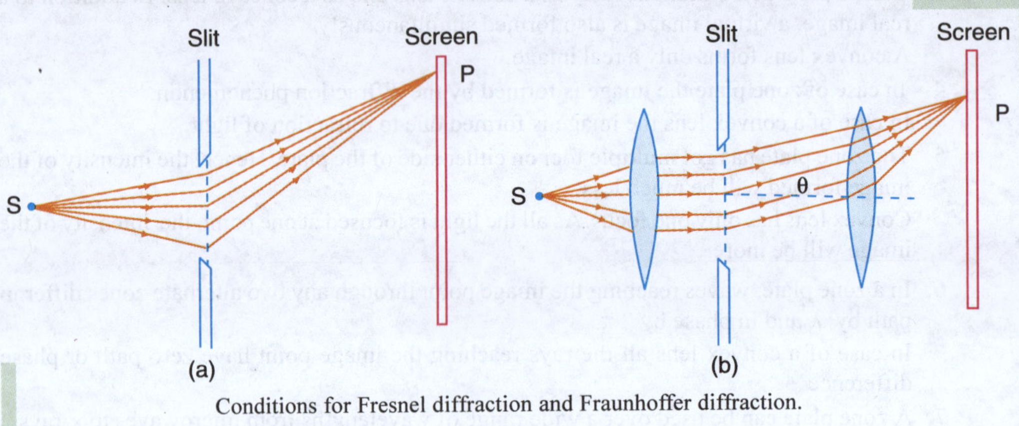

Conditions for Fresnel diffraction and Fraunhoffer diffraction.

Fig. 17.9

2. Fraunhoffer diffraction: In this type of diffraction, the source of light and the screen are effectively at infinite distances from the obstacle. Fraunhoffer diffraction pattern can be easily observed in practice. The conditions required for Fraunhoffer diffraction are achieved using two convex lenses, one to make the light from the source parallel and the other to focus the light after diffraction on to the screen (Fig. 17.9b). The diffraction is thus produced by the interference between parallel rays. The incident wave front as such is plane and the secondary wavelets, which originate from the unblocked portions of the wave front, are in the same phase at every point in the plane of the obstacle. This problem is simple to handle mathematically because the rays are parallel. The incoming light is rendered parallel with a lens and diffracted beam is focused on the screen with another lens.

Fresnel class of diffraction phenomenon is treated in this chapter.

17.8. DIFFRACTION AT A CIRCULAR APERTURE

Let AB be a small aperture (say a pin hole) and S is a point source of monochromatic light. XY is a screen perpendicular to the plane of the paper and P is a point on the screen. SP is perpendicular to the screen O is the center of the aperture and r is the radius of the aperture.

Let the distance of a source from the aperture be a (SO = a) and the distance of the screen from the aperture be b (OP=b). QOQ_1 is the incident spherical wavefront and with reference to the point P, O is the pole of the wavefront (Fig. 17.10). To consider the intensity at P, half period zones can be constructed with P as center and radii $b+\frac{\lambda}{2}, b+\frac{2\lambda}{2}$ etc. on the exposed wavefront AOB. Depending on the distance of P from the aperture (i.e. the distance b) the number of half period zones that can be constructed may be odd or even. If the distance a is such that only one half period zone can be constructed, then the intensity at P will be proportional to m_1^2 (where m_1 is the amplitude due to the first zone at P). On the other hand, if the whole of the wavefront is exposed to the point P, the resultant amplitude is $\frac{m_1}{2}$ or the intensity at P will be proportional to $\frac{m_1^2}{4}$. The position of the screen can be altered so as to construct 2, 3 or more half period zones for the same area of the aperture. If only two zones are exposed, the resultant amplitude at $P = m_1 - m_2$ (minimum) and if 3 zones are exposed, the amplitude = $m_1 - m_2 + m_3$ (maximum) and so on. Thus by continuously altering the value of b, the point P becomes alternately bright and dark depending on whether odd or even number of zones are exposed by the aperture.

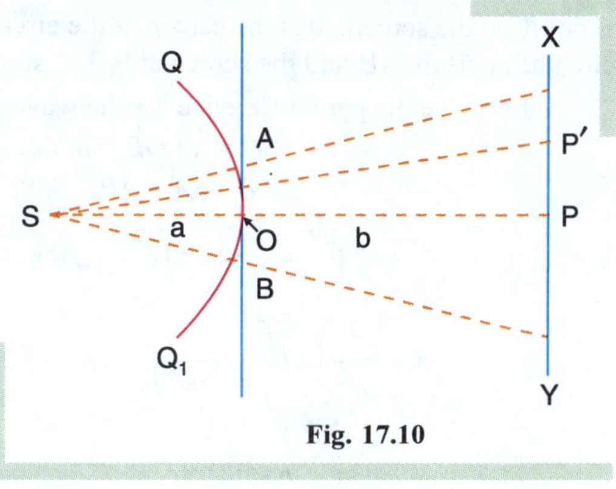

Fig. 17.10

Now let us consider a point P′ on the screen XY (Fig.17.11). Let S to P′ be joined. The line SP′ meets the wavefront at O′. O′ is the pole of the wavefront with reference to the point P′. Now let us construct half period zones with the point O′ as the pole of the wavefront. The upper half of the wavefront is cut off by the obstacle. If the first two zones are cut off by the obstacle between the points O′ and A and if only the 3rd, 4th and 5th zones are exposed by the aperture AOB, then intensity at P′ will be maximum. Thus if odd number of half period zones are exposed, point P′ will be of maximum intensity and if even number of zones are exposed, point P′ will be of minimum intensity. As the distance of P′ from P increases the intensity of maxima and minima gradually decreases. It is because with the point P′ far removed from P, the most effective central half period zones are cut off by the obstacle between the points O′ and A. With the outer zones the obliquity increases with reference to the point P′ and hence the intensity of maxima and minima also will be less. If the point P′ happens to be of maximum intensity, then all the points lying on a circle of radius PP′ on the screen also will be of maximum intensity. Thus with a circular aperture, the diffraction pattern will be concentric bright and dark rings with the centre P bright or dark depending on the distance b. The width of the rings continuously decreases.

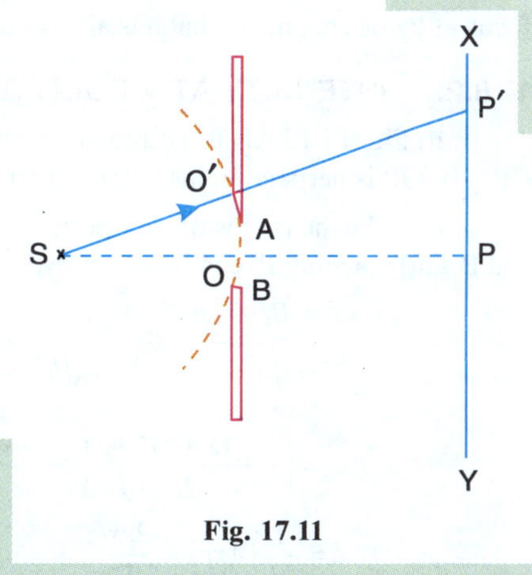

Fig. 17.11

17.8.1. MATHEMATICAL TREATMENT OF DIFFRACTION AT A CIRCULAR APERTURE

In Fig. 17.12 S is a point source of a monochromatic light, AB is the circular aperture and P is point on the screen. O is the center of the circular aperture. The line SOP is perpendicular to the circular aperture AB and the screen at P. The screen is perpendicular to the plane of the paper.

Let δ be the path difference for the wave reaching P along the paths SAP and SOP.

$$SO = a\ ;\ OP = b;\ OA = r$$

$$\delta = SA + AP - SOP$$

$$= \left(a^2+r^2\right)^{1/2} + \left(b^2+r^2\right)^{1/2} - (a+b) \quad = a\left(1+\frac{r^2}{a^2}\right)^{1/2} + b\left(1+\frac{r^2}{b^2}\right)^{1/2} - (a+b)$$

$$= a\left(1+\frac{r^2}{2a^2}\right) + b\left(1+\frac{r^2}{2b^2}\right) - (a+b) = \frac{r^2}{2}\left(\frac{1}{a}+\frac{1}{b}\right) \qquad (17.11a)$$

$$\therefore \qquad \frac{1}{a}+\frac{1}{b} = \frac{2\delta}{r^2}$$

If the position of the screen is such that n full number of half period zones can be constructed on the aperture, then the path difference, $\delta = \frac{n\lambda}{2}$ or $2\delta = n\lambda$,

Substituting the value of 2δ in (17.11a), we get ,

$$\frac{1}{a}+\frac{1}{b} = \frac{n\lambda}{r^2} \qquad (17.11b)$$

The point P will be of maximum or minimum intensity depending on whether n is odd or even. If the source is at infinite distance (for an incident plane wave front), then $a = \infty$ and

$$\frac{1}{b} = \frac{1}{f} = \frac{n\lambda}{r^2} \qquad (17.12)$$

If n is odd, P will be a bright point. The idea of focus at P does not mean that it is always a bright point.

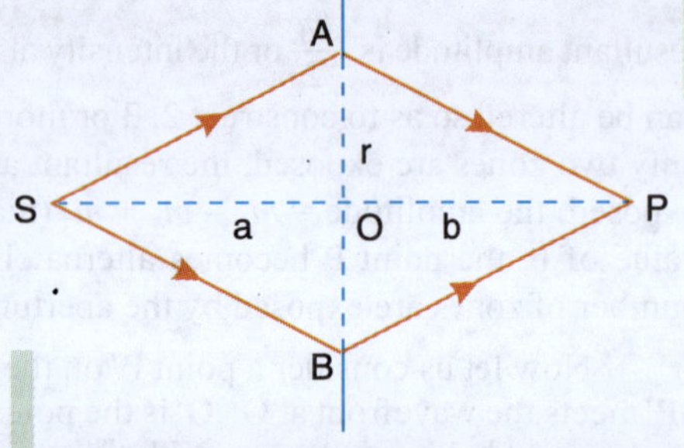

Fig. 17.12

17.8.2. INTENSITY AT A POINT AWAY FROM THE CENTRE

In Fig. 17.13 AB is a circular aperture and P and P′ are two points on the screen. PP′ = x and OP = b. OP is perpendicular to the screen.

Let r be the radius of the aperture. The path difference between the secondary waves from A and B and reaching P′ can be given by,

$$\delta = BP' - AP'$$

$$= \sqrt{b^2+(x+r)^2} - \sqrt{b^2+(x-r)^2} \quad = b\left(1+\frac{(x+r)^2}{2b^2}\right) - b\left(1+\frac{(x-r)^2}{2b^2}\right)$$

$$= \left(b+\frac{(x+r)^2}{2b}\right) - \left(b-\frac{(x-r)^2}{2b}\right) \quad = \frac{1}{2b}\left[(x+r)^2-(x-r)^2\right]$$

$$\therefore \qquad \delta = \frac{1}{2b}(4xr) = \frac{2rx}{b} \qquad (17.13)$$

The point P′ will be dark if the path difference,

$\delta = 2n\frac{\lambda}{2}$ (2n means even number of zones).

$$2n\frac{\lambda}{2} = \frac{2rx_n}{b}$$

or
$$x_n = \frac{nb\lambda}{2r} \qquad (17.14)$$

where x_n gives the radius of nth dark ring.

Similarly, if $\delta = \dfrac{(2n+1)\lambda}{2}$,

then $$\frac{(2n+1)\lambda}{2} = \frac{2rx_n}{b}$$

or $$x_n = \frac{(2n+1)\, b\,\lambda}{4r} \quad (17.15)$$

where x_n gives the radius of the nth bright ring.

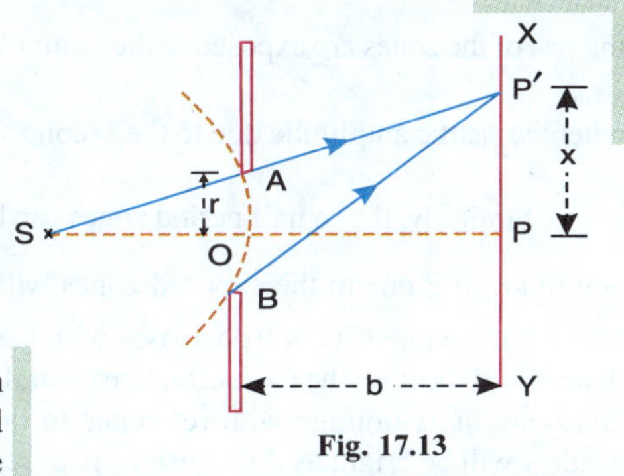

Fig. 17.13

The objective of a telescope consists of an achromatic convex lens and a circular aperture is fixed in front of the lens. Let the diameter of the aperture be D (= 2r). While viewing distant objects, the incident wave front is plane and the diffraction pattern consists of a bright centre surrounded by dark and bright rings of gradually decreasing intensity. The radii of the dark rings are given by

$$x_n = \frac{nb\lambda}{2r} = \frac{nb\lambda}{D} \quad (17.16)$$

The radius of the first dark ring is, $x_1 = \dfrac{b\lambda}{D}$

For an incident plane wavefront, $b = f$ the focal length of the objective.

$$\therefore \quad x_1 = \frac{f\lambda}{D}$$

The value of x_1 measures the distance of the first secondary minimum from the central bright maximum. However, according to Airy's theory, the radius of the first dark ring is given by

$$x_1 = \frac{1.22 f\lambda}{D} \quad (17.17)$$

It is interesting to note that the size of the central image depends on λ, the wavelength of light, *f*, the focal length of the lens and D, diameter of the lens aperture.

17.9. DIFFRACTION AT AN OPAQUE CIRCULAR DISC

S is a point source of monochromatic light. CD is an opaque disc and MN is the screen. P is a point on the screen such that SAP is perpendicular to the screen. The screen is perpendicular to the plane of the paper. XY is the incident spherical wave front. EF is the geometrical shadow. With reference to the point P, the wave front can be divided into half period zones taking the centre of the disc (A) as the pole (Fig.17.14). If one half period zone can be constructed on the surface of the disc,

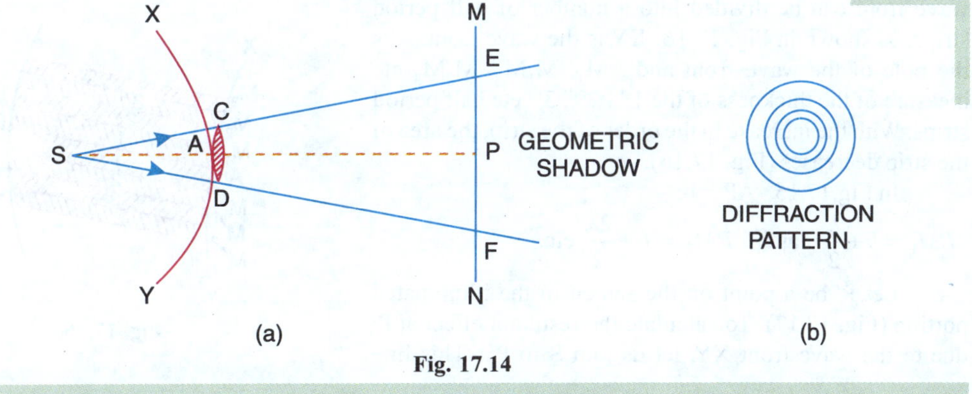

Fig. 17.14

the rest of the zones are exposed to the point P and the resultant amplitude at $P = \frac{m_2}{2}$ approximately, where m_2 is the amplitude due to the second zone $\left[m_2 - m_3 + m_4 - .... = \frac{m_2^2}{2} \text{ approximately} \right]$.

Similarly, if two half period zones can be constructed on the surface of the disc, the resultant amplitude at P due to the exposed zones will be $\frac{m_3}{2}$ and so on. Thus, the point P will always be bright but the intensity at P decreases with increase in the diameter of the disc. That is, with a large diameter of the disc, the most effective central zones will be cut off by the disc and the exposed outer zones are more oblique with reference to the point P. Thus (at P) the center of the geometrical shadow will be bright as if the disc were absent. The diffraction pattern consists of a central bright spot surrounded by alternate bright and dark rings, as shown in Fig. 17.14 (b).

17.10. DIFFRACTION PATTERN DUE TO A STRAIGHT EDGE

Let S be narrow slit illuminated by a source of monochromatic light of wavelength, λ. The length of the slit is perpendicular to the plane of the paper. AD is the straight edge and the length of the edge is parallel to the length of slit (Fig.17.15). XY is the incident cylindrical wavefront. P is a

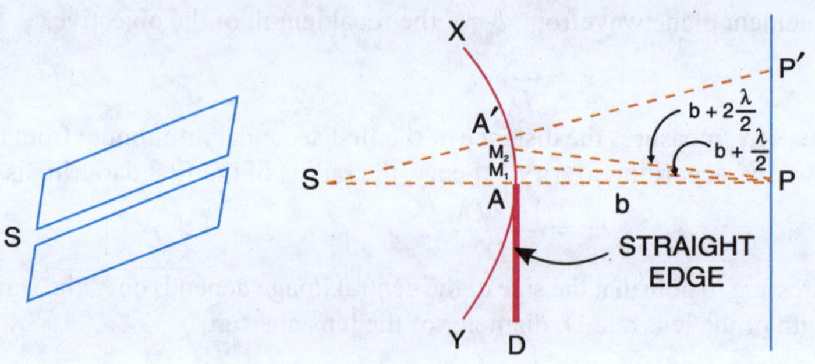

Fig. 17.15

point on the screen and SAP is perpendicular to the screen. The screen is perpendicular to the plane of the paper. Below the point P is the geometrical shadow and above P is the illuminated portion. Let the distance AP be b. With reference to the point P, the wave front can be divided into a number of half period strips, as shown in Fig. 17.16. XY is the wave front, A is the pole of the wave front and AM_1, M_1M_2, M_2M_3 etc measure of the thickness of the 1st, 2nd, 3rd etc half period strips. With the increase in the order of the strip, the area of the strip decreases (Fig. 17.16).

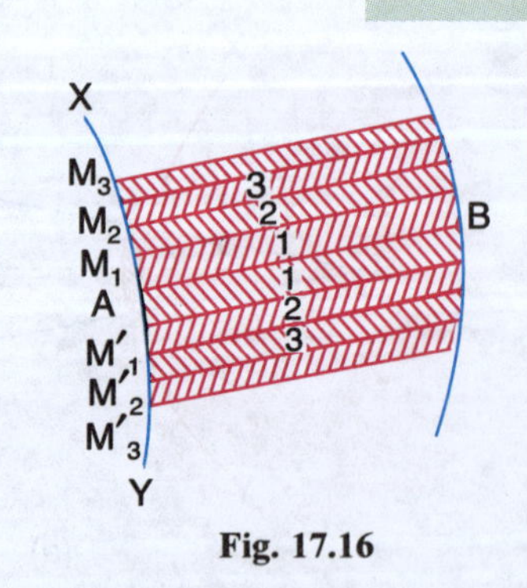

Fig. 17.16

In Fig.17.15, AP = b,

$$PM_1 = b + \frac{\lambda}{2} \quad \text{and} \quad PM_2 = b + \frac{2\lambda}{2} \text{ etc.}$$

Let P′ be a point on the screen in the illuminated portion (Fig. 17.17). To calculate the resultant effect at P′ due to the wave front XY, let us join S to P′. This line meets the wave front at B. B is the pole of the wave front with reference to the point P′ and the intensity at P′ will depend mainly on the number of half period strips enclosed between the points A and B. The effect at P′ due to the wave front above B is same at all points on the screen whereas it is different at different points due to the wave front between B and A. The point P′ will be of maximum intensity, if the number of half period strips enclosed between B and A is odd and the intensity at P′ will be minimum if the number of half period strips enclosed between B and A is even.

17.10.1. POSITIONS OF MAXIMUM AND MINIMUM INTENSITY

Let the distance between the slit and the straight edge be a and the distance between the straight edge and the screen be b (Fig. 17.17). Let PP′ be x.

The path difference, $\delta = AP' - BP'$

$$= (b^2 + x^2)^{1/2} - [SP' - SB]$$

$$= (b^2 + x^2)^{1/2} - \left(\sqrt{(a+b)^2 + x^2} - a\right)$$

$$= b\left[1 + \frac{x^2}{2b^2}\right] - (a+b)\left[1 + \frac{x^2}{2(a+b)^2}\right] + a$$

$$= \frac{x^2}{2}\left(\frac{1}{b} - \frac{1}{a+b}\right) = \frac{x^2}{2}\left(\frac{a+b-b}{b(a+b)}\right)$$

$$\therefore \quad \delta = \frac{x^2}{2} \cdot \frac{a}{b(a+b)}$$

Fig. 17.17

The point P′ will be of maximum intensity if $\delta = (2n+1)\frac{\lambda}{2}$

$$\therefore \quad (2n+1)\frac{\lambda}{2} = \frac{ax_n^2}{2b(a+b)}$$

$$x_n^2 = \frac{(2n+1)(a+b)b\lambda}{a}$$

or $$x_n = \sqrt{\frac{(2n+1)(a+b)b\lambda}{a}} \qquad (17.18)$$

where x_n is the distance of the nth bright band from P.

Similarly, P′ will be of minimum intensity if $\delta = 2n\frac{\lambda}{2}$.

$$\therefore \quad 2n\frac{\lambda}{2} = \frac{ax_n^2}{2b(a+b)} \quad \text{or} \quad x_n = \frac{\sqrt{2n(a+b)b\lambda}}{a}$$

where x_n is the distance of the nth dark band from P. Thus, diffraction bands of varying intensity (roughly corresponding to maxima and minima) are observed above the geometrical shadow i.e., above P and the bands disappear and uniform illumination occurs if P′ is far away from P.

17.10.2. INTENSITY AT A POINT INSIDE THE GEOMETRICAL SHADOW (STRAIGHT EDGE)

If P′ is a point below P (Fig. 17.18) and B is the new pole of the wave front with reference to the point P′, then the half period strips below B are cut off by the obstacle and only the uncovered half period strips above B will be effective in producing the illumination at P′. As P′ moves farther from P, more number of half period strips above B is also cut off and the intensity gradually falls. Thus within the geometrical shadow, the intensity gradually falls off depending on the position of P′ with respect to P.

The intensity distribution on the screen due to a straight edge is shown in Fig. 17.19. S is the source, AD is the straight edge and MN is the screen. In the illuminated portion PM, alternate bright and dark bands of gradually diminishing intensity will be observed and the intensity falls off gradually in the region of the geometrical shadow. Thus according to the wave theory, the shadows cast by obstacles in the path of light are not sharp and hence rectilinear propagation of light is only approximately true. In general, there is gradual fading of intensity in the region of the geometrical shadow and with monochromatic light bright and dark bands (diffraction bands) are observed in the illuminated portion of the screen. However, with white light coloured bands will be observed and the bands of shorter wavelength are nearer the point P.

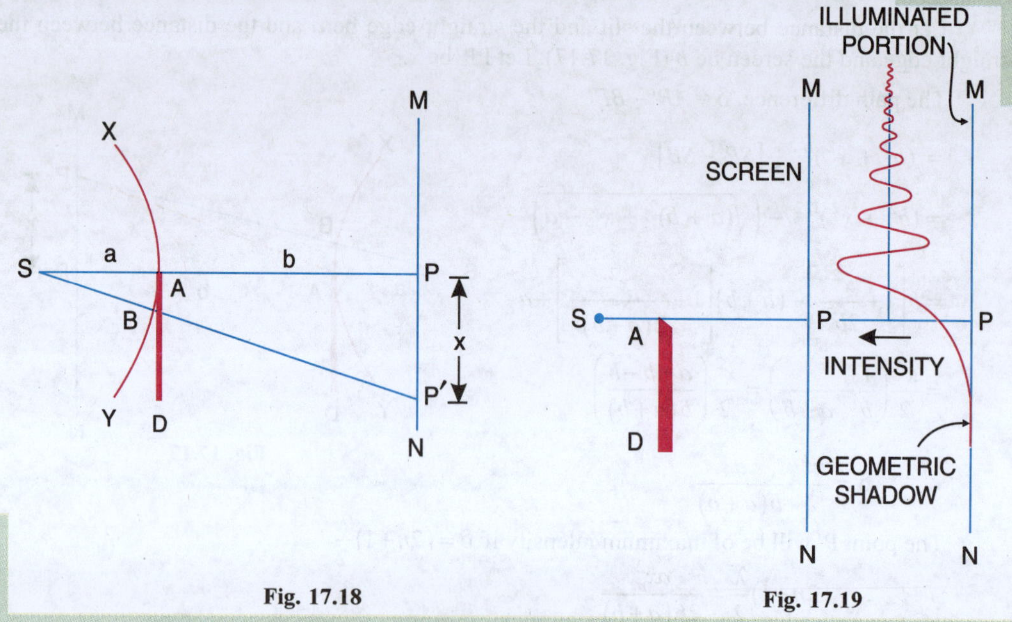

Fig. 17.18

Fig. 17.19

17.11. DIFFRACTION PATTERN DUE TO A NARROW SLIT

S is a narrow slit illuminated by monochromatic light. The length of the slit is perpendicular to the plane of the paper. AB is a rectangular aperture parallel to the slit, MN is the screen and P is a point on the screen such that SOP is perpendicular to the plane of the paper; XY is the incident cylindrical wave front (Fig. 17.20). On the screen, EF is the illuminated portion and above E and below F is the region of the geometrical shadow.

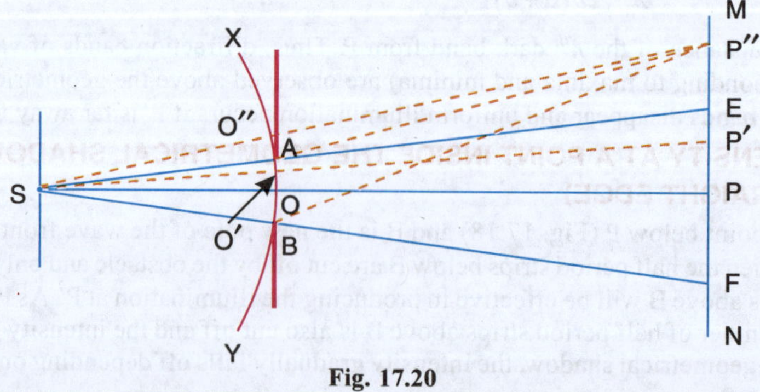

Fig. 17.20

If the slit AB is wide, then with reference to the point P, the cylindrical wave front can be divided into a large number of half period strips and the resultant amplitude at P will be $\frac{m_1}{2}$ where m_1 is the amplitude due to the first half period strip. Thus, the point P will be illuminated. Even points very near to P will be equally illuminated. If the wave front is divided with reference to points nearer P, the number of half period strips above and below the new pole in the exposed portion of the wavefront will be quite large and hence this results in uniform illumination.

Now, let us consider a point P′ nearer to the edge of the geometrical shadow (see Fig.17.20). Let us join S to P′. Here O′ is the pole of the wave front with reference to the point P′. If the wave front is divided into half period strips, the number of half period strips between O′ and B will be quite large and the illumination at P′ due to the lower portion of the wave front will be the same at all points near the edge of the geometrical shadow. But the intensity at P due to the exposed portion of the wave front between A and O′ will depend on the number of half period strips present. If the number of half period elements is odd, the point P′ will be of maximum intensity and if it is even the point will be of minimum intensity.

Let P″ be a point in the region of the geometrical shadow. Let us join S to P″. Here O″ is the pole of the wave front with reference to the point P″. If the wave front is divided into half period elements, then the upper half of the wave front between X and O″ is cut off by the obstacle and only a portion between A and B is exposed to the point P″. If the number of half period elements exposed by AB is odd, then P″ will be of maximum intensity and if it is even, it will be of minimum intensity. But as the most effective central half period strips between O″ and A are cut off, the intensity falls off rapidly in the region of the geometrical shadow and maxima and minima cannot be distinguished. The intensity distribution due to a wide aperture is shown in Fig. 17.21 (b).

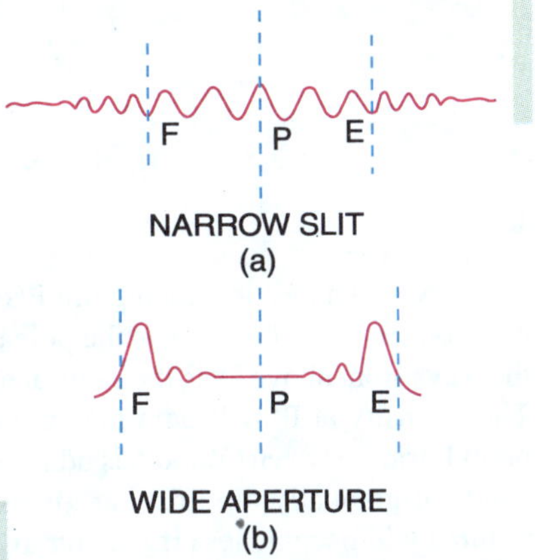

Fig. 17.21

On the other hand if the slit is narrow, the intensity at the point P will depend on the number of half period strips that can be constructed on the exposed wave front between A and B. If the number of half period strips is odd, the intensity at P will be maximum and if it is even, the intensity at P will be minimum (Fig. 17.20). Thus, the point P can be bright or dark. If we consider a point P′ in the illuminated portion EF of the screen, the intensity at P will depend on number of half period strips that can be constructed between A and O′ where O′ is the pole of the wave front with reference to the point P′. If the number of half periods strips between A and O′ is odd, P′ will be a point of maximum intensity. Thus, between E and F alternate bright and dark bands will be observed and the point P may be bright or dark.

Now consider a point P″ in the region of the geometrical shadow (Fig.17.20). O″ is the pole of the wavefront with reference to the point P″ and the intensity at P″ will depend on the number of half period strips exposed by the slit AB. The upper half of the wavefront above O″ is obstructed by the obstacle and even the most effective central half period strips between O″ and A are cut off by the obstacle. Thus, at P″ which is far away from E, the maxima and minima become indistinguishable. There is no marked transition between the diffraction bands observed in the geometrical shadow and the illuminated portion. In the intensity distribution on the screen due to a narrow slit (say less than the wavelength of light), a broad central maximum will be observed in the illuminated portion and the intensity variation cannot be distinguished. The intensity gradually falls off in the region of geometrical shadow.

17.12. DIFFRACTION DUE TO A NARROW WIRE

In Fig. 17.22, S is a narrow slit illuminated by monochromatic light, AB is the diameter of the narrow wire and MN is the screen. The length of the wire is parallel to the illuminated slit and perpendicular to the plane of the paper. The screen is also perpendicular to the plane of the paper. XY is the incident cylindrical wave front. P is a point on the screen such that SOP is perpendicular to the screen. EF is the region of the geometrical shadow and above E and below F, the screen is illuminated.

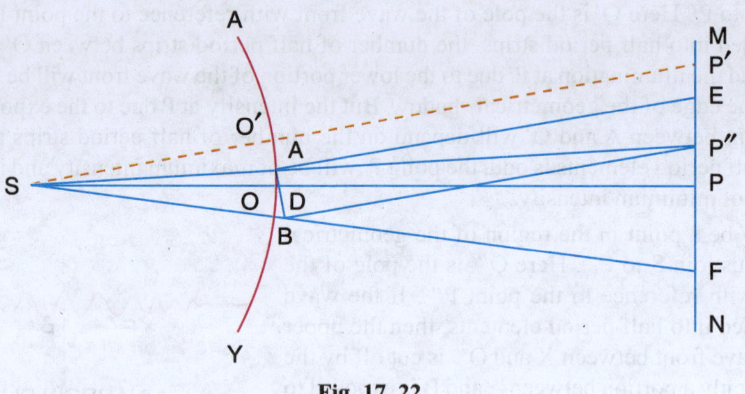

Fig. 17. 22

Now, let us consider a point P′ on the screen in the illuminated portion. Let us join S to O′, a point on the wave front. O′ is the pole of the wave front with reference to P′. The intensity P′ due to the wave front above O′ is the same at all points and the effect due to the wave front BY is negligible. The intensity at P′ will be a maximum or a minimum depending on whether the number of half period strips between O′ and A is odd or even. Thus, in the illuminated portion of the screen, diffraction bands of gradually diminishing intensity will be observed. The distinction between maxima and minima will become less if P′ is far away from the edge E of the geometrical shadow. Maxima and minima cannot be distinguished if the wire is very narrow, because in that case the portion BY of the wavefront also produces illumination at P.

Next let us consider a point P″ in the region of the geometrical shadow. Interference bands of equal width will be observed in this region due to the fact that the points A and B, of the incident wave front, are similar to two coherent sources. The point P″ will be of maximum or minimum intensity, depending on whether the path difference (BP″ – AP″) is equal to even or odd multiples of $\frac{\lambda}{2}$. The fringe width β is given by

$$\beta = \frac{D\lambda}{d}$$

where D is the distance between the wire and the screen, λ is the wave length of light and d is the distance between the two coherent sources. In this case $d = 2r$ where $2r$ is diameter of the wire $(AB = 2r)$.

$$\therefore \quad \beta = \frac{D\lambda}{2r} \qquad (17.19)$$

$$\therefore \quad r = \frac{D\lambda}{2\beta} \qquad (17.20)$$

$$\text{or} \quad \lambda = \frac{2r\beta}{D} \qquad (17.21)$$

Here, β the fringe width corresponds to the distance between any two consecutive maxima. Thus, from equations (17.20) and (17.21), knowing the values of *r* or λ ; λ or r can be determined. In Fig. 17.23 the bands marked "*a*" represent the interference bands in the region of the geometrical shadow and the bands marked "*b*" and "c"represent the diffraction bands in the illuminated portion. The intensity distribution due to a narrow wire is shown in Fig. 17.24 (*a*). The center of the geometrical shadow is bright.

(b) (a) (c)

Fig. 17.23

On the other hand, if the wire is very thick, the interference bands cannot be noticed. From equation (17.19), $\beta = \frac{D\lambda}{2r}$; where β is the fringe width. As the diameter of the wire increases the fringe width decreases and if the wire is sufficiently thick, the width of the interference fringes decreases considerably and they cannot be distinguished. The intensity falls off rapidly in the geometrical shadow. The diffraction pattern in the illuminated portion will be similar to that of a thin wire Fig. 17.24(b). Coloured fringes will be observed with white light.

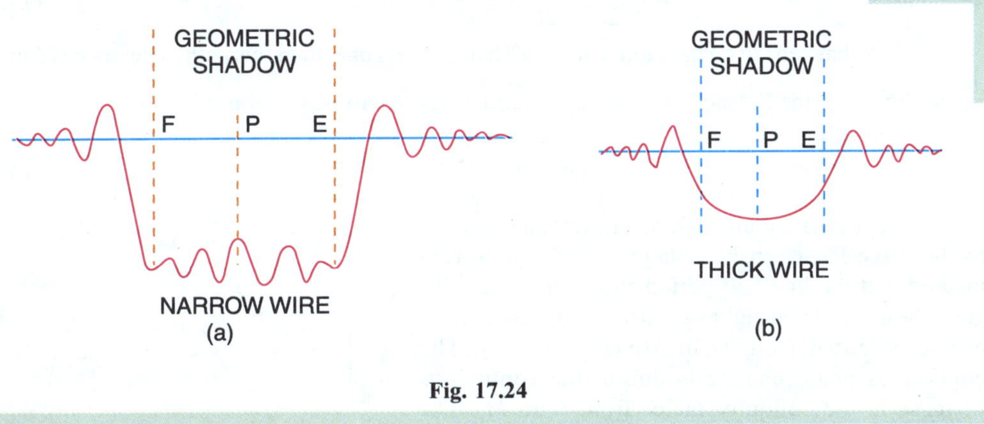

Fig. 17.24

17.13. CORNU'S SPIRAL

To find the effect at a point due to an incident wave front Fresnel's method consists in dividing the wavefront into half period strips or half period zones. The path difference between secondary

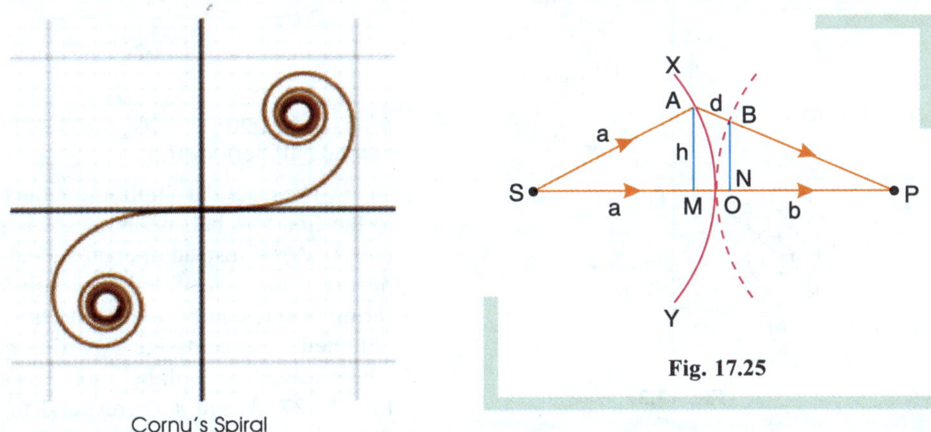

Cornu's Spiral

Fig. 17.25

waves from two corresponding points of neighboring zones is equal to $\frac{\lambda}{2}$.

In Fig. 17.25, S is a point source of light and *XY* is the incident spherical wave front. With reference to the point *P*, *O* is the pole of the wave front. Let *a* and *b* be the distances of the points *S* and *P* from the pole of the wave front. With *P* as centre and radius *b*, let us draw a sphere touching the incident wavefront at *O*. The path difference between the waves travelling in the directions *SAP* and *SOP* is given by

$$d = SA + AP - SOP = SA + AP - (SO + OP) = a + AB + b - (a + b) = AB$$

For large distances of *a* and *b*, AM and BN can be taken to be approximately equal and the path difference *d* can be written as

$$d = AB = MO + ON$$

But, from the property of a circle,

$$MO = \frac{AM^2}{2SO} = \frac{h^2}{2a} \text{ and } ON = \frac{BN^2}{2OP} = \frac{h^2}{2b} \quad \text{(approximately)}$$

$$\therefore \quad d = \frac{h^2}{2a} + \frac{h^2}{2b} = \frac{h^2}{2ab}(a + b) \quad (17.22)$$

If AM happens to be the radius of the nth half period zone, then this path difference is equal to $\frac{n\lambda}{2}$ according to the Fresnel's method of constructing the half period zones.

$$\therefore \quad \frac{h^2}{2ab}(a+b) = \frac{n\lambda}{2} \quad (17.23)$$

The resultant amplitude at an external point due to the wave front can be obtained by the following method. Let the first half period strip of the Fresnel's zones be divided into eight sub- strips and these vectors are represented from *O* to M_1 (Fig. 17.26). The continuous phase change is due to the continuous increase in the obliquity factor from *O* to M_1. The resultant amplitude at the external point due to the first half period strip is given by OM_1 (= m_1). Similarly, if the process is continued, we obtain the vibration curve M_1M_2. The portion M_1M_2 corresponds to the second half period strip.

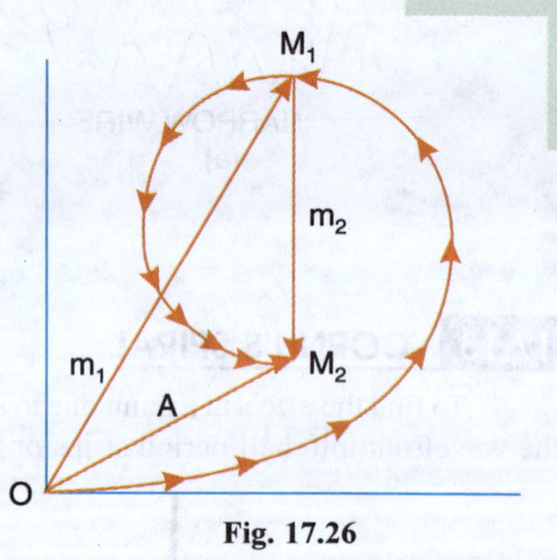

Fig. 17.26

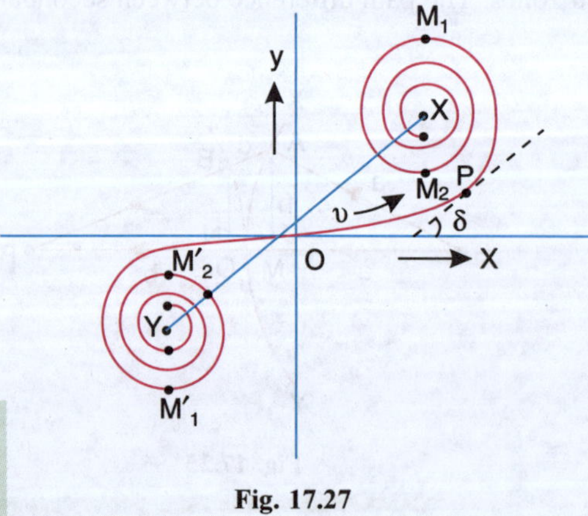

Fig. 17.27

The resultant amplitude at the point due to the first two half period strips is given by OM_2 (= A). If instead of eight sub-strips, each period zone is divided into sub-strips of infinitesimal width, a smooth curve will be obtained. The complete vibration curve for whole wave front will be a spiral as shown in Fig. 17.27. *X* and *Y* correspond to the two extremities of the wave front and M_1 and M_2

etc refer to the edge of the first, second, etc. half period strips. Similarly $M_1{}'$ and $M_2{}'$ etc refer to the edge of the first, second, etc half period strips of the lower portion of the wave front. This is called **Cornu's spiral**. The characteristic of this curve is that for any point P on the curve, the phase lag δ is directly proportional to the square of the distance υ. The distance is measured along the length of the curve from the point O. For a path difference of λ, the phase difference 2π. Hence, for a path difference of d, the phase difference δ is given by,

$$\delta = \frac{2\pi}{\lambda} d.$$

Substituting the value of d from equation (17.22) we get,

$$\delta = \frac{\pi}{2}\left[\frac{2h^2(a+b)}{ab\lambda}\right] \tag{17.24}$$

$$\delta = \frac{\pi}{2}\upsilon^2 \tag{17.25}$$

The value of υ is given by

$$\upsilon^2 = \frac{2h^2(a+b)}{ab\lambda}$$

or

$$\upsilon = h\sqrt{\frac{2(a+b)}{ab\lambda}} \tag{17.26}$$

Cornu's spiral can be used for any diffraction problem irrespective of the values of a, b and λ.

17.13.1. FRESNEL'S INTEGRALS

For any point on the Cornu's spiral, the x and y co-ordinates are given by two integrals known as *Fresnel's integrals*. Let us consider the point P on the spiral (Fig 17.27). The distance of the point P along the curve from the origin is υ. The tangent to the curve at P makes an angle δ with the x-axis. δ corresponds to the phase change from O to P. For a small displacement $d\upsilon$ of the point along the curve, let the corresponding changes in the co-ordinates of the point be dx and dy.

Then, $dx = d\upsilon \cos\delta$ and $dy = d\upsilon \sin\delta$

Substituting the value of δ from equation (17.25) of §17.13, we get

$$dx = \cos\left(\frac{\pi\upsilon^2}{2}\right)d\upsilon \tag{17.27}$$

and

$$dy = \sin\left(\frac{\pi\upsilon^2}{2}\right)d\upsilon \tag{17.28}$$

The coordinates x and y of the Cornu's spiral are given by,

$$x = \int dx = \int_0^{\upsilon} \cos\left(\frac{\pi\upsilon^2}{2}\right)d\upsilon \tag{17.29}$$

and

$$y = \int dy = \int_0^{\upsilon} \sin\left(\frac{\pi\upsilon^2}{2}\right)d\upsilon \tag{17.30}$$

These are called **Fresnel's integrals.**

17.13.2. MAXIMA AND MINIMA IN DIFFRACTION PATTERNS (CORNU'S SPIRAL)

The various diffraction patterns as discussed in the earlier articles and the positions of maxima and minima can be easily explained with the help of Cornu's spiral.

In Fig. 17.28, O is the origin of co-ordinates, OX is the vibration curve for the upper half of the wave front and OY refers to the vibration curve for the lower half of the wave front. If the whole wave front is unobstructed, the resultant amplitude at a point is given by XY.

If a cylindrical wave front is incident on a straight edge, the amplitude at a point P on the edge of the geometrical shadow (refer to the discussion on diffraction at a straight edge) is given by XY. If points above P in the illuminated portion are considered, gradually more of the lower half of the wave front is also exposed to the screen and amplitude vector passes through maxima and minima. Xb′, Xd′ etc refer to maximum amplitudes and Xc′ refers to the minimum amplitude. Thus, in the illuminated portion alternate bright and dark bands parallel to the length of the slit are observed on the screen. If points below P and in the region of (the geometrical shadow are considered) the lower half of the wave front and a portion of the upper half of the wave front are cut off and the tail of the amplitude vector moves to the right of O. The amplitude gradually decreases and becomes zero when the tail approaches X. Thus, in the region of the geometrical shadow the intensity falls off gradually. Quantitative values of intensity for different points on the screen can be obtained by finding the amplitude A for different values of υ. The square of the amplitude measures the intensity at that point. The point M_1, M_2 etc correspond to the edges of the first, second etc half period strips in the upper half of the wave front and the points M_1', M_2' etc refer to the lower half of the wave front. The points b′, d′ etc on the spiral, corresponding to maximum intensity, occur a little before the points M_1', M_3' etc are reached.

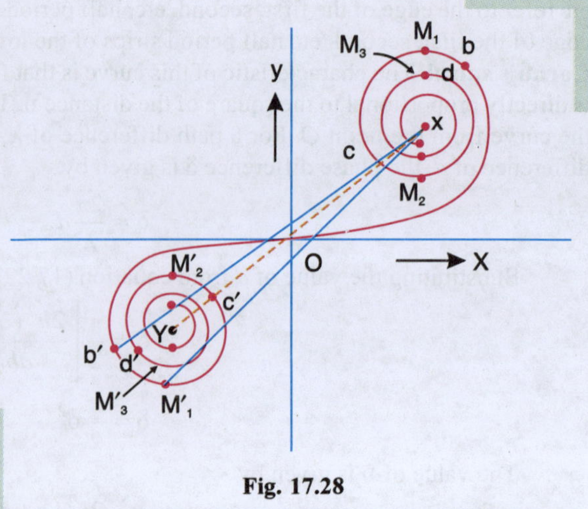

Fig. 17.28

The coordinates x and y of the Cornu's spiral are given by

$$x = \int dx = \int_0^{\upsilon} \cos\left(\frac{\pi \upsilon^2}{2}\right) d\upsilon \tag{17.31}$$

$$y = \int dy = \int_0^{\upsilon} \sin\left(\frac{\pi \upsilon^2}{2}\right) d\upsilon \tag{17.32}$$

The values of these integrals can be calculated for different values of υ. The graph is as shown in Fig. 17.28. The two integrals represent the horizontal and vertical components of the resultant amplitude. The intensity is proportional to the square of the resultant amplitude.

$$I_p = k\left[x^2 + y^2\right] \tag{17.33}$$

When the whole of the wave front is exposed to the point, $\upsilon \to \infty$, the values of the integrals will be,

$$x = \int_0^{\infty} \cos\left(\frac{\pi \upsilon^2}{2}\right) d\upsilon = \frac{1}{2}$$

$$y = \int_0^{\infty} \sin\left(\frac{\pi \upsilon^2}{2}\right) d\upsilon = \frac{1}{2}$$

Thus, for the point X in Fig 17.28, the x and y co-ordinates are [1/2, 1/2]. Similarly, for the point Y on the lower half of the spiral, the coordinates are [–1/2, –1/2].

At the origin, i.e. when $\upsilon = 0$, $x = 0$ and $y = 0$, the spiral passes through the origin and it is symmetrical with the origin. At any point on the spiral, the tangent to the curve makes an angle ϕ with the x-axis and $\tan \phi = \frac{dy}{dx}$.

$$\tan = \phi \frac{\sin\left(\frac{\pi \upsilon^2}{2}\right) d\upsilon}{\cos\left(\frac{\pi \upsilon^2}{2}\right) d\upsilon} = \tan\left(\frac{\pi \upsilon^2}{2}\right)$$

or
$$\phi = \left(\frac{\pi \upsilon^2}{2}\right) \tag{17.34}$$

When $\upsilon = 0, \quad \phi = 0.$

It means the curve is parallel to the x-axis at the origin.

The element of length $d\upsilon$ along the spiral is given by,

$$d\upsilon = \sqrt{(dx)^2 + (dy)^2} \tag{17.35}$$

Differentiating equation (17.34), we get

$$d\phi = \frac{2\pi \upsilon}{2} d\upsilon \quad \text{or} \quad \frac{d\upsilon}{d\phi} = \frac{1}{\pi \upsilon} \tag{17.36}$$

Hence $\frac{d\upsilon}{d\phi}$ measures the radius of curvature of the spiral at the point under consideration.

From equation (17.36), $\frac{d\upsilon}{d\phi} \propto \frac{1}{\upsilon}$

It shows that with the increase in the value of υ, the radius of curvature of the curve gradually decreases and takes the shape of a spiral. Finally for $\upsilon \to \infty$, the curve ends in a point (X or Y).

17.14. CORNU'S SPIRAL (ALTERNATIVE METHOD)

A narrow slit illuminated by light gives rise to a cylindrical wave front. Let S be a narrow slit perpendicular to the plane of the paper and illuminated by monochromatic light of wavelength λ (Fig 17.29).

For the cylindrical wave front XY, the slit is the axis of the cylindrical surface. The effect of the wave front at a point P in the plane is the same as that at all points along a line passing through the point P and parallel to the length of the slit. Let y be the displacement at all points on the wavefront XY at time t such that

$$y = a \sin \omega t = a \sin \frac{2\pi t}{T}$$

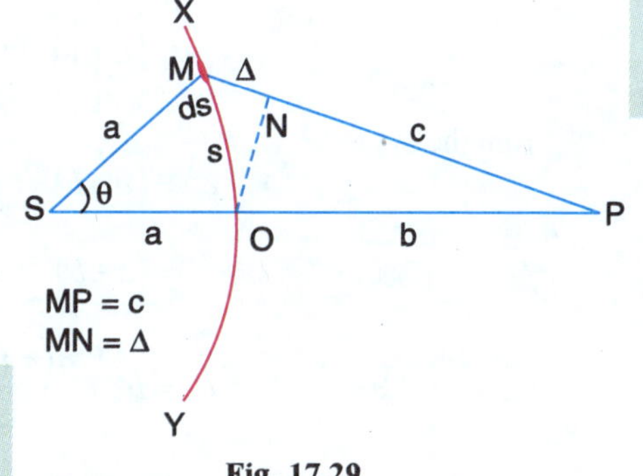

Fig. 17.29

According to Huygens principle, every point on the primary wave front is a source of secondary disturbance and the resultant intensity at P can be obtained by combining the effect of all the secondary wavelets. Let us consider a small element ds of the wave front at M. The distance of the element measured along the curve is s. For the secondary disturbances from O to P, i.e., for a distance OP = b, the phase difference is $\frac{2\pi b}{\lambda}$. The disturbance at the point P due to a small element ds is given by

$$dy = K \sin 2\pi \left[\frac{t}{T} - \frac{b}{\lambda}\right] ds \tag{17.37}$$

Similarly, the disturbance at P due to a small element at M is given by

$$dy = K \sin 2\pi \left[\frac{t}{T} - \frac{c}{\lambda}\right] ds \tag{17.38}$$

where MP = $c = b + \Delta$

$$dy = K \sin 2\pi \left[\frac{t}{T} - \frac{b}{\lambda} - \frac{\Delta}{\lambda}\right] ds \tag{17.39}$$

$$dy = K \sin \left[2\pi\left(\frac{t}{T} - \frac{b}{\lambda}\right) - \frac{2\pi\Delta}{\lambda}\right] ds$$

or $$dy = K\left[\sin 2\pi\left(\frac{t}{T} - \frac{b}{\lambda}\right)\cos \frac{2\pi\Delta}{\lambda}\right] ds - K\left[\cos 2\pi\left(\frac{t}{T} - \frac{b}{\lambda}\right)\sin \frac{2\pi\Delta}{\lambda}\right] ds \tag{17.40}$$

Integrating the above equation between the limits O and s, we get

$$y = K \sin 2\pi\left(\frac{\text{t}}{\text{T}} - \frac{b}{\lambda}\right)\int_0^{\text{s}} \cos\frac{2\pi\Delta}{\lambda}\, ds - K \cos 2\pi\left(\frac{\text{t}}{\text{T}} - \frac{b}{\lambda}\right)\int_0^{\text{s}} \sin\frac{2\pi\Delta}{\lambda}\, ds \tag{17.41}$$

Let $$K\int_0^s \cos\frac{2\pi\Delta}{\lambda} ds = R\cos\phi \quad \text{and} \quad K\int_0^s \sin\frac{2\pi\Delta}{\lambda} ds = R\sin\phi$$

Substituting these values in equation (17.41), we obtain

$$y = R \sin 2\pi\left(\frac{t}{T} - \frac{b}{\lambda}\right)\cos\phi - R\cos 2\pi\left(\frac{t}{T} - \frac{b}{\lambda}\right)\sin\phi$$

$$= R\sin\left[2\pi\left(\frac{t}{T} - \frac{b}{\lambda}\right) - \phi\right]$$

Also $R^2 = R^2\sin^2\phi + R^2\cos^2\phi$

$$= K^2\left[\left\{\int_0^s \sin\frac{2\pi\Delta}{\lambda} ds\right\}^2 + \left\{\int_0^s \cos\frac{2\pi\Delta}{\lambda} ds\right\}^2\right]$$

From the ΔSMP

$$c^2 = (a+b)^2 + a^2 - 2a(a+b)\cos\theta$$

$$= (a+b)^2 + a^2 - 2a(a+b)\left[1 - \frac{\theta^2}{2\,!}\right] = b^2 + a(a+b)\theta^2$$

But $$\theta = \frac{s}{a} \qquad \therefore\ c^2 = b^2 + \frac{a(a+b)s^2}{a^2} = b^2 + \frac{(a+b)}{a}s^2$$

$$\therefore \qquad c = b\left[1+\frac{(a+b)}{a}\frac{s^2}{b^2}\right]^{1/2} = b+\frac{(a+b)}{2ab}s^2$$

$$MP - NP = \Delta = c - b$$

$$= b+\frac{(a+b)}{2ab}s^2 - b$$

or
$$\Delta = \frac{(a+b)s^2}{2ab} \tag{17.42}$$

Let $\frac{2\pi\,\Delta}{\lambda} = \frac{\pi\,\upsilon^2}{2}$, where υ is a new variable depending on the values of a, b, λ and s. Substituting the value of Δ from equ. (17.42), we obtain

$$\frac{2\pi}{\lambda}\left[\frac{(a+b)s^2}{2ab}\right] = \frac{\pi\,\upsilon^2}{2}$$

or
$$s^2 = \upsilon^2\left[\frac{ab\lambda}{2(a+b)}\right]$$

$$\therefore \qquad s = \upsilon\sqrt{\frac{ab\lambda}{2(a+b)}} \tag{17.43}$$

or
$$\upsilon = s\sqrt{\frac{2(a+b)}{ab\lambda}} \tag{17.44}$$

From equation (17.43),

$$ds = d\upsilon\sqrt{\frac{ab\lambda}{2(a+b)}}$$

Substituting the value of ds and $\frac{2\pi\,\Delta}{\lambda}$ in the following relation

$$R^2 = K^2\left[\left\{\int_0^s \cos\frac{2\pi\,\Delta}{\lambda}ds\right\}^2 + \left\{\int_0^s \sin\frac{2\pi\,\Delta}{\lambda}\right\}^2\right]$$

we get

$$R^2 = \frac{K^2 ab\lambda}{2(a+b)}\left[\left(\int_0^\upsilon \cos\frac{\pi\,\upsilon^2}{2}d\upsilon\right)^2 + \left(\int_0^\upsilon \sin\frac{\pi\,\upsilon^2}{2}d\upsilon\right)^2\right]$$

or
$$R^2 = K_1\left[X^2 + Y^2\right]$$

where $X = \int_0^\upsilon \cos\frac{\pi\,\upsilon^2}{2}\,d\upsilon$ and $Y = \int_0^\upsilon \sin\frac{\pi\,\upsilon^2}{2}\,d\upsilon$

$$\int_0^\upsilon \cos\frac{\pi\,\upsilon^2}{2}\,d\upsilon = M\cos\left(\frac{\pi\,\upsilon^2}{2}\right) + N\sin\left(\frac{\pi\,\upsilon^2}{2}\right)$$

and
$$\int_0^\upsilon \sin\left(\frac{\pi\,\upsilon^2}{2}\right)d\upsilon = M\sin\left(\frac{\pi\,\upsilon^2}{2}\right) - N\sin\left(\frac{\pi\,\upsilon^2}{2}\right)$$

Here $M = \frac{\pi^{o}\upsilon}{1} - \frac{\pi^{2}\upsilon^{5}}{1.3.5} + \frac{\pi^{4}\upsilon^{9}}{1.3.5.7.9}$ and $N = \frac{\pi\,\upsilon^{3}}{1.3} - \frac{\pi^{3}\upsilon^{7}}{1.3.5.7} + \frac{\pi^{5}\upsilon^{11}}{1.3.5.7.9.11}$

For a given value of s (the extent of the wavefront contributing for the intensity at a point on the screen) the corresponding value of υ is evaluated from the relation,

$$\upsilon = \sqrt{\frac{2(a+b)}{ab\lambda}}.$$

From the value of υ thus obtained, the values of X and Y corresponding to the X and Y coordinates of a point on the spiral are obtained from Fresnel's integrals. For the properties of Cornu's spiral refer to §17.13.2. Cornu's spiral is useful in understanding the Fresnel's diffraction patterns due to obstacles such as straight edge, thin wire, thick wire, narrow slit, wide aperture etc placed in the path of light.

Application of Cornu's spiral to Fresnel's diffraction at a straight edge is dealt in §17.15. Proceeding in the same way, the intensity distribution on the screen due to Fresnel's diffraction at a narrow wire, narrow slit, thick wire, wide aperture etc can also be obtained.

17.15. DIFFRACTION AT A STRAIGHT EDGE

S is a narrow slit illuminated by monochromatic light of wavelength λ. The length of the slit is perpendicular to the plane of the paper. *AD* is a straight edge and the length of the edge is parallel length of the slit (Fig.17.30). *XY* is the incident cylindrical wave front. Cornu's spiral helps to obtain qualitatively the intensity distribution at a point on the screen.

Diffract spatially filtered beam from 60-mW helium-neon laser around edge of razor blade mounting on rotating stage.

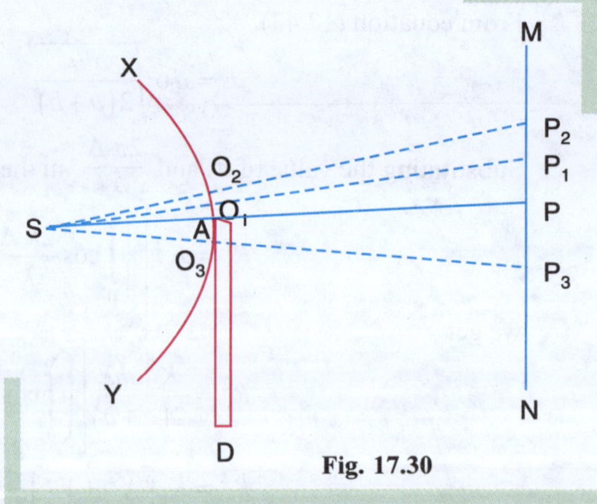

Fig. 17.30

Let us consider the points P, P_1 and P_2 in the illuminated portion and point P_3 in the geometrical shadow region of the screen *MN*. A, O_1, O_2 and O_3 are the poles of the wave front for the points P, P_1, P_2 and P_3 respectively.

Intensity at the point P:

The pole of the wave front is the origin of coordinates. The lower half of the wave front is cut off by the obstacle and the intensity at P is due to the upper half of the wave front between A and X. This intensity is equal to $I_o/4$, where I_o is the intensity due to the whole wave front (see Fig.17.31).

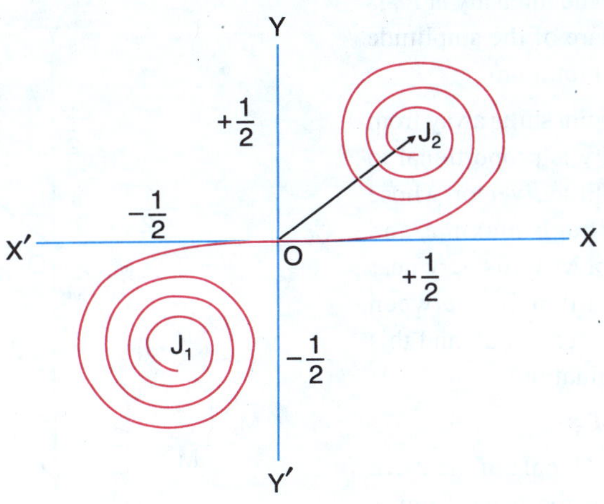

Fig. 17.31

$$I = K_1[X^2 + Y^2]$$

$$I = K_1\left[\left(\int_{-\infty}^{+\infty} \cos \frac{\pi \upsilon^2}{2} d\upsilon\right)^2 + \left(\int_{-\infty}^{+\infty} \sin \frac{\pi \upsilon^2}{2} d\upsilon\right)^2\right]$$

$$= K_1[1+1] = 2K_1$$

Let I_o be the intensity due to the whole wave front.

Then $\qquad I_o = 2K_1$

or $\qquad K_1 = I_o/2.$

The intensity due to the upper half of the wave front alone is given by

$$I = \frac{I_o}{2}\left[\left(\int_0^{\infty} \cos \frac{\pi \upsilon^2}{2} d\upsilon\right)^2 + \left(\int_0^{\infty} \sin \frac{\pi \upsilon^2}{2} d\upsilon\right)^2\right] = \frac{I_0}{2}\left[\frac{1}{4}+\frac{1}{4}\right] = \frac{I_0}{4}$$

This intensity is proportional to the square of the amplitude vector OJ_2.

Intensity at the point P_1:

For the point P_1, O_1 is the pole of the wave front and let the exposed portion of the wave front between A and O_1 correspond to the spiral OM_1 (see Fig. 17.32). The intensity in this case is proportional to the square of the vector M_1J_2. Thus, the point P_1 will be of maximum intensity.

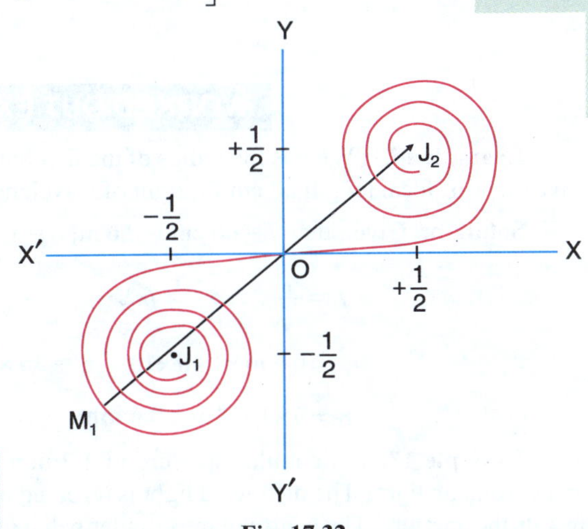

Fig. 17.32

Intensity at the point P_2:

For the point P_2, O_2 is the pole of the wave front and let the exposed portion of the wave front between A and O_2 correspond to the length of the spiral

OM_2 (see Fig. 17.33). The intensity at P_2 is proportional to the square of the amplitude vector M_2J_2 which is a minimum.

Thus, when the point shifts away from P, the resultant intensity is proportional to the square of the amplitude vector whose magnitude passes through maxima and minima. When the point M shifts very near J_1, the difference in intensity between maxima and minima is very small and this results in uniform illumination.

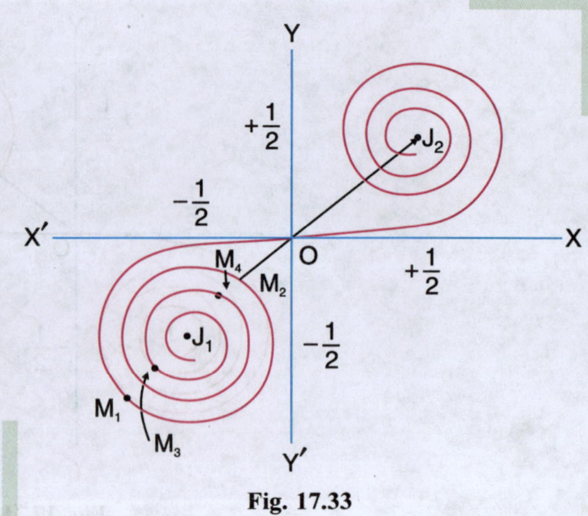

Fig. 17.33

Intensity at the point P_3:

For the point P_3, O_3 pole of the wave front. The lower half of the wave front is cut off by the obstacle and the portion of the wave front between A and O_3 is also obstructed corresponding to the length of the spiral OM (see Fig. 17.34). The intensity at P_3 is proportional to the square of the amplitude MJ_2. As the point P_3 shifts more and more into the region of geometrical shadow, the point M shifts more and more towards J_2. Thus, the magnitude of MJ_2 gradually decreases. In other words, the intensity falls off gradually in the region of the geometrical shadow. The intensity distribution due to Fresnel diffraction at a straight edge is given in Fig. 17.19.

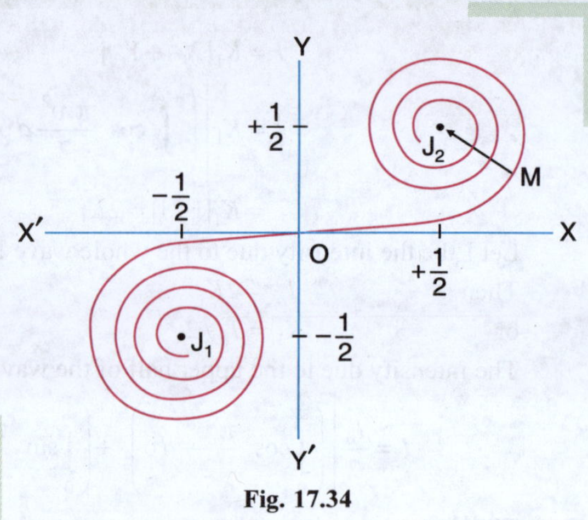

Fig. 17.34

WORKED OUT EXAMPLES

Example 17.1: What is the radius of the first half period zone in a zone plate behaving like a convex lens of focal length 60 cm for light of wavelength 6000 Å? **(Purvanchal, 2000)**

Solution: Given that $f = 60$ cm = 0.6 m, $n = 1$, and $\lambda = 6000$ Å $= 6 \times 10^{-7}$m

$$f = \frac{r_n^2}{n\lambda} \quad \therefore\ r_n^2 = fn\lambda$$

$$\therefore \quad r_1^2 = 0.6\ \text{m} \times 1 \times 6 \times 10^{-7}\ \text{m} = 36 \times 10^{-8}\ \text{m}^2$$

or $\quad r_1 = 6 \times 10^{-4}\ \text{m} = \mathbf{0.6\ mm}$.

Example 17.2: A circular aperture of 1.2 mm diameter is illuminated by plane waves of monochromatic light. The diffracted light is received on a distant screen which is gradually moved towards the aperture. The centre of the circular patch of light first becomes dark when the screen is 30 cm from the aperture. Calculate the wavelength of light.

Solution: Given that diameter = 1.2 mm = 0.12 cm

$$\text{Radius} = r = 0.06 \text{ cm}$$

$$b = 30 \text{ cm}$$

Here

$$\left(b^2 + r^2\right) = (b+\lambda)^2$$

$$30^2 + (0.06)^2 = (30+\lambda)^2$$

$$\lambda = \frac{(0.06)^2 \text{ cm}^2}{2 \times 30 \text{ cm}} \quad (\lambda^2 \text{ neglected})$$

$$= 0.00006 \text{ cm} = \mathbf{6000\ Å}$$

QUESTIONS

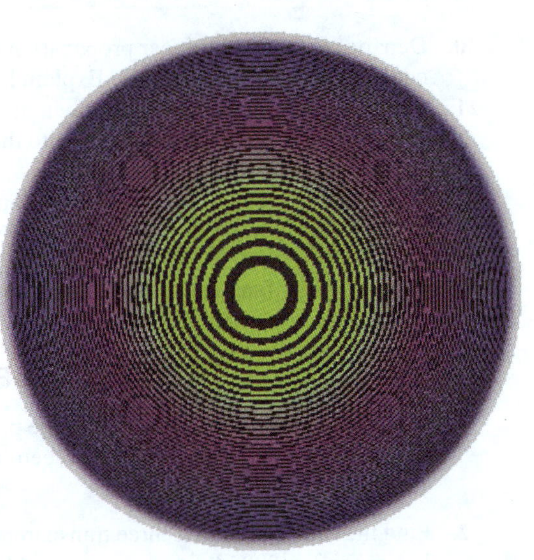

1. What is a zone plate? Give its theory. Show that a zone plate has multiple foci. Compare the zone plate with a convex lens. **(Madhurai Kamaraj, 2003; Agra, 2005)**
2. Give an account of the phenomenon and the related theory of diffraction due to a straight edge.
3. Describe and explain the diffraction pattern formed by a narrow wire illuminated by monochromatic light from a narrow slit parallel to the wire. How do you use the pattern to measure the thickness of a wire?
4. Describe and explain the phenomenon of diffraction due to a straight edge. Explain why the bands are neither equidistant nor equally illuminated. **(Gulbarg, 2005)**
5. A sharp razor blade is held vertically in a beam of monochromatic light coming from a narrow slit parallel to the edge of the blade. Discuss the position and intensities observed on a screen placed behind the blade. How will you use it to determine the wavelength of light? **(Agra, 2000)**
6. Discuss the properties of Cornu's spiral and explain its relationship with Fresnel's half period zones. Show how the spiral can be used to obtain the intensity distribution in the Fresnel diffraction pattern due to a straight edge. **(Nagpur, 2004)**
7. Explain the difference between Fraunhoffer and Fresnel diffraction. **(Meerut, 2005; Madhurai Kamaraj, 2003)**
8. Discuss the phenomenon of diffraction at a straight edge to obtain the position of maximum and minimum intensity from the center of screen. **(Nagpur, 2005)**
9. Explain the construction of Fresnel's half period zones on a plane wave front. **(Nagpur, 2005)**

10. Define Diffraction phenomenon of light. What are its type? **(Nagpur, 2005)**
11. Explain the construction of a zone plate. **(Nagpur, 2004)**
12. Show that area of each half period zone of a plane wavefront is equal to $\pi b \lambda$; where b is distance of external point on the screen from wavefront. **(Nagpur, 2004)**
13. Explain the construction of Fresnel's half period zones. **(Nagpur, 2004)**
14. What is a zone plate? How does it differ from a convex lens? **(Nagpur, 2004)**
15. Give the theory of diffraction at a straight edge and show that the Bands produced are not equally spaced. **(Kovempu, 2005)**
16. Describe with necessary theory the Fresnel class of diffraction due to straight edge. How will you use it to determine the wavelength of light? **(Meerut, 2005)**
17. Differentiate between interference and diffraction of light.
18. Can wave theory explain rectilinear propagation of light? Explain in brief. **(Lucknow, 2004)**
19. Radio waves diffract around buildings although light waves do not. Explain why? **(Lucknow, 2004)**
20. Departures from rectilinear propagation of light are limited to a small region near the edges of geometrical shadow. Explain in brief. **(Lucknow, 2004)**
21. What are half period zones? Show that the amplitude due to a large wave front at a point in front, it is just half that due to the first half period zone acting along. **(Garhwal, 2005)**
22. What are Fresnel half period zones? Show that the resultant amplitude at a point due to whole wavefront is equal to half of the amplitude due to first half zone only at that point. **(Punjab, 2005)**
23. Define diffraction phenomenon of light. What are its types? **(RTMNU, 2010)**

PROBLEMS FOR PRACTICE

1. The diameter of the first ring of a zone plate is 1.1 mm. If plane waves ($\lambda = 6000$ Å) fall on the plate, where should the screen be placed so that light is focused to a bright spot? **(Kurukshetra, 2002) [Ans: 0.5 m]**
2. Find the radii of the first three transparent zones of a zone plate whose first focal length is 1 m for $\lambda = 5893$Å. **(Agra, 2001) [Ans: 0.77 mm, 1.3 mm, 1.7 mm]**
3. The diameter of the central zone of a zone plate is 3 mm. If a point source of light ($\lambda = 6000$ Å) is placed at a distance of 5 m from it, calculate the position of the first image. **(Garhwal, 2001) [Ans: 15 m]**
4. A parallel beam of monochromatic light of $\lambda = 5000$ Å is incident normally on a plate having a circular hole of diameter 1mm. The screen is at the farthest position for which the axial point is almost black. The screen is moved towards the plate so that the axial point is again seen black. How far is the screen moved from the first position to the second? **(Allahabad, 2001) [Ans: 0.12 m]**
5. A narrow slit illuminated by light of wave length 4500 Å is placed at a distance of 20 cm from a straight edge. Calculate the distance between the second and third diffraction maxima formed on a screen placed at a distance of 60 cm from straight edge. **(Gulbarga, 2005)**

18

CHAPTER

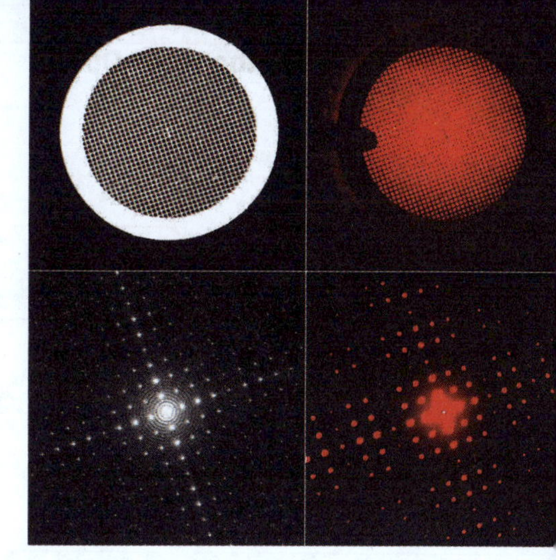

Fraunhoffer Diffraction

18.1. INTRODUCTION

To obtain a Fraunhoffer diffraction pattern, the incident wave front must be plane and the diffracted light is collected on the screen with the help of a lens. Thus, the source of light should either be at a large distance from the slit or a collimating lens must be used.

At a Glance

18.2. FRAUNHOFFER DIFFRACTION AT A SINGLE SLIT

In Fig. 18.1 S is a narrow slit perpendicular to the plane of the paper and illuminated by monochromatic light. L_1 is the collimating lens and AB is a slit of width *a*. XY is the incident spherical wave front. The light passing through the slit AB is incident on the lens L_2 and the final refracted beam is observed on the screen MN. The screen is perpendicular to the plane of the paper. The line SP is perpendicular to the screen. L_1 and L_2 are achromatic lenses.

A plane wave front is incident on the slit AB and each point on this wave front is a source of secondary disturbance. The secondary waves traveling in the direction parallel to OP *viz.* AQ and BV come to focus at P and a bright central image is observed. The secondary waves from points equidistant from O and situated in the upper and lower halves OA and OB of the wave front travel the same distance in reaching P and hence the path difference is zero. The secondary waves reinforce one another and P will be a point of maximum intensity.

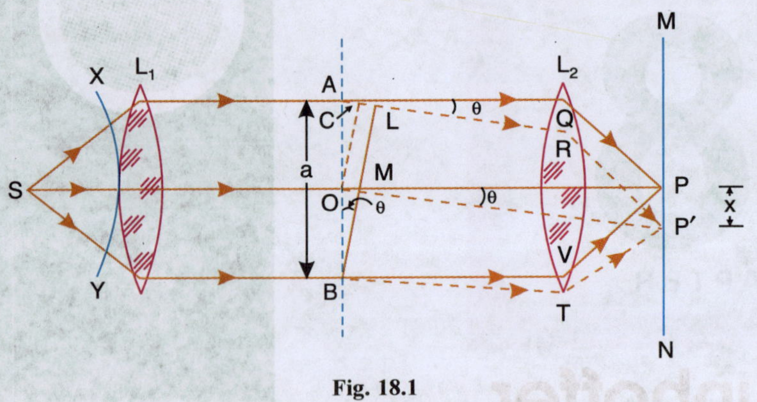

Fig. 18.1

Now, consider the secondary waves traveling in the direction AR, inclined at an angle è to the direction OP. All the secondary waves traveling in this direction reach the point P′ on the screen. The point P′ will be of maximum or minimum intensity depending on the path difference between the secondary waves originating from the corresponding points of the wave front. Let us draw OC and BL perpendicular to AR.

Then, in ΔABL

$$\sin\theta = \frac{AL}{AB} = \frac{AL}{a}$$

or
$$AL = a\sin\theta \tag{18.1}$$

where a is the width of the slit and AL is the path difference between the secondary waves originating from A and B. If this path difference is equal to λ the wavelength of the light used, then P′ will be a point of minimum intensity. The whole wave front can be considered to be of two halves OA and OB and if the path difference between the secondary waves from A and B is λ, then the path difference between the secondary waves from A and O will be λ / 2. Similarly, for every point in the upper half OA, there is a corresponding point in the lower half OB, and the path difference between the secondary waves from these points is λ / 2 .Thus, destructive interference takes place and the point P′ will be of minimum intensity. If the direction of the secondary waves is such that $AL = 2\lambda$, then also the point where they meet the screen will be of minimum intensity. This is so because the secondary waves from the corresponding points of the lower half differ in path by λ / 2. And this again gives the position of minimum intensity. In general,

$$a\sin\theta_n = n\lambda$$

$$\sin\theta_n = \frac{n\lambda}{a} \tag{18.2}$$

where θ_n gives the direction of the nth minimum. Here n is an integer. If, however, the path difference is odd multiples of λ / 2, the directions of the secondary maxima can be obtained. In this case,

$$a\sin\theta_n = (2n+1)\,\lambda/2$$

$$\sin\theta_n = \frac{(2n+1)\lambda}{2a} \tag{18.3}$$

where $n = 1,2,3,\ldots$

Thus, the diffraction pattern due to a single slit consists of a central bright maximum at P followed by secondary maxima and minima on both the sides, as shown in Fig. 18.2. P corresponds to the position of the central bright maximum and the points on the screen for which the path difference between the points A and B is λ, 2λ etc correspond to the position of secondary minima. The secondary maxima are of much less intensity. The intensity falls off rapidly from the point P outwards.

If the lens L_2 is very near the slit or the screen is far away from the lens L_2, then,

$$\sin\theta = \frac{x}{f} \tag{18.4}$$

where f is the focal length of the lens L_2.

But
$$\sin\theta = \frac{\lambda}{a} \tag{18.5}$$

$$\therefore \quad \frac{x}{f} = \frac{\lambda}{a} \quad \text{or} \quad x = \frac{f\lambda}{a}$$

where x is the distance of the secondary minimum from the point P.

Thus, the width of the central maximum W = 2x

or
$$W = \frac{2f\lambda}{a} \tag{18.6}$$

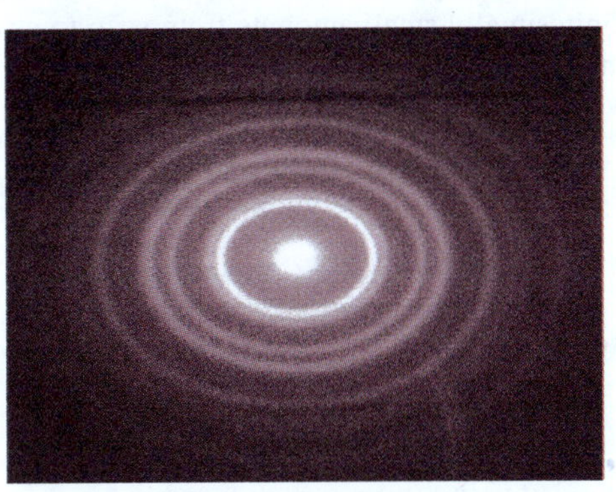

Diffraction pattern.

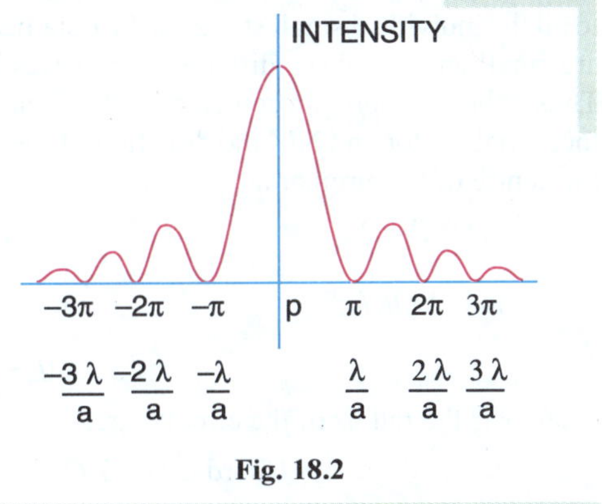

Fig. 18.2

The width of the central maximum is proportional to the wavelength of the light. With red light (longer wavelength), the width of the central maximum is more than with violet light (shorter wavelength). With a narrow slit, the width of central maximum is more. The diffraction pattern consists of alternate bright and dark bands with monochromatic light. With white light, the central maximum is white and the rest of the diffraction bands are coloured. From equation (18.5) if the width a of the slit is large, sin θ is small and hence θ is small. The maxima and minima are very close to the central maximum at P. But with a narrow slit, a is small and hence θ is large. This results in a distinct diffraction maxima and minima on both the sides of P.

18.2.1. INTENSITY DISTRIBUTION IN DIFFRACTION PATTERN DUE TO A SINGLE SLIT

The intensity variation in the diffraction pattern due to a single slit can be investigated as follows. The incident plane wave front on the slit AB (Fig.18.1) can be imagined to be divided into a large number of infinitesimally small strips. The path difference between the secondary waves emanating from the extreme points A and B is $a \sin\theta$ where a is the width of the slit and $\angle ABL = \theta$. For a parallel beam of incident light, the amplitude of vibration of the waves from each strip can be taken to be the same. As one considers the secondary waves in a direction inclined at an angle θ from the point B upwards, the path difference changes and hence the phase difference also increases. Let α be the phase difference between the secondary waves from the points B and A of the slit

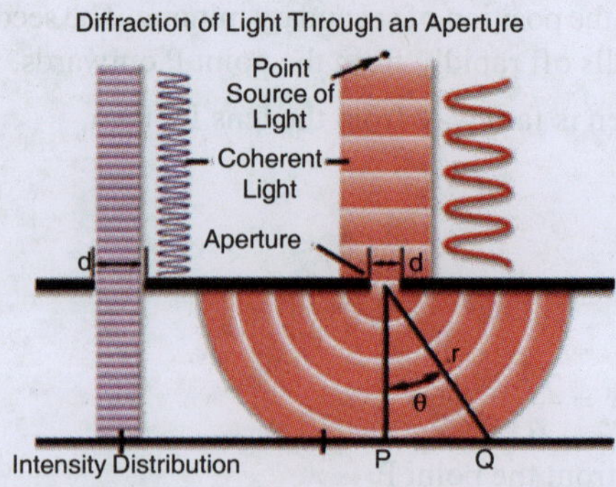

Intensity distribution of light diffracted by single slit experiment

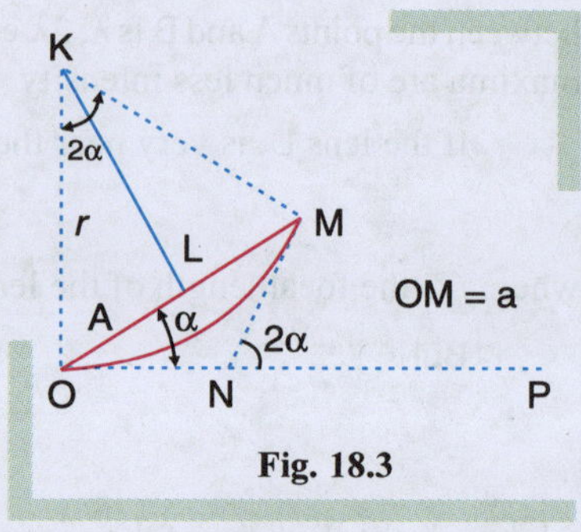

Fig. 18.3

(see Fig. 18.1). As the wave front is divided into a large number of strips, the resultant amplitude due to all the individual small strips can be obtained by the vector polygon method. Here, the amplitudes are small and the phase difference increases by infinitesimally small amounts from strip to strip. Thus, vibration polygon coincides with the circular arc OM (see Fig. 18.3). OP gives the direction of the initial vector and NM the direction of the final vector due to the secondary waves from A. K is the centre of the circular arc.

$$\angle MNP = 2\alpha \qquad \therefore \quad \angle OKM = 2\alpha$$

In ΔOKL $\qquad \sin\alpha = \dfrac{OL}{r};$

$$\therefore \quad OL = r\sin\alpha$$

where r is the radius of the circular arc.

$$\therefore \qquad \text{Chord } OM = 2\,OL = 2\,r\sin\alpha \qquad (18.7)$$

The length of the arc OM is proportional to the width of the slit.

$$\text{Length of the arc } OM = Ka$$

where K is a constant and a is the width of the slit.

Also, $\qquad 2\alpha = \dfrac{\text{Arc } OM}{\text{radius}} = \dfrac{Ka}{r}$

or $$2r = \frac{Ka}{\alpha} \qquad (18.8)$$

Substituting the value of $2r$ in equation (18.7), we get

$$\text{Chord OM} = \frac{Ka}{\alpha}.\sin\alpha$$

But, $OM = A$ where A is the amplitude of the resultant vibration.

$$\therefore \qquad A = (Ka)\frac{\sin\alpha}{\alpha}.$$

or $$A = A_0\frac{\sin\alpha}{\alpha}. \qquad (18.9)$$

Thus, the resultant amplitude of vibration at a point on the screen is given by $A_0\dfrac{\sin\alpha}{\alpha}$ and the intensity I at the point is given by

$$I = A^2 = A_0^{\,2}\frac{\sin^2\alpha}{\alpha^2} = I_0\left(\frac{\sin\alpha}{\alpha}\right)^2 \qquad (18.10)$$

The intensity at any point on the screen is proportional to $\left(\frac{\sin\alpha}{\alpha}\right)^2$. A phase difference of 2π corresponds to a path difference of λ. Therefore, a phase difference of 2α is given by

$$2\alpha = \frac{2\pi}{\lambda} a \sin\theta \tag{18.11}$$

where $a \sin\theta$ is the path difference between the secondary waves from A and B (Fig.18.1).

$$\alpha = \frac{\pi}{\lambda} a \sin\theta \tag{18.12}$$

Thus, the value of α depends on the angle of the diffraction θ. The value of $\left(\frac{\sin^2\alpha}{\alpha^2}\right)$ for different values of θ gives the intensity at the point under consideration. Fig.18.2 represents the intensity distribution. It is a graph of $\left(\frac{\sin^2\alpha}{\alpha^2}\right)$ (along the Y-axis), as a function of α or $\sin\theta$ (along the X-axis).

18.2.2. FRAUNHOFFER DIFFRACTION AT A SINGLE SLIT (CALCULUS METHOD)

Let a monochromatic parallel beam of light be incident on the slit AB of width a. The secondary waves traveling in the same direction as the incident light come to focus at the point P. The secondary waves traveling at an angle θ come to the focus at P′ (Fig.18.4). Let us consider the screen to be at a distance r from the slit. The center of the slit O is the origin of coordinates. Further, let us consider a small element dz of the wave front with the coordinates (o,z). The coordinates of the point P′ are (x_o, z_o) (see Fig.18.5). The distance of the element dz from the point P′ is ρ.

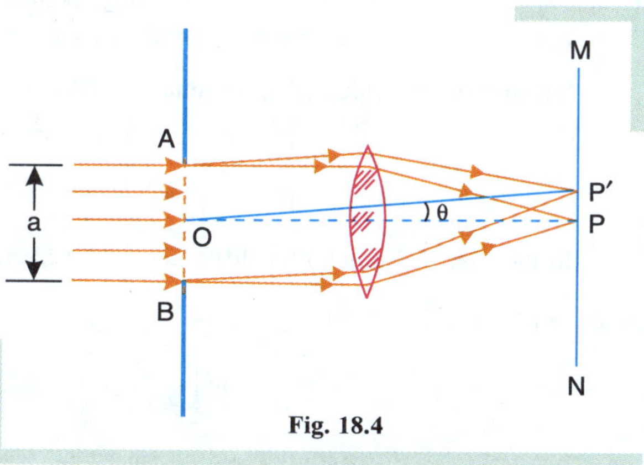

Fig. 18.4

The displacement at the point P′ due to the element dz at any instant is given by

$$dy = K dz \sin 2\pi\left(\frac{t}{T} - \frac{\rho}{\lambda}\right) \tag{18.13}$$

The resultant displacement at P′ due to the whole wave front is obtained by integrating the above equ.(18.13).

$$y = K \int_{-a/2}^{+a/2} \sin 2\pi\left(\frac{t}{T} - \frac{\rho}{\lambda}\right) dz \tag{18.14}$$

Also,

$$\rho^2 = x_o^2 + (z_0 - z)^2 \tag{18.15}$$

$$r^2 = x_o^2 + z_o^2$$

$$\therefore \quad x_o^2 = r^2 - z_o^2$$

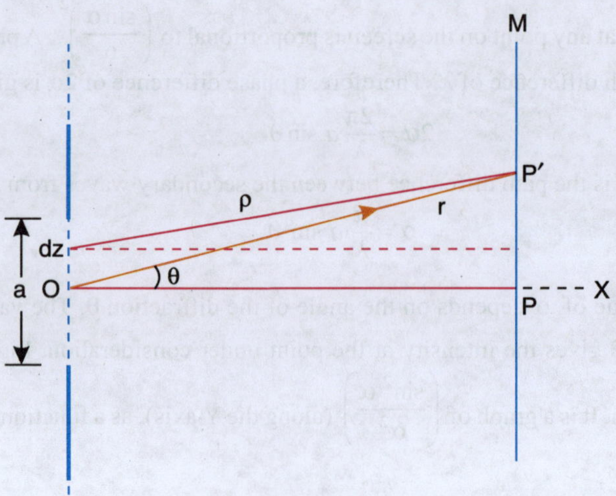

Fig. 18.5

Substituting the value of x_0^2 in equation (18.15), we get

$$\rho^2 = r^2 - z_o^2 + (z_0 - z)^2$$

$$\therefore \quad \rho^2 = r^2\left[1 - \frac{2zz_0}{r^2} + \frac{z^2}{r^2}\right] \tag{18.16}$$

In the case of Fraunhoffer diffraction, the screen is at a very large distance from the slit, therefore $r >> z$ and $\frac{z^2}{r^2}$ is negligible.

$$\therefore \quad \rho^2 = r^2\left[1 - \frac{2zz_0}{r^2}\right]$$

or

$$\rho = r\left[1 - \frac{zz_0}{r^2}\right] = \left[r - \frac{zz_0}{r}\right]$$

But $\frac{z_0}{r} = \sin\theta$

$$\therefore \quad \rho = r - z\sin\theta \tag{18.17}$$

Substituting this value of ρ in equation (18.14), we get

$$y = K\int_{-a/2}^{+a/2} \sin 2\pi\left(\frac{t}{T} - \frac{r}{\lambda} + \frac{z\sin\theta}{\lambda}\right)dz$$

$$y = -\frac{K\lambda}{2\pi\sin\theta}\left[\cos 2\pi\left[\frac{t}{T} - \frac{r}{\lambda} + \frac{a\sin\theta}{2\lambda}\right] - \cos 2\pi\left[\frac{t}{T} - \frac{r}{\lambda} - \frac{a\sin\theta}{2\lambda}\right]\right]$$

$$= -\frac{K\lambda}{2\pi\sin\theta}\left[2\sin 2\pi\left(\frac{t}{T} - \frac{r}{\lambda}\right)\sin 2\pi\left(-\frac{a\sin\theta}{2\lambda}\right)\right]$$

$$= \frac{K\lambda}{\pi\sin\theta}\sin 2\pi\left(\frac{t}{T} - \frac{r}{\lambda}\right)\sin\left(\frac{\pi a\sin\theta}{\lambda}\right)$$

Let $\frac{\pi a \sin\theta}{\lambda} = \alpha$

$$\therefore \quad y = Ka\left(\frac{\sin\alpha}{\alpha}\right)\sin 2\pi\left(\frac{t}{T}-\frac{r}{\lambda}\right) \tag{18.18}$$

The amplitude at P′ is $Ka\left(\frac{\sin\alpha}{\alpha}\right)$ and the intensity at P′ is

$$I' = K^2a^2\left(\frac{\sin^2\alpha}{\alpha^2}\right) = I_o\left(\frac{\sin^2\alpha}{\alpha^2}\right) \tag{18.19}$$

Here $I_o = K^2a^2$ and is the value of intensity at P, for $\alpha = 0$, as $\frac{\sin\alpha}{\alpha} = 1$ when $\alpha \to 0$.

(*i*) Central Maximum:

For the point P on the screen (Fig.18.5) $\theta = 0$; and hence $\alpha = 0$.

The value of $\frac{\sin\alpha}{\alpha}$ when $\alpha \to 0$ is equal to 1. Hence, the intensity at $P = I_0\left(\frac{\sin\alpha}{\alpha}\right)^2 = I_0$ which is a maximum.

(*ii*) Secondary Maxima:

The directions of secondary maxima are given by the equation,

$$\sin\theta_n = \frac{(2n+1)\lambda}{2a}$$

Substituting this value of θ_n in equation (18.12), we get

$$\alpha = \frac{\pi}{\lambda}\cdot\frac{a(2n+1)\lambda}{2a} = \frac{(2n+1)\pi}{2} \tag{18.20}$$

Substituting $n = 1,2,3$ etc in equation (18.20), the values of α are given by

$$\alpha = \frac{3\pi}{2}, \frac{5\pi}{2}, \frac{7\pi}{2} \text{ etc.}$$

(*a*) For the first secondary maximum, $\alpha = \frac{3\pi}{2}$

and
$$I = I_0\left(\frac{\sin\alpha}{\alpha}\right)^2 = I_0\left[\frac{\sin\frac{3\pi}{2}}{\frac{3\pi}{2}}\right]^2 = I_0\left[\frac{-1}{\frac{3\pi}{2}}\right]^2 = \frac{4I_0}{9\pi^2} = \frac{I_0}{22}$$

(*b*) For the secondary maximum, $\alpha = \frac{5\pi}{2}$ and

$$I = I_0\left(\frac{\sin\alpha}{\alpha}\right)^2 = I_0\left[\frac{\sin\frac{5\pi}{2}}{\frac{5\pi}{2}}\right]^2 = I_0\left[\frac{1}{\frac{5\pi}{2}}\right]^2 = \frac{4I_0}{25\pi^2} = \frac{I_0}{61}$$

Thus, the secondary maxima are of *decreasing intensity* and the directions of these maxima are obtained from the equation given above.

The intensity at P′ is given by, $I' = I_0\left(\frac{\sin\alpha}{\alpha}\right)^2$

$$dI' = I_o\left[\frac{\alpha^2 2\sin\alpha\ \cos\alpha - (\sin^2\alpha)2\alpha}{\alpha^4}\right]d\alpha$$

For I' to be a maximum, $\frac{dI'}{d\alpha} = 0$

$$\therefore \quad \alpha^2(2\sin\alpha \ \cos\alpha) - (\sin^2\alpha)2\alpha = 0$$

$$\therefore \quad \tan\alpha = \alpha$$

If graphs are plotted for $y = \alpha$, and $y = \tan\alpha$, it will be found that the secondary maxima are not exactly midway between two minima. The positions of the secondary maxima are slightly towards the central maximum (Fig.18.6)

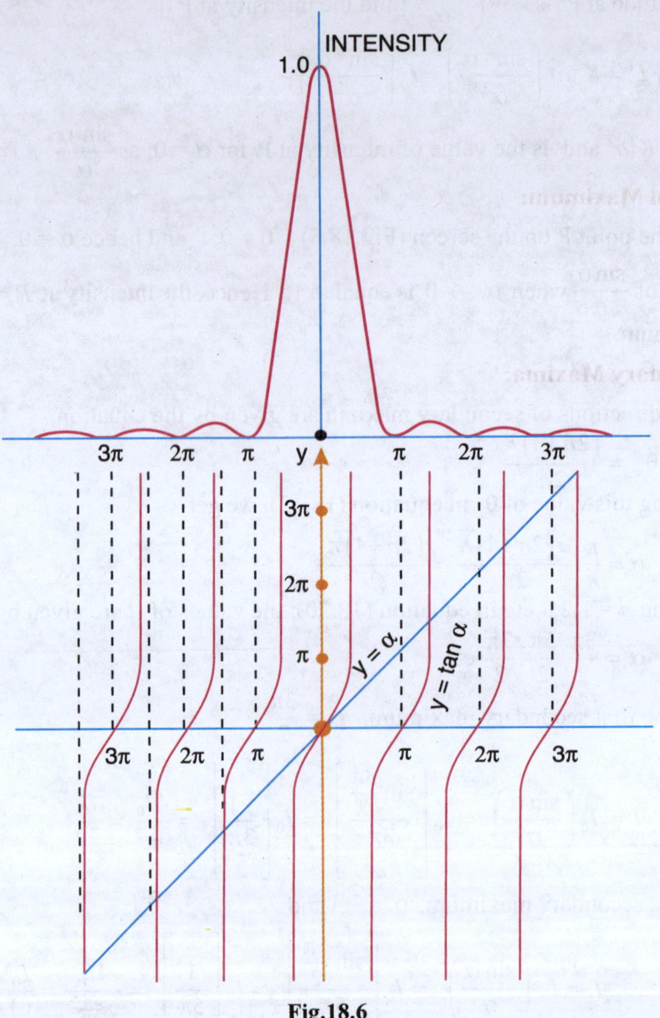

Fig.18.6

(iii) Secondary Minima:

The directions of the secondary minima are given by the equation

$$a\sin\theta = n\lambda$$

Substituting the value of $a\sin\theta$ in equation (18.12),we get

$$\alpha = \frac{\pi}{\lambda}.n\lambda = n\pi \qquad (18.21)$$

Substituting $\quad n = 1,2,3$ etc in equation (18.21), we obtain

$$\alpha = \pi, 2\pi, 3\pi \text{ etc.}$$

When these values of α are substituted in the equation (18.19) for intensity, we get

$$I = 0$$

In Fig.18.6, the positions of the secondary minima are shown for values of

$$\alpha = \pi,\ 2\,\pi,\ 3\,\pi \text{ etc.}$$

$\frac{\lambda}{a}, \frac{2\lambda}{a}, \frac{3\lambda}{a}$ etc. refer to the values of sin θ for these positions.

18.3. FRAUNHOFFER DIFFRACTION AT A CIRCULAR APERTURE

In Fig. 18.7, AB is a circular aperture diameter *d*. C is the centre of the aperture and P is a point on the screen. CP is perpendicular to the screen. The screen is perpendicular to the plane of the paper. A plane wave front is incident on the circular aperture. The secondary wave traveling in the direction CO comes to the focus at *P*. Therefore, *P* corresponds to the position of the central maximum. Here, all the secondary waves emanating from points equidistant from *O* travel the same distance before reaching P and hence they all reinforce one another. Now, let us consider the secondary waves traveling in a direction inclined at an angle θ with the direction CP. All these secondary waves meet at P_1 on the screen. Let the distance PP_1 be *x*. The path difference between the secondary waves emanating from the points *B* and *A* (extremities of diameter) is *AD*.

From the Δ*ABD*, $AD = d \sin\theta$

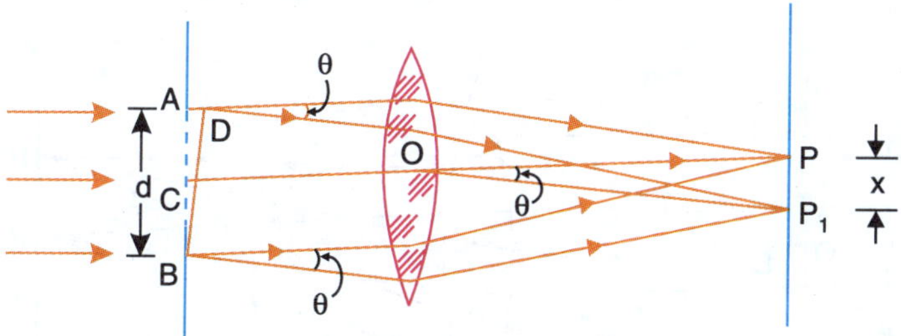

Fig. 18.7

Arguing as in article 18.2, the point P_1 will be of minimum intensity if this path difference is equal to integral multiples of λ, i.e.,

$$d \sin\theta_n = n\,\lambda \qquad (18.22)$$

The point P_1 will be of maximum intensity if the path difference is equal to odd multiples of λ/2 *i.e.*,

$$d\, sin\, \theta_n = \frac{(2n+1)\lambda}{2} \qquad (18.23)$$

If P_1 is the point of minimum intensity , then all the points at the same distance from *P* as P_1 and lying on a circle of radius *x* will be of minimum intensity. Thus, the diffraction pattern due to a circular aperture consists of a central disc called the Airy's disc, surrounded by alternate dark and bright concentric rings called the Airy's rings. The intensity of the dark rings is zero and that of the bright rings decreases gradually outwards from *P*. Further, if the collecting lens is very near the slit or when the screen is at a large distance from the lens,

$$\sin\theta = \theta = \frac{x}{f} \qquad (18.24)$$

Also, for the first secondary minimum, $d \sin\theta = \lambda$

$$\sin\theta = \theta = \frac{\lambda}{d} \qquad (18.25)$$

From equation (18.24) and (18.25)

$$\frac{x}{f} = \frac{\lambda}{d}$$

or
$$x = \frac{f\lambda}{d} \tag{18.26}$$

where x is the radius of the Airy's disc. But actually, the radius of the first dark ring is slightly more than that given by equation (18.26). According to Airy, it is given by

$$x = 1.22\frac{f\lambda}{d} \tag{18.27}$$

The discussion on the intensity distribution of the bright and dark rings is similar to the one given for a rectangular slit. With increase in diameter of the aperture, the radius of the central bright ring decreases.

18.4. FRAUNHOFFER DIFFRACTION AT DOUBLE SLIT

In Fig. 18.8, AB and CD are two rectangular slits parallel to one another and perpendicular to the plane of the paper. The width of each slit is a and the width of the opaque portion is b. L is a collecting lens and MN is a screen perpendicular to the plane of the paper. P is a point on the screen such that OP is perpendicular to the screen. Let a plane wave front be incident on the surface of XY. All the secondary waves traveling in a direction parallel to OP come to focus at P. Therefore, P corresponds to the position of the central bright maximum.

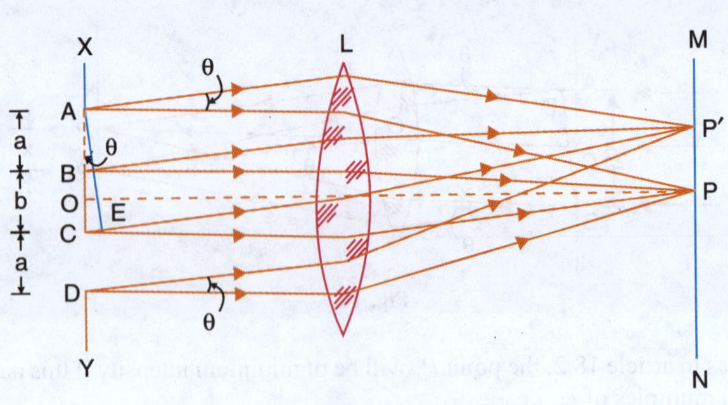

Fig. 18.8

In this case, the diffraction pattern has to be considered in two parts (*i*) the interference phenomenon due to the secondary waves emanating from the corresponding points of the two slits and (*ii*) the diffraction pattern due to the secondary waves from the two slits individually. For calculating the positions of interference maxima and minima, the diffracting angle is denoted as θ and for the diffraction maxima and minima it is denoted as ϕ. Both the angles θ and ϕ refer to the angle between the direction of the secondary waves and the initial direction of the incident light.

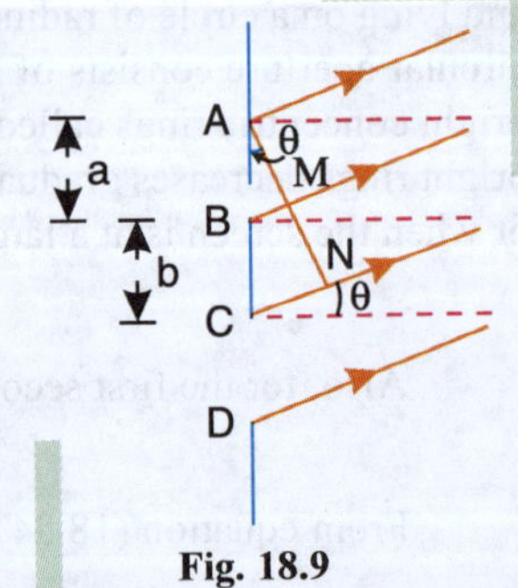

Fig. 18.9

(*i*) Interference maxima and minima:

Let us consider the secondary waves traveling in a direction inclined at an angle θ with the initial direction.

In the Δ CAN (Fig. 18.9) $\quad \sin\theta = \dfrac{CN}{AC} = \dfrac{CN}{a+b}$

or
$$CN = (a+b)\sin\theta$$

If this path difference is equal to odd multiples of $\lambda/2$, θ gives the direction of minima due to interference of the secondary waves from the two slits.

$$\therefore \qquad CN = (a+b)\sin\theta_n = (2n+1)\frac{\lambda}{2} \qquad (18.28)$$

Putting $n = 1, 2, 3$, etc, the values of θ_1, θ_2, θ_3, etc, corresponding to the directions of minima can be obtained.

From equation (18.28)

$$\sin\theta_n = \frac{(2n+1)\lambda}{2(a+b)} \qquad (18.29)$$

On the other hand, if the secondary waves travel in a direction θ' such that the path difference is even multiples of $\lambda/2$, then θ' gives the direction of the maxima due to interference of light waves emanating from the two slits.

$$\therefore \qquad CN = (a+b)\sin\theta'_n = 2n\frac{\lambda}{2}$$

or

$$\sin\theta'_n = \frac{n\lambda}{(a+b)} \qquad (18.30)$$

Putting $n = 1, 2, 3$ etc, θ'_1, θ'_2, θ'_3 etc corresponding to the directions of the maxima can be obtained. From equation(18.29), we get

$$\sin\theta_1 = \frac{3\lambda}{2(a+b)}$$

and

$$\sin\theta_2 = \frac{5\lambda}{2(a+b)}$$

$$\therefore \qquad \sin\theta_2 - \sin\theta_1 = \frac{\lambda}{(a+b)} \qquad (18.31)$$

Thus, the angular separation between any two consecutive minima (or maxima) is equal to $\frac{\lambda}{a+b}$. The angular separation is inversely proportional to $(a+b)$, the distance between the two slits.

***(ii)* Diffraction maxima and Minima:**

Let us consider the secondary waves traveling in a direction inclined at an angle ϕ with the initial direction of the incident light.

If the path difference BM is equal to λ the wavelength of the light used, then ϕ will give the direction of the diffraction minimum (Fig. 18.9). That is, the path difference between secondary waves emanating from the extremities of a slit (i.e., points A and B) is equal to λ. Considering the wave front on AB to be made up of the two halves, the path difference between the corresponding points of the upper and lower halves is equal to $\lambda/2$. The effect at P′ due to the wave front incident on AB is zero. Similarly, for the same direction of the secondary waves, the effect at P′ due to the wave front incident on the slit CD is also zero. In general,

$$a\sin\phi_n = n\lambda.$$

Putting $n = 1, 2, 3$, etc, the values of ϕ_1, ϕ_2, ϕ_3 etc, corresponding to the directions of diffraction minima can be obtained.

18.4.1. FRAUNHOFER DIFFRACTION AT DOUBLE SLIT (CALCULUS METHOD)

The intensity distribution due to Fraunhoffer diffraction at double slit (two parallel slits) can be obtained by integrating the expression for dy (vide single slit) for both the slits.

$$y = K\left[\int_{-a/2}^{a/2} \sin\left\{2\pi\left(\frac{t}{T}-\frac{r}{\lambda}+\frac{z\sin\theta}{\lambda}\right)\right\}dz + \int_{d-a/2}^{d+a/2} \sin\left\{2\pi\left(\frac{t}{T}-\frac{r}{\lambda}+\frac{z\sin\theta}{\lambda}\right)\right\}dz\right] \qquad (18.32)$$

$$\therefore \qquad y = Ka\left(\frac{\sin\alpha}{\alpha}\right)\sin 2\pi\left(\frac{t}{T}-\frac{r}{\lambda}\right) - \frac{K\lambda}{2\pi\sin\theta}\left[\cos 2\pi\left[\frac{t}{T}-\frac{r}{\lambda}+\frac{z\sin\theta}{\lambda}\right]\right]_{d-a/2}^{d+a/2}$$

$$\therefore \quad y = Ka\left(\frac{\sin\alpha}{\alpha}\right)\sin 2\pi\left(\frac{t}{T}-\frac{r}{\lambda}\right)$$

$$-\frac{K\lambda}{2\pi\sin\theta}\left[\cos 2\pi\left[\frac{t}{T}-\frac{r}{\lambda}+\frac{d\sin\theta}{\lambda}+\frac{a\sin\theta}{2\lambda}\right]-\cos 2\pi\left[\frac{t}{T}-\frac{r}{\lambda}-\frac{d\sin\theta}{\lambda}-\frac{a\sin\theta}{2\lambda}\right]\right]$$

$$y = Ka\left(\frac{\sin\alpha}{\alpha}\right)\sin 2\pi\left(\frac{t}{T}-\frac{r}{\lambda}\right)+\frac{K\lambda}{\pi\sin\theta}\left[\sin 2\pi\left[\frac{t}{T}-\frac{r}{\lambda}+\frac{d\sin\theta}{\lambda}\right]\sin\left(\frac{\pi a\sin\theta}{\lambda}\right)\right]$$

But $\alpha = \dfrac{\pi a\sin\theta}{\lambda}$

$$\therefore \quad y = Ka\left(\frac{\sin\alpha}{\alpha}\right)\left[\sin 2\pi\left(\frac{t}{T}-\frac{r}{\lambda}\right)+\sin 2\pi\left[\frac{t}{T}-\frac{r}{\lambda}+\frac{d\sin\theta}{\lambda}\right]\right]$$

$$y = 2Ka\left(\frac{\sin\alpha}{\alpha}\right)\sin 2\pi\left(\frac{t}{T}-\frac{r}{\lambda}+\frac{d\sin\theta}{\lambda}\right).\cos\frac{\pi d\sin\theta}{\lambda}$$

Let $\dfrac{\pi d\sin\theta}{\lambda} = \beta.$

$$\therefore \quad y = 2Ka\left(\frac{\sin\alpha}{\alpha}\right)\cos\beta\,\sin 2\pi\left(\frac{t}{T}-\frac{r}{\lambda}+\frac{d\sin\theta}{\lambda}\right) \qquad (18.33)$$

The intensity at the point P′ is given by

$$I = 4K^2a^2\left(\frac{\sin^2\alpha}{\alpha^2}\right)\cos^2\beta$$

$$\text{But } I_0 = K^2a^2 \;\therefore\; I = 4I_0\left(\frac{\sin^2\alpha}{\alpha^2}\right)\cos^2\beta \qquad (18.34)$$

The intensity of the central maximum = $4I_0$ when $\alpha = 0$ and $\beta = 0$. In Fig. 18.11, the dotted curve represents the intensity distribution due to diffraction pattern due to double slit and the thick line curve represents the intensity distribution duo to the interference between the light from both the slits. The pattern consists of interference maxima within each diffraction maximum.

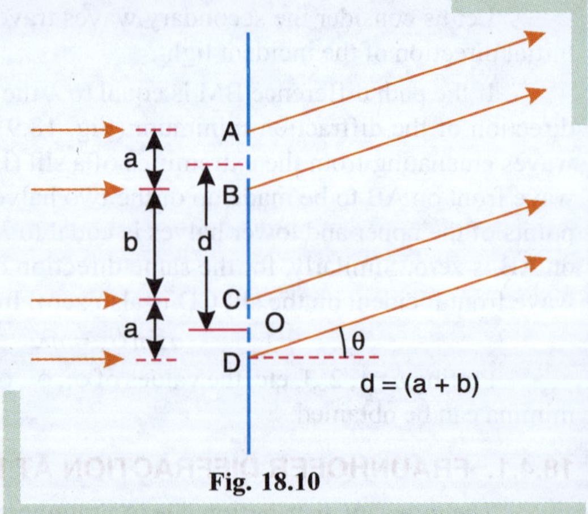

Fig. 18.10

Intensity distribution due to the Fraunhoffer diffraction at two parallel slits is shown in Fig. 18.11. The full line represents equally spaced interference maxima and minima and the dotted curve represents the diffraction maxima and minima. In the region originally occupied by the central maximum of the single slit diffraction pattern, equally spaced interference maxima and minima are observed. The intensity of the central interference maximum is four times the intensity of the central maximum of the single slit diffraction pattern. The intensity of other interference maxima on the two sides of the central maximum gradually decreases. In the region of the secondary maxima due to diffraction at a single slit, equally spaced interference maxima of low intensity are observed. The intensity distribution shown in Fig. 18.11 corresponds to $2a = b$ where a is the width of each slit and b is the opaque spacing between the two slits (see Fig. 18.10). Thus, the pattern due to diffraction at a double slit consists of a diffraction pattern due to the individual slits of width a each and the interference maxima and minima of equal spacing. The spacing of the interference maxima and minima is dependent on the values of a and b.

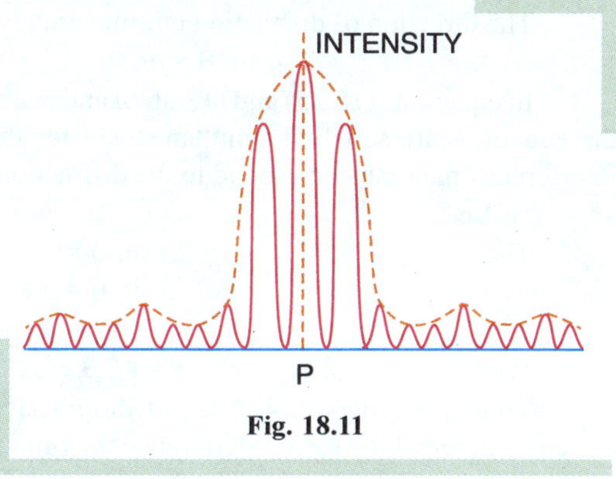

Fig. 18.11

18.4.2. DISTINCTION BETWEEN SINGLE SLIT AND DOUBLE SLIT DIFFRACTION PATTERNS

The single slit diffraction pattern consists of a central bright maximum with secondary maxima and minima of gradually decreasing intensity. The double slit diffraction pattern consists of equally spaced interference maxima and minima with in the central maximum. The intensity of the central maximum in diffraction pattern due to a double slit is four times that of the central maximum in the diffraction pattern due to diffraction at a single slit. In the above arrangement, if one of the slits is covered with opaque screen, the pattern observed is similar to the one observed with a single slit.

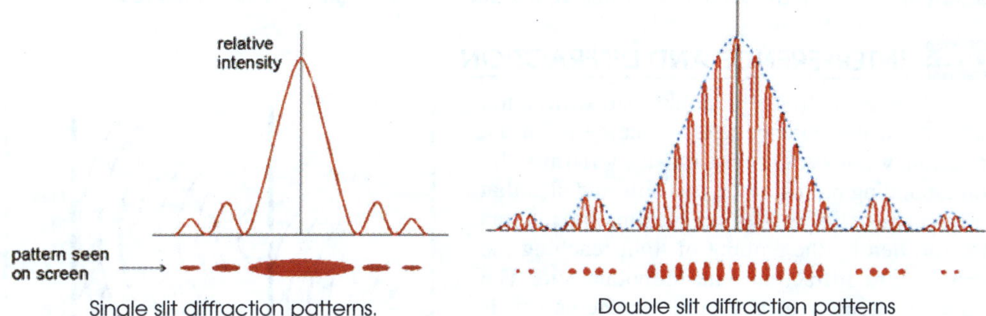

Single slit diffraction patterns. Double slit diffraction patterns

The spacing of diffraction maxima and minima depends on a, the width of the slit and the spacing of the interference maxima and minima depends on the value of a and b where b is opaque spacing between the two slits. The intensities of the interference maxima are not constant but decrease to zero on either side of the central maximum. These maxima reappear two or three times before the intensity becomes too low to be observed.

18.4.3. MISSING ORDERS IN A DOUBLE SLIT DIFFRACTION PATTERN

In the diffraction pattern due to a double slit discussed earlier, the slit width is taken as a and the separation between the slits as b. If the slit width a is kept constant, the diffraction pattern remains the same. Keeping a constant, if the spacing b is altered the spacing between the interference maxima changes. Depending on the relative values of a and b certain orders of interference maxima will be missing in the resultant pattern.

The directions of interference maxima are given by the equation,

$$(a + b) \sin \theta = n \lambda \tag{18.35}$$

The direction of diffraction minima are given by the equation,

$$a \sin \theta = p\lambda \tag{18.36}$$

In equations (18.35) and (18.36) n and p are integers. If the value of a and b are such that both the equations are satisfied simultaneously for the same value of θ, then the positions of certain interference maxima correspond to the diffraction minima at the same position on the screen.

(*i*) Let $a = b$

Then $2a \sin \theta = n\lambda$

and $a \sin \theta = p\lambda$

$\therefore$ $n/p = 2$ or $n = 2p$

If $p = 1,2,3$ etc., then $n = 2, 4, 6$ etc.

Thus, the orders 2, 4, 6 etc of the interference maxima will be missing in the diffraction pattern. There will be three interference maxima in the central diffraction maximum.

(*ii*) If $2a = b$, then $3a \sin \theta = n\lambda$

and $a \sin \theta = p\lambda$

$\therefore$ $n/p = 3$ or $n = 3p$

If $p = 1, 2, 3$, etc, $n = 3, 6, 9$, etc.

Thus the orders 3, 6, 9 etc of the interference maxima will be missing in the diffraction pattern. On both sides of the central maximum, the number of interference maxima is 2 and hence there will be five interference maxima in the central diffraction maximum. The position of the third interference maximum also corresponds to the first diffraction minimum

(*iii*) If $a + b = a$ i.e., $b = 0$

The two slits join and all the orders of the interference maxima will be missing. The diffraction pattern observed on the screen is similar to that due to a single slit of width equal to $2a$.

18.5. INTERFERENCE AND DIFFRACTION

It is clear from the double slit diffraction pattern that interference takes place between the secondary waves originating from the corresponding points of the two slits and also that the intensity of the interference maxima and minima is controlled by the amount of light reaching the screen due to diffraction at the individual slits. The resultant intensity at any point on the screen is obtained by multiplying the intensity function for the interference and the intensity function for the diffraction at the two slits. The values of the intensity functions are taken for the same direction of the secondary waves. But the interference of all the secondary waves originating from the whole wave front is termed as diffraction. Hence the pattern obtained on the screen may be called an interference pattern or a diffraction pattern. The term interference may be used for those cases in which the resultant amplitude at a point is obtained by the superimposition of two or more beams. Diffraction can be defined as the phenomenon in which the resultant amplitude at any point on the screen is obtained by integrating the effect of infinitesimally small number of elements in to which the whole wave front can be divided. Thus, the resultant diffraction pattern obtained with a double slit can be taken as a combination of the effect of both interference and diffraction.

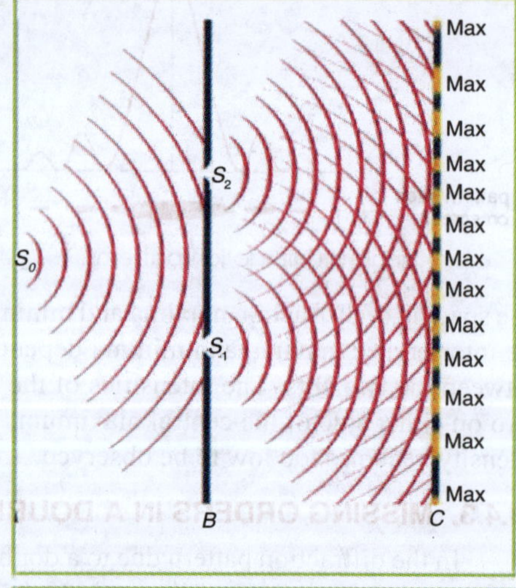

Interference and Diffraction.

18.6. FRAUNHOFFER DIFFRACTION AT N SLITS

Fraunhoffer diffraction at two slits consists of diffraction maxima and minima given by $\frac{\sin^2\alpha}{\alpha^2}$ and the sharp interference maxima and minima, in each diffraction maximum governed by $\cos^2\beta$ term.

To derive an expression for the intensity distribution due to diffraction at N slits, the expression for dy has to be integrated for N slits.

For a single slit, $dy = K\int_{-a/2}^{+a/2} \sin 2\pi\left(\frac{t}{T}-\frac{r}{\lambda}+\frac{z\sin\theta}{\lambda}\right)dz$

Let $\sin 2\pi\left(\frac{t}{T}-\frac{r}{\lambda}+\frac{z\sin\theta}{\lambda}\right)$ be equal to $\phi(z)$ (i.e. function of z).

For N slits

$$dy = K\left[\int_{\frac{-a}{2}}^{\frac{+a}{2}} \phi(z)dz + \int_{d-\frac{a}{2}}^{d+\frac{a}{2}} \phi(z)dz + \int_{2d-\frac{a}{2}}^{2d+\frac{a}{2}} \phi(z) + ...+ \int_{(N-1)d-\frac{a}{2}}^{(N-1)d+\frac{a}{2}} \phi(z)dz\right] \tag{18.37}$$

On simplification

$$Y = Ka\frac{\sin\alpha}{\alpha}\left[\begin{array}{l}\sin 2\pi\left(\frac{t}{T}-\frac{r}{\lambda}\right)+\sin 2\pi\left(\frac{t}{T}-\frac{r}{\lambda}+\frac{d\sin\theta}{\lambda}\right)+\sin 2\pi\left(\frac{t}{T}-\frac{r}{\lambda}+\frac{2d\sin\theta}{\lambda}\right)+ \\ \quad ....+\sin 2\pi\left(\frac{t}{T}-\frac{r}{\lambda}+\frac{(N-1)d\sin\theta}{\lambda}\right).\end{array}\right]$$

Here $\alpha = \frac{\pi a\sin\theta}{\lambda}$. For a general trigonometric summation

$$\sum_{p=0}^{p=n}\sin(x+pm) = \frac{\sin\left(x+\frac{nm}{2}\right)\sin\left[\left(\frac{n+1}{2}\right)m\right]}{\sin\left(\frac{m}{2}\right)}$$

Here $x = 2\pi\left(\frac{t}{T}-\frac{r}{\lambda}\right)$; $m = \frac{2\pi d\sin\theta}{\lambda} = 2\left[\frac{\pi d\sin\theta}{\lambda}\right] = 2\beta$

where $\beta = \frac{\pi d\sin\theta}{\lambda}$ and $n = (N-1)$

$$\therefore \quad y = \frac{Ka\left(\frac{\sin\alpha}{\alpha}\right)\left[\sin\left(x+\frac{(N-1)m}{2}\right)\sin\left(\frac{Nm}{2}\right)\right]}{\sin\left(\frac{m}{2}\right)}$$

$$y = Ka\left(\frac{\sin\alpha}{\alpha}\right)\frac{(\sin N\beta)}{\sin\beta}\left[\sin 2\pi\left\{\frac{t}{T}-\frac{r}{\lambda}+\frac{(N-1)d\sin\theta}{2\lambda}\right\}\right] \tag{18.38}$$

The intensity at a point P′ is given by

$$I = K^2a^2\left(\frac{\sin^2\alpha}{\alpha^2}\right)\left(\frac{\sin^2 N\beta}{\sin^2\beta}\right) \tag{18.39}$$

The maximum intensity, when $\alpha = 0$ and $\beta = 0$, is $I_0 = K^2a^2$

$$\therefore \qquad I = I_0\left(\frac{\sin^2\alpha}{\alpha}\right)\left(\frac{\sin^2 N\beta}{\sin^2\beta}\right) \qquad (18.40)$$

The expression $\frac{\sin^2\alpha}{\alpha^2}$ represents the diffraction pattern due to a single slit. The additional factor $\frac{\sin^2 N\beta}{\sin^2\beta}$ represents the interference effects due to the secondary waves from the N slits.

The numerator will be zero when , $N\beta = 0, \pi, 2\pi, 3\pi$.....etc = $k\pi$

Denominator is also zero when, $\beta = 0, \pi, 2\pi, 3\pi$.....etc.

Since the quotient $\frac{0}{0}$ is indeterminate, $N\beta = k\pi$ gives the condition for minimum intensity for all values of k other than $k = 0, N, 2N, 3N$ etc.

The directions of principal maxima correspond to the values of $k = 0, N, 2N$ etc.

$$N\beta = \frac{N\pi d\sin\theta}{\lambda} \quad \text{or} \quad k\pi = \frac{N\pi\, d\sin\theta}{\lambda}$$

For directions of principal maxima, $k = 0, 1N, 2N, 3N$, etc. $= nN$

When $\qquad n = 1, 2, 3$ etc., $\qquad n\pi N = \frac{N\pi\, d\sin\theta}{\lambda}$

$$\therefore \qquad d\sin\theta = n\lambda \qquad (n = 1, 2, 3, \ldots \text{ etc})$$

If the width of the slit is a and the width of the opaque spacing is b, $d = (a + b)$

and $\qquad (a + b)\sin\theta = n\lambda$

Putting $n = 1, 2,3$ etc, the directions of principal maxima $\theta_1, \theta_2, \theta_3$ ....etc can be determined.

For values of k in between 0 and N between N and 2N, etc., there are (N–1) secondary minima and (N–2) secondary maxima.

The intensity distribution due to diffraction and N slits is shown in Fig. 18.14.

18.6.1. INTENSITY OF PRINCIPAL MAXIMA

In a diffraction grating there are about 6000 narrow slits in one cm. For values of $\beta = k\pi$ and $\beta = 0, \pi, 2\pi, \; 3\pi$ etc.

$$\frac{\sin N\beta}{\sin\beta} = \frac{0}{0}$$

which is indeterminate.

To find the value of this limit, the numerator and the denominator are differentiated. Thus, we get

$$\underset{\beta\to n\pi}{Lt}\ \frac{\sin N\beta}{\sin\beta} = \underset{\beta\to n\pi}{Lt}\ \frac{\frac{d}{d\beta}(\sin N\beta)}{\frac{d}{d\beta}(\sin\beta)} = \underset{\beta\to n\pi}{Lt}\ \frac{N\cos N\beta}{\cos\beta} = \pm N$$

Thus, the resultant amplitude is proportional to N and resultant intensity is proportional to N^2.

$$I = N^2 I_0\left(\frac{\sin^2\alpha}{\alpha^2}\right) \qquad (18.41)$$

These maxima are intense and are called **principal maxima**.

18.7. PLANE DIFFRACTION GRATING

A diffraction grating is an extremely useful device and in one of its forms it consists of a very large number of narrow slits side by side. The slits are separated by opaque spaces. When a wave front is incident on a grating surface, light is transmitted through the slits and obstructed by the opaque portions. Such a grating is called a **transmission grating**. The secondary waves from the positions of the slits interfere with one another, similar to the interference of waves in Young's experiment. Joseph Fraunhoffer used the first grating which consisted of a large number of parallel fine wires stretched on a frame. Now, gratings are prepared by ruling equidistant parallel lines on a glass surface. The lines are drawn with a fine diamond point. The space in between any two lines is transparent to light and the lined portion is opaque to light. Such surfaces act as transmission gratings. If, on the other hand, the lines are drawn on a silvered surface (plane or concave) then light is reflected from the positions of the mirror in between any two lines and such surfaces act as *reflection gratings*.

Transmission grating.

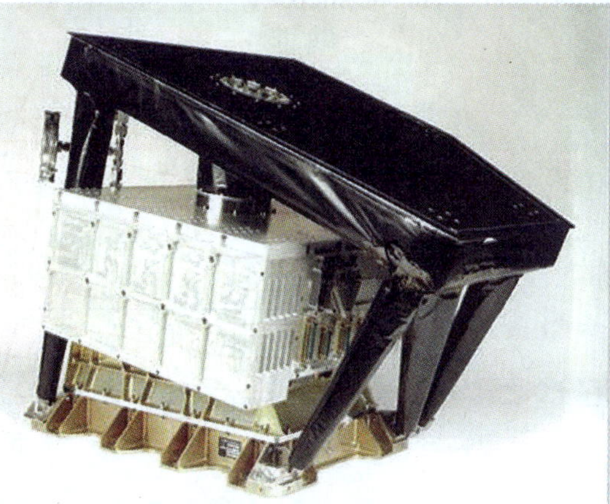

Reflection grating images.

If the spacing between the lines is of the order of the wavelength of light, then an appreciable deviation of the light is produced. Gratings used for the study of the visible region of the spectrum contain 10,000 lines per cm. Gratings, with originally ruled surfaces are only few. For practical purposes, replicas of the original grating are prepared. On the original grating surface a thin layer of collodion solution is poured and the solution is allowed to harden. Then, the film of collodion is removed from the grating surface and then fixed between two glass plates. This serves as a plane transmission grating. A large number of replicas are prepared in this way from a single original ruled surface.

18.7.1. THEORY OF PLANE TRANSMISSION GRATING

In Fig.18.12, XY is the grating surface and MN is the screen, both perpendicular to the plane of the paper. The slits are all parallel to one another and perpendicular to the plane of the paper. Here

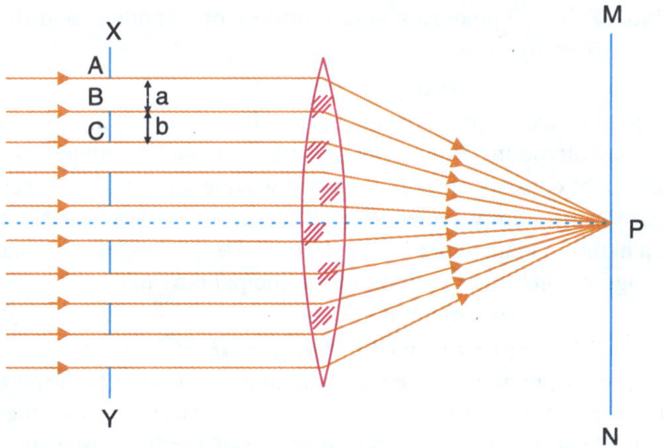

Fig. 18.12

AB is a slit and BC is an opaque portion. The width of each slit is a and the opaque spacing between any two consecutive slits is b. Let a plane wave front be incident on the grating surface. Then all the secondary waves traveling in the same direction as that of the incident light will come to focus at the point P on the screen. The screen is placed at the focal plane of the collecting lens. The point P where all the secondary waves reinforce one another corresponds to the position of the central bright maximum.

Now, consider the secondary waves traveling in a direction inclined at an angle θ with the direction of the incident light (Fig.18.13). The collecting lens also is suitably rotated such that the axis of the lens is parallel to the direction of the secondary waves. These secondary waves come to focus at point P_1 on the screen. The intensity at P_1 will depend on the path difference between the secondary waves originating from the corresponding points A and C of two neighbouring slits. In Fig. 18.13, AB = a and BC = b. The path difference between the secondary waves starting from A and C is equal to AC sin θ.

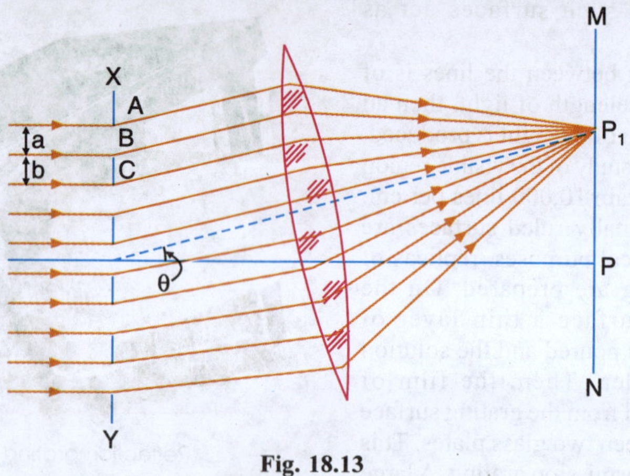

Fig. 18.13

But $$AC = AB + BC = a + b$$

Path difference $$= AC \sin\theta = (a + b)\sin\theta$$

The point P_1 will be of maximum intensity if this path difference is equal to integral multiples of λ where λ is the wavelength of light. In this case, all the secondary waves originating from the corresponding points of the neighbouring slits reinforce one another and the angle θ gives the direction of maximum intensity. In general

$$(a + b)\sin\theta_n = n\lambda \tag{18.42}$$

where θ_n is the direction of the nth principal maximum. Putting $n = 1, 2, 3$, etc, the angles θ_1, θ_2, θ_3, etc corresponding to the directions of the principal maxima can be obtained.

If the incident light consists of more than one wavelength, the beam gets dispersed and the angles of diffraction for different wavelengths will be different. Let λ and λ + $d\lambda$ be two nearby wavelengths present in the incident light and θ and (θ + $d\theta$) be the angles of diffraction corresponding to these two wavelengths. Then, for the first order principal maxima

$$(a + b)\sin\theta = \lambda$$

and $$(a + b)\sin(\theta + d\theta) = \lambda + d\lambda$$

Thus, in any order, the number of principal maxima corresponds to the number of wavelengths present. A number of parallel slit images corresponding to the different wavelengths will be observed on the screen. In equ.(18.42), $n = 1$ gives the direction of the first order image, $n = 2$ gives the direction of the second order image and so on. When white light is used, the diffraction pattern on the screen consists of a white central bright maximum and on both sides of this maximum a spectrum corresponding to the different wavelengths of light present in the incident beam will be observed in each order.

Secondary maxima and minima:

The angle of diffraction θ_n corresponding to the direction of the nth principal maximum is given by the equation

$$(a+b)\sin\theta_n = n\lambda$$

In this equation, $(a+b)$ is called the **grating constant**. For a grating with 15,000 lines/inch, the value of

$$(a+b) = \frac{2.54}{15000}\text{ cm}$$

Now, let the angle of diffraction be increased by a small amount $d\theta$ such that the path difference between the secondary waves from the points A and C increases by λ / N (See Fig. 18.13). Here N is the total number of lines on the grating surface. Then, the path difference between the secondary waves from the extreme points of the grating surface will be $(\lambda/N)N = \lambda$. Assuming the whole wave front to be divided into two halves, the path difference between the corresponding points of the two halves will be $\lambda/2$ and all the secondary waves cancel one another's effect. Thus, $(\theta_n + d\theta)$ will give the direction of the first secondary minimum after the nth primary maximum. Similarly, if the path difference between the secondary waves from the points A and C is $2\lambda / N$, $3\lambda / N$ etc, for gradually increasing values of $d\theta$, these angles correspond to the directions of 2nd, 3rd etc secondary minima after the nth primary maximum. If the value is $2\lambda / N$, then the path difference between the secondary waves from the extreme points of the grating surface is $(2\lambda / N)N = 2\lambda$ and considering the wave front to be divided into 4 portions, the concept of the 2nd secondary minimum can be understood. The number of secondary minima in between any two primary maxima is (N–1) and the number of secondary maxima is (N–2).

Four rings of transmission gratings.

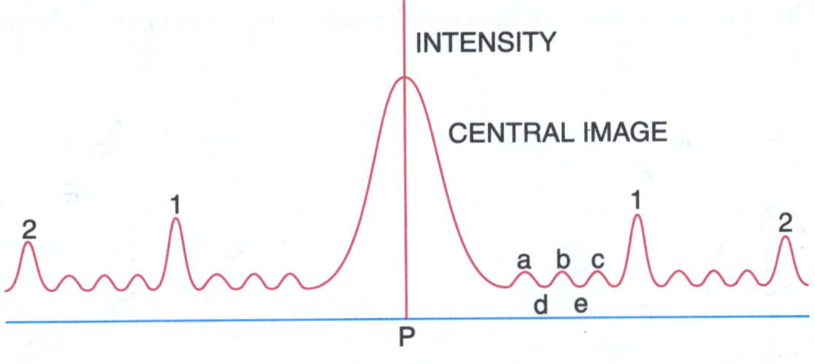

Fig. 18.14

The intensity distribution on the screen is shown in Fig.18.14. P corresponds to the position of the central maxima and 1, 2, etc, on the two sides of P represent the 1st, 2nd, etc principal maxima. *a, b*, etc are secondary maxima and *d, e* etc are the secondary minima. The intensity as well as angular spacing of the secondary maxima and minima is so small in comparison to the principal maxima that they cannot be observed. It results in uniform darkness between any two principal maxima.

18.7.2. WIDTH OF PRINCIPAL MAXIMA

The direction of the nth principal maximum is given by

$$(a+b)\sin\theta_n = n\lambda$$

Let $\theta_n + d\theta$ and $\theta_n - d\theta$ give the directions of the first secondary minima on the two sides of the nth primary maxima (see Fig. 18.15). Then

$$(a+b)\sin[\theta_n \pm d\theta] = n\lambda \pm \frac{\lambda}{N} \quad (18.43)$$

where N is the total number of the lines on the grating surface.

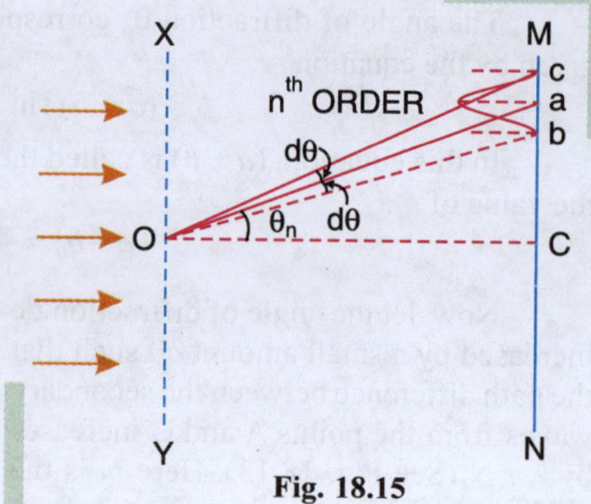

Fig. 18.15

Dividing (18.43) by (18.42), we get

$$\frac{(a+b)\sin[\theta_n \pm d\theta]}{(a+b)\sin\theta_n} = \frac{n\lambda \pm \frac{\lambda}{N}}{n\lambda}$$

$$\frac{\sin[\theta_n \pm d\theta]}{\sin\theta_n} = 1 \pm \frac{1}{nN}$$

Expanding this equation, we get $\dfrac{\sin\theta_n.\cos d\theta \pm \cos\theta_n \sin d\theta]}{\sin\theta_n} = 1 \pm \dfrac{1}{nN}$ (18.44)

For small values of $d\theta$; $\cos d\theta = 1$ and $\sin d\theta = d\theta$.

$$\therefore \quad 1 \pm \cot\theta_n\, d\theta = 1 \pm \frac{1}{nN} \quad \text{or} \quad \cot\theta_n\, d\theta = \frac{1}{nN} \quad \therefore \quad d\theta = \frac{1}{nN\cot\theta_n} \quad (18.45)$$

In equation (18.45), $d\theta$ refers to half the angular width of the principal maximum. The half width $d\theta$ is (*i*) inversely proportional to N, the total number of lines and (*ii*) inversely proportional to $n\cot\theta_n$. The value of $n\cot\theta_n$ is more for higher orders because the increase in the value of $\cot\theta_n$ is less than the increase in the order. Thus, the half width of the principal maximum is less for higher orders. Also, the larger the number of lines on the grating surface, the smaller is the value of $d\theta$. Further, the value of θ_n is higher for longer wavelengths and hence the spectral lines are sharper towards the violet than the red end of the spectrum.

18.7.3. OBLIQUE INCIDENCE

Let a parallel beam of light be incident obliquely on the grating surface at an angle of incidence *i* (Fig. 18.16). Then the path difference between the secondary waves passing through the points A and C = FC + CE

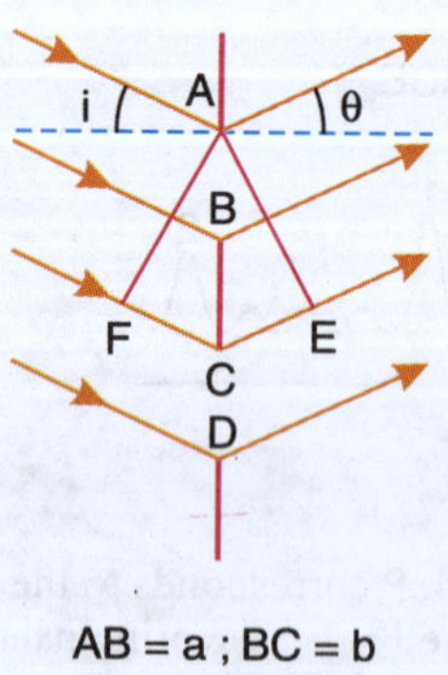

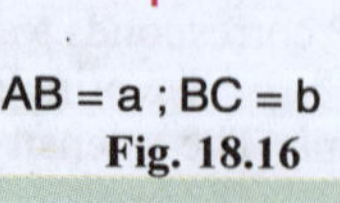

Fig. 18.16

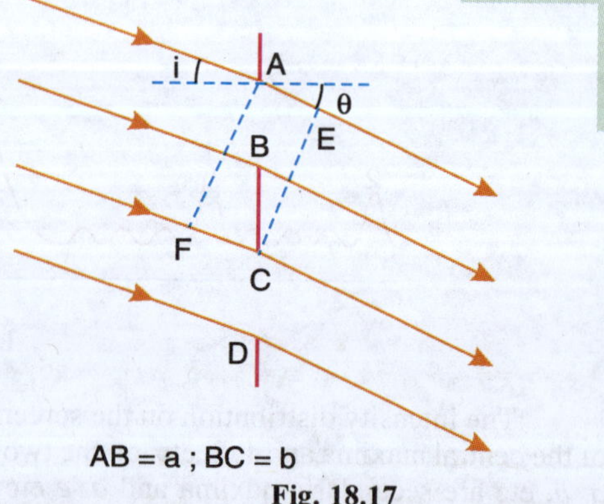

Fig. 18.17

From the Δ^{le} AFC it is seen that $FC = (a + b) \sin i$

and from Δ^{le}ACE $CE = (a + b) \sin \theta$

$$\therefore \quad FC + CE = (a + b) [\sin \theta + \sin i] \tag{18.46}$$

The equation (18.46) holds good if the beam is diffracted upwards. Fig.18.17 illustrates the diffraction of the beam downwards. In this case the path difference

$$= (a + b)[\sin \theta - \sin i] \tag{18.47}$$

For the nth primary maximum $(a + b) [\sin \theta_n + \sin i] = n\lambda$ (18.48)

or

$$(a+b)\left[2\sin\frac{\theta_n+i}{2}.\cos\frac{\theta_n-i}{2}\right] = n\lambda$$

or

$$\sin\frac{\theta_n+i}{2} = \frac{n\lambda}{2(a+b)\cos\frac{\theta_n-i}{2}} \tag{18.49}$$

The deviation of the diffraction beam $= \theta_n + i$.

For the deviation $\theta_n + i$ to be a minimum, $\sin\frac{\theta_n+i}{2}$ must be minimum. This is possible if the value of $\cos\frac{\theta_n-i}{2}$ is maximum, i.e., $\frac{\theta_n-i}{2} = 0$ or $\theta_n = i$.

Thus, the deviation produced in the diffracted beam is a minimum when the angle of incidence is equal to the angle of diffraction. Let D_n be the angle of minimum deviation.

Then $D_m = \theta_n + i$

But $\theta_n = i$

$$\therefore \quad \theta_n = \frac{D_m}{2} \text{ and } i = \frac{D_m}{2}$$

$$(a + b)\left[\sin\frac{D_m}{2} + \sin\frac{D_m}{2}\right] = n\lambda$$

or

$$2(a + b)\sin\frac{D_m}{2} = n\lambda \tag{18.50}$$

Equation (18.50) refers to the principal maximum of the nth order for a wavelength λ.

18.7.4. ABSENT SPECTRA WITH A DIFFRACTION GRATING

In the equation $(a + b)\sin\theta = \lambda$, if $(a + b) < \lambda$, then $\sin\theta > 1$. But this is not possible. Hence the first order spectrum is absent. Similarly, the second, the third, etc order spectra will be absent if $(a + b) < 2\lambda$, $(a + b) < 3\lambda$ etc. In general, if $(a+ b) < n\lambda$, then the nth order spectrum will be absent.

The condition for absent spectra can be obtained from the following considerations. For the nth order principal maximum

$$(a + b)\sin\theta_n = n\lambda \tag{18.51}$$

Further, if the value of a and θ_n are such that

$$a\sin\theta_n = \lambda \tag{18.52}$$

then, the effect of the wave front from any particular slit will be zero. Considering each slit to be made up of two halves, the path difference between the secondary waves from the corresponding points will be $\lambda / 2$ and they cancel one another's effect. If the two conditions given by equations (18.51) and (18.52) are simultaneously satisfied, then dividing (18.51) by (18.52), we get

$$\frac{(a+b)\sin\theta_n}{a\sin\theta_n} = \frac{n\lambda}{\lambda}$$

or

$$\frac{a+b}{a} = n \tag{18.53}$$

In equation (18.53), the values of n =1, 2, 3 etc refer to the order of the principal maxima that are absent in the diffraction pattern.

(*i*) If $\frac{a+b}{a} = 1; b = 0$

In this case, the first order spectrum will be absent and the resultant diffraction pattern is similar to that due to the single slit.

(*ii*) If $\frac{a+b}{a} = 2; \quad a = b$

i.e., the width of the slit is equal to the width of the opaque spacing between any two consecutive slits. In this case, the second order spectrum will be absent.

18.7.5. OVERLAPPING OF SPECTRAL LINES

If the light incident on the grating surface consists of a large range of wavelengths, then the spectral lines of shorter wavelength and of higher order overlap on the spectral lines of longer wavelength and of lower order. Let the angle of diffraction θ be the same for (*i*) the spectral line of wavelength λ_1 in the first order, (*ii*) the spectral line of wavelength λ_2 in the second order and (*iii*) the spectral line of wavelength λ_3 in the third order. Then

$$(a + b) \sin \theta = 1.\lambda_1 = 2\lambda_2 = 3\,\lambda_3 = \ldots\ldots$$

The red line of wavelength 7000 Å in the third order, the green line of wavelength 5250 Å in the fourth order and the violet line of wavelength 4200 Å in the fifth order are all formed at the same position of the screen because,

$$\begin{aligned}(a + b) \sin \theta &= 3 \times 7000 \times 10^{-8}\text{cm}\\ &= 4 \times 5250 \times 10^{-8}\text{ cm}\\ &= 5 \times 4200 \times 10^{-8}\text{cm}\end{aligned}$$

For the visible region of the spectrum, there is no overlapping of the spectral lines. The range of wavelengths for the visible part of the spectrum is 4000 Å to 7200 Å. Thus, the diffracting angle for the red end of the spectrum in the first order is less than the diffracting angle for the violet end of the spectrum in the second order. If, however, the observations are made with a photographic plate, the spectrum recorded may extend up to 2000 Å in the ultra violet region. In this case, the spectral line corresponding to a wavelength of 4000 Å in the first order and a spectral line of wavelength 2000 Å in the second order overlap. Suitable filters are used to absorb those wavelengths of the incident light which will overlap with the spectral lines in the region under investigation.

18.7.6. DETERMINATION OF WAVELENGTH OF A SPECTRAL LINE USING THE TRANSMISSION GRATING

In the laboratory, the grating spectrum of a given source of light is obtained by using a spectrometer. Initially all the adjustments of the spectrometer are made and it is adjusted for parallel rays by Schuster's method. The slit of the collimator is illuminated by monochromatic light (say light from sodium lamp) and the position of the telescope is adjusted such that the image of the slit is obtained at the position of the vertical cross-wire in the field of view of the telescope. Now the axes of the collimator and the telescope are in the

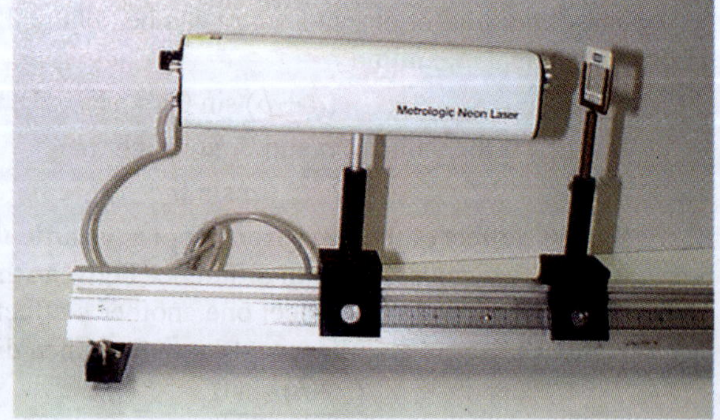

Determination of wavelength of a spectral line using the transmission grating.

same line. The position of the telescope is noted on the circular scale and 90° is added to this reading. The telescope is turned to this position. In this position the axis of the telescope is perpendicular to the axis of collimator. The position of the telescope is fixed. The given transmission grating is mounted at the centre of the prism table such that the grating surface is perpendicular to the prism table. The prism table is suitably rotated such that the image of the slit reflected from the grating surface is obtained in the centre of field of view of the telescope. This means that the parallel rays of light from the collimator are incident at an angle 45° on the grating surface because the axis of the collimator and the telescope are perpendicular to each other. The reading of the prism table is noted and adding 45° to this reading, the prism table is suitably rotated to the new position so that the grating surface is normal to the incident light.

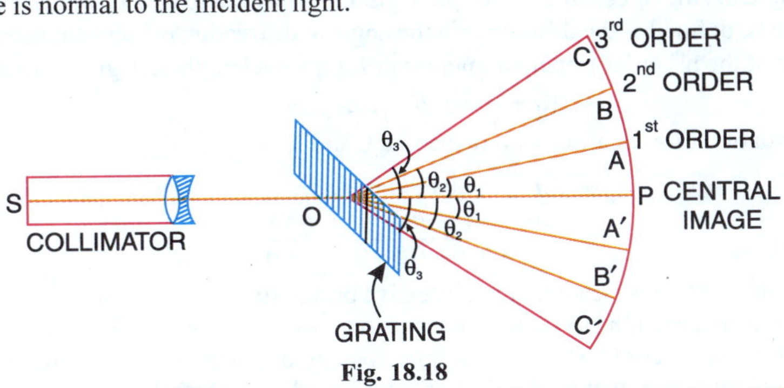

Fig. 18.18

If the wavelength of the sodium light is to be determined, then the angles of diffraction θ_1 and θ_2 corresponding to the first and second order principal maxima are determined (Fig. 18.18). OA, OB etc., give the directions of the telescope corresponding to the first and second order images. A′, B′ etc refer to the positions of these images towards the left of the central maximum. The angles AOA′ and BOB′ are measured and half of these angles measure θ_1 and θ_2. Then

$$(a + b) \sin \theta_1 = 1\lambda \tag{18.54}$$

and

$$(a + b) \sin \theta_2 = 2\lambda \tag{18.55}$$

Then the value of λ is calculated from equations (18.54) and (18.55) and the mean value is taken. $(a + b)$ is the grating element and it is equal to the reciprocal of the number of lines per cm. If the number of lines on the grating surface is 15,000 per inch then

$$(a + b) = \frac{2.54}{15000} \text{ cm}$$

If the source of light emits radiations of different wavelengths, then the beam gets dispersed by the grating and in each order a spectrum of the constituent wavelengths is observed. To find the wavelength of any spectral line, the diffracting angles are noted in the first and second orders and using the equations given above, the wavelength of the spectral line can be calculated. Overlapping spectral orders can be avoided by using suitable colour filters so that the wavelengths beyond the range of study are eliminated.

With a diffraction grating, the wavelength of the spectral line can be determined very accurately. The method involves only the accurate measurement of the angles of diffraction.

Taking $\lambda = 6000 \text{ Å} = 6000 \times 10^{-8}$ cm and $(a + b) = \frac{2.54}{15000}$ cm

We get from the equations,

$$(a + b) \sin \theta_1 = 1\lambda$$

and

$$(a + b) \sin \theta_2 = 2\lambda$$

$$\theta_1 = 20° - 45'$$

and $$\theta_2 = 45° - 7'$$

As the angles are large they can be measured accurately with a properly calibrated spectrometer. The number of lines per inch (or cm), is given on the grating by the manufacturing company and hence $(a + b)$ can be calculated. As the method does not involve measurements of very small distances (as in the case of interference experiments) an accurate value of λ can be obtained.

18.7.7. DISPERSIVE POWER OF GRATING

Dispersive power of a grating is defined as the ratio of the difference in the angle of diffraction of any two neighbouring spectral lines to the difference in wavelength between the two spectral lines. It can also be defined as the difference in the angle of diffraction per unit change in wavelength. The diffraction of the nth order principal maximum for a wavelength λ, is given by the equation

$$(a + b) \sin \theta = n \lambda$$

Differentiating this equation with respect to θ and λ, we get

$$(a + b) \cos \theta \, d\theta = n d\lambda$$

or $$\frac{d\theta}{d\lambda} = \frac{n}{(a+b)\cos\theta} = \frac{nN'}{\cos\theta} \quad (18.56)$$

From equ.(18.56) it is clear that the dispersive power of the grating is (*i*) directly proportional to the order of the spectrum, n (*ii*) directly proportional to the number of lines per cm, N′ and (*iii*) inversely proportional to cos θ. Thus, the angular spacing of any two spectral lines is double in the second order spectrum than that in the first order. Secondly, the angular dispersion of the lines is more with a grating having a larger number of lines per cm. Thirdly, the angular dispersion is a minimum when $\theta = 0$. If the value of θ is not large, the value of cos θ can be taken as unity and the influence of this factor can be neglected. Then it is clear that the angular dispersion of any two spectral lines is directly proportional to the difference in wavelength of the spectral lines. A spectrum of this type is called a **normal spectrum.**

If the linear spacing of two spectral lines of wavelengths λ and $\lambda + d\lambda$ is dx in the focal plane of the telescope objective or photographic plate, then

$$dx = f \, d\theta$$

where f is the focal length of the objective. The linear dispersion is

$$\frac{dx}{d\lambda} = f\frac{d\theta}{d\lambda} = \frac{fnN'}{\cos\theta} \quad (18.57)$$

or $$dx = \frac{fnN'}{\cos\theta} \cdot d\lambda$$

The linear dispersion is useful in studying the photographs of a spectrum.

18.7.8. PRISM AND GRATING SPECTRA

For dispersing a given beam of light and for studying the resultant spectrum, a diffraction grating is mostly used instead of a prism. The grating and prism spectra differ in the following points.

(*i*) With a grating, a number of spectra of different orders can be obtained on the two sides of the central maximum whereas with a prism only one spectrum is obtained.

(*ii*) The spectra obtained with a grating are comparatively purer than those with a prism.

(*iii*) Knowing the grating element $(a + b)$ and measuring the diffraction angle, the wavelength of any spectral line can be measured accurately. But in case of a prism, the angles of deviation are not directly related to the wavelength of the spectal line. The angles of deviation are dependent on the refractive index of the material of the prism, which depends on the wavelength of light.

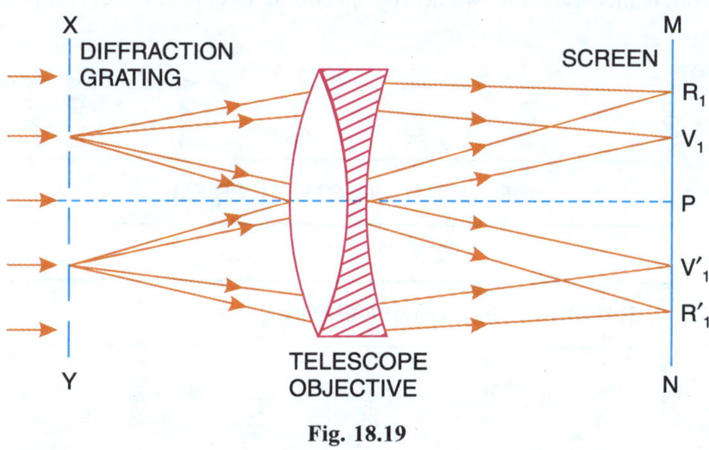

Fig. 18.19

(*iv*) With a grating, the diffraction angle for violet end of the spectrum is less than for red. In Fig. 18.19, V_1R_1 and $V_1'\,R_1'$ refer to the first order spectra on the two sides of the central maximum P. With a prism (Fig.18.20), the angle of deviation for the violet rays of light is more than for the red rays of light.

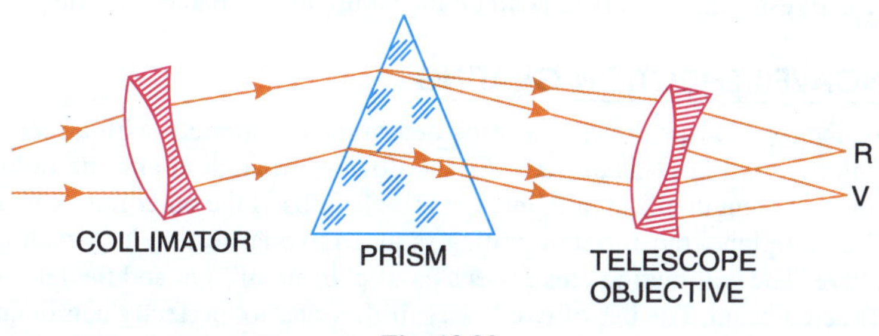

Fig. 18.20

(*v*) The intensities of the spectral lines with a grating are much less than with a prism. In a grating spectrum, most of the incident light energy is associated with the undispersed central bright maximum and the rest of the energy is distributed in the different order spectra on the two sides of the central maximum. But in a prism most of the incident light energy is distributed in a single spectrum and hence brighter spectral lines are obtained.

(*vi*) The dispersive power of a grating is given by

$$\frac{d\theta}{d\lambda} = \frac{nN'}{\cos\theta}$$

and this is constant for a particular order. Thus, the spectral lines are evenly distributed. Hence, the spectrum obtained with a grating is said to be **rational** (Fig. 18.21). The refractive index of the material of a prism changes more rapidly at the violet end than at the red end of the spectrum. The dispersive power of a prism is given by $\frac{d\mu}{\mu - 1}$ and this has higher value in the violet region than in the red region. Hence, there will be more spreading of the spectral lines towards the violet and the spectrum obtained with a prism is said to be **irrational** (see Fig. 18.21).

(*vii*) The resolving power of a grating is given by *n*N whereas the resolving power of a prism is given by $t\,\frac{d\mu}{d\lambda}$ where *t* is the base of the prism. The resolving power of a grating is much higher than that of a prism. Hence the same two nearby spectral lines appear better resolved with a grating than with a prism.

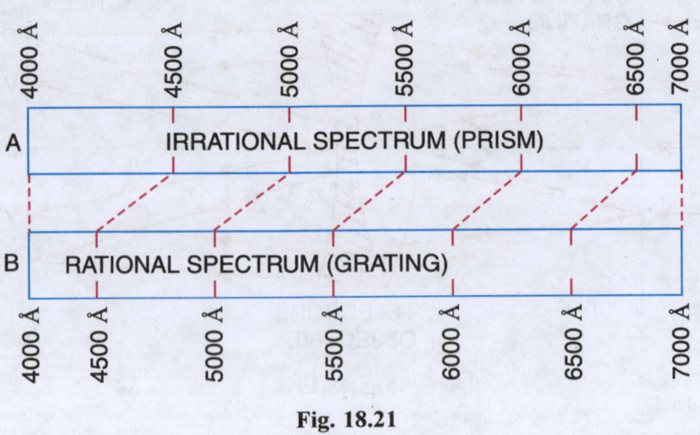

Fig. 18.21

(*viii*) Lastly, the spectra obtained with different gratings are identical because the dispersive power and the resolving power of a grating do not depend on the nature of the material of the grating. But the spectra obtained with prisms made of different materials are never identical because both dispersive and resolving powers depend on the nature of the material of the prism.

18.8. CONCAVE REFLECTION GRATING

The wavelength of a spectral line can be determined accurately with a plane transmission grating, knowing the grating constant $(a + b)$, the diffraction angle θ and the order n. From the knowledge of the wavelength of a single line, the wavelengths of the other lines can be obtained by comparison. Use of a plane transmission grating requires two lenses, the collimating lens and the telescope objective. The collimating lens gives a parallel beam of light and the telescope objective focuses the diffracted beam. The use of two lenses, if they are not perfectly achromatic, makes the spectrum more complex due to the chromatic aberration present in the lenses. Rowland developed the *concave reflection grating*, the use of which dispenses the use of the lenses. The rulings are made on a concave reflecting surface instead of a plane surface. The concave mirror is highly polished metal surface and it will diffract the incident beam and also focuses it at the same time. In a concave

Reflection Grating Arrays

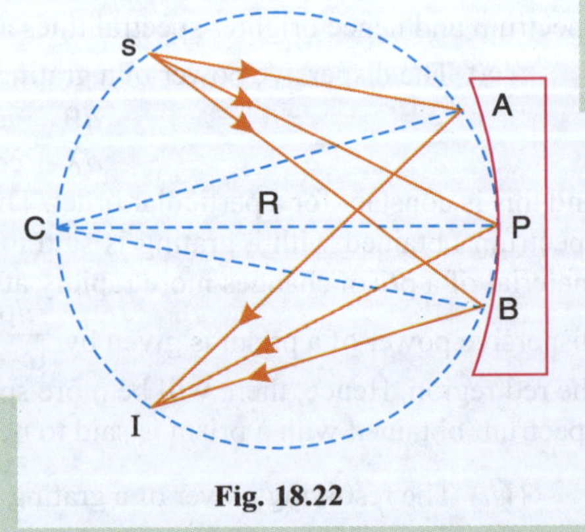

Fig. 18.22

reflection grating, the effect of chromatic aberration is completely eliminated and it can be conveniently used in those regions of the spectrum for which the glass lenses are not transparent.

In Fig. 18.22, APB is the surface of a concave reflection grating in which the rulings are perpendicular to the plane of the paper. C is the centre of curvature of the surface, that is CP = R. The dotted circle is the Rowland circle, which has a diameter R. The circle touches the grating surface at P. If a source of light S is placed at any point on the circumference of the Rowland circle PSCI, the dispersed spectral images of the slit are obtained at points such as I on the circumference of the same circle. By keeping the source of light at C, the spectra can be observed at other points on the circumference of the circle.

18.8.1. THEORY OF CONCAVE REFLECTION GRATING

In Fig. 18.23 GG′ is the concave reflection grating and C is the centre of the curvature of the grating surface. CC′ is the radius of curvature of the surface (CC′ = R). The dotted circle represents the Rowland circle of diameter CC′. S is a narrow slit perpendicular to the plane of the paper and illuminated by light. A and C are two corresponding points on the grating. Here C′A and CC′ are normals. SA and SC are the incident rays and AI and CI are the diffracted rays. Let the angles of incidence and diffraction be i and θ.

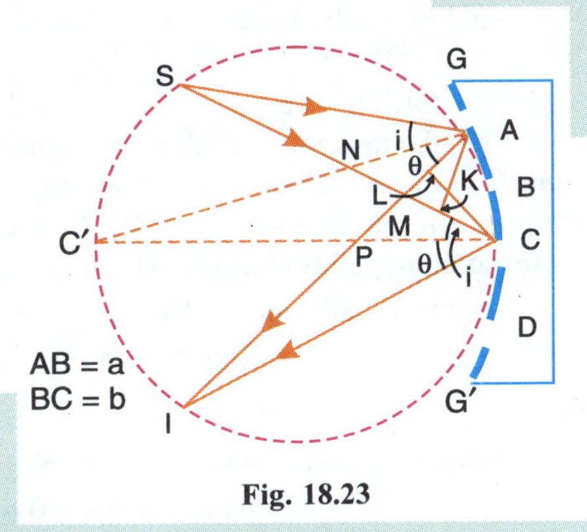

Fig. 18.23

$\angle SAC' = i$ and $\angle C'AI = \theta$

As the points A and C are very close, $\angle SAC' = \angle SCC' = i$ and $\angle C'AI = \angle C'CI = \theta$

From Δ^{les} SMA and IMC, we find that

$$\angle ASC = \angle AIC \tag{18.58}$$

From Δ^{les} ASN and CC′N, we have $\angle ASC = \angle AC'C$ (18.59)

From equations (18.58) and (18.59) $\angle ASC = \angle AIC = \angle AC'C$ (18.60)

Hence, the points A, S, C′, I and C lie on the same circle.

To calculate the path difference between the secondary waves emanating from the points A and C, draw perpendiculars AK and CL. As angles at S and I are small, SA = SK and IC = IL.

The path difference = (SC + CI) – (SA + AI)

= (SK + KC + CI) – (SA + AL + LI)

But SK = SA; LI = CI

∴ Path difference = KC – AL (18.61)

In the $\Delta^{le} KAC$, $\angle KAC = i$

and in the Δ^{le} ALC, $\angle LCA = \theta$ because C′A and CC′ are two radii with reference to the concave surface.

But AB = a; BC = b and AC = $a + b$

$$\therefore \quad KC = AC \sin i = (a + b) \sin i$$

and

$$AL = AC \sin \theta = (a + b) \sin \theta$$

$$\therefore \quad \text{Path difference} = KC - AL = (a + b)(\sin i - \sin \theta)$$

For the nth order spectrum to lie at I,

$$(a + b)(\sin i - \sin \theta) = n\lambda \tag{18.62}$$

If the points S and I lie on the same side of C′, it can be shown that the path difference

$$(a + b)(\sin i + \sin \theta) = n\lambda \tag{18.63}$$

The secondary waves from the corresponding points A and C reinforce at I and the same holds good for other pairs of corresponding points in AB and CD and also for the whole grating surface.

Equation (18.63) must hold good for diffraction at every element of the grating surface. In other words, it is essential that the path difference for any pair of diffracted rays from the corresponding points of the grating surface must be constant for a particular wavelength. If this condition is satisfied, then all the rays corresponding to a particular wavelength in a given order will come to focus at a single point on the conference of Rowland circle.

Differentiating equation (18.63), we get

$$(a + b)(\cos i\, di + \cos\theta\, d\theta) = 0$$

or $$(\cos i\, di - \cos\theta\, d\theta) = 0 \qquad (18.64)$$

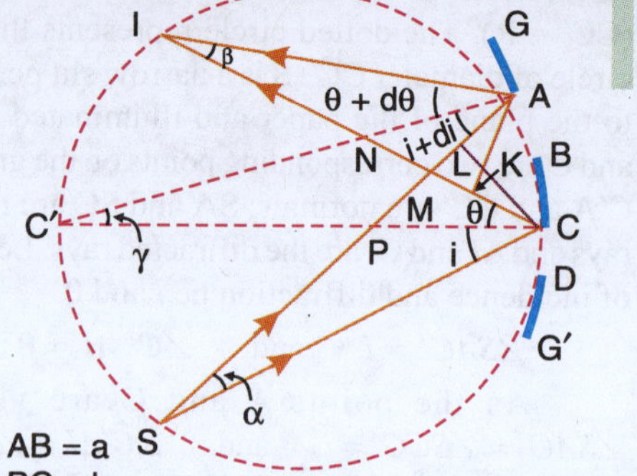

In Fig. 18.24, let

$\angle ASC$ be α; $\angle AIC$ be β

and $\angle AC'C$ be γ.

Let the rays SC and SA be incident at angles i and $i + di$ and let the corresponding angles of diffraction be θ and $\theta + d\theta$. A and C are corresponding points (Fig.18.24).

Fig. 18.24

From Δ^{les} SPC and C′PA

$$i + \angle ASC = i + di + \angle AC'C$$

or $$i + \alpha = i + di + \gamma \qquad (18.65)$$

Similarly from Δ^{les} CNC′, and ANI

$$\theta + \gamma = \theta + d\theta + \beta$$

or $$d\theta = \gamma - \beta \qquad (18.66)$$

Let the radius of curvature CC′ be R and let the distance SA and AI be s_1 and s_2.

$$AC = AB + BC = a + b$$

Substituting the values of di and $d\theta$ of equations (18.65) and (18.66) in equation (18.64), we get

$$(\alpha - \gamma)\cos i - (\gamma - \beta)\cos\theta = 0 \qquad (18.67)$$

CL is perpendicular to SA and AK is perpendicular to IC.

$\therefore$ $\angle ACL = i$ and $\angle KAC = \theta$ approximately.

In the Δ^{le} ACL, LC = AC cos ACL = AC cos i

$$= (a + b)\cos i \qquad (18.68)$$

Similarly in the Δ^{le} AKC, AK = AC cos KAC = AC cos θ

$$= (a + b)\cos\theta \qquad (18.69)$$

As the points A and C are very near, LC = SC.α = SA.α = $s_1\alpha$

and AK = IA . β = $s_2\beta$

Substituting these values of LC and AK in equations (18.68) and (18.69),we get

$$s_1\alpha = (a + b)\cos i$$

$$s_2\beta = (a + b)\cos\theta$$

Also, $$AC = (a + b) = R\gamma$$

From the above equations we obtain $\alpha = \dfrac{(a+b)\cos i}{s_1}$; $\beta = \dfrac{(a+b)\cos\theta}{s_2}$ and $\gamma = \dfrac{(a+b)}{R}$

Substituting these values of α, β and γ in equation (18.67), we obtain

$$(\alpha-\gamma)\cos i-(\gamma-\beta)\cos\theta=0$$

$$\left[\frac{(a+b)\cos i}{s_1}-\frac{(a+b)}{R}\right]\cos i-\left[\frac{(a+b)}{R}-\frac{(a+b)\cos\theta}{s_2}\right]\cos\theta=0$$

or

$$\left[\frac{\cos i}{s_1}-\frac{1}{R}\right]\cos i-\left[\frac{1}{R}-\frac{\cos\theta}{s_2}\right]\cos\theta=0$$

$$\frac{\cos^2 i}{s_1}-\frac{\cos i}{R}-\frac{\cos\theta}{R}+\frac{\cos^2\theta}{s_2}=0$$

$$\frac{\cos^2\theta}{s_2}=\frac{s_1[\cos i+\cos\theta]-R\cos^2 i}{Rs_1}$$

$$\therefore \quad s_2=\frac{Rs_1\cos^2\theta}{s_1(\cos i+\cos\theta)-R\cos^2 i} \qquad (18.70)$$

In equation (18.70), if R cos $i = s_1$, then on simplification, we get

$$s_2 = R\cos\theta$$

Thus, if the point S lies on the circumference of the circle of diameter R, then I also lies on the same circle.

18.9. PASCHEN MOUNTING

The common form of mounting used for a concave reflection grating is shown in Fig.18.25. It is called the *Paschen mounting.*

In this mounting, the slit S is set on the circumference of the Rowland circle as shown. The slit is perpendicular to the plane of the paper. GG′ is the concave refection grating and OC′ is the

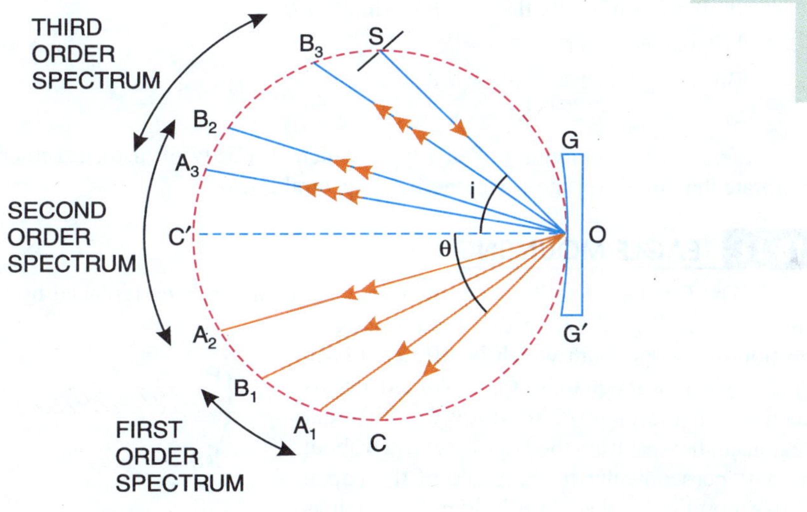

Fig. 18.25

diameter of the Rowland circle. The spectra of different orders are imaged on the circumference of the circle. In the figure, C is the central image; A_1B_1 is the first order spectrum. A_2B_2 is second order spectrum and A_3B_3 is the third order spectrum. With this mounting, several orders of the spectrum can be photographed simultaneously. The photographic plates are held in a frame which can give the plates the proper curvature coincident with the Rowland circle. For any particular order, the dispersion is minimum when $\theta = 0$. The disadvantage in this mounting is that the spacing of the spectral lines in the different regions of the spectrum is not proportional to the difference in wavelength between the lines.

18.10. ROWLAND MOUNTING

The principle of Rowland mounting is illustrated in Fig.18.26. G is the concave grating and P is the plate holder. The grating and plate holder are mounted at the ends of a beam of length R equal to the radius of curvature of the grating surface. This beam GP can slide along two rails SX and SY. G, P and G′, P′, represent two positions of the beam. The slit S is set at the point of intersection of the rails SX and SY. With an arrangement of this type, the region of the spectrum imaged at P can be altered by sliding the beam. Sliding the beam alters the angle of incidence i. The spectrum obtained with this arrangement is nearly normal because the angle θ is nearly zero. For any position of the beam the spectrum is imaged at P.

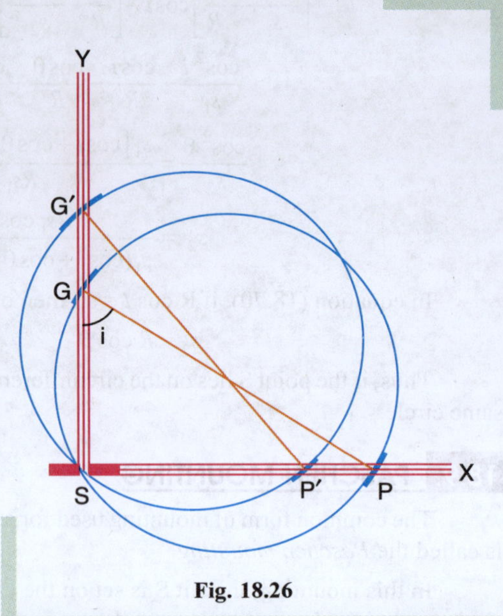

Fig. 18.26

From the equation,

$$(a + b)(\sin i - \sin \theta) = n\lambda$$

if $\theta = 0$; $\sin \theta = 0$ then, $(a + b) \sin i = n\lambda$

For a given order $\sin i \propto \lambda$

But $\sin i \propto SP$

$\therefore$ $SP \propto \lambda$ (18.71)

Thus, with a mounting of this type, which is mostly of historical interest, it is possible to calibrate the rail SP for the wavelengths of spectral lines.

18.11. EAGLE MOUNTING

The Rowland and Paschen mountings have largely been replaced by the **Eagle mounting** illustrated in Fig.18.27. In this mounting, that portion of the spectrum which is diffracted back at an angle almost equal the angle of incidence, is focused on the plate P. To study the different regions of the spectrum the grating is turned about an axis perpendicular to the plane of the paper. Correspondingly, the plate holder P, which is hinged on one side of S is turned such that P and S lie on the Rowland circle. For studying spectra in the ultra violet region, the Eagle mounting is commonly used in vacuum spectrographs.

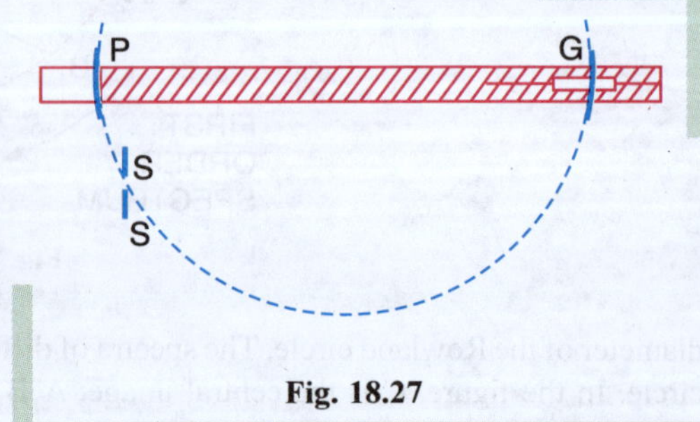

Fig. 18.27

18.12. LITTROW MOUNTING

Littrow mounting is illustrated in Fig.18.28. Large plane reflection gratings are mounted this way. G is a plane reflection grating, P is a photographic plate, S is a slit and L is a large achromatic lens. The principle of Littrow mounting is similar to that of Eagle mounting.

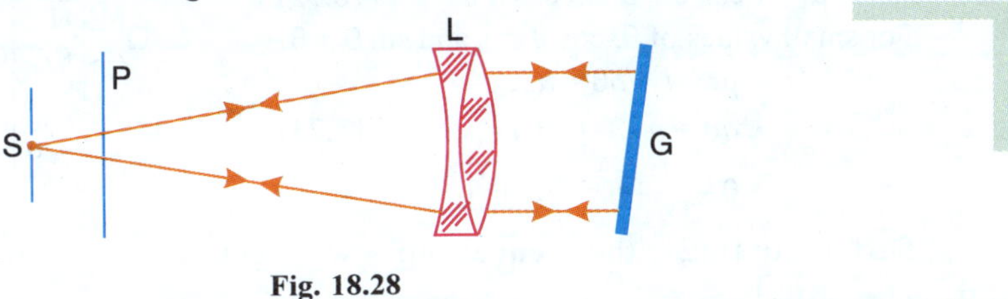

Fig. 18.28

The achromatic lens L serves two functions. It renders the incident light from the slit parallel as well as it focuses the diffracted beam on P. Thus, it serves both as collimating lens and a telescope objective.

18.13. ECHELON GRATING

An echelon transmission grating consists of a number of optically worked glass plates arranged in the form of steps, as shown in Fig. 18.29. All the plates are cut from a single optically worked glass plate. Each plate overlaps on the next by the same distance i.e., the step-width is the same throughout and is of the order of 1mm. A parallel beam of monochromatic light incident normally is diffracted through a small angle.

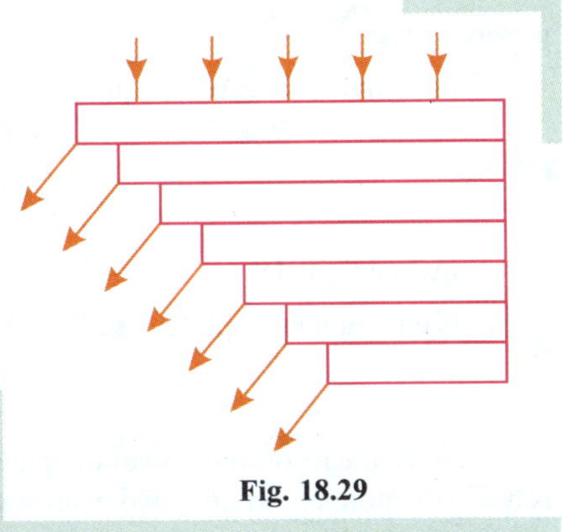

Fig. 18.29

One can observe the fifth or sixth order spectrum with a concave grating and only the second or third order spectrum with a plane diffraction grating. The resolving power of a ruled grating is dependent on the number of lines on the grating surface and the order of the spectrum. It is difficult and very tedious to draw a large number of equidistant parallel lines on a grating surface. With an echelon grating, designed by Michelson, one can observe the spectrum of a very high order and hence the resolving power of the echelon grating is very high. But, with the increase in the order of the spectrum, the intensity of the spectral lines decreases considerably. Also, the angular spacing $d\theta$ is very small and hence there will be many over lapping orders. An echelon grating, therefore, is not suitable to study the spectrum as such but essentially helps in detecting the true monocromatism of a beam of light.

18.13.1. THEORY OF ECHELON GRATING

In Fig. 18.30 the step width $AB = CD = a$ and the thickness of each plate is 't'. A parallel beam of monochromatic light is incident on the surface of the top plate. Let μ be the refractive index of the material of the plate.

The path difference between the secondary waves from the corresponding points A and C $= BC - AM$

But $AM = AQ - MQ$

$= AQ - NP$

$\therefore$ Path difference $= BC - AQ + NP$

$BC = \mu t$, $AQ = t \cos\theta$ and $NP = a \sin\theta$

where θ is the angle between the direction of the diffracted beam and the incident direction of light.

$\therefore$ Path difference = $\mu t - t\cos\theta + a\sin\theta$

If this path difference is equal to integral multiples of λ, all the secondary waves traveling in this direction reinforce with one another.

$$\mu t - t\cos\theta + a\sin\theta = n\,\theta \qquad (18.72)$$

For small values of θ, $\cos\theta = 1$ and $\sin\theta = \theta$.

$$\therefore \quad \mu t - t + a\theta = n\lambda$$

or
$$a\theta = n\lambda - \mu t + t \qquad (18.73)$$

or
$$\theta = \frac{1}{a}[n\lambda - \mu t + t]$$

Differentiating the equation, we get

$$\frac{d\theta}{d\lambda} = \frac{1}{a}\left[n - t\frac{d\mu}{d\lambda}\right] \qquad (18.74)$$

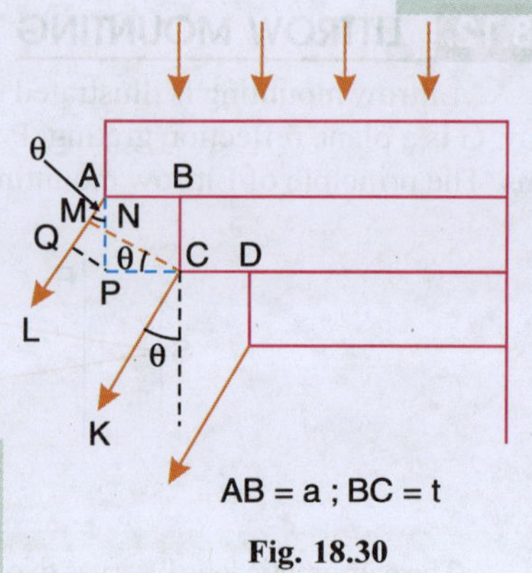

Fig. 18.30

In equ. (18.74), $\frac{d\theta}{d\lambda}$ is the dispersive power of the grating and $\frac{d\mu}{d\lambda}$ is the rate of change of refractive index of the material of the echelon with respect to wavelength.

For small values of θ, equation (18.72) can be written as

$$(\mu - 1)\,t = n\lambda$$

or
$$n = \frac{(\mu - 1)\,t}{\lambda} \qquad (18.75)$$

In equation (18.75), if $t = 1$ cm, $\lambda = 5000$ Å and $\mu = 1.5$ then $n = 10{,}000$.

If the number of plates used is 40, the resolving power of the grating

$$= n\mathrm{N} = 10{,}000 \times 40$$
$$= 4 \times 10^5$$

Thus, the resolving power of an echelon grating is very high and if the incident beam of light is not truly monochromatic, two nearby spectral lines will appear well resolved. The high resolving power of an echelon grating helps in the study of hyperfine structure e.g., the splitting of spectral lines in Zeeman effect.

WORKED OUT EXAMPLES

Example 18.1: In Fraunhoffer diffraction pattern due to a narrow slit a screen is placed 2m away from the lens to obtain the pattern. If the slit width is 0.2 mm and the first minima lie 5 mm on either sides of the central maximum, find the wavelength of light.

Solution: Here, $a = 0.2$ mm $= 0.02$ cm , $x = 5$ mm $= 0.5$ cm, $D = 2$ m $= 200$ cm

In the case of Fraunhofer diffraction at a narrow rectangular aperture,

$$a\sin\theta = n\lambda$$

Here $n = 1$ $\qquad \therefore\ a\sin\theta = \lambda$

$$\sin\theta = \frac{x}{D} \quad \therefore \lambda = \frac{ax}{D}$$

or
$$\lambda = \frac{0.02\text{ cm} \times 0.5\text{cm}}{200\text{cm}} = 5 \times 10^{-5}\text{ cm} = \mathbf{5000\ Å}$$

Example 18.2: Diffraction pattern of a single slit of width 0.5 cm is formed by a lens of focal length 40 cm. Calculate the distance between the first dark and the next bright fringe from the axis. Wavelength of light used is 4890 Å.

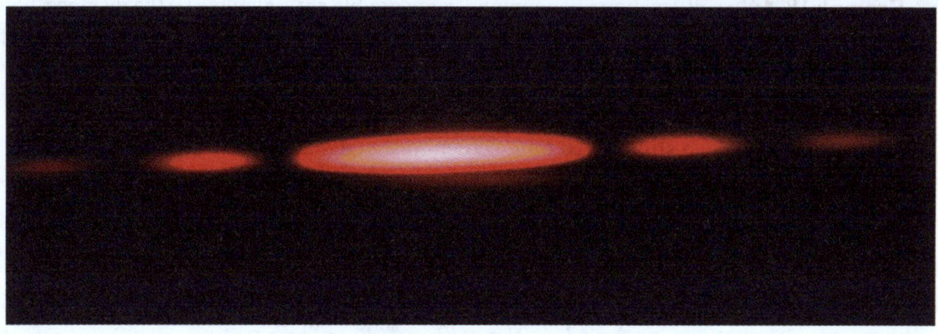

Diffraction pattern of a single slit.

Solution: Here, $a = 0.5$ cm $= 5 \times 10^{-3}$ m, $f = 40$ cm $= 0.4$ m,

$\lambda = 4890$ Å $= 4890 \times 10^{-10}$ m

For a minima, $a \sin\theta = n\lambda$. Also $\sin\theta = \frac{x_1}{f}$

As $n = 1$, we get $\frac{x_1}{f} = \frac{\lambda}{a}$

or $$x_1 = \frac{f\lambda}{a} = \frac{0.4 \times 4890 \times 10^{-10}}{5 \times 10^{-3}} \text{ m}$$

∴ $$x_1 = 3.912 \times 10^{-5} \text{ m.}$$

Now for a maximum, $a \sin\theta = \frac{(2n+1)\lambda}{2}$. Also $\sin\theta = \frac{x_2}{f}$

As $n = 1$, we get $\frac{x_2}{f} = \frac{3\lambda}{2a}$ or $x_2 = \frac{3\lambda f}{2a}$

∴ $$x_2 = \frac{3 \times 4890 \times 10^{-10} \times 0.4}{2 \times 5 \times 10^{-3}} \text{ m} = 5.868 \times 10^{-5} \text{ m}$$

∴ $$x_2 - x_1 = 5.868 \times 10^{-5} \text{ m} - 3.912 \times 10^{-5} \text{ m}$$

$$= 1.956 \times 10^{-5} \text{ m} = \mathbf{1.9 \times 10^{-2} \text{ mm.}}$$

Example 18.3: Deduce the missing orders for a double slit Fraunhofer diffraction pattern, if the slit widths are 0.16 mm and they are 0.8 mm apart.

Solution: Here $a = 0.16$ mm $= 0.016$ cm and $b = 0.8$ mm $= 0.08$ cm

Equation for interference maxima is, $(a + b) \sin\theta = n\theta$

Equation for diffraction minima is, $a \sin\theta = p\lambda$

∴ $$\frac{(a+b)}{a} = \frac{n}{p}$$

∴ $$\frac{n}{p} = \frac{(0.016 + 0.080)}{0.016} = 6$$

∴ $$n = 6p$$

For the values of $p = 1, 2, 3$ etc.

$n = 6, 12, 18$ etc.

Thus the orders 6, 12, 18 etc. of the interference maxima will be missing from the diffraction pattern.

Example18. 4: In a diffraction phenomenon using double slit, calculate (*i*) the distance between the central maximum and the first minimum of the fringe envelope and (*ii*) the distance between any two consecutive double slit dark fringes.

Given data : Wavelength of light = 5000 Å, slit width = 0.02 mm,

Spacing between two slits = 0.10 mm, screen to slits distance = 100 cm

Solution: Here, $a = 0.02$ mm $= 2 \times 10^{-5}$ m, $b = 0.1$ mm $= 10^{-4}$ m, $(a + b) = 1.2 \times 10^{-4}$ m, $\lambda = 5000$ Å, $d = 100$ cm $= 1$ m

(*i*) The angular separation between the central maximum and the first minimum is,

$$\sin\theta_1 = \theta_1 = \frac{\lambda}{2(a+b)} \quad \text{and} \quad \theta_1 = \frac{x_1}{D}$$

$$\therefore \quad \frac{x_1}{D} = \frac{\lambda}{2(a+b)} \quad \therefore \; x_1 = \frac{\lambda D}{2(a+b)}$$

$$\therefore \quad x_1 = \frac{5\times10^{-7}\text{ m}\times 1\text{m}}{2\left(1.2\times10^{-4}\text{ m}\right)} = 2.08\times10^{-3}\text{ m} = \mathbf{2.08\ mm}$$

The separation between the central maximum and the first minimum is **2.08 mm.**

(*ii*) The angular separation between two consecutive dark fringes,

$$\sin\theta_1 - \sin\theta_2 = \theta_1 - \theta_2 = \theta = \frac{3\lambda}{2(a+b)} - \frac{\lambda}{2(a+b)}$$

$$\theta = \frac{\lambda}{(a+b)}; \quad \text{Also } \theta = \frac{x_2}{D}$$

$$\therefore \quad x_2 = \frac{\lambda D}{(a+b)} \quad \therefore \; x_2 = \frac{5\times10^{-7}\text{ m}\times1\text{m}}{1.2\times10^{-4}\text{ m}}$$

$$= 4.16\times10^{-3}\text{ m} = \mathbf{4.2\ mm}$$

Example 18.5: A parallel beam of light of wavelength 5460 Å is incident at an angle of 30º on a plane transmission grating which has 6000 lines /cm. Find the highest order spectrum that can be observed.

Solution : Here $\theta = 30º$, $\lambda = 5460$ Å $= 5460 \times 10^{-10}$ m,

$$(a+b) = \frac{1}{6\times10^3}\text{ cm} = \frac{1}{6\times10^5}\text{ m}$$

Now, $$(a+b)[\sin\theta_n + \sin i] = n\lambda$$

But here $\theta_n = i$

$$\therefore \quad (a+b)\,[2\sin i\,] = n\,\lambda$$

$$\therefore \quad n = \frac{(a+b)[2\sin i]}{\lambda} = \frac{1}{6\times10^5\times5460\times10^{-10}} = 3.05$$

or $$n = \mathbf{3}$$

Example18.6: Calculate the possible order of spectra with a plane transmission grating having 18,000 lines per inch when light of wavelength 4500 Å is used.

Solution: Order of spectra, $n = \dfrac{d\sin\theta}{\lambda}$. The highest order occurs when $\sin\theta = 1$;

$$\therefore \quad n = \frac{d}{\lambda} = \frac{1}{N\lambda} = \frac{1}{(7.09\times10^5\text{ lines/m})\,(4500\times10^{-10}\text{ m})} = \mathbf{3}$$

Example18.7: Light which is a mixture of two wavelengths 5000 Å and 5200 Å is incident normally on a plane transmission grating having 10000 lines per cm. A lens of focal length 150 cm is used to observe the spectrum on the screen. Calculate the separation in cm of the two lines in the first order spectrum.

Solution: Here $(a + b) = 10^{-4}$ cm, $\lambda_1 = 5000$ Å $= 5 \times 10^{-5}$ cm;

$\lambda_2 = 5200$ Å $= 5.2 \times 10^{-5}$ cm and $n = 1$

Now $$\sin \theta_1 = \frac{n\lambda_1}{a+b} = \frac{1 \times 5 \times 10^{-5}\,\text{cm}}{10^{-4}\,\text{cm}}$$

$\therefore$ $$\theta_1 = \sin^{-1}(0.5) = 30^\circ$$

Similarly, $$\sin \theta_2 \frac{n\lambda_2}{a+b} = \frac{1 \times 5.2 \times 10^{-5}\,\text{cm}}{10^{-4}\,\text{cm}}$$

$$= 0.52$$

$$\theta_2 = \sin^{-1}(0.52) = 31.3^\circ$$

Further, $$\tan \theta_1 = \frac{x_1}{f} \text{ and } \tan \theta_2 = \frac{x_2}{f}$$

$\therefore$ $$(x_2 - x_1) = f[\tan \theta_2 - \tan \theta_1]$$

$$= 150\,[0.0687 - 0.5774]$$

$$= \mathbf{4.7\ cm}$$

QUESTIONS

1. Describe and explain the Fraunhofer pattern obtained with a narrow slit and illuminated by a parallel beam of monochromatic light.
2. Explain the construction and the mode of action of a diffraction grating and derive an expression for its resolving power.
3. Give the theory of a concave grating. Give a short account of the important methods of mounting the gratings with their respective advantages and disadvantages.
4. Describe Rowland's grating and a method of mounting it for measuring wavelength of light. Deduce the formula used.
5. In what respect is an echelon grating superior to an ordinary ruled grating? Obtain expressions for these in the case of a plane transmission grating.
6. In what respect is an echelon grating superior to an ordinary ruled grating? Give an account of its theory and its practical applications.
7. Derive an expression for the angular dispersion of a plane diffraction grating.
8. Distinguish between Fresnel and the Fraunhofer classes of diffraction. Compare the diffraction phenomenon in the following cases:
   (*a*) Diffraction by a slit and a narrow wire,
   (*b*) Diffraction by a circular hole and a circular obstacle. **(Agra, 2005)**
9. Give the theory of a diffraction grating. Describe in detail how you would use a transmission grating for measuring the wavelength of light.
10. Explain giving suitable examples, what is meant by Fraunhofer class of diffraction. Describe in detail how you would use this phenomenon for determining the wavelength of light.
11. Discuss the diffraction of light by a narrow slit.
12. Discuss Fraunhofer diffraction due to a single slit. Extend the theory to the case of a plane transmission grating. Explain what is meant by diffraction spectra of different orders and state the conditions under which the grating spectra of even order are absent.
13. Discuss Fraunhofer diffraction pattern of a straight edge. How does this pattern differ from that due to a straight wire?

14. Give an account of the diffraction effects produced by a slit. Explain what happens when the slit width is gradually increased and also when the screen is gradually moved away from the slit.
15. Explain the formation of spectra by a plane diffraction grating. What are its chief characteristics?
16. Discuss Fraunhofer diffraction due to a single slit. Draw a curve indicating distribution of light in the diffraction pattern. Is there any fundamental difference between interference and diffraction? Give reasons.
17. Describe and explain the phenomenon of diffraction due to a straight edge. Explain why the bands are neither equidistant nor equally illuminated.
18. Explain how plane transmission grating spectra can be used to determine the wavelength of sodium light. Compare the grating spectra with prism spectra. **(Nagpur, 2004; Kovempu, 2005)**
19. Explain the formation of spectra by a plane diffraction grating. What particular spectra would be absent if the width of the transparencies and opacities of the grating are equal?
20. Give the theory of the formation of the spectra of the various orders on the Rowland's circle by a concave refection grating.
21. Explain why a grating is designed to produce only two orders. Why the grating spectrum is called a normal spectrum?
22. Discuss the theory of a plane diffraction grating for oblique incidence. **(Nagpur, 2004)**
23. In a Fraunhofer diffraction due to a single slit, show that the intensity of the first secondary maximum is roughly 4.5% of that of the principal maximum.
24. Derive an expression for the intensity at a point in the Fraunhoffer type of diffraction produced by two nearby parallel narrow slits. Draw a diagram to indicate the intensity distribution in this case. **(Lucknow, 2004)**
25. What do you understand by diffraction of waves? Explain why the diffraction of sound is readily observed than the diffraction of light.
26. Explain why gratings with larger number of lines are preferred?
27. In a diffraction grating how are the spectral lines affected when the rulings are made closer.
28. Give the theory of a concave reflection grating. Deduce the condition to obtain focused spectra. Explain why a concave grating is preferred over a plane grating.
29. (*i*) Define grating element.
    (*ii*) Derive an expression for the resolving power of a plane transmission grating.
    (*iii*) Distinguish between grating spectra and prism spectra. **(Nagpur, 2004, 2005)**
30. (*i*) What is plane diffraction grating?
    (*ii*) Discuss its theory for normal incident of light.
    (*iii*) Describe in detail how plane transmission grating is used to determine the wavelength of light. **(Nagpur, 2004)**
31. Explain Fraunhoffer Diffraction through a single slit. Draw the intensity distribution curve. **(Bangalore, 2005)**
32. Give the construction and theory of a plane diffraction grating of the transmission type and obtain expressions for principal maxima, minima and secondary maxima. **(Garhwal, 2005)**
33. Discuss Fraunhofer diffraction at a circular aperture. **(RTMNU, Nagpur, 2010)**
34. Explain the construction of a zone plate. **(RTMNU, 2010)**
35. Describe the participle, construction and action of the Zone plate. **(GNDU, Amritsar, 2010)**
36. (*i*) What is plane transmission grating? Derive grating equation.
    (*ii*) How is it used to determine the wavelength of light? **(RTMNU, 2010)**

PROBLEMS FOR PRACTICE

1. Find the half angular width of the central bright maximum in the Fraunhoffer diffraction pattern of a slit of width 12×10^{-5} cm when the slit is illuminated by monochromatic light of wavelength 6000 Å. **[Ans. 30°]**
2. In a single slit diffraction pattern the distance between the first minima on either side of the central zero maximum is 4.4 mm as observed on a screen at a distance of 0.7 m. The wavelength of light used is 5890Å. Calculate the slit width.
3. A single slit illuminated by red light of 6000Å wavelength gives first order Fraunhoffer diffraction minima that subtends angle of 4° with the axis. How wide is the slit?
4. Light of wavelength 6000 Å is incident on a slit of width 0.30 mm. The screen is placed 2 m away from the slit. Find (a) the position of the first dark fringe and (b) the width of the central bright fringe. **[Ans. $\pm 4 \times 10^{-3}$ m , 8 mm]**
5. Light of wavelength 6000 Å is incident on a narrow slit. The screen is placed 2 m away from the slit. If the first minima lie 5 mm on either side of the central maximum, calculate the slit width. **[Ans. 0.24 mm]**
6. What is the highest order of spectrum which may be observed with monochromatic light of wavelength 5000 Å by means of grating with 5000 lines/cm? **(Nagpur, 2005)**
7. A narrow slit illuminated by light of wave-length 5890 Å. Calculate the distance between first and second dark band if a = 0.1 m and b = 0.5 m. **(Nagpur, 2005)**
8. Calculate the minimum number of lines per cm in a 2.5 cm wide grating which will just resolve the sodium lines in the second order. Given: λ_1= 5890 Å and λ_2 = 5896 Å.
9. A parallel beam of light incident normally on a grating having 4250 lines/cm and second order deviation is observed through 30°. Calculate the wavelength of light used. **(Nagpur, 2004)**
10. In Fraunhofer Diffraction Pattern due to a single slit, the screen is at a distance of 100 cm from slit of width 0.1 mm. The slit is illuminated by monochromatic light of wavelength 5893 Å. Calculate the distance of second minimum from central maximum. **(Nagpur, 2004)**
11. A parallel beam of sodium light is incident normally on a plane transmission grating having 6 x 10^5 lines per meter length. The first order spectrum is observed to be deviated through an angle of 20.7° from the normal. Calculate the wave length of light used. **(Bangalore, 2005)**
12. A plane grating has 15,000 lines per inch. Find the angle of separation of the 5048 Å and 5016 Å lines of helium in the second order spectrum. **(Kovempu, 2005)**
13. What is longest wavelength that can be observed in the fourth order for a transmission grating having 5000 lines per centimeter? **(Lucknow, 2004)**
14. How many orders will be visible if the wave length of light be 5644 A.U. and the number of lines per inch on the transmission grating are 15000? **(Lucknow, 2004)**
15. A grating has 4000 lines per cm. How many orders of the entire visible spectrum (4000 – 7000 Å) will be visible? **(Garhwal, 2005)**
16. A diffraction grating at normal incidence gives a line of wavelength 6000 Å in a certain order of spectrum. This line coincides with another line of wavelength 4800Å in next higher order of spectrum. If angle of diffraction is $\sin^{-1}$ [2/3], obtain the value of grating element, number of lines per cm. and order of spectrum. **(RTMNU, 2010)**
17. A parallel beam of light of wavelength 5000Å is incident normality slit of width 0.02 cm. A convex lens of focal length 20 cm is placed to observe the Fraunhofer diffraction. Calculate the distance between first two minima. **(RTMNU, 2010)**

19

CHAPTER

Resolving Power

At a Glance

- Resolving Power
- Rayleigh's Criterion
- Limit of Resolution of The Eye
- Limit of Resolution of a Convex Lens
- Resolving Power of Optical Instruments
- Criterion for Resolution According to Lord Rayleigh
- Resolving Power of a Telescope
- Resolving Power of a Microscope
- Ways of Increasing Resolution
- Magnification Versus Resolution
- Resolving Power of A Prism
- Resolving Power of a Plane Transmission Grating
- Michelson's Stellar Interferometer

19.1. RESOLVING POWER

When two objects or their images are very close to each other, they appear as one and it may not be possible for the eye to see them as separate. If the objects are not seen separately, then we say that the details are *not resolved* by the eye. Optical instruments are used to assist the eye in resolving the objects or images. The method adapted to seeing the close objects as separate objects is called **resolution**. The ability of an optical instrument to produce distinctly separate images of two objects located very close to each other is called its *resolving power*. We use the term 'resolving power' in two different senses. In case of microscopes and telescopes, we talk of *geometrical resolution* where the geometrical positions between two nearby objects are to be resolved and in case of spectroscopes we refer to *spectral resolution* where differences of wavelengths of light in a given source are to be resolved. *Resolving power is normally defined as the reciprocal of the smallest angle subtended at the objective of optical instrument by two point objects, which can just be distinguished as separate.*

19.2. RAYLEIGH'S CRITERION

When a beam of light from a point object passes through the objective of a telescope, the lens acts like a circular aperture and produces a diffraction pattern instead of a point image. This diffraction pattern is a bright disc surrounded by alternate dark and bright rings (see Fig. 19.1).

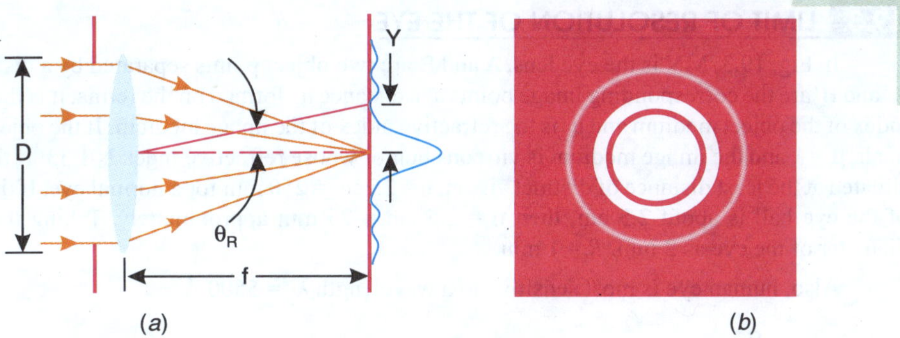

(*a*) The image of a point source of monochromatic light, formed by a lens of diameter *D*, is a diffraction pattern.

(*b*) The intensity pattern is a maximum at the center of the Airy disc.

Fig. 19.1

It is known as *Airy's disc*. If there are two point objects lying close to each other, then two diffraction patterns are produced, which may overlap on each other and it may be difficult to distinguish them as separate (see Fig. 19.2a).

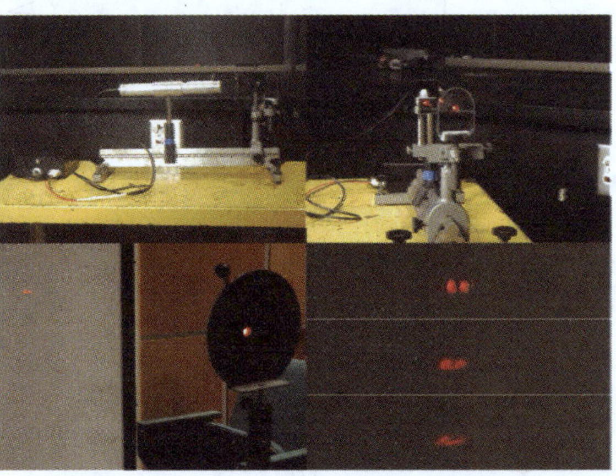

Rayleigh's criterion is usually applied to circular apertures.

To obtain the measure of the resolving power of an objective lens Rayleigh suggested that the two images of such point-objects lying close to each other may be regarded as separated if the central maximum of one falls on the first minimum of the other. In other words, when the central bright image of one falls on the first dark ring of the other, the two images are said to be just resolved (see Fig.19.2b). This is equivalent to the condition that the distance between the centers of the patterns shall be equal to the radius of the central disc. This is called the **Rayleigh criterion** for resolution and is also known as **Rayleigh's limit of resolution.**

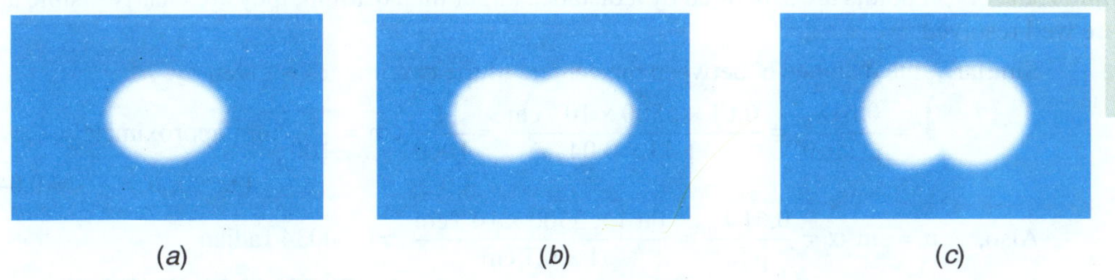

(*a*) The angular separation of the sources is so small that the images are not resolved.

(*b*) The images are just resolved.

(*c*) The images are well resolved.

Fig. 19.2

19.3. LIMIT OF RESOLUTION OF THE EYE

In Fig. 19.3, MN is the eye lens, A and B are two object points separated by a distance h and A′ and B′ are the corresponding image points at a distance h′ formed on the retina. μ is the refractive index of the object medium and μ′ is the refractive index of the image medium. If the object is placed in air, μ = 1 and the image medium is vitreous humor whose refractive index is 1.33. If the object is situated at the least distance of distinct vision, u = 25 cm = 250 mm for a normal eye. If the diameter of the eye ball is about 2.5 cm, then υ = 2.5 cm = 25 mm approximately. Taking the pupillary diameter of the eye as 2 mm, R = 1 mm.

Also, human eye is most sensitive to a wavelength λ_o = 5500 Å.

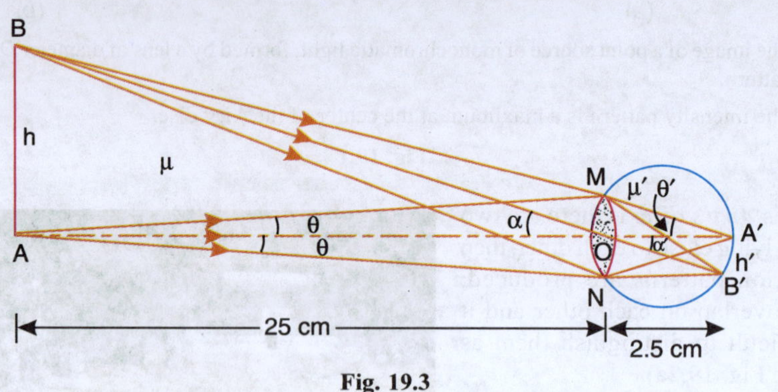

Fig. 19.3

From the ΔAMO for small angles of θ,

$$\sin\theta = \tan\theta = \frac{R}{u} = \frac{1}{250} = 0.004$$

Numerical aperture = μ sin θ

= 1 × 0.004 = 0.004

Applying Rayleigh's criterion, the minimum distance (h) between two just resolvable object points of equal intensity is given by,

$$h = \frac{0.61\lambda_o}{\mu \sin\theta} = \frac{0.61 \times 5500 \times 10^{-8}\ \text{cm}}{0.004} = \frac{1}{100}\ \text{cm} = \frac{1}{10}\ \text{mm approximately.}$$

It means that if the object is situated at the least distance of distinct vision from the eye (25 cm), the minimum separation between two nearby object points should be of the order of 0.1 mm. If the object points are separated by a distance larger than 0.1 mm, they are clearly visible and are well resolved.

Similarly, the distance h′ between the centers of the two images is given by,

$$h' = \frac{0.61\lambda_o}{\mu' \sin\theta'} = \frac{0.61 \times 5500 \times 10^{-8}\text{cm}}{1.33 \times 0.04} = \frac{1}{1000}\ \text{cm} = \frac{1}{100}\ \text{mm approximately.}$$

$$\left(\because \sin\theta' = \frac{1}{25} = 0.04\right)$$

Also, $$\alpha = \sin\alpha = \frac{0.61\,\lambda_o}{\mu R} = \frac{0.61 \times 5500 \times 10^{-8}\text{cm}}{1 \times 0.1\ \text{cm}} = 0.00034\ \text{radian}$$

= 1 minute of an arc (approx.)

The value of h′ (= 10^{-2}mm) is approximately equal to the distance between the cones in the fovea and thus the retinal structure is strikingly in accordance with the limit of resolution of the eye. Further, two point objects appear to be just resolved if the angle subtended by them at the eye is 1 minute of an arc. If the diameter of the pupil of the eye is smaller than 2 mm, the numerical aperture decreases and hence the value of h increases, i.e. two points will appear to be just resolved if the distance between the two is larger. Thus the resolving ability of the eye is decreased.

19.4. LIMIT OF RESOLUTION OF A CONVEX LENS

In Fig. 19.4 L is a convex lens. A and B are two object points and A′ and B′ are the corresponding image points. The distance between the object points is h and the distance between the image points is h′. The distance of the object points from the lens is u and the distance of the image points is v. μ and μ' are the refractive indices of the object and image media.

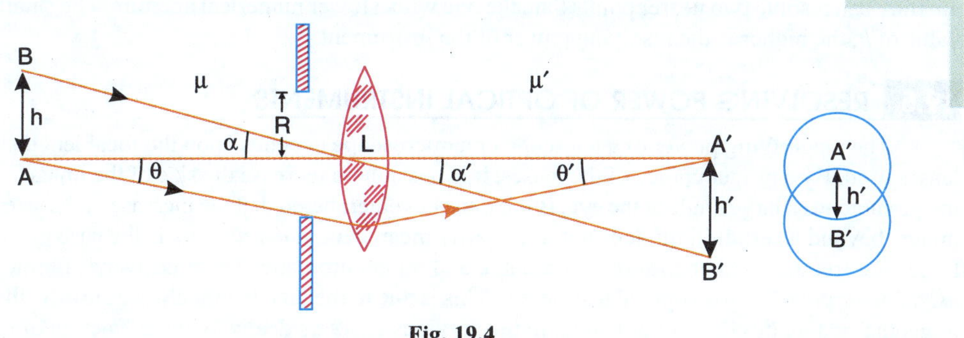

Fig. 19.4

R is the radius of the aperture kept in front of the lens (D is the diameter of the aperture). In the side figure, A′ and B′ are the centers of the central bright discs of the patterns of A and B. Let λ and λ' be the wavelengths of light in the object and image media and λ_o be the wavelength of light in vacuum. Then,

$$\lambda = \frac{\lambda_o}{\mu} \text{ and } \lambda' = \frac{\lambda_o}{\mu'}$$

According to Rayleigh criterion, if the two images are just resolved, distance between the centers of the two discs (h') is equal to the radius of either disc. If this condition is satisfied, then,

$$\sin\alpha = \frac{1.22\lambda}{D} = \frac{1.22\,\lambda_o}{\mu D} = \frac{1.22\,\lambda_o}{2\mu R} = \frac{0.61\,\lambda_o}{\mu R} \quad (19.1)$$

Similarly,
$$\sin\alpha' = \frac{0.61\,\lambda_o}{\mu' R} \quad (19.2)$$

From eq. (19.1),
$$\mu R \sin\alpha = 0.61\lambda_o \quad (19.3)$$

But in Fig. 19.4, when the angles α and θ are small, we write

$$\sin\alpha = \tan\alpha = \frac{h}{u}$$

and
$$\sin\theta = \tan\theta = \frac{R}{u} \text{ or } R = u\sin\theta$$

Substituting the values of $\sin\alpha$ and R in equation (19.3),

$$\mu u \sin\theta . \frac{h}{u} = 0.61\,\lambda_o$$

or
$$h = \frac{0.61\,\lambda_o}{\mu\sin\theta}$$

Similarly it can be shown that, $h' = \dfrac{0.61\,\lambda_o}{\mu\sin\theta'}$

Thus, according to the Rayleigh's criterion of resolution, the linear distance between two just resolved point objects is given by,

$$h = \frac{0.61\,\lambda_o}{\mu \sin\theta} \qquad (19.4)$$

and the distance between corresponding image points is given by,

$$h' = \frac{0.61\,\lambda_o}{\mu' \sin\theta'} \qquad (19.5)$$

The quantity $\mu \sin\theta$ in equation (19.4) is called the **numerical aperture** (N.A.) of the optical instrument. From equation (19.4), the distance between two just resolvable object points is inversely proportional to the numerical aperture of the instrument. An optical instrument with higher numerical aperture can resolve two nearer points than the one with a lower numerical aperture. The smaller the value of h, the higher is the resolving power of the instrument.

19.5. RESOLVING POWER OF OPTICAL INSTRUMENTS

The magnifying power of a telescope or a microscope depends upon the focal lengths of the lenses used. By a proper choice of the lenses, it is possible to increase the size of the image, i.e. the image subtends a large angle at the eye. But it must be remembered that the increase in the size of the image, beyond a certain limit does not necessarily mean gain in *detail*. This is the case even if the lenses are free from all aberrations, chromatic and monochromatic. There is always a limit to the *useful magnification* of an optical instrument. This is due to the fact that for a wave surface, the laws of geometrical optics do not hold good. In the previous chapters dealing with diffraction of light, it has been shown that the image of a point source is not a point but it is a diffraction pattern. With a circular aperture kept in the path of incident light, the diffraction pattern of a point source of light consists of a central bright disc surrounded by alternatively dark and bright rings. This is called **Airy's disc.**

If the lens diameter or the size of the aperture is large, the diffraction pattern of a point source of light is small. If there are two nearby point sources, the diffraction discs of the two patterns may overlap and the two images may not be distinguished. An optical instrument like a telescope or a microscope is said to have resolved the two point sources when the two diffraction patterns are well separated from each other or when the diffraction patterns are small so that in both the cases, the two images are seen as separate ones. **The ability of an optical instrument, expressed in numerical measure, to resolve the images of two nearby points is termed as its resolving power.**

A telescope gives us geometric resolution between two far away objects subtending a very angle. There we define resolving limit of a telescope = θ_m; where θ_m is the smallest angle resolved.

A microscope resolves the linear distance between two close objects. There we define, resolving limit of a microscope = x_m; where x_m is the smallest distance resolved.

In the case of a prism or a grating spectrograph, the term resolving power is referred to the ability of the prism or grating to resolve two nearby spectral lines so that the two lines can be viewed or photographed as separate lines.

Spectral resolving power = $\dfrac{\lambda}{\Delta\lambda_m}$ where $\Delta\lambda_m$ is the smallest difference of wavelength which can be resolved by the instrument at wavelength λ.

19.6. CRITERION FOR RESOLUTION ACCORDING TO LORD RAYLEIGH

To express the resolving power of an optical instrument as a numerical value, Lord Rayleigh proposed an arbitrary criterion. According to him, two nearby images are said to be resolved if the position of the central maximum of one coincides with the first secondary minimum of the other and vice versa. The same criterion can be conveniently applied to calculate the resolving power of a telescope, microscope, grating, prism, etc.

In Fig. 19.5, A and B are the central maxima of the diffraction patterns of two spectral lines of wavelengths λ_1 and λ_2. The difference in the angle of diffraction is large and the two images can be seen as separate ones. The angle of diffraction corresponding to the central maximum of the image B is greater than the angle of diffraction corresponding to the first minimum at the right of A. Hence the two spectral lines will appear to be well resolved (see Fig. 19.2c also).

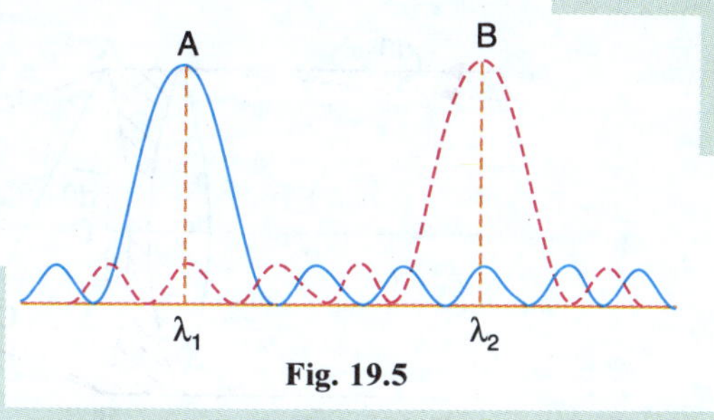

Fig. 19.5

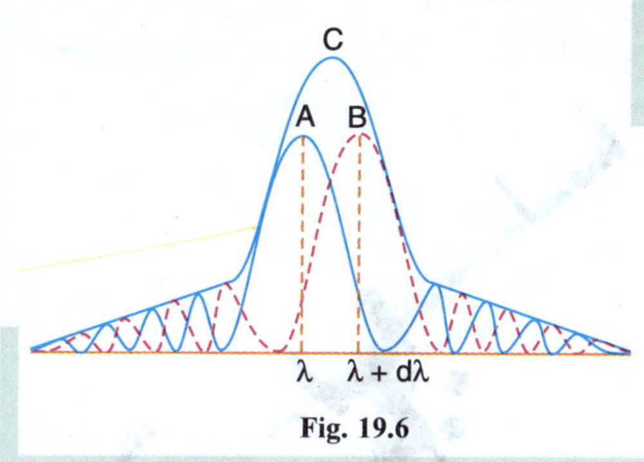

Fig. 19.6

In Fig. 19.6, the central maximum corresponding to the wavelengths λ and $\lambda + d\lambda$ are very close. The angle of diffraction corresponding to the first minimum of A is greater than the angle of diffraction corresponding to the central maximum of B. The two images overlap and they can not be distinguished as separate images. The resultant intensity curve gives the maximum as at C and the intensity of this maximum is higher than the individual intensities of A and B. Thus when the spectrograph is turned from A to B, the intensity increases, becomes maximum at C and then decreases. In this case, the two spectral lines are **not resolved** see Fig. 19.2(a) also.

In Fig. 19.7, the position of the central maximum of A (wavelength λ) coincides with the position of the first minimum of B (wavelength $\lambda + d\lambda$). Similarly, the position of the central maximum of B coincides with the position of the first minimum of A. Further, the resultant intensity curve shows a **dip** at C i.e. in the middle of the central maximum of A and B (here it is assumed that the two spectral lines are of the same intensity). The intensity at C is approximately 20 % less than that at A or B. If a spectrograph is turned from the position corresponding to the central image of A to the one corresponding to the image of B, there is noticeable decrease in intensity between the two central maxima. The spectral lines can be distinguished from one another and according to Rayleigh they are said to be **just resolved** (see Fig. 19.2 b also). Rayleigh's condition can also be stated as follows. **Two images are said to be just resolved if the radius of the central disc of either pattern is equal to the distance between the centers of the two patterns.**

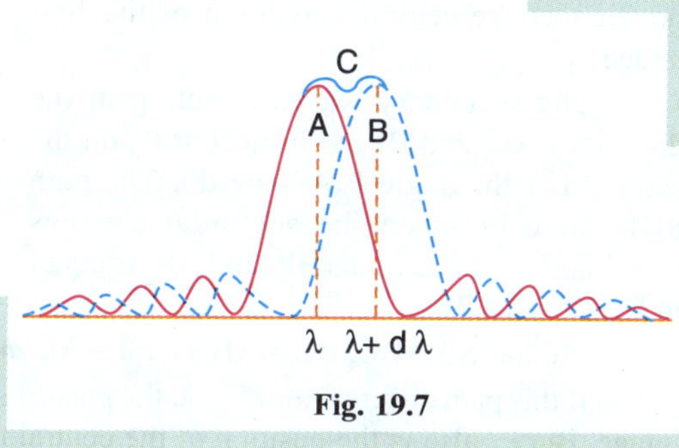

Fig. 19.7

19.7. RESOLVING POWER OF A TELESCOPE

Let a be the diameter of the objective of the telescope (Fig. 19.8). Consider the rays of light from two neighboring points of a distant object. The image of each point object is a Fraunhofer diffraction pattern.

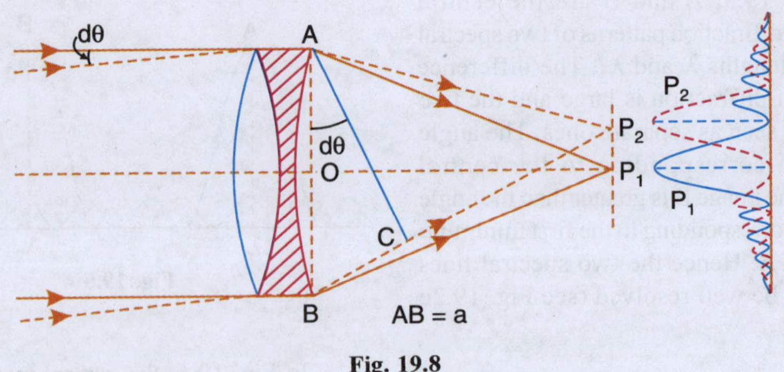

Fig. 19.8

Let P_1 and P_2 be the positions of the central maxima of the two images. According to Rayleigh, these two images are said to be resolved if the position of the central maximum of the second image coincides with the first minimum of the first image and vice versa. The path difference between the secondary waves traveling in the directions AP_1 and BP_1 is zero and hence they reinforce with one another at P_1. Similarly, all the secondary waves from the corresponding points between A and B will have zero path difference. Thus P_1 corresponds to the position of the central maximum of the first image.

Telescope.

The secondary waves traveling in the directions AP_2 and BP_2 will meet at P_2 on the screen. Let the angle P_2AP_1 be dθ. The path difference between the secondary waves traveling in the directions BP_2 and AP_2 is equal to BC (Fig. 19.8).

From the Δ ABC, $BC = AB \sin d\theta = AB.d\theta = a.d\theta$ (for small angles)

If this path difference $a.d\theta = \lambda$, the position of P_2 corresponds to the first minimum of the first image. But P_2 also is the position of the central maximum of the second image. Thus, Rayleigh's condition of resolution is satisfied if,

$$a.d\theta = \lambda \quad \text{or} \quad d\theta = \frac{\lambda}{a} \tag{19.6}$$

The whole aperture AB can be considered to be made of two halves AO and OB. The path difference between the secondary waves from the corresponding points in the two halves will be $\frac{\lambda}{2}$. All the secondary waves destructively interfere with one another and hence P_2 will be the first minimum of the first image. The equation $d\theta = \frac{\lambda}{a}$ holds good for rectangular apertures. For circular apertures, this equation, according to Airy, can be written as,

$$d\theta = \frac{1.22\,\lambda}{a} \tag{19.7}$$

where λ is the wavelength of light and a is the aperture of the telescope objective. The aperture is equal to the diameter of the metal ring which the objective lens is mounted. Here $d\theta$ refers to the **limit of resolution** of the telescope. **The reciprocal of $d\theta$ measures the resolving power of the telescope.**

$$\therefore \quad \frac{1}{d\theta} = \frac{a}{1.22\lambda} \tag{19.8}$$

$d\theta$ is also the angle subtended by the two distant object points at the objective.

From equation (19.8), it is clear that a telescope with large diameter of the objective has higher resolving power,

Thus, resolving power of a telescope can be defined as the reciprocal of the angular separation that two distant object points must have, so that their images will appear just resolved according to Rayleigh's criterion.

If f is the focal length of the telescope objective, then,

$$d\theta = \frac{r}{f} = \frac{1.22\,\lambda}{a}$$

or

$$r = \frac{1.22\,f\lambda}{a} \tag{19.9}$$

where r is the radius of the central bright image. The diameter of the first dark ring is equal to the diameter of the central image. The central bright disc is the Airy's disc.

From equation (19.9), if the focal length of the objective is small, the wavelength is small and the aperture is large, then the radius of the central bright disc is small. The diffraction patterns will appear sharper and the angular separation between the two just resolvable point objects will be smaller. Correspondingly, the resolving power of the telescope will be higher.

Let two distant stars subtend an angle of 1 second of an arc at the objective of the telescope.

1 second of an arc = 4.85×10^{-6} radian. Let the wavelength of light be 5500 Å. Then, the diameter of the objective required for just resolution can be calculated from the equation,

$$d\theta = \frac{1.22\,\lambda}{a}$$

or

$$a = \frac{1.22\,\lambda}{d\theta} = \frac{1.22 \times 5500 \times 10^{-8}\text{cm}}{4.85 \times 10^{-8}} = 13.9 \text{ cm (approximately)}$$

The resolving power of a telescope increases with increase in the diameter of the objective. With the increase in the diameter of the objective, the effect of the spherical aberration becomes appreciable. So, in the case of large telescope objectives, the central portion of the objective is covered with a stop so as to minimize the effect of spherical aberration. This, however, does not affect the resolving power of the telescope.

19.7.1. RELATION BETWEEN MAGNIFYING POWER AND RESOLVING POWER OF A TELESCOPE

The magnifying power of a telescope is given by,

$$M = \frac{D}{d} \tag{19.10}$$

where D is the diameter of the objective (entrance pupil) and d is the diameter of the exit pupil. The magnification of a telescope is said to be normal if d, the diameter of the exit pupil is equal to d_e, the diameter of the pupil of the eye. Therefore, the normal magnification of the telescope is given by,

$$M = \frac{D}{d_e} \tag{19.11}$$

Further, the limit of resolution of a telescope is given by, $d\theta = \frac{1.22\,\lambda}{D}$ (19.12)

And the limit of resolution of eye is given by, $d\theta' = \frac{1.22\,\lambda}{d_e}$ (19.13)

From eq. (19.12) and (19.13), we get

$$\frac{d\theta'}{d\theta} = \frac{\text{Limit of resolution of the eye}}{\text{Limit of resolution of the telescope}} = \frac{1.22\,\lambda}{de} \Big/ \frac{1.22\,\lambda}{D} = \frac{D}{de} = M$$

Thus, the product of normal magnifying power of a telescope and its limit of resolution is equal to the limit of resolution of the unaided eye.

Taking a pupil diameter of the eye as 2 mm and wavelength of light as 5500 Å, the angular separation ($d\theta'$) between two distant object points resolvable by the eye is given by,

$$d\theta' = \frac{1.22\,\lambda}{d_e} = \frac{1.22 \times 5500 \times 10^{-8}\ \text{cm}}{0.2\ \text{cm}}$$

$$= \frac{1.22 \times 5500 \times 10^{-8}\ \text{cm}}{0.2\ \text{cm}} \times \frac{180}{\pi} \times 60 \text{ minutes} = 1 \text{ minute of an arc (approx.)}$$

Similarly, the angular separation between two distant stars just resolvable by a telescope objective of diameter 254 cm is approximately $\frac{1}{20}$ th second of an arc.

i.e. $d\theta = \frac{1}{20}$ th second of an arc

$d\theta'$ = one minute of an arc = 60 seconds of an arc

Normal magnifying power of a telescope objective of diameter 254 cm

$$= \frac{d\theta'}{d\theta} = \frac{60}{1/20} = 1200$$

If the normal magnifying power is 1200, full advantage of the high resolving power of the telescope can be taken.

If two telescope objectives have the same focal length, the magnifying power will be the same in the two cases. But, the telescope with an objective of larger aperture has high resolving power than the one with a smaller aperture. With increase in size of the diameter of the objective of a telescope, the resolving power increases. Also with a large diameter objective, the radius of the central disc of the diffraction pattern is smaller and consequently the image obtained is sharp and more intense.

19.8. RESOLVING POWER OF A MICROSCOPE

The minimum distance by which two points in the object are separated from each other so that their images as produced by the microscope are just seen separate is called the *limit of resolution.* The reciprocal of limit of resolution is known as the *resolving power.*

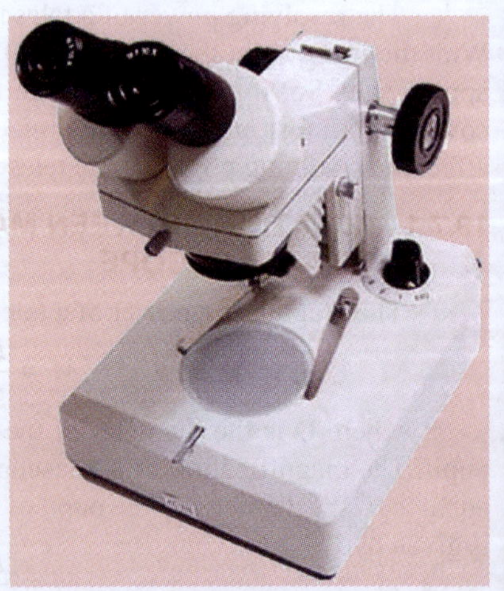

Microscope.

In Fig. 19.9, MN is the aperture of the objective of a microscope and A and B are two object points at a distance *d* apart. *A'* and *B'* correspond to diffraction patterns due to A and B. *A'* and *B'* are surrounded by alternate dark and bright diffraction rings. The two images are said to be just resolved if the position of the central maximum of B' also corresponds to the first minimum of the image of A'.

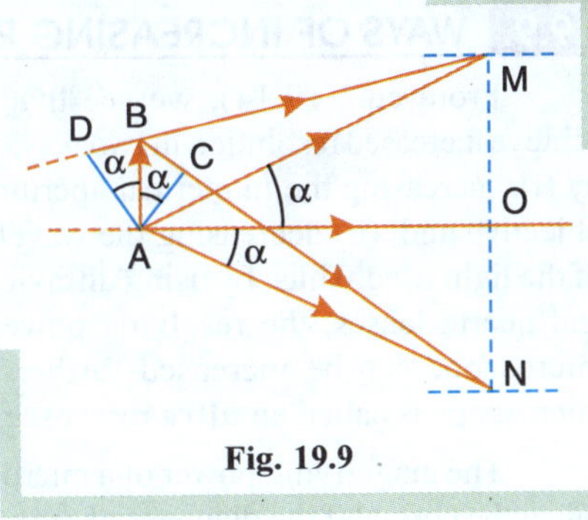

Fig. 19.9

The path difference between the extreme rays from the point B and reaching A' is given by,

$(BN + NA') - (BM + MA')$

But $NA' = MA'$

$\therefore$ Path difference = BN – BM

In Fig. 19.10, AD is perpendicular to DM and AC is perpendicular to BN.

$\therefore$ BN– BM = (BC + CN) – (DM – DB)

But CN = AN = AM = DM

$\therefore$ Path difference = BC + DB

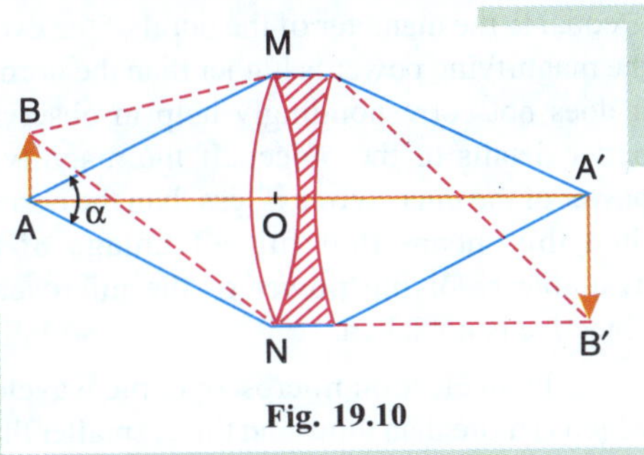

Fig. 19.10

From the Δ^S ACB and ADB,

$$BC = AB \sin\alpha = d \sin\alpha$$

and

$$DB = AB \sin\alpha = d \sin\alpha$$

$\therefore$ Path difference = $2d \sin\alpha$

If this path difference $2\,d \sin\alpha = 1.22\,\lambda$, then, A' corresponds to the first minimum of the image B' and the two images appear just resolved.

$\therefore \quad 2\,d \sin\alpha = 1.22\,\lambda$

or

$$d = \frac{1.22\,\lambda}{2 \sin\alpha} \tag{19.14}$$

Equation (19.14) derived above is based on the assumption that the object points A and B are self luminous. But actually, the objects viewed with a microscope are not self luminous but are illuminated with light from a condenser. It is found that the resolving power depends upon the mode of illumination. According to Abbe, the least distance between two just resolvable object points is given by,

$$d = \frac{\lambda_o}{2\mu \sin\alpha} \tag{19.15}$$

where λ_o is the wavelength of light in vacuum and μ is the refractive index of the medium between the object and the objective. The space between the objective and the object is filled with oil (cedar wood oil) in microscopes of high resolving power. This has two advantages. Firstly the loss of light by reflection at first lens surface is decreased and secondly the resolving power of the microscope is increased.

The expression $\mu \sin\alpha$ in eq. (19.15) is called the **numerical aperture** of the objective of the microscope and is a characteristic of the particular objective used. The highest value of numerical aperture obtained in practice is about 1.6. Taking the effective wavelength of white light as 5500 Å and $\mu \sin\alpha = 1.6$,

$$d = \frac{5500 \times 10^{-8}\text{cm}}{2 \times 1.6} = 1.72 \times 10^{-5}\ \text{cm}$$

where d is the linear distance between two just resolvable object points.

19.9. WAYS OF INCREASING RESOLUTION

From equ.(19.14), we see that we can achieve increased resolution in two ways, namely by (i) increasing the numerical aperture of the objective and (ii) decreasing the wavelength λ of the light used. Thus, by using ultraviolet light and quartz lenses, the resolving power of the microscope can be increased further. Such a microscope is called an **ultra microscope.**

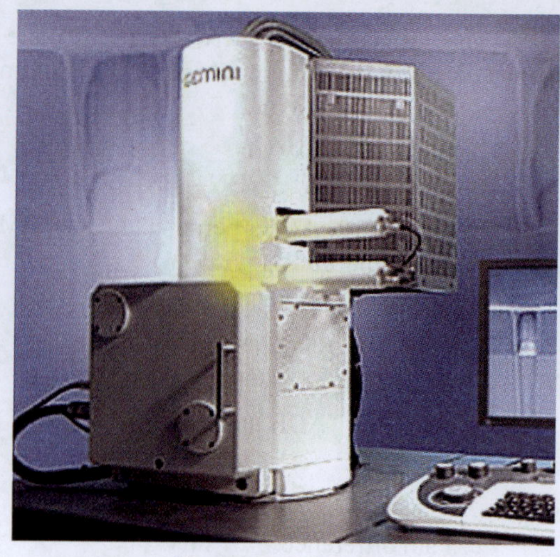

Electron microscope

The magnifying power of a microscope is said to be normal if the diameter of the exit pupil is equal to the diameter of the pupil of the eye. If the magnifying power is higher than the normal, it does not correspondingly help in observing better details of the object. If the magnifying power of the microscope is less than the normal, then this means that full advantage of the available resolving power of the microscope objective is not taken.

In an electron microscope, the wavelength of electron beam used is of the order of 0.12 Å, which is more than thousand times smaller than the wavelength of visible light. Hence, the resolving power of an electron microscope is much higher than that of an ordinary microscope. However, the numerical aperture of an electron microscope is smaller than that of an ordinary microscope.

19.10. MAGNIFICATION VERSUS RESOLUTION

The student should now be careful to notice the significance of resolution. In any optical instrument, the first magnification is done by the objective (because it faces the object). If the diffraction patterns of the different parts of the object are not well resolved at this stage, they can *never* be resolved further by any amount of magnification at the later stages. The patterns may get *magnified but not resolved*. High magnification alone will not reveal any further details of the object. Thus, for a given optical system, there will be a limit to useful magnification.

For the standard observer, the near point is about 25cm from the eye, and for many purposes one can assume that at this distance the limit of resolution is about 0.01 cm. It means that objects can be resolved by the eye when they subtend an angle of about one minute. Thus, for two optical images to be resolved by the observer, they must have an angular separation not less than this. Therefore, the magnifying power of the instrument should be such that the physically resolved images by the instrument subtend an angle of at least one minute at the observer's eye. Thus, for any visual instrument, there is a *minimum magnifying power* below which the eye will be unable to resolve the objects even though the images as formed by the instrument itself may clearly be resolved. This minimum magnifying power is usually called the **necessary magnifying power**. A magnifying power excessively greater than this is of no advantage, since the eye can never resolve images that the instrument does not resolve. This additional magnification is a disadvantage because it renders visible the diffuse nature of the Airy discs and one gets an impression that image quality is poor. Such additional magnification, greatly in excess of the necessary value, is called **empty magnification**.

19.11. RESOLVING POWER OF A PRISM

The term resolving power applied to the spectrographic devices (using a grating or a prism) signifies the ability of the instrument to form two separate spectral images of two neighboring wavelengths, λ and $\lambda + d\lambda$ in the wavelength region λ.

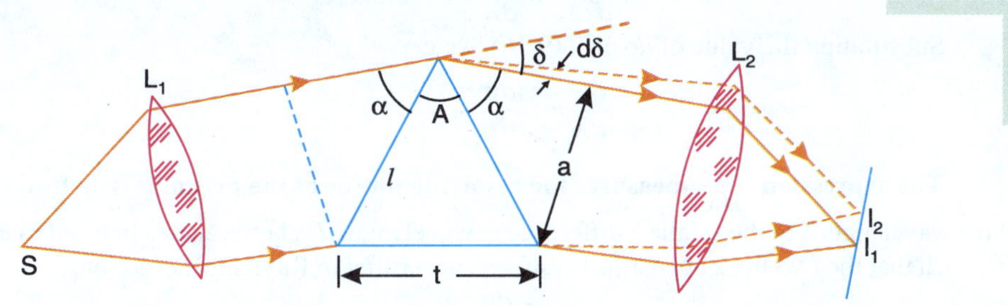

Fig. 19.11

In Fig. 19.11**,** S is a source of light, L_1 is a collimating lens and L_2 is the telescope objective. As two wavelengths λ and $\lambda + d\lambda$ are very close, if the prism is set in minimum deviation position would hold good for both the wavelengths. The final image I_1 corresponds to the principal maximum for the wavelength λ and I_2 corresponds to the principal maximum for the wavelength $\lambda + d\lambda$. I_1 and I_2 are formed at the focal plane of the telescope objective L_2. The face of the prism limits the incident beam to a rectangular section of width a. Hence, the Rayleigh criterion can be applied in the case of a rectangular aperture.

In the case of diffraction at a rectangular aperture, the position of I_2 will correspond to the first minimum of the image I_1 for wavelength λ_1 provided

$$a.d\delta = \lambda \text{ or } d\delta = \frac{\lambda}{a} \quad (19.16)$$

Here δ is the angle of minimum deviation for wavelength λ.

From the Fig. 19.11, $\alpha + A + \alpha + \delta = \pi$

$$\therefore \quad \alpha = \left[\left(\frac{\pi}{2}\right) - \left(\frac{A+\delta}{2}\right)\right]$$

$$\therefore \sin\alpha = \sin\left[\left(\frac{\pi}{2}\right) - \left(\frac{A+\delta}{2}\right)\right] = \cos\left(\frac{A+\delta}{2}\right)$$

But $\sin\alpha = \frac{a}{l}$; $\quad \therefore \cos\left(\frac{A+\delta}{2}\right) = \frac{a}{l}$ (19.17)

Also, $$\sin\frac{A}{2} = \frac{t}{2l} \quad (19.18)$$

In the case of a prism, $$\mu = \frac{\sin\frac{A+\delta}{2}}{\sin\frac{A}{2}}$$

$$\therefore \quad \sin\frac{A+\delta}{2} = \mu\sin\frac{A}{2} \quad (19.19)$$

Here μ and δ are dependant on wavelength of light λ.

Differentiating eq. (19.19) with respect to λ, we get

$$\frac{1}{2}\cos\left(\frac{A+\delta}{2}\right)\frac{d\delta}{d\lambda} = \frac{d\mu}{d\lambda}\left(\sin\frac{A}{2}\right) \quad (19.20)$$

Substituting the values of $d\delta$ from equations (19.17) and (19.18), we obtain

$$\frac{1}{2}\left(\frac{a}{l}\right)\frac{d\delta}{d\lambda} = \frac{d\mu}{d\lambda}\left(\frac{t}{2l}\right)$$

or $$a.\frac{d\delta}{d\lambda} = t.\frac{d\mu}{d\lambda} \qquad (19.21)$$

Substituting the value of $d\delta$ eq. (19.16), we get

$$\frac{\lambda}{d\lambda} = t \cdot \frac{d\mu}{d\lambda} \qquad (19.22)$$

The expression $\frac{\lambda}{d\lambda}$ measures the resolving power of the prism. It is defined as the ratio of the wavelength λ to the smallest difference in wavelength $d\lambda$, between this line and a neighboring line such that the two lines appear just resolved, according to Rayleigh's criterion.

So, resolving power of a prism $= t \cdot \frac{d\mu}{d\lambda}$ (19.23)

It means that the resolving power (i) is directly proportional to the length of the base of the prism and (ii) rate of change of refractive index with respect to wavelength for that particular material.

19.12. RESOLVING POWER OF A PLANE TRANSMISSION GRATING

The resolving power of a grating is defined as the ratio of the wavelength λ of any spectral line to the smallest difference in wavelength $d\lambda$, between this line and a neighboring line such that the two lines appear just resolved, according to Rayleigh's criterion.

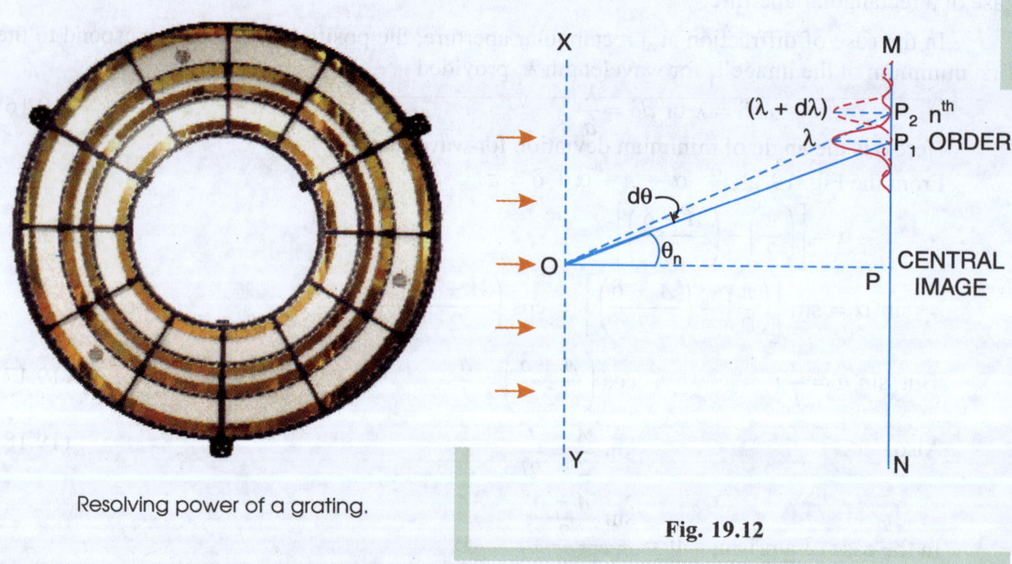

Resolving power of a grating.

Fig. 19.12

So resolving power of a grating $= \frac{\lambda}{d\lambda}$

In Fig.19.12, XY is the grating surface and MN is the field of view of the telescope, P_1 is n^{th} primary maximum of a spectral line of wavelength λ at an angle of diffraction θ_n. P_2 is the nth primary maximum of a second spectral line of wavelength $\lambda + d\lambda$ at a diffracting angle $\theta_n + d\theta$. P_1 and P_2 are the spectral lines in the nth order. These two spectral lines according to Rayleigh, will appear just resolved if the position of P_2 also corresponds to the first minimum of P_1.

The direction of the nth primary maximum for a wavelength λ is given by,

$$(a+b)\sin\theta_n = n\lambda \tag{19.24}$$

The direction of the nth primary maximum for a wavelength $\lambda + d\lambda$ is given by,

$$(a+b)\sin(\theta_n + d\theta) = n(\lambda + d\lambda) \tag{19.25}$$

These two lines will appear just resolved if the angle of diffraction $(\theta_n + d\theta)$ also corresponds to the direction of the first secondary minimum after the n_{th} primary maximum at P_1 (corresponding to the wavelength λ).This is possible if the extra path difference introduced is $\frac{\lambda}{N}$, where N is the total number of lines on the grating surface.

$$\therefore \qquad (a+b)\sin(\theta_n + d\theta) = n\lambda + \frac{\lambda}{N} \tag{19.26}$$

Equating the right hand sides of equations (19.25) and (19.26),

$$n(\lambda + d\lambda) = n\lambda + \frac{\lambda}{N} \quad \text{or} \quad nd\lambda = \frac{\lambda}{N} \quad \text{or} \quad \frac{\lambda}{d\lambda} = nN \tag{19.27}$$

The quantity $\frac{\lambda}{d\lambda} = nN$ measures the resolving power of a grating. Thus, the resolving power of a grating is independent of the grating constant. The resolving power is directly proportional to (i) the order of the spectrum and (ii) the total number of lines on the grating surface. For a given grating, the distance between the spectral lines is double in the second order spectrum than that in the first order spectrum.

The dispersive power of a grating is given by, $\frac{d\theta}{d\lambda} = \frac{n}{(a+b)\cos\theta} = \frac{nN'}{\cos\theta}$; and the resolving power of a grating is given by, $\frac{\lambda}{d\lambda} = nN$, where n is the order of the spectrum and N is the total number of lines on the grating surface. N' is the number of lines per cm on the grating surface. Here θ gives the direction of the nth principal maximum corresponding to a wavelength λ. From the above equation, it is clear that the dispersive power increases with increase in the number of lines per cm and the resolving power increases with increases in the total number of lines on the grating surface (i.e. the width of the grating surface).

High dispersive power refers to wide separation of the spectral lines whereas high resolving power refers to the ability of the instrument to show nearby spectral lines as separate ones.

19.13. MICHELSON'S STELLAR INTERFEROMETER

The smallest angular separation (θ) between two distant point sources for viewing the two images of the sources as separate with a telescope, is given by ,

$$\theta = \frac{1.22\,\lambda}{D} \tag{19.28}$$

where λ is the wavelength of light and D is the diameter of the objective of the telescope. Let the telescope objective be covered with a screen which is pierced with two parallel slits. Let the slit separation (d) be almost equal to the diameter of the telescope; a suitable value for $d = \frac{D}{1.22}$. Now let the telescope be directed towards a distant double star so that the line joining the two stars is perpendicular to the length of either slit. Interference fringes due to the double slit will be observed in the focal plane of the objective. The condition for the first appearance of fringes is given by,

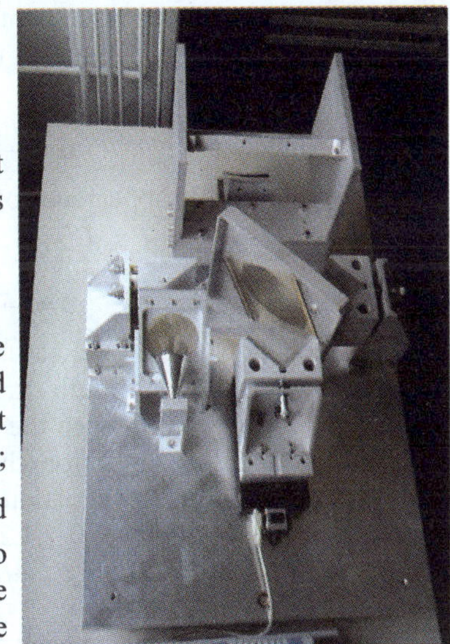

Michelson's Stellar Interferometer.

$$\alpha = \frac{\lambda}{2d} = \frac{1.22\,\lambda}{2D} = \frac{\theta}{2} \qquad (19.29)$$

where α is the angular separation between the two stars when the first disappearance of the fringes takes place. Similarly, for values of α given by the multiples of $\frac{\lambda}{2d}$, disappearance of the fringes can be observed. If the double slit is avoided and the observations are made directly, the multiples can be ruled out. The angular separation α is half the angle θ, where θ is the minimum angle of resolution of the telescope objective.

The method employing the double slit interference is used to measure the angular separation between two stars.

Michelson in 1920 successfully used this method to find the diameters of stars. The arrangement is known as Michelson's stellar interferometer (Fig. 19.13).

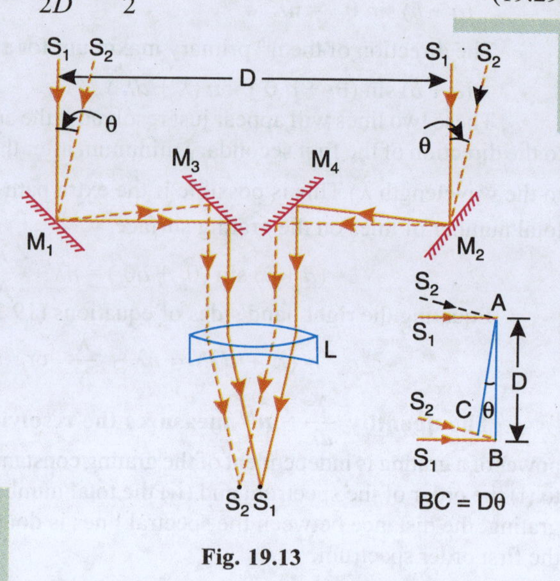

Fig. 19.13

It consists of four mirrors M_1, M_2, M_3 and M_4 arranged as shown in the figure. L is the objective of the telescope and the two slits are kept in the paths of light reflected from the mirrors M_3 and M_4. Let S_1 and S_2 be the ends of a diameter of the star. The paths of the rays of light from these two points S_1 and S_2 are shown in the figure. The mirrors M_1 and M_3 are parallel. The mirrors M_1 and M_2 are mounted on a girder and by sliding these mirrors, the distance D between the mirrors can be altered. The silvered faces of M_1 and M_3 (and M_3 and M_4) face each other. Interference fringes will be observed in the field of view of the telescope. The path difference between the rays of light from M_1 to L and M_2 to L is zero.

In the side figure, A is the point of incidence of the rays of light on the mirror M_2 and B is the point of incidence of the rays of light on the mirror M_1. The path difference between the rays traveling from S_2 (one end of the diameter of the disc of the star) and reaching A and B is equal to the distance BC.

From the ΔABC, $\quad \theta = \frac{BC}{D}$ or $BC = D\theta$.

For the first disappearance of the fringes, this path difference must be equal to 1.22 λ.

or $\quad D\theta = 1.22\,\lambda$

or $$\theta = \frac{1.22\,\lambda}{D} \qquad (19.30)$$

In equation (19.30), θ measures the angular diameter of the star. In one of the experiments of Michelson, using a 250 cm reflecting telescope at Mount Wilson observatory, the disappearance of the fringes was observed when the distance between the mirrors M_1 and M_2 was 306.5 cm. If the average wavelength of light from the star is assumed to be 5750 Å, the angular diameter of the star can be calculated from the equation

$$\theta = 1.22\lambda/D$$

$$\lambda = 5750\text{ Å} = 5750 \times 10^{-8}\text{ cm},\ D = 306.5\text{ cm}$$

$$\theta = \frac{1.22 \times 5750 \times 10^{-8}}{306.5} \times \frac{180}{\pi} \times 60 \times 60 \text{ second of an arc}$$

$$= 0.04718 \text{ second of an arc.}$$

WORKED OUT EXAMPLES

Example 19.1: Find the separation of two points on the moon that can be resolved by a 500 cm telescope. The distance of the moon is 3.8×10^5 km. The eye is most sensitive to light of wavelength 5500 Å.

Solution: The limit of resolution of a telescope is given by $d\theta = \dfrac{1.22\,\lambda}{a}$

Here $\lambda = 5500$ Å $= 5500 \times 10^{-8}$ cm, a = 500 cm

$$\therefore \quad d\theta = \frac{1.2 \times 5500 \times 10^{-8}\ \text{cm}}{500\ \text{cm}}$$

$$\therefore \quad d\theta = 13.42 \times 10^{-8}$$

Let the distance between the two points be x.

$$\therefore \quad d\theta = \frac{x}{R}$$

Here $R = 3.8 \times 10^{10}$ cm

$$x = R.\,d\theta = 3.8 \times 10^{10}\ \text{cm} \times 13.42 \times 10^{-8}$$

= 51.0 m.

Example 19.2 : Two pin holes 1.5 mm apart are placed in front of a source of light of wavelength 5.5×10^{-5} cm and seen through a telescope with its objective stopped down to a diameter of 0.4 cm. Find the maximum distance from the telescope at which the pin holes can be resolved.

Solution: Here, $\lambda = 5.5 \times 10^{-5}$ cm; a = 0.4 cm; x = 1.5 mm = 0.15 cm

Now, $$d\theta = \frac{1.22\,\lambda}{a}$$

$$d\theta = \frac{x}{d} \quad \therefore\ d = \frac{x\,a}{1.22\,\lambda} = \frac{0.15 \times 0.4}{1.22 \times 5.5 \times 10^{-5}}\ \text{cm} = 894.2\ \text{cm}$$

= 8.9 m

Example 19.3: Calculate the useful magnifying power of a telescope of 10 cm objective, assuming that the limit of resolution of the eye is 2 minutes of an arc. Wavelength of light used is 6000 Å.

Solution: Here diameter of the objective D = 10 cm, λ = 6000 Å = 6×10^{-5} cm.

Limit of resolution of the telescope, $d\theta = \dfrac{1.22\,\lambda}{D}$

$$= \frac{1.22 \times 6 \times 10^{-5}\ \text{cm}}{10} = 7.32 \times 10^{-6}\ \text{radian}$$

Limit of resolution of the eye, $d\theta' = 2$ minutes of an arc

$$= \frac{2}{60} \times \frac{22}{7 \times 180} = 582 \times 10^{-6}\ \text{radian}$$

Useful magnifying power of the telescope $= \dfrac{d\theta}{d\theta'} = \dfrac{582 \times 10^{-6}}{7.32 \times 10^{-6}} = \mathbf{79.5}$

Example 19.4: Sodium light of wavelength 5890 Å is used to view an object under a microscope. The aperture of the objective has a diameter of 0.9 cm.

(*a*) Calculate the limiting angle of resolution

(*b*) Using violet light, what is the maximum limit of resolution for this microscope?

Solution: (*a*) Limiting angle of resolution, $\theta_m = 1.22\left(\dfrac{\lambda}{d}\right)$

Here, $\lambda = 5890$ Å $= 5.89 \times 10^{-7}$ m, $d = 0.9$ cm $= 9 \times 10^{-3}$ m

$$\theta_m = 1.22\left(\frac{5.89 \times 10^{-7}}{9 \times 10^{-3}}\right) = \mathbf{7.98 \times 10^{-5}\ radian}$$

(*b*) The wavelength of violet light is 4000 Å.

The maximum limit of resolution for the microscope corresponds to the smallest angle.

Here, $\lambda = 4000$ Å $= 4 \times 10^{-7}$ m

$$\theta_m = 1.22\left(\frac{\lambda}{d}\right)$$

$$\therefore \quad \theta_m = 1.22\left(\frac{4 \times 10^{-7}}{9 \times 10^{-3}}\right) = \mathbf{5.42 \times 10^{-5}\ radian}$$

Example 19.5: Calculate the minimum thickness of the base of a prism which will just resolve the D_1 and D_2 lines of sodium. The refractive index of glass is 1.6545 for $\lambda = 6563$Å and 1.6635 for $\lambda = 5270$Å.

Solution: Resolving power of the prism $= \dfrac{\lambda}{d\lambda} = t\dfrac{d\mu}{d\lambda}$

$$\frac{d\mu}{d\lambda} = \left[\frac{1.6635 - 1.6545}{(6563 - 5270) \times 10^{-10}\,\text{m}}\right]$$

and $$\frac{\lambda}{d\lambda} = \frac{5893}{6}$$

now $$t = \frac{\lambda/d\lambda}{d\mu/d\lambda}$$

$$\therefore \quad t = \frac{5893 \times 1293 \times 10^{-10}\ \text{m}}{6 \times 0.0090} = \mathbf{14\ mm}$$

Example 19.6: Light is incident normally on a grating of total ruled width 5×10^{-3} m with 2500 lines in all. Calculate the angular separation of the two sodium lines in the first order spectrum. Can they be seen distinctly?

Solution: (*i*) Here, $N = 2500$, width of ruling $= 5 \times 10^{-3}$ m, $n = 1$,

$$\lambda_1 = 5890 \times 10^{-10}\ \text{m}, \quad \lambda_2 = 5896 \times 10^{-10}\ \text{m}$$

Number of lines per meter $= \dfrac{2500}{5 \times 10^{-3}} = 5 \times 10^5$

$$\therefore \quad (a + b) = \frac{1}{5 \times 10^5} = 2 \times 10^{-6}\ \text{m}$$

For the first order ($n = 1$),

$$\sin\theta_1 = \frac{n\lambda_1}{(a+b)} = \frac{\lambda_1}{(a+b)} = \frac{5890 \times 10^{-10}\ \text{cm}}{2 \times 10^{-6}\ \text{cm}} = 0.2945$$

$$\therefore \quad \theta_1 = 17°8'$$

and $$\sin\theta_2 = \frac{n\lambda_2}{(a+b)} = \frac{\lambda_2}{(a+b)} = \frac{5896 \times 10^{-10}\ \text{cm}}{2 \times 10^{-6}\ \text{cm}} = 0.2948$$

$$\therefore \quad \theta_2 = 17°9'$$

(*ii*) The resolving power of the grating, $\dfrac{\lambda}{d\lambda} = \dfrac{5890 \times 10^{-10}\ \text{cm}}{6 \times 10^{-10}\ \text{cm}} = 982$

As the total number of lines on the grating is 2500 which is more than 982, the lines can be seen distinctly.

QUESTIONS

1. What is resolving power? Explain Rayleigh criterion. **(Meerut, 2005)**
2. Define resolving power and dispersive power of a plane diffraction grating. Obtain expression for these in the case of a plane diffraction grating. **(Meerut, 2005)**
3. Derive an expression for an angular dispersion of a plane diffraction grating.
4. What is meant by resolving power of an optical instrument? Distinguish between resolving power and magnifying power of a microscope.
5. Explain clearly the importance of a very large number of sharp lines drawn over a large area on the grating.
6. Derive an expression for resolving power of a telescope and discuss the relationship between the resolving power and the magnifying power. **(Bangalore, 2005)**
7. Obtain expression for the resolving power of a plane diffraction grating. **(Garhwal, 2005)**
8. Distinguish between the resolving power and the dispersive power of a plane diffraction grating. **(Kovempu, 2005)**
9. Write short notes on : (*i*) Resolving power of a plane diffraction grating
   (*ii*) Resolving power and magnifying power of a microscope.
10. Explain what do you understand by the resolving power of an optical instrument? State Rayleigh's criterion for resolution and apply it to deduce an expression for the resolving power of a plane diffraction grating. **(Nagpur, 2004)**
11. Explain Rayleigh criterion for resolution and determine the resolving power of the telescope. **(Nagpur, 2004)**
12. Explain the concept of resolving power of an optical instrument. What is Rayleigh's criterion of limit of resolution? Derive an expression for the resolving power of a plane diffraction grating.
13. Explain the Rayleigh criterion for resolution and discuss it in relation to the resolving power of a microscope. **(Nagpur, 2005)**
14. Which optical instrument will you prefer to resolve two closely spaced spectral lines? Why ? **(Nagpur, 2005)**
15. What do you understand by resolving power of an optical instrument? Find an expression for resolving power of a plane transmission grating. **(Lucknow, 2004)**
16. Distinguish between dispersive and resoling power of a grating. **(Punjab, 2005)**
17. What is Airy's diffraction pattern? **(RTMNU, Nagpur, 2010)**
18. Derive the equation for resolving power of a microscope. **(RTMNU, 2010)**
19. Derive the expression for resolving power of a grating. **(RTMNU, Nagpur, 2010)**
20. Write a note on resolving power of a grating. **(A.P.University, 2010)**

PROBLEMS FOR PRACTICE

1. Calculate the resolving power of a Telescope whose objective lens has a diameter of 508 cm and $\lambda = 6000$ Å. **(Nagpur, 2005)**
2. Find the number of lines a grating should have to resolve the second order doublet having wavelength difference 6×10^{-10} m at 5893×10^{-10} m. **(Nagpur, 2005)**
3. Find the limit of resolution of a Laboratory Microscope having numerical aperture 1.2 when used for light of wavelength $\lambda = 6 \times 10^{-7}$ m. **(Nagpur, 2004)**
4. Calculate the aperture of the objective of a telescope which can resolve two stars separated by an angular distance of 4.84×10^{-6} m. radians. The wavelength of light is 5000 Å. **(Nagpur, 2004)**
5. Calculate the number of lines that a grating must have to resolve D_1 and D_2 lines of sodium in the second order. ($\lambda_1 = 5890$ Å and $\lambda_2 = 5896$ Å). **(Nagpur, 2004)**
6. A telescope of aperture 3 cm, is focused on a window 80 m away fitted with a wire mesh of spacing 2 mm. Will the telescope able to observe the wire- mesh?
7. Calculate the minimum spectral width for two wavelengths which can just be resolved in first order with a grating having 1100 lines when a light of wavelength 680 mm is used. **(Punjab, 2005)**

20

CHAPTER

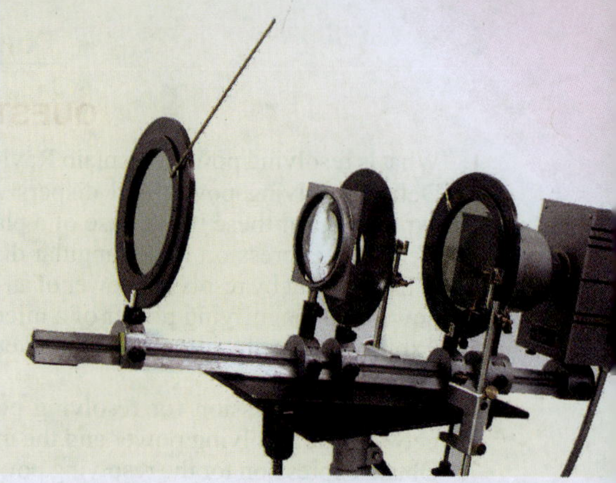

Polarization

20.1. INTRODUCTION

Interference and diffraction phenomena proved that light is a wave motion and enabled the determination of the wavelength. However, they do not give any indication regarding the character of the waves. Whether the light waves are longitudinal or transverse, or whether the vibrations are linear or circular cannot be deduced from the above two phenomena, as all kinds of waves under suitable conditions exhibit interference and diffraction. In 1816 Arago and Fresnel showed that light waves vibrating in mutually perpendicular planes do not interfere. In 1817 Thomas Young postulated that light waves are *transverse waves* and explained

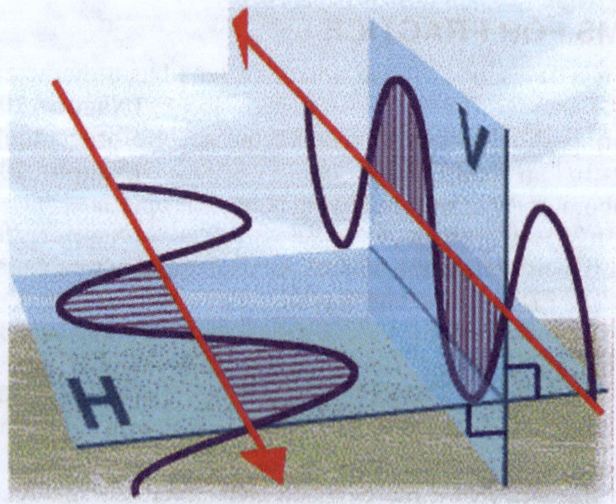

Polarization

At a Glance

- Introduction
- Unpolarized and Polarized Light
- Types of Polarization
- Polaroid Sheets
- Malus' Law
- Double Refraction in Calcite Crystal
- Effect of Polarizer on Light of Different Polarizations
- Experimental Determination of Principal Refractive Indices
- Phase Difference Between E-ray and O-ray
- Retarders
- Production of Circularly Polarized Light
- Babinet Compensator
- Double Image Polarizing Prisms
- Optical Activity
- Specific Rotation
- Experimental Verification of Fresnel's Theory
- Biquartz
- Electro-optic and Magneto-optic Effects
- Magneto-optic Effects
- Anisotropy Induced by Mechanical Strain
- Photoelasticity

the absence of interference between light waves polarized in mutually perpendicular planes. Thus, the existence of polarization property is a direct consequence of light being a transverse wave. Light coming from common light sources is unpolarized. It can be transformed into different types of polarization using optical devices. The state of polarization cannot be perceived by an unaided human eye. An understanding of polarization is essential for understanding the propagation of electromagnetic waves guided through wave-guides and optical fibres. Polarized light has many important applications in industry and engineering. One of the most important applications is in liquid crystal displays (LCDs), which are widely used in wristwatches, calculators, TV screens, etc.

20.2. POLARIZATION

Waves are basically of two types: (*i*) longitudinal waves and (*ii*) transverse waves.

(*i*) A wave in which particles of the medium oscillate to and fro along the direction of propagation is called a **longitudinal wave**. Waves produced on a spring and sound waves are examples of longitudinal waves. The longitudinal wave consists of alternate compressions and rarefactions, as shown in Fig.20.1 (c) & (d).

(*ii*) A wave in which every particle of the medium oscillates up and down at right angles to the direction of wave propagation is called a **transverse wave**. Ripples on water surface and waves on a rope are examples of transverse waves. The wave propagates in the form of alternating crests and troughs, as shown in Fig.20.1 (a) & (b).

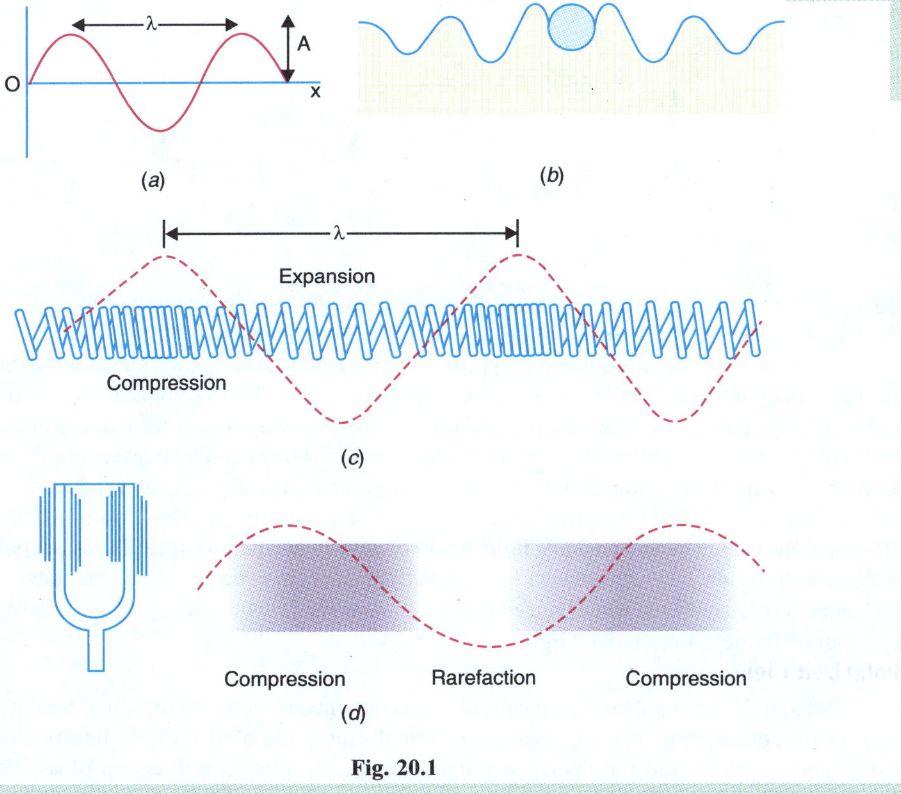

Fig. 20.1

In a longitudinal wave, all directions perpendicular to the wave propagation are equivalent. On the other hand, a preferential direction normal to the wave propagation exists in a transverse wave.

The preferential direction in a transverse wave is the direction of vibration of the particles and it differs from all other directions. The existence of a preferential direction for a transverse wave leads to the characteristic phenomenon known as polarization. Polarization is not found with longitudinal waves as they do not possess a directional property. Thus, polarization is specific to transverse waves.

Light waves are transverse waves consisting of electric and magnetic fields vibrating perpendicular to each other and to the direction of propagation. The vibrating electric field vector and the direction of propagation of the wave constitute a plane. There is an infinite number of such planes around the direction of propagation. In an ideal light wave, the vibrations of electric vector are confined to a single plane. In practice, light sources emit a mixture of light waves whose planes of vibration are randomly oriented about the direction of propagation. Such random orientation of vibration planes gives rise to symmetry about the wave propagation direction. As a result, the transverse nature of the wave gets concealed. The process of removing the symmetry and bringing in one-sidedness in the light wave is called **polarization**.

20.3. UNPOLARIZED AND POLARIZED LIGHT

Light wave is a transverse electromagnetic wave made up of mutually perpendicular, fluctuating electric and magnetic fields. Fig.20.2 (*a*) shows the electric field in the *xy*- plane, the magnetic field in the *xz*-plane and the propagation of the wave in the *x*- direction. The right-hand part of the diagram shows the variation of the electric field in space as the wave propagates. Traditionally, light wave is described by the **electric field vector**, ***E***, and accordingly, only the electric field vector is shown in Fig.20.2 (*b*).

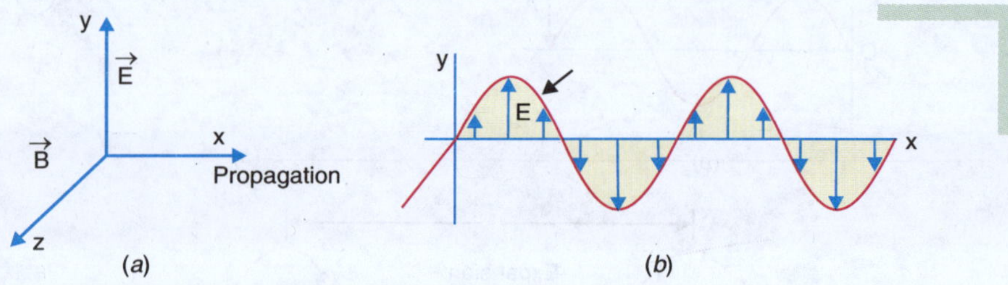

Fig. 20.2: A light wave is described by the electric field vector, *E*

As the electric field is a vector, it points in a particular direction in space. The **polarization** of an electromagnetic wave refers to the orientation of its electric field vector *E*. If we could view a light wave coming from an ordinary source towards us, then we would observe that the direction of *E* is randomly varying with time on a very fast scale. The light from an incandescent bulb, for example, emits a mixture of light waves with electric field components that change randomly on a scale of 10^{-14} s, almost as fast as the optical frequency itself. As a result, the **direction of oscillation of the electric field vector in an ordinary light beam occurs in all the possible planes perpendicular to the beam direction,** as illustrated in Fig.20.3. A light wave, in which **E**-vector oscillates in more than one plane, is referred to as **unpolarized light**. Light emitted by the sun, by an incandescent lamp, or by a candle flame is unpolarized light.

Polarized Light

Polarized light is not produced naturally. It is obtained by converting natural light into polarized light using optical elements. The process of transforming unpolarized light into polarized light is polarization. A polarized light wave is a light wave with a definite direction of oscillation of the *E*- vector, which occurs in a *single plane* or in *some specific way*. For example, the wave in Fig.20.2 (b) is a polarized wave. **Polarized light** is the light that contains waves that only fluctuate in one specific plane.

We designate the plane created by the direction of oscillation of the electric field vector ***E*** and the direction of propagation of the beam as the **plane of polarization of light wave**. Thus, the *xy*-plane is the plane of polarization in Fig.20.2 (*b*).

20.4. NATURAL LIGHT IS UNPOLARIZED LIGHT

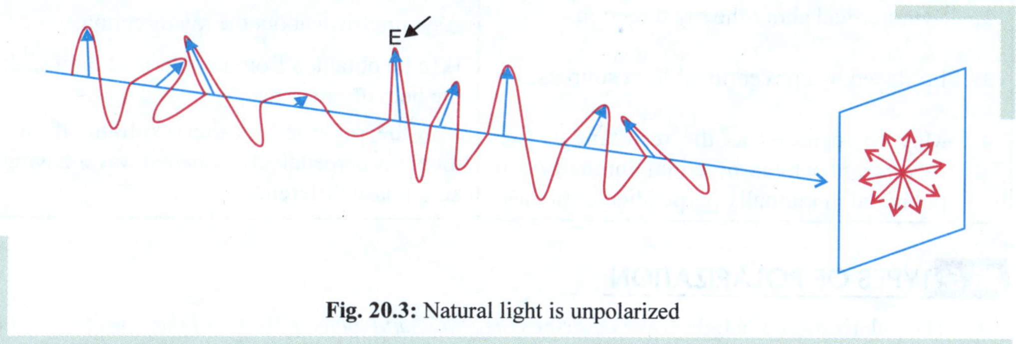

Fig. 20.3: Natural light is unpolarized

We mentioned earlier that light from the ordinary light sources is unpolarized. Let us find out the reason. We know that atoms emit light. Any light source consists of a very large number of atomic emitters. Each atom radiates, at a specific instant, a **wave packet** (also known as a **wave train**) that lasts for about 10^{-8} s. Light radiated by a source is a mixture of wave packets emitted by different atoms at different instances (see Fig.20.3). Individual wave packets will be polarized, but each wave packet has its own polarization, and they are not correlated in any way. There is no continuity of plane of polarization and the plane of polarization varies from wave packet to wave packet in a completely random manner. The polarization of the waves determined at any particular spot will fluctuate randomly, and very rapidly, with no preferred direction. Any direction is equally likely, and the usual graphical representation of ordinary light is shown in the right-hand figure in Fig.20.4 (*a*). Fig.20.4 (*a*) is a fictitious diagram. It only implies that the natural light consists of electric field vectors of many possible orientations lying at different angles between 0 and 360° and hence symmetrically distributed about the direction of propagation.

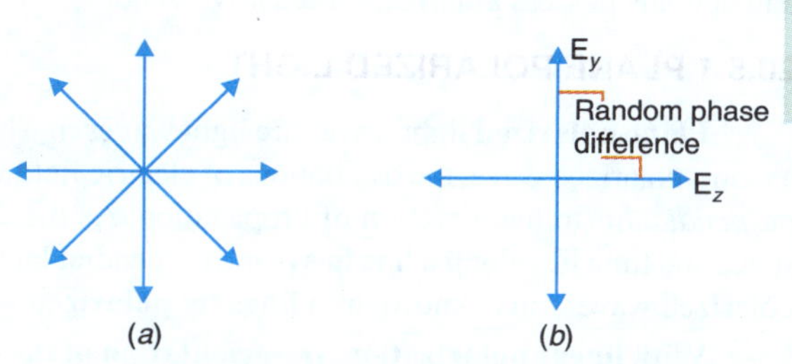

Fig.20.4: (*a*) Pictorial representation of Natural light. (*b*) Unpolarized light is viewed as a combination of *incoherent* vertically and horizontally polarized waves.

The concept of unpolarized light is rather difficult to visualize. In general, it is looked upon as consisting of an average of half of its vibrations horizontally polarized and half of its vibrations vertically polarized. Due to the random distribution of the optical vectors, the amplitude of the vertical and horizontal component vectors are taken as equal; however, the **two components are incoherent**, *i.e.* have a randomly changing phase difference. In view of this, unpolarized is pictorially represented by two electric vectors as shown in Fig.20.4 (*b*).

A comparison of Unpolarized and Polarized light

	Unpolarized light	Polarized light
1.	Consists of waves with planes of vibration equally distributed in all directions about the ray direction.	Consists of waves having their electric vector vibrating in a single plane normal to ray direction.
2.	Symmetrical about the ray direction	Asymmetrical about the ray direction.
3.	Produced by conventional light sources.	Is to be obtained from unpolarized light with the help of polarizers.
4.	May be regarded as the resultant of two *incoherent* waves of equal intensity but polarized in mutually perpendicular planes.	May be regarded as the resultant of two mutually perpendicular *coherent* waves having zero phase difference.

20.5. TYPES OF POLARIZATION

The polarization of a light wave describes *the shape and locus of the tip of the E vector* (in the plane perpendicular to the direction of propagation) *at a given point in space as a function of time*. Depending upon the locus of the tip of the *E* vector, light may exhibit three different states of polarization. They are

(*i*) plane or linear polarization,

(*ii*) elliptical polarization and

(*iii*) circular polarization.

Apart from these, the light may also be partially polarized.

An unaided human eye cannot identify the state of polarization of light. However, some insects and animals possess polarization sensitive vision.

20.5.1 PLANE POLARIZED LIGHT

Plane polarized light waves are light waves in which the oscillations occur in a single plane. In a plane-polarized wave, the oscillations of electric field vector *E* are strictly confined to a single plane *perpendicular* to the direction of propagation. As the direction of the field vector at some point in space and time lies along a line in a plane perpendicular to the direction of wave propagation, a plane-polarized wave is also known as a **linearly polarized** wave.

With linear polarization, the orientation of the *E*-vector stays constant at a point in space. That is, the direction of *E* does not vary with time, but its magnitude varies sinusoidally with time. If the field is pointing either up or down, we call it **vertical polarization**, and if it's pointing either right or left, we call it **horizontal polarization**. Electric fields are not restricted to pointing exactly along vertical or horizontal axes, but can be at any arbitrary angle to those axes. Linearly polarized light, polarized at any arbitrary angle, may be regarded as a combination of horizontally and vertically polarized light, with appropriate amplitude, and which are oscillating **in phase** or **180° out**

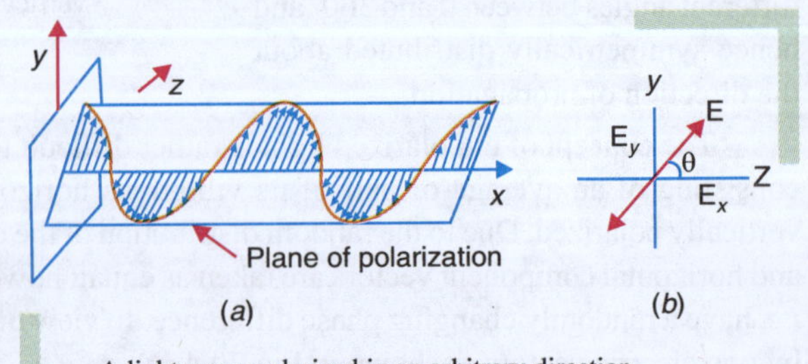

A light wave polarized in an arbitrary direction

Fig.20.5

of phase. The **key point** is that the two component waves are **coherent**.

Let ***x*** be the direction of travel of the light and, *y* and *z* be directions in the plane of the electric field. The electric field makes a constant angle to the *z*-direction, as shown in the Fig.20.5 (*a*). The wave in Fig.20.5 (*a*) is the resultant wave due to superposition of two **coherent** linearly polarized waves, oscillating in phase, as shown in Fig.20.6.

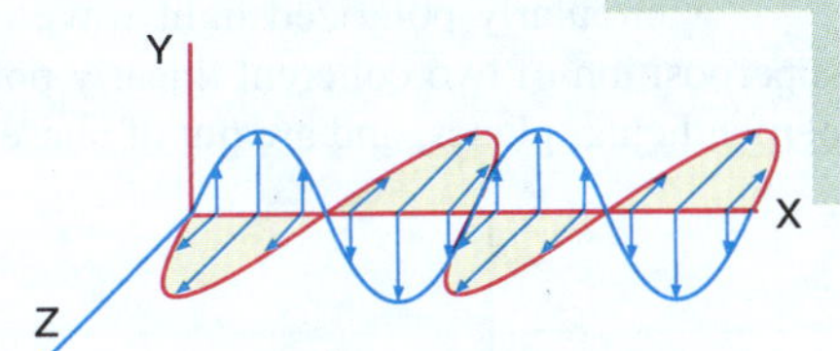

A linearly polarized wave is viewed as a combination of *coherent* vertically and horizontally polarized waves.

Fig.20.6

Representation of linearly polarized light in diagrams

Linearly polarized light is represented in diagrams as shown in Fig.20.7. When the electric field vector oscillates horizontally in a direction perpendicular to the plane of the paper, the light wave is represented by dots (Fig.20.7 *a*). When the electric field vector oscillates vertically in the plane of the paper, the light wave is represented by arrows, as shown in Fig.20.7 (*b*).

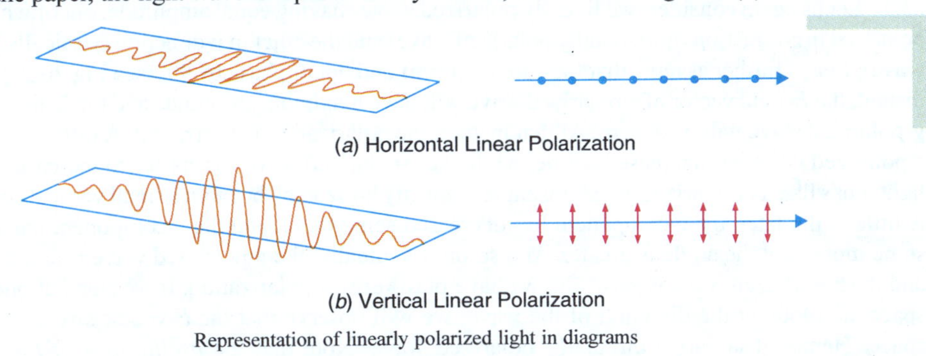

(*a*) Horizontal Linear Polarization

(*b*) Vertical Linear Polarization

Representation of linearly polarized light in diagrams

Fig. 20.7

20.5.2 CIRCULARLY POLARIZED LIGHT

A light wave is said to be ***circularly polarized***, if in the course of wave propagation, the magnitude of the electric vector *E* stays constant but it rotates at a constant rate about the direction of propagation and sweeps a circular helix in space, as shown in Fig.20.8. This is a picture in terms of the **space variation** of *E***.** Alternately, if we could see the wave advancing towards our eyes, we would find that the tip of the *E* vector tracing a circle in space, completing one revolution within one wavelength. Hence, the state of polarization is called *circular polarization*. In, circularly polarized light, there is no preference to specific direction of oscillation.

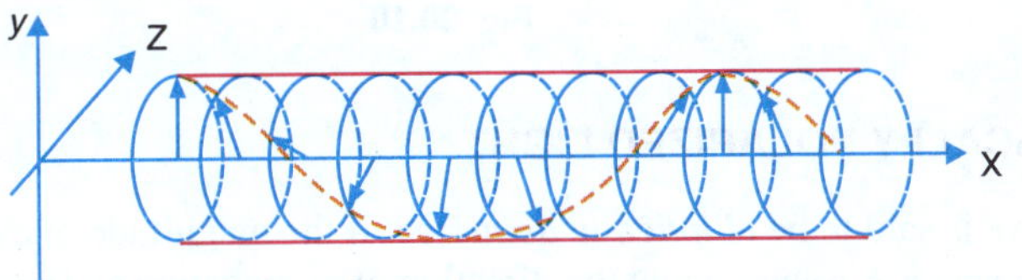

Space variation of *E*; *E* vector sweeps a circular helix in space

Fig. 20.8

A circularly polarized light wave may be regarded as the resultant wave produced due to superposition of two coherent linearly polarized waves of *equal amplitude* oscillating in mutually perpendicular planes, and are out of phase by 90°.

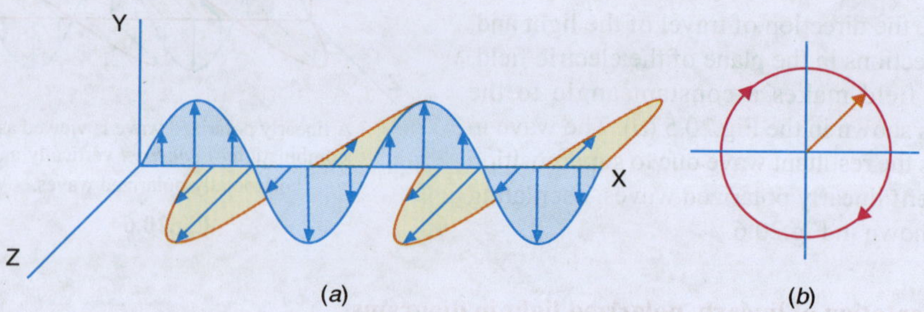

Circularly polarized wave is a combination of horizontally polarized and vertically polarized waves that are out of phase by 90° and having equal amplitudes.

Fig. 20.9

Let us again consider two linearly polarized waves having equal amplitude, out of which one is polarized in *z*-direction (horizontally polarized wave) and the other in *y*-direction (vertically polarized wave). Let us further assume that they are **coherent** and out of phase by 90° (see Fig.20.9a). At some instant, the *E*-field vector of *z*-polarized wave will have maximum amplitude and the *E*-field vector of y-polarized wave will be at zero. At that instant, the polarization is horizontal. A little time later, the *z*-polarized wave has decreased a little, while the *y*-polarized wave has started to increase. Then, the light looks like it is polarized at a slight angle— mostly horizontal, but with a small vertical component. A little while later, the *z*-component has decreased some more, and the *y*-component has increased some more, and the angle is greater. And so on. Eventually, the *y*-polarized wave is at a maximum, and the *z*-polarized wave is zero, and we have pure vertical polarization. If we stand at one point in space, and look at the direction of the wave, we will observe that the *E*-vector sweeps a circle in space. Hence, it is called circularly polarized wave. Note that the *oscillations of the resultant E-vector do not take place in a single plane.*

If the rotation of the tip of *E* is clockwise as seen by an observer looking back towards the source, then the wave is said to be **right-circularly polarized** (Fig.20.10 *a*). If the tip of *E* rotates anti-clockwise, then the wave is said to be **left-circularly polarized** (Fig.20.10. *b*). This is a description of circular polarization in terms of the **time variation** of *E*.

(a) Right circular polarization

(b) Left circular polarization

Fig. 20.10

20.5.3 ELLIPTICALLY POLARIZED LIGHT

A light wave is said to be ***elliptically polarized***, if the magnitude of electric vector *E* changes with time and the vector *E* rotates about the direction of propagation and sweeps a flattened helix in space, as shown in Fig.20.11. This is a description of elliptically polarized light in terms of the **space variation** of ***E***. Alternately, if we imagine that we are looking at the light wave advancing towards us, we would observe that the tip of the *E* vector traces an ellipse in space. Hence, it is called elliptically polarized light. This is a description of elliptical polarization in terms of the **time variation** of *E*.

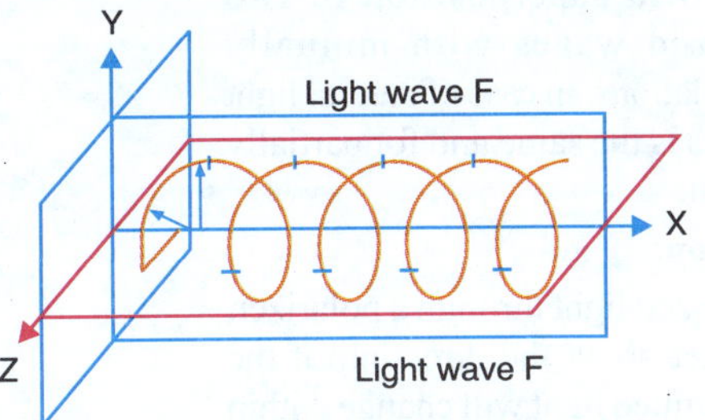

Space variation of **E**; **E** vector sweeps a flattened helix in space

Fig. 20.11

An elliptically polarized light wave may be regarded as the resultant wave produced due to superposition of two **coherent** linearly polarized waves of different amplitudes, oscillating in mutually perpendicular planes and are out of phase. If waves of differing amplitude are related in phase by 90°, or if the relative phase difference is other than 90° then the resultant light wave is elliptically polarized.

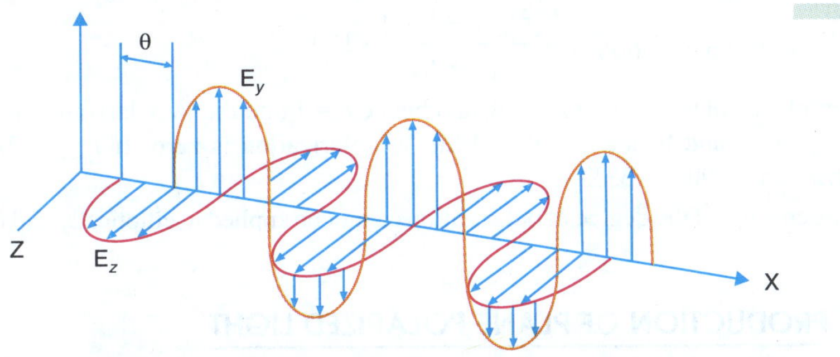

Elliptically polarized wave is a combination of horizontally polarized and vertically polarized waves that are of different amplitudes and out of phase by θ

Fig. 20.12

Let us consider two linearly polarized waves having different amplitudes, out of which one is polarized in *x*-direction and the other in y-direction. Let us further assume that they are **coherent** and out of phase by an arbitrary angle θ (see Fig.20.12). The *E*-field vector of *Z*-polarized wave will have maximum amplitude at times when the *E*-field vector of *y*-polarized wave is a minimum and vice versa (Fig.20.12). Then the oscillations of the resultant *E*-vector do not take place in a single plane. The magnitude of resultant *E*-vector varies at each point in space and the overall rotation of the *E*-vector has the appearance of a flattened helix.

When we are looking back towards the source, if the rotation of *E* vector occurs clockwise, it is said to be a **right-elliptically-polarized** wave. If it rotates anti-clockwise, it is said to be a **left-elliptically polarized** wave.

20.5.4 PARTIALLY POLARIZED LIGHT

Usually, light is neither totally polarized nor unpolarized but a mixture of the two types. It can be viewed as a mixture of plane polarized light and unpolarized light. Partially polarized light is represented as shown in Fig.20.13.

Partially polarized light, like natural light, can be represented in the form of a superposition of two incoherent plane-polarized waves with mutually perpendicular planes of oscillations. In case of natural light the amplitude of these waves is the same and for partially polarized light, it is different.

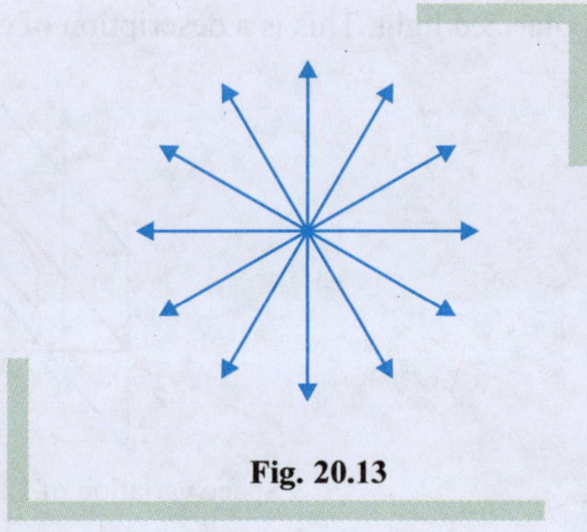

Fig. 20.13

Degree of Polarization:

If we pass partly polarized light through a polarizer, and if we rotate the polarizer about the direction of the ray, the intensity of the transmitted light will change within the limits from I_{max} to I_{min}. The transition from one of these values to the other will occur upon rotation through an angle of 90°. We define the degree of polarization with the help of the following expression.

$$P = \frac{I_{max} - I_{min}}{I_{max} + I_{min}} \quad \text{... (20.1)}$$

% polarisation

$$\% \text{ polarization} = \frac{(I_{max} - I_{min})}{(I_{max} + I_{min})} \times 100 \quad \text{... (20.2)}$$

For plane polarized light $I_{min} = 0$, and hence $P = 1$ and the % polarization is 100%. For natural light, $I_{max} = I_{min}$, and hence $P = 0$ and the % polarization is zero. If $I_{max} = 2I_{min}$, $P = 0.33$ and % polarisation = 100/3 = 33%.

The concept of the degree of polarization cannot be applied to elliptically and circularly polarized light.

20.6. PRODUCTION OF PLANE POLARIZED LIGHT

We now study the methods of producing plane-polarized light. Plane polarized light may be produced from unpolarized light using the following five optical phenomena:

(*i*) reflection, (*ii*) refraction, (*iii*) scattering, (*iv*) selective absorption (dichroism), and (*v*) double refraction. Out of these five, the phenomena of selective absorption and double refraction are helpful in practical production of plane polarized light.

20.6.1 POLARIZATION BY REFLECTION FROM DIELECTRIC SURFACES

E.L.Malus, French engineer discovered in 1808 polarization of light by reflection. He noticed that when natural light is incident on a smooth surface, at a certain angle the reflected beam is plane polarized. The extent to which polarization occurs is dependent upon the angle at which the light is incident on the surface and upon the material, which the surface is made of. Metallic surfaces reflect light with a variety of vibrational directions; such reflected light is unpolarized. However, light that is specularly reflected from dielectric surfaces, such as asphalt roadways, water etc, is linearly polarized. If the extent of linear polarization is large, a person perceives **glare** from such surfaces. On bright sunny days, the glare caused by sunlight on a roadway or a field of snow, may be almost blinding to the human eye.

When light wave is incident on a boundary between two dielectric materials, part of it is reflected, and part of it is transmitted.

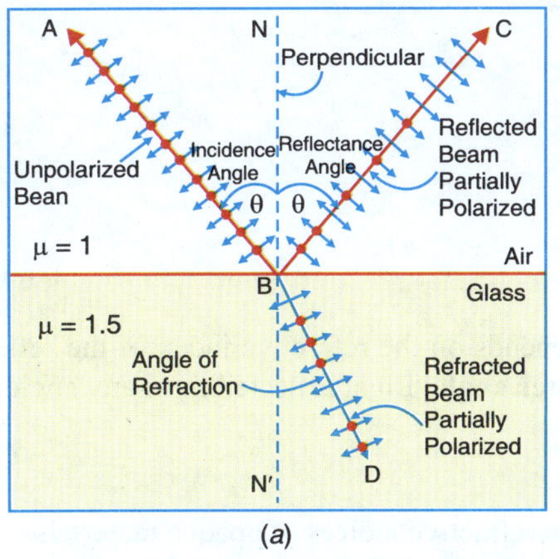

(a)

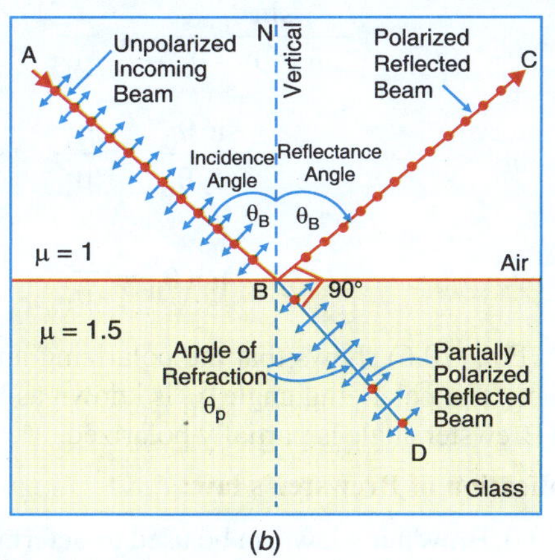

(b)

(*a*) Reflection and Refraction at the surface between two media (*b*) Brewster Law

Fig. 20.14

Fig.20.14 (*a*) shows an unpolarized light beam AB incident on a glass surface. The incident ray AB and the normal NBN′ define the plane of incidence. The electric vector E of the ray AB can be resolved into two components, one perpendicular to the plane of incidence and the other lying in the plane of incidence. The perpendicular component is represented by **dots** and is called the ***s*-component**. The parallel component is represented by the **arrows** and is called the ***p*-component**. In case of completely unpolarized light the two components are of equal magnitude. At a particular angle θ_B, the reflection coefficient for *p*-component goes to zero and the reflected beam does not contain any *p*-component (see Fig.20.14 *b*). It contains only *s*-component and is totally *plane polarized.* The angle θ_B is called the **polarizing angle** or **Brewster's angle.**

This particular method of polarizing light is not advantageous, as the intensity of the reflected beam is very small.

20.6.1.1 Brewster's Law

Sir David Brewster performed a series of experiments on the polarization of light by reflection at a number of surfaces. He found that the polarizing angle depends upon the refractive index of the medium. In 1892, Brewster proved that **the tangent of the angle at which polarization is obtained by reflection is numerically equal to the refractive index of the medium.** If θ_B is the angle and m is the refractive index of the medium, then

$$\mu = \tan \theta_B \quad ...(20.3)$$

This is known as Brewster's law.

If natural light is incident on a smooth surface at the polarizing angle, it is reflected along *BC* and refracted along *BD*, as shown in Fig.20.14 (*b*). Brewster found that the maximum polarization of reflected ray occurs when it is at right angles to the refracted ray. It means that $\theta_B + r = 90^\circ$.

$$\therefore \quad r = 90^\circ - \theta_B \quad ...(20.4)$$

According to Snell's law,

$$\frac{\sin \theta_B}{\sin r} = \frac{\mu_2}{\mu_1} \quad ...(20.5)$$

where μ_2 is the absolute refractive index of reflecting surface and μ_1 is the refractive index of the surrounding medium. It follows from equ.(20.4) and equ.(20.5) that

$$\frac{\sin \theta_B}{\sin (90° - \theta_B)} = \frac{\mu_2}{\mu_1}$$

or $$\frac{\sin \theta_B}{\cos \theta_B} = \frac{\mu_2}{\mu_1}$$

∴ $$\tan \theta_B = \frac{\mu_2}{\mu_1} \qquad \text{... (20.6)}$$

Equ.(20.6) shows that the polarizing angle depends on the refractive index of the reflecting surface. The polarizing angle q_B is known as **Brewster angle**. Light reflected from any angle other than Brewster angle is partially polarized.

Application of Brewster's law:

(*i*) Brewster's law can be used to determine the refractive indices of opaque materials.

(*ii*) It helps us in calculating the polarizing angle necessary for total polarization of reflected light for any material if its refractive index is known. However, the law is not applicable for metallic surfaces.

(*iii*) In gas lasers it is common to arrange two glass windows at the two ends of the laser tube. The windows are arranged at Brewster angle to the axis of the laser tube and hence they are called **Brewster windows**. The light beam traveling between the mirrors of the laser is reflected many times from these mirrors. Since the mirrors are at Brewster angle, all the light that is polarized perpendicular to the beam plane is emitted out of the laser cavity at early stage. In the gas tube remains only radiation polarized in the beam plane. The advantage of this arrangement is that the beam has no reflection losses, since only the transmitted polarized beam is traveling between the mirrors. The radiation out of these lasers is polarized as can be seen in Fig.20.15.

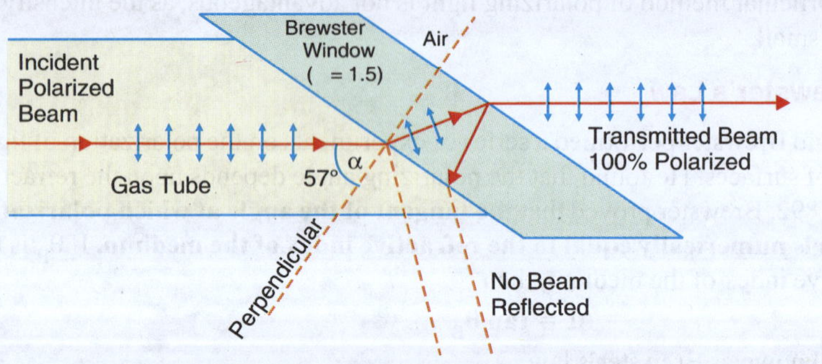

Brewster angle window at the end of gas laser

Fig. 20.15

(*iv*) Another application utilises the Brewster angle for transmitting a light beam into or out of an optical fibre without reflection losses.

20.6.2 POLARIZATION BY REFRACTION - PILE OF PLATES

When unpolarized light is incident at Brewster angle on a smooth glass surface, the reflected light is totally polarized, while the refracted light is partially polarized. If natural light is transmitted through a single plate, the transmitted beam is only partially polarized. If a stack of glass plates is used instead of a single plate, reflections from successive surfaces occur leading to the filtering of the s-component in the transmitted ray. Ultimately, the transmitted ray consists of p-component alone. It is found that a stack of about 15 glass plates is required for this purpose. The glass plates are supported in a tube of suitable size and inclined at an angle of about 33° to the axis of the tube, as shown in Fig.20.16. Such an arrangement is called a **pile of plates**. Unpolarized light enters the tube and is incident on the plates at Brewster angle and the transmitted light will be totally polarized parallel to the plane of incidence.

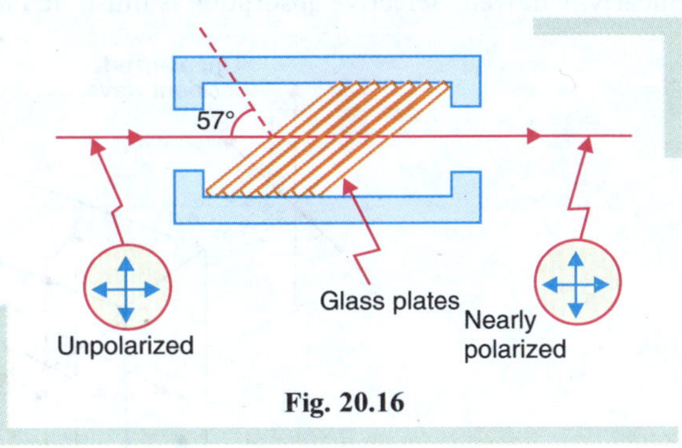

Fig. 20.16

20.6.3 POLARIZATION BY SCATTERING

If a narrow beam of natural light is incident on a transparent medium containing a suspension of ultramicroscopic particles, the light scattered is partially polarized. The incident light causes electrons in the scattering medium to vibrate. A vibrating electron emits most light in a direction perpendicular to its vibration and none along the direction of its vibration. The electric field of the emitted radiation is parallel to the direction of electron vibration. Hence light scattered through about 90° with respect to the incident direction is strongly polarized. The direction of vibration of E vector in the scattered light will be perpendicular to the plane defined by the direction of propagation and the direction of observation, *i.e.,* the plane of the paper, as illustrated in Fig.20.17.

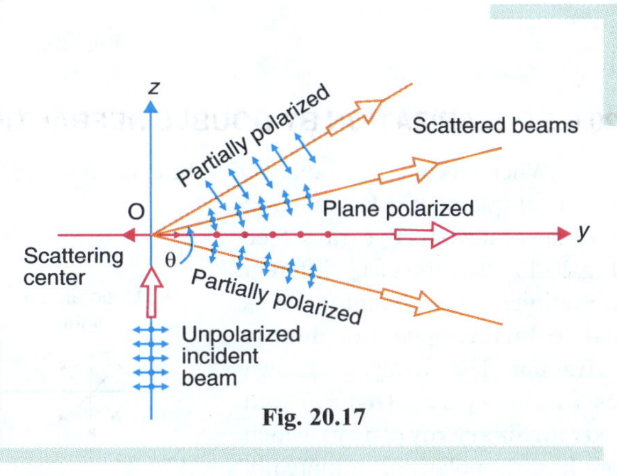

Fig. 20.17

The light from a blue sky is quite strongly polarised, particularly at 90° from the sun. It is not completely polarised because a significant amount of sunlight has undergone multiple-scattering, i.e. has been scattered more than once. Light scattered twice through a total angle of 90° would be less polarised than light scattered once.

20.6.4 POLARIZATION BY SELECTIVE ABSORPTION

A number of crystalline materials absorb more light in one incident plane than another, so that light progressing through the material become more and more polarized as they proceed. This difference in the absorption for the light rays is known as **selective absorption** or **dichroism**. Biot discovered this phenomenon in 1815. When natural light passes through a crystal such as tourmaline, it is split into two components, which are polarized in mutually perpendicular planes. The crystal absorbs light that is polarized in a direction parallel to a particular plane in the crystal but freely transmits the light component polarized in a direction perpendicular to that plane. If the crystal is of proper thickness, one of the components is totally absorbed and the other component emerging from the crystal is linearly polarized. Selective absorption is illustrated in Fig.20.18.

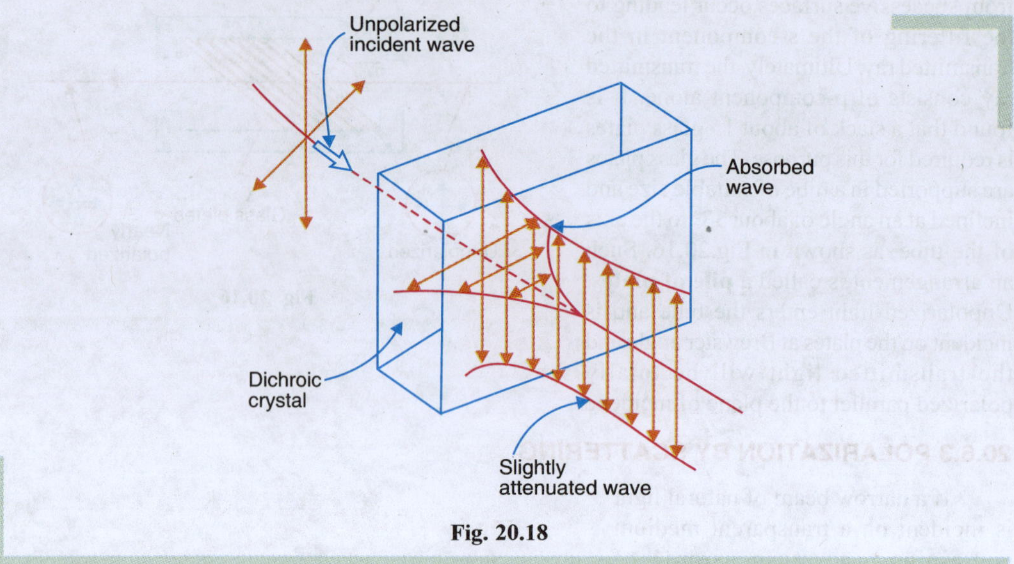

Fig. 20.18

20.6.5 POLARIZATION BY DOUBLE REFRACTION

When a beam of unpolarized light is incident on the surface of an anisotropic crystal such as calcite or quartz, it is found that it will separate into two rays (see Fig.20.19) that travel in different directions. This phenomenon is called **birefringence** or **double refraction**. The two rays are known as **ordinary ray (o-ray)** and **extraordinary ray (e-ray)**, which are linearly polarized in mutually perpendicular directions. A single linearly polarized ray is obtained in practice through elimination of one of the two polarized rays.

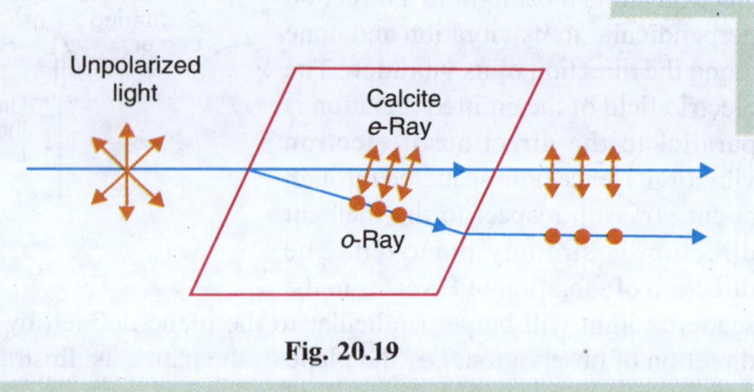

Fig. 20.19

20.7. POLAROID SHEETS

In 1928 E. H. Land invented Polaroid sheets, which utilize the phenomenon of selective absorption. The sheets are fabricated as follows. A clear plastic sheet of long chain molecules of PVA (polyvinyl alcohol) is heated and then stretched in a given direction to many times its original length. During the stretching process the PVA molecules become aligned along the direction of stretching. The sheet is then laminated to a rigid sheet of plastic to stabilise its size. It is then exposed to iodine vapour. The iodine atoms attach themselves to the straight long chain PVA molecules and consequently form long parallel conducting chains. The iodine atoms provide electrons, which can move easily along the aligned chains, but not perpendicular to them. When natural light is incident on the sheet, the electromagnetic vibrations that are in a direction parallel to the alignment of the iodine atoms are strongly absorbed because of the dissipative effects of the electron motion in the chains. Consequently, only those vibrations in a direction perpendicular to the direction of molecular chains are transmitted. Thus, the light transmitted through the polaroid sheet is polarized. A sheet fabricated according to this process is known as H-sheet.

These sheet polarizers are inexpensive and can be made in large sizes. Polaroid sheets are widely used in sunglasses, camera filters etc to eliminate the unwanted glare from objects.

Polaroid sheets are extensively used for the production and detection of linearly polarized light.

20.8. POLARIZER AND ANALYZER

A *polarizer* is an optical element, which utilizes the phenomenon of selective absorption or double refraction, and transforms unpolarized light into polarized light. Plane polarized light is obtained by eliminating one of the two components in the unpolarized light. When natural light is incident on a polarizer, the *E*-field component that is parallel to the chains of iodine atoms induces current in the conducting chains and is therefore strongly absorbed. Consequently, the light transmitted contains only the component that is perpendicular to the direction of molecular chains. The direction of *E*-vector in the transmitted beam corresponds to the **transmission axis** of the Polaroid sheet.

Effect of polarizer on natural light

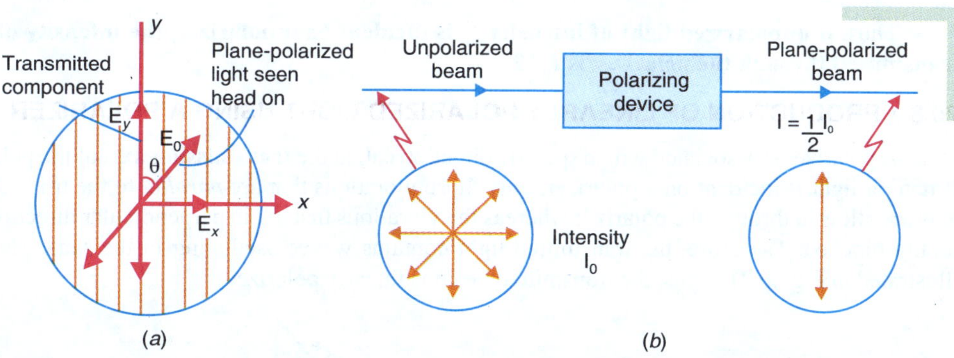

(*a*) Action of polarizer on linearly polarized wave
(*b*) The intensity of an unpolarized beam reduces to half after passing through a polarizer.

Fig. 20.20

Let us now understand the action of polarizer on the incident unpolarized light. Let us consider unpolarized light incident on a polarizer with the electric vector E_0 making an angle θ with respect to the *transmission axis* of the polarizer. The electric vector E_0 may be resolved into its component vectors lying parallel and perpendicular to the transmission axis of the polarizer (see Fig.20.20 *a*), that is E_y, parallel to the transmission axis and E_z, perpendicular to the transmission axis of the polarizer. The polarizer transmits the parallel component while blocking the perpendicular component. Thus, it is the parallel component E_y that is transmitted by the polarizer. But

$$E_y = E_0 \cos\theta \qquad \text{... (20.7)}$$

and hence, the intensity of the transmitted component is given by

$$I \propto E_y^2 = E_0^2 \cos^2\theta \qquad \text{... (20.8)}$$

In unpolarized light all the values of θ are equally probable. Therefore, the fraction of light transmitted through the polarizer equals the average value of $\cos^2\theta$.

$$\therefore \quad I = E_0(\cos^2\theta) = I_0(\cos^2\theta)$$

$$= \frac{I_0}{2\pi}\int_0^{2\pi}\cos^2\theta\, d\theta$$

$$= \frac{I_0}{2\pi}\int_0^{2\pi}\left(\frac{1+\cos 2\theta}{2}\right)d\theta$$

$$= \frac{I_0}{4\pi}\left[(\theta)_0^{2\pi} + \left\{\frac{\sin 2\theta}{2}\right\}_0^{2\pi}\right]$$

$$= \frac{I_0}{4\pi}(2\pi + 0)$$

$$= \frac{1}{2}I_0$$

$$\therefore \quad I = \frac{1}{2}E_0^2 = \frac{1}{2}I_0 \qquad \text{... (20.9)}$$

Thus, **if unpolarized light of intensity I_0 is incident on a polarizer, the intensity of light transmitted through the polarizer is $I_0 / 2$.**

20.8.1 PRODUCTION OF LINEARLY POLARIZED LIGHT USING A POLARIZER

A polarizer is associated with a specific direction called the **transmission axis** of the polarizer. If natural light is incident on a polarizer, only those vibrations that are *parallel* to the transmission axis are allowed through the polarizer whereas the vibrations that are in perpendicular directions are totally blocked. Therefore, the transmitted light contains waves oscillating in the same plane, as illustrated in Fig.20.21. Thus, the transmitted beam is linearly polarized.

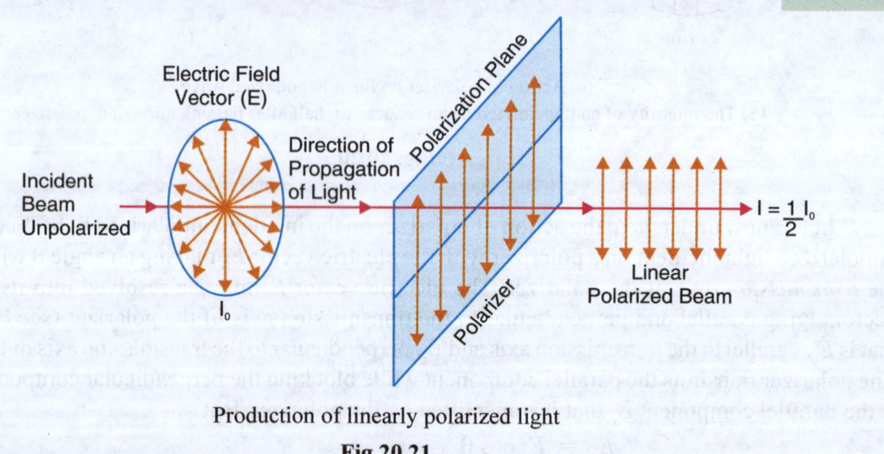

Production of linearly polarized light

Fig.20.21

According to equ.(20.9), *when unpolarized light of intensity I_0 is incident on a polarizer, the intensity of light transmitted by the polarizer is $I_0 / 2$.*

20.8.2 DETECTION OF LINEARLY POLARIZED LIGHT

An *analyzer* is an optical element, which is used to identify the plane of vibration of plane polarized light.

To examine light coming from some direction either after emission or reflection etc, we use a Polaroid sheet. The Polaroid sheet used to determine the plane of polarization of light is known as an **analyzer**. There is no difference between a polarizer and analyzer in fabrication but they differ in their roles. Both the polarizer and analyzer are characterized by a transmission axis.

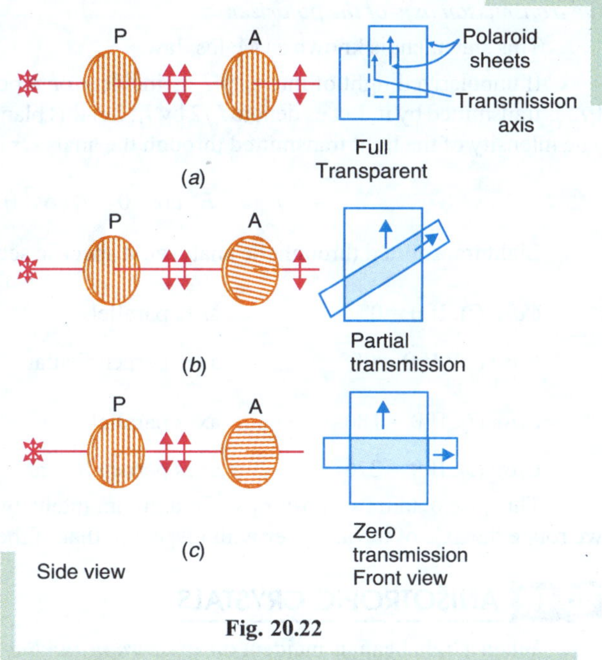

Fig. 20.22

When the transmission axis of the analyzer *A* is set up parallel to that of polarizer *P*, light transmitted by the polarizer, passes unhindered through the analyzer (Fig. 20.22 *a*).

If the transmission axes are set at an angle q, light is partially transmitted (Fig.20.22 *b*). As the angle rotates from 0 to 90 degrees, the amount of light that is transmitted decreases. When the axes are perpendicular to each other, the polarized light from P is extinguished by the analyzer A (Fig.20.22 *c*). The polarizer and analyzer are said to be **crossed** in this configuration.

When we rotate the axis of the analyzer with respect to that of the polarizer, we obtain two positions of maximum intensity and two positions of zero intensity in *one full rotation*. Conversely, if we obtain two positions of maximum intensity and two positions of zero intensity in one full rotation of the analyzer, we conclude that the light incident on the analyzer is plane-polarized light.

20.9. MALUS' LAW

The amount of light transmitted through a polarizer at an arbitrary angle [Fig.20.22 (*b*)] is given by **Malus's Law**.

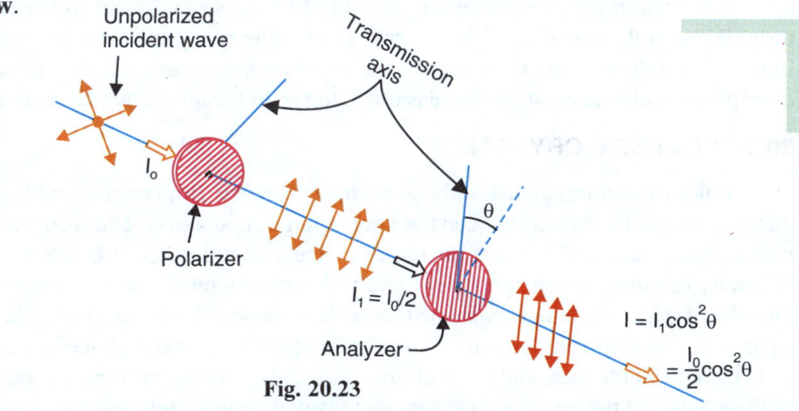

Fig. 20.23

In 1809 Malus found that the *intensity of polarized light transmitted through a polarizer is proportional to the square of cosine of the angle between the plane of polarization of the light and the transmisson axis of the polarizer.*

This statement is known as Malus' law.

If unpolarized light of intensity I_o is incident on a polarizer, plane polarized light of intensity $I_o/2$ is transmitted by it. Let us denote $I_o/2$ by I_1. Let this plane polarized light pass through an analyzer. The intensity of the light transmitted through the analyzer is given by

$$I = E_1^2 \cos^2\theta = I_1 \cos^2\theta = \frac{1}{2} I_0 \cos^2\theta \qquad \text{... (20.10)}$$

Light transmitted through the analyzer at specific settings are as follows.

Case (*i*): If $\theta = 0°$	axes parallel	$I = I_1 = \frac{1}{2} I_0$
Case (*ii*): If $\theta = 90°$	axes perpendicular	$I = 0$
Case (*i*): If $\theta = 180°$	axes parallel	$I = I_1 = \frac{1}{2} I_0$
Case (*i*): If $\theta = 270°$	axes perpendicular	$I = 0$

Thus, we obtain two positions of maximum intensity and two positions of zero intensity when we rotate the axis of the analyzer with respect to that of the polarizer.

20.10. ANISOTROPIC CRYSTALS

When a light beam is incident on an *isotropic medium* such as a glass slab, it refracts as a single ray. An optically isotropic material is one in which the index of refraction is the same in all directions. Glass, water and air are examples of isotropic materials. The atoms in a crystal are arranged in a regular periodic manner. If the arrangement of atoms differ in different directions within a crystal, then the physical properties vary with the direction. The thermal conductivity, electrical conductivity, velocity of light and hence refractive index etc properties depend on the crystallographic direction along which the property is measured. Then we say that the crystal is **anisotropic**. In such anisotropic crystals the force of interaction between the electron cloud and the lattice is different in different crystallographic directions. The natural frequency of the electron cloud is likewise dependent on the direction in which the electrons are caused to vibrate by the incident light wave. This results in different velocities in different directions and the index of refraction is different in different directions within the crystal.

The anisotropic crystals are divided into two classes: uniaxial and biaxial crystals. In case of **uniaxial crystals**, one of the refracted rays is an ordinary ray and the other is an extraordinary ray. In **biaxial crystals** both the refracted rays are extraordinary rays. Calcite, tourmaline and quartz are examples of uniaxial crystals whereas mica, topaz and aragonite are examples of biaxial crystals.

20.10.1 CALCITE CRYSTAL

Calcite is a common naturally occurring substance. Both marble and limestone are made up of many small calcite crystals bonded together. A large crystal of calcite is colourless and transparent. It was at one time found in great quantities in Iceland and hence it is also known as **Iceland spar**. Naturally occurring calcite crystals (Fig.20.24*a*) has rhombohedral cleavage which means it breaks into blocks with parallelogram-shaped faces. It is bounded by six faces (Fig.20.24*b*), each of which is a parallelogram with angles equal to 101°55'and 78°5'. The rhombohedron has only two corners A and H where all the face angles are obtuse (101°55'). These two corners appear as the *blunt corners* of the crystal. At the rest of six corners there is one obtuse angle and two acute angles.

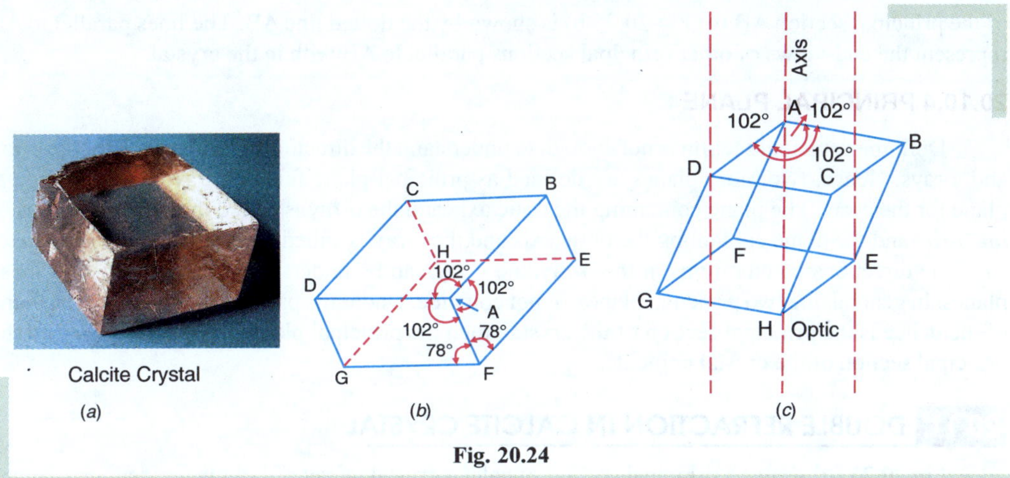

Fig. 20.24

20.10.2 OPTIC AXIS

A line bisecting any one of the blunt corners (A or H) and making equal angles with each of the three edges meeting there, is the **optic axis** (see Fig.20.24 *c*). In fact any line parallel to this line is also an optic axis. Thus, the optic axis is a direction and not a specific line in the crystal. Hence an optic axis can be drawn through every point in the crystal, that is, any line parallel to the line above will also be the optic axis. It is to be noted that the optic axis is not obtained by joining the two blunt corners. Only in a special case, when the three edges of the crystal are equal, the line joining the two blunt corners A and H coincide with the crystallographic axis of the crystal and it gives the direction of the optic axis. The optic axis is actually the axis of symmetry of the crystal. A ray of light propagating along optic axis does not suffer double refraction, because the structure of the crystal is symmetric about that direction.

The optic axis is the direction in a uniaxial crystal along which the e-ray and the o-ray travel with the *same* speed and consequently double refraction *does not* take place along this direction. The corresponding refractive index is the refractive index for ordinary light, say μ_o.

20.10.3 PRINCIPAL SECTION

A plane containing the optic axis and perpendicular to a pair of opposite faces of the crystal is called the **principal section** of the crystal for that pair of faces (Fig.20.25 *a*). Thus, there are three principal sections passing through any point within the crystal, one corresponding to each pair of opposite faces. A principal section always cuts surfaces of calcite crystal in a parallelogram having angles 71° and 109° (see Fig.20.25 *b*). Fig.20.25 (*c*) shows a face of the crystal in which the end-view

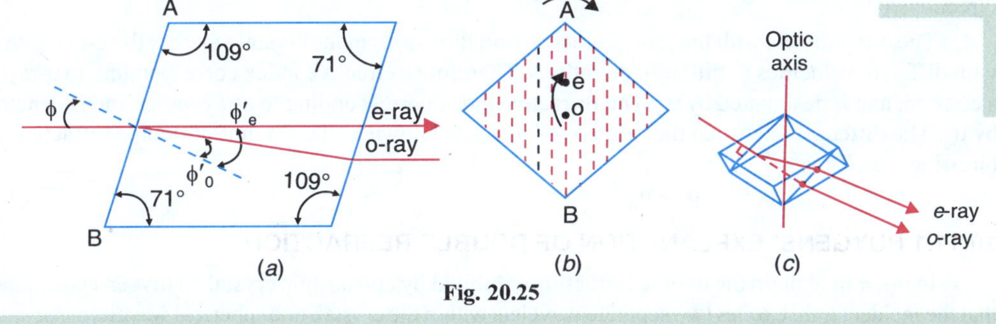

Fig. 20.25

of the principal section AB (of Fig.20.25 b) is shown by the dotted line AB. The lines parallel to AB represent the end-views of other principal sections parallel to AB with in the crystal.

20.10.4 PRINCIPAL PLANE

Defining principal section is not enough to understand the directions of vibrations for the o-ray and e-rays. Hence, two more planes are defined as principal plane for the o-ray and the principal plane for the e-ray. The plane containing the optic axis and the o-ray is called the *principal plane of the o-ray* and the plane containing the optic axis and the e-ray is called the *principal plane of the e-ray*. The directions of vibrations in the o-ray and e-ray can be understood with reference to these planes. In general, the two principal planes do not coincide. Under the particular case, when the plane of incidence is the principal section of the crystal, then the principal planes of o- and e-rays and the principal section of the crystal coincide.

20.11. DOUBLE REFRACTION IN CALCITE CRYSTAL

Fig.20.25 (*a*) shows a principal section of calcite crystal. A ray of light is incident on the face AB of the crystal and it travels along the principal section. The ray is split into two rays, namely o-ray (fast ray) and e-ray (slow ray). The o-ray travels through the crystal without deviation while the e-ray is refracted at some angle. As the opposite faces of the crystal are parallel, the rays emerge out parallel to the incident ray. Within the crystal the o-ray *always* lies in the plane of incidence whereas e-ray does not lie in the plane of incidence. e- ray lies in the plane of incidence only when the plane of incidence is a principal section.

If a mark (dot or cross) is made on a paper and then the calcite crystal (AB face) is placed on it, two images are seen through the crystal, as illustrated in Fig.20.25 (*b*). The images are produced by the o-ray and e-ray. The intensities of the images are lesser than that of the original mark. The line joining them lies in the principal section. If now the crystal is rotated slowly about an axis passing through the o-image, the e-image moves round in a circle while the o-image remains stationary. It shows that the velocity of propagation of o-ray is the same in all directions, while that of e-ray changes with direction. O-ray obeys the laws of refraction and the e-ray does not follow the ordinary laws of refraction.

The e-ray and o-ray are linearly polarized. The e-ray has its vibrations (i.e., the optical vector) *parallel* to the principal section whereas the vibrations (optical vector) in o-ray are *perpendicular* to the principal section, as indicated in Fig.20.19. The vibration directions can be established by examining the rays through a polarizer. As the polarizer is held in the path of the rays and rotated slowly, the intensity of one of the images, say the o-image, increase while that of the e-image decreases. In one position, the intensity of the o-image will be a maximum while the e-image is extinguished. Further rotation through 90° from this particular position, causes the o-image to disappear and e-image intensity to become a maximum. It proves that the e- and o-rays are linearly polarized in mutually perpendicular directions.

The o-ray travels with the same velocity in all directions in the crystal whereas the e-ray travels with different velocities in different directions. Therefore, refractive index corresponding to o-ray is a constant and is designated by μ_o. The refractive index corresponding to e-ray varies and is denoted by μ_e. The difference between the refractive indices is known as the amount of double refraction or birefringence. Thus,

$$\Delta_\mu = \mu_e - \mu_o \qquad \text{... (20.11)}$$

20.11.1 HUYGENS' EXPLANATION OF DOUBLE REFRACTION

In order to explain the double refraction exhibited by anisotropic crystals, Huygens postulated that the incident light excites two separate wavelets within the crystal, one spherical wavelet associated with the ordinary waves and one ellipsoidal wavelet associated with the extraordinary waves. For example, the plane wavefront, in Fig.20.26, incident normally on the crystal surface generates spherical wavelets as well as ellipsoidal wavelets. The spherical wavelets propagate equally quickly in all directions. The wave surface corresponding to o-ray is therefore spherical. The tangent to these waves lies straight ahead and, by successive application of the principle, the plane wave propagates straight ahead with speed υ. The ellipsoidal wavelets propagate at different speeds in different directions. The wave surface corresponding to e-ray is therefore an ellipsoid of revolution about the optic axis. The common tangent to these ellipsoids after a little time is the new wavefront. The line from the point of generation of each ellipsoid to the tangent point on that ellipsoid is off at an angle and defines the direction of travel of the extraordinary wavefronts. The wavefronts are not perpendicular to their direction of travel.

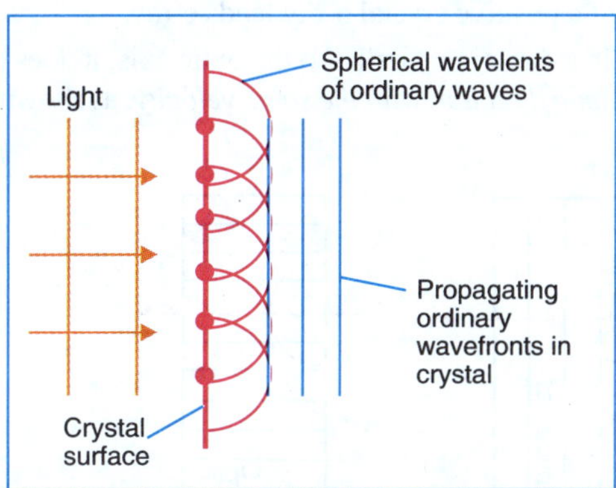

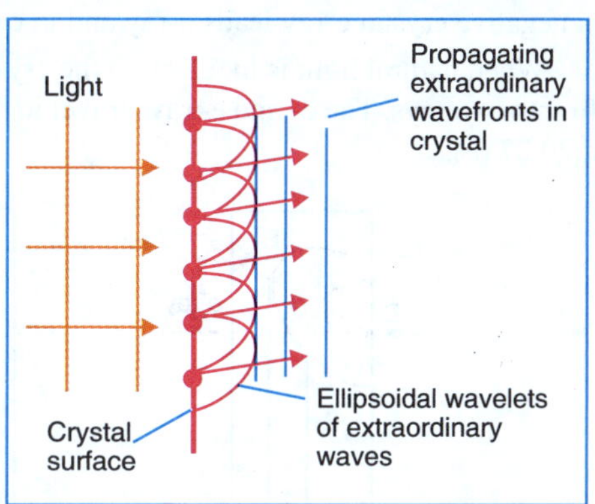

Fig. 20.26

The two wave surfaces touch each other at the two points where they are intersected by the optic axis. As light propagates through the crystal, the two wave surfaces travel in different directions in the crystal. Ultimately, two refracted rays emerge from the crystal.

20.11.2 ORDINARY AND EXTRA-ORDINARY RAYS

We now compare the properties of o- and e-rays:

(*i*) o-ray obeys the laws of refraction and the e-ray does not follow the ordinary laws of refraction.

(*ii*) Both o-ray and e-ray are plane polarized. They are polarized in mutually perpendicular planes. The electric vector of o-ray vibrates perpendicular to the principal section of o-ray while the vibrations of e-ray take place parallel to the principal section of e-ray.

(*iii*) O-ray travels with the same speed in all directions within the crystal. The e-ray travels with different speeds along different directions in the crystal. However, the speed of e-ray will be equal to that of o-ray along the optic axis direction.

(*iv*) Because o-ray travels with the same velocity in all directions, the refractive index corresponding to it has a constant value. On the other hand, the refractive index for e-ray varies from direction to direction. The principal refractive index for o-ray is defined as follows:

$$\mu_0 = \frac{c}{\upsilon_0} = \frac{\text{velocity of light in a vacuum}}{\text{velocity of o-ray in the crystal}} \quad ...(20.12)$$

The principal refractive index for e-ray in ***positive crystals*** is defined as follows:

$$\mu_o = \frac{c}{(\upsilon_e)_{min}} = \frac{\text{velocity of light in a vacuum}}{\text{minimum velocity of e-ray in the crystal}} \quad \text{... (20.13)}$$

The principal refractive index for e-ray in ***negative crystals*** is defined as follows:

$$\mu_e = \frac{c}{(\upsilon_e)_{max}} = \frac{\text{velocity of light in a vacuum}}{\text{maximum velocity of e-ray in the crystal}} \quad \text{... (20.14)}$$

(*v*) When natural light is incident on an anisotropic crystal at an angle to the optic axis, it splits into o-and e-rays, which travel in *different directions* with *different velocities* (20.27 *a*).

When natural light is incident in a direction perpendicular to the optic axis, o-ray and e-ray propagate in the *same direction* in the crystal but with *different velocities*, as shown in Fig.20.27 (*b*). In a negative crystal e-ray leads o-ray and in case of a positive crystal o-ray leads e-ray.

When natural light is incident on the crystal in a direction parallel to the optic axis, it does not split into two rays. The o- and e- rays travel in the *same direction with the same velocity*, as shown in Fig.20.27 (*c*).

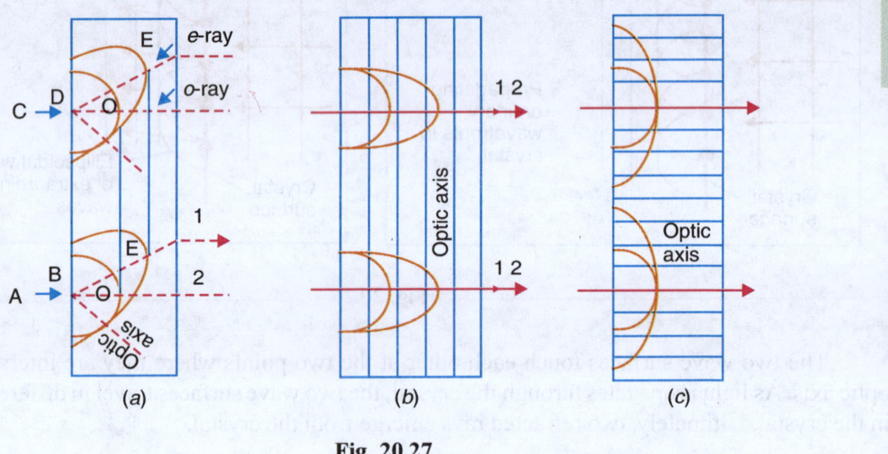

Fig. 20.27

(*vi*) The distinction of o-ray and e-ray exists only within the crystal. Once they emerge from the crystal, they travel with the same velocity. The rays outside the crystal differ only in their direction of travel and plane of polarization. The designation of o- ray and e-ray has no meaning outside the crystal.

20.11.3 POSITIVE AND NEGATIVE CRYSTALS

Because of two different wave fronts, two different types of uniaxial crystals exist. In one type of crystals, the spherical wave front of o-ray is enclosed by the ellipsoidal wave front of e-ray. Such crystals are known as **negative crystals**. They are called negative crystals because the refractive index corresponding to the e-ray is less than that corresponding to o-ray. Calcite crystal is an example of negative type crystals. In the other case, the extraordinary wave front lies within the ordinary wave front and such crystals are called **positive crystals**. They are positive because the refractive index for the extraordinary ray is greater than that of o-ray. Quartz crystal is an example of positive crystals.

We compare here the characteristics of the positive and negative crystals.

(*i*) In positive uniaxial crystals, the ellipsoid of revolution corrrsponding to the e-ray is totally contained within the sphere corresponding to the o-ray.

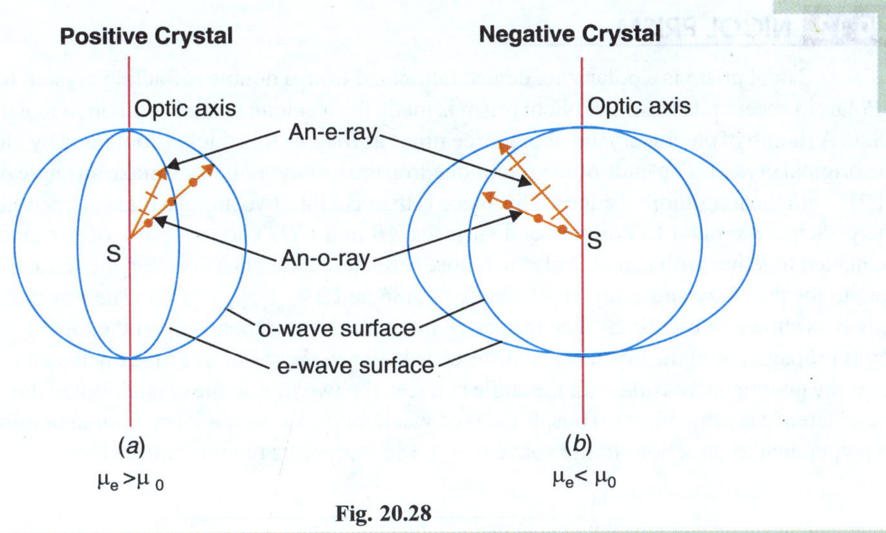

Fig. 20.28

In negative uniaxial crystals, the ellipsoid of revolution for e-ray lies completely outside the sphere corresponding to o-ray. The two cases are depicted in Fig. 20.28.

(*ii*) In positive crystals the e-ray velocity has a maximum value along the optic axis and a minimum value in a direction perpendicular to the optic axis.

On the other hand, in negative crystals the velocity of e-ray has a minimum value parallel to the optic axis and a maximum value in a direction perpendicular to the optic axis.

(*iii*) In positive crystals, e-ray travels slower than o-ray in all directions except along the optic axis.

$\upsilon_e = \upsilon_o$ — parallel to optic axis

$\upsilon_e < \upsilon_o$ — other directions

In negative crystals, o-ray travels slower than e-ray in all directions except along the optic axis.

$\upsilon_e = \upsilon_o$ — parallel to optic axis

$\upsilon_e > \upsilon_o$ — other directions

(*iv*) In positive crystals the principal refractive index for e-ray is larger than the principal refractive index for o-ray.

$$\mu_e > \mu_o$$

In negative crystals the principal refractive index for o-ray is larger than the principal refractive index for e-ray.

$$\mu_e < \mu_o$$

(*v*) **Birefringence** or amount of double refraction of a crystal is defined as

$$\Delta\mu = \mu_e - \mu_o \qquad \text{... (20.15 } a\text{)}$$

As $\mu_e > \mu_o$ in these crystals, $\Delta\mu$ is a positive quantity for positive crystals.

$\Delta\mu$ is a negative quantity for negative crystals as $\mu_e < \mu_o$ in these crystals and the birefringence of a negative crystal is given by

$$\Delta\mu = \mu_o - \mu_e \qquad \text{... (20.15 } b\text{)}$$

20.12. NICOL PRISM

Nicol prism is a polarizing device fabricated from a double refracting crystal. It is similar to a Polaroid sheet in its action. A Nicol prism is made from calcite crystal. William Nicol designed it in 1820. A rhomb of calcite crystal about three times as long as it is thick, is obtained by cleavage from the original crystal. The ends of the rhombohedron are ground until they make an angle of 68°instead of 71° with the longitudinal edges. This piece is then cut into two along a plane perpendicular both to the principal axis and to the new end surfaces *AB* and *CD*. The two parts of the crystal are then cemented together with canada balsam, whose refractive index lies between the refractive indices of calcite for the o-ray and e-ray. μ_o=1.66, μ_e=1.486 and $\mu_{\text{canada balsam}}$=1.55. The position of the optic axis is as shown in Fig.20.29. The refractive index for e-ray depends upon the direction in which e-ray is propagating in the crystal. The difference between the refractive index between o-ray and that for e-ray goes on increasing with the angle between the two rays in the crystal. When this angle is 90°, the difference is a maximum. Thus, for a fixed value for μ_o, the μ_e has its maximum or minimum value in perpendicular direction. In the above μ_e = 1.486 represents the minimum value.

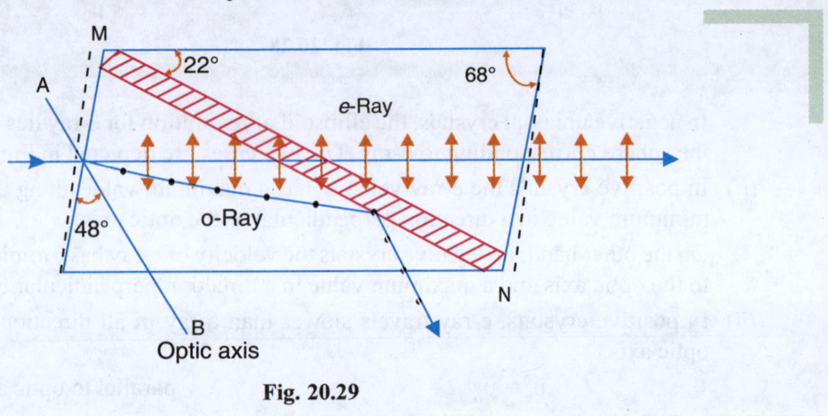

Fig. 20.29

Unpolarized light is made to fall on the crystal as shown in Fig.20.29 at an angle of about 15°. The ray after entering the crystal suffers double refraction and splits up into o-ray and e-ray. The two rays with their directions of vibrations are as shown in the Fig.20.29. The values of the refractive indices and the angles of incidence at the canada balsam layer are such that the e-ray is transmitted while the o-ray is internally reflected. The face where the o-ray is incident is blackened so that the o-ray is completely absorbed. Then we get only the plane-polarized e-ray coming out of the Nicol. Thus, the Nicol works as a polarizer.

For studying the optical properties of transparent substances, two Nicols are used - one as a polarizer and the other as an analyzer.

When two Nicol prisms P and A are placed adjacent to each other as shown in Fig.20.30, one of them acts as a polarizer and the other acts as an analyser. If unpolarized ray of light is incident on the Nicol prism P, a linearly polarized e-ray emerges from P with its vibration direction lying in the principal section of P. The state of the polarization of the light emerging from polarizer P can be examined with another polarizer A, which for convenience is called an analyser. Let now this ray be incident on the second Nicol prism A, whose principal section is parallel to that to P. The vibration direction of the ray will be in the principal section of A and hence it is transmitted unhindered through the analyser A.

If the Nicol prism A is gradually rotated, the intensity of the e-ray decreases in accordance with Malus law. When its principal section becomes perpendicular to that of the Nicol prism

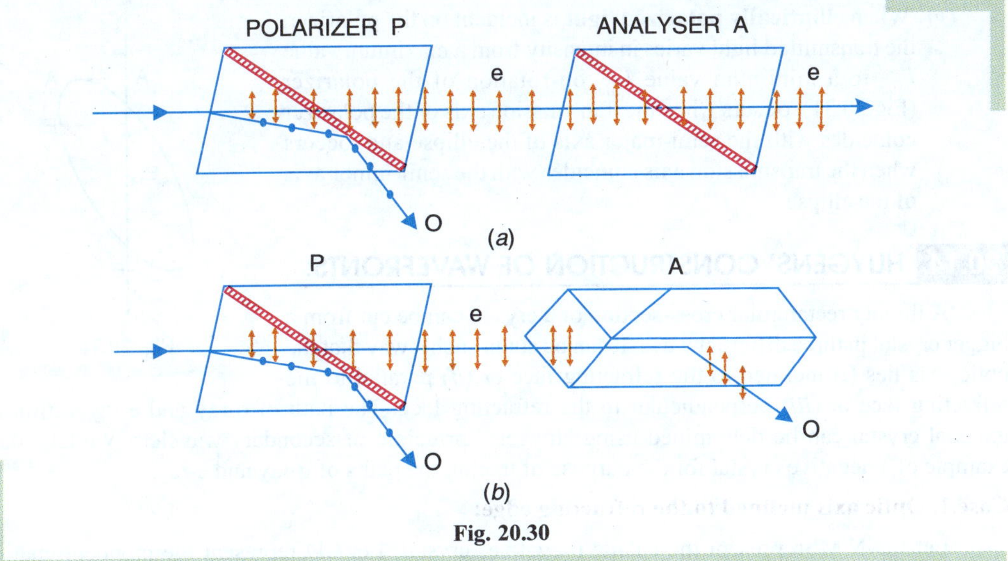

Fig. 20.30

P (Fig.20.30 b), the vibrations of the ray, emerging from P and incident on A, will be perpendicular to the principal section of A. In this position the ray behaves as o-ray inside the prism A and is totally internally reflected by the Canada balsam layer. Hence no light is transmitted by the prism A. In this configuration, the two Nicol prisms P and A are said to be **crossed.** If the Nicol prism A is further rotated through another 90°, the intensity of light emerging from A will go on increasing. The intensity will become a maximum when its principal section is again parallel to that of the prism P. Thus, the prism P produces linearly polarised light while the prism A detects it. Hence the prism P is called a polarizer and the prism A an analyser.

20.13. EFFECT OF POLARIZER ON LIGHT OF DIFFERENT POLARIZATIONS

The action of a polarizer, whether a Nicol prism or a Polaroid sheet, on light of different types of polarization is as follows:

(*i*) If **unpolarized light** is incident on a polarizer, it transmits half the intensity of light incident on it. The transmitted light stays constant in intensity on rotation of the polarizer.

(*ii*) If **partially polarized light** is incident on the polarizer, the intensity of the transmitted light depends on the direction of the transmission axis of the polarizer. The intensity varies from a maximum value I_{max} to a minimum value I_{min}. Two positions of I_{max} and two positions of I_{min} occur in one full rotation of the polarizer.

(*iii*) If **plane polarized light** is incident on the polarizer, the intensity of the transmitted light varies from zero to a maximum twice in one full rotation of the polarizer.

(*iv*) When **circularly polarized light** is incident on the polarizer, the transmitted light remains constant in intensity on rotation of the polarizer. This is interpreted as follows: the circular vibrations may be resolved into two mutually perpendicular linear vibrations of equal amplitude. When the circularly polarized light is incident on the polarizer, the vibration parallel to its transmission axis passes through the polarizer while the perpendicular component is not allowed. This happens in all positions of the polarizer in its rotation. Therefore, the intensity of the transmitted light stays constant in one full rotation of the polarizer.

(*v*) When **elliptically polarized light** is incident on the polarizer, the transmitted light varies in intensity from a maximum value I_{max} to a minimum value I_{min} on rotation of the polarizer (Fig.20.31). occurs when the transmission axis of the polarizer coincides with the semi-major axis of the ellipse and occurs when the transmission axis coincides with the semi-minor axis of the ellipse.

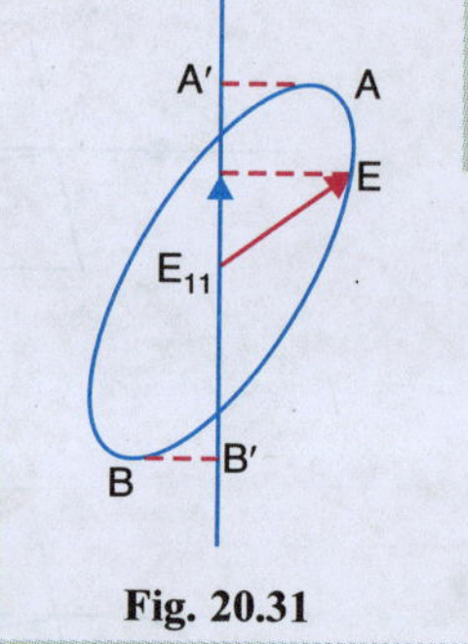

Fig. 20.31

20.14. HUYGENS' CONSTRUCTION OF WAVEFRONTS:

A thinner rectangular cross-section of a crystal can be cut from a bigger crystal in three different ways. It can be cut in such a way that the optic axis lies (*i*) inclined to the refracting face or (*ii*) parallel to the refracting face or (*iii*) perpendicular to the refracting face. The path of o-ray and e-ray within a uniaxial crystal can be determined using Huygens' principle of secondary wavelets. We take the example of a negative crystal for the purpose of tracing the paths of o-ray and e-ray.

Case.1. Optic axis inclined to the refracting edge:

Let MNN′M′ represent the calcite (negative) crystal. Let CD represent the monochromatic plane wave front incident normal to the crystal surface. The optic axis is ***inclined to the refracting edge.*** According to Huygens the phenomenon of double refraction involves two types of propagation of light waves. The wave known as o-wave has a spherical wave front whereas the e-wave has an ellipsoidal wave front. The two wave fronts touch each other along the optic axis of the crystal. In a negative crystal the e-ray velocity is greater than the o-ray velocity and hence the ellipsoidal surface lies outside the spherical surface.

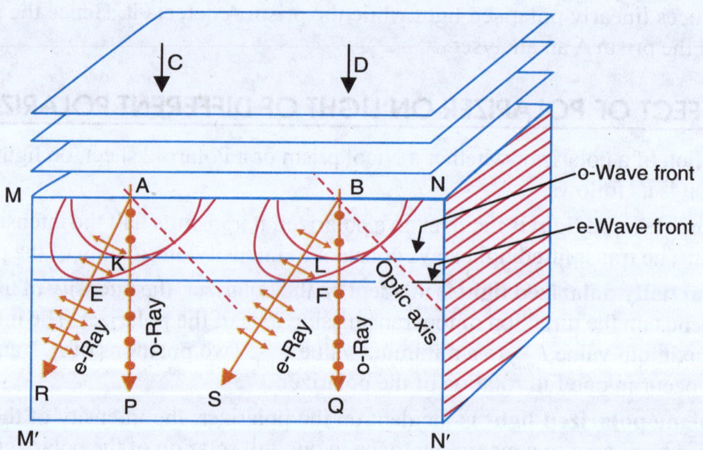

Unpolarized light beam is incident at normal on a calcite crystal slice. Optic axis is inclined to the incident ray. Huygens wavelets showing double refraction, and difference in velocities of *o*-ray and *e*-ray.

Fig. 20.32

Let us consider that parallel beam of light CD falls **normally** on the surface of the negative crystal. As soon as the parallel beam strikes the crystal boundary, each point on the wave front becomes a source of secondary disturbance. The points A and B are chosen for the purpose of illustration.

According to Huygens principle the points A and B produce elliptical and spherical wavelets. The position of the two wave fronts after a lapse of '*t*' seconds can be determined as follows. A circle of radius '$\upsilon_o t$' is drawn taking A as centre. Similarly, a circle of the same radius is drawn with B as centre. The ellipsoidal wave front can be drawn if the major and minor axes are known. In a negative crystal the major axis is equal to '$2\upsilon_e t$' and minor axis is '$2\upsilon_o t$'. The circle and the ellipse touch each other, as required, along the optic axis. If we now draw the common tangents to the secondary wavelets, they represent the plane wave fronts corresponding to the two rays. Thus KL is the tangent to the spherical wavelets and EF is the tangent to the ellipsoidal wave fronts. If now we join the point of origin of wavelets to the points of tangency, the direction of propagation of the rays will be known. Thus, if we join A to K and B to L, we obtain the direction of propagation of o-ray. Similarly, the lines AP and BQ show the direction of propagation of e-ray. We find that in this case the o- ray and e-ray travel along different directions with different velocities.

Case 2(*a*): Optic axis in the plane of incidence and parallel to the refracting edge:

Let MNN′M′ represent the refracting face of the calcite (negative) crystal. Let CD represent the monochromatic plane wavefront incident normal to the crystal surface. Let the optic axis be ***parallel*** to the refracting edge MN and lie in the plane of incidence. See Fig.20.33 (a). According to Huygens the phenomenon of double refraction involves propagation of light waves of two types. The wave known as o-wave has a spherical wavefront whereas the e-wave has an ellipsoidal wavefront. The two wavefronts touch each other along the optic axis of the crystal. In a negative crystal the e-ray velocity is greater than the o-ray velocity and hence the ellipsoidal surface lies outside the spherical surface.

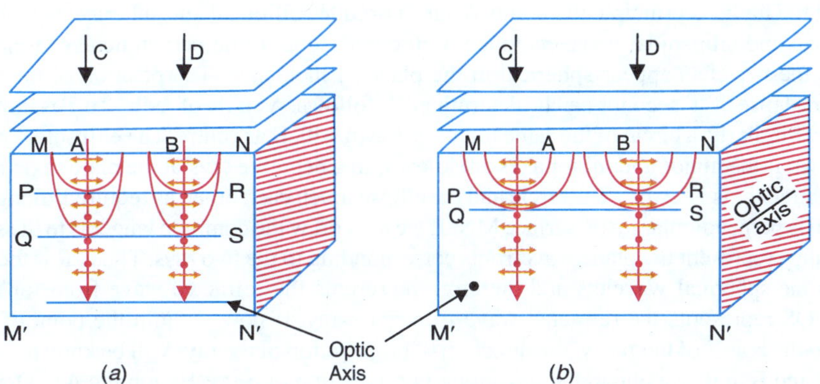

Unpolarized light beam incident normally on a calcite slab at right angles to the optic axis. Double refraction does not occur but *o*-ray and *e*-ray travel with different velocities in the crystal. (*a*) Optic axis in the plane of incidence and parallel to the crystal surface; (*b*) Optic axis perpendicular to the plane of incidence and parallel to the crystal surface.

Fig. 20.33

Let us consider that parallel beam of light CD falls **normally** on the surface of the negative crystal. As soon as the parallel beam strikes the crystal boundary, each point on the wavefront becomes a source of secondary disturbance. The points A and B are chosen for the purpose of illustration. According to Huygens' principle the points A and B produce elliptical and spherical wavelets. The two wavelets touch each other along AM and BN, which is the direction of optic axis, as required. The wave fronts at any instant can be drawn as tangential surfaces to the two wavelets. Note that the two wave fronts here are parallel to each other. The position of the two wavefronts after a lapse of '*t*' seconds can be determined as follows. A circle of radius '$\upsilon_o t$' is drawn taking A as centre. Similarly, a circle of the same radius is drawn with B as centre. The ellipsoidal wavefront can be drawn if the major and minor axes are known. In a negative crystal the major axis is equal to '$2\upsilon_e t$' and minor axis is '$2\upsilon_o t$'. PR is the tangent to the spherical wavelet and therefore represents the refracted wavefront for the o-ray. Similarly, QS represents the refracted wavefront for e-ray. If now we join the point of origin of wavelets to the points of tangency, the direction of propagation of the rays will be known. Thus, if we join A to P and B to R, we obtain the direction of propagation of o-ray. By joining A to Q and B to S we get the direction of propagation of e-ray. We find that in this case the o- ray and e-ray travel along the same direction but have different velocities.

Case 2 (*b*): Optic axis perpendicular to the plane of incidence and parallel to the refracting edge:

Let MN be the refracting edge of the calcite (negative) crystal. Let CD represent the monochromatic plane wave front incident normal to the crystal surface. The optic axis is ***parallel to the refracting edge but is perpendicular to the plane of incidence*** (Fig.20.33 b). According to Huygens the phenomenon of double refraction involves two types of propagation of light waves. The wave known as o-wave has a spherical wave front whereas the e-wave has an ellipsoidal wave front. The two wave fronts touch each other along the optic axis of the crystal. In a negative crystal the e-ray velocity is greater than the o-ray velocity and hence the ellipsoidal surface lies outside the spherical surface.

Let us consider that parallel beam of light CD falls **normally** on the surface of the negative crystal. As soon as the parallel beam strikes the crystal boundary, each point on the wave front becomes a source of secondary disturbance. The points A and B are chosen for the purpose of illustration. According to Huygens principle the points A and B produce elliptical and spherical wavelets. Since the spherical and ellipsoidal wavelets touch each other along a line perpendicular to the plane of incidence, the wavelets appear spherical in the plane of incidence. The position of the two wave fronts after a lapse of 't' seconds can be determined as follows. A circle of radius '$\upsilon_o t$' is drawn taking A as centre. Similarly, a circle of the same radius is drawn with B as centre. The ellipsoidal wave front can be drawn if the major and minor axes are known. In a negative crystal the major axis is equal to '$2\upsilon_e t$' and minor axis is '$2\upsilon_o t$'. The circle and the ellipse touch each other, as required, along the optic axis, which is perpendicular to the surface MN. If we now draw the common tangents to the secondary wavelets, they represent the plane wavefronts corresponding to the two rays. Thus PR is the common tangent to the spherical wavelets and therefore represents the refracted wave front for the o-ray. Similarly, QS represents the refracted wave front for e-ray. If now we join the point of origin of wavelets to the points of tangency, the direction of propagation of the rays will be known. Thus, if we join A to P and B to R, we obtain the direction of propagation of o-ray. By joining A to Q and B to S we get the direction of propagation of e-ray. We find that in this case the o- ray and e-ray travel along the same direction but have different velocities.

Case 3 : Optic axis perpendicular to the refracting edge and lying in the plane of incidence:

Let MNN′M′ represent the calcite (negative) crystal. Let CD represent the monochromatic plane wave front incident normal to the crystal surface. The optic axis is ***perpendicular to the refracting edge and lies in the plane of incidence***. According to Huygens the phenomenon of double refraction involves two types of propagation of light waves. The wave known as o-wave has a spherical wave front whereas the e-wave has an ellipsoidal wave front. The two wave fronts touch each other along the optic axis of the crystal. In a negative crystal the e-ray velocity is greater than the o-ray velocity and hence the ellipsoidal surface lies outside the spherical surface.

Let us consider that parallel beam of light CD falls **normally** on the surface of the negative crystal. As soon as the parallel beam strikes the crystal boundary, each point on the wavefront becomes a new point source of light. The points A and B are chosen for the purpose of illustration. According to Huygens' principle the points A and B produce elliptical and spherical wavelets. Since the two wavelets must touch each other along the optic axis, they touch at points P and Q (Fig.20.34). Note that the section of the ellipsoid lies outside the section of the circle. The minor axis of the ellipse and the radius of the circle are equal. The position of the two wave fronts after a lapse of 't' seconds can be determined as follows. A circle of radius '$\upsilon_o t$' is drawn taking A as centre. Similarly, a circle of the same radius is drawn with B as centre. The ellipsoidal wave front can be drawn if the major and minor axes are known. In a negative crystal the major axis is equal to '$2\upsilon_e t$' and minor axis is '$2\upsilon_o t$'. The circle and the ellipse touch each other, as required, along the optic axis, which is perpendicular to the refracting edge MN. If we now draw the common tangents to the secondary wavelets, they represent the plane wavefronts corresponding to the two rays. Thus PQ is the common tangent to the spherical wavelets as well as the ellipsoidal wave fronts. If now we join the point of origin of wavelets to the points of tangency, the direction of propagation of the rays will be known. Thus, if we join A to P and B to Q, we obtain the direction of propagation of o-ray as well as e-ray. We find that in this case the o- ray and e-ray travel along the same direction with the same velocity.

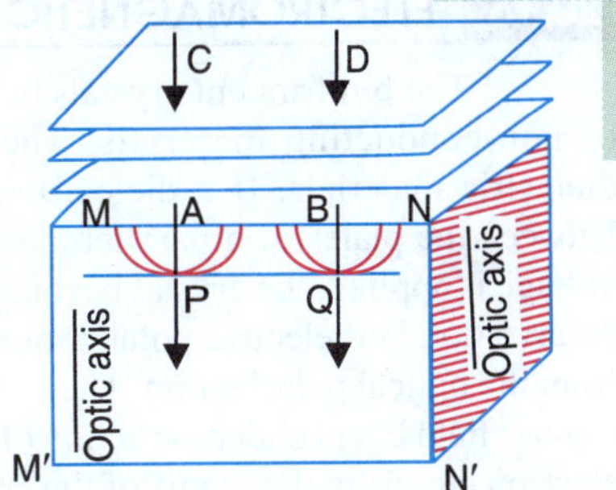

Unpolarized light beam incident normally on a calcite slab parallel to the optic axis. The o-ray and e-ray travel along the same velocity.

Fig. 20.34

20.11. EXPERIMENTAL DETERMINATION OF PRINCIPAL REFRACTIVE INDICES

For determining the refractive index for the extraordinary ray a calcite crystal is cut in the form of a prism with the optic axis perpendicular to the refracting edge of the prism and perpendicular to the base BC (Fig.20.35). It can also be cut with the optic axis parallel to the refracting edge of the prism. The prism is placed on the spectrometer table and is adjusted for the minimum deviation position for the extraordinary rays. The angle of minimum deviation δ_m is determined and the principal refractive index for the extraordinary ray is calculated from the relation (20.16a).

$$\mu_e = \frac{\sin\left(\frac{A+\delta_m}{2}\right)}{\sin\frac{A}{2}} \qquad \text{... (20.16 } a\text{)}$$

For a given wavelength, the ordinary and the extraordinary rays are separated while passing through the prism. Therefore, the angle of minimum deviation δ_m' for the ordinary ray can be measured and thus its refractive index can be calculated from the relation (20.16b).

$$\mu_0 = \frac{\sin\left(\frac{A+\delta_m'}{2}\right)}{\sin\frac{A}{2}} \qquad \text{... (20.16 } b\text{)}$$

(a) (b)

Fig. 20.35

20.16. ELECTROMAGNETIC THEORY OF DOUBLE REFRACTION:

The birefringent crystals belong to the group of non-conducting materials. They are known as dielectric materials. If a dielectric crystal is placed between the plates of a parallel plate capacitor and a voltage is applied, the crystal becomes polarized. This polarization is dielectric polarization and is different from the optical polarization. The voltage produces an electric field E and under the action of the electric field, electron clouds in the atoms of the crystal distort with the result that the centres of action of negative charge and positive charge get separated by a small distance. It means that electrical dipoles are produced throughout the crystal. Negative charges of the dipoles lie on the side of the positively charged plate of the capacitor and positive charges lie on the side of the negatively charged plate. Thus electric charges of opposite nature are induced on the opposite faces of the crystal in response to the applied electric field (see Fig.20.36). This is known as **dielectric polarization.**

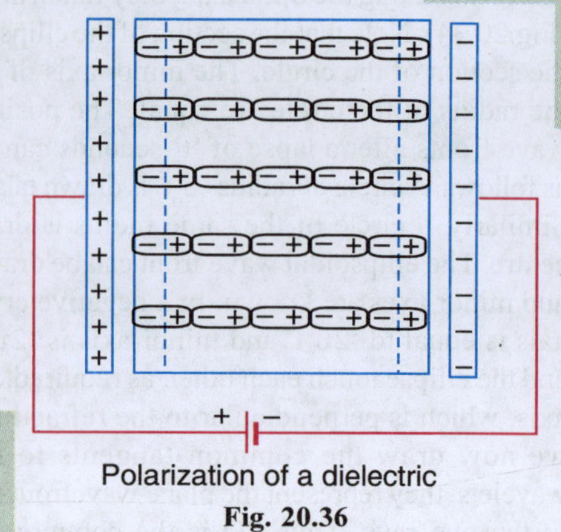

Polarization of a dielectric

Fig. 20.36

The electric field, polarization and displacement in the crystal are related by

$$D = \varepsilon_o E + P \quad \text{... (20.17)}$$

where $$D = \varepsilon_o \varepsilon_r E \quad \text{... (20.18)}$$

and $$P = \chi \varepsilon_0 E \quad \text{... (20.19)}$$

ε_r and χ are known as *relative permittivity* and *electric susceptibility* of the crystal. The relative permittivity is related to the refractive index of the material through the relation

$$\mu = \sqrt{\varepsilon_r} = \sqrt{\varepsilon / \varepsilon_o} \quad \text{... (20.20)}$$

We know that refractive index is related to the velocity of light in the medium.

$$\mu = \frac{c}{\upsilon} \quad \text{... (20.21)}$$

It is not necessary for dielectric polarization that the crystal be placed in between the plates of a capacitor. The electric field of a light wave propagating in the crystal also can cause dielectric polarization. According to the above relations the dielectric polarization caused by the light wave is related to the velocity (direction of propagation) of light wave with in the crystal.

In case of isotropic crystals, the vectors *D, E* and *P* are parallel and ε_r, μ and χ are scalars. In anisotropic crystals the induced polarization is in a direction different from that of the field. As a result, ε_r, μ and χ are tensor quantities. The components of polarization are given by

$$\begin{aligned} P_x &= \varepsilon_o \left(\chi_{11} E_x + \chi_{12} E_y + \chi_{13} E_z\right) \\ P_y &= \varepsilon_o \left(\chi_{21} E_x + \chi_{22} E_y + \chi_{23} E_z\right) \\ P_z &= \varepsilon_o \left(\chi_{31} E_x + \chi_{32} E_y + \chi_{33} E_z\right) \end{aligned} \quad \text{(20.22)}$$

The above equations can be simplified by choosing the coordinate axes in such a way that the off-diagonal elements vanish. Then

$$\begin{aligned} P_x &= \varepsilon_o \chi_{11} E_x \\ P_y &= \varepsilon_o \chi_{22} E_y \end{aligned} \quad \text{(20.23)}$$

$$P_z = \varepsilon_o \chi_{33} E_z$$

Similarly, we can express $D_x = \varepsilon_{11} E_x$,

$$D_y = \varepsilon_{22} E_y \qquad \text{... (20.24)}$$

$$D_z = \varepsilon_{33} E_z$$

These directions are called the **principal axes** of the crystal and the corresponding diagonal terms $\varepsilon_{11}, \varepsilon_{22}$ and ε_{33} the *principal permittivities*. The variation in the permittivity and the corresponding variation in refractive index and hence in wave velocity is the origin of double refraction. μ_x, μ_y and μ_z are the three *principal refractive indices* corresponding to the three principal permittivities. **The subscripts *x*, *y* and *z* relate to the direction of polarization of the light waves and not to their direction of propagation.** The principal permittivities and refractive indices are related as

$$\mu_x = \sqrt{\varepsilon_{11} / \varepsilon_o}\,, \quad \mu_y = \sqrt{\varepsilon_{22} / \varepsilon_o} \quad \text{and} \quad \mu_z = \sqrt{\varepsilon_{33} / \varepsilon_o} \qquad \text{... (20.25)}$$

20.16.1 INDEX ELLIPSOID OR OPTICAL INDICATRIX:

The different speeds of light in different directions in a crystal are conveniently represented by a three-dimensional figure that shows the refractive index for light waves in their direction of polarization. The representation is known as the **indicatrix**. *The optical indicatrix is spherical or ellipsoidal surface having the three different refractive indices as axes.*

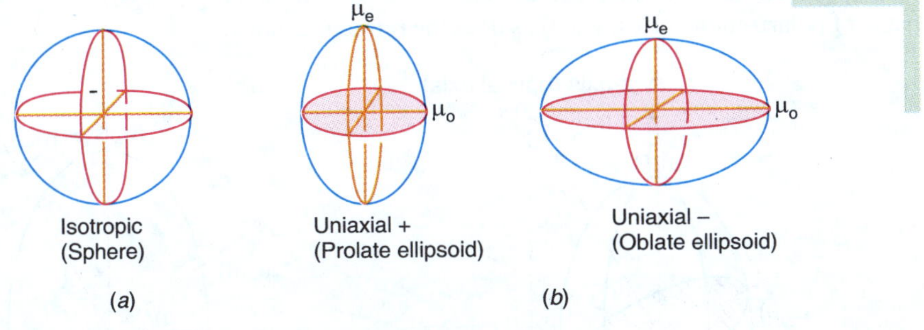

Fig. 20.37

In case of isotropic crystals, the axes are all of the same length and consequently the refractive indices in three directions are the same. Therefore, the indicatrix is a *sphere*, as shown in Fig. 20.37 (a). In case of uniaxial crystals, a unique axis exists normal to the plane of two equal axes. The velocity of propagation of light along this unique axis (z-direction) is independent of the wave polarisation and this direction is the *optic axis* direction. The refractive index along this direction is μ_e. In a plane perpendicular to the optic axis, the refractive index is independent of direction, that is $\mu_x = \mu_y$ denoted by μ_o. Therefore, in uniaxial crystals the indicatrix is an **ellipsoid** having a circular cross-section in the *xy*-plane. The ellipsoid is prolate if $\mu_e > \mu_o$ and oblate if $\mu_e < \mu_o$ (see Fig. 20.37b).

The form of optical indicatrix can be obtained as follows. The energy density in a dielectric medium is given by

$$W = \tfrac{1}{2} D \cdot E \qquad \text{... (20.26)}$$

$$W = \tfrac{1}{2}\left[\frac{D_x^2}{\varepsilon_{11}} + \frac{D_y^2}{\varepsilon_{22}} + \frac{D_z^2}{\varepsilon_{33}}\right] \qquad \text{... (20.27)}$$

or $$\frac{1}{2}\left[\frac{D_x^2}{W\varepsilon_{11}}+\frac{D_y^2}{W\varepsilon_{22}}+\frac{D_z^2}{W\varepsilon_{33}}\right]=1$$

or $$\left[\frac{D_x^2}{2W\mu_x^2\varepsilon_o}+\frac{D_y^2}{2W\mu_y^2\varepsilon_o}+\frac{D_z^2}{2W\mu_z^2\varepsilon_o}\right]=1$$

Writing $x^2 = D_x^2/2\varepsilon_o W$, $y^2 = D_y^2/2\varepsilon_o W$ and $z^2 = D_z^2/2\varepsilon_o W$, we get

$$\frac{x^2}{\mu_x^2}+\frac{y^2}{\mu_y^2}+\frac{z^2}{\mu_z^2}=1 \qquad \text{... (20.28)}$$

This is the equation of an ellipsoid with semi-axes μ_x, μ_y and μ_z. In the case of a uniaxial crystal the indicatrix has circular symmetry about the z-axis. In this case,

$$\frac{x^2}{\mu_0^2}+\frac{y^2}{\mu_0^2}+\frac{z^2}{\mu_e^2}=1 \qquad \text{... (20.29)}$$

Let us now consider light propagating in a direction ***r***, at an angle θ to the optic axis. Because of the circular symmetry, we can choose that the y-axis should coincide with the projection of ***r*** on the *xy*-plane. The plane normal to *r* intersects the ellipsoid in the shaded ellipse. The two allowed directions of polarization are parallel to the axes of the ellipse and thus correspond to *OP* and *OQ* (Fig.20.38). They are thus perpendicular to *r* as well as to each other. The two waves polarized along these directions have refractive indices given by $OP = \mu_o$ and $OQ = \mu_e(\theta)$. In case of the e-ray, the plane of polarization varies with θ as does the refractive index.

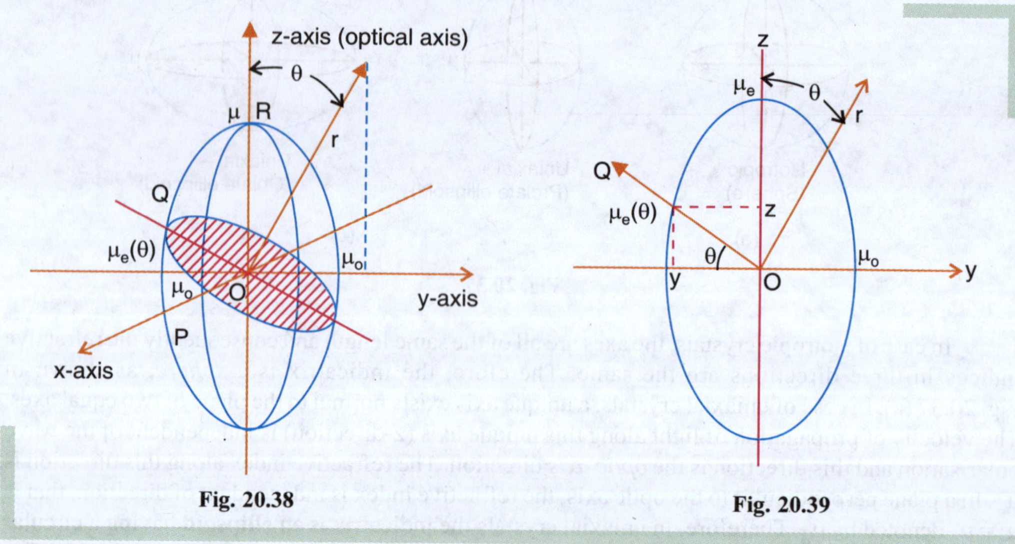

Fig. 20.38

Fig. 20.39

We can determine the relationship between $\mu_e(\theta)$, μ_e and μ_o with the help of the Fig.20.39. The figure shows the intersection of the indicatrix with the yz-plane. From the diagram, it is seen that

$$\mu_e^2(\theta) = z^2 + y^2 \qquad \text{... (20.30)}$$

and $$z = \mu_e(\theta)\sin\theta \qquad \text{... (20.31)}$$

∴ $$\mu_e^2(\theta) = \mu_e^2(\theta)\sin^2\theta + y^2$$

or $$y^2 = \mu_e^2(\theta)\cos^2\theta \qquad \text{... (20.32)}$$

The equation of the ellipse shown in Fig. 20.39 is

$$\frac{y^2}{\mu_o^2}+\frac{z^2}{\mu_e^2}=1 \quad (20.23)$$

Substituting equ. (20.31) and equ. (20.32) into equ. (20.33), we obtain

$$\frac{1}{\mu_e^2(\theta)}=\frac{\cos^2\theta}{\mu_o^2}+\frac{\sin^2\theta}{\mu_e^2} \quad (20.34)$$

Thus, for $\theta = 0^o$, that is propagation along the optic axis, $\mu_e(0°) = n_0$ while for $\theta = 90^o$, $\mu_e(90°) = \mu_e$. The two polarizations which can be propagated correspond to the maximum and minimum refractive indices given by the index ellipsoid. For propagation parallel to the optic axis (that is *z*-direction), there is no birefringence as the section of the ellipsoid perpendicular to this direction is a circle. For propagation perpendicular to the optic axis, the birefringence will be a maximum, the permitted polarizations will be parallel to the y-axis with refractive index μ_o and parallel to the *z*-axis with refractive index μ_e.

20.17. PHASE DIFFERENCE BETWEEN e-RAY AND o-RAY

We have seen that natural light incident on the surface of an anisotropic crystal undergoes double refraction and produces two plane polarized waves (Fig.20.40a). Let us consider the particular case of a slice of a calcite crystal (a negative uniaxial crystal) where the optic axis is parallel to refracting face of the crystal. The two waves travel along the same direction in the crystal but with different velocities. As a result, when the waves emerge from the rear face of the crystal, an optical path difference would have developed between them (Fig.20.40b). The optical path difference can be calculated as follows:

Let *d* be the thickness of the crystal.

The optical path for o-ray within the crystal $= \mu_o d$

The optical path for e-ray within the crystal $= \mu_e d$

∴ The optical path difference between e-ray and o-ray $\Delta = (\mu_o - \mu_e)\, d$ ... (20.35)

Consequently, a phase difference arises between the two waves. It is given by

$$\delta = \frac{2\pi}{\lambda}(\mu_o - \mu_e)d \quad ... (20.36)$$

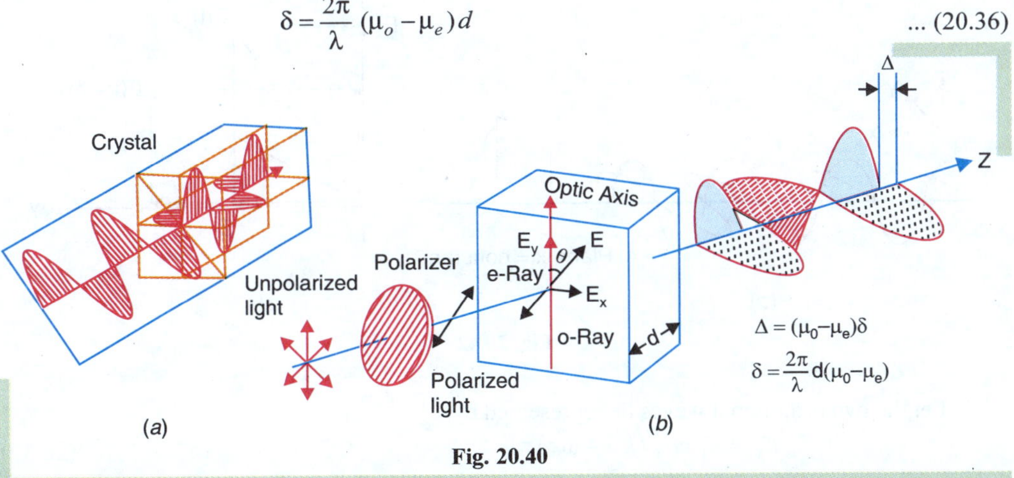

Fig. 20.40

As the two component waves are derived from the same incident wave, the two waves are in phase at the front face and have emerged from the crystal with a constant phase difference and hence it may be expected that the waves are in a position to interfere with each other. However, as the planes of polarization of o-ray and e-ray are perpendicular to each other, interference cannot take place between e- and o-rays. The waves instead combine with each other to give elliptically polarized wave.

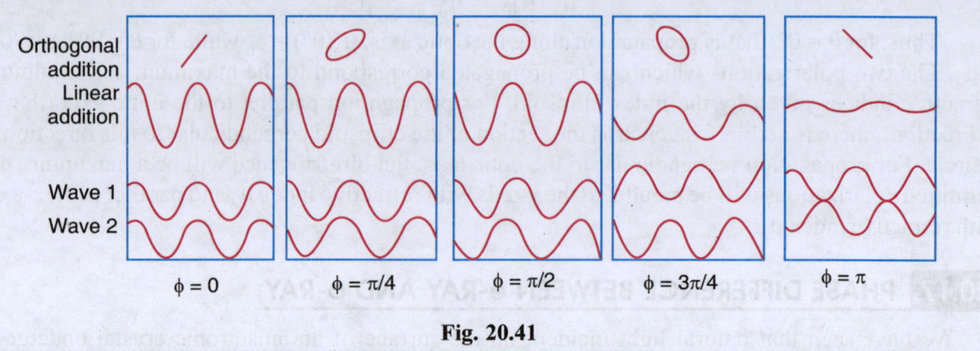

Fig. 20.41

The most important point to be noted here is as follows: The superposition of two coherent waves having a *common plane* of polarization yields a wave polarized in the same plane. The superposition leads to linear addition of waves, i.e., **interference**. The resultant vibration is *linear* and occurs in the same direction as that of the two superposing vibrations (see Fig.20.41). On the other hand, if the two coherent waves are polarized in orthogonal planes, the resultant vibrational motion takes place in two dimensions either in the form of an ellipse, a circle or a straight line depending on the phase difference between the waves (Fig.20.41). This is known as *orthogonal addition.*

20.18. SUPERPOSITION OF WAVES LINEARLY POLARISED AT RIGHT ANGLES

Let us now look at the result of superposition of two waves linearly polarized at right angles to each other (see Fig.20.42 a). Let us consider two light waves travelling in the same direction, *x* ; one wave is polarized in the *xy*-plane and the other is polarized in *yz*-plane. We are interested to know the state of polarization of the resultant wave at the plane *x* = *constant*.

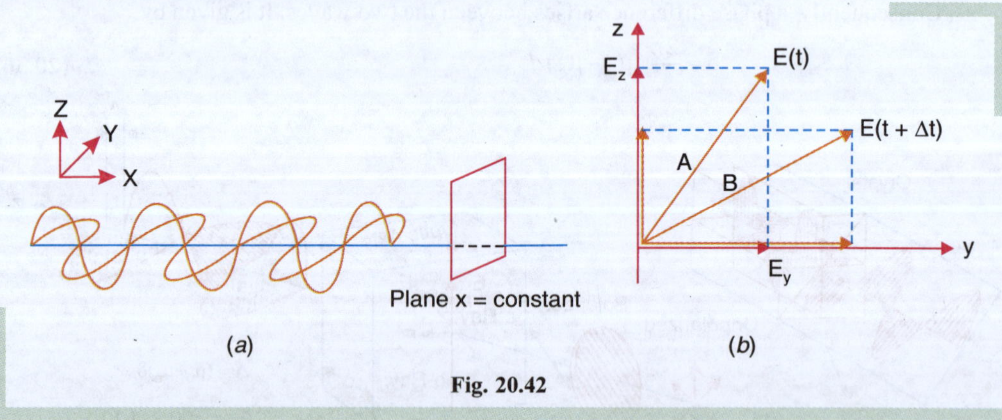

Fig. 20.42

Let the two orthogonal waves be represented by

$$E_y = E_1 \cos\left(k\,x - \omega\,t\right) \qquad (20.37)$$

$$E_z = E_2 \cos\left(k x - \omega t + \delta\right) \quad \text{... (20.38)}$$

The waves are of the same frequency $\nu = \omega/2\pi$. δ is the phase difference between the waves. At a given time t, the optical vectors E_y and E_z produce a resultant optical vector of magnitude, say A and at a slightly later time, $t + \Delta t$, they give rise to a resultant vector of magnitude B which points in a different direction (see Fig.20.42b). With progress of time the tip of the resultant optical vector moves along a curve in the yz-plane. We apply the *principle of superposition* to find the equation of the curve traced by the tip of the resultant vector of the two vectors.

According to the principle of superposition

$$\begin{aligned} E &= E_y + E_z \\ &= E_1 \cos(kx - \omega t) + E_2 \cos(kx - \omega t + \delta) \quad \text{... (20.39)} \end{aligned}$$

The equation of the curve may be found by eliminating 't' from the equations. We can write the expansion of equ. (20.39) as

$$\begin{aligned} E_z &= E_2 \cos(kx - \omega t) \cos\delta - E_2 \sin(kx - \omega t) \sin\delta \quad \text{... (20.40)} \\ &= E_2 \cos(kx - \omega t) \cos\delta \pm [1 - \cos^2(kx - \omega t)]^{1/2} E_2 \sin\delta \end{aligned}$$

We find from equ.(20.37) that

$$\cos(kx - \omega t) = E_y/E_1$$

$$\therefore \quad E_z = E_2 \frac{E_y}{E_1} \cos\delta \pm \sqrt{1 - \frac{E_y^2}{E_1^2}}\; E_2 \sin\delta \quad \text{... (20.41)}$$

Rearranging the terms, we get

$$\left[E_z - \frac{E_2}{E_1} E_y \cos\delta\right] = \pm\sqrt{1 - \left(\frac{E_y}{E_1}\right)^2}\; E_2 \sin\delta$$

On squaring both the sides, we obtain

$$E_z^2 + \frac{E_y^2 E_2^2}{E_1^2} \cos^2\delta - \frac{2E_y E_z E_2}{E_1} \cos\delta = E_2^2 \sin^2\delta - \frac{E_y^2 E_2^2}{E_1^2} \sin^2\delta$$

Rearranging the terms, we get

$$E_z^2 + \frac{E_y^2 E_2^2}{E_1^2} (\cos^2\delta + \sin^2\delta) - \frac{2E_y E_z E_2}{E_1} \cos\delta = E_2^2 \sin^2\delta \quad \text{... (20.42)}$$

$$E_z^2 + \frac{E_y^2 E_2^2}{E_1^2} - \frac{2E_y E_z E_2}{E_1} \cos\delta = E_2^2 \sin^2\delta$$

Dividing both the sides by E^2_2 and rearranging the terms, we obtain

$$\frac{E_y^2}{E_1^2} + \frac{E_z^2}{E_2^2} - \frac{2E_y E_z}{E_1 E_2} \cos\delta = \sin^2\delta \quad \text{... (20.43)}$$

Equation (20.43) is the general equation of an *ellipse*. Hence, the tip of the resultant vector traces an ellipse in the yz-plane. The ellipse is constrained within a rectangle having sides $2E_1$ and $2E_2$. The major axis makes an angle α with the y-axis (see Fig.20.43).

$$\tan 2\alpha = 2E_1 E_2 \cos\delta / \left(E_1^2 - E_2^2\right)$$

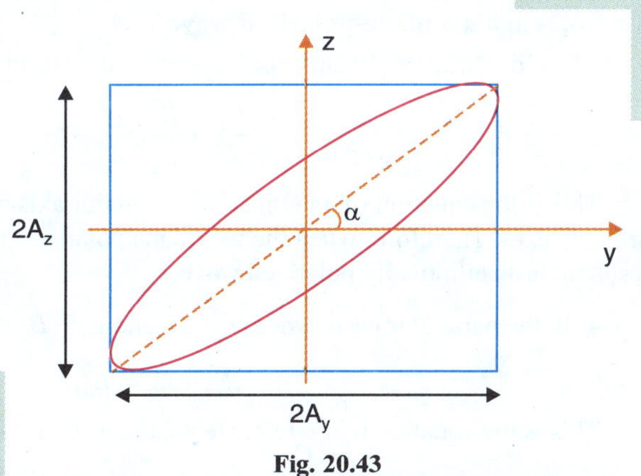

Fig. 20.43

Special cases:

1. When $\delta = 0$, or $\pm 2m\pi$, the two waves are in phase.
 $\cos\delta = 1$ and $\sin\delta = 0$ and the equ. (20.43) reduces to

$$\frac{E_y^2}{E_1^2}+\frac{E_z^2}{E_2^2}-\frac{2E_yE_z}{E_1E_2}=0$$

$$\left[\frac{E_y}{E_1}-\frac{E_z}{E_2}\right]^2=0$$

$$\frac{E_y}{E_1}-\frac{E_z}{E_2}=0$$

$$\therefore \qquad E_z=\frac{E_2}{E_1}E_y \qquad \text{... (20.44)}$$

The above equation represents a straight line, having a slope (E_2/E_1). Therefore, the equation represents a wave having its plane of polarisation making an angle $\tan^{-1}(E_2 - E_1)$ with respect to the y-axis. It means that **the resultant of two plane-polarised waves, which are in phase (*i.e.,* coherent waves), is again a plane-polarised wave.**

2. When $\delta = \pi$, or $\pm(2m+1)\pi$, the two waves are in opposite phase.
 $\cos\delta = -1$ and $\sin\delta = 0$ and the equ. (20.43) reduces to

$$\frac{E_y^2}{E_1^2}+\frac{E_z^2}{E_2^2}-\frac{2E_yE_z}{E_1E_2}=0$$

$$\left[\frac{E_y}{E_1}+\frac{E_z}{E_2}\right]^2=0$$

$$\frac{E_y}{E_1}+\frac{E_z}{E_2}=0$$

$$\therefore \qquad E_z=-\frac{E_2}{E_1}E_y \qquad \text{... (20.45)}$$

This equation represents a straight line of a slope $(-E_2/E_1)$. Therefore, the equation represents a wave having its plane of polarization making an angle $\tan^{-1}(-E_2/E_1)$ with respect to the y-axis. It means that **the resultant of two plane-polarized waves, which are in opposite phase (*i.e.,* coherent waves), is again a plane-polarized wave.**

3. If $\delta = \pi/2$, or $\pm(2m+1)\pi/2$, then $\cos\delta = 0$ and $\sin\delta = 1$. Equ.(20.43) reduces to

$$\frac{E_y^2}{E_1^2}+\frac{E_z^2}{E_2^2}=1 \qquad \text{... (20.46)}$$

This is the equation of an ellipse whose major axis and minor axis coincide with *y*- and *z*-co-ordinate axes. Therefore, when the two plane polarized waves are out of phase by 90º, their resultant is an elliptically polarized wave.

4. In the particular case, when $\delta = \pi/2$ and $E_1 = E_2 = E_o$, equ.(20.43) reduces to

$$E_y^2+E_z^2=E_o^2 \qquad \text{... (20.47)}$$

This is the equation of a circle. Hence the resultant light is circularly polarized.

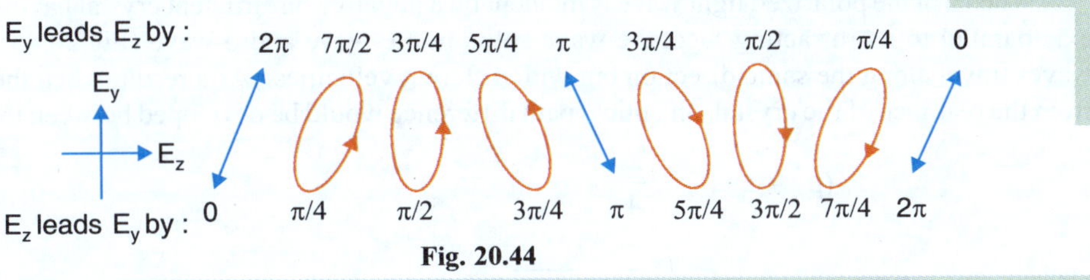

Fig. 20.44

Fig.20.44 shows more generally how the E vector changes with time for various values of δ.

20.18.1 SUPERPOSITION OF e-RAY AND O-RAY:

Now we are in a position to understand what happens when e-ray and o- ray overlap on each other after emerging from an anisotropic crystal plate. It is obvious that they cannot produce interference fringes as in a double slit experiment, since the o-ray are not vibrating in the same plane. On the other hand, they combine to produce different states of polarization depending upon their optical path difference.

1. When the optical path difference is 0 or an even or odd multiple of λ/2, the resultant light wave is linearly polarized.
2. When the optical path difference is λ/4, the resultant light wave is **elliptically polarized.**
3. In the particular instance when the wave amplitudes are equal and the optical path difference is λ/4, the resultant light wave is **circularly polarized**.

20.19. RETARDERS

A **retarder** is a uniform plate of birefringent material whose optic axis lies in the plane of the plate. Retarders are called quarter-wave plates, half-wave plates and full-wave plates depending on their action. They divide the incident wave into two polarized waves that travel perpendicular to the plate at different speeds. A phase retardation of one wave relative to the other is therefore introduced as the waves cross the thickness *d* of the plate. They are used to produce circularly or elliptically polarised light and to analyse polarised light into its elliptical components.

20.19.1 QUARTER WAVE PLATE

A *quarter wave plate* is a thin plate of birefringent crystal having the optic axis parallel to its refracting faces and its thickness adjusted such that it introduces a quarter-wave (λ/4) path difference (or a phase difference of 90°) between the e-ray and o-ray propagating through it.

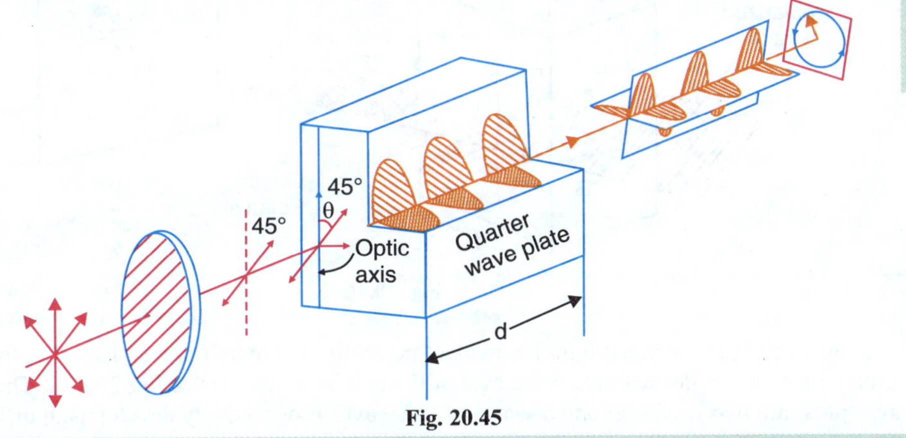

Fig. 20.45

When a plane polarized light wave is incident on a negative birefringent crystal having the optic axis parallel to its refracting face, the wave splits into e-wave and o-wave (Fig.20.45). The two waves travel along the same direction but with different velocities. As a result, when they emerge from the rear face of the crystal, an optical path difference would be developed between them. Thus,

$$\therefore \quad (\mu_o - \mu_e)d = \frac{\lambda}{4} \qquad \text{... (20.48)}$$

$$d = \frac{\lambda}{4(\mu_o - \mu_e)} \qquad \text{... (20.49)}$$

A quarter wave plate introduces a phase difference δ, between e-ray and o-ray given by $\delta = (2\pi/\lambda)\Delta = \pi/2 = 90°$.

A quarter-wave plate is used for producing elliptically or circularly polarized light. It converts plane-polarized light into elliptically or circularly polarized light depending upon the angle that the incident light vector makes with the optic axis of the quarter wave plate.

20.19.2 HALF WAVE PLATE

A *half wave plate* is a thin plate of birefringent crystal having the optic axis parallel to its refracting faces and its thickness chosen such that it introduces a half-wave ($\lambda/2$) path difference (or a phase difference of 180°) between e-ray and o-ray.

When a plane polarized light wave is incident on a birefringent crystal having the optic axis parallel to its refracting faces, it splits into two waves: o- and e-waves. The two waves travel along the same direction inside the crystal but with different velocities. As a result, when they emerge from the rear face of the crystal, an optical path difference would be developed between them.

$$\therefore \quad (\mu_o - \mu_e)d = \frac{\lambda}{2} \qquad \text{... (20.50)}$$

$$\therefore \quad d = \frac{\lambda}{2(\mu_o - \mu_e)} \qquad \text{... (20.51)}$$

A half wave plate introduces a phase difference δ, between e-ray and o-ray given by $\delta = (2\pi/\lambda)\Delta = \pi = 180°$.

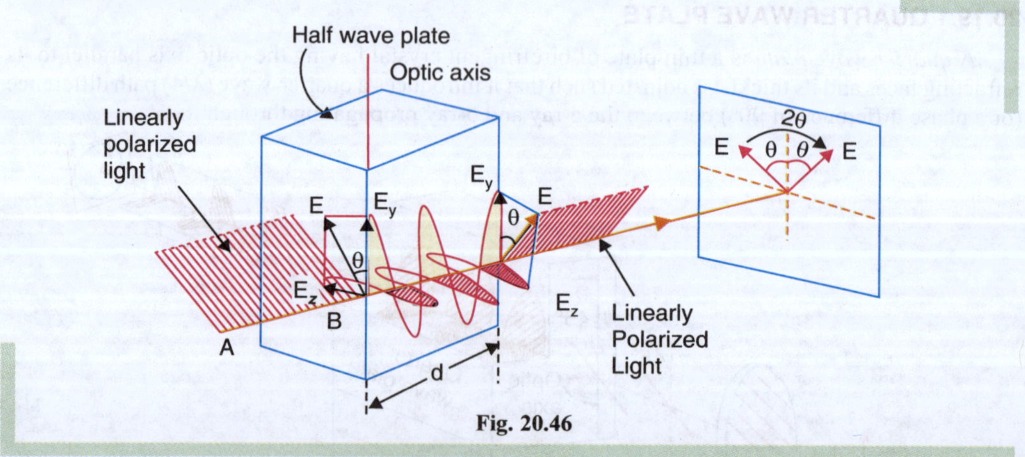

Fig. 20.46

Now let a plane polarized light be incident normally on the half-wave plate. Let the electric vector E make an angle with the optic axis of the half wave plate (See Fig.20.46). The incident wave splits into two waves, e- and o-waves. The waves progressively develop path difference as they travel through the crystal and they emerge with a phase difference of 180°. When the two waves combine, they yield a plane-polarized wave, which has its plane of polarization rotated through an angle of 2θ.

Therefore, ***a half-wave plate rotates the plane of polarization of the incident plane polarized light through an angle 2θ.*** The half wave plate will invert the handedness of elliptical or circular polarized light, changing right to left and vice versa.

Now we are in a position to understand what happens when e-ray and o- ray overlap on each other after emerging from an anisotropic crystal plate. It is obvious that they cannot produce interference fringes as in a double slit experiment. On the other hand, they combine to produce different states of polarization depending upon their optical path difference.

1. When the optical path difference is 0 or an even or odd multiple of λ/2, the resultant light wave is linearly polarized.
2. When the optical path difference is λ/4, the resultant light wave is **elliptically polarized.**
3. In the particular instance when the wave amplitudes are equal and the optical path difference is λ/4, the resultant light wave is **circularly polarized.**

20.20. PRODUCTION OF ELLIPTICALLY POLARIZED LIGHT

A quarter wave plate and a polarizer are the optical devices necessary to produce elliptically polarized light from unpolarized light.

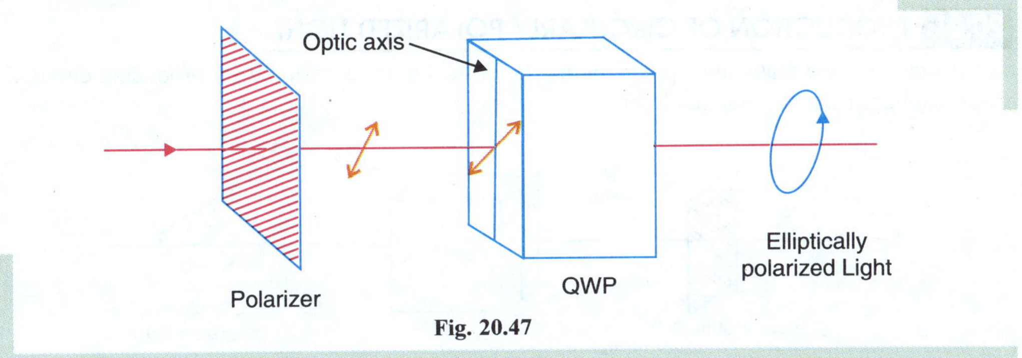

Fig. 20.47

Unpolarized light is first converted to plane polarized light by allowing it to pass through a polarizer (a polaroid sheet or a Nicol prism), as shown in Fig.20.47. The plane-polarized light is then made incident on a quarter wave plate. The quarter wave plate or the polarizer is rotated such that the electric vector E of plane polarized light wave makes an angle θ (≠45°) with the optic axis of the quarter wave plate. The incident ray divides into o-ray and e-ray of amplitudes E sin θ and E cos θ. The rays travel along the same direction in the crystal with different velocities. The two rays are polarized in orthogonal planes. They are in phase at the front face but progressively get out of phase as they travel through the crystal. When they emerge out of the crystal they will have a path difference of λ /4 or a phase difference of 90°. When they combine, they produce elliptically polarized light.

20.20.1 DETECTION OF ELLIPTICALLY POLARIZED LIGHT

The light beam is allowed to pass through an analyzer (a polaroid sheet or a Nicol prism). If on rotating the analyzing Polaroid sheet or Nicol, the intensity of the emerging beam varies from a maximum to a minimum value, but never reaching zero, then the incident light is elliptically polarized.

A similar result would be obtained if the incident light were partially polarized. The two cases may be distinguished by inserting a quarter wave plate in the path of light before it falls on the analyzer.

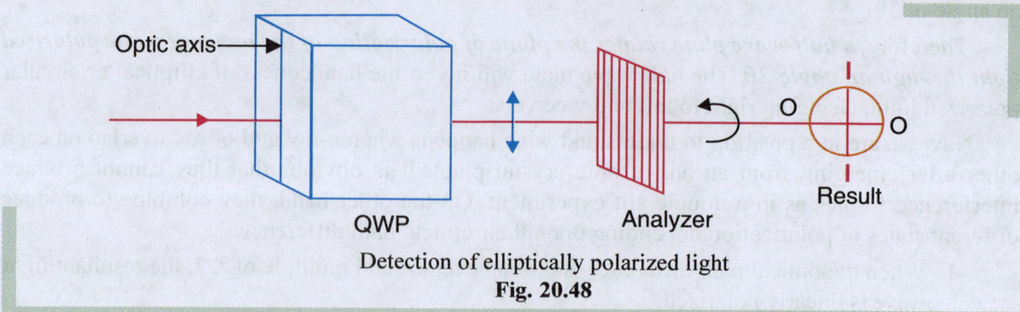

Detection of elliptically polarized light

Fig. 20.48

If the original light is elliptically polarized, it may be considered as resultant of two coherent plane polarized waves that is e-ray and o-ray, which are out of phase by 90°. If the light passes through the quarter wave plate, an additional phase difference of 90° is introduced between the e-ray and o-ray. Therefore, the total phase difference becomes 180° between the e-ray and o-ray. On emerging from the quarter plate, the e-and o-rays combine to produce plane-polarized light. If the light coming out of quarter wave plate is examined with an analyzer, light will be extinguished twice in one full rotation of the polarizer as shown in Fig.20.48.

20.21. PRODUCTION OF CIRCULARLY POLARIZED LIGHT

A quarter wave plate and a polarizer are the optical devices required for producing circularly polarized light from unpolarized light.

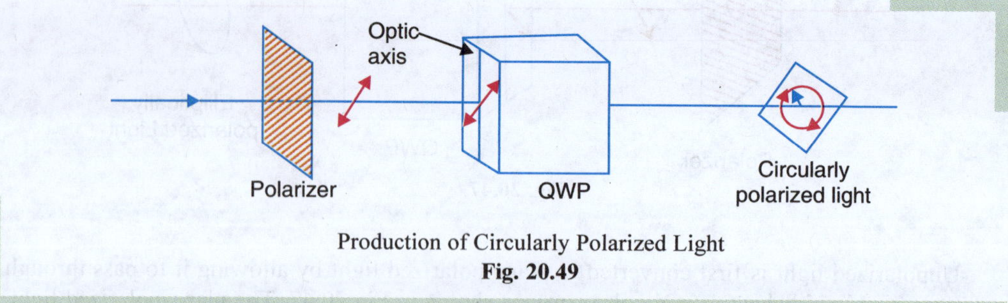

Production of Circularly Polarized Light

Fig. 20.49

Unpolarized light is first converted to plane polarized light by allowing it to pass through a polarizer (a polaroid sheet or a Nicol prism), as shown in Fig.20.49. Plane polarized light is then made incident on a quarter wave plate. The polarizer and the quarter wave plate are rotated such that the electric vector E of the plane-polarized wave makes an angle of 45° with the optic axis of the quarter wave plate. The plane polarized wave incident on the quarter wave plate splits into two rays, o-ray and e-ray of equal amplitude ($E_1 \cos 45^\circ = E_2 \sin 45^\circ$). The two rays travel in the same direction inside the crystal but with different velocities. The two rays are in phase at the front face of the crystal but progressively get out of phase as they travel through the crystal. As they emerge from the rear face of the crystal, they will have a path difference of $\lambda/4$ or phase difference of 90°. The two rays are linearly polarized in mutually perpendicular directions. When they combine, they produce circularly polarized light.

20.21.1 DETECTION OF CIRCULARLY POLARIZED LIGHT

The light beam is allowed to pass through an analyzer (a polaroid sheet or a Nicol prism). If on rotating the analyzing polaroid sheet or Nicol, the intensity of the emerging beam remains uniform, then the incident light is circularly polarized. A similar result would be obtained if the incident light is ordinary unpolarized light. The two cases may be distinguished by inserting a quarter wave plate in the path of light before it falls on the analyzer. If the original light is circularly polarized, it may be considered as resultant of two coherent plane-polarized waves, that is e-ray and o-ray, which are out of phase by 90°. If the light passes through the quarter wave plate, an additional phase difference of 90° is introduced between the e-ray and o-ray. Therefore, the total phase difference becomes 180° between the e-ray and o-ray. On emerging from the quarter plate, the e- and o-rays combine to produce plane-polarized light. Therefore, if the light coming out of quarter wave plate is examined with an analyzer, light will be extinguished twice in one full rotation of the polarizer as shown in Fig.20.50 or otherwise the incident light is unpolarized.

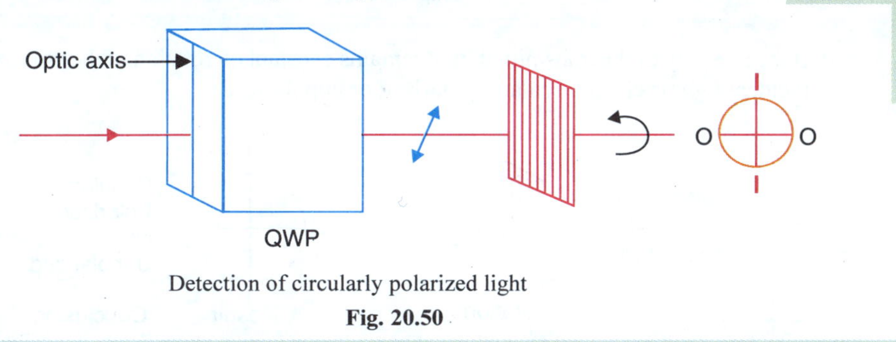

Detection of circularly polarized light

Fig. 20.50

20.22. ANALYSIS OF POLARIZED LIGHT

In practice light may exhibit any one of the three types of polarization, or may be unpolarized or a mixed type. The unaided eye cannot distinguish the different types of polarization. However, using a polarizer and a quarter wave plate, the actual type of polarization of a light beam can be ascertained. The following steps are used in the analysis of the type of polarization.

The light of unknown polarization is allowed to fall normally on a polarizer. The polarizer is slowly rotated through a full circle and the intensity of the transmitted light is observed. If the intensity of the transmitted light is extinguished twice in one full rotation of the polarizer, then the incident light is *plane polarized.*

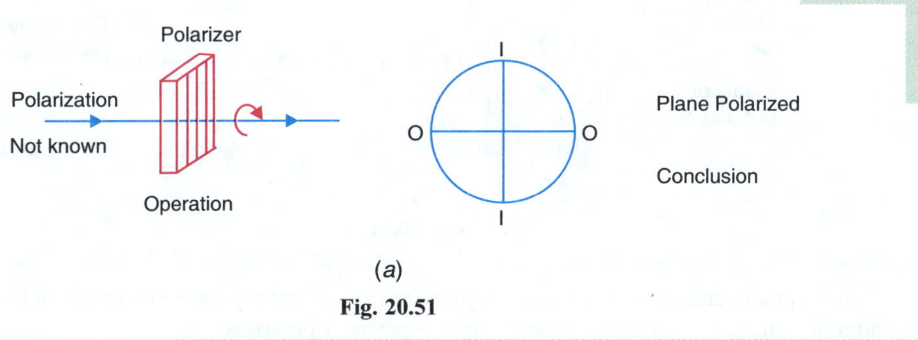

(*a*)

Fig. 20.51

(*i*) If the intensity of the transmitted light varies between a maximum and a minimum value but does not become extinguished in any position of the polarizer, then the incident light is either elliptically polarized or partially polarized.

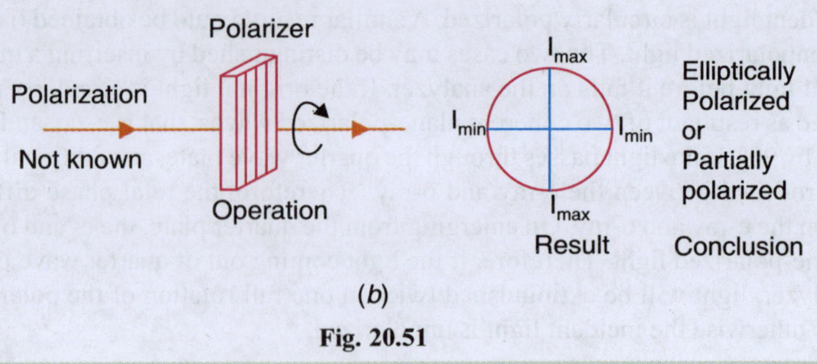

(*b*)

Fig. 20.51

(*ii*) If the intensity of the transmitted light remains constant on rotation of the polarizer, then the incident light is either circularly polarized or unpolarized.

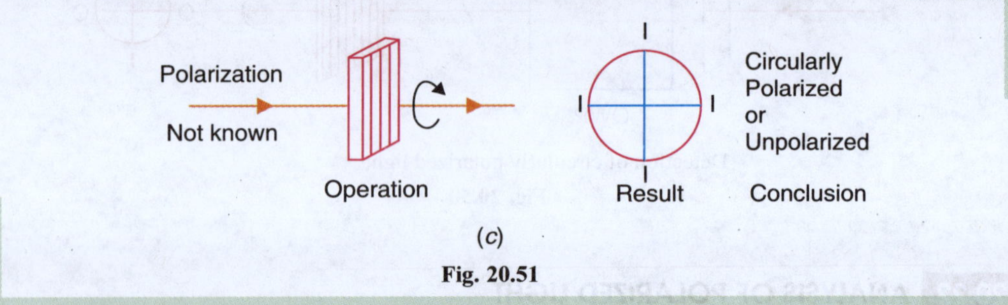

(*c*)

Fig. 20.51

To distinguish between elliptically polarized and partially polarized or between the circularly polarized and unpolarized light, we take the help of quarter wave plate. The light is first made to be incident on the quarter wave plate and then it passes through the polarizer.

(*iii*) If the incident light is elliptically polarized, the quarter wave plate converts it into a plane polarized beam. When this polarized light passes through the polarizer, it would be extinguished twice in one full rotation of the polarizer.

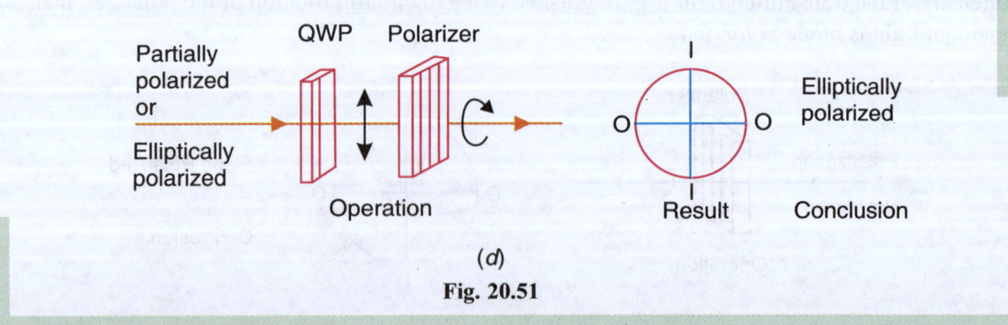

(*d*)

Fig. 20.51

On the other hand, if the transmitted light intensity varies between a maximum and a minimum without becoming zero, then the incident light is partially polarized.

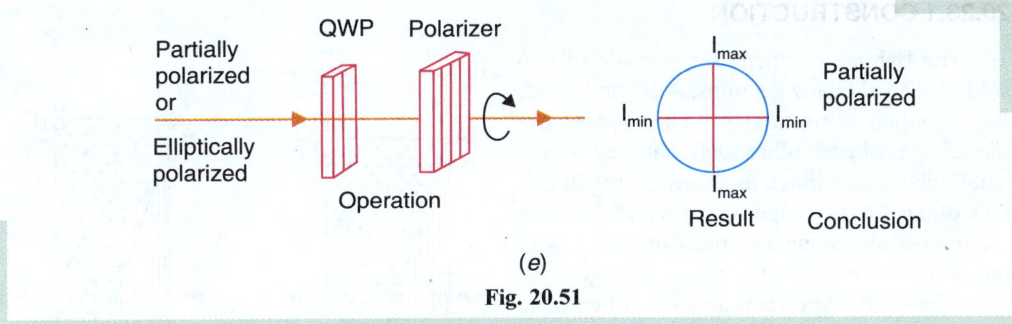

(e)

Fig. 20.51

(*iv*) If the incident light is circularly polarized, the quarter wave plate converts it into plane polarized light. When this plane polarized light passes through the polarizer, it would be completely extinguished twice in one full rotation of the polarizer.

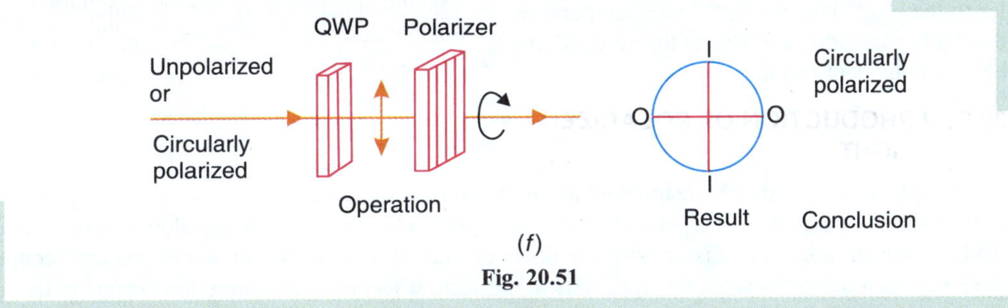

(f)

Fig. 20.51

On the other hand, if the intensity of the transmitted light stays constant, then the incident light is unpolarized.

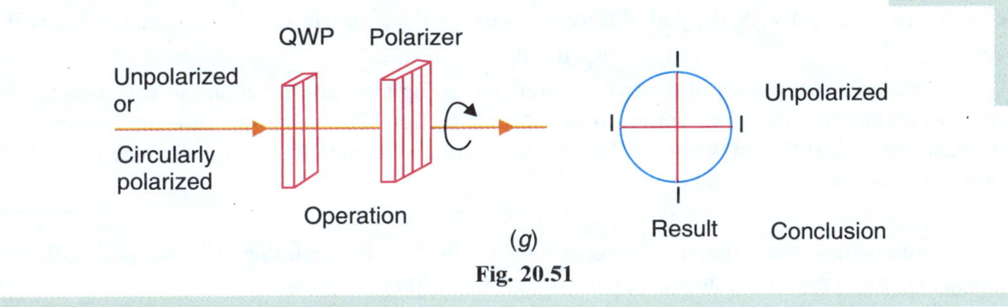

(g)

Fig. 20.51

20.23. BABINET COMPENSATOR

A *compensator* is an optical device whose function is to compensate a path difference. It is used in conjunction with a polarizer and analyzer combination to investigate elliptically polarized light. The compensator helps in determining the axis of the ellipse and the ratio of their lengths. Elliptically polarized light may be considered as the resultant of two coherent plane polarized vibrations occurring in mutually orthogonal directions and with an initial path difference of $\lambda/4$. When the elliptically polarized light passes through the device such that it introduces a further path difference of $\lambda/4$. The total path difference between the perpendicular vibrations becomes $\lambda/2$ and the vibrations recombine after emerging from the device to produce plane polarized light. From the analysis of this plane polarized light, the information regarding the incident elliptically polarized light can be obtained.

20.23.1 CONSTRUCTION

The Babinet compensator is made of two wedge-shaped quartz sections, ABC and ADC, having equal acute angles. The wedges are placed against each other such that they form a small rectangular block as shown in Fig.20.52. One of the quartz wedges is fixed and the other can be displaced along their plane of contact with the help of a micrometer screw arrangement. Thus, the combination acts as a plate of variable thickness.

The optic axis of the first section is parallel to its refracting edge AB and the optic axis of the second section is in a direction perpendicular to the edge. The two optic axes are perpendicular to each other and also to the perpendicular to the incident beam.

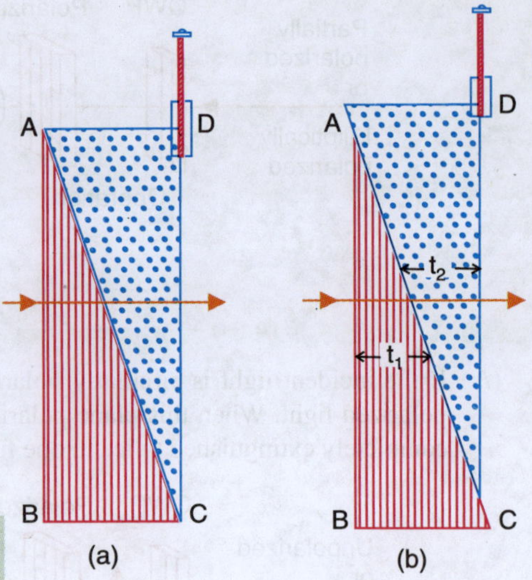

Fig. 20.52

20.23.2 PRODUCTION OF POLARIZED LIGHT

Let plane polarized light be incident normally on the face AB of the compensator. It splits into e-ray and o-ray parallel and perpendicular to AB respectively. The e-ray travels slower than o-ray in the first section, since quartz is a positive uniaxial crystal. When these rays enter the second section, the e-ray becomes o-ray since the optic axis in the second section is in a direction normal to that in the first prism. Similarly, o-ray becomes e-ray. Thus, the two rays exchange their velocities in passing from one section to the other section. The net effect is that one section cancels the effect of the other.

If d_1 is the thickness of the first section, and μ_e and μ_o are the refractive indices of quartz for e- and o-rays respectively, the path difference between the e- and o- rays in the first section will be

$$\Delta_1 = [\mu_e - \mu_o]\, d_1,$$

As the principal planes of the two sections are at right angles, the e- and o- rays change their roles in going from the first section to the second section. The velocities of e-ray and o-ray are interchanges and if the thickness of the second section is d_2, then the path difference between the rays in the second prism will be

$$\Delta_2 = [\mu_o - \mu_e]\, d_2.$$

As the compensator is thin, the separation of the rays is negligible. The net path difference between the two rays after emerging from the crystal will be

$$\Delta = \Delta_1 + \Delta_2$$

$$\Delta = (\mu_e - \mu_0)d_1 + (\mu_0 - \mu_e)d_2$$

$$= (\mu_e - \mu_0)(d_1 - d_2) \qquad \text{... (20.52)}$$

The net phase difference is

$$\delta = \frac{2\pi}{\lambda}(\mu_e - \mu_0)\,(d_1 - d_2) \qquad \text{... (20.53)}$$

For a ray passing through the centre of the compensator where $d_1 = d_2$ the net path difference and hence the phase difference is zero. It means that the effect of one wedge is exactly cancelled by the other. This is true for all wavelengths and the incident vibration is transmitted as such. Plane polarized light incident on the compensator will emerge as plane polarized light with its plane of vibration parallel to that of the incident light.

Any desired thickness difference ($d_2 - d_1$) can be achieved at the centre of the compensator by moving the second section relative to the first section. Thus, any desired value of phase difference can be obtained between the e and o- rays. Therefore, the light emerging will be either plane or circular or elliptically polarized light, depending on the phase difference.

Thus the compensator has the same effect as that of a wave plate of varying thickness. The advantage of compensator is that it can be arranged to suit any wavelength while a quarter wave plate is designed to suit only one particular wavelength.

20.23.3 ANALYSIS OF ELLIPTICALLY POLARIZED LIGHT:

Using the compensator, one can determine the characteristics of elliptically polarized light.

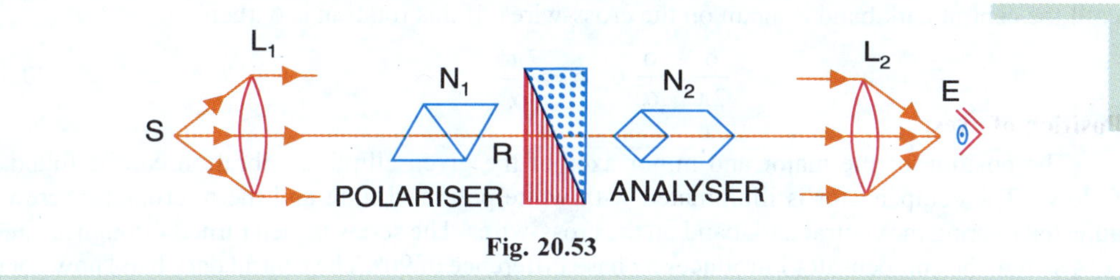

Fig. 20.53

Let the compensator be placed between crossed polarizer P and analyser A, as shown in Fig.20.53. Let the transmission axis of polarizer be oriented at 45° with respect to the optic axis of wedge ABC of the compensator. At midpoint R the light emergent from the compensator is plane polarized in the same plane as transmitted by P and therefore it will be extinguished by the analyser A. Similarly, at distances from the midpoint for which the retardation is 1λ,2λ,3λ,.....mλ, and the emergent light is plane polarized in the same plane as transmitted by P and hence will be extinguished by the analyser. So the field of view is crossed by a series of equidistant parallel dark bands.

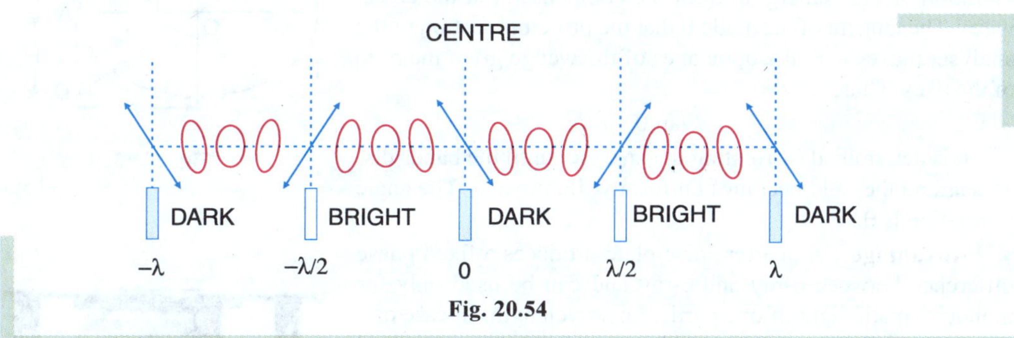

Fig. 20.54

At positions between them, where the path difference corresponds to an odd multiple of λ/2, i.e.,λ/2, 3λ/2, 5λ/2, ...(2m+1)λ/2, the transmitted light is plane polarized. The analyser transmits the light completely and those regions will be bright. In all other cases, the emerging light is elliptically polarized with varying parameters of the ellipse, as shown in Fig.20.54. If white light is used, the central band will be dark while others will be coloured.

By using white light source, the compensator is adjusted such that the central dark band is under cross wire and the micrometer reading is noted. The micrometer screw is turned through an angle such that the compensator introduces a phase difference of π/2 at cross wire. Then elliptically polarized light is made to be incident on the compensator. The central dark band undergoes a shift with respect to the cross wire. The compensator is rotated through an angle α in its own plane until the central dark band is on the cross wire. The axes of the incident elliptically polarized light are parallel to the optic axes of the wedges of the compensator.

Phase difference:

The elliptical vibration can be regarded as made of two mutually perpendicular linear vibrations, which are having a phase difference, δ. δ can be determined as follows.

First the compensator is illuminated with white plane-polarised light and the micrometer is adjusted to bring the central dark band on the cross-wires. The white light is then replaced by elliptically polarised light. The central band shifts to a point where the original phase difference δ between the two component vibrations of elliptical polarised light is exactly balanced by the phase difference introduced by the compensator. This phase difference is determined by rotating the screw until the central dark band is again on the cross-wires. If this rotation is φ, then

$$\frac{\delta}{2\pi} = \frac{\phi}{\alpha} \text{ or } \delta = \frac{2\pi\phi}{\alpha} \quad \text{... (20.54)}$$

Position of axes:

The position of the major and minor axes of the given elliptical vibration can be found as follows. The compensator is illuminated with white polarised light and the micrometer screw is adjusted to bring the central dark band on the cross-wires. The screw is then turned through an angle $\alpha / 4$ so that the compensator introduces a phase difference of 90°. The central dark band now not on the cross wires. The elliptically polarised light is made incident on the compensator. Then the compensator is rotated in its own plane until the central dark band again comes on to the cross-wires. The axes of the incident light are parallel to the optic axes of the wedges of the compensator.

Ratio of the axes:

Referring to the Fig.20.55, OA and OB represent the optic axes of the two wedgews of the compensator. OC is the direction of the principal section of the analyser. DE is the direction of vibration of light emerging from the compensator at the cross-wires. The tangent of the angle θ that the principal section of the analyser makes with the optic axes of the wedge gives the ratio of the axes. Thus,

$$\tan \theta = a/b$$

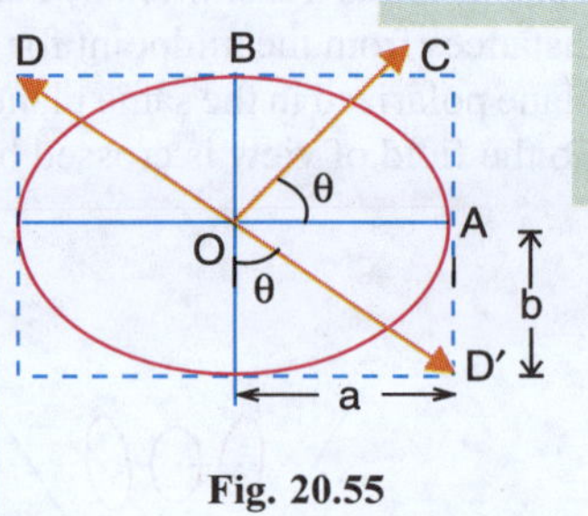

Fig. 20.55

θ is determined by rotating the analyser until the bands disappear and the field becomes uniformly illuminated. The angle of rotation is θ.

Advantages: A quarter wave plate produces a fixed phase difference between o-ray and e-ray and can be used only for monochromatic light of one particular wavelength. In case of a compensator, the phase difference between the rays can be varied continuously and hence a compensator, made of a combination of wedges, can be used for light of any given wavelength.

20.24. FRESNEL'S RHOMB

Fresnel designed a rhomb of glass whose angles are 54° and 126° as shown in Fig.20.56.Its functioning is based on the fact that a phase difference of $\pi / 4$ is introduced between the component vibrations when light is totally internally reflected back at glass - air- interface when the angle of incidence is 54°.

Fresnel's Rhomb.

A ray of light enters normally at one end of the rhomb and is totally internally reflected at the point B along BC. The angle of incidence at B is 54°, which is more than the critical angle of glass. Let the incident light be plane polarised and let the vibrations make an angle of 45° with the plane of incidence. Its components (i) parallel to the plane of incidence and (ii) perpendicular to the plane of incidence are equal. These components after reflection at the point B undergo phase difference of $\pi/4$ or a path difference of $\lambda/8$. A further phase difference of $\pi/4$ or a path difference $\lambda/8$ is introduced between the components when the ray Bc is totally internally reflected back along CD. Therefore, the final emergent ray DE has two components, vibrating at right angles to each other and they have a path difference of $\lambda/4$. Therefore, the emergent ray is circularly polarised.

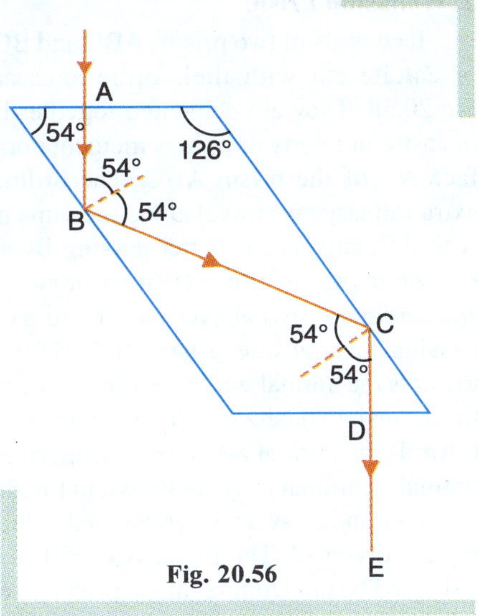

Fig. 20.56

If the light entering the Fresnel's rhomb is circularly polarised, a further path difference of $\lambda/4$ is introduced between the component vibrations. The total path difference between the component vibrations is $\lambda/2$. Therefore, the emergent light is plane polarised and its vibrations make an angle of 45° with the plane of incidence.

When an elliptically polarised light is passed through a Fresnel's rhomb, a further path difference of $\lambda/4$ is introduced between the component vibrations. The total path difference between the component vibrations is $\lambda/2$ and the emergent light is plane polarised.

Thus, Fresnel's rhomb works very much similar to a quarter wave plate. A quarter wave plate is used for a particular wavelength light whereas a Fresnel's rhomb can be used for light of all wavelengths.

20.25. DOUBLE IMAGE POLARIZING PRISMS

Nicol prism cannot be used with ultraviolet light on account of the canada balsam layer which absorbs UV rays. Sometimes it is also desirable to have both the ordinary and extraordinary rays widely separated. For this purpose, two prisms, namely Rochon prism and Wollaston prism, are used.

1. Rochon Prism:

It consists of two prisms ABC and BCD (of quartz or calcite) cut with their optic axes as shown in Fig.20.57. The prism ABC is cut such that the optic axis is parallel to the face AB and the incident light. The prism BCD has the optic axis perpendicular to the plane of incidence. Light incident normally on the face AC of the prism passes without deviation up to the boundary BC. In the prism BCD, the ordinary ray passes without deviation. If the prisms are made of quartz, the extraordinary ray is deviated as shown in Fig.20.57.

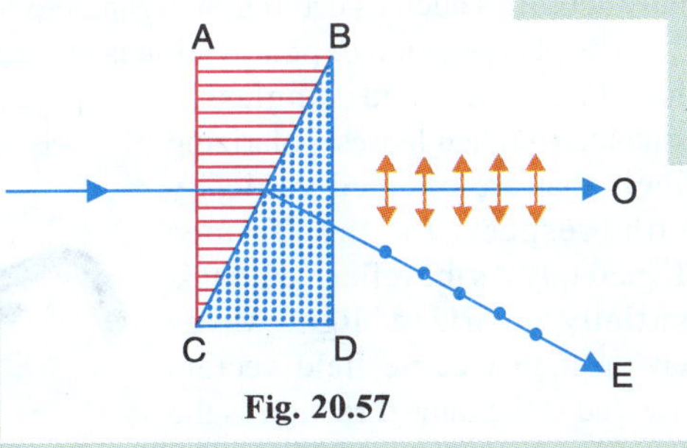

Fig. 20.57

In the case of calcite, the extraordinary ray will be deviated to the other side. The prisms ABC and BCD are cemented together by glycerine or castor oil. Here, the ordinary emergent beam is achromatic whereas the extraordinary beam is chromatic.

2. Wollaston Prism:

It consists of two prisms ABC and BCD of quartz or calcite cut with their optic axes as shown in Fig.20.58. They are cemented together by glycerine or castor oil. A ray of light is incident normally on the face AC of the prism ABC. The ordinary and the extraordinary rays travel along the same direction but with different speeds. After passing BC the ordinary ray behaves as the extraordinary ray and the extraordinary ray behaves as the ordinary ray while passing through the prism BCD. One ray is bent towards the normal while the other is bent away from the normal. In quartz $\mu_e > \mu_o$. Therefore, the ordinary ray while passing the boundary BC is refracted towards the normal as an extraordinary ray while the extraordinary ray is refracted away from the normal as ordinary ray, as shown in Fig.20.58. If the prisms are made from calcite, the directions of the o-ray and e-ray are interchanged. While coming out of the face BD of the prism, the o-ray and e-ray are diverged. The prism is useful in determining the percentage of polarisation in a partially polarised beam. Double image prisms are used in spectrophotometers and pyrometers.

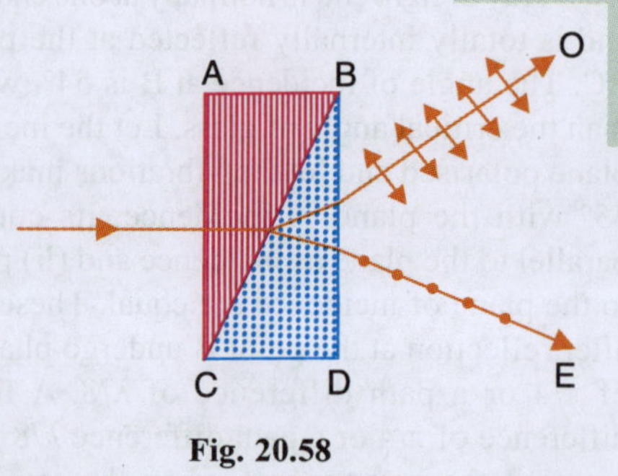

Fig. 20.58

20.26. APPLICATIONS OF POLARIZED LIGHT

The phenomenon of polarization has many practical applications in daily life. We discuss here some of the interesting applications here.

1. Sunglasses: Light reflected from nonmetallic surfaces such as water, snow clad mountains, asphalt roads etc is partially polarized. At angles nearer to the Brewster's angle, the reflected light contains a large concentration of vibrations in a plane parallel to the reflecting surface (Fig.20.59). Such a highly polarized light causes **glare** in one's eyes and makes it difficult to view the objects. When the amount of glare is large, daily activities such as driving on a road etc would become very difficult to perform.

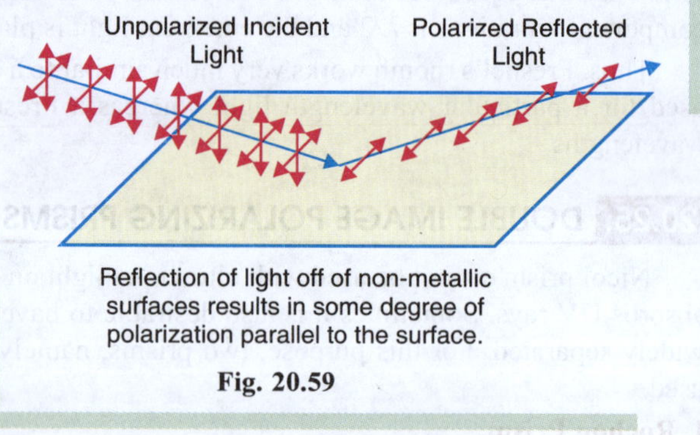

Reflection of light off of non-metallic surfaces results in some degree of polarization parallel to the surface.

Fig. 20.59

The phenomenon of polarization is utilized in making sunglasses, which will drastically reduce the glare. Polarized sunglasses contain, over their lenses, polarizing filters that are oriented vertically with respect to the frames (Fig.20.60). As the reflected light is partially polarized, light waves having their electric field vectors oriented in the same direction as the polarizing lenses (and perpendicular to the reflecting surface) are passed through. On the other hand, light waves having their electric field

Fig. 20.60

vectors oriented parallel to the reflecting surface (and perpendicular to the filters in the lenses) are blocked by the lenses. Thus, polarized sunglasses eliminate the glare from an illuminated surface.

2. Photography: Polarization by scattering occurs as light passes through our atmosphere. The scattered light often produces a glare in the skies. In photography, this partial polarization of scattered light produces a washed-out sky. The problem is overcome by the use of a polarizing filter fitted to the camera. As the filter is rotated, the partially polarized light is blocked and the glare is reduced. Thus, a vivid blue sky as the backdrop of a beautiful foreground is captured using polarizing filters.

3. Stereoscopic Movies: The phenomenon of polarization is used in making and viewing stereoscopic movies. Stereoscopic movies are three-dimensional movies, which give the same effect of depth as seen on a stage. The three-dimensional impression is obtained through binocular vision. Three-dimensional movies are actually two movies being shown at the same time through two projectors. For making a stereoscopic movie, two views of the same scene are shot simultaneously from two slightly different camera locations. One view corresponds to that seen by the right eye and the other corresponds to the view seen by the left eye. Each individual movie is then projected from different sides of the audience onto a screen through a polarizing filter. The polarizing filter used for the projector on the left may have its polarization axis aligned horizontally while the polarizing filter used for the projector on the right would have its polarization axis aligned vertically. Consequently, there are two slightly different movies being projected onto a screen; each movie is cast by light, which is polarized with an orientation perpendicular to the other movie. The audience then wears glasses, which have two Polaroid filters. Each filter has a different polarization axis - one is horizontal and the other is vertical. The result of these arrangements of projectors and filters is that the left eye sees the movie, which is projected from the right projector while the right eye sees the movie that is projected from the left projector. This gives the viewer a perception of depth.

4. Optical Microscopy: Polarization of light is also very useful in many aspects of optical microscopy. Microscopes may be configured to use crossed polarizers, in which case the first polarizer, described as the **polarizer**, is placed below the sample in the light path and the second polarizer, known as the **analyzer**, is placed above the sample, between the objective and the eyepieces. If the microscope stage is left empty, the analyzer blocks the light polarized by the polarizer and no light is visible. However, when a **birefringent**, or doubly refracting, sample is placed on the stage between the crossed polarizers, the microscopist can visualize various aspects of the sample. This is because the birefringent sample rotates the light, allowing it to successfully pass through the analyzer.

5. LCDs: Another interesting use of light polarization is the liquid crystal display (**LCD**) utilized in applications such as wristwatches, computer screens, timers, and clocks. These devices are based upon the interaction of rod-like liquid crystalline molecules with an electric field and polarized light waves.

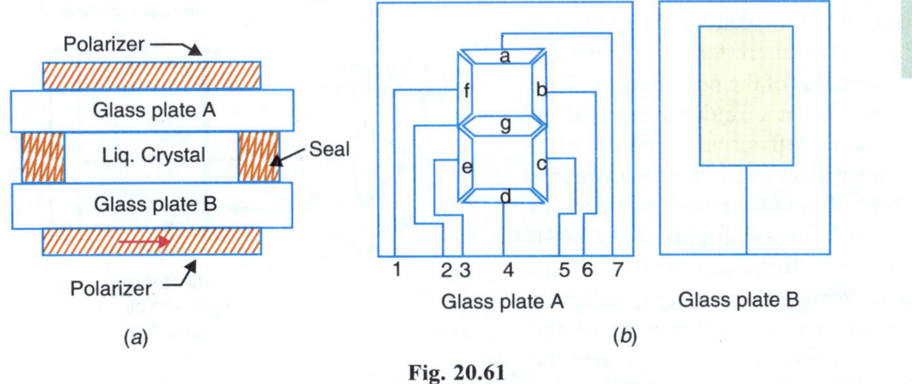

Fig. 20.61

An LCD consists of a liquid crystal material, which is double refracting, of about 10 μm thick suitably supported between two thin glass plates having transparent conducting coatings on their inner surfaces. The conducting coating is etched in the form of a digit or character, as shown in Fig.20.61. The assembly of glass plates with liquid crystal material in between is sandwiched between two crossed-polarizer sheets.

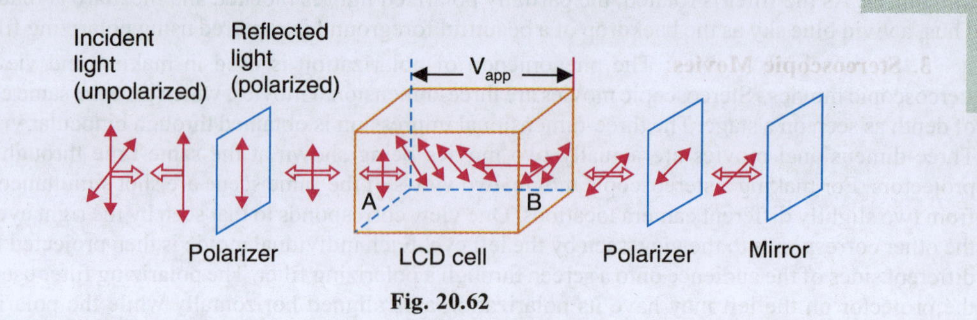

Fig. 20.62

During the fabrication of LCDs, the liquid crystal molecules are aligned in such a way that their long axes undergo a 90° rotation, as illustrated in Fig.20.62. It is called a twisted molecular arrangement. When natural light is incident on the assembly, the front polarizer converts it into linearly polarized light. As the linearly polarized light propagates through the LCD, the optical vector is rotated through 90° by the twisted molecular arrangement. Therefore, it passes unhindered through the rear polarizer whose transmission axis is perpendicular to that of the front polarizer. A reflecting coating at the back of the rear polarizer sends back the light, which emerges unobstructed by the front polarizer. Consequently, the display appears uniformly illuminated. When a voltage is applied to the device, the molecules between the electrodes untwist and align along the field direction. As a result, the optical vector does not undergo rotation as it passes through that region. The rear polarizer blocks the light and therefore a dark digit or character is seen in that region, as illustrated in Fig.20.63.

Fig. 20.63

6. Enhancing visibility of digital displays: Circular polarizers are used to enhance the visibility of digital displays. They can cut out the extraneous light reflected from the face of the display, improving the contrast of the display. They use the fact that mirror reflection changes the handedness of the polarization of the light. Right circular polarization becomes left circular polarization upon reflection. Let us consider a sheet of circular polarizer placed in front of a digital display (Fig.20.64). External light that falls on the polarizer becomes circularly polarized before it reaches the front of the display. The reflected light gets the handedness of its polarization

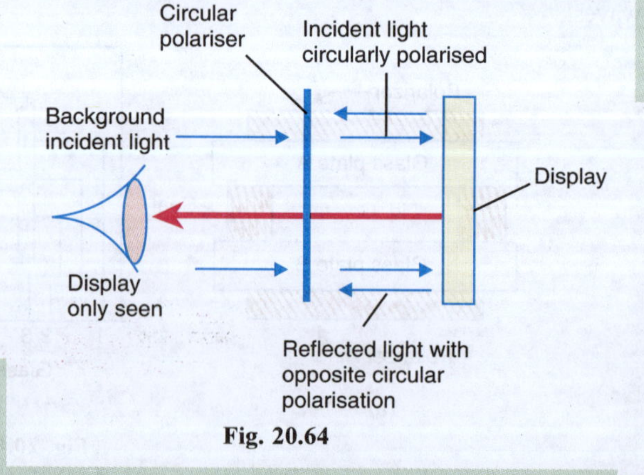

Fig. 20.64

reversed, and its return path is blocked by the polarizer. Light generated by the display passes through the polarizer and hence is seen without the background reflecting light.

7. Photoelasticity: Photoelasticity is an experimental method to determine stress distribution in various engineering components. The method is mostly used in cases where mathematical methods become quite cumbersome. Photoelasticity is especially useful for the study of objects with irregular boundaries and stress concentrations, such as pieces of machinery with notches or curves, structural components with slits or holes, and materials with cracks.

Principle: The method is based on the property of double refraction, which is exhibited by photoelastic materials on the application of stress. Double refraction or birefringence is a property by virtue of which a ray of light passing through a birefringent material splits into two beams (e- and o-rays). The two beams travel along the same path in the material and their speed at each point in the material is directly related to the state of stress at that point. Because the velocities of light propagation are different in each direction, there occurs a phase shifting of the light waves. Therefore, light emerges out of the component as two beams vibrting out of phase with one another and when they are combined, produce interference pattern.

The stressed component is examined under monochromatic polarized light in a polariscope. The polarizer in the polariscope produces polarized light. When the analyzer in the polariscope recombines the waves, interference pattern is observed. Regions of stress where the wave phases cancel appear dark, and regions of stress where the wave phases add appear bright. Therefore, in models of complex stress distribution, bright and dark fringe patterns (isochromatic fringes) are projected from the model. As these fringes are related to the stresses, the magnitude and direction of stresses at any point can be determined by examination of the fringe pattern. When the component is unloaded, the photoelastic fringe pattern disappears.

When white light is used in place of monochromatic light, coloured fringes are observed. White light is often used for demonstration, and monochromatic light is used for precise measurements.

The above method is suitable when the component is transparent. In the case of opaque components, a thin sheet of photoelastic material is suitably bonded to the surface of the component. When the component is loaded, the surface strain in the component is transmitted to the photoelastic sheet producing stress in it. The resulting fringe pattern is observed by illuminating the component with polarized light and viewing it through an analyzer. More commonly, a transparent scale model of the part is made out of a material, which is optically sensitive to stress such as epoxy, glyptol or polyester resins. The model is then subjected to the forces that the actual component would experience in use. The birefringence varies from point to point over the surface of the model. When viewed with crossed polarizers, a complicated fringe pattern is seen which provides a visual means of observing overall stress characteristics of an object. The patterns can be projected on a screen or photographic film.

20.27. OPTICAL ACTIVITY

Certain crystals and solutions possess a natural ability to rotate the plane of polarization about the direction of propagation. It is known as *optical activity*. In case of crystals this ability arises due to the twisted arrangement of atomic layers with respect to one another. In liquids and solutions the optical activity is due to certain structural symmetry in their molecules. The optical activity found in bigger organic molecules provides a number of clues which help us understand biological activity.

Many crystalline materials exhibit birefringence naturally, without application of any voltage. The birefringence is present all the time. Examples of such crystals are quartz and calcite. There are also a number of crystals that are not birefringent naturally but in which application of a voltage or magnetic field induces birefringence. The induced optical activity leads to the ability to control light beams in a variety of ways and is the basis of a number of applications such as light-beam modulators, Q-switches, and deflectors.

20.28. OPTICAL ROTATION

When a beam of plane polarized light propagates through a quartz crystal along the optic axis, the plane of polarization steadily turns about the direction of the beam. The optical rotation can be detected as follows. If two polaroid sheets or Nicol prisms are held in crossed configuration and if a beam of unpolarized light is viewed through them, the field of view appears to be completely dark. Now let a quartz crystal, cut with its faces perpendicular to the optic axis, be inserted between the polarizers such that light is incident normally on the crystal. The field of view now appears lit up indicating that the light is not cut off by the analyzer. In order to cut off the transmitted light, we find that the analyzer is to be rotated through a certain angle. The experiment establishes that the plane polarized light produced by the polarizer remains plane polarized while passing through the quartz crystal but the plane of polarization is rotated through an angle. This angle is the angle through which the analyzer is rotated in order to cut off the light totally. The optical rotation, i.e., rotation of the plane of polarized light is shown in Fig.20.65(*a*).

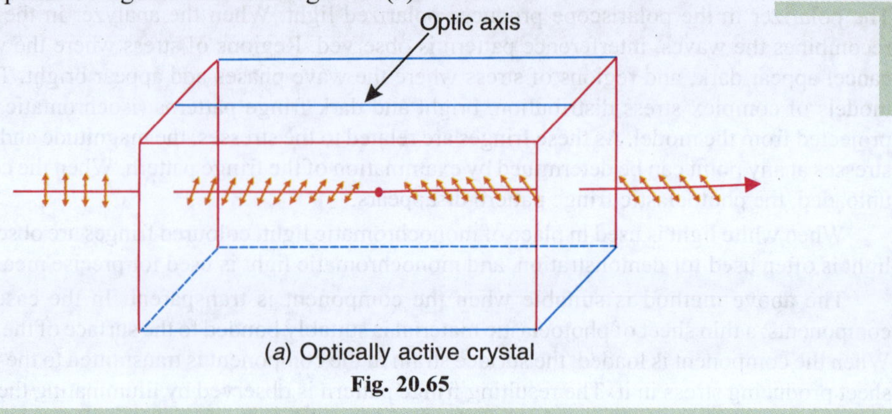

(*a*) Optically active crystal

Fig. 20.65

The ability to rotate the plane of polarization of plane polarized light by certain substance is called **optical activity**. Substances, which have the ability to rotate the plane of the polarized light passing through them, are called ***optically active*** substances. Quartz and cinnabar are examples of optically active crystals while aqueous solutions of sugar, tartaric acid are optically active solutions.

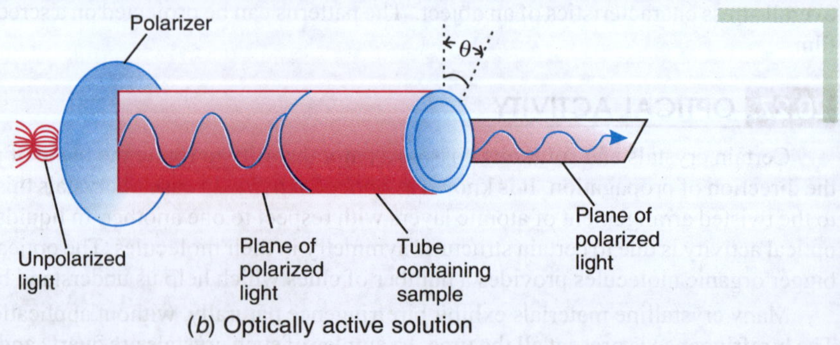

(*b*) Optically active solution

Rotation of the plane-polarized light. The plane is rotated by an angle after the light passes through an optically (*a*) active crystal or (*b*) solution.

Fig. 20.65

Optically active substances are classified into two types.

(*i*) **Dextrorotatory substances:** Substances which rotate the plane of polarization of the light toward the right are known as right-handed or dextrorotatory .

(*ii*) **Laevorotatory substances:** Substances which rotate the plane of polarization of the light toward the left are known as left-handed or laevorotatory .

20.29. SPECIFIC ROTATION

A measure of the optically activity of a sample is the rotation produced for a 1 mm slab for a solid, or a 100 mm path length for a liquid. This measure is called the *specific rotation*. Liquids usually rotate the light much less than solids. Solutions of solids will obviously show an effect that depends on the concentration of active material and, to a small extent, both on temperature and the solvent.

If an optically active material is kept between two crossed polarizers, the field of view becomes bright. In order to get darkness once again, the analyzer has to be rotated through an angle. The angle through which the analyzer is rotated equals the angle through which the plane of polarization is rotated by the optically active substance. This angle depends on

(*a*) The thickness of the substance,

(*b*) Density of the material or concentration of the solution,

(*c*) Wavelength of light, and

(*d*) The temperature.

The amount of rotation θ caused by crystalline materials is given by

$$\theta = \alpha l \qquad \text{... (20.55)}$$

where α is called the **rotational constant**.

In solutions the amount of rotation θ is given by

$$\theta = s\,c\,l \qquad \text{... (20.56)}$$

where c is the concentration and s is called the **specific rotation**.

The specific rotation for a given wavelength of light at a given temperature is defined conventionally as the rotation produced by one decimetre long column of the solution containing 1 gm of optically active material per c.c .of solution.

$$[S]_\lambda^t = \frac{\theta}{l \times C} = \frac{\text{Rotation in degrees}}{\text{Length in decimetres} \times \text{conc. in gm/c.c.}} = \frac{10\theta}{l(cm)C} \qquad \text{... (20.57)}$$

20.30. FRESNEL'S EXPLANATION

A linearly polarised light can be considered as a resultant of two circularly polarised vibrations rotating in opposite directions with the same angular velocity. Fresnel assumed that plane-polarised light on entering a crystal along the optic axis is resolved into two circularly polarised vibrations rotating in opposite directions with the same angular frequency. In an optically inactive crystal like calcite, the two circularly polarised vibrations travel with the same angular velocity. On the other hand, in an optically active crystal like quartz, the two circularly polarised vibrations travel with the different angular velocities.

Fig.20.66 (a) shows plane polarized light entering a calcite crystal along the optic axis AB and split up into two circular motions rotating in opposite directions. They are represented by OL and OR. OL is the circularly polarised vector rotating in the anticlockwise direction and OR is the circularly polarised vector rotating in the clockwise direction. If OL and OR start simultaneously from OA and rotate with the same angular velocity, then at any subsequent time, the resultant of OL and OR will lie along OA. Hence on emerging from the crystal, the two circular waves combine to

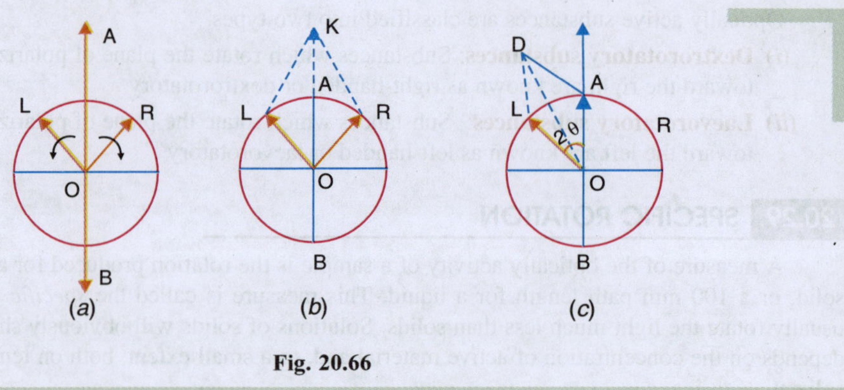

Fig. 20.66

produce a linear vibration along the initial direction, OA (Fig.20.66 b). Therefore, crystals like calcite do not rotate the plane of vibration.

In case of quartz, the linearly polarised light, the component having clockwise rotation travels faster than the anticlockwise component. When the components emerge out of the crystal, they are at an angle 2è. The resultant of these two vectors OR and OL is now along OD (Fig.20.66 c). Before entering the crystal, the plane of vibration is along OA and after emerging from the crystal, it is along OD which makes an angle è with the optic axis OA. After passing through the crystal, the circular components combine to produce a linear vibration whose direction is now along OD. Therefore, the plane of vibration has rotated through an angle è. The angle through which the plane of vibration is rotated depends on the thickness of the crystal.

Left and right circularly polarised light travel through the crystal at different speeds. Because one travels more slowly than the other, a phase shift builds up between them. When right and left-handed light are combined together at any point, the result is always linear polarization; but the angle of the polarisation depends on the phase shift *between* the two circular polarisations. If μ_R and μ_L are the refractive indices corresponding to the right and left citrcularly polarized light, the angle of polarisation, è, is given by:

$$\theta = \frac{\pi}{\lambda}(\mu_R - \mu_L)d \qquad \text{... (20.58)}$$

where d is the thickness of the material. As a result, the direction of polarisation spirals around z in space.

The cause of this special behaviour for circularly polarised light is that the optically active molecules are chiral, meaning that they have a helical twist in them. Any arrangement of atoms with a helical structure can form left-handed helices or right-handed helices. One is a mirror image of the other. Now there is a curious property of helices not possessed by rotating circles. If we look at a wheel rotating clockwise from one side, the same wheel appears to go around anticlockwise from the other side. The 'wise' description just depends on our viewpoint. Helices are different. If we turn round a right-handed helix (in which the spiral appears to go around clockwise as we look into the helix), then it is still a right-handed helix. So, helical molecules may interact with circularly polarised light differently, depending on whether their chirality is the same handedness as the circular polarisation or opposite handedness. This is the fundamental reason for the two refractive indicesand . It is interesting that the cause of the phenomenon lies at a molecular level and does *not* have anything to do with spatial ordering of molecules. This is why the phenomenon occurs with liquids as well as solids.

20.30.1. ANALYTICAL TREATMENT

Let us assume that linearly polarised light is incident normally on a quartz plate cut perpendicular to the optic axis. Let the vibrations in the incident light be represented by

$$y = a \cos \omega t$$

On entering the crystal these vibrations are broken up into two equal and opposite circular vibrations which are represented by

$$x_1 = \frac{a}{2} \sin \omega t \quad \text{and} \quad y_1 = \frac{a}{2} \cos \omega t \qquad \text{— clockwise circular polarisation}$$

and $$x_2 = -\frac{a}{2} \sin \omega t \quad \text{and} \quad y_2 = \frac{a}{2} \cos \omega t \qquad \text{— anti-clockwise circular polarisation}$$

These circular components travel through the crystal with different velocities. When they emerge from the crystal, there is a phase difference δ between them. In case of quartz crystal, which is a right-handed optically active crystal, the clockwise component travels faster. The circular components are then represented by

$$x_1 = \frac{a}{2} \sin (\omega t + \delta) \quad \text{and} \quad y_1 = \frac{a}{2} \cos (\omega t + \delta) \qquad \text{—clockwise component}$$

and $$x_2 = -\frac{a}{2} \sin \omega t \quad \text{and} \quad y_2 = \frac{a}{2} \cos \omega t \qquad \text{— anti-clockwise component}$$

The resultant displacements along the two axes are

$$X = x_1 + x_2 = \frac{a}{2}\left[\sin(\omega t + \delta) - \sin \omega t\right] = a \cos\left[\omega t + \frac{\delta}{2}\right] \sin \frac{\delta}{2} \qquad \text{...(20.59)}$$

and $$Y = y_1 + y_2 = \frac{a}{2}\left[\cos(\omega t + \delta) + \cos \omega t\right] = a \cos\left[\omega t + \frac{\delta}{2}\right] \cos \frac{\delta}{2} \qquad \text{... (20.60)}$$

the resultant vibrations along the X-axis and Y-axis have the same phase. Therefore, the resultant vibration is plane polarised and it makes an angle $\delta/2$ with the original direction. Therefore, the plane of polarisation is rotated through an angle $\delta/2$ on passing through the crystal.

Dividing equ.(20.59) by equ.(20.60), we obtain

$$\frac{X}{Y} = \tan \frac{\delta}{2} \qquad \text{... (20.61)}$$

If μ_R and μ_L are the refractive indices of quartz in the direction of optic axis for clockwise and anticlockwise circularly polarised light and d is the thickness of the crystal plate, then the phase difference δ is given by

$$\delta = \frac{2\pi}{\lambda}(\mu_L - \mu_R) d \qquad \text{... (20.62)}$$

Hence the plane of polarisation is rotated through

$$\theta = \frac{\delta}{2} = \frac{\pi d}{\lambda}(\mu_L - \mu_R) \qquad \text{... (20.63)}$$

20.31. EXPERIMENTAL VERIFICATION OF FRESNEL'S THEORY

Fresnel showed that linearly polarised light on entering an optically active crystal is resolved into two circularly polarised vibrations. He arranged alternatively a number of negative and positive optically active quartz prisms to form a rectangular block, as shown in Fig.20.67. The optic axis is

Fig. 20.67

parallel to the base of each prism. When the linearly polarised light is incident normally on the first crystal surface AB, the two component circular vibrations travel along the same direction but with different speeds. When the light beam passes through the first oblique surface BC, the R-component, which is faster in the first prism, becomes slower in the second prism. The opposite is the situation for L-component. It means that the second prism BCD acts as a denser medium for the R-component and rarer medium for the L-component. As a result, the R-component bends towards the base in the second prism while L-component bends away from the base. Therefore, the two components are separated apart while they travel through prism BCD. Again at the boundary of the next prism, the speeds are interchanged and the R-component bends away from the base and the L-component bends towards the base. The net result is that the two beams are separated more and more as they pass through the successive prisms. When they ultimately emerge out, they are widely separated.

When the two waves are analysed by a quarter wave plate and a Nicol prism, both are found to be circularly polarised in opposite directions. Thus, Fresnel's hypothesis is verified.

20.32. LAURENT'S HALF SHADE POLARIMETER

A polarimeter is an instrument used for determining the optical rotation of solutions. When used for determining the optical rotation of sugar it is called a saccharimeter.

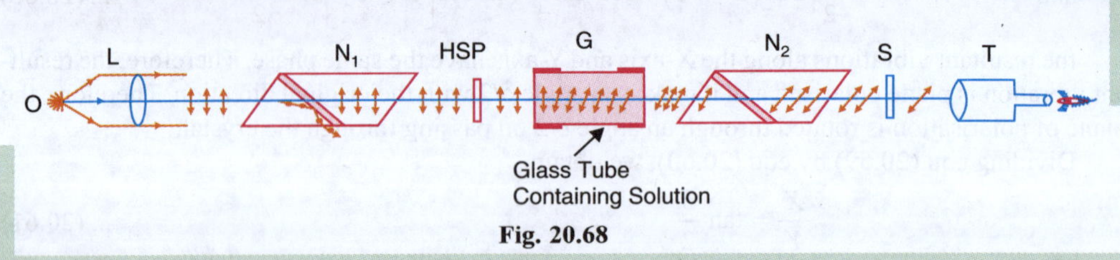

Fig. 20.68

Construction: A polarimeter consists of a glass tube for holding the solution under test held between crossed Nicol prisms (Fig.20.68). Beyond the polarizing Nicol prism a half-shade plate is located which is used for accurately adjusting the two Nicols for crossed position. Light from a monochromatic source is rendered parallel by the lens L and is incident on the polarizer, N_1. The light transmitted by the polarizer is plane polarized. The polarized beam then passes through the half-shade plate and a glass tube G containing the solution. The light emerging from the solution will be incident on the analyzer N_2. The light is observed through a telescope T. The analyzing Nicol N_2 can be rotated about the axis of the tube and the rotation can be measured with the help of a graduated circular scale.

Working: To find the specific rotation of a solution, the analyzer is first adjusted such that field of view is completely dark. Then the glass tube is filled with the solution and is held in position. The field of view now becomes illuminated. The field of view can be again be made dark by rotating the analyzer through a certain angle which gives the optical rotation of the solution. The practical difficulty in this method is in determination of the exact position for which complete darkness is achieved. The difficulty is overcome by using what is known as a Laurent's half-shade device.

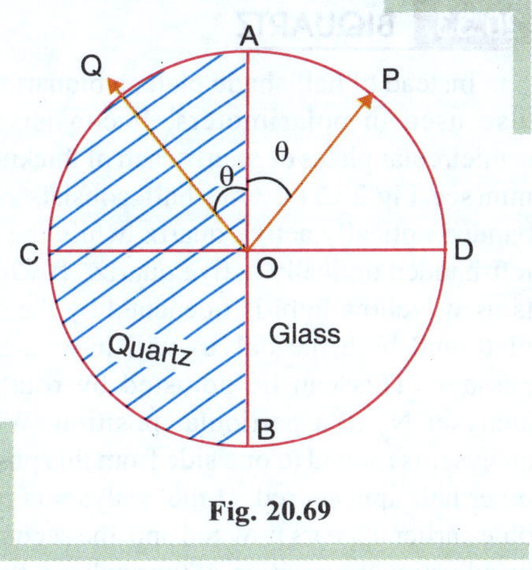

Fig. 20.69

It consists of a semicircular half wave plate ACB of quartz cemented to a semicircular Plate ADB of glass (Fig.20.69). The optic axis of the wave plate is parallel to the line of separation AB. The half wave plate introduces a phase difference of 180° between e-ray and o-ray passing through it. The thickness of the glass plate is such that it transmits the same amount of light as done by the quartz half wave plate. One half of the incident light passes through the quartz plate ACB and the other half through the glass plate ADB. When the light after passing through the polarizer is incident normally on the half shade plate and has vibrations along OP. On passing through the glass, half the vibrations will remain along OP but on passing through the quartz half, the vibrations will split into e- and o-rays. The o-vibrations are along OD and e-vibrations are along OA. The half wave plate introduces a phase difference of π rad between the two vibrations. The vibrations of o-ray will occur along OC instead of OD on emerging from the plate. Therefore the resultant vibration will be along OQ whereas the vibrations of the beam emerging from glass plate will be along OP. In effect, the half wave plate turns the plane of polarization of the incident light through an angle 2θ.

If the principal plane of the Nicol N_2 is aligned parallel to OP, the plane polarized light emerging from the glass tube will pass through the glass plate of the half shade plate and that part appears brighter. On the other hand light coming out of the quartz plate is partially obstructed and the corresponding field of view appears less bright. If the principal plane of N_2 is parallel to OQ the quartz half will appear brighter than the glass half. Thus, the two halves of the plate are unequally illuminated. When the principal plane of N_2 is parallel to AB, the two halves appear equally bright and when it is parallel to CD, the two halves are equally dark.

To find the specific rotation of a solution, the analyzer is first in the position for equal darkness without solution in the tube G. The reading on the circular scale is noted. Next, the tube is filled with the optically active solution of known concentration. The field of view is now partially illuminated. The analyzer is rotated till the field of view becomes equally dark. The reading on the circular scale is noted again. The difference between the two scale readings gives the angle of rotation of the plane of polarization caused by the solution. Knowing the values of *θ, l, and c*, the specific rotation is obtained using the formula (20.57). Or otherwise, knowing the value of the specific rotation, the concentration of the solution can be determined with the help of the equation (20.57).

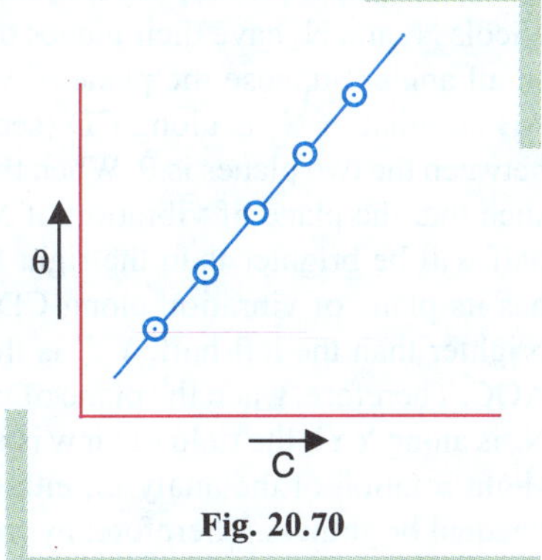

Fig. 20.70

In the actual experiment, different concentrations of solutions are taken and the corresponding angles of rotation are determined. A graph is plotted between concentration C and the angle of rotation θ. The graph is a straight line (see Fig.20.70). Using the value of the slope in equ.(20.56), the specific rotation of the optically active substance is calculated.

20.33. BIQUARTZ

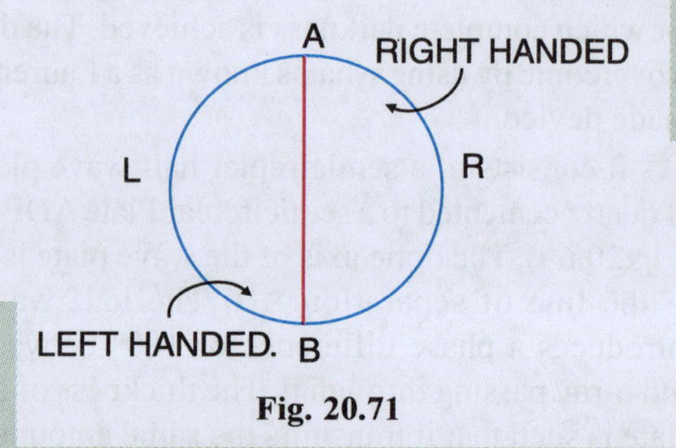

Fig. 20.71

Instead of half shade plate, a biquartz plate is also used in polarimeters. It consists of two semicircular plates of quartz each of thickness 3.75 mm(see Fig.20.71). One half consists of right-handed optically active quartz, while the other is left-handed optically active quartz. If white light is used, yellow light is quenched by the biquartz plate and both the halves will have the tint of passage. This can be adjusted by rotating the analyser N_2 to a particular position. When the analyser is rotated to one side from this position, one half of the field of view appears blue, while the other half appears red. If the analyser is rotated in the opposite direction, the first half which was blue earlier appears now red and the second half which was earlier red appears blue now. Therefore, by adjusting the position of the analyser, the field of view appears equally bright with tint of passage.

20.34. LIPPICH POLARIMETER

Laurent's polarimeter suffers from the defect that it can be used only for light of a particular wavelength for which the half wave plate has been designed. Lippich developed a polarimeter that can be used for light of any wavelength.

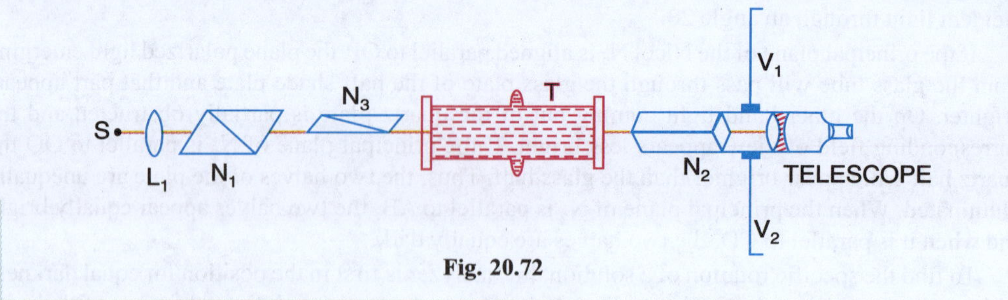

Fig. 20.72

The schematic of the polarimeter is shown in Fig.20.72. It consists of two Nicol prismsN_1 and N_2. Behind N_1, there is a Nicol prism N_3 that covers half the field of view. The Nicols N_1 and N_3 have their planes of vibration inclined at a small angle. Suppose the plane of vibration of N_1 is along AB and that of N_3 is along CD (see Fig.20.73). The angle between the two planes is θ. When the analyser N_2 is rotated such that the plane of vibration of N_2 is along AB, the left half will be brighter than the right half. If the analyser N_2 has its plane of vibration along CD, the right half will be brighter than the left half. YY' is the bisector of the angle AOC. Therefore, when the plane of vibration of the analyser N_2 is along YY', the field of view is equally illumined. For a slight rotation of the analyser, either to the right or to the left, the field of view appears to be of unequal brightness. Therefore, by rotating N_2, the position for equal brightness of the field of view is obtained. To determine the specific rotation of the optically active substance, the procedure is the same as discussed in § 20.32.

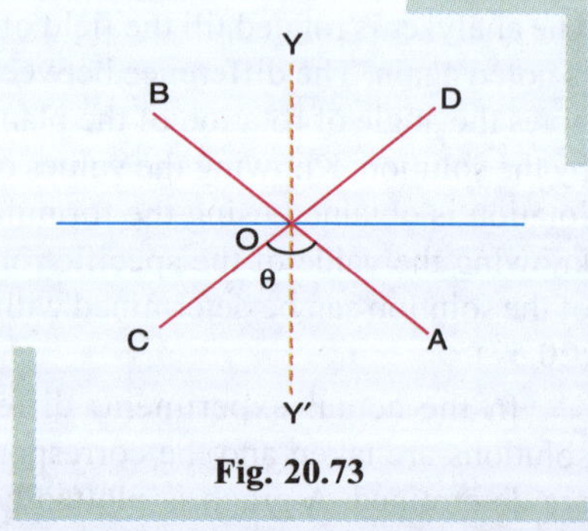

Fig. 20.73

Three-field system: in the improved form, Lippich polarimeter has a three-field system. the defect in the two field system is that if the eye is off the axis, even for the position of equal brightness of the field of view, one side appears brighter than the other. Just behind N_1, there are two Nicol prisms N_3 and N_4, as shown in Fig.20.74.

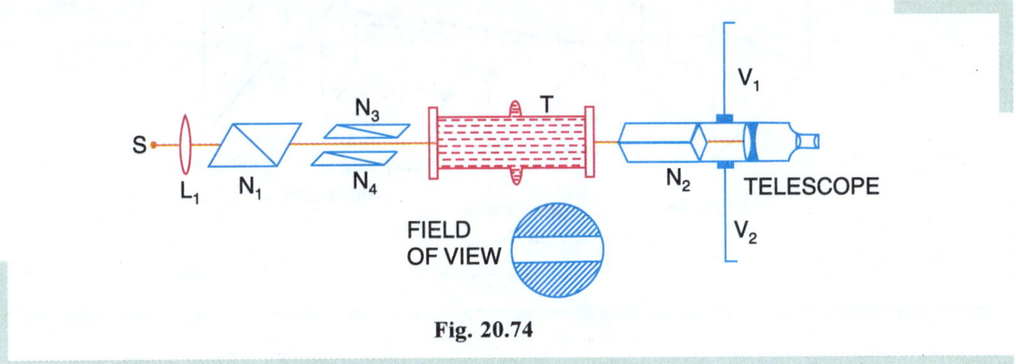

Fig. 20.74

The planes of vibration of N_3 and N_4 are parallel to each other and make a small angle with the plane of vibration of N_1. For a particular position of the analyser N_2, the field has three parts. The central portion is illuminated by light, which has passed through N_1, and N_2 while the other two portions, which are equally bright, are illuminated by light passing through N_1, N_2 and one of the prisms N_3 and N_4.

20.35. ELECTRO-OPTIC AND MAGNETO-OPTIC EFFECTS

Isotropic transparent materials such as glass do not exhibit double refraction under ordinary circumstances. However, they acquire the optical properties of a uniaxial crystal under the action of external forces. Consequently, they exhibit double refraction. The appearance of double refraction under the influence of an external agent is known as **artificial double refraction** or **induced birefringence**. The direction of the optical axis in such materials will be collinear with the direction of the external force. The action of the external force is to cause distortion of the molecular arrangement within the material and thereby transform the isotropic substance into an *anisotropic* substance. The induced birefringence disappears as soon as the external force ceases to act.

The materials which experience a change in their optical behaviour under the action of an electric field are called electro-optic materials and the resulting optical effects are known as **electro-optic effects**. Similarly, the materials that get influenced by a magnetic field are called magneto-optic materials and the resulting optic effects are known as **magneto-optic effects**. The electro-optic and magneto-optic materials play a very important role in modern technology.

20.36. ELECTRO-OPTIC EFFECTS

20.36.1 KERR EFFECT

Optical anisotropy induced in an isotropic liquid under the influence of an electric field is known as the Kerr effect. John Kerr discovered it in 1875.

A Kerr cell is required for studying the effect. It consists of a sealed glass cell filled with a liquid comprising of asymmetric molecules. Two plane electrodes of specific length l are arranged in it with their faces strictly parallel to each other. When a voltage is applied to them a uniform electric field is produced in the cell. The Kerr cell is placed between a crossed polarizer system. When the electric field is applied, the molecules of the liquid tend to align along the field direction. As the

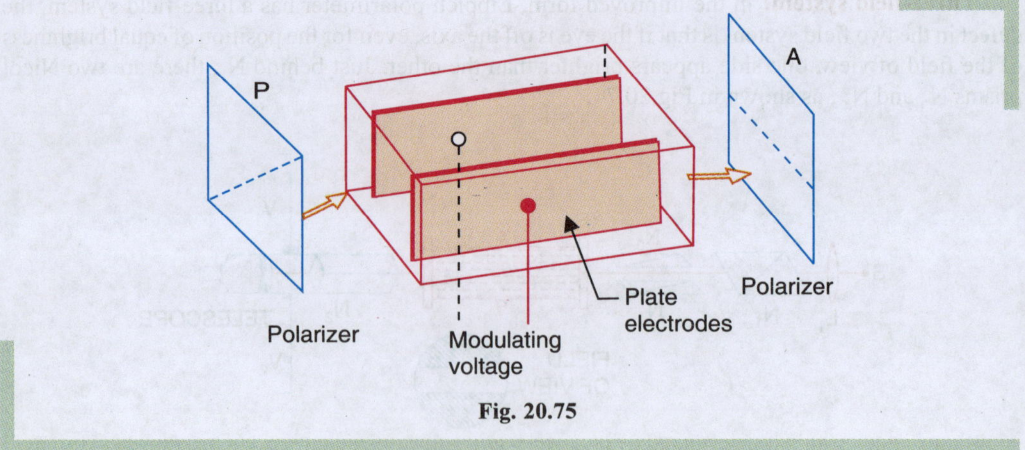

Fig. 20.75

molecules are asymmetric, the alignment causes anisotropy and the liquid becomes double refracting. The induced birefringence is proportional to the square of the applied electric field and to the wavelength of incident light. Thus,

$$\Delta\mu = K\lambda E^2 \tag{20.64}$$

where K is known as the Kerr constant.

Among the liquids, nitrobenzene ($C_6H_5NO_2$) is found to have the highest value for the Kerr constant. therefore, Kerr cells use nitrobenzene.

Kerr cell is used as an electro-optic shutter in high-speed photography, as a light chopper in the measurement of the speed of light.

20.36.2 POCKELS EFFECT

F. Pockels discovered in 1893 that the application of an electric field to piezoelectric crystals makes them birefringent. Normally, piezoelectric crystals are birefringent but in certain directions do not exhibit double refraction. When an electric field is impressed along these directions, double refraction is induced along these directions also.

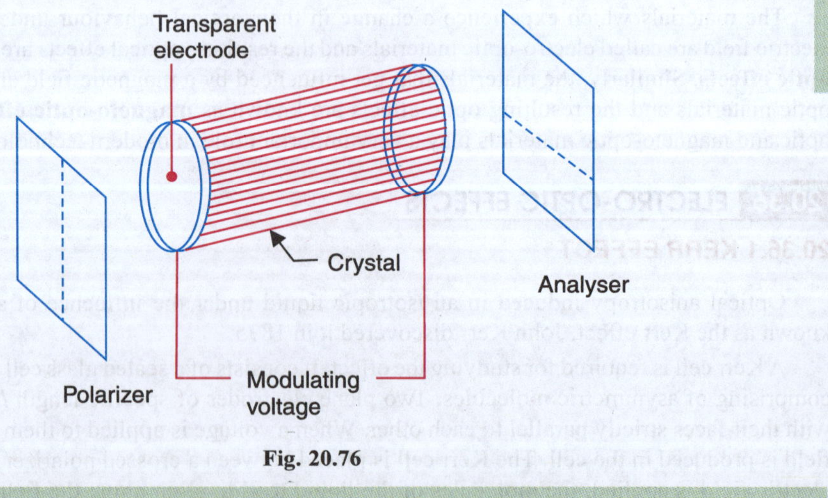

Fig. 20.76

A Pockels cell consists of a piezoelectric crystal, for example lithium niobate placed between crossed polarisers. Transparent electrodes (thin conducting coatings of tin oxide or indium) are deposited on opposite sides of the crystal. The crystal is oriented in such a way that its optic axis lies along the direction of the electric field applied between the electrodes. The transparent electrodes ensure free propagation of light through the crystal. A Pockels set up is shown in Fig.20.76.

The birefringence induced in the crystal is proportional to the strength of the applied field. Thus,

$$\Delta\mu = kE \tag{20.65}$$

where k is a constant characteristic of the material. Equ.(20.65) shows that Pockels effect is a linear effect.

The total birefringence of the cell is initially made equal to $\lambda/2$. When the electric field is increased, the beam is transmitted or hindered depending on the phase difference between the o-ray and e-ray. The device switches on and off periodically. Pockels cells are used in fast switching applications and in fibre optics. It can be used to obtain amplitude, frequency or phase modulation.

A Pockels cell is simple in construction and requires a small voltage of the order of 1.5 kV where as a Kerr cell is complicated in construction and requires higher voltages of the order of 15 kV. The piezoelectric crystals of ammonium dihydrophosphate (ADP) and potassium dihydrophosphate (KDP) are widely used in Pockels cell.

Kerr and Pockels cells are widely used as electro-optic shutters in Q-switching of lasers.

20.37. MAGNETO-OPTIC EFFECTS

20.37.1 COTTON-MOUTON EFFECT

The Cotton-Mouton effect is a magneto-optic effect. An isotropic material acquires the optical behaviour of a uniaxial crystal under the action of an external magnetic field. The set up is shown in Fig.20.77.

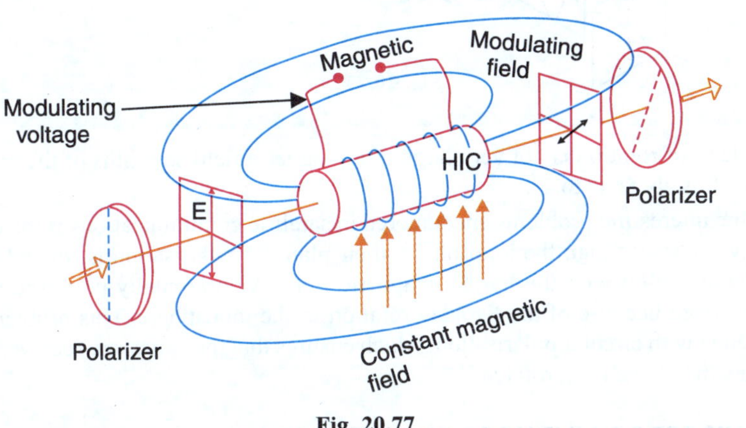

Fig. 20.77

The induced birefringence is governed by the relation

$$\Delta\mu = C\lambda B^2 \tag{20.66}$$

where C is a constant characteristic of the material. The magnitude of the induced birefringence is usually very small.

20.37.2 FARADAY EFFECT

Optically inactive substances acquire the ability of rotating the plane of polarisation of light when they are subjected to a magnetic field, parallel to propagation direction. Michael Faraday discovered this effect and hence it is called Faraday effect. This effect occurs in most optically transparent dielectric materials (including liquids) when they are subject to strong magnetic fields. The set up for observing Faraday effect is shown in Fig.20.78.

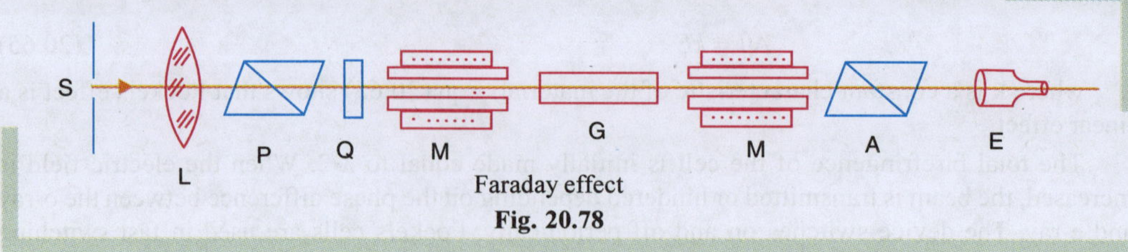

Faraday effect

Fig. 20.78

The angle of rotation θ of the plane of polarisation is proportional to the length of the path of light in the material and to the strength of the applied magnetic field. Thus,

$$\theta = VlH \tag{20.67}$$

where V is known as Verdet constant.

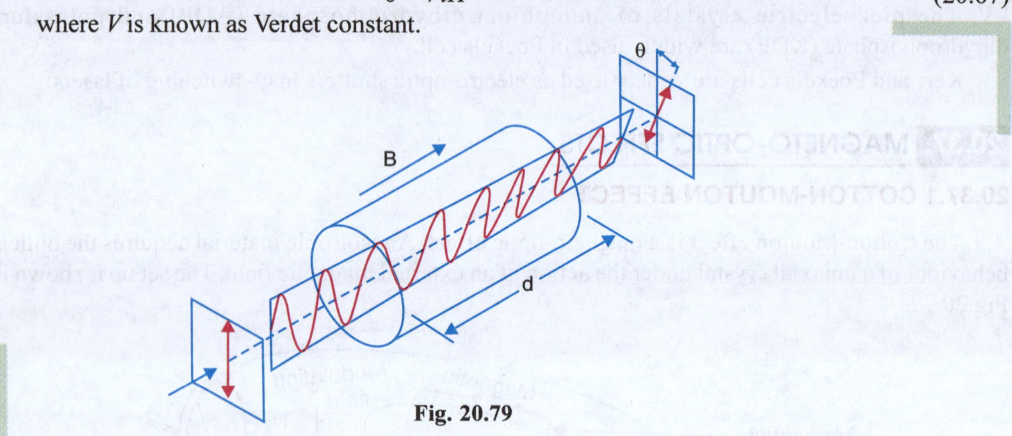

Fig. 20.79

The angle θ of rotation is not very large. For magnetic field strengths of the order of 10^6 A/m and $l = 0.1$ m, θ is about 1° to 2°.

One of the interesting problems encountered in satellite communications is the Faraday effect. As radio waves pass through the ionosphere, their plane of polarisation is rotated by the ionised particles in conjunction with the Earth's magnetic field. A horizontally polarised wave becomes vertically polarised because of the Faraday rotation in the ionosphere. This problem is solved by using an antenna with circular polarisation, which ensures that the waves are received satisfactorily, no matter how they have been rotated.

20.38. ANISOTROPY INDUCED BY MECHANICAL STRAIN

Materials such as glass become double refracting when they are subjected to mechanical strain. Experiments show that the induced birefringence is directly proportional to the stress σ experienced at a given point of the material. Thus,

$$\Delta\mu = k\sigma \tag{20.68}$$

where k is the proportionality constant characteristic of the material. This effect was discovered by Sir David Brewster in 1816.

When the material is held between crossed polarisers P and A, as in Fig.20.80, the field of view appears dark as long as the external force is not applied. As soon as the force is applied, coloured contours will be seen. Dark regions indicate the absence of strain in those areas. Each coloured contour shows the areas that are identically deformed. Such contours enable us assess the distribution of the stresses in the material.

The mechanically induced birefringence is used to study stresses in girders, beams etc. the model of the object under investigation is made of transparent plastic material and is then loaded. Using crossed polarizer system, the stresses produced at different positions are analysed and estimated. This method of analysis is known as photo-elastic analysis and is widely employed in civil and mechanical engineering practices.

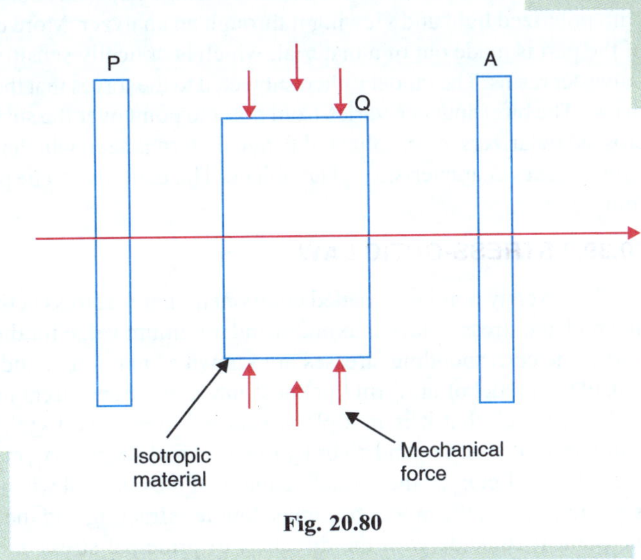

Fig. 20.80

20.39. PHOTOELASTICITY

Photoelasticity is an experimental method to determine stress distribution in various engineering components. The method is mostly used in cases where mathematical methods become quite cumbersome. Photoelasticity is especially useful for the study of objects with irregular boundaries and stress concentrations, such as pieces of machinery with notches or curves, structural components with slits or holes, and materials with cracks.

Principle: The method is based on the property of double refraction, which is exhibited by photoelastic materials on the application of stress. Double refraction or birefringence is a property by virtue of which a ray of light passing through a birefringent material splits into two beams (e- and o-rays). The two beams travel along the same path in the material and their speed at each point in the material is directly related to the state of stress at that point. Because the velocities of light propagation are different in each direction, there occurs a phase shifting of the light waves. Therefore, light emerges out of the component as two beams vibrating out of phase with one another and when they are combined, produce interference pattern.

The stressed component is examined under monochromatic polarized light in a polariscope. The polarizer in the polariscope produces polarized light. When the analyzer in the polariscope recombines the waves, interference pattern is observed. Regions of stress where the wave phases cancel appear dark, and regions of stress where the wave phases add appear bright. Therefore, in models of complex stress distribution, bright and dark fringe patterns (isochromatic fringes) are projected from the model. As these fringes are related to the stresses, the magnitude and direction of stresses at any point can be determined by examination of the fringe pattern. When the component is unloaded, the photoelastic fringe pattern disappears.

When white light is used in place of monochromatic light, coloured fringes are observed. White light is often used for demonstration, and monochromatic light is used for precise measurements.

The above method is suitable when the component is transparent. In the case of opaque components, a thin sheet of photoelastic material is suitably bonded to the surface of the component. When the component is loaded, the surface strain in the component is transmitted to the photoelastic sheet producing stress in it. The resulting fringe pattern is observed by illuminating the component with polarized light and viewing it through an analyzer. More commonly, a transparent scale model of the part is made out of a material, which is optically sensitive to stress such as epoxy, glyptol or polyester resins. The model is then subjected to the forces that the actual component would experience in use. The birefringence varies from point to point over the surface of the model. When viewed with crossed polarizers, a complicated fringe pattern is seen which provides a visual means of observing overall stress characteristics of an object. The patterns can be projected on a screen or photographic film.

20.39.2 STRESS-OPTIC LAW

At any point in a loaded component there is stress acting in every direction. The directions in which the stresses have maximum and minimum value for the point are known as *principal directions*. The corresponding stresses are known as maximum and minimum principal stresses. Let us consider a model of uniform thickness made of a transparent high polymer material. Let the model be loaded such that it is in a plane state of stress (see Fig.9.11). Then the state of stress can be characterized by σ_x, σ_y and τ_{xy} or by the principal stresses σ_1, σ_2 and their orientation with respect to a set of axes. Let n_o be the refractive index of the material when it was not stressed. When the model is put in a stress, the model becomes double refracting and the directions of polarization of light at the point P coincide with the direction of principal stress axis at that point. If n_1 and n_2 are the refractive indices for vibration corresponding to these two directions, then

$$n_1 - n_0 = c_1\sigma_1 + c_2\sigma_2 \tag{20.69}$$

$$n_2 - n_0 = c_1\sigma_2 + c_2\sigma_1 \tag{20.70}$$

where c_1 is called the direct stress optic coefficient and c_2 the transverse stress optic coefficient. Since the stresses vary uniformly, σ_1, σ_2 and θ are continuously distributed functions over the model in the *xy*-plane. The directions of the polarizing axes as well as the values of n_1 and n_2 vary uniformly over the -face of the model.

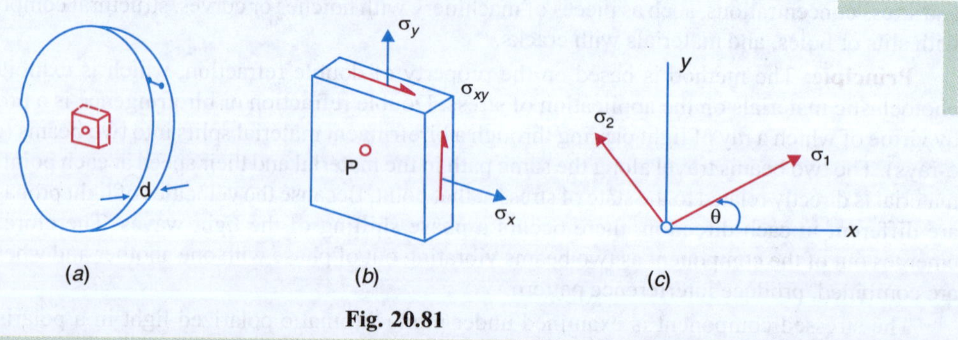

Fig. 20.81

If linearly polarized light is incident normally at any point P of the model, the incident light gets resolved along σ_1 and σ_2; and these two vibrating components travel through the thickness of the model with different velocities. When they emerge there will be a certain amount of relative phase difference between these two components. The phase difference is given by

$$\delta = \frac{2\pi d}{\lambda}(n_1 - n_2) \tag{20.71}$$

$$\delta = \frac{2\pi d}{\lambda}(c_2 - c_1)(\sigma_1 - \sigma_2)$$

$$\delta = \frac{2\pi d}{\lambda} C\,(\sigma_1 - \sigma_2) \qquad (20.72)$$

where $C = c_2 - c_1$ is the relative or **differential stress-optic coefficient** expressed in terms of brewsters (1 brewster = 10^{-12} m^2/N).

Equ.(20.72) shows that in a transparent and isotropic model in which the stresses are two-dimensional, the phase difference between the two wave components traveling through the model is directly proportional to the difference of the principal stresses.

When the two wave components are brought together, interference takes place and we get a fringe pattern, which depends on relative retardation, given by equ.(20.72). Thus, studying the fringe pattern one can determine the state of stress at various points in the material.

The number of wavelengths of relative path difference is given by

$$N = \frac{\delta}{2\pi} = \frac{d}{\lambda} C\,(\sigma_1 - \sigma_2) \qquad (20.73)$$

Equs.(20.72) and (20.73) are called stress-optic relations or **stress-optic law**. These equations relate the state of stress at a point to the optical behaviour of the model.

In practice one computes the values of $(\sigma_1 - \sigma_2)$ from the observed values of δ or N. Then,

$$(\sigma_1 - \sigma_2) = \frac{N\lambda}{dC} = \frac{N}{d} F \qquad (20.74)$$

F is called the **material fringe value**. If d = 1cm and N = 1 wavelength, then F gives the value of $(\sigma_1 - \sigma_2)$. It produces a relative phase difference of 2π radians on a model of unit thickness. This is a property of the model material and wavelength of light used. The quantity

$$\frac{F}{d} = f \qquad (20.75)$$

is called the **model fringe value.**

At those points in a stressed model where $\sigma_1 = \sigma_2$, the fringe order is zero and black dots appear at these points. Such points are called *isotropic points*. If $\sigma_1 = \sigma_2 = 0$, then also the fringe order is zero at these points and black dots appear. Such points are called *singular points*.

20.39.3 DEFINITION OF ISOCLINICS AND ISOCHROMATICS

Isoclinics are the locus of the points in the specimen along which the principal stresses are in the same direction.

Isochromatics are the locus of the points along which the difference in the first and second principal stress remains the same. Thus they are the lines which join the points with equal maximum shear stress magnitude.

20.39.4 PHOTO ELASTIC BENCH

Photo elastic bench is an optical instrument used to analyse the stress distribution in a model subjected to load. The instrument utilizes the properties of polarized light in its operation.

Principle

When a transparent material is stressed, it becomes double refracting. On examining the stressed material between crossed polarizers using light, interference fringes are observed. The fringes are used to test and measure the stress and strain produced in the material.

Construction

A photoelastic bench mainly consists of the following parts.

(*i*) Polariscope
(*ii*) Loading frame
(*iii*) Light source and
(*iv*) Camera.

The block diagram of the arrangement of a photoelastic bench is shown in Fig.20.82.

For photoelastic analysis, two types of polariscopes are used.

(*a*) Plane polariscope (*b*) Circular polariscope.

In the plane polariscope, plane-polarized light is used and in the circular polariscope, circularly polarized light is used.

20.39.4.1 PLANE POLARISCOPE

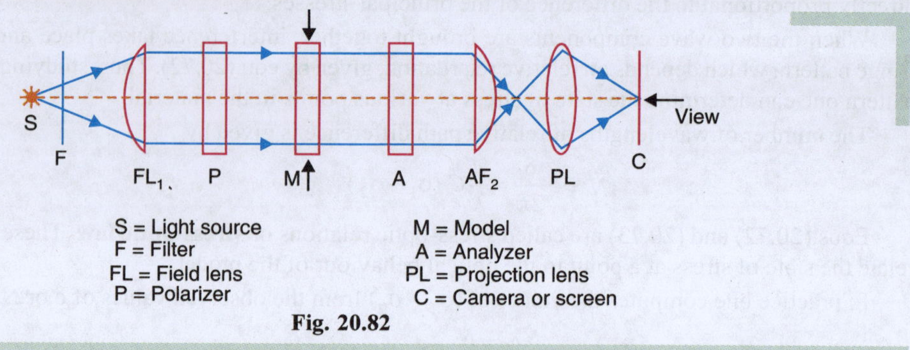

Fig. 20.82

The basic arrangement of a lens type polariscope is shown in Fig.20.82.

Working

An incandescent lamp serves as a white light source. The first field lens L_1 gives a parallel beam of light in the field of view. The unpolarized light emerging from the field lens is then passed through the polarizer P and gets plane-polarized.

The model M made of a photoelastic material is loaded in a loading frame by which various types of loads can be applied. The plane polarized light emerging out of P passes through the model in stressed condition and splits into two beams. These beams are plane polarized in mutually perpendicular planes and are incident on the analyzer A. The analyzer combines these two beams coming from the model. The polarizer and analyzer are generally coupled together by a flexible coupling to achieve their simultaneous rotation.

The second field lens L_2 makes the parallel beam of light to converge on the projection lens, L, which finally projects the interference fringes on to the screen or camera C.

The fringe pattern in a plane polariscope setup consists of both the isochromatics and the isoclinics. The isoclinics change with the orientation of the polariscope while there is no change in the isochromatics.

20.39.4.2 CIRCULAR POLARISCOPE

The circular polariscope contains two quarter-wave plates extra in addition to all the elements of a plane polariscope (Fig.20.83). The first quarter-wave plate is kept between the polarizer P and

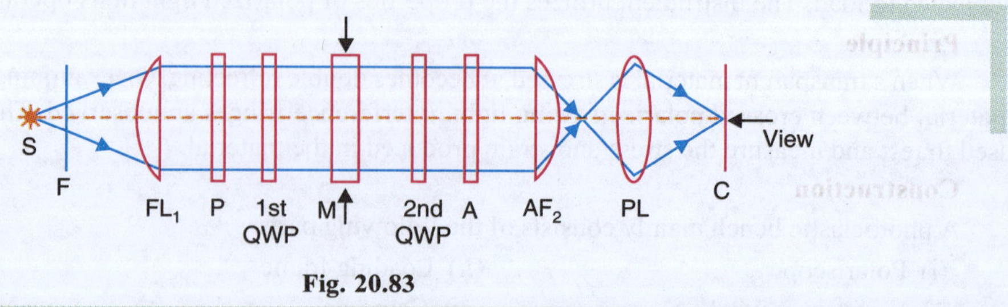

Fig. 20.83

the model M, while the second one is held between the model and the analyzer A. The first QWP converts plane polarized light into circularly polarized light and the second QWP converts circularly polarized light into plane-polarized light.

The basic advantage of a circular polariscope over a plane polariscope is that in a circular polariscope setup we only get the isochromatics and not the isoclinics. This eliminates the problem of differentiating between the isoclinics and the isochromatics.

20.39.5 ISOCLINICS AND ISOCHROMATICS

Isoclinics and isochromatics are the two different types of fringes observed in photoelastic studies. Isoclinic fringes occur whenever either principal stress direction coincides with the axis of polarization of the polarizer. They provide information about the *directions of the principal stresses* in the component. Isoclinic fringes can be removed by using a circular polarizer. Isochromatic fringes are lines of constant **principal stress difference** ($\sigma_1 - \sigma_2$). With monochromatic light, they appear as dark and bright fringes and with white light illumination, they appear as coloured fringes. In a plane polariscope, the two types of fringes are found superposed on each other and can be distinguished by rotating the component. Isoclinic fringes vary in intensity as they pass through the extinction positions, whereas isochromatic fringes remain unchanged.

Let us consider a model suitably held in a plane polariscope. Let the polarizer and analyzer combination be held in crossed configuration, which is called as the *dark field* set up. When the model is stressed, it becomes double refracting. At the point where the ray passes, the polarizing axes coincide with the principal stress axes σ_1, σ_2 at that point.

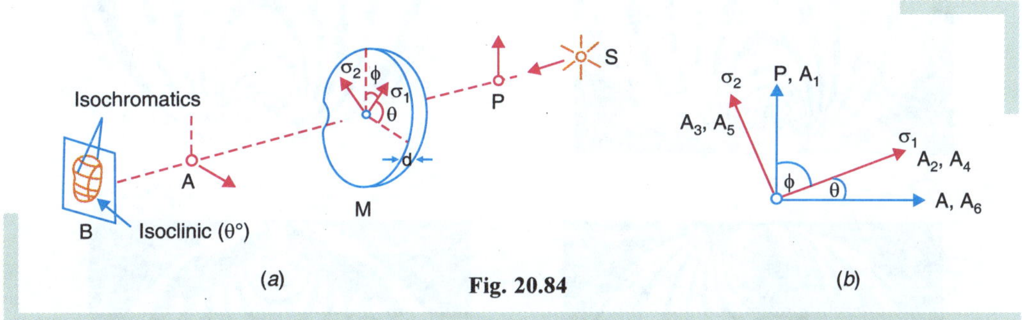

Fig. 20.84

In general, the polarizer makes an angle ϕ with the σ_1-axis. If ϕ happens to be zero or 90°, the direction of the polarizer coincides with either σ_1 (or σ_2), then plane polarized light incident on the model at that point will emerge as plane polarized light. Since the analyzer is kept crossed with respect to the polarizer, no light comes out of the analyzer. Consequently, at all those points of the model where the directions of the principal stresses happen to coincide with the particular orientation of the polarizer-analyzer combinations, the light coming out of the analyzer is zero. If the polarizer-analyzer combination happens to coincide with the directions of σ_1, σ_2 stresses at one point of the model, then in general, there will be a locus of points in the model along which this condition is satisfied.

The locus of points where the directions of principal stresses coincide with a particular orientation of the polarizer-analyzer combination is known as the **isoclinic**. For example, if the polarizing element is kept vertical and the analyzer horizontal, then on the screen a dark band will be seen which is the locus of points where σ_1, and σ_2 directions happen to be vertical and horizontal. If one measures angles from the vertical reference axis, this isoclinic will be called the 0° isoclinic. If now the polarizer is turned through 30°and the analyzer is also rotated through an equal amount, then the previously observed 0° isoclinic vanishes and a new dark band is observed on the screen. This is the

30° isoclinic and it represents the locus of points in the model where the principal stress axes are oriented at 30° and 30° + π/2 with respect to the vertical.

Let us now consider another situation. Suppose at a particular point of the model, the values of and are such that they cause a relative phase difference of 2 *m*π where *m* is an integer. If the phase difference is 2 *m*π, the model behaves like a full wave plate at that particular point. Therefore, at all these points of the model where the values of and are such as to cause a phase difference of 2 *m*π, the intensity of light will be zero. On the screen a series of dark bands corresponding to these locus of points are observed. These dark bands or fringes are called **isochromatics**. An isochromatic is a locus of points where the values of and are such as to cause a phase difference of 2 *m*π, when the background is dark. The locus of points, where the values of are such as to cause zero radian of phase difference, is called the *zero-order* fringe. The locus of points, where the values of are such as to cause 2π radian of phase difference, is called the *first-order* fringe. Similarly, one can observe second order fringe, third order fringe and so on. Fig.20.85 shows typical isoclinic and isochromatics for a stressed circular disc.

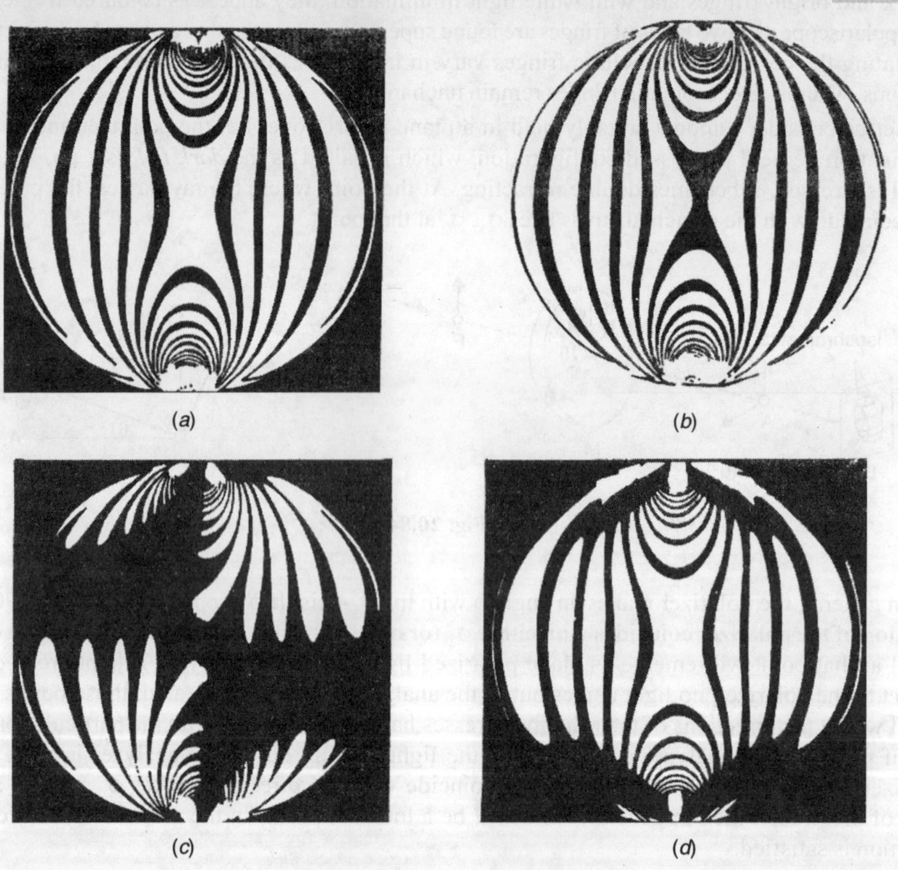

Isoclinecs and Isochromatics for a diametrically stressed circular disc (*a*) Isochromatics – in dark field set up (*b*) Isochromatics – in bright field view (*c*) 15° - Isoclinic (*d*) 45° - Isoclinic

Fig. 20.85

WORKED OUT PROBLEMS

Example 20.1: It is desired to use a plate of glass to obtain polarized light. If the refractive index of glass is 1.5, what is the polarizing angle?

Solution: Polarizing angle, $\theta_p = \tan^{-1}\mu = \tan^{-1}(1.5) = \mathbf{56.31°}$.

Example 20.2: Sunlight is reflected from a calm lake. The reflected light is 100% polarized at a certain instant. What is the angle between the sun and the horizon at that instant? The refractive index of water is 1.33.

Solution: Since the reflected is 100% polarized, the angle of incidence is equal to the Brewster angle. By Brewster's law, $\mu = \tan\theta_B$. Therefore,

$$1.33 = \tan\theta_B. \qquad \therefore \theta_B = 53.06°$$

The angle between the sun and the horizon = 90°–53.06° = 36.94° = 36° 54'.

Example 20.3: *The critical angle of incidence for total reflection in case of water is 48°. What is its polarization angle? What is the angle of refraction corresponding to the polarization angle?*

Solution: The refractive index, $\mu = \dfrac{1}{\sin\theta_C} = \dfrac{1}{\sin 48°} = 1.346.$

Now, $\mu = \tan\theta_B$ or $1.346 = \tan\theta_B$ $\therefore \theta_B = 53.4° = 53°21'$

If r is the angle of refraction, $r = 90° - 53°21' = 36°39'$.

Example 20.4: Unpolarized light falls on two polarizing sheets placed one on top of the other. What must be the angle between the characteristic directions of the sheets if the intensity of the transmitted light is one-third intensity of the incident beam?

Solution:

Intensity of the light transmitted through the first polarizer $I_1 = I_o/2$,

where I_o is the intensity of the incident unpolarized light.

Intensity of the light transmitted through the second polarizer is $I_2 = I_1\cos^2\theta$

where θ is the angle between the characteristic directions of the polarizer sheets.

But $\quad I_2 = I_o/3 \quad$ (given)

$$\therefore \quad I_2 = I_1\cos^2\theta = I_o\cos^2\theta/2 = I_o/3$$

$$\therefore \quad \cos^2\theta = 2/3$$

$$\therefore \quad \cos\theta = 0.8165$$

$$\therefore \quad \boldsymbol{\theta = 35.3°}$$

Example 20.5: If the plane of vibration of the incident beam makes an angle of 30° with the optic axis, compare the intensities of extraordinary and ordinary light.

Solution: Intensity of the extraordinary ray $I_e = E^2\cos^2\theta$

Intensity of the ordinary ray $I_o = E^2\sin^2\theta$

$$\frac{I_e}{I_o} = \frac{\cos^2\theta}{\sin^2\theta} = \frac{\cos^2 30^o}{\sin^2 30^o} = \mathbf{3}$$

Example 20.6: *Light of intensity I_o is incident on a polarizer. What is the intensity of the resultant beam if: (i) incident light is unpolarized? (ii) incident light is plane polarized with its electric field making an angle of 30° with the axis of the polarizer?*

Solution:

If incident beam is unpolarized, then the intensity of the resultant beam will be $I_o/2$.

When the incident light is plane polarized, according to Malus' law

$$I = I_o\cos^2\theta = I_o\cos^2 30° = I_o(0.866)^2 = (3/4)I_o$$

Example 20.7: Plane-polarized light of wavelength 5400 Å is incident perpendicularly on a quartz plate cut with faces parallel to optic axis. Find the thickness of quartz plate, which introduces phase difference of 60° between e- and o-rays.

Solution:

The path difference between the waves is given by $\Delta = (\mu_e - \mu_o)d$

The phase difference between the waves is given by $\delta = \dfrac{2\pi}{\lambda}\Delta$

$$\Delta = \frac{60^o}{360^o}\cdot\lambda = \frac{\lambda}{6}$$

$$\therefore \quad d = \frac{\lambda}{6(\mu_e - \mu_o)}$$

$$d = \frac{5400\ \text{Å}}{6(1.553-1.544)} = \frac{0.54}{0.054}\,\mu\text{m} = \mathbf{10\ \mu m}$$

Example 20.8: Plane-polarized light of wavelength 6000 Å is incident perpendicularly on a calcite plate of thickness 0.04 mm. Calculate the phase retardation that it will introduce between the e-ray and o-ray. Given that $\mu_o = 1.642$ and $\mu_e = 1.478$.

Solution: The phase difference between the waves is given by $\delta = \dfrac{2\pi}{\lambda}\Delta = \dfrac{2\pi}{\lambda}d\,(\mu_e - \mu_o)$

$$\therefore \quad \delta = \frac{2\times 3.143}{6000\times 10^{-10}\ \text{m}} \times 4\times 10^{-5}\ \text{m}\,(1.642-1.478) = \mathbf{68.7\ rad.}$$

Example 20.9: At a given point the electric fields of component waves of a polarized wave are

$$E_x = 10\cos\omega t \text{ and } E_y = 20\cos(\omega t + \pi)$$

Determine the type of polarization and the direction of polarization.

Solution: At any instant, *t*, we have $E_x = 10\cos\omega t$ and $E_y = -20\cos(\omega t + \pi)$.

$$\therefore \quad \cos\omega t = \frac{E_x}{10} = -\frac{E_y}{20} \quad \text{or} \quad E_y = -2E_x$$

Example 20.10: Find out the state of polarization represented by the following equations.

$$E_x = E_0 \sin(\omega t - kz)$$

and

$$E_y = E_0 \sin(\omega t - kz)$$

Solution : $E_x = E_0 \sin(\omega t - kz)$

$E_y = E_0 \sin(\omega t - kz)$

Squaring the above equations, we get

$$E_x^2 = E_0^2 \sin^2(\omega t - kz)$$

$$E_y^2 = E_0^2 \sin^2(\omega t - kz)$$

Adding the above equations, we get

$$E_x^2 + E_y^2 = E_0^2$$

The equation represents a circle. Therefore, the light is circularly polarized.

It means that the light is plane polarized in *xy*-plane with slope (–2). As $\tan^{-1}(-2) = -63.4°$, the light is polarized at an angle of –63.4° with the *x*-axis.

Example 20.11: Plane-polarized light passes through a double refracting crystal of thickness 40 μm and emerges out as circularly polarized light. If the birefringence of the crystal is 0.00004, find the wavelength of the incident light.

Solution: A quarter-wave plate changes plane-polarized light into circularly polarized light. Its thickness is given by $d = \dfrac{\lambda}{4(\mu_e - \mu_o)}$.

$$\therefore \qquad \lambda = 4d\,(\mu_e - \mu_o) = 4 \times 40 \times 10^{-6}\,\text{m} \times 0.00004 = 6400\ \text{Å}.$$

Example 20.12: A beam of plane polarized light is changed into circularly polarized light by passing it through a slice of cysrtal 0.003 cm thick. Calculate the birefringence of the crystal assuming this to be the minimum thickness that will produce the effect, ($\lambda = 6 \times 10^{-5}$cm).

Solution: Plane polarized light is converted into circularly polarized light by a suitably oriented quarter wave plate. Its thickness is given by

$$d = \frac{\lambda}{4(\mu_e - \mu_o)} = \frac{\lambda}{4\Delta\mu}$$

$$\therefore \qquad \Delta\mu = \frac{\lambda}{4d} = \frac{6 \times 10^{-5}\,\text{cm}}{4 \times 0.003\text{cm}} = \mathbf{0.005}$$

Example 20.13: *Plane polarized light is incident on a piece of quartz cut parallel to the axis. Find the least thickness for which the ordinary and extraordinary rays combine to form plane polarized light.*

Solution: When plane polarized light is incident on a half wave plate, the emergent beam will also be plane polarized. The least thickness of the quartz plate is given by

$$\therefore \qquad d = \frac{\lambda}{2(\mu_o - \mu_e)} = \frac{5 \times 10^{-5}\,\text{cm}}{2(1.5533 - 1.5442)} = \mathbf{27\mu m}$$

Example 20.14: *A half-wave plate is fabricated for a wavelength of 3800 Å. For what wavelength does it work as a quarter-wave plate?*

Solution: The thickness of a half-wave plate is $d = \dfrac{\lambda_1}{2(\mu_e - \mu_o)}$. The same plate is required to act as a quarter-wave plate. Therefore, we can write that $d = \dfrac{\lambda_2}{4(\mu_e - \mu_o)}$.

$$\therefore \qquad d = \frac{\lambda_1}{2(\mu_e - \mu_o)} = \frac{\lambda_2}{4(\mu_e - \mu_o)}$$

$$\therefore \qquad \lambda_2 = 2\lambda_1 = 2 \times 3800\text{Å} = 7600°.$$

Example 20.15: Calculate the least thickness of a calcite plate which would convert plane polarized light into circularly polarized light. Given $\mu_o = 1.658$, $\mu_e = 1.486$ and wavelength of light is 5890Å.

Solution:

Plane polarized light gets converted into circularly polarized light by a suitably oriented quarter wave plate. Its thickness is given by

$$\therefore \qquad d = \frac{\lambda}{4(\mu_o - \mu_e)} = \frac{5890}{4(1.658 - 1.486)}\,\text{Å}$$

or $\qquad d = \mathbf{0.856\ \mu m.}$

Example 20.16: Plane polarized light passes through a calcite plate with its optic axis parallel to the faces. Calculate the least thickness of the plate for which the emergent beam will be plane polarized. Given $\mu_o = 1.6584$, $\mu_e = 1.4864$ and wavelength of light is 5000Å.

Solution:

When a plane-polarized beam is incident on a half wave plate, the emergent beam will also be plane polarized but the plane of polarization undergoes a rotation through an angle 2è.

$$d = \frac{\lambda}{2(\mu_o - \mu_e)} = \frac{5\times10^{-7}\text{ m}}{2(1.6584 - 1.4864)} = \mathbf{1.45\ \text{ìm.}}$$

Example 20.17: The rotation in the plane of polarization in a certain substance is 10°/cm. Calculate the difference between the refractive indices for right and left circularly polarized light in the substance. Given λ = 5893 Å.

Solution: $$q = \frac{\pi}{\lambda}[\mu_R - \mu_L]d \qquad \therefore \quad [\mu_R - \mu_L] = \frac{\theta\lambda}{\pi d}$$

It is given that $\frac{\theta}{d} = 10° = \frac{10\times\pi}{360°} = \frac{\pi}{36}$ radian/cm and $\lambda = 5893$ Å $= 5893\times10^{-8}$cm.

$$\therefore \quad \mu_R - \mu_L = \frac{\pi}{36}.\frac{5893\times10^{-8}\text{ cm}}{\pi} = \mathbf{1.6\times10^{-6}.}$$

Example 20.18: *The indices of refraction of quartz for right-handed and left-handed circularly polarized waves of wavelength 7620 Å travelling in the direction of optic axis have the following values.*

$$\mu_R = 1.53914 \text{ and } \mu_L = 1.53920$$

Calculate the rotation of the plane of polarization of light in degrees produced by a plate of 0.5 mm thick.

Solution: $$\theta = \frac{\pi}{\lambda}[\mu_R - \mu_L]d = \frac{3.14\times0.5\times10^{-3}\text{ m}}{7620\times10^{-10}\text{ m}}(1.53920 - 1.53914)$$

$$= 0.1236 \text{ radian} = \frac{0.1236\times180°}{\pi} = \frac{0.1236\times180°}{3.14} = \mathbf{7°5'.}$$

Example 20.19: A 200 mm long tube containing 48 cm³ of sugar solution produces an optical rotation of 11° when placed in a saccharimeter. If the specific rotation of sugar solution is 66°, calculate the quantity of sugar contained in the tube in the form of a solution.

Solution: It is given that θ = 11°, l = 200 mm = 20 cm, S = 66°, and V = 48 cm³.

$$C = \frac{10\theta}{1S} = \frac{10\times11°}{20\text{cm}\times66°} = 0.0833\text{ g/cm}^3$$

Mass of sugar in solution M = CV = 0.0833 g/cm³ × 48 cm³ = **4 grams.**

Example 20.20: A 20 cm long tube containing sugar solution is placed between crossed Nicols and illuminated by light of wavelength of 6000 Å. If the specific rotation is 60° and optical rotation is 12°, what is the strength of the solution?

Solution: The specific rotation at a given temperature and for a given wavelength is given by

$$C = \frac{10\theta}{1S} = \frac{10\times12°}{20\text{cm}\times60°} = \mathbf{0.1}$$

Therefore, it is 10% solution of sugar, i.e., 1 gm sugar is dissolved in 10 gm of water.

Example 20.21: 20 cm length of a certain optically active solution causes right-handed rotation of 40° and 30 cm of another solution causes left-handed rotation of 24°.What will be the optical rotation produced by 30 cm length of the mixture of the above solutions in volume ratio 1:2. It is given that the solutions do not react chemically.

Solution: As the length of the mixtures is 30 cm and the solutions are in the volume ratio 1:2, we may assume that 10 cm length is of the first solution and 20 cm length is of the second solution.

The optical rotation produced by the first solution $= 40° \times \dfrac{10 \text{ cm}}{20 \text{ cm}} = 20°$ (right-handed) $= -20°$.

The optical rotation produced by the first solution $= 24° \times \dfrac{20 \text{ cm}}{30 \text{ cm}} = 16°$ (left-handed) $= +16°$

∴ Total optical rotation $= -20° + 16° = -4°$.

∴ The resultant optical rotation is **4° right-handed.**

Example 20.22: *In an experiment using a photo-elastic bench, the difference between principal stresses is 8×10^9 Nm^{-2}. The photelastic material has relative stree optic coefficient of 2 brewsters. Calculate the difference between refractive indices along the principal stresses.*

Solution: Difference between principal stresses, $\sigma_1 - \sigma_2 = 8 \times 10^9 \text{Nm}^{-2}$

Stress optic coefficient, C = 2 brewsters $= 2 \times 10^{-12} \text{ m}^2 \text{ N}^{-1}$

Difference between refractive indices, $(n_2 - n_1)$ $= C \times (\sigma_1 - \sigma_2)$

$= 2 \times 10^{-12} \text{ m}^2 \text{ N}^{-1} \times 8 \times 10^9 \text{Nm}^{-2}$

$= \mathbf{0.016.}$

QUESTIONS

1. Define and explain polarization.
2. What do you understand by polarization of light? Distinguish between polarized and unpolarized light?
3. Define the terms plane of vibration and plane of polarization. Discuss the methods by which plane polarized light can be produced.
4. Explain polarization of light on the basis of electromagnetic theory. **(Nagpur 2005)**
5. Describe the process of production of plane polarized light by reflection. State Brewster's law and give its significance.
6. How would you obtain plane polarized light by reflection? **(Nagpur 2004)**
7. State Brewster's law. How can this law be used to produce plane-polarized light?
8. State and explain Brewster's law.
9. What is Brewster's law? Give any two applications of it.
10. Show that when light is incident on a transparent material at the Brewster angle, the reflected and refracted rays are at right angles.
11. Discuss some of the applications of Brewster's law.
12. State Brewster's law. **(RTMNU, Nagpur 2010)**
13. How can the refractive index of a smooth opaque dielectric material be determined?
14. Write a note on Law of Malus. **(A.P.University, 2010)**
15. State and explain the law of Malus. **(Garhwal 2005)**
16. Unpolarized light falls on two polarizing sheets so oriented that no light is transmitted from the combination. If a third polarizing sheet is placed between them, can light be transmitted? Explain.
17. Describe the fabrication of a Polaroid.

18. How can plane-polarized light be detected?
19. Unpolarized light falls on two polarizing sheets so oriented that no light is transmitted from the combination. If a third polarizing sheet is placed between them, can light be transmitted? Explain.
20. What is dichroism?
21. Explain double refraction in uniaxial crystals. **(Nagpur 2005)**
22. What is meant by double refraction? Explain it. **(Madurai Kamaraj 2003)**
23. Explain the phenomenon of double refraction either in a calcite crystal or quartz crystal. **(Agra 2005)**
24. What are ordinary and extra-ordinary rays? **(Nagpur 2004)**
25. (*i*) What do you understand by double refraction? Explain on the basis of Huygens theory, propagation of light in a double refracting uniaxial crystal.

    (*ii*) What do you understand by rotatory polarization and rotatory dispersion? **(Garhwal 2005)**
26. What is an optic axis?
27. Give the analytical treatment of elliptically polarized light. **(Punjab 2005)**
28. A black dot is marked on a white paper. It is then viewed through a calcite crystal from the top.

    (*i*) How many images are expected to be seen and why?

    (*ii*) Is it possible to see one image only? If the answer is yes, explain the reason for it.
29. Explain the terms: (i) Double refraction (ii) optic axis (iii) positive and negative crystals.
30. What are positive and negative crystals? **(Bangalore 2005)**
31. Distinguish between:

    (*i*) Ordinary ray and extraordinary ray.

    (*ii*) Positive and negative crystals

    (*iii*) Uniaxial and biaxial crystals
32. (*a*) Describe a Nicol prism and explain how it acts as an analyzer.

    (*b*) Explain the phenomenon of double refraction in uniaxial crystals.
33. Distinguish between positive and negative crystals. **(Nagpur 2005)**
34. What is double refraction? Give the Huygen's theory of double refraction in uniaxial crystals. **(Kovempu 2005)**
35. Explain the principle, construction and working of a Nicol prism with a neat diagram.
36. Give the construction and working of Nicol prism.
37. How is Nicol prism used as an analyzer?
38. How Nicol prism can be used as polarizer and analyzer? Explain in detail with the help of diagram
39. Describe the construction of a Nicol prism and show how it can be used as a polarizer and analyser.
40. Linearly polarized light incident on a polarizer sheet such that its electric vector E makes an angle è w.r.t the transmission axis of the polarizer. What will be the transmitted intensity?
41. An unpolarized light beam is incident on a polarizer. Show that the transmitted intensity will be the half the incident intensity.
42. Explain the propagation of ordinary and extraordinary wave fronts in a calcite crystal for normal incidence with optic axis:

    (*i*) Parallel to the direction of propagation

(*ii*) Normal to the direction of propagation

43. Using Huygens principle, construct refracted beams and wave fronts in calcite crystal when:

(*i*) Incident ray is normal to crystal surface, optic axis parallel to the crystal surface and lies in the plane of incidence

(*ii*) Incident ray makes an angle and with normal while optic axis is as above.

43. Discuss the stare of polarization when the x and y components of the electric field are given by the following equations:-

(*i*) $E_x = E_0\cos(wt + kz), E_y = \frac{E_0}{2}\cos(wt + kz + \pi)$

(*ii*) $E_x = E_0\sin(wt + kz),\ E_y = E_0\cos(wt + kz)$.

44. Show that the light waves represented by equations,

$E_x = j\ E_1\cos(wt - kz)$ and $E_y = j\ E_2\cos(wt - kz + \varphi\ )$

45. What are retardation plates? State their uses. **(Nagpur 2004)**

46. Describe how quarter wave and half wave plates are made. Explain their uses in the study of different types of polarization. **(Garhwal 2005)**

47. Derive the equation for minimum thickness of quarter plate for light of wavelength ë. **Nagpur 2004)**

48. Describe in brief the phenomenon of birefringence. Discuss briefly Huygens' theory of double refraction.

49. Determine the path and wave fronts of ordinary and extra ordinary rays inside a double refracting crystal on the basis of Huygens' theory when the optic axis is inclined to the incident surface.

50. If, in a doubly refracting crystal, optic axis and incident ray are parallel to each other, find the direction of emergent ordinary and extra-ordinary rays. **(Lucknow 2004)**

51. Give the construction and theory of half wave plate.

52. Define a quarter wave plate and a half wave plate.

53. (*a*) What is meant by plane polarized, circularly polarized and elliptically polarized light?

(*b*) Show that the plane polarized and circularly polarized lights are special cases of elliptically polarized light.

50. What is half wave plate? Explain its action on plane polarized light incident on it with it its electric vector E making an angle è with the optic axis of the half wave plate.

51. What do you understand by an ordinary ray and extraordinary ray? How do these differ from each other?

52. What is a quarter wave plate?

53. What is a quarter-wave plate? Deduce its thickness for a given wavelength in terms of its refractive indices.

54. What is a quarter wave plate? Explain how it is used in producing elliptically and circularly polarized light.

55. If a quarter wave plate and half wave plate are given to you, how would you proceed to distinguish them from each other?

56. Distinguish between a quarter wave plate and a half wave plate. **(Madurai Kamaraj 2003)**

57. A transparent plate is given. Using two Nicol prisms how would you find whether the given plate is a quarter wave plate, a half wave plate or a simple glass plate.

58. What would be the phase difference introduced by quarter-wave plate and half-wave plate? **(GNDU, Amritsar, 2010)**

59. (*a*) Describe the phenomenon of double refraction.

(*b*) A quarter wave plate is rotated between two crossed polaroids. If an unpolarized bean is incident on the first Polaroid, discuss the variation of the intensity of the emergent beam as the quarter wave plate is rotated. What would happen if we have a half-wave plate instead of a quarter –wave plate? **(GNDU, Amritsar, 2010)**

60. What are the optical devices require producing circularly polarized light from unpolarized light? Explain how they are used to produce circular and elliptical polarized light.

61. Explain how we can convert circularly polarized light into plane polarized light. **(RTMNU, Nagpur 2010)**

62. What is elliptically polarized light? How is it produced?

63. Explain the construction and working of quarter wave plate. How is it used to produce circularly polarized light? **(Gulbarga 2005)**

64. How would you produce and detect the following with the help of a nicol prism and a quarter wave plate:

(*i*) plane polarized (*ii*) circularly polarized (*iii*) elliptically polarized light.

65. What is meant by circularly and elliptically polarized light? How would you produce these lights?

66. What will be the state of polarization of the transmitted light when

(*a*) a beam of plane polarized light is passed through a quarter wave plate

(*b*) a beam of circularly polarized light is passed through a quarter wave plate

(*c*) a beam of elliptically polarized light is passed through a quarter wave plate

(*d*) a beam of unpolarized light is passed through a quarter wave plate

67. Distinguish between plane polarized, circularly polarized and elliptically polarized light. How can circularly and elliptically polarized lights be produced? Explain briefly how will you determine whether a beam of light is unpolarized, plane polarized, circularly polarized or elliptically polarized? **(Agra 2005)**

68. Distinguish between (*i*) circularly polarized light and unpolarized light and (*ii*) elliptically polarized and partially polarized light?

69. Explain production of plane polarized and circularly polarized light.

70. Explain how you will distinguish between unpolarized light and circularly polarized light.

71. Describe how, with the help of a Nicol prism and a quarter wave plate, plane polarized light, circularly polarized light and elliptically polarized light are produced and detected. **(Meerut 2005)**

72. Explain production of elliptically and circularly polarized light.

73. Explain how circularly polarized and elliptically polarized light are produced and detected.

74. How can we experimentally distinguish between plane polarized, circularly polarized and elliptically polarised light?

75. Explain some of the applications of polarized light.

76. What is Babinet's compensator? Explain how it can be used to analyse elliptically polarised light?

77. What is a Babinet's compensator? What are its advantages over a quarter wave plate?

78. Discuss double image polarising prisms: (i) Rochon's prism and (ii) Wollaston's prism.

79. Explain origin of optical rotation in liquid. **(Nagpur 2005)**

80. Define specific rotatory power. **(Madurai Kamaraj 2003)**

81. (*i*) What is 'specific rotation' in optical activity?(*ii*) On what factors does it depend? **(Nagpur 2004)**

82. What is meant by optical rotation? State any two laws of optical rotation. **(Nagpur 2004)**

83. Write a note on optical activity. **(Kovempu 2005)**

84. Describe the construction and working of a Laurent's half shade polarimeter. **(Garhwal 2005)**

85. Explain Fresnel's theory of rotation of the plane of polarisation.

86. Give the construction and theory of Lippich polarimeter. How will you determine the specific rotation of a given solution with its help?

87. Explain Fresnel's theory of rotation of the plane of polarisation. How would you increase the sensitiveness of a pair of crossed Nicol prisms?

88. Define specific rotation. How it is experimentally determined using Laurent's half shade polarimeter? **(A.P.University, 2010)**

89. State the laws of rotatory polarisation. Give Fresnel's hypothesis for rotatory polarisation and derive a formula for the rotation of quartz. Give the experimental verification of this formula.

90. Define specific rotation. Explain the action of biquartz plate in a biquartz polarimeter. **(Meerut 2005)**

91. What is meant by artificial double refraction?

92. What are electro-optic effects?

93. What are magneto-optic effects?

94. Explain Kerr effect? **(Bangalore 2005)**

95. What is Pockels effect?

96. What is Cotton-Mouton effect?

97. Discuss Faraday effect?

98. Explain photoelasticity. What is photoelastic effect?

99. State the stress-optic law and obtain an expression for the same.

100. What are isoclinic and isochromatic fringes?

101. Describe the working of a plane polariscope

102. Explain the working of a circular polariscope.

PROBLEMS FOR PRACTICE

1. Calculate the thickness of (i) a quarter wave plate and (ii) a half wave plate given that μ_e=1.553, μ_o=1.544 and λ = 5000 Å. **[Ans: 0.14 μm; 0.28 μm]**

2. Plane polarised light is incident on a piece of quartz cut parallel to the axis. Find the least thickness for which the o-ray and the e-ray combine to form plane polarised light. Given that μ_e=1.5533, μ_o=1.5442 and $\lambda = 5\times10^{-5}$ cm. **[Ans: 0.028 μm]**

3. Calculate the thickness of a double refracting plate capable of producing a path difference of $\lambda / 4$ between e- and o- waves. **[Ans: 0.15 μm]**

4. Calculate the thickness of a quarter wave plate whose refractive indices for e-ray and o-ray are 1.5533 and 1.5433 respectively for sodium light of wavelength 5890Å.

5. Determine the specific rotation of the given sample of sugar solution if the plane of polarisation is turned through 13.2°. the length of the tube containing 10% sugar solution is 20 cm. **[Ans: 66°]**

6. On introducing a polarimeter tube 25 cm long and containing sugar solution of unknown strength, it is found that the plane of polarisation is rotated through 10°. Find the strength of the sugar solution in g/cm³. Given that the specific rotation of sugar solution is 60° per decimetre per unit concentration. **[Ans:0.067 g/cc]**

7. How will you orient the polarizer and analyser so that a beam of natural light is reduced to (*i*) 0.5 (*ii*) 0.25 (*iii*) 0.75 and (*iv*) 0.125 of its original intensity?

[Ans: (i) 45° (ii) 60° (iii) 30° (iv)69°]

8. Consider a positive crystal with refractive indices for e-ray 1.31 and for o-ray 1.309. What should be the minimum thickness of that crystal so that it can act as a quarter wave plate for light of wavelength 6000 Å?

9. Ice is a positive crystal with induces of refraction of 1.309 and 1.310. W hat should be the minimum thickness of ice so that it can act as a quarter wave plate for light of wavelength 6000Å?

10. Light of intensity I_0 is incident on a polarizer. What is the intensity of the resultant beam if:
   (*i*) Incident light is unpolarized?
   (*ii*) Incident light is plane polarized with its electric field making an angle of 30° with the axis of the polarizer?

11. Find the thickness of a quarter wave plane that can convert plane polarized light into elliptically polarized light. Use the following data: $\lambda = 5890$Å, $\mu_o = 1.658$ and $\mu_e = 1.486$.

12. Calculate the thickness of
   (*i*) Quarter wave plate and
   (*ii*) Half wave plate.
   Given that $\mu_e = 1.553$ $\mu_o = 1.544$, $\lambda = 5000$ Å

13. Calculate the thickness of double refracting plate capable of producing a path difference of $\lambda/4$ between extraordinary and ordinary waves.
   Given λ=5890Å, μ_0=1.530, μ_e= 1.540

14. Calculate the thickness of a plate which would convert plane polarized into circularly polarized light. Given: λ=5890Å, μ_0=1.658, μ_e= 1.486.

15. A half wave plate is designed from a crystal for $\lambda = 600nm$. If $(\mu_o - \mu_e) = 0.0057$, calculate the thickness of the plate. **[Ans: 52 μm]**

16. Calculate the thickness of a mica sheet required for making a quarter wave plate for $\lambda = 500nm$. The refractive indices for o-ray and e-ray in mica are 1.586 and 1.592.

[Ans: 20 μm]

17. Two Nicol prisms are so arranged that the amount of light transmitted through them is maximum. What will be the percentage reduction in intensity of the transmitted light when the analyzer is rotated through (a) 30° (b) 90°? **[Ans: 25%, 100%]**

18. Plane polarized light of is incident on a quartz crystal parallel to the optic axis. Find the least thickness for which the o-ray and e-rays combine to form plane-polarized light. Given: μ_o=1.5442, μ_e= 1.5533. **[Ans: 27.5 μm]**

19. A 10 cm long tube containing 10% sugar solution produces optical rotation of 13.2°. Find the specific rotation of sugar under given experimental conditions. **[Ans: 66°]**

20. Calculate the specific rotation if the plane of polarization is turned through 26.4°, traversing 20 cm length of 20% sugar solution. **[Ans: 66°]**

21. A length of 25 cm of a solution containing 50 gm of solute per litre causes a rotation of the plane of polarization of light by 5°. Find the rotation of plane of polarization by a length of 75 cm of a solution containing 100 gm of solute per litre. **[Ans: 30°]**

22. For quartz the refractive indices for right-handed and left-handed vibrations are 1.55810 and 1.55821 respectively for λ = 4000Å. Find the amount of optical rotation produced at λ = 4000Å by a plate of quartz 2 mm thick and with its faces perpendicular to the optic axis.

[Ans: 98°]

23. A solution of camphor in alcohol in a tube of 20 cm long is found to rotate the plane of vibration of light by 27°. What is the mass of the camphor in unit volume of solution? The specific rotation of camphor is +54°. **[Ans: 0.25gm/cc]**

24. A plate of quartz cut with its faces perpendicular to the optic axis is found to annul exactly the rotation of plane of polarization of sodium light produced by a 30 cm length of 18% solution of lactose. Find the thickness of the quartz palte. Given that specific rotation of lactose is 52.53° and that 1 mm quartz rotates the plane of polarization of sodium light by 21.71°. **[Ans: 1.31 mm]**

25. A sugar solution in a tube of length 20 cm produces optical rotation of 13°. The solution is then diluted to one-third of its previous concentration. Find the optical rotation produced by 30 cm long tube containing the diluted solution. **[Ans: 6.5°]**

26. 80 gm of impure sugar when dissolved in a litre of water gives an optical rotation of 99° when placed in a tube of length 20 cm. if the specific rotation of sugar is 66°, find the percentage purity of the sugar sample. **[Ans: 93.75%]**

27. The rotation in the plane of polarization in a certain substance, at 5893 Å, is 10°/cm. Calculate the difference between the refractive indices for right and left circularly polarized light in the substance. **[Ans: 8.186×10^{-7}]**

28. For a given wavelength 1 mm of quartz cut perpendicular to the optic axis rotates the plane of polarization by 20°. Find for what thickness will no light of this wavelength be transmitted when the quartz pieces is interposed between parallel Nicols. **[Ans: 4.5 mm]**

QUANTUM OPTICS

Einstein postulated that electromagnetic radiation is made up of photons indicating thereby that the electromagnetic field itself is quantized. Photons are the basic and discrete units of energy. Each photon has an energy $E = h\nu$ and momentum $p = h\nu/c = (h/2\pi)k$. Photons are stable, electrically neutral and massless elementary particles. They always travel with the speed of light '*c*'and exist only at that speed.

A photon has the mass $m = h\nu/c^2$ which is the mass of the electromagnetic field and is not associated with the rest mass because photons at rest do not exist. Photons always travel with the speed of light whether in vacuum or in a medium. Note that the velocity of photons in a medium is different from the velocity of the propagation of the wave front of light in the medium. Photons are emitted when atoms make a transit from a higher energy state to a lower energy state. They are also emitted upon the acceleration or retardation of charged particles, in the decay of small particles, and in the annihilation process of electron-positron pairs.

Photons have spin angular momentum of 1η and obey Bose-Einstein statistics. Therefore, they are called **bosons.** Bosons do not obey Pauli exclusion principle and therefore any number of photons can occupy the same state. When a very large number of photons occupy the same state, the inherent discreteness of the light beam disappears and the beam appears to be a bunch of continuous electromagnetic waves. A monochromatic plane wave may be regarded as a stream of photons having high population density, all photons occupying the same state.

There is an intimate relationship between the wave and corpuscular nature of photons. Photons exhibit interference and diffraction phenomenon which are the result of their redistribution in space. The intensity of light reaching any point of space is a measure of the number of photons striking this point. The quantum properties of light are due to the fact that the energy, momentum and mass of electromagnetic radiation is concentrated in photons.

Quantum optics deals with the discrete nature of the emission, propagation and interaction of light with a substance. It is chiefly concerned with the generation of coherent light and detection of light. Lasers, light emitting diodes (LEDs) and various types of photo-detectors are quantum optic devices.

Light emitting diodes are very low consumption devices that reduce the overall building electrical load.

21 CHAPTER

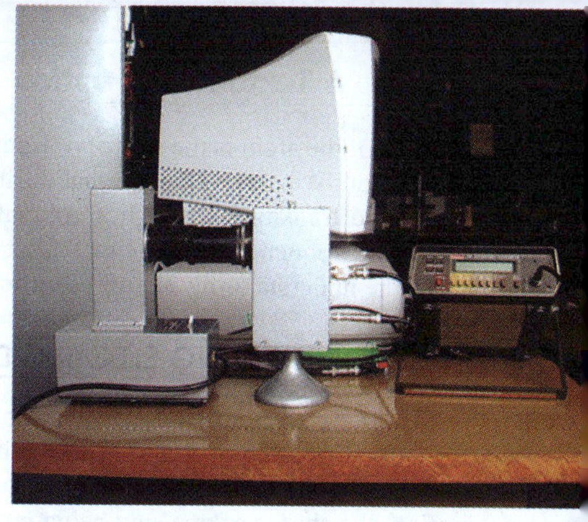

Mechanism of Light Emission

21.1. INTRODUCTION

Radiation comes in a broad range of frequencies grouped into radiowaves, microwaves, infrared rays, visible rays, ultraviolet rays, *x*-rays and γ-rays. They are all essentially electromagnetic waves and therefore, the mechanism of their emission should be the same. In all cases electromagnetic waves originate in the oscillations of electric charge. The simplest device, which produces electromagnetic waves, is an *oscillating dipole.* A dipole consists of two opposite charges. When these opposite charges vibrate to and fro along a straight line, they constitute an oscillating dipole. In terms of the simple Bohr's theory light emission from atoms is explained as due to electron's transition from an excited state to the ground state. According to quantum-mechanical considerations, the electric dipole

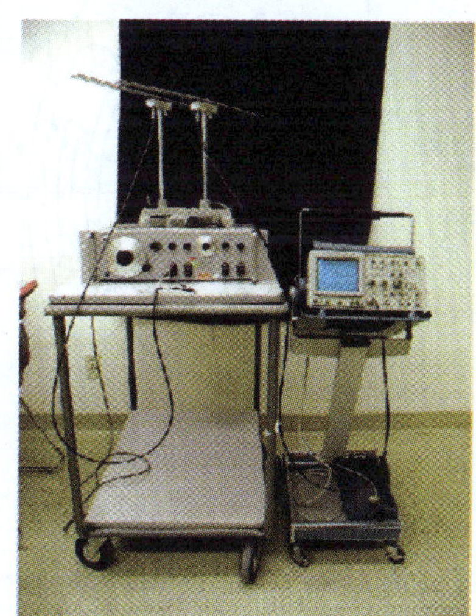

Oscillating dipole.

At a Glance

- Introduction
- Oscillating Electric Dipole
- Thermal Radiation
- The Ultraviolet Catastrophe
- The Planck's Radiation Law
- The Photon
- Photoelectric Effect
- Compton Effect
- Spectrum and Spectral Lines
- Atomic Structure
- De Broglie Hypothesis
- Heisenberg Uncertainty Principle
- Wave Functions
- Schrödinger Wave Equation
- The Wave Mechanical Model of Atom
- The Structure of the Atom
- Wave Mechanical Explanation of Photon Emission
- Properties of Spectral Lines
- Luminescence
- Scattering

moment of the atom is the source of light and one can calculate the probability of electron making a transition. We may visualize that during the act of light emission the orbital electron makes its downward transition through a gradually damped oscillatory motion occurring at the specific resonance frequency. In this chapter, we briefly describe some of the important developments that led to the understanding of atomic structure and mechanism of light emission.

21.2. OSCILLATING ELECTRIC DIPOLE

Let us consider an oscillating dipole which consists of equal positive and negative charges $+q$ and $-q$ located at A and B respectively (see Fig. 21.1*a*). The charges vibrate along the line AB with simple harmonic motion, the charge of one sign being out of phase with that of the other by 180°. Let us further assume that the oscillations occur at constant amplitude and there is no damping.

If A_1 is the amplitude of the vibration of the charge $+q$, the displacement of the charge at any instant is

$$z_1 = A_1 \cos \omega t$$

The displacement of the negative charge vibrating with the same frequency but out of phase by 180° with the first is given by

$$z_2 = A_2 \cos (\omega t + \pi) = -A_2 \cos \omega t$$

where A_2 is the amplitude of vibration of the second charge. At any instant the distance between the charges is

$$z_1 - z_2 = (A_1 + A_2) \cos \omega t.$$

The electric moment of the dipole is

$$q\,(z_1 - z_2) = p_o \cos \omega t \text{ where } p_o = q\,(A_1 + A_2)$$

Z, P, r, B, θ, O, Y, A, X

(*a*)
Oscillating dipole
Fig. 21.1

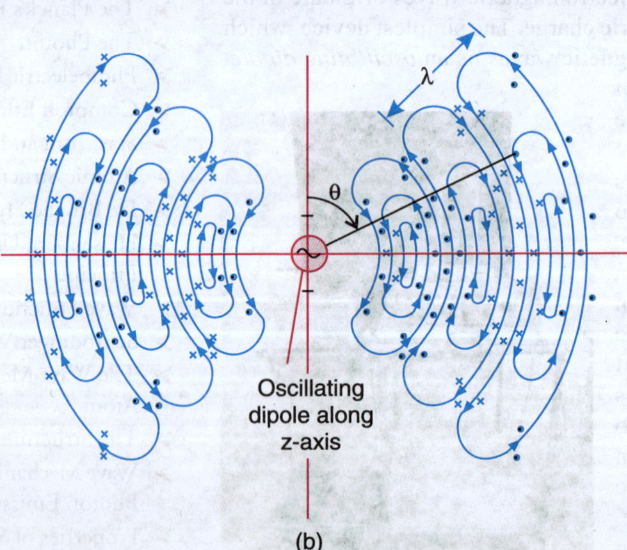

(*b*) Representation of the electric field and the magnetic field in a plane containing an oscillating electric dipole.

Fig. 21.1

According to Maxwell's equations, the electric intensity at any point around the dipole is made up of two components. Close to the dipole the field at any instant may be treated as if static. At a distance r from the centre of the dipole large compared with AB

$$E_r = \frac{2\,p_o \cos\theta}{r^3} \cos \omega t \text{ and } E_\theta = \frac{p_o \sin\theta}{r^3} \cos \omega t$$

The magnetic field near the dipole is

$$H = -\frac{\omega p_o \sin\theta}{cr^2} \sin \omega t$$

at right angles to the rz-plane. The lines of magnetic force are therefore circles about the z-axis.

At a great distance from the dipole E_r has a negligible magnitude compared with E_θ, and the electric field intensity is given by

$$E_\theta = -\frac{4\pi^2\, p_o \sin\theta}{\lambda^2 r} \cos(\omega t - kr) \qquad (21.1)$$

and the magnetic intensity is given by

$$H_\Phi = -\frac{4\pi^2\, p_o \sin\theta}{\lambda^2 r} \cos(\omega t - kr) \qquad (21.2)$$

in a direction at right angles to the rz-plane, *i.e.*, in the direction of increasing azimuth Φ measured about the z-axis. Therefore, the lines of magnetic force are circles around the z-axis. Both E_θ and H_Φ are zero along the axis of the dipole on account of the factor $\sin\theta$ and are maximum in the equatorial plane. The radiation is emitted from the dipole in greatest intensity in directions at right angles to the line of oscillation.

The Poynting flux is

$$S = \frac{c}{4\pi} E_\theta H_\Phi \frac{4\pi^3 c p_o^2 \sin^2\theta}{\lambda^4 r^2} \cos^2(\omega t - kr) \qquad (21.3)$$

in the direction of radius vector.

The energy radiated per unit time is given by

$$I = \int_o^\pi S \cdot 2\pi r^2 \sin\theta \, d\theta = \frac{32\pi r^4 c p_o^2}{3\lambda^4} \cos^2(\omega t - kr)$$

The mean rate of radiation is

$$I_{ave} = \frac{16\pi^4 c\, p_o^2}{3\lambda^4} = \frac{16\pi^4 \nu^4 p_o^2}{3\,c^3} \qquad (21.4)$$

21.3. THERMAL RADIATION

In majority of the cases, light is produced as a consequence of their temperature. Therefore, they are said to be thermal sources and the radiation emitted by the source by virtue of its temperature is called **thermal radiation**. Thermal radiation is electromagnetic in nature and its energy is smoothly distributed over all wavelengths. Therefore, a thermal source produces **continuous spectrum.** The intensity and the predominant wavelength of radiation vary with the temperature of

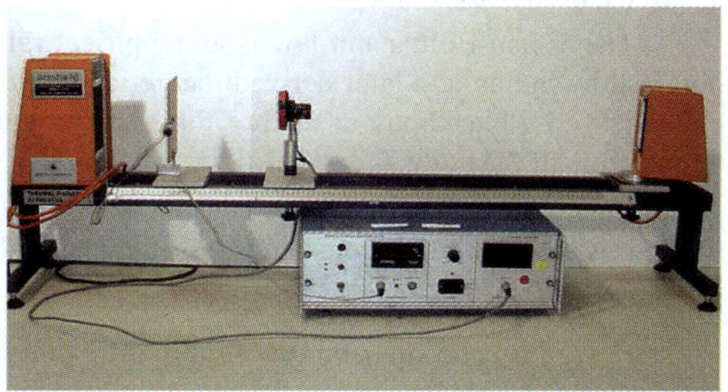

Thermal Radiation Apparatus.

the body. At low temperatures the radiation mainly lies in the infra red region. As the temperature of the body is increased, the component of maximum intensity shifts to a higher and higher frequency. For example, the filament of an incandescent bulb appears dark at room temperature and as current is passed, it gets heated up. As the current increases through the filament, it appears initially red, orange gradually and then yellow and finally it emits white light. At temperatures above 1000°C, a heated body is capable of giving out energy in the form of waves of all possible wavelengths. Normally the amount of radiation emitted by a hot body depends on factors such as the properties of its surface.

21.3.1. KIRCHHOFF'S LAW

Bodies radiate energy either in the invisible form or in the visible form. Similarly, they absorb light to varying degrees. A body that is a good light-absorber appears black. In 1859, Gustav Kirchhoff formulated the following law of radiation.

If a body is in thermal equilibrium, the energy it absorbs equals the energy it gives away in the form of radiation.

The law is expressed mathematically as

$$\frac{E(\nu,T)}{A(\nu,T)} = \varepsilon(\nu,T) \tag{21.5}$$

where $E(\nu,T)$ is the emissive power of the body, $A(\nu,T)$ is its absorptive power and $\varepsilon(\nu,T)$ is the emissive power of a perfect blackbody.

A **perfect blackbody** is a body that absorbs all the radiation that is incident on it, no matter what the radiation frequency is. Conversely, when a perfect blackbody is heated it emits radiation at all frequencies. Thus, it is a good radiator as well as a good absorber. Transparent and reflective materials are poor radiators and poor absorbers. **An objet that is a good radiator at a given wavelength is also a good absorber at the same wavelength.** This is another form of Kirchhoff's law.

Gustav Kirchhoff.
(1828-1887)

The emissive power represents the ability of a body to radiate efficiently and is also associated with its ability to absorb radiation. Only for a true blackbody, $\varepsilon = 1$. For other bodies, the emissivity usually lies between 0.2 and 0.9, and for highly polished metals ε may be as low as 0.01.

21.3.2. LAWS OF BLACKBODY RADIATION

(*i*) Stefan-Boltzmann law:

The Stefan-Boltzmann law is an empirical relationship obtained by Stefan and later derived theoretically by Boltzmann. It states that the total radiation emitted from a blackbody at temperature T is proportional to the fourth power of the absolute temperature of the body T^4.

$$I = \sigma T^4 \tag{21.6}$$

where σ is called Stefan's constant having a numerical value of 5.67×10^{-8} W/ m^2-K^4.

(*ii*) Wien's law

If we plot the distribution of radiant energy as a function of wavelength at different temperatures, we obtain a set of curves as shown in Fig. 21.2. All curves have a peak and the peak is displaced towards the shorter wavelengths as the temperature rises. Wien's displacement law states that the peak wavelength, λ_{max} at which the maximum emission occurs for any given temperature is inversely proportional to the absolute temperature of the body. Thus,

$$\lambda_{max} = \frac{2.8978 \times 10^{-3}}{T} \; mK \qquad (21.7)$$

The shift to shorter wavelengths agrees with common experience.

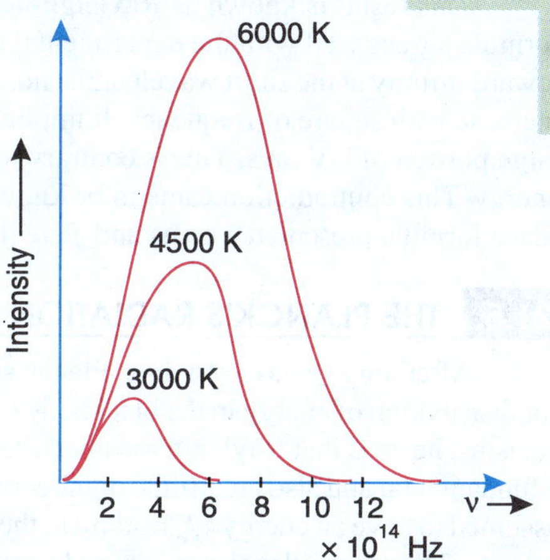

Spectral energy distribution of black body radiatiom. Area under a curve represents the total radiation.

Fig. 21.2

21.4. THE ULTRAVIOLET CATASTROPHE

In practice, a perfect black body is made by taking a hollow sphere (cavity) and drilling a small hole in it (see Fig. 21.3). The hole acts a perfect absorber. Light entering the cavity undergoes multiple reflections at the walls and gets trapped inside the cavity. Consequently, the hole appears perfectly dark. Conversely, when the cavity is heated, the radiation produced in the cavity comes out through the aperture and contains all the wavelengths. Therefore, the hole acts as a perfect emitter. The experimental results showed that at a given temperature the radiation energy density initially increases with frequency, then peaks at around a particular frequency and after that decreases finally to zero at very high frequencies, as shown in Fig. 21.2. The spectral distribution of that radiation is a function of temperature alone and the material as such plays no role.

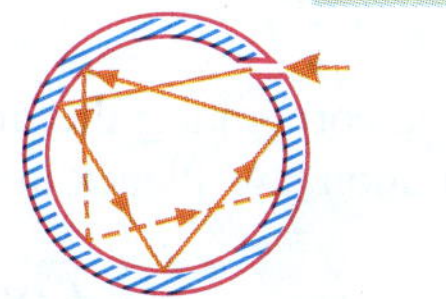

A black body: A black body absorbs all light and reflects none of the light incident on it. A spherical cavity blackened inside and completely closed except a narrow aperture serves as an ideal black body. Light entering the cavity is trapped inside by multiple reflections from the walls. When heated, the black body would emit more from a unit area than any other body at a given temperature.

Fig. 21.3

Various efforts were made to calculate theoretically the frequency distribution of thermal radiation. At that period of time electron was discovered but the inner structure of the atom was unknown. The generally accepted model was that a blackbody was made up of a huge number of atoms and the atoms are regarded as small harmonic oscillators. The random thermal motion of atoms within the walls generates electromagnetic waves, which is the *thermal radiation* emitted from the walls of the cavity. The radiation emitted by the atoms is reflected back and forth by the cavity walls to form a system of standing waves for each frequency present. There would be many modes of vibration present in the cavity space. Finally, when thermal equilibrium is attained, the average rate of emission of radiant energy by atomic oscillators in the walls equals the rate of absorption of radiant energy by the interior walls. According to Boltzmann's *principle of equipartition of energy*, each simple harmonic oscillator has an average thermal energy of 'kT' at thermal equilibrium. Rayleigh assumed that each of the standing waves ought to have energy 'kT' and derived an expression for the energy density of radiation distribution within the cavity.

$$E(v) = \frac{8\pi\, v^2\, kT}{c^3} \qquad (21.8)$$

This result is known as **Rayleigh-Jeans law**. The energy density calculated with the above formula agrees well with the experimental results at long wavelength end of the spectrum but goes toward infinity at the short wavelength end. It predicted that the intensity of thermal radiation should increase with square of frequency. It implied that the radiation emitted by a hot body should have a large portion of UV rays. This is contrary to our experience and violates the law of conservation of energy. This contradiction came to be known as **ultraviolet catastrophe**. The failure of Rayleigh-Jeans formula presented a crisis and gave the first indication of inadequacy of classical physics.

21.5. THE PLANCK'S RADIATION LAW

After long years of struggle, Planck succeeded in 1900 to obtain the correct mathematical law for distribution of energy in the blackbody radiation. He recognized that the reason for the ultraviolet catastrophe was that Rayleigh assumed that the standing waves in the cavity of the body consist of a fundamental and also an infinite number of harmonic modes of vibration. Each of these modes was assumed to have an energy kT, leading to the absurd result that the total energy in the cavity would be infinite. Therefore, Planck put forward a revolutionary postulate that an oscillating atom can absorb or reemit energy only in quantities that are integer multiples of '$h\nu$'. All other values of energy are forbidden. The indivisible discrete unit of energy $h\nu$ is called an **energy quantum.**

In general, the possible values of the energy of an oscillator of frequency ν are

$$E = n\,h\nu \qquad (n = 1,2,3,\ldots) \tag{21.9}$$

where n is called the **quantum number** of the oscillator, h is a constant now known as Planck's constant.

Incorporating the discreteness in the emission of energy, and using Maxwell-Blotzmann distribution law Planck obtained the energy density in blackbody radiation as

$$E(\nu) = \frac{8\pi h\nu^3}{c^3}\left[\frac{1}{e^{h\nu/kT} - 1}\right] \tag{21.10}$$

This is **Planck's radiation law**. It is seen from the formula that the exponential term in the denominator prevents E (ν) from rising to infinity at small wavelength values. Planck's formula gives complete fit with the experimental observations at all wavelengths and for all temperatures.

The existence of discrete energy values (Fig. 21.4) represented a departure from the classical physics and our everyday experience. If we take a mass–spring harmonic oscillator, it can receive any amount of energy form zero to some maximum value (Fig. 21.5). Thus, in the realm of classical physics energy always appears to occur with continuous values and energy exchange between bodies involves any arbitrary amounts of energy.

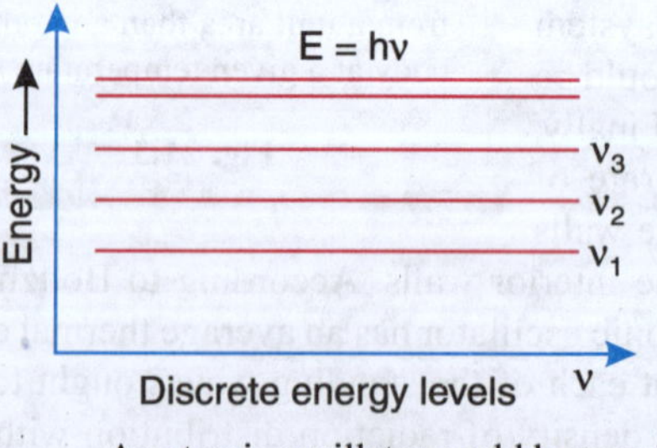

An atomic oscillator could have only discrete amount of energy, according to Planck's hypothesis.

Fig. 21.4

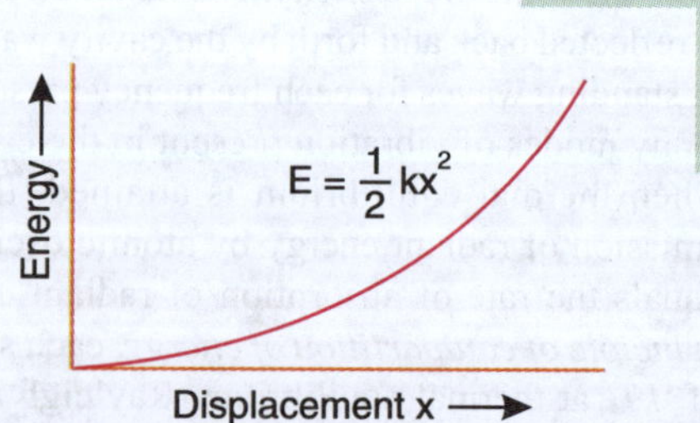

A classical oscillator such as a spring can have any amount of energy from zero to some maximum value; and the energy distribution is continuous.

Fig. 21.5

Planck's hypothesis of discrete quanta of energy was the origin of the quantum physics. Thus, the investigations on the continuous spectrum emitted by a blackbody gave birth to quantum physics.

21.6. THE PHOTON

Max Planck introduced the concept of discontinuous emission and absorption of radiation by bodies but he treated the propagation of radiation through space as occurring in the form of continuous waves as demanded by electromagnetic theory. Einstein refined the Planck's hypothesis and invested the quantum with a clear and distinct identity. The energy quanta are named *photons*.

1. According to Einstein the quantization of energy, which is present in the emission and absorption processes, is retained as the energy propagates through space. A light beam is regarded as a stream of photons travelling with a velocity 'c'.
2. An electromagnetic wave having a frequency contains identical photons, each having an energy $h\nu$. The higher the frequency of the wave, the higher is the energy content of each photon. Thus, γ-ray and x-ray photons are more energetic compared to optical photons while the photons of r.f. frequencies are the feeblest.
3. The intensity of a monochromatic light beam I is related to the concentration of photons present in the beam.

$$I = N\,h\nu \tag{21.11}$$

4. When photons encounter matter, they impart all their energy to the particles of matter and vanish.

Two experiments, namely the photoelectric effect and Compton effect, have provided the convincing experimental evidence for the existence of photon.

21.7. PHOTOELECTRIC EFFECT

Photoelectric effect is the phenomenon in which electrons are knocked off a metal surface when light is incident on it (see Fig. 21.6). The metal is said to be a photosensitive material and the liberated electrons are called *photoelectrons*.

Electronic emission increases with the intensity of the radiation falling on the metal surface, since more energy is available to release electrons. But a characteristic frequency dependence is also observed. For each substance there is a minimum frequency ν_o of light and for light of frequency less than ν_o, photoelectrons are not produced, however intense the incident light may be. The minimum

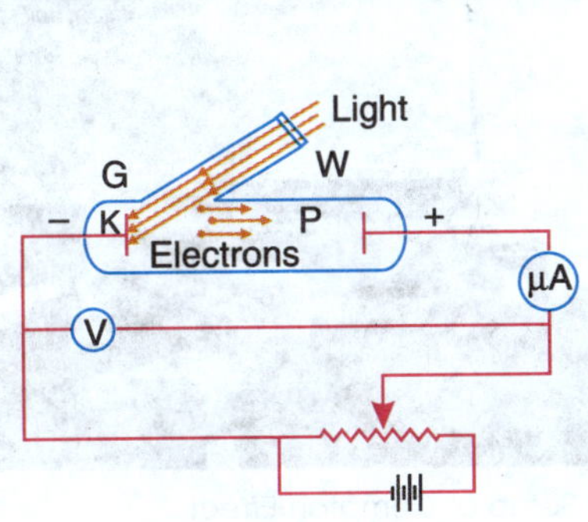

Schematic of the arrangement to study photoelectric effect.

Fig. 21.6

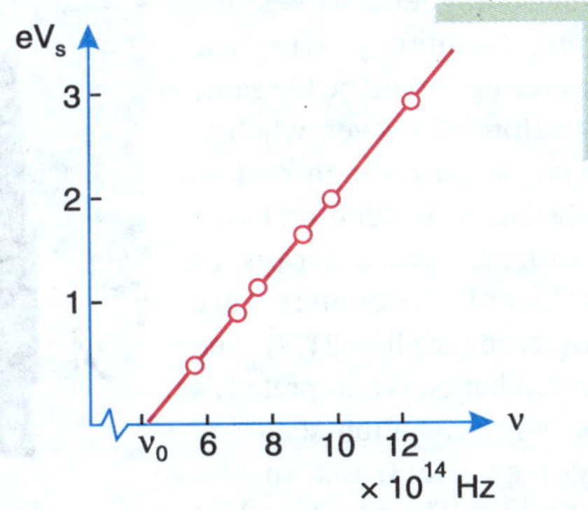

A plot of K.E.$_{max}$ of photoelectrons with frequency. The kinetic energy is equal to zero at threshold frequency ν_0. Above threshold, the K.E. of photoelectrons is directly proportional to the frequency of the incident light.

Fig. 21.7

frequency ν_o is called **threshold frequency**. The photoelectrons that are emitted in the process are found to have a range of energies. The maximum kinetic energy of the photoelectrons varies with the frequency of the incident light (see Fig. 21.7)and is independent of the intensity of light.

These features could not be explained on the basis of electromagnetic wave theory of light. Einstein used Planck's quantum concept to explain the photoelectric effect. Einstein regarded light as a stream of photons. In a beam of monochromic light of frequency ν, all photons have the same energy $h\nu$. When the photons encounter electrons in a metal, they give up their energy to the electrons. However, the electrons are held within the metal by a **potential barrier** at the surface. In order to escape from the metal, an electron should have enough energy to overcome the potential barrier at the surface. It requires an input energy that is equal to the work function, W_o. This work function is specific for the metal. When an electron absorbs a photon, the energy gained by it is equal to $h\nu$. If $h\nu \geq W_o$, the electron will escape from the metal. Thus, out of the total energy $h\nu$, a small portion, W_o, is spent on surmounting the potential barrier and the balance energy $(h\nu - W_o)$ is given to the electron as kinetic energy. Applying the law of conservation of energy to this phenomenon, we find that

$$h\nu = W_o + \frac{1}{2} m\upsilon^2_{max} \qquad (21.12)$$

The above equation is known as **Einstein's photoelectric equation.** It explains all the features of the photoelectric effect.

From this equation it is seen that a photoelectron is emitted from the metal only if the energy of the incident photon $h\nu \geq W_o$. If $h\nu < W_o$, electrons cannot surmount the barrier at the metal surface and emerge from the metal, however intense the incident light may be. The minimum energy that a photon should have to dislodge an electron from the metal, without imparting it any kinetic energy (K.E. = 0), is given by $h\nu_o = W_o$. Hence, photoelectric effect can occur only when the frequency of the incident light is equal to or greater than the minimum value ν_o, which is given by $\nu_o = W_o / h$.

21.8. COMPTON EFFECT

In 1922 Arthur H. Compton studied scattering of x-rays. From classical standpoint, X-rays are just electromagnetic waves and their frequency after the scattering should be the same as before. However, when x-rays passed through certain substances, in addition to the incident x-rays, x-rays of different frequency are observed (see Fig. 21.8). The new radiation is interpreted as the x-ray radiation scattered by free electrons in the substance. The wavelength of the scattered x-rays is longer than that of the incident x-rays. Further, the wavelength of the scattered x-rays is different for each direction of scattering (see Fig. 21.9).

Experimental Setup of Compton Effect.

Compton visualized the scattering of x-rays by an electron as a collision between a photon and the electron and during the collision process there occured an exchange of energy and momentum.

By using the laws of conservation of energy and momentum, we can calculate the change in the frequency of photon as a function of the scattering angle. We obtain the following relation in terms of the wavelength.

$$\Delta\lambda = \lambda_f - \lambda_i = \frac{h}{m_o c}(1 - \cos\theta) \qquad (21.13)$$

where θ is the scattering angle. This simple formula was found to be in a perfect agreement with experimental results (Fig. 21.9).

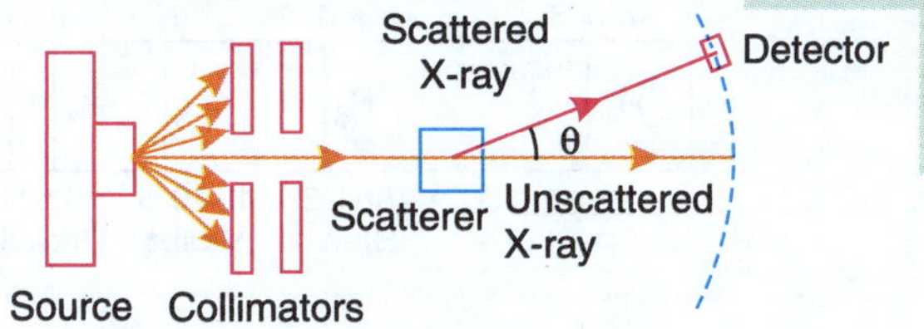

Schematic of the experimental arrangement for the study of Compton effect.

Fig. 21.8

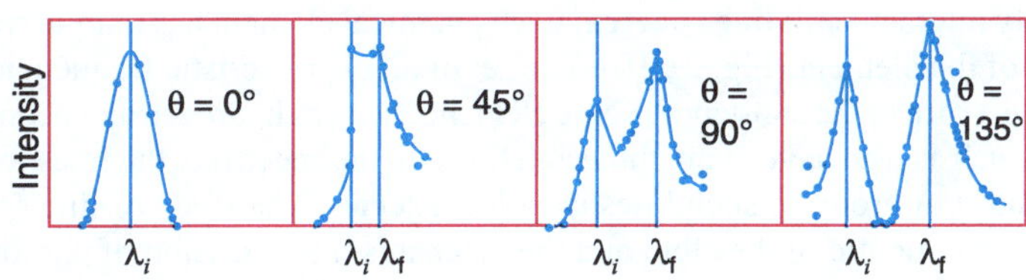

The variation of intensity X-ray as a function of wavelength, at different angles. λ_i corresponds to the unmodified (primary) component and λ_f to modified component. Note that the Compton shift $\Delta\lambda = \lambda_i - \lambda_i$ increases with increasing scattering angle θ.

Fig. 21.9

Photoelectric effect and Compton effect established beyond doubt the discrete nature of light and the existence of photons. While the photoelectric effect showed that the photons transport energy, the Compton effect indicated that photons carry momentum. Henceforth photons were granted the full-pledged status of particles.

21.9. SPECTRUM AND SPECTRAL LINES

The next question we are concerned is about how the photons are emitted. In 1666 Newton discovered that different coloured rays of light were refracted at different angles when sunlight coming through a slit was passed through a prism. When a lens was kept in the path of the band of colours, a series of coloured images of the slit were found on a screen. Newton called these series of coloured slits a **spectrum**. In general, solid materials and gases at high pressure produce **continuous spectra**, whose investigations led to quantum hypothesis by Planck.

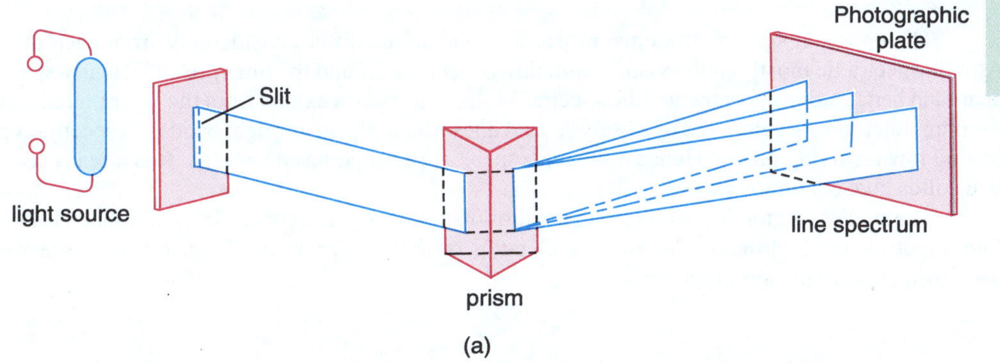

(a)

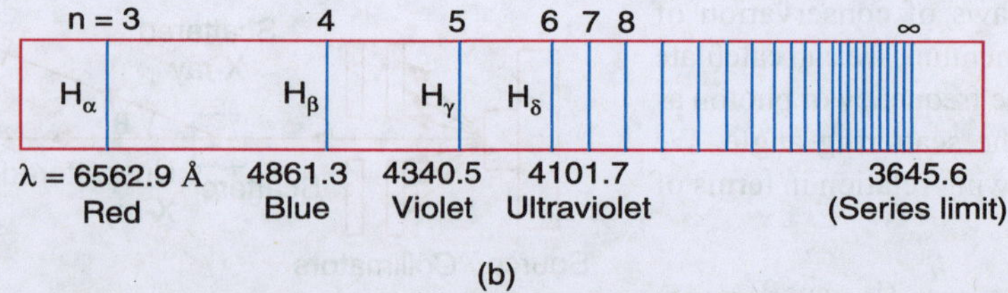

(*a*) Light emitted by an element in gaseous form passed through a prism or grating and recorded on a photographic film. (*b*) A series of s emitted by hydrogen atom. The lines marked as H_α, H_β, H_γ and H_δ belong to B and lie in the visible region.

Fig. 21.10

In contrast, vapours and gases at low pressure, where the atoms or molecules are far apart and do not significantly interact, emit **line spectra.** Each chemical element has unique spectral lines that are characteristic of the element (Fig. 21.10). The set of all characteristic frequencies is designated as the **emission spectrum** of the substance. Some elements, such as hydrogen and neon, have relatively few lines; others, such as iron have many thousand lines. In the spectrum, the lines follow each other in an **orderly** fashion. A group of such lines is called a **series**. The study of line spectra led to the understanding of the structure of the atom and the mechanism of emission of photons by atoms.

In 1859, Gustav Kirchhoff summed up his observations on line spectra as follows:

1. Light from a hot body on passing through a prism gives a continuous spectrum with no lines.
2. The same light when passed through a cool gas yields continuous spectrum but with certain wavelengths of light removed.
3. If the light emitted by a hot gas is viewed through a prism, one observes a series of bright lines at specific wavelengths against an otherwise dark background.

These are known as **Kirchhoff's three laws of spectroscopy**. If the same gas is used in 2 and 3 the absorption and emission lines appear at the same wavelengths in both the cases.

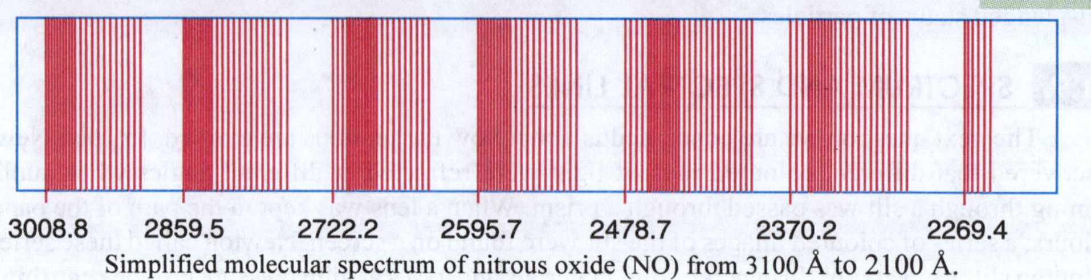

Simplified molecular spectrum of nitrous oxide (NO) from 3100 Å to 2100 Å.

Fig. 21.11

The emission spectra of atoms, molecules, and solids differ considerably from each other. Atomic spectra lie mostly in the visible and ultraviolet regions and the lines are sufficiently spaced apart and hence atomic spectra are line spectra. Molecular spectra extend from the far infra red up to the ultraviolet and are composed of closely spaced lines which appear as bright bands in a spectroscope having low resolving power. Hence molecular spectra are called **band spectra.** It is already learnt that solids give continuous spectra.

The studies on atomic spectra led to the formulation of atomic models. In 1885 Johann Balmer found that for the hydrogen atom the wavelengths of the regular array of spectral lines are well described by a simple empirical formula,

$$\lambda = 3646\left[\frac{n^2}{n^2 - 4}\right] \text{Å} \quad n = 3, 4, 5.... \tag{21.14}$$

which was later rearranged by *J.* Rydberg in the following form.

$$\frac{1}{\lambda} = R\left[\frac{1}{2^2} - \frac{1}{n^2}\right] \tag{21.15}$$

where R is a constant and known as Rydberg constant and has the value

$$R = 1.0973732 \times 10^7 \text{m}^{-1}. \tag{21.16}$$

The reciprocal of wavelength $1/\lambda$ is called the **wave number**.

21.10. ATOMIC STRUCTURE

J.J. Thomson discovered electron in 1897. He put forward the first ever model of the atom. Thomson represented it as a positively charged sphere with electrons immersed in it here and there. When unperturbed, both electrons and the positive charge were believed to be at rest. When in motion the electrons lose their energy in the form of radiation. Basing on the results of α-scattering experiments conducted by him, Rutherford proposed 'planetary model' of the atom. The atom consists of a relatively small positive nucleus at the centre and electrons orbiting well away from it. But an electron moving in an orbit around the nucleus is in a state of accelerated motion and according to Maxwell, any accelerated charge generates electromagnetic waves. Consequently, the orbiting electron loses continually its energy until it approaches the nucleus and falls into it. This leads to the prediction that an atom exists for only about 10^{-8}s; but in fact atoms have remarkable stability. Secondly, if an electron is rotating around the nucleus, it radiates light whose frequency is the same as that of the rotation. As the electron spirals and falls onto the nucleus the light frequency continually increases and therefore the radiation given out by Rutherford atom must be continuous. However, the atomic spectra were in fact a set of discrete lines. In 1913, Niels Bohr removed the deficiencies of Rutherford model by incorporating some adhoc quantum hypotheses into it.

Bohr explained the origin of line spectra in general terms on the basis of two central ideas. One is the concept of photon and the other is the concept of energy levels of atoms. Bohr combined these two concepts nicely to explain how light is emitted and why the spectral lines are arranged in an orderly fashion.

21.10.1. BOHR'S MODEL OF ATOM

Bohr fused the quantum idea with the purely mechanical model of Rutherford and introduced the following three revolutionary ad hoc postulates.

1. Electrons revolve about a nucleus only in certain special orbits called stationary orbits, though an infinite number of orbits are mechanically allowed. While moving in the permitted orbits, electrons do not emit or absorb electromagnetic radiation though they are in accelerated motion. Hence the atom is stable.

2. The allowed electron orbits are those for which the angular momentum is an integral multiple of $\hbar$, where h is Planck's constant. The angular momentum of electron is $I\omega = (mr^2)(\upsilon/r) = m\upsilon r$. Accordingly

$$m\upsilon r = n\hbar \quad (n = 1,2,3,...) \tag{21.17}$$

where n is called the **principal quantum number**.

3. The atom radiates energy only when the electron jumps from one of the upper allowed orbits, with energy say E_2, to a lower allowed orbit, with energy say E_1. The change in energy during the transition is given by

$$E_2 - E_1 = h\nu \tag{21.18}$$

Note that **the frequency of the emitted light is not determined by the motion of the electron but is governed by changes in its motion.**

Bohr obtained values of the energies of various states of hydrogen atom by assuming that an electron with velocity 'υ' revolves in a circular orbit of radius 'r' around the nucleus (see Fig. 21.12). For a dynamically stable orbit, the centripetal force experienced by the electron equals the force of electrical attraction between the nucleus and the electron. Thus,

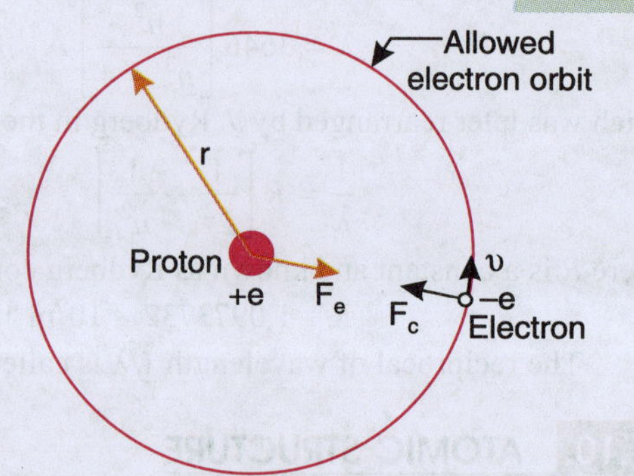

Bohr's model of the hydrogen atom. The radius of the orbit is quantized in this model.

Fig. 21.12

$$\frac{m\upsilon^2}{r} = \frac{e^2}{4\pi\varepsilon_o r^2} \qquad (21.19)$$

Using (21.17) and (21.19) to eliminate υ, we obtain for the radius of the orbits as

$$\therefore \quad r = \frac{\varepsilon_o h^2}{\pi\, me^2} n^2 = r_o\, n^2 \qquad (21.20)$$

where r_o is the radius of the first Bohr orbit ($n = 1$) given by

$$r_o = \frac{\varepsilon_o\, h^2}{\pi\, me^2} = 0.53\,\text{Å}. \qquad (21.21)$$

The total energy for a specific electron orbit is obtained by adding the kinetic energy and the potential energy of the electron with respect to the nucleus. Thus,

$$E = \frac{1}{2}\, m\upsilon^2 - \frac{e^2}{4\pi\varepsilon_o r} \qquad (21.22)$$

where the first term represents the kinetic energy of electron due to its angular motion and the second term is the potential energy due to electrical attraction. Substituting the value of r from (21.20) and eliminating υ using (21.17), we get

$$E = -\frac{e^2}{8\pi\,\varepsilon_o r} \qquad (21.23)$$

Using the value of r into the above equation (21.23), we find that electron can have only certain discrete negative values of energy E_n given by

$$E_n = -\frac{me^4}{8\varepsilon_o^2\, h^2}\cdot\frac{1}{n^2} = -\frac{E_o}{n^2} \qquad (21.24)$$

where $$E_o = \frac{me^4}{8\varepsilon_o^2\, h^2} = 13.6\; eV \qquad (21.25)$$

21.10.2. ENERGY LEVELS

We find from (21.24) that the electron in a hydrogen atom can have any one of a series of negative energies, which are referred to as **energy states.** It is easier to represent the energy states of an atom in the form of **energy levels**, rather than in the form of orbits (see Fig. 21.13). The words energy state and energy level are used interchangeably. (To be more specific, the hydrogen atom has two states in its ground level corresponding to – 13.60 eV, eight states in its – 3.40 eV level, that is the first excited level and so on).

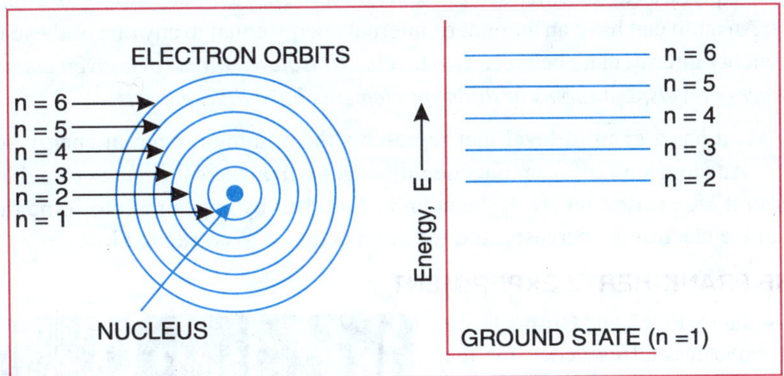

Electron orbits and the corresponding energy levels of the hydrogen atom

Fig. 21.13

An energy-level diagram of hydrogen is shown in Fig. 21.14.

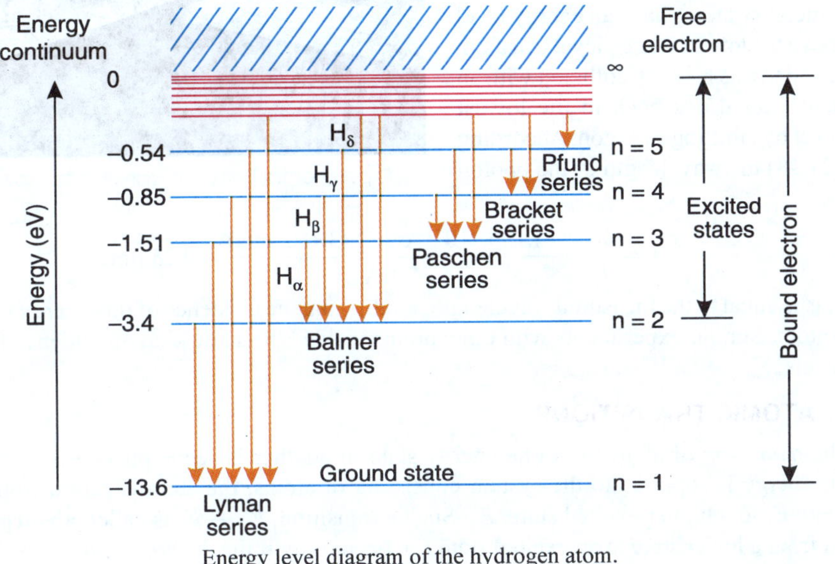

Energy level diagram of the hydrogen atom.

Fig. 21.14

Each horizontal line represents an allowed energy state. The vertical arrows between levels represent various discrete transitions. The lowest energy level is the level in which the electron revolves in the innermost orbit (n = 1); that level is called the **ground state**. For hydrogen, the ground sate is at – 13.6 eV. When the hydrogen gas is heated or subjected to an electric discharge, the atoms are raised to higher, **excited states**. For example, the first excited level is at – 3.4 eV (n = 2) and so on. The actual value of the discrete negative energy indicates the amount of energy required to remove the electron completely away from its orbit around the nucleus, and it is referred to as **ionization** energy.

The above picture can be extended to atoms of all elements. Each atom has a set of possible energy levels. An atom can have an amount of internal energy equal to any one of these levels, but it cannot have energy intermediate between two levels. *All isolated atoms of a given element have the same set of energy levels, but atoms of different elements have different sets.*

Every atom has a **ground level** that represents the minimum internal energy state that the atom can have. An atom shut off from outside influences will always lie in the ground level. All the levels higher to it are **excited levels.** If the atom is disturbed by exposing it to some radiation, the total energy of the electron is increased and the atom is raised to an excited level.

21.10.3. THE FRANK-HERTZ EXPERIMENT

In 1914, James Frank and Gustav Hertz found direct experimental evidence for the existence of atomic energy levels. Frank and Hertz studied the motion of electrons through mercury vapour under the action of electric field. They found that when an electron has kinetic energy 4.9 eV or greater, the vapour emitted UV light of wavelength 0.25 mm. Suppose mercury atoms have an energy level 4.9 eV above the lowest energy level. An atom can be raised to this level by collision with an electron; it later drops back to the lowest energy level by emitting a photon. According to equ. (21.18) the wavelength of the photon should be

Experimental Setup of Frank-Hertz Experiment.

$$\lambda = \frac{hc}{E} = \frac{(4.136\times10^{-15}\ \text{eV.s})\ (3\times10^{8}\ \text{m/s})}{4.9\,\text{eV}} = 0.25\ \mu\text{m}.$$

This is equal to the measured wavelength, confirming the existence of this energy level of the mercury atom. Similar experiments with other atoms yielded the same kind of evidence for atomic energy levels.

21.10.4. ATOMIC TRANSITIONS

The *transition* of atom from one energy state to another is accomplished by a transfer of energy. If energy is supplied to the system consisting of atoms, the atom is raised from a lower energy state E_1 to a higher excited state, E_2. Such a transition, $E_1 \rightarrow E_2$ is called **absorption**. The transition from a lower state to an excited state can occur only if the difference in energy is exactly equal to the photon energy $h\nu$. The atom does not stay in an excited state indefinitely. It usually returns on its own to the lower state after about 10 *ns* by emitting a photon. The downward transition $E_2 \rightarrow E_1$ is called **emission**. During the emission process, as the atom returns from a higher energy state E_2 to a lower energy state E_1, it emits a quantum of energy, $h\nu$. If the photon energy is $h\nu = \frac{hc}{\lambda}$, then conservation of energy gives

$$h\nu = \frac{hc}{\lambda} = E_2 - E_1 \tag{21.26}$$

Vertical lines in Fig. 21.14 show the jumps that electrons can make from one level to another. The location of these transitions, or **emission lines**, can be shown on a wavelength graph. The ensuing pattern corresponds exactly to that recorded by the spectrographic plates. Thus, **position** is one of the important properties of spectral lines.

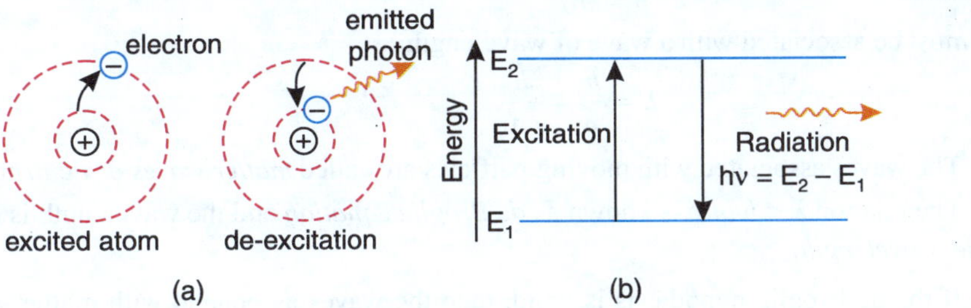

(*a*) Absorbing energy, electron jumps from an inner orbit to an outer orbit. When the electron returns to the inner orbit, it emits the same energy in the form of a photon. (*b*) energy level representation of excitation and de-excitation of the atom

Fig. 21.15

The average time spent by the atom in an excited level is called the **lifetime** of the level. The lifetime is usually of the order of 10^{-8}s. Besides the short-lived excited states, there are states with average lifetimes greater than 10 milliseconds or even as long as several seconds. They are called **metastable states**.

21.10.5. LIMITATIONS OF BOHR THEORY

Bohr's theory employed a semi-classical model where quantum postulates are introduced into a mechanical model. But it explained with remarkable success the origin of line spectra in case of hydrogen atom, which is a one-electron atom. However the theory could not give correct predictions for two electron atoms. The Bohr hypothesis established the relation of wavelengths to energy levels, but it did not provide any general principles for predicting the energy levels of a particular atom. It also did not explain why angular momentum is quantized and why and when atoms make transitions. Further, it cannot determine what the intensity of the spectral line will be. Inspite of such limitations, Bohr's theory contributed immensely to the understanding of atomic structure. A more general understanding of atomic structure and energy levels is provided by the quantum mechanics.

21.11. DE BROGLIE HYPOTHESIS

We learnt earlier that a light beam consists of electromagnetic waves and in this chapter, we have learnt that it consists of a stream of photons. Thus, we find that light has dual nature and behaves as waves sometimes and as particles at other times.

As a photon travels with the velocity *c*, we can express its momentum as

$$p = \frac{E}{c} = \frac{h\nu}{c} = \frac{h}{\lambda} \tag{21.27}$$

Thus, the wavelength and momentum *p* of a photon are related to each other through the expression

$$\lambda = \frac{h}{p} \tag{21.28}$$

The quantities ν and λ are wave properties and the quantities *E* and *p* are particle properties. The relations (21.27) and (21.28) demonstrate that the wave and particle natures of a photon are intimately tied up to each other.

In 1924, de Broglie suggested that the wave–particle dualism need not be a special feature of light. The relation (21.28) between the momentum and the wavelength of a photon must be a universal relation applying to photons and material particles alike. A particle of mass '*m*' moving with a velocity υ carries a momentum

$$p = m\upsilon$$

and it must be associated with a wave of wavelength

$$\lambda = \frac{h}{p} = \frac{h}{m\upsilon} \tag{21.29}$$

The waves associated with moving particles are called *matter waves or de Broglie waves*.

The relation $\lambda = h/m\upsilon$ is known as *de Broglie equation* and the wavelength is called the *de Broglie wavelength.*

If the de Broglie hypothesis is valid, then the waves associated with matter should suffer diffraction. A beam of electrons diffracted by a crystal should show interference phenomena. Davisson and Germer obtained, in 1927, the first experimental evidence that gave convincing proof of the wave nature of matter.

21.11.1. DE BROGLIE'S JUSTIFICATION OF BOHR'S POSTULATE

One of the postulates that Bohr used in formulating a model of atom is that the angular momentum L of the electron revolving in a stationary orbit is quantized. Thus,

$$L = n\hbar \quad \text{-- Bohr's postulate} \tag{21.30}$$

The above postulate of Bohr follows directly from the concept of matter waves. If a stretched string is fastened at both ends and is made to vibrate, standing waves are formed provided the length of the string is an integral number of half–wavelengths of the disturbance. If the string is formed into a circular loop, the condition for standing waves is that the circumference of the loop should be an integral number of whole wavelengths of the disturbance. Thus, if r is the radius of the circular loop,

$$2\pi r = n\lambda \qquad n = (1,2,3, \text{-----}) \tag{21.31}$$

We may regard the stationary electron orbits in an atom to be analogous to the circular loop of string. We conclude that stationary electron wave pattern can form in the orbit if only an integral number of electron wavelengths fit into the orbit, as shown in Fig. 21.16.

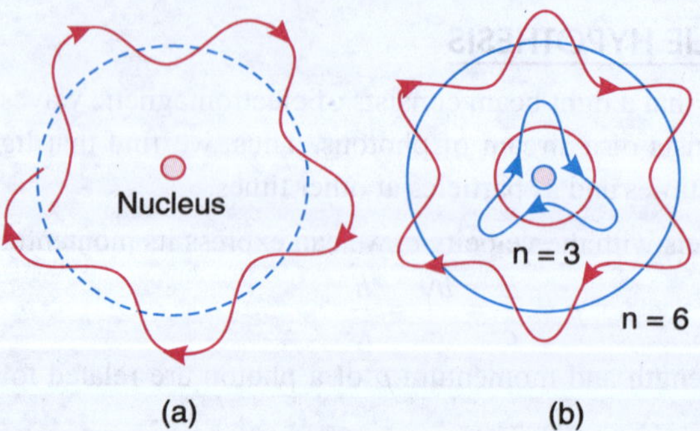

(*a*) When the circumference of the circular orbit it not equal to an integral multiple of wavelengths, the electron wave does not form standing waves and the orbit is not allowed.
(*b*) The allowed orbits accommodate an integral multiple of electron waves.

Fig. 21.16

The above equation (21.31) can be applied for electron waves, taking λ as de Broglie wavelength of electron waves.

The de Broglie wavelength of electron wave is given by

$$\lambda = \frac{h}{m\upsilon}$$

where υ is the speed of the electron in the orbit. Using de Broglie wavelength into equ.(21.31), we obtain

$$2\pi r = \frac{nh}{m\upsilon}$$

$$\therefore \quad m\upsilon r = \frac{nh}{2\pi}$$

But the quantity '$m\upsilon r$' is the angular momentum,L, of electron in the orbit of radius r. Thus,

$$L = m\upsilon r$$

It ,therefore, follows that

$$L = n\hbar \tag{21.32}$$

which is precisely Bohr's postulate. De Broglie thus demonstrated that the quantization of angular momentum is a direct consequence of wave nature of electron.

Let us calculate the wavelength of the electron in the first orbit of hydrogen atom. The electron speed υ in the orbit is given by

$$\upsilon = \frac{e}{\sqrt{4\pi\varepsilon_o mr}}$$

$$\therefore \quad \lambda = \frac{h}{e}\sqrt{\frac{4\pi\varepsilon_o r}{m}} \tag{21.33}$$

Taking $r = 5.3 \times 10^{-11}$ m, we get $\lambda = 3.3$ Å. The circumference of the orbit is $2\pi r = 3.3$ Å. Hence, the first orbit of the electron in a hydrogen atom corresponds to one complete electron wave joined on itself. The de Broglie hypothesis thus offered a new meaning to the principal quantum number "n". n is the number of de Broglie wavelengths that fit into the circumference of Bohr allowed orbits.

21.12. HEISENBERG UNCERTAINTY PRINCIPLE

The wave nature of moving particles leads to some inevitable consequences. Classically, the state of a particle is defined by its position and momentum. At each instant, the position and momentum of a classical particle can be measured to a very high accuracy. In case of a quantum particle, there are uncertainties associated with its location and momentum.

At about the same time that Davisson and Germer conducted the experiments of wavelike properties of electrons, Heisenberg put forth his **uncertainty principle**. He suggested that the product of uncertainty in the location of the quantum particle and the uncertainty in its momentum, $\Delta x\, \Delta p$, would always be of the order of Planck's constant h. Thus,

$$\Delta x \cdot \Delta p \approx \hbar \tag{21.34}$$

This is known as Heisenberg uncertainty principle for position and momentum, which may be stated as follows:

"*It is not possible to know simultaneously and with exactness both the position and the momentum of a microparticle*".

This uncertainty principle expresses a fundamental limitation in nature that also limits the precision of our measurements.

The uncertainty principle asserts that it is physically impossible to know simultaneously the exact position ($\Delta x = 0$) and exact momentum ($\Delta p_x = 0$) of a microparticle. According to it, the more precisely we know the position of the particle, the less precise is our information about its momentum. The momentum of a particle cannot be precisely specified without our loss of knowledge of the position of the particle at that time. Similarly, a particle cannot be precisely localized in a particular direction without our loss of knowledge of momentum in that particular direction. We can at best specify that certain momentum of the particle is more probable than other or that the particle is more likely to be here than there. Thus, the uncertainly principle implies that we can never define the path of an atomic particle with the absolute precision indicated in classical mechanics. To describe the quantum particle the concept of *energy* becomes important since it is related to the *state* of the system rather than to its path.

In addition to the uncertainty relation between co-ordinates and momentum of a moving particle, there is an uncertainty relation between energy and time. Suppose that we want to determine the energy of a particle and the time at which the particle has such energy. If ΔE and Δt are the uncertainties in the values of these quantities, then

$$\Delta E \cdot \Delta t \approx \hbar \tag{21.35}$$

Thus, it is impossible to know simultaneously and with exactness the energy of a particle and the time at which it has that energy.

In view of the wavelike properties of particles and uncertainties in determining the physical parameters, a theory taking into consideration of the **probability distribution** of parameters such as position, momentum, energy in place of *precise values* is required. Efforts in this direction were done independently by Schrodinger and Heisenberg who used different approaches namely wave mechanics and matrix mechanics but which were found to be equivalent.

21.13. WAVE FUNCTIONS

In quantum mechanics, various atomic parameters are described with the help of wave functions ψ. In general, these wave functions ψ are mathematical functions that depend on the variables necessary to specify the particular features of the particle. A wave function is a complex quantity and is not an *observable* quantity; hence it has no direct physical significance. For example, let the location of an electron be described by the wave function $\psi(x, y, z, t)$. This wave function by itself cannot indicate us the location of the electron. It is known that the intensity of a wave motion is proportional to the square of the amplitude of the wave. Therefore, the square of the absolute value $|\psi|^2$ of the wave function gives where the intensity of the field $|\psi|^2$ is large and thus the regions of space where the particle is more likely to be found at time t. The function ψ is, therefore, called the *probability amplitude*. Thus,

Probability, P, of finding the particle in an infinitesimal volume dV (= $dx\, dy\, dz$) is proportional to $|\psi(x, y, z)|^2\, dx\, dy\, dz$ at time t.

or
$$P \propto |\psi(x, y, z)|^2\, dV \tag{21.36}$$

The probability is a real quantity.

The probability of finding the particle between the limits $x = a$ and $x = b$, $y = g$ and $y = h$ and $z = p$ and $z = q$, is then

$$P \propto \int_a^b \int_g^h \int_p^q \psi^* (x, y, z)\, \psi(x, y, z)\, dx\, dy\, dz \tag{21.37}$$

The probability that the electron is located somewhere must be unity.

$$\therefore \qquad \iiint \psi^* \psi\, dx\, dy\, dz = 1 \tag{21.38}$$

21.14. SCHRÖDINGER WAVE EQUATION

In 1926 Erwin Schrödinger developed a wave equation that describes the behaviour of atomic particles. Let us assume that a particle of mass 'm' is in motion along the x-direction. Let the wave function ψ be the dependent variable of the de Broglie wave which is a function of the coordinates x and t. Analogous to the classical wave, we may expect that, ψ will be a function of $(x - \omega t)$. As $\upsilon = \omega / k$, the wave function may be written as a function of $(kx - \omega t)$. Using the relation $p = \hbar k$ and $E = \hbar\omega$, we can write

$$\psi = f\left(\frac{px - Et}{\hbar}\right) \tag{21.39}$$

The more general wave would be a sum of a sine and cosine waves. Taking help of Euler's identity, we write the above equation in an exponential form as follows:

$$\psi = A\, exp\left[\frac{i}{\hbar}(px - Et)\right] \tag{21.40}$$

We assume that the energy and momentum of the particle are constant. Differentiating the above equation with respect to x, we get

$$\frac{\partial \psi}{\partial x} = A\frac{ip}{\hbar}\, exp\left[\frac{i}{\hbar}(px - Et)\right]$$

$$\therefore \qquad \frac{\partial \psi}{\partial x} = \frac{ip}{\hbar}\psi$$

Rearranging the terms in the above equation, we get

$$p\psi = \frac{\hbar}{i}\frac{\partial \psi}{\partial x} = -i\hbar\frac{\partial \psi}{\partial x} \tag{21.41}$$

Differentiating the eq.(21.40) with respect to t gives

$$\frac{\partial \psi}{\partial t} = -\frac{iE}{\hbar}A\, exp\left[\frac{i}{\hbar}(px - Et)\right] = -\frac{iE}{\hbar}\psi$$

Rearranging the terms in the above equation, we get

$$E\psi = -\frac{\hbar}{i}\frac{\partial \psi}{\partial t} = i\hbar\frac{\partial \psi}{\partial t} \tag{21.42}$$

The partial derivatives with respect to x and t are connected by means of the relation between the energy and momentum. The classical expression for the kinetic energy in terms of the momentum is

$$E_k = \frac{m\upsilon^2}{2} = \frac{(m\upsilon)^2}{2m} = \frac{p^2}{2m} \tag{21.43}$$

The total energy and momentum are related by the expression

$$\frac{p^2}{2m} + V = E \tag{21.44}$$

where V is the potential energy of the particle. Multiplying the eq.(21.44) with ψ, we obtain

$$\frac{p^2}{2m}\psi + V\psi = E\psi$$

Using the relations (21.41) and (21.42) into the above equation, we get

$$-\frac{\hbar^2}{2m}\frac{\partial^2 \psi}{\partial x^2} + V\psi = i\hbar\frac{\partial \psi}{\partial t} \tag{21.45}$$

The above equation is known as the **time-dependent Schrödinger wave equation.**

Knowing the form of V, equn.(21.45) can be solved for the wave function ψ. In a number of cases the potential energy V of a particle does not depend on time; it varies with the position of the

Erwin Schrödinger.

particle only and the field is said to be *stationary*. In the stationary problems Schrödinger equation can be simplified by separating out time and position-dependent parts. Accordingly, we can write the wave function as a product of x, $\psi(x)$ and a function of t, $\varphi(t)$.

We, therefore, write that

$$\psi(x, t) = \psi(x)\,\varphi(t) \tag{21.46}$$

Equation (21.45) may be written as

$$-\frac{\hbar^2}{2m}\varphi\frac{\partial^2\psi}{\partial x^2} + V\psi\varphi = i\hbar\,\psi\frac{\partial\phi}{\partial t}$$

Dividing the above equation with $\psi\varphi$, we get

$$-\frac{\hbar^2}{2m}\frac{1}{\psi}\frac{d^2\psi}{dx^2} + V = i\hbar\frac{1}{\varphi}\frac{d\varphi}{dt} \tag{21.47}$$

If we assume that the potential energy V is a function of x only, the entire left hand side of eq.(21.47) is a function of x only while the right hand side is a function of t only. Since x and t are independent variables, both the function of x and t must be equal to a constant. The constant that each side must equal is called the *separation constant* E. Thus,

$$-\frac{\hbar^2}{2m}\frac{1}{\psi}\frac{d^2\psi}{dx^2} + V = E \tag{21.48}$$

and

$$i\hbar\frac{1}{\varphi}\frac{d\varphi}{dt} = E$$

Eq.(21.48) may be rewritten as

$$-\frac{\hbar^2}{2m}\frac{d^2\psi}{dx^2} + V\psi = E\psi$$

or

$$\frac{d^2\psi}{dx^2} + \frac{2m}{\hbar^2}(V - E)\psi = 0$$

$\therefore$

$$\frac{d^2\psi}{dx^2} + \frac{8\pi^2 m}{h^2}(V - E)\psi = 0 \tag{21.49}$$

The above equation is called the **time–independent Schrödinger equation.**

21.14.1. ALLOWED WAVE FUNCTIONS AND ENERGIES

The time-independent wave equation is the pertinent equation for studying properties of atomic systems in stationary conditions. Wave mechanical methods of solving the problem of particle motion are essentially based on ψ functions. Appropriate wave equation is formulated by incorporating the particle mass 'm' and potential energy function V for the region in which the particle is located. The next step consists of solving the differential equation for solutions, namely for the ψ functions which will satisfy the differential equation. By solving the Schrodinger equation, we obtain the possible set of ψ functions. In case of bound particles the acceptable solutions for the differential equation are possible only for certain specified values of energy. These energy values will be the only possible results of precise measurements of the total energy of the particle. These discrete values of energy $E_1, E_2, \ldots.. E_n$ are called **eigen values** or **allowed values** of the energy of the particle. The solutions $\psi_1, \psi_2, \ldots\ldots \psi_n$ corresponding to the eigen energy values E_n are called the **eigen functions**. The quantization of energy thus appears as a natural element of the wave equation.

Thus, using the Schrodinger equation, it is possible to determine first the electron energies and then the wave functions. These wave functions can then be used to determine the probability distribution function $\psi^*\psi$ of the electron, for various discrete energies as it revolves around the nucleus.

21.15. THE WAVE MECHANICAL MODEL OF ATOM

We now apply the time-independent Schrödinger equation to the hydrogen atom and obtain the eigen values of energy and eigen functions related to the electronic states of the hydrogen atom. The Schrodinger equation for the electron in three dimensions is

$$\nabla^2 \psi + \frac{8\pi^2 m}{h^2}(E - V)\,\psi = 0$$

The potential V is that of a positive charge and negative charge separated by a distance 'r' which is given by

$$V = -\frac{e^2}{4\pi\varepsilon_o r} \tag{21.50}$$

$$\therefore \quad \nabla^2 \psi + \frac{8\pi^2 m}{h^2}\left(E + \frac{e^2}{4\pi\varepsilon_o r}\right)\psi = 0 \tag{21.51}$$

It is difficult to solve this partial differential equation. It can be easily solved if it is expressed in spherical polar coordinates. The equation takes the following form in spherical polar coordinates.

$$\frac{1}{r^2}\left[\frac{\partial}{\partial r}\left(r^2\frac{\partial}{\partial r}\right) + \frac{1}{\sin\theta}\frac{\partial}{\partial\theta}\left(\sin\theta\frac{\partial}{\partial\theta}\right) + \frac{1}{\sin^2\theta}\frac{\partial^2}{\partial\phi^2}\right]\psi + \frac{8\pi^2 m}{h^2}\left(E + \frac{e^2}{4\pi\varepsilon_o r}\right)\psi = 0 \tag{21.52}$$

The above equation can be easily separated into three independent equations, each involving only in one variable. After carrying out appropriate substitutions and separating the variables, we obtain the following three total differential equations.

Azimuthal wave equation:

$$\frac{d^2\Phi}{d\phi^2} + m_l^2\,\Phi = 0 \tag{21.53}$$

Polar wave equation:

$$\frac{1}{\sin\theta}\frac{d}{d\theta}\left(\sin\theta\frac{d\Theta}{d\theta}\right) + \left[l(l+1) - \frac{m_l^2}{\sin^2\theta}\right]\Theta = 0 \tag{21.54}$$

Radial wave equation:

$$\frac{1}{r^2}\frac{d}{dr}\left(r^2\frac{dR}{dr}\right) + \left[\frac{8\pi^2 m}{h^2}\left(E + \frac{e^2}{4\pi\varepsilon_o r}\right) - \frac{l(l+1)}{r^2}\right]R = 0 \tag{21.55}$$

The azimuthal wave equation (21.53) is a simple differential equation having the following solutions.

$$\Phi(\phi) = Ae^{im_l\phi}$$

The wave function Φ must have a single value at a given point in space. It can happen only when

$$m_l = 0, \pm 1, \pm 2, \pm 3, \ldots\ldots\ldots \tag{21.56}$$

The polar wave equation (21.54) has a complicated solution in terms of associated Legender functions. The solution exists only when the constant l is an integer equal to or greater than $|m_l|$. This condition implies that for a given value of l, m_l can have values

$$m_l = 0, \pm 1, \pm 2, \pm 3, \ldots\ldots \pm l. \tag{21.57}$$

The radial wave equation (21.55) also has a complicated solution in terms of polynomials called the associated Laguerre functions. In case of the electron bound to atom, the radial wave equation can be solved only under the following conditions.

1. E has one of the negative values E_n given by

$$E_n = -\frac{me^4}{8\varepsilon_o^2 h^2} \cdot \frac{1}{n^2} \tag{21.58}$$

2. n is an integer equal to or greater than $(l + 1)$. This condition means that

$$l = 0,1,2,3, \ldots\ldots\ldots\ldots, \quad (n-1). \tag{21.59}$$

The energy eigen values specified by equ.(21.58) are precisely the same as those obtained by Bohr on the basis of the semi-classical model. The three quantum numbers needed to describe the three-dimensional motion of the electron in the hydrogen atom are n, l and m_l. They must satisfy the following conditions:

(*i*) Principal quantum number, $n \geq 1$; $n = 1, 2, 3, \ldots\ldots$

(*ii*) Orbital quantum number, $l \leq (n-1)$; $l = 0, 1, 2, 3, \ldots\ldots\ldots\ldots, (n-1)$.

(*iii*) Orbital magnetic quantum number, $|m_l| \leq l$; $m_l = 0, \pm 1, \pm 2, \pm 3, \ldots\ldots \pm l$.

Using the above conditions, we can write down all the allowed eigen functions describing the various possible quantum states for the hydrogen atom by using the notation for each state.

21.15.1. WAVE FUNCTIONS

The wave function ψ_1 of the ground state ($n = 1$) of the hydrogen atom is relatively simple to find out. As ψ_1 has to be completely spherically symmetric, we assume the solution for it.

$$\psi_1 = C_1 e^{-a r} \tag{21.60}$$

α is a constant which turns out to be the reciprocal of the value of the first Bohr radius, r_o. For a spherically symmetric volume element $4\pi r^2\, dr$, the normalization constant C_1 has a value

$$C_1 = \frac{1}{\pi^{1/2}\, r_o^{3/2}}$$

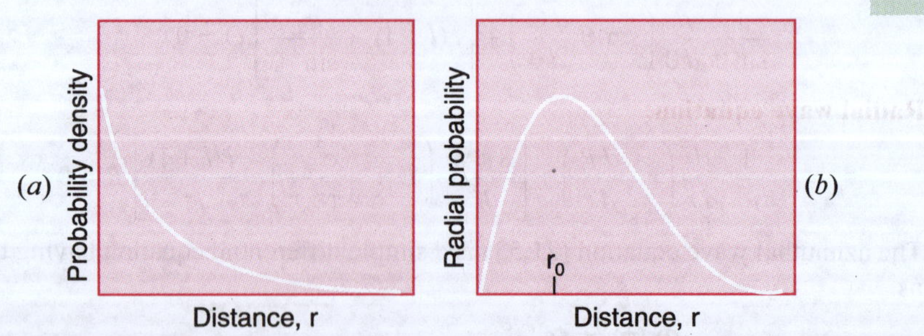

Probability curves for an electron in the $n = 1$ state of the hydrogen atom.

(*a*) Probability of finding the electron per unit volume versus distance form the nucleus.

(*b*) Probability of finding the electron at a given distance versus distance from the nucleus.

Fig. 21.17

Therefore, the wave function for the ground state of hydrogen is

$$\psi_1 = \left(\frac{1}{\sqrt{\pi}\, r_o^{3/2}}\right) e^{-r/r_o} \tag{21.61}$$

The wave function does not tell us where we actually would find the electron. The probability of finding an electron at any radial distance with in a spherically symmetric shell of thickness dr and volume $4\pi r^2\, dr$ is given by

$$\psi_1^* \psi_1\, 4\pi r^2\, dr = 4\pi r^2\, C_1^2 e^{-2r/r_o}\, dr \tag{21.62}$$

Curve 21.17 (*b*) is a radial probability curve. The total probability of finding the electron at a given distance from the nucleus is plotted against distance. The curve shows a maximum at a distance r_o. The total probability of finding the electron at all points of distance r from the nucleus is greatest when $r = r_o$. This value is the same as the value determined by the Bohr theory for the radius of the first orbit. In wave mechanics, r_o is the distance at which the electron is likely to be found most often. It is not possible to draw a shape that bounds a region in which the probability of finding the electron is 100%. The electron probability is distributed over the entire volume of the atom. However, a surface can be drawn that connects points of equal probability and that encloses a volume in which the probability of finding the electron is high. An electron in an atom may therefore be visualized as an electron cloud (see Fig. 21.17*c*). There is a region where the electron probability density falls steeply to a low value and it corresponds to the boundary of the atom. The concept of orbit is replaced in quantum mechanics with that of an orbital. An **orbital** is the region around the nucleus in which the probability of finding the electron is the highest. Depending upon the configuration and size of the boundary surface, the shape and size of the orbital is determined.

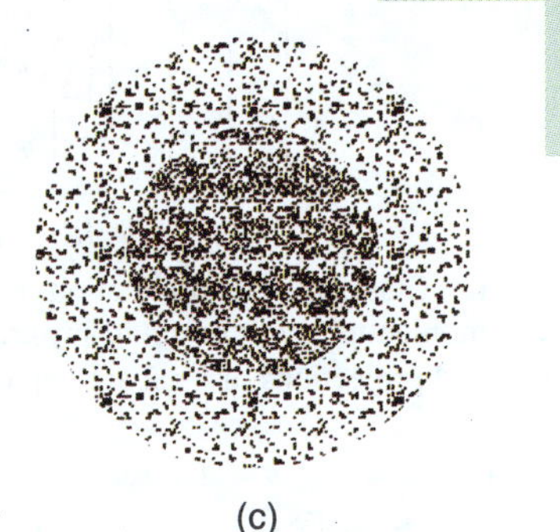

(c)

(*c*) Boundary surface representation of an electron in the $n = 1$ state of the hydrogen atom. (Volume encloses 90% of the electron density.)

Fig. 21.17

21.15.2. ORBITAL ANGULAR MOMENTUM

It is found that like total energy E, the angular momentum of electron is both conserved and quantized. Quantum theory shows that the magnitude L of the electron orbital angular momentum L is given by

$$L = \sqrt{l\,(l+1)}\,\hbar \tag{21.63}$$

The law of conservation angular momentum implies that an electron in a definite stationary state can have a definite angular momentum. However (= $m\upsilon r$) contains simultaneously both p (= $m\upsilon$) and r. The uncertainty principle forbids a particle to have a definite momentum p and a definite coordinate r at the same time. The quantum theory shows that the conserved angular momentum of the electron can be characterized by the magnitude L and one of its rectangular components, L_z. The component L_z is also quantized which implies the quantization of direction of L. Thus, space quantization emerges as a natural element of quantum mechanical solution of electron motion in the atom.

$$L_z = m_l \hbar \tag{21.64}$$

The choice of L_z among the three components L_x, L_y, L_z is simply a matter of convention where a magnetic field is assumed parallel to z-direction for the sake of reference. When L_z is

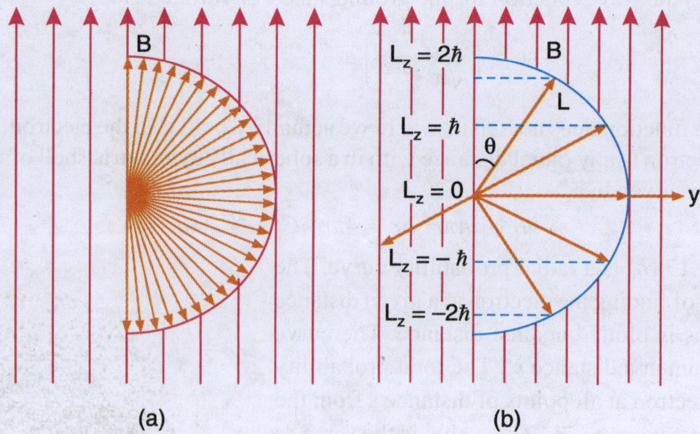

(*a*) Classically, all orientations of electron orbits in space are allowed. Correspondingly, the angular momentum *L* can orient in any arbitrary direction in space. (*b*) In quantum mechanics the orientation and projections of $l = 2$ are depicted in the diagram.

Fig. 21.18

chosen to have a definite value, then the other components L_x and L_y do not have well defined values; one can only determine the probabilities of specific values of L_x and L_y.

The simultaneous quantization of *L* and L_z implies that the vector *L* can never be fixed in space pointing any specific direction. In fact, the direction of *L* constantly changes and it precesses around the *z*-axis tracing out a cone in space. The tilt angle of *L* is determined by m_l.

$$\cos \theta = \frac{L_z}{L} = \frac{m_l}{\sqrt{l\,(l+1)}} \qquad (21.65)$$

21.15.3. INTRINSIC ANGULAR MOMENTUM

Electron is known to possess spin angular momentum. Schrodinger equation does not give any hint regarding electron spin. When relativity is included into the quantum mechanical treatment of electron motion in an atom, the intrinsic non-orbital angular momentum of the electron emerges in a natural way. The relativistic wave equation was first formulated in 1928 by P.A.M.Dirac (1902-1984). According to Dirac's theory, the only value that the spin quantum numbers can have is 1/2. The magnitude *S* of the intrinsic angular momentum of the electron is given by

$$S = \sqrt{s\,(s+1)}\,\hbar \qquad (21.66)$$

The direction of this intrinsic angular momentum is also quantized. The space quantization of intrinsic angular momentum is described by the spin magnetic quantum number m_s. The components S_z along a magnetic field in the *z*-direction is given by

$$S_z = m_s \hbar \qquad (21.67)$$

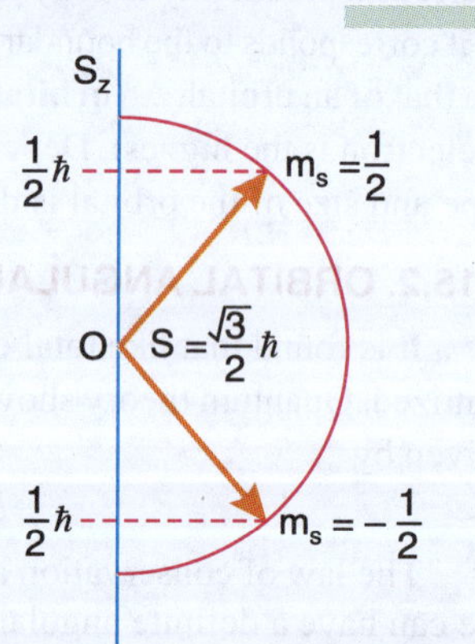

The spin angular momentum also exhibits space quantization. This figure shows the two allowed projections of the spin angular momentum vector for a single electron.

Fig. 21.19

21.15.4. SELECTION RULES

The fact that each level in a hydrogen-like atom is composed of several angular momentum states is important from the point of view of light emission. The electron transitions are restricted by **selection rules**. The selection rules for electric dipole transitions are

$$\Delta l = \pm 1 \text{ and } \Delta m_1 = 0, \pm 1$$

These selection rules are imposed by the law of conservation of angular momentum since the emitted or absorbed photon carries a spin angular momentum of 1. Therefore, the angular momentum of the atom must change by one unit to compensate for the angular momentum carried by the emitted or absorbed photon.

21.16. THE STRUCTURE OF THE ATOM

An atom of a certain kind cannot be said to have a definite size but from a practical point of view a definite size is attributed to it. Thus, hydrogen atom is supposed to have a diameter of about 1 Å.

The electrons in an atom occupy energy or orbitals characterized by the quantum numbers n, l and m_l. A set of these numbers defines the **state** of the electron. The states (or orbitals) are organized into subshells and shells. The possible orbitals corresponding to a particular value of n are said to constitute a **shell.** All the electrons in an atom which have the same principal quantum number n belong to the electron shell. The electrons having $n = 1$ are said to form the K shell of the atom; electrons in the $n = 2$ state form the L shell and so on. The designation of the electron shells is as follows,

Principal quantum number, n	1	2	3	4	5	6
Designation of the electron shell	K	L	M	N	O	P

A shell characterized by the principal quantum number n accommodates a maximum of $2n^2$ electrons. The value of n increases as the distance of a shell from the nucleus increases. The number of electrons that a shell can accommodate increases in accordance with the value of n. Thus, a K shell with $n = 1$ accommodates 2 electrons, an L shell with $n = 2$ accommodates 8 electrons and so on. The maximum capacities of the shells are given below.

Electron shell	K $n = 1$	L $n = 2$	M $n = 3$	N $n = 4$	O $n = 5$	P $n = 6$
Maximum capacity, $2n^2$	2	8	18	32	50	72

Shells are built from subshells which accommodate electrons of the same value of the orbital quantum number l. Thus, electrons that share a certain value of l in a shell said to occupy the same subshell. The number of subshells in a shell is equal to the value of n. Thus a K shell has only one subshell corresponding to $l = 0$, an L shell has two subshells and so on. In atomic physics, states with a particular value of l have a particular name. A state with $l = 0$ is called an s state; a state with $l = 1$ is called a p state and so on. The designations are listed in a table below.

Orbital quantum number, l	0	1	2	3	4	5
Designation of the state or subshell	s	p	d	f	g	h

For a given set of values of n and l, there are $(2l + 1)$ possible values of m_l. It means that for a given value of n and l, there are $(2l + 1)$ orbitals or electron **states.** Thus, if $l = 0$, there is only one electron state; if $l = 1$, three electronic states etc. The electronic states correspond to the different orientations of the orbitals.

An energy state described by the three quantum numbers n, l and m_l can be occupied by only two electrons which have opposite spin directions. This is known as **Pauli's exclusion principle.** This principle indicates why electrons do not crowd into the lowest energy state. The electron spin is characterized by quantum number m_s. The state of the electron described by the four quantum numbers n, l, m_l and m_s is called the **quantum state.** Thus each energy state consists of two quantum states.

The terms energy level and energy state are used often in the same meaning. It may be noted that an energy level is not equivalent with an quantum state. An energy level is determined by the value of the principal quantum number n, and such a level corresponds to n^2 quantum states. Thus, for $n = 2$, there are $2^2 = 4$ different quantum states. The energy of an electron is mainly determined by the value of n, but to some extent by the quantum numbers l and m_l also.

The occupancy of various subshells and shells in a complex atom is governed by three basic rules;

1. **Minimum energy condition.** Electrons tend to occupy the lowest available energy state such that their total energy is a minimum. They go to higher energy states only when the lower energy states are not vacant.
2. **Pauli's exclusion principle.** An orbital in an atom described by the quantum numbers n, l and m_l can be occupied by only two electrons having opposite spin directions.
3. **Hund's rule.** The order of filling of the orbitals of subshells obeys Hund's rule. According to this rule the *total spin number of the electrons of a shell must be maximum*. It means that the orbitals of a subshell are filled first with one electron each and then with the second electron respectively.

Wolfgang Pauli (1900–1958)

For instance in nitrogen atom, there are three electrons in $2p$ subshell. Three electrons occupy the three orbitals p_x, p_y and p_z, instead of occupying only p_x and p_y orbitals.

The sequence of the energy states in order of increasing energy of the orbitals of a multi electron atom is in the following order.

$$1s < 2s < 2p < 3s < 3p < 4s < 3d < 4p < 5s < 4d < 5p < 6s < 4f < 5d < 6p.$$

As each orbital characterized by the set of quantum numbers n, l, m_l can accommodate two electrons having m_s values $+\frac{1}{2}$ and $-\frac{1}{2}$, the electron capacity of a subshell is a maximum of $2(2l + 1)$ electrons.

Therefore, the maximum capacity N of a shell is given by,

$$\begin{aligned} N &= \sum_{l=0}^{n-1}(2(2l+1) = 2\,[1 + 3 + 5 + ... + 2(n-1) + 1] \\ &= 2\,[1 + 3 + 5 + ... + (2n-1)] \\ &= 2 \times \frac{n}{2}\,[1 + (2n-1)] \\ &= 2n^2 \end{aligned} \tag{21.68}$$

The following table shows at a glance the capacities of individual orbitals and the shell in total.

TABLE-1

n	shell	$l = 0$	Electron capacity	$l = 1$	Electron capacity	$l = 2$	Electron capacity	$l = 3$	Electron capacity	$l = 4$	Electron capacity	$l = 5$	Electron capacity	Max. capacity of shell, $2n^2$.
1	*K*	1*s*	2	-	-	-	-	-	-	-	-	-	-	2
2	*L*	2*s*	2	2*p*	6	-	-	-	-	-	-	-	-	8
3	*M*	3*s*	2	3*p*	6	3*d*	10	-	-	-	-	-	-	18
4	*N*	4*s*	2	4*p*	6	4*d*	10	4*f*	14	-	-	-	-	32
5	*O*	5*s*	2	5*p*	6	5*d*	10	5*f*	14	5*g*	18	-	-	50
6	*P*	6*s*	2	6*p*	6	6*d*	10	6*f*	14	6*g*	18	6*h*	22	72

The electron configuration of an atom is described using nl^x notation, where x denotes the number of electrons occupying the subshell. For example, we write the electron configuration of the sodium atom as $1s^2\,2s^2\,2p^6\,3s^1$.

A shell or subshell that contains its full quota of electrons is said to be **closed** or **completely filled.** Thus, the *K* shell and *L* shell of sodium atom are closed while the *M* shell is partially filled. The 3*s* subshell of *M* shell contains only one electron and is half filled.

The completely filled shells and subshells of atoms are stable and are not readily disturbed. The electrons in the outer shell will be in a position to interact with similar electrons in adjacent atoms. The number of such electrons determines the **valency** of the atom and the chemical behavior of the element.

21.17. WAVE MECHANICAL EXPLANATION OF PHOTON EMISSION

A stationary state of electron is denoted by an eigen function ψ. The time dependence of the wave function is given by the exponential factor $e^{-iEt/\hbar} = e^{-i\omega t}$. Then

$$\Psi(x, y, z, t) = \psi(x, y, z)\, e^{-i\omega t} \tag{21.69}$$

The probability of locating electron at a point x, y, z can now be found as

$$\Psi^*\Psi\, dx\, dy\, dz = \psi^*\, e^{i\omega t}\, \psi\, e^{-i\omega t} dx\, dy\, dz = \psi^*\, \psi\, dx\, dy\, dz \tag{21.70}$$

The probability is not dependent on time and hence it is justified to refer the state as a stationary state.

If an external influence causes the atomic electron to jump from state n to m, the probability that the electron will be in an element of ΔV in the atom is no longer given by $|\psi_n|^2 dV$. The electron will reside part of the time in state n and part of the time in state m. It implies that the charge density at any point is no longer constant but varies with time. Thus, the charge oscillates and the electron is said to acquire electric dipole moment which varies periodically with time. When an electron is in transit moving from one stationary state to another stationary state, the wave function must be a linear combination of both the states. Thus,

$$\Psi = C_1\, \psi_m e^{-iE_1 t/\hbar} + C_2 \psi_n\, e^{-iE_2 t/\hbar} \tag{21.71}$$

This state is called a **coherent state**. The energy is not well defined in a coherent state. The probability distribution function of this state may be represented by

$$\Psi^*\Psi = C_1^* C_1 \psi_m^* \psi_m + C_2^* C_2\, \psi_n^* \psi_n + C_1^* C_2\, \psi_m^* \psi_n e^{-i\omega_{mn} t} + C_2^* C_1\, \psi_n^* \psi_m e^{i\omega_{nm} t} \tag{21.72}$$

Equ. (21.72) shows that the superposition of two stationary states is not a stationary state, as it contains oscillating terms. The oscillating terms represent an oscillating charge distribution with an oscillating electric dipole moment. Such an oscillating charge distribution will radiate electromagnetic waves with a frequency

$$\omega_{nm} = \frac{E_n - E_m}{\hbar} \tag{21.73}$$

Thus, the quantum mechanical description of an atom undergoing transition consists of an electron oscillating between the initial and final states at a frequency equal to

$$\nu_{nm} = \frac{E_n - E_m}{h} \tag{21.74}$$

This corresponds to the transition frequency postulated by Bohr.

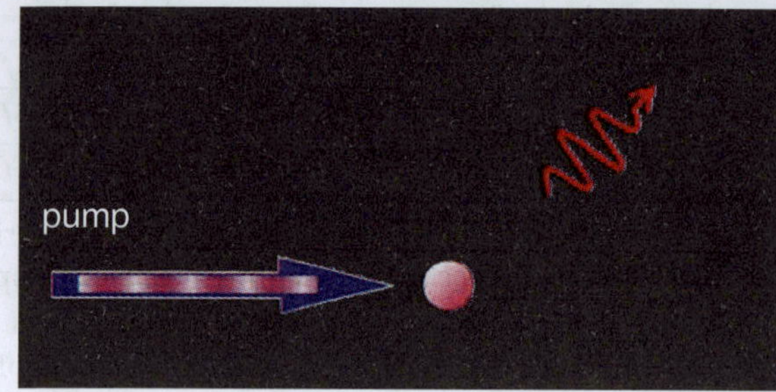

Spontaneous emission.

Therefore, the oscillating electron is equivalent to an **electric dipole** which emits light radiation. This process of light emission is known as **spontaneous emission**.

The power radiated by an atom is

$$P_r = \frac{dE}{dt} = A_{nm}\, h\nu \tag{21.75}$$

where A_{nm} is the *transition probability per second* and $h\nu$ is the energy of photon emitted. By equating the above expression to the classical expression (21.4), we can find the transition probability. Thus,

$$A_{nm}\, h\nu = \frac{16\pi^4 \nu^4 (e x_o)^2}{3\varepsilon_o c^3}$$ where $(e x_o)$ is the classical dipole moment

or $$A_{nm} = \frac{16\pi^4 \nu^3 M_{nm}{}^2}{3h\varepsilon_o c^3} \tag{21.76}$$

M_{nm} is the quantum mechanical dipole matrix element. If M_{nm} is non-vanishing, the transition is allowed. If $M_{nm} = 0$, the transition is forbidden. The units of equation are reciprocal seconds, that is transitions /sec. The reciprocal of A_{nm} gives the **lifetime** of state *n* for transition to state *m*. For any quantum state it is possible to derive a set of "transition rules" or "selection rules" that indicate in what way the quantum numbers of the system must change during a transition. A_{nm} is known as the **Einstein's coefficient of spontaneous emission**.

The transition frequencies computed correspond to what are known as **electric-dipole transitions.** Equ.(21.76) helps us to predict the relative intensities of spectral lines also.

Einstein showed that light emission can take place through another process which is known as **stimulated emission**. Stimulated emission is a *resonance phenomenon* – an atom at the excited state drops to the ground state only when a photon of the same frequency interacts with the system. Stimulated emission will be discussed in Chapter 22. For high frequency transitions (*IR*, *UV*, visible upwards) spontaneous emission is by far the most likely. Conversely, for low-frequency changes

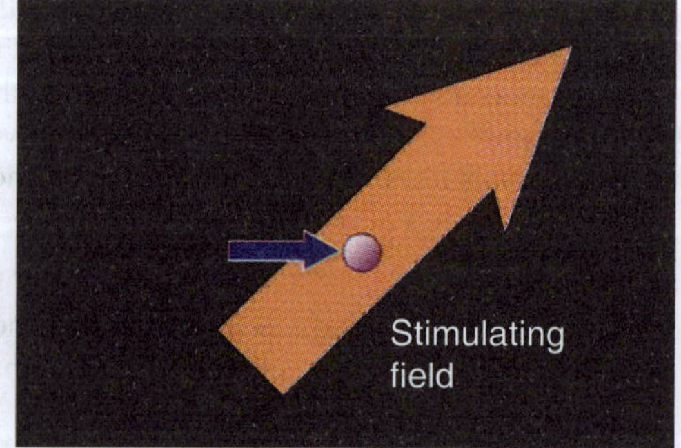

Stimulated emission.

(microwave and magnetic resonance) spontaneous emission is unlikely and if the right conditions are obtained, stimulated emission will occur.

21.18. PROPERTIES OF SPECTRAL LINES

We record the spectral lines with the help of spectrometers, which have narrow slits. The spectral lines are therefore linear and possess three important properties; namely **position** measured in terms of frequency, finite **width,** which is a *range* of frequencies and **intensity**, the brightness of the line.

Spectral lines.

21.18.1. WIDTH OF SPECTRAL LINES

Real atoms or molecules do not emit radiation at precise frequencies; each emission is more or less broadened by various processes, and so each line is really a small package of slightly different frequencies. Therefore, even if we make the slit of the spectrometer infinitely narrow, there is nonetheless a minimum line width. The natural shape of an emission line appears under ideal conditions when the emitting atom is at rest and is not subjected to the external forces during the process of emission. In practice, several factors contribute to the line broadening. They are broadly classified into two categories: (*i*) homogeneous and (*ii*) inhomogenenous broadening. If the broadening mechanism affects each individual atom in the sample to the same extent, then the broadening is said to be *homogeneous*. In such a case all of the atoms in the sample will have the centre frequency ν_o and the same line shape. Natural broadening and collision broadening belong to this category. On the other hand, if different atoms in the sample have slightly different frequencies for the same transition, the overall response of the sample broadens out and the broadening is said to be *inhomogeneous*. Doppler broadening and broadening due to crystal defects belong to this category.

1. Collision broadening: Atoms or molecules in liquid and gaseous phases are in continual motion and collide frequently with each other. When a radiation-emitting atom undergoes a collision, the emission conditions change and it is equivalent to an interruption. In the process, the phase of the emitted wave suffers random variations. Thus, each collision act leads to a random phase change, which means shortening of wave trains, as shown in Fig. 21.20. It results in the broadening of the spectral line. This broadening is called *collision broadening* since it is a consequence of the collisions of atoms.

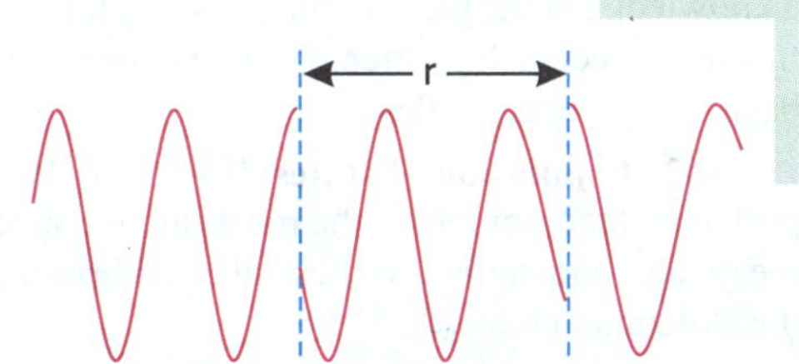

Random collision with other atoms cause abrupt changes in phase of the wave emitted by an atom.

Fig. 21.20

2. Doppler broadening: Doppler effect occurs when a source and an observer are in relative motion. The frequency as measured by the observer increases if the source and observer approach each other and decreases when they recede. In a gas due to the chaotic nature of thermal motion, all the directions of the molecules' velocities relative to a spectrograph are equally probable. Therefore, the radiation received by the spectrograph contains all the frequencies in the interval from $\nu_o(1-\upsilon/c)$ to $\nu_0(1+\upsilon/c)$, where ν_o is the frequency emitted by the molecules and υ is the velocity of thermal motion. Hence the spectral line is broadened. The broadening of the line caused by the Doppler effect is called *Doppler broadening*. In general, for liquids collision broadening is the most important factor, whereas for gases where collision broadening is less pronounced, the Doppler effect often determines the natural line width.

3. Natural broadening: Even in an isolated, stationary atom the energy levels are not indefinitely sharp. According to Heisenberg uncertainty principle, if a system exists in an energy state for a limited time Δt seconds, then the energy of that state will be uncertain to an extent ΔE where

$$\Delta E \times \Delta t \approx h/2\pi$$

Thus, we see that the ground state of a system is sharply defined since, left to itself, the system will remain in that state for an infinite time. Thus $\Delta t = \infty$ and $\Delta E = 0$. In contrast, the lifetime of an excited state is about 10^{-8}s, which gives a value for ΔE of about 10^{-34}J.s $\div$ 10^{-8} s $= 10^{-26}$ J. A transition between an excited state and the ground state will thus have an energy uncertainty of ΔE, and a corresponding uncertainty in the associated radiation frequency of $\Delta E/h$, which we can write as

$$\Delta\nu = \frac{\Delta E}{h} \approx \frac{1}{2\pi\Delta t} \qquad (21.77)$$

Natural width of energy levels.

Fig. 21.21

If an excited electronic state lifetime is 10^{-8}s, then $\Delta\nu = 10^8$ Hz. Thus, the *natural broadening* is relatively small in magnitude and it is often masked by other mechanisms.

21.18.2. THE INTENSITY OF SPECTRAL LINES

The spectral line intensities are dependent on two factors: the *transition probability*, the likelihood of a system in one state changing to another state; and the *population of state*, the number of atoms or molecules initially in the state from which the transition occurs.

1. Transition probability: The detailed calculation of absolute transition probabilities involves a knowledge of the precise quantum mechanical wave functions of the two states between which the transition occurs. It is often possible to decide whether a particular transition is allowed or forbidden basing on selection rules.

2. Population of states: If we have two levels from which transitions to a third are equally probable, then obviously the most intense spectral line will arise from the level which initially has the greater population. At thermal equilibrium the population of a set of energy levels is governed by the Boltzmann law

$$\frac{N_2}{N_1} = e^{-\Delta E/kT} \qquad (21.78)$$

where N_1 is the population in the lower energy state E_1 and N_2 in the higher energy state E_2, $\Delta E = E_2 - E_1$, T is the temperature in K, and k is Boltzmann's constant.

21.19. LUMINESCENCE

Luminescence.

A body that emits light on account of high temperature is called *incandescent*. There are bodies that emit light due to causes other than high temperature and they are said to be **luminescent**. The emission of light from a substance, when it is stimulated by the input of energy from a source of suitable radiation, is known as **luminescence**. Materials exhibiting luminescence are broadly called as *phosphors*. There are various ways of exciting substances and so there are the following kinds of luminescence.

(*i*) **Photoluminescence** is the glow of bodies as a result of their irradiation by visible or *uv* light, *x*-rays or γ-rays. Tube lights use this phenomenon to produce visible light. The internal surface of the tube is coated with phosphors and these substances emit visible light of lower frequency under the action of *uv* light of higher frequency.

(*ii*) **Electroluminescence** is caused due to the passage of an electric current through a substance or the action of an electric field. The glow of a gas discharge in the tubes of advertisement is an example of this kind of luminescence.

(*iii*) **Cathodoluminescence** is the glow of materials due to the bombardment of the material by electrons or other charged particles.

(*iv*) **Chemiluminescence** is caused at the expense of energy produced in a chemical reaction. The glow of many living organisms such as bacteria and insects is due to chemiluminescene.

Many substances emit light when a beam of light is made to fall on them. When the light is emitted only as long as the exciting radiation is maintained, it is called **fluorescence.** For fluorescence, the time interval between the acts of excitation and emission is 1 to 10 ns. For example, kerosene emits a faint bluish fluorescence when illuminated by daylight.

Substances, which continue to emit light for some time after the exciting radiation is removed, are called **phosphorescent** substances. Phosphorescence is displayed for example by zinc sulphide. A coat of zinc sulphide applied to a sheet of glass makes a fluorescent screen, which will emit a faint glow for minutes after the exciting radiation is switched off.

Fluorescence occurs when the atoms of the fluorescent material absorb a portion of the incident radiation of shorter wavelength and re-emit it in the form of radiation of longer wavelength.

Luminescence in solids is closely related to impurities and lattice defects. Crystal luminescence is more important and practically useful. Whatever may the form of energy given to the luminescent material, the final stage in the process is an electronic transition between two energy levels, E_1 and E_2 with the emission of light of frequency ν, where

$$\nu = \frac{E_2 - E_1}{h}$$

Invariably E_1 and E_2 are part of two groups of energy levels, so that, instead of a single emission frequency, a band of frequencies is usually observed.

Fig. 21.22 illustrates some processes that take place in luminescent solids. When an electron is excited from the valence band into the conduction band, a hole is left in the valence band. In a perfectly regular lattice, the electron usually returns to the valence band, although it may take some time to do so. However, if the lattice has some impurity that introduces energy levels in the forbidden region, an electron in a low lying impurity level may fill the hole in the valence band while the electron in the conduction band may fall into one of the vacant high impurity energy levels (See Fig. 21.22 *b*). These transitions generally involve photons of small energy that do not fall into the visible region. Finally, the electron may fall from higher energy impurity level to the lower energy impurity level, emitting radiation of lower energy or longer wavelength than the incident radiation; this constitutes the luminescence (see Fig. 21.22 *c*).

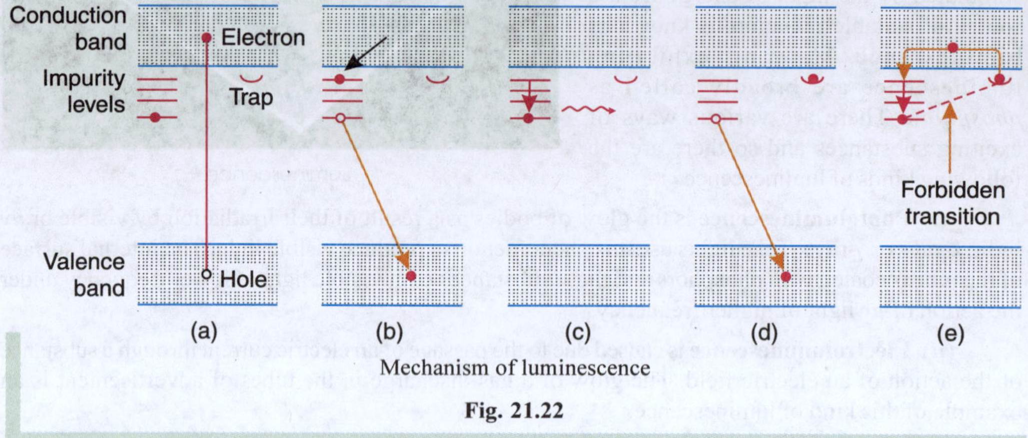

Mechanism of luminescence

Fig. 21.22

In some instances, instead of going through the process depicted in Fig. 21.22 (*b*), the electron in the conduction band may fall into an energy level called a *trap*. From there electron cannot jump to the ground state impurity energy level. The transition is forbidden. In such a case the electron is in a state similar to an atom in a metastable state. The trapped electron waits till it is returned to the conduction band through some other mechanism. Then it follows the steps (*b*) and (*c*). Due to the time delay involved, the process is *phosphorescence*. These substances are called *phosphors*. One such substance is zinc sulphide. Phosphors are widely used in coating the screen of CRTs used in CROs, TVs and video terminals of computers.

21.20. SCATTERING

When the electrical polarizability of molecule changes during its motion, light scattering takes place. This phenomenon also gives rise to light emission but does not involve electron transitions from a higher energy level to a lower energy level. Rayleigh scattering is discussed in Chapter 13. We now describe the Raman scattering.

21.20.1. RAMAN EFFECT

C.V. Raman (1888-1970)

In 1928, Sir C.V. Raman discovered that when a beam of monochromatic light was passed through an organic liquid such as benzene, the scattered light was found to contain a strong line of frequency ν_o equal to the frequency of the incident light and a few weak lines on either side of the incident line $\nu' = \nu_o \pm \nu_M$. This is referred to as **Raman Scattering**. The scattering of light with change of frequency is known as **Raman effect**. The spectrum formed due to Raman effect is called **Raman spectrum** and the spectral lines obtained are called **Raman lines.** In order to observe Raman effect, the incident light should be monochromatic and very intense. Raman scattering is always accompanied by Rayleigh scattering. Raman lines at frequencies less than that of the incident frequency ($\nu_o - \nu_M$) are known as **Stokes lines** and those with frequencies greater than that of the incident frequency ($\nu_o + \nu_M$) are known as **anti-Stokes lines.** Experiments show that Stokes lines are far more intense than anti-Stokes lines. Overall, however, the total radiation scattered at any but the incident frequency is extremely small, and sensitive apparatus is needed for its detection and study.

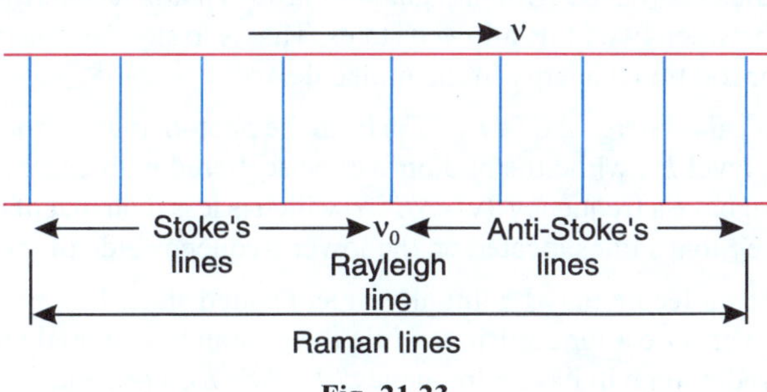

Fig. 21.23

If the frequency of the incident light is varied, the weak lines are once again observed on either side of the Rayleigh line with the same difference in frequency. It is evident that the frequency difference ν' between the incident and the scattered light in Raman effect is determined by the nature of the scattering molecules and is independent of the frequency of incident light.

21.20.2. CLASSICAL EXPLANATION

When a sample of molecules is subjected to a beam of light of frequency ν, the induced dipole undergoes oscillations of frequency ν;

$$\mu = \alpha E = \alpha E_o \sin 2\pi\nu t \qquad (21.79)$$

If in addition, the molecule undergoes some internal motion, such as vibration or rotation, which changes the *polarizability* periodically, then the oscillating dipole will have superimposed upon it the vibrational or rotational oscillation. As an example, let us assume a vibration of frequency ν_{vib}, which changes the polarizability. Then

$$\alpha = \alpha_o + \beta \sin 2\pi\nu_{vib} t \qquad (21.80)$$

where α_o is the equilibrium polarizability and β represents the rate of change of polarizability with the vibration. Using (21.80) into (21.79), we get

$$\mu = \alpha E = [\alpha_o + \beta \sin 2\pi\nu_{vib} t]\, E_o \sin 2\pi\nu t \qquad (21.81)$$

Expanding the above equation and using the trigonometric relation, we get

$$\mu = \alpha_o E_o \sin 2\pi\nu t + \frac{1}{2}\beta E_o \left[\cos 2\pi(\nu - \nu_{vib})t - \cos 2\pi(\nu + \nu_{vib})t\right] \qquad (21.82)$$

It means that the oscillating dipole has frequency components $\nu \pm \nu_{vib}$, as well the exciting frequency ν. These additional frequencies are responsible for producing the Raman lines. Thus, for Raman scattering, *a molecular rotation or vibration must cause some change in a component of molecular polarizability.*

21.20.3. QUANTUM THEORY

The explanation of Raman scattering in terms of the quantum theory is very simple. When light is incident on a solid, liquid, or gas, the photons can be imagined to undergo collisions with molecules.

(*i*) If the collision is perfectly *elastic*, there will be no transfer of energy from the photon to the molecule or from the molecule to the photon. The photon is scattered without any change of energy. Therefore, the frequency of the scattered photon is the same as that of the incident photon. A detector placed to collect energy at right angles to an incident beam will thus receive photons of energy $h\nu$, that is the radiation of frequency ν. This explains the Rayleigh line in the Raman spectrum.

(*ii*) However, it may happen that energy is exchanged between photon and molecule during the collision; such collisions are called *inelastic collisions*. The molecule can gain or lose amounts of energy only with the accordance with the quantum laws. That is its energy changes must be the difference in energy between two of its allowed states. That is to say, ΔE must represent a change in the vibrational and /or rotational energy of the molecule.

If the molecule gains energy $\Delta E\ (= E_2 - E_1)$ from the photon, it goes from a lower energy level E_1 to a higher energy level E_2, while the photon will be scattered with energy $(h\nu - \Delta E)$. Then, the scattered radiation will have a frequency $(\nu - \Delta E / h)$ which is less than that of the incident radiation. The resulting lines are Stokes lines located on the lower frequency side of the Raman spectrum.

Conversely, the molecule may be initially in an excited state. It may lose energy ΔE to the photon and go to a lower state after collision, while the photon is scattered with energy $(h\nu + \Delta E)$. Hence, the scattered radiation will have a frequency $(\nu + \Delta E / h)$. The resulting lines are anti-Stokes lines, which are situated on the higher frequency side of the Raman spectrum.

At ordinary temperatures, there are more molecules in the lower energy state. Therefore, transitions are more likely from lower energy state to upper energy state. Hence, energy is absorbed by the molecules from the photons. The reverse process of energy being given to photons by molecules is less likely. Therefore, Stokes lines are more intense than the anti-Stokes lines. The intensity of anti-Stokes lines increases with temperature, as a rise in temperature increases the relative population of molecules in the higher energy state.

WORKED OUT EXAMPLES

Example 21.1: About 0.1% of electrical energy supplied to a laboratory mercury vapour lamp of 80 watt is converted into *UV* light of wavelength 2500 Å. Calculate the number of *UV* photons emitted per second by the lamp.

Solution:

$$\text{Number of } UV \text{ photons emitted per second} = \frac{\text{Energy converted into } UV \text{ light}}{\text{Energy carried by one } UV \text{ photon}}$$

$$= \frac{80\ J/s \times (0.1/100) \times (6.24 \times 10^{18}\ eV/J)}{12400\ eV/2500}$$

$$= \mathbf{10^{17}\ photons/s.}$$

Example 21.2: *X*-rays of 0.5 Å are scattered by free electrons in a block of carbon through 90°. Find the velocity of recoil electrons.

Solution:

$$\text{K.E. of the recoil electrons} = h\nu_i - h\nu_f = 12400 \left(\frac{1}{\lambda_i} - \frac{1}{\lambda_f}\right) eV\ (1.602 \times 10^{-19}\ J/eV)$$

$$\lambda_f = \lambda_i + \frac{h}{m_o c}(1 - \cos\theta) = (0.5 + 0.02426)\,\text{Å} = 0.5243\ \text{Å}$$

$$\therefore \quad K.E. = 12400 \left(\frac{1}{0.5} - \frac{1}{0.5243}\right)(1.602 \times 10^{-19})\,J$$

$$= \mathbf{1.84 \times 10^{-16} J}$$

$$\therefore \quad \upsilon = \left[\frac{2(K.E.)}{m}\right]^{1/2} = \left[\frac{2(1.84 \times 10^{-16} J)}{9.11 \times 10^{-31} Kg.}\right]^{1/2} = \left\{4.04 \times 10^{14} \frac{kg.m^2/s^2}{kg}\right\}^{1/2}$$

$\therefore \quad \mathbf{\upsilon = 2 \times 10^7}$ **m/s.**

Example 21.3 : Calculate de Broglie wavelength of an electron moving with velocity 10^7 m/s.

Solution:

$$\lambda = \frac{h}{m\upsilon} = \frac{6.626 \times 10^{-34} J.s}{9.11 \times 10^{-31} kg \times 10^7 m/s} = \frac{6.626 \times 10^{-34}}{9.11 \times 10^{-24}} \frac{kg.m^2/s^2.(s^2)}{kg.m} = 7.28 \times 10^{-11} m$$

$\therefore \mathbf{\lambda = 0.72}$ **Å**

Example 21.4: Compute the minimum uncertainty in the location of a mass of 2.0 gm moving with a speed of 1.5 m/s and the minimum uncertainty in the location of an electron moving with a speed of 0.5×10^8 m/s. Given that the uncertainty in the momentum p for both is $\Delta p = 10^{-3}p$.

Solution: $p = m\upsilon = 2 \times 10^{-3}$ kg $\times$ 1.5 m/s $= 3 \times 10^{-3}$ kg m/s.

$$\therefore \quad \Delta p = 10^{-3}\ p = 3 \times 10^{-6} \text{ kg m/s}$$

$$\therefore \quad \Delta x = \frac{h}{2\pi.\Delta p} = \frac{6.63 \times 10^{-34} Js}{2\pi \times 3 \times 10^{-6} \text{kg.m/s}} = \mathbf{3.5 \times 10^{-19}} \text{ Å}$$

For an electron, $p = m\upsilon = (9.11 \times 10^{-31}$ kg) $(0.5 \times 10^8$ m/s) $= 4.55 \times 10^{-25}$ kg. m/s

$$\therefore \quad \Delta p = 10^{-3}\ p = 4.55 \times 10^{-26} \text{ kg.m/s}$$

$$\therefore \quad \Delta x = \frac{h}{2\pi.\Delta p} = \frac{6.63 \times 10^{-34} Js}{2\pi \times 4.55 \times 10^{-26} \text{ kg.m/s}} = \mathbf{23} \text{ Å}$$

QUESTIONS

1. State the basic postulates of Planck's quantum theory. How did Einstein modify this theory in the light of experimental results?
2. What is photoelectric effect? How does Einstein's equation explain the features of photoelectric effect?
3. What is meant by work function of a material?
4. What is Compton effect? Show that Compton shift is independent of wavelength of incident x-rays.
5. Explain de Broglie's concept of matter waves.
6. Using the concept of matter waves, obtain the Bohr's condition for quantization of angular momentum.
7. State uncertainty principle. Write its mathematical form for the following pairs of variables:
   (*i*) Position and momentum
   (*ii*) Energy and time
   (*iii*) Angular position and angular momentum.
8. What is Heisenberg Uncertainty principle? Explain how it is the out come of the wave description of a particle.
9. Write down Schrodinger's time dependent and time independent wave equations for matter waves. Explain, why:

(i) The wave function ψ must be single valued and continuous function of position.
(ii) The integral of $|\psi|^2$ over all space must equal unity.

10. What is the physical significance of wave function 'ψ'?
11. Explain the physical significance of quantum numbers n, l and m_l.
12. Discuss the space quantization of angular momentum.
13. Explain the term probability density.
14. What is luminescence? Explain the difference between fluorescence and phosphorescence.
15. What is Raman effect? What are Stokes and anti-Stokes lines in Raman spectrum?
16. How is Raman effect explained on the basis of quantum theory?
17. Why are the anti-Stokes lines fainter than Stokes lines?
18. Explain the mechanisms responsible for broadening the resonant response of a transition. **(Punjab, 2005)**
19. What is the cause of Doppler broadening ? **(Punjab, 2005)**

PROBLEMS FOR PRACTICE

1. The threshold frequency for photoelectric emission in copper is 1.1×10^{15} Hz. Find the maximum energy in eV when light of frequency 1.2×10^{15} Hz is incident on the copper surface. **[Ans: 0.41 eV]**
2. Find the energy of neutron having de Broglie wavelength of 10^{-14} m. Given the rest mass of neutron is 1.6×10^{-27} kg. **[Ans: 8.5 MeV]**
3. Calculate de Broglie wavelength of an electron whose kinetic energy is 500 eV. **[Ans: 0.55 Å]**
4. A particle of mass 10^{-6} gm has a speed of 1 m/s. If the speed is uncertain by 0.01% what is the minimum uncertainty in the position of the particle? **[Ans: 1.45×10^{-3} m]**
5. The first member of Balmer series of hydrogen has a wavelength of 6536 Å. Calculate the wavelength of its second member. **[Ans: 4861 Å]**

22

CHAPTER

Lasers

22.1. INTRODUCTION

Laser is one of the outstanding inventions of the 20th century. Laser is a **photonic device**, which is actually responsible for the resurgence of interest in optical technology and for the birth of a new field, namely *photoncis*. The word 'LASER' is the acronym for **L**ight **A**mplification through **S**timulated **E**mission of **R**adiation. However, laser is *not* a simple amplifier of light but is actually a generator of light. In fact, the device should have been called a LOSER signifying **L**ight **O**scillation through **S**timulated **E**mission of **R**adiation; this name was avoided because of its bad connotation and the name laser has been preferred. Though it is a source of light, laser differs vastly from the traditional light sources. It is not used for illumination purposes as we use other light sources. Laser is more akin to radio and microwave transmitters and produces a highly directional coherent monochromatic light beam. Einstein gave the theoretical basis for the development of laser in 1916, when he predicted the possibility of stimulated emission. In 1954, C.H.Townes and his co-workers put Einstein's prediction for practical realization. They developed a microwave amplifier based on stimulated emission of radiation. It was called a **maser**. Shortly thereafter, in 1958, A.Schawlow and C.H.Townes extended the principle of masers to light and T.H.Maiman built the first laser device in 1960. In 1961, A.Javan and associates developed the first gas laser, the helium-neon laser. Laser is a high technology device and is the most sought after tool in a wide variety of fields such as metalworking, entertainment, communications, surgery, and ophthalmology and weapon guidance in wars.

At a Glance

22.2 ATTENUATION OF LIGHT IN AN OPTICAL MEDIUM

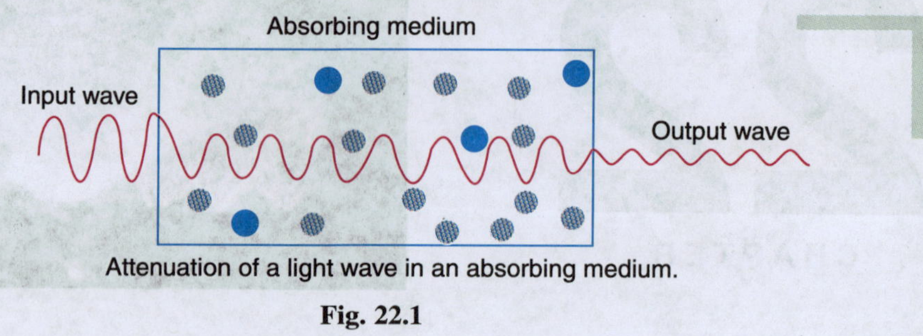

Attenuation of a light wave in an absorbing medium.

Fig. 22.1

When light travels through a medium, a gradual reduction in its intensity occurs mainly because of the processes of absorption and scattering of light in the medium. (*i*) Light *absorption* occurs because part of the incident light is transformed into the energy of motion of the atoms in the medium; and (*ii*) light is scattered when it encounters obstacles of sizes smaller than a wavelength. The reduction in intensity with distance in a medium is called **attenuation of light**. The following relation governs the attenuation of light in a transparent medium.

$$I = I_o e^{-\alpha x} \quad (22.1)$$

where x is the distance in the medium, I_o is the value of intensity at $x = 0$, and α is the *coefficient of attenuation* or *absorption coefficient* of the material at frequency ν.

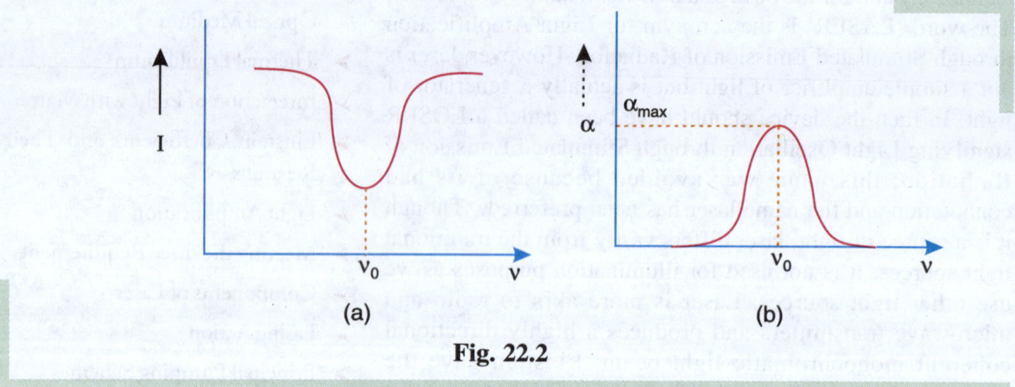

Fig. 22.2

Fig.22.2(a) shows the variation of intensity I with frequency ν at a fixed distance x in the medium. Fig.22.2 (b) shows the variation of the attenuation coefficient α with frequency ν. In general α is a positive quantity and we say that the material has a *positive coefficient of absorption*.

22.3 THERMAL EQUILIBRIUM

A material medium is composed of *identical atoms,* which are characterized by a specific system of energy levels. These energy levels are common to all atoms in the medium. We can therefore say that a certain number of atoms occupy a certain energy level. The number of atoms per unit volume that occupy a given energy level is called the **population** of that energy level. We make a simple assumption here that a particular medium has atoms, which are characterized by only two energy levels (see Fig.22.3). Let them be E_1 the *ground level* and E_2 the *excited level*. Atoms are distributed differently in these two energy levels. Let the populations at the levels E_1 and E_2 be N_1 and N_2 respectively.

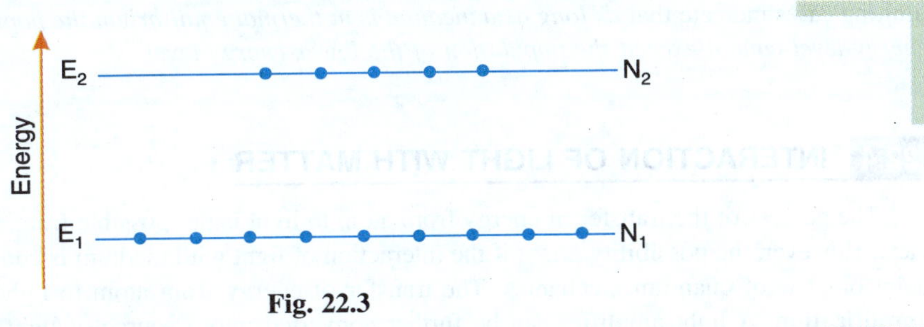

Fig. 22.3

At thermal equilibrium, the population at the energy levels can be found with the help of Boltzmann law.

$$N_1 = e^{-E_1/kT} \tag{22.2}$$

$$N_2 = e^{-E_2/kT} \tag{22.3}$$

The relative population N_2 / N_1 is given by

$$\frac{N_2}{N_1} = e^{-(E_2 - E_1)/kT} \tag{22.4}$$

Equ.(22.4) shows that the relative population N_2 / N_1 is dependent on two factors: (i) the temperature T and (ii) the energy difference $(E_2 - E_1)$. We conclude that the fraction of excited atoms would be large if the temperature is very high or if the energy levels are close.

Effect of temperature

Let us now make an estimate of the population in the energy level N_2. Let us take the example of hydrogen gas and assume it to be a monatomic gas. In case of hydrogen atoms,

$$E_1 = -13.6\,\text{eV} \text{ and } E_2 = 3.39\,\text{eV}.$$

$$\therefore \quad (E_2 - E_1) = 10.21\,\text{eV}.$$

At room temperature, T = 300 K, kT = 0.025eV and $\frac{N_2}{N_1} = e^{-\frac{10.21\,\text{eV}}{0.025\,\text{eV}}} = e^{-408.4} \approx 0$ or $N_2 \approx 0$.

It means that at room temperature all atoms are in the ground state. Now, suppose the temperature is raised say to 6000K,

$$\frac{N_2}{N_1} = e^{-\frac{10.21\,\text{eV}}{0.516\,\text{eV}}} = e^{-19.79} \cong \frac{4}{10^{10}}$$

It means that roughly 4 atoms out of 10^{10} atoms are excited to higher energy level at 6000K. This is a very small number. We thus find that even at very high temperatures the fraction of excited atoms is very small. The number of atoms in the lower energy level is *always larger* than that in the higher energy level. Under normal conditions higher the energy of the energy level, lesser is its population. Hence $N_1 > N_2$. Such distribution of atoms among the energy levels is termed **normal distribution** (also see Fig. 22.10a).

Now, if we assume the limiting case of $T \rightarrow \infty$, then $\frac{N_2}{N_1} = e^0 = 1$. Therefore, $N_2 = N_1$. On the other hand, if the energy difference $(E_2 - E_1) \to 0$, then also $\frac{N_2}{N_1} = e^0 = 1$. Again, $N_2 = N_1$. Both these limiting cases indicate that *as long as a medium is in thermal equilibrium the population of higher energy level cannot exceed the population of the lower energy level.*

22.4 INTERACTION OF LIGHT WITH MATTER

The process of the transfer of energy from atom to light is not possible from classical point of view. However, the possibility arises if the interaction of light with medium is considered from the point of view of quantum mechanics. The transfer of energy from atom to light results in **light amplification**. A light amplifier can be further converted into a source of light having superior characteristics compared to traditional light sources. A **laser is a monochromatic coherent light source** that depends on quantum processes for its operation. It is therefore, necessary to appreciate the quantum processes involved in the development of a laser.

The radiation incident on a material is viewed as a stream of photons, where each photon carries an energy $E = h\nu$. We assume that the two energy levels of the atoms in the material have an energy difference $(E_2 - E_1) = h\nu$. When photons travel through the medium, three different processes are likely to occur. They are *absorption, spontaneous emission and stimulated emission*. We study these in detail.

22.4.1 ABSORPTION

Suppose an atom is in the lower energy level E_1. If a photon of energy $h\nu = (E_2 - E_1)$ is incident on the atom, it imparts its energy to the atom and disappears. Then we say that the atom absorbed an incident photon. As a result of absorption of adequate energy, the atom jumps to the excited state E_2 (Fig.22.4). The transition is called an **absorption transition**. It is also referred to as **induced absorption**. We may express the process as

$$A + h\nu = A^*$$

where A is an atom in the lower state and A^* is an excited atom.

In each absorption transition event, an atom in the medium is excited and one photon is subtracted from the incident light beam, which results in attenuation of light in the medium.

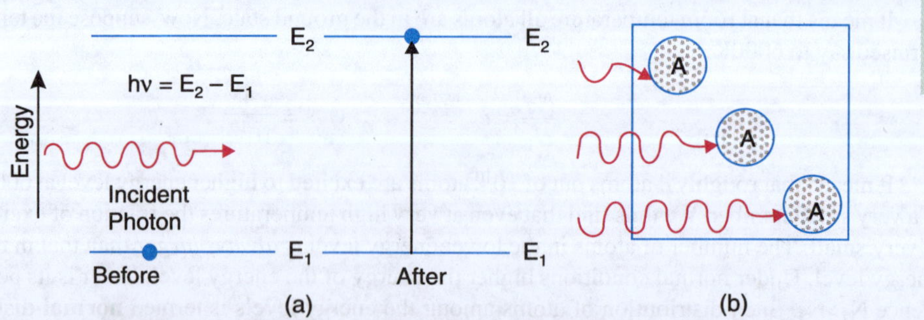

Abosorption process (*a*) Induced absorption (*b*) Material absorbs photons.

Fig.22.4

The probability that an absorption transition occurs is proportional to the photon density ρ(ν).

$$P_{12} \propto \rho(\nu)$$

or
$$P_{12} = B_{12}\rho(\nu) \tag{22.5}$$

where B_{12} is the constant of proportionality. B_{12} is known as the *Einstein coefficient for induced absorption.* It indicates the probability of occurrence of an induced transition from level 1→2. B_{12} is a constant characteristic of the atom and represents the properties of the energy states E_1 and E_2.

The number of atoms per unit volume that undergo absorption transitions per second is called the *rate of absorption transition.* It is denoted by

$$R_{abs} = -\frac{dN_1}{dt} \tag{22.6}$$

where N_1 is the population of atoms at E_1, and $(-dN_1/dt)$ represents the rate of decrease of population at the lower level E_1. The rate of absorption can also be represented by the rate of increase of population at the upper level E_2, as

$$R_{abs} = \frac{dN_2}{dt} \tag{22.7}$$

The number of absorption transitions occurring in the material at any instant will be equal to the product of the number of atoms at the energy level E_1 and the probability P_{12} for the absorption transition. When the atoms are more at the lower energy level, then more atoms can jump into the excited state. Similarly, when more photons are incident on the assembly of atoms, then more atoms can get excited to the higher energy level. Then the rate of absorption transitions is given by

$$R_{abs} = B_{12}\rho(\nu)N_1 \tag{22.8}$$

Induced absorption involves the excitation of atom to the fixed higher level only. As a result of the absorption, N_1decreases and N_2 increases. But under normal conditions N_2 cannot be greater than N_1 (see§22.3). Therefore, as light propagates through the medium, it gets absorbed. However, N_2 can be made greater than N_1 using special techniques.

22.4.2 SPONTANEOUS EMISSION

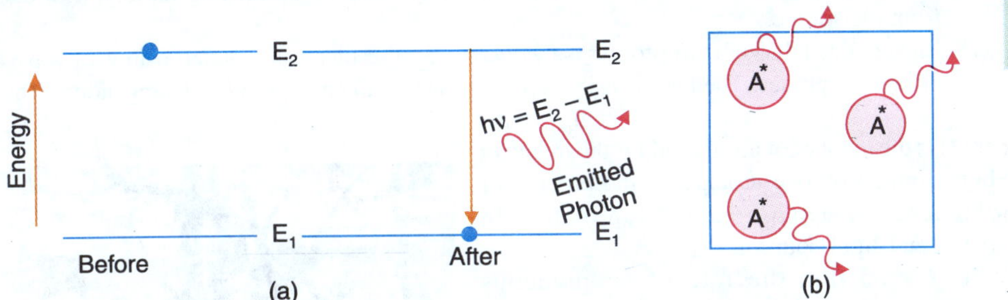

Spontaneous emission (*a*) emission process (*b*) Material emits photons haphazardly.

Fig. 22.5

When an atom at lower energy level is excited to a higher energy level, it cannot stay in the excited state for a relatively longer time. In a time of about 10^{-8}s, the atom reverts to the lower energy state by releasing a photon of energy hν, where $h\nu = (E_2 - E_1)$. The emission of photon occurs on its own and without any external impetus given to the excited atom (Fig.22.5). Emission of a photon by an atom *without any external impetus* is called **spontaneous emission**. We may write the process as

$$A^* \rightarrow A + h\nu$$

The probability that a spontaneous transition occurs depends only on the properties of energy states E_2 and E_1 and is independent of the photon density. It is equal to the lifetime of level E_2. Thus,

$$(P_{21})_{Spont.} = A_{21} \tag{22.9}$$

where A_{21} is a constant and known as the *Einstein coefficient for spontaneous emission.* A_{21} represents the probability of a spontaneous transition from level $2 \rightarrow 1$. A_{21} is a constant characteristic of the atom. *$1/A_{21}$ is a measure of the lifetime of the upper state against spontaneous transition to the lower state.*

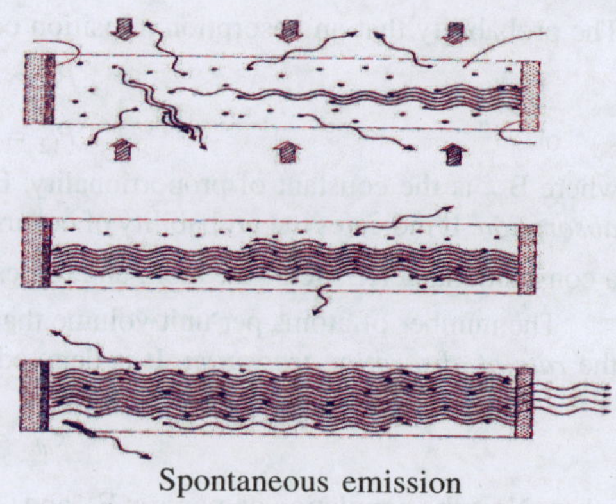

Spontaneous emission

Therefore, the rate of spontaneous transitions is given by

$$R_{sp} = A_{21} N_2 \tag{22.10}$$

It is to be noted that the process of spontaneous emission is independent of the incident light energy.

It follows from quantum mechanical considerations that spontaneous transition takes place from a given state to states lying lower in energy. Thus, spontaneous transition is not possible from level E_1 to level E_2. Therefore, the probability of spontaneous transition from E_1 to E_2 is zero.

$$\therefore \quad A_{12} = 0 \tag{22.11}$$

Let us now look at the salient features of spontaneous emission of light.

Characteristics of spontaneous emission:

(*i*) The process of spontaneous emission is essentially *probabilistic* in nature and is not amenable for control from outside.

(*ii*) The instant of transition, direction of propagation, the initial phase and the plane of polarisation of each photon are all *random*.

(*iii*) The light resulting through this process is *not monochromatic*.

(*iv*) As different atoms in the source emit photons in different directions, light spreads in all directions around the source. The light intensity goes on decreasing rapidly with distance from the source.

(*v*) Light emitted through this process is *incoherent*, as it results from a superposition of wave trains of random phases. The net intensity is proportional to the number of radiating atoms. Thus,

$$I_{total} = N\,I \tag{22.12}$$

where N is the number of atoms and I is the intensity of light emitted by one atom. It is the process of spontaneous emission that dominates in conventional light sources.

It is seen that the rate of spontaneous transitions is determined only by the population N_2 at the higher energy level whereas the rate of absorption transitions is determined by the population N_1 at the lower energy level and the energy density $\rho(\nu)$ in the incident light. If absorption and spontaneous emission were the only processes operative, then obviously the number of atoms absorbing radiation per second would be more than the number of atoms emitting light per

Photoluminescence spectroscopy measures stimulated light emission spectra.

second. Eventually, we may end up with a non-equilibrium situation where all the atoms in the medium are excited. But this condition is not observed in practice. It means that equilibrium is maintained. Therefore, in order to account for the state of equilibrium between light and matter, Einstein pointed out that *if a photon can stimulate an atom to move from a lower energy level E_1 to a higher energy level E_2 by means of absorption transition, then a photon should also be able to stimulate an atom from the same upper level E_2 to the lower level E_1*. This alternative mechanism of photon emission depends on the photon density present in the medium and is known as **stimulated emission**.

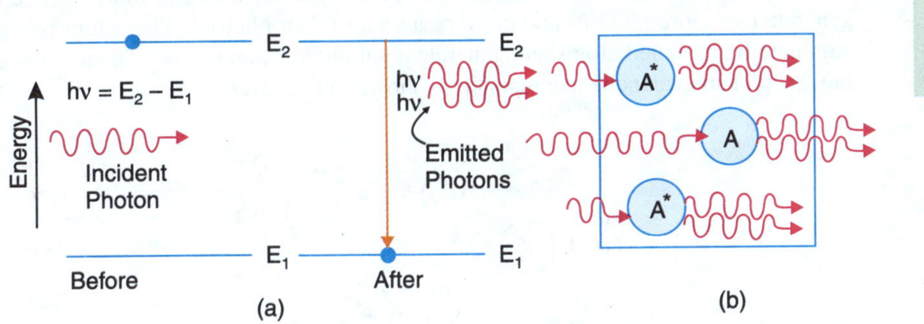

Stimulated emission (*a*) emission process. (*b*) Material emits photons in a coordinated manner.
Fig. 22.6

An atom in the excited state need not "wait" for spontaneous emission of photon. Well before the atom can make a spontaneous transition, it may interact with a photon with energy $h\nu = E_2 - E_1$, and make a downward transition. The photon is said to **stimulate** or **induce** the excited atom to emit a photon of energy $h\nu = (E_2 - E_1)$. The passing photon does not disappear and in addition to it there is a *second photon* which is emitted by the excited atom (see Fig. 22.6). The phenomenon of forced photon emission by an excited atom *due to the action of an external agency* is called **stimulated emission** or **induced emission**. The process may be expressed as

$$A^* + h\nu \rightarrow A + 2h\nu$$

The probability that a stimulated transition occurs is given by

$$(P_{21})_{stimulated} \propto \rho(\nu)$$

or

$$(P_{21})_{stimulated} = B_{21}\rho(\nu) \tag{22.13}$$

where B_{21} is the constant of proportionality and is known as the *Einstein coefficient for stimulated emission.* It is a constant characteristic of the atom and represents the properties of the energy states E_1 and E_2. B_{21} represents the probability for induced transition from level $2 \rightarrow 1$. The number of stimulated transitions occurring in the material at any instant will be equal to the product of the number of atoms at the energy level E_2 and the probability P_{21} for the stimulated transition. Thus, the rate of stimulated emission of photons is given by

$$R_{st} = B_{21}\rho(\nu)N_2 \tag{22.14}$$

In stimulated emission each incident photon encounters a previously excited atom, and the optical field of the photon interacts with the electron. The result of the interaction is a kind of resonance effect, which induces each atom to emit a second photon with the same frequency, direction, phase, and polarization as the incident photon.

Let us now look at the salient features of the stimulated emission of light.

Characteristics of Stimulated emission:

(*i*) The process of stimulated emission is *controllable* from outside.

(*ii*) The photon induced in this process propagates in the *same direction* as that of stimulating photon.

(*iii*) The induced photon has features identical to that of the inducing photon. It has the *same frequency, phase and plane of polarisation* as that of the stimulating photon.

(*iv*) **Multiplication of Photons:** The outstanding feature of this process is the *multiplication of photons*. For one photon interacting with an excited atom, there are two photons emerging. The two photons travelling in the same direction interact with two more excited atoms and generate two more photons and produce a total of four photons. These four photons in turn stimulate four excited atoms and generate eight photons, and so on. The number of photons builds up in an avalanche like manner, as shown in Fig. 22.7.

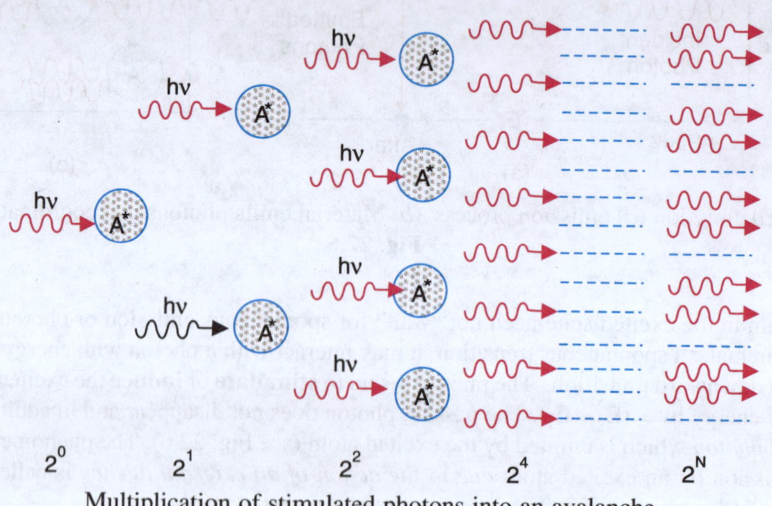

Multiplication of stimulated photons into an avalanche

Fig. 22.7

(*v*) **Light amplification:** All the light waves generated in the medium are due to one initial wave and all of the waves are in phase. Thus, the waves are coherent and *interfere constructively* (See Fig. 22.8).

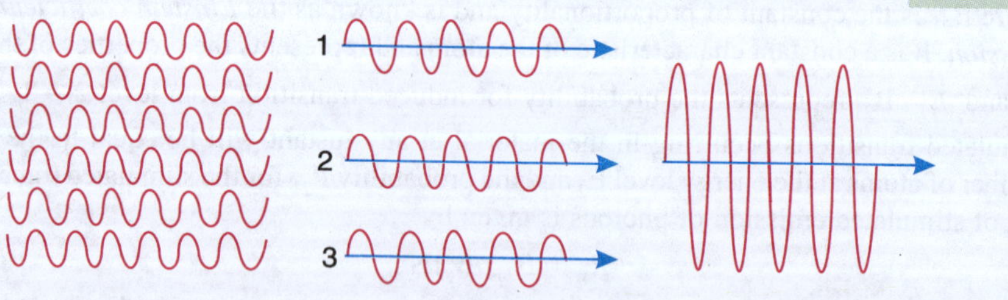

Coherent radiation (*a*) coherent waves (*b*) component wave form (*c*) Resultant wave

Fig. 22.8

(*vi*) The net intensity of light will be proportional to the square of the number of atoms radiating light. Thus,

$$I_{total} = N^2 I \tag{22.15}$$

The light emitted through the process of stimulated emission is of very high intensity and we say light *is amplified.*

At any reasonable temperature there cannot be enough atoms in excited states for any appreciable amount of stimulated emission from these states to occur.

Rather absorption is much more probable.

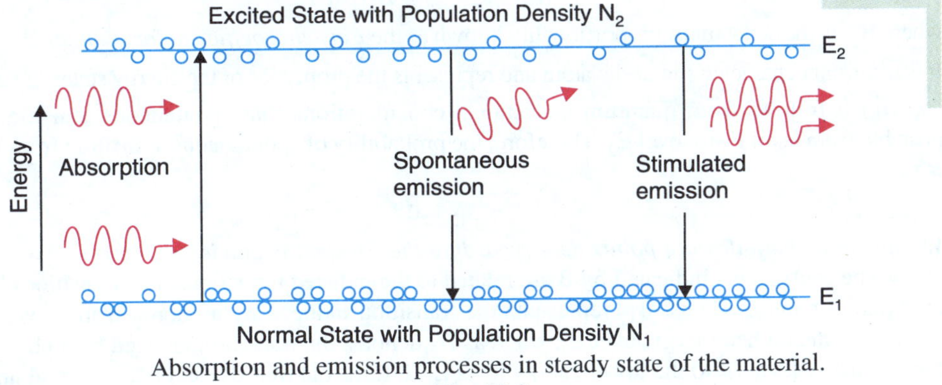

Absorption and emission processes in steady state of the material.

Fig. 22.9

The three processes, namely, absorption, spontaneous emission and stimulated emission occur simultaneously in a medium. Under steady state condition the absorption and emission processes balance each other. Thus,

$$R_{abs} = R_{sp} + R_{st} \tag{22.16}$$

$$\therefore \quad B_{12} N_1 \rho(\nu) = A_{21} N_2 + B_{21} N_2 \rho(\nu) \tag{22.17}$$

If we consider a medium in thermal equilibrium, there would be more atoms in the lower level than at higher level. That is $N_1 >> N_2$. As the probability for absorption transition is equal to the probability for stimulated transition, a photon traveling through the medium is *more likely to get absorbed* than to stimulate an excited atom to emit a photon. Therefore, usually the process of absorption dominates the process of stimulated emission. Similarly, an atom that is at the excited state is more likely to jump to the lower level on its own than being stimulated by a photon. It is due to the fact that the photon density in the incident beam is not sufficient to interact with the excited atoms; and the photons interact with atoms at lower level because of the large population available at that level. Owing to this, the spontaneous emission dominates the stimulated emission.

22.5 EINSTEIN COEFFICIENTS AND THEIR RELATIONS

22.5.1 EINSTEIN COEFFICIENTS

We summarize here the Einstein coefficients, which are the proportionality constants introduced in the above discussions.

(*i*) The probability that an absorption transition occurs is given by

$$P_{12} = B_{12}\rho(\nu)$$

where B_{12} is the constant of proportionality known as the *Einstein coefficient for induced absorption.* It is a constant characteristic of the atom and represents the properties of the energy states E_1 and E_2.

(*ii*) The probability that a spontaneous transition occurs is given by

$$(P_{21})_{Spontaneous} = A_{21}$$

where A_{21} is a constant known as the *Einstein coefficient for spontaneous emission.* A_{21} is a constant characteristic of the atom and is known as the *radiative rate* measured in units of s^{-1}. *$1/A_{21}$ is the lifetime of the upper state against spontaneous decay to the lower state.*

(*iii*) The probability that a stimulated transition occurs is given by

$$(P_{21})_{stimulated} = B_{21}\rho(\nu)$$

where B_{21} is the constant of proportionality known as the *Einstein coefficient for stimulated emission.* It is a constant characteristic of the atom and represents the properties of the energy states E_1 and E_2.

(*iv*) It follows from quantum mechanical considerations that spontaneous transition is not possible from level E_1 to level E_2. Therefore, the probability of spontaneous transition from E_1 to E_2 is zero.

$$\therefore \qquad A_{12} = 0 \tag{22.18}$$

We may note the following points here regarding the Einstein coefficients.

- The coefficients indicated by B are related to the induced transitions, i.e., transitions induced by external photons. Thus, B_{12} represents the transition induced by a photon from lower energy level E_1 to the higher energy level E_2, whereas B_{21}denotes the transition induced by a photon from higher energy level E_2 to the lower energy level E_1. It turns out that B_{12} and B_{21} are equal under the special condition that the quantum states E_1 and E_2 are single energy levels (*i.e.*, nondegenerate levels).
- The coefficient indicated by A is related to the spontaneous transition, *i.e.*, transition occurred on its own without the assistance of external agent. Since a spontaneous transition cannot take place from lower energy state E_1 to the higher energy state E_2, we do not have the coefficient A_{12}. In other words,

$$A_{12} = 0.$$

22.5.2 EINSTEIN RELATIONS

Under thermal equilibrium, the mean population N_1 and N_2 in the lower and upper energy levels respectively must remain constant. This condition requires that the number of transitions from E_2 to E_1 must be equal to the number of transitions from E_1 to E_2 (Fig. 22.9). Thus,

The numbe of atoms absorbing photons per second per unit volume = The number of atoms emitting photons per unit volume

The number of atoms absorbing photons per second per unit volume = $B_{12}\,\rho(\nu)\,N_1$

The number of atoms emitting photons per second per unit volume = $A_{21}N_2 + B_{21}\,\rho(\nu)\,N_2$

As the number of transitions from E_1 to E_2 must equal the number of transitions from E_2 to E_1, we have

$$B_{12}\,\rho(\nu)\,N_1 = A_{21}\,N_2 + B_{21}\,\rho(\nu)\,N_2 \tag{22.19}$$

$$\rho(\nu)\,[B_{12}\,N_1 - B_{21}\,N_2] = A_{21}\,N_2$$

$$\therefore \qquad \rho(\nu) = \frac{A_{21}N_2}{[B_{12}N_1 - B_{21}N_2]} \tag{22.20}$$

By dividing both the numerator and denominator on the right hand side of the above equation with $B_{12}N_2$, we obtain.

$$\rho(\nu) = \frac{A_{21}/B_{12}}{\left[\frac{N_1}{N_2} - \frac{B_{21}}{B_{12}}\right]} \tag{22.21}$$

But $$\frac{N_2}{N_1} = e^{-(E_2 - E_1)/kT}$$

As $$E_2 - E_1 = h\nu,$$

$$\frac{N_2}{N_1} = e^{-h\nu/kT} \quad \text{or} \quad \frac{N_1}{N_2} = e^{h\nu/kT}$$

$$\therefore \quad \rho(\nu) = \frac{A_{21}}{B_{12}}\left[\frac{1}{e^{h\nu/kT} - B_{21}/B_{12}}\right] \tag{22.22}$$

To maintain thermal equilibrium, the system must release energy in the form of electromagnetic radiation. It is required that the radiation be identical with black body radiation and be consistent with Planck's radiation law for any value of T. According to Planck's law

$$\rho(\nu) = \left(\frac{8\pi h\nu^3\mu^3}{c^3}\right)\left[\frac{1}{e^{h\nu/kT} - 1}\right] \tag{22.23}$$

where μ is the refractive index of the medium and c is the velocity of light in free space. Energy density ρ(ν) given by equ.(22.22) will be consistent with Planck's law (22.23), only if

$$\frac{A_{21}}{B_{12}} = \frac{8\pi h\nu^3\mu^3}{c^3} \tag{22.24}$$

and $$\frac{B_{21}}{B_{12}} = 1 \quad \text{or} \quad B_{12} = B_{21} \tag{22.25}$$

The above equations are known as the **Einstein relations**. The coefficients B_{12}, B_{21} and A_{21} are known as **Einstein coefficients**. It follows that the coefficients are related through

$$B_{12} = B_{21} = \frac{c^3}{8\pi h\nu^3\mu^3} A_{21} \tag{22.26}$$

The relation (22.25) shows that the coefficients for both absorption and stimulated emission are numerically equal. The equality implies that when an atom with two energy levels is placed in the radiation field, the probability for an upward (absorption) transition is equal to the probability for a downward (stimulated) transition.

The relation (22.26) shows that the ratio of coefficients of spontaneous versus stimulated emission is proportional to the third power of frequency of the radiation. This is why it is difficult to achieve laser action in higher frequency ranges such as x-rays.

22.6 LIGHT AMPLIFICATION

If we consider a medium in thermal equilibrium, there would be more atoms in the lower level than at higher level. That is $N_1 >> N_2$. As the probability for absorption transition is equal to the probability for stimulated transition, a photon travelling through the medium is more likely to get absorbed than to stimulate an excited atom to emit a photon. Therefore, usually the process of absorption dominates the process of stimulated emission. Similarly, an atom that is at the excited state is more likely to jump to the lower level on its own than being stimulated by a photon. Further, the photon density in the incident beam is not sufficient to interact with the excited atoms. Owing to this, the spontaneous emission dominates the stimulated emission.

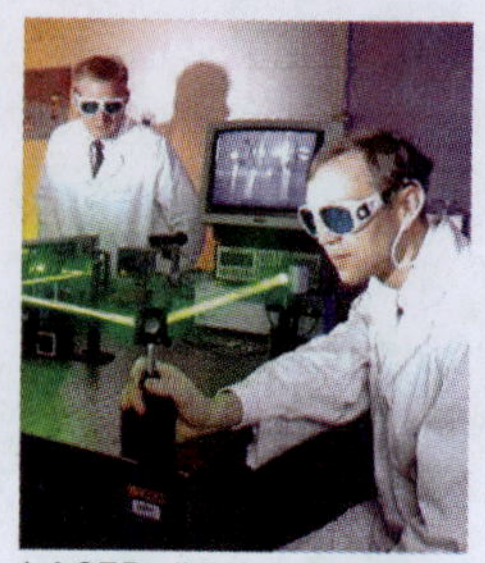
LASER (Light Amplification by Stimulated Emission of Radiation)

Light amplification requires that stimulated emission occur almost exclusively. In practice, absorption and spontaneous emission always occur together with stimulated emission. The laser operation is achieved when stimulated emission exceeds in a large way the other two processes. Let us now look at the conditions under which the number of stimulated transitions can be made larger than the other two transitions.

22.6.1 CONDITION FOR STIMULATED EMISSION TO DOMINATE SPONTANEOUS EMISSION

The ratio of equ.(22.14) to equ. (22.10) gives

$$R_1 = \frac{\text{Stimulated transitions}}{\text{Spontaneous transitions}} = \frac{B_{21}\rho(\nu)N_2}{A_{21}N_2} = \frac{B_{21}}{A_{21}}\rho(\nu) \tag{22.27}$$

Equ.(22.27) indicates that stimulated transitions will dominate the spontaneous transitions if the radiation density $\rho(\nu)$ is very large and the value of the ratio B_{21}/A_{21} is also large.

(*i*) Using equ.(22.23) into equ. (22.27), we get

$$R_1 = \left(\frac{B_{21}}{A_{21}}\right)\left[\frac{8\pi\, h\nu^3\mu^3}{c^3}\cdot\frac{1}{e^{h\nu/kT}-1}\right]$$

But

$$\frac{B_{21}}{A_{21}} = \frac{c^3}{8\pi\, h\nu^3\mu^3}$$

$\therefore$

$$R_1 = \left(\frac{c^3}{8\pi\, h\nu^3\mu^3}\right)\left[\frac{8\pi\, h\nu^3\mu^3}{c^3}\cdot\frac{1}{e^{h\nu/kT}-1}\right]$$

or

$$R_1 = \left[\frac{1}{e^{h\nu/kT}-1}\right] \tag{22.28}$$

If we assume $\nu = 5\times10^{14}$Hz and T=300 K, the value of R_1 comes to 10^{-58}.

The above result shows that in the optical region stimulated emission is negligible compared to spontaneous emission.

(*ii*) Equ.(22.27) indicates that stimulated transitions will dominate the spontaneous transitions if the radiation density $\rho(\nu)$ is very large. Thus, the presence of a large number of photons in the active medium is required. However, it will lead to more absorption transitions. Hence, large photon density alone will not guarantee more stimulated emissions.

(*iii*) **Requirement of states of larger lifetimes**

Equ.(22.27) further indicates that stimulated transitions will dominate the spontaneous transitions if the value of the ratio B_{21}/A_{21} is also large. To increase the probability of stimulated emissions, the lifetime of atoms at the excited state should be larger. In other words, it is necessary that the excited state has a longer lifetime (remember that $1/A_{21}$ represents the lifetime of the excited state).

22.6.2 CONDITION FOR STIMULATED EMISSION TO DOMINATE ABSORPTION TRANSITIONS

It may be noted that the presence of a large number of photons will lead to more absorption transitions rather than stimulated emissions. Hence large photon density alone will not guarantee more stimulated emissions.

The ratio of equ.(22.14) to equ. (22.8) yields

$$R_2 = \frac{\textit{Stimulated}\text{ transition}}{\text{Absorption transition}} = \frac{B_{21}\rho(\nu)N_2}{B_{12}\rho(\nu)N_1} \tag{22.29}$$

$$\text{As } B_{21} = B_{12}, \quad R_2 = \frac{N_2}{N_1} \tag{22.30}$$

The above condition indicates that the stimulated transitions will overwhelm the absorption process if N_2 is greater than N_1. It means that there should be more atoms present in the higher energy level than in the lower energy level for stimulated emissions to dominate over the spontaneous emissions.

At thermodynamic equilibrium, $N_2 << N_1$ and the population at the ground level far exceeds that of the excited level. When light propagates through the medium, a photon may hit an excited atom leading to stimulated emission, or be absorbed on hitting an atom in the ground state. As more number of atoms is there at the ground level, a photon has a much higher probability of being absorbed than of stimulating an atom at the excited level. Therefore, the absorption transitions will be larger than stimulated transitions and the medium will absorb the incident light. If, on the other hand, the number the opposite situation prevails, photons are more likely to cause stimulated transitions than absorption transitions.

Two sum up, three conditions are to be satisfied to make stimulated transitions overwhelm the other transitions: (i) the population at excited level should be greater than that at the lower energy level, (ii) the ratio B_{21}/A_{21} should be large and (iii) a very high radiation density should be present in the medium. A medium amplifies light only when these *three conditions* are fulfilled.

To achieve high percentage of stimulated emissions, (i) an artificial situation known as **population inversion** is to be created in the medium. (ii) A larger value of B_{21}/A_{21} is achieved by choosing a **metastable energy level** as the higher level. As spontaneous transitions are forbidden from a metastable state, the ratio B_{21}/A_{21} will be larger. (iii) $\rho(\nu)$ is made larger by enclosing the emitted radiation in an **optical resonant cavity** formed by two parallel mirrors. The radiation is reflected many times till the photon density reaches to a very high value and a favourable condition is created for large stimulated emissions.

22.7 MEETING THE THREE REQUIREMENTS

22.7.1 POPULATION INVERSION

When the material is in thermal equilibrium condition, the population ratio is governed by the Boltzmann distribution law according to the following equation:

$$\frac{N_2}{N_1} = e^{-[E_2-E_1]/kT} \tag{22.31}$$

It means that the population N_2 at the excited level E_2 will be far smaller than the population N_1 at the level E_1. The condition in which there are more atoms in the lower energy level and relatively lesser number of atoms in the higher energy level is called **normal condition** or **thermal equilibrium** (Fig.22.10a). Thus, under thermal equilibrium, $N_1 >> N_2$.

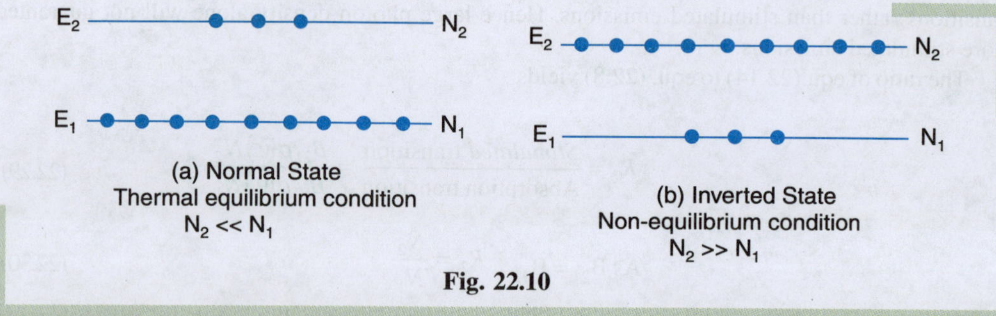

Fig. 22.10

To achieve a high percentage of stimulated emission, a majority of atoms should be at the higher energy level than at the lower level. We may somehow enhance the number of atoms in the excited level, such that the population ratio N_2 / N_1 **momentarily** increases without change in temperature. This is *a non-equilibrium condition* and is known as *inverted population* condition. **Population inversion** is the non-equilibrium condition of the material in which population of the upper energy level N_2 momentarily exceeds the population of the lower energy level N_1 (Fig.22.10 b). That is,

$$N_2 >> N_1 \tag{22.32}$$

From equ.(22.31) it is seen that N_2 can exceed N_1 only if the temperature *were* negative. In view of this, the state of population inversion is sometimes referred to as a ***negative temperature state***. It does not mean that we can attain temperatures below absolute zero. The terminology underlines the fact that the state of population inversion is a non-equilibrium state. It should be borne in mind that the population inversion is attained at normal temperatures.

The system shown in Fig.22.10(a) has two energy levels. At thermal equilibrium, photon absorption and emission processes take place side by side, but because $N_1 > N_2$, the system absorbs photons rather than emit photons. Now suppose that the system is supplied with energy from an external source till N_2 exceeds N_1. Then, the system is said to have attained the state of population inversion. The population inversion took place between the levels E_2 and E_1.

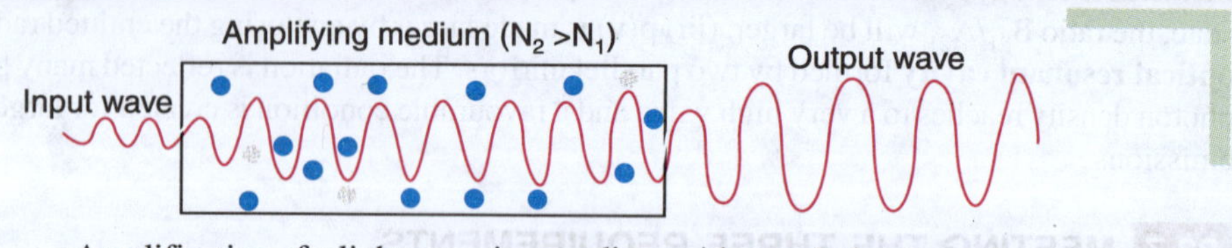

Amplification of a light wave in a medium with population inversion.

Fig. 22.11

When the system is in the population inversion condition, a few randomly emitted photons trigger stimulated emission of photons and those stimulated photons induce more stimulated emissions and so on. Consequently light gets amplified (Fig.22.11) and a cascade of light is produced. However, in this process atoms from E_2 level make downward transitions and as soon as the population at lower level becomes equal or larger than that at the excited level, population inversion comes to an end. Energy is again to be supplied to the system to take it into the state of population inversion.

The non-equilibrium condition is attained by employing pumping techniques to transfer large number of atoms from lower energy level to higher energy level.

22.7.2 METASTABLE STATES

An atom can be excited to a higher level by supplying energy to it. Normally, excited atoms have short lifetimes and release their energy in a matter of nanoseconds (10^{-9}s) through spontaneous emission. It means that atoms do not stay long enough at the excited state to be stimulated. As a result, even though the pumping agent continuously raises the atoms to the excited level, they undergo spontaneous transitions and rapidly return to the lower energy level. Population inversion cannot be established under such circumstances. In order to establish the condition of population inversion, the excited atoms are required to 'wait' at the upper energy level till a large number of atoms accumulate at that level. In other words, it is necessary that the excited state has a longer lifetime. A metastable state is such a state. Because of restrictions imposed by conservation of angular momentum, an electron excited to a metastable state cannot return to the ground state by emitting a photon, as it is generally expected to do. Such a state in which single-photon emission is impossible, has an unusually long time and is called a **metastable state**. Atoms excited to the metastable states remain excited for an appreciable time, which is of the order of 10^{-6} to 10^{-3}s. This is 10^3 to 10^6 times the lifetimes of the ordinary energy levels.

New optical pumping apparatus.

Therefore, the metastable state allows accumulation of a large number of excited atoms at that level. The metastable state population can exceed the population at a lower level and establish the condition of population inversion in the lasing medium. It would be impossible to create the state of population inversion without a metastable state. Metastable state can be readily obtained in a crystal system containing impurity atoms. These levels lie in the forbidden band gap of the host crystal. Population inversion readily takes place as the lifetimes of these levels are large, and secondly, there is no competition in filling these levels, as they are localized levels.

There could be no population inversion and hence no laser action, if metastable states do not exist.

22.7.3 CONFINING RADIATION WITHIN THE MEDIUM

According to equ.(22.27) a high radiation density $\rho(\nu)$ is required to be present in the active medium so that stimulated emission dominates spontaneous emission. If laser medium is enclosed in between a pair of optically plane parallel mirrors, photon density builds up to a very high value through repeated reflections of photons which remain within the medium. Such an arrangement is known as an **optical resonant cavity** or **optical resonator**.

22.8 COMPONENTS OF LASER

The essential components of a laser are (i) an active medium, (ii) a pumping agent and (iii) an optical resonator (see Fig. 22.12).

Fig. 22.12. Components of a Laser

22.8.1 ACTIVE MEDIUM

Atoms are in general characterized by a large number of energy levels. However, all types of atoms are not suitable for laser operation. Even in a medium consisting of different species of atoms, only a small fraction of atoms of a particular type have energy level system suitable for achieving population inversion. Such atoms can produce more stimulated emission than spontaneous emission and cause amplification of light. Those atoms, which cause laser action, are called *active centers*. The rest of the medium acts as host and supports active centers. The medium hosting the active centers is called the **active medium**. It is also called **laser medium**. *An active medium is a medium which when excited reaches the state of population inversion and promotes stimulated emissions leading to light amplification.*

22.8.2 PUMP

For achieving and maintaining the condition of population inversion, we have to raise continuously the atoms in the lower energy level to the upper energy level. It requires energy to be supplied to the system. **Pumping** is the process of supplying energy to the laser medium with a view to transfer it into the state of population inversion. Because N_1 is originally very much larger than N_2, a large amount of input energy is required to momentarily increase N_2 to a value comparable to N_1. **Pump** is the agency which supplies the energy.

There are a number of techniques for pumping a collection of atoms to an inverted state. Optical pumping, electrical discharge and direct conversion are some of the methods of pumping. In *optical pumping*, a light source such as a flash discharge tube is used to illuminate the active medium. This method is adopted in solid state lasers. In *electrical discharge* method, the electric field causes ionization of the medium and raises it to the excited state. In semiconductor diode lasers, a *direct conversion* of electrical energy into light energy takes place.

22.8.3 OPTICAL RESONANT CAVITY

Laser is a light source and it is analogous to an electronic oscillator. An electronic oscillator (Fig.22.13) is essentially an amplifier supplied with a positive feed back. A part of the output of the amplifier is taken and fed back at its input. When the amplifier is switched on, electrical noise signal of appropriate frequency present at the input will

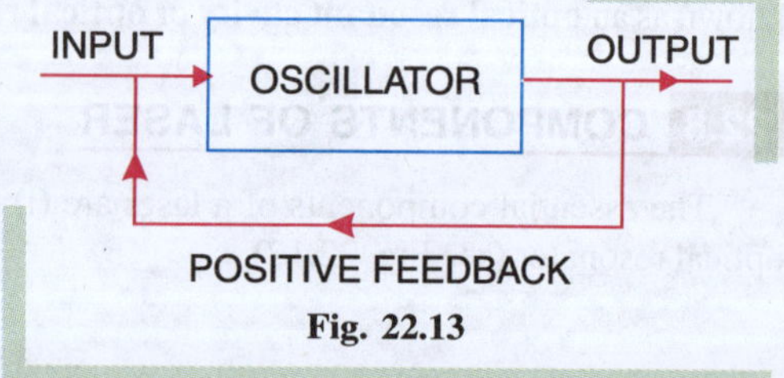

Fig. 22.13

be amplified; the output is fed back to the input and amplified again and so on. A stable output is quickly reached when the oscillator acts as a source of a particular frequency.

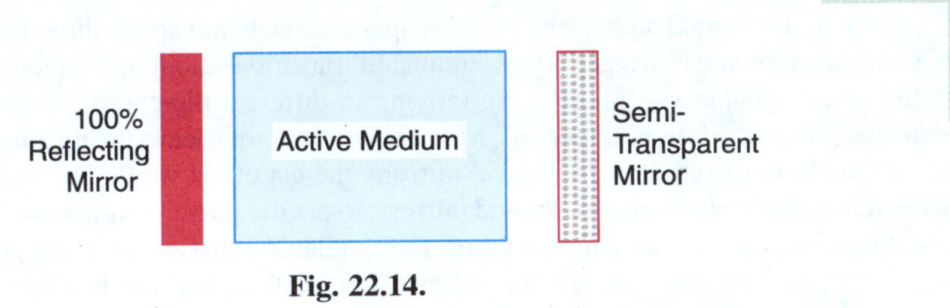

Fig. 22.14.

In laser the active medium is the amplifying medium. It is converted into an oscillator through the feed back mechanism established by an optical resonator. A pair of optically plane parallel mirrors (see Fig.22.14) constitutes an optical resonant cavity. It is known as a Fabry-Perot resonator. One of these mirrors is fully reflecting and reflects all the light that is incident on it. The other mirror is made partially reflecting such that more than 90% of incident light is reflected from it and a small fraction is transmitted through it as the laser beam.

In laser, the role of noise is played by chance photons emitted spontaneously. The photons emitted along the optic axis of the resonant cavity travel through the medium and trigger stimulated emissions. They are reflected by the end mirror and reverse their path. The photons are thus fed back into the medium and travel toward the opposite end mirror causing more stimulated emissions The photons are once more reflected at the mirror and travel toward the opposite mirror. Substantial light amplification takes place because the light beam is reflected several times at the mirrors and gains strength in each passage, Ultimately, when the amplification balances the losses in the cavity, the laser beam emerges out from the front – end mirror. *In the absence of resonator cavity, there would be no amplification of light.*

22.9 LASING ACTION

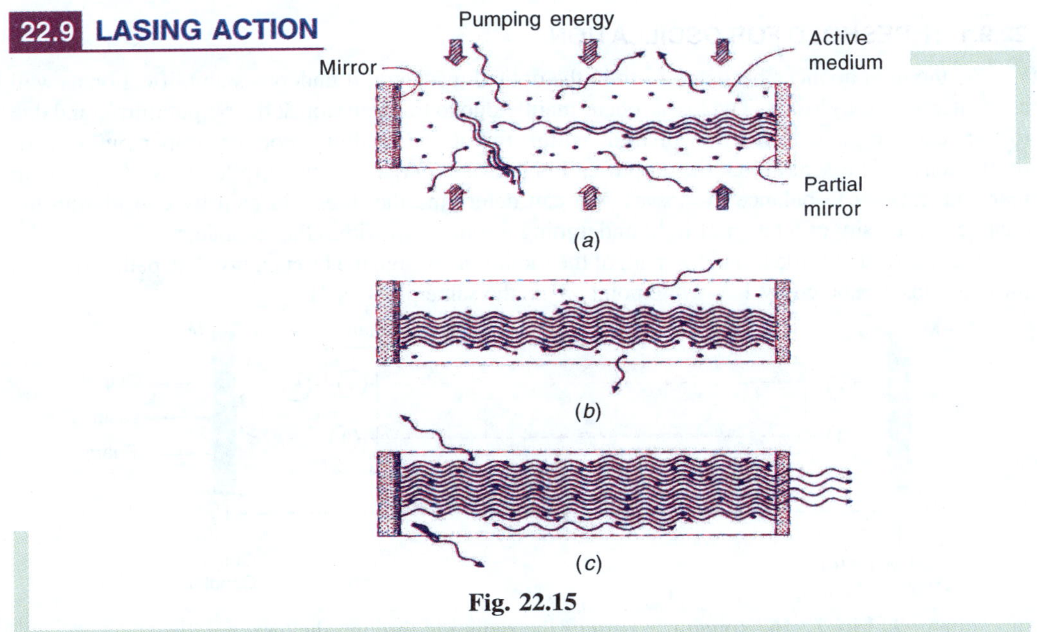

Fig. 22.15

Fig. 22.15 shows the action of an optical resonator. The active centers in the medium are in the ground state initially. Through suitable pumping mechanism, the medium is taken into the state of population inversion.

Some of the excited atoms emit photons spontaneously in various directions (Fig.22.15a). Each spontaneous photon can trigger many stimulated transitions along the direction of its propagation. As the initial spontaneous photons are moving in different directions, the photons stimulated by them also travel in different directions. Many of such photons leave the medium without reinforcing their strength. In the absence of the end mirrors, the net effect would have been the production of incoherent light. Now because of the end mirrors, a specific direction is imposed on photons. Photons travelling along the axis are amplified through stimulated emission while the photons emitted in any other direction will pass through the sides of the medium and are lost forever. Thus, a specific direction is selected for further amplification of light.

A majority of photons travelling along the axis are reflected back on reaching the end mirror. They travel towards the opposite mirror and on their way stimulate more and more atoms and build up the photon strength, as shown in Fig. 22.15 (b). The photons that strike the opposite mirror are reflected once more into the medium, as shown in Fig. 22.15 (b). The photons travel once more through the medium generating more photons and more amplification. The photons are then reflected again at the mirror and travel through the medium. As the photons are reflected back and forth between the mirrors, stimulated emission sharply increases and the amplification of light is augmented. The mirrors thus provide positive feed back of light into the medium so that stimulated emission acts are sustained and the medium operates as an oscillator.

At each reflection at the front-end mirror, light is partially transmitted through it. The transmitted component constitutes a *loss of energy* from the resonator. When the losses at the mirrors and within the medium balance the gain, the laser oscillations build up. A steady and strong laser beam will emerge from the front-end mirror, as shown in Fig.22.15 (c).

As the end mirrors reflect light strongly into the laser medium, the light levels are very high with in the optical cavity. Consequently, the condition that $\rho(\nu)$ should be large within the medium is satisfied.

22.9.1 THRESHOLD FOR OSCILLATION

As the light bounces back and forth in the optical resonator, it undergoes amplification as well as it suffers various losses. The losses occur mainly due to transmission at the output mirror and due to the scattering and diffraction of light within the active medium. For the proper build up of oscillations, it is essential that the amplification between two consecutive reflections of light from rear end mirror can balance the losses. We can determine the threshold gain by considering the change in intensity of a beam of light undergoing a round trip within the resonator.

Fig. 22.16 shows the round trip path of the radiation through the laser cavity. The path is divided into sections numbered by 1-5, while point "5" is the same point as "1".

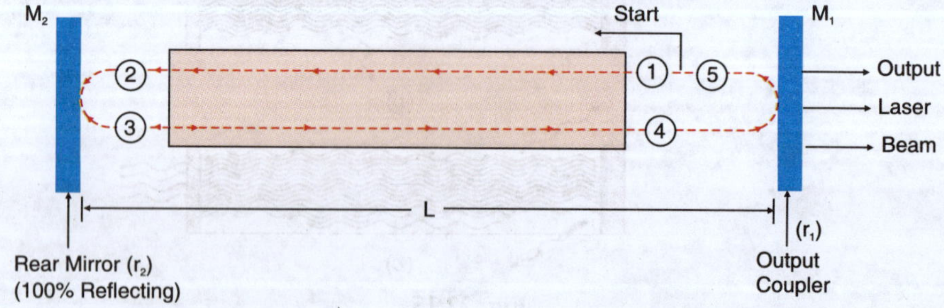

Fig. 22.16. Round trip path of the radiation through the laser cavity

Let us assume that the laser medium fills the space between the mirrors M_1 and M_2 (see Fig.22.16), which have reflectivity r_1 and r_2 respectively. Let the mirrors be separated by a distance L. Further, let the intensity of the light beam be I_o at M_1. Then, in travelling from mirror M_1 to mirror M_2, the beam intensity increases from I_o to I(L), which is given by

$$I(L) = I_o e^{(\gamma-\alpha_s)L} \tag{22.33}$$

After reflection at M_2, the beam intensity will be $r_2 I_o e^{(\gamma-\alpha_s)L}$ and after a complete round trip the final intensity will be

$$I(2L) = r_1 r_2 I_o e^{(\gamma-\alpha_s)2L} \tag{22.34}$$

The amplification obtained during the round trip is

$$G = \frac{I(2L)}{I_o} = r_1 r_2 e^{(\gamma-\alpha_s)2L} \tag{22.35}$$

The product $r_1 r_2$ represents the losses at the mirrors whereas α_s includes all the distributed losses such as scattering, diffraction and absorption occurring in the medium. The losses are balanced by gain, when $G \geq 1$ or $I(2L) = I_o$. It requires that

$$r_1 r_2 e^{2(\gamma-\alpha_s)L} \geq 1 \tag{22.36}$$

or

$$e^{2(\gamma-\alpha_s)L} \geq \frac{1}{r_1 r_2}$$

Taking logarithms on both sides, we get

$$2L(\gamma - \alpha_s) \geq -ln\, r_1 r_2$$

$$\gamma - \alpha_s \geq -\frac{1}{2L} ln\, r_1 r_2$$

$$\therefore \quad \gamma \geq \alpha_s - \frac{1}{2L} ln\, r_1 r_2 \tag{22.37}$$

or

$$\gamma \geq \alpha_s + \frac{1}{2L} ln \frac{1}{r_1 r_2} \tag{22.38}$$

Equ.(22.38) is known as the ***condition for lasing***. It shows that the initial gain must exceed the sum of the losses in the cavity. This condition is used to determine the threshold value of pumping energy for lasing action.

γ, the ***amplification*** of the laser will be dependent on how hard the laser medium is pumped. As the pump power is slowly increased, a value of γ_{th} called **threshold value** is reached and the laser starts oscillating. The threshold value γ_{th} is given by

$$\gamma_{th} = \alpha_s + \frac{1}{2L} ln \frac{1}{r_1 r_2} \tag{22.39}$$

Equ.(22.39) states the condition when the net gain would be able to counteract the effect of losses in the cavity and is known as the **threshold condition for lasing**. The value of γ must be atleast γ_{th} for laser oscillations to commence.

22.10 PRINCIPAL PUMPING SCHEMES

Atoms in general are characterized by a large number of energy levels. Among them only three or four levels will be pertinent to the pumping process. Therefore, only those levels are depicted in the pumping scheme diagrams. Two important pumping schemes are widely employed. They are known as three-level and four-level pumping schemes.

22.10.1 THREE-LEVEL PUMPING SCHEME

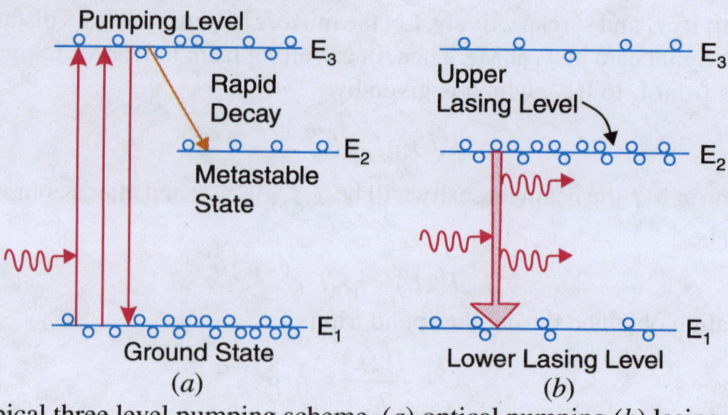

A typical three level pumping scheme–(*a*) optical pumping (*b*) lasing action

Fig. 22.17

A typical three-level pumping scheme is shown in Fig.22.17. The state E_1 is the ground level; E_3 is the pump level and E_2 is the metastable upper lasing level. When the medium is exposed to pump frequency radiation, a large number of atoms will be excited to E_3 level. However, they do not stay at that level but rapidly undergo downward transitions to the metastable level E_2 through non-radiative transitions. The atoms are trapped at this level as spontaneous transition from the level E_2 to the level E_1 is forbidden. The pumping continues and after a short time there will be a large accumulation of atoms at the level E_2. When more than half of the ground level atoms accumulate at E_2, the population inversion condition is achieved between the two levels E_1 and E_2. Now a chance photon can trigger stimulated emission.

22.10.2 FOUR-LEVEL PUMPING SCHEME

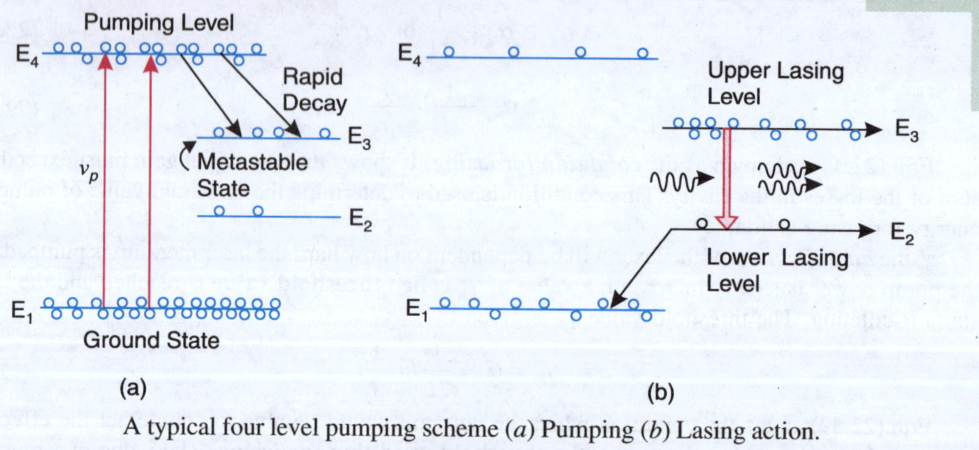

A typical four level pumping scheme (*a*) Pumping (*b*) Lasing action.

Fig. 22.18.

A typical four-level pumping scheme is shown in Fig.22.18. The level E_1 is the ground level, E_4 the pumping level, E_3 the metastable upper lasing level and E_2 the lower lasing level. E_2, E_3 and E_4 are the excited levels. When light of pump frequency ν_p is incident on the lasing medium, the active centers are readily excited from the ground level to the pumping level E_4. The atoms stay at the E_4 level for only about 10^{-8}s, and quickly drop down to the metastable level E_3. As spontaneous transitions from the level E_3 to level E_2 cannot take place, the atoms get trapped at the level E_3. The population at the level E_3 grows rapidly. The level E_2 is well above the ground level such that $(E_2 - E_1) > kT$. Therefore, at normal temperature atoms cannot jump to level E_2 on the strength of thermal energy. As a result, the level E_2 is virtually empty. Therefore, population inversion is attained between the levels E_3 and E_2. A chance photon of energy $h\nu = (E_3 - E_2)$ emitted spontaneously can start a chain of stimulated emissions, bringing the atoms to the lower laser level E_2. From the level E_2, the atoms subsequently under go non–radiative transitions to the ground level E_1 and will be once again available for excitation.

22.10.3 COMPARISON OF FOUR LEVEL LASER WITH THE THREE LEVEL LASER

1. In the three–level pumping scheme, the terminal level of laser transition is simultaneously the ground level. Therefore, in order to achieve population inversion more than half of the ground level atoms have to be pumped up to the upper lasing level, such that $N_2 > N_1/2$. As the number of atoms in the ground level is very large, high pump power is required in order to promote $N_1/2$ atoms and establish the required population inversion. On the other hand, in the four-level pumping scheme, the terminal level of laser transition is virtually empty and population inversion condition is readily established even if a smaller number of atoms arrive at the upper lasing level. Therefore, relatively small pumping power is required to establish population inversion in four level pumping schemes.
2. In case of three level pumping scheme, once stimulated emission commences, the population inversion condition reverts to normal population condition.. Lasing ceases as soon as the excited atoms drop to the ground level. Lasing occurs again *only* when the population inversion is re-established. The light output therefore is a *pulsed output.*

   In case of four level scheme, the condition of population inversion can be held without interruption and light output is obtained continuously. Thus, the laser operates in *continuous wave* (cw) mode.

22.10.4 NECESSITY OF BROAD ABSORPTION BAND AT PUMPING LEVEL

In a laser medium, active centers are excited through absorption of energy from a pumping source. We desire that the energy given through the pumping agent is utilized to the largest possible extent in exciting the ground level atoms. This is denoted by *pumping efficiency*. In case of optical pumping, a flash discharge from a lamp serves as the pumping agent. Normally, light sources such as the flash discharge emit light over a wide frequency range. More number of atoms can be excited to the higher level if large number of the frequency components is utilized instead of a single frequency. That can happen *only* when there is a band of close spaced energy levels at the pumping level.

Thus, for a larger pumping efficiency, the pump level should be a broad band rather than a narrow discrete level.

22.11 ROLE OF RESONANT CAVITY

(*i*) The primary function of optical resonator is to provide a positive feedback of light into the lasing medium so that the stimulated emission acts are sustained and the laser acts as a generator of light.

(*ii*) A chance photon spontaneously emitted by an excited atom acts as the input and induces stimulated emission. To sustain stimulated emission acts and to increase the light intensity in a cumulative way, a positive feedback of light must be provided. The mirrors by way of reflecting the incident photons provide the feedback.

(*iii*) Laser oscillation is initiated by the photons spontaneously radiated by some of the excited atoms. Each spontaneous photon can trigger many stimulated transitions along the path of its travel. As the initial spontaneous photons are emitted in various directions, the secondary stimulated photons will also travel in various directions. The result would be production of incoherent light. In order to build up oscillations, a specific direction has to be defined for photon propagation in the lasing medium.

The optical resonator sets its optic axis as the most favourable direction for build–up of light beam.

(*iv*) In order to make the stimulated emissions dominate spontaneous emissions, a high optical energy density Q is necessary to be present in the active medium. The mirrors constituting the cavity confine more than 90% of the emitted photons to be within the laser medium such that a very high optical energy density is always present in the lasing medium.

(*v*) The optical cavity is very much similar to a resonating column. Just as standing waves form in a resonating column, standing waves of optical frequencies are formed in the optical cavity. If L is the length of the cavity, the longest wavelength that will produce a standing wave pattern is $\lambda = 2L$. In general the cavity supports the wavelengths

$$\lambda_m = \frac{2nL}{m} \quad (m = 1,2,3,4,\text{——}) \qquad (22.40)$$

where n is the refractive index of the laser medium, and nL is the optical path. Waves of other wavelengths attenuate quickly. Thus, the optical cavity selects and amplifies only certain frequencies. Hence, the value of L should be properly selected.

(*vi*) Active centers may have a number of lasing transitions instead of only one transition. The mirrors of optical cavity suppress the undesired transitions. The reflectivity of the mirrors is further made less for the undesired photons, which therefore get absorbed at the mirrors.

22.12 MODES OF THE LASER BEAM

We have noted that part of the laser light in the laser cavity emerges through the output mirror. The optical waves within an optical resonant cavity are characterized by their **resonant modes**, which are discrete resonant conditions determined by the **dimensions of the cavity** [equ.(22.40)]. The laser beam radiated from the laser cavity is thus not arbitrary. Only the waves oscillating at modes that match the oscillation modes of the laser cavity can be produced. The laser modes governed by the axial dimensions of the resonant cavity are called the **longitudinal modes**, and the modes determined by the cross-sectional dimensions of the laser cavity are called **transverse modes**.

22.12.1 LONGITUDINAL MODES

An electromagnetic wave which move inside the laser cavity from right to left, is reflected by the left mirror, and move to the right until it is reflected from the right mirror, and so on. Thus, two waves of the same frequency and amplitude are moving in opposite directions, which is the condition for creating a standing wave. In order to create a *standing wave*, the wave must start with the same phase at the mirror.

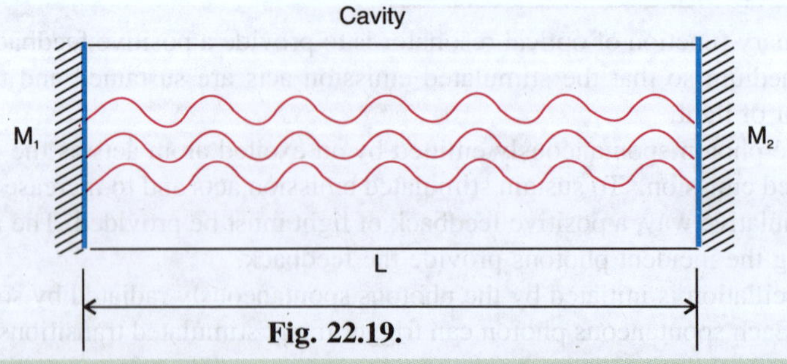

Fig. 22.19.

The condition for standing wave pattern is that the path length traveled by a wave between two consecutive reflections at an end mirror should equal an integral multiple of the wavelength. It means that

$$2\,n\,L = m\,\lambda \qquad (m = 1, 2, 3, \text{---}) \tag{22.41}$$

The above equations indicate that only those waves, which satisfy the above condition, can exist inside the cavity in a steady state. Waves of other wavelengths interfere destructively and are quickly attenuated. Because of its length which is very large compared to light wavelength, optical resonator supports simultaneously several standing waves of multiple wavelengths. These wavelengths are called **longitudinal** or **axial modes**. Therefore, m in equ.(22.41) is called the *mode number*. Thus, longitudinal modes are standing waves along the optical axis of the laser.

The frequencies of the longitudinal modes are given by

$$\nu_m = \frac{c}{n\lambda_m} = m\left(\frac{c}{2nL}\right) \tag{22.42}$$

Fig. 22.20 is a graphic representation of allowed frequencies inside a laser cavity.

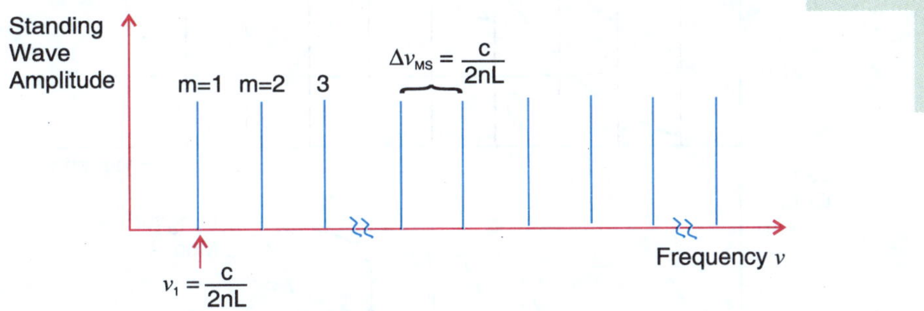

Fig. 22.20. Allowed Longitudinal modes inside a Laser Cavity

Theoretically, the cavity can resonate at a very large number of frequencies (see Fig. 22.20.) that satisfy the equation (22.42). For example, if we take $L = 0.5$ m, and $\lambda = 5000$ Å, we obtain $m = 2 \times 10^6$. It means that the cavity supports 2×10^6 longitudinal modes.

All these allowed frequencies will not be emitted from the laser, since there are more limiting conditions. The modes, which we calculate from (22.42), belong to the *passive cavity* where the active medium is not present. Out of these frequencies, only those frequencies (modes) that have amplification above lasing threshold to overcome absorption, will be emitted out of the laser.

Mode Separation:

The frequency separation between the neighbouring modes is constant and is given by

$$\Delta\nu = \nu_{m+1} - \nu_m = \frac{c}{2nL} \tag{22.43}$$

Equ.(22.43) can be expressed in terms of $\Delta\lambda$ as follows.

$$\Delta\lambda = \lambda^2 \cdot \frac{\Delta\nu}{c}$$

Using the relation (22.43) into the above equation, we find the wavelength separation

$$\Delta\lambda = \frac{\lambda^2}{2nL} \tag{22.44}$$

22.12.2 GAIN CURVE AND LASER OPERATING FREQUENCIES

When the active medium is placed between the two mirrors, the cavity becomes an **active cavity**. Then, the standing waves supported by the cavity are the light waves emitted by the stimulated atoms of the medium. Ideally, the emission should occur at a single frequency, as a group of identical atoms radiate at the same frequency. However, in practice, because of the various line broadening mechanisms, there will be a spread of frequencies about the central frequency of the emission line. Fig. 22.21 shows the shape of a typical emission line. It is also called **gain curve** or **gain profile** because it indicates the range of frequencies over which stimulated emission can provide sufficient gain.

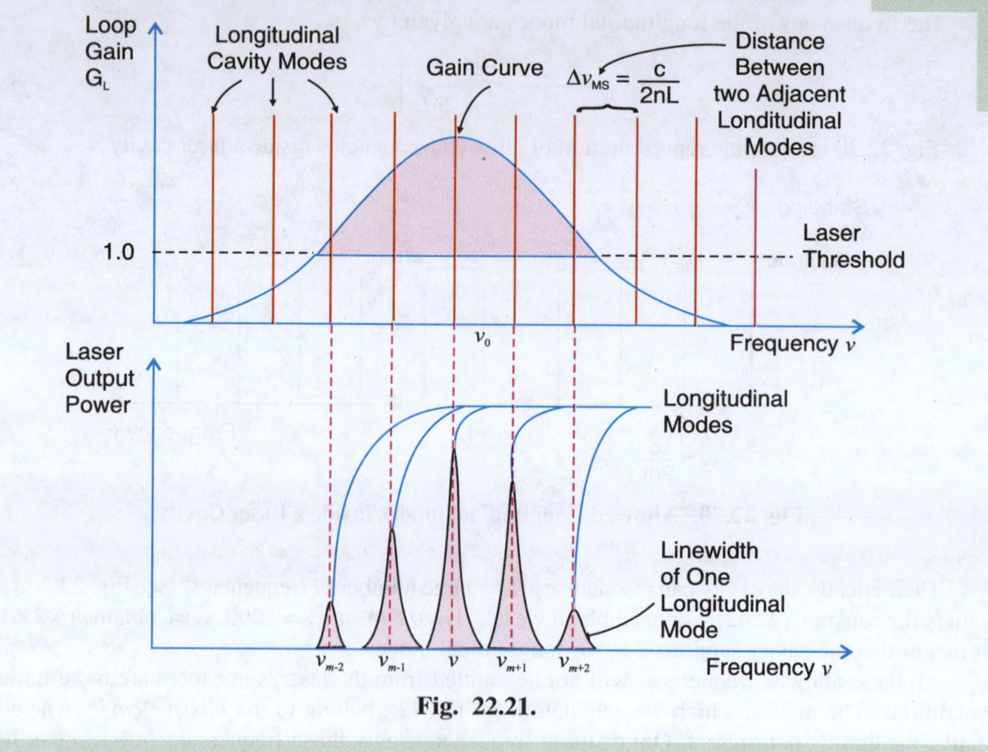

Fig. 22.21.

The laser operating frequencies are determined together by the resonant frequencies of the cavity and by the laser emission line width. If an output has to exist at a particular frequency, the cavity must be resonant at that specific frequency and there must be sufficient gain at that frequency. Laser oscillation can take place only when the gain is large enough to maintain resonance. For example, dashed lines are shown in Fig. 22.21 (a) which corresponds to threshold level. It is seen that frequencies between ν_{m+1} and ν_{m+2} can be amplified, while all other frequencies are cut-off. Thus, there are only a few frequencies that resonate. As a result, the output of a laser consists of a few closely spaced frequencies as shown in Fig. 22.21 (b). Thus, the laser emission line transforms into a series of narrow spectral lines corresponding to cavity modes. If $\delta\lambda$ is the line width of the emission line, then the number of modes that would be ultimately present is given by

$$N = \frac{\delta\lambda}{\Delta\lambda} \tag{22.45}$$

22.13 TRANSVERSE MODES

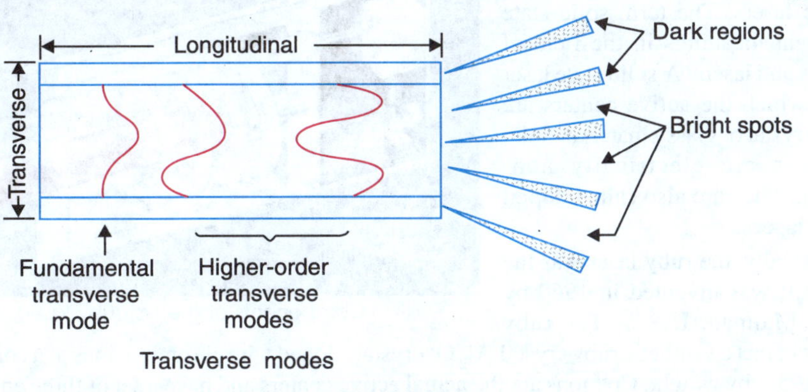

Fig. 22.22.

The longitudinal dimension L of the resonant cavity governs the axial modes. The laser modes (see Fig. 22.22) governed by the cross-sectional dimension of the optical cavity are called **transverse electromagnetic (TEM) modes**. The TEM modes are generally few in numbers and they are easy to see. If the laser beam is spread out by a negative lens and focussed on to a screen, several bright patches are seen on the screen. The patches are separated by intervals called nodal lines. The transverse modes characterize the intensity distribution across the cross-section of the laser beam. In general, the allowed modes are designated as TEM_{mn}, where m and n are integers. The integers m and n represent the number of intensity minima in two orthogonal directions of the laser beam.

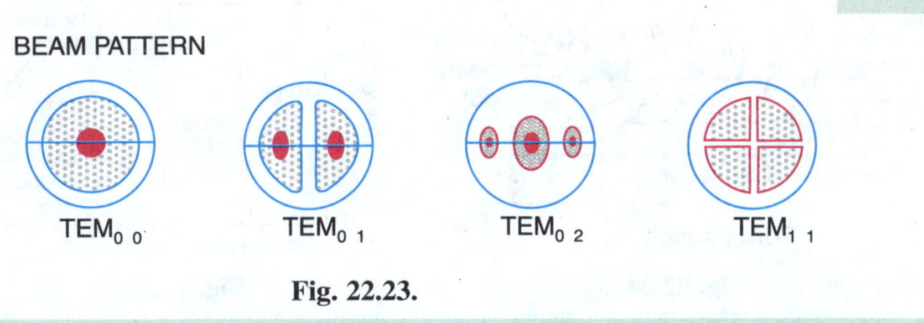

Fig. 22.23.

The lowest order transverse mode is TEM_{00}. It is the simplest mode and has a smooth cross-section profile with a peak in the middle (see Fig. 22.23.). TEM_{01} beam has a single minimum dividing the beam into two bright spots. A TEM_{11} beam has two perpendicular minima dividing the beam into four quadrants, and so on. Operation of a laser in multimode form provides considerably more power than in single mode operation.

22.14 TYPES OF LASERS

There are several ways in which we can classify lasers into different types. We prefer here to classify the lasers on the basis of the material used as active medium. Accordingly, they are broadly divided into four categories, namely solid state lasers, gas lasers, liquid lasers, and semiconductor diode lasers. Most lasers emit light in the red or IR regions. Lasers work in continuous mode or in a pulsed mode.

22.14.1 RUBY LASER

Ruby laser belongs to the class of solid state lasers. The term solid state has different meanings in the field of electronics and lasers. A solid state laser is one in which the active centers are fixed in a crystal or glassy material. Solid state lasers are electrically non-conducting. They are also called **doped insulator lasers**.

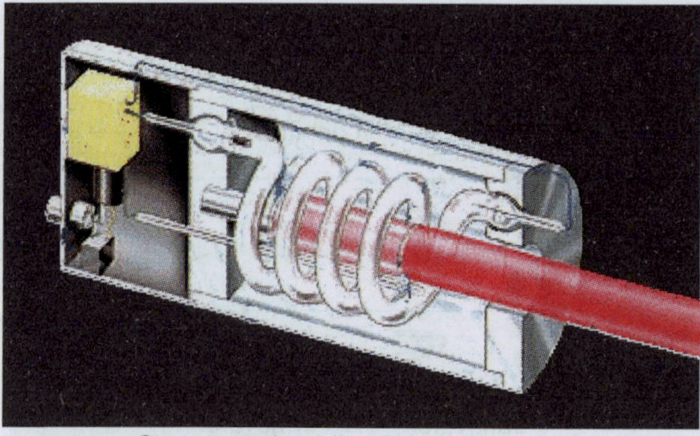

Components of the first ruby laser.

Historically, the ruby laser was the first laser. It was invented in 1960 by Theodore Maiman, U.S.A. The ruby laser rod is in fact a synthetic ruby crystal, Al_2O_3 crystal, doped with chromium ions at a concentration of about 0.05% by weight. Cr^{3+} ions are the actual active centers and have a set of three energy levels suitable for realizing lasing action whereas aluminum and oxygen atoms are inert.

Construction: The schematic of a ruby laser is shown in Fig.22.24. Ruby rod is taken in the form of a cylindrical rod of about 4 cm in length and 0.5 cm in diameter. Its ends are grounded and polished such that the end faces are exactly parallel and are also perpendicular the axis of the rod. One face is silvered to achieve 100% reflection while the other is silvered to give 10% transmission and 90% reflection. The silvered faces constitute the Fabry-Perot resonator. The laser rod is surrounded by a helical photographic flash lamp filled with xenon. Whenever activated by the power supply the lamp produces flashes of white light.

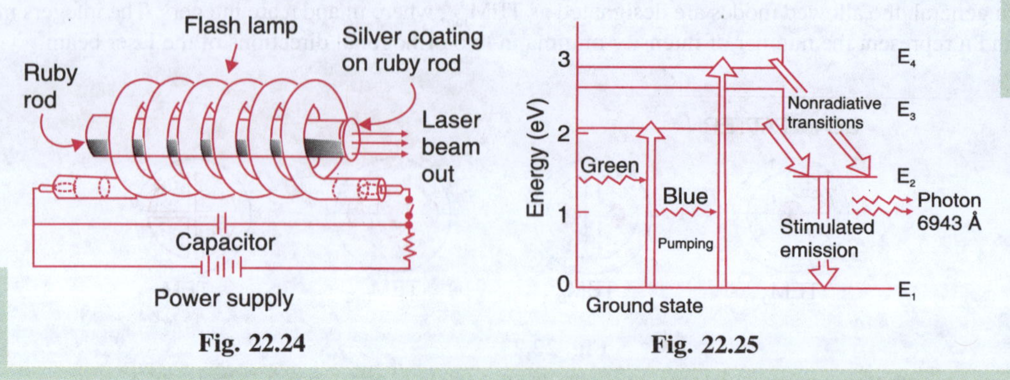

Fig. 22.24 Fig. 22.25

Working: Ruby laser uses a three – level pumping scheme. The energy levels of Cr^{3+} ions in the crystal lattice are shown in Fig. 22.25. There are two wide energy bands E_3 and E_3' and a pair of closely spaced levels at E_2. When the flash lamp is activated, the xenon discharge generates an intense burst of white light lasting for a few milliseconds. The Cr^{3+} ions are excited to the energy bands E_3 and E_3' by the green and blue components of white light. the energy levels in these bands have a very small lifetime ($\approx 10^{-9}$ s). Hence the excited Cr^{3+} ions rapidly lose some of the energy to the crystal lattice and undergo non–radiative transitions. They quickly drop to the levels E_2. The pair of levels at E_2 are metastable states having a lifetime of approximately 1000 times more than the lifetime of E_3 level. Therefore, Cr^{3+} ions accumulate at E_2 level. When more than half of the Cr^{3+} ion population accumulates at E_2 level, the state of population inversion is established between E_2 and E_1 levels. A chance photon emitted spontaneously by a Cr^{3+} ion initiates a chain of stimulated emissions by other Cr^{3+} ions in the metastable state. Red photons of wavelength 6943 Å travelling along the axis of the ruby rod are repeatedly reflected at the end mirrors and light amplification takes place. A strong intense beam of red light emerges out of the front-end mirror.

Note that the green and blue components of light play the role of pumping agents and are responsible for causing population inversion. The spontaneous photons of $\lambda = 6943$ Å, corresponding to red colour, act as the input of the oscillator which actually gets amplified. The xenon flash lasts for a few milliseconds. However, the laser does not operate throughout this period. Its output occurs in the form of irregular pulses of microsecond duration. It is because the stimulated transitions occur faster than the rate at which population inversion is maintained in the crystal. Once stimulated transitions commence, the metastable state E_2 gets depopulated very rapidly and at the end of each small pulse, the population at E_2 has fallen blow the threshold value required for sustained emission of light. As a result the lasing ceases and laser becomes inactive. The next pulse appears after the population inversion is once again restored. The process repeats.

22.14.2 ND: YAG LASER

Nd: YAG laser is one of the most popular types of solid state laser. It is a four-level laser. Yttrium aluminium garnet, $Y_3Al_5O_{12}$, commonly called YAG is an optically isotropic crystal. Some of the Y^{3+} ions in the crystal are replaced by neodymium ions, Nd^{3+}. Doping concentrations are typically of the order of 0.725% by weight. The crystal atoms do not participate in the lasing action but serve as a host lattice in which the active centres, namely Nd^{3+} ions reside.

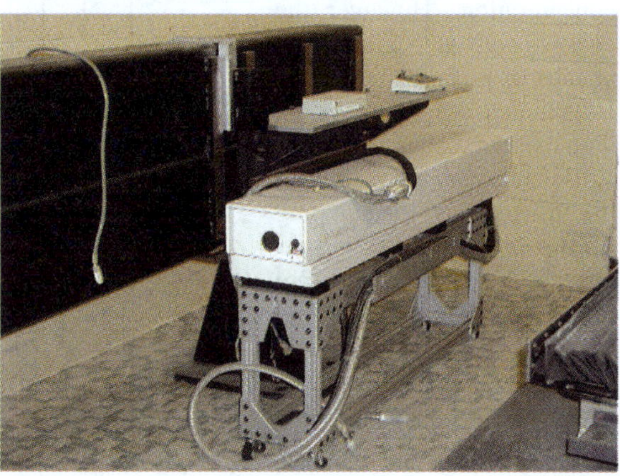

YAG Laser.

Construction: Fig. 22.26 illustrates a typical design of Nd: YAG laser. The system consists of an elliptically cylindrical reflector housing the laser rod along one of its focus line and a flash lamp along the other focus line. The light leaving one focus of the ellipse will pass through the other focus after reflection from the silvered surface of the reflector. Thus the entire flash lamp radiation gets focused on the laser rod. The YAG crystal rods are typically of 10 cm in

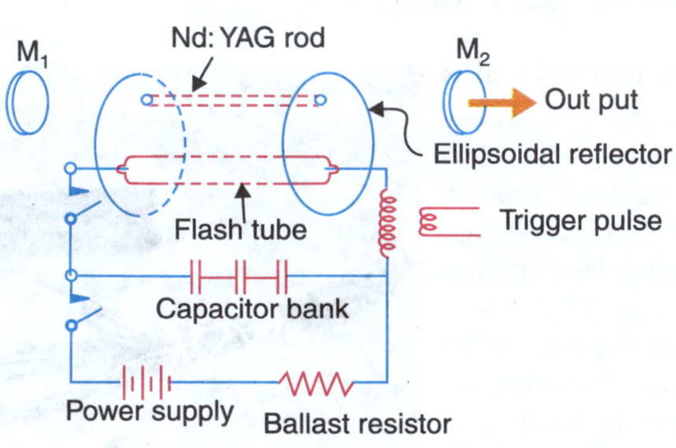

Fig. 22.26

length and 12 mm in diameter. The two ends of the laser rod are polished and silvered and constitute the optical resonator.

Working: A simplified energy level diagram for the neodymium ion in YAG crystal is shown in Fig. 22.27. The energy level structure of the free neodymium atom is preserved to a certain extend because of its relatively low concentration. However, the energy levels are split and the structure is complex. It is essentially a four-level system with the terminal laser level E_2 sufficiently far removed from the ground level. The pumping of the Nd^{3+} ions to upper states is done by a krypton arc lamp. The optical pumping with light of wavelength range to 5000 to 8000Å excites the ground state Nd^{3+} ions to the multiple energy levels at E_4. The metastable level E_3 is the upper laser level, while the E_2 forms the lower laser level. The upper laser level E_3 will be rapidly populated, as the excited Nd^{3+} ions quickly make downward transitions from the upper energy bands. The lower laser level E_2 is far above the ground level and hence it can not be populated by Nd^{3+} ions through thermal transitions from the ground level. Therefore, the population inversion is readily achieved between the E_3 level and E_2 level. The laser emission occurs in infrared (IR) region at a wavelength of about 10,600Å (1.06 μm). As the laser is a four-level laser, the population inversion can be maintained in the face of continuous laser emission. Thus Nd: YAG laser can be operated in CW mode. An efficiency of better than 1% is achieved. Nd: YAG lasers find many industrial applications such as resistor trimming, machining operations like welding, hole drilling etc. They are also used in surgery.

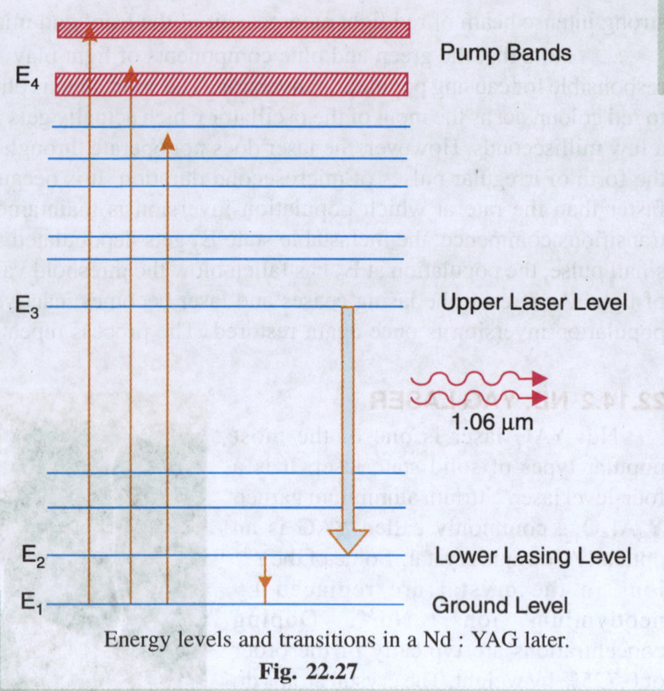

Energy levels and transitions in a Nd : YAG later.

Fig. 22.27

22.14.3 HELIUM-NEON LASER

Gas lasers are the most widely used lasers. They range from the low power helium-neon laser used in college laboratories to very high power carbon dioxide laser used in industrial applications. These lasers operate with rarefied gases as the active media and are excited by an electric discharge. In gases, the energy levels of atoms involved in the lasing process are narrow and as such require sources with sharp

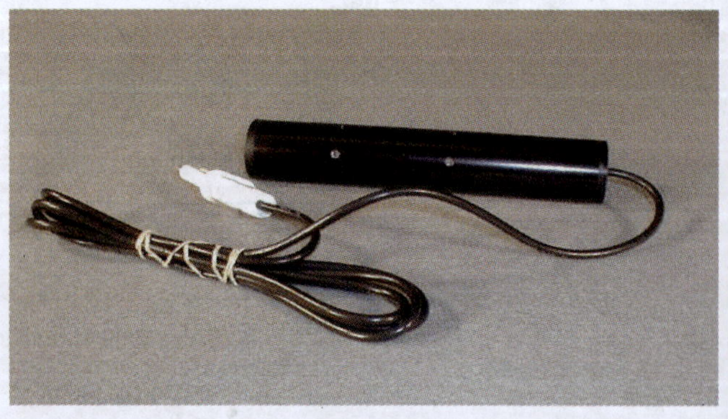
Helium-Neon Laser.

wavelength to excite atoms. Finding an appropriate optical source for pumping poses a problem. Therefore optical pumping is not used in gas lasers. The most common method of exciting gas laser medium is by passing an electric discharge through the gas. Electrons present in the discharge transfer energy to atoms in the laser gas by collisions.

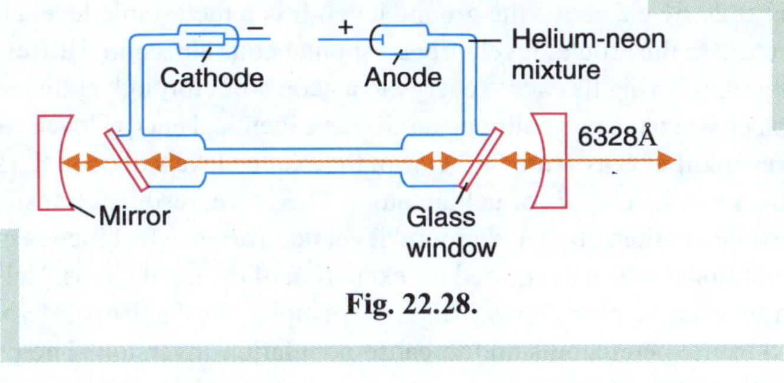

Fig. 22.28.

The first gas laser was He-Ne laser which was invented in 1961 by Ali Javan, William R. Bennett, Jr. and Donald R.Herriott.

Construction: The schematic of a He-Ne laser is shown in Fig.22.28. Helium – Neon laser consists of a long discharge tube filled with a mixture of helium and neon gases in the ratio 10: 1. Neon atoms are the active centers and have energy levels suitable for laser transitions while helium atoms help in exciting neon atoms. Electrodes are provided in the discharge tube to produce discharge in the gas. They are connected to a high voltage power supply. The tube is hermetically sealed by inclined windows arranged at its two ends. On the axis of the tube, two mirrors are arranged externally which form the Fabry-Perot optical resonator. The distance between the mirrors is adjusted to be m λ /2 such that the resonator supports standing wave pattern.

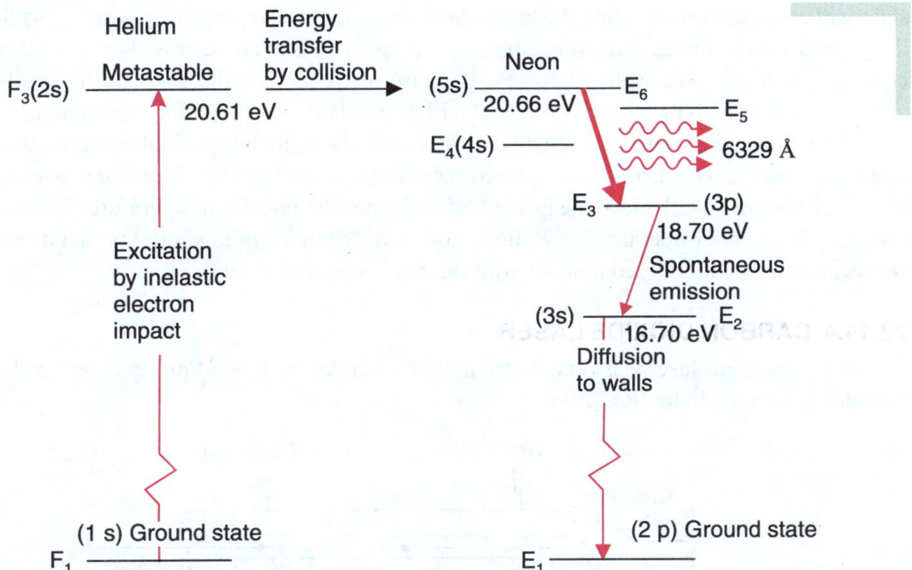

Energy level diagram for a helium-neon laser. Only the relevant energy levels are shown.

Fig. 22.29

Working: Helium – Neon laser employs a four – level pumping scheme. The energy levels of helium and neon are shown in Fig.22.29. When the power is switched on, a high voltage of about 10 kV is applied across the gas. It is sufficient to ionize the gas. The electrons and ions produced in the process of discharge are accelerated towards the anode and cathode respectively. The energetic electrons excite helium atoms through collisions. One of the excited levels of helium F_3 (2s) is at 20.61 eV above the ground level. It is a metastable level and the excited helium atom cannot return to the ground level through spontaneous emission. However, it can return to the ground level by transferring its excess energy to a neon atom through collision. Such an energy transfer can take place when the two colliding atoms have identical energy levels. Such an energy transfer is known as **resonant energy transfer**. One of the excited levels of neon E_6 (5s) is at 20.66eV, which is nearly at the same level as F_3 of helium atom. Therefore, resonant transfer of energy can occur between the excited helium atom and ground level neon atom. The kinetic energy of helium atoms provides the additional 0.05 eV required for excitation of the neon atoms. Helium atoms drop to the ground state after exciting neon atoms. This is the pumping mechanism in He-Ne laser. The role of helium atoms is to excite neon atoms and to cause population inversion. The probability of energy transfer from helium atoms to neon atoms is more, as there are 10 helium atoms per 1 neon atom in the gas mixture. The probability of reverse transfer of energy from neon to helium atom is negligible.

The upper state of neon atom E_6 is a metastable state. Therefore, neon atoms accumulate in this upper state. The E_3 (3p) is sparsely populated at ordinary temperatures, and a state of population inversion is readily established between E_6 and E_3 levels. Random photons emitted spontaneously prompt stimulated emission and lasing occurs. The transition $E_6 \rightarrow E_3$ generates a laser beam of red colour of wavelength 6328 Å. Other possible transitions produce 3.39 μm and 1.15 μm laser beams respectively. These transitions are not shown in Fig. 22.29.

From the level E_3 the neon atoms drop to E_2 (3s) level spontaneously. E_2 level is however a metastable state. Consequently, neon atoms tend to accumulate at E_2 level. It is necessary that these atoms are brought to the ground state E_1(2p) quickly; otherwise the number of atoms at the ground state will go on diminishing and the laser ceases to function. The only way of bringing the atoms to the ground state is through collisions. If the discharge tube is made narrow, the probability of atomic collisions with the tube walls increases. Because of frequent collisions with the walls, the neon atoms rapidly drop to the ground level and will be available for excitation once again.

If the diameter of the discharge tube is increased, the probability of collisions of atoms with the walls decreases and the neon atoms tend to accumulate at energy level E_2. In due course of time, the atoms are no more available at the ground level for further excitation. Therefore, the laser ceases to operate. He-Ne laser operates in cw mode and is widely used in laboratories as a monochromatic source. It is also widely used in laser printing, bar code reading, etc.

22.14.4 CARBON DIOXIDE LASER

The carbon gas laser is a very useful and efficient laser. It is a four-level molecular laser and operates at 10.6 μm in far IR region.

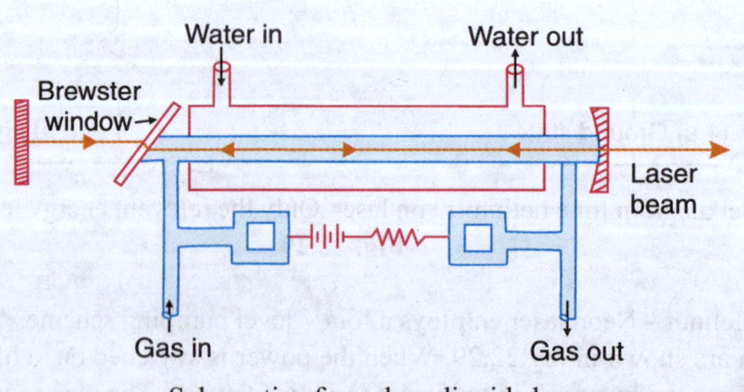

Schematic of a carbon dioxide laser

Fig. 22.30

Construction: The schematic of typical CO_2 laser is shown in Fig. 22.30. It is basically a discharge tube having a bore of cross section of about 1.5 mm^2 and a length of about 260 mm. The discharge tube is filled with a mixture of carbon dioxide, nitrogen and helium gases in 1:4:5 proportions respectively. Other additives such as water vapour are also added. The active centres are CO_2 molecules lasing on the transitions between the vibrational levels of the electronic ground state.

Energy levels of CO_2 molecule

The electron energy levels of an isolated atom are discrete and narrow. However, in case of molecules the energy spectrum is complicated due to many additional features. Each electron energy level is associated with nearly equally spaced vibrational levels and each vibrational level in turn has a number of rotational levels.

Carbon Dioxide Laser.

CO_2 molecule is a linear molecule consisting of a central carbon atom with two oxygen atoms attached one on either side. It undergoes three independent vibrational oscillations known as the **vibrational modes.** These vibrational degrees of freedom are quantized. At any one time, a CO_2 molecule can vibrate in a linear combination of three fundamental modes. The energy states of the molecule are then represented by three quantum numbers (m n q). These numbers represent the amount of energy associated with each mode. For example, the number (020) indicates that the molecule in this energy state is in the pure bending mode with two units of energy. Each vibrational state is associated with rotational states corresponding to the rotation of CO_2 molecule about its centre of mass. The separations between vibrational – rotational states are much smaller on the energy scale compared to the separations between electron energy levels. The nitrogen molecule N_2 is also characterized by similar vibrational levels. Fig. 22.31 shows the vibrational modes and rotations of CO_2 molecule.

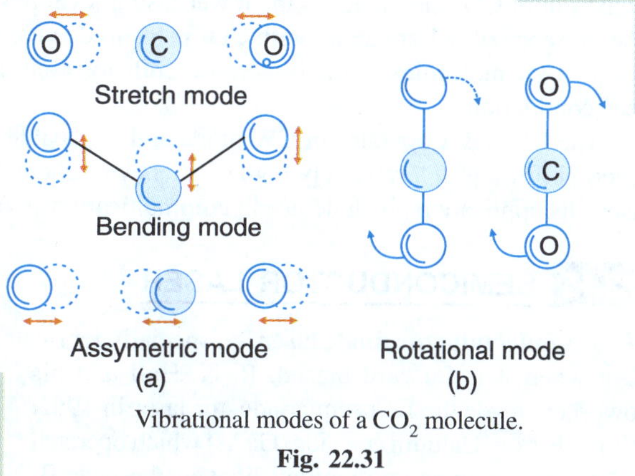

Vibrational modes of a CO_2 molecule.

Fig. 22.31

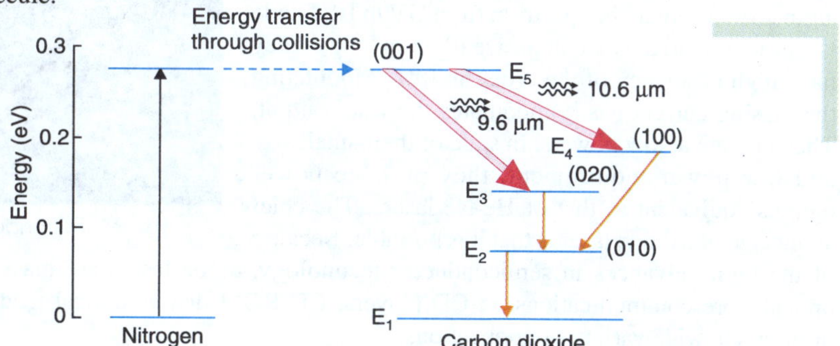

Energy levels of nitrogen and carbon dioxide molecules and transitions between the levels

Fig. 22.32.

Working: Fig.22.32 shows the lowest vibrational levels of the ground electron energy state of CO_2 molecule and an N_2 molecule. The excited state of an N_2 molecule is metastable and it is identical in energy to (001) vibrational level of CO_2 molecule, indicated as E_5 in Fig. 22.32. As current passes through the mixture of gases, the N_2 molecules get excited to the metastable state. The excited N_2 molecules cannot spontaneously lose their energy and consequently, the number of N_2 molecules at the level keeps on increasing. The N_2 molecules return to ground state through inelastic collisions with ground state CO_2 molecules. Thereby the CO_2 molecules are excited to E_5 level. Some of the CO_2 molecules are also excited to the upper level E_5 through collisions with electrons. The excitation of CO_2 molecules through collisions with excited N_2 molecules is similar to that of helium atoms by neon atoms in helium-neon laser. The E_5 level is the upper lasing level while the (020) and (100) states marked as E_3 and E_4 levels act as the lower lasing levels. As the population of CO_2 molecules builds up at E_5 levels, population inversion is achieved between E_5 level and the levels at E_4 and E_3. The laser transition between $E_5 \rightarrow E_4$ levels produces far IR radiation at the wavelength 10.6 μm (1,06,000Å). The lasing transition between $E_5 \rightarrow E_3$ levels produces far IR radiation at 9.6 μm (96,000Å) wavelength. E_3 and E_4 levels are also metastable states and the CO_2 molecules at these levels fall to the lower level E_2 through inelastic collisions with normal (unexcited) CO_2 molecules. This process leads to accumulation of population at E_2 level. And also, as the gaseous mixture heats up, the E_2 level, which is close to the ground state, E_1 tends to be populated through thermal excitations. Thus, the de-excitation of CO_2 molecules at the lower lasing level poses a problem and inhibits the laser action. The presence of helium along with CO_2 helps to decrease the population density at E_2 level. It de-excites CO_2 molecules through inelastic collisions and aids cooling the gaseous mixture through heat conduction.

The CO_2 laser operates in CW mode and is capable of generating high powers of the order of several kilowatts at a relatively high efficiency of about 40%. Therefore, it is the most widely used laser. Its applications include use in communications, weaponry and laser fusion.

22.15 SEMICONDUCTOR LASER

A **semiconductor diode laser** is a specially fabricated pn junction device, which emits coherent light when it is forward biased. R. N. Hall and his coworkers made the first semiconductor laser in 1962. It is made from Gallium arsenide (GaAs) which operated at low temperatures and emitted light in the near IR region. Semiconductor lasers working at room temperature and in continuous wave mode are produced by 1970. Now pn-junction lasers are made to emit light almost anywhere in the spectrum from UV to IR. Diode lasers are remarkably small in size (0.1mm long). They have high efficiency of the order of 40%. Modulating the biasing current easily modulates the laser output. They operate at low powers. In spite of their small size and low power requirement, they produce power outputs equivalent to that of He-Ne lasers. The chief advantage of a diode laser is that it is portable. Because of the rapid advances in semiconductor technology, diode lasers are mass produced for use in optical fibre communications, in CD players, CD-ROM drives, optical reading, high speed laser printing etc wide variety of applications.

Semiconductor Laser.

A **semiconductor** is a material with electrical properties intermediate to those of a conductor and an insulator. The allowed energy values of the valence electrons in semiconductors occur within two well-defined energy bands separated by an energy gap known as **band gap**. A pure semiconductor crystal has exactly enough electrons to fill all the states in the lower band, namely **valence band**. However, when a covalent bond is just broken, an electron is just set free. Then we say that the electron jumped into the upper band, namely **conduction band**. The electron jumping to the conduction band leaves behind a vacancy in the valence band. The vacancy is called a **hole** and is assigned a positive charge and a mass equivalent to that of an electron. In a pure semiconductor, for each covalent bond broken an electron and a hole are generated. Therefore, the number of electrons in the conduction band and the number of holes in the valence band are equal. When a conduction electron falls into the valence band, it recombines with a hole there. The electron rejoins the broken covalent bond and therefore both the electron and hole disappear. The recombination energy is released in the form of heat in silicon and germanium crystals. In some crystals it is released in the form of light.

Doping with small amounts of impurities can drastically increase the electrical conductivity of a pure semiconductor. When the dopant is a pentavalent element, each dopant atom contributes an electron to the conduction band without creating a hole simultaneously in the valence band. Hence the addition of the pentavalent element increases the number of conduction electrons which become the **majority carriers** in the silicon crystal. As negatively charged electrons are current carriers in this crystal, it is called a **n-type semiconductor**. On the other hand, a trivalent dopant atom produces a hole in the valence band without the simultaneous generation of electron in the conduction band. Hence the addition of the trivalent element increases the number of holes which become the *majority carriers* in the silicon crystal. As positively charged holes are current carriers in this crystal, it is called a **p-type semiconductor**.

There is a reference level in the energy band diagram of each type of semiconductor. The reference level is called the **Fermi level**. The Fermi level E_{Fp} is nearer to the top of the valence band in the p-type material and the Fermi level E_{Fn} is nearer to the bottom of the conduction band in the n-type material. When the p-type and n-type materials are joined at the atomic level to form a pn-junction device, equilibrium is attained only when equalization of Fermi levels takes place. The energy levels in p-region move up and those in n-region move down till the Fermi levels (E_{Fp} and E_{Fn}) in both the regions come to the same level. The mutual displacement of the energy levels on both sides of the junction causes a bending of the energy bands around the junction.

Achieving population inversion in a semiconductor:

Population inversion is required for producing stimulated emission. The way in which population inversion is achieved in semiconductors is very different from the way it is established in other types of lasers. A semiconductor cannot be regarded as two-level atomic system. It consists of electrons and holes distributed in the respective energy bands. Therefore, laser action in semiconductors involves energy bands rather than discrete levels. Secondly, in other types of lasers, population inversion is obtained by exciting electrons in spatially isolated atoms. In semiconductors, electrons are not associated with specific atoms but are injected into the conduction band from the external circuit. Therefore, the conduction band plays the role of excited level while the valence band plays the role of ground level. Population inversion requires the presence of a large concentration of electrons in the conduction band and a large concentration of holes in the valence band. A simple way to achieve population inversion is to use a semiconductor in the form of a pn-junction diode formed from heavily doped p- and n-type semiconductors.

22.15.1 PN-JUNCTION LASER

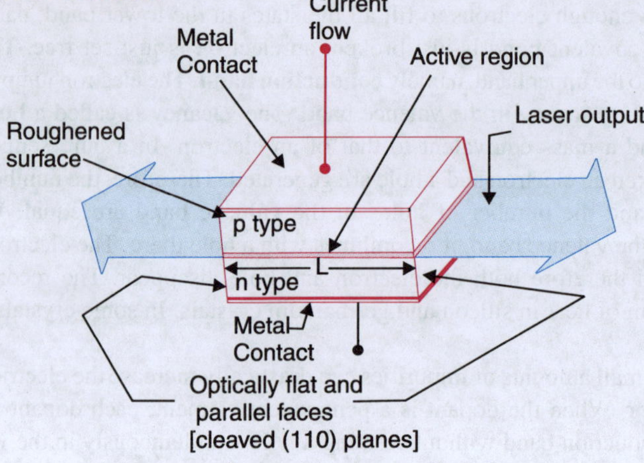

Schematic of a semiconductor diode laser

Fig. 22.33

Construction : Fig.22.33 shows the schematic of a semiconductor laser. A simple diode makes use of the same semiconductor material, say, GaAs on both sides of the junction. Starting with a heavily doped n-type GaAs material, a p-region is formed on its top by diffusing zinc atoms into it. A heavily zinc doped layer constitutes the heavily doped p-region. The diode is extremely small in size. Typical diode chips are 500 μm long and about 100 μm wide and thick. The top and bottom faces are metallized and metal contacts are provided to pass current through the diode. The front and rear faces are polished parallel to each other and perpendicular to the plane of the junction. The polished faces constitute the Fabry-Perot resonator. In practice there is no necessity to polish the faces. A pair of parallel planes cleaved at the two ends of the pn junction provides the required reflection to form the cavity. The two remaining sides of the diode are roughened to eliminate lasing action in that direction. The entire structure is packaged in small case which looks like the metal case used for discrete transistors.

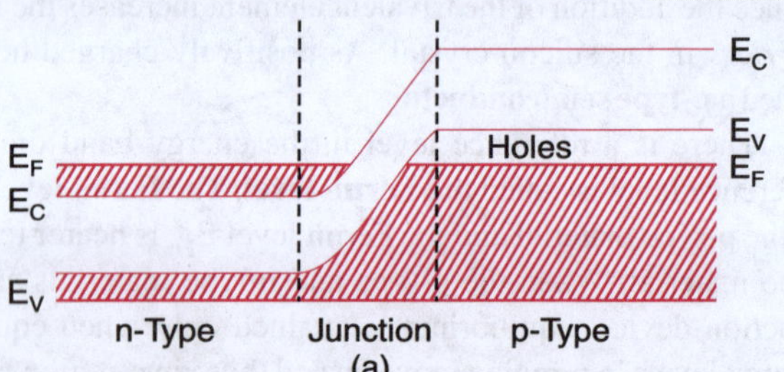

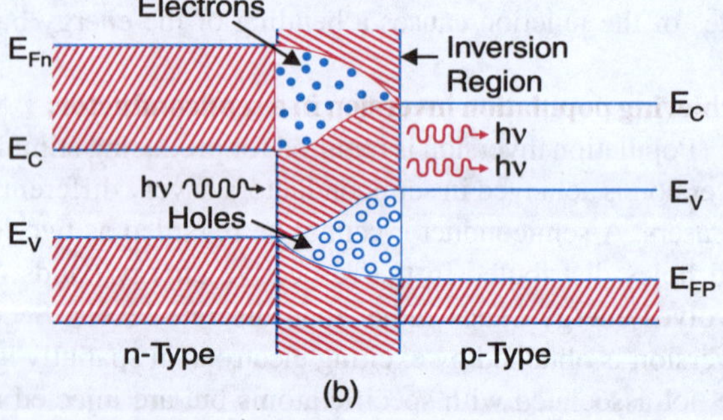

Energy band structure of a semiconductor diode (*a*) Heavily doped pn-junction without bias. (*b*) Heavily doped pn-junction forward biased above threshold value

Fig. 22.34

Working : The energy band diagram of a heavily doped pn junction is shown in Fig.22.34 (a). Because of very high doping on n-side, the donor levels are broadened and extend into the conduction band. The Fermi level also is pushed into the conduction band. Electrons occupy the portion of the conduction band lying below the Fermi level. Similarly, on the heavily doped p-side the Fermi level lies within the valence band and holes occupy the portion of the valence band that lies above the Fermi level. At thermal equilibrium, the Fermi level is uniform across the junction.

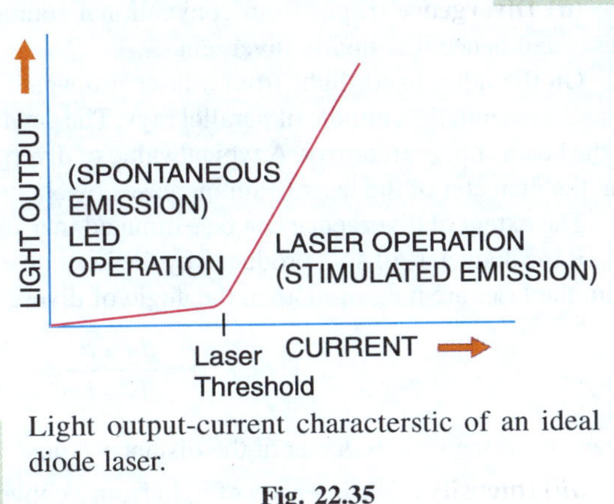

Light output-current characterstic of an ideal diode laser.

Fig. 22.35

When the junction is forward-biased, electrons and holes are injected into the junction region in high concentrations. In other words, carriers are *pumped* by the dc voltage source. At low forward current level, the electron-hole recombination causes spontaneous emission of photons and the junction acts as an LED. As the forward current through the junction is increased the intensity of the light increases linearly. However, when the current reaches a threshold value (see Fig.22.35), the carrier concentrations in the junction region will rise to a very high value. As a result, the region (region'd' in Fig.22.34 b) contains a large concentration of electrons within the conduction band and *simultaneously* a large number of holes within the valence band. Holes represent absence of electrons. Thus, the upper energy levels in the narrow region are having a high electron population while the lower energy levels in the same region are vacant. Therefore, the condition of population inversion is attained in the narrow junction region. This narrow zone in which population inversion occurs is called an **inversion region** or **active region**. Chance recombination acts of electron and hole pairs lead to emission of spontaneous photons. The spontaneous photons propagating in the junction plane stimulate the conduction electrons to jump into the vacant states of valence band. This stimulated electron-hole recombination produces coherent radiation. GaAs laser emits light at a wavelength of 9000 Å in IR region.

22.16 LASER BEAM CHARACTERISTICS

The important characteristics of a laser beam are:

(*i*) directionality
(*ii*) negligible divergence
(*iii*) high intensity
(*iv*) high degree of coherence and
(*v*) high monochromaticity.

(*i*) Directionality: The conventional light sources emit light uniformly in all directions. When we need a narrow beam in a specific direction, we obtain it by placing a slit in front of the source of light.

In case of laser, the active material is in a cylindrical resonant cavity. Any light that is travelling in a direction other than parallel to the cavity axis is eliminated and only the light that is travelling parallel to the axis is selected and reinforced. Light propagating along the axial direction emerges from the cavity and becomes the laser beam. Thus, a laser emits light only in one direction.

(ii) Divergence: Light from conventional sources spreads out in the form of spherical wave fronts and hence it is highly divergent.

On the other hand, light from a laser propagates in the form of plane waves. The light beam remains essentially a bundle of parallel rays. The small divergence that exists is due to the diffraction of the beam at the exit mirror. A typical value of divergence of a He-Ne laser is 10^{-3} radians. It means that the diameter of the laser beam increases by about 1 mm for every meter it travels.

The extent of divergence can be estimated in a simple way as follows:

If the diameters of spot produced by the laser on a screen which is held at two different distances from the laser are measured, then the angle of divergence is given by

$$\phi = \frac{d_2 - d_1}{l_2 - l_1} \tag{22.46}$$

where d_1 is the spot diameter at the distance l_1 and d_2 is the spot diameter at the distance l_2.

(iii) Intensity : The intensity of light from a conventional source decreases rapidly with distance as it spreads out in space. Laser emits light in the form of a narrow beam with its energy concentrated in a small region of space. Therefore, the beam intensity would be tremendously large and stays constant with distance. The intensity of a laser beam is approximately given by

$$I = \left[\frac{10}{\lambda}\right]^2 \mathrm{P} \tag{22.47}$$

where P is the power radiated by the laser.

To obtain light of same intensity from a tungsten bulb, it would have to be raised to a temperature of 4.6×10^6 K.

(iv) Coherence : The light that emerges from a conventional light source is a jumble of short wave trains which combine with each other in a random manner. The resultant light is incoherent. Coherence length is one of the parameters used as a measure of coherence. In case of a laser a large number of identical photons are emitted through stimulated emissions and therefore they will be in phase with each other. The resultant light exhibits a high degree of coherence.

The coherence length of light from a sodium lamp, which is a traditional monochromatic source, is of the order of 0.3 mm. On the other hand the coherence length of light emitted by an ordinary helium-neon laser is about 100 m.

(v) Monochromaticity: If light coming from a source has only one frequency (single wavelength) of oscillation, the light is said to be *monochromatic* and the source a *monochromatic source.* Light from traditional monochromatic sources spreads over a wavelength range of 100 Å to 1000 Å. On the other hand, the light from lasers is highly monochromatic and contains a very narrow range of a few angstroms (< 10 Å).

22.19. APPLICATIONS

Lasers find application in almost every field. They are used in mechanical working, industrial electronics, entertainment electronics, communications, information processing, and even in wars to guide missiles to the target. Lasers are used in CD players, laser printers, laser copiers, optical floppy discs, optical memory cards etc. We discuss here a few of the applications in industry.

The large intensity that is possible in the focused output of a laser beam and its directionality makes laser an extremely useful tool for a variety of industrial applications.

Welding: Welding is the joining of two or more pieces into a single unit. If we consider welding of two metal plates, the metal plates are held in contact at their edges and a laser beam is made to move along the line of contact of the plates. The laser beam heats the edges of the two plates to their melting points and causes them to fuse together where they are in contact. The main advantage of the laser welding is that it is a contact-less process and hence there is no possibility of introduction of impurities into the joint. In the process, the work-pieces do not get distorted, as the total amount of input is very small compared to conventional welding processes. The heat-effected zone is relatively small because of rapid cooling. Laser welding can be done even at difficult to reach place. CO_2 lasers are used in welding thin sheets and foils.

Drilling: The principle underlying drilling is the vaporization of the material at the focus of the beam. With lasers, one can drill holes as small as 10 μm in diameter. For drilling, the energy must be supplied in such a way that rapid evaporation of material takes place without significant radial diffusion of heat into the work piece. The vaporized material is removed with the help of a gas jet. Pulsed ruby and neodymium lasers are commonly used for drilling holes of small l / D ratio, where l is the thickness of the work and D is the hole diameter.

Hardening: Heat treatment is the process, which is done for sometime to harden metals and certain other materials. Heat treatment is common in the tooling and automotive industry. Heat-treating converts the surface layer to a crystalline state that is harder and more resistant to wear. In general CO_2 lasers of about 1 kW output power operating in cw mode are used for heat treatment. As metals are more reflecting at IR frequencies, a heat absorbing coating such as graphite or zinc phosphate is applied on the surface of the work piece to help it absorb laser energy more efficiently. Laser heat treatment requires a low amount of energy input to the work piece. Laser processing advantageous as it can provide selective treatment of the desirable areas. Heat treatment is used to strengthen cylinder blocks, gears, camshafts etc in the automobile industry. As the method is a non-contact method, stress is not induced in the work-pieces.

Electronics Industry: Electronics industry uses lasers in the manufacture of electronic components and integrated circuits. Lasers have been used to perforate and divide silicon slices having several hundred circuits. They are also used for the isolation of faulty components in a large integrated circuit by disconnecting the conducting paths by evaporation. Trimming of thick and thin film resistors using lasers is a very common application.

Measurement of atmospheric pollutants: Laser is a very useful tool for the measurement of the concentrations of various atmospheric pollutants such as N_2, CO, SO_2 etc gases and particulate matter such as dust, smoke and flyash. Conventional methods of pollution measurements require that samples of pollutants are to be collected for chemical analysis. Therefore, these methods cannot give real-time data. In contrast, laser methods permit measurements by remotely sensing the composition of atmosphere without the necessity of sample collection or chemical processing.

In one of the laser techniques, the light scattered by pollutants is studied. A pulsed laser is used as the source of light and the light scattered back is detected by a photodetector. The distance to particulate matter and the concentrations of particulate matter is obtained in this method. The distance is inferred from the time that light takes to travel up to the pollutant region and to return back. This technique is known as LIDAR which stands for light detection and ranging. The principle is very much similar to that of RADAR. The method helps in determining the concentration of particulate matter as a function of distance. However, this method cannot provide any information regarding the nature of the scattering particles. It is mainly useful in knowing the distribution of atmospheric pollutants in different vertical sections and in monitoring their variations. Environmental agencies measure concentrations of harmful gases such as SO_2 and NO_2 using this method.

Another technique uses study of absorption of light beam by pollutants. The existence of specific gases in the atmosphere is detected using absorption spectroscopy techniques. A laser beam is transmitted through polluted sample and the attenuation of intensity of light due to absorption in the sample is detected and recorded. Each chemical absorbs light of characteristic wavelengths and from the absorption spectrum, its existence can be inferred.

A third method uses Raman effect to detect the pollutants. The Raman effect involves scattering of light by gas molecules accompanied by a shift in the wavelength of light. Raman shifts are characteristic of each molecular species. Hence, analysis of backscattered laser light reveals the constituents of the gas sample. The ozone concentration high in the atmosphere is determined using this technique.

WORKED-OUT EXAMPLES

Example 22.1: Find the ratio of populations of the two states in a He-Ne laser that produces light of wavelength 6328Å at 27°C.

Solution: The ratio of population is given by $\frac{N_2}{N_1} = e^{-(E_2-E_1)/kT}$

$$E_2 - E_1 = \frac{12400}{6328} eV = 1.96 eV$$

$$\therefore \frac{N_2}{N_1} = \exp\left[\frac{-1.96\,\text{eV}}{(8.61\times10^{-5}\,eV)(300K)}\right] = e^{-75.88} = 1.1\times10^{-33}$$

Example 22.2: The wavelength of emission is 6000Å and the coefficient of spontaneous emission is 10^6/s. Determine the coefficient for the stimulated emission.

Solution: The coefficient for stimulated emission is given by

$$B_{21} = \frac{c^3}{8\pi\, h\nu^3 \mu^3} A_{21} = \frac{\lambda^3}{8\pi\, h} A_{21} \text{ (Taking } \mu = 1)$$

$$\therefore \quad B_{21} = \frac{\left(6000\times10^{-10}\right)^3 m^3 \left(10^6 / s\right)}{8\pi\times6.626\times10^{-34}\, Js} = 1.3\times10^{19}\,\text{m/kg}.$$

Example 22.3: At what temperature are the rates of spontaneous and stimulated emission equal? Assume $\lambda = 5000$Å.

Solution: If the rates of spontaneous and stimulated emission are equal, then

$$R_1 = \left[\frac{1}{e^{h\nu/kT} - 1}\right] = 1 \quad \text{or} \quad e^{h\nu/kT} = 2$$

As $\lambda = 5000$Å, $\nu = c/\lambda = 6\times10^{14}$Hz and

$$\frac{h\nu}{kT} = \frac{6.626\times10^{-34}\, Js\left(6\times10^{14}/s\right)}{\left(1.38\times10^{-23}\, J/K\right)T} = \frac{28.8\times10^3}{T} K$$

$$e^{h\nu/kT} = \exp\left[\frac{28.8\times10^3}{T} K\right] = 2$$

or

$$\frac{28.8\times10^3}{T} K = \ln 2 = 0.693$$

$$\therefore \quad T = \frac{28.8\times10^3}{0.693} K = \mathbf{41{,}558\ K.}$$

Example 22.4: The length of a laser tube is150 mm and the gain factor of the laser material is 0.0005/cm. If one of the cavity mirrors reflects 100% light that is incident on it, what is the required reflectance of the other cavity mirror?

Solution:
$$\gamma_{th} = \frac{1}{2L}\ln\frac{1}{r_1 r_2}$$

$$\therefore \quad r_2 = \frac{1}{r_1 e^{2L\gamma}} = \frac{1}{1\times e^{2\times15\times0.0005}} = 0.985$$

It means that the second mirror should have a reflectance of 98.5%.

Example 22.5: The half-width of the gain profile of a He-Ne laser material is 2×10^{-3} nm. If the length of the cavity is 30 cm, how many longitudinal modes can be excited? The emission wavelength of He-Ne laser is 6328 Å.

Solution: The separation between successive longitudinal modes is given by

$$\Delta\lambda = \frac{\lambda^2}{2L} = \frac{\left(6328\times10^{-10}\,m\right)^2}{2\left(30\times10^{-2}\,m\right)} = 0.66\times10^{-3}\text{ nm}$$

Number of modes
$$N = \frac{\delta\lambda}{\Delta\lambda} = \frac{2\times10^{-3}\,nm}{0.66\times10^{-3}\,nm} = \mathbf{3.}$$

QUESTIONS

1. Explain with neat diagram absorption, spontaneous emission and stimulated emission of radiation.
2. What is population inversion? Explain why laser action cannot occur without population inversion between atomic levels?
3. What do you understand by a negative temperature state? How can it be achieved?
4. Discuss the four-level (pumping) scheme for laser action.
5. What do you understand by an optical resonant cavity? Explain.
6. Why is the optical resonator required in lasers? Illustrate your answer with neat sketches.
7. What are the essential components of a laser? Explain their functions briefly.
8. Describe the working of solid state ruby laser.
9. Explain the principle and working of a He-Ne laser.
10. In helium – neon laser lasing is through neon gas. What is then the role of helium gas?
11. In helium – neon laser why is it necessary to use narrow tubes?
12. What is the reason for monochromaticity of laser beam?
13. With the help of energy band diagram discuss the working of a semiconductor laser.
14. Explain in brief the characteristics of a laser beam.
15. What is population inversion? Describe the construction and working of He-Ne laser. **(GNDU, Amritsar, 2010)**
16. What do you mean by non-radiative decay? **(GNDU, Amritsar, 2010)**
17. What is the role of He in He-Ne Laser ? **(GNDU, Amritsar, 2010)**
18. Write the rate equation for three level laser system. **(GNDU, Amritsar, 2010)**
19. Why heterojunction semiconductor lasers are preferred? **(GNDU, Amritsar, 2010)**
20. Discuss the absorption and amplification of a parallel beam light passing through a medium. **(GNDU, Amritsar, 2010)**
21. Explain the concept of the Longitudinal and Transverse modes. **(GNDU, Amritsar, 2010)**

22. Explain the terms Doppler and line broadening. **(GNDU, Amritsar, 2010)**
23. Discuss the working, construction and applications of CO_2 Laser with neat labeled diagrams. **(GNDU, Amritsar, 2010)**
24. Write the requirements for producing high inversion density for Q-Switching. **GNDU, Amritsar, 2010)**
25. Draw the electronic energy levels diagrams showing the output wavelength / frequency of Ruby and Nd:YAGI Lasers ? **(GNDU, Amritsar, 2010)**

PROBLEMS FOR PRACTICE

1. A pulsed laser is constructed with a ruby crystal as the active element. The ruby rod contains typically a total of 3×10^{19} Cr^{3+}ions. If the laser emits light at 6943 Å wavelength, find
   a. the energy of emitted photon (in eV)
   b. the total energy available per laser pulse(assuming total population inversion)

   [**Ans :** (a)E = 1.79 eV; (b) 8.6 J]
2. Find the relative populations of the two states in a ruby laser that produces a light beam of wavelength 6943Å at 300K. [**Ans :** 8×10^{-31}]
3. Find the ratio of populations of the two states in a He-ne laser that produces light of wavelength 6328Å at27°C. [**Ans :** 1.1×10^{-33}]
4. The He-Ne system is capable of lasing at several different IR wavelengths, the prominent one being 3.3913μ m. Determine the energy difference (in eV) between the upper and lower levels for this wavelength. [**Ans :** E = 0.37 eV]
5. The CO_2 laser is one of the most powerful lasers. The energy difference between the two laser levels is 0.117 eV. Determine the frequency and wavelength of the radiation.

   [**Ans :** λ=10.5 μ m. $\nu = 2.9 \times 10^{13}$Hz]
6. A laser beam can be focused on an area equal to the square of its wavelength. For a He-Ne laser, the wavelength of emitted light is 6328Å.If the laser radiates energy at the rate of 1 mW, find out the intensity of the focused beam. [**Ans :** $I = 2.5\times10^{15}$ W/m^2]
7. Compute the Doppler broadening for the 6328A laser transition in the He-Ne Laser, assuming a single isotope of Ne^{20} and that the laser operate at a discharge-bore temperature of 373 K. **(GNDU, Amritsar, 2010)**

23 CHAPTER

Holography

23.1. INTRODUCTION

Images of objects are generally obtained using photographic method. In this method a lens focuses the light reflected from a three-dimensional object onto a photographic film where a two-dimensional image of the object is formed. A negative is first obtained by developing the film and then a positive is obtained through printing. The positive print is a two-dimensional record of light intensity received from the object. It, thus, contains information about the square of the amplitude of the light wave that produced the image but information about the phase of the wave is not recorded and is lost.

In 1948 Dennis Gabor outlined a two-step lensless imaging process. It is radically a new technique of photographing the objects and is known as **wave front reconstruction**. The technique is also called

A white light reflection hologram.

At a Glance

- Introduction
- Principle of Holography
- Coaxial Holography
- Off-axis Holography
- Holograms
- Important Properties of Hologram
- Classification of Holograms Applications
- Medical Applications of Holography

holography. The word 'holography' is formed by combining parts of two Greek words: 'holos', meaning "whole", and 'graphein' meaning "to write". Thus holography means writing the complete image. Holography is actually a recording of interference pattern formed between two beams of coherent light coming from the same source. In this process both the amplitude and phase components of light wave are recorded on a light sensitive medium such as a photographic plate. The recording is known as a **hologram.** Holography required an intense coherent light source. Laser was not available when Gabor formulated the idea of holography. Holographic technique became a practical proposition only after the invention of lasers. Leith and Upatnicks prepared laser holograms for the first time. In this chapter we discuss the fundamental concept of holography.

23.2. PRINCIPLE OF HOLOGRAPHY

Holography is a two-step process. First step is the **recording** of hologram where the object is transformed into a photographic record and the second step is the **reconstruction** in which the hologram is transformed into the image. Unlike in the conventional photography, lens is not required in either of the steps. A hologram is the result of interference occurring between two waves, an object beam which is the light scattered off the object and a coherent background, the reference beam, which is the light reaching the photographic plate directly. In Gabor's original experiments, the reference beam and object beams were *coaxial*. Further advance was made by Leith and Upatnieks, who used the reference beam at an *offset angle*. That made possible the recording of holograms of three-dimensional objects.

The off-axis arrangement for generating and viewing holograms is described here.

23.2.1 RECORDING OF THE HOLOGRAM

In the off-axis arrangement a broad laser beam is divided into two beams, namely a **reference beam** and an **object beam** by a beam splitter (Fig.23.1). The reference beam goes directly to the photographic plate. The second beam of light is directed onto the object to be photographed. Each point of the object scatters the incident light and acts as the source of spherical waves. Part of the light, scattered by the object, travels towards the photographic plate. At the photographic plate the innumerable spherical waves from the object combine with the plane light wave from the reference

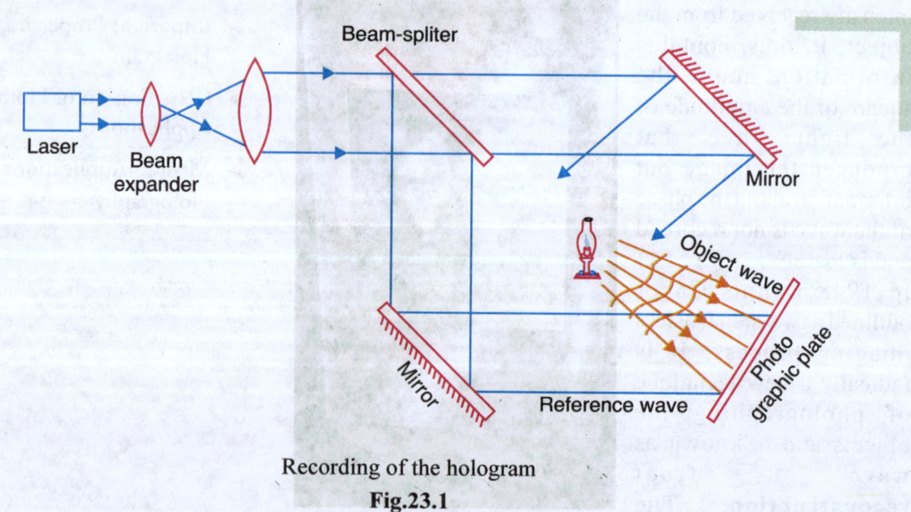

Recording of the hologram

Fig.23.1

beam. The sets of light waves are coherent because they are from the same laser. They interfere and form interference fringes on the plane of the photographic plate. These interference fringes are a series of zone-plate like rings, but these rings are also superimposed, making a complex pattern of lines and swirls (Fig.23.2). The developed negative of these interference fringe-patterns is a hologram. Thus, the hologram does not contain a distinct image of the object but carries a record of both the intensity and the relative phase of the light waves at each point.

The image on the film for a transmission hologram

Fig.23.2

23.2.2. RECONSTRUCTION OF THE IMAGE

Whenever required, the object can be viewed. For **reconstruction** of the image, the hologram is illuminated by a parallel beam of light from the laser (Fig.23.3). Most of the light passes straight through, but the complex of fine fringes acts as an elaborate diffraction grating. Light is diffracted at a fairly wide angle. The diffracted rays form two images: a virtual *image* and a *real image.* The virtual image appears at the location formerly occupied by the object and is sometimes called as the true image. The real image is formed in front of the hologram. Since the light rays pass through the point where the real image is, it can be photographed. The virtual image of the hologram is only for viewing. Observer can move to different positions and look around the image to the same extent that he would be able to, were he looking directly at the real object. This type of hologram is known as a *transmission* hologram since the image is seen by looking through it. The three dimensional image is seen suspended in midair at a point which corresponds to the position of the real object which was photographed.

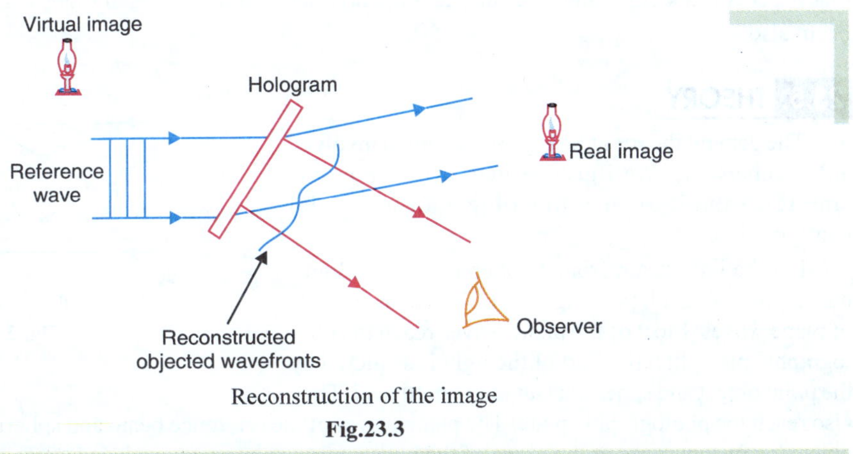

Reconstruction of the image

Fig.23.3

23.3. COAXIAL HOLOGRAPHY

The original technique adopted by Gabor for recording hologram was a **coaxial arrangement** (Fig.23.4), where he made use of a mercury discharge lamp and *collinear* object and reference beams. In this arrangement both virtual and real images are on the same axis. The real image is

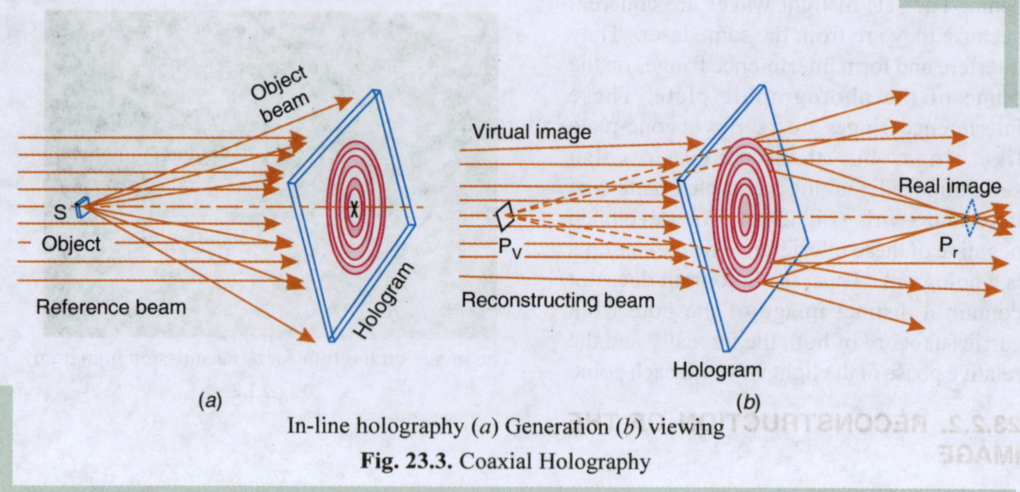

In-line holography (*a*) Generation (*b*) viewing

Fig. 23.3. Coaxial Holography

located in front of the virtual image. Thus an observer focusing on one image, always sees it accompanied by the out-of-focus twin image. As such it is inconvenient for viewing or photographing the image.

23.4. OFF-AXIS HOLOGRAPHY

In 1962, Leith and Upatnieks demonstrated a technique which made it possible to separate the twin images. In this technique a separate coherent reference wave is allowed to fall on the hologram plate during the recording process, at *an offset angle* to the beam from the object. The exposed plate is developed by normal photographic procedures so that the amplitude transmittance of the plate after development is proportional to the exposure. The advantage of the off-axis configuration is that it generates virtual and real images angularly separated from each other and from the direct beam also.

23.5. THEORY

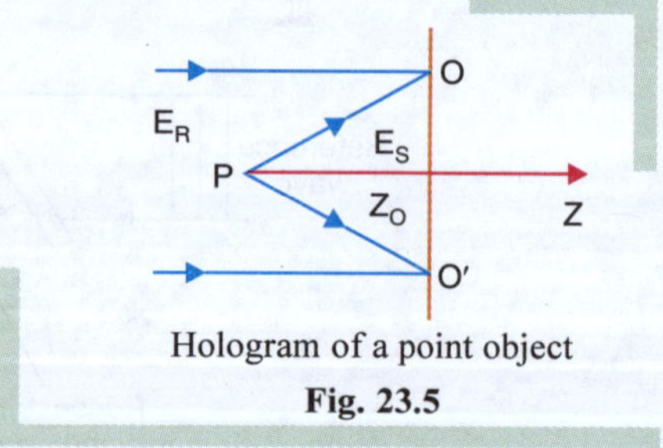

Hologram of a point object

Fig. 23.5

The general theory of holography is much involved and cumbersome. We illustrate here it by taking the simple example of a point object in a coaxial configuration.

Let the light beam from a coherent source illuminate a point object P (see Fig.23.5). The beam consists of plane waves. Most of the plane waves reach the photographic plate directly. Part of the light is scattered by the point object and spherical waves are produced. They also reach the photographic plate. The plane waves of the reference beam and spherical waves of the object beam superpose at the plane of the photographic plate and produce interference.

We may write the optical field arriving at point O on the photographic plate as

$$E = E_R + E_S \tag{23.1}$$

where E_R is the field due to the reference beam and E_S is the field scattered from the object. The scattered field E_S is not simple, both amplitude and phase vary greatly with position. The scattered wave fronts are spherical and concentric around the point of origin. We represent the field of the scattered wave front by

$$E_S = \frac{E_o}{r_o}\exp\left[i\left(k\, r_o - \omega\, t\right)\right] \qquad (23.2)$$

and the field E_R by the plane wave

$$E_R = E_R = E_r \exp\left[i\left(k\, z_o - \omega\, t\right)\right] \qquad (23.3)$$

where $r_o = PO$ and z_o is the distance from P to the plate. The intensity at O is

$$I = |E_R - E_S|^2$$

$$= \left|E_R\right|^2 + \frac{\left|E_S\right|^2}{r_o^2} + \frac{E_S E_r^*}{r_o}\exp\left[ik\left(r_o - z_o\right)\right] + \frac{E_S^* E_r}{r_o}\exp\left[ik\left(z_o - r_o\right)\right] \qquad (23.4)$$

We combine the last two terms of the above equation and write it as

$$I = \left|E_R\right|^2 + \frac{\left|E_S\right|^2}{r_o^2} + K\frac{\cos\left[k\left(r_o - z_o\right) + \varphi\right]}{r_o} \qquad (23.5)$$

The total intensity I is a function of cosine term and shows a series of maxima and minima. Thus, the interference of the spherical wave E_S and plane wave E_R produces a set of circular interference fringes. If we assume that the plate response is proportional to the intensity I, the power transmission of the plate, T^2 is given by

$$T^2 = 1 - \alpha I \qquad (23.6)$$

P Virtual image Z₀ Z₀ P′ Real image Z

Fig. 23.6

where α is a constant. Equ.(24.6) can be approximated as

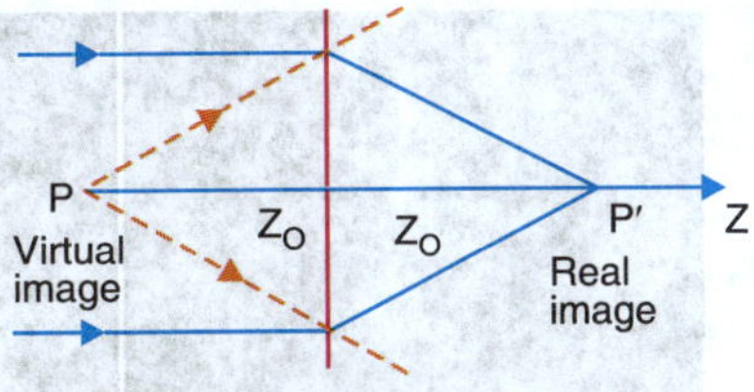

$$T = 1 - \frac{1}{2}\alpha I \qquad (23.7)$$

When the hologram is illuminated by the reference beam, the field of the transmitted wave may be written as

$$E = TE_2 = \left[1 - \frac{\alpha}{2} I\right] \mathrm{E_r} \exp\left[i\left(kz_o - \omega\, t\right)\right]$$

$$= \left[1 - \frac{\alpha}{2}\left|E_R\right|^2 - \frac{\alpha}{2}\frac{\left|E_S\right|^2}{r_o^2}\right] \mathrm{E_R} \exp\left[i\left(kz_o - \omega\, t\right)\right] -$$

$$\frac{\alpha}{2}\frac{E_S E_r^*}{r_o}\exp\left[ik\left(r_o - z_o\right)\right] \mathrm{E_r} \exp\left[i\left(kz_o - \omega\, t\right)\right]$$

$$-\frac{\alpha}{2}\frac{E_S^* E_r}{r_o}\exp\left[ik\left(z_o - r_o\right)\right] \mathrm{E_r} \exp\left[i\left(kz_o - \omega\, t\right)\right]$$

$$= \left[1 - \frac{\alpha}{2}\left|E_r\right|^2 - \frac{\alpha}{2}\frac{\left|E_S\right|^2}{r_o^2}\right] E_r \exp\left[i\left(kz_o - \omega\, t\right)\right] -$$

$$\frac{\alpha E_S \left|E_r\right|^2}{2r_o}\exp\left[i\left(kr_o - \omega\, t\right)\right] - \frac{\alpha}{2}\frac{E_S^* E_r^2}{2}\exp\left[i\left(2kz_o - kr_o - \omega\, t\right)\right] \qquad (23.8)$$

The first term in (23.8) represents the incident plane wave with some attenuation.

The second term represents a spherical wave identical with that emitted by the object except for a constant factor. The wave surface when projected back appears to have come from an apparent object located at the place where the original objected was located. This is the *virtual image* of the object.

The third term represents also a spherical wave, which is identical to the original wave but converges at a point P'. A *real image* is produced at P' which can be photographed without a lens. The hologram thus produces both a real image P and a virtual image P'.

23.6. HOLOGRAMS

Holograms are true three-dimensional images. This is evidenced by the fact that one can move his head while viewing the image and see it in a different perspective. It reveals part of the image which was hidden at another viewing angle.

For example, three images are shown below (Fig.23.7). They are from the *same* hologram but are obtained by looking through the hologram at *different angles*. Note that the pawn appears in different perspective in front of the king behind it.

Fig. 23.7

23.6.1 ORTHOSCOPIC AND PSEUDOSCOPIC IMAGES

A hologram reconstructs two images, one real and the other virtual which are exact replicas of the object. However, the two images differ in appearance to the observer. The virtual image is produced at the same position as the object and has the same appearance of depth and the parallax as the original three dimensional object. The virtual image appears as if the observer is viewing the original object through a window defined by the size of the hologram. This image is known as **orthoscopic image**.

The real image is also formed at the same distance from the hologram, but in front of it. In the real image, however the scene depth is inverted. This is due to the fact that the corresponding points on the two images (virtual and real) are located at the same distances from the hologram. The real image is known as **pseudoscopic image** and does not give a pleasing sensation as we do not come across objects with inverted depths in normal life.

23.6.2 HOLOGRAPHY AND PHOTOGRAPHY

Recording a hologram is different from taking a photograph.

i. In hologram, it is necessary that a coherent light source like a laser is used to illuminate the object.
ii. There is a second beam of coherent light which strikes the film on which the hologram is to be recorded. This is called the reference beam. The reference beam and the object beam overlap at the surface of the film and they form an interference pattern.
iii. A high-resolution photographic film is used for recording the fine patterns.

iv. Because the lines that make up the hologram are usually less than a micron, holographic recording is more sensitive to movement and vibrations compared to photographic recording. This is actually a very serious restriction. Therefore, holography technique is restricted to the laboratory where vibration isolation is created.

The fundamental difference between a **hologram** and **an ordinary photograph** is like this.

i. In a photograph the information is stored in an orderly fashion: each point in the object relates to a conjugate point in the image. In a hologram there is no such relationship; light from every object point goes to the entire hologram. This has two main advantages. As the observer moves sideways in viewing the hologram, the image is seen in three dimensions.

ii. If the hologram were shattered or cut into small pieces, each fragment would still reconstruct the whole scene, not just part of the scene.

23.7. IMPORTANT PROPERTIES OF HOLOGRAM

1. In an ordinary photograph each region contains separate and individual part of the original object. Therefore, destruction of a portion of a photographic image leads to an irreparable loss of information corresponding to the destroyed part. On the other hand, in a hologram each part contains information about the entire object. From even a small pat of the hologram the entire image can be reconstructed if only with a reduced clarity and definition of the image. Therefore, a hologram is a reliable medium for data storage.
2. It is not useful to record several images on a single photographic film. Such a record cannot give information about any of the individual images. On the other hand, several images can be recorded on a hologram. Therefore the information holding capacity of a hologram is extremely high. While a 6×9 mm photograph can hold one printed page, a hologram of the same size can store up to 300 such pages.
3. On a hologram information is recorded in the form of interference pattern. The type of the pattern obtained depends on the reference beam used to record the hologram. The information can be decoded only by a coherent wave identical to that of the reference wave. The reference wave can be chosen appropriately. Consequently without the knowledge of the shape of the reference wave front the information encoded in the form of interference pattern on the hologram cannot be deciphered.
4. The reconstruction of the image of the hologram can be done with reference beam of any wavelength if it is coherent and identical to the original reference beam. If the wavelength λ of the reconstructing beam is greater than that λ_o of the reference beam, the reconstructed image will be a magnified image. The magnification will be proportional to the ratio of the two wavelengths.

23.8. CLASSIFICATION OF HOLOGRAMS

Holograms may be classified in a number of different ways depending on their thickness, method of recording, method of reconstruction etc.

i. Thin Holograms or Plane Holograms

Holograms may be thin (plane) or thick (volume) (see Fig.23.8). A hologram may be regarded as *thin* if its emulsion thickness is much less than the fringe spacings. Otherwise it is called a *thick hologram.* Thick hologram is also known as *volume hologram.*

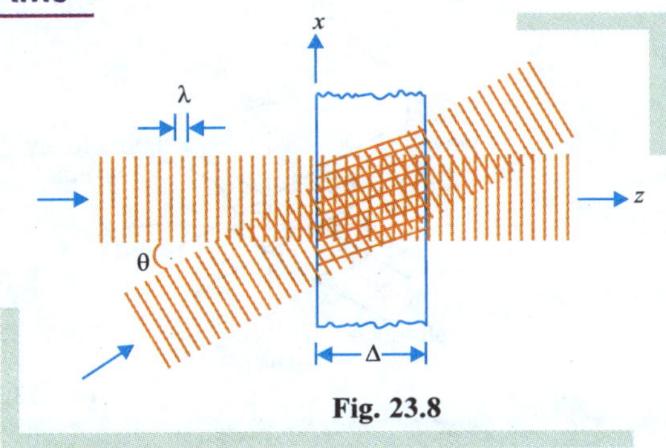

Fig. 23.8

Thin holograms produce several orders (i) zero order which is the directly transmitted reference beam, (ii) the first order diffraction producing virtual image, (iii) the minus first order diffraction equal in intensity to the first order producing the conjugate image and (iv) higher orders of decreasing intensity.

ii. Volume hologram

In 1962 Yuri Denisyuk used a process similar to Lippmann colour photography. In this method the object wave is reflected from the object and propagates backward and overlaps the incoming reference wave. The two waves form standing wave pattern. The fringes are recorded by the photo-emulsion throughout its entire thickness to form a **volume hologram**. The hologram may be regarded as a three-dimensional grating. In the volume holograms there is an interdependence of the wavelength and the scattering angle, because the scattering follows Bragg's law, $2d\sin\theta = m\lambda$. Therefore, by successively changing the incident angle or wavelength, a number of holograms can be stored in the medium. Different and mutually incoherent laser beams may be used to produce different component holograms of the object and when they are illuminated, a multicolored image is seen. A consequence of Bragg condition is that the volume hologram reconstructs the virtual image at the original position of the object if the reconstruction beam exactly coincides with the reference beam. However, the conjugate image and higher order diffractions are absent.

iii. Amplitude and Phase Holograms

Holograms recorded in photographic emulsions change both the amplitude and the phase of the illuminating wave. The shape of the recorded fringe planes depend on the relative phase of the interfering beams. Consequently, the reconstructed wave is reflected from the hologram according to the density of the silver deposited with the amplitude variation proportional to the amplitude of the object wave. Similarly the phase of the reconstruction wave is modulated in proportional to the phase of the object wave. Thus both amplitude and phase of the object wave are reproduced. An **absorption type** hologram produces a change in the amplitude of the reconstruction beam. The **phase type hologram** produces phase changes in the reconstruction beam due to a variation in the refractive index or thickness of the medium. Phase holograms have the advantage over amplitude holograms that no energy is dissipated within the hologram medium and have higher diffraction efficiency.

iv. Transmission Holograms

For preparing a transmitting hologram, both the reference beam and object beam are made incident from the same side of the recording medium, as shown in Fig.23.9 (a). These two beams form interference pattern and the spacing between the maxima is given by

$$\Lambda = \frac{\lambda}{2n\sin\alpha} \tag{23.9}$$

where n is the index of refraction of the recording medium and is the angle shown in Fig.23.9 (b). In the reconstruction process, the interference fringes act like reflective layers to the incident

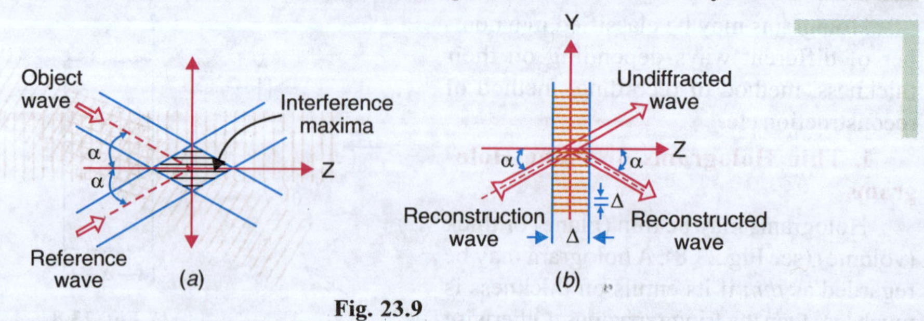

Fig. 23.9

reference wave. These reflecting layers are perpendicular to the plane of the hologram. The incident reconstruction wave is reflected toward the other side of the hologram. The reconstructed beam appears as if the beam is transmitted through the hologram.

v. Reflection Holograms

The distinction between transmitting and reflection holograms is due to the angle between the reference and object beams. In case of reflection hologram, the object beam is introduced at almost 180° with respect to the reference beam. Further, reflection holograms are often thicker than transmission holograms. There is more physical space for recording interference fringes. The interference pattern consists of vertical strips at a spacing given by

$$\Lambda = \frac{\lambda}{2n \sin \alpha} \tag{23.10}$$

The reflecting layers formed in the recording medium are parallel to the surface of the hologram. One can think of holograms that are made this way as having multiple **layers** that are only about half a wavelength deep. In the reconstruction process, some of the incident light reflects back toward the light source, and some continues to the next layer, where the process repeats. The light from each layer interferes with the light in the layers above it. This is known as the **Bragg effect**, and it is a necessary part of the reconstruction of the object beam in reflection holograms. The reconstructed wave is on the same side of the hologram and thus appears as if light is reflected from the hologram.

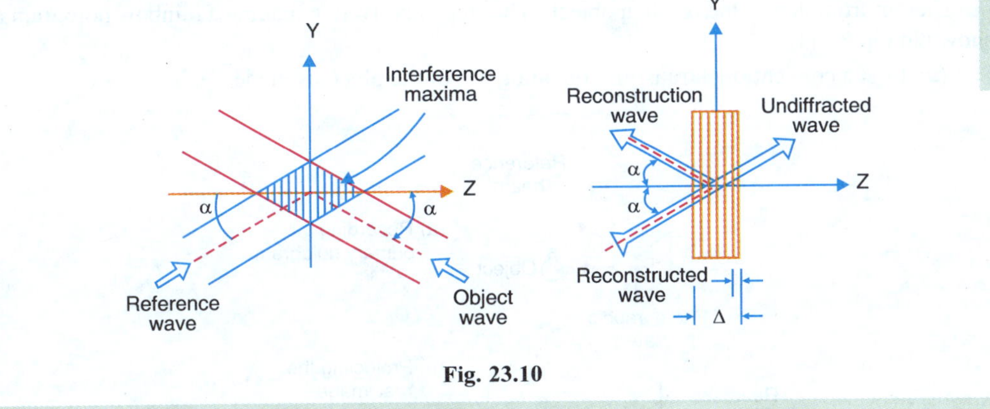

Fig. 23.10

The Bragg effect can also change the way the hologram reflects light, especially in holograms that one can view in white light. At different viewing angles, the Bragg effect can be different for different wavelengths of light. This means that you might see the hologram as one color from one angle and another color from another angle. The Bragg effect is also one of the reasons why most novelty holograms appear green even though they were created with a red laser.

Reflection holograms are more useful since a laser is not required for reconstruction. They can be viewed in white light.

vi. White-Light Reflection Holography

This method is developed by Stroke and Labeyrie. In this scheme, the hologram is generated using coherent light but in the reconstruction process an ordinary white-light beam having a wave front similar to the original coherent waves is used. Using coherent sources at different wavelengths, several holograms are stored in a single film. When the hologram is illuminated by ordinary white light, a multicolored image is seen in reflection.

vii. Colour Holography

Colour holograms are basically multiplexed holograms which produce multicolour images. They can be recorded with three wavelengths. When reconstructed with the recording wavelengths the hologram produces overlapping images in three colours producing a multicolour image. The behaviour of the reconstructed image depends on whether the hologram has been recorded in a thin medium or in a thick medium. Colour holograms recorded in a thin recording medium suffer from cross-talk. Volume holograms effectively eliminate cross-talk images utilizing Bragg effect. Both transmission and reflection volume colour holograms can be recorded in thick media. The transmission volume holograms are reconstructed with the laser beams used to record it, while the volume colour reflection holograms are reconstructed with white light due to their inherent wavelength discrimination ability.

viii. Rainbow Holograms

Rainbow hologram is a new type of transmission hologram capable of reconstructing a bright, sharp and monochromatic image when viewed in white light. They are made by a double holographic process where an ordinary hologram such as a transmission hologram is used as the object and a second hologram is made through a slit. A horizontal slit limits the vertical perspective of the first image so that there is no vertical parallax in the resultant rainbow hologram. This slit process removes the coherence requirement on the viewing light so that full advantage can be taken of the image brightness obtained from ordinary room light, while maintaining the three-dimensional character of the image. A true-color hologram image can be observed when the hologram is viewed in the correct plane. If the viewer moves off this plane, different shades of color can still be seen, but the color will be different from that of the original object. The steps involved in making a rainbow hologram are shown in Fig.23.11.

(*a*) First a conventional transmission hologram of the object is made.

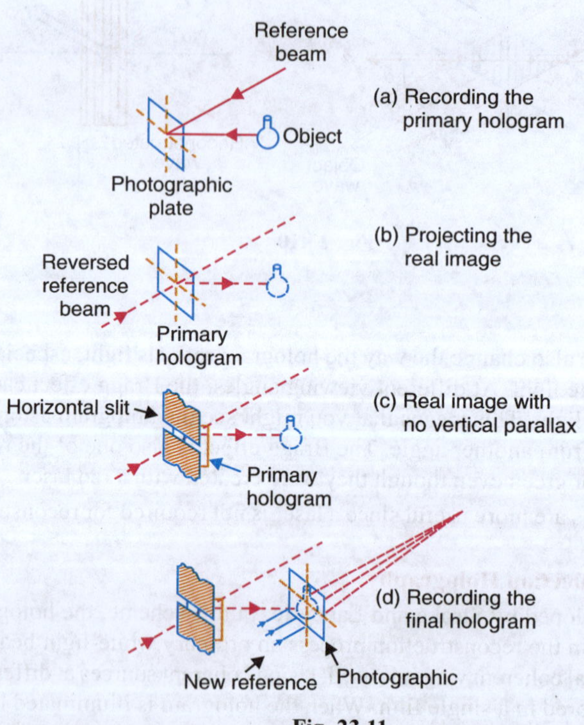

Fig. 23.11

(*b*) The hologram is then illuminated by the conjugate of the original reference beam. It generates a diffracted wave, which is the real image of the object.

(*c*) A horizontal slit is placed over the primary hologram to eliminate vertical parallax.

(*d*) A second and final hologram is recorded which is used for reconstructing the object.

ix. 360° Holograms

This type of hologram is made on a 360° circular film. A photographic film is mounted on a cylindrical surface surrounding the object to be holographed. A divergent laser beam is made to be incident on the object from the top (Fig.23.12). A convex mirror at the bottom illuminates the object.

When the cylindrical hologram is illuminated, the virtual image is observed at the centre of the cylinder and can be viewed from all sides.

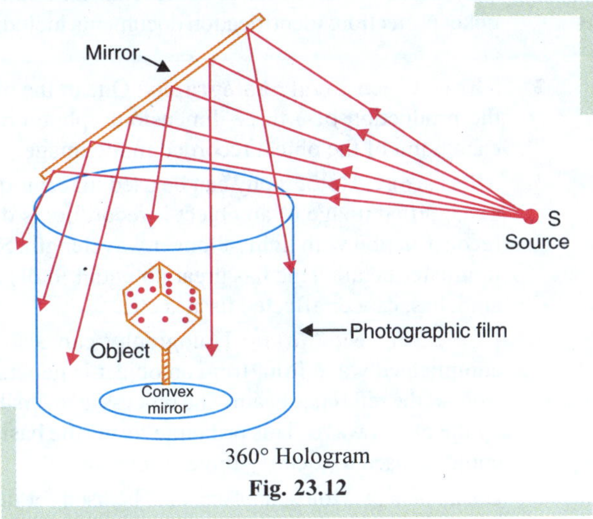

360° Hologram

Fig. 23.12

x. Copying Holograms

Many of the commercial products bear a holographic logo and trade mark as a mark of the identity and authenticity of the manufacturer. It requires bulk production of holograms. Large number of copies of the original master hologram can be made using different ways.

Light reflected or transmitted from the master hologram and a direct beam are made to interfere on a copy plate (Fig.23.13). This results in a copy hologram. Excellent copies can be made in this way.

Alternately, an embossing method can be used for making copies. A phase hologram where the thickness of the emulsion varies is used in this method. This master hologram has a surface relief structure that is metallized. The metallized hologram is used for impressing the pattern onto thin sheets of plastic.

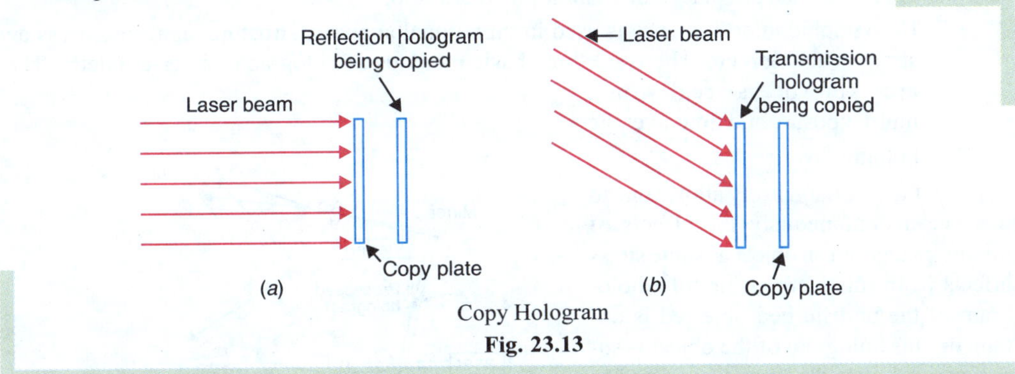

Copy Hologram

Fig. 23.13

23.9. APPLICATIONS

Holography can be used for a broad range of applications in different fields. It is not possible to describe all of them here. Only some typical applications are discussed here.

1. **Security**: One of the most popular areas for the use of holograms is the security and product authentication. The presence of holograms indicates the authenticity of these items. They provide a powerful obstacle to counterfeiting. The security holograms have proven to be unsurpassed when added to documents, anti-counterfeiting, tamper-proofing, customizing ticket protection, identification documents including credit and phone cards, drivers licenses etc.
2. **Three-dimensional photography**: One of the most obvious applications of holography is the production of a three-dimensional photograph, with the distance and orientation of each point of the object recorded in the image.
3. **Microscopy**: Holography can be used in techniques of microscopy. It is possible to obtain a magnified image of an object if recording is done with light of smaller wavelength and reconstruction with light of longer wavelength. Smaller areas in an object can be examined in greater detail. This has great potential in observing micro-objects such as blood cells, amoebas, cancer affected tissues etc.
4. **Character recognition**: Holography can also be used for character recognition. The complicated wave front from an object is generated from a hologram by the simple wave front of the reference beam. The process is reversible so that reference wave can be generated by the object wave. This principle forms the basis of holographic pattern recognition. This could be used to identify fingerprints etc.
5. **Data storage**: Holograms can also be used for data storage devices and hence are of much use in computer technology. A large amount of information such as 10^{12} letters/digits can be stored in a cubic cm of a volume hologram. These memories have long lifetime because a small mechanical damage to the portion of a hologram will not erase the stored information.
6. **Photolithography:** Holography is used in the production of photographic masks used to produce microelectronic circuits.
7. **Holographic projection** is used to display flight information at the pilot's eye level in an airplane cockpit.
8. **Holographic interferometry:** One of the most important applications of holography has been in interferometry. **Holographic interferometry** is an optical technique to visualize in a dark environment small deformations (200 nm to 100 μm) of objects. It is applied to objects, which are placed in a vibration-free set-up.

   Holographic interferometry is used in vibrational analysis, structural analysis, stress and strain evaluation etc. There are three basic methods of holographic interferometry. They are known as real time, multiexposure and time-average holography.

Real time holography allows one to observe instantaneously the effects of minute changes in an object as some stress affects it. In this method, first the hologram of the undisturbed objected is obtained. This hologram of the object is superimposed over the object subjected to some small stress (Fig.23.14). The distortions that appear as a fringe pattern are analysed.

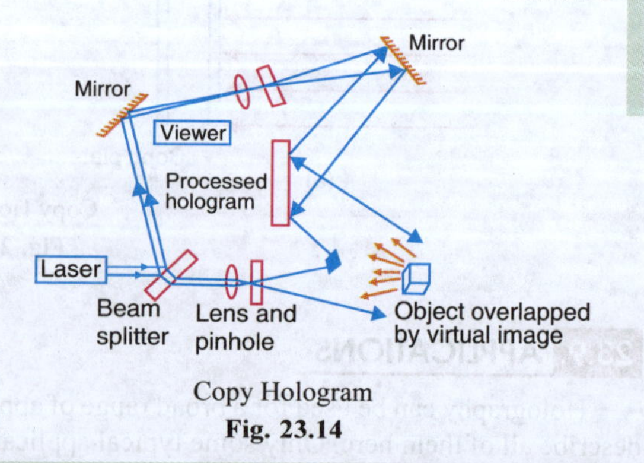

Copy Hologram

Fig. 23.14

Multiexposure holography creates a hologram by using two or sometimes more exposures. The first exposure shows the object in an undisturbed state. Subsequent exposures are made on the same image while the object is subjected to some stress. The resulting image depicts the difference between the two states.

The technique where two exposures are made is known as **double exposure holographic interferometry.** In the technique, a hologram of the undisturbed object is first recorded on the photographic plate with an exposure to a reference wave. Then, before the hologram is removed or developed, the object is stressed and is recorded on the same photographic plate through a second exposure along with the same reference wave. After this double exposure, the hologram is developed. If the hologram is now illuminated by a reconstruction wave, there would emerge from the hologram two object waves - one corresponding to the unstressed object and the other corresponding to the stressed object. These two object waves interfere to produce interference pattern. Thus on viewing through the hologram, the object as well as the surface covered by the interference fringes caused by surface irregularities can be seen. These fringes reveal the distribution of strain in the object.

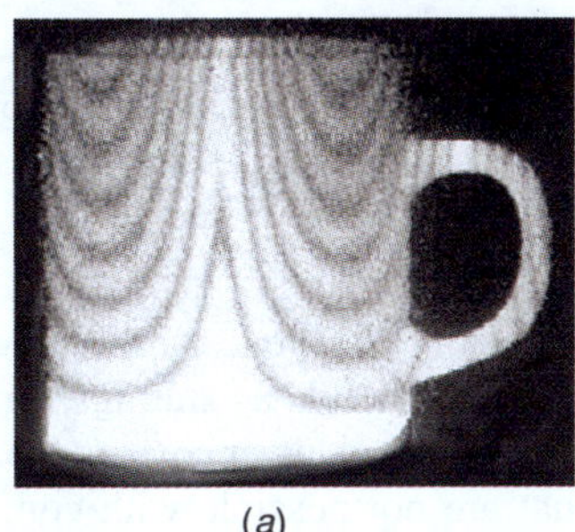

(*a*)

(*b*)

The Zebra-like pattern of fringes on (*a*) a vibrating coffee cup (*b*) on a violin shows the nodal pattern of vibrations.

Fig. 23.15

The third technique is known as the **time –average holographic interferometry** which is used in studying vibrating objects. In this technique, a hologram of the vibrating object is prepared by exposing the photographic plate to a reference wave, for a relatively long period of time such that the vibrating object has undergone a number of oscillations during that time. The resulting hologram consists of a standing wave pattern caused by a superposition of the number of images corresponding to the successive states of vibration of the object. The bright areas in the hologram correspond to undeflected regions whereas the contour lines indicate the regions of constant amplitude of vibration. This technique is very much useful in the vibrational analysis of any vibrating system or vibrating component of a machine.

Nowadays, the above techniques are widely used in nondestructive testing. Although holography can solve many problems, it still is a relatively expensive procedure.

9. Acoustical holography

It is easy to produce coherent sound waves. Therefore, holograms can be made using ultrasonic waves initially and then visible light can be used for reconstruction of the visual image. Light waves cannot propagate considerable distances in dense liquids and solids whereas sound waves can propagate through them. Therefore, a three dimensional acoustical hologram of an opaque object can be made. By viewing such hologram in visible light the internal structure of the object can be observed. Such techniques will be highly useful in the fields of medicine and technology. In one of the techniques, two submerged coherent sound wave generators emit the reference and the signal waves scattered by an object (Fig.23.16). On a calm surface of water, these two contributions produce

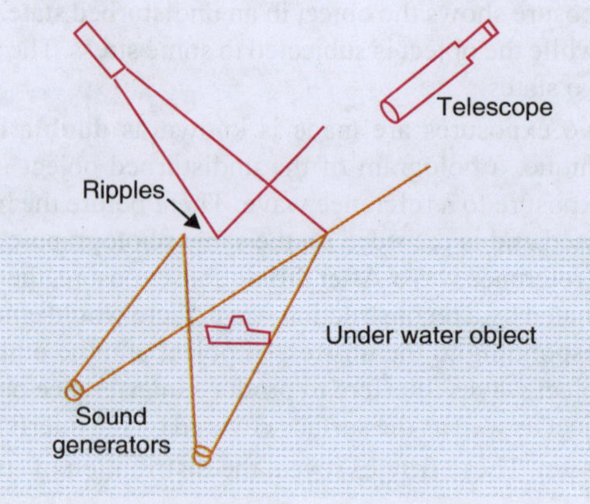

Fig. 23.16

ripples. The ripple pattern is the hologram. The pattern is photographed and reconstructed as and when required. As sound waves can propagate through dense liquids and solids, acoustical holography has an advantage in locating underwater submarines etc and study of internal body organs.

10. Holographic optical elements

Traditional optical elements operate on the principle of refraction. A diffractive optical element (DOE) operates on the principle of diffraction. DOEs can function as gratings, lenses or any other type of optical element. Large optical apertures, lightweight and lower cost are the main features of DOEs. They can offer unique optical properties that are not possible with conventional optical elements. They can be fabricated in a wide range of materials namely, aluminum, silicon, silica, plastics, etc providing the user greater flexibility in selecting the material for a particular application. DOEs can be used to perform more specialized functions, like making the panel instruments of a car visible in the windshield for increased safety.

Some of the important features of DOEs are as follows.

- Several different optical elements can share the same substrate without interfering with one another. Thus, a single DOE can be used as a lens, beam splitter and spectral filter simultaneously.
- Diffractive elements are very light, as they are formed in thin films of a few um thickness only.
- Because DOEs can generate unique optical functions that are not possible by conventional reflective or refractive optical elements, they provide greater flexibility in system configuration.
- At least one surface of a conventional glass lens is curved, whereas for a diffractive lens there is no such requirement. A diffractive element can be fabricated on any arbitrary shape of the substrate.
- They can be made to operate over a narrow wavelength band.
- The fabrication and replication of DOEs are relatively easy and cheap because no precision shaping of a surface is required.

23.10. MEDICAL APPLICATIONS OF HOLOGRAPHY

Holographic technique is also used in various medical applications like opthalmology, endoscopy, otology, orthopedics and many more. Recent improvements in hologram recording techniques and the availability of tools for the interpretation of holographic interferograms and the success of holographic techniques in imaging through tissues, ophthalmology, dentistry, urology, otology, pathology, and orthopedics shows that holography may emerge as a powerful tool for medical applications. We discuss here some of the applications.

1. Holographic Endoscope

Endoscopic holography provides a powerful tool for non-contact, high resolution, 3D imaging and nondestructive measurements inside natural cavities of human body. It combines the features of holography and endoscopy. The holographic endoscopy is of two types. In one of the forms, the hologram is recorded inside the endoscope, while the other form uses an external recording device.

(a). Internal Hologram Recording Endoscope

The endoscope accommodates a miniaturized holographic setup inside the instrument and records a reflection hologram. It mainly consists of three parts; a film cartridge, a diaphragm and a single mode optical fibre (core diameter 4 μm) cable. The three parts are assembled in three adjustable stainless steel tubes. The film is placed normal to the endoscope. The holograms are viewed under a powerful microscope allowing for the observation of individual cells. Due to large hologram aperture, the image with a low speckle noise and high lateral resolution is obtained. A lateral resolution of 7 μm has been obtained in the reconstructed image that shows that the technique can be used for cellular structure analysis and may even substitute biopsy in tumour diagnosis. Specific dyes can be used to enhance the contrast of the tissue before recording the holograms as has been used extensively in gynecology and gastrointestinal tract.

(b). External Hologram recording Endoscope

In the external hologram recording endoscope, a conventional endoscope is used. The system records the hologram outside the endoscope using an external reference beam. An endoscope with extremely small outer diameter can be used. In order to obtain a high signal-to-noise ratio, the holographic endoscope must use gradient-index (GRIN) rod lenses. The speckle noise is reduced by illuminating and imaging the object by the same GRIN lens. An electro-optic crystal can be used as the photographic storage device in the holographic endoscope to provide in-situ recording, reconstruction, and erasure.

Holographic endoscope has been used with success for early recognition of cancerous indurations in the wall of urinary bladder.

2. Holography in Ophthalmology

Recording of a three dimensional image of the eye was one of the earliest applications of holography in the field of ophthalmology. Any retinal detachment or intraocular foreign body can be detected. Holography can also be applied for the measurement of corneal topography and crystalline lens changes and for the study of surface characteristics of both the nerve head and the cornea. Current methods of determining the shape of the central surface miss the central part and its periphery. The major advantage of holographic technique is the ultra high precision (sub-um range) with which such measurements are possible. The elastic expansion of the cornea can also be measured by holographic interferometry. This information is vital for corneal surgery. The studies made so far show that holography has potential to investigate changes on the cornea, crystalline lens changes, and surface characteristics of both the nerve head and the retina.

3. Diffractive bifocal Intraocular Lens

A very useful application of diffractive optics is in the correction of refractive errors for old persons who have been operated for cataract by the use of a bifocal intraocular lens. Such persons have difficulty in changing the focus of their eyes for near distant and far distant objects. Bifocal lenses are implanted in place of the natural eye lenses. The bifocal lens is a combination of a conventional refractive lens and a diffractive lens, the former focussed to infinity and the later for near distance vision. The efficiency of the diffractive lens is set at 50%, thus both the near and the far foci are accommodated over the whole visual field. The diffractive lens is fabricated on the rear of the conventional lens. When the eyes are focussed for a distant object, a blurred image is superimposed due to the presence of diffractive lens and vice versa, which obviously reduces the image quality. In most cases, the blurred image is discarded by the human visual perception and retinal processing system.

4. Holography in Orthopedics

Holography offers an excellent tool for the contactless study of orthopedic structures, specifically external fixtures to measure strains on fixation pins and rods. Such studies are important in osteosynthesis with external fixture used for long bone fractures, to prevent dislocations of both fractured ends that are mainly caused by decrease in strength of the fixation pins.

QUESTIONS

1. What is meant by holography? Why is it called wave front reconstruction?
2. Describe a how hologram is generated and image is reconstructed using off-axis configuration?
3. What is a hologram? How does it differ from an ordinary photograph? Describe in short how a hologram is generated and viewed.
4. Describe the recording and reconstruction processes in Holography with the help of suitable diagrams.
5. Write the principle of holography. With neat sketches explain in brief recording of a hologram and reconstruction of images.
6. What is holography? How does it differ from ordinary photographic technique? How do interference and diffraction phenomena related to construction reconstruction of the hologram? Explain in detail with suitable diagram.
7. What is Holography ? Describe its principle and the process of recording and reconstruction of a hologram. **(GNDU, Amritsar, 2010)**
8. How a Hologram is different from a photograph? **(GNDU, Amritsar, 2010)**
9. What is Holography? Write its applications. **(A.P.University, 2010)**
10. What is the advantage of off-axis configuration over the coaxial configuration?
11. Explain the principle of holography?
12. Explain some of the important properties of hologram?
13. How are the holograms classified? Explain.
14. Write the differences between holography and photography?
15. Discuss some of the important applications of holography?
16. Explain holographic interferometery.
17. Explain acoustic holography.
18. Describe some of the important medical applications of holography.
19. What are DOEs (diffractive optical elements)? Describe their salient features and applications.

PHOTONICS

Photonics is the technology of harnessing light and other forms of radiant energy. In photonics photons play the same role as that played by electrons in electronics. The subject of photonics is therefore also referred to as **optoelectronics** or **optronics**. Photons travel with a speed far larger than that of electrons, and light beams can cross each other without affecting each other. Optical signals have large bandwidth and can accommodate a large number of channels per a given volume. Photonics is mainly concerned with lasers, fibre-optics, harmonic generation using nonlinear media, electro-optic devices, imaging, optical computing etc.

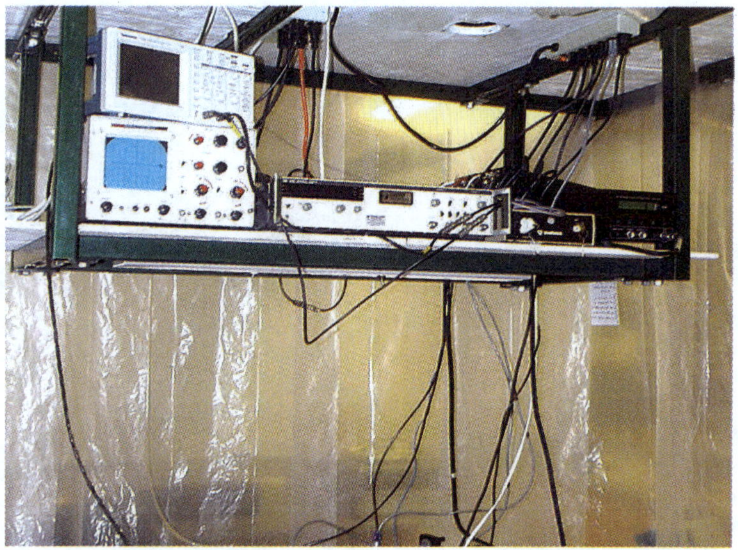

View of the optoelectronics for the wavemeter.

24 CHAPTER

Fibre Optics

24.1. INTRODUCTION

In 1870 John Tyndall, a British physicist demonstrated that light can be guided along the curve of a stream of water. Owing to total internal reflections light gets confined to the water stream and the stream appears luminous. A luminous water stream is the precursor of an optical fibre. In the 1950's, the transmission of images through optical fibres was realized in practice. Hopkins and Kapany developed the flexible fibrescope, which was used by the medical world in remote illumination and viewing the interior of human body. It was Kapany who coined the term fibre optics. By 1960, it had been established that light could be guided by a glass fibre. In 1966 Charles Kao* and George Hockham proposed the transmission of information over glass fibre, but the fibres available at that time heavily attenuated light propagating through them. In 1970 Corning Glass Works produced low-loss glass fibres. The invention of solid state lasers in 1970 made optical communications practicable. Commercial communication systems based on optical fibres made their appearance by 1977. Apart from the use as communicational channel, optical fibres are widely used in other areas. Fibroscopes made of optical fibres are widely used in a variety of forms in medical diagnostics. Sensors for detecting electrical, mechanical, thermal energies are made using optical fibres.

Fibre optics is a technology in which signals are converted from electrical into optical signals, transmitted through a thin glass fibre and reconverted into electrical signals.

At a Glance

*(received Nobel Prize in Physics in 2009)

24.2 OPTICAL FIBRE

Optical Fibre.

Definition: *An optical fibre is a cylindrical wave guide made of transparent dielectric, (glass or clear plastic), which guides light waves along its length by total internal reflection.* It is as thin as human hair, approximately 70 μm or 0.003 inch diameter. (Note that a thin strand of a metal is called a *wire* and a thin strand of dielectric materials is called a *fibre*).

Fig. 24.1: Illustration of a transparent fibre guiding light along its length.

Principle: The propagation of light in an optical fibre from one of its ends to the other end is based on the principle of *total internal reflection*. When light enters one end of the fibre, it undergoes successive total internal reflections from sidewalls and travels down the length of the fibre along a zigzag path, as shown in Fig.24.1. A small fraction of light may escape through sidewalls but a major fraction emerges out from the exit end of the fibre, as illustrated in Fig. 24.1. Light can travel through fibre even if it is bent.

Structure:

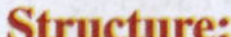

Fig. 24.2: Side view and cross sectional view of a typical optical fibre

A practical optical fibre is cylindrical in shape (Fig. 24.2a) and has in general three coaxial regions (Fig. 24.2b).

(*i*) The innermost cylindrical region is the light guiding region known as the **core**. In general, the diameter of the core is of the order of 8.5 μm to 62.5 μm.

(*ii*) It is surrounded by a coaxial middle region known as the **cladding**. The diameter of the cladding is of the order of 125 μm. The refractive index of cladding (n_2) is always lower than that of the core (n_1). Light launched into the core and striking the core-to-cladding interface at an angle greater than critical angle will be reflected back into the core. Since the angles of incidence and reflection are equal, the light will continue to rebound and propagate through the fibre.

(*iii*) The outermost region is called the **sheath** or a **protective buffer coating**. It is a plastic coating given to the cladding for extra protection. This coating is applied during the manufacturing process to provide physical and environmental protection for the fiber. The buffer is elastic in nature and prevents abrasions. The coating can vary in size from 250 μm or 900 μm. To sum up

- Core is the inner light-carrying member.
- Cladding is the middle layer, which serves to confine the light to the core.
- Buffer coating surrounds the cladding, which protects the fibre from physical damage and environmental effects.

24.2.1 NECESSITY OF CLADDING

The actual fibre is very thin and light entering a bare fibre will travel along the fibre through repeated total internal reflections at the glass-air boundary. However, bare fibres are used only in certain applications. For use in communications and some other applications, the optical fibre is provided with a cladding. *The cladding maintains uniform size of the fibre, protects the walls of the fibre from chipping, and reduces the size of the cone of light that will be trapped in the fibre*.

➢ It is necessary that the diameter of an optical fibre remains constant throughout its length and is surrounded by the same medium. Any change in the thickness of the fibre or the medium outside the fibre (when the fibre gets wet due to moisture etc) will cause loss of light energy through the walls of the fibre.

➢ A very large number of reflections occur through the fibre and it is necessary that the condition for total internal reflection must be accurately met over the entire length of the fibre. If the surface of the glass fibre becomes scratched or chipped, the normal to the edge will no longer be uniform. As a result, the light traveling through the fibre will get scattered and escapes from the fibre. This also causes loss of light energy.

➢ Part of light energy penetrates the fibre surface. The intensity of the light decreases exponentially as we move away from the surface, as the light is able to penetrate only a very small distance outside the fibre. However, anytime the fibre touches something else, the light can leak into the new medium or be scattered away from the fibre. This effect causes a significant leakage of the light energy out of the fibre. Even a small amount of dust on the surface would cause a fair amount of leakage.

➢ If bare optic fibres are packed closely together in a bundle, light energy traveling through the individual fibres tends to get coupled through the phenomenon of *frustrated total internal reflection*. Cladding of sufficient thickness prevents the leakage of light energy from one fibre to the other.

The fiber is provided with a cladding in order to prevent loss of light energy due to the above reasons.

➢ The cladding causes a reduction in the size of the cone of light that can be trapped in the fibre. Light entering the fibre at larger angles will strike the fibre walls at smaller angles (higher modes) and ultimately travel a longer distance. Such higher modes of a light signal will take longer time to reach the end of the fibre than the lower modes. Therefore, a pulse sent through optical fibre spreads out. The spreading would be larger, the larger the cone of acceptance. Such pulse spreading limits the rate of data transmission through the fibre. As fibers with a cladding have smaller cone of acceptance, they carry information at a much higher bit rate than those without a cladding.

Thus, the cladding performs the following important functions:

- Keeps the size of the fibre constant and reduces loss of light from the core into the surrounding air.
- Protects the fiber from physical damage and absorbing surface contaminants.
- Prevents leakage of light energy from the fibre through evanescent waves.
- Prevents leakage of light energy from the core through frustrated total internal reflection.
- Reduces the cone of acceptance and increases the rate of transmission of data.
- A solid cladding, instead of air, also makes it easier to add other protective layers over the fibre.

24.2.2 OPTICAL FIBRE SYSTEM

An optical fibre is used to transmit **light signals** over long distances. It is essentially a **light-transmitting medium**, its role being very much similar to a coaxial cable or wave-guide used in microwave communications. Optical fibre requires a **light source** for launching light into the fibre at its input end and a **photodetector** to receive light at its output end. As the diameter of the fibre is very small, the light source has to be dimensionally compatible with the fibre core. Light emitting diodes and laser diodes, which are very small in size, serve as the light sources. The electrical input signal is in general of digital form. It is converted into an optical signal by varying the current flowing through the light source. Hence, the intensity of the light emitted by the source is modulated with the input signal and the output will be in the form of light pulses. The light pulses constitute the signal that travels through the optical fibre. At the receiver end, semiconductor photodiodes, which are very small in size, are used for detection of these light pulses. The photodetector converts the optical signal into electrical form. Thus, a basic *optical fibre system* consists of a LED/laser diode, optical fibre cable and a semiconductor photodiode.

24.2.3 OPTICAL FIBRE CABLE

Optical fibre cables are designed in different ways to serve different applications. More protection is provided to the optical fibre by the "cable" which has the fibres and strength members inside an outer covering called a "jacket". We study here two typical designs:a single fibre cable or a multifibre cable.

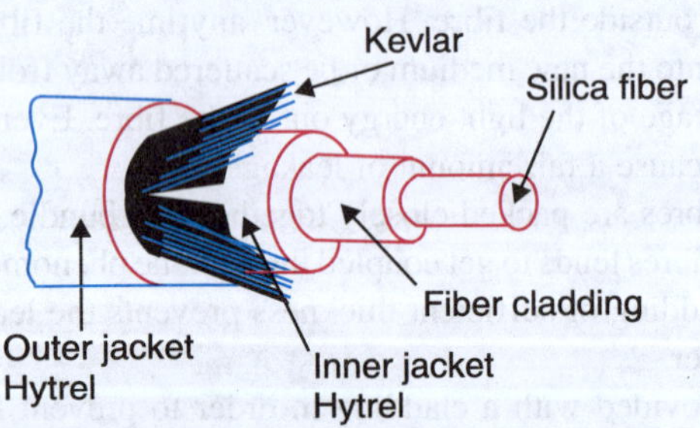

Fig. 24.3: Single fibre cable

- **Single Fibre Cable:** Around the fibre a tight buffer jacket of Hytrel is used (see Fig.24.3). The buffer jacket protects the fibre from moisture and abrasion. A strength member is arranged around the buffer jacket in order to provide the necessary toughness and tensile strength. The strength member may be a steel wire, polymer film, nylon yarn or Kevlar yarn. Finally, the fibre cable is covered by a Hytrel outer jacket. Because of this arrangement fibre cable will not get damaged during bending, rolling, stretching or pulling and transport and installation processes. The single fibre cable is used for indoor applications.

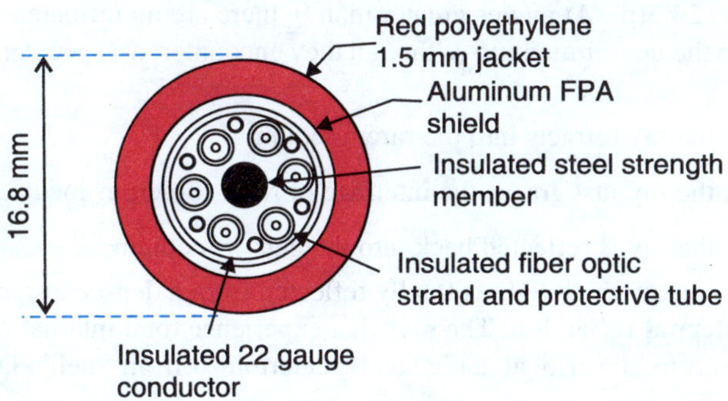

Fig. 24.4: Cross-sectional view of a typical multi fibre cable

• **Multifibre Cable:** A multifibre cable consists of a number of fibres in a single jacket. Each fibre carries light independently. The cross-sectional view of a typical telecommunication cable is shown in Fig.24.4. It contains six insulated optical fibre strands and has an insulated steel cable at the center for providing tensile strength. Each optical fibre strand consists of a core surrounded by a cladding, which in turn is coated with insulating jacket. The fibres are thus individually buffered and strengthened. Six insulated copper wires are distributed in the space between the fibres. They are used for electrical transmission, if required. The assembly is then fitted with in a corrugated aluminium sheath, which acts as a shield. A polyethylene jacket is applied over the top.

24.3 TOTAL INTERNAL REFLECTION

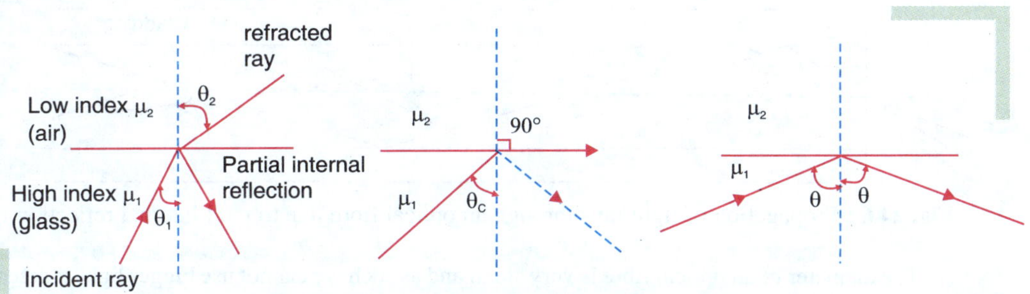

Fig. 24.5: Phenomenon of total internal reflection

A medium having a lower refractive index is said to be an optically **rarer medium** while a medium having a higher refractive index is known as an optically **denser medium**. When a ray of light passes from a denser medium to a rarer medium, it is bent away from the normal in the rarer medium (see Fig.24.5a). Snell's law for this case may be written as

$$\sin\theta_2 = \left(\frac{\mu_1}{\mu_2}\right)\sin\theta_1 \tag{24.1}$$

where θ_1 is the angle of incidence of light ray in the denser medium and θ_2 is the angle of refraction in the rarer medium. Also $\mu_1 > \mu_2$. When the angle of incidence, θ_1 in the denser medium is increased, the transmission angle, θ_2 increases and the refracted rays bend more and more away from the normal. At some particular angle θ_c the refracted ray glides along the boundary surface so that θ_2 =

90°, as seen in Fig. 24.5(b). At angles greater than θ_c there are no refracted rays at all. The rays are reflected back into the denser medium as though they encountered a specular reflecting surface (Fig. 24.5c). Thus,

- If $\theta_1 < \theta_c$, the ray refracts into the rarer medium
- If $\theta_1 = \theta_c$, the ray just grazes the interface of rarer-to-denser media
- If $\theta_1 > \theta_c$, the ray is reflected back into the denser medium.

The phenomenon in which light is totally reflected from a denser-to-rarer medium boundary is known as **total internal reflection**. The rays that experience total internal reflection obey the laws of reflection. Therefore, the critical angle can be determined from Snell's law.

When $\theta_1 = \theta_c, \quad \theta_2 = 90^o$.

Therefore, from equ.(24.1), we get

$$\mu_1 \sin\theta_c = \mu_2 \sin 90^o = \mu_2$$

$$\therefore \quad \sin\theta_c = \frac{\mu_2}{\mu_1} \tag{24.2}$$

When the rarer medium is air, $\mu_2 = 1$ and writing $\mu_1 = \mu$, we obtain

$$\sin\theta_c = \frac{1}{\mu} \tag{24.3}$$

24.4 PROPAGATION OF LIGHT THROUGH AN OPTICAL FIBRE

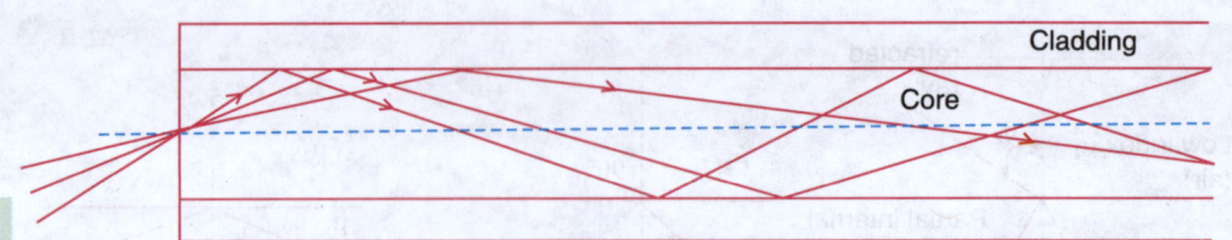

Fig. 24.6 : Propagation of light rays through an optical fibre due to total internal reflection.

The diameter of an optical fibre is very small and as such we cannot use bigger light sources for launching light beam into it. Light emitting diodes (LEDs) and laser diodes are the optical sources used in fibre optics. Even in case of these small sized sources, a focusing lens has to be used to concentrate the beam on to the fibre core. Light propagates as an electromagnetic wave through an optical fibre. However, light propagation through an optical fibre can as well be understood on the basis of *ray model*. According to the ray model, light rays entering the fibre strike the core-clad interface at different angles. As the refractive index of the cladding is less than that of the core, majority of the rays undergo total internal reflection at the interface and the angle of reflection is equal to the angle of incidence in each case. Due to the cylindrical symmetry in the fibre structure, the rays reflected from an interface on one side of the fibre axis will suffer total internal reflections at the interface on the opposite side also. Thus, the rays travel forward through the fibre via a series of total internal reflections and emerge out from the exit end of the fibre (Fig.24.6). Since each reflection is a total internal reflection, there is no loss of light energy and light confines itself within the core during the course of propagation. Because of the negligible loss during the total internal reflections, optical fibre can carry the light waves over very long distances. Thus, the optical fibre acts essentially as a wave-guide and is often called a **light guide** or **light pipe**. At the exit end of the fibre, the light is received by a photo-detector.

Total internal reflection at the fibre wall can occur and light propagates down the fibre, only if the following two conditions are satisfied.

1. The refractive index of the core material, n_1, must be slightly greater than that of the cladding, n_2.
2. At the core-cladding interface (Fig.24.7), the angle of incidence ϕ between the ray and the normal to the interface must be greater than the critical angle ϕ_C defined by

$$\sin\phi_C = \frac{n_2}{n_1} \tag{24.4}$$

It is to be noted here that only those rays, that are incident at the core-clad interface at angles greater than the critical angle will propagate through the fibre. Rays that are incident at smaller angles are refracted into the cladding and are lost.

24.4.1 CRITICAL ANGLE OF PROPAGATION

Let us consider a step index optical fibre into which light is launched at one end. The end at which light enters the fibre is called the **launching end**. Fig. 23.7 depicts the conditions at the launching end. In a step-index fibre, the refractive index changes abruptly form the core to the cladding. Now, we consider two rays entering the fibre at two different angles of incidence.

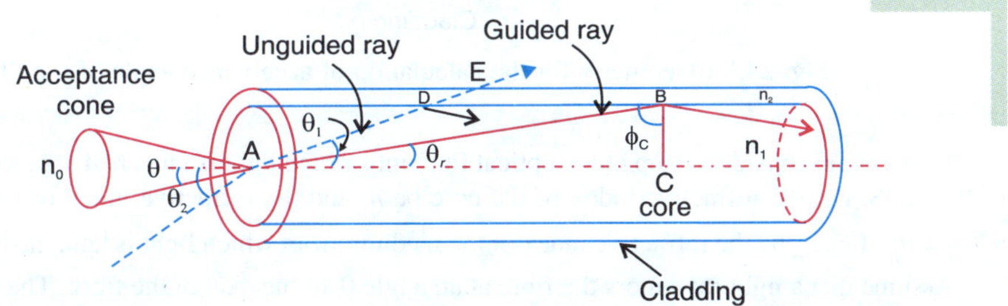

Fig. 24.7: Light rays incident at an angle smaller than critical propagation angle will propagate through the fibre.

The ray shown by the broken line is incident at an angle θ_2 with respect to the axis of the fibre. This ray undergoes refraction at point A on the interface between air and the core. The ray refracts into the fibre at an angle θ_1 ($\theta_1 < \theta_2$). The ray reaches the core-cladding interface at point D. At point D, refraction takes place again and the ray travels in the cladding. Finally, at point E, the ray refracts once again and emerges out of fibre into the air. It means that the ray does not propagate through the fibre.

Let us next consider the ray shown by the solid line in Fig.24.7. The ray incident at an angle θ undergoes refraction at point A on the interface and propagates at an angle θ_c in the fibre. At point B on the core-cladding interface, the ray undergoes total internal reflection, since $n_1 > n_2$. Let us assume that the angle of incidence at the core-cladding interface is the *critical angle* ϕ_c, where ϕ_c is given by

$$\phi_C = \sin^{-1}(n_2 / n_1) \tag{24.4a}$$

A ray incident with an angle larger than ϕ_c will be confined to the fibre and propagate in the fibre. A ray incident, at the core-cladding boundary, at the critical angle is called a **critical ray**. The critical ray makes an angle θ_c with axis of the fibre. It is obvious that rays with propagation angles larger than θ_c will not propagate in the fibre. Therefore, the angle θ_c is called the **critical propagation angle.** From the Δ^{le} ABC, it is seen that

$$\frac{AC}{AB} = \sin \phi_C . \qquad \text{Also, } \frac{AC}{AB} = \cos \theta_c$$

From the relation (24.4a), $\sin \phi_c = n_2 / n_1$.

$$\cos \theta_c = n_2 / n_1 \qquad (24.5)$$

$$\therefore \qquad \theta_c = \cos^{-1}(n_2/n_1) \qquad (24.6)$$

Thus, only those rays which are refracted into the cable at angles $\theta_r < \theta_c$ will propagate in the optical fibre.

24.4.2 ACCEPTANCE ANGLE

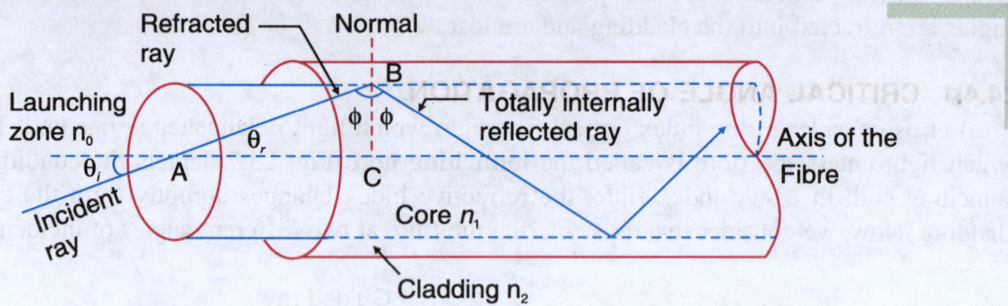

Fig. 24.8 : Geometry for the calculation of acceptance angle of the fibre.

Let us again consider a step index optical fibre into which light is launched at one end, as shown in Fig. 24.8. Let the refractive index of the core be n_1 and the refractive index of the cladding be $n_2 (n_2 < n_1)$. Let n_o be the refractive index of the medium from which light is launched into the fibre.

Assume that a light ray enters the fibre at an angle θ_i to the axis of the fibre. The ray refracts at an angle θ_r and strikes the core-cladding interface at an angle ϕ. If ϕ is greater than critical angle ϕ_c, the ray undergoes total internal reflection at the interface, since $n_1 > n_2$. As long as the angle ϕ is greater than ϕ_c, the light will stay within the fibre.

Applying Snell's law to the launching face of the fibre, we get

$$\frac{\sin \theta_i}{\sin \theta_r} = \frac{n_1}{n_0} \qquad (24.7)$$

If θ_i is increased beyond a limit, ϕ will drop below the critical value ϕ_c and the ray escapes from the sidewalls of the fibre. The largest value of θ_i occurs when $\phi = \phi_c$.

From the Δ^{le}ABC, it is seen that

$$\sin \theta_r = \sin (90^\circ - \phi) = \cos \phi \qquad (24.8)$$

Using equation (24.8) into equation (24.7), we obtain

$$\sin \theta_i = \frac{n_1}{n_o} \cos \phi$$

When $$\phi = \phi_c, \quad \sin \left[\theta_{i_{max}}\right] = \frac{n_1}{n_o} \cos \phi_c \qquad (24.9)$$

But $$\sin \phi_C = \frac{n_2}{n_1}$$

$$\therefore \qquad \cos\phi_C = \frac{\sqrt{n_1^2 - n_2^2}}{n_1} \qquad (24.10)$$

Substituting the expression (24.10) into (24.9), we get

$$\sin\left[\theta_i(\text{max})\right] = \left[\frac{\sqrt{n_1^2 - n_2^2}}{n_0}\right] \qquad (24.11)$$

Quite often the incident ray is launched from air medium, for which n_o=1. Designating θ_i *(max)*= θ_o, equation (24.11) may be simplified to

$$\sin\theta_0 = \sqrt{n_1^2 - n_2^2}$$

$$\therefore \qquad \theta_o = \sin^{-1}\left[\sqrt{n_1^2 - n_2^2}\right] \qquad (24.12)$$

The angle θ_o is called the **acceptance angle** of the fibre. *Acceptance angle is the maximum angle that a light ray can have relative to the axis of the fibre and propagate down the fibre.* Thus, only those rays that are incident on the face of the fibre making angles less than θ_o will undergo repeated total internal reflections and reach the other end of the fibre. Obviously, larger acceptance angles make it easier to launch light into the fibre.

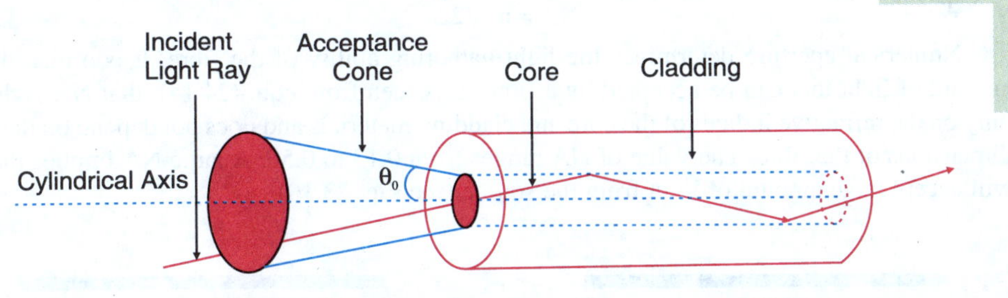

Fig. 24.9 Acceptance Cone

In three dimensions, the light rays contained within the cone having a full angle $2\theta_o$ are accepted and transmitted along the fibre (see Fig. 24.9). Therefore, the cone is called the **acceptance cone.**

Light incident at an angle beyond θ_o refracts through the cladding and the corresponding optical energy is lost.

24.5 FRACTIONAL REFRACTIVE INDEX CHANGE

The fractional difference Δ between the refractive indices of the core and the cladding is known as *fractional refractive index change*. It is expressed as

$$\Delta = \frac{n_1 - n_2}{n_1} \qquad (24.13)$$

This parameter is always positive because n_1 must be larger than n_2 for the total internal reflection condition. In order to guide light rays effectively through a fibre, $\Delta \ll 1$. Typically, Δ is of the order of 0.01.

24.6 NUMERICAL APERTURE

The main function of an optical fibre is to accept and transmit as much light from the source as possible. The light gathering ability of a fibre depends on the numerical aperture. The acceptance angle and the fractional refractive index change determine the numerical aperture of fibre.

The numerical aperture (NA) is defined as the sine of the acceptance angle. Thus,

$$NA = \sin \theta_o$$

where θ_o is the acceptance angle.

But

$$\sin \theta_o = \sqrt{n_1^2 - n_2^2}$$

$$\therefore \quad NA = \sqrt{n_1^2 - n_2^2} \tag{24.14}$$

$$n_1^2 - n_2^2 = (n_1 + n_2)(n_1 - n_2) = \left(\frac{n_1 + n_2}{2}\right)\left(\frac{n_1 - n_2}{n_1}\right) 2n_1$$

Approximating $\frac{n_1 + n_2}{2} \approx n_1$, we can express the above relation as $\left(n_1^2 - n_2^2\right) = 2n_1^2 \Delta$. It gives

$$NA = \sqrt{2n_1^2 \Delta}$$

$$\therefore \quad NA = n_1 \sqrt{2\Delta} \tag{24.15}$$

Numerical aperture determines the light gathering ability of the fibre. It is a measure of the amount of light that can be accepted by a fibre. It is seen from equ. (24.14) that NA is dependent only on the refractive indices of the core and cladding materials and does not depend on the physical dimensions of the fibre. The value of NA ranges from 0.13 to 0.50. A large NA implies that a fibre will accept large amount of light from the source (see Fig. 23.10).

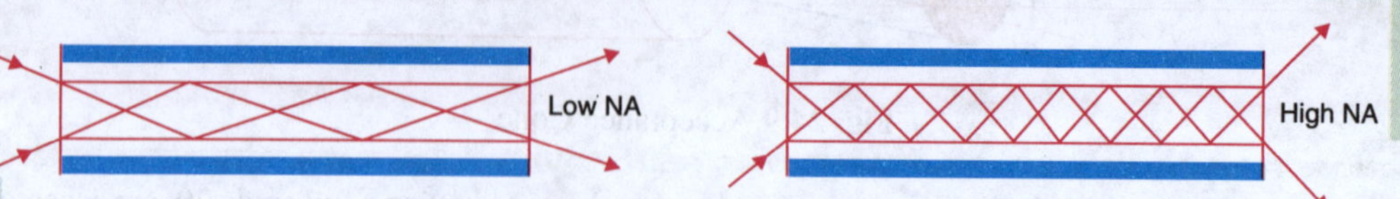

Fig. 24.10: Illustration of the propagation of light through low and high numerical aperture fibres.

24.7 SKIP DISTANCE AND NUMBER OF TOTAL INTERNAL REFLECTIONS

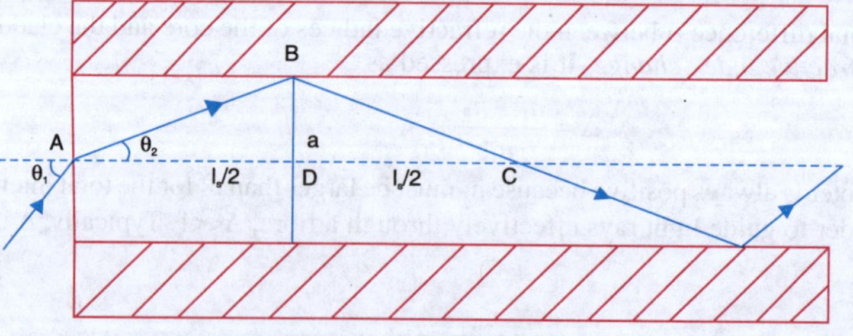

Fig. 24.11 : Skip Distance l_S

We shall now calculate the number of total internal reflections that a light ray undergoes as it travels through an optical fibre of length L. Let a be the radius of the fibre. Let a light ray be incident on one end of the fibre at an angle θ_1 to the axis and refract into the fibre at an angle θ_2, as shown in Fig. 24.11. The ray undergoes the first reflection at B. The distance $AC = l_s$ is known as the **skip distance** and represents the distance between two successive reflections of the ray. In the Δ^{le}ABD,

$$AD = \tfrac{1}{2} AC = \frac{l_s}{2}, \; BD = a \text{ and } \angle BAD = \theta_2 .$$

$$\tan\theta_2 = \frac{BD}{AD} = \frac{a}{l_s/2}$$

or
$$l_s = \frac{2a}{\tan\theta_2} \tag{24.16}$$

In terms of the incidence angle θ_1, the above equation may be rewritten as

$$l_s = 2a\left[\frac{\cos\theta_2}{\sin\theta_2}\right] \quad \text{or} \quad l_s^2 = (2a)^2\left[\frac{1}{\sin^2\theta_2} - 1\right]$$

Using equ.(24.7) into the above expression, we obtain

$$l_s^2 = (2a)^2\left[\left(\frac{n_1}{n_o \sin\theta_1}\right)^2 - 1\right]$$

or
$$l_s = 2a\left[\left(\frac{n_1}{n_o \sin\theta_1}\right)^2 - 1\right]^{1/2} \tag{24.17}$$

In case of air , $n_o = 1$ and
$$l_s = 2a\left[\left(\frac{n_1}{\sin\theta_1}\right)^2 - 1\right]^{1/2} \tag{24.17a}$$

The number of total internal reflections in the total fibre length is given by

$$N = \frac{\textit{Total length of the cable}}{\textit{The distance travelled during one reflection}} = \frac{L}{l_s}$$

$\therefore$
$$N = \frac{L\tan\theta_2}{2a} \tag{24.18a}$$

Also,
$$N = \frac{L}{2a\left[\left(\frac{n_1}{\sin\theta_1}\right)^2 - 1\right]^{1/2}} \tag{24.18b}$$

For example, if $n_1 = 1.50$, $\theta_1 = 30^o$ and $a = 25\,\mu m$, equ.(24.18b) gives $l_s = 141\,\mu m$. Alternately, computing the value of θ_2 and using it in equ.(24.18a), we obtain $l_s = 141\,\mu m$. Therefore, if light travels through a length of 1 m of the optical fibre of the above specifications, it is reflected 7092 times.

24.8 MODES OF PROPAGATION

Light propagates as an electromagnetic wave through an optical fibre and its propagation is governed by Maxwell's equations. Complete understanding of propagation of light waves through optical fibres requires a thorough understanding of solution of these equations in the context of optical fibres. When a plane electromagnetic wave propagates in free space, it travels as a transverse electromagnetic wave. The electric field and magnetic field components associated with the wave are perpendicular to each other and also perpendicular to the direction of propagation. It is known as a TEM wave. When the light ray is guided through an optical fibre, it propagates in different types of modes. Each of these guided modes consists of a variety of electromagnetic field configurations, such as transverse electric (TE), transverse magnetic (T M) and hybrid modes. Hybrid modes are combination of transverse electric and magnetic modes.

In simple terms these *modes can be visualised as the possible number of allowed paths of light* in an optical fibre (see Fig.24.6). The paths are all zigzag paths excepting the axial direction. Though the rays having propagation angles between $\theta = 0^o$ and $\theta = \theta_c$ will be in a position to undergo total internal reflections, all of them will not however propagate along the optical fibre. Only a certain ray directions are allowed. As a zigzag ray gets repeatedly reflected at the walls of the fibre, phase shift occurs. Consequently, the waves travelling along certain zigzag paths will be in phase and undergo constructive interference, while the waves coursing along certain other paths will be out of phase and diminish due to destructive interference. The light ray paths along which the waves are in phase inside the fibre are known as **modes**. Each mode is a pattern of electric and magnetic field distributions that is repeated along the fibre at equal intervals. The number of modes propagating in a fibre increases as θ_c or Δ increases. Increasing the core refractive index increases the number of propagating modes. On the other hand, increasing the clad refractive index decreases the number of propagating modes. The number of modes that a fibre will support depends on the ratio d / λ, where d is the diameter of the core and λ is the wavelength of the wave being transmitted. The zero order ray travels along the axis is known as the *axial ray*.

Note that each mode carries a portion of the light from the input signal.

Types of modes:

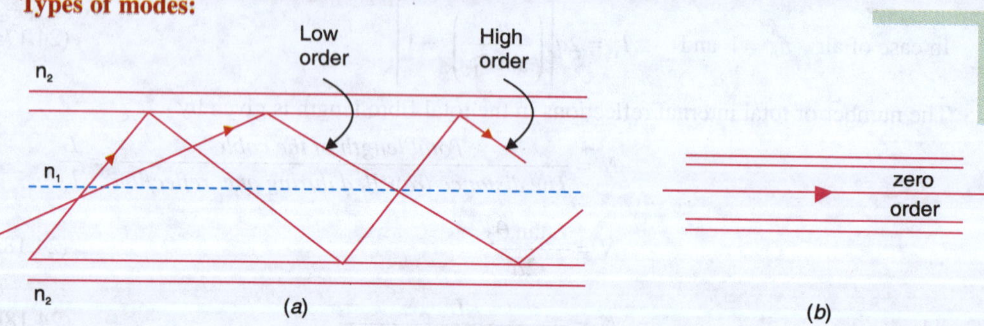

Fig.24.12 : (a) Low and High-order ray paths in a multimode fibre. (b) Axial ray in a single mode fibre

In a fibre of fixed thickness, the modes that propagate at angles close to the critical angle ϕ_C (i.e., critical propagation angle θ_c) are **higher order modes**, and modes that propagate with angles larger than the critical angle (i.e., lower than the critical propagation angle) are **lower order modes** (see Fig.24.12). In case of lower order modes, the fields are concentrated near the center of the fibre. In case of higher order modes, the fields are distributed more towards the edge of the wave-guide and tend to send light energy into the cladding. This energy is lost ultimately. The higher order modes have to traverse longer paths and hence take larger time than the lower order modes to cover a given length of the fibre. Thus, the higher order modes arrive at the output end of the fiber later than the lower order modes.

24.9 TYPES OF RAYS

The rays that propagate through an optical fibre can be classified into two categories: (*i*) meridional rays and (*ii*) skew rays.

1. Meridional ray: A ray that propagates through the fibre undergoing total internal reflection is called meridional ray. It passes through the longitudinal axis of the fibre core (Fig. 24.13a). The propagation of meridional rays is possible only in the TM or TE modes.

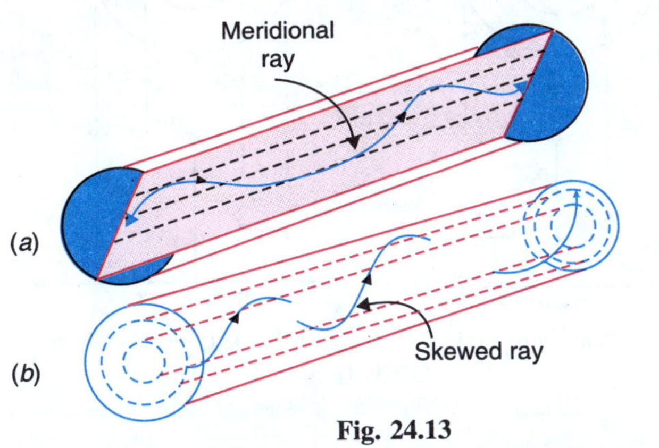

Fig. 24.13

2. Skew ray: The ray that describes angular helical path along the fibre is called a skew ray (see Fig.24.13b). These rays do not pass through the axis of the core. These rays are propagated in either hybrid EH or HE modes. Some of these modes produce losses through leakage of radiation. In real situations, the skew rays constitute a substantial portion of the total number of guided rays. They tend to propagate only in the annular region near the outer surface of the core and do not fully utilize the core as the medium. However, they are complementary to the meridional rays and increase the light gathering capacity of the fibre.

24.10 CLASSIFICATION OF OPTICAL FIBRES

Optical fibres are differently classified into various types basing on different parameters.

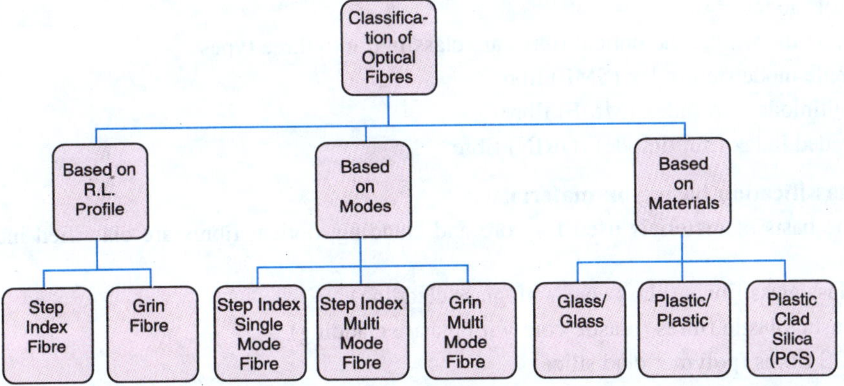

A. Classification basing on refractive index profile:

Refractive index profile of an optical fibre is a plot of refractive index drawn on one of the axes and the distance from the core axis drawn on the other axis (see Fig. 24.14). Optical fibres are classified into the following two categories on the basis of refractive index profile.

1. Step index fibres and 2. **Graded index (GRIN) fibres.**

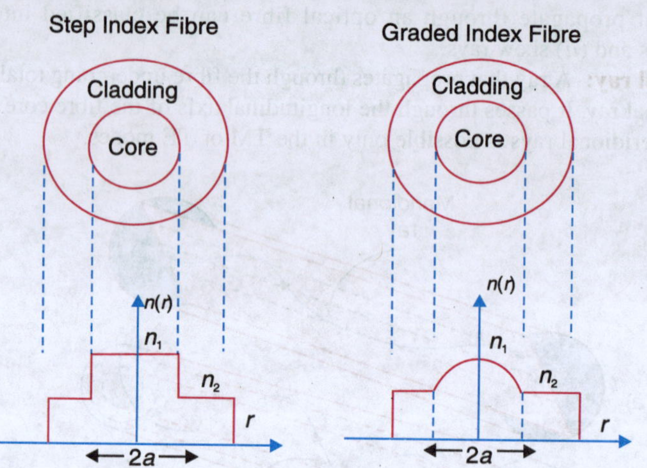

Fig. 24.14 : Classification of optical fibres based on R.I. profile (*a*) Step index fibre (*b*) GRIN fibre

Step index refers to the fact that the refractive index of the core is constant along the radial direction and abruptly falls to a lower value at the cladding and core boundary (see Fig.24.14a). In case of GRIN fibres, the refractive index of the core is not constant but varies smoothly over the diameter of the core (see Fig. 24.14b). It has a maximum value at the center and decreases gradually towards the outer edge of the core. At the core-cladding interface the refractive index of the core matches with the refractive index of the cladding. The refractive index of the cladding is constant.

B. Classification basing on the modes of light propagation:

On the basis of the modes of light propagation, optical fibres are classified into two categories as **1. Single mode fibres (SMF)** and 2. **Multimode fibres (MMF).**

A **single mode fibre** (SMF) has a smaller core diameter and can support only one mode of propagation. On the other hand, a **multimode fibre** (MMF) has a larger core diameter and supports a number of modes.

Thus, on the whole, the optical fibres are classified into three types:

- Single mode step-index (SMF) fibre
- Multimode step-index (MMF) fibre
- Graded index (multimode) (GRIN) fibre.

C. Classification basing on materials:

On the basis of materials used for core and cladding, optical fibres are classified into three categories.

1. Glass/glass fibres (glass core with glass cladding)
2. Plastic/plastic fibres (plastic core with plastic cladding)
3. PCS fibres (polymer clad silica)

24.11 THE THREE TYPES OF FIBRES

We now study the detailed structure and characteristics of the three types of optical fibres.

24.11.1 SINGLE MODE STEP INDEX FIBRE

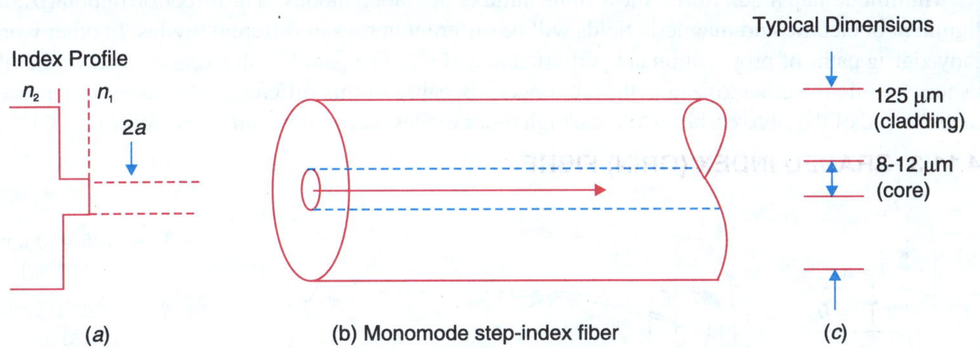

Fig. 24.15: Single mode step index fibre (a) R.I. profile (b) ray paths (c) typical dimensions

Structure

A single mode step index fibre has a very fine thin core of diameter of 8 μm to 12 μm (see Fig.24.15 c). It is usually made of germanium doped silicon. The core is surrounded by a thick cladding of lower refractive index. The cladding is composed of silica lightly doped with phosphorous oxide. The external diameter of the cladding is of the order of 125 μm. The fibre is surrounded by an opaque protective sheath. The refractive index of the fibre changes abruptly at the core-cladding boundary, as shown in Fig. 24.15 (a). The variation of the refractive index of a step index fibre as a function of radial distance can be mathematically represented as

$$\begin{aligned} n(r) &= n_1 \; [r < a \quad \text{inside core}\;] \\ &= n_2 \; [r > a \quad \text{in cladding}] \end{aligned} \qquad (24.19)$$

Propagation of light in SMF

Light travels in SMF along a single path that is along the axis (Fig.24.15b). Obviously, it is the zero order mode that is supported by a SMF. Both Δ and NA are very small for single mode fibres. This relatively small value is obtained by reducing the fibre radius and by making Δ, the relative refractive index change, to be small. The low NA means a low acceptance angle. Therefore, light coupling into the fibre becomes difficult. Costly laser diodes are needed to launch light into the SMF.

24.11.2 MULTIMODE STEP INDEX FIBRE

Structure

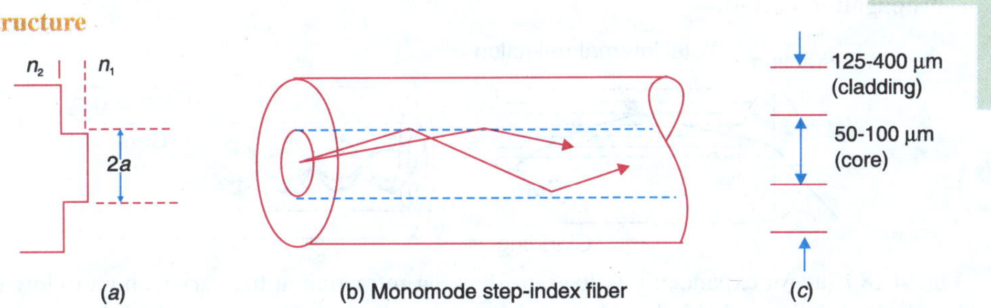

Fig. 24.16: Multimode step index fibre (*a*) R.I. Profile (*b*) Ray paths (*c*) typical dimensions.

A multimode step index fibre is very much similar to the single mode step index fibre except that its core is of larger diameter. The core diameter is of the order of 50 to 100 μm, which is very large compared to the wavelength of light. The external diameter of cladding is about 150 to 250 μm (Fig. 24.16 c).

Propagation of light in MMF

Multimode step index fibres allow finite number of guided modes. The direction of polarization, alignment of electric and magnetic fields will be different in rays of different modes. In other words, many zigzag paths of propagation are permitted in a MMF. The path length along the axis of the fibre is shorter while the other zigzag paths are longer. Because of this difference, the lower order modes reach the end of the fibre earlier while the high order modes reach after some time delay (Fig.24.16b).

24.11.3 GRADED INDEX (GRIN) FIBRE

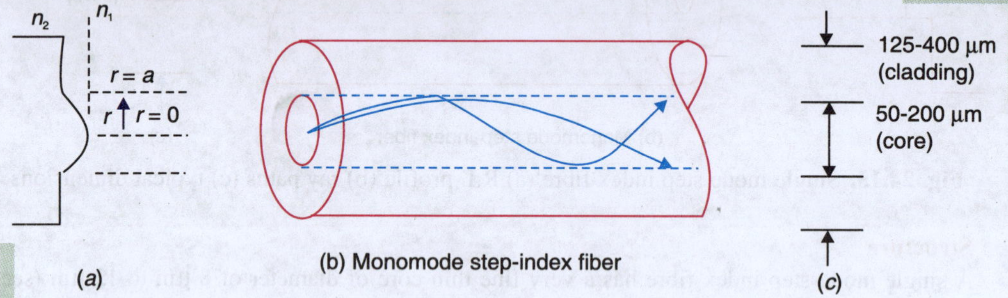

Fig. 24.17 : GRIN fibre (*a*) R.I. Profile (*b*) Ray paths (*c*) typical dimensions

A graded index fibre is a multimode fibre with a core consisting of concentric layers of different refractive indices. Therefore, the refractive index of the core varies with distance from the fibre axis. It has a high value at the centre and falls of with increasing radial distance from the axis. A typical structure and its index profile are shown in Fig. 24.17 (a). Such a profile causes a periodic focussing of light propagating through the fibre. The size of the graded index fibre is about the same as the step index fibre. The variation of the refractive index of the core with radius measured from the center is given by

$$n(r) = \begin{cases} n_1\sqrt{1 - \left[2\Delta\left(\dfrac{r}{a}\right)^{\alpha}\right]}, & r < a \text{ inside core} \\ n_2, & r > a \text{ in cladding} \end{cases} \qquad (24.20)$$

where n_1 is maximum refractive index at the core axis, a the core radius, and α the grading profile index number which varies from 1 to ∞. When $\alpha = 2$, the index profile is parabolic and is preferred for different applications.

Propagation of light

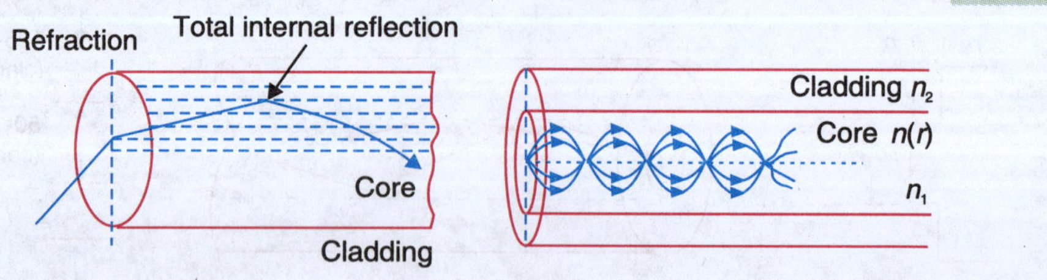

Fig.24.18 : (a) An expanded ray diagram showing refraction at the various high to low index interfaces within graded index fibre, giving an overall curved ray path. (b) Light transmission in a graded index fibre.

As a light ray goes from a region of higher refractive index to a region of refractive index, it is bent away from the normal. The process continues till the condition for total internal reflection is met. Then the ray travels back towards the core axis, again being continuously refracted (Fig.24.18a). The turning around may take place even before reaching the core-cladding interface. Thus, continuous refraction is followed by total internal reflection and again continuous refraction towards the axis. In the graded index fibre, rays making larger angles with the axis traverse longer path but they travel in a region of lower refractive index and hence at a higher speed of propagation. Consequently, all rays traveling through the fibre, irrespective of their modes of travel, will have almost the same optical path length and reach the output end of the fibre at the same time (see Fig. 24.18b).

In case of GRIN fibres, the acceptance angle and numerical aperture decrease with radial distance from the axis. The numerical aperture of a graded index fibre is given by

$$NA = \sqrt{n^2(r) - n_2^2} \approx n_1 (2\Delta)^{\frac{1}{2}} \sqrt{1 - \left(\frac{r}{a}\right)^2}$$

$$= n_1 \sqrt{2\Delta\left[1 - \left(\frac{r}{a}\right)^2\right]} \qquad (24.21)$$

24.12 MATERIALS

Optical fibres are fabricated from glass or plastic which are transparent to optical frequencies. Step index fibres are produced in three common forms – (*i*) a glass core cladded with a glass having a slightly lower refractive index, (*ii*) a silica glass core cladded with plastic and (*iii*) a plastic core cladded with another plastic. Generally, the refractive index step is the smallest for all glass fibres, a little larger for the plastic clad silica (PCS) fibres and the largest for all plastic construction.

24.12.1 ALL GLASS FIBRES

The basic material for fabrication of optical fibres is silica (SiO_2). It has a refractive index of 1.458 at $\lambda = 850\, nm$. Materials having slightly different refractive index are obtained by doping the basic silica material with small quantities of various oxides. If the basic silica material is doped with germania (GeO_2) or phosphorous pentoxide (P_2O_5), the refractive index of the material increases. Such materials are used as core materials and pure silica is used as cladding material in these cases. When pure silica is doped with boria (B_2O_3) or fluorine, its refractive index decreases. These materials are used for cladding when pure silica is used as core material. Examples of fiber compositions are

- SiO_2 core – $B_2O_3.SiO_2$ cladding
- $GeO_2.SiO_2$ core – SiO_2 cladding

The glass optical fibres exhibit very low losses and are used in long distance communications.

24.12.2 ALL PLASTIC FIBRES

In these fibres, perspex (PMMA) and polysterene are used for core. Their refractive indices are 1.49 and 1.59 respectively. A fluorocarbon polymer or a silicone resin is used as a cladding material. A high refractive index difference is achieved between the core and the cladding materials. Therefore, plastic fibres have large NA of the order of 0.6 and large acceptance angles up to 77°. The main advantages of the plastic fibres are low cost and higher mechanical flexibility. The mechanical flexibility allows the plastic fibres to have large cores, of diameters ranging from 110 to 1400 μm. However, they are temperature sensitive and exhibit very high loss. Therefore, they are used in low cost applications and at ordinary temperatures (below 80°C). Examples of plastic fiber compositions are

- Polysterene core $n_1 = 1.60$ NA = 0.60
  - Methyl methacrylate cladding $n_2 = 1.49$
- Polymethyl methacrylate core $n_1 = 1.49$ NA = 0.50
  - cladding made of its copolymer $n_2 = 1.40$

24.12.3 PCS FIBRES

The plastic clad silica (PCS) fibres are composed of silica cores surrounded by a low refractive index transparent polymer as cladding. The core is made from high purity quartz. The cladding is made of a silicone resin having a refractive index of 1.405 or of perfluoronated ethylene propylene (Teflon) having a refractive index of 1.338. Plastic claddings are used for step-index fibres only. The PCS fibres are less expensive but have high losses. Therefore, they are mainly used in short distance applications.

24.13 V-NUMBER

Let us consider a narrow beam of monochromatic light launched on the front end of a step-index fibre, at an angle less than the acceptance angle of the fibre. Let the wavelength of the light be λ_o and the diameter of the fiber be *d*. It appears to us from the ray concept that all the rays contained in the beam propagate along the fibre, such that there can be infinite modes of propagation. However, in practice, only a limited number of modes of propagation are possible in an optical fibre. To understand the reason for this behaviour, we have to recall that phase changes occur as the light waves travel forward. The phase shift takes place due to two reasons – (*i*) due to optical path length traversed and (*ii*) due to total internal reflection at the core-cladding interface.

(*i*) When a wave travels a distance l in a medium of refractive index n_1, it undergoes a phase change δ_1 given by

$$\delta_1 = k\, n_1 l = \frac{2\pi\, l\, n_1}{\lambda} \tag{24.22}$$

where k is the propagation constant.

(*ii*) Whenever, a wave with component normal to the reflecting surface undergoes total internal reflection, the phase shift, δ_2, is given by

$$\delta_2 = 2\tan^{-1}\frac{\sqrt{n_1^2\cos^2\phi - n_2^2}}{n_1\sin\phi} \tag{24.23}$$

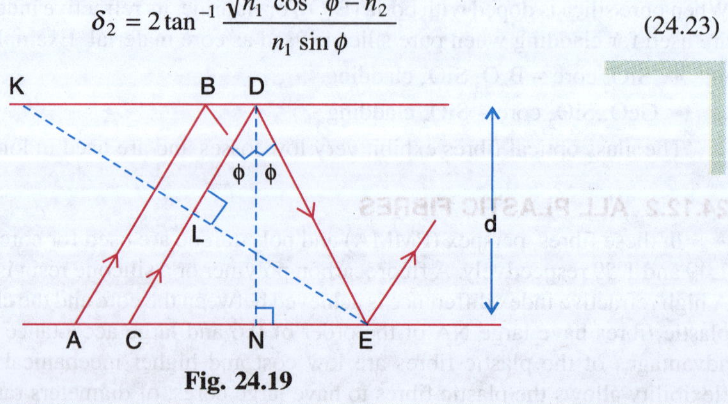

Fig. 24.19

In Fig. 24.19, AB and CD are parallel rays in an incident beam. The line KL is perpendicular to the propagation path of the rays AB, CD and hence represents a plane wavefront. The points K and L lying on the same wavefront will be in phase with each other. As the point E, which is on the reflected ray DE, lies on the wavefront KL, the points L and E must be in phase with each other. However, moving from the point L to E along the ray, we find that there occurs a phase shift given by

$$\delta = (LD + DE)\frac{2\pi\, n_1}{\lambda_o} - 2\delta_2 \tag{24.24}$$

The factor 2 in the above equation takes into account the two total internal reflections at D and E. In the $\Delta^{le}\, DNE$

$$\frac{DN}{DE} = \cos\phi \text{ . Therefore, } \quad DE = \frac{DN}{\cos\phi} = \frac{d}{\cos\phi}$$

Further, in the $\Delta^{le}\, LDE$

$$\frac{LD}{DE} = \cos 2\phi \text{ . Therefore, } \quad LD = DE\cos 2\phi$$

$$LD + DE = DE\,(1 + \cos 2\phi) = 2\, DE\cos^2\phi.$$

Therefore, $$LD + DE = 2\frac{d}{\cos\phi}\cos^2\phi = 2d\cos\phi$$

Using the above expression into equ.(24.23), we obtain

$$\delta = \frac{4d\pi\, n_1 \cos\phi}{\lambda_o} - 2\delta_2$$

Now the condition for the wave associated with the ray CD to propagate along the optical fibre is that the phase of the twice reflected wave must be the same as that of the incident wave. That is, the wave must interfere constructively with itself. If this phase condition is not satisfied, the wave would interfere destructively with itself and just die out. It means that the total phase shift must be equal to an integer multiple of 2π radians. Thus

$$\frac{4d\pi\, n_1 \cos\phi}{\lambda_o} - 2\delta_2 = 2\pi\, m$$

or $$m = \frac{2d\, n_1 \cos\phi_m}{\lambda_o} - \frac{\delta_2}{\pi} \tag{24.25}$$

where m is an integer that determines the allowed ray angles for propagation of the wave and ϕ_m is the value of ϕ corresponding to a particular value of m.In order to sustain total internal reflection,

$$\sin\phi_m \geq \frac{n_2}{n_1}$$

∴ $$\cos\phi_m \leq \frac{\sqrt{n_1^2 - n_2^2}}{n_1}$$

∴ $$m \leq \frac{2d\sqrt{n_1^2 - n_2^2}}{\lambda_o} - \frac{\delta_2}{\pi} \tag{24.26}$$

or $$m \leq \frac{2V}{\pi} - \frac{\delta_2}{\pi} \tag{24.27}$$

where V is given by

$$V = \frac{\pi\, d}{\lambda_o}\sqrt{n_1^2 - n_2^2} \tag{24.28}$$

V-number is more generally called normalized **frequency** of the fibre. Each mode has a definite value of V-number below which the mode is cut off. Equ.(24.28) can be written as

$$V = \frac{\pi\, d}{\lambda_o}(NA) \qquad (24.29)$$

or

$$V = \frac{\pi\, d}{\lambda_o} n_1 \sqrt{2\Delta} \qquad (24.30)$$

The maximum number of modes N_m supported by an SI fibre is given by

$$N_m = \tfrac{1}{2} V^2 \qquad (24.31)$$

Thus, for $V = 10$, N_m is 50. When the normalized frequency V is less than 2.405, the fibre can support only one mode, which propagates along the axial length of the fibre, and the fibre becomes a single mode fibre. It means that for single mode transmission in a MMF, V must be less than 2.405. The wavelength at which the fibre becomes single mode is called **cutoff wavelength**, λ_c of the fibre. Using equ.(24.29), we can write

$$\lambda_c = \frac{\pi\, d}{2.405}(NA) \qquad (24.32)$$

It is seen from the above equation that single mode property can be realized in a multimode fibre by decreasing the core diameter and/or decreasing Δ such that V < 2.405.

In case of GRIN fibres, for larger values of V,

$$N_m \cong \frac{V^2}{4} \qquad (24.33)$$

24.14 FABRICATION

A number of techniques are available to produce all glass fibres. In one of the methods, known as the double crucible method, fibres are directly produced from the melt.

Double Crucible Technique

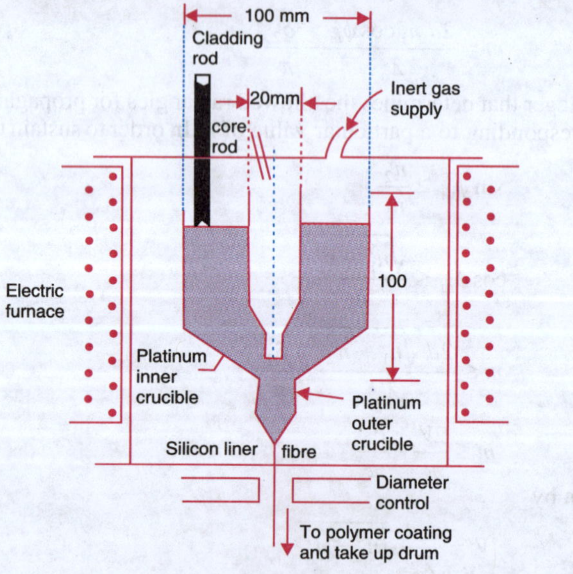

Fig. 24.20

The double crucible consists of two concentric platinum crucibles having thin orifices at the bottom. Raw material for the core-glass is placed in the inner crucible and the raw material for cladding is fed to the outer crucible. The double crucible arrangement is mounted vertically (Fig.24.20) in a furnace. The furnace is maintained at a suitable temperature to take the raw material into molten state. The fibres are drawn through the thin orifices at the bottom of the crucibles. As both the materials are drawn simultaneously, a filament of core glass surrounded by a tube of cladding glass is obtained in the process. The thickness of the fibre is monitored and the fibre is then coated with a polymer. Subsequently, it is passed through a plastic extrusion die to form a plastic sheath over the fibre.

24.15 LOSSES IN OPTICAL FIBRE

As a light signal propagates through a fibre, it suffers loss of amplitude and change in shape. The loss of amplitude is referred to as *attenuation* and the change in shape as *distortion*.

24.15.1 ATTENUATION

When an optical signal propagates through a fibre, its power decreases exponentially with distance. The loss of optical power as light travels down a fiber is known as **attenuation**. The attenuation of optical signal is defined as *the ratio of the optical output power from a fibre of length L to the input optical power.* If P_i is the optical power launched at the input end of the fibre, then the power P_o at a distance L down the fibre is given by

$$P_o = P_i \, e^{-\alpha L} \qquad (24.34)$$

where α is called the **fibre attenuation coefficient** expressed in units of km^{-1}. Taking logarithms on both the sides of the above equation, we obtain

$$\alpha = \frac{1}{L} ln \frac{P_i}{P_o} \qquad (24.35)$$

In units of dB / km, α is defined through the equation

$$\therefore \quad \alpha_{dB/km} = \frac{10}{L} \log \frac{P_i}{P_o} \qquad (24.36)$$

In case of an ideal fibre, $P_o = P_i$ and the attenuation would be zero.

24.15.2 DIFFERENT MECHANISMS OF ATTENUATION

There are several loss mechanisms responsible for attenuation in optical fibres. They are broadly divided into two categories: *intrinsic* and *extrinsic* attenuation. Intrinsic attenuation is caused by substances inherently present in the fiber, whereas extrinsic attenuation is caused by external forces such as bending.

A. Intrinsic Attenuation

Intrinsic attenuation results from materials inherent to the fiber. It is caused by impurities present in the glass. During manufacturing, there is no way to eliminate all impurities. When a light signal hits an impurity in the fiber, either it is scattered or it is absorbed. Intrinsic attenuation can be further characterized by two components:

- **Material absorption**
- **Rayleigh scattering**

Absorption by material

Material absorption occurs as a result of the imperfection and impurities in the fiber and accounts for 3-5% of fiber attenuation. The most common impurity is the hydroxyl (OH-) molecule, which remains as a residue despite stringent manufacturing techniques. These radicals result from the presence of water remnants that enter the fiber-optic cable material through either a chemical reaction in the manufacturing process or as humidity in the environment. The natural impurities in the glass absorb light signal, and convert it into vibrational energy or some other form of energy. Hydroxyl radical ions(OH), and transition metals such as copper, nickel, chromium, vanadium and manganese have electronic absorption in and near visible part of the spectrum. Their presence causes heavy losses.

Even a highly pure glass absorbs light in specific wavelength regions. Strong electronic absorption occurs at UV wavelengths, while vibrational absorption occurs at IR wavelengths.

Losses due to impurities can be reduced by better manufacturing processes. In improved fibres, metal ions are practically negligible. The largest loss is caused by OH ions. These cannot be sufficiently reduced. The absorption of light either through intrinsic or impurity process constitutes a transmission loss because that much energy is subtracted from the light propagating through the fibre. The absorption losses are found to be at minimum at around 1.3 μm.

Unlike scattering, absorption can be limited by controlling the amount of impurities during the manufacturing process.

Rayleigh Scattering

Rayleigh scattering accounts for the majority (about 96%) of attenuation in optical fiber. The local microscopic density variations in glass cause local variations in refractive index. These variations, which are inherent in the manufacturing process and cannot be eliminated, act as obstructions and scatter light in all directions (Fig.24.21). This is known as *Rayleigh scattering*. The Rayleigh scattering loss greatly depends on the wavelength. It varies as $1/\lambda^4$ and becomes important at lower wavelengths. Thus, Rayleigh scattering sets a lower limit, on the wavelengths that can be transmitted by a glass fibre at 0.8 μm, below which the scattering loss is very high.

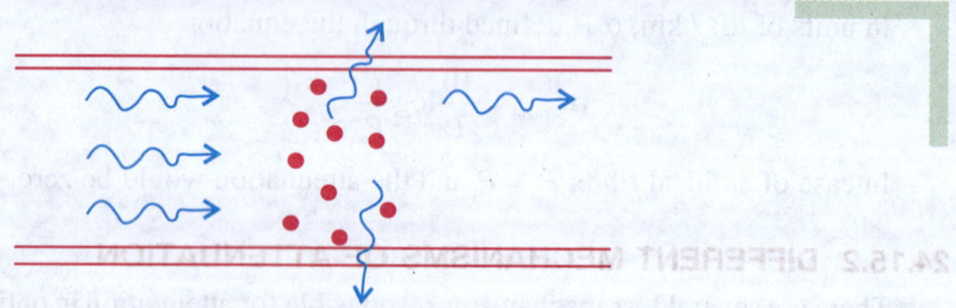

Fig. 24.21: Rayleigh scattering, showing attenuation of an incident stream of photons due to localized variations in refractive index.

Any wavelength that is below 800 nm is unusable for optical communication because attenuation due to Rayleigh scattering is high. At the same time, propagation above 1700 nm is not possible due to high losses resulting from infrared absorption.

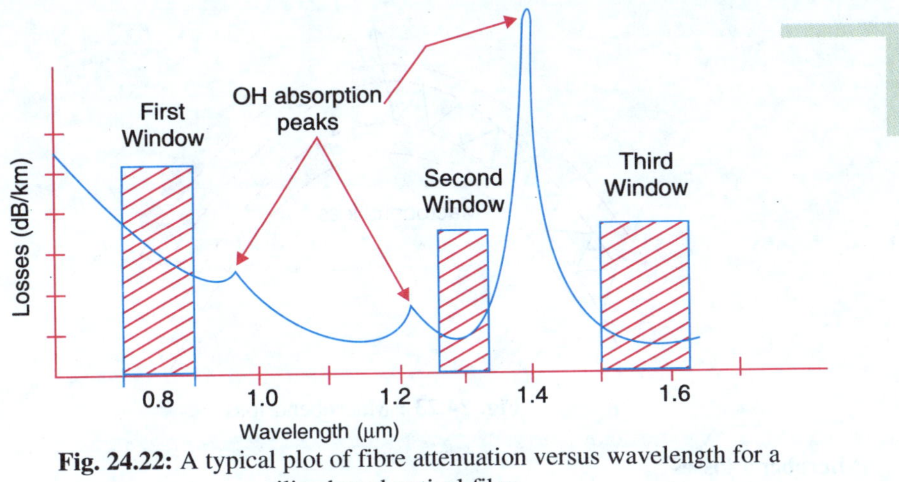

Fig. 24.22: A typical plot of fibre attenuation versus wavelength for a silica based optical fibre.

Fig. 24.22 shows the variation of attenuation with wavelength measured for a typical fiber-optic cable.

For better performance, the choice of wavelength must be based on minimizing loss and minimizing dispersion. Such windows are selected for communication purposes. It is seen from the attenuation curve that it has a minimum at around a particular band of optical wavelengths. The band of wavelengths at which the attenuation is a minimum is called **optical window** or **transmission window** or **low-loss window**. There are three **principal windows**. These correspond to wavelength regions in which attenuation is low and matched to the capability of a transmitter to generate light efficiently and a receiver to carry out detection.

λ (nm)	**Approx. loss (dB/km)**
820 - 880	2.2
1200 - 1320	0.6
1550 - 1610	0.2

From the above data it is seen that the range 1550 to 1610 is most preferable. From the point of view of dispersion, the low intramodal dispersion wavelength of about 1300nm is most suitable.

B. Extrinsic Attenuation or Bending losses

Extrinsic attenuation is caused by two external mechanisms: **macrobending** or **microbending**. Both of them cause a reduction of optical power. If a bend is imposed on an optical fiber, strain is placed on the fiber along the region that is bent. The bending strain affects the refractive index and the critical angle of the light ray in that specific area. As a result, the condition for total internal reflection is no longer satisfied. Hence, light traveling in the core can refract out, and loss occurs.

Macrobend losses

A **macrobend** is a large-scale bend that is visible. When a fibre is bent through a large angle, strain is placed on the fiber along the region that is bent. The bending strain will affect the refractive index and the critical angle of the light ray in that specific area. As a result, light traveling in the core can refract out, and loss occurs. (Fig. 24.23). To prevent macrobends, optical fiber has a *minimum bend radius* specification that should not be exceeded. This is a restriction on how much bend a fiber can withstand before experiencing problems in optical performance or mechanical reliability.

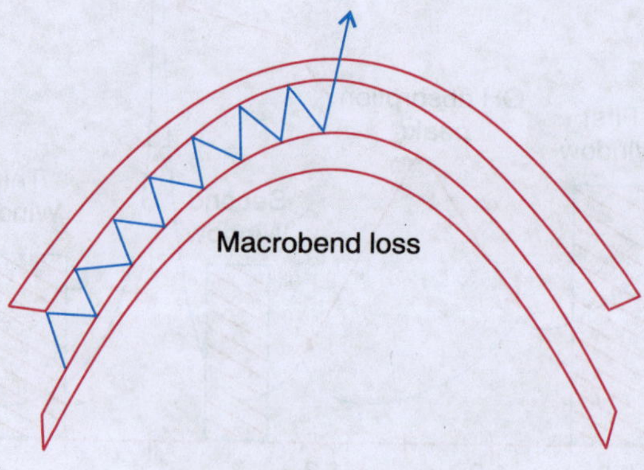

Fig. 24.23 : Macrobend loss

Microbend losses

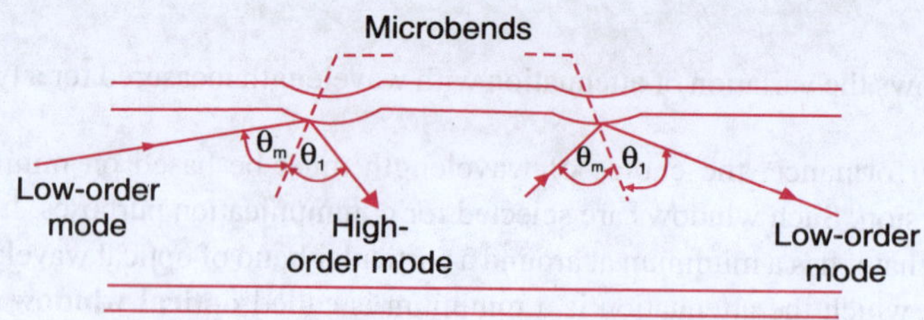

Fig. 24.24: Microbend losses

Microbend is a small-scale distortion. It is localized and generally indicative of pressure on the fiber. Microbending might be related to temperature, tensile stress, or crushing force. Microbending is caused by imperfections in the cylindrical geometry of fiber during the manufacturing process or installation processes. The bend may not be clearly visible upon inspection. Structural variations in the fibre, or fibre deformation, cause radiation of light away from the fibre (Fig. 24.24). Microbending may occur, for example, due to winding of optical fibre cable over spools. Light rays get scattered at the small bends and escape into the cladding. Such losses are known as microbend losses.

24.16 DISTORTION

In an optical fibre communication system, the information (signal) is coded in the form of discrete pulses of light, which are transmitted through the fibre. The light pulses are of a given width, amplitude and interval. The number of pulses that can be sent per unit time will determine the information capacity of the fibre. More information can be sent by optical cable when distinct pulses can be transmitted in more rapid succession. The pulses travel through the transmitting medium (i.e., optical fibre) and reach the detector at the receiving end. For the information to be retrieved at the detector, it is necessary that the optical pulses are well resolved in time. However, the light pulses broaden and spread into a wider time interval because of the different times taken by different rays propagating through the fibre. This phenomenon is known as **distortion** or **pulse dispersion**. Hence, even though two pulses may be well resolved at the input end, they may overlap on each other at the output end, as shown in Fig.24.25. It is obvious that the pulse broadening depends on the length of the travel of the pulses through the fibre. Hence, dispersion is expressed in units of **ns/km** (time/distance).

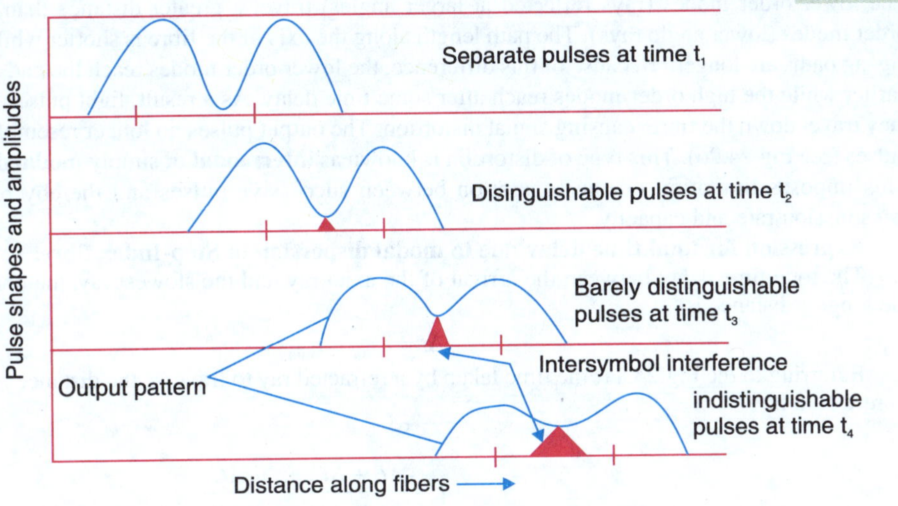

Fig. 24.25 : Distortion of the pulses traveling along a fibre

The following three different dispersion mechanisms determine the distortion of the signal in an optical fibre. They are

- Intermodal dispersion and
- Intramodal dispersion.

Intramodal dispersion is again divided into the following two types.

- Material dispersion
- Waveguide dispersion

24.16.1 INTERMODAL DISPERSION

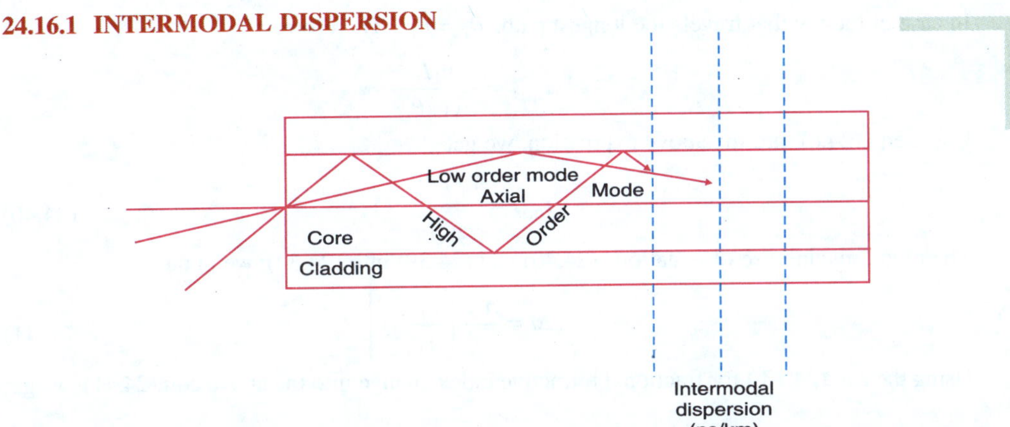

Fig. 24.26: Lower order modes reach the end of the fibre earlier while the high order modes reach after some time delay

Intermodal dispersion occurs as a result of the differences in the group velocities of the modes. For example, let us consider the propagation of a pulse through a multimode fibre. The power associated with the single pulse gets distributed into the various modes or paths guided by the fibre.

The lower order modes (rays reflected at larger angles) travel a greater distance than the higher order modes (lower angle rays). The path length along the axis of the fibre is shorter while the other zigzag paths are longer. Because of this difference, the lower order modes reach the end of the fibre earlier while the high order modes reach after some time delay. As a result, light pulses broaden as they travel down the fibre, causing signal distortion. The output pulses no longer resemble the input pulses (see Fig.24.26). This type of distortion is known as **intermodal** or simply **modal dispersion**. This imposes limitation on the separation between successive pulses and thereby reduces the transmission rate and capacity.

Expression for total time delay due to modal dispersion in Step-Index fibre:

The total time delay between the arrival of the axial ray and the slowest ray, the one traveling the longest distance is

$$\Delta t = t_{max} - t_{min} \tag{24.37}$$

Referring to the Fig.24.11, the time taken by a refracted ray to traverse the distance ABC of the fibre would be

$$t' = \frac{AB + BC}{\upsilon} = \frac{n_1 AC}{c \cos\theta_r}$$

where $\upsilon = c / n_1$ is the speed of the light in the core. Since the ray path will repeat itself, the time taken by a ray to traverse a length L of the fibre is

$$t = \frac{n_1 L}{c \cos\theta_r} \tag{24.38}$$

The above relation shows that the time taken by a ray in the fibre core is a function of the angle θ_r. For the axial ray $\theta_r = 0$ and hence

$$t_{min} = \frac{n_1 L}{c} \tag{24.39}$$

In case of the ray that travels the longest path, $\theta_r = \theta_C$. Therefore,

$$t_{max} = \frac{n_1 L}{c \cos\theta_C}$$

Using equ.(24.6) into the above expression, we get

$$t_{max} = \frac{n_1^2 L}{n_2 c} \tag{24.40}$$

Therefore, making use of equations (24.40) and (24.39) into (24.37), we obtain

$$\Delta t = \frac{n_1 L}{c}\left[\frac{n_1}{n_2} - 1\right] \tag{24.41}$$

Using the equ.(24.13) for fractional refractive index change into the above equ.(24.41), we get

$$\Delta t = \frac{n_1 L}{c}\left[\frac{\Delta}{1 - \Delta}\right] \tag{24.42}$$

We can also express the relation (24.41) in the following form.

$$\Delta t = \frac{n_1 L}{c}\left[\frac{n_1 - n_2}{n_2}\right] = \frac{n_1 L}{c}\left[\frac{n_1 - n_2}{n_2}\right]\left[\frac{n_1 + n_2}{n_1 + n_2}\right]$$

$$= \frac{n_1 L}{c}\left[\frac{\left(n_1^2 - n_2^2\right)}{n_2\left(n_1 + n_2\right)}\right] = \frac{n_1 L}{c}\frac{\left(n_1^2 - n_2^2\right)}{2n_1 n_2}$$

or $$\Delta t = \frac{L}{2n_2 c}(NA)^2 \tag{24.43}$$

It is seen from the equ.(24.43) that the time delay is proportional to the square of the value of NA. Therefore, a large NA fibre allows more modes of propagation of light, which will result in greater modal dispersion. A smaller NA limits the number of modes, hence reduces dispersion. It is further seen that the intermodal dispersion does not depend upon the spectral width of the source. It follows that a light pulse from an ideal monochromatic source would still get broadened.

24.16.2 INTRAMODAL DISPERSION

Intramodal dispersion is the spreading of light pulse within a single mode. The two main causes of intramodal dispersion are (*a*) material dispersion and (*b*) waveguide dispersion.

a. Material Dispersion: Glass is a dispersive medium. A light pulse is a wave packet, composed of a group of components of different wavelengths. The different wavelength components will propagate at different speeds along the fibre (Fig. 24.27). The short wavelength components travel slower than long wavelength components, eventually causing the light pulse to broaden. This type of distortion is known as **material dispersion**. It is often called the **chromatic dispersion**. Obviously, the spectral width of the source determines the extent of material dispersion.

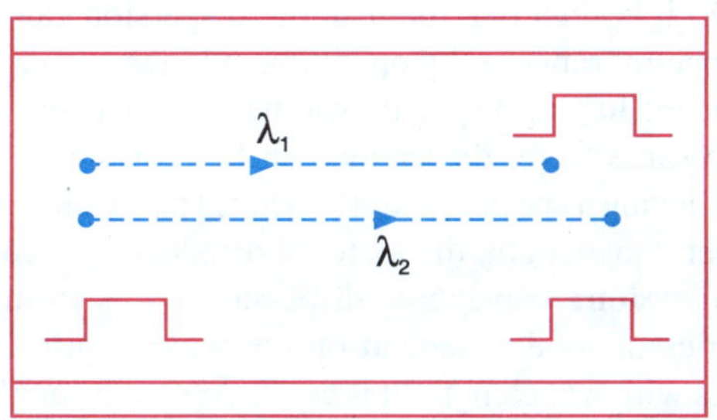

Fig. 24.27 : The different wavelength components propagate at different speeds along the fibre

Expression for time delay due to material dispersion:

Let us consider a plane wave propagating in fiber core. It is represented by $\psi \propto \exp(kx - \omega t)$. The wave number k is given by

$$k = \frac{2\pi}{\lambda} = \frac{2\pi}{\lambda_o}\cdot\frac{\lambda_o}{\lambda} = \frac{2\pi}{\lambda_o}\cdot n = \frac{2\pi\nu}{c}\cdot n$$

or $$k = \frac{\omega n}{c} \tag{24.44}$$

and $$\omega = 2\pi\nu = \frac{2\pi}{\lambda}c \tag{24.45}$$

A wave packet of finite spread of wavelengths travels with group velocity υ_g is given by

$$\upsilon_g = \frac{d\omega}{dk}$$

$$\therefore \quad \frac{1}{v_g} = \frac{dk}{d\omega} = \frac{d}{d\omega}\left(\frac{\omega n}{c}\right) = \frac{n}{c} + \frac{\omega}{c}\frac{dn}{d\omega} = \frac{1}{c}\left[n + \omega\frac{dn}{d\omega}\right]$$

But

$$\frac{dn}{d\omega} = \frac{dn}{d\lambda}\cdot\frac{d\lambda}{d\omega} = -\frac{\lambda^2}{2\pi c}\cdot\frac{dn}{d\lambda}$$

$$\therefore \quad \frac{1}{v_g} = \frac{1}{c}\left[n - \frac{\omega\lambda^2}{2\pi c}\frac{dn}{d\lambda}\right] = \frac{1}{c}\left[n - \lambda\frac{dn}{d\lambda}\right]$$

As the signal propagates through the fibre, each spectral component can be assumed to travel independently and to undergo a time delay per unit length in the direction of propagation, which is given by

$$t_{mat} = \frac{L}{v_g} = \frac{L}{c}\left[n - \lambda\frac{dn}{d\lambda}\right] \tag{24.46}$$

The pulse spread Δt_{mat} for a source of spectral width $\Delta\lambda$ is found by differentiating the equ.(24.46) with respect to λ and then multiplying by $\Delta\lambda$. Thus,

$$\Delta t_{mat} = \frac{dt_{mat}}{d\lambda}\Delta\lambda = -\frac{L\lambda}{c}\frac{d^2n}{d\lambda^2}\Delta\lambda = D_{mat}(\lambda)L\lambda \tag{24.47}$$

where $D_{mat}(\lambda)$ is the mateial dispersion.

$$D_{mat}(\lambda) = -\frac{\lambda}{c}\frac{d^2n}{d\lambda^2} \tag{24.48}$$

From the equ.(24.48) it is seen that the material dispersion can be reduced either by choosing sources with narrower spectral range or by operating at longer wavelengths. To cite an example, an LED operating at 820 nm and having a spectral width of 38 nm results in dispersion of about 3 ns/km in a certain fibre. In the same fibre, dispersion can be reduced to 0.3 ns/km using a laser diode operating at 1140 nm and having a spectral width of 3 nm. Thus, using a more and more monochromatic source operating at higher wavelength, the material dispersion is reduced.

b. Wave-guide Dispersion: Waveguide dispersion arises from the guiding properties of the fibre. The group velocities of modes depend on the wavelength. Hence, the effective refractive index for any mode varies with wavelength. It is equivalent to the angle between the ray and the fibre axis varying with wavelength which subsequently leads to a variation in the transmission times for the rays and hence dispersion (see Fig.24.28). Waveguide dispersion is generally small in MMF, but it is important in SMF.

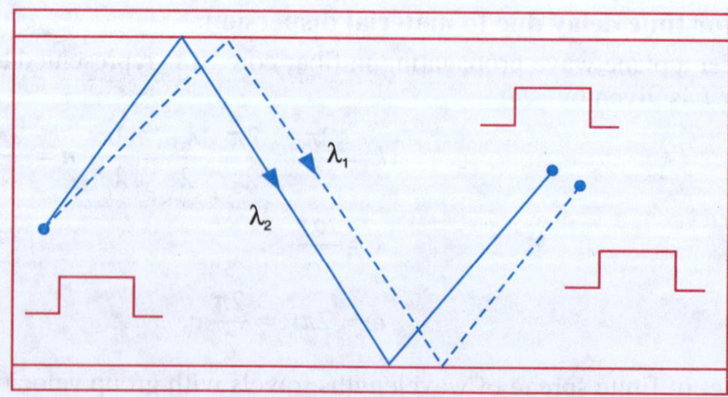

Fig. 24.28: Wave guide dispersion

The intermodal distortion can be reduced if graded index fibre is used. In case of a graded index fibre, the refractive index is larger at the center and it gradually decreases away from the center. A pulse traveling along the axis of the fibre, travels along a shorter path but it takes longer time to reach the end of the fibre since it is traveling through a medium of higher refractive index. On the other hand, the pulse traveling away from the axis travels a longer distance but takes lesser time since it is traveling through a medium of lower refractive index. As a result both the pulses reach the end of the fibre simultaneously. Thus, using a GRIN fibre can reduce the problem of intermodal dispersion. Low NA fibres exhibit smaller dispersion. Dispersion may be restricted by a careful selection of low NA fibre and a narrow spectral width fibre.

In a MMF, all three pulse spreading mechanism exist simultaneously. In case of SMF, only material and wave-guide dispersion exist.

24.16.3 TOTAL DISPERSION

All the above three dispersions contribute pulse spreading during signal transmission through an optical fibre. The total dispersion introduced by an optical fibre is given by the root mean square value of all the three dispersions. Thus,

$$(\Delta t)_T = \sqrt{(\Delta t)^2_{intermodal} + (\Delta t)^2_{mat} + (\Delta t)^2_{wg}} \tag{24.49}$$

24.17 BANDWIDTH

It is learnt in the above section that various dispersion mechanisms cause broadening of the information signal in time domain. If the pulses spread more, they can interfere with the adjacent pulses resulting in *Inter Symbol Interference* or ISI in short and there can be so much of ISI that it becomes impossible to distinguish between the individual pulses. Therefore, for a given broadening, the pulses have to be separated by a minimum time interval in order to avoid overlapping of the pulses. This would determine the ultimate information-carrying capacity of the system. When the pulse separation is increased, the data transfer rate decreases. Thus, broadening of pulses puts an upper limit on the rate of pulse transmission. To a first approximation, it may be taken that the bandwidth in hertz is equal to the digital bit rate. Thus, $B_T = \frac{1}{\tau} = B$, where τ is the input pulse duration. In other words, the maximum allowable transmission rate is called **bandwidth**. In practice, the fibre bandwidth is expressed in terms of MHz.km, a product of frequency and distance. This is known as **bandwidth-distance** product, which specifies the usable bandwidth over a definite distance. With the increase in distance, different dispersion effects would increase in the optical fibre and as a result the usable bandwidth reduces. The attenuation per kilometer and the bandwidth-kilometer product are the important performance parameters of optical fibres.

24.18 CHARACTERISTICS OF THE FIBRES

A. Step-index single-mode fibre

- It has a very small core diameter, typically of about 10 μm.
- Its numerical aperture is very small.
- It supports only one mode in which the entire light energy is concentrated.
- A single mode step index fibre is designed to have a V number between 0 and 2.4.
- Because of a single mode of propagation, loss due to intermodal dispersion does not exist. With careful choice of material, dimensions, and wavelength, the total dispersion can be made extremely small.

- The attenuation is least.
- The single mode fibres carry higher bandwidth than multimode fiber.
- It requires a monochromatic and coherent light source. Therefore, laser diodes are used along with single mode fibres.

Advantages

- No degradation of signal
- Low dispersion makes the fibre suitable for use with high data rates. Single-mode fiber gives higher transmission rate and up to 50 times more distance than multimode.
- Highly suited for communications.

Disadvantages

- Manufacturing and handling of SMF are more difficult.
- The fibre is costlier.
- Launching of light into fibre is difficult.
- Coupling is difficult.

Applications

- Used as under water cables

B. Step-index multi-mode fibre

- It has larger core diameter, typically ranging between 50-100 μm.
- The numerical aperture is larger and it is of the order of 0.3.
- Larger numerical aperture allows more number of modes, which causes larger dispersion. The dispersion is mostly intermodal.
- Attenuation is high.
- Incoherent sources like LEDs can be used as light sources with multimode fibres.

Advantages

- The multimode step index fibre is relatively easy to manufacture and is less expensive.
- LED or laser source can be used.
- Launching of light into fibre is easier.
- It is easier to couple multi-mode fibres with other fibres.

Disadvantages

- Has smaller bandwidth.
- Due to higher dispersion data rate is lower and transmission is less efficient.
- It is less suitable for long distance communications.

Applications

- Used in data links.

C. Graded-index multi-mode fibre

- Core diameter is in the range of 50-100 μm.
- Numerical aperture is smaller than that of step-index multimode fibre.
- The number of modes in a graded index fibre is about half that in a similar multimode step-index fibre.
- Has medium attenuation.
- Intermodal dispersion is zero, but material dispersion is present.
- Has better bandwidth than multimode step-index fibre.

Advantages

- Either an LED or a laser can be used as the source of light with GRIN fibres.

Disadvantages

- The manufacture of graded index fibre is more complex. Hence, it is the most expensive fibre.
- Coupling fibre to the light source is difficult.

Applications

- Used in telephone links.

24.19 SPLICING

It is often required to join two optical fibres together to form a continuous optical waveguide. The method and technique for connecting the fibres depends on whether a permanent joint is required or easily disconnected joint is required. The permanent bonding technique is called **splice** technique and easily disconnected joint techniques are called **connectors**. Splicing is analogous to soldering in metal wires. It consists of fusing of two fibre ends and bonding them together in an alignment structure. The generally accepted splicing method is arc fusion splicing, which melts the fiber ends together with an electric arc. For quicker fastening jobs, a "mechanical splice" is used.

1. Fusion splicing

Fusion splicing is the act of joining two optical fibres end-to-end using heat. The goal is to fuse the two fibers together in such a way that light passing through the fibers is not scattered or reflected back by the splice, and so that the splice and the region surrounding it are almost as strong as the virgin fiber itself.

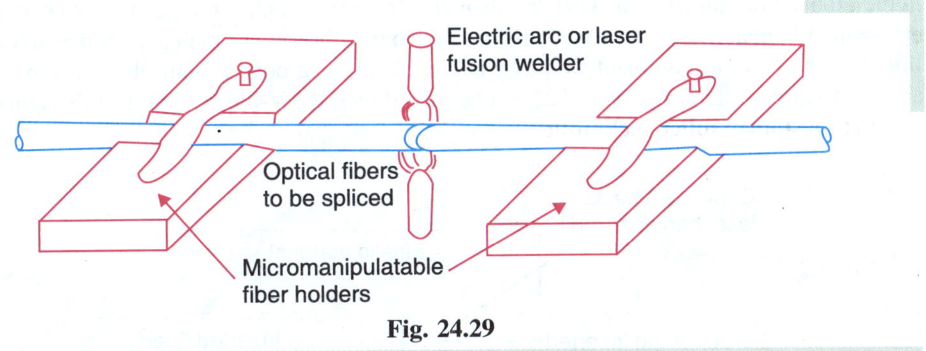

Fig. 24.29

Fusion splicing is done with a specialized instrument that typically operates as follows: The two cable ends are fastened inside a splice enclosure that will protect the splices (Fig.24.29), and the fiber ends are stripped of their protective polymer coating (as well as the sturdier outer jacket, if present). The ends are *cleaved* (cut) with a precision cleaver to make them perpendicular, and are placed into special holders in the splicer. The splice is usually inspected via a magnified viewing screen to check the cleaves before and after the splice. The splicer uses small motors to align the end faces together, and emits a small spark between electrodes at the gap to burn off dust and moisture. Then the splicer generates a larger spark that raises the temperature above the melting point of the glass, fusing the ends together permanently. The splices offer controlled alignment of fiber optic cables to achieve losses as low as 0.05 dB.

2. Mechanical splicing

A **mechanical splice** is a junction of two or more optical fibres that are aligned and held in place by a self-contained assembly. The fibers are not permanently joined, just precisely held together so that light can pass from one to another. They are easily applied in the field, require little or no tooling and offer losses of about 0.2 dB.

Mechanical fiber splices are designed to be quicker and easier to install, but there is still the need for stripping, careful cleaning and precision cleaving. The fiber ends are aligned and held together by a precision-made sleeve, often using a clear index-matching gel that enhances the transmission of light across the joint. Such joints typically have higher optical loss and are less robust than fusion splices, especially if the gel is used. All splicing techniques involve the use of an enclosure into which the splice is placed for protection afterward.

***(i)* V- groove splice technique**

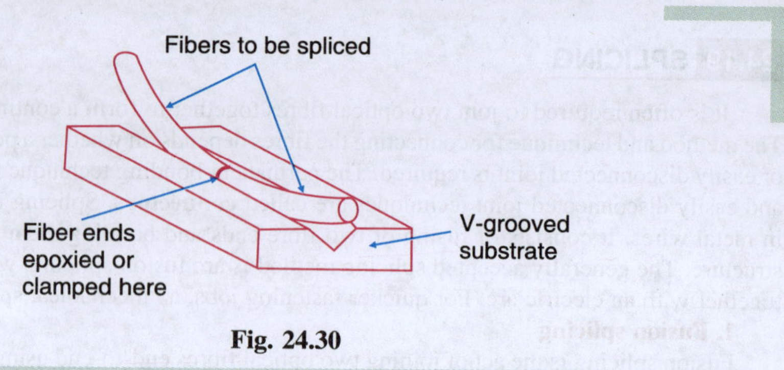

Fig. 24.30

The V-block is the simplest mechanical splice. The bared fibres to be joined are placed in the groove (Fig. 24.30). Angular alignment is particularly well controlled. The two fibres can slide in the groove until they touch. They are then epoxied permanently into position, so end-separation errors are minimal. If the epoxy is index matched to the fibre, even small gaps can be tolerated with little loss. Lateral misalignment would be negligible in the groove if both fibres had the same core and cladding diameters. A cover plate can be placed over the V-block to protect the splice further.

***(ii)* Elastomer splice technique**

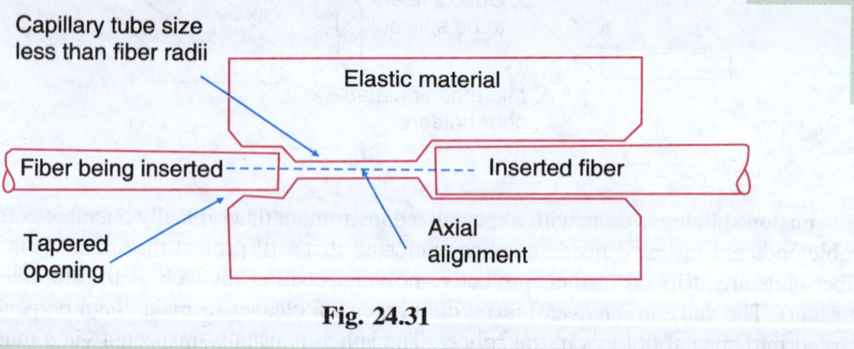

Fig. 24.31

Another splice is essentially a precision sleeve made with elastomeric materials. The elastomer is an elastic material usually made into a cylinder with an opening along its axis. The groove is a little smaller than the fibre but accepts and centers it by expanding slightly when the fibre is inserted (Fig. 24.31). The fibres are inserted from both the ends of the cylinder and touch near its midpoint. The slice can be epoxied for permanent connection. An external splice holder is used for full protection of the splice.

24.20 APPLICATIONS

Transmission of light via an optical fibre has a wide variety of applications. We discuss here some of the applications. Broadly, optical fibres have three different applications, apart from other miscellaneous applications.

a. They are used for illumination and short distance transmission of images.

b. They are used as wave-guides in telecommunications.

c. They are used in fabricating a new family of sensors.

24.20.1 ILLUMINATION AND IMAGE TRANSMISSION

A large number of fibres whose ends are bound together, ground and polished, form flexible bundles. One of the ends of the bundle acts as an input end while the other acts as an output end. If the relative positions of the fibre terminations at both the ends are not the same, and if no attempt is made to align the fibres in an orderly array, the bundle is said to be an *incoherent bundle*. In such a case, there would not be any correlation in the positions of the fibre terminations at one end of the bundle with that at the other end of the bundle. The primary function of such bundles is simply to conduct light from one region to another. Such **flexible light carriers** are relatively easy to make and inexpensive. They are used for illumination purpose.

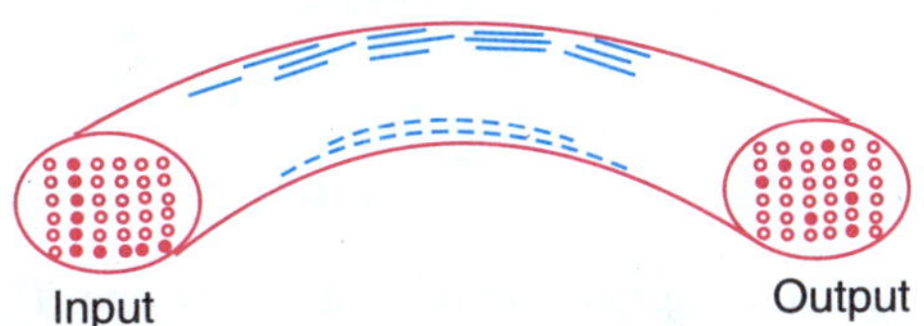

(a) Incoherent bundle-image of letter L scrambled as dark spots at the out put

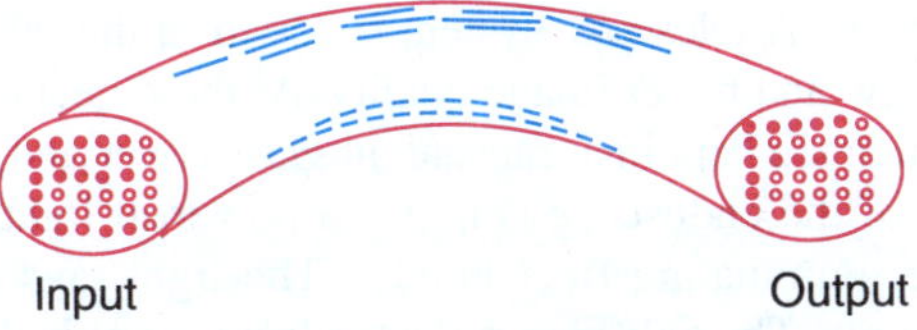

(a) Coherent bundle-image of letter E

Fig. 24.32 : Fibre Optic bundles

When the fibres are carefully arranged so that their terminations occupy the same relative positions in both of the bound ends of the bundle, the bundle is said to be *coherent*. Such a bundle is capable of transmitting undistorted images to a distant place. When one end of such a **flexible image carrier** is placed face down flat on an illuminated surface, a point-by-point image of the surface appears at the other end (Fig. 24.32).

When the fibres are carefully arranged so that their terminations occupy the same relative positions in both of the bound ends of the bundle, the bundle is said to be *coherent*. Such a bundle is capable of transmitting undistorted images to a distant place. When one end of such a **flexible image carrier** is placed face down flat on an illuminated surface, a point-by-point image of the surface appears at the other end.

Endoscopes

The most important application of the coherent bundles is in diagnostic field as an optical endoscope. An endoscope is an optical instrument which facilitates visual inspection of internal parts of a human body. It is also called a fiberoscope. It requires about 10,000 fibres forming a bundle of 1 mm diameter and it can resolve objects with a separation of 70 μm. By allowing direct viewing of what was formerly hidden, a fiberoscope has become a vital diagnostic tool for industry and medicine. The broncho-fiberoscope, gastrointestinal fiberoscope, laparoscope etc are the endoscopes used in medical diagnosis.

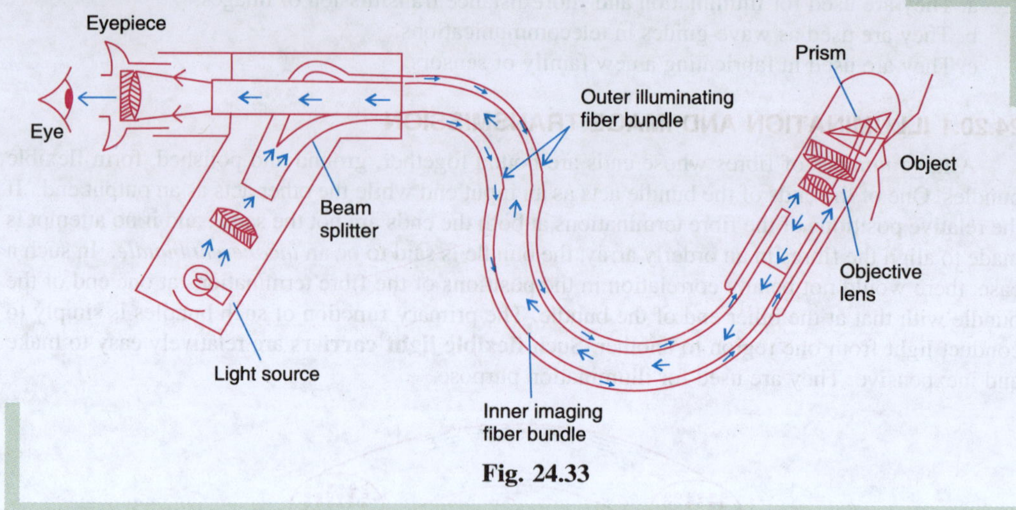

Fig. 24.33

Fig. 24.33 shows the schematic diagram of a flexible endoscope. The endoscopes are designed using low quality, large diameter and short silica fibres. There are two fiber bundles in an endoscope. One of them is used to illuminate the interior of the body and the other is used to collect the reflected light from the illuminated area. A telescope system is added in the internal part of endoscope for obtaining a wider field of view and better image quality. At the object end, there is an assembly of objective lens and prism which are kept in a transparent glass cover and at the viewing end, there is an eye lens. The input end of the endoscope contains a powerful light source. The light rays are focused and coupled to the illuminating fiber bundle. The light rays are finally incident on the surface of the object under study. The light rays reflected from the object surface are received by the objective lens through a prism and are transmitted through the imaging fibre bundle to the viewing end of the scope. Here the eye piece reconstructs the image of the object and one can view the image of the surface of the object. Endoscope pictures can be recorded on a videotape recorder.

24.20.2 OPTICAL COMMUNICATIONS

Traditionally, electronic communications were carried out by sending electrical signals through copper cables, coaxial cables or waveguides. In recent years optical fibres are being used, where light signals replace electrical signals. A basic communications system consists of a transmitter, a receiver and an information pathway. Normally, the information to be communicated is a non-electrical message, which is to be converted first into an electrical form. The conversion is done by a transducer. For example, a microphone converts sound waves into currents. Similarly, a video camera converts images into currents. These electrical messages are of low frequency and cannot be transmitted directly. Therefore, they are superposed on a carrier wave of very high frequency. The process of imposing a message signal on a carrier wave is called modulation. Two different techniques of modulation are available. In analog modulation a continuous wave carries the message. In digital modulation message is transmitted in discrete form using binary digits. The message travels along the transmission channel and is received at the receiver. The receiver demodulates the modulated wave and separates out the message and feeds to a transducer such a loud speaker. The bandwidth requirement of the message and the bandwidth of the carrier determine the number of messages that can be simultaneously transmitted on an information channel. For example, a bandwidth of 4 kHz is required for voice transmission while 6 MHz bandwidth is required for TV signal transmission. When signals are transmitted in analog form the carrier should have double the above bandwidth. The normal TV communications has a bandwidth of 250 MHz and therefore, it can simultaneously transmit 20 TV programmes. However, instead of microwaves if light waves are used as carrier wave, the bandwidth will be about 10^8 MHz and can therefore transmit about 10^6 TV programs at a time. Thus, the use of the light waves expands our communication capabilities tremendously.

24.20.3 MEDICAL APPLICATIONS

Fibre optic technology is used in medical diagnostics as well as in medical procedures. The fibre optic endoscope is used to inspect internal organs for diagnostic purposes.

In ophthalmology, a laser beam guided by optical fibres is used to reattach detached retina and to correct defective vision.

In cardiology, optical energy transmitted through a optical fibre is used to evaporate built-up plaque that is blocking an artery.

In the treatment of cancer also the optical fibre technology is used. The process involves injection of special chemicals that penetrate only the cancerous cells. Infra red energy transmitted via the fibre illuminates the affected area and is absorbed by the special chemical in the cancerous cells. The heat generated destroys the cancerous cells.

24.20.4 MILITARY APPLICATIONS

An aircraft, a ship or a tank needs tons of copper wire for wiring of the communication equipment, control mechanisms, instrument panel illumination etc. Use of optical fibre in place of copper reduces weight and further maintains true communication silence to the enemy. For example, a shipboard radar system requires about 250 m of coaxial cable, with a weight of 7 tons and a diameter of 45 cm. These cables can be replaced by optical fibre weighing 20 kg and measuring 2.5 cm in diameter.

Fibre guided missiles are used in recent wars. Sensors mounted on the missile transmit video information through the optical fibre to a ground control van and receive commands from the van again. The control van continuously monitors the course of the missile and if necessary corrects its course to ensure that the missile precisely hits the target.

24.21 FIBRE OPTIC COMMUNICATION SYSTEM

A fibre optic communication system is very much similar to a traditional communications system and has three major components. A **transmitter** converts electrical signal to light signals, an **optical fibre** transmits the signals and a **receiver** captures the signals at the other end of the fibre and converts them to electrical signals.

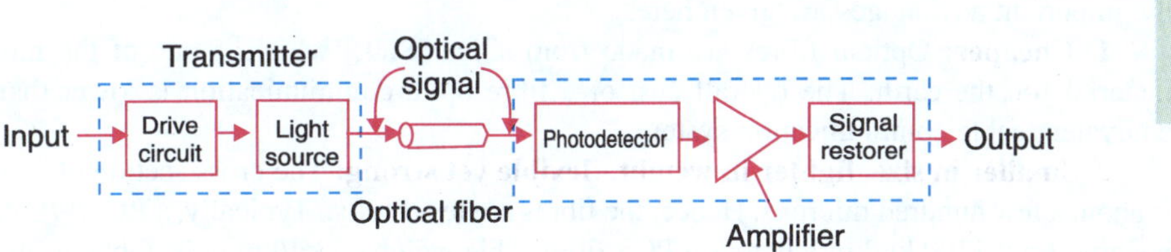

Fig. 24.34: Illustration of a typical fibre optic communication link.

The block diagram Fig.24.34 illustrates a typical communications system. The transmitter consists of a light source supported by necessary drive circuits. A transducer converts a non-electrical message into an electrical signal and is fed to a light source. The light source is a miniature source, either a light emitting diode or a semiconductor laser. In either case, light is emitted in the IR range with a wavelength of 850 nm (0.85 μm), 1300 nm (1.3 μm) or 1550 nm (1.55 μm). The light waves are modulated with the signal. By varying the intensity of the light beam from the laser diode or LED, analog modulation is achieved. By flashing the laser diode or LED on and off at an extremely fast rate, digital modulation is achieved. A pulse of light represents the number 1 and the absence of light at a specified time represents zero. A message can be transmitted by a particular sequence of these 1s or 0s. If the receiver is programmed to recognize such digital patterns, it can reconstruct the original message. Though the digital modulation requires more complicated equipment such as encoders and decoders and also more bandwidth than analog modulation, it allows greater transmission distance with the same power. This is a great advantage and hence digital modulation has become popular and widely used nowadays. The transmitter feeds the analog or digitally modulated light wave to the transmission channel, namely optical fibre link. The optical signal travelling through the fibre will get attenuated progressively and distorted due to dispersion effects. Therefore, repeaters are to be used at specific intervals to regenerate the signal. At the end of the fibre, an output coupler directs the light from the fibre onto a semiconductor photodiode, which converts the light signals to electrical signals. The photodetector converts the light waves into electrical signals which are then amplified and decoded to obtain the message. The output is fed to a suitable transducer to convert it into an audio or video form.

Applications

Optical fibre communications systems can be broadly classified into two groups: (*i*) local and intermediate range systems where the distances involved are small and (*ii*) long-haul systems where cables span large distances.

(*i*) Local area networks: The local area network (LAN) is a computer oriented communication system. LAN operates over short distances of about 1 to 2 km. It is multiuser oriented system. In LAN, a number of computer terminals are interconnected over a common channel allowing each computer to use data and programs from any other. An optical data bus offers a great reduction in cost and increases enormously the information handling capacity.

(*ii*) Long-haul communication: One of the most important applications of fibre optic communication is long-haul communication. Long-haul communication systems are used for long distances, 10 km or more. Telephone cables connecting various countries come under this category. A rather sophisticated long-haul network is the NSFNET which links six supercomputer centres throughout U.S.A.

24.22 MERITS OF OPTICAL FIBRES

Optical fibres have many advantageous features that are not found in conducting wires. Some of the important advantages are given here.

1. Cheaper: Optical fibres are made from silica (SiO_2) which is one of the most abundant materials on the earth. The overall cost of a fibre optic communication is lower than that of an equivalent cable communication system.

2. Smaller in size, lighter in weight, flexible yet strong: The cross section of an optical fibre is about a few hundred microns. Hence, the fibres are less bulky. Typically, a RG-19/U coaxial cable weighs about 1100 kg/km whereas a PCS fibre cable weighs 6 kg/km only. Optical fibres are quite flexible and strong.

3. Not hazardous: A wire communication link could accidentally short circuit high voltage lines and the sparking occurring thereby could ignite combustible gases in the area leading to a great damage. Such accidents cannot occur with fibre links since fibres are made of insulating materials.

5. Immune to EMI and RFI: In optical fibres, information is carried by photons. Photons are electrically neutral and cannot be disturbed by high voltage fields, lightening, etc. Therefore, fibres are immune to externally caused background noise generated through electromagnetic interference (EMI) and radiofrequency interference (RFI).

6. No cross talk: The light waves propagating along the optical fibre are completely trapped within the fibre and cannot leak out. Further, light cannot couple into the fibre from sides. In view of these features, possibility of cross talk is minimized when optical fibre is used. Therefore, transmission is more secure and private.

7. Wider bandwidth: Optical fibres have ability to carry large amounts of information. While a telephone cable composed of 900 pairs of wire can handle 10,000 calls, a 1mm optical fibre can transmit 50,000 calls.

8. Low loss per unit length: The transmission loss per unit length of an optical fibre is about 4 dB/km. Therefore, longer cable-runs between repeaters are feasible. If copper cables are used, the repeaters are to be spaced at intervals of about 2 km. In case of optical fibres, the interval can be as large as 100 km and above.

24.22.1 DISADVANTAGES

Installation and maintenance of optical fibres require a new set of skills. They require specialized and costly equipment like optical time domain reflectometers etc. All this means heavy investment.

24.23 FIBRE OPTIC SENSORS

Fibre optic sensors are transducers, which generally consist of a light source coupled with an optical fibre and a light detector held at the receiver-end. The fibres used could be either multimode or single mode type. The sensors can be used to measure pressure, temperature, strain, the acoustic field, magnetic field, etc physical parameters. The advantages of these sensors are that they are lighter, occupy lesser volume and are cheaper.

The optical fibre merely carries the light beam in some of the sensors and in others the fibre itself acts as the sensor. We study here a few typical examples of the sensors.

24.23.1 TEMPERATURE SENSORS

(*a*) Intensity modulated sensor

Principle: In this type of sensor, temperature is measured by the modulation of intensity of the reflected light from a target, a silicon layer. The operation of the temperature sensor is based on the 1 μm wavelength light-absorption characteristics of silicon as a function of temperature. Depending on the temperature, the amount of light absorbed by the silicon layer varies. The change in intensity of the reflected light is proportional to the change in temperature.

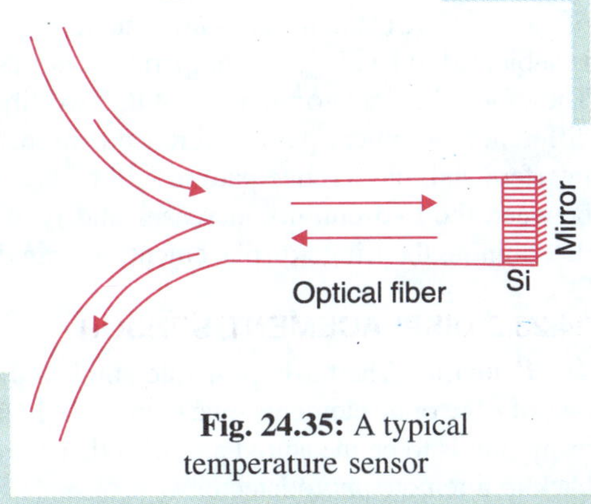

Fig. 24.35: A typical temperature sensor

Construction: Fig.24.35 illustrates a temperature sensor with a multimode fibre. The fibre is coated at one end with a thin silicon layer.

The silicon layer is in turn coated with a reflective coating at the back. The silicon layer acts as the sensing element.

Working: The light from a light source is launched into the fibre from one of the ends of one of its branches (see Fig. 24.35). It passes first through the fibre and then through the silicon layer. The mirror coating at the other end of the silicon layer reflects the light back which again travels through the silicon layer. The reflected light emerges out through another branch of multimode fibre and is collected by a photodetector. The amount of the reflected light is converted into voltage by the photodetector. The absorption of light by the silicon layer varies with temperature and the variation modulates the intensity of the light received at the detector. Temperature measurements can be made with a sensitivity of 0.001°C.

(*b*) Phase modulated sensor

Principle: This temperature sensor is based on phase variation resulting due to the variation of refractive index of the optical fibre under the influence of temperature.

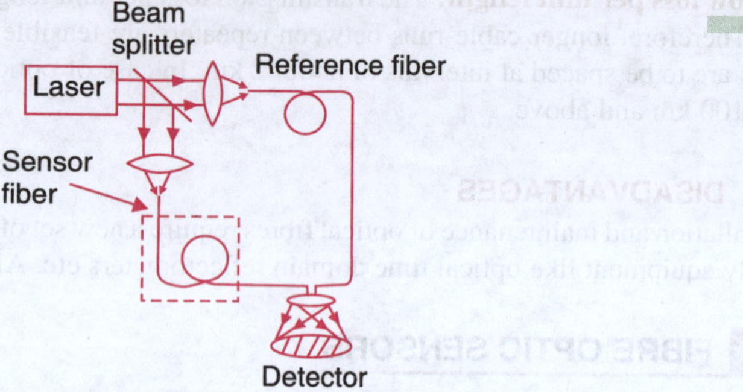

Fig. 24.36: Temperature sensor using phase variations

Construction: Fig.24.36 shows a single mode fibre sensor arranged in what is known as the Mach-Zehnder arrangement. A light source produces light. A beam splitter divides the light into two parts and sends light through the sensing fibre and the reference fibre. Light passing out of the two fibre elements is fed to a detector, which measures the difference in phase of the two light waves. Accurate measurements of the temperature may be obtained from these patterns.

Working: The light from the source is divided into two parts by the beam splitter. One part is allowed through sensor fibre, and the other part is passed through the reference fibre. Light rays entering the fibres are coherent and have the same phase. Prior to heating, the optical path lengths of the two fibre elements are same and hence both the outputs will be in phase. When the sensor fibre is subjected to heating, the temperature causes a change in the refractive index of the optical fibre. Therefore, the light coming out of the two fibres at the other end will have phase difference due to difference in optical path difference caused by the heating. When the rays are superposed, they interfere and interference pattern will be observed. As temperature increases, the phase difference between the two outputs increases and is observed as a displacement of the fringe pattern. By determining the fringe displacement, we can determine the magnitude of temperature.

24.23.2 DISPLACEMENT SENSOR

Principle: The basic principle employed in displacement sensor consists of using an adjacent pair of fibre optic elements, one to carry light from a remote source to an object whose displacement or motion is to be measured and the other to receive the light reflected from the object and carry it back to a remote photodetector.

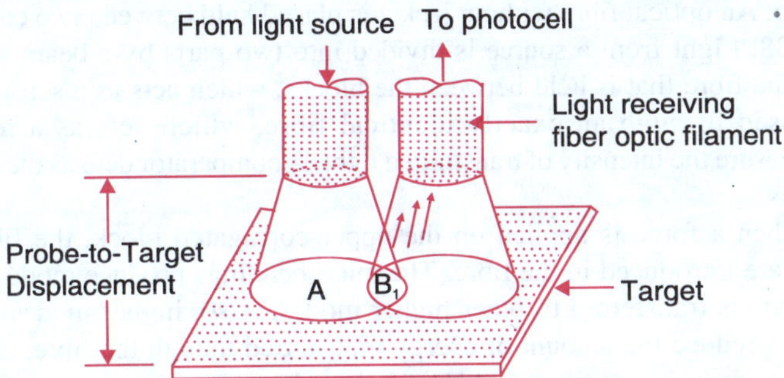

Fig. 24.37: Object moving away from probe, causes increase in reflected light intensity.

Construction: Fig.24.37 shows the arrangement of a displacement sensor. Two separate optical fibres are positioned adjacent to each other. One of them transmits light coming from a light source. The other fibre receives light reflected from the object under study and passes it on to a photodetector.

Working: Light from the transmitting fibre element is incident on the object under study. The light receiver fibre element is positioned adjacent to the transmitting fibre. If the gap between the object and the fibre elements is zero, the light from the transmit fibre would be directly reflected back into itself and little or no light would go into the receive fibre. When the object moves away, the gap increases and some of the reflected light is captured by the receive fibre which in turn is carried to the photodetector. As the gap increases, a distance will be reached at which maximum reflected light is received by the photodetector. Further increase in the gap will result in a decrease in the light at the receiver fibre face and corresponding drop in the signal output from the photodetector. By proper calibration, we can obtain the displacement of the object in terms of the strength of the output signal of the photodetector.

24.23.3 FORCE SENSOR

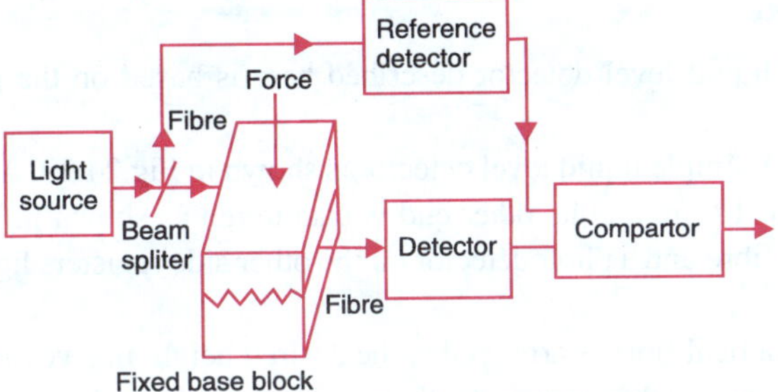

Fig. 24.38: Force sensor using microbend losses

Principle: This sensor is based on variations of light intensity. When an optical fibre is pressed, a small change occurs in light propagation direction due to microbending of the fibre. As a result, energy from one mode is transferred to another mode through mode coupling. In addition, higher order modes are likely to change into radiation modes. All these effects cause a loss in intensity of the light transmitted through the fibre. Therefore, the change in intensity of the transmitted light is proportional to the force applied on the optical fibre.

Construction: An optical fibre without jacket is placed held between two corrugated blocks, as shown in Fig.24.38. Light from a source is divided into two parts by a beam splitter. One part is allowed through the fibre that is held between the blocks, which acts as a sensor element, and the other part is passed through an exactly identical fibre, which acts as a reference element. Photodetectors measure the intensity of transmitted light. A comparator detects the difference between the light intensities.

Working: When a force is applied on the upper corrugated block, the fibre is pressed and microbend losses are introduced in the fibre. The microbendings produce mode coupling such that energy of one mode is transferred to other higher modes. Also, higher modes are converted into leaky modes which reduce the amount of energy transmitted though the fibre. The changes in the light intensity due to these losses are detected by a photodetector and compared with that of the light coming out of the reference element. The change in intensity is related to the force and hence is a measure of the applied force.

24.23.4 LIQUID LEVEL DETECTOR

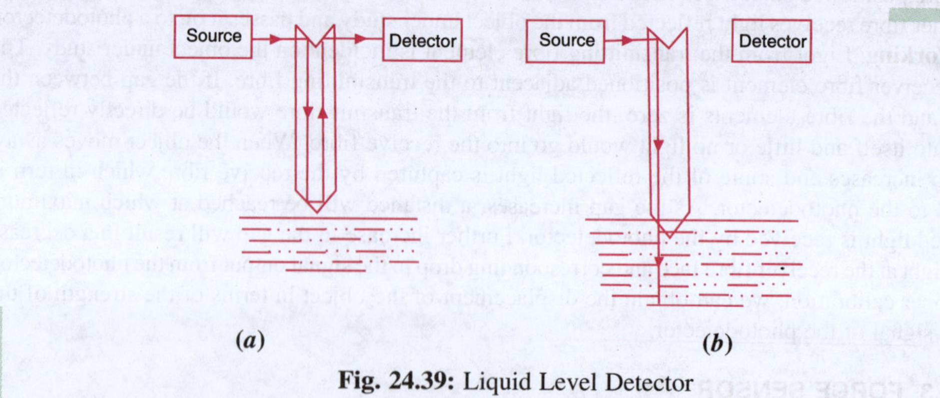

Fig. 24.39: Liquid Level Detector

Principle: The liquid level detector described here is based on the principle of total internal reflection.

Construction: A simple liquid level detector is shown in Fig.24.39. A notch is made at one end of a multimode optical fibre and its other end is chamfered as shown in Fig.24.39. A light source sends light on to the fibre and a photodetector on the other side registers light emerging out from the fibre.

Working: The optical fibre is arranged at the desired height in a vessel. The refractive index of the fibre is chosen to be less than that of the liquid whose level is to be detected. Light from the light source is made to be incident on one of the inclined faces of the notch. The light turns through 90° and travels through the fibre. On reaching the chamfered end of the fibre, it gets internally reflected at the fibre-air boundary, if the liquid is below the desired level. Then, it is again turned through 90° at the opposite face, travels back through the fibre to be turned once again through 90° and is detected at the detector (Fig. 24.39 a).

When the liquid rises and touches the fibre end, total internal reflection ceases and the light is transmitted into the liquid. Hence, the photodetector does not receive any light (Fig.24.39 b). Thus, an indication of the liquid level is obtained at the detector.

WORKED-OUT EXAMPLES

Example 24.1: In an optical fibre, the core material has refractive index 1.43 and refractive index of clad material is 1.4. Find the propagation angle.

Solution: $$\cos\theta_C = \frac{n_2}{n_1} = \frac{1.40}{1.43} = 0.979$$

Therefore, propagation angle $\theta_C = \cos^{-1}(0.979) = 11.8^o$

Example 24.2: In an optical fibre, the core material has refractive index 1.6 and refractive index of clad material is 1.3. What is the value of critical angle? Also calculate the value of angle of acceptance cone.

Solution: Critical angle is given by $\sin\phi_c = \frac{n_2}{n_1} = \frac{1.3}{1.6} = 0.8125$

$$\therefore \phi_c = \mathbf{54.3^o}$$

Acceptance angle $$\theta_0 = \sin^{-1}\left[\sqrt{n_1^2 - n_2^2}\right] = \sin^{-1}\left[\sqrt{1.6^2 - 1.3^2}\right]$$

$$= \sin^{-1}(0.87)$$

$$= \mathbf{60.5^o}$$

Angle of acceptance cone $=2\theta_o = \mathbf{121^o}$

Example 24.3: Calculate the numerical aperture and acceptance angle of an optical fibre from the following data:

$$\mu_1(core) = 1.55 \text{ and } \mu_2(cladding) = 1.50$$

Solution: $$NA = \sqrt{n_1^2 - n_2^2} = \sqrt{1.55^2 - 1.50^2} = \sqrt{0.153} = \mathbf{0.391.}$$

Acceptance angle $$\theta_0 = \sin^{-1}\left[\sqrt{n_1^2 - n_2^2}\right] = \sin^{-1}\left[\sqrt{1.55^2 - 1.50^2}\right] = \mathbf{23.02^o}$$

Example 24.4: What is the numerical aperture of an optical fibre cable with a clad index of 1.378 and a core index of 1.546?

Solution: $$NA = \sqrt{n_1^2 - n_2^2} = \sqrt{1.546^2 - 1.378^2} = \sqrt{0.491} = 0.70$$

Example 24.5: A fibre cable has an acceptance angle of 30° and a core index of refraction of 1.4. Calculate the refractive index of the cladding.

Solution: $$\sin\theta_o = \sqrt{n_1^2 - n_2^2}$$

$$\therefore \sin^2\theta_o = n_1^2 - n_2^2$$

$$n_2^2 = n_1^2 - \sin^2\theta_o = (1.4)^2 - \sin^2 30^o = 1.96 - 0.25$$

$$= 1.71$$

$$\therefore \quad n_2 = \mathbf{1.308}$$

Example 24.6: Calculate the angle of acceptance of a given optical fibre, if the refractive indices of the core and the cladding are 1.563 and 1.498 respectively.

Solution:
$$\sin\theta_o = \sqrt{n_1^2 - n_2^2} = \sqrt{(1.563)^2 - (1.498)^2} = 0.4461$$

$$\theta_o = \sin^{-1}(0.4461) = \mathbf{26.49^o}$$

Example 24.7: Calculate the fractional index change for a given optical fibre if the refractive indices of the core and the cladding are 1.563 and 1.498 respectively.

Solution: Fractional index change $\Delta = \frac{n_1 - n_2}{n_1} = \frac{1.563 - 1.498}{1.563} = \frac{0.065}{1.563} = \mathbf{0.0415}$

Example 24.8: Calculate the refractive indices of the core and the cladding material of a fiber from the following data:

Numerical aperture (NA) = 0.22 and $\Delta = 0.012$

where Δ is the fractional refractive index change.

Solution:
$$NA = n_1\sqrt{2\Delta}$$

$$0.22 = n_1\sqrt{2\times 0.012} = 0.155\ n_1 .$$

$$\therefore \quad n_1 = \frac{0.22}{0.155} = \mathbf{1.42}$$

$$\Delta = \frac{n_1 - n_2}{n_1} \quad \therefore \quad \frac{1.42 - n_2}{1.42} = 0.012 \quad \therefore \quad n_2 = 1.42 - 1.42\times 0.012 = \mathbf{1.403}$$

Example 24.9: Find the fractional refractive index and numerical aperture for an optical fibre with refractive indices of core and cladding as 1.5 and 1.49 respectively.

Solution:
$$\Delta = \frac{n_1 - n_2}{n_1} = \frac{1.5 - 1.49}{1.5} = \mathbf{0.0067}$$

$$NA = n_1\sqrt{2\Delta} = 1.5\sqrt{2\times 0.0067} = \mathbf{0.174}$$

Example 24.10: A step-index fibre is made with a core of refractive index 1.52, a diameter of 29 μm and a fractional difference index of 0.0007. It is operated at a wavelength of 1.3 μm. Find the V-number and the number of modes that the fibre will support.

Solution:
$$V = \frac{\pi d}{\lambda_o} n_1\sqrt{2\Delta} = \frac{3.143\times 29\times 10^{-6}\,m}{1.3\times 10^{-6}\,m}\times 1.52\sqrt{2\times 0.0007} = \mathbf{4.049}$$

$\therefore$ Number of modes, $N = \frac{1}{2}V^2 = \frac{1}{2}(4.049)^2 = \mathbf{8\ modes}$

Example 24.11: A step-index fibre is with a core of refractive index 1.55 and cladding of refractive index 1.51. Compute the intermodal dispersion per kilometer of length of the fibre and the total dispersion in a 15 km length of the fibre.

Solution:
$$\Delta t = \frac{n_1 L}{c}\left[\frac{n_1}{n_2} - 1\right] = \frac{1.55\times 10^3\,m}{3\times 10^8\,m/s}\left[\frac{1.55}{1.51} - 1\right] = \mathbf{138\ ns/km.}$$

Total dispersion for 15 km length = Δt ×15 km = (138 ns/km)×15 km = **2.07 μs**.

Example 24.12: Optical power of 1 mW is launched into an optical fibre of length 100 m. If the power emerging from the other end is 0.3.mW, calculate the fibre attenuation.

Solution: Attenuation, $\alpha = \frac{10}{L}\log\frac{P_i}{P_o} = \frac{10}{0.1km}\log\frac{1mW}{0.3mW}$ **= 52.3 dB/km**

Example 24.13: What is the attenuation in dB/km, if 15% of the power fed at the launching end of a $\frac{1}{2}$ km fibre is lost during propagation?

Solution: Attenuation, $\alpha = \frac{10}{L}\log\frac{P_i}{P_o} = \frac{10}{0.5km}\log\frac{1}{0.15} = 16.48$ dB/km

QUESTIONS

1. Explain the phenomenon of total internal reflection of light. How is it used in fiber optic communications?
2. What is meant by critical propagation angle of an optical fibre? Obtain an expression for the critical propagation angle.
3. What is meant by critical angle of an optical fibre? Obtain an expression for the critical angle.
4. What is an optical fibre? What is the principle involved in its working?
5. Explain the following terms:
   (*i*) critical angle
   (*ii*) Acceptance cone
   (*iii*) Numerical aperture
6. Deduce an expression for acceptance angle of an optical fibre.
7. Using ray theory, derive the condition for transmission of light within an optical fibre.
8. With the help of a ray diagram, show how optical fibres can guide light waves.
9. Derive an expression for angle of acceptance of fibre in terms of refractive index of the core and the cladding of an optical fibre. What is meant by acceptance cone?
10. What is meant by acceptance of angle for an optical fibre? Show how it is related to numerical aperture.
11. What do you understand by the terms acceptance angle and acceptance cone? Derive an expression for acceptance angle in terms of refractive indices of the core and the cladding.
12. Derive an expression for acceptance angle and numerical aperture for step index fibre with the help of suitable diagram.
13. Explain what you understand by acceptance angle and numerical aperture.
14. Derive an expression for N.A. for S.I. fibre in terms of refractive index of the core and relative refractive index difference between the core and the cladding.
15. Derive an expression for numerical aperture of a step-index fibre in terms of Δ.
16. Define the relative refractive index difference of an optical fibre. Show how it is related to numerical aperture.
17. Classify the fibres on the basis of refractive index profile, on the basis of modes and on the basis of materials.
18. Explain what is step-index, graded index, single mode, multimode fibre.
19. Explain what is step-index, graded index, monomode and multimode fibre. Draw relevant sketches.
20. Explain the difference between the step-index fibre and graded-index fibre.
21. Compare a single mode step index fibre with a multimode step index fibre.

22. What are the different types of attenuation losses in an optical fibre? Discuss the absorption losses.
23. Describe various mechanisms of attenuation in optical fibres.
24. Draw the diagram for an optical fibre link and explain the function of each block.
25. List the main components of optical communication system. Describe the basic optical communication system.
26. Explain optical communication through block diagram. For long distance communication whether (*i*) mono-mode or multimode and (*ii*) step index or graded index fibre, which are preferable and why?
27. Discuss the advantages and disadvantages of optical fibres over conventional communication transmission media.
28. Explain with basic principle, the construction and working of any one type of optical fibre sensor.
29. Discuss any one application of an optical fibre as a sensor.

PROBLEMS FOR PRACTICE

1. An optical fibre has a core material of refractive index of 1.55 and cladding material of refractive index 1.50. The light is launched into the fibre from air. Calculate its numerical aperture.
2. The numerical aperture of an optical fibre is 0.39. If the difference in the refractive indices of the material of its core and the cladding is 0.05, calculate the refractive index of material of the core.
3. An optical fibre has an acceptance angle 26.80°. Calculate its numerical aperture. **(Ans: 0.4508)**
4. An optical fibre refractive indices of core and cladding are 1.53 and 1.42 respectively. Calculate its critical angle. **(Ans: 68.14°)**
5. Consider a fibre having a core of index 1.48, a cladding of index 1.46 and has a core diameter of 30 μm. Show that all rays making an angle less than 9.43° with the axis will propagate through the fibre.
6. A step-index fibre is made with a core of index 1.54, a cladding of index 1.50 and has a core diameter of 50 μm. It is operated at a wavelength of 1.3 μm. Find the V- number and the number of modes that the fibre will support. **(Ans: 42.15, 888)**
7. Using a step index fibre with $n_1 = 1.48$ and $n_2 = 1.46$ and the core radius $a = 30$ μm. Calculate the number of total internal reflections that will occur on its propagation in a length of 1 km fibre.
8. A step-index fibre has a core refractive index of 1.44 and the cladding refractive index of 1.41. Find (i) the numerical aperture,(ii) the relative refractive index difference, and the acceptance angle. **(Ans: 0.292, 0.021, 33.96°)**
9. An optical fibre has a numerical aperture of 0.20 and a cladding refractive index of 1.59. Find the acceptance angle for the fibre in water which has a refractive index of 1.33. **(Ans: 8°39')**
10. Compute the cut-off parameter and the number of modes supported by a fibre which has a core refractive index of 1.54 and the cladding refractive index of 1.50. The radius of the core is 25 μm and operating wavelength is 1300 nm. **(Ans: 42.15, 888)**
11. Find the numerical aperture and acceptance angle of a fibre of core index 1.4 and $\Delta = 0.02$. **(Ans: 0.28, 32.52°)**

12. Compute the total dispersion in 10 km length of a step index fibre, which has a core refractive index of 1.55 and the relative refractive index difference of 0.026. **(Ans: 138ns)**

13. Consider a bare step index fibre having a refractive index of 1.46. The radius of the fibre is 50 μm. Compute the pulse dispersion per km. **(Ans: 2238ns)**

14. Compute the cut-off parameter and the number of modes supported by a fibre, which has a core refractive index of 1.47 and the cladding refractive index of 1.45. The radius of the core is 50 μm and operating wavelength is 850 nm. **(Ans: 44.64, 996)**

15. A step-index fibre has a normalized frequency V = 26.6 at 1300 nm wavelength. If the core radius is 25 μm, calculate the numerical aperture. **(Ans: 0.22)**

16. Find the core radius necessary for single mode operation at 820 nm of a step index fibre, which has a core refractive index of 1.480 and the cladding refractive index of 1.478. **(Ans: 4.08 μm)**

17. A signal of 100mW is injected into a fibre. The outcoming signal from the other end is 40 mW. What is the loss in dB? **(Ans: 3.98 dB)**

18. A communication system uses a 10 km fibre having a fibre loss of 2.5 dB/km. Find the input power if the output power is 1.265 μW. **(Ans: 400 μW)**

19. A fibre length 100m has power input 10 μW and power output 8.8 μW. Find the power loss in dB/km. **(Ans: 5.55 dB/km)**

20. When the mean optical power launched into a 8 km length fibre is 120 μW, the mean optical power at the fibre output is 3 μW. Determine

(*i*) the overall signal attenuation in dB through the fibre,

(*ii*) the signal attenuation per km for the fibre,

(*iii*) the overall signal attenuation for a 10 km optical link using the same fibre,

(*iv*) the numerical input/output ratio. **(Ans: 16 dB, 2.0 dB/km, 20 dB, 100)**

CHAPTER

Non-Linear Optics

At a Glance

- Introduction
- Wave Propagation and Momentum Conservation
- Linear Medium
- Nonlinear Polarization
- Second Harmonic Generation
- Phase Matching
- Sum and Difference Frequency Generation
- Parametric Oscillation
- Self-Focussing of Light
- Stimulated Raman Scattering

25.1. INTRODUCTION

Lasers generate coherent radiation at many wavelengths ranging from meter wavelength region to the soft x-rays region. However, it is not possible to produce light covering all wavelengths of interest in spite of the fact that a large number of active materials are available and lasers can be built using them. It becomes, therefore, necessary to transform the frequency of light generated by lasers into light of desired frequency. Nonlinear optical media help us generate frequencies that were not available, through frequency conversion techniques. Harmonic generation, sum and difference frequency generation and parametric oscillations are some of the important nonlinear processes utilized in laser light frequency transformations. Stimulated Raman scattering is another important process that is used in generating new wavelengths. The processes such as second harmonic generation, sum and difference frequency conversion, parametric oscillation are associated with *passive media*, i.e., media that do not make evident their own characteristic frequencies. Stimulated Raman scattering arises in *active media* that impose their characteristic frequencies on the light wave. In this chapter we acquaint ourselves with the methods of nonlinear processes.

25.2. WAVE PROPAGATION AND MOMENTUM CONSERVATION

A plane wave propagating through a vacuum in a general direction is represented by the equation (12.21).

$$\boldsymbol{E} = \boldsymbol{E}_o e^{i(\omega t - \boldsymbol{k}.r)} \tag{25.1}$$

where $\boldsymbol{k}$ is the vector that defines the direction of propagation of the wave. $\boldsymbol{k}$ can be written as

$$\boldsymbol{k} = \boldsymbol{i}k_x + \boldsymbol{j}\,k_y + \boldsymbol{k}\,k_z \tag{25.2}$$

When a light wave interacts with an isotropic material, it will be affected in the same way whatever may be the direction of the beam with respect to the material. Hence, the refractive index of the material is the same in every direction and the velocity of the beam passing through the material is also the same in every direction. Since the magnitude of **k** is related to the velocity, the components k_x, k_y and k_z would all be the same in case of isotropic crystals. In anisotropic crystals, the light beam experiences different refractive indices in different directions and propagates at different velocities in different directions. Therefore, the components of **k** will have different values in different directions.

If we consider the photon description of light beam, we know that the momentum and wavelength of a photon are related through the following expression.

$$p = \frac{h}{\lambda} = \frac{2\pi\, h}{2\pi\lambda} = \hbar k$$

Momentum p is a vector quantity and hence we write the above equation in vector form as

$$\mathbf{p} = \hbar\,\mathbf{k} \tag{25.3}$$

where **k** is the vector which defines the propagation direction of the wave associated with the photon. Thus, the significance of **k** in equ.(25.1) and (25.3) is the same.

When a beam of light passes from air into a crystal, the momentum of the wave must be conserved at the interface of the two media. It requires that the normal component and parallel component of the momentum must be separately conserved at the boundary. Therefore, the transverse component of the momentum must be continuous across the boundary. There cannot be an abrupt jump or change in the transverse momentum of the wave as the beam enters the medium. In view of the relation (25.3), it means that the transverse component of **k** should not change abruptly at the boundary.

25.3. LINEAR MEDIUM

The optical media are basically dielectric materials. They do not allow electric current to pass through when an electric field is applied across them. Instead, they get polarized. The electric field exerts forces on the valence electrons. These forces are quite small, and induce **electric dipoles** in the medium. These dipoles orient in the direction of the electric field and the dielectric is said to be *polarized.* Dielectrics are polarized, for example, when they are placed between the charged plates of a capacitor. The polarization vector, P, denotes the extent to which the dielectric is polarized. The electric polarization is parallel with and directly proportional to the applied field, **E**. It is given by

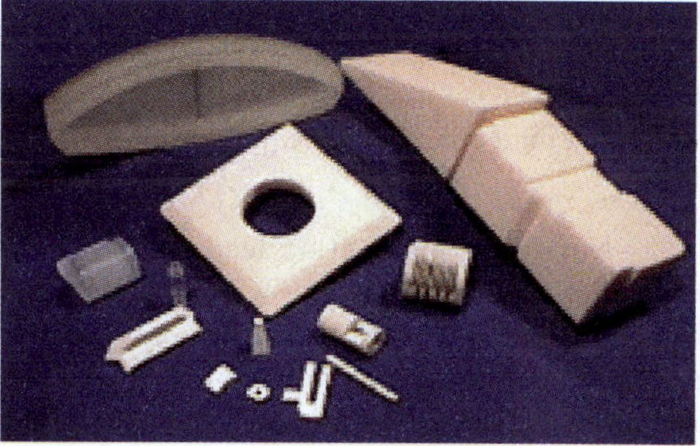

Linear dielectrics.

$$\mathbf{P} = \varepsilon_0 \chi\, \mathbf{E} \tag{25.4}$$

where ε_o is the permitivity of free space and χ [= $(\varepsilon/\varepsilon_o) - 1$] is a dimensionless constant known as the **electric susceptibility** of the medium. The stronger the electric field, the greater will be the polarization and a plot of P versus E is a straight line. Materials, in which such kind of linear relationship holds, are known as **linear dielectrics**. Light waves are electromagnetic waves and when they propagate through a dielectric, the electric field of the waves polarizes the dielectric. Hence the optical parameters of the dielectric are closely related to the dielectric polarization. The refractive index of the medium is given by,

$$\mu = \sqrt{\varepsilon} = \sqrt{1 + 4\pi\chi} \qquad (25.5)$$

As long as the intensity of the light propagating in the dielectric medium is small, the parameters χ and μ are constant quantities and are independent of the intensity of light. Ordinary light sources generate light of field strengths of the order of 10^5 V/m which are very small compared to atomic fields, and therefore cannot affect the optical parameters of the medium.

25.4. NONLINEAR POLARIZATION

When high electric field strength is used, it is expected that P cannot increase linearly indefinitely with E and will become saturated. Therefore, we may anticipate nonlinear behaviour of P at very high field strengths. Lasers produce light of field strengths of the order of 10^7 to 10^{11} V/m, which are of the order of the atomic field strengths. Therefore, the intense light of lasers is in a position to cause nonlinearity of P and influence the optical parameters of the medium. When the electric field E in the light is very large, the parameters χ, ε and μ become the functions of E. Since the directions of **P** and **E** coincide in an isotropic medium, we can express χ as a power series in the field strength as,

$$\chi(E) = \chi_1 + \chi_2 E + \chi_3 E^2 + \ldots \qquad (25.6)$$

and as $\quad P = \varepsilon_o \chi(E) E$

$\therefore \quad P = \varepsilon_o (\chi_1 E + \chi_2 E^2 + \chi_3 E^3 + \ldots) \qquad (25.7)$

where χ_1 is linear susceptibility and is much greater than the coefficients of the nonlinear terms χ_2, χ_3 and so on. The nonlinear terms contribute noticeably only at very high-amplitude electric fields.

The second order nonlinear polarization is given by

$$P_2 = \varepsilon_o \chi_2 E^2 \qquad (25.8)$$

and third order nonlinear polarization by

$$P_3 = \varepsilon_o \chi_3 E^3. \qquad (25.9)$$

Nonlinear polarization leads to nonlinear optical effects. Materials, in which polarization exhibits nonlinear dependence on the field strengths, are called **nonlinear media**. Fig. 25.1 shows the nonlinear variation of electric polarization with the electric field strength in a nonlinear medium.

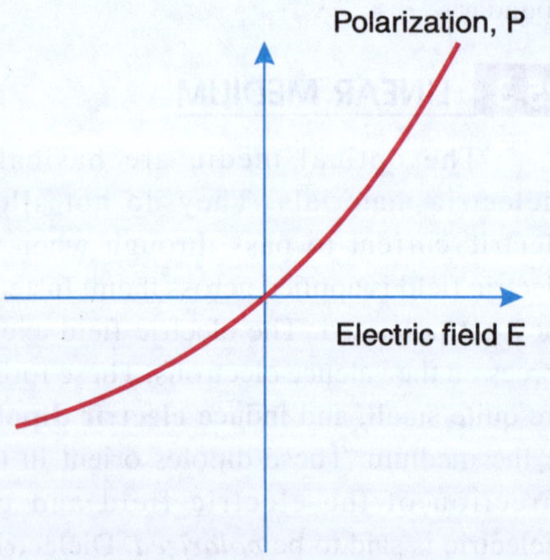

A typical plot of variation of electric polarization with electric field in a nonlinear medium.

Fig. 25.1

In optically isotropic materials, the coefficients of even powers of E in Eq. (25.7) are zero. But in case of anisotropic materials, coefficients of both odd and even powers of E exist. Strictly speaking, any medium becomes *nonlinear* provided the electric field of the incident radiation is very high. With the advent of intense lasers, the nonlinear polarization has assumed importance and made it possible the phenomena of frequency conversion and other non-linear effects.

Now if we take into the consideration of electric field oscillations of a light wave, we can write for the electric field of the wave as

$$E = E_0 \sin \omega t$$

The polarization caused by this wave can be expressed as

$$P = \varepsilon_o (\chi_1 E_o \sin \omega t + \chi_2 E_o^2 \sin^2 \omega t + \chi_3 E_o^3 \sin^3 \omega t + \ldots.) \quad (25.10)$$

The above equation can be re-written as

$$P = \varepsilon_o [\chi_1 E_o \sin \omega t + \tfrac{1}{2}\chi_2 E_o^2 (1 - \cos 2\omega t) + \tfrac{1}{4}\chi_3 E_o^3 (3 \sin \omega t - \sin 3\omega t) + \ldots.] \quad (25.11)$$

Thus, as the electric field in the incident wave oscillates, the array of dipoles produces an electromagnetic wave, which may be called a **polarization wave**. When the incident light beam is not intense, the polarization wave will be in phase and have the same frequency as the incident light wave. The higher order terms in equ.(25.11) indicate that for higher beam intensities, polarization waves of higher frequencies also will be produced. We have to consider the phases of the waves at these other frequencies in order to determine the frequencies that get enhanced.

25.5. SECOND HARMONIC GENERATION

In case of anisotropic crystals equ. (25.7) may be rewritten as,

$$P = \varepsilon_0 [\chi_1 E + \chi_2 E^2] \quad (25.12)$$

$$= P_1 + P_{nl}$$

where P_1 and P_{nl} are linear polarization and nonlinear polarization components respectively.

Fig. 25.2 shows the response of the nonlinear medium, in which a symmetrical electric field produces an asymmetrical polarization. Fourier analysis of the polarization shows that it consists of components having frequencies ν and 2ν as well as a d.c. component. Let us consider a plane monochromatic light wave of frequency ν, travelling at the velocity υ_1 in a nonlinear medium. The electric field of the wave may be of the form,

$$E = E_o \sin [2\pi\nu (t - x/\upsilon_1)] + \chi\, Eo^2 \sin^2 [2\pi\nu (t - x/\upsilon_1)]$$

$$= E_o \sin [2\pi\nu (t - x/\upsilon_1) + \frac{1}{2}\chi_1 Eo^2 - \frac{1}{2} Eo^2 \cos [4\pi\nu (t - x/\upsilon_1)] \quad (25.13)$$

Eq. (25.13) contains a term in $4\pi v$ which corresponds to a wave of polarization propagating in the same direction and at the same speed of light wave with twice higher frequency. The magnitude of the term in $4\pi v \{= 2(2\pi v)\}$ approaches the magnitude of the first term $\chi_1 E_o$ at high values of E_o. Secondly, under certain conditions, this polarization wave may cause emission of a new light wave. It means that a strong light of frequency v propagating in a nonlinear medium can generate a new light wave at frequency $2v$. It is a second harmonic and hence the phenomenon is known as *second harmonic generation.* At some initial point the incident wave generates the second harmonic wave and at that point the two waves are coherent. As the incident wave propagates through the crystal, it continues to generate second harmonic waves. All these waves combine constructively only if they satisfy the phase matching condition (25.14).

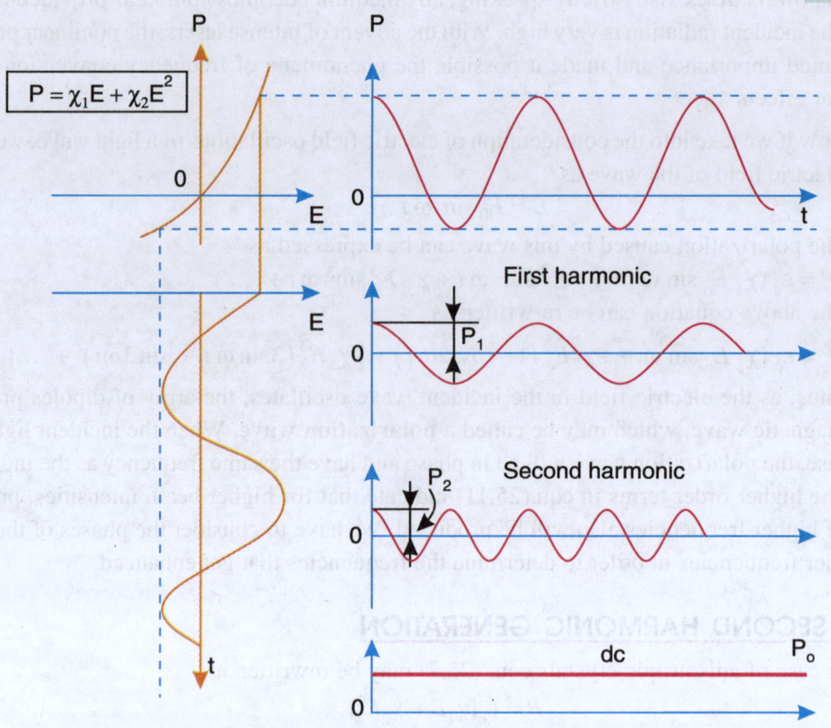

Fourier analysis of the asymmetrical polarization wave into fundamental, second harmonic and d.c. components

Fig. 25.2

The input wave of frequency ν is normally called the **pump wave**. The ratio of the intensity of the generated second harmonic wave to that of the pump wave is dependent on the **phase mismatch factor**, Δk, which is given by

$$\Delta k = k(2\nu) - 2k(\nu) = 2\nu\,[\mu(2\nu) - \mu(\nu)]$$

In order to obtain an intense second harmonic wave, which is directed like the pump laser beam, one has to achieve the condition $\Delta k = 0$. This implies that

$$\mu(2\nu) = \mu(\nu) \tag{25.14}$$

Equ.(25.14) is known as the **phase matching condition**. It is also known as **index matching condition.**

This condition is satisfied only when the pump and generated waves are of orthogonal polarization. Therefore, the process of second harmonic generation requires a uniaxial double refracting crystal exhibiting high nonlinear susceptibility χ_2. It must be cut in the shape of a rectangular parallelepiped with the axis along the direction of phase matching. The incident light must be plane polarized, with the plane of polarization perpendicular to the plane of principal section of the crystal. When these conditions are satisfied, a wave of frequency ν produces another wave of frequency 2ν, which is the second harmonic. The direction of propagation of the second harmonic coincides with that of the fundamental wave. The plane of polarization of second harmonic will be perpendicular to that of the fundamental wave (see Fig. 25.6 also).

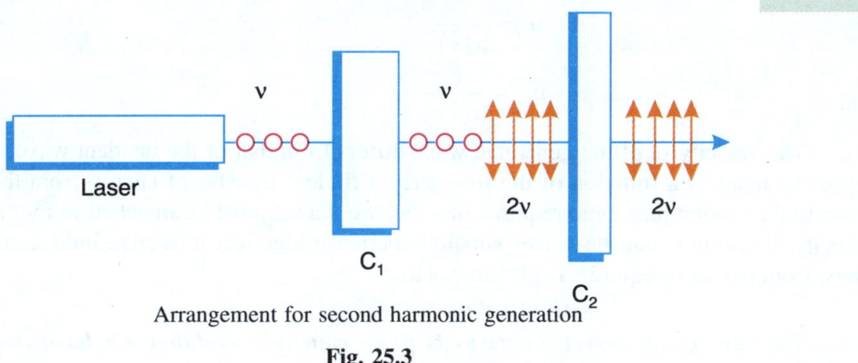

Arrangement for second harmonic generation

Fig. 25.3

A commonly used material for second harmonic generation is KDP which is a negative double refracting crystal. In negative crystals, if the fundamental light is a linearly polarized *ordinary wave*, the resulting second harmonic will be an *extraordinary wave*. The reverse is the situation in case of positive crystals. If the fundamental light is a linearly polarized *extraordinary wave*, the resulting second harmonic will be an *ordinary wave*.

The second harmonic generation may also be described from the standpoint of photon concept. According to this picture (see Fig. 25.4), two photons of the fundamental frequency ν, each having energy hν combine to produce an energy of 2hν. The energy levels located at hν and 2 hν are known as virtual levels and shown by dashed lines, as they are not allowed states of the material and as such are not populated. In this process, two photons of frequency ν are destroyed and one photon of frequency 2ν is simultaneously created.

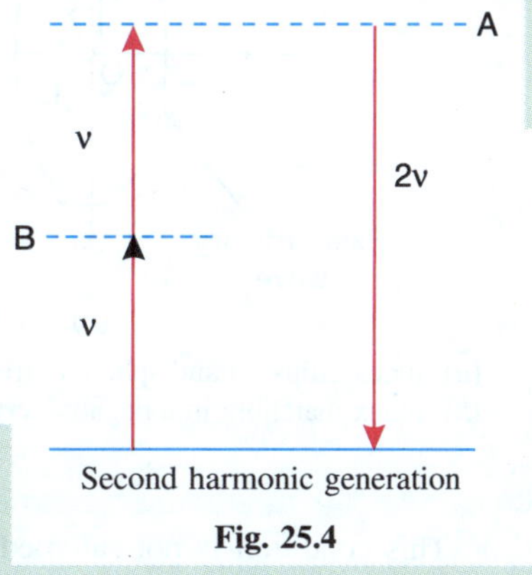

Second harmonic generation

Fig. 25.4

Second harmonic generation was first demonstrated by Franken and his co-workers in 1961. They focused a 3kW ruby laser pulse (λ = 6943 Å) onto a quartz crystal and obtained a very low intensity output at a wavelength of 3471.5Å. Second harmonic generation helps in extending the range of laser wavelengths into the blue and UV part of the spectrum, which are not rich in naturally occurring laser lines. Second harmonic generators are now available in market for use with visible and IR lasers.

25.6. PHASE MATCHING

It was noted earlier that light waves are electromagnetic waves and the electric field of the waves polarizes the medium through which they propagate. The polarization response of the medium consists in the orientation of the electric dipoles in the direction of the externally applied electric field. As the electric field in the incident wave oscillates, the array of dipoles produces *polarization wave*. For higher beam intensities, polarization waves of higher frequencies also will be produced. To understand the condition under which the enhancement of a particular frequency occurs, let us take the case of second harmonic generation. In this case, the polarization wave propagates at the velocity υ_1, the second harmonic propagates at the velocity v_2 in the medium. The velocities are given by,

$$\upsilon_1 = \frac{c}{\mu(\nu)} \tag{25.15a}$$

and
$$\upsilon_2 = \frac{c}{\mu(2\nu)} \tag{25.15b}$$

The velocity υ_2 of the generated wave differs from that of the incident wave,υ_1 because the refractive index is a function of the frequency. Efficient transfer of energy from the polarization wave to the second harmonic requires that the two waves must be matched in momentum, that is velocity. It requires that the waves should experience identical refractive indices in the medium. Thus, from the set of equ.(25.15) it follows that,

$$\mu(\nu) = \mu(2\nu) \tag{25.16}$$

The above condition is the same as the *phase matching condition* or *index-matching* condition (25.14).

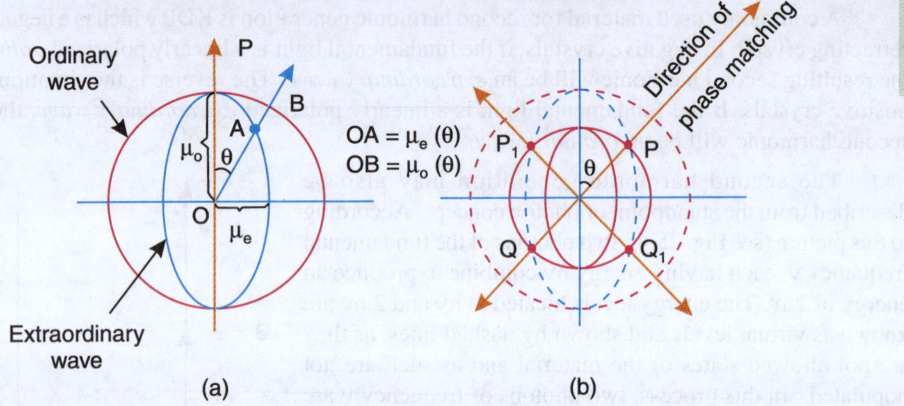

(*a*) Index ellipsoid and sphere corresponding to a e-ray and o-ray in a negative uniaxial crystal.
(*b*) Index matching in a negative crystal.

Fig. 25.5

This condition is not satisfied in general cases. It is satisfied *only* in anisotropic crystals known as *uniaxial crystals*. When a light ray is incident on a *uniaxial crystal,* it splits up into two rays, namely, an *ordinary ray* (o-ray) and an *extra ordinary ray* (e-ray). The refractive index of an e-ray depends upon the direction of propagation. The refractive index surface is a *sphere* in case of o-ray and an *ellipsoid* in case of e-ray. The two refractive index surfaces in case of a negative *uniaxial crystal* are shown in Fig. 25.5(a). Fig. 25.5(b) shows the refractive index surfaces for the frequency *v* and for the frequency 2 *v*. It is seen that the refractive index surface of the o-ray with frequency *v* and that of the e-ray with frequency 2*v* intersect; one of the points of intersection is at point P. At the point of intersection the velocities υ_1 and υ_2 are equal. It is obvious that the phase-matching condition is satisfied for light waves propagating in the direction PQ. Thus, the direction PQ is *the direction of phase- matching*. Along PQ,

Uniaxial birefringent lithium niobate crystal.

$$\mu_o(\nu) = \mu_e(2\nu) \tag{25.17}$$

25.6.1. ANGLE TUNING FOR PHASE MATCHING

There are two methods for achieving phase matching in double refracting crystals. They are known as *angle tuning* and *temperature tuning*. The refractive index of extraordinary ray μ_e varies with the angle between the direction of propagation and the optic axis. The variation is given by

$$\mu_e(\theta) = \left[\frac{\cos^2\theta}{\mu_o^2} + \frac{\sin^2\theta}{\mu_e^2}\right]^{-1/2} \tag{25.18}$$

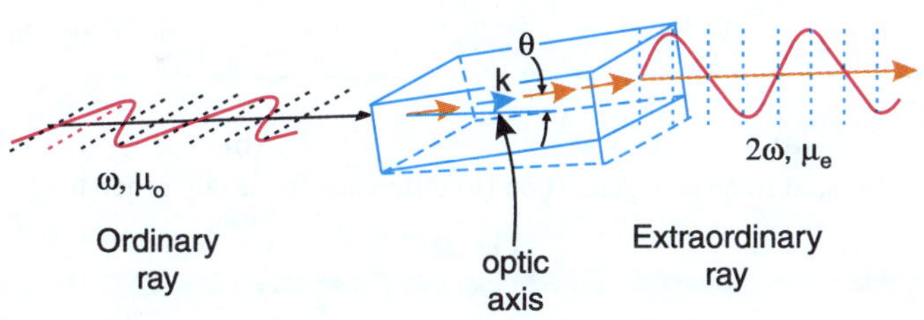

Phase matching using angle tuning

Fig. 25.6

By rotating the crystal in the plane of the optic axis and the incident beam (Fig.25.6), it is possible to select the value of μ_e that will satisfy the phase matching condition (25.16).

25.7. SUM AND DIFFERENCE FREQUENCY GENERATION

Nonlinear crystals make it also possible to generate beams of sum and difference frequencies when two light waves of different frequencies interact within them. The interaction of waves with frequencies ν_1 and ν_2 generate light waves with frequencies $\nu_1 + \nu_2$ and $\nu_1 - \nu_2$.

Let us consider two coherent waves of frequencies ν_1 and ν_2 propagating simultaneously in a nonlinear medium. Let the waves be represented by $E_1 \sin 2\pi\nu_1 t$ and $E_2 \sin 2\pi\nu_2 t$, respectively. The superposition of the waves leads to,

$$E = E_1 \sin 2\pi\nu_1 t + E_2 \sin 2\pi\nu_2 t \tag{25.19}$$

Using Eq. (25.19) in Eq.(25.12), we obtain the following expression for the nonlinear term in the polarization of the medium.

$$\begin{aligned} P_{nl} &= \varepsilon_o \chi_2 E^2 \\ &= \varepsilon_0\, \chi_2 [E_1 \sin 2\pi\nu_1 t + E_2 \sin 2\pi\nu_2 t]^2 \\ &= \varepsilon_o \chi^2 E_1^2 \sin^2 2\pi\nu_1 t + \varepsilon_o \chi_2 E_2^2 \sin^2 2\pi\nu_2 t + 2\varepsilon_o \chi_2 E_1 E_2 \sin(2\pi\nu_1 t)\sin(2\pi\nu_2 t) \\ &= \varepsilon_o \chi_2 E_1^2 (1 - \cos 4\pi\nu_1 t) + \varepsilon_o \chi_2 E_2^2 (1 - \cos 4\pi\nu_2 t) \\ &\quad + \varepsilon_o \chi_2 E_1 E_2 \{\cos 2\pi(\nu_1 - \nu_2)t - \cos 2\pi(\nu_1 + \nu_2)t\} \end{aligned}$$

or

$$\begin{aligned} P_{nl} &= \varepsilon_o \chi_2 \left(E_1^2 + E_2^2\right) - \varepsilon_o \chi_2 E_1^2 \cos 2\pi(2\nu_1)\, t - \varepsilon_o \chi_2 E_1^2 \cos 2\pi(2\nu_2)\, t \\ &\quad + \varepsilon_o \chi_2 E_1 E_2 \{\cos 2\pi(\nu_1 - \nu_2)\, t - \cos 2\pi(\nu_1 + \nu_2)\, t\} \end{aligned} \tag{25.20}$$

The Eq. (25.20) shows that the interaction of light waves of frequencies $\nu_1$and ν_2 propagating in a nonlinear medium generates new light waves of sum $(\nu_1 + \nu_2)$, difference $(\nu_1 - \nu_2)$ and double frequencies $2\,\nu_1$ and $2\,\nu_2$. It is to be noted that all the above mentioned frequencies are not generated simultaneously when two waves of frequencies ν_1 and ν_2 propagate through the crystal. If any of these additional frequencies are to be generated, the phase matching must be implemented and it can be implemented only to one of the frequencies at a time.

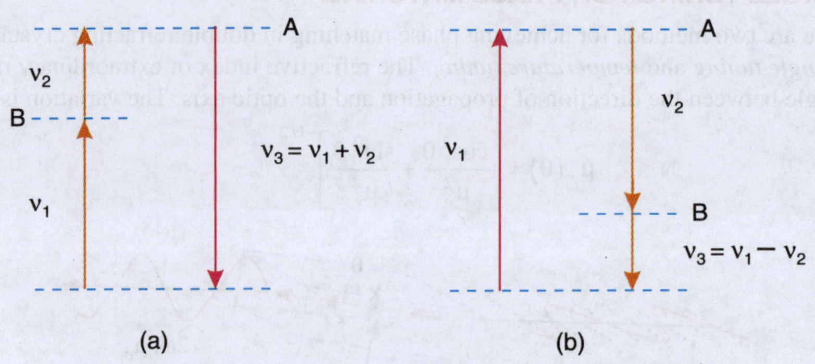

(*a*) sum frequency generation (*b*) difference frequency generation

Fig. 25.7

The processes of sum and difference frequency generation are depicted in Fig. 25.7. In sum frequency generation the frequencies ν_1 and ν_2 are destroyed and a new photon $\nu_3 = (\nu_1 + \nu_2)$ is created, as shown in Fig. 25.7 (*a*). In difference frequency generation, the photon of higher frequency ν_1 is destroyed and two new photons of frequencies ν_2 and $\nu_3 = (\nu_1 - \nu_2)$ are created, as shown in Fig. 25.7 (*b*). Note that ν_2 is already present as one of the input beams; it indicates that ν_2 is amplified in the process.

In sum frequency generation, ν_1 is taken as a weak infrared wave that sums with a strong pump wave ν_2 to generate a high frequency wave at $\nu_3 = (\nu_1 - \nu_2)$. Thus, the light of wavelength of 10.6 μm produced by a CO_2 laser is made to interact with a laser beam of 1.06 μm wavelength produced by Nd: YAG laser to yield a light beam of wavelength of 0.964 μm. Such upconversion is done because it is difficult to detect the radiation at 10 μm while the radiation at 0.964 μm can be easily detected.

25.8. PARAMETRIC OSCILLATION

The production of coherent radiation at frequencies in the UV region is very much essential. Apart from the technique of the harmonic generation, tuning of frequency by parametric oscillation is another technique available for this purpose. The method of parametric excitation of oscillations is widely used in electronics, where use of nonlinear capacitors is made. The parametric generation of light involves use of nonlinear medium and thus bears close similarity with the electronic process.

In sum frequency generation, light at frequencies ν_1 and ν_2 add together to produce the frequency $(\nu_1 + \nu_2)$. On the other hand, if a strong beam at frequency $\nu_1 = (\nu_2 + \nu_3)$ alone is applied to a suitable nonlinear material such as lithium niobate, two beams at lower frequencies ν_1 and ν_3 can be generated. For exciting parametric oscillations, the nonlinear medium is placed between two mirrors, which form a resonant cavity comparable to a laser cavity. The wave of frequency $\nu_1 = (\nu_2 + \nu_3)$ is the *pump wave* and it should be highly intense such that it can induce nonlinear behaviour of the crystal. Secondly the process requires that the phase-matching condition be satisfied. The wave of higher frequency, say ν_2, is called the *signal wave* and the lower frequency wave, say at ν_3, is called the *idler wave*. Both the *signal and idler* waves are weak and always present in the crystal in the form of *noise*, which arises due to spontaneous photons. If the pump frequency $(\nu_2 + \nu_3)$ is fixed, then the two frequencies ν_2 and ν_3 are free to spread over a wide range of values. This

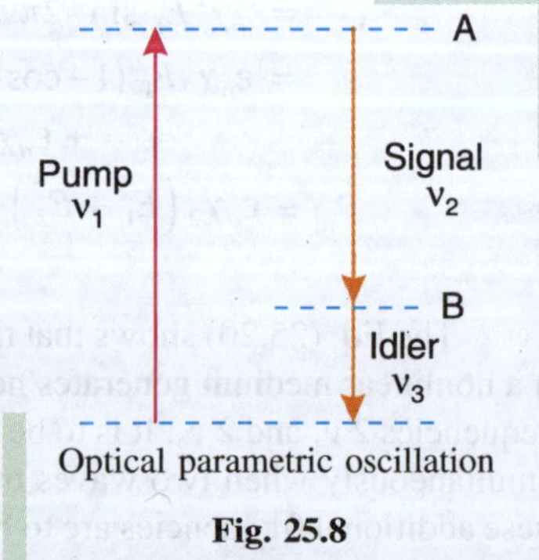

Optical parametric oscillation

Fig. 25.8

effect is known as *parametric amplification.* By varying the orientation of the crystal, the values of ν_2 and ν_3 can be varied. If the mirrors of the resonant cavity are reflective at ν_2 and ν_3, and not at the frequency $(\nu_2 + \nu_3)$, resonance of these waves in the cavity enhances the interaction in the nonlinear crystal and builds up the waves. Thus the system acts as a parametric oscillator. Tuning can be achieved by varying the phase-matching conditions through mechanical or temperature control of the cavity.

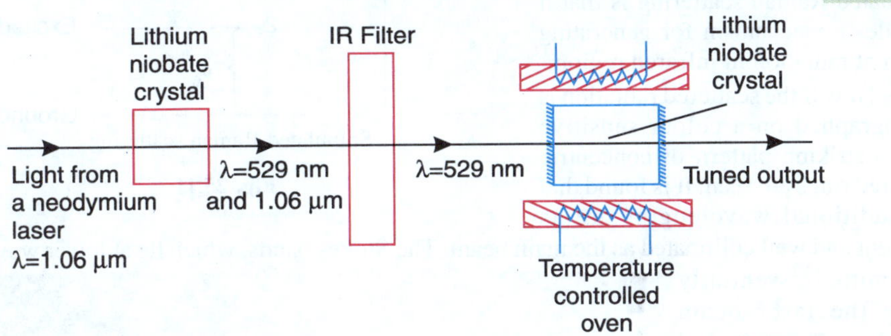

A typical arrangement for observing parametric oscillation in a nonlinear medium

Fig. 25.9

Parametric oscillation was first observed by Giordmaine and Miller in 1965. The schematic of the set up used by them is shown in Fig. 25.9. The output was tuned by changing the temperature of the lithium niobate crystal. A temperature range of about 11° C produced wavelengths in the range 9680 Å to 11540 Å. The conversion efficiency was low as 1%. Currently, conversion efficiency is improved from 40% to 50%.

25.9. SELF-FOCUSSING OF LIGHT

When a light beam of very high intensity passes through an optical medium, an internal force will result. It in turn alters the density, changes permittivity and thereby the refractive index. If the beam has a Gaussian cross-section, the intensity would be the greatest on the axis of the rod and hence the index of refraction would be greater on the axis than off the axis. These induced refractive index variations will cause the medium to function as a lens, as illustrated in Fig. 25.10. Consequently, the beam contracts and results in the process known as **self-focussing** of the beam.

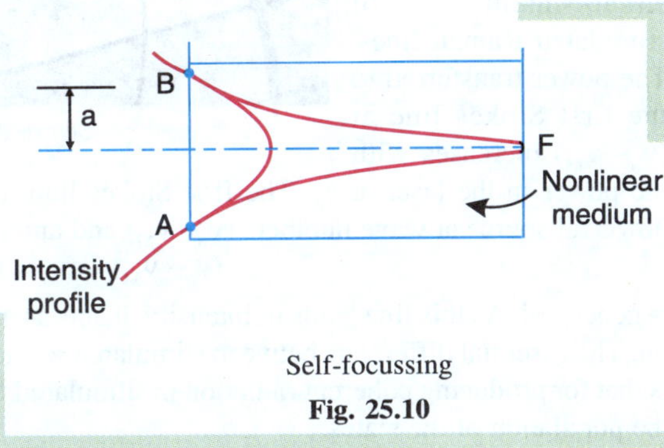

Self-focussing

Fig. 25.10

25.10. STIMULATED RAMAN SCATTERING

Stimulated Raman effect is a process in which an optical medium absorbs a higher energy photon and emits a lower energy photon and simultaneously absorbs the balance of the original photon energy and goes to an excited state. The process is illustrated in Fig. 25.11. A photon of frequency ν_1 is incident upon the material. A molecule in the material absorbs the photon and goes to a virtual level A, from where it emits a photon of lower frequency ν_2 and simultaneously absorbs the energy $h\nu_\upsilon = h(\nu_1 - \nu_2)$. As a result the molecule is left at the excited level B. The normal Raman scattering is incoherent whereas the stimulated Raman scattering is coherent.

The main advantage of stimulated Raman scattering is that it provides a mechanism for generating coherent radiation in infrared region.

Now if the scattered radiation is photographed on a colour sensitive film, a striking pattern of concentric coloured rings are seen. It is found that the additional wavelengths are as coherent and well collimated as the main beam. The Stokes bands, which lie at lower wave numbers are emitted essentially along the laser beam direction. The coloured rings correspond to successive anti-Stokes lines at higher wave numbers which are emitted only along directions which make specific small angles with the laser beam direction. Parametric amplification is responsible for the enhancement of stimulated Raman lines. The power transferred to the first Stokes line at $(\nu_o - \nu_M)$ increases with the power in the laser at ν_o. The first Stokes line rapidly becomes intense enough and acts as a powerful source at wave numbers $(\nu_o - \nu_M)$ and another Stokes line at

A — Virtual level; ν_1; ν_2; 2 — Excited level, B; ν_υ; Ground level

Stimulated Raman scattering

Fig. 25.11

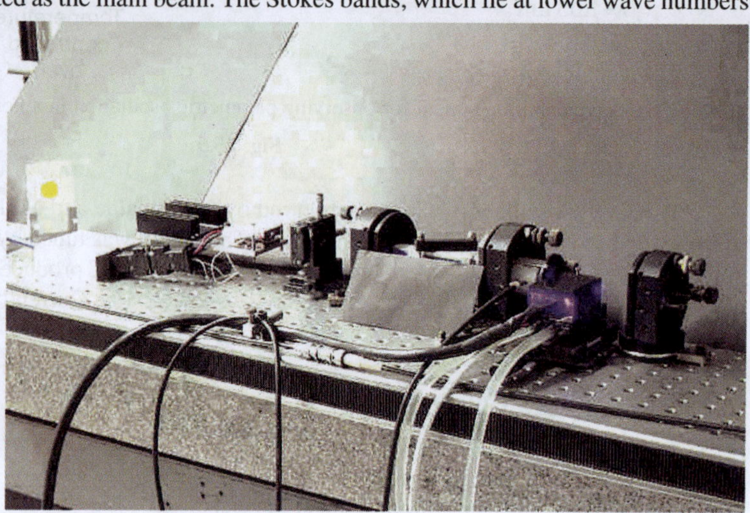

Stimulated Raman Scattering (SRS)

$$(\nu_o - \nu_M) - \nu_M = (\nu_o - 2\,\nu_M)$$

is generated. As this line gains in intensity, it acts as another source giving rise to a third line and so on. The essential difference between stimulated Raman scattering and stimulated emission in lasers is that for producing coherent radiation in stimulated Raman scattering there is no need for inverting the population of the states.

QUESTIONS

1. What is a nonlinear medium?
2. What is nonlinear polarization and why it assumed importance after the advent of lasers?
3. Explain why phase matching condition must be satisfied to generate coherent radiation at new wavelengths?
4. How is second harmonic generation achieved?
5. What are the different methods of tuning to obtain index-matching?
6. What is meant by parametric oscillation? How a nonlinear medium can be used as a parametric oscillator?
7. Explain what is meant by self-focussing?
8. Explain stimulated Raman scattering? How can it be explained using quantum concept?

26 CHAPTER

Atom Laser

At a Glance

- Introduction
- Bose-Einstein Condensation
- Methods of Cooling Atoms
- Laser Doppler Cooling
- Evaporative Cooling
- Basic Atom Laser
- Atom Laser Applications

26.1. INTRODUCTION

An atom laser is analogous to an optical laser. Optical laser generates a coherent beam of electromagnetic waves whereas an atom laser produces a coherent beam of matter waves. Each atom is associated with a de Broglie wave. When a dilute gas is cooled to temperatures near to zero, the de Broglie waves merge together to form a coherent matter wave. Therefore, an atom laser requires a thermal cloud of ultra-cold atoms which constitutes active medium of the laser. Hence, the main challenge in making an atom laser is the cooling of atoms to near zero temperature. In 1975 T.W. Hansch and A.L. Schawlow, USA, proposed first the technique of laser cooling. They suggested that neutral atoms might be cooled to millikelvin temperatures by the action of a near resonant laser beam that excites the atoms followed by spontaneous emission. Steven Chu, Claude Cohen-Tannoudji and William D.Phillips succeeded in cooling and trapping atoms with laser light and were awarded in 1997 the Nobel prize in Physics for their work. Cooling to millikelvin temperatures is not enough to create coherent matter waves. Atoms are to be cooled to nanokelvin temperatures, when they form a Bose-Einstein condensate and create the state of coherent matter waves. In 1995 Cornell, Wieman and their colleagues succeeded for the first time in cooling a dilute gas of rubidium atoms (Rb^{87}) to nanokelvin temperatures. They were awarded the Nobel prize in physics in the year 2001 for this work. The discovery of Bose-Einstein condensation has heralded the invention of atom laser. An atom laser will have a major impact on the fields of atom optics, atom lithography and precision measurements.

26.2. BOSE-EINSTEIN CONDENSATION

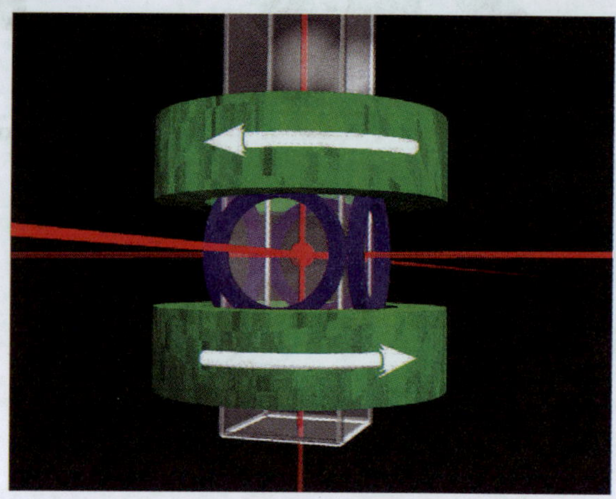

Bose-Einstein Condensation Trapping and Cooling Apparatus.

It is necessary to understand what a Bose-Einstein condensate is and how it is produced, before we go into the details of atom laser.

In 1924 the Indian physicist Satyendra Nath Bose derived the Planck law for black-body radiation by treating the photons as a gas of identical particles. Photons have an intrinsic angular momentum, or "spin", of the Planck constant h divided by 2π. Einstein generalized Bose's theory to other particles of integral spin. The theory is now known as **Bose-Einstein statistics**. Particles that have a spin that is an integer multiple of $h/2\pi$ obey Bose-Einstein statistics and are called **bosons**. It was shown that more than one boson can occupy the same quantum state. Einstein predicted that at sufficiently low temperatures all the atoms in an ideal gas of identical atoms might be condensed to a single lowest quantum state of the system. This large number of atoms locked together in the same state is called **Bose-Einstein condensate** (BEC). Bose-Einstein condensation is an exotic quantum phenomenon that was observed in dilute atomic gases. Bose-Einstein condensation happens only for "bosons". The low temperature necessary for obtaining BEC is of the order of nanokelvin.

26.3. METHODS OF COOLING ATOMS

Let us consider a gas inside a container at room temperature. The atoms of gas move at random in all directions and attain thermal equilibrium with the container. They move with velocities,υ, which areof the order of a few hundred meters/second. According to the kinetic theory the absolute temperature is proportional to $\left\langle \upsilon^2 \right\rangle$. Temperature is an expression of kinetic motion of the particles. Hence the kinetic energy of a particle may define the kinetic temperature.

The successful route to form Bose-Einstein condensate of the gas involves two steps. The first step consists in cooling down the atoms at low density to millikelvin temperatures by the action of a laser beam of an appropriate frequency. In the second step, the cold atoms are further cooled by using the technique of evaporative cooling. These techniques provided a new route to ultracold temperatures that does not involve cryogenics.

26.4. LASER DOPPLER COOLING

The basic principle of this method is as follows. If a photon falls on an atom, which has a resonance frequency equal to that of the photon, the atom absorbs the photon. If an atom absorbs a photon coming from the opposite direction, the momentum of the photon is transferred to the atom and the atom is pushed back and loses its velocity. In order to ensure deceleration of the atom, the atoms should absorb only the oppositely moving radiation. This is achieved by using Doppler effect.

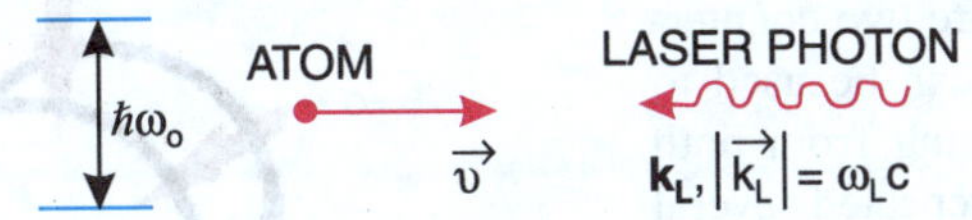

Doppler-resonant absorption of a red-detuned laser beam by a moving two-level atom

Fig. 26.1

An atom with a lower energy level E_1 and an upper energy level E_2 loses or gains energy by emitting or absorbing light of frequency ν_L such that $E_2 - E_1 = h\nu_L$ (see Fig. 26.1). The photons associated with a plane light wave of frequency ν_L , propagating along a direction with unit vector $\mathbf{k_L}$ have an energy $E = h\nu_L$ and a momentum $\mathbf{p_L} = \left(\frac{h\nu_L}{c}\right) \mathbf{k_L}$ each. Absorption of a single photon by an atom results in a momentum transfer of $\boldsymbol{p_L}$ to the atom. In one second, the atom can absorb and emit a large number of times and hence the net momentum transfer rate to the tiny atom become substantial. After about ten thousand absorption and subsequent emission cycles, an atom which was originally moving at about 700 m/s can be slowed down to near zero speed.

Fig. 26.2 shows a beam of atoms, each of mass m traveling with a velocity $\boldsymbol{\upsilon}$, colliding with a counterdirected beam of laser photons having a propagation vector $\mathbf{k_L}$. The laser frequency ν_L is selected such that it is just beneath the resonant frequency v_o of the atoms. Because of its motion, any particular atom sees an oncoming photon with a frequency that is Doppler-shifted upward by an amount $v_L \upsilon / c$. When the laser frequency is tuned so that $v_o = v_L (1 + \upsilon / c)$, collisions with photons will resonate the atoms. In the process, each photon transfers its momentum of $\hbar\mathbf{k_L}$ to the absorbing atom, whose speed there upon reduced by an amount $\Delta\upsilon$ where $m\,\Delta\upsilon = \hbar\mathbf{k_L}$. The cloud of atoms is not very dense, and each excited atom can drop back to its ground state with spontaneous emission of a photon of energy hv_o. The emission is randomly directed and so although the atom recoils, the average amount of momentum regained by it over thousands of cycles tends to zero. The change in momentum of the atom per photon absorption-emission cycle is therefore effectively $\hbar k_L$ and it slows down. By contrast, an atom moving in the opposite direction, away from the light source sees

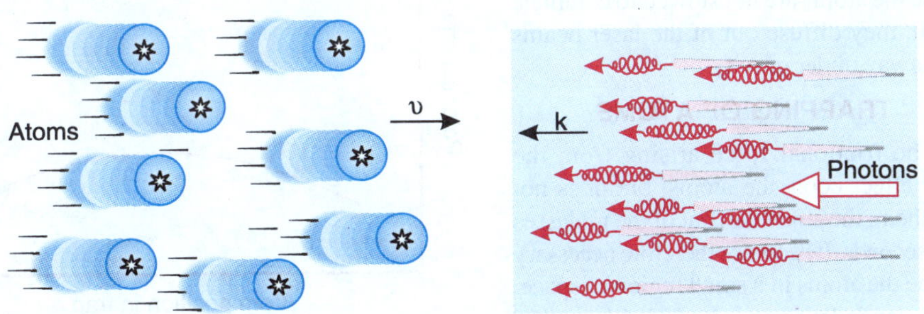

A stream of atoms colliding with a laserbeam in the process of laser cooling.

Fig. 26.2

photons to have a frequency v_L (1– υ/c), far enough away from v_0 that there can be little or no absorption and therefore no momentum gain. If there are two counter-propagating waves, they can be used to decelerate the atoms coming from both directions. The process is repeated several times and slows down the atoms. Hence, the gas is cooled. Thus, the absorption slows down the atom where as the emission does not have an effect and on the average there is a reduction in momentum. However, the lowest temperature that the laser cooling technique can reach is limited. The process of Doppler cooling can lower the temperature to the order of 100 μK for alkali atoms. Alkali atoms are well suited to laser-based methods because their optical transitions can be excited by available lasers and because they have a favourable internal energy-level structure for cooling to low temperatures. Once the atoms are cooled to μK temperature, we have to confine them. If the cooling laser is turned off, the atoms will start moving apart with their residual velocities, and will hit the walls and eventually fall down. The cold atoms usually have a velocity of a few centimeters per second.

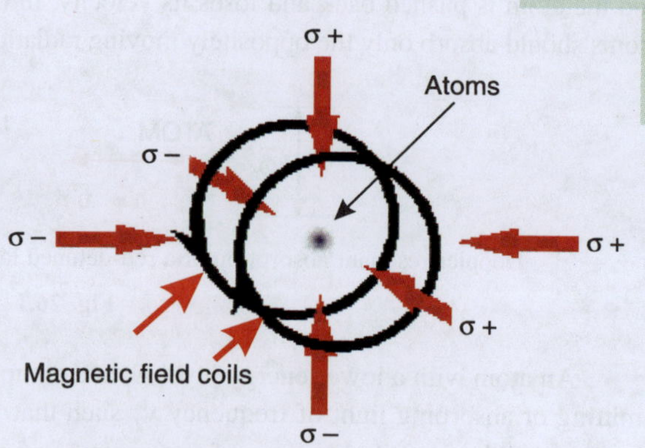

Six laser beams and a pair of magnetic field coils forming a magneto-optical trap can cool more than a billion atoms to micro-kelvin temperatures.

Fig. 26.3

In practice six laser beams are arranged – two counter-propagating beams each, (one along the positive direction and the other along the negative direction) along each of the three Cartesian axes, as shown in Fig. 26.3. An atom, in whichever direction it moves, always encounters a beam in opposite direction and the net effect is that the atom experiences a strong viscous force in the three dimensional space in the field of photons. This six laser beam configuration used to cool a sample of atoms from a vapour is given the name **optical molasses,** since the light bath always opposes the motion of atoms as if they were submerged in molasses and causes viscous effect on the atomic motion. The optical molasses could produce a temperature much lower than Doppler cooling limit. However there is no position dependent force and since the atoms are not slowed to complete standstill, they diffuse out of the laser beams and fall away under gravity.

26.4.1. TRAPPING OF ATOMS

The frictional force arising from the Doppler effect cools the atoms, but does not capture them; the atoms fall out of the molasses in a few seconds. It becomes therefore necessary to confine the atoms in a small region of space. Position dependence is introduced by using (σ^+ / σ^-) polarized laser beams and applying a weak magnetic field. This arrangement is known as a *magneto-optical trap* (MOT). The trap may be a spherical quadrupole magnetic field,

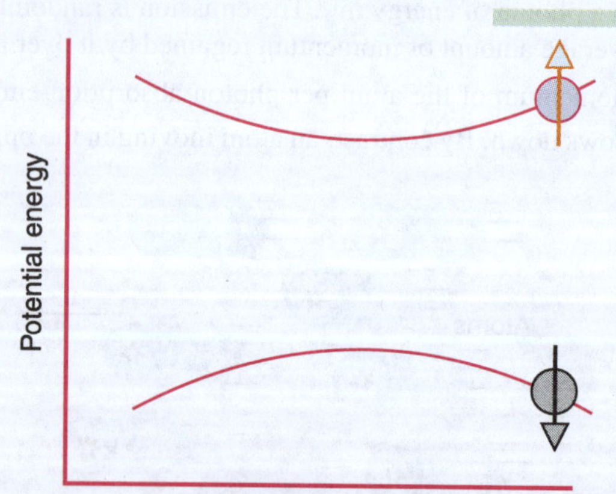

An atom with spin parallel to the magnetic field is attracted to the energy minimum and an atom with anti-parallel spin is repelled.

Fig. 26.4

produced by two anti-Helmholtz coils carrying currents in opposite directions (see Fig. 26.3). Such an arrangement gives rise to a zero-field region between the two coils, at the intersection of the six laser beams and the field increases in all directions away from this point. This field minimum attract molecules in the low-field seeking states, that is molecules with their magnetic moments oriented anti-parallel to the magnetic fields, and repel strong field seeking molecules (see Fig. 26.4).

Alkali atoms have a magnetic moment because they have an unpaired electron. The magnetic moment is in a direction opposite to that of the electron spin. The magnetic moment interacts with the applied weak magnetic field. If the magnetic moment is parallel to the external magnetic field, the atom is attracted to the local minimum of the field (Fig. 26.4) and can be trapped. It is also necessary that the atoms must be thermally isolated from their surroundings, since at ultracold temperatures atoms stick to all surfaces. After the atoms are trapped and cooled with lasers, all light is extinguished and a potential is built up around the atoms with an inhomogeneous magnetic field. Thus, the magnetic field can act as a little bowl and which confines the atoms to a small region of space, as shown in Fig. 26.5 (a). In addition to holding the atoms at a point in space, the trapping force compresses them into a dense cold cloud.

26.5. EVAPORATIVE COOLING

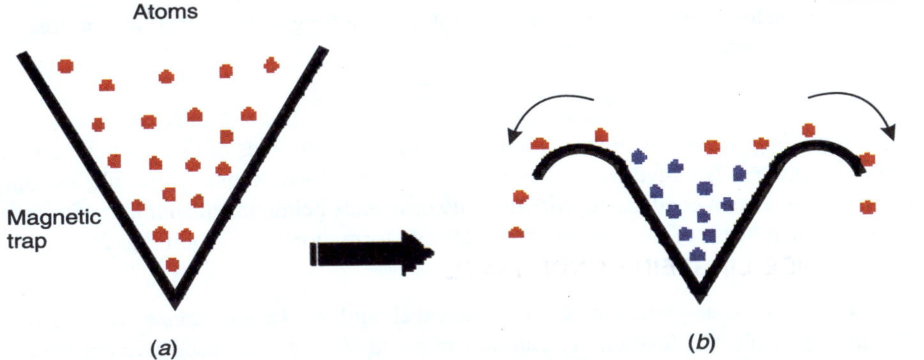

Evaporative cooling. (*a*) Atoms held in a magnetic trap. (*b*) Cooling works by slowly lowering the sides to release the hot atoms at the top.

Fig. 26.5

The atoms are further cooled to ultracold temperatures using the technique of evaporative cooling. The basic physics of evaporating cooling is very simple. It is known that a cup of hot milk or coffee cools down by evaporation. In the same way, high-energy atoms are allowed to escape from the sample so that the average energy of the remaining atoms is reduced.

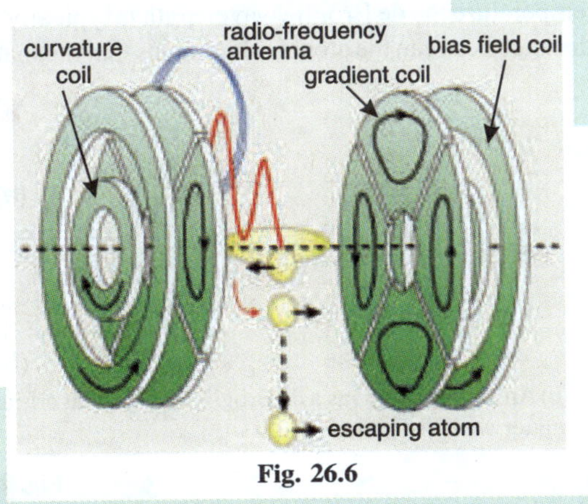

Fig. 26.6

It was shown by H.F. Hess that the idea of evaporative cooling could be applied to the atoms confined in a magnetic bowl. In this method the atoms are cooled first in the presence of the magneto-optic trap. Then the laser beams are turned off. Then a magnetic field, of the same type but much stronger than that used in MOT, is applied to perform the role of a bowl and confine the cold atoms. Now, a radio frequency oscillating field is applied to induce transitions in the atom between spin up state (attracted to the magnetic trap) and the spin down state (repelled by the magnetic trap), so that the atoms will undergo a spin-flip transition. If the rf field is tuned to the higher side of the magnetic bowl the atoms having higher energy will jump the well and fall out of the bowl (Fig.26.6). Lowering the rf frequency will induce some more atoms to leave the bowl. When the atoms with higher energy leave the bowl, the average energy of the remaining atoms becomes lower and atoms get colder in the process.

Cooling to progressively colder temperatures resulting in the formation of a Bose-Einstein condensate (far right).

Fig. 26.7

Continuing to cool involves continuously lowering the sides of the magnetic trap (see Fig. 26.5 b). The process is halted when the temperature goes down to nearly 100 nK and only a few atoms are left to form a condensate, with the only drawback being that the number of trapped atoms is reduced. The process of Bose-Einstein condensate formation is shown in Fig. 26.7.

26.5.1. BOSE-EINSTEIN CONDENSATE

Each atom has an associated de Broglie wavelength λ_B. In accordance with the Heisenberg uncertainty principle, the position of an atom is smeared out over a distance given by the thermal de Broglie wavelength,

$$\lambda_B = \left[\frac{h^2}{2kmT}\right]^{1/2}$$

where k is the Boltzmann constant, m is the atomic mass and T is the temperature of the gas. At room temperature the de Broglie wavelength is typically about ten thousand times smaller than the average distance between the atoms. This means that the matter waves of the individual atoms are uncorrelated

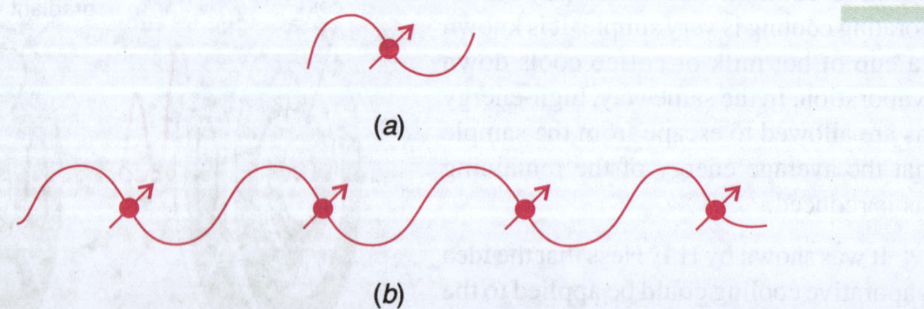

(*a*) An atom behaves as a de Broglie wave; (*b*) at critical temperature Bose-Einstein condensate or a coherent matter wave is formed.

Fig. 26.8

or "disordered" and the gas can thus be described by *classical Boltzmann statistics.* As the gas is cooled, however, the wavelength increases, the smearing increases, and eventually there is more than one atom in each cube of dimension λ_B. The wave functions of adjacent atoms then "overlap"(Fig. 26.8), causing the atoms to lose their identity, and become one '*super-atom*'.

Bose-Einstein statistics dramatically increase the chances of finding more than one atom in the same state. Thus, at very low temperatures, all the atoms fall down to the lowest state, as they do not have enough energy to go to higher energy states. The result is *Bose-Einstein condensation*, a macroscopic occupation of the ground state of the gas. Although small, typically 0.1 mm across, a Bose-Einstein condensate can be seen with weakly magnifying lenses and a video camera. Thus, it is a microscopic object. The transition to BEC corresponds to a transition from a set of disordered atoms to coherent matter waves. Indeed, the transition from disordered to coherent matter waves can be compared to the change from incoherent to laser light. In a Bose condensate all the atoms occupy the same quantum state and can be described by the same wave function. The condensate therefore has many unusual properties not found in other states of matter.

26.6. BASIC ATOM LASER

An optical laser consists of an active medium held in a cavity resonator (see Fig. 26.9). When the active medium is pumped, say with an optical source, into the state of population inversion, stimulated emission is triggered and an intense beam of light emerges out of the partially reflecting mirror of the cavity. An atom laser also should consist of similar principal parts.

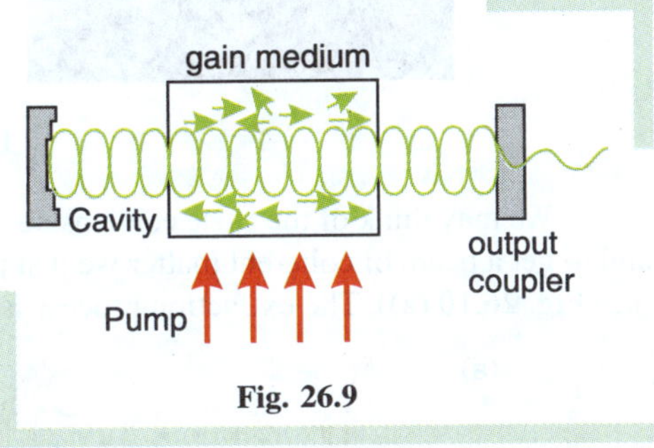

Fig. 26.9

1. Active medium: In the case of atom laser, the active medium is a thermal cloud of ultracold atoms.

2. Cavity: The cavity resonator is a magnetic trap in which the atoms are confined by magnetic mirrors. In a magnetic trap, for instance, once the atoms have been cooled and trapped by lasers, the light is switched off and an inhomogeneous magnetic field provides a confining potential around the atoms. The trap is analogous to the optical cavity formed by the mirrors in a conventional laser.

3. Pumping: Pumping is done by the evaporative cooling. The evaporation process creates a cloud which is not in thermal equilibrium and relaxes toward colder temperatures. This results in growth of the condensate.

4. Stimulated emission: In an atom laser, the presence of a Bose-Einstein condensate with N atoms enhances the probability that an atom joins the condensate increasing the size of the condensate to N+1 atoms. The process of condensing atoms into the ground state of a magnetic trap is analogous to stimulated emission into a single mode of an optical laser.

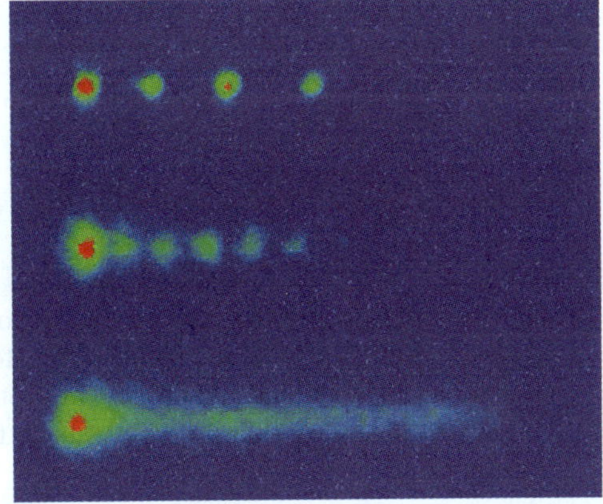

Basic Atom Laser.

5. Laser threshold: An optical laser starts working when the losses in the cavity are balanced by the gain in the medium. In atom laser the critical temperature for Bose-Einstein condensation resembles to the laser threshold. When the critical temperature for BEC is reached, atoms predominantly go into the lowest energy state of the system.

6. Output-coupler: An important feature of a laser is an output coupler to extract a fraction of the coherent field in a controlled way. In the case of a conventional laser the output coupler is a partially transmitting mirror.

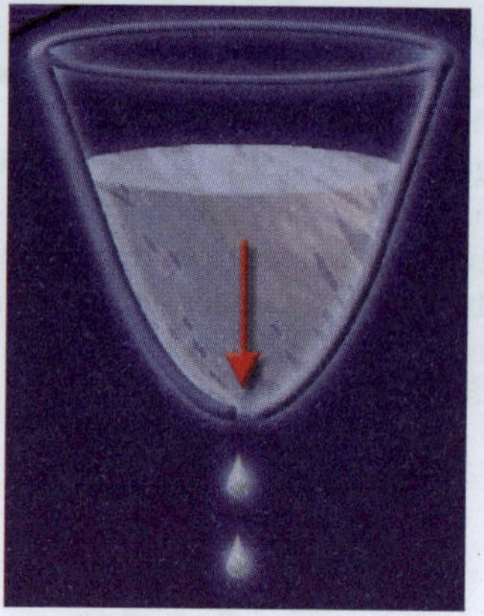

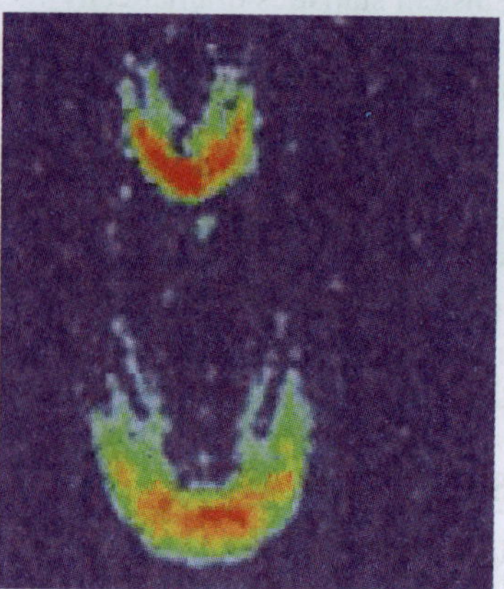

Fig. 26.10

We may think of the Bose condensate as being held in a container much like water in a bowl and to get a beam of coherent matter we just puncture the container and Bose condensate leaks out (see Fig. 26.10 (a)). The extraction process is called **output-coupling.**

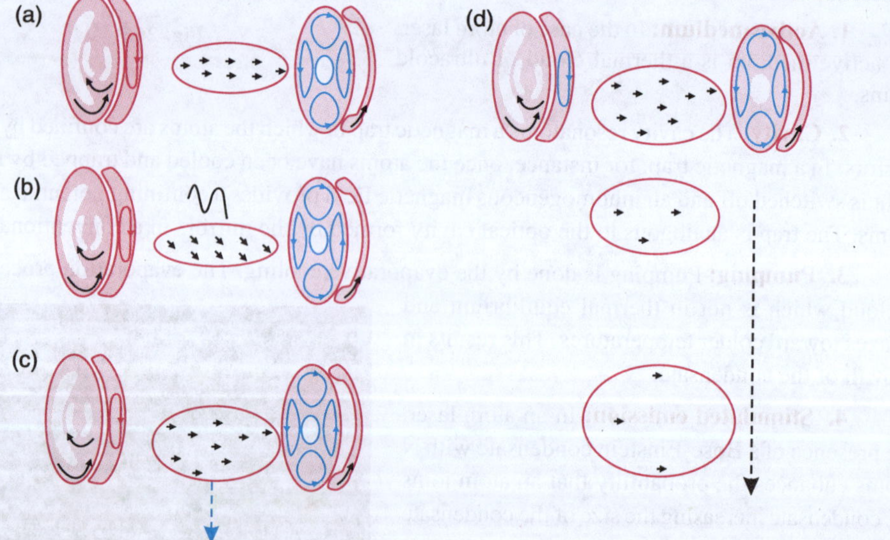

The rf output coupler (*a*) shows a Bose condensate trapped in a magnetic trap (*b*) A short pulse of rf radiation tilts the spins of the atoms (*c*) Quantum-mechanically, a tilted spin is a superposition of spin up and down. The cloud is split into a trapped cloud and an out-coupled cloud. (*d*) Several output pulses can be extracted, which spread out and are accelerated by gravity.

Fig. 26.11

Output coupling for atoms can be achieved by transferring them from states that are confined to states that are not, typically by changing an internal degree of freedom, such as the magnetic states of the atoms. Only the atoms that had their magnetic moments pointing in the opposite direction to the magnetic field were trapped. The "reflectivity" of the magnetic mirrors is altered by applying a short radio-frequency pulse to "flip" the spins of some of the atoms and therefore release them from the trap (see Fig. 26.11). The extracted atoms then are accelerated away from the trap under the force of gravity. By changing the amplitude of the radio-frequency field, the extracted fraction could be varied between 0% and 100%.

A pulsed output beam of atoms is shown in Fig. 26.12.

7. Output of laser: The output of an optical laser is a well collimated beam of light (electromagnetic waves).For an atom laser, the output is a beam of atoms (matter waves).

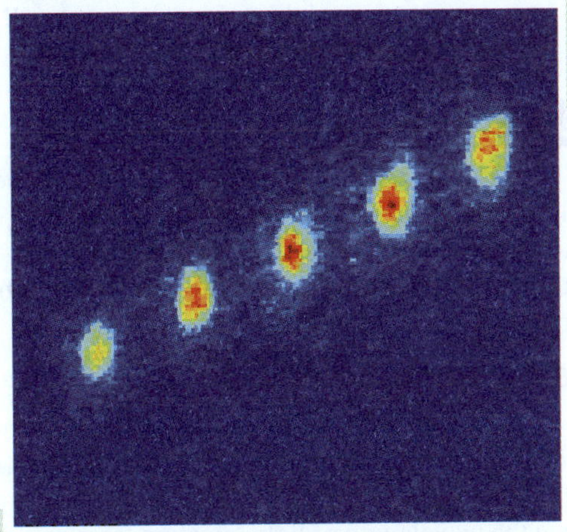

A pulsed atom-laser output beam.

Fig. 26.12

8. Nature of output: Optical lasers can work either in a pulsed mode or continuous wave (cw) mode. So far, the atom laser has been realized in the pulsed mode only. Most continuous-wave optical lasers are truly continuous in the sense that they are continually fed energy or "pumped" so that they can supply photons indefinitely. A truly continuous source of coherent matter waves could be similarly achieved only if the condensate could be replenished continually. Schemes for steady-state condensate formation are being explored currently.

9. Modes: Optical lasers radiate in several modes, i.e., at several nearby frequencies. The cavity in an optical laser is typically many wavelengths long and as a result can support several different frequencies or modes. Atom laser operates in a single mode, since the Bose condensate actually involves mode competition and the atoms typically occupy the lowest energy state of the trap.

26.6.1. IMPORTANT FEATURES OF A LASER BEAM

1. Monochromaticity: An important property of laser light is that it is monochromatic. By analogy, in a Bose-Einstein condensate all the atoms have the same energy and hence the same de Broglie wavelength. If this property can be maintained when the atoms are released from the condensate, we will have a highly **monochromatic** source of matter waves.

2. Coherence: A crucial feature of a laser is the **coherence** of its output - in other words, the presence of a macroscopic wave.

It is likely that the **coherence length** of the lasers is in fact much larger than the size of the condensate.

3. Intensity: Light produced by an optical laser is highly intense due to the constructive interference of coherent waves. Coherent matter waves are more intense than ordinary atoms.

4. Directionality: The spread of optical beam issuing out of a laser is governed by the diffraction effect at the output aperture. In case of atom laser, the beam is limited by Heisenberg uncertainty.

26.6.2. DIFFERENCES BETWEEN AN ATOM LASER AND AN OPTICAL LASER

1. In optical lasers population inversion is essential for realizing laser action. Population inversion does not occur in an atom laser.
2. In case of an optical laser, a large number of photons are generated through stimulated emission. In case of atom laser, atoms are not created. The atoms in the lower state are increased while the number of atoms in the upper states decreases.
3. Atoms are heavier than photons. They are therefore accelerated by gravity and a matter wave beam will fall like a beam of ordinary atoms.
4. A light beam from an optical laser can travel very far and does not require a vacuum for its operation. Unlike photons, atoms will not travel very far in air, so that the atom laser must be used in a vacuum.

26.7. ATOM LASER APPLICATIONS

The possibility of producing a coherent beam of atoms that could be collimated to travel long distances, or brought to a tiny focus like an optical laser, opens up a whole host of applications. Atom lasers may have a major impact on the fields of atom optics, atom lithography, precision atomic clocks and other measurements of fundamental standards. We study here three typical applications of atom laser.

26.7.1. HOLOGRAPHY

One application for which the coherence of an atom laser is critical is atom holography. Just as conventional holography uses the diffraction of a photon beam to reconstruct a 3-D image, atom holography uses the diffraction of atoms. As the de Broglie wavelength of the atoms is much smaller than the wavelength of light, an atom laser could create much higher resolution holographic images. Atom holography might be used to project complex integrated-circuit patterns, just a few nanometres in scale, onto semiconductors.

The first atom holograms were demonstrated in 1996 by Fujio Shimuzu and colleagues at the University of Tokyo, using laser-cooled atoms. In the case of laser-cooled gases, the level of coherence needed to create a hologram is achieved by selecting a small portion of the atoms. The problem would be simplified if a source, in which most of the atoms are in the same quantum state, is used. An atom laser is such a source and could provide a much more intense and fully coherent beam of atoms.

Holography is a two-step process. First, a hologram - a sort of diffraction grating containing information about the object - is produced. Then a beam of light (or atoms) is diffracted by the hologram to form the image. In optical holography, the hologram is often made by interfering a laser beam with light that has been reflected from an object. The resulting diffraction pattern is recorded on photographic film.

In the atom-holography experiments, an image has been created by diffracting a coherent beam of atoms through a grating that was manufactured using electron-beam lithography. The image may be recorded on a "microchannel plate" - a detector that is sensitive to atoms. So far, atom holography has been able to produce 2-D images.

26.7.2. ATOM INTERFEROMETRY

Another important application is atom interferometry. In an atom interferometer an atomic wave packet is coherently split into two wave packets that follow different paths before recombining. The interference pattern created when the two wave packets recombine tells us something about the phase difference between the two paths. Atom interferometers that are more sensitive than optical interferometers could be used to test quantum theory, and may even be able to detect changes in space-time. This is because the de Broglie wavelength of the atoms is smaller than the wavelength of light, and the atoms have mass. Atom lasers would allow the use of devices with unequal path lengths, such as Michelson interferometers. Such devices may provide a way to measure lengths over large distances with unprecedented precision.

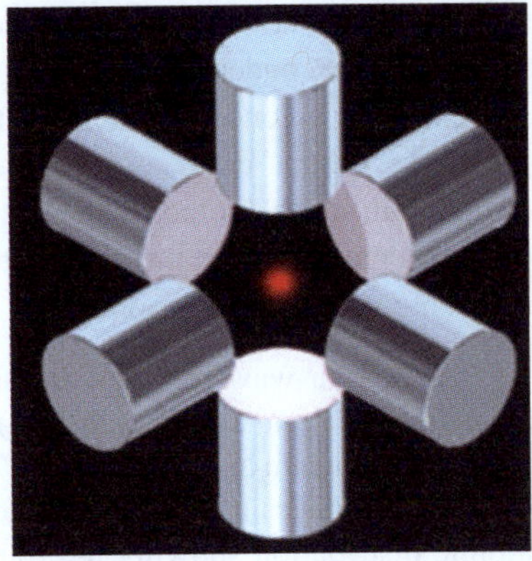

Atom laser may be useful as an input source for an atom interferometer.

26.7.3. NONLINEAR ATOM OPTICS

Just as the invention of the laser enabled the field of nonlinear optics to flourish, intense sources of coherent matter waves have opened up a similar field in atom optics. Until recently, most atom-optics experiments could be thought of as single-particle phenomena where the interactions between particles could be neglected. In conventional nonlinear optics, photons interact with each other through some mediating material, such as a transparent crystal. A common nonlinear optical phenomenon is "four-wave mixing". Typically three waves, of frequency ω_1, ω_2 and ω_3, are sent into a nonlinear crystal. The exchange of energy and momentum between the waves, mediated by the nonlinear crystal, results in the production of a fourth wave with frequency $\omega_4 = \omega_1 + \omega_2 - \omega_3$ (see Fig. 26.13a). A quantum mechanical description of this process shows that two photons from separate beams annihilate in the crystal and produce two new photons. The energy and momentum of one of these photons adds to the third beam, while the other photon corresponds to a new, fourth beam.

In 1998 an analogous process with matter waves was predicted. It is predicted that if three condensates of appropriate momenta collided, the term in the nonlinear Schrödinger equation that describes the interactions between the atoms would give rise to a fourth. At the atomic level, this process can be described as a collision between two atoms from separate matter-wave beams. One of the atoms is stimulated so that it scatters in the direction of the third incident matter-wave beam. By the conservation of momentum, the other atom goes off to make a fourth, separate beam.

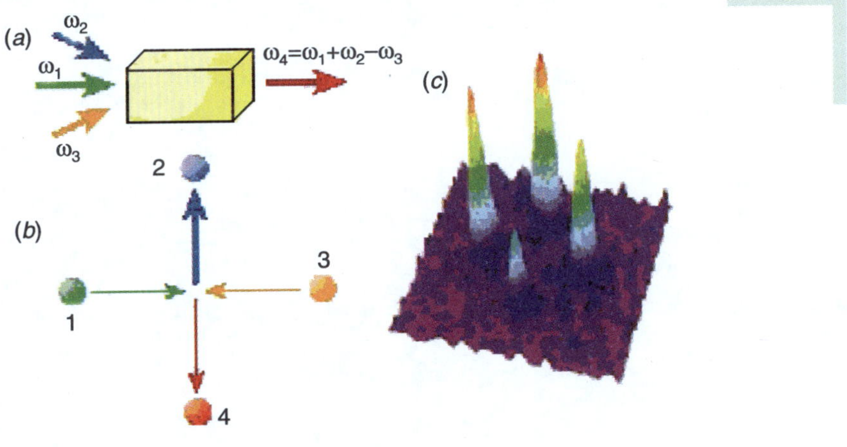

Fig. 26.13

The actual experiment did not use three separate condensates. Instead, lasers were used to divide a single condensate into three different momentum states via a process called Bragg diffraction. Starting with a condensate at rest, two separate pulses of interfering laser beams were applied to create the Bragg "diffraction grating" that divided the atoms roughly equally into three different momentum states, including the state of the initial condensate. When these pulses were applied fast enough - that is before the different momentum states had a chance to separate - atoms in a fourth momentum state were produced.

The four-wave mixing process arises from collisions between pairs of atoms from two matter-wave beams (1 and 2 in Fig. 26.13 b). One pair of atoms scatters in the direction of the third, incident matter-wave beam and amplifies it (3). By the conservation of energy and momentum, the other pair of atoms produces a fourth, separate beam (4). Fig. 26.13 (*c*) shows an image of the experimental atomic distribution showing the fourth (small) wave packet generated by the matter-wave mixing process.

QUESTIONS

1. Explain in brief the method of cooling a dilute gas to nanokelvin temperatures.
2. What is meant by Bose-Einstein condensation? What is its importance?
3. Explain the technique of Doppler cooling? What is the temperature limit that can be attained through this method?
4. What is an optical molasses? What is its role in cooling atoms?
5. What is a magneto-optic trap? Why is it required in cooling of atoms?
6. Describe the technique of evaporative cooling.
7. Explain the principle of an atom laser and describe its important components.
8. Describe the similarities and differences between an optical laser and an atom laser.
9. Describe any two important applications of atom laser.
10. Explain how atom laser is used in holography?
11. Explain the four-wave mixing of matter waves.

Appendix

Nobel Laureates in Physics

Year	Nobel Laureate		Recognized for
1901	Wilhelm Konrad Rontgen	1845-1923	discovery of X-rays
1902	Hendrik Antoon Lorentz Pieter Zeeman	1853-1928 1865-1943	their researches into the influence of magnetism upon radiation phenomena
1903	Antoine Henri Becquerel Pierre Curie Marie Sklodowskan Curie	1852-1908 1859-1906 1867-1934	his discovery of spontaneous radioactivity; their joint researches on the radiation phenomena discovered by Prof. Henri Becquerel
1904	John William Strutt, (Lord Rayleigh)	1842-1919	investigations of the densities of the most important gases and his discovery of argon
1905	Philipp Eduard Anton von Lenard	1862-1947	his work on cathode rays
1906	Joseph John Thomson	1856-1940	his theoretical and experimental investigations on the conduction of electricity by gases
1907	Albert Abraham Michelson	1852-1931	his optical precision instruments and metrological investigations carried out with their aid
1908	Gabriel Lippmann	1845-1921	his method of reproducing colors photographically based on the phenomena of interference
1909	Guglielmo Marconi Karl Ferdinand Braun	1874-1937 1850-1918	their contribution to the development of wireless telegraphy
1910	Johannes Diderik van der Waals	1837-1932	his work on the equation of state of gases and liquids
1911	Wilhelm Wien	1864-1928	his discoveries regarding the laws governing the radiation of heat
1912	Nils Gustaf Dalen	1869-1937	his invention of automatic regulators for use in conjunction with gas accumulators for illuminating lighthouses and buoys
1913	Heike Kamerlingh Onnes	1853-1926	his investigations of the properties of matter at low temperatures which led, inter alia to the production of liquid helium.
1914	Max Theodor Felix von Laue	1879-1960	his discovery of the diffraction of Rontgen rays by crystals
1915	William Henry Bragg William Lawrence Bragg	1862-1942 1890-1971	their services in the analysis of crystal structure by means of X rays
1917	Charles Glover Barkla	1877-1944	his discovery of the characteristic X rays of the elements
1918	Max Planck	1858-1947	his discovery of energy quanta
1919	Johannes Stark	1874-1957	his discovery of the Doppler effect in canal rays and the splitting of spectral lines in electric fields

Year	Nobel Laureate		Recognized for
1920	Charles Edouard Guillaume	1861-1938	the service he has rendered to precision measurements in Physics by his discovery of anomalies in nickel steel alloys
1921	Albert Einstein	1879-1955	his services to Theoretical Physics, and especially for his discovery of the laws of the photoelectric effect
1922	Niels Henrik David Bohr	1885-1962	the investigation of the structure of atoms, and of the radiation emanating from them
1923	Robert Andrews Millikan	1868-1953	his work on the elementary charge of electricity and on the photoelectric effect.
1924	Karl Manne Georg Siegbahn	1888-1979	his discovery and research in the field of X-ray spectroscopy
1925	James Franck Gustav Hertz	1882-1964 1887-1975	their discovery of the laws governing the impact of an electron upon an atom
1926	Jean Baptiste Perrin	1870-1942	his work on the discontinuous structure of matter, and especially for his discovery of sedimentation equilibrium
1927	Arthu Holly Compton Charles Thomson Rees Wilson	1892-1962 1869-1959	his discovery of the effect named after him, his method of making the paths of electrically charged particles visible by condensation of vapour
1928	Owen Willans Richardson	1879-1959	his work on the thermionic phenomenon, and especially for the discovery of the law named after him
1929	Prince Louis Victor de Broglie	1892-1987	his discovery of the wave nature of electrons
1930	Sir Chandrasekhara Venkata Raman	1888-1970	his work on the scattering of light and for the discovery of the effect named after him
1932	Werner Heisenberg	1901-1976	the creation of quantum mechanics, the application of which has, among other things, led to the discovery of the allotropic forms of hydrogen
1933	Erwin Schrödinger Pal Adrien Maurice Dirac	1887-1961 1902-1984	the discovery of new productive forms of atomic theory
1934	(no award)		
1935	James Chadwick	1891-1974	his discovery of the neutron
1936	Victor Franz Hess Carl David Anderson	1883-1964 1905-1991	the discovery of cosmic radiation his discovery of the position
1937	Clinton Joseph Davisson George Paget Thomson	1881-1958 1892-1975	their experimental discovery of the diffraction of electron by crystals
1938	Enrico Fermi	1901-1954	his demonstrations of the existence of new radioactive elements produced by neutron irradiation, and for his related discovery of nuclear reactions brought about by slow neutrons
1939	Ernest Orlando Lawrence	1901-1958	the invention and development of the cyclotron and for results obtained with it, especially for artificial radioactive elements

Year	Nobel Laureate		Recognized for
1943	Otto Stern	1888-1969	his contribution to the development of the molecular rays method and his discovery of the magnetic moment of the proton
1944	Isidor Issac Rabi	1898-1988	his resonance method for recording the magnetic properties of atomic nuclei
1945	Wolfgang Pauli	1900-1958	his discovery of the Exclusion Principle (Pauli Principle)
1946	Percy Williams Bridgman	1882-1961	the invention of an apparatus to produce extremely high pressures, and for the discoveries he made therewith in the field of high-pressure physics
1947	Sir Edward Victor Appleton	1892-1965	his investigations of the physics of the upper atmosphere, especially for the discovery of the so-called Appleton layer
1948	Patrick Maynard Stuart Blackett	1897-1974	his development of the Wilson cloud chamber method, and his discoveries therewith in nuclear physics and cosmic radiation
1949	Hideki Yukawa	1907-1981	his prediction of the existence of mesons on the basis of theoretical work on nuclear forces
1950	Cecil Frank Powell	1903-1969	his development of the photographic method of studying nuclear processes and his discoveries regarding mesons made with this method
1951	Sir John Douglas Cockcroft Ernest Thomas Sinton Walton	1897-1967 1903-	their pioneer work on the transmutaion of atomic nuclei by artificially accelerated atomic particles
1952	Felix Bloch Edward Mills Purcell	1905-1983 1912-	their development of new methods for nuclear magnetic precision measurements and discoveries in connection therewith
1953	Frits Zernike	1888-1966	his demonstration of the phase contrast method, especially for his invention of the phase contrast microscope
1954	Max Born	1882-1970	his fundamental research in quantum mechanics especially for his statistical interpretation of the wave function
	Walther Bothe	1891-1958	the coincidence method and his discoveries made therewith
1955	Willis Eugene Lamb	b. 1913	his discoveries concerning the fine structure of the hydrogen spectrum
	Polykarp Kusch	b. 1911	his precision determination of the magnetic moment of the electron
1956	William Bradford Shockley John Bardeen Walter Houser Brattain	1910-1989 1908-1991 1902-1987	their researches on semiconductors and their discovery of the transistor effect
1957	Chen Ning Yang Tsung Dao Lee	b. 1922 b. 1926	their penetrating investigation of the parity laws which has led to important discoveries regarding elementary particles

Year	Nobel Laureate		Recognized for
1958	Pavel Alekseyevich Cerenkov	b. 1904	the discovery and interpretation of the Cerenkov effect
	Ilya Mikhaylovich Frank	1908-1990	
	Igor Yevgenyevich Tamm	1895-1971	
1959	Emillio Gino Segre	1905-1990	their discovery of the antiproton
	Own Chamberlain	b. 1920	
1960	Donald Arthur Glaser	b. 1926	the invention of the bubble chamber
1961	Robert Hofstadter	1915-1990	his pioneering studies of electron scattering in atomic nuclei and for his thereby achieved discoveries concerning the structure of the nucleons
	Rudolf Ludwig Mossbauer	b. 1929	his research concerning the resonance absorption of γ rays and his discovery in this connection of the effect which bears his name
1962	Lev Davidovich Landau	1908-1968	his pioneering theories of condensed matter, especially liquid helium
1963	Eugene Paul Wigner	b. 1902	his contributions to the theory of the atomic nucleus and the elementary particles, particularly through the discovery and application of fundamental symmetry principles
	Maria Goeppert Mayer	1906-1972	their discoveries concerning nuclear shell structure
	J. Hans D. Jensen	1907-1973	
1964	Charles Hard Townes	b. 1915	fundamental work in the field of quantum electronics which has led to the construction of oscillators and amplifiers based on the maser-laser principle
	Nikolay Gennadiyevich Basov	b. 1922	
	Alekanndr Mikhazlovich Prokhorov	b. 1916	
1965	Schinichiro Tomonaga	1906-1979	their fundamental work in quantum electrodynamics, with profound consequences for the physics of elementary particles
	Julian Schwinger	b. 1918	
	Richard Phillips Feynman	1918-1988	
1966	Alfred Kastler	1902-1984	the discovery and development of optical methods for studying Hertzian resonance in atoms
1967	Hans Albrecht Bethe	b. 1906	his contributions to the theory of nuclear reactions, especially his discoveries concerning the energy production in stars
1968	Luis Walter Alvarez	1911-1988	his decisive contribution to elementary particle physics, in particular the discovery of a large number of resonance states, made possible through his development of the technique of using the hydrogen bubble chamber and data analysis
1969	Murray Gell-Mann	b. 1929	his contributions and discoveries concerning the classification of elementary particles and their interactions.
1970	Hannes Olof Gosta Allven	b. 1908	fundamental work and discoveries in magnetohydro-dynamics with fruitful applications in different parts on plasma physics

Year	Nobel Laureate		Recognized for
	Louis Eugene Felix Neel	b. 1904	fundamental work and discoveries concerning anti-ferromagnetism and ferrimagnetism which have led to important applications in solid-state physics
1971	Dennis Gabor	1900-1979	his discovery of the principles of holography
1972	John Bardeen Leon Neil Cooper John Robert Schrieffer	1908-1991 b. 1930 b. 1931	their development of a theory of superconductivity
1973	Leo Esaki Ivar Giaever Brian David Josephson	b. 1925 b. 1929 b. 1940	his discovery of tunneling in semiconductors; his discovery of tunneling in superconductors; his theoretical prediction of the properties of a supercurrent through a tunnel barrier
1974	Anthony Hewish	b. 1924	the discovery of pulsars
	Sir Martin Ryle	1918-1984	his pioneering work in radio astronomy
1975	Aage Bohr Ben Mottelson Leo James Rainwater	b. 1922 b. 1926 1917-1986	the discovery of the connection between collective motion and particle motion and the development of the theory of the structure of the atomic nucleus based on this connection
1976	Burton Richter Samuel Chao Chung Ting	b. 1931 b. 1936	their (independent) discovery of an important fundamental particle
1977	Phillip Warren Anderson Sir Nevill Francis Mott John Hasbrouck Van Vleck	b. 1923 b. 1905 1899-1980	their fundamental theoretical investigations of the electronic structure of magnetic and disordered system
1978	Pyotr Leonidovich Kapitsa	1894-1984	his basic inventions and discoveries in low temperature physics
	Arno Allan Penzias Robert Woodrow Wilson	b. 1926 b. 1936	their discovery of cosmic microwave background radiation
1979	Sheldon Lee Glashow Abdus Salam Steven Weinberg	b. 1932 b. 1926 b. 1933	their unified model of the action of the weak and electromagnetic forces and for their prediction of the existence of neutral currents.
1980	James Watson Cronin Val L. Fitch	b. 1931 b. 1923	the discovery of violations of fundamental symmetry principles in the decay of neutral K mesons
1981	Nicolaas Bloembergen Arthur Leonard Schawlow Kai M. Siegbahn	b. 1920 b. 1921 b. 1918	their contribution to the development of laser spectroscopy; His contribution to high resolution electron microscopy.
1982	Kenneth Geddes Wilson	b. 1936	his method of analyzing the critical phenomena inherent in the changes of matter under the influence of pressure and temperature
1983	Subrahmanyan Chandrasekhar	b. 1910	his theoretical studies of the structure and evolution of stars
	William A. Fowler	b. 1911	his studies of the formation of chemical elements in the universe
1984	Carlo Rubbia Simon van der Meer	b. 1934 b. 1925	their decisive contributions to the large project, which led to the discovery of the field particles W and Z, communicators of the weak interaction
1985	Klaus von Klitzing	b. 1943	his discovery of the quantized Hall resistance
1986	Ernst August Friedrich Ruska	1906-1988	his invention of the electron microscope

Year	Nobel Laureate		Recognized for
	Gerd Binning Heinrich Rohrer	b. 1947 b. 1933	their invention of the scanning-tunneling electron microscope
1987	Karl Alex Muller Johannes Georg Bednorz	b. 1927 b. 1950	their discovery of a new class of superconductors
1988	Leon Max Lederman Malvin Schwartz Jack Steinberger	b. 1922 b. 1932 b. 1921	the neutrino-beam method and the demonstration of the doublet structure of the leptons through the discovery of the muon neutrino
1989	Norman Foster Ramsey, Jr.	b. 1915	his invention of the separated-oscillatory-field method and its use in the hydrogen maser and other atomic clocks
	Hans Georg Dehmelt Wilhelm Paul	b. 1922 b. 1913	their development of the ion trap technique
1990	Jerome I, Friedman Henry W. Kendall Richard E. Taylor	b. 1930 b. 1926 b. 1929	their experiment which led to breakthrough in our understanding of matter
1991	Pierre-Gilles de Gennes	b. 1932	his work on polymers and liquid crystals
1992	Georges Charpak		
1993	Joseph H. Taylor Russel A. Hulse		discovery of binary pulsar
1994	Betran N. Brockhouse Clifford G. Shull		their development of neutron scattering technique
1995	Martin L. Perl Frederick Reines		
1996	David Lee Robert Richardson Douglas Osheroff		discovery of superfluidity in ^{3}He isotope of helium.
1997	Steven Chu Claude Cohen-Tannoudji William D. Phillips		development of methods to cool and trap atoms with laser light
1998	Robert C. Laughlin Horst L. Stoermer Daniel C. Tsui		discovery of how subatomic particles can behave like a fluid.
1999	Gerardus Hooft Martinus J.G. Veltman		theoretical work on the structure and motion of subatomic particles.
2000	Zhores I. Alfernov Herbert Kroemer and Jack Kilby		development of technology used in satellite communications and cellular phones. invention and development of the ICs, and invention of the pocket calculator.
2001	Erick A. Cornell Carl E. Wieman Wolfgang Ketterle		the achievement of Bose-Einstein condensation in dilute gases of alkali atoms and for early fundamental studies of the properties of the condensates.
2002	Raymond Davis Jr. Masatoshi Koshiba; Riccardo Giacconi		their research into cosmic neutrinos and; his construction of instruments needed to investigate cosmic X-rays radiation which is absorbed in the Earth's atmosphere
2003	Alexei A. Abrikosov Vitaly L. Gineburg Anthony J. Leggett		their work in quantum physics concerning superconductivity and superfluidity.
2004	David Gross David Politzer Frank Wilczek		a unified description of all forces of Nature – from the tiniest distances within the atomic nucleus to the vast distances of the universe.

Year	Nobel Laureate		Recognized for
2005	Roy Glauber John Hall Theodor Haensch		their contribution to the developoment of laser-based precision spectroscopy including the optical frequency combtechnique.
2006	John Mathen George Smoot		spearheading a satellite programme that added weight to the Big Bang Theory of the origin of the Universe.
2007	Albert Fert Peter Gruenberg		discoveries allowing the miniaturisation of hard disks in electronic devices from laptops to ipods.
2008	Yoichiro Nambu	b. 1921	for the discovery of the mechanism of spontaneous broken symmetry in Subatomic Physics.
	Makoto Kobayashi Toshihide Maskawa	b. 1944 b. 1940	for the discovery of the origin of the broken symmetry.
2009	Charles Kuen Kao		for groundbreaking achievements concerning the transmission of light in fibres for optical communication
	Willard S.Boyle George. E. Smith		for the invention of an imaging semiconductor circuit - the CCD sensor.
2010	Andre Geim Konstantin Novoselov		for groundbreaking experiments regarding the two-dimensional material graphene.
2010	Soul perlmutter Adman G. Reiss Brian P. Schmidt		for the discovery of the accelerating expansion of the universe through observations of distant supernovae.